S0-DQX-123

LACRIMAL GLAND, TEAR FILM, AND DRY EYE SYNDROMES 2

Basic Science and Clinical Relevance

ADVANCES IN EXPERIMENTAL MEDICINE AND BIOLOGY

Recent Volumes in this Series

A Continuation Order Plan is available for this series. A continuation order will bring delivery of each new volume immediately upon publication. Volumes are billed only upon actual shipment. For further information please contact the publisher.

LACRIMAL GLAND, TEAR FILM, AND DRY EYE SYNDROMES 2

Basic Science and Clinical Relevance

Edited by

David A. Sullivan
Darlene A. Dartt

The Schepens Eye Research Institute and
Harvard Medical School
Boston, Massachusetts

and

Michele A. Meneray

Louisiana State University Medical Center
New Orleans, Louisiana

PLENUM PRESS • NEW YORK AND LONDON

Library of Congress Cataloging-in-Publication Data

Lacrimal gland, tear film, and dry eye syndromes 2 : basic science and clinical relevance / edited by David A. Sullivan, Darlene A. Dartt, and Michele A. Meneray.
p. cm. -- (Advances in experimental medicine and biology ; v. 438.)
"Proceedings of the Second International Conference on the Lacrimal Gland, Tear Film, and Dry Eye Syndromes, held November 16-19, 1996, at the Southhampton Princess Resort, Bermuda"--T.p. verso.
Includes bibliographical references and index.
ISBN 0-306-45812-8
1. Lacrimal apparatus--Physiology--Congresses. 2. Tears--Congresses. 3. Dry eye syndromes--Congresses. I. Sullivan, David D. II. Dartt, Darlene A. III. Meneray, Michele A. IV. International Conference on the Lacrimal Gland, Tear Film, and Dry Eye Syndromes (2nd : 1996 : Southampton, Bermuda Islands) V. Series.
[DNLM: 1. Lacrimal Apparatus--congresses. 2. Tears--physiology--congresses. 3. Dry Eye Syndromes--congresses. W1 Ad559 v. 438 1998]
QP188.T4L332 1998
612.8'47--dc21
DNLM/DLC
for Library of Congress 98-17987
CIP

Proceedings of the Second International Conference on the Lacrimal Gland, Tear Film, and Dry Eye Syndromes, held November 16 – 19, 1996, at the Southampton Princess Resort, Bermuda

ISBN 0-306-45812-8

A Division of Plenum Publishing Corporation
233 Spring Street, New York, N.Y. 10013

http://www.plenum.com

10 9 8 7 6 5 4 3 2 1

Printed in the United States of America

PREFACE

During the past two decades, a significant international research effort has been directed toward understanding the composition and regulation of the preocular tear film. This effort has been motivated by the recognition that the tear film plays an essential role in maintaining corneal and conjunctival integrity, protecting against microbial challenge, and preserving visual acuity. In addition, research has been stimulated by the knowledge that alteration or deficiency of the tear film, which occurs in countless individuals throughout the world, may lead to desiccation of the ocular surface, ulceration and perforation of the cornea, an increased incidence of infectious disease, and, potentially, pronounced visual disability and blindness.

To promote further progress in this field of vision research, the Second International Conference on the Lacrimal Gland, Tear Film and Dry Eye Syndromes: Basic Science and Clinical Relevance was held at the Southampton Princess Resort in Bermuda November 16–19, 1996. This conference was organized and directed by David A. Sullivan, Ph.D., codirected by Darlene A. Dartt, Ph.D., and Michele A. Meneray, Ph.D., and sponsored by the Schepens Eye Research Institute (Boston, MA), an affiliate of Harvard Medical School. The meeting was designed to assess critically the current knowledge and "state of the art" research on the structure and function of lacrimal tissue, tears, and the ocular surface in both health and disease. The goal of this conference was to promote an international exchange of information that would be of value to basic scientists involved in eye research, to physicians in the ophthalmological community, to pharmaceutical companies with an interest in the treatment of lacrimal gland, tear film, or ocular surface disorders, and to representatives of the U.S. Food and Drug Administration and the National Eye Institute.

To help achieve this objective, over 230 scientists, physicians, and industry representatives from 21 countries, including Australia, Austria, Belgium, Brazil, Canada, Denmark, England, Finland, France, Germany, Hungary, India, Italy, Japan, Scotland, Spain, Sweden, Switzerland, The Netherlands, the United States, and Uruguay, registered as active participants in this conference. In addition, this volume, which contains summaries of the conference's keynote, oral, and poster presentations, was created to provide an educational foundation and scientific reference for research on the tear film and dry eye syndromes.

The editors commend and thank Leona Greenhill for her excellent copy editing, as well as Benjamin D. Sullivan for his outstanding book indexing and superlative technical advice. In addition, the editors would like to thank the following individuals for their organizational, advisory, administrative, technical, and/or editorial help: members of the Organization and Program Committee, including Drs. Anthony J. Bron, Robert I. Fox, Aize Kijlstra, Peter Laibson, Michael A. Lemp, Austin K. Mircheff, J. Daniel Nelson, Stephen C. Pflugfelder, Bernard Rossignol, John M. Tiffany, Kazuo Tsubota, John L. Ubels, Benjamin Walcott, and Steven E. Wilson; members of the International Advisory Committee, including Mark Abelson, Roger W. Beuerman, B. Britt Bromberg, Barbara Caffery, P. Noel Dilly, Marshall Doane, Claes H. Dohlman, Peter C. Donshik, Henry F. Edelhauser, R. Linsy Farris, Gary N. Foulks, Philip Fox, Mitchell H. Friedlaender, Roderick J. Fullard, Anne-Marie Gachon, Jeffrey Gilbard, Ilene K. Gipson, Martin J. Göbbels, Jack Greiner, Jean-Pierre Guillon, Brien Holden, Frank J. Holly, Brett Jessee, Renee Kaswan, Donald R. Korb, Gordon W. Laurie, Donald L. MacKeen, Rolf Marquardt, William D. Mathers, Mitchell D. McCartney, James P. McCulley, Charles McMonnies, Paul C. Montgomery, Juan Murube del Castillo, Sudi Patel, Jan Ulrik Prause, Al Reaves, Miguel Refojo, Brenda L. Reis, Maurizio Rolando, Hans-Walter Roth, Robert A. Sack, Ichiro Saito, Elcio H. Sato, Oliver D. Schein, Michael E. Stern, Norman Talal, Alan Tomlinson, O.P. van Bijsterveld, N.J. van Haeringen, Dwight Warren, Graeme Wilson, and Koji Yamamoto, and Ms. Sue Dauphin and Jean S. Kahan; and individuals in Acton, MA, Southampton Parish, Bermuda, Boston, MA, Littleton, MA, Louisville, KY, New York, NY, and Research Triangle Park, NC, including Lawrence Bair, Alan Bergl, Tony Best, Gail Burke, Mary Gallagher, Marie Groome, Robin Hodges, Marcia M. Jumblatt, PhD, Kevin J. Klein, Kathleen Krenzer, OD, PhD, Margie Leak, Catherine Louis, Karen Madeiros, Margaret Rocco, Eduardo M. Rocha, MD, James W. Putney, PhD, Lilia Aikawa da Silveira, Amy G. Sullivan, Rose M. Sullivan, and Driss Zoukhri, PhD.

David A. Sullivan

ACKNOWLEDGMENTS

The editors sincerely thank the following companies and foundations, whose generous financial contributions significantly offset the educational expenses and publication costs associated with the Second International Conference on the Lacrimal Gland, Tear Film and Dry Eye Syndromes: Basic Science and Clinical Relevance.

Allergan, Inc.
Tear Products Division, Irvine, CA, USA

Allergan, Inc.
New Products Marketing Division, Irvine, CA, USA

Alcon Ophthalmic
Marketing Division, Alcon Laboratories, Fort Worth, TX, USA

CIBA Vision Ophthalmics
Pharmaceutical Research and Development, Duluth, GA, USA

Alcon Ophthalmic
Research & Development Division, Alcon Laboratories, Fort Worth, TX, USA

Santen Pharmaceutical Co., Ltd.
Research and Development Division, Osaka, Japan

Taisho Pharmaceutical Co., Ltd.
Research and Development Division, Tokyo, Japan

Pfizer Inc.
Consumer Health Care Division, New York, NY, USA

Vistakon
Materials Research & Development Division, Jacksonville, FL, USA

Bausch & Lomb
Global Clinical Research Division, Rochester, NY, USA

Bausch & Lomb
Pharmaceutical Division, Tampa, FL, USA

Alcon Foundation
Fort Worth, TX, USA

Eagle Vision
Memphis, TN, USA

Advanced Instruments, Inc.
Business Development Division, Norwood, MA, USA

SPECIAL ACKNOWLEDGMENT

The editors express their grateful appreciation to L. Alexandra Wickham, whose exceptional administrative, technical, and editorial assistance was truly invaluable in making the Second International Conference on the Lacrimal Gland, Tear Film and Dry Eye Syndromes: Basic Science and Clinical Relevance, as well as these proceedings, a reality.

CONTENTS

Conference Address

Lacrimal Gland: Cellular and Molecular Biology

Lacrimal Gland and Ocular Surface: Signal Transduction, Membrane Traffic, and Fluid and Protein Secretion

Mucins: Origin, Biochemistry, and Regulation

Meibomian Gland and Tear Film Lipids: Structure, Function, and Control

Tear Film Stability, Evaporation, and Biophysics

Cytokines, Growth Factors, Proto-Oncogenes, and Apoptosis

Inflammation and Immunity

Tear Film Components and Influence on the Ocular Surface

Classification, Diagnosis, Clinical Features, and Epidemiology of Dry Eye Syndromes

Pathogenesis of Dry Eye Syndromes

Management and Therapy of Dry Eye Syndromes

The editors dedicate this book to Drs. Anthony J. Bron, Claes H. Dohlman, Anne-Marie F. Gachon, Michael A. Lemp, Bernard Rossignol, and John M. Tiffany for their pioneering efforts and outstanding achievements in basic and clinical research on the lacrimal gland, tear film, and/or dry eye syndromes.

1

A CLINICIAN LOOKS AT THE TEARFILM

J. Daniel Nelson

Departments of Ophthalmology, HealthPartners-Ramsey Clinic and
Healthpartners-St. Paul Ramsey Medical Center
St. Paul, Minnesota

1. INTRODUCTION

Today as we open the Second International Conference on the Lacrimal Gland, Tear Film, and Dry Eye Syndromes, I want us to let go of preconceived ideas, long-held dogma, and bias and to look to the future as we pursue basic knowledge about pathophysiology and regulation of the lacrimal and meibomian glands and the ocular surface, the causes of dry eye, and new treatments. Remember, the reason we are pursuing this knowledge is to help those who suffer from dry eye, our patients, and, perhaps, someday offer a cure for each of the different types of dry eye.

The principal function of the tear film is to maintain a smooth, clear, refractive optical surface in a very hostile external environment. Any adverse effect on corneal regularity and clarity will interfere with vision and may cause pain. It is pain and decreased vision that result in disability to dry eye patients. It is pain and decreased vision that affect the quality of life of dry eye sufferers.

Defining the term dry eye has always been a problem. To the clinician, it is defined by objective findings (often an abnormal Schirmer's test). To the patient, it is defined by symptoms. At an NEI-sponsored clinician/industry workshop on dry eye, the terms dry eye and keratoconjunctivitis sicca (KCS) were agreed to be the same, and the following definition of dry eye was agreed upon:

"Dry eye is a disorder of the tear film due to tear deficiency or excessive tear evaporation which causes damage to the interpalpebral ocular surface and is associated with symptoms of ocular discomfort."[1]

Dry eye is classified into two broad categories, the lacrimal gland deficient dry eye and the evaporative dry eye. Dry eye due to lacrimal gland deficiency is further broken down into dry eye associated with Sjögren's syndrome and dry eye not associated with Sjögren's syndrome. It may be more appropriate to classify lacrimal gland deficiency as immune and non-immune related. Immune causes include Sjögren's syndrome, graft-vs.-host disease, acquired immune deficiency syndrome (AIDS), and sarcoidosis. Non-im-

Lacrimal Gland, Tear Film, and Dry Eye Syndromes 2
edited by Sullivan *et al.*, Plenum Press, New York, 1998

mune causes include primary lacrimal gland deficiency (PLD) and secondary lacrimal deficiency. PLD refers to a group of patients with no discernible underlying cause for the dry eye. Secondary lacrimal deficiency includes congenital alacrima, sensory and secretomotor nerve damage, radiation-induced lacrimal gland atrophy, and drug-induced hyposecretion. Dry eye due to evaporation is further broken down, with major classifications being dry eye due to blepharitis, lid abnormalities, and contact lens wear. The term dry eye therefore refers to dry eye due to all causes.

Dry eye affects all ages and both sexes and is responsible for the purchase of millions of dollars worth of artificial tear products. All of us have experienced transient dry eye symptoms at one time or another. Airplane flights, contact lens wear, sleepless nights, cosmetics, dry weather, air pollution can stress even the normal tear film. The question is—what makes the tear film in dry eye patients so different from normal individuals? Why is the tear film of the normal eye so resilient to the external environment? Other questions remain to be answered. Why are women more affected than men in most instances? What effect does aging have? Is there a genetic predisposition? Blepharitis, contact lenses, aging, immunologic disease, infectious disease, and systemic medications are commonly associated with dry eye. Is there a common thread tying these causes together?

Our knowledge of dry eye has been limited by only a rudimentary understanding of the normal pathophysiology of tear secretion and the tear film. Many of our long-held assumptions concerning the tear film need to be challenged. You are all here because you hold the keys to answering these questions. Today, as a clinician who treats many patients with dry eye, I would like to cover several areas as they relate to the tear film. I will discuss the role of androgens and neural innervation, lacrimal fluid and mucin, as they relate to the tear film. I will offer observations, speculations, and conclusions from my perspective as a clinician scientist.

2. THE ROLE OF ANDROGENS

Primary lacrimal gland deficiency (PLD) is more common in women than men. It has been suspected that estrogen is the key factor because of this gender bias and because there are changes in the ocular surface cell morphology that parallel the menstrual cycle. However, PLD can be associated with lactation, pregnancy, post-menopause, oral contraceptives, and supplemental estrogens.[2–4] Post-menopausal and lactating women have decreased levels of circulating estrogens; women on oral contraceptives or who are pregnant have increased levels. The common feature of these two groups is low bioavailability of androgens. Remember that older men, who are also prone to PLD, have lower levels of androgens. Elevated levels of circulating estrogens decrease androgens by a positive feedback loop to the hypothalamic-pituitary-gonadal axis. Increased estrogen levels also cause the liver to increase production of sex hormone binding globulin (SHBG), which binds estrogens and androgens, decreasing androgen as well as estrogen levels. Aging, which is associated with PLD, is accompanied by decreased androgen levels.[5]

Although protein and DNA levels are similar in both sexes prior to puberty, adult female lacrimal glands have lower amounts of protein and a lower DNA content than the adult male.[6] It is known that androgens are essential for maintaining the sexual dimorphism of the lacrimal gland secretory immune system. Androgens increase secretory factor and IgA secretion by the lacrimal gland.[7] In rats, castration decreases their secretion, and androgen replacement increases their secretion.[7,8] An intact pituitary-gonadal axis is necessary for normal functioning of the secretory immune system.[9] However, the effect of androgen treatment on tear secretion is species dependent.[10,11]

In female rats, hypophysectomy decreases lacrimal function. If these rats are then given dihydrotestosterone (DHT), a potent androgen that can not be converted to estrogens, lacrimal function improves.[12] In rabbits, ovarianectomy reduces lacrimal gland function, which is improved with DHT treatment.[13] In addition, androgens may suppress apoptosis.[13]

It is thus apparent that androgens play the key role in supporting lacrimal gland function. However, DHT has no effect on increasing cholinergic receptor content in either the hypophysectomized rat or the ovarianectomized rabbit.[12,13] Prolactin, which is synthesized and secreted by the lacrimal gland[14–17] and has receptors in the lacrimal gland,[16] also plays a role in regulating lacrimal gland function. It is likely that the content of cholinergic receptors is uniquely dependent on prolactin.[12] In addition, cholinergic receptor production is reduced by clomiphene (an anti-estrogen), suggesting that pituitary prolactin production is dependent on estrogen. Estrogens and prolactin may also be involved in the pathophysiology of Sjögren's syndrome (SS).[18,19] Although the issue of hormonal regulation of the lacrimal gland is complex and more knowledge is needed, it is likely that the influence of estrogen on the lacrimal glands is indirect, by suppressing ovarian androgen production and increasing pituitary prolactin production.

At the 1996 American Academy of Ophthalmology meeting, Mamalis et al.[20] suggested that low testosterone levels were more prevalent in women with dry eyes and correlated with the severity of dry eye. Androgen levels are reported to be decreased in SS and in systemic lupus erythematosus (SLE).[21–23] In mouse models of SS, androgen treatment suppressed inflammation in lacrimal glands.[10,24] It may be that there is a critical level of androgen support needed to maintain lacrimal gland function.

In summary, androgens and prolactin play a direct role in lacrimal gland regulation and estrogens an indirect role. Androgens are likely in the starring role, with prolactin and estrogens in strong supporting roles. However, there is significant complexity and much remains undiscovered. Although we have gained significant knowledge that hormones act to influence lacrimal gland function, little is known about regulatory feedback systems from the lacrimal gland or ocular surface. As with other hormonally regulated systems in the body, one would anticipate feedback mechanisms that influence hormonal action on lacrimal gland function.

Recently androgen receptor mRNA has been found in the meibomian glands, as well as in the bulbar and palpebral conjunctiva of humans, rats, and rabbits.[25] At this conference, Wickham et al. will present information demonstrating that sex steroid mRNAs and/or proteins are present in numerous ocular tissues. Sullivan et al. will show that the meibomian gland is an androgen target organ and that androgens modify lipid production. It is likely that androgens also play a significant role in meibomian gland disease and the evaporative dry eye. This may explain why meibomian gland disease is often found in conjunction with both PLD and SS.

3. THE ROLE OF NEURONAL INNERVATION

Stimuli at the ocular surface increase lacrimal gland fluid secretion via parasympathetic and sympathetic branches of the autonomic nervous system. There are also efferent and afferent sensory fibers in the lacrimal gland. A number of neuropeptides have been detected in the lacrimal glands of various species, including enkephalins, neuropeptide Y, calcitonin gene-related peptide (CGRP), substance P, and VIP.[26–32] Lacrimal epithelial cells have muscarinic, α and β adrenergic, ACTH, VIP, 5-hydroxytryptamine, neuropeptide Y, and dopamine receptors, along with opiate receptors activated by enkephalins. These nerves are in close apposition to vascular, myoepithelial, and/or epithelial cells and are critical to the control of tear production.[33] In SS, extensive lacrimal gland inflammation may cause a loss of neuronal innervation, which has been shown to occur in the salivary gland in areas of large lymphocytic infiltration.[34,35] In addition, IL-1β has been shown to suppress release of neurotransmitters from cholinergic and adrenergic nerves during intestinal inflammation.[36,37] Lacrimal gland cytokines may similarly suppress release of neurotransmitters. Finally, methionine-enkephalin suppresses parasympathetic and peptidergic stimulation of epithelial cell secretion in the lacrimal gland.[38] Therefore, inflammation alone, independent of gland destruction, may interfere with neuronal regulation in the lacrimal gland epithelial cells, with resulting decreased tear secretion. This may explain why patients with SS may have decreased tear secretion, while retaining significant portions of intact lacrimal gland tissue.

Although we know that stimuli at the ocular surface result in secretion of lacrimal fluid, there is some evidence that the ocular surface may influence phenotypic expression in the lacrimal gland. It has been reported that trauma to the ocular surface resulted in increased mRNA levels and synthesis of EGF along with EGF receptors in the lacrimal gland.[39]

On the ocular surface, efferent parasympathetic and sympathetic, but not sensory, nerves are located adjacent to goblet cell clusters. It has been proposed that parasympathetic (containing VIP) and sympathetic (containing norepinephrine) nerve activation could directly stimulate goblet cell mucus secretion.[40] Topical or subconjunctival lidocaine inhibits wound-induced stimulation of goblet cell mucous secretion.[41] The likely purpose of this neural innervation is to provide rapid goblet cell mucous secretion to protect the ocular surface.

Rat meibomian glands receive innervation from nerve fibers containing tyrosine hydroxylase, dopamine, beta hydroxylase, VIP, peptide histidine isoleucine, substance P, or calcitonin gene-related peptide (CGRP).[42–47] Recent studies in rhesus and cynomolgus monkeys also demonstrate numerous nerve fibers near the meibomian gland acini with immunoreactivity to neuropeptide Y and VIP, tyrosine hydroxylase, CGRP, and substance P.[48] It is likely that meibomian gland secretion is under neural control.

In summary, the lacrimal gland, goblet cells, and meibomian glands, all of which contribute to the tears or tear film, are under neural control. Neural regulation of lacrimal secretion can be interrupted by lymphocytic infiltration, cytokine production, or methionine-enkephalin production. What effects neural interruption has on the goblet cells or meibomian glands is not known. It may be that conjunctival and lid inflammation also decrease secretion of meibomian glands and goblet cells or alter their function.

4. THE ROLE OF TEARS

Most believe that tears water the eye, providing moisture and protection from a hostile environment. Several clinical observations in dry eye patients are worth mentioning. First, frequently the use of artificial lubricants, while they may decrease symptoms, have little effect on rose bengal or fluorescein staining or on squamous metaplasia of the ocular surface. Second, the ocular surface disease that occurs in SS is not correlated with the amount of tear secretion. Third, in the aqueous-deficient dry eye, while frequent use of artificial lubricants does not reverse rose bengal staining, punctal occlusion, which increases retention of tears, significantly decreases staining.[49] These observations suggest that there may be more to tears than just water.

Epidermal growth factor (EGF),[50,51] endothelin-I,[51] basic fibroblastic growth factor (FGF-β),[52] transforming growth factor - α (TGF-α),[53] transforming growth factor-β (TGF-β),[54] and hepatic growth factor (HGF)[55] are secreted by the lacrimal gland, as are thyroid hormone[56] and retinol.[57] These factors may exert paracrine effects on the ocular surface and paracrine and autocrine effects within the lacrimal gland. Not only is TGF-β2 secreted by the lacrimal gland,[54] but high levels of TGF-β2 are also present in tears.[58] Could these and other hormones and growth factors be involved in the homeostasis of the ocular surface and in corneal epithelial wound healing? It has been suggested that TGF-β may play an important regulatory role in suppressing ocular surface inflammation, promoting normal growth and differentiation of ocular surface epithelia, and promoting key elements of the wound-healing cascade.[54,58] Prolactin, synthesized by the lacrimal gland ductal epithelium and secreted in tears,[14,15,59] may play a role in corneal epithelial wound healing. It is known that cholinergic receptors, when stimulated, accelerate wound healing.[60–62] Prolactin, by its positive effect on the expression of cholinergic receptors, might indirectly promote wound healing.

As the lacrimal gland does secrete various hormones and cytokines, and as at least a few of these products make their way to tears, it is likely that it is the composition—and not the amount—of tears that is important. Tears may actually serve as the delivery system of these substances to the ocular surface much like loggers used rivers to deliver logs to sawmills. The normal homeostasis of the ocular surface is likely dependent on tear secretion. Defining dry eye by tests that measure the volume or rate of secretion of tears may be immaterial. The cytokine and hormone composition of tears may be more relevant to defining and classifying normal and dry eyes. Our efforts in defining and diagnosing the causes of dry eye should be aimed at looking at the composition of the tear film.

It is also likely that cytokines are produced in or by the lacrimal gland in response to inflammation. Research has shown that mRNA for IL-α, IL-1β, IL-2 receptor, IL-6, TGF-β, and TNF-α can be detected in lacrimal glands of autoimmune female MRL/lpr mice.[63] These factors may also find their way into the tear fluid and be delivered to the ocular surface. Once delivered, they may induce ocular surface inflammation seen, for example, in early SS. This may explain why some SS patients have significant ocular surface inflammation with relatively normal Schirmer test values.

In summary, we need to think of lacrimal tears as a drug delivery system and not just water for the eyes. In health, tears deliver substances to maintain the health of the ocular surface. In lacrimal gland disease, they deliver inflammatory cytokines that may lead to surface inflammation. It does beg the issue, if lacrimal gland fluid is only a delivery system, does it contribute anything else to the tear film? Is our concept of an aqueous-dominated tear film the appropriate model?

5. THE ROLE OF MUCIN

A hypothesis made in 1986 by Kaura went as follows:

> "There is clearly a need in the external eye for both a mucous gel and a viscous mucous solution. The gel is required for lubrication between the lids and globe to bind pathogens and entrap foreign bodies. The viscous fluid would facilitate exchange between its components and the corneal epithelium and add to the stability of its aqueous component."[64]

Further work since has demonstrated that the entire ocular surface, cornea and conjunctiva, produce mucins,[65] and that the stratified squamous epithelia of the conjunctiva and cornea synthesize mucin (MUC1).[66] It has been assumed that rose bengal was a vital dye and stained dead and dying epithelium. It is now known that rose bengal staining occurs whenever there is poor protection or coverage of the surface epithelium with mucin.[67] Rose bengal staining, therefore, represents a deficiency of the preocular tear film protection.[68] Finally, it is proposed that the tear film consists mostly of mucus.[69,70] Using two methods, interferometry and confocal microscopy, the tear film was estimated to be ~40 μm in thickness. With the application of acetylcysteine, the tear film thinned to ~11 μm, the estimated thickness of the aqueous layer.

In the stomach, the gastric mucosa is protected from autodigestion by a mucin-gel formed by the interaction of bicarbonate and mucin. Bicarbonate is a necessary component of the gel, as mucin alone does not provide sufficient protection.[71] Based on these findings, it is reasonable to propose that we must discard our classical concept of an aqueous-dominated preocular tear film. As Dilly proposed at this meeting in 1994,

"(The tear film) is nearer 30–40 μm (thick) with the extra thickness contributed by the mucin layer.... It contains many vital biological active substances that can probably be slow released.... The aqueous phase contains dissolved mucins and provides the cleavage plane for lid movements.... It is probably more accurate to describe the mucus and aqueous layers of the tear film as phases with more or less mucus, respectively."[72]

It is likely that the tear film is mucin dominated, with a ~30 μm thick, hydrated mucin gel, with overlying aqueous and lipid layers. With this view of the tear film, one would not expect the frequent use of artificial tears to reverse rose bengal staining. This suggests that rather than watering the eye, therapy needs to be aimed at restoring the protective mucin gel.

6. FUTURE TREATMENTS

Based on our discussion, new treatments may be aimed at using androgens to improve lacrimal and meibomian gland function or neural peptides to improve meibomian gland and goblet cell function. Prolactin may prove useful as a wound-healing agent. Cytokines or growth factors may be used to maintain normal ocular surface homeostasis when lacrimal gland secretion is reduced or absent. Agents that restore or maintain the bicarbonate-mucin gel may prove quite useful.

7. CONCLUSIONS

Where are we? To speculate a little, it is likely that the lacrimal gland, meibomian glands, and ocular surface are regulated both by hormones, androgens and prolactin in particular, and by neural factors. It is likely that substances secreted in lacrimal gland fluid—the tears—are involved in homeostasis and in disease of the ocular surface. Finally, the tear film is relatively thick, likely a bicarbonate-mucin gel rather than an aqueous-dominated film. What's new? We will hear that during this conference. Listen closely. Think critically. Talk to your colleagues. Above all, be open to new ideas. Ideas that may run contrary to your long-held beliefs.

REFERENCES

1. Lemp MA. Report of the National Eye Institute/Industry Workshop on Clinical Trials in Dry Eyes. *CLAO J.* 1995;21:221–231.
2. Serrander A-M, Peek K. Changes in contact lens comfort related to the menstrual cycle and menopause: A review of articles. *J Am Optom Assoc.* 1993;64:162–166.
3. Brennan N, Efrom N. Sumptomatology of HEMA contact lens wear. *Optom Vis Sci.* 1989;66:834–838.
4. Verbeck B. Augenbefunde und Soffwechselverhalten bei Einnahme von Ovulationshemmern. *Klin Monatsbl Augenheilkd.* 1973;162:612–621.
5. Zumoff B, Rosenfeld R, Strain G, Levin J, Fukushima D. Sex differences in twenty-four mean plasma concentrations of dehydroisoandrosterone (DHA) and dehydroisoandrosterone sulfate (DHAS) and the DHA to DHS ratio in normal adults. *J Clin Endocrinol Metabol.* 1980;51:330–333.
6. Azzarolo A, Mazaheri A, Mircheff A, Warren D. Sex-dependent parameters related to electrolyte, water and glycoprotein secretion in rabbit lacrimal glands. *Curr Eye Res.* 1993;12:795–802.
7. Sullivan D, Allansmith M. The effect of aging on the secretory immune system of the eye. *Immunology.* 1988;63:403–410.
8. Rocha F, Wickham L, Pena J, et al. Influence of gender and the endocrine environment on the distribution of androgen receptors in the lacrimal gland. *J Steroid Biochem Mol Biol.* 1993;46:737–749.
9. Sullivan D, Allansmith M. Hormonal influence on the secretory immune system of the eye: Endocrine interactions in the control of IgA and secretory component levels in tears of rats. *Immunology.* 1987;60:337–343.
10. Vendramini A, Soo C, Sullivan D. Testosterone-induced suppression of autoimmune disease in lacrimal tissue of a mouse model (NZB/NZW F1) of Sjögren's syndrome. *Invest Ophthalmol Vis Sci.* 1991;32:3002–3006.
11. Sato E, Sullivan D. Comparative influence of steroid hormones and immunosuppressive agents on autoimmune expression in lacrimal glands of a female mouse model of Sjögren's syndrome. *Invest Ophthalmol Vis Sci.* 1994;35:2632–2642.
12. Azzarolo A, Bjerrum K, Maves C, et al. Hypophysectomy-induced regression of female rat lacrimal glands: Partial restoration and maintenance by dihydrotestosterone and prolactin. *Invest Ophthalmol Vis Sci.* 1995;36:216–226.
13. Azzarolo A, Kaswan R, Mircheff A, Warren D. Androgen prevention of lacrimal gland regression after ovarianectomy of rabbits. ARVO Abstracts. *Invest Ophthalmol Vis Sci..* 1994;35:S1793.
14. Mircheff AK, Warren DW, Wood RL, Tortoriello PJ, Kaswan RI. Prolactin localization, binding, and effects on peroxidase release in rat exorbital lacrimal gland. *Invest Ophthalmol Vis Sci.* 1992;33:641–650.
15. Markoff E, Lee D, Fellows JL, Nelson J, Frey WI. Human lacrimal glands synthesize and release prolactin. *Endocrinol* 1995;152(Suppl):440A.
16. Zhang J, Whang G, Gierow J, Warren D, Mircheff A, Wood R. Prolactin receptors are present in rabbit lacrimal gland. ARVO Abstracts. *Invest Ophthalmol Vis Sci.* 1995;36:S991.
17. Zhang J, Yang T, Platler B, Kuda A, Mircheff A, Wood R. Endocytosis and degradation of prolactin by reconstituted lacrimal acini. ARVO Abstracts. *Invest Ophthalmol Vis Sci.* 1996;37:S856.
18. Carlsten H, Tarkowski A, Holmdahl R, Nilsson L. Oestrogen is a potent disease accelerator in SLE-prone MRL lpr/lpr mice. *Clin Exp Immunol.* 1990;80:467–473.
19. Ahmed S, Aufdemorte T, Chen J, Montoya A, Olive D, Talal N. Estrogen induces the development of autoantibodies and promotes salivary gland lymphoid infiltrates in normal mice. *J Autoimmun.* 1989;2:543–552.
20. Mamalis N, Harrison D, Hiura G, et al. Dry eyes and testosterone deficiency in women. American Academy of Ophthalmology meeting, Chicago, 1996:132.
21. Xu G, Shang H, Zhu F. Measurement of serum testosterone level in female patients with dry eye. Proceedings of the International Congress of Ophthalmology, 1994.
22. Lahita R, Bradlow H, Ginzler E, Pang S, New M. Low plasma androgens in women with systemic lupus erythematosus. *Arthritis Rheum.* 1987;30:241–248.
23. Lavalle C, Loyo E, Paniagua R, et al. Correlation study between prolactin and androgens in male patients with systemic lupus erythematosus. *J Rheumatol.* 1987;14:268–272.
24. Ariga H, Edwards J, Sullivan D. Androgen control of autoimmune expression in lacrimal glands of MRL/Mp-lpr/lpr mice. *Clin Immunol Immunopathol.* 1989;53:499–508.
25. Sullivan D, Wickham L, Toda I, Gao J. Identification of androgen, estrogen, and progesterone receptor mRNAs in rat, rabbit, and human ocular tissues. ARVO Abstracts. *Invest Ophthalmol Vis Sci.* 1995;35:S651.

26. Dartt D, Baker A, Vaillant C, Rose P. Vasoactive intestinal polypeptide stimulation of protein secretion from rat exorbital lacrimal gland acini. *Am J Physiol.* 1984;247:G502-G509.
27. Nikkinen A, Lehtosalo J, Uusitalo H, Palkama A, Panula P. The lacrimal glands of the rat and guinea pig are innervated by nerve fibers containing immunoreactivities for substance P and vasoactive intestinal polypeptide. *Histochemistry.* 1984;81:23–27.
28. Walcott B, Sibony P, Keyser K. Neuropeptides and the innervation of the avian lacrimal gland. *Invest Ophthalmol Vis Sci.* 1989;30:1666–1674.
29. Matsumoto Y, Tanabe T, Ueda S, Kawata M. Immunohistochemical and enzymehistochemical studies of peptidergic, aminergic, and cholinergic innervation of the lacrimal gland of the monkey (*Macaca fuscata*). *J Auton Nerv Syst.* 1991;37:207–217.
30. Cripps M, Bennett D. Proenkephalin A derivatives in lacrimal gland: Occurrence and regulation of lacrimal function. *Exp Eye Res.* 1992;54:829–834.
31. Singh J, Adeghate E, Burrows S, Howarth F, Donath T. Protein secretion and the identification of neurotransmitters in the isolated pig lacrimal gland. *Adv Exp Med Biol.* 1994;350:57–60.
32. Williams R, Singh J, Sharkey K. Innervation and mast cells of the rat lacrimal gland: The effects of age. *Adv Exp Med Biol.* 1994;350:1–9.
33. Dartt D. Physiology of tear production. In: Lemp M, Marquardt R, eds. *The Dry Eye.* Berlin: Springer-Verlag; 1992:65–99.
34. Konttinen Y, Sorsa T, Hukkanen M, et al. Topology of innervation of labial salivary glands by protein gene product 9.5 and synaptophysin immunoreactive nerves in patients with Sjögren's syndrome. *J Rheumatol.* 1992;19:30–37.
35. Kontinen Y, Hukkanen M, Kemppenen P, et al. Peptide-containing nerves in labial salivary glands in Sjögren's syndrome. *Arthritis Rheum.* 1992;35:815–820.
36. Collins S, Hurst S, Main C, et al. Effect of inflammation of enteric nerves: Cytokine-induced changes in neurotransmitter content and release. *Ann NY Acad Sci.* 1992;664:415–424.
37. Main C, Blennerhasset P, Collins S. Human recombinant interleukin 1β suppresses acetylcholine release from rat myenteric plexus. *Gastroenterology.* 1993;104:1648–1654.
38. Cripps M, Patchen-Moor K. Inhibition of stimulated lacrimal secretion by [D-ALa^2] metenkephalinamide. *Am J Physiol.* 1989;20:G151-G156.
39. Steinemann T, Thompson H, Maroney K, Palmer C, Henderson L, Malter J. Changes in epidermal growth factor receptor and lacrimal gland EGF concentration after corneal wounding. *Invest Ophthalmol Vis Sci.* 1990(Suppl);31:55.
40. Dartt D, McCarthy D, Mercer H, Kessler T, Chung E-H, Zieske J. Localization of nerves adjacent to goblet cells in rat conjunctiva. *Curr Eye Res.* 1995;14:993–1000.
41. Kessler T, Mercer H, Zieske J, McCarthy D, Dartt D. Stimulation of goblet cell mucous secretion by activation of nerves in rat conjunctiva. *Curr Eye Res.* 1995;14:985–992.
42. Meyer-Bothling U, Bron A, Osborne N. Immunohistochemical evidence for an adrenergic innervation of the rat meibomian glands. ARVO Abstracts. *Invest Ophthalmol Vis Sci.* 1994;35:S1789.
43. Simons E. Smith P. Sensory and autonomic innervation of the rat eyelid: Neuronal origins and peptide phenotypes. *J Chem Neuroanat.* 1994;7:35–47.
44. Hartschuh W, Weihe E, Reinecke M. Peptidergic (neurotensin, VIP, substance P) nerve fibers in the skin: Immunohistochemical evidence of an involvement of neuropeptides in nociception, pruritus and inflammation. *Br J Dermatol.* 1983;109(suppl 25):14–17.
45. Elsas T, Edvinsson L, Sundler F, Uddman R. Neuronal pathways to the rat conjunctiva revealed by retrograde tracing and immunohistochemistry. *Exp Eye Res.* 1994;58:117–126.
46. Luhtala J, Uusitalo H. The distribution and origin of substance P immunoreactive nerve fibers in the rat conjunctiva. *Exp Eye Res.* 1991;53:641–646.
47. Luhtala J, Palkama A, Uusitalo H. Calcitonin gene related peptide immunoreactive nerve fibers in the rat conjunctiva. *Invest Ophthalmol Vis Sci.* 1991;32:640–645.
48. Chung C, Tigges M, Stone R. Peptidergic innervation of primate meibomian gland. *Invest Ophthalmol Vis Sci.* 1996;37:238–245.
49. Gilbard J, Rossi S, Azar D, Heyda K. Effect of punctal occlusion by Freeman silicone plug insertion on tear osmolarity in dry eye disorders. *CLAO J.* 1989;15:216–218.
50. Ohashi Y, Motokura M, Kinoshita Y, Mano T, Watanabe H, Kinoshita S. Presence of epidermal growth factor in tears. *Invest Ophthalmol Vis Sci.* 1989;30:1879–1882.
51. Takashima Y, Takagi H, Reinach P, Takahashi M, Yoshimura N. Detection of ET-I lacrimal gland gene expression and ET-I like immunoreactivity in rabbit tears. ARVO Abstracts. *Invest Ophthalmol Vis Sci.* 1994;35:S1791.

52. Wilson S, Lloyd S, Kennedy R. Basic fibroblastic growth factor (FGFβ) and epidermal growth factor (EGF) messenger RNA production in human lacrimal gland. *Invest Ophthalmol Vis Sci.* 1991;32:2816–2820.
53. van Setten G, Macauley S, Humphreys-Beher M, Chegini N, Schultz G. Detection of transforming growth factor-α in human tears. *Invest Ophthalmol Vis Sci.* 1996;37:166–173.
54. Yoshino K, Garg R, Monroy D, Zhonghua J, Pflugfelder S. Production and secretion of transforming growth factor beta (TGF-β) by the human lacrimal gland. *Curr Eye Res.* 1996;15:615–624.
55. Li Q, Weng J, Mohan R, Bennett G, Schwall R, Wang Z-F. Hepatocyte growth factor (HGF) and HGF receptor in lacrimal gland, tears, and cornea. *Invest Ophthalmol Vis Sci.* 1996;37:727–739.
56. Liotet S, Glomaud J. Significance of T_3 and T_4 assays in tears. In: Holly F, ed. *The Preocular Tear Film in Health, Disease, and Contact Lens Wear.* Lubbock, TX: The Dry Eye Institute; 1986:792–797.
57. Ubels J, Rismondo V, Osgood TB. The relationship between secretion of retinol and protein by the lacrimal gland. *Invest Ophthalmol Vis Sci.* 1989;30:952–960.
58. Kokawa N, Sotozono C, Nishida K, Kinoshita S. High total TGF-β2 levels in normal human tears. *Curr Eye Res.* 1996;15:341–343.
59. Wood R, Park K-H, Gierow J, Mircheff A. Immunogold localization of prolactin in acinar cells of lacrimal gland. *Adv Exp Med Biol.* 1994;350:75–77.
60. Lindt G, Cavanagh H. Nuclear muscarinic acetylcholine receptors in corneal cells from rabbit. *Invest Ophthalmol Vis Sci.* 1993;34:2943–2952.
61. Colley A, Cavanagh H, Law M. Effects of topical carbamylcholine on corneal epithelial resurfacing. *Metab Pediatr Syst Ophthalmol.* 1987;10:71–72.
62. Colley A, Law M, Drake L, Cavanagh H. Activity of DNA and RNA polymerases in resurfacing rabbit corneal epithelium. *Curr Eye Res.* 1987;6:477–487.
63. Ono M, Huang Z, Wickham L, Gao J, Sullivan D. Analysis of androgen receptors and cytokines in lacrimal glands of a mouse model of Sjögren's syndrome. ARVO Abstracts. *Invest Ophthalmol Vis Sci.* 1994;35:S1793.
64. Kaura R. Ocular mucus: A hypothesis for its role in the external eye based on physiological considerations and on the properties and functions known to be common with other mucous secretions. In: Holly F, ed. *The Preocular Tear Film: In Health, Disease, and Contact Lens Wear.* Lubbock, TX: Dry Eye Institute; 1986:743–747.
65. Watanabe H, Fabricant M, Tisdale A, Spurr-Michaud S, Lindberg K, Gipson I. Human corneal and conjunctival epithelia produce a mucin-like glycoprotein for the apical surface. *Invest Ophthalmol Vis Sci.* 1995;36:337–344.
66. Inatomi T, Spurr-Michaud S, Tisdale A, Gipson I. Human corneal and conjunctival epithelia express MUC1 mucin. *Invest Ophthalmol Vis Sci.* 1995;36:1818–1827.
67. Feenstra R, Tseng S. What is actually stained by rose bengal? *Arch Ophthalmol.* 1992;110:984–993.
68. Feenstra R, Tseng S. Comparison of fluorescein and rose bengal staining. *Ophthalmology.* 1992;99:605–617.
69. Prydal J, Artal P, Woon H, Campbell F. Study of human precorneal tear film thickness and structure using laser interferometry. *Invest Ophthalmol Vis Sci.* 1992;33:2006–2011.
70. Prydal J, Campbell F. Study of precorneal tear film thickness and structure by interferometry and confocal microscopy. *Invest Ophthalmol Vis Sci.* 1992;33:1996–2005.
71. Slomiany BL, Slomiany A. Role of mucus in gastric mucosal protection. *J Physiol Pharmacol.* 1991;42:147–161.
72. Dilly P. Structure and function of the tear film. *Adv Exp Med Biol.* 1994;350:239–347.

2

INFLUENCE OF GENDER, SEX STEROID HORMONES, AND THE HYPOTHALAMIC-PITUITARY AXIS ON THE STRUCTURE AND FUNCTION OF THE LACRIMAL GLAND

David A. Sullivan, L. Alexandra Wickham, Eduardo M. Rocha, Robin S. Kelleher, Lilia Aikawa da Silveira, and Ikuko Toda

Schepens Eye Research Institute and
Department of Ophthalmology, Harvard Medical School
Boston, Massachusetts

1. GENDER-RELATED DIFFERENCES IN THE LACRIMAL GLAND

Throughout the twentieth century it has become increasingly apparent that males and females are different, and not just in terms of physical characteristics. Scientists have discovered that fundamental, gender-related differences exist in almost every cell, tissue and organ of the body, including those associated with respiration, digestion, metabolism, circulation, renal function, and neural and endocrine activity. Indeed, during a recent five year period, at least 8,159 scientific reports were published that addressed the basic and/or clinical influence of gender on health and disease (Table 1).

Of interest, very few of these publications address the eye (Table 1). This lack of research is somewhat surprising, given that gender appears to exert a significant impact on the eye, including such tissues as the conjunctiva, cornea, anterior chamber, lens and retina, and may influence the development and/or progression of keratoconjunctivitis sicca (KCS; dry eye syndromes), vernal keratoconjunctivitis, glaucoma, macular holes and macular degeneration.[1] However, there is one ocular tissue with innate, gender-related differences that has received considerable attention, and that is the lacrimal gland.

Over the past 55 years, research investigators have found that that striking, gender-associated variations are present in the structure, function and pathology of the lacrimal gland in a number of species, including mice, rats, hamsters, guinea pigs, rabbits and/or humans. Thus, depending upon the species, distinct differences may occur between lacri-

Table 1. Number of medical research articles related to gender published during the period 1990 to 1995

Gender 8159					
Disease	1372	Muscle	170	Vision	27
Health	1088	Kidney	120	Foot	26
Sex	924	Sleep	118		
Blood	815			Lens	14
Heart	579	Eye	73	Cornea	8
Brain	278	Speech	70	Optic nerve	6
Lung	203			Anterior chamber	2
Liver	192	Mouth	32	Conjunctiva	0
Skin	144	Ear	29	Meibomian gland	0

Data in this Table were obtained from a "Medline" computer search.

mal tissues of males and females in the mean area and density of acinar complexes, the membrane contours, cytoplasmic appearance and vesicular content of acinar epithelial cells, the position, size, and shape of acinar cell nuclei, the quantity of intranuclear inclusions, the prominence of nucleoli, the frequency of intercellular channels, the number of capillary endothelial pores, the population density of lymphocytes, the expression of secretory immunity, the response to α-adrenergic stimulation, the levels of numerous mRNAs (e.g. for cytokines, proto-oncogenes, apoptotic factors, hormone receptors and other proteins), the production, amount, activity and/or affinity of various proteins, enzymes and receptors, the secretion of specific proteins, and the susceptibility to viral replication, focal adenitis, atrophy and autoimmune disease (e.g. Sjögren's syndrome) (Table 2).[2–50] In addition, gender-related differences in lacrimal gland morphology appear to become more profound during aging[7,9,10,35] and, depending upon the species, may be paralleled by a relative decrease[9,10] or increase[25] in DNA content, as well as a reduction in acinar cell turnover,[10] in male lacrimal tissue.

In effect, gender influences not only the anatomy and physiology of the lacrimal gland, but also such indices as the expression of this tissue's mucosal immune system, which is designed to protect the ocular surface against viral and bacterial infection,[51] and the incidence of various aqueous tear deficiencies (e.g. lacrimal gland dysfunction in Sjögren's syndrome).[23,52] Particularly intriguing are two observations: first, that the transcription of certain genes (e.g. cystatin-related protein [CRP], pancreatic lipase), and the translation of the corresponding proteins (e.g. CRP, also a 20 kDa protein), may be completely gender-restricted.[26,31,50] Thus, some genes appear to be expressed only in lacrimal tissues of males or females, but not of both sexes; and second, that fundamental, gender-related differences may exist in the susceptibility of lacrimal gland epithelial cells to programmed cell death. Recent findings concerning the expression of proto-oncogenes and apoptotic factor mRNAs in lacrimal tissues (e.g. high Fas antigen, low bcl-2 of males) suggest that epithelial cells in lacrimal glands of males may have a higher rate of apoptosis and turn-over than those of females.[40,41]

However, as additional considerations, it is important to note that: (1) most of the gender variations in the lacrimal gland have been documented in non-human species. Relatively few studies have been performed to assess the comparative structural profile of lacrimal tissues in human males and females. Moreover, no investigations appear to have been conducted exploring gender's possible impact on the functional activity of the human lacrimal gland. This absence of 'human' information may be due to the difficulty in obtaining sufficient amounts of viable, experimental tissue; (2) no consistent gender-re-

Table 2. Gender-related differences in the tear film and in the morphology, biochemistry, immunology, molecular biology and secretion of the lacrimal gland

Male	Female	Species
Morphology		
Large and irregular acini with wide lumina	Smaller, more regular acini with narrow lumina	Mouse, rat, guinea pig and/or rabbit
Glandular epithelial cells with cloudy, light granular and basophilic cytoplasm	Epithelial cells with clearer and less structured cytoplasm with heavy basophilic staining around nucleus (lighter toward periphery)	
Centrally located nucleus varying considerably in size and shape	Basally-situated nucleus showing more regularity in size and shape	
Distinct nuclear polymorphism		
Elevated number of polyploid nuclei		
	Acinar cell contours more conspicuous	
Nuclei frequently harbour prominent nucleoli		
Cell borders are either indistinct or not evident	Cell borders are clear and lobulated	
Basal vacuoles and enhanced quantity of intranuclear inclusions in acinar cells	Numerous, large cytoplasmic vesicles	
Sparse intercellular channels	Frequent intercellular channels	
Specialized structure of Golgi fields		
Capillary endothelia display few pores	Capillary endothelia typically show pores	
Increased labelling index of epithelial cells suggesting decreased cell turnover during aging		
Larger acinar area		Human
Greater extent of harderianization		Rat
Marked sexual dimorphism during aging		Rat
More frequent lobular fibrosis and focalatrophy in elderly	More frequent diffuse fibrosis and diffuseatrophy in elderly	Human
Biochemistry		
Higher number and affinity of β-adrenergic binding sites	Greater amounts of melatonin and N-acetyltransferase	Hamster
Higher total quantity of β-adrenergic receptors	Higher specific activity of Na^+,K^+-ATPase, cholinergic receptors, acid and alkaline phosphatase, and galactosyltransferase	Rabbit (note: different results at younger ages)
Higher activity of hydroxyindole-o-methyltransferase		Hamster
Greater level of androgen receptor protein and higher activity of carbonic anhydrase	Greater content of leucine aminopeptidase after puberty	Rat
	Higher peroxidase activity and 20kDa protein levels	Hamster
Immunology		
Higher synthesis of secretory component (SC)		Rat
Greater production of immunoglobulin A (IgA)		Rat
Increased number of IgA-containing cells after puberty		Rat
	Higher susceptibility to cytomegalovirus invasion and/or replication	Rat
	Greater incidence of focal adenitis, and higher yet in females > 45 years old	Human
	Greater incidence of autoimmune disease	Mouse, Human

(continued)

Table 2. (*Continued*)

Male	Female	Species
Molecular biology		
Greater levels of α2u-globulin, SC, cystatin-related protein, TGF-β1 mRNAs	Higher content of androgen receptor mRNA	Rat
Greater amounts of Fas antigen and mouse urinary protein mRNA	Greater levels of bcl-2, c-myc, c-myb, p53, androgen receptor, IL-1β, TNF-α and pancreatic lipase mRNA (some differences are strain-dependent)	Mouse
Secretion and tears		
Higher secretion and tear levels of SC, IgA and cystatin-related protein		Rat
Greater phenylephrine-induced secretion of peroxidase and total protein by gland *in vitro*		Rat
Higher tear levels of total protein, IgA and 42 kDa and 46 kDa proteins	Higher 90 kDa protein levels	Mouse
	Higher 20 kDa protein levels	Hamster
Greater amounts of EGF, TGF-α and gender-specific tear protein	Lower non-invasive tear break-up time, and an an increase in tear osmolality during aging	Human
Higher tear osmolality (particularly < 41 years old)		Human

The research findings cited in this table were obtained from the following references: 2-50, 54-58. This table was modified from Table 3 in reference 23.

lated differences exist, at least in rats, in the total amount of lacrimal gland protein;[9] (3) no consistent differences appear to occur between 'healthy' males and females in the production, total volume or total protein content of tears;[15,29,53,54] and (4) although the absolute weight of lacrimal tissue is often greater in males,[16,21,40,41] the lacrimal gland/body weight ratio may be lower, equal or higher in females.[16,40,41]

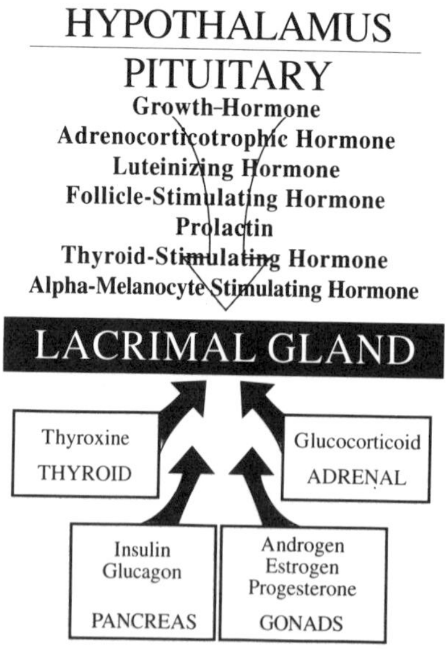

HYPOTHALAMUS
PITUITARY
Growth-Hormone
Adrenocorticotrophic Hormone
Luteinizing Hormone
Follicle-Stimulating Hormone
Prolactin
Thyroid-Stimulating Hormone
Alpha-Melanocyte Stimulating Hormone
LACRIMAL GLAND
Thyroxine
THYROID
Glucocorticoid
ADRENAL
Insulin
Glucagon
PANCREAS
Androgen
Estrogen
Progesterone
GONADS

Figure 1. Hormones from the hypothalamic-pituitary-gonadal axis that influence the structure and/or function of the lacrimal gland.

Nevertheless, despite these considerations, gender does exert a significant influence on lacrimal tissue. The underlying basis for this sexual dimorphism in the lacrimal gland is not completely known, but has been attributed most often to the influence of hormones from the hypothalamic-pituitary-gonadal axis (Figure 1). The nature and extent of these endocrine effects, with a particular focus on sex steroids (e.g. androgens, estrogens, progestins) and pituitary hormones, are reviewed below.

2. HORMONAL REGULATION OF THE LACRIMAL GLAND

2.1. Androgens

2.1.1. Influence of Orchiectomy and Androgen Administration on the Lacrimal Gland. Perhaps the most potent hormones known to control the lacrimal gland are androgens. These sex steroids modulate the morphology, biochemistry, physiology, immunology, molecular biology and/or secretion of lacrimal tissue in mice, rats, hamsters, guinea pigs, rabbits and/or humans.[23,59] Moreover, androgen action may be responsible for many of the gender-related differences in the lacrimal gland.[23,60]

Insight into the nature and magnitude of androgen influence may be gained by briefly summarizing the results of investigations into the effects of castration and/or androgen replacement on the lacrimal gland. Thus, during the past 50 years scientists have reported that orchiectomy, or treatment with androgen receptor antagonists, cause degenerative changes in lacrimal tissue, including decreased growth and activity, disappearance of glandular elements, an attenuation in acinar cell size, diminished cytoplasmic basophilia, a loss of nuclear polymorphism, reduced nuclear volume, an alteration in nuclear

Table 3. Influence of orchiectomy and androgen treatment on the structure, function and secretion of the lacrimal gland in mice, rats, hamsters, guinea pigs and/or rabbits

Orchiectomy	*Androgen treatment*
• Alteration in glandular appearance, • Proliferation of interfollicular connective tissue • Decreased glandular tissue • Diminished cytoplasmic basophilia • Reduced size of acinar cells and nuclei • Loss of cellular and nuclear polymorphism • Fewer basophilic glandular cells • Diminished or increased harderianization • Reduced alkaline phosphatase activity • Increased N-acetyltransferase and hydroxyindole-o-methyl-transferase activity • Alterations in levels of various mRNAs and proteins • Increased acinar epithelial cell susceptibility to cytomegalovirus infection • Reduction in growth (80 days after orchiectomy) • Change in structure to neutral (40 days after orchiectomy) or female type morphology • Alteration in fluid or specific protein secretion	• Reversal of the influence of orchiectomy on glandular structure, function and secretion • Induction of acinar cell and parenchymal hyperactivity • Enlargement of glandular vesicles • Generation of numerous glycoprotein-secreting cells • Production of mucus and highly polymerized carbohydrates • Appearance of PAS-positive material in acinar cells and central lumina • Suppression of glandular inflammation • Transformation of the glandular acino-serous structure into a "vesicular mucus" structure

Please note that, as mentioned in the text, several of the above listed changes may be species- and/or strain-dependent. The data in this table were obtained from the following references: 2,6,12,14,18-20,23,24,26,28-34,36,37,39-43,49,60-91,95,97-101,221.

shape, a proliferation of connective tissue, an increased susceptibility to cytomegalovirus infection, variations in the levels of immune proteins, enzymes and mRNA species, changes in fluid and specific protein secretion, and a transformation of histological appearance into a neutral or 'female' type (Table 3).[2,6,12,18–20,23,26,28–33,36,37,42,43,49,60–67] Conversely, researchers have also reported that androgen administration reverses the influence of orchiectomy on lacrimal gland structure and function, and may lead to parenchymal and acinar epithelial cell hyperactivity, the generation of numerous glycoprotein secreting cells, an enlargement of glandular vesicles, the appearance of PAS-positive material in acinar cells and central lumina, the synthesis of mucus and highly polymerized carbohydrates, the modulation of mRNA and protein levels, the suppression of inflammation, an alteration of the glandular acino-serous structure into a "vesicular mucus" shape, and a change in tear protein and fluid output (Table 3).[2,6,14,18–20,24,26,29–34,36,37,39–42,49,64–91,95,221]

2.1.2. Mechanism(s) of Androgen Action on the Lacrimal Gland. The mechanism(s) underlying these androgen actions, which may be species- and strain-dependent,[49] appears to be mediated primarily through a hormone interaction with saturable, high-affinity and steroid-specific binding sites within epithelial cell nuclei; these receptors are members of the steroid/thyroid hormone/retinoic acid family of ligand-activated transcription factors.[92] Classically, the monomeric, activated hormone-receptor complex associates with a response element in the regulatory region of specific target genes, then dimerizes with another sex steroid-bound complex and, in combination with appropriate co-activators, tissue-specific and basal promoter elements, modulates gene transcription and eventually protein synthesis.[92,93] In support of this 'classical' mechanism (Figure 2), studies have

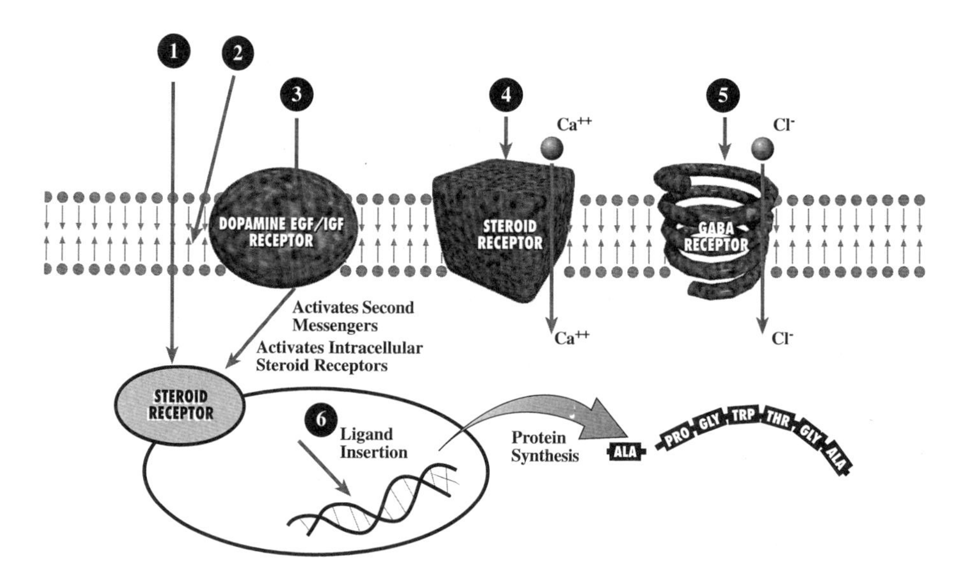

Figure 2. Classical and non-classical mechanisms of sex steroid hormone action. The numbers represent the following pathways: (1) classical hormone binding to intranuclear receptors; (2) non-specific hormone effect on membrane fluidity; (3) activation of steroid receptors by growth factors; (4) hormone association with specific membrane receptors; (5) interaction with membrane neurotransmitter receptors; and (6) insertion of the hormone between DNA base pairs. This figure is adapted from Figure 4 in reference 105, with permission from the Journal of Steroid Biochemistry & Molecular Biology. A detailed description of these pathways and their impact on cellular processes is described in that article (i.e. 105).

demonstrated that: (1) androgen receptor mRNA exists in lacrimal tissues of mice, rats, hamsters, guinea pigs, rabbits and humans;[1,40,42,94] (2) androgen receptor protein is present within epithelial cell nuclei of lacrimal glands in all species examined, including mice, rats and hamsters;[36,42,95] (3) rat lacrimal glands contain a single class of saturable, high-affinity and steroid-specific androgen binding sites, which have a dissociation constant and stereochemical selectivity analogous to those found in numerous cells and tissues throughout the body;[94,96] (4) androgen-receptor complexes in lacrimal tissue adhere to DNA;[96] and (5) androgens regulate the accumulation of various mRNA species, and induce or suppress the synthesis of many proteins, in the lacrimal gland (Table 4).[14,20,26,30–32,34,36,37,39–43,61,66,67,75,76,90,91,95,97–101,221] The functional nature of these androgen binding sites is further indicated by the observation that androgen activity in rat lacrimal tissue or acinar epithelial cells may be inhibited by antagonists of,[18,100] or mutations within,[67] androgen receptors, as well as by inhibitors of transcription[100] and translation.[30] Of interest, androgens control the level of their own receptors in the lacrimal gland, by increasing the amount of receptor protein and decreasing the content of receptor mRNA.[36,40,42,95] This type of autoregulation is also found in other androgen target organs (e.g. prostate[102–104]).

It should be noted that androgens may also act on lacrimal tissue through non-classical pathways (Figure 2). These pathways, which are very rapid (i.e. occuring in seconds or minutes), involve alterations of membrane fluidity, regulation of neurotransmitter receptors and interaction with stereospecific plasma membrane receptors.[105] Androgens might also influence the lacrimal gland in humans by associating with sex-hormone binding

Table 4. Influence of androgens on the levels of specific mRNAs and proteins in the lacrimal glands of mice, rats, hamsters or rabbits

	mRNA	Protein
Increase		
	Secretory component	Secretory component
	α-2u globulin	Immunoglobulin A
	Mouse major urinary protein	Cystatin-related protein
	Transforming growth factor-β*	Transforming growth factor-β *
	Bax	C3 component of prostatic binding protein
		Androgen receptor
		Na^+,K^+-ATPase activity (total)
		Acid phosphatase activity (total)
		Alkaline phosphatase activity (total)
		β-Adrenergic receptor activity (total)
		Seminal vesicle secretion VI protein †
Decrease		
	Androgen receptor	Leucine aminopeptidase
	Estrogen receptor	Cholinergic receptor activity (total)
	Interleukin-1β	20 kDa protein
	Tumor necrosis factor-α	
	bcl-2	
	c-myb (MRL/lpr mouse)	
	PRL-inducible protein/ gross cystic disease fluid protein-15	

*The androgen-induced increases in the levels of transforming growth factor-β mRNA and protein were found in the lacrimal glands of rats and mice, respectively; †Extrapolated from data which showed that the amount of this protein decreased following orchiectomy; Data in this Table are from References: 14,20,26,30-32,34,36,37,39-43,61,66,67,75,76, 90,91,95,97-101,221.

globulin, which in turn may interact with a cell surface receptor. This binding would result in stimulation of adenylyl cyclase, production of cAMP, activation of protein kinase A, phosphorylation of a cAMP response element binding protein and eventually modulation of a gene's cAMP response element, thereby influencing transcription.[106] Whether these 'non-classical' pathways are operative in the lacrimal gland, though, remains to be established. Such routes could theoretically account for androgen effects on certain cell membrane phenomena, such as the stimulation of Na^+,K^+-ATPase activity.[89] It is of interest that this latter androgen action also occurs in the prostate,[107] but whether the hormone's effect on Na^+,K^+-ATPase activity in lacrimal tissue is paralleled by a corresponding increase or decrease in the expression the enzyme's α-subunit protein[e.g. 108] has yet to be determined.

Very recently, the mRNA for the alpha subunit of androgen-binding protein, a low molecular weight protein, has been identified in lacrimal glands of male and female mice.[222] Whether this mRNA is translated, and plays a role in androgen action, is unknown.

2.1.3. Potential Clinical Relevance of Androgen Effects on the Lacrimal Gland. It is possible that the impact of androgens on the lacrimal gland may have significant clinical relevance in the prevention and/or treatment of disease. Examples might be the influence of androgens on mucosal immunity and autoimmune disease in the eye. Thus, as shown in experimental animals, androgens control the ocular secretory immune system, which defends the anterior surface of the eye against microbial infection and toxic challenge.[51] This immune function is mediated primarily through secretory IgA (SIgA) antibodies, which originate (as polymeric IgA) from plasma cells in the lacrimal gland,[109,110] are transported across epithelial cells into tears by secretory component (SC),[91] and serve to inhibit viral invasion, prevent bacterial colonization, interfere with parasitic infestation and suppress antigen-induced damage on the ocular surface.[51] In rats, androgens induce a striking increase in the synthesis and secretion of SC by lacrimal gland acinar epithelial cells,[20,30,43,97–101] elevate the concentration of IgA in lacrimal tissue[91] and significantly enhance the transfer and accumulation of SC and IgA in tears.[29,30,32,87,88,91] Androgens also augment the lacrimal gland output of IgA, presumably (at least in part) through SC regulation, into tears in several strains of mice.[73] Given that the promoter region of the human SC gene has recently been shown to contain several putative androgen receptor binding sites,[111] it may be that androgens may also increase SC production in the human lacrimal gland. If so, these hormones might increase SIgA transport into human tears and thereby promote the maintenance of corneal and conjunctival integrity and the preservation of visual acuity.

Another clinically relevant aspect of androgen-lacrimal gland interactions may be the effect of these hormones on autoimmune disease in lacrimal tissue. Recent research has demonstrated that androgen therapy dramatically suppresses the inflammation in, and stimulates the functional activity of, lacrimal glands in mouse models of Sjögren's syndrome.[68–73] This syndrome is an insidious and currently incurable autoimmune disorder, that occurs almost exclusively in females (> 90%), and is associated with an extensive lymphocyte accumulation in the lacrimal gland, significant changes in the expression of cytokines, adhesion molecules and proto-oncogenes, epithelial cell destruction and/or dysfunction, a precipitous decrease in tear secretion and severe dry eye.[23,38–41,52,112–117] The immunosuppressive effect of androgens appears to be mediated in part through a hormone interaction with receptors in epithelial cell nuclei, which then cause an altered expression of cytokines (e.g. IL-1β, TNF-α) and apoptotic factors (e.g. proto-oncogenes bcl-2 and Bax) in the lacrimal gland.[23,39–41,95] Of particular importance, this ability of androgens to diminish inflammation and enhance function in autoimmune lacrimal tissue suggests that the targeted delivery (e.g. topical) of these hormones might potentially provide a safe and effective treatment for the lacrimal gland dysfunction and aqueous tear deficiency in Sjögren's syndrome patients.[23] In support of this hypothesis, investigators of three different, albeit uncontrolled, clinical studies reported that systemic androgen administration led to a precipitous decline in dry eye signs and symptoms and to a considerable rise in tear flow in individuals with Sjögren's syndrome.[84–86] However, whether the topical application of androgens to Sjögren's syndrome patients will elicit analogous ocular effects remains to be elucidated. If so, such an approach would compensate, at least in part, for the reduced androgen levels in these patients,[118] and also avoid the dangerous side effects of systemic exposure to exogenous androgens.[23,119]

2.1.4. Qualifications and Considerations Regarding Androgen-Lacrimal Gland Interactions. Several qualifications and considerations should be mentioned concerning the nature and extent of androgen influence on the lacrimal gland:

2.1.4.1. Androgens and the Human Lacrimal Gland. Whether androgens regulate the human lacrimal gland in a manner similar to that of other species is unclear. As of today, no studies have been published regarding the possible direct effects of androgens on the structure or function of human lacrimal tissue.

2.1.4.2. Androgen Influence on Lacrimal Gland Structure. In contrast to the reported effects of androgens on lacrimal gland structure (see section 2.1.1.), other studies have found that neither orchiectomy, nor the administration of androgens, has any impact on this tissue's growth, acinar cell or nuclear characteristics or histological appearance.[3,120] This apparent discrepancy in findings may be attributed, at least in part, to variations in experimental design, including differences in the age, sex, and endocrine status of animals, as well as in the dosage and time course of androgen treatment, and even in the methods of analysis. However, a more important factor seems to be species. Thus, research has shown that androgens have no consistent effect on the growth, or acinar complex features, of lacrimal tissues in various species: androgen treatment does increase the lacrimal gland weight/body weight (LGW/BW) ratio, as well as acinar area in intact female mice,[49,68,69,71,72] but rarely alters these variables in lacrimal glands of orchiectomized or ovariectomized rats, guinea pigs or rabbits.[49] One exception has been observed with castrated male or female guinea pigs, wherein androgen exposure for one week significantly reduced the LGW/BW ratio (Table 5).[49] It may be that prolonged androgen treatment may enlarge the lacrimal gland, as has been found after 7 to 15 weeks of androgen administration in rabbits,[78,79] but shorter term androgen exposure, which is sufficient to regulate many functional changes in lacrimal tissue,[14,20,26,29–34,36,37,39–43,66–76,87–91,95,97–101] does not elicit a species-independent change in lacrimal gland weight.[49] Overall, it appears that androgens do modulate certain structural aspects of lacrimal tissue, such as membrane contours or nuclear characteristics (see section 2.1.1.). However, androgen regulation of the lacrimal gland is unlike that of the ventral prostate, which is completely dependent upon androgens for size maintenance and undergoes involution and programmed cell death following androgen removal.[121]

2.1.4.3. Androgen Impact on fluid and Total Protein Secretion by the Lacrimal Gland. Although androgens clearly regulate the lacrimal gland output of specific proteins,[29,30–32,73,87,88,91] these hormones do not elicit a consistent, species-independent action

Table 5. Effect of androgen treatment on the lacrimal gland weight/body weight ratio and the area and density of acinar complexes

Species		LGW/BW *	Acinar area**	Acinar density	Reference
Mouse (Female)†					
	MRL/lpr	↑	↑	↓	49,68,71,72
	NZB/NZW F1	↑			69
	BALB/c	↑			49
	C57BL/6	↑			49
	C3H/lpr	↑			38
	C3H/HeJ	↑			41
Rat‡					
	Male	—	—	—	49
	Female	—			49
Guinea Pig§					
	Male	—↓	—	—	49
	Female	—↓	—	—	49
Rabbit¶					
	Male	—	—	—	49
	Female	—	—	—	49

*The abbreviation "LGW/BW" represents the lacrimal gland weight/body weight ratio; **"Acinar" refers to the acinar complex; †Mice were intact and treated with androgens for 17 to 54 days; ‡Rats were castrated and administered androgens for 1 to 17 days; §Guinea pigs were castrated and exposed to testosterone for 4 or 7 days. A decrease in the LGW/BW was observed after 7, but not 4, days of hormone treatment; ¶Male and female rabbits were castrated and treated with testosterone for 4 or 7 days.

on fluid or total protein secretion.[49] Rather, androgens induce time-, species- and strain-dependent effects, leading to a non-uniform increase, decrease or no impact on the volume and total protein level of tears in mice, rats, guinea pigs and rabbits (Table 6).[49,65,69,71,73,122]

In humans, systemic androgen administration to patients with KCS or Sjögren's syndrome appears to increase the tear volume.[78,79,84–86] This response may be due to the: [a] the anti-inflammatory effect of androgens on the lacrimal gland (e.g. in Sjögren's syndrome patients), which may alleviate the immune-related damage to acinar and ductal epithelial cells and permit at least basal aqueous tear secretion;[23] and [b] the lipid-stimulating action of androgens on the meibomian gland, which may enhance tear film stability and reduce tear film evaporation (Figure 3).[123] This explanation seems to make sense physiologically, given that many patients with aqueous tear deficiency (e.g. due to lacrimal gland defects in Sjögren's syndrome, systemic lupus erythematosus or rheumatoid arthritis) or evaporative dry eye (e.g. due to meibomian gland dysfunction during menopause, aging or anti-androgen treatment) have diminished serum androgen concentrations.[223]

It has also been suggested that the diminished androgen levels that occur during menopause, pregnancy, lactation or the use of estrogen-containing oral contraceptive may predispose women to the development of a type of non-immune dry eye, termed primary lacrimal gland deficiency (PLD).[89,122,124,125] In effect, this hypothesis suggests that androgens are critical for the maintenance of fluid output by the human lacrimal gland, and that androgen insufficiency may lead to the generation of PLD.[89,122,124–126] However, if this hypothesis is correct, then it may reflect a hormone action that is gender-specific (i.e. females, not males). A recent study has shown that extended (e.g. years) exposure to anti-androgen treatment for prostatic indications had no effect on the aqueous tear production of these patients (i.e. Schirmer's test), as compared to that of age-matched, untreated controls.[123] These 'anti-androgen' patients, though, did suffer from meibomian gland dysfunction, decreased tear film breakup time and functional dry eye.[123] Thus, androgen defi-

Table 6. Effect of androgen administration on the volume and total protein content of tears

Species	Tear volume	Tear protein	Reference
Mouse (Female)*			
MRL/lpr	—	↑	49,71,73
NZB/NZW F1	↑	↑	49,69,73
BALB/c	—	↑	49
C57BL/6	↓		49
Rat†			
Male	↓	—	49,65
Female	—	—	49
Guinea Pig‡			
Male	—	—	49
Female	—	—	49
Rabbit			
Male§1	—	—	49
Male§2	↑		83
Female§3	↓		122
Female§1	—	—	49
Female§4	↑		83,122
Human¶			
Male	↑	—	78,79,84,86
Female	↑	—	78,79,84,85

*Mice were intact and treated with androgens for 17 to 54 days; † Rats were castrated and administered androgens for 4 or more days. Acute treatment (24 hours) of orchiectomized rats with physiological levels of testosterone had no effect on either the volume or total protein content of tears, when these values were standardized to pretreatment levels; ‡ Guinea pigs were castrated and exposed to testosterone for 4 or 7 days; §1 Male and female rabbits were castrated and treated with testosterone for 4 or 7 days; §2 Male and female rabbits were intact and administered high doses of androtestone for over 6 weeks; §3 Female rabbits were castrated and treated with dihydrotestosterone (DHT) for an unreported number (several?) of days; §4 Female rabbits were either intact and administered high doses of androtestone for over 6 weeks,[83] or castrated and treated with DHT for an unreported number (several?) of days, as well as pilocarpine;[122] ¶ Humans were intact and treated with androgens for varying intervals of time.

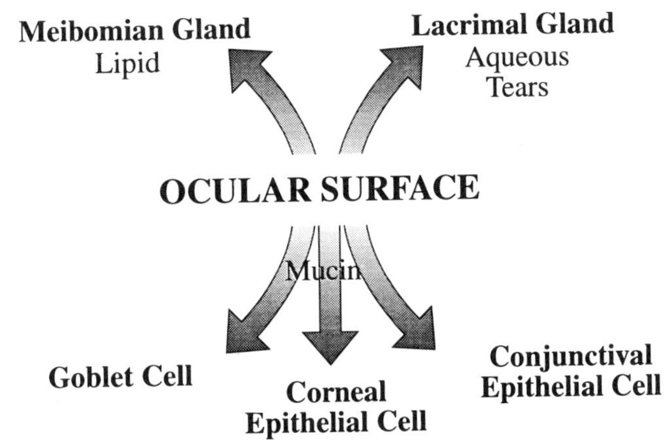

Figure 3. Potential clinical impact of androgens on the lacrimal and meibomian glands.[223] As explained in section 2.1.4.3., the topical administration of androgens may serve as a possible treatment for aqueous-deficient (e.g. due to lacrimal gland dysfunction in Sjögren's syndrome)[23] and evaporative (e.g. due to meibomian gland dysfunction during menopause and aging)[123] dry eye syndromes.

ciency in males, who show no evidence of immune pathology (e.g. autoimmune disease), does not appear to promote the development of PLD.

2.1.4.4. Role of Androgens in the Etiology of Autoimmune Disease in the Lacrimal Gland. A recent report suggests that androgen insufficiency may trigger an autoimmune process in the lacrimal gland, resulting in lymphocyte infiltration and pathological characteristics similar to those encountered in Sjögren's syndrome.[124] In support of this hypothesis, androgen deficiency does seem to be a risk factor for the development of Sjögren's syndrome,[23,52,117,127] as well as its corresponding KCS,[118] and decreased androgen levels are apparently associated with an immigration of certain immune cells into rabbit lacrimal tissue.[128] However, whether low concentrations of serum androgens are the cause, or lead to, the extensive, multifocal lacrimal gland inflammation found in Sjögren's syndrome is questionable. As shown in other studies, no lymphocyte accumulation, and no gross morphological changes, occur in lacrimal glands of male or female rats, guinea pigs or rabbits for periods of up to one month after castration.[49] Discrete lymphocyte infiltrates may be found in rat lacrimal glands 40 days after orchiectomy, but these small foci, when present, remain following androgen treatment.[19] Thus, although androgens suppress immunopathological lesions in lacrimal tissues of autoimmune (e.g. Sjögren's syndrome) animals,[23] current evidence does not support the hypothesis that attenuated androgen levels alone may induce such overwhelming inflammation in the lacrimal gland.

2.1.4.5. Modulation of Androgen Action in the Lacrimal Gland. Androgen action on the lacrimal gland may be enhanced or inhibited by neurotransmitters, cytokines, secretagogues, autocoids and changes in the endocrine environment (Table 7). For example, research has demonstrated that vasoactive intestinal peptide (VIP), the β-adrenergic agent, isoproterenol, IL-1α, IL-1β and tumor necrosis factor-α (TNF-α) all increase the androgen-stimulated synthesis and secretion of SC by rat acinar epithelial cells.[98–100] These neuroimmune effects on SC production may be mediated through the regulation of intracellular adenylate cyclase and cyclic AMP (cAMP) activity. In support of this hypothesis, VIP, adrenergic agonists, IL-1 and TNF-α are known to increase the generation of cellular cAMP.[129,130] Moreover, exposure of lacrimal gland acinar cells to cyclic AMP analogues (e.g. 8-bromoadenosine 3':5'-cyclic monophosphate), cyclic AMP inducers (e.g. cholera toxin and prostaglandin E_2) or phosphodiesterase inhibitors (e.g. 3-isobutyl-1-methylxanthine) promotes SC synthesis in the absence, and/or presence, of androgens.[99] It may be that cAMP may directly affect androgen dynamics, given that a functional cAMP response

Table 7. Factors that influence androgen action on the lacrimal gland

Vasoactive intestinal peptide	Insulin
β-Adrenergic agonist	Glucocorticoids
cAMP Analogues (e.g. 8-bromoadenosine 3':5'-cyclic monophosphate)	Estrogens
cAMP inducers (e.g. cholera toxin, prostaglandin E_2)	Progestins
Phosphodiesterase inhibitors	Retinoic acid (?)
Cholinergic agonist	Prolactin
Interleukin-1α	Hypothalamic-pituitary factors
Interleukin-1β	Thyroid gland factors
Tumor necrosis factor-α	Adrenal gland factors
Extracellular calcium	High-density lipoprotein

Information in this Table was obtained from the following references: 20,42,87,89,91,97-100,133

element exists in a promoter of the androgen receptor gene, and that 8-bromoadenosine 3':5'-cyclic monophosphate may upregulate androgen receptor mRNA levels.[131,226] In contrast, the cholinergic agonist, carbamyl choline, acutely (i.e. hours) enhances, but chronically (i.e. days) reduces, androgen-induced SC production by rat lacrimal gland acinar epithelial cells.[99,100] The transient effect of this cholinergic agent may be mediated through the mobilization of intracellular calcium, the activation of protein kinase C and the rapid enhancement of cellular secretion.[100,129] The mechanism underlying the long-term inhibitory action of carbachol, which may be prevented by atropine,[100] is unknown, but may involve an attenuation of cAMP levels[132] or an alteration of SC gene activity.[100] In addition to these compounds, insulin, glucocorticoids, extracellular calcium, high-density lipoprotein, and factors from the pituitary, thyroid and adrenal glands may also modify androgen action on SC synthesis and secretion by lacrimal tissue.[20,87,91,97,98,133]

Other actions of androgens on the lacrimal gland may possibly be modulated by progestins or corticosteroids, given that these hormones increase or decrease, respectively, the content of androgen receptor protein in lacrimal tissue.[42] Similarly, retinoic acid may also influence androgen effects, considering that this compound may elevate or depress androgen receptor levels in other cells.[134,135] Lastly, a high dose of prolactin appears to reduce the stimulatory impact of androgen treatment on Na^+,K^+-ATPase and acid phosphatase activities in rat lacrimal tissue.[89]

2.1.4.6. Androgen Regulation of Responsive Genes. The extent of androgen regulation of responsive genes in non-ocular tissues is critically dependent upon the presence of appropriate transcription factors and co-activators.[92] However, with the exception of one recent report,[136] it appears that almost no information is available concerning these androgen regulatory elements in the lacrimal gland or in any other part of the eye. To date, it is known that transcription factors that bind to the promoter region of androgen target genes are present in the lacrimal gland, and that these proteins are different than found in the prostate.[136] This finding is intriguing, given that a series of androgen-controlled prostatic and seminal vesicle genes and proteins have recently been identified in the lacrimal gland[31,61,67,75,76,90] and may possibly be controlled in a site-specific manner (e.g. lacrimal tissue different than prostate). As concerns potential 'post'-gene effects, it remains to be determined whether androgens definitively influence such processes as post-transcriptional remodeling, mRNA stability, translational efficiency or protein longevity, which may occur in other exocrine tissues. (e.g. [137])

2.1.4.7. Androgen Receptor Mutations. Many defects are known to occur in sex steroid receptors, including splicing and point mutations, deletions, and termination codons, and these alterations may result in significant changes in the affinity or specificity of ligand binding, nuclear translocation, receptor dimerization, DNA association and transcriptional activation.[138,139] In fact, well over 100 different mutations have been identified in androgen receptors, which may lead to partial or complete insensitivity to androgens in various tissues.[139] How these androgen receptor variants may influence the lacrimal gland, or perhaps contribute to disease in this tissue, has yet to be clarified.

2.1.4.8. Androgen Metabolism. The ability of androgens to act on the lacrimal gland may be significantly altered by changes in metabolism. For example, if aromatase activity is enhanced in lacrimal tissue, then testosterone could be converted to 17β-estradiol.[140] Alternatively, if 5α-reductase activity is stimulated, then testosterone could be metabolized to the potent androgen, 5α-dihydrotestosterone.[141] As one further example, the adrenal androgen precursor, dehydroepiandrosterone, may be converted in tissues to either androgens or estrogens depending upon the prevailing metabolic pathways.[e.g. 142] At present, it is known that both Types 1 and 2 5α-reductase mRNA are present in human and rabbit lacrimal tissues,[143] and that this enzyme is apparently not required for testosterone's induction of SC synthesis in acinar epithelial cells of the rat lacrimal gland.[97] However, no further information appears to be available concerning possible androgen metabolic routes in the lacrimal gland. Such information is important for our understanding of androgen-lacrimal tissue interactions, given that these enzymes may change androgen action,[144] and that the activity of these metabolic enzymes may be augmented in autoimmune disease (e.g. increased oxidation of testosterone, thereby decreasing androgen levels), inflammation (e.g. promotion of aromatase function by pro-inflammatory cytokines, or stimulation of 5α-reductase activity by other factors) and stress (e.g. potentiation of androgen aromatization to estrogens by glucocorticoids).[e.g. 127,142,145–147]

2.2. Estrogens

2.2.1. Estrogen Effects on Lacrimal Gland Structure and Function. The precise nature and extent of estrogen action on the lacrimal gland is unclear, as well as controversial. A number of investigators have proposed that estrogens may be critical factors in the sexual di-

Table 8. Reported effects of ovariectomy, anti-estrogen treatment or estrogen administration on the structure, function and secretion of the lacrimal gland in rats, hamsters, rabbits and/or humans

Ovariectomy or anti-estrogen treatment	*Estrogen treatment*
• Lacrimal gland regression • Increased connective tissue • Decreased glandular tissue • Acinar cell disruption and vacuolization (8 weeks after surgery) • Reduced DNA content • Diminished total and membrane bound protein • Attenuated total activity of cholinergic receptors, β-adrenergic receptors and Na ,K -ATPase • Lymphocyte infiltration (20 weeks after surgery) • Alteration of morphological appearance to a "male" type (30 days after surgery) • Change in structural profile to a "neutral" type (40 days after surgery) • Development of KCS • Increased level of 20 kDa protein • Enhanced phenylephrine (α-adrenergic agonist)-induced peroxidase and total protein secretion • Reduced acinar epithelial cell susceptibility to cytomegalovirus infection	*'Positive'* • Restores glandular appearance to "female" type • Augments DNA and RNA levels • Induction of leucine aminopeptidase • Increased total activity of β-adrenergic receptors *'Neutral'* • No effect on the weight, morphology, peroxidase activity, total protein content or specific protein secretion of lacrimal tissue *'Negative'* • Lacrimal gland regression • Reduced estrogen receptor mRNA levels • Loss of PAS-staining • Suppression of 20 kDa protein content • Decreased acid and alkaline phosphatase activities • Attenuated total activities of cholinergic receptors and Na, K -ATPase • Diminished tear output

It is possible that several of the above listed changes may be species-dependent. The data in this table were obtained from the following references: 3,5,19,24,26,28,29,31,32,41-43,71,76,77,97,124,148-157.

morphism of the lacrimal gland, and that these hormones may exert a significant influence on the structure, function and/or secretion of this tissue.[3,19,77] In support of this hypothesis, ovariectomy, or other forms of estrogen deficiency (e.g. anti-estrogen exposure), have been reported to cause regressive changes in the lacrimal gland, including diminished glandular tissue, cell vacuolization, acinar cell disruption, a decreased quantity of membrane-associated protein, an attenuated total activity of β-adrenergic receptors, and a change in the morphological appearance to a neutral or "male" type (Table 8).[3,19,28,124,148–150] Estrogen loss has also been linked to the development of KCS and postmenopausal dry eye syndromes.[151–153] Conversely, estrogen treatment reportedly restores the lacrimal gland profile to that of intact females,[3,19,24,28] increases the content of β-adrenergic receptors in lacrimal tissue[149,154] and may antagonize certain androgen effects on lacrimal gland structure[81] (Table 8). However, a number of other researchers have reported that estrogen deficiency or administration has no influence on the weight, total protein content, peroxidase activity, morphological pattern or specific protein secretion of lacrimal tissue[5,24,29,31,32,76,97,155,156] and yet others have found that estrogenic compounds elicit glandular regression, a loss of PAS-staining, suppression of 20 kDa protein levels, a decline in acid and alkaline phosphatase activities and a decreased tear output[24,26,71,124,154,156,157] (Table 8). Indeed, one group of investigators has reported that both the administration[154] and the withdrawal (i.e. anti-estrogen exposure or ovariectomy)[148,149] of estrogens causes a reduction in the total activities of Na^+,K^+-ATPase and/or cholinergic receptors in lacrimal tissue of female rabbits. Thus, the actual impact of estrogens on the lacrimal gland requires further clarification. As concerns humans, no studies appear to have been performed that address the direct effect of estrogens on lacrimal tissue structure or function.

It should be noted that ovariectomy has also been reported to significantly increase the phenylephrine (i.e. α-adrenergic agonist)-induced secretion of peroxidase and total protein by the rat lacrimal gland *in vitro*, but this effect is apparently unrelated to estrogen deficiency.[5]

2.2.2. Mechanism of Estrogen Action on the Lacrimal Gland. It is possible that many of the proposed estrogen effects on the lacrimal gland may be mediated through 'classical' mechanisms (i.e. binding to intranuclear receptors; Figure 2). Estrogen receptor mRNA has been identified in lacrimal tissues of rats, rabbits and humans[1,42] and putative estrogen binding sites have been detected in a pooled cytosol preparation from rabbit lacrimal glands.[149] Moreover, recent studies have shown that the administration of 17β-estradiol to ovariectomized rats reduces the levels of estrogen receptor mRNA in lacrimal tissue,[42] an autoregulatory action that, as in other tissues,[158] would presumably occur through nuclear estrogen receptors. However, other investigations have been unable to find any evidence for the presence of saturable, high-affinity and specific binding sites for estrogens in male or female rat lacrimal glands.[94,155] Analyses have revealed the existence of apparent low-affinity receptors for estrogens in rat lacrimal tissue, but these measurements may actually reflect low-affinity association of estrogens to androgen binding sites.[94]

Another possibility is that estrogen effects on the lacrimal gland may be mediated through 'non-classical' pathways, such as influencing membrane fluidity or binding to membrane receptors (Figure 2). Alternatively (or in addition), estrogens may act through indirect processes, such as the control of the levels of certain pituitary hormones (e.g. prolactin, luteinizing or follicle-stimulating hormones), which apparently may modulate lacrimal gland structure and/or function.[89,159] Of interest, testosterone and dexamethasone both decrease the amount of estrogen receptor mRNA in lacrimal tissues of ovariectomized rats,[42] but the physiological meaning of this effect remains to be determined. Overall, further research is needed in order to delineate whether estrogens actually do possess the ability to act directly on the lacrimal gland, and if so, through what mechanism.

2.2.3. Estrogen Influence on the Tear Film. The effect of estrogens on the tear film is also controversial. Some studies have reported that the topical or systemic administration of estrogens to women with dry eye syndromes elicits a subjective amelioration of symptoms, a decrease in foreign body sensation and redness, an increase in Schirmer test values, an improvement in tear film break-up time, as well as enhanced corneal clarity and visual acuity[153,156,160–162] (Table 9). In contrast, other studies have reported that estrogens, or estrogen-containing oral contraceptives, have either no influence[65,163] or induce negative actions on the tear film, including lacrimal gland hyposecretion, decreased tear volume, diminished tear film break-up time, attenuated mucous production, foreign body sensation, contact lens intolerance and dryness in humans[71,156,157,164–167] (Table 9). In addition, research indicates that estrogens may be involved in the pathogenesis and/or progression of Sjögren's syndrome and systemic lupus erythematosus,[52,127,168–172] thereby possibly promoting the pathological manifestations in the lacrimal gland and the decreased aqueous tear output. Thus, it would seem that the utility of estrogens, which have been proposed for the treatment of dry eye syndromes,[153,162] remains to be definitively established. It would be helpful if future studies with estrogens clearly identified which ocular sites are specific targets for estrogen action, as well as clarify the exact cellular processes that may be controlled. This type of experimental approach with androgen research has resulted in the identification of ocular targets for androgens, information concerning the mechanisms of hormone action, as well as consideration of androgens as potential treatments for lacrimal[23] and meibomian gland dysfunction.[123,223]

2.3. Progestins

To date, little is known about the influence of progestins on lacrimal tissue. Progesterone may interfere with certain androgen-induced effects on lacrimal gland structure,[81] but not necessarily function.[88] Progesterone has also been demonstrated to increase the number of androgen receptor-containing acinar epithelial cells and decrease the level of

Table 9. Influence of topical or systemic estrogens, or estrogen-containing contraceptives, in dry eye patients

Positive	
	Induce subjective amelioration
	Promote clearer cornea and better vision
	Increase Schirmer test values
	Improve tear film breakup time
	Decrease foreign body sensation and redness
Neutral	
	No effect
Negative	
	Induce lacrimal hyposecretion
	Decrease tear volume
	Suppress mucous production
	Diminish tear film breakup time
	Elicit foreign body sensation, ocular "scratchiness" feeling
	Promote contact lens intolerance and dryness

Information in this Table was obtained from the following references: 65,71,153,156,157,160-167. For comparison, additional investigators have reported that birth control pills increase the blink rate in women,[224] and that topical estradiol may help relieve symptoms of postmenopausal chronic conjunctivitis.[225]

androgen receptor mRNA.[42] The mechanism of these progestin actions is unknown, but may be mediated through intranuclear receptors, given that progesterone receptor mRNA has been found in lacrimal glands of rats, rabbits and humans.[1,42] Of interest, if progesterone, as well as estrogen, receptor mRNAs are translated, then their receptor proteins may possibly be susceptible to activation by neurotransmitters, growth factors, secretagogues and pharmacological agents. Thus, several non-steroidal agents, including dopamine, cAMP and cyclosporine A for the progesterone receptor, and epidermal growth factor, insulin growth factor and IL-2 for the estrogen receptor, may cause these receptors to become transcriptionally active in the absence of the estrogen or progesterone ligand.[105,173,174]

2.4. Hormones from the Hypothalamic-Pituitary Axis

2.4.1. Effect of Hypophysectomy, Anterior Pituitary Ablation or Interruption of the Hypothalamic-Pituitary Axis on the Lacrimal Gland. Hormones originating from the hypothalamic-pituitary axis appear to exert a significant impact on the growth, differentiation, structural architecture and/or functional activity of the lacrimal gland.[89,159,175–178] In addition, these hormones may play a direct or indirect (e.g. control of sex steroid levels) role in promoting the sexual dimorphism of lacrimal tissue.[159]

The overall influence of this axis on the lacrimal gland is readily apparent in dwarf mice with deficient pituitary function, or in animals following hypophysectomy, selective anterior pituitary ablation or interruption of the hypothalamic-pituitary connection (Table 10). These conditions, which cause a functional castration, are reportedly associated with a loss of gender-related differences in lacrimal tissue,[159] as well as glandular atrophy, acinar cell contraction, nuclear pycnosis, cytoplasmic vacuolar metamorphosis, an increase in sulphate uptake, a decline in tissue protein, total RNA and mRNA content, a decrease in fluid and protein secretion and a diminished tear volume.[14,20,49,65,87,89,91,97,120,133,159,179–181] The extent of these changes, at least insofar as the reduction in size[180] and the decreased level of specific, androgen-regulated mRNAs[20] are concerned, are significantly greater in males than in females.

The reduction in lacrimal gland weight in hypophysectomized male rats appears to be primarily attributable to a decline in body weight, given that the LGW/BW ratios in these animals are not consistently affected.[49] In addition, removal of the entire or anterior pituitary in male rats, although causing a significant drop in the area of acinar complexes, typically induces a significant increase in acinar complex density.[49]

The effects of hypophyseal dysfunction on the lacrimal gland are apparently due to a deficiency in anterior, but not posterior, pituitary hormones.[78,79,83,159] In addition, it is

Table 10. Interruption of the hypothalamic-pituitary axis: effect on the lacrimal gland

Loss of sexual dimorphism
Glandular atrophy
Acinar cell contraction
Nuclear pycnosis
Cytoplasmic vacuolar metamorphosis
Increase in sulphate uptake
Decline in total RNA, mRNA and tissue protein content
Decrease in fluid and protein secretion

Information in this Table was obtained from the following references: 14,20,49,65,87,89,91,97,120,133,159,179-181

likely that the ability of pituitary extracts to increase lacrimal gland weight in guinea pigs[182] is also attributable to hormones from the anterior pituitary. Of interest, treatment of KCS patients with an anterior pituitary extract improved their dry eye,[78,79] whereas the administration of posterior pituitary hormones to rabbits had no effect on lacrimal secretion.[83]

2.4.1.1. Influence of Hypophysectomy or Anterior Pituitary Ablation on Androgen-Lacrimal Gland Interactions. An intact hypothalamic-pituitary axis appears to be an essential requirement for certain aspects of the androgen regulation of the lacrimal gland. An example of this influence may be seen in the androgen control of the ocular secretory immune system. Thus, hypophysectomy or ablation of the anterior pituitary significantly decreases the number of IgA plasma cells in lacrimal tissue, reduces the acinar cell production of SC mRNA and protein, elicits a marked decline in the levels of tear IgA and SC and almost completely inhibits androgen action on ocular mucosal immunity *in vivo*.[20,87,91,133] Hypophysectomy also leads to a significant attenuation in the acinar cell capacity to synthesize SC following androgen exposure *in vitro*.[97] The mechanism(s) underlying this interference with androgen effects on the lacrimal gland may be quite complex,[97,99] but appears to involve, at least, a marked diminution in the amount of androgen receptor protein in acinar cell nuclei,[36] as well as a decrease in androgen-related transcriptional and post-transcriptional events.[20] This latter explanation is prompted by the observation that androgens increase the levels of SC mRNA in lacrimal glands of hypophysectomized rats,[20] but this action is not translated into a corresponding rise in the production and/or secretion of SC protein *in vivo*.[87,91,133] Of interest, the ability of androgens to augment mRNA amounts (e.g. major urinary protein) in lacrimal tissues of hypophysectomized animals has also been found in mice,[14] but whether this message is translated is not known.

In addition to the above findings, it has been reported that androgen exposure may enhance the biochemical correlates of fluid and protein secretion in female rat lacrimal glands following hypophysectomy.[89] However, a number of other studies have shown that acute or chronic testosterone treatment has no consistent effect on the volume or protein concentration in tears of orchiectomized animals that had undergone prior anterior pituitary ablation, hypophysectomy or pituitary transplant to the kidney capsule.[49,65,87] Furthermore, investigations have demonstrated that androgens do not influence the weight, LGW/BW ratio, morphological appearance or fluid secretory activity of lacrimal tissue in pituitary-deficient animals.[49,65,87,159,183] The possibility exists that some of these discrepant findings may be due to gender differences, given that the nature of the hypothalamic-pituitary control of the lacrimal gland appears to be influenced by gender.[20,159]

Of interest, a recent study demonstrated that androgen therapy augmented the area of acinar complexes in lacrimal tissue of hypophysectomized rats and helped delay the loss of lacrimal gland size following pituitary transplant to the kidney capsule.[49] In addition, treatment of hypophysectomized rats with a combination of testosterone and insulin significantly increased the lacrimal gland weight and LGW/BW ratio.[49] The mechanism(s) responsible for these hormone actions remains to be determined.

It is important to note that the impairment of lacrimal gland responses to androgens in pituitary-operated animals does not reflect a generalized lack of tissue responsiveness throughout the body. Testosterone administration to rats with anterior pituitary ablation elicited a significant, 20-fold rise in both the seminal vesicle weight (SVW) and SVW/BW ratio.[49] Thus, the effect of the hypothalamic-pituitary axis on androgen target organs appears to vary according to the site.

2.4.2. Pituitary Hormone Actions on the Lacrimal Gland. Hormones from the hypothalamic-pituitary axis that are known to influence the lacrimal gland include α-melanocyte stimulating hormone (α-MSH), adrenocorticotropic hormone (ACTH), luteinizing hormone, follicle-stimulating hormone, growth hormone, thyroid-stimulating hormone and prolactin. These hormones from the anterior pituitary, as well as those modulated by this tissue (e.g. sex steroids, thyroxine, insulin, glucocorticoids), may modulate the growth, morphological differentiation, protein secretion and fluid output of the lacrimal gland.[78,79,89,124,159,175–179,182] More specifically, the effects of these hormones on lacrimal tissue are as follows:

2.4.2.1. Prolactin. Prolactin administration has been shown to increase the acinar cell diameter, nuclear size and lacrimal gland weight of male and female dwarf mice,[159] to enhance the Na^+,K^+-ATPase (total and specific) and muscarinic cholinergic receptor (total) activity in lacrimal glands of hypophysectomized female rats,[89] and to decrease the carbachol-induced peroxidase release in lacrimal tissue epithelial cells.[186] In contrast, other studies have demonstrated that prolactin has no effect on lacrimal gland weight (i.e. in hypophysectomized male rats),[49] SC production[99] or basal peroxidase release[186] by acinar epithelial cells, the tear levels of SC, IgA or total protein,[133] or the tear volume.[49]

Of interest, recent research has found that prolactin is not only present, but also transcribed and translated, in lacrimal gland acinar epithelial cells,[184–187] which also express prolactin receptors.[188] Moreover, prolactin is secreted by lacrimal tissue into tears.[187,189] It is unclear, though, what factors may regulate the production and output of this 'intra-lacrimal' prolactin. The content of lacrimal gland prolactin is not influenced by treatment with prolactin, prolactin antagonists (i.e. bromocriptine) or estrogen,[190] and the concentration of prolactin secreted by lacrimal tissue into tears is not affected by cholinergic agonists (i.e. pilocarpine).[189] However, given observations in other tissues, it may be that the synthesis of,[191] and receptor levels for,[192,227,228] prolactin in the lacrimal gland may be regulated by androgens. Conversely, as found in the prostate[193] or Harderian gland,[194] it is possible that prolactin may modulate the levels of androgen receptors in lacrimal tissue.

Overall, it remains to be determined what the species-independent role of prolactin may be in lacrimal tissue or in the tear film. However, considering this hormone's pro-inflammatory effects,[195,227] and its proposed role in the pathogenesis of Sjögren's syndrome,[196] it is quite possible that the local production of prolactin may serve to promote autoimmune disease in the lacrimal gland.

2.4.2.2. α-MSH and ACTH. The hormones α-MSH and ACTH appear to be involved in the control of protein secretion by the lacrimal gland. α-MSH, which has receptors present on acinar epithelial cells of the rat lacrimal gland,[176,178,197] stimulates the production of cAMP[178] and induces protein release[175,176] by acinar cells. Similarly, ACTH appears to augment peroxidase and total protein secretion by lacrimal acini through a cAMP-dependent mechanism.[175,177] ACTH may also be synthesized by, or accumulate within, myoepithelial cells.[184] However, these hormones do not affect the output of all proteins, given that neither α-MSH nor ACTH influences the secretion of SC by acinar epithelial cells.[99] With regard to possible other actions of these hormones, investigators have reported that α-MSH has no impact on tear IgA levels,[133] the relative (i.e. LGW/BW ratio) or absolute size of the lacrimal gland,[49,183] or the volume of tears.[49] One study, though, suggests that in combination with testosterone treatment, α-MSH may enhance lacrimal gland weight in orchiectomized rats.[183] As concerns ACTH, this peptide appears to have no effect on acinar cell or nuclear diameters in lacrimal tissues of pituitary-deficient mice.[159] An ACTH insensitivity syndrome (e.g. Allgrove) has been described, which results in adrenal insufficiency, glucocorticoid deficency and alacrima,[198–201] but the specific cause of the decreased tear output is not known.

2.4.2.3. Luteinizing, Follicle-Stimulating, Thyroid-Stimulating and Growth Hormones. Luteinizing hormone, follicle-stimulating hormone and thyroid-stimulating hormone have been reported to enhance the weight and morphological differentiation of lacrimal glands in male, but not female, pituitary dwarf mice.[159] The reason for these gender-related differences in peptide hormone action is unclear. Moreover, whether these pituitary hormone effects are direct (e.g through lacrimal gland receptors), and/or are mediated through the control of other hormones (e.g. sex steroids, thyroxine), has yet to be determined.

Growth hormone (GH) also increases the weight, but causes no change in the morphology, of lacrimal glands in both male and female dwarf mice.[159] This GH action, though, is not necessarily duplicated in other species. Thus, other studies have reported that GH has no influence on the weight of lacrimal tissue in hypophysectomized rats,[49,120] or in orchiectomized and thyroidectomized rats.[63] Growth hormone also has no effect on the levels of mouse urinary protein mRNA in lacrimal glands of hypophysectomized mice,[14] on the amounts of tear SC, IgA or total protein in hypophysectomized rats,[49,133] or on the secretion of SC by acinar epithelial cells *in vitro*.[99]

2.4.3.4. Other Hormones and Factors from the Hypothalamic-Pituitary Axis. Little information is known concerning the effects of other hormones or factors from the hypothalamic-pituitary axis on the lacrimal gland. Research has shown that arginine vasopressin, oxytocin and rat hypothalamic extract have no significant influence on the basal and/or androgen-induced production of SC by rat lacrimal gland acinar epithelial cells *in vitro*,[99] and that hypophysin does not alter tear secretion in rabbits.[83] Of interest, vasopressin may be produced or taken up by lacrimal epithelial cells,[202] but the local effect of this hormone on lacrimal tissue remains to be determined.

2.5. Thyroxine, Insulin, Glucocorticoids, Melatonin, Aldosterone, Human Chorionic Gonadotropin, Glucagon and Retinoic Acid

Additional hormones, including thyroxine, insulin, glucocorticoids and glucagon, are known to influence structural, functional and/or secretory aspects of the lacrimal gland. Thyroxine increases lacrimal gland weight in orchiectomized and thyroidectomized rats,[63] and enhances lacrimal tissue regeneration and secretion in rabbits.[83] However, this hormone, with or without parallel glucocorticoid therapy, does not appear to affect the size, activity and/or secretion of the lacrimal gland in pituitary-deficient animals.[14,49,159] A possibility exists that thyroxine may be produced by the lacrimal gland, secreted into tears and promote alterations in corneal curvature (e.g. development of keratoconus),[203] but further research is required to support this speculation.

Insulin appears to play an important role in maintaining the structure and function of the lacrimal gland, as evidenced by the impact of diabetes. This condition leads to a decrease in lacrimal tissue weight, a decline in the density of IgA-containing cells in lacrimal tissue, a reduction in the concentrations of IgA and SC in tears,[91] an attenuation in the extent of the immune response to androgens,[91] a drop in tear volume[91] and rise in the incidence of KCS.[204] Insulin deficiency also causes a significant decline in the basal and androgen-induced output of SC by acinar epithelial cells, as well as a striking decrease in acinar cell endurance, *in vitro*.[98] Insulin therapy, in turn, appears to restore tear IgA levels to normal in diabetic patients.[205]

The mechanism of action of insulin on the lacrimal gland may involve an initial binding of this hormone to specific surface receptors on acinar epithelial cells.[98] This hormone's effect on androgen activity, though, does not seem to be mediated through the control of androgen receptor content or affinity.[36,49] Of interest, some of the effects of pituitary deficiency on lacrimal tissue may be due to the loss of insulin, given that serum insulin levels are significantly diminished in hypopituitarism or following hypophysectomy.[206–208] However, insulin administration is unable to reverse the influence of pituitary removal on lacrimal gland weight or tear volume.[49]

Physiological levels of glucocorticoids stimulate the synthesis of the C3 component of prostatic binding protein[76] and promote the androgen-induced production of SC,[98] CRP and C3[76] by rat lacrimal gland acinar epithelial cells. Low amounts of corticosteroids are also critical for the maintenance of acinar cells *in vitro*.[98] Higher concentrations of glucocorticoids augment CRP and C3 synthesis,[76] have no effect on basal SC output,[97] and inhibit the androgen-related SC response.[197] In other studies, it has been shown that glucocorticoids suppress the uptake of sulfate into newborn rat lacrimal tissue *in vivo*.[82]

Glucocorticoids have also been proposed as a possible therapy for lacrimal gland dysfunction during the early stages of Sjögren's syndrome.[209] The utility of this treatment is questionable, though, given that chronic corticosteroid treatment of Sjögren's syndrome patients does not appear to mitigate the severity of lacrimal gland disease.[210,211] In fact, glucocorticoids (e.g. cortisol, dexamethasone) have been shown to significantly decrease tear volume in autoimmune mice,[71] orchiectomized rats[29] and 'normal' humans.[212] This corticosteroid effect may involve feedback action on the pituitary gland, because dexamethasone administration has no influence on the tear volume, or lacrimal tissue weight, in hypophysectomized rats.[49] The mechanism by which glucocorticoids act on the lacrimal gland may involve specific receptors, given that glucorticoid receptor mRNA has been identified in human lacrimal tissue.[213]

With regard to other hormones, melatonin is synthesized by the lacrimal gland,[12] but its effect on this tissue is unclear. Melatonin, aldosterone or human chorionic gonadotropin do not influence SC production by rat lacrimal gland acinar epithelial cells, whether in the presence of absence of androgens.[97,99] Glucagon attenuates the uptake of sultate by newborn rat lacrimal tissue,[82] but the mechanism underlying this hormone action is unknown. Lastly, the lacrimal glands of rats, rabbits and/or humans harbor cellular retinoic acid- and retinol-binding proteins,[214,215] contain nuclear retinoic acid receptors (subtypes α, β and/or γ)[216] and metabolize and secrete retinoids.[217,218] Retinoic acid reduces acinar cell growth *in vitro*[219] and may increase aqueous tear output *in vivo*,[220] but the full range of possible effects of this hormone on lacrimal tissue remains to be elucidated.

3. FUTURE DIRECTIONS

It is apparent that gender and the endocrine system play a major role in the architectural profile, functional activity and disease susceptibility of the lacrimal gland. However, many questions remain concerning the full nature and extent of this gender and hormonal influence. For example, what are the processes, aside from sex steroid action, that may underly the fundamental, gender-related differences in lacrimal tissue? Does gender impact the pattern of innervation, the capacity to release neurotransmitters, and the sensitivity to neural stimulation, as occurs in other exocrine or peripheral tissues?[e.g. 230–232] Do androgens regulate the differentiation of the lacrimal gland, in a manner analogous to that of the salivary gland?[233] Are androgen actions in lacrimal tissue enhanced by regulators of protein phosphorylation (e.g. activators of protein kinases A and C, inhibitors of protein phosphatase-1 and 2A),[234] suppressed by calcium-binding proteins (e.g. calreticulin)[235] or duplicated by growth factor (e.g. insulin-like growth factor I, epidermal growth factor and keratinocyte growth factor) interaction with androgen receptors?[236] Are some androgen actions in epithelial cells mediated through a plasma membrane-associated signaling system, and if so, is this pathway (e.g. involving sex hormone binding globulin and protein kinase A), as in prostatic cells, also activated by estrogens?[237] Do estrogens have direct effects on the lacrimal gland? What is the cellular distribution of receptors for, and the functional role of, peptide hormones in lacrimal tissue? Is it possible to develop cultures of immortalized acinar, ductal and myoepithelial cells from the lacrimal gland, in order to better understand the hormonal regulation of, and the communication systems between, these cells?

Clearly, there are numerous questions, and those mentioned above are but a few. One of the most intriguing, unresolved issues, at least from an endocrine standpoint, is clarification of the nature of feedback regulation from the ocular surface to the lacrimal gland, and from the front of the eye and/or lacrimal tissue to endocrine organs. In other words, hormone actions typically occur as part of a control system, whereby information is relayed back from the effector site to the endocrine source to communicate whether the response is optimal, or whether more or less hormone is required (e.g. gonadotropin control of sex steroid synthesis and output). Thus, for example, if hormones control the lacrimal gland secretion of various proteins (e.g. IgA antibody, growth factors) into the tear film, what are the signals from the ocular surface that let the lacrimal gland know that the protein amount is correct or should be altered? Moreover, what are the signals from the ocular surface and/or lacrimal gland that relay information to endocrine organs about the appropriateness of hormone effects in lacrimal tissue? The lacrimal gland apparently does have a two-way communication system with endocrine sites (e.g. pituitary, testis, thyroid, pancreas),[238] but the nature of the lacrimal signals involved in these inter-tissue relationships is unknown.

Overall, future research may result in a better understanding of the influence of gender and the endocrine system on the anatomy and physiology of the lacrimal gland. In addition, such an effort may lead to the development of new and unique therapeutic strategies to treat diverse lacrimal gland and associated tear film disorders.

ACKNOWLEDGMENTS

The authors would like to express their appreciation to Benjamin D. Sullivan (Acton, MA) for his expertise in computer graphics (i.e. producing Figures 1, 2 and 3). This research review was supported by NIH grant EY05612. In addition, many of our studies that are reported herein were supported by NIH grant EY05612, Allergan, the Massachusetts Lions' Eye Research Fund, the Sjogren's Syndrome Foundation and the Uehara Memorial Foundation (Japan).

REFERENCES

1. Wickham LA, Gao J, Toda I, Rocha EM, Sullivan DA. Identification of androgen, estrogen and progesterone receptor mRNAs in rat, rabbit and human ocular tissues. Submitted for publication, 1997.

2. Cavallero C. Relative effectiveness of various steroids in an androgen assay using the exorbital lacrimal gland of the castrated rat. *Acta Endocrinol. (Copenh.)* 1992;55:119–130.
3. Gabe M. Conditionnement hormonal de la morphologie des glandes sus-parotidiennes chez le Rat albinos. *Compt rend Séanc Soc Biol.* 1955;149:223–225.
4. Cornell-Bell AH, Sullivan DA, Allansmith MR. Gender-related differences in the morphology of the lacrimal gland. *Invest Ophthalmol Vis Sci.* 1985;26:1170–1175.
5. Cripps MM, Bromberg BB, Welch MH. Gender-dependent lacrimal protein secretion. *Invest Ophthalmol Vis Sci Suppl.* 1986;27:25.
6. Ducommun P, Ducommun S, Baquiche M. Comparaison entre l'action du 17-ethyl-19-nor-testosterone et du propionate de testosterone chez le rat adulte et immature. *Acta Endocrin.* 1959;30:78–92.
7. Walker R. Age changes in the rat's exorbital lacrimal gland. *Anat Rec.* 1958;132:49–69.
8. Luciano L. Die feinstruktur der tränendrüse der ratte und ihr geschlechtsdimorphismus. *Zeitschrift für Zellforschung* 1967;76:1–20.
9. Paulini K, Beneke G, Kulka R. Age- and sex-dependent changes in glandular cells. I. Histologic and chemical investigations on the glandular lacrimalis, glandular intraorbitalis, and glandula orbitalis external of the rat. *Gerontologia* 1972;18:131–146.
10. Paulini K, Mohr W, Beneke G, Kulka R. Age- and sex-dependent changes in glandular cells. II. Cytomorphometric and autoradiographic investigations on the glandular lacrimalis, glandular intraorbitalis, and glandula orbitalis external of the rat. *Gerontologia* 1972;18:147–156.
11. Pangerl A, Pangerl B, Jones DJ, Reiter RJ. b-Adrenoreceptors in the extraorbital lacrimal gland of the syrian hamster. Characterization with [^{125}I]-iodopindolol and evidence of sexual dimorphism. *J Neural Transm.* 1989;77:153–162.
12. Mhatre MC, van Jaarsveld AS, Reiter RJ. Melatonin in the lacrimal gland: first demonstration and experimental manipulation. *Biochem Biophys Res Comm.* 1988;153:1186–1192.
13. Sullivan DA, Hann LE, Yee L, Allansmith MR. Age- and gender-related influence on the lacrimal gland and tears. *Acta Ophthalmologica* 1990;68:188–194.
14. Shaw PH, Held WA, Hastie ND. The gene family for major urinary proteins: expression in several secretory tissues of the mouse. *Cell* 1983;32:755–761.
15. Tier H. Über Zellteilung und Kernklassenbildung in der Glandula orbitalis externa der Ratte. *Acta path microbiol scand Suppl.* 1944;50:1–185.
16. Sullivan DA, Allansmith MR. The effect of aging on the secretory immune system of the eye. *Immunology* 1988;63:403–410.
17. Hann LE, Allansmith MR, Sullivan DA. Impact of aging and gender on the Ig-containing cell profile of the lacrimal gland. *Acta Ophthalmologica* 1988;66:87–92.
18. Hahn JD. Effect of cyproterone acetate on sexual dimorphism of the exorbital lacrimal gland in rats. *J Endocr.* 1969;45:421–425.
19. Baquiche M. Le dimorphisme sexuel de la glande de Loewenthal chez le rat albinos. *Acta anat.* 1959;36:247–280.
20. Gao J, Lambert RW, Wickham LA, Banting G, Sullivan DA. Androgen control of secretory component mRNA levels in the rat lacrimal gland. *J Ster Biochem Mol Biol.* 1995;52:239–249.
21. Lorber M, Vidic B. Weights and dimensions of 16 lacrimal glands from human cadavers. *Invest Ophthalmol Vis Sci Suppl.* 1994;35:1790.
22. Waterhouse JP. Focal adenitis in salivary and lacrimal glands. *Proc R Soc Med.* 1963;56:911–918.
23. Sullivan DA, Wickham LA, Krenzer KL, Rocha EM, Toda I. Aqueous tear deficiency in Sjögren's syndrome: Possible causes and potential treatment. In: Pleyer U, Hartmann C and Sterry W, eds. *Oculodermal Diseases - Immunology of Bullous Oculo-Muco-Cutaneous Disorders.* Buren, The Netherlands: Aeolus Press; 1997:95–152.
24. Lauria A, Porcelli F. Leucine aminopeptidase (LAP) activity and sexual dimorphism in rat exorbital gland. *Basic Appl Histochem.* 1979;23:171–177.
25. Azzarolo AM, Mazaheri AH, Mircheff AK, Warren DW. Sex-dependent parameters related to electrolyte, water and glycoprotein secretion in rabbit lacrimal glands. *Curr Eye Res.* 1993;12:795–802.
26. Ranganathan V, De PK. Androgens and estrogens markedly inhibit expression of a 20-kDa major protein in hamster exorbital lacrimal gland. *Biochem Biophys Res Comm.* 1995; 208:412–417.
27. Kenney MC, Brown DJ, Hamdi H. Proteinase activity in normal human tears: male-female dimorphism. *Invest Ophthalmol Vis Sci.* 1995;36:S4606.
28. Krawczuk-Hermanowiczowa O. Effects of sexual glands on the lacrimal gland. II. Changes in rat lacrimal glands after castration. *Klin oczna* 1983;85:15–17.
29. Sullivan DA, Bloch KJ, Allansmith MR. Hormonal influence on the secretory immune system of the eye: Androgen regulation of secretory component levels in rat tears. *J Immunol.* 1984;132:1130–1135.

30. Sullivan DA, Bloch KJ, Allansmith MR. Hormonal influence on the secretory immune system of the eye: androgen control of secretory component production by the rat exorbital gland. *Immunology* 1984;52:239–246.
31. Winderickx J, Vercaeren I, Verhoeven G, Heyns W. Androgen-dependent expression of cystatin-related protein (CRP) in the exorbital lacrimal gland of the rat. *J Steroid Biochem Molec Biol.* 1994;48:165–170.
32. Sullivan DA, Allansmith MR. Hormonal influence on the secretory immune system of the eye: androgen modulation of IgA levels in tears of rats. *J. Immunol.* 1985;134:2978–2982.
33. Calmettes L, Déodati F, Planel H, Bec P. Influence des hormones génitales sur la glande lacrymale. *Bull Soc franc Ophtal.* 1956;69:263–270.
34. Gubits RM, Lynch KR, Kulkarni AB, Dolan KP, Gresik EW, Hollander P, Feigelson P. Differential regulation of a 2u globulin gene expression in liver, lachrymal gland, and salivary gland. *J Biol Chem.* 1984;259:12803–12809.
35. Obata H, Yamamoto S, Horiuchi H, Machinami R. Histopathologic study of the human lacrimal gland. Statistical analysis with special reference to aging. *Ophthalmol.* 1995; 102:678–686.
36. Rocha FJ, Wickham LA, Pena JDO, Gao J, Ono M, Lambert RW, Kelleher RS, Sullivan DA. Influence of gender and the endocrine environment on the distribution of androgen receptors in the lacrimal gland. *J Ster Biochem Mol Biol.* 1993;46:737–749.
37. Huang Z, Gao J, Wickham LA, Sullivan, DA. Influence of gender and androgen treatment on TGF-β1 mRNA levels in the rat lacrimal gland. *Invest Ophthalmol Vis Sci.* 1995; 35:S991.
38. Toda I, Rocha EM, Siveira LA, Wickham LA, Sullivan DA. Gender-related difference in the extent of lymphocyte infiltration in lacrimal and salivary glands of mouse models of Sjögren's syndrome. Submitted for publication, 1997.
39. Rocha EM, Toda I, Wickham, LA, Silveira LA, Sullivan DA. Effect of gender, androgens and cyclophosphamide on cytokine mRNA levels in lacrimal tissues of mouse models of Sjögren's syndrome. Submitted for publication, 1997.
40. Toda I, Wickham LA, Sullivan DA. Influence of gender and androgen treatment on the mRNA expression of proto-oncogenes and apoptotic factors in lacrimal and salivary tissues of the MRL/lpr mouse model of Sjögren's syndrome. Submitted for publication, 1997.
41. Toda I, Wickham LA, Sullivan DA. Gender and androgen-related influence on the expression of proto-oncogene and apoptotic factor mRNAs in lacrimal glands of autoimmune and non-autoimmune mice. Submitted for publication, 1997.
42. Wickham LA, Gao J, Toda I, Sullivan DA. Endocrine regulation of androgen receptor protein and sex steroid mRNA levels in the rat lacrimal gland. Submitted for publication, 1997.
43. Huang Z, Lambert RW, Wickham A, Sullivan DA. Analysis of cytomegalovirus infection and replication in acinar epithelial cells of the rat lacrimal gland. *Invest Ophthalmol Vis Sci.* 1996;37:1174–1186.
44. Bromberg BB, Welch MH, Beuerman RW, Chew S-J, Thompson HW, Ramage D, Githens S. Histochemical distribution of carbonic anhydrase in rat and rabbit lacrimal gland. *Invest Ophthalmol Vis Sci.* 1993;34:339–348.
45. De PK. Sex differences in content of peroxidase, a porphyrin containing enzyme, in hamster lacrimal gland. In: Abstracts of the International Conference on the Lacrimal Gland, Tear Film and Dry Eye Syndromes: Basic Science and Clinical Relevance (Bermuda); 1992:40.
46. Sashima M, Hatakeyama S, Satoh M, Suzuki A. Harderianization is another sexual dimorphism of rat exorbital lacrimal gland. *Acta Anat.* 1989;135:303–306.
47. Parhon CI, Babes A, Petrea I, Istrati F, Burgher E. Structura si Dimorfismul sexual al glandelor parotide la Sobolanul Alb. *Bul Stiint Sect de Stünte med.* 1955;7:3.
48. Sullivan DA, Colby E, Hann LE, Allansmith MR, Wira CR. Production and utilization of a mouse monoclonal antibody to rat IgA: Identification of gender-related differences in the secretory immune system. *Immunol Invest.* 1986;15:311–318.
49. Sullivan DA, Block L, Pena JDO. Influence of androgens and pituitary hormones on the structural profile and secretory activity of the lacrimal gland. *Acta Ophthalmol Scand.* 1996;74:421–435.
50. Remington SG, Lima PH, Nelson JD. Pancreatic lipase mRNA isolated from female mouse. *Invest Ophthalmol Vis Sci.* 1997;38:S149.
51. Sullivan DA. Ocular mucosal immunity. In: Ogra PL, Mestecky J, Lamm ME, Strober W, McGhee J, Bienenstock J, eds. *Handbook of Mucosal Immunology, 2nd Edition.* Orlando, FL: Academic Press; 1997: in press.
52. Sullivan DA. Sex hormones and Sjögren's syndrome. J Rheumatology Supplement. In: Parke AL, Tanzer JM, Donshik P, editors. Proceedings of the VI International Symposium on Sjögren's Syndrome (Connecticut, 10/97); 1997: in press.
53. Henderson JW, Prough WA. Influence of age and sex on flow of tears. *Arch Ophthalmol.* 1950;43:224–231.

54. Craig JP, Tomlinson A. Age and gender effects on the normal tear film. *Adv Exp Med Biol.* 1997; in press.
55. Barton K, Nava A, Monroy DC, Pflugfelder SC. Cytokines and tear function in ocular surface disease. *Adv Exp Med Biol.* 1997: in press.
56. van Setten G, Schultz G. Transforming growth factor-alpha is a constant component of human tear fluid. *Graefe's Arch Clin Exp Ophthalmol.* 1994;232:523–526.
57. Bodelier VMW, van Haeringen NJ. Gender related differences in tear fluid protein profiles of mice. In: Abstracts of the IVth International Congress of the International Society of Dacryology (Stockholm, Sweden); 1996:24.
58. Craig JP, Tomlinson A. Effect of age on tear osmolality. *Optom Vis Sci.* 1995;72:713–717.
59. Sullivan DA. Hormonal influence on the secretory immune system of the eye. In: Freier S, editor. *The Neuroendocrine-Immune Network.* Boca Raton, FL: CRC Press; 1990:199–238.
60. Sullivan DA, Sato EH. Immunology of the lacrimal gland. In Albert DM, Jakobiec, FA, editors. *Principles and Practice of Ophthalmology: Basic Sciences.* Philadelphia, PA: WB Saunders Company; 1994:479–486.
61. Aumüller G, Arce EA, Heyns W, Vercaeren I, Dammshäuser, Seitz J. Immunocytochemical localization of seminal proteins in salivary and lacrimal glands of the rat. *Cell Tissue Res.* 1995;280:171–181.
62. Dzierzykray-Rogalska I, Chodynicki S, Wisniewski L. The effect of gonadectomy on the parotid salivary gland and Loeventhal's gland in white mice. *Acta Med Polona* 1963;2:221–228.
63. Carriere R. The influence of the thyroid gland on polyploid cell formation in the external orbital gland of the rat. *Am J Anat.* 1964;115:1–16.
64. Cavallero C, Morera P. Effect of testosterone on the nuclear volume of exorbital lacrimal glands of the white rat. *Experentia* 1960;16:285–286.
65. Sullivan DA, Allansmith MR. Hormonal modulation of tear volume in the rat. *Exp Eye Res.* 1986;42:131–139.
66. Myal Y, Iwasiow B, Yarmill A, Harrison E, Paterson JA, Shiu RP. Tissue-specific androgen-inhibited gene expression of a submaxillary gland protein, a rodent homolog of the human prolactin-inducible protein/GCDFP-15 gene. *Endocrinol.* 1994;135:1605–1610.
67. Winderickx J, Hemschoote K, De Clercq N, Van Dijck P, Peeters B, Rombauts W, Verhoeven G, Heyns W. Tissue-specific expression and androgen regulation of different genes encoding rat prostatic 22-kilodalton glycoproteins homologous to human and rat cystatin. *Molec Endocrin.* 1990;4:657–667.
68. Ariga H, Edwards J, Sullivan DA. Androgen control of autoimmune expression in lacrimal glands of MRL/Mp-lpr/lpr mice. *Clin Immunol Immunopath.* 1989;53:499–508.
69. Vendramini AC, Soo CH, Sullivan DA. Testosterone-induced suppression of autoimmune disease in lacrimal tissue of a mouse model (NZB/NZW F1) of Sjögren's Syndrome. *Invest Ophthalmol Vis Sci.* 1991;32:3002–3006.
70. Sato EH, Ariga H, Sullivan DA. Impact of androgen therapy in Sjögren's syndrome: Hormonal influence on lymphocyte populations and Ia expression in lacrimal glands of MRL/Mp-lpr/lpr mice. *Invest Ophthalmol Vis Sci.* 1992;33:2537–2545.
71. Sato EH, Sullivan DA. Comparative influence of steroid hormones and immunosuppressive agents on autoimmune expression in lacrimal glands of a female mouse model of Sjögren's syndrome. *Invest Ophthalmol Vis Sci.* 1993;35:2632–2642.
72. Rocha FJ, Sato EH, Sullivan BD, Sullivan DA. Effect of androgen analogue treatment and androgen withdrawal on lacrimal gland inflammation in a mouse model (MRL/Mp-lpr/lpr) of Sjögren's syndrome. *Reg Immunol.* 1994;6:270–277.
73. Sullivan DA, Edwards J. Androgen stimulation of lacrimal gland function in mouse models of Sjögren's syndrome. *J Ster Biochem Mol Biol.* 1997;60:237–245.
74. Quintarelli G, Dellovo MC. Activation of glycoprotein biosynthesis by testosterone propionate on mouse exorbital glands. *J Histochem Cytochem.* 1965;13:361–364.
75. Vercaeren I, Winderickx J, Devos A, Peeters B, Heyns W. An effect of androgens on the length of the poly(A)-tail and alternative splicing cause size heterogeneity of the messenger ribonucleic acids encoding cystatin-related protein. *Endocrin.* 1992;131:2496–2502.
76. Vanaken H, Claessens F, Vercaeren I, Heyns W, Peeters B, Rombauts W. Androgenic induction of cystatin-related protein and the C3 component of prostatic binding protein in primary cultures from the rat lacrimal gland. *Mol Cell Endocrinol.* 1996;121:197–205.
77. Krawczuk-Hermanowiczowa O. Effect of sex hormones on the lacrimal gland. III. Effects of testosterone and oestradiol and of both these hormones jointly on the morphological appearance of the lacrimal gland in castrated rats. *Klin oczna* 1983;85:337–339.
78. Radnót VM, Németh B. Wirkung der Testosteronpräparate auf die Tränendrüse. *Ophthalmologica* 1955;129:376–380.
79. Radnot M, Nemeth B. Testosteronkészítmények hatása a könnymirigyre. *Orvosi Hetilap* 1954;95:580–581.

80. Cavallero C, Offner P. Relative effectiveness of various steroids in an androgen assay using the exorbital lacrimal gland of the castrated rat. II. C19-steroids of the 5α-androstane series. *Acta Endocrinol. (Copenh.)* 1967;55:131–135.
81. Cavallero C. The influence of various steroids on the Lowenthal lachrymal glands of the rat. *Acta Endocrinol Suppl. (Copenh.)* 1960;51:861.
82. Cavallero C, Chiappino G, Milani F, Casella E. Uptake of ^{35}S Labelled sulfate in the exorbital lacrymal glands of adult and newborn rats under different hormonal treatment. *Experentia (Basel)* 1960;16:429
83. Nover A. The influence of testosterone and hypophysine on lacrimal secretion. *Arzneimittel-Forschg.* 1957;7:277–278.
84. Appelmans M. La Kerato-conjonctivite seche de Gougerot-Sjogren. *Arch 'Ophtalmologie* 1948;81:577–588.
85. Brückner R. Uber einem erfolgreich mit perandren behandelten fall von Sjogren'schem symptomen komplex. *Ophthalmologica* 1945;110:37–42.
86. Bizzarro A, Valentini G, Di Marinto G, Daponte A, De Bellis A, Iacono G. Influence of testosterone therapy on clinincal and immunological features of autoimmune diseases associated with Klinefelter's syndrome. *J Clin End Metab.* 1987;64:32–36.
87. Sullivan DA, Allansmith MR. Hormonal influence on the secretory immune system of the eye: endocrine interactions in the control of IgA and secretory component levels in tears of rats. *Immunology* 1987;60:337–343.
88. Sullivan DA, Hann LE, Vaerman JP. Selectivity, specificity and kinetics of the androgen regulation of the ocular secretory immune system. *Immunol Invest.* 1988; 17:183–194.
89. Azzarolo Am, Bjerrum K, Maves CA, Becker L, Wood RL, Mircheff AK, Warren DW. Hypophysectomy-induced regression of female rat lacrimal glands: Partial restoration and maintenance by dihydrotestosterone and prolactin. *Invest Ophthalmol Vis Sci.* 1995;36:216–226.
90. Claessens F, Vanaken H, Vercaeren I, Verrijdt G, Haelens A, Schoenmakers E, Alen P, Peeters B, Verhoeven G, Rombauts W, Heyns W. Androgen-regulated transcription in the epithelium of the rat lacrimal gland. *Adv Exp Med Biol.* 1997: in press.
91. Sullivan DA, Hann LE. Hormonal influence on the secretory immune system of the eye: endocrine impact on the lacrimal gland accumulation and secretion of IgA and IgG. *J Steroid Biochem.* 1989;34:253–262.
92. Clark JH, Schrader WT, O'Malley BW. Mechanisms of action of steroid hormones. In: Wilson JD, Foster DW, eds. *Williams Textbook of Endocrinology*. Philadelphia: WB Saunders; 1992:35–90.
93. Rundlett SE, Wu X-P, Miesfeld RL. Functional characterizations of the androgen receptor confirm that the molecular basis of androgen action is transcriptional regulation. *Mol Endocr.* 1990;4:708–714.
94. Sullivan DA, Edwards JA, Wickham LA, Pena JDO, Gao J, Ono M, Kelleher RS. Identification and endocrine control of sex steroid binding sites in the lacrimal gland. *Curr Eye Res.* 1996;15:279–291.
95. Ono M, Rocha FJ, Sullivan DA. Immunocytochemical location and hormonal control of androgen receptors in lacrimal tissues of the female MRL/Mp-lpr/lpr mouse model of Sjögren's syndrome. *Exp Eye Res.* 1995;61:659–666.
96. Ota M, Kyakumoto S, Nemoto T. Demonstration and characterization of cytosol androgen receptor in rat exorbital lacrimal gland. *Biochem Internat.* 1985;10:129–135.
97. Sullivan DA, Kelleher RS, Vaerman JP, Hann LE. Androgen regulation of secretory component synthesis by lacrimal gland acinar cells in vitro. *J Immunol.* 1990;145:4238–4244..
98. Hann LE, Kelleher RS, Sullivan D.A. Influence of culture conditions on the androgen control of secretory component production by acinar cells from the lacrimal gland. *Invest Ophthalmol Vis Sci.* 1991;32:2610–2621.
99. Kelleher RS, Hann LE, Edwards JA, Sullivan DA. Endocrine, neural and immune control of secretory component output by lacrimal gland acinar cells. *J Immunol.* 1991;146:3405–3412.
100. Lambert RW, Kelleher RS, Wickham LA, Vaerman JP, Sullivan DA. Neuroendocrinimmune modulation of secretory component production by rat lacrimal, salivary and intestinal epithelial cells. *Invest Ophthalmol Vis Sci.* 1994;35:1192–1201.
101. Wickham A, Huang Z, Lambert RW, Sullivan DA. Effect of sialodacryoadenitis virus exposure on acinar epithelial cells from the rat lacrimal gland. *Ocular Immunol Immunopathol.* 1997: in press.
102. Tan J, Joseph DR, Quarmby VE, Lubahn DB, Sar M, French FS, Wilson EM. The rat androgen receptor: primary structure, autoregulation of its messenger ribonucleic acid, and immunocytochemical localization of the receptor protein. *Mol Endocr.* 1988;2:1276–1285.
103. Quarmby VE, Yarbrough WG, Lubahn DB, French FS, Wilson EM. Autologous down-regulation of androgen receptor messenger ribonucleic acid. *Mol. Endocr.* 1990;4:22–28.
104. Krongrad A, Wilson CM, Wilson JD, Allman DR, McPhaul MJ. Androgen increases androgen receptor protein while decreasing receptor mRNA in LNCaP cells. *Mol Cell Endocr.* 1991;76:79–88.

105. Brann DW, Hendry LB, Mahesh VB. Emerging diversities in the mechanism of action of steroid hormones. *J Steroid Biochem Mol Biol.* 1995;52:113–133.
106. Lewin DI: From outside or in, sex hormones tweak prostate cells. *J NIH Res.* 1996;8:29–30.
107. Farnsworth WE. Prostate plasma membrane receptor: a hypothesis. *Prostate* 1991;19:329–352.
108. Kurihara K, Maruyama S, Hosoi K, Sato S, Ueha T, Gresik EW. Regulation of Na^+,K^+-ATPase in submandibular glands of hypophysectomized male mice by steroid and thyroid hormones. *J Histochem Cytochem.* 1996;44:703–711.
109. Sullivan DA, Allansmith MR. Source of IgA in tears of rats. *Immunol.* 1994;53:791–799.
110. Peppard JV, Montgomery PC. Studies on the origin and composition of IgA in rat tears. *Immunol.* 1987;62:194–198.
111. Verrijdt G, Swinnen J, Peeters B, Verhoeven G, Rombauts W, Claessens R. Characterization of the human secretory component gene promoter. *Biochim Biophys Acta* 1997;1350:147–154.
112. Talal N, Moutsopoulos HM, Kassan SS, editors. *Sjögren's Syndrome. Clinical and Immunological Aspects.* Berlin: Springer Verlag; 1987.
113. Fox RI, editor. *Sjögren's Syndrome.* Rheum Dis Clin NA. 1992:vol 18.
114. Homma M, Sugai S, Tojo T, Miyasaka N, Akizuki M, editors. *Sjögren's Syndrome.* State of the Art. Amsterdam: Kugler Press; 1994.
115. Saito I, Terauchi K, Shimuta M, Nishiimura S, Yoshino K, Takeuchi T, Tsubota K, Miyasaka N. Expression of cell adhesion molecules in the salivary and lacrimal glands of Sjögren's syndrome. *J Clin Lab Anal.* 1993; 7:180–187.
116. Fox RI, Saito I. Sjögren's syndrome: immunologic and neuroendocrine mechanisms. *Adv Exp Med Biol.* 1994; 350:609–621.
117. Sullivan DA. Possible mechanisms involved in the reduced tear secretion in Sjögren's syndrome. In: Homma M, Sugai S, Tojo T, Miyasaka N and Akizuki M, eds. *Sjögren's Syndrome. State of the Art.* Amsterdam: Kugler Press; 1994:13–19.
118. Xu G, Shang H, Zhu F. Measurement of serum testosterone level in female patients with dry eye. Proceedings of the International Congress of Ophthalmology Meeting (Abstract); 1994.
119. Lahita R. Sex hormones, Sjögren's syndrome and the immune response. *The Moisture Seekers Newsletter* 1991;8:1.
120. Ebling FJ, Ebling E, Randall V, Skinner J. The effects of hypophysectomy and of bovine growth hormone on the responses to testosterone of prostate, preputial, harderian and lachrymal glands and of brown adipose tissue in the rat. *J Endocr.* 1975;66:401–406.
121. Kyprianou N, Isaacs JT. Activation of programmed cell death in the rat ventral prostate after castration. *Endocrinology* 1988;122:552–562.
122. Warren DW, Azzarolo AM, Huang ZM, Platler BW, Kaswan RL, Gentschein E, Stanczyk FL, Becker L, Mircheff AK. Androgen support of lacrimal gland function in the female rabbit. *Adv Exp Med Biol.* 1997: in press.
123. Sullivan DA, Krenzer KL, Ullman MD, Wickham LA, Toda I, Bazzinotti D, Dana MR. Androgen control of the meibomian gland. Submitted for publication, 1997.
124. Mircheff AK. Understanding the causes of lacrimal insufficiency: implications for treatment and prevention of dry eye syndrome. In: *Research to Prevent Blindness Science Writers Seminar.* New York: Research to Prevent Blindness; 1993:51–54.
125. Azzarolo AM, Wood RL, Mircheff AK, Olsen E, Huang ZM, Zolfagari R, Warren DW. Time course of apoptosis in lacrimal gland after rabbit ovariectomy. *Adv Exp Med Biol.* 1997: in press.
126. Mamalis N, Harrison DY, Hiura G, Hanover R, Meikle AW, Warren DW, Mazer NA. Dry eyes and testosterone deficiency in women. In: Abstracts of the Centennial Annual Meeting of the American Academy of Ophthalmology 1996; p 132.
127. Homo-Delarche F, Fitzpatrick F, Christeff N, Nunez EA, Bach JF, Dardenne M. Sex steroids, glucocorticoids, stress and autoimmunity. *J Ster Biochem Mol Biol.* 1991;40:619–637.
128. Azzarolo AM, Olsen E, Huang ZM, Mircheff AK, Wood RL, Warren DW. Ovariectomy induces apoptosis and necrosis in rabbit lacrimal glands. Prevention by androgen. *Invest Ophthalmol Vis Sci.* 1997;38:S1155.
129. Dartt DA, Sullivan DA. Wetting of the ocular surface. In Albert DM, Jakobiec, FA, eds. *Principles and Practice of Ophthalmology.* Philadelphia, PA: WB Saunders Company; 1997: in press.
130. Mauduit P, Herman G, Rossignol B. Protein secretion induced by isoproterenol or pentoxifylline in lacrimal gland: Ca^{2+} effects. *Am J Physiol.* 1984; 246:C37-C44.
131. Mizokami A, Yeh SY, Chang C. Identification of 3',5'-cyclic adenosine monophosphate response element and other cis-acting elements in the human androgen receptor gene promoter. *Mol Endocrinol.* 1994;8:77–88.

132. Jumblatt JE, North GT, Hackmiller RC. Muscarinic cholinergic inhibition of adenylate cyclase in the rabbit iris-ciliary body and ciliary epithelium. *Invest Ophthalmol Vis Sci.* 1990;31:1103–1108.
133. Sullivan DA. Influence of the hypothalamic-pituitary axis on the androgen regulation of the ocular secretory immune system. *J Steroid Biochem.* 1988;30:429–433.
134. Zhuang YH, Blauer M, Ylikomi T, Tuohimaa P. Spermatogenesis in the vitamin A-deficient rat: possible interplay between retinoic acid receptors, androgen receptor and inhibin alpha-subunit. *J Ster Biochem Mol Biol.* 1997;60:67–76.
135. Hall RE, Tilley WD, McPhaul MJ, Sutherland RL. Regulation of androgen receptor gene expression by steroids and retinoic acid in human breast-cancer cells. *Int J Cancer* 1992;52:778–784.
136. Devos A, Claessens F, Heyns W, Rombauts W, Peeters B. Distinctive interactions of nuclear proteins on the highly homologous promoter regions of two differentially androgen-regulated cystatin-related protein genes. In: Abstracts of the 12th International Symposium of the Journal of Steroid Biochemistry & Molecular Biology (Berlin, Germany); 1995;68P.
137. Sheflin LG, Brooks EM, Keegan BP, Spaulding SW. Increased epidermal growth factor expression produced by testosterone in the submaxillary gland of female mice is accompanied by changes in poly-A tail length and periodicity. *Endocrinol.* 1996;137:2085–2092.
138. MacLean HE, Warne GL, Zajac JD. Defects of androgen receptor function: from sex reversal to motor neuron disease. *Mol Cell Endocr.* 1995;112:133–141.
139. Brinkmann AO, Jenster G, Ris-Stalpers C, van der Korput JAGM, Brüggenwirth HT, Boehmer ALM, Trapman J. Androgen receptor mutations. *J Steroid Biochem Mol Biol.* 1995;53:443–448.
140. Martel C, Melner MH, Gagné D, Simard J, Labrie F. Widespread tissue distribution of steroid sulfatase, 3β-hydroxysteroid dehydrogenase/$Æ^5$-$Æ^4$ isomerase (3β-HSD), 17β-HSD 5α-reductase and aromatase activities in the rhesus monkey. Mol Cell Endocrinol. 1994;104:103–111.
141. Luu-The V, Sugimoto Y, Puy L, Labrie Y, Solache IL, Singh M, Labrie F. Characterization, expression, and immunohistochemical localization of 5α-reductase in human skin. *J Invest Dermatol.* 1994;102:221–226.
142. Lahita RG. The connective tissue diseases and the overall influence of gender. *Int J Fertil.* 1996; 41:156–165.
143. Rocha EM, Wickham LA, Silveria LA, Krenzer KL, Toda I, Sullivan DA. Identification of androgen receptor protein and 5α-reductase mRNA in human ocular tissues. Submitted for publication, 1997.
144. Labrie F, Bélanger A, Simard J, Luu-The V, Labrie C. DHEA and peripheral androgen and estrogen formation: Intracrinology. *Ann NY Acad Sci.* 1995;774:16–28.
145. Sooriyamoorthy M, Gower DB, Eley BM. Androgen metabolism in gingival hyperplasia induced by nifedipine and cyclosporin. *J Periodont Res.* 1990;25:25–30.
146. Purohit A, Ghilchik MW, Duncan L, Wang DY, Singh A, Walker MM, Reed MJ. Aromatase activity and interleukin-6 production by normal and malignant breast tissues. *J Clin Endocr Metab.* 1995;80:3052–3058.
147. Macdiarmid F, Wang D, Duncan LJ, Purohit A, Ghilchick MW, Reed MJ. Stimulation of aromatase activity in breast fibroblasts by tumor necrosis factor alpha. *Mol Cell Endocr.* 1994;106:17–21.
148. Azzarolo AM, Kaswan RL, Mircheff AK, Warren DW. Androgen prevention of lacrimal gland regression after ovariectomy of rabbits. *Invest. Ophthalmol Vis Sci Suppl.* 1994;35:1793.
149. Jacobs M, Buxton D, Kramer P, Lubkin V, Dunn M, Herp A, Weinstein B, Southren AL, Perry H. The effect of oophorectomy on the rabbit lacrimal system. *Invest Ophthalmol Vis Sci Suppl.* 1986;27:25.
150. Huang SM, Azzarolo AM, Mircheff AK, Esrail R, Grayson G, Heller K, Zimmerman K, Feldon S, Warren DW. Does estrogen directly affect lacrimal gland function. *Invest Ophthalmol Vis Sci.* 1995;36:S651.
151. Coles N, Lubkin V, Kramer P, Weinstein B, Southren L, Vittek J. Hormonal analysis of tears, saliva, and serum from normals and postmenopausal dry eyes. *Invest Ophthalmol Vis Sci Suppl.* 1988;29:48.
152. Krasso I. Die behandlung der erkrankungen des vorderen bulbusabschnittes mit buckys grenzstrahlen. *Ztschr f Augenh.* 1930;71:1–11.
153. Lubkin V, Kramer P, Nash R, Bennett G. Evaluation of safety and efficacy of topical 17β-estradiol, 0.1% and 0.25%, in postmenopausal dry eye syndrome. In: Abstracts of the Second International Conference on the Lacrimal Gland, Tear Film and Dry Eye Syndromes: Basic Science and Clinical Relevance (Bermuda); 1996:160.
154. Azzarolo AM, Mircheff AK, Kaswan R, Warren DW. Hypothesis for an indirect role of estrogens in maintaining lacrimal gland function. *Invest Ophthalmol Vis Sci Suppl.* 1993;34:1466.
155. Laine M, Tenovuo J. Effect on peroxidase activity and specific binding of the hormone 17b-estradiol and rat salivary glands. *Arch Oral Biol.* 1983;8:847–852.
156. Prijot E, Bazin L, Destexhe B. Essai de traitment hormonal de la keratocon-jonctivite seche. *Bull Soc Belge Ophtalmol.* 1972;162:795–800.
157. Verbeck B. Augenbefunde und stoffwechselverhalten bei einnahme von ovulationshemmern. *Klin Mbl Augenheilk* 1973;162:612–621.

158. Saceda M, Lippman ME, Lindsey RK, Puente M, Martin MB. Role of an estrogen receptor-dependent mechanism in the regulation of estrogen receptor mRNA in MCF-7 cells. *Mol Endocrinol.* 1989;3:1782–1787.
159. Martinazzi M, Baroni C. Controllo ormonale delle ghiandola lacrimale extraorbitale nel topo con nanismo ipofisario. *Folia Endocrinol.* 1963;16:123–132.
160. Ostachowicz M, Jettmar A, Laukienicki A. Próba leczenia zespolu Sjögrena hormonami plciowymi zenskimi. *Wiad Lek.* 1973;11:1075–1077.
161. Lubkin V, Nash R, Kramer P. The treatment of perimenopausal dry eye syndrome with topical estradiol. *Invest Ophthalmol Vis Sci Suppl.* 1992;33:1289.
162. Akramian J, Wedrich A, Nepp J, Sator M. Estrogen therapy in keratoconjunctivitis sicca. *Adv Exp Med Biol.* 1997: in press.
163. Valde G, Ghini M, Gammi L, Passarini M, Schiavi L. Effets des contraceptifs oraux triphases sur la secretion lacrymale. *Ophtalmologie* 1988;2:129–130.
164. Christ T, Marquardt R, Stodtmeister R, Pillunat LE. Zur Beeinflussung der tränenfilmaufreibzeit durch hormonale kontrazeptiva. *Fortschr Ophthalmol.* 1986;83:108–111.
165. Brennan NA, Efron N. Symptomatology of HEMA contact lens wear. *Optom Vis Sci.* 1989;66:834–838.
166. Gurwood AS, Gurwood I, Gubman DT, Brzezicki. Idiosyncratic ocular symptoms associated with the estradiol transdermal estrogen replacement patch. *Optom Vis Sci.* 1995;72:29–33.
167. Medical economics data of Medical Economics Co, Inc. *Physicians Desk Reference.* Montvale, NJ, 1993:895–898.
168. Carlsten H, Tarkowski A, Holmdahl R, Nilsson LA. Oestrogen is a potent disease accelerator in SLE-prone MRL lpr/lpr mice. *Clin exp Immunol.* 1990;80:467–473.
169. Ahmed SA, Aufdemorte TB, Chen JR, Montoya AI, Olive D, Talal N. Estrogen induces the development of autoantibodies and promotes salivary gland lymphoid infiltrates in normal mice. *J Autoimmunity* 1989;2:543–552.
170. Homo-Delarche F, Durant S. Hormones, neurotransmitters and neuropeptides as modulators of lymphocyte functions. In: Rola-Pleszczynski M, editor. *Handbook of Immunopharmacology.* London: Academic Press Ltd; 1994:169–240.
171. Cutolo M, Sulli A, Seriolo B, Masi AT. Estrogens, the immune response and autoimmunity. *Clin Exp Rheumatol.* 1995;13:217–226.
172. Ahmed SA, Talal N. Importance of sex hormones in systemic lupus erythematosus. In: Wallace D, Hahn B, editors. *Dubois' Lupus Erythematosus.* Philadelphia: Lea & Febiger 1993; 148–156.
173. Newton CJ, Arzt E, Stalla GK. Involvement of the estrogen receptor in the growth response of pituitary tumor cells to interleukin-2. *Biochem Biophys Res Commun.* 1994;205:1930–1937.
174. Jung-Testas I, Lebeau MC, Catelli MG, Baulieu EE. Cyclosporin A promotes nuclear transfer of a cytoplasmic progesterone receptor mutant. *Comptes Rendus de l Academie des Sciences - Serie Iii, Sciences de la Vie* 1995;318:873–878.
175. Jahn R, Padel U, Porsch PH, Soling HD. Adrenocorticotrophic hormone and alpha-melanocyte stimulating hormone induce secretion and protein phosphorylation in the rat lacrimal gland by activation of a cAMP-dependent pathway. *Eur J Biochem.* 1982;126:623–629.
176. Leiba H, Garty NB, Schmidt-Sole J, Piterman O, Azrad A, Salomon Y. The melanocortin receptor in the rat lacrimal gland: a model system for the study of MSH (melanocyte stimulating hormone) as a potential neurotransmitter. *Eur J Pharmacol.* 1990;181:71–82.
177. Cripps MM, Bromberg BB, Patchen-Moor K, Welch MH. Adrenocorticotropic hormone stimulation of lacrimal peroxidase secretion. *Exp Eye Res.* 1987;45:673–683.
178. Entwistle ML, Hann LE, Sullivan DA, Tatro JB. Characterization of functional melanotropin receptors in lacrimal glands of the rat. *Peptides* 1990;11:477–483.
179. Martinazzi M. Effetti dell'ipofisectomia sulla ghiandola lacrimale extraorbitale del ratto. *Folia Endocrinol.* 1962;150:120–129.
180. Minami A, Kamel T. Sur la glande lacrymale extérieure chez le rat et ses modifications aprés hypophysectomie. *Compt rend Séanc Soc Biol.* 1959;153:269–273.
181. Wegelius O, Friman C. Two different pituitary-controlled sulphation mechanisms in the rat. *Acta Medica Scand Suppl.* 1964;412:221–228.
182. Pochin EE. The mechanism of experimental exophthalmos caused by pituitary extracts. *Ciba Foundation Colloquia on Endocrinology* 1952;4:316–326.
183. Ebling FJ, Ebling E, Randall V, Skinner J. The synergistic action of alpha-melanocyte stimulating hormone and testosterone on the sebaceous, prostate, preputial, harderian and lachrymal glands, seminal vesicles and brown adipose tissue in the hypophysectomized-castrated rat. *J Endocr.* 1975;66:407–412.

184. Frey WH, Nelson JD, Frick ML, Elde RP. Prolactin immunoreactivity in human tears and lacrimal gland: Possible implications for tear production. In: Holly FJ, editor. *The Preocular Tear Film: In Health, Disease and Contact Lens Wear.* Lubbock, TX: Dry Eye Institute; 1986:798–807.
185. Wood RL, Park K-H, Gierow JP, Mircheff AK. Immunogold localization of prolactin in acinar cells of lacrimal gland. *Adv Exp Med Biol.* 1994;350:75–77.
186. Mircheff AK, Warren DW, Wood RL, Tortoriello PJ, Kaswan RL. Prolactin localization, binding, and effects on peroxidase release in rat exorbital lacrimal gland. *Invest Ophthalmol Vis Sci.* 1992; 33:641–650.
187. Markoff E, Lee DW, Fellows JL, Nelson JD, Frey WH. Human lacrimal glands synthesize and release prolactin. *Endocrinol Suppl.* 1995;152:440A.
188. Zhang J, Whang G, Gierow J, Warren D, Mircheff A, Wood R. Prolactin receptors are present in rabbit lacrimal gland. *Invest Ophthalmol Vis Sci.* 1995;36:S991.
189. Warren DW, Platler BW, Azzarolo AM, Huang ZM, Wang G, Wood RL, Mircheff AK. Pilocarpine stimulates prolactin secretion by rabbit lacrimal glands. *Invest Ophthalmol Vis Sci.* 1995;36:S651.
190. Azzarolo AM, Kaswan RL, Huang ZM, Platler BW, Mircheff AK, Warren DW. Is lacrimal gland prolactin content regulated? *Invest Ophthalmol Vis Sci.* 1995;36:S651.
191. Narukawa S, Kanzaki H, Inoue T, Imai K, Higuchi T, Hatayama H, Kariya M, Mori T. Androgens induce prolactin production by human endometrial stromal cells in vitro. *J Clin Endocrinol Metab.* 1994;78:165–168.
192. Nevalainen MT, Martikainen P, Valve EM, Ping W, Nurmi M, Härkönen PL. Prolactin regulation of rat and human prostate. In: Abstracts of the 12th International Symposium of the Journal of Steroid Biochemistry & Molecular Biology (Berlin, Germany) 1995;53P.
193. Reiter E, Bonnet P, Sente B, Dombrowicz D, de Leval J, Closset J, Hennen G. Growth hormone and prolactin stimulate androgen receptor, insulin-like growth factor-I (IGF-I) and IGF-I receptor levels in the prostate of immature rats. *Mol Cell Endocrinol.* 1992;88:77–87.
194. McBlain WA, Hoffman RA, Buzzell GR. Androgen receptor in the harderian glands of the golden hamster: characterization and the effects of androgen deprivation, the pituitary, and gender. *J Exp Zoology* 1994;268:442–451.
195. Reber PM. Prolactin and immunomodulation. *Amer J Med.* 1993; 95:637–644.
196. McMurray R, Keisler D, Kanuckel K, Izui S, Walker SE. Prolactin influences autoimmune disease activity in the female B/W mouse. *J Immunol.* 1991;147:3780–3787.
197. Hann LE, Tatro J , Sullivan DA. Morphology and function of lacrimal gland acinar cells in primary culture. *Invest Ophthalmol Vis Sci.* 1989;30:145–158.
198. Khong PL, Peh WC, Low LC, Leong LL. Variant of the Triple A syndrome. *Australasian Radiol.* 1994; 38:222–224.
199. Heinrichs C, Tsigos C, Deschepper J, Drews R, Collu R, Dugardeyn C, Goyens P, Ghanem GE, Bosson D, Chrousos GP. et al. Familial adrenocorticotropin unresponsiveness associated with alacrima and achalasia: biochemical and molecular studies in two siblings with clinical heterogeneity. *Eur J Pediatrics* 1995;154:191–196.
200. Tsigos C, Arai K, Latronico AC, DiGeorge AM, Rapaport R, Chrousos GP. A novel mutation of the adrenocorticotropin receptor (ACTH-R) gene in a family with the syndrome of isolated glucocorticoid deficiency, but no ACTH-R abnormalities in two families with the triple A syndrome. *J Clin Endocrinol & Metab.* 1995;80:2186–2189.
201. Chavez M, Moreno C, Perez A, Garcia F, Solis J, Cargone A, Astete M, Contardo C. Sindrome de Allgrove (acalasia-alacrima-insuficiencia adrenal): Reporte de un caso. *Revista de Gastroenterologia del Peru* 1996;16:153–157.
202. Djeridane Y. Immunohistochemical evidence for the presence of vasopressin in the rat harderian gland, retina and lacrimal gland. *Exp Eye Res.* 1994;59:117.
203. Kahan IL, Varsanyi-Nagy M, Toth M, Nadrai A. The possible role of tear fluid thyroxine in keratoconus development. *Exp Eye Res.* 1990;50:339–343.
204. Ramos-Remus C, Suarez-Almazor M, Russell AS. Low tear production in patients with diabetes mellitus is not due to Sjogren's syndrome. *Clin Exp Rheumatol.* 1994;12:375–380.
205. Stolwijk TR, Kuizenga A, van Haeringen NJ, Kijlstra A, Oosterhuis JA, van Best JA. Analysis of tear fluid proteins in insulin-dependent diabetes mellitus. *Acta Ophthalmologica* 1994; 72:357–362.
206. Gause I, Isaksson O, Lindahl A, Eden S. Effect of insulin treatment of hypophysectomized rats on adipose tissue responsiveness to insulin and growth hormone. *Endocrinol.* 1985;116:945.
207. Heinze E, Kleine, W, Voigt KH. Insulin release in rats 1 and 5 days after hypophyectomy. *Horm Res.* 1981;14:243.
208. Van Lan V, Yamaguchi N, Garcia MJ, Ramey ER, Penhos JC. Effect of hypophysectomy and adrenalectomy on glucagon and insulin concentrations. *Endocrinol.* 1974;94:671–675.

209. Tabbara KF, Frayha RA. Alternate-day steroid therapy for patients with primary Sjögren's syndrome. *Ann Ophthalmol.* 1983;15:358–361.
210. Nasu M, Matsubara O, Yamamoto H. Post-mortem prevalence of lymphocytic infiltration of the lacrymal gland: a comparative study in autoimmune and non-autoimmune diseases. *J Pathol.* 1984;143:11–15.
211. Fox PC, Datiles M, Atkinson JC, Macynski AA, Scott J, Fletcher D, Valdez IH, Kurrasch RHM, Delapenha R, Jackson W. Prednison and piroxicam for treatment of primary Sjögren's syndrome. *Clin Exp Rheumatol.* 1993;11:149–156.
212. Singh G, Kaur J. Iatrogenic dry eye: late effect of topical steroid formulations. *J Indian Med Assoc.* 1992;90:235–237.
213. Wilson SE, Lloyd SA, Kennedy RH. Fibroblast growth factor receptor-1, interleukin-1 receptor, and glucocorticoid receptor messenger RNA production in the human lacrimal gland. *Invest Ophthalmol Vis Sci.* 1993;34:1977–1982.
214. Perkovich CL, Ubels JL, Lee SY, Soprano DR. Cellular retinol-binding protein and cellular retinoic acid-binding protein in the lacrimal gland. *Exp Eye Res.* 1993;56:513–519.
215. Yamaguchi K, Gaur VP, Young RW, Sweatt AJ. Cellular retinoic acid-binding protein in rat lacrimal gland. *Invest Ophthalmol Vis Sci.* 1991;32:3273–3276.
216. Ubels JL, Dennis MH, Rigatti BW, Vergnes JP, Beatty R, Kinchington PR. Nuclear retinoic acid receptors in the lacrimal gland. *Curr Eye Res.* 1995;14:1055–1062.
217. Ubels JL, Osgood TB. 13-cis retinoyl-beta-glucuronide in lacrimal gland fluid of rabbits treated with 13-cis retinoic acid. *J Ocular Pharmacol.* 1990;6:321–327.
218. Rismondo V, Ubels JL. Isotretinoin in lacrimal gland fluid and tears. *Arch Ophthalmol.* 1987;105:416–420.
219. Ubels JL, Dennis M, Lantz W. The influence of retinoic acid on growth and morphology of rat exorbital lacrimal gland acinar cells in culture. *Curr Eye Res.* 1994;13:441–449.
220. Rismondo V, Ubels JL, Osgood TB. Tear secretion and lacrimal gland function of rabbits treated with isotretinoin. *J Am Acad Dermatol.* 1988;19:280–285.
221. Vercaeren I, Vanaken H, Devos A, Peeters B, Verhoeven G, Heyns W. Androgens transcriptionally regulate the expression of cystatin-related protein and the C3 component of prostatic binding protein in rat ventral prostate and lacrimal gland. *Endocrinol.* 1996;137:4713–4720.
222. Lima PH, Georges SA, Remington SG, Nelson JD. mRNA in mouse lacrimal gland with homology to salivary androgen-binding protein. *Invest Ophthalmol Vis Sci.* 1997;38:S148.
223. Sullivan DA. Gender, sex steroids and dry eye. In: *Research to Prevent Blindness Science Writers Seminar.* New York: Research to Prevent Blindness, 1997: in press.
224. Yolton DP, Yolton RL, Lopez R, Bogner B, Stevens R, Rao D. The effects of gender and birth control pill use on spontaneous blink rates. *J Amer Optom Assoc.* 1994;65:763–770.
225. Öhman L. Topical estrogen treatment of postmenopausal chronic conjunctivitis. In: Abstracts of the IVth International Congress of the International Society of Dacryology (Stockholm, Sweden); 1996:91.
226. Stubbs AP, Lalani EN, Stamp GW, Hurst H, Abel P, Waxman J. Second messenger up-regulation of androgen receptor gene transcription is absent in androgen insensitive human prostatic carcinoma cell lines, PC-3 and DU-145. *FEBS Letters* 1996;383:237–240.
227. Wilder RL. Adrenal and gonadal steroid hormone deficiency in the pathogenesis of rheumatoid arthritis. *J Rheumatol Suppl.* 1996;44:10–12.
228. Ormandy CJ, Clarke CL, Kelly PA, Sutherland RL. Androgen regulation of prolactin-receptor gene expression in MCF-7 and MDA-MB-453 human breast cancer cells. *Int J Cancer* 1992;50:777–782.
229. Smirnova OV, Petraschuk OM, Kelly PA. Immunocytochemical localization of prolactin receptors in rat liver cells: I. Dependence on sex and sex steroids. *Mol Cell Endocrinol.* 1994;105:77–81.
230. Wright LL, Luebke JL. Somatostatin-, vasoactive intestinal polypeptide- and neuropeptide Y-like immunoreactivity in eye- and submandibular gland-projecting sympathetic neurons. *Brain Res.* 1989;494:267–275.
231. Tumer N, Mortimer ML, Roberts J. Gender differences in the effect of age on adrenergic neurotransmission in the heart. *Exp Gerontol.* 1992;27:301–307.
232. Li Z, Duckles SP. Influence of gender on vascular reactivity in the rat. *J Pharm Exp Therapeutics* 1994;268:1426–1431.
233. Durban EM, Nagpala PG, Barreto PD, Durban E. Emergence of salivary gland cell lineage diversity suggests a role for androgen-independent epidermal growth factor receptor signaling. *J Cell Sci.* 1995;108:2205–2212.
234. Ikonen T, Palvimo JJ, Kallio PJ, Reinikainen P, Janne OA. Stimulation of androgen-regulated transactivation by modulators of protein phosphorylation. *Endocrinol.* 1994;135:1359–1366.

235. Dedhar S, Rennie PS, Shago M, Hagesteijn CY, Yang H, Filmus J, Hawley RG, Bruchovsky N., Cheng H, Matusik RJ et al. Inhibition of nuclear hormone receptor activity by calreticulin. *Nature* 1994;367:480–483.
236. Culig Z, Hobisch A, Cronauer MV, Hittmair A, Radmayr C, Bartsch G, Klocker H. Activation of the androgen receptor by polypeptide growth factors and cellular regulators. *World J Urol.* 1995;13:285–289.
237. Nakhla AM, Romas NA, Rosner W. Estradiol activates the prostate androgen receptor and prostate-specific antigen secretion through the intermediacy of sex hormone-binding globulin. *J Biol Chem.* 1997;272:6838–6841.
238. Kaku K. Experimental studies on the function of the lacrimal gland. II. Report histological findings of various endocrine glands in lacrimotectomized rats. *Acta Soc Ophthalmol Jpn.* 1955;59:975–979.

3

ANDROGEN-REGULATED TRANSCRIPTION IN THE EPITHELIUM OF THE RAT LACRIMAL GLAND

F. Claessens, H. Vanaken, I. Vercaeren, G. Verrijdt, A. Haelens, E. Schoenmakers, P. Alen, A. Devos, B. Peeters, G. Verhoeven, W. Rombauts, and W. Heyns

Division of Biochemistry, and
Laboratory for Experimental Medicine and Endocrinology
University of Leuven
Campus Gasthuisberg, Leuven, Belgium

1. INTRODUCTION

Androgens are male sex hormones produced by the testes and, to a lesser extent, by the adrenals and ovaries. The responses evoked by androgens can be very diverse depending on the tissue under investigation: sexual accessory glands depend on androgens for organogenesis, maintenance, and cellular differentiation, whereas in other organs such as kidney, liver, salivary gland, and lacrimal gland, a more limited number of genes are influenced.[1]

Androgens are steroid hormones that bind to specific receptors in the target cells. The receptor for androgens is a transcription factor belonging to the large gene family of nuclear receptors.[2] The androgen receptor (AR) contains a ligand-binding domain, a DNA-binding domain, and a large N-terminal domain. After complexation of the receptor with the hormone, the complex is translocated to specific DNA sites in the chromatin called androgen response elements (AREs).[3] Subsequently, the transcription of the nearby located genes can be activated.

The best-studied target organ for androgens is the rat ventral prostate. It secretes several proteins under androgenic control, the most abundantly expressed being the C1, C2, and C3 components of prostatic (steroid) binding protein (PBP), proline-rich polypeptides, and cystatin-related proteins (CRP).[4] Much to our surprise, we found that some of the androgen-inducible gene products originally described in rat ventral prostate are also expressed in the rat lacrimal gland.[5] Recently, we were able to demonstrate expression of the C3 component of PBP and CRP in primary cultures of epithelial cells from lacrimal glands.[6]

Lacrimal Gland, Tear Film, and Dry Eye Syndromes 2
edited by Sullivan *et al.*, Plenum Press, New York, 1998

Other genes have also been reported to be androgen-responsive in the lacrimal gland, the best-studied being secretory component (SC), which is the polymeric Ig receptor, and the androgen receptor itself.[7,8] These genes are also responsive to androgens in rat ventral prostate.[9] Moreover, tear lipocalin originally described in human tears has recently been demonstrated to be expressed in prostate epithelium as well.[10]

We have isolated the genes that code for the C1, C2, and C3 components of PBP, for the CRPs of the rat, and the human SC gene, in order to define the molecular mechanisms of their androgen responsiveness.[11–14] This initially resulted in the description of AR-binding sites in the promoter regions and in the first intron of the C1, C2, and C3(1) genes.[11] Subsequently, AREs were described in these genes by *in vitro* binding studies with receptor fragments produced in *E. coli* [15,16] and with full-size androgen receptors prepared in a vaccinia expression system.[17] Functionality of these AREs was demonstrated in co-transfection experiments.[18] Moreover, these AREs are part of more complex androgen responsive units (ARUs) which can be located upstream, but also within the introns and exons of the target genes.[19,20] Such ARUs contain binding sites not only for the androgen receptor, but also for other transcription factors, which might be involved, for example, in the tissue specificity of the expression of these genes.

2. MATERIALS AND METHODS

Restriction and other enzymes were obtained from Pharmacia (Uppsala, Sweden), Boehringer (Mannheim, Germany) and Promega (Madison, WI, USA). Standard molecular biology techniques were used.[21]

The cell line T-47D, a human mammary carcinoma line that contains functional androgen receptors, was obtained from the ATCC (Rockville, MD, USA). Transient transfections, induction by hormones, and CAT assays were performed as described earlier.[18] Mutations or substitutions were made as described.[9]

3. RESULTS

3.1. Androgen-Induced Expression in the Epithelial Cells of the Lacrimal Gland

The androgen-induced expression of SC in primary cultures of epithelial cells of male rat lacrimal glands has been well studied.[7,8] In media samples from these cultures, provided by Dr. D.A. Sullivan, we were able to detect expression of CRP, although it was not clearly androgen-modulated (results not shown). In primary cultures of epithelial cells from lacrimal glands taken from female rats, however, we can clearly demonstrate by Western blotting (Fig. 1) that expression of not only SC and CRP, but also of C3 is strongly androgen-responsive. The complexity of the androgen regulation of transcription of the C3 gene is illustrated in the next paragraphs.

3.2. The C3(1) Intronic ARU as Enhancer Trap

The first introns of the C1 and C3(1) genes were shown to contain an AR-binding site by DNA cellulose competition assays.[12] By DNase I footprinting, the boundaries and exact sequence of the binding sites were established.[22] To analyze the functional relevance

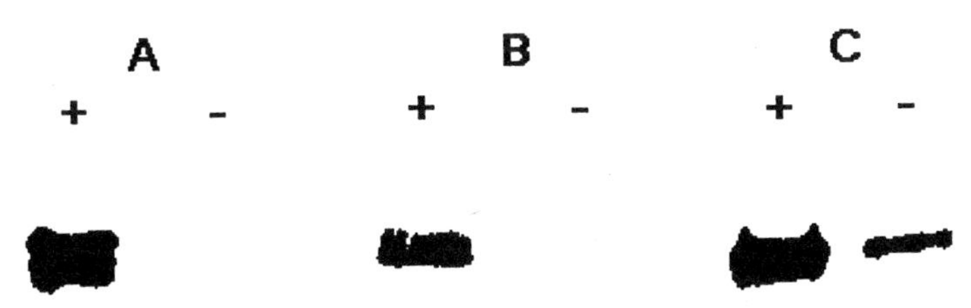

Figure 1. Androgen-regulated proteins expressed by epithelial cells of rat lacrimal gland. Medium of epithelial cells of rat lacrimal gland cultured as in ref. 6 was TCA-precipitated and analyzed by Western blot with antibodies specific for (A) C3, (B) CRP and (C) SC. Cells were cultured in medium supplemented with (-) ethanol or (+) 10^{-7} M R1881, an androgen agonist.

of the AR-binding sites, we substituted the C3 (1) ARE for the C1 androgen receptor-binding site as well as for the consensus glucocorticoid response element (GRE) or the other ARE-like sequence, called core I, found in the C3(1) intron. As shown in Fig. 2, the constructs containing the C3(1)ARE or the GRE consensus are clearly androgen inducible; constructs containing the C1 binding site or core I are not responsive.

3.3. The C3(1) ARE Is Functionally Modulated by the Prostate-Specific Footprint

In the C3 (1) intronic ARU, binding sites for a nuclear factor 1 (NF1), an octamer transcription factor (OTF), and a prostate-specific factor (PSF) are present immediately upstream of the ARE[19] (Fig. 2). To examine the implications of the immediate flanking sequences on the androgen response mediated by the C3(1) ARE, we made the constructs depicted in Fig. 3, containing parts of the ARU, cloned upstream of a reporter gene consisting of the thymidine kinase promoter driving transcription of a chloramphenicol acetyltransferase gene (tk-CAT). Although the PSF was observed only with nuclear extracts from rat ventral prostate, from the functional data it is clear that it can function as a cis-acting repressor sequence in T-47D cells.

Another remarkable cis-acting element was detected when fragments containing the ARE and its immediate downstream sequence were included in the reporter constructs. Indeed, this sequence exhibits a clear positive effect on the androgen response mediated by

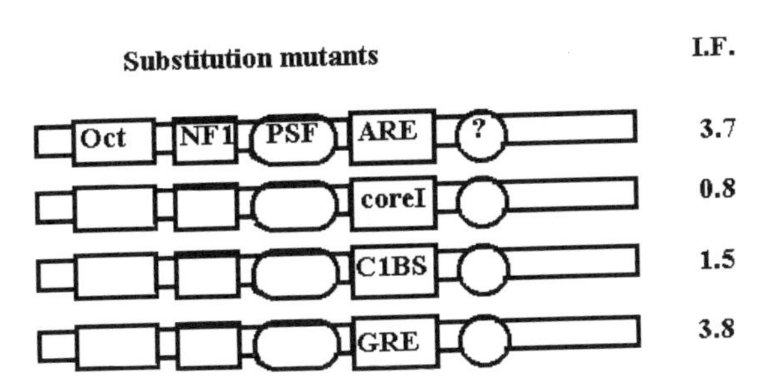

Figure 2. The C3(1) ARU depends on a strong response element. (Left) The original C3(1) intron and the substitution mutants as cloned in front of a tk-CAT reporter gene. Constructs were transfected in T-47D cells. (Right) The induction factors (I.F.) as the CAT-activity in extracts from cells (transfected with the corresponding constructs) grown in the presence of 10^{-7} M 5α DHT relative to CAT activity obtained in non-induced cells. I.F. given here is the mean of at least three independent measurements.

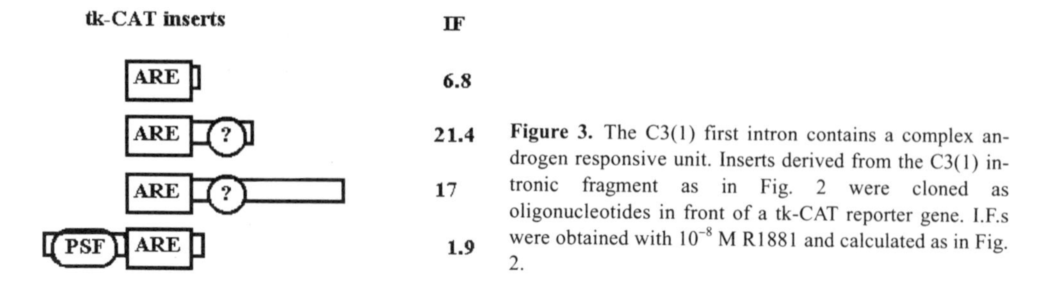

Figure 3. The C3(1) first intron contains a complex androgen responsive unit. Inserts derived from the C3(1) intronic fragment as in Fig. 2 were cloned as oligonucleotides in front of a tk-CAT reporter gene. I.F.s were obtained with 10^{-8} M R1881 and calculated as in Fig. 2.

the ARE. Although we did not detect a footprint on this region in our earlier experiments, it contains an OTF consensus sequence of the type 5'-TAATGARAT-3.[23] However, in gel retardations with purified DNA-binding domain of the OTF-1 factor (a kind gift of Dr. P. van der Vliet, University of Utrecht, Netherlands), we could not detect any binding (results not shown).

4. DISCUSSION

Several androgen-regulated prostate-specific genes are expressed in other androgen-target tissues like salivary and lacrimal glands.[24] To explain this puzzling pattern of expression, we probably need more information on the function of these proteins and on the mechanisms of regulation of the corresponding genes. Based on the homology among the prostatic binding protein components C1, C2, and C3, the rabbit uteroglobin, and the human CC10 protein, they can be catalogued in the same gene family. Putative functions of proteins could be the downregulation of the immune response or the transport of small lipids.[25] In this respect, it is noteworthy that a C3-like mRNA has been detected in human mammary gland (and not in the prostate); it was called mammaglobin.[26] It will be interesting to verify whether this gene is also active in the lacrimal gland, and whether the steroid regulation of its expression is also conserved.

Recently, we cloned the human gene encoding secretory component.[14] The upstream sequence of the SC gene contains several interesting elements, which could explain not only its androgen and glucocorticoid responsiveness, but also its regulation by cytokines and peptide hormones.
[7–9,27] Moreover, an ARU could be delineated in the far upstream region.[28] The human tPA gene, expressed under androgen control in the prostate and after injury in many tissues, also contains in its far upstream sequence a complex enhancer that is inducible by both androgens and glucocorticoids.[29] These data clearly illustrate the importance of androgens as modulators of gene expression in the lacrimal gland.

Here, we report in more detail on the mechanisms of transcriptional regulation of the C3(1) gene which is expressed in an androgen-inducible way in rat lacrimal gland and prostate as well.[5] Androgen receptor-binding regions in the C1, C2, C3, and CRP1 genes have been documented elsewhere.[12,20] These binding sites can be located within the promoter, within the further upstream region, or even within the genes. Moreover, the relative affinity of these sites for the DNA-binding domain of the androgen receptor can be very low, an observation that probably explains their low activity in transient transfection experiments.[12] It is clear that sequences located immediately adjacent to these AREs can either positively or negatively influence the ARE activity, thus potentiating the weaker

AREs or attenuating the stronger AREs (Figs. 2 and 3). The C3(1) ARU seems to depend on the presence of a strong ARE. Indeed, we observed that the C3(1) ARE and the GRE consensus have a high affinity for the AR-DBD,[30] and were able to confer a clear androgen responsiveness to a reporter gene, even when cloned as such in front of a reporter gene (see also Fig. 3). The low affinity of both the C1 and the C3(1) core I sequences for the AR-DBD[12] together with the presence of the repressing PSF element (Fig. 3) might explain why the substitution mutants did not display androgen responsiveness (Fig. 2). Possibly the repressive nature of the PSF in T-47D cells reflects the inactivity of the C3(1) gene in nonprostate and nonlacrimal tissues. Due to the lack of cell lines homologous to prostate or lacrimal gland epithelium, the exact nature of the PSF-binding factor is difficult to study. In conclusion, the androgen responsiveness of the C3(1) gene seems explained, at least in part, by the presence of a complex androgen responsive enhancer in its first intron.

ACKNOWLEDGMENTS

The authors thank all members of the technical staff for their excellent assistance. We are grateful to Dr. D. A. Sullivan for practical and theoretical advice during our work on the lacrimal gland and secretory component. We thank Dr. J. P. Vaerman for providing the anti-rat SC antibodies. F.C. holds a postdoctoral fellowship of the Belgian National Fund of Scientific Research (N.F.W.O.). G.V. and P.A. hold grants, from I.W.T. Supported by grants from F.G.W.O., G.O.A. and Belgian Cancer Fund.

REFERENCES

1. Walters MR. Steroid hormone receptors and the nucleus. *Endocrine Rev.* 1985;6:512–543.
2. Brinkmann AO, Faber PW, Van Rooij HCJ, et al. The human androgen receptor: domain structure, genomic organisation and regulation of expression. *J Steroid Biochem*. 1989;34:307–310.
3. Evans RM. The steroid and thyroid hormone receptor superfamily. *Science*. 1988;240:889–895.
4. Heyns W. Androgen-regulated proteins in rat ventral prostate. *Andrologia*. 1990;22 (suppl.1):67–73.
5. Winderickx j, Vercaeren I, Verhoeven, G, Heyns W. Androgen-dependent expression of cystatin-related protein (CRP) in the exorbital lacrimal gland of the rat. *J Steroid Biochem Mol Biol*. 1994;48:165–170.
6. Vanaken H, Claessens F, Vercaeren I, Heyns W, Peeters B, Rombauts W. Androgen regulation of secretory component synthesis by lacrimal gland acinar cells in vitro. *Mol Cell Endocrinol*. 1996;121:197–205.
7. Sullivan DA, Kelleher RS, Vaerman JP, Hann LE. Androgen regulation of secretory component synthesis by lacrimal gland acinar cells in vitro. *J Immunol*. 1990;145:4238–4244.
8. Gao J, Lambert RW, Wickham LA, Banting G, Sullivan DA. Androgen control of secretory component mRNA levels in rat lacrimal gland. *J Steroid Biochem Mol Biol*. 1995;52:239–249.
9. Stern JE, Gardner S, Quirk D, Wira CR. Secretory immune system of the male reproductive tract: effects of dihydrotestosterone and estradiol on IgA and secretory component levels. *J Reprod Immunol*. 1992;22:73–85.
10. Holzfeind P, Merschak P, Rogattsch H, et al. Expression of the gene for tear lipocalin/von Ebner's gland protein in human prostate. *FEBS Lett*. 1996;395:95–98.
11. Claessens F, Dirckx L, Delaey, B, et al. The androgen-dependent rat prostatic binding protein: comparison of the sequences in the 5'part and upstream region of the C1 and C2 genes and analysis of their transcripts. *J Mol Endocrinol*. 1989;3:93–103.
12. Claessens F, Rushmere NK, Davies P, Celis L, Peeters B, Rombauts WA. Sequence-specific binding of androgen-receptor complexes to prostatic binding protein genes. *Mol Cell Endrocrinol*.1990;74:203–212.
13. Devos A, De Clercq N, Vercaeren I, Heyns W, Rombauts W, Peeters B. Structure of rat genes encoding androgen-regulated cystatin-related proteins (CRPs): a new member of the cystatin superfamily. *Gene*. 1993;125:159–167.

14. Verrijdt G, Swinnen J, Peeters B, Verhoeven G, Rombauts W, Claessens F. Characterization of the human secretory component gene promoter. *Biochim Biophys Acta.* 1997.
15. De Vos P, Claessens F, Winderickx J et al. Interaction of androgen response elements with the DNA-binding domain of the androgen receptor expressed in E. coli. *J Biol Chem.* 1991;266:3439–3443.
16. De Vos P, Claessens F, Peeters B, Rombauts W, Heyns W, Verhoeven G. Interaction of androgen and glucocorticoid receptor DNA-binding domains with their response elements. *Mol Cell Endocrinol.* 1993;90:R11-R16.
17. De Vos P, Schmitt J, Verhoeven G, Stunnenberg HG. Human androgen receptor expressed in HeLa cells activates transcription in vitro. *Nucleic Acids Res.* 1994;22:1161–1166.
18. Claessens F, Celis L, Peeters B, Heyns W, Verhoeven G, Rombauts W. Functional charaterization of an androgen response element in the first intron of the C3(1) gene of prostatic binding protein. *Biochem Biophys Res Comm.* 1989;164:833–840.
19. Celis L, Claessens F, Peeters B, Heyns W, Verhoeven G, Rombauts W. Proteins interacting with an androgen-responsive unit in the C3(1) gene intron. *Mol Cell Endocrinol.* 1993;94:165–172.
20. Devos A., Claessens F, Alen P, Heyns W, Rombauts W, Peeters B. Identification of a functional androgen response element in the exon 1-coding sequence of the Cystatin-related protein gene crp2. *Mol Endocrinol.* 5, in press
21. Sambrook J, Fritsch EF, Maniatis T. Molecular Cloning: a Laboratory Manual. 2nd ed Cold Spring Harbor, NY: Cold Spring Harbor Laboratory;1989.
22. Claessens F, Celis L, De Vos P, et al. Intronic androgen response elements of prostatic binding protein genes. *Biochem Biophys Res Commun.* 1993;191:688–694.
23. Verrijzer CP, Van der Vliet PC. POU domain transcription factors. *Biochim Biophys Acta.* 1993;1173:1–21.
24. Aumuller G, Arce EA, Heyns W, Vercaeren I Dammshauser I, Seitz J. Immunocytochemical localization of seminal proteins in salivary and lacrimal glands of the rat. *Cell Tissue Res.* 1995;280:171–181.
25. Hagen G, Wolf M, Katyal SL, Singh G, Beato M, Suske G. Tissue-specific expression, hormonal regulation and 5'-flanking gene region of the rat Clara cell 10 kDa protein: comparison to rabbit uteroglobin. *Nucleic Acids Res.* 1990;18:2939–2946.
26. Watson MA, Flemming TP. *Cancer Res.* 1996;56:860–865.
27. Sullivan DA. In: Freier S, ed. The Neuroendocrine-Immune Network. Boca Raton, FL. CRC Press; 1990:199–238.
28. Verrijdt G, Claessens F, Swinnen J, Peeters B,Rombauts W. Characterisation of the human secretory component gene promoter. *Biochem Soc Trans.* 1997; 25:186S.
29. Bulens F, Merchiers P, Ibanez-Tallon I et al. Identification of a multihormonal responsive enhancer far upstream from the human tissue-type plasminogn activator gene. *J Biol Chem.* 1997.
30. Claessens F, Alen P, Devos A, Peeters B, Verhoeven G, Rombauts W. *J Biol Chem.* 1996;271:19013–19016.

4

GENE CLONING OF BM180, A LACRIMAL GLAND ENRICHED BASEMENT MEMBRANE PROTEIN WITH A ROLE IN STIMULATED SECRETION

Anil C. Asrani, Angela J. Lumsden, Rajesh Kumar, and Gordon W. Laurie

Department of Cell Biology
University of Virginia
Charlottesville, Virginia

1. INTRODUCTION

1.1. Basement Membrane-Acinar Cell Interface

Dry eye is the ocular manifestation of a collective secretory deficiency by lacrimal acinar cells.[1] Lacrimal acinar cells are polarized tear protein factories, with tear protein-filled secretory granules packed in the apical cytoplasm adjacent to the acinar lumen into which tear proteins are released. These cells are wrapped into acini by a poorly characterized basement membrane[2] sheet, the complex cellular adhesiveness of which is the molecular basis for exocrine cell polarity.[3] Abnormality of the basement membrane-cell interface is the primary lesion in several autoimmune and genetic diseases.[4] In Goodpasture's syndrome, for example, nephrotoxic autoantibodies appear as a consequence of infection-associated exposure of a sequestered epitope in the $\alpha3$ chain of collagen IV. In Alport's syndrome, function-altering point mutations in the collagen IV $\alpha3$ chain have been detected.

Could a similar defect in the lacrimal acinar cell-basement membrane interface contribute to dry eye? One possibility is that a key adhesive component in the interface could be inactivated by circulating autoantibodies. The resultant altered adhesion could diminish cell polarity, thereby displacing not only secretory granules, but also cytoskeletal-associated signalling mediators including phospholipase C-γ, diacylglycerol kinase, phosphatidylinositol phosphate kinase, and phosphatidylinositol kinase.[5] Alternatively, cell polarity could remain normal, but stimulus-secretion coupling is altered. The latter proposal would require the existence of an adhesive component capable of promoting the differentiation and maintenance of stimulus-secretion coupling, a concept proposed by Jamieson[6] based on descriptive and functional studies in the pancreas.

1.2. Functional Identification of BM180

A search for basement membrane components required for regulated tear protein secretion was carried out using overnight primary cultures of lacrimal acinar cells adhering to basement membrane fractions in the presence or absence of anti-basement membrane monoclonal antibodies. This led to the identification of BM180, the presence of which appears to be required for normal regulated, but not constitutive, tear protein secretion.[7] Subsequent studies determined that BM180 shares some homology to the plant protein gliadin,[8] against which autoantibodies have been reported in Sjögren's syndrome.[9] cDNA cloning of BM180 has been initiated using α-gliadin cDNA as a probe.

2. METHODS

2.1. Secretion Studies

Secretion experiments[7] were conducted using overnight cultures of freshly isolated lacrimal acinar cells,[10] plated on the following: (i) 10 mM EDTA extract of mouse EHS tumor basement membrane, BMS[11]; (ii) a laminin-rich, high-molecular-weight peak of BMS; and (iii) a partially BM180-enriched, low-molecular-weight peak of BMS. BMS fractions were prepared by gel filtration.[7] Tear protein peroxidase activity was used as a marker[12] to assess both constitutive secretion and carbachol/VIP-stimulated[13] regulated secretion.

2.2. Screening of cDNA Libraries

Newborn mouse and human fetal brain cDNA libraries (in λZAPII and λUniZapXR, respectively) and a human genomic library (in λDASHII) were purchased from Stratagene (La Jolla, CA, USA). Full-length α-gliadin cDNA[8] was obtained from Dr. Tom Okita (Washington State University, Pullman, WA), and labeled with ^{32}P-dCTP by random priming. Libraries were titered and screened with ^{32}P-gliadin cDNA. Positives were picked and rescreened two additional times to ensure clonality. Inserts were excised as pBluescript, sized by restriction digestion and partially sequenced using an ABI semiautomatic sequencer (Biomolecular Research Facility, University of Virginia).

3. RESULTS

3.1. Role of BM180 in Regulated Tear Secretion

Plating on BMS was found to suppress the degradation of the stimulus-secretion coupling pathway commonly observed in overnight lacrimal acinar cell cultures. Indirect evidence using blocking monoclonal antibody 3E12 and an addback strategy suggested that the BMS activity was, in part, attributable to a synergistic relationship between laminin-1 and novel protein BM180. Adding back increasing amounts of BM180-enriched, low-molecular-weight fraction of BMS to a constant amount of laminin-1 resulted

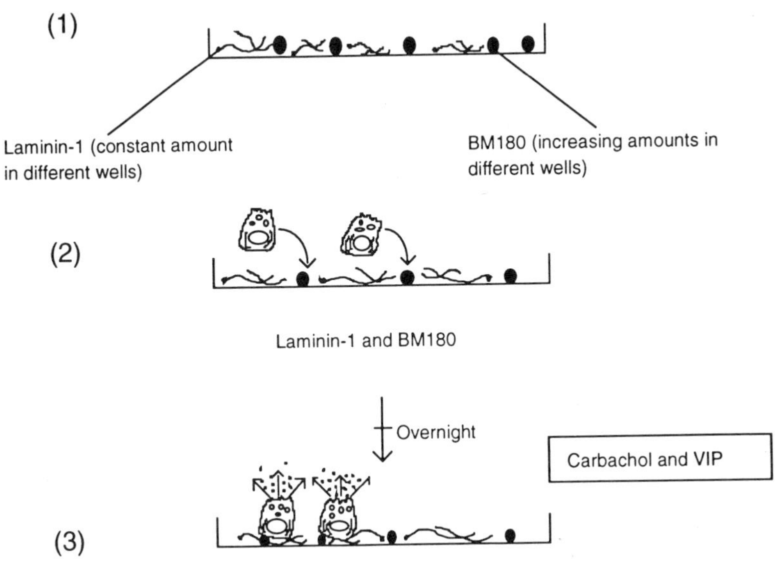

Figure 1. Schematic representation of the addback assay. Laminin-1 and increasing amounts of BM180-enriched peak 2 are coated simultaneously. Lacrimal acinar cells are plated overnight and stimulated the next day with carbachol and VIP. Peak 2 promotes a dose-dependent increase in regulated, but not constitutive, secretion.

in a functional reconstitution of BMS, thereby increasing regulated, but not constitutive, secretion in a dose-dependent manner.[7] The addback approach is presented in Fig. 1.

N-terminal sequencing of BM180 revealed homology to the plant protein α-gliadin, a result reinforced by Western blot and ELISA using BMS and anti-gliadin monoclonal and polyclonal antibodies.[8] Similar cross-species homology exists between the selectin family of lymphocyte-homing proteins and plant lectins.[14]

3.2. Preliminary Model of BM180

A preliminary model of BM180 has been developed using these data (Fig. 2). BM180 is a trimer of 60 kD monomers linked by interchain disulfide bonds; it is distinguished by the N-terminal sequence VRVPVPQLQPQNP. BM180 also shares an epitope with anti-gliadin monoclonal antibody KG9, the antigen of which is amino acids 206–217 of α-gliadin. Lack of cross-reactivity with other anti-α-gliadin monoclonal antibodies (WB8 and WC2)[8] suggests that the N-terminal homology is quite limited.

3.3. Screening of cDNA Libraries

Mouse and human cDNA libraries and a human genomic library were probed with ^{32}P-labeled α-gliadin cDNA under wash conditions of 0.1X sodium citrate, sodium chloride (SSC), 0.1% SDS at 42–50°C. This strategy led to the identification of positive clones in all three libraries (Table 1). Homology of the clones to α-gliadin was confirmed by Southern hybridization. Sequencing through inserts using the nested deletion approach,

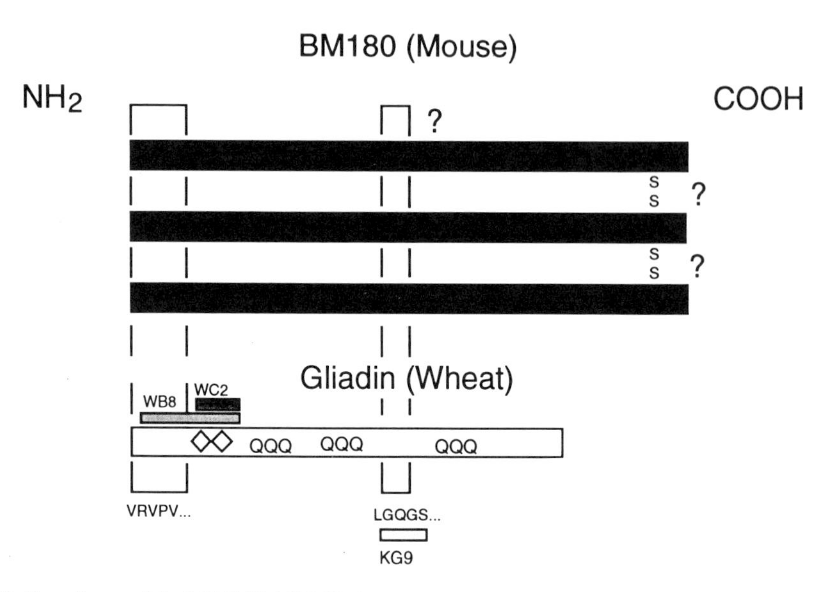

Figure 2. Tentative model of BM180. BM180 shares limited sequence and immunological similarity with α-gliadin. KG9, WB8, and WC2 are monoclonal antibodies prepared against α-gliadin. KG9, but not WB8 or WC2, appears to cross-react with BM180. Site of interchain disulfide bonding and location of the KG9 epitope in BM180 are hypothetical.

Table 1. cDNA cloning—in progress

Probe/Approach (Stringency)	cNDA Library/mRNA	cDNA Clones Name	Size (Kb)	Novel	Sequence identities (%/bp)
1. α-Gliadin cDNA (0.1xSSC, 42-50°C)	Mouse Newborn Brain (dT/Random)	2a	2.0	Y	E-B Virus (66/55) SSA/Ro (64/45)
		4a	2.7	Y	Neuroligin (69/313)
		7a	3.0	Y	cAMP kin (83/618)
	Human Fetal Brain (dT)	1a	2.3	Y	Hunt. Dis (86/78)
		2a	2.5	Y	CpG clone (80/129)
		3a	2.5	Y	Retino sus (81/95)
		4a	2.9	Y	Hunt. Dis (74/558)
		5a	5.0	Y	C. eleg (76/37)
	Human Genomic	2			
		7			
		8			
		12			
2. 5' Oligonucleotide (2xSSC, 55°C)	Mouse newborn brain (dT/Random)	Screening underway			
	Human fetal brain (dT)	Screening underway			

Sequence identities: E-B Virus = Epstein-Barr Virus. SSA/Ro = Human SSA/Ro autoantigen. Neuroligin = Rat Neuroligin. cAMP kin = Mouse cAMP dependent protein kinase. Hunt. Dis = Huntington's Disease region Human DNA cosmid. CpG clone = X chromosome contains CpG clone. Retino sus = Human retinoblastoma susceptibility. C. eleg = C. elegans cosmid DHII.

followed by FastA or Blast analysis, revealed all sequences to be novel. Several interesting homologies have been identified at the nucleotide level. For example, mouse clone 2a contains short regions of partial homology to human SS-A/Ro autoantigen and Epstein-Barr virus[15] genes (Table 1). The relevance of these homologies remains to be determined. However, based on lack of hybridization of all clones with an oligonucleotide corresponding to the BM180 N-terminus, the BM180 coding sequence apparently remains unrepresented among these clones. This observation may indicate either that 5' terminal sequence is incomplete, or that other cloning strategies must be devised that will more efficiently select for BM180 cDNA.

4. SUMMARY AND CONCLUSIONS

BM180 is a novel basement membrane component with a role in regulated tear secretion by lacrimal acinar cells. BM180 bears some sequence similarity to α-gliadin, a plant protein against which antibodies have been reported in patients with Sjögren's syndrome. A precedent for plantlike sequence in the mammalian genome is provided by selectins, which possess a plant lectinlike domain involved in inflammatory cell homing. Cloning the mouse and human BM180 gene will aid molecular investigation of lacrimal acinar cell-BM180 interactions and may lead to a new molecular understanding of mechanisms contributing to dry eye.

REFERENCES

1. Sullivan DA, Wickham A, Krenzer KL, Rocha EM, Toda I. Aqueous tear deficiency in Sjögren's syndrome. In: Pleyer U, Hartmann C, Sterry W, eds. *Oculodermal Diseases - Immunology of Bullous Oculo-Muco-Cutaneous Disorders*. Berlin: Aeolus Press; 1997. pp. 95–152.
2. Timpl R. Structure and biological activity of basement membrane proteins. *Eur J Biochem.* 1989;120:487–502.
3. Nelson WJ, Wang AZ, Ojakian GK. Steps in the morphogenesis of a polarized epithelium. *J Cell Sci.* 1990;95:153–165.
4. Hudson BG, Reeders ST, Tryggvason K. Type IV collagen: Structure, gene organization, and role in human diseases. Molecular basis of Goodpasture and Alport syndromes and diffuse leiomyomatosis. *J Biol Chem.* 1993;268:26033–26036.
5. Payrastre B, vanBergen en Henegouwen PM, Brenton M, et al. Phosphoinositide kinase, diacylglycerol kinase, and phospholipase C activities associated to the cytoskeleton: Effect of epidermal growth factor. *J Cell Biol.* 1991;115:121–128.
6. Jamieson JD. Plasmalemmal glycoproteins and basal lamina: Involvement in pancreatic morphogenesis. *Prog Clin Biol Res.* 1982;91:413–427.
7. Laurie GW, Glass JD, Ogle RA, Stone CM, Sluss JR, Chen L. BM180: A novel basement membrane protein with a role in stimulus-secretion coupling by lacrimal acinar cells. *Am J Physiol.* 1996;270:C1743-C1750.
8. Laurie GW, Ciclitira PJ, Ellis HJ, Pogany G. Immunological and partial sequence identity of mouse BM180 with wheat α-gliadin. *Biochem Biophys Res Commun.* 1995;217:10–15.
9. Teppo A-M, Maury CPJ. Antibodies to gliadin, gluten and reticulin glycoprotein in rheumatic diseases: Elevated levels in Sjögren's syndrome. *Clin Exp Immunol.* 1984;57:73–78.
10. Hann LE, Kelleher RS, Sullivan DA. Influence of culture conditions on the androgen control of secretory component production by acinar cells from the rat lacrimal gland. *Invest Ophthalmol Vis Sci.* 1991;32:2610–2621.
11. Matter ML, Laurie GW. A novel laminin E8 cell adhesion site required for lung alveolar formation in vitro. *J Cell Biol.* 1994;124:1083–1090.
12. Bromberg BB, Cripps MM, Welch MH. Peroxidase secretion by lacrimal glands from juvenile F344 rats. *Invest Ophthalmol Vis Sci.* 1989;30:562–568.

13. Dartt DA, Baker AK, Rose PE, Murphy SA, Ronco LV, Unser MF. Role of cyclic AMP and Ca^{2+} in potentiation of rat lacrimal gland protein secretion. *Invest Ophthalmol Vis Sci.* 1988;29:1732–1738.
14. Ley K, Bullard DC, Arbones ML, et al. Sequential contribution of P-selectin to leukocyte rolling in vivo. *J Exp Med.* 1995;181:669–675.
15. Fox RI, Luppi M, Kang HI, Pisa P. Reactivation of Epstein-Barr virus in Sjögren's syndrome. *Springer Semin Immunopathol..* 1991;13:217–231.

5

SENSORY DENERVATION LEADS TO DEREGULATED PROTEIN SYNTHESIS IN THE LACRIMAL GLAND

Doan H. Nguyen,[1] Roger W. Beuerman,[1] Michele A. Meneray,[2] and Dmitri Maitchouk[1]

[1]Department of Ophthalmology
Louisiana State University Eye Center
New Orleans, Louisiana
[2]Department of Physiology
Louisiana State University Medical Center
New Orleans, Louisiana

1. INTRODUCTION

The secretory function of the lacrimal gland is influenced by the release of classic and peptide neurotransmitters from sensory, sympathetic, and parasympathetic nerve terminals in the gland. These neuronal pathways constitute an integrated system that regulates the integrity of the front of the eye, particularly the optical qualities of the cornea. In everyday experience, this relationship is evidenced by the rapid tearing response to corneal stimulation. The release of neuromodulators resulting from activation of these pathways leads to receptor activation, mobilization of intracellular second messengers, exocytosis, and synthesis of secretory proteins. Consequently, the processes of cell signaling and synthesis of new secretory material must be regulated in coordination with secretion to maintain the normal exocrine function of the gland.

The importance of the sensory innervation that links the front of the eye with the lacrimal gland was examined by unilateral ablation of the trigeminal ganglion (TG).[1] We found that this ablation resulted in an enhanced secretory response in the denervated lacrimal gland following activation by cholinergic and adrenergic agonists. In addition, electron microscopy showed an increase in secretory granules and altered acinar structure.[2] These findings were unexpected because ablation of the TG leads to alterations of the corneal epithelium presumed to be exacerbated by decreased lacrimal secretion.[1]

Using differential mRNA display,[3] we identified a gene that is downregulated following sensory denervation. This gene belongs to a serine/threonine family of eIF-2α kinases that play a major regulatory role in protein synthesis at the level of translational initiation.[4] The inability to control the rate-limiting step of translation can lead to deregulated protein synthesis and malignant transformation.[5] In the lacrimal gland, removal of sensory input is hypothesized to result in a stimulatory shift in protein translation. Substance P, which is localized to sensory nerves of trigeminal origin, may be important for maintaining lacrimal function by modulating the normal balance of protein translation to secretion by maintaining normal levels of eIF-2α kinase.

2. MATERIALS AND METHODS

2.1. Sensory Denervation

Ablation of TG cells subserving the lacrimal nerve was carried out as previously described.[1] Then the eyes were taped shut until the rabbit recovered from the anesthesia. All animals ate and drank water normally after the procedure.

2.2. Matrix Assisted Laser Desorption/Ionization (MALDI) Mass Spectrometry Analysis of Tear Proteins

Tear proteins were collected using glass capillaries. The samples were placed in 0.5 ml microfuge tubes, sealed tight with Parafilm, and stored frozen at -70°C until analysis. Samples were mixed with a matrix solution containing α-cyano-4-hydroxycinnamic acid (Aldrich Chemical Co., Milwaukee, WI, USA), placed on a metal sample plate, air dried, and placed into a mass spectrometer.

2.3. In Vitro Translation of Total mRNA

Total mRNA was isolated from the intact and denervated gland using the Quick Prep mRNA isolation kit (Pharmacia, Piscataway, NJ, USA). The translatability of mRNA was examined using a commercially available rabbit reticulocyte lysate (Biorad, Hercules, CA). Procedures for translation were carried out as recommended by the manufacturer. Equal amounts of mRNA were heated to 70°C for 5 min and placed on ice. The reticulocyte lysate translation assay was set up in a final volume of 25 μl containing ^{35}S methionine (0.8 mCi/ml). For analysis of ^{35}S methionine incorporation, 2 μl were TCA-precipitated, and total radioactivity was counted.

2.4. In Vitro Translation of Isolated Acinar Cells

Acinar cells from intact and denervated glands were isolated with few modifications of reported procedures.[1,6] All steps were done on ice and all solutions were RNase free. The procedures of Carroll and Lucas-Lenard[6] were used. The post-mitochondrial supernatant was collected by centrifugation. For each translation reaction, the supernatant was adjusted with 0.4 mCi/ml ^{35}S methionine to a final volume of 25 μl. Translation was carried out by incubating at 30°C for 30 min. SDS/PAGE was analyzed as described above.

2.5. In Vitro Kinase Phosphorylation Assay

The kinase buffer was prepared as described by Carroll and Lucas-Lenard.[6] The post-mitochondrial supernatant (5–20 μg) was placed in buffer containing 40 μCi/ml $\gamma^{32}P$

ATP (1500–3000 Ci/mmole) to a final volume of 14 μl. The reactions were heated to 42°C for 3 min, excess eIF-2α synthetic peptide[7] was added, and the mixture was incubated at 30°C for 30 min. The peptide (ILLSELSRRIR) corresponding to the phosphorylation site of eIF-2α was made by LSU CORE Laboratories. The reactions were quickly ice-chilled, and 5 μl was separated on 10–20% Tris-Tricine Gel (Biorad). The gel was fixed, stained in 0.025% Coomassie brilliant blue, and destained in 10% acetic acid. The gel was dried and autoradiographed at room temperature.

3. RESULTS

3.1. MALDI Mass Spectrometry

Analysis of tears from the intact and denervated glands showed qualitative differences in protein profiles (Fig. 1). The numbers of peaks in the two conditions were about the same, although several peaks were distinct. This suggests that the secretory mechanism was still operative in the denervated gland. The changes in peak profile may be related to the alteration of the protein synthetic machinery, with resultant differences in the secretion of specific proteins.

3.2. Translation of Total mRNA

In vitro translation of isolated total mRNA from the intact and denervated glands, analyzed by SDS/PAGE, showed increases in both intensity and band patterns (Fig. 2A). Closer examination at the 35 kD standard showed a band in the intact reaction products (lanes 6 and 7) that was not present in the denervated reaction products (lanes 3, 4, and 5). From the top of the gel to the 97 kD standard, there were labeled products in the denervated reactions (lanes 3, 4, and 5) and very few to none in the intact reactions (lanes 6 and 7). The fact that mRNA from both the intact and the denervated glands could be translated suggests that denervation does not lead to increased degradation of total cellular mRNA. On the other hand, there may be selective degradation or upregulation of subsets of mRNAs. Scintillation counting of TCA precipitate showed significantly greater ^{35}S methionine incorporation in the denervated assay (Fig. 2B), which agreed with the SDS/PAGE findings. This suggests that total cellular mRNA is neither lost nor degraded but that the massive accumulation of protein granules may result from an increased rate of translation. The increased incorporation may also be the result of the absence or low level of inhibitor of eIF-2α kinase.

3.3. Acinar Translation Assay

We determined that two lacrimal glands are required for preparation of the translation lysate (Fig. 3A). SDS/PAGE showed an overall greater band intensity of labeled products in the denervated lysate. Furthermore, there were no distinct bands in either sample, suggesting that there was no overall shift in gene transcription, but that translation was more efficient in the denervated gland. Analysis of ^{35}S methionine incorporation showed more incorporation in the denervated sample. These results further suggest that translation in the denervated gland may be more efficient, leading to a higher level of protein synthesis.

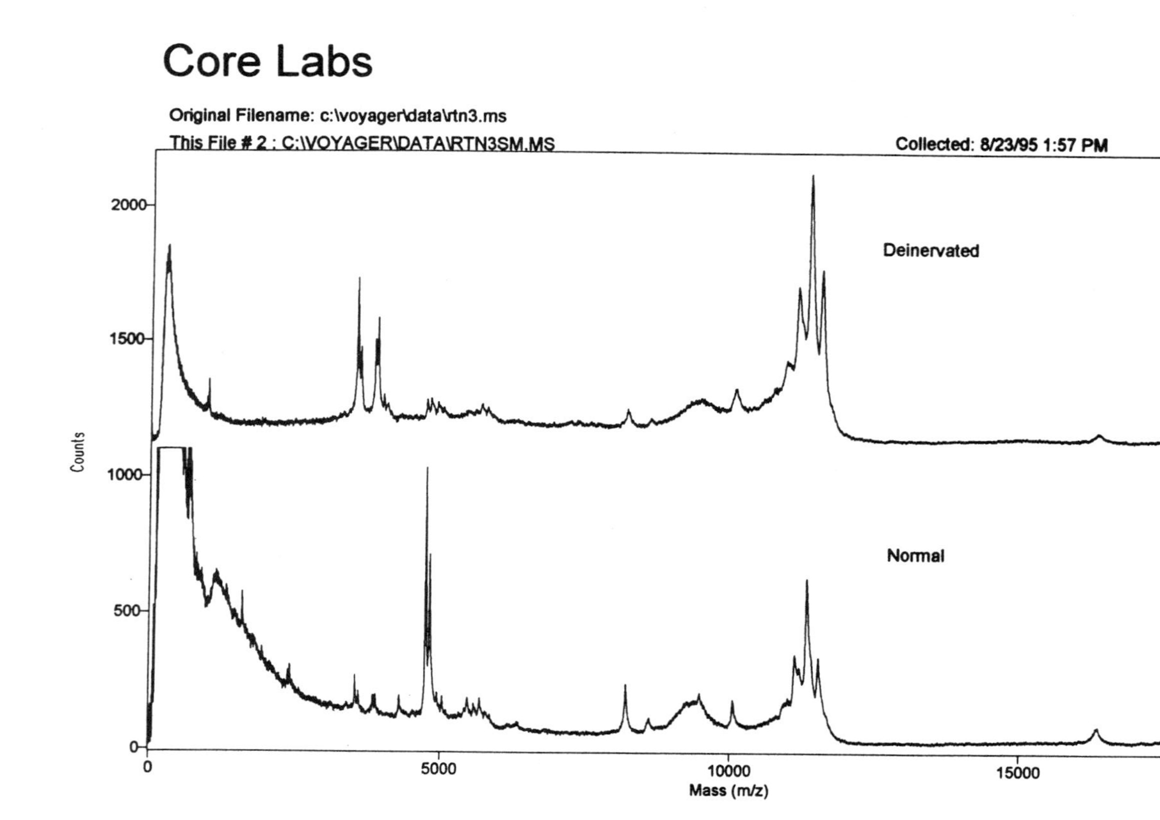

Figure 1. Mass spectra comparing proteins in the tears of the two eyes from one rabbit after unilateral trigeminal ganglion ablation. The results suggest a qualitative change in the nature of the tear proteins after sensory denervation.

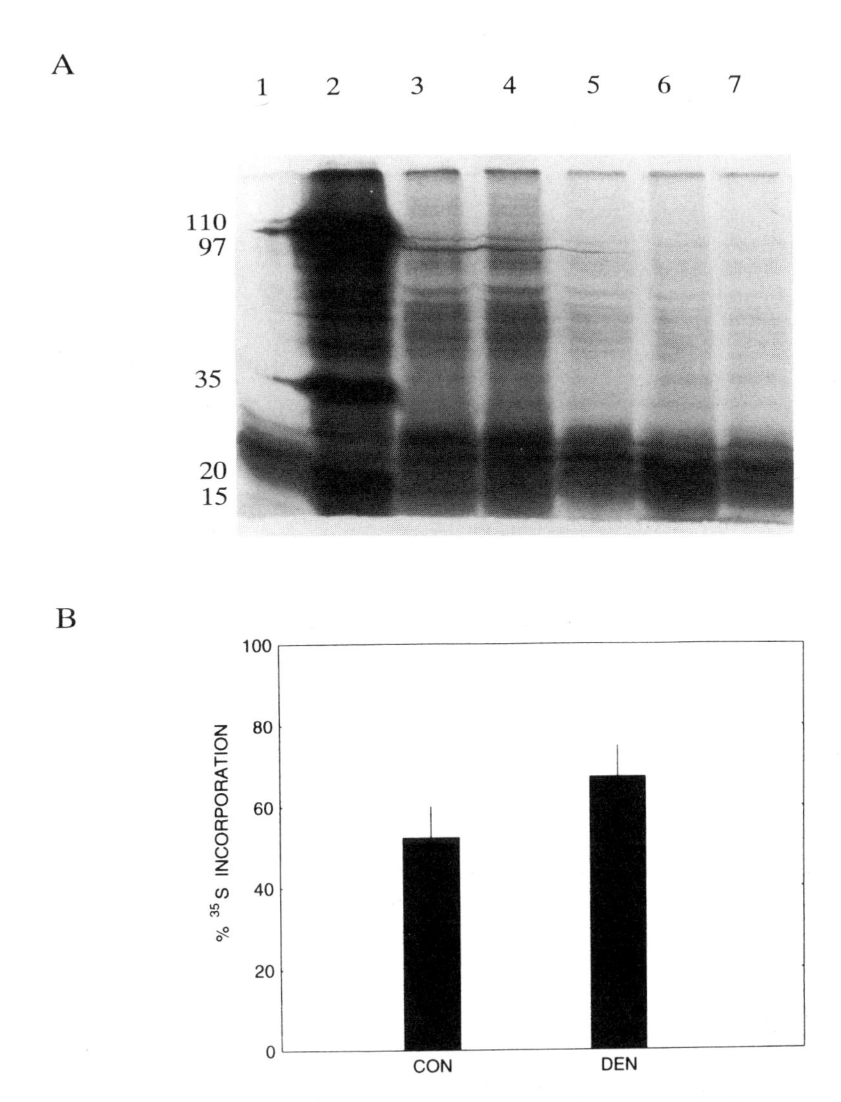

Figure 2. In vitro reticulocyte protein assay of control and denervated lacrimal mRNA. (A) The amount of protein synthesized following sensory denervation (Ln 3, 4, 5) was greater than the amount synthesized by the intact control (Ln 6, 7). Ln 1, no mRNA. Ln 2, BMV standard (positive control). Ln 3, denervated mRNA (22.5 μg/ml final concentration). Ln 4, denervated mRNA (12.5 μg/ml). Ln 5, denervated mRNA (5 μg/ml). Ln 6, intact control mRNA (22.5 μg/ml). Ln 7, intact control mRNA (12.5 μg/ml). (B) Histogram of ^{35}S incorporation of TCA precipitates, expressed as control (con) or denervated (den), above background (no mRNA). The amounts of incorporation were significantly greater in the denervated than in the control mRNA ($p<0.05$, n=6).

3.4. Kinase Phosphorylation Assay

The eIF-2α peptide (mol wt 1500), which contains the phosphorylation site (Ser 51) of eIF-2, has been shown to be a specific substrate for eIF-2α kinases in vitro.[7] To examine the amounts of eIF-2α kinase in each lysate preparation, the lysates were used at 1–20 μg. Because eIF-2α kinase is activated via autophosphorylation, the lysate was heat-

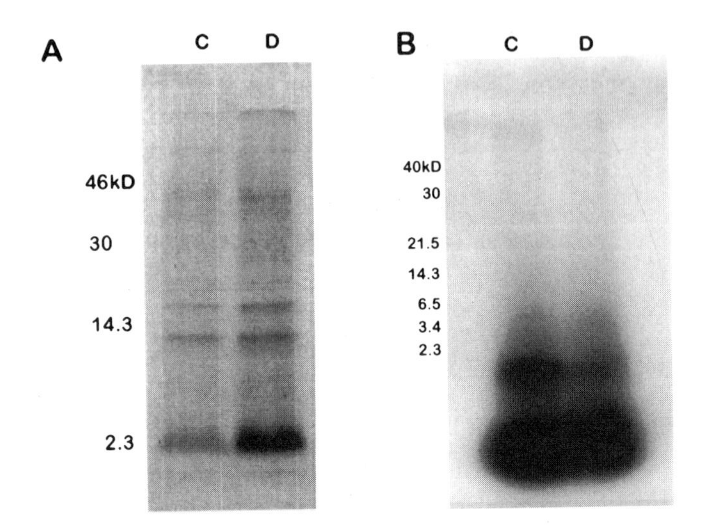

Figure 3. (A) SDS/PAGE analysis of translated products using acinar lysate from the intact (c) and denervated (d) glands. The results suggest greater translational efficiency (band intensity) in the denervated than in the intact gland. (B) Analysis of eIF-2α kinase activity in the intact (c) and denervated (d) glands using a synthetic peptide substrate for eIF-2α kinase. The lower level of phosphorylation in the denervated gland lysate suggests decreased levels/activity of eIF-2α kinase. A lower level of phosphorylated eIF-2α kinase should result in increased efficiency of translation, as shown in (A).

treated to activate eIF-2α kinase before addition of the peptide. The results showed that the amount of phosphorylation of the peptide was low in the denervated sample (Fig. 3B). This suggests that the level of eIF-2α kinase may be low or that its activity may be affected in the denervated gland.

4. DISCUSSION

Selective sensory denervation leads to the alteration of the corneal epithelium, presumed to be exacerbated by decreased reflex lacrimal protein secretion.[1] A change in secretomotor signals may, over time, result in reduction in the volume of the gland.[8] Lacrimal gland disruption, reduction of the size and number of secretory vesicles, and an increase in inflammatory cells have been reported in aged animals.[9] These changes are accompanied by the loss of neuropeptide and neurotransmitter immunoreactive fibers. In addition, decreased sensory function of corneal afferents is a well-recognized change with aging.[10] In the salivary gland, reinnervation reverses the reduced glandular size and constituents. Consequently, decreased innervation may result in loss of lacrimal function that may underlie symptoms of neuroparalytic cornea, including delayed corneal wound healing and increased incidence of ulceration.[11]

In normal glands, neurostimulation leads to release of secretory proteins and a rapid, compensatory increase in protein synthesis critical for maintenance of lacrimal exocrine function. We hypothesized authors that decreased stimulation should lead to depletion of secretory granules. In contrast to parasympathectomy, which results in depletion of secretory proteins, sensory denervation was followed by massive accumulation of immature secretory granules suggestive of enhanced protein synthesis.[12] Tears collected prior to gland removal and analyzed by MALDI mass spectrometry showed qualitative differences in protein profiles.

Translation of total mRNA using a rabbit reticulocyte lysate showed greater band intensity and greater incorporation of radioactive methionine. The fact that mRNAs from both the intact and the denervated glands were translatable suggests that denervation did not result in the overall degradation of total cellular mRNA. On the other hand, denervation may affect the stability of particular classes of mRNA. Because messages for secretory proteins are present in abundance and are relatively stable, their expression is controlled at the post-transcriptional level.[13] Secretory proteins may represent subsets of mRNA that are selectively translated and stimuli dependent. Initiation of their synthesis is then under the control of translation factors and regulatory proteins.

The rate of protein synthesis is controlled at the level of translational initiation. This is the major control step for the activation or inhibition of protein synthesis in response to cellular signal and is catalyzed by a family of eIF-2α kinases. These kinases are activated by various stress stimuli such as heat shock, amino acid deprivation, viral infection, and calcium mobilizing agonists.[14] Activated eIF-2α kinase, in turn, phosphorylates the alpha subunit of initiation factor 2 (eIF-2αβγ) and leads to inhibition of protein synthesis at the level of translational initiation. The inhibitory response is rapid, with a shift to the synthesis of specific proteins such as the heat shock proteins.[14] Using differential mRNA display, we have identified the loss or much reduced level of this kinase following sensory denervation. This finding, along with that reported by Meneray et al.,[2] suggests that the control of initiation is lost in the denervated gland.

We have reconstituted the translational machinery of the lacrimal acinar cells by preparing active acinar lysate that can be used to examine protein translation in vitro. The results from SDS/PAGE and ^{35}S methionine incorporation suggest that protein synthesis is more efficient in the denervated gland. An increased efficiency of translation suggests that the denervated acinar cells are geared toward active protein synthesis and implicates a shift toward polysome formation, indicative of increased translational initiation events. Examination of the activity of eIF-2α kinase in both conditions using a synthetic peptide as target in an in vitro kinase assay showed lower levels of phosphorylation in the denervated gland, indicating a reduced level or activity of eIF-2α kinase. Since mRNA levels are not affected, these two results suggest a decreased level of translational control.

The reduction in eIF-2α kinase, which controls translational initiation, has important implications in lacrimal physiology, as patterns of protein synthesis are influenced by alterations in translational efficiency. Resynthesis of proteins is characterized by a shift from monomeric ribosomes to polysomes, suggestive of increased translational initiation events.[15] Because secretion and resynthesis of proteins are dynamic processes in exocrine tissues, the transitional shift in the polysomal state may be influenced by receptor activation and second messenger mobilization. In general, it is believed that activation of the cAMP/PKC pathway stimulates protein synthesis, whereas the IP_3/Ca^{2+} pathway is inhibitory.[16] The inhibitory effect of calcium flux is believed to be at the level of translational initiation via the activation of an eIF-2α kinase activity.[17] This process may be altered as a result of the reduced level of eIF-2α kinase in the denervated gland. This suggests that either decreased reflex-stimulated secretion or decreased localized influence of substance P may lead to changes in the level of intracellular calcium and interfere with the calcium-dependent mechanism for controlling protein synthesis. Sensory innervation may affect lacrimal structure and exocrine function by modulating the level of eIF-2α kinase via the regulation of protein translation. Finally, the activity of eIF-2α kinase may be an important link for receptor activation to secretion by modulating protein translation.

ACKNOWLEDGMENTS

This work was supported by U.S. Public Health Service grants EY04074 and EY02377 from the National Eye Institute, National Institutes of Health, Bethesda, MD, and an unrestricted grant from Research to Prevent Blindness, Inc., New York, NY.

REFERENCES

1. Beuerman RW, Schimmelpfennig B. Sensory denervation of the rabbit cornea affects epithelial properties. *Exp Neurol*. 1980;69:196–201.
2. Meneray MA, Bennett JD, Nguyen D, Beuerman RW. Effect of sensory denervation on the morphology and physiologic responsiveness of rabbit lacrimal gland. Exp. Eye Res. Submitted.
3. Liang P, Pardee AB. Differential display of eukaryotic messenger RNA by means of the polymerase chain reaction. *Science*. 1992;257:967–971.
4. Chen JJ, Throop MS, Gehrke L, et al. Cloning of the cDNA of the heme-regulated eukaryotic initiation factor eIF-2α kinase of rabbit reticulocytes: Homology to yeast GCN2 protein kinase and human double-stranded RNA dependent eIF-2α kinase. *Proc Natl Acad Sci USA*. 1991;88:7729–7733.
5. Sonnenberg N. Translation factors as effectors of cell growth and tumorigenesis. *Curr Opin Cell Biol*. 1993;5:955–960.
6. Carroll R, Lucas-Lenard J. Preparation of cell-free translation system with minimal loss of initiation factor eIF-2/eIF-2B activity. *Anal Biochem*. 1993;312:17–23.
7. Mellor H, Proud CG. A synthetic peptide substrate for initiation factor-2 kinases. *Biochem Biophys Res Commun*. 1991;178:430–437.
8. Schneyer CA. Autonomic regulation of secretory activity and growth responses of rat parotid gland. In: Thorn NA, Petersen OH, eds. *Secretory Mechanism of Exocrine Gland*. Copenhagen; Munksgaard: 1974:42–55.
9. Williams RM, Singh J, Sharkey KA. Innervation and mast cells of rat exorbital lacrimal gland: The effects of age. *J Auton Nerv Syst*. 1994;47:95–108.
10. Boberg-Ans J. On the corneal sensitivity. *Acta Ophthalmol (Copenh)*. 1956;34:149–162.
11. de Hass EBH. Desiccation of cornea and conjunctiva after sensory denervation. *Arch Ophthalmol*. 1962;67:439–452.
12. Ruskell GL. Changes in nerve terminals and acini of the lacrimal gland and changes in secretion induced by autonomic denervation. *Z Zellforsch*. 1969;94:261–281.
13. Liu Y, Woon PY, Lim SC, Jeyaseelan K, Thiyagarajah P. Cholinergic stimulation of amylase gene expression in the rat parotid gland: Inhibition by two distinct post-transcriptional mechanisms. *Biochem J*. 1995;305:637–642.
14. Proud CG. Protein phosphorylation in translational control. Curr. Top. Cell. Reg. 1992;32:243–369.
15. Perkins PS, Pandol SJ. Cholecystokinin-induced changes in polysome structure regulate protein synthesis in pancreas. *Biochim Biophys Acta*. 1992;1136:265–271.
16. Kanagasuntheram P, Lim SC. Calcium-dependent inhibition of protein synthesis in rat parotid gland. *Biochem J*. 1978;176:23–29.
17. Brostrom CO, Brostrom MA. Calcium-dependent regulation of protein synthesis in intact mammalian cells. *Annu Rev Physiol*. 1990;52:577–590.

ACINAR CELL BASAL-LATERAL MEMBRANE–ENDOMEMBRANE TRAFFIC MAY MEDIATE INTERACTIONS WITH BOTH T CELLS AND B CELLS

Austin K. Mircheff,[1,3] Tao Yang,[1,3] Jian Zhang,[2] Hongtao Zeng,[1,3]
J. Peter Gierow,[4] Dwight W. Warren,[2] and Richard L. Wood[2]

[1]Department of Physiology and Biophysics
[2]Department of Cell and Neurobiology
[3]Department of Ophthalmology
University of Southern California School of Medicine
Los Angeles, California
[4]Department of Natural Sciences
University of Kalmar, Kalmar, Sweden

1. INTRODUCTION

The lacrimal glands are normally populated by significant numbers of IgA-producing plasma cells and T lymphocytes, with CD8 cells in twofold excess over CD4 cells. The numbers of these cells increase during normal aging, and there has been speculation that they may exacerbate the lacrimal dysfunction of primary lacrimal deficiency. Lymphocytic infiltration increases much more dramatically in Sjögren's syndrome, an immune system disorder in which CD4 cells dominate the T lymphocyte population, and the antibody-producing population includes large numbers of B cells. The B cell infiltrates are polyclonal and often produce IgG and IgM against intracellular antigens, such as the ribonuclear proteins Ro and La and a variety of Golgi proteins. Because of the severity of the ocular surface disease and the debilitating and potentially life-threatening nature of the systemic manifestations of Sjögren's syndrome, it is important to identify the mechanisms that lead to CD4 cell and B cell proliferation in the lacrimal glands.

We have previously shown that a remarkably rapid recycling traffic occurs between the basal-lateral (ie, interstitium-facing) plasma membranes (BLM) and endomembrane compartments of lacrimal gland acinar cells. The half-time for internalization of BLM constituents that could be labeled with sulfo-NHS-biotin was less than 30 sec. Moreover, this traffic was accelerated several-fold by stimulation with 10 μM carbachol.[1,2] We have speculated that BLM–endomembrane traffic might underlie acinar cells' ability to provoke CD4 cell and B cell activation in at least two different ways.

First, it is likely that the ongoing traffic secretes a spectrum of proteins, such as growth factors and extracellular matrix constituents, into the interstitium. The proteins normally secreted via this pathway must be subject to usual mechanisms of immune surveillance, including internalization, processing, and presentation by professional antigen-presenting cells, but they must also be subject to either central or peripheral tolerance. However, the observation that secretomotor stimulation accelerates BLM–endomembrane recycling traffic suggested that sustained or excessive stimulation might induce changes in compartmentation and targetting that alter the spectrum of secreted proteins and thereby threaten to overcome peripheral tolerance.

Second, acinar cells can be induced to express Class II MHC molecules, and they readily do so when placed in primary culture. When MHC II molecules are expressed, they participate in the BLM–endomembrane recycling traffic.[3] Thus, it would seem that if proteases, such as cathepsin B, are present in the endomembrane compartments traversed by MHC II, then proteins that enter these compartments might be subject to catheptic processing and MHC II-mediated presentation analogous to the processing and presentation functions of the professional antigen presenting cells.

As one approach to evaluating the merits of these hypothetical mechanisms, we have used methods of subcellular fractionation analysis to dissect the BLM–endomembrane pathway, to investigate whether its compartmental organization is altered by secretomotor stimulation, and to survey the subcellular distributions of MHC II and cathepsin B. A parallel approach, which exploits a newly discovered autologous mixed cell reaction between lacrimal gland cells and lymphocytes, is discussed by Kaslow et al. in these proceedings.

2. METHODS

We employed a relatively new preparation, acinus-like structures that acinar cells reconstitute when they are placed in primary culture in a serum-free medium supplemented with soluble laminin. These structures have identifiable lumina, as well as cell-cell contacts that prevent permeation of extracellular horseradish peroxidase (HRP).[4]

Cells in the reconstituted acini were allowed take up markers of fluid-phase endocytosis, such as HRP, and markers of receptor-mediated endocytosis, such as [^{125}I]-EGF and [^{125}I]-PRL, at 4°C, 18°C, or 37°C. After chilling to 4°C and washing to remove extracellular medium, cells were lysed in a Balch cell press. Lysates were analyzed according to an analytical fractionation scheme that combined differential centrifugation, isopycnic centrifugation on sorbitol density gradients, and partitioning in aqueous dextran-polyethyleneglycol two-phase systems. These methods have been described in detail.[4]

3. RESULTS

3.1. Delineating Plasma Membrane and Endomembrane Compartments

The density gradient distributions of HRP taken up at 4°C and 37°C and of several intrinsic biochemical markers are depicted in Fig. 1. These distributions suggest that BLM

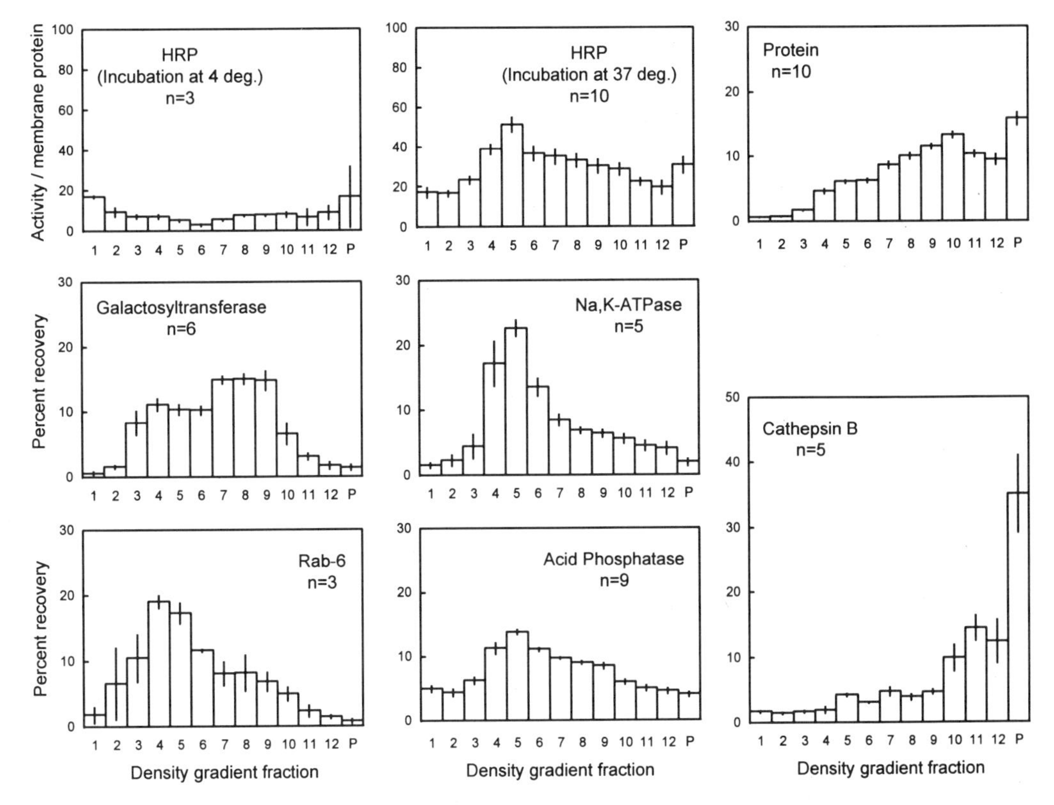

Figure 1. Density gradient distributions of HRP and intrinsic membrane markers. For the other intrinsic markers, protein is expressed as percent recovery. Fraction 1 contains the lowest density membranes, and fraction P contains membranes which formed a pellet beneath the density gradient. (Redrawn from ref. 4 with permission of the publisher.)

constituents, such as Na,K-ATPase, travel between a small surface-expressed pool and a much larger intracellular pool. Moreover, the intracellular pool is distributed among several distinct compartments.

When cells had been incubated with HRP at 4°C, a temperature that suppresses membrane recycling traffic, the plasma membranes were evident as a peak of high-specific activity HRP labeling in the first two fractions of the density gradient (Fig. 1). Lower specific activity labeling in the remaining density gradient fractions was probably due to non-specific phenomena, such as permeation into damaged cells and adsorption to intracellular membranes, since this component of the labeling pattern was reproduced when HRP was added to the cell lysates (not shown).

A striking feature of the data in Fig. 1 is that the BLM, identified by high specific activity HRP labeling in fractions 1 and 2, did not account for the major peak of Na,K-ATPase, an enzyme traditionally used as a marker for the BLM in fractionation analyses of other epithelial cell types. Rather, the membranes containing the Na,K-ATPase peak became labeled only when cells were incubated with HRP at 37°C, indicating that they were derived from one or more endomembrane compartments, not from the BLM. Clues about the identity of these endomembrane compartments were provided by the observation that they accounted for a minor peak in the distribution of the Golgi enzyme, galactosyltransferase, and for the major peak in the distribution of Rab6, a low-molecular-weight GTP-binding protein that has been localized to the Golgi complex and trans-Golgi network in several different cell types.

Further analysis of density gradient fractions 3–6 by phase partitioning resolved two distinct membrane populations, one substantially enriched in galactosyltransferase and the other enriched in acid phosphatase.[4]

Preliminary data, not shown, indicate that both populations contain significant amounts of Rab6, suggesting that they may have been derived from distinct domains of the trans-Golgi network. Accordingly, we have provisionally designated them TGNLC1 and TGNLC2.

The density distribution of endocytosed HRP suggests that fluid phase traffic reaches the Golgi complex (marked by the peak of galactosyltransferase in density gradient fractions 7–9), the lysosomes (marked by the major peak of cathepsin B catalytic activity in the pellet that formed beneath the density gradient), and a presumptive prelysosomal compartment (marked by the minor peak of cathepsin B activity in fractions 10–12). This interpretation is consistent with EM-cytochemical studies of endocytosis across lacrimal gland acinar cells in vivo, which have demonstrated endocytic traffic to the Golgi complex and to lysosome-like structures.[5,6] It has also been confirmed by HRP EM cytochemistry of the reconstituted acinus preparation.[4]

Phase partitioning analyses of density gradient fractions 7+8, 9+10, and 11+12 have resolved several membrane populations that overlapped the distributions of the Golgi population and the prelysosomal population. These included a population, provisionally designated Hex1,[5] which is believed to contain secretory vesicle membrane fragments (marked by a peak of beta-hexosaminidase activity in fractions 9–11; Hamm-Alvarez et al., these proceedings), as well as acid phosphatase- and Na,K-ATPase-containing populations that also became labeled by the endocytic markers and so appeared to represent distinct compartments of the basal-lateral membrane recycling endosome (BLMRE).

Traffic to the TGNLC1 and TGNLC2 is suppressed when the incubation temperature is reduced to 18°C, indicating that these compartments are separated from earlier compartments of the endocytic pathway by a temperature-sensitive step (ie, an "18°C block"). In contrast, the Golgi, prelysosomal, and lysosomal membrane populations all were labeled to similar extents by HRP and [^{125}I]-PRL during incubations at 18°C and at 37°C (not shown). Thus, these compartments are located proximal to the 18°C block. Only small amounts of the endocytosed [^{125}I]-PRL reach the TGNLC1 and TGNLC2, and only small amounts of endocytosed [^{125}I]-EGF reach the prelysosomes and lysosomes, even when cells were incubated at 37°C. Thus, it appears that important sorting processes occur after endocytosis to the BLMRE.

3.2. Stimulation-Induced Changes in Subcellular Compartmentation and BLM Composition

We found that stimulating cells for 20 min with carbachol at a concentration of 10 mM accelerates traffic between the BLMRE and the BLM[1,2,6] and increases traffic to the Golgi complex and the lysosomes.[4] As reviewed elsewhere in these proceedings (Hamm-Alvarez et al.), stimulation also triggers a net shift of beta-hexosaminidase activity from Hex1 to the Golgi complex. This shift has been interpreted as the result of adsorption of secretory product to secretory vesicle membrane constituents that are retrieved and returned to the Golgi complex for recycling.[7] Because vesicular traffic mediates rapid fluid phase communication among the Golgi complex, the TGNLC1, TGNLC2, and BLMRE, we suggest that an increased return of secretory proteins to the Golgi complex during sustained stimulation might change the spectrum of proteins that acinar cells secrete to the interstitium via the BLMRE.

More recent experiments indicate that stimulating cells with carbachol for 20 min also causes a net shift of galactosyltransferase from the Golgi complex to the TGNLC and BLM, and net shifts of prepro- and pro-forms of cathepsin B from the Golgi complex and prelysosomes to the TGNLC and BLM. The mechanisms underlying these changes in compartmentation are not yet known. Since the levels at which galactosyltransferase and the immature cathepsin B forms are expressed in the BLM of resting cells are quite low, it is possible that their enhanced surface display could be an additional occurrence which perturbs peripheral tolerance, ie, by increasing the likelihood of encounters with B cell surface immunoglobulins.

3.3. Co-Localization of MHC II and Cathepsin B

Subcellular fractionation analyses such as those described in Fig. 1 indicate that cathepsin B and cathepsin D are present, along with MHC II, in several, if not all, of the compartments of the BLM–endomembrane traffic. Thus, it would seem that cellular proteins that enter these compartments may be subject to catheptic processing and that resulting peptides may bind MHC II in transit to the BLM.

4. DISCUSSION

There seems little reason to doubt that peptide–MHC II complexes reaching the BLM are accessible to CD4 cell antigen receptors. Whether or not CD 4 cells that recognize these complexes will be activated or rendered tolerant depends on the nature of the available accessory signals, including surface-expressed and secreted factors arising from acinar cells, from autonomic and sensory nerve endings, and from connective tissue and other interstitial cells.

If, as the experimental observations outlined above indicate, acinar cells may both aberrantly display intracellular proteins at their surface membranes and also process and present intracellular proteins via MHC II, then the possibility arises that they can engage in triadic relationships with B cells and CD4 cells.

Proteins aberrantly displayed at the acinar cell BLM would stimulate B cell surface immunoglobulins, peptides bound to acinar cell MHC II would stimulate the antigen receptors on CD4 cells, B cells would provide accessory stimulation to CD4 cells, and CD4 cells would produce cytokines that favor B cell activation. Such interactions would be consistent with the unique character of the lymphocytic infiltrates in Sjögrens's syndrome, as well as with the production of autoantibodies to intracellular, particularly Golgi-related, proteins. They could also account for the production of autoantibodies to ribonuclear protein antigens if, as suggested by others, these antigens appear in the cytoplasm and at the BLM.[8]

One of the many remaining questions is whether or not levels of secretomotor stimulation sufficient to alter intracellular traffic and compartmentation and to threaten peripheral tolerance are likely to occur in vivo. It seems quite plausible to suggest that they are, in view of the principle that the lacrimal glands are effectors in a servomechanism that has evolved to keep the ocular surface well-lubricated and free of irritants. This servomechanism should respond to accelerated tear breakup and evaporation during exposure to dry and polluted environments by increasing secretomotor output. As reported elsewhere in these proceedings (Azzarolo et al.; Warren et al.), certain physiological states, such as the post-menopause, pregnancy, lactation, and oral contraceptive use, entail decreases in the levels of available androgens and, therefore, are likely to trigger atrophy of the lacrimal glands. To continue producing fluid at an adequate rate, the surviving lacrimal tissue must function at a greater fraction of its maximal capacity, and this greater level of function would, presumably, be driven by a greater level of secretomotor stimulation. Thus, the environmental and physiological circumstances in which lacrimal insufficiency is most frequently encountered are circumstances that would appear to favor initiation of autoimmune responses in the lacrimal glands.

ACKNOWLEDGMENTS

This work was supported by NIH grants EY05801 (AKM), EY09405 (DWW), and EY10550 (RLW) and by a Sjögren's Syndrome Foundation postdoctoral research fellowship (JPG).

REFERENCES

1. Lambert RW, Maves CA, Gierow JP, Wood RL, Mircheff AK. Plasma membrane internalization and recycling in rabbit lacrimal acinar cells. *Invest Ophthalmol Vis Sci.* 1993;34:305–316.
2. Gierow JP, Lambert RW, Mircheff AK. Fluid phase endocytosis by isolated rabbit lacrimal acinar cells. *Exp Eye Res.* 1995;60:511–525.
3. Mircheff AK, Wood RL, Gierow JP. Traffic of major histocompatibility complex Class II molecules in rabbit lacrimal gland acinar cells. *Invest Ophthalmol Vis Sci.* 1994;35:3943–3951.
4. Gierow JP, Yang T, Bekmezian A, et al. Na,K-ATPase in lacrimal gland acinar cell endosomal system: Correcting a case of mistaken identity. *Am J Physiol.* 1996;271:C1685-C1698.
5. Farquhar MG. Membrane recycling in secretory cells:Implications for traffic of products and specialized membranes within the Golgi complex. *Methods Cell Biol.* 1988;23:399–427.
6. Oliver, C. Endocytic pathways at the lateral and basal cell surface of exocrine acinar cells. *J Cell Biol.* 1982;95:154–161.
7. Hamm-Alvarez SF, Da Costa S, Yang T, Gierow JP, Mircheff AK. Cholinergic stimulation of lacrimal acinar cells promotes redistribution of membrane-associated kinesin and the secretory enzyme, beta-hexosaminidase, and activation of soluble kinesin. *Exp Eye Res.* 1997;64:141–156.
8. Bachmann M, Althoff H, TrÖster H, Selenk C, Falke D, Müller WEG. Translocation of the nuclear autoantigen La to the cell surface of herpes simplex virus type 1 infected cells. *Autoimmunity* 1992;12:37–45.

7

TISSUE EXPRESSION OF TEAR LIPOCALIN IN HUMANS

Catherine Ressot,[1] Hervé Lassagne,[1] Jean-Louis Kemeny,[2] and Anne-Marie Françoise Gachon[1]

[1]Laboratoire de Biochimie Médicale
Faculté de Médecine
Clermont-Ferrand Cedex, France
[2]Laboratoire d'Anatomie-Pathologique
CHU de Clermont-Ferrand
Clermont-Ferrand Cedex, France

1. INTRODUCTION

Human tear fluid is a very complex mixture. The main lacrimal gland secretes a group of six proteins, the tear lipocalins, characterized by low molecular weight and an acidic pI.[1] These secreted proteins are processed to differing extents at their N-terminal end, resulting in heterogeneity evidenced by multiple spots in two-dimensional electrophoresis.[2,3] These tear proteins belong to the large lipocalin family[4,5] of which the archetype is retinol binding protein (RBP). This is demonstrated by their amino-acid structure (176 amino-acid residues, some highly conserved amino-acid positions),[3] by the corresponding cDNA structure[6] and by gene structure and location.[7–10] However, there is a low degree of overall homology (20–37% for amino acids and 23–30% for nucleic acids) between tear lipocalin and the other human lipocalins.

No specific ligand has been clearly described for tear lipocalin (TL): retinol has been proposed, but its affinity for TL is much lower than for RBP.[11] Lipocalins are associated with a broad array of ligands in the tear film; among them, stearic and palmitic acids, co-eluting in column fractions containing tear lipocalins.[12] The role of TL may be approached through an exploration of its tissue expression. We have obtained[6] the sequence of a cDNA clone from human lacrimal glands (accession number X 67647, H. sapiens mRNA for tear lipocalin). This sequence has been demonstrated to be the same as one from von Ebner's glands, minor salivary glands located on the dorsal surface of the tongue.[13] To investigate the overall expression pattern, 18 different human tissues and various ocular tissues were explored; a polyclonal antibody was used for direct immunodetection of the protein. A radiolabeled cDNA and an anti-sense RNA were used for in situ hybridization (ISH).

2. MATERIALS AND METHODS

Lacrimal gland, salivary glands (parotid, submaxillary, labial, and lingual von Ebner's), trachea, mammary gland, endometrium, testis, pancreas, kidney, adrenal medulla, thyroid gland, spleen, liver, colon, striated muscle, and brain were collected either fresh or within 3 h after death, frozen in isopentane cooled by liquid nitrogen, and stored at -80°C until needed. An entire eyeball was submitted to more gradual freezing to prevent ice crystal formation. The quality of the structures was checked by toluidine blue staining.

Antibodies were prepared as previously described[1] and tested for specificity by Western blot. Immunocytochemical labeling was performed on successive sections. Some sections were used for ISH. After blocking unspecific binding with 1% bovine serum albumin (BSA), the sections were incubated in a 1/2500 dilution of tear lipocalin antiserum for 2 h at room temperature (RT) and treated again with 1% BSA for 15 min. Antigens were detected using a 1/200 dilution of either a fluorescein- or a peroxidase-conjugated goat anti-rabbit IgG antibody (Biosys). Control sections were obtained using pre-immune serum or without the first antibody.

For hybridization experiments, RNA and cDNA probes were constructed. We used a previously sequenced clone 16 cDNA, isolated from a personal human lacrimal gland cDNA library.[3] This cDNA represents a nearly full-length cDNA (17 nucleotides are missing at the non-translated 5' side). A simplified restriction map of the clone 16, used for ISH experiments, is presented in Fig. 1.

An α ^{35}S CTP-labeled anti-sense RNA probe was generated from clone 16 by in vitro transcription (Stratagene), using PstI and T7 polymerase. The presence of a PstI site inside the insert affords a 165-base antisense riboprobe, purified using a spin column (Clontech Chroma Spin 100). A labeled cDNA was generated by the technique of random priming, using ^{35}S dATP following the manufacturer's recommendations (Appligene). The sections were formaldehyde fixed, proteinase K treated (40 µg/ml for 12 min at 37°C), and acetylated in 1/250 triethanolamine acetic anhydride for 30 min at RT.[14] As negative or positive controls, the following probes were prepared: (i) a radiolabeled sense RNA of clone 16, using EcoRV and T3 polymerase; before purification on the spin column, the approx. 800-base-long probe obtained was submitted to alkaline hydrolysis according to Wilkinson et al.,[15] yielding 200 to 300 base-long fragments; (ii) the ubiquitously expressed β-actin (gift of Dr. M. Fiszmann, Institut Pasteur, Paris); (iii) the selectively brain-

Figure 1. Simplified map of the tear lipocalin cDNA insert (clone 16) in pBluescript SK.

expressed subcommissural organ cDNA (gift of Dr A. Meiniel, INSERM 384, Clermont-Ferrand). Controls for specificity of the labeling consisted of an RNase treatment (100 μg/ml in 2X SSC) for 1 h at 37°C, followed by hybridization with the radioactive cDNA.

For RNA probes, the sections were hybridized overnight at 40°C in a moist chamber and then washed once extensively in 4X SSC formamide 50% for 30 min at 25°C, twice with 1X SSC for 30 min at RT and 0.1X SSC for 10 min at 37°C. To remove non-hybridized RNA, the slides were then treated with RNAase for 15 min at 37°C, and washed in 0.01X SSC for 3 h at 37°C.

For DNA probes, the sections were hybridized in the same conditions as for RNA probes and washed in 1X SSC at RT for 30 min, twice with 0.1X SSC for 30 min at 40°C, and 0.1X SSC for 2.5 h.

After dehydration, the sections were covered with nuclear emulsion (Ilford K5) diluted 1/1 in distilled water. After 15 to 21 days of exposure in Clay-Adams boxes, the sections were developed using Kodak D19 and then fixed. After toluidine blue counterstaining to define tissue architecture, the sections were observed with a light microscope equipped with a dark field.

For each tissue, the presence of single-stranded and double-stranded nucleic acid was checked by acridine orange staining on a separate section.

Immunochemistry, using either fluorescein or DAB, showed no staining or fluorescence in the conjunctival septa of the different sections. No labeling, corresponding to hybridization signals, was observed in mammary gland, endometrium, testis, pancreas, thyroid gland, kidney, adrenal medulla, spleen, liver, or colon, or muscular and nervous structures. Among secretory structures, specific labeling was observed in lacrimal glands, von Ebner's glands, submaxillary glands and the secretory units embedded in the submucosa of the trachea. Counterstaining with toluidine blue indicated that the protein was present only in the serous acini. The labeled acini seemed to be all of serous type, some acini remained unlabeled, and the labeling was faint. Positive staining was also observed, with DAB, on the outermost layer of the corneal epithelium but not in the other parts of the eyeball; in this case, the toluidine blue staining was performed on a separate section. Control sections, incubated with preimmune serum or without the first antibody, showed no specific immunofluorescence or peroxidase staining.

3. RESULTS AND DISCUSSION

In situ hybridization results indicated no specific labeling in the conjunctival septa or in the muscular or nervous structures. All the tissues found negative using immunochemistry were negative using ISH. Location of the labeling was the same for ISH experiments as for immunodetection experiments. Intense labeling of the acini structures was obtained with the following tissues: lacrimal glands, von Ebner's glands, submaxillary glands, and secretory units located in the submucosa of the trachea (Fig. 2). Both the ^{35}S clone 16 cDNA probe and ^{35}S anti-sense RNA gave hybridization signals; the conjunctival septa were always unlabeled. However, no labeling was observed on the first layer of the corneal epithelium (a positive signal had been obtained using immunocytochemistry), indicating the absence of tear lipocalin mRNA. No labeling was observed in the positive tissues previously described either after RNase treatment or using a bovine central nervous system probe. Furthermore, an actin ^{35}S cDNA probe produced homogeneous labeling on the whole sliced tissues.

Tear lipocalin was never expressed in muscular or nervous tissues. The specific labeling observed after in situ hybridization of both radiolabeled cDNA and anti-sense RNA

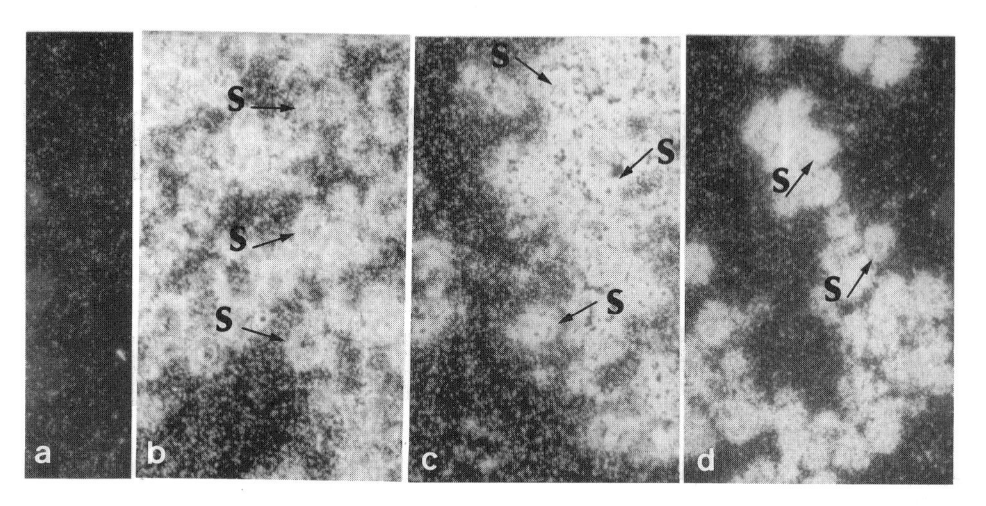

Figure 2. ISH and dark-field images (X400) (a) ISH with a sense probe; only a few silver grains correspond to the background. (b, c, d) ISH with an anti-sense probe: lacrimal gland (b), submaxillary gland (c), and secretory units of tracheal submucosa (d). ISH is performed with the ^{35}S CTP labelled anti-sense RNA probe. Silver grains, indicating the presence of specific transcripts, are observed on the acini cells (S) of each positive tissue.

points to the presence of specific transcripts in serous acini of specific secretory tissues, namely lacrimal, von Ebner's, submaxillary, and tracheal secretory units. Furthermore, well-characterized cDNA probes exhibited either a completely different pattern of labeling (β-actin probe) or no labeling (subcommissural organ probe), when compared to tear lipocalin cDNA or tear lipocalin anti-sense RNA results. The localization of the tear lipocalin proteins and mRNA in the same structures is supported on successive sections, by the tear lipocalin pattern of DAB or immunofluorescence, indicating the presence of the secretory products, and that of the radioactive labeling, indicating the presence of transcripts. Concerning the immunodetected tear lipocalin in the first outermost layer of the corneal epithelium, the negative ISH signal seems to indicate that tear lipocalin, coming from the lacrimal gland, is adsorbed onto only the first epithelial layer of the cornea.

Not all the secretory structures express tear lipocalin, however. No hybridization signal was noted in the other secretory exocrine glands tested: mammary gland, parotid, labial glands, pancreas, testis. No hybridization signal was noted in endocrine glands such as thyroid, pancreas, adrenal medulla, and testis.

4. CONCLUSIONS

For a long time, TL has been described as specific to lacrimal glands and von Ebner's glands. We demonstrate here that TL is also synthesized (presence of mRNA) in the submaxillary gland and the trachea. Tear lipocalin is detected by immunocytochemistry on the outermost layer of the corneal epithelium, but its corresponding mRNA is not detected in any corneal layer by in situ hybridization. This indicates that TL, secreted by the lacrimal gland, is trapped only in the thick glycocalyx of the epithelium and not synthesized by corneal epithelium. TL function remains uncertain; it has been suggested[13,16] that this protein, also expressed in minor salivary glands, von Ebner's glands, is probably not implicated in taste perception as proposed by Schmale et al.[17] A broad array of lipids co-elute with tear lipocalin fractions[12] and binding studies indicate that retinol is a possible ligand.[11] The presence of TL in submaxillary gland and in the secretory units of trachea questions the involvement of TL in protection of mucous membranes of the upper airway.

ACKNOWLEDGMENTS

We would like to thank Dr. A. Meiniel and Dr. R. Meiniel (U 384 INSERM), histology specialists, for free access to their facilities and helpful advice and discussions. Dr. Delatour (Pathological Anatomy Laboratory), Prof. Chipponi (Hepato-Gastro-Enterology Department), Prof. Mondié and Prof. Chazal (Face and Brain Surgery Department) are thanked for their help in tissue collection. This work was supported by a grant from the Ministère de l'Enseignement et de la Recherche (MESR: N° 93Y 0113/1) and by University funding (BQR).

REFERENCES

1. Gachon AM, Verrelle P, Bétail G, Dastugue B. Immunological and electrophoretic studies of human tear proteins. *Exp Eye Res.* 1979;29: 539–553.
2. Fullard RJ, Kissner DM. Purification of the isoforms of tear specific prealbumin. Curr Eye Res. 1991;10:613–628.
3. Delaire A, Lassagne H, Gachon AMF. New members of the lipocalin family in human tear fluid. *Exp. Eye Res.* 1992;55:645–647.
4. Flower DR. The lipocalin protein family: a role in cell regulation. *FEBS Lett.* 1992;354:7–11.
5. Flower DR, Sansom C, Beck ME, Attwood TK.The first prokaryotic lipocalins. *Trends Biochem Sci.* 1995;20:498–499.
6. Lassagne H, Gachon AMF. Cloning of a human lacrimal lipocalin secreted in tears. *Exp Eye Res.* 1993;56:605–609.
7. Glasgow BJ, Heinzmann C, Kojis T, Sparkes R, Mohandas T, Bateman JB. Assignment of tear lipocalin gene to human chromosome 9q34–9qter. *Curr Eye Res.* 1993;12:1019–1023.
8. Lassagne H, Ressot C, Mattei MG, Gachon AMF. Assignment of the human tear lipocalin gene (LCN1) to 9q34 by in situ hybridization. *Genomics* 1993;18:160–161.
9. Lassagne H, Nguyen VC, Mattei MG, Gachon AMF. Assignment of LCN1 to human chromosome 9 is confirmed. *Cytogenet Cell Genet.* 1995;71:104.
10. Holzfeind P, Redl B. Structural organization of the gene encoding the human lipocalin tear prealbumin and synthesis of the recombinant protein in *Escherichia coli. Gene* . 1994;139:177–183.
11. Redl B, Holzfeind P, Lottspeich F. cDNA cloning and sequencing reveals human tear prealbumin to be a member of the lipophilic-ligand carrier protein superfamily. *J Biol Chem.* 1992;28:20282–20287.
12. Glasgow BJ, Abduragimov AR, Farahbakhsh ZT, Faull KF, Hubbell WL. Tear lipocalin bind a broad array of lipid ligands. *Curr Eye Res.* 1995;14:363–372.
13. Bläker M, Kock K, Ahlers C, Buck F, Schmale H. Molecular cloning of human von Ebner's gland protein, a member of the lipocalin superfamily highly expressed in lingual salivary glands. *Biochim Biophys Acta.* 1993;1172:131–137.
14. Meiniel R, Creveaux I, Dastugue B, Meinicl A. Specific transcripts analyzed by in situ hybridization in the subcommissural organ of bovine embryos. *Cell Tissue Res.* 1995;279:101–107.
15. Wilkinson DG, Bailes JA, Champion JE, McMahon AP. A molecular analysis of mouse development from 8 to 10 days post-coïtum detects changes only in embryonic globin expression. *Development.* 1987;99:493–500.
16. Gachon AMF. Lipocalins: Do we taste with our tears? *Trends Biochem Sci.* 1993;18:206–207.
17. Schmale H, Ahlers C, Bläker M, Kock K, Spielman AI. Perireceptor events in taste. In: *The Molecular Basis of Smell and Taste Transduction*, Chichester: Wiley (Ciba Foundation Symposium) 1993;179:167–185.

8

THE EXORBITAL LACRIMAL GLANDS OF THE RAT ARE TENSED IN SITU

Mortimer Lorber

Department of Physiology and Biophysics
Georgetown University School of Medicine
Washington, D.C.

1. INTRODUCTION

The main lacrimal gland of rodents is a paired, extraorbital organ. Each exists subcutaneously anterior to and slightly below the ear. This study will demonstrate that this exorbital lacrimal gland exists under tension. This characteristic became manifest when we noted that the gland's width and length diminished as its attachments were removed during excision. Were it not tensed, no such dimensional changes would have occurred as it was freed (Fig. 1).

Surgeons have long known that incised skin gapes and transected vessels retract. Such adjustments indicate release of pre-existing tension. Experimentally, those observations have been confirmed for both skin[1] and blood vessels.[2] Spaghetti when cut does not retract—only tensed structures do. In contrast to organs that change appreciably in size, such as the lung which exhibits elastic recoil[3] and the spleen which has stretch receptors,[4] the exorbital lacrimal gland is a parenchymal organ that functions while essentially stable in size but, nonetheless, is physically tensed.

2. EXPERIMENTAL PROCEDURES

2.1. Measurements

Under pentobarbital anesthesia, the left exorbital lacrimal glands of eight adult female rats (222 ± 5 g) (mean ± 1 SD) were exposed by removing the overlying skin and fascia. Measurements were made using a dividers to span an organ's greatest width and then length. The dividers' points were inserted in the jaws of a digital micrometer to obtain the values. Those dimensions were determined six times during the sequence of pro-

Lacrimal Gland, Tear Film, and Dry Eye Syndromes 2
edited by Sullivan *et al.*, Plenum Press, New York, 1998

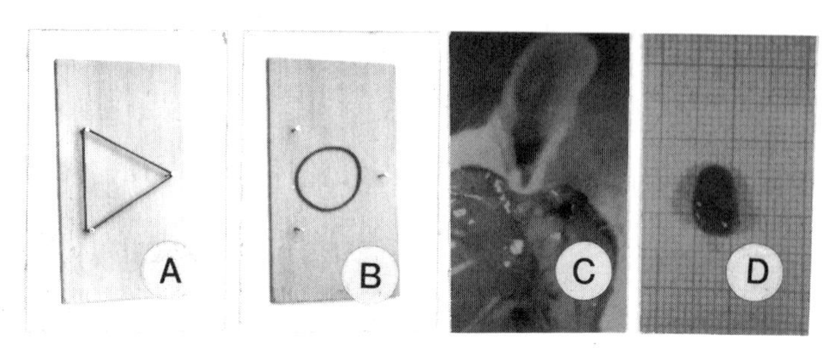

Figure 1. Questions and answers. (A) How do you know the band is stretched? (B) Because now it is smaller. (C) How do you know the gland is stretched? (D) Because now it is smaller.

cedures: initially, following removal of the remaining fascia along the organ's borders and deep surface, and then after ligation and transection of the duct, the artery, and the vein individually. Finally, the excised organ was measured. Only then could its thickness be measured by calibrated calipers.

2.1.1. Estimation of in Situ Organ Thickness. Exorbital lacrimal gland shape approximates a geometric figure. Because the gland's deep surface is flat rather than convex as is a true ellipsoid, the organ can be termed a hemi-ellipsoid. Consequently, in applying the formula for the volume of an ellipsoid,[5] $4\Pi abc/3$, organ thickness represents not "c", but "1/2 c". Because gland width and length were measured at every step of the sequence and because volume was constant, the modified formula could be used to estimate organ thickness *in situ* from its measurement in the excised organ. For simplicity, the usual asymmetry of the gland's rostral and caudal ends has been ignored in relating it to an ideal geometric figure.

2.2. Statistics

Statistical analysis was by ANOVA for dependent variables. Tukey tests were then used for pairwise comparisons. A difference that achieved a p value of < 0.05 was considered significant.

3. RESULTS

Fig. 2 shows the sequential changes in contour as one exorbital lacrimal gland is freed and excised.

The detailed measurements at every step of organ removal appear in Table 1.

Initially, the eight glands had a mean width of 8.78 ± 1.33 mm (mean ± SD) and length of 12.69 ± 1.08 mm. Width did not diminish significantly until excision, when it was 7.59 ± 0.55 mm ($p < 0.05$) (Table 1). This narrowing of 1.19 mm was a 13.6% reduction. In contrast, shortening began when the fascia, the first attachment to be separated, was removed, a 1.11 mm (8.8%) reduction ($p < 0.05$). On excision, mean length was 11.23 ± 0.48 mm, 12.7% less than initially ($p < 0.01$). Thus, as the organs were freed, the decreases in width and length were not simultaneous.

Because the duct, artery, and vein had been doubly ligated before transection, loss of their fluids did not occur. Gland volume must have remained constant other than for water

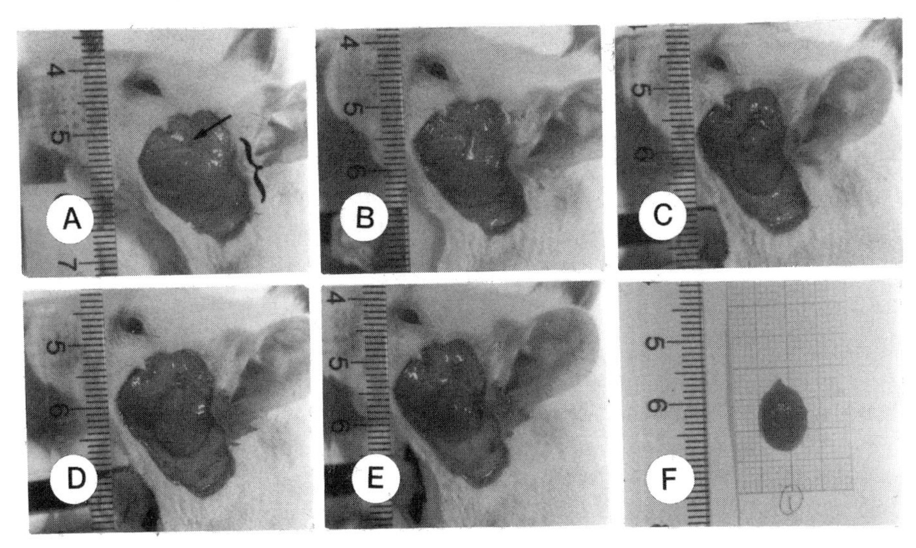

Figure 2. Changes in rat exorbital lacrimal gland dimensions. (A) Before severance of attachments. The arrow points to the duct, and the bracket indicates the gland's location. (B) Fascia removed. (C) Duct transected. (D) Artery transected. (E) Vein transected. (F) Organ excised.

loss by evaporation. In light of that constant volume, as its width and length decreased organ thickness had to increase. Although only the excised organs' thicknesses could be measured, yet by modifying the formula for an ellipsoid (Section 2.1.1) their mean thickness at each preceding step could be estimated (Table 1).

Although the measurements demonstrate that the glands exist under tension, a truer index of the forces *in situ* is obtained by expressing the initial measurements in terms of the excised values. It is only when the glands had become liberated from their attachments that their inherent shape and dimensions became manifest. In terms of the excised values being 100%, *in situ* mean lacrimal gland width was 115.7%, mean length 113.0%, and estimated mean thickness 76.4%. The percentage changes in width and length are even greater than had been noted previously.

Table 1. Measurement data of eight rat exorbital lacrimal glands

	1 Organ *in situ*	2 Fascia Removed	3 Duct Transected	4 Artery Transected	5 Vein Transected	6 Organ Excised
Width	8.78±1.33	8.60±0.90	8.80±1.14	8.54±1.35	8.49±0.91	7.59±0.55
Length	12.69±1.08	11.58±0.68	11.64±0.76	11.37±0.74	10.93±0.62	11.23±0.48
Thickness	2.11*	2.36*	2.30*	2.42*	2.54*	2.76±0.45

Statistics: Width - col. 6 vs col. 1 or 3, $p < 0.01$; col. 6 vs col. 2, 4, or 5, $p < 0.05$
Length - col. 1 vs col. 2, 3, or 4, $p < 0.05$; col. 1 vs col. 5 or 6, $p < 0.01$

All values are millimeters. Width, length, and excised thickness values are mean ± 1 SD.
*Estimated mean value

4. DISCUSSION

Unfortunately, the possible tensile force exerted by the subcutaneous tissue superficial to the lacrimal gland could not be evaluated because of having to remove it to see the organ's borders clearly. It is credible that at least some functions would be affected by tensile forces that are able to modify organ dimensions greatly. Such forces, even if progressively damped after first being applied to the lacrimal gland's surface, would without doubt also be transmitted to the organ's interior, thereby altering the contours of both its extra- and intracellular structures. This must affect function. For example, tensile forces, by elongating vessel and duct branches, would diminish their diameters. Thus, in accord with Poiseuille's law,[6] which states that flow of a fluid changes directly with the fourth power of its vessel's radius, the flow of both blood and lacrimal secretion would be much less in those radicles than if the organ were not tensed *in situ.*

It is also to be expected that connective tissue cells and fibers, as well as parenchymal cells, would elongate in accord with the force applied. These stresses would not stop at a cell's plasma membrane but would also affect the dimensions of its membrane-bound organelles, which in rat exorbital lacrimal gland acinar cells contain numerous enzymes[7] whose production might be modified by the distortion. Likewise, deformation of fibrillar cytoskeletal structures, such as the intermediate filaments that span the distance between the plasma membrane and the nucleus,[8] would disturb nuclear shape. Major functional consequences might result.

Changes in function have taken place when tension was increased experimentally. Stretching the skin of living rats[9] caused the epidermal cells and their nuclei to elongate, presumably by traction on the basement membrane which has supportive and elastic properties.[10] In addition, dermal vessels narrowed and dermal collagen straightened. These deformations impaired migration of basal cells toward the skin surface and affected the cell cycle. Thymidine uptake increased in areas where the applied tensile force was weak, and diminished or halted where it was stronger. In tissue culture, as individual cells adhere to the matrix their cytoskeletal organization changes, causing the cells to change shape, that is, to spread. Subsequently, the cell cycle progresses.[11] However, in rat skin *in vivo*, extrinsic tensile force caused anchored cells to change shape. That undoubtedly altered their cytoskeletons, causing the nuclei to distort. The cell cycle and the spatial relationships of cells were then affected so their migration was curtailed. Because the nuclear matrix localizes transcription factors and the folding patterns of both DNA and the chromosomes in the area of a gene, contributing to proper and timely gene expression,[12] nuclear distortion might well influence those normal abilities.

As the skin was stretched intercellular distances increased.[9] It would be expected that attachment sites and channels parallel to tensile forces would elongate. Thus, channel diameters would diminish, as when the ends of a thin rubber tube are pulled. However, the major diameter of any channel perpendicular to the force would expand, as occurs when the sides of that rubber tube are distracted. Changes in diameter might affect "mechanogated" channels that open or close during mechanical stimulation.[13] If so, such dimensional adjustments might affect the composition and flow of secretion. That would be the case if the flow of molecules from cell to cell through gap junctions were altered. Likewise, their passage through paracellular channels[14] lying between duct or acinar lumens and the basolateral membranes of the cells might be affected. It is through the latter membranes that ion flux in rat exorbital lacrimal gland acinar cells occurs.[15]

Tensile forces might cause the spatial orientations and foldings of molecules to vary in direction or length and distort bonds connecting them. These adjustments could alter the geometry of the lacrimal receptors for hormones,[16] which might influence the organ's sexual dimorphism.[17] Immunological consequences, such as development of altered immunoglobulin production[18] or of autoimmunity,[19] are possible if the configuration of proteins and distances between their surface molecules change.

Tensile forces comparable to those presently shown in the rat might be exerted on the human lacrimal gland by its fascial attachments to the orbital periosteum,[20] such as the ligament of Soemmering, as well as by other structures that suspend it.[21] Relaxation of the suspensory connective tissue is recognized as the basis of prolapse of the palpebral lobe.[22] Should lacrimal gland tension exist in man, it might affect physiological and pathological processes. Perhaps as yet unrecognized variations in magnitude of the gland's tension might either predispose to or mitigate against the development of disorders of mitosis (neoplasia), secretion (dry eye), and immune reactivity (Sjögren's syndrome). Might there be sex differences or age changes in this biophysical parameter that could influence such occurrences?

REFERENCES

1. Wilkes GL, Brown LA, Wildnauer RH. The biomechanical properties of skin. *Crit Rev Biomed Eng.* 1973;1:453–495.
2. Tickner EG, Sacks AH. A theory for the static elastic behavior of blood vessels. *Biorheology* 1967;4:151–163.
3. Setnikar L, Agostini E, Toglietti A. Entita, caratteristiche e origine della depressione pleurica. *Arch Sci Biol.* 1957;41:312–325.
4. Calaresu FR, Tobey JC, Heidemann SR, Weaver LC. Splenic and renal sympathetic responses to stimulation of splenic receptors in cats. *Am J Physiol.* 1984;247:R856-R865.
5. Frame JS. Ellipsoid and spheroid. In: *McGraw-Hill Encyclopedia of Science & Technology*. 7th ed. New York: McGraw-Hill; 1992:315–316.
6. Guyton AC, Hall JE. *Textbook of Medical Physiology*. 9th ed. Philadelphia: W.B. Saunders Co.; 1996:167.
7. Mircheff AK, Lu CC. A map of membrane populations isolated from rat exorbital gland. *Am J Physiol.* 1984;247:G651-G661.
8. Goldman R, Goldman A, Green K, Jones J, Lieska N, Yang H-Y. Intermediate filaments: Possible functions as cytoskeletal connecting links between the nucleus and the cell surface. *Ann N Y Acad Sci.* 1985;455:12–17.
9. Lorber M, Milobsky SA. Stretching of the skin *in vivo*: A method of influencing cell division and migration in the rat epidermis. *J Invest Dermatol.* 1968;51:395–402.
10. Welling LW, Zupka MT, Welling DJ. Mechanical properties of basement membrane. *News in Physiol Sci.* 1995;10:30–35.
11. Zhu X, Assoian RK. Integrin-dependent activation of MAP kinase: A link to shape-dependent cell proliferation. *Mol Biol Cell.* 1995;6:273–282.
12. Stein GS, Stein JL, van Wijnen AJ, Lian JB. The maturation of a cell. *Am Scientist* 1996;84:28–37.
13. Hamill OP, McBride DW Jr. Mechanoreceptive membrane channels. *Am Scientist* 1995;83:30–37.
14. Griepp EB, Robbins ES. Epithelium. In: Weiss L, ed. *Cell and Tissue Biology*. Baltimore: Urban & Schwarzenberg; 1988:131–142.
15. Yiu SC, Lambert RW, Tortoriello PJ, Mircheff AK. Secretagogue-induced redistributions of Na, K-ATPase in rat lacrimal acini. *Invest Ophthalmol Vis Sci.* 1991;32:2976–2984.
16. Warren DW, Azzarolo AM, Becker L, Bjerrum K, Kaswan RL, Mircheff AK. Effects of dihydrotestosterone and prolactin on lacrimal gland function. *Adv Exp Med Biol.* 1994;350:99–104.
17. Sullivan DA, Sato EH. Immunology of the lacrimal gland. In: Albert DM, Jakobiec FA, eds. *Principles and Practice of Ophthalmology: Basic Sciences.* Philadelphia: W.B. Saunders Co.; 1994:479–486.
18. Gudmundsson OG, Sullivan DA, Bloch KJ, Allansmith MR. The ocular secretory immune system of the rat. *Exp Eye Res.* 1985;40:231–238.
19. Mircheff AK, Gierow JP, Wood RL. Autoimmunity of the lacrimal gland. *Int Ophthalmol Clin.* 1994;34:1–18.
20. Whitwell J. Denervation of the lacrimal gland. *Br J Ophthalmol.* 1958;42: 518–525.

21. Bergin DJ. Anatomy of the eyelids, lacrimal system, and orbit. In: McCord CD Jr, Tanenbaum M, Nunnery WR, eds. *Oculoplastic Surgery*. 3rd ed. New York: Raven Press; 1995;73–77.
22. Smith B, Petrelli R. Surgical repair of prolapsed lacrimal glands. Arch Ophthalmol. 1978;96:113–115.

9

ABERRANT LACRIMAL GLAND DEVELOPMENT IN AN ANOPHTHALMIC MUTANT STRAIN OF RAT

Prabir K. De

Centre for Cellular and Molecular Biology
Hyderabad, India

1. INTRODUCTION

Anophthalmia (congenital absence of eyeball) is a rare congenital defect seen in laboratory animals of different species[1–6] as well as in humans.[7] Several instances of hereditary anophthalmia have been reported,[3–6] where both unilateral and bilateral anophthalmic phenotypes have been found. Anophthalmia is also known to result as a teratogenic effect of drugs.[7,8] Such teratogenic anophthalmia reported for laboratory rats was found to be bilateral as well as unilateral.[8] Our institution has recently isolated a strain of mutant anophthalmic rats that were derived from a parent colony of inbred CFY rats.[6]

There is no information in the literature regarding the status of lacrimal glands in reported strains of mutant anophthalmic rodents. Moreover, in instances of drug-induced (teratogenic) anophthalmia, the state of the lacrimal glands have not been investigated. Rodents have three conspicuous lacrimal glands. The exorbital lacrimal gland is present beneath the skin below the ear. The intraorbital lacrimal and Harderian glands are located in the orbit of the eye. The secretory enzyme peroxidase is a well-known marker of lacrimal gland secretory activity.[9] We investigated whether there are any obvious changes in gland weight, peroxidase activity, or SDS-PAGE protein profile of lacrimal glands in our strain of mutant anophthalmic rats. Comparisons were made with normal rats of the parent colony from which the mutant rats were derived.

2. MATERIALS AND METHODS

This study used a strain of anophthalmic mutant CFY rats isolated and maintained in our Centre's animal facility. These mutant rats have been described.[6] The mutant rats were

Lacrimal Gland, Tear Film, and Dry Eye Syndromes 2
edited by Sullivan *et al.*, Plenum Press, New York, 1998

of unilateral or bilateral anophthalmic phenotype and were offspring of bilaterally anophthalmic parents. Other than their anophthalmic phenotype, the mutant rats were otherwise outwardly indistinguishable from normal-sighted CFY rats of the parent colony. Unless otherwise stated, all investigations were done on adult (3–4 months) anophthalmic female rats. Glands of the affected side of unilaterally anophthalmic rats were compared with those from the normal side. For bilaterally anophthalmic rats, comparisons were made with normal female rats of the parent colony. Rats were weighed before sacrifice. A pair of glands of each type (Harderian, and exorbital and intraorbital lacrimal) were excised after sacrifice. Individual glands of a pair from a rat were weighed and frozen separately. Glands were homogenized (5% w/v) in 20 mM potassium phosphate buffer (pH 8.0) and centrifuged (28000*g*, 30 min) at 4°C. Supernatants were assayed for peroxidase activity,[9] and 140 μl of supernatants were run in 10% SDS-PAGE and gels stained as described.[10]

3. RESULTS

In unilaterally anophthalmic rats, the mean weight of exorbital lacrimal glands of the normal side was not significantly different from that of glands of either side of normal rats. However, the exorbital lacrimal glands of the affected sides varied considerably in weight, which ranged from 1.5 to 93% of respective normal-side gland weight. Of 35 affected-side glands examined, weights of 14 (40%) were within 10% of the weight of their respective normal-side glands. Such drastically affected-side exorbital glands lacked peroxidase activity and appeared as a tiny white mass apparently lacking glandular tissue. The remaining 21 (60%) affected side exorbital glands, which clearly contained glandular tissue (but were affected in weight to varying extents), had a mean peroxidase activity (units/mg tissue) of 12.7 ± 3.4 (mean ± SD; n = 21) which was significantly ($p < 0.001$) less than the mean value of 20.3 ± 3.7 (n=21) for normal-side glands. SDS-PAGE profiles of these 21 affected-side exorbital glands were, however, indistinguishable from those of normal-side glands or glands from normal rats. The protein profiles of the 14 affected-side exorbital glands (which lacked glandular tissue and peroxidase activity) were indistinguishable from each other, but when compared with those of normal-side glands showed an absence or markedly reduced levels of almost all protein bands seen in the latter, except a major 67 kDa protein whose level appeared unaltered.

Exorbital lacrimal glands from either side of bilaterally anophthalmic rats (when compared with glands from normal rats) were found to be similarly affected in weight, peroxidase activity, and protein profile as seen for affected-side glands of unilaterally anophthalmic rats. Thus, out of 48 single exorbital glands examined from 24 adult bilaterally anophthalmic rats, 31 glands were within 8 mg in weight, and such glands lacked glandular tissue and peroxidase activity and had altered protein profile as seen for the drastically weight-affected glands of the affected side of unilaterally anophthalmic rats. Moreover, of these 31 glands from bilaterally anophthalmic rats, 18 were from 9 rats, ie, both side glands were drastically affected in weight. Thus, in bilaterally anophthalmic rats considerable variation was frequently seen between individual exorbital gland weights within a pair. We found that the mean fold of difference in exorbital gland weight within a pair of glands was 4.2 ± 4.1 (mean ± SD; n=50; range = 1–15.2) for bilaterally anophthalmic rats, which was significantly ($p < 0.001$) higher than that of normal rats (1.04 ± 0.04; n=45; range = 1–1.6). The marked variation in individual exorbital gland weights on the affected side of unilaterally anophthalmic rats as well as in bilaterally anophthalmic rats was intriguing, since, at least on gross outward examination, no apparent differences in the anophthalmic phenotype was seen. Essentially similar variation in exorbital lacrimal gland weights were seen when 14-day-old, bilaterally anophthalmic rats were examined. In such rats a large percentage of exorbital lacrimal glands had markedly low weights compared with 14-day-old normal rats.

Intraorbital lacrimal glands of bilaterally anophthalmic rats had slightly higher weights when compared with glands from normal rats. The mean weight of a pair of intraorbital glands from bilaterally anophthalmic rats was 16.7 ± 5.0 mg (mean ± SD; n=19 pairs), which was significantly ($p < 0.05$) higher than that of normal rats (14.17 ± 2.14 mg; n=19 pairs). The mean peroxidase activity (units/mg tissue) for a pair of intraorbital lacrimal glands from bilaterally anophthalmic rats was 12.4 ± 3.4 (mean ± SD; n=19); this was significantly ($p < 0.001$) less than the mean value of 20.6 ± 4.0 (n=19) for glands from normal rats. However, SDS-PAGE protein profiles of intraorbital glands from bilaterally anophthalmic rats were never found to have any detectable difference when compared with profiles of glands from normal rats. Likewise, no detectable difference was seen when SDS-PAGE protein profiles of intraorbital glands of the affected side from unilaterally anophthalmic rats were compared with profiles of respective normal-side glands.

Mean weight of Harderian glands from bilaterally anophthalmic rats did not differ significantly from that of normal rats. Moreover, no significant difference was found between mean Harderian gland weights of the affected and normal sides of unilaterally anophthalmic rats. Protein profiles of Harderian glands from bilaterally anophthalmic rats were indistinguishable from those from normal rats. Moreover, SDS-PAGE protein profiles of Harderian glands of the affected and normal sides of unilaterally anophthalmic rats were always found to be indistinguishable from each other. No peroxidase activity was detectable in Harderian glands of normal rats or of unilaterally or bilaterally anophthalmic rats.

4. CONCLUSIONS

Our investigations show that the mutation in the present strain of anophthalmic rats affects the development of both the eyeball and the exorbital lacrimal gland. Since the low weight of exorbital lacrimal glands of anophthalmic rats was also apparent at age 14 days, it can be concluded that the low gland weights seen in adults were due to an early developmental defect. The reason for the marked variation in exorbital lacrimal gland weight in anophthalmic rats is unknown. Careful histological examination must be done to check whether rat-to-rat differences exist in the anophthalmic phenotype in the mutant rats used in this study. Our observations make it imperative to know the state of the lacrimal glands in other known strains of mutant anophthalmic rats,[3] as well as in rats born anophthalmic due to teratogenic effect of drugs.[8]

REFERENCES

1. Eaton ON. A hereditary eye defect in guinea pigs. *J* Hered. 1937;28:353–358.
2. Komich RJ. Anophthalmos: An inherited trait in a new stock of guinea pigs. *Am J Vet Res.* 1971;32:2099–2105.
3. Ibuka N. Circadian rhythms in sleep-wakefulness and wheel-running activity in a congenitally anophthalmic rat mutant. *Physiol Behav.* 1987;39:321–326.
4. Chase HB. Studies on an anophthalmic strain of mice III. Results of crosses with other strains. *Genetics.* 1942;27:339–348.

5. Knopp BH, Polivanov S. Anophthalmic albino: A new mutation in the Syrian hamster. *American Naturalist.* 1958;92:317–318.
6. Harinarayana Rao S, Sesikeran B. Congenital anophthalmia in CFY rats: A newly identified autosomal recessive mutation. *Lab Anim Sci.* 1992;42:623–625.
7. Spagnolo A, Bianchi F, Calabro A, et al. Anophthalmia and benomyl in Italy: a multicenter study based on 940, 615 newborns. *Reprod Toxicol.* 1994;8:397–403.
8. Vorhees CV, Acuff-Smith KD. Prenatal methamphetamine-induced anophthalmia in rats. *Neurotoxicol Teratol.* 1990;12:409.
9. De PK. Tissue distribution of constitutive and induced soluble peroxidase in rat: purification and characterization from lacrimal gland. *Eur J Biochem.* 1992;206:59–67.
10. Ranganathan V, De PK. Androgens and estrogens markedly inhibit expression of a 20 kDa major protein in hamster exorbital lacrimal gland. *Biochem Biophys Res Commun.* 1995;208: 412–417.

10

HORMONAL INFLUENCES ON SYRIAN HAMSTER LACRIMAL GLAND

Marked Repression of a Major 20 kDa Secretory Protein by Estrogens, Androgens, and Thyroid Hormones

Prabir K. De and Velvizhi Ranganathan

Centre for Cellular and Molecular Biology
Hyderabad, India

1. INTRODUCTION

Lacrimal gland secretory activity is important for maintaining a healthy ocular mucosa.[1,2] Lacrimal glands in rodents and other species including humans show histomorphological and biochemical sexual dimorphisms.[3–6] Moreover, in humans, lacrimal gland secretory insufficiency leading to dry eyes (keratoconjunctivitis sicca) is known to predominantly afflict women.[5] The hormonal regulation of this gland and its secretory activity has been investigated mainly in the rat, where marked histomorphological sex differences and higher levels of secretory component and IgA in male gland and tears have been attributed to the inductive effect of androgens on protein synthesis.[7] Such effects are likely to be mediated via androgen receptors in the lacrimal gland.[7] Unlike androgens, estrogens are believed to have no effect on the lacrimal gland.[7]

Prolactin has been shown to have some trophic effects on the rat lacrimal gland,[8] although it had no effect on IgA or secretory component levels.[9] Moreover, thyroidectomy did not affect IgA or secretory component levels in rat tears.[10]

We recently investigated the sex-hormonal effects on the Syrian hamster lacrimal gland by examining protein profiles of lacrimal glands taken from hamsters in different hormonal states.[6] The expression of a hamster lacrimal 20 kDa major protein was markedly inhibited by both androgens and estrogens.[6] We report here our further studies in which we investigated the expression of this protein in hamsters during pregnancy and lactation. The effect of tamoxifen (an estrogen receptor antagonist) administration on the inhibitory effect of estrogen and the effect of thyroid hormones and bromocriptine (a prolactin release inhibitor) on the expression of this protein were studied. The hormonal regu-

Lacrimal Gland, Tear Film, and Dry Eye Syndromes 2
edited by Sullivan *et al.*, Plenum Press, New York, 1998

lation of the lacrimal 20 kDa protein was also checked in Western blots using the purified protein's antisera. Some properties of the purified protein are also described.

2. MATERIALS AND METHODS

Syrian hamsters of both sexes were bilaterally gonadectomized at 2 months of age. After 30 days, different groups were separately injected daily (sc) with various hormones (Sigma) for 15 days, and then sacrificed along with vehicle-injected gonadectomized and intact controls.[6] Groups of intact and gonadectomized hamsters of both sexes were injected with 500 μg bromocriptine for 45 days and then sacrificed along with controls. All investigations were done on groups of animals (*n*=5–6). After sacrifice, lacrimal glands were excised, weighed, and homogenized (2.5% w/v), and supernatants were prepared as described.[6]

Equal volumes of supernatants (140 μl) were run in SDS-PAGE and gels stained as described.[6] Western blot of lacrimal extracts was done[11] using rabbit antisera against hamster lacrimal 20 kDa protein. Lacrimal 20 kDa protein was purified from supernatants of lacrimal glands taken from ovariectomized hamsters. Supernatant was passed consecutively through columns of Biogel-HTP and concanavalin A-Sepharose. The unbound proteins in the final flowthrough in which the 20 kDa protein was present were concentrated and then resolved in a Sephadex G-75 column and fractions collected. Several fractions that were clearly homogeneous for the 20 kDa protein were pooled and used for immunization and characterization studies.

3. RESULTS

A 20 kDa major protein was seen in SDS-PAGE profiles of lacrimal glands of female hamsters but not in males or 15-day-pregnant females. Females ovariectomized for 45 days and 15-day-lactating females had several-fold higher levels of this lacrimal 20 kDa protein, which then constituted ~ 20 % of total soluble proteins. Males gonadectomized for 45 days had high levels of this lacrimal 20 kDa protein similar to ovariectomized females. Daily administration for 15 days of either estradiol (3.6 μg), testosterone (50 μg), diethylstilbestrol (3.6 μg), dihydrotestosterone (50 μg), or thyroxine (60 μg), but not progesterone (100 μg) or dexamethasone (100 μg), to 30-day-gonadectomized males or females obliterated this lacrimal 20 kDa protein. Bromocriptine (500 μg) administered for 45 days to gonadectomized or intact hamsters of either sex had no effect on the levels of this lacrimal 20 kDa protein when compared with vehicle-treated controls. Daily postpartum administration of estradiol (3.6 μg), but not bromocriptine (500 μg), to lactating hamsters for 15 days obliterated the lacrimal 20 kDa protein. The post-gonadectomy increase of lacrimal 20 kDa protein was time-dependent, and similar maximum levels were found in males and females by 30 days and 10 days respectively. Daily administration of different doses of estradiol or testosterone for 15 days to 30-day-orchiectomized males showed that although a minimum of 3.6 μg estradiol could cause complete inhibition of lacrimal 20 kDa protein, it required 50 μg testosterone for the same effect. Daily administration of different doses of thyroid hormones to ovariectomized females for 15 days showed no inhibition of lacrimal 20 kDa protein with a dose of 2.2 μg thyroxine, whereas doses of 20 μg and 60 μg showed ~ 80% and complete inhibition respectively. Triiodothyronine was more potent than thyroxine and showed ~ 80% inhibition at 0.75 μg dose. Daily administration of 500 μg tamoxifen, along with complete inhibitory doses (see above) of either estradiol, testosterone, or thyroxine, for 15 days to ovariectomized hamsters prevented only estradiol's inhibitory effect on lacrimal 20 kDa protein.

Western blots using antisera against hamster lacrimal 20 kDa protein failed to detect any immunoreactive 20 kDa protein in lacrimal glands and tears of male hamsters or from gonadectomized hamsters treated for 15 days with inhibitory doses of estradiol, testosterone, or thyroid hormone. Strong 20 kDa cross-reaction was seen in lacrimal glands and tears from 45-day- gonadectomized male or female hamsters. Cross-reacting 20 kDa protein was also seen in lacrimal gland and tears of intact female hamsters.

Purified lacrimal 20 kDa protein did not stain for glycoproteins and did not bind concanavalin A-Sepharose, and its mobility in SDS-PAGE was unaffected by *N*-glycosidase F or *O*-glycosidase treatment. It had no lysozyme-like activity or any detectable antibacterial activity in in vitro tests using *E. coli* as test organism. The protein had no protease activity using azocasein as substrate. The lacrimal 20 kDa protein had a pI of 4.2 as determined in the Phast system using Phast gels (Pharmacia), and in its amino acid analysis aspartic and glutamic acids were the two most abundant amino acid residues recovered (13.7 and 14.7% nanomoles respectively).

4. DISCUSSION AND FUTURE DIRECTIONS

Our results show that hamster lacrimal major 20 kDa protein is a secretory protein, and its expression is markedly inhibited by androgens, estrogens, and thyroid hormones. The effects of androgens and estrogens are obviously at physiological levels, but this may not be true for thyroid hormones. Our inhibitory doses of thyroid hormones may be supraphysiological, and additionally, there is no report of sex- or gonadectomy-related differences in thyroid hormone levels in hamsters, which could account for the different level of expression of 20 kDa protein in these states. The effect of hypothyroid state on the expression of this protein in both sexes remains to be investigated. Although the inhibitory effect of androgen might be mediated via its receptor present in lacrimal gland,[7] the effect of estrogen (which is prevented by tamoxifen) is likely to be via estrogen receptors presumably present in hamster lacrimal gland. The high level of lacrimal 20 kDa protein during lactation and its absence during pregnancy could be respectively due to very low and high level of endogenous estrogen known to be present during these states. The absence of this protein in intact males and not in females is likely due to its complete inhibition by endogenous levels of testosterone in males and incomplete inhibition by endogenous estrogens in females.[6] Results also rule out any influence of prolactin, progesterone, or glucocorticoids on the expression of this major lacrimal protein.

Future investigations should check the effect of estrogen, androgen, and thyroid hormone on the expression of lacrimal 20 kDa protein using lacrimal cell cultures. The effect of hypophysectomy and the possible influence of other pituitary hormones on the expression of this lacrimal protein need investigation, as well as do the presence of estrogen and thyroid hormone receptors in this tissue. Cloning and sequencing the cDNA of this major lacrimal protein might give a clue to its possible function and allow studies on the molecular mechanism of its multihormonal repression. Research by other laboratories is directed at identifying steroidal compounds that might be useful in ameliorating symptoms of dry eye in humans by enhancing lacrimal secretory activity and preventing autoimmune destruction of lacrimal gland. The hamster lacrimal 20 kDa protein might be an excellent indicator useful for testing potential candidate steroids for their effects on the lacrimal gland and also for studying the regulation of the lacrimal gland in detail.

REFERENCES

1. Gudmundsson OG, Sullivan DA, Bloch KJ, Allansmith MR. The ocular secretory immune system of the rat. *Exp Eye Res.* 1985;40:231–239.
2. De PK. Tissue distribution of constitutive and induced soluble peroxidase in rat: purification and characterization from lacrimal gland. *Eur J Biochem.* 1992;206:59–67.
3. Cornell-Bell AH, Sullivan DA, Allansmith MR. Gender-related differences in the morphology of the lacrimal gland. *Invest Ophthalmol Vis Sci.* 1985;26:1170–1175.
4. Mhatre MC, VanJaarsveld AS, Reiter RJ. Melatonin in the lacrimal gland: first demonstration and experimental manipulation. *Biochem Biophys Res Commun.* 1988;153:1186–1192.
5. Azzarolo AM, Mazaheri AH, Mircheff AK, Warren DW. Sex-dependent parameters related to electrolyte, water and glycoprotein secretion in rabbit lacrimal glands. *Curr Eye Res.* 1993;12:795–802.
6. Ranganathan V, De PK. Androgens and estrogens markedly inhibit expression of a 20 kDa major protein in hamster exorbital lacrimal gland. Biochem Biophys Res Commun. 1995;208:412–417.
7. Sullivan DA, Edwards JA, Wickham LA, et al. Identification and endocrine control of sex steroid binding sites in the lacrimal gland. *Curr Eye Res.* 1996;15:279–291.
8. Azzarolo AM, Bjerrum K, Maves CA, et al. Hypophysectomy-induced regression of female rat lacrimal glands: partial restoration and maintenance by dihydrotestosterone and prolactin. *Invest Ophthalmol Vis Sci.* 1995;36:216–226.
9. Sullivan DA. Influence of hypothalamic-pituitary axis on the androgen regulation of the ocular secretory immune system. *J Steroid Biochem.* 1988;30:429–433.
10. Sullivan DA, Allansmith MR. Hormonal influence on the secretory immune system of the eye: endocrine interactions in the control of IgA and secretory component levels in tears of rats. Immunology. 1987;60:337–343.
11. Ranganathan V, De PK. Western blot of proteins from Coomassie stained polyacrylamide gels. *Anal Biochem.* 1996;234:102–104.

11

ANDROGEN SUPPORT OF LACRIMAL GLAND FUNCTION IN THE FEMALE RABBIT

Dwight W. Warren,[1,2,3] Ana Maria Azzarolo,[1] Zuo Ming Huang,[1]
Barbara W. Platler,[2] Renee L. Kaswan,[5] Elizabeth Gentschein,[4]
Frank L. Stanczyk,[4] Laren Becker,[1] and Austin K. Mircheff[2,3]

[1]Department of Cell and Neurobiology
[2]Department of Physiology and Biophysics
[3]Department of Ophthalmology
[4]Department of Obstetrics and Gynecology
University of Southern California
Los Angeles, California
[5]College of Veterinary Medicine
University of Georgia
Athens, Georgia

1. INTRODUCTION

Dry eye is a major reason for visits to an ophthalmologist's office. The most probable cause for dry eye is primary lacrimal deficiency (PLD).[1] PLD is usually detected in women, most frequently after menopause, during pregnancy or lactation, or when taking estrogen-containing oral contraceptives. These various endocrine states exhibit a complete range of plasma estrogen levels from very low to very high. Thus, plasma estrogen concentrations do not appear to be a common variable in PLD. However, plasma free androgen levels are potentially decreased in all of these states. We have previously demonstrated that ovariectomy of female rabbits[2] and hypophysectomy of female rats[3] result in a decrease in biochemical correlates of lacrimal gland function. Treatment of these endocrinectomized animals with the potent androgen dihydrotestosterone (DHT) restores the decreases in the biochemical markers of secretion,[2,3] specifically, lacrimal gland protein, DNA, Na,K-ATPase, and ß-adrenergic receptors. Androgens have been shown to be responsible for the male-like morphological and functional characteristics of the gland, including larger acini,[4] greater secretion of IgA,[5] and greater production of polymeric IgA receptor, measured as secretory component (SC).[6] When female rats are treated with androgens, the morphology of the lacrimal gland changes and resembles the male lacrimal gland.[7] However, the major neurotransmitter receptor coupled to secretion in the lacrimal

Lacrimal Gland, Tear Film, and Dry Eye Syndromes 2
edited by Sullivan *et al.*, Plenum Press, New York, 1998

gland, the muscarinic cholinergic receptor, is regulated by circulating levels of prolactin, not androgens.[3]

It was our hypothesis that ovariectomy of female rabbits causes regression of the biochemical markers of lacrimal gland function and that this regression correlates with a decrease in lacrimal fluid flow rates. Additionally, we hypothesized that treatment of ovariectomized rabbits with DHT restores or even increases lacrimal fluid flow rates.

2. METHODS

All animal experiments were approved by the University of Southern California Institutional Animal Care and Use Committee. Sexually mature female New Zealand white rabbits (4–4.5 kg) were anesthetized with an intramuscular injection of 40 mg/kg ketamine and 10 mg/kg xylazine and ovariectomized through a midline abdominal incision. Control animals received sham operations of a midline abdominal incision. Some of the ovariectomized rabbits received daily treatment of 1 mg/kg DHT.

Unanesthetized rabbits were placed in a restraint cage, and tear fluid was collected from the lateral canthus for 5 min using a 5 or 10 μl calibrated, flame-polished micropipet. The tear fluid collected was considered the basal level of tear fluid flow rate. Some rabbits were anesthetized with 40 mg/kg ketamine and 10 mg/kg xylazine and the lacrimal gland ducts cannulated using polyethylene tubing (Intramedic PE-10). Pilocarpine, a cholinergic agonist, was injected through the ear vein at a maximally stimulating dose of 0.3 mg/kg.[8] Lacrimal gland fluid was collected in preweighed micro centrifuge tubes for three consecutive collections and measured gravimetrically. The results are expressed in μl/min.

At the end of the stimulated fluid collection period, blood was collected from the central ear artery. The rabbits were then killed by injecting a lethal dose of sodium pentobarbital. The superior and inferior lobes of the lacrimal gland were obtained, snap frozen in liquid nitrogen, and stored at -70°C until processed for measurement of protein, DNA, Na,K-ATPase, ß-adrenergic receptor, and muscarinic cholinergic receptor content as previously described.[2] The total serum levels of testosterone and androstenedione were determined using previously described radioimmunoassay procedures.[9]

Comparisons between groups were made with Student's t tests for unpaired samples or analysis of variance followed by Duncan's New Multiple Range Test.

3. RESULTS

Ovariectomy significantly decreased serum levels of the androgens testosterone and androstenedione by 88.5% and 35.9% respectively. Ovariectomy caused a remarkable regression of the lacrimal gland. Ovariectomy significantly decreased the major biochemical markers of secretion, including total gland protein by 22% and DNA by 35%. Other markers of biochemical secretion (Na,K-ATPase, muscarinic cholinergic receptors, and ß-adrenergic receptors) decreased, but the decrease did not reach the $p<0.05$ level of significance when ANOVA and Duncan's New Multiple Range Test were used to determine significant differences between groups.

DHT treatment prevented the decrease in total protein and DNA content. Additionally, DHT treatment significantly increased ß-adrenergic receptor (23% over the ovariectomized group) and Na,K-ATPase (29% over the ovariectomized group) content. However, DHT treatment failed to increase muscarinic cholinergic receptor content.

Maximally stimulated lacrimal gland fluid flow rate, collected by cannulation of the lacrimal excretory ducts, showed a significant decrease after ovariectomy (7.72 ± 0.41 μl/min in ovariectomized rabbits, and 9.16 ± 0.77 μl/min in sham-operated controls). DHT treatment increased the stimulated lacrimal gland fluid flow rate compared to ovariectomized rabbits (13.26 ± 1.48 μl/min, and 7.72 ± 0.41 μl/min, respectively). No changes were observed in basal tear flow rate, measured by collection from the tear meniscus by capillary pipettes, after ovariectomy (1.96 ± 0.12 μl/min) compared to controls (1.84 ± 0.07 μl/min). Paradoxically, DHT significantly decreased the basal tear flow rate (1.02 ± 0.04 μl/min) compared to ovariectomized controls (1.96 ± 0.12 μl/min).

4. DISCUSSION

In the mammalian ovary, androgen production is an essential intermediate for estrogen synthesis. Thus, ovariectomy decreases the production of androgens as well as estrogens. In this study, ovariectomy resulted in a decrease in circulating androgens but did not cause a reduction to zero levels. The remaining androgen levels are most likely of adrenal gland origin.

Although ovariectomy caused the regression of biochemical markers of lacrimal gland function, these changes were not as dramatic as in the hypophysectomized rat model.[3] These two animal models differ in several respects, and we can visualize at least two possible explanations for the quantitative difference in responses. One explanation could be that hypophysectomy, in addition to removing the gonadotropins luteinizing hormone (LH) and follicle-stimulating hormone (FSH), thus eliminating the stimulation for production of the ovarian androgens, also removes adrenocorticotropic hormone (ACTH), thus eliminating stimulation of adrenal gland hormone production, including the adrenal androgens, predominantly dehydroepiandrosterone (DHEA) and its sulfate (DHEAS), but also testosterone and androstenedione. Ovariectomy, on the other hand, removes only ovarian androgens without altering adrenal androgen production. Thus, the greater decrease in total circulating levels of androgens in the hypophysectomized model could account for the greater regression observed after hypophysectomy.

Another factor could be that hypophysectomy also removes prolactin (PRL) in addition to other pituitary hormones. PRL receptors have been found in both male and female rat lacrimal gland acinar cells.[10] PRL has also been implicated in regulating certain aspects of lacrimal gland function,[3] predominantly stimulating an increase in Na,K-ATPase and muscarinic cholinergic receptor content. Thus, the combined lack of androgens, PRL, and perhaps other hormones, such as thyroid hormone or glucocorticoids, in the hypophysectomized rat model might impair the gland's function more than a lack of androgens alone, as is probably the case in the ovariectomized rabbit model.

DHT treatment simultaneous with ovariectomy was used to verify the importance of androgen in maintaining lacrimal gland secretion. DHT was used instead of testosterone because it cannot be converted to estrogens, although the dose used (1 mg/kg) is clearly pharmacologic in female rabbits. DHT given at the time of ovariectomy prevented the marked regression of the lacrimal gland and maintained or increased the expression of the biochemical markers of secretory capacity, confirming the previous results observed in the hypophysectomized rat model.[3] The observation that DHT had no effect on muscarinic cholinergic binding sites is in accord with the previous results obtained with hypophysectomized rats. Increasing the content of this receptor appears to be dependent on increased levels of circulating PRL.

Ovariectomy appears to decrease the maximally stimulated lacrimal fluid flow rate compared to control, and DHT significantly increases this value over that in ovariectomized rabbits. Thus, maximal secretory capacity of the lacrimal gland seems to be androgen dependent.

Ovariectomy did not change the basal tear fluid flow rate, and a significant decrease was observed after DHT treatment, as compared to the ovariectomized group. Sullivan and Allansmith[11] found similar results, showing that castration of rats increased basal levels of tear volume, while chronic testosterone treatment after castration decreased basal tear volume. These results might seem inconsistent. One possible explanation relates to our incomplete understanding of the variety of signals that are integrated to determine the rate of lacrimal gland fluid production. It is of particular interest that androgens increase ß-adrenergic receptor binding sites with no effect on muscarinic cholinergic receptor binding sites. Botelho et al.[12] have shown that sympathetic nerve impulses inhibit secretion of lacrimal gland fluid when the gland is stimulated with a cholinergic agonist. In addition, a large body of evidence has recently accumulated in vascular and nonvascular smooth muscle that indicate that cross talk between the cAMP and the polyphosphoinositide (PPI) signaling cascade plays an important role in the functional antagonism between the sympathetic and parasympathetic nervous system.[13] Thus, agents that elevate intracellular cAMP conentrations, such as ß-adrenergic agonist, inhibit agonist-stimulated hydrolysis of the PPI system, such as muscarinic cholinergic agonist.[13] Therefore, we might surmise that an increase in ß-adrenergic receptor binding sites without a concomitant increase in muscarinic cholinergic receptor binding sites, as is the case in the DHT-treated, ovariectomized rabbits, could shift the balance between stimulatory and inhibitory intracellular signals generated in response to basal autonomic output, resulting in a decrease of basal lacrimal gland fluid production and precorneal tear fluid volume, even though the levels of cellular elements responsible for fluid production in the lacrimal gland are increased. When maximal stimulation with a cholinergic agonist occurs, the stimulatory effect of the cholinergic pathway can override the inhibition of the cAMP-driven pathways.

The present experiments support the thesis that lacrimal gland secretory function depends for the most part on the action of androgens. While the volume of evidence for this thesis is growing in animal models, there has been no confirmation that androgens are also responsible for maintenance of function in the human lacrimal gland. Recently, Mamalis et al.[14] have demonstrated in women that low testosterone levels were significantly more prevalent in dry eye patients than in controls. Additionally, subjective severity of dry eyes was negatively correlated with testosterone levels ($p<0.05$). Thus, if additional human studies confirm these initial investigations on the relationship between androgens and dry eye, the animal models of dry eye will be validated and support further studies on the cellular and molecular mechanisms of regulation of lacrimal fluid secretion.

ACKNOWLEDGMENTS

The authors thank Barbara Platler for outstanding technical help. Supported by NIH grants EY09405, EY05801, and EY10550.

REFERENCES

1. Lemp MA. Report of the National Eye Institute/industry workshop on clinical trials in dry eyes. *CLAO J.* 1995;21:221–232.

2. Azzarolo AM, Mircheff AK, Kaswan RL, et al. Androgen support of lacrimal gland function. *Endocrine.* 1997;6:39–45.
3. Azzarolo AM, Bjerrum K, Maves CA, et al. Hypophysectomy-induced regression of female rat lacrimal glands: Partial restoration/maintenance by dihydrotestosterone and prolactin. *Invest Ophthalmol Vis Sci.* 1995;36:216–226,
4. Cavallero C. Relative effectiveness of various steroids in an androgen assay using the exorbital lacrimal gland of the castrated rat. *Acta Endocrinol.* 1967;55:119–130.
5. Sullivan DA, Allansmith MR. Hormonal influence on the secretory immune system of the eye: Androgen modulation of IgA levels in tears of rats. *J Immunol.* 1985;134:2978–2982.
6. Sullivan DA, Bloch KJ. Allansmith MR. Hormonal influence on the secretory immune system of the eye: Androgen regulation of secretory component levels in rat tears. *J Immunol.* 1984;132:1130–1135.
7. Sullivan DA, Bloch KJ, Allansmith MR. Hormonal influence on the secretory immune system of the eye: Androgen control of secretory component production by the rat exorbital gland. *Immunology.* 1984;52:239–246.
8. Rismondo V, Osgood TB, Leering P, Hattenhauer MG, Ubels JL, Edelhauser HF. Electrolyte composition of lacrimal gland fluid and tears of normal and vitamin A-deficient rabbits. *CLAO.* 1989;15:222–228.
9. Stanczyk FZ, Shoupe D, Nunez V, Macias-Gonzales P, Vijod MA, Lobo RA. A randomized comparison of non-oral estradiol delivery in postmenopausal women. *Am J Obstet Gynecol.* 1988;159:1540–1546.
10. Mircheff AK, Warren DW, Wood RL, Tortoriello PJ, Kaswan RL. Prolactin localization, binding, and effects on peroxidase release in rat exorbital lacrimal gland. *Invest Ophthalmol Vis Sci.* 1992;33:641–650.
11. Sullivan DA, Allansmith MR. Hormonal modulation of tear volume in the rat. *Exp Eye Res.* 1986;42:131–139.
12. Botelho SY, Martinez EV, Pholpramool C, et al. Modification of stimulated lacrimal gland flow by sympathetic nerve impulses in rabbit. *Am J Physiol.* 1996;230:80–84.
13. Abdel-Latif AA. Cross talk between cyclic AMP and the polyphosphoinositide signaling cascade in iris sphincter and other nonvascular smooth muscle. *Soc Exp Biol Med.* 1996;211:163–177.
14. Mamalis N, Harrison DY, Hiura G, et al. Are "dry eyes" a sign of testosterone deficiency in women? Abstract presented at the 10th International Congress of Endocrinology, June 12–15, 1996, San Francisco, CA, USA.

12

IDENTIFICATION AND HORMONAL CONTROL OF SEX STEROID RECEPTORS IN THE EYE

L. Alexandra Wickham, Eduardo M. Rocha, Jianping Gao, Kathleen L. Krenzer, Lilia Aikawa da Silveira, Ikuko Toda, and David A. Sullivan

Schepens Eye Research Institute and Department of Ophthalmology
Harvard Medical School
Boston, Massachusetts

1. INTRODUCTION

During the past several decades, it has become quite evident that androgens, estrogens and progestins may exert a significant influence on the structure and/or function of a variety of ocular tissues, including the lacrimal gland, meibomian gland, conjunctiva, goblet cells, cornea, anterior chamber, lens and/or retina.[1,2] The nature of these sex steroid effects appears to involve modulation of such ocular parameters as tissue morphology, gene expression, protein synthesis, lipid production, mucous secretion, aqueous tear output, tear film stability, immunological activity, corneal curvature, aqueous humor outflow and visual acuity.[1,2] In addition, these hormones have been proposed as topical therapies for such conditions as dry eye syndromes (both aqueous-deficient and evaporative), corneal wound healing and high intraocular pressure.[1,3,4] However, despite these findings, very little information exists concerning the precise target cells for sex steroid action, the specific ocular processes controlled by these hormones, or the mechanisms (e.g. classical vs. non-classical) underlying potential sex steroid-eye interactions.

To advance our understanding of sex steroid effects in the eye, the purpose of the following studies was to: (1) determine whether tissues of the anterior and posterior segments contain sex steroid receptors, which might make them susceptible to hormone action following topical application; (2) examine whether these tissues contain the mRNA for Types 1 and/or 2 5α-reductase, an enzyme that converts testosterone to a very potent metabolite, dihydrotestosterone; and (3) explore the regulation of these receptors in a specific ocular tissue, the lacrimal gland.

Lacrimal Gland, Tear Film, and Dry Eye Syndromes 2
edited by Sullivan *et al.*, Plenum Press, New York, 1998

2. METHODS

2.1. Animals and Surgical Procedures

Young adult male and/or female Sprague-Dawley rats (Zivic-Miller Laboratories, Allison Park, PA), BALB/c mice (Taconic Laboratories, Germantown, NY), golden Syrian hamsters (Charles River Breeding Laboratories, Wilmington, MA), Hartley guinea pigs (Elm Hill Breeding Labs, Chelmsford, MA) and New Zealand White rabbits (Pine Acre Rabbitry, Brattleboro, VT) were purchased and housed in constant temperature rooms with light/dark intervals of 12 hours duration. When indicated, surgical procedures, including orchiectomy and ovariectomy, were performed on rats by surgeons at Zivic-Miller. All animals were allowed to recover for at least 7 days prior to further experimentation. For hormone regulation studies, castrated rats were administered subcutaneous pellets (Innovative Research of America, Sarasota, FL) in the subscapular space. These pellets contained vehicle, testosterone (50 mg), 17β-estradiol (0.1 mg), progesterone (50 mg) or dexamethasone (50 mg), and were designed to release a constant amount of hormone over a 21 day period. All studies with experimental animals adhered to the The Association for Research in Vision and Ophthalmology Resolution on the Use of Animals in Research.

2.2. General Procedures

Lacrimal (exorbital) glands, lids, conjunctivae, corneas, iris/ciliary bodies, lens and retina/uvea, retina/choroid, prostatic and uterine tissues were obtained following animal sacrifice, during surgical procedures on human males and females, or from Eye Bank Donor material. Human retinal pigment epithelial cells were a gift from Kathleen Dorey, PhD (Boston, MA). Total human prostate RNA was purchased from Clontech Laboratories (Palo Alto, CA), and two human cell lines were obtained from American Type Culture Collection (ATCC; Rockville, MD), including LNCaP epithelial cells (human prostate carcinoma, clone FGC, ATCC #CRL 1740) and MCF-7 mammary epithelial cells (human breast adenocarcinoma, ATCC HTB 22). These cell lines were cultured according to the recommendations of ATCC. Samples were processed for molecular biological or immunohistochemical techniques. Statistical analysis of the data was conducted with the unpaired, two-tailed Student's t test or the Mann Whitney U test.

2.3. Histological Procedures

The distribution of androgen receptor protein in ocular tissues was determined by using immunohistochemical procedures, as previously described.[6–8] In brief, samples were embedded in OCT compound, cut into 6 μm sections and fixed in acetone, then 4% paraformaldehyde (Sigma Chemical Company, St. Louis, MO). Sections were blocked with a 2% 'normal' goat serum solution (Vector Laboratories, Burlingame, CA), exposed to purified rabbit polyclonal antibody (gift from Dr. Gail S. Prins, Chicago, IL) to the androgen receptor, and incubated sequentially with avidin D, biotin and biotinylated goat anti-rabbit IgG (Vector). After second antibody treatment, sections were exposed to Vectastain Elite ABC reagent (Vector), developed with an acetate buffer containing 3-amino-9-ethylcarbazole (Sigma), N, N-dimethylformamide and hydrogen peroxide, postfixed in 2% paraformaldehyde and preserved in Crystal Mount (Biomeda, Foster City, CA). To confirm the specificity of the rabbit polyclonal antibody, control studies were performed with each tissue and involved antibody preincubation with 0.1% gelatin/PBS or excess amounts of androgen receptor peptides (from Dr. Prins) "1–21" (receptor segment used to generate the original antibody[5]) or "462–478" (derived from a distant and non-reactive section of the receptor[5]), according to reported methods.[6–8] Under these conditions, peptide "1–21," completely inhibited immunohistochemical staining, whereas neither vehicle nor peptide "462–478" interfered with first antibody binding.

To quantitate the density of androgen-receptor containing cells, tissues were evaluated with a Zeiss light microscope at 160 x magnification and the quantity of immunologically stained nuclei were counted in a defined, geometric area in non-adjacent sections, as previously described.[6,8]

2.4. Molecular Biological Procedures

To identify and/or quantitate the mRNA for androgen, estrogen and progesterone receptors, as well as Types 1 and 2 5α-reductase, in ocular tissues, reverse transcription polymerase chain reactions (RT-PCR) were utilized, as previously described.[1,7,8] Briefly, total RNA was isolated from tissues and cells by using modified acid guanidinium-thiocyanate-phenol-chloroform extraction procedures[9] or TRI-Reagent (Molecular Research Center, Inc., Cincinnati, OH). The RNA preparations were analyzed spectrophotometrically at 260 nm to determine their concentration and examined on 1.2% agarose (Gibco/BRL, Grand Island, NY) gels to confirm RNA integrity. cDNAs were transcribed from RNA samples (5 μg) by employing AMV or MMLV reverse transcriptase, oligo dT and either the First-Strand cDNA Synthesis kit from Invitrogen (San Diego, CA) or the Advantage RT-for-PCR Kit from Clontech Laboratories Inc. (Palo Alto, CA), according to modifications of the manufacturer's protocol. PCR amplification of the cDNAs was performed with a Perkin Elmer Cetus GeneAmp PCR System 9600 (Perkin Elmer, Norwalk, CT) by utilizing Taq DNA polymerase (Gibco/BRL), dNTPs (Perkin-Elmer), Invitrogen PCR buffer and 0.4 μM of each 5' and 3' primer corresponding to rat, rabbit and/or human steroid hormone receptors, 5α-reductases and β-actin cDNAs, as previously reported.[1,7,8] Primers and oligonucleotide probes were designed as outlined,[1,7,8] synthesized by National Biosciences, Inc. (Plymouth, MN) or Gibco/BRL Custom Primers, Inc. (Palo Alto, CA), or purchased from Clontech. cDNA probes were gifts from Drs. Elizabeth Wilson (Chapel Hill, NC), Geoffrey Greene (Chicago, IL) and Lan Hu (Boston, MA). The PCR conditions, number of cycles and types of controls (including the use of RNA from LNCaP and MCF-7 cells, and prostatic and uterine tissues) have been described.[1,7,8] After amplification, PCR products were electrophoresed on 1.5% agarose gels, which contained a 100 bp DNA molecular weight ladder (Gibco) and were post-stained with ethidium bromide in order to verify the anticipated fragment sizes. Amplified cDNA products were transfered by positive pressure to GeneScreen nylon membranes (Dupont/NEN, Boston, MA), fixed by UV cross-linking and incubated with specific ^{32}P-labeled probes, which had been phosphorylated by random priming (Random Primer Labeling System, Dupont/NEN, Boston, MA) or end-labelling (T4 Polynucleotide Kinase, New England BioLabs, Beverly, MA) techniques. Following an overnight hybridization, the Southern blots were processed for autoradiography by utilizing X-OMAT AR film (Kodak, Rochester, NY) at -80°C. To measure mRNA levels, band densities in autoradiographs were analyzed with a computer-assisted image analysis system, as previously reported,[1] and absorbance results were normalized to the corresponding level of β-actin mRNA. To confirm the identity of Types 1 and 2 5α-reductase PCR products, DNA sequence analysis was utilized.[1]

Table 1. Identification of androgen, estrogen and progesterone receptor mRNAs in ocular tissues of rats, rabbits and humans

	Androgen Receptor mRNA			Estrogen Receptor mRNA			Progesterone Receptor mRNA		
Ocular Tissue	Rat	Rabbit	Human	Rat	Rabbit	Human	Rat	Rabbit	Human
Lacrimal gland	+	+	+	+	+	+	+	+	+
Lid	+			+			+		
Conjunctiva		+	+*		+	+		+	+
Cornea	+	+	+*	+	+*	+	+	+	+
Iris/ciliary body		+			+			+	
Lens	+	+		+	+*		–	+	
Retina/uvea	+			+			+*		
Retina/choroid		+			+*			+	
Retinal pigment epithelial cells			+			+			+

Tissues (rats: n = 11-16 samples/tissue type; rabbits: n = 6-14 samples/tissue type; human: n = 2-5 samples/tissue type) were processed for the analysis of sex steroid receptor mRNAs, as outlined in the Methods section. The symbols refer to detectable (+) and undetectable (–). * The number of positive tissues showed gender-related differences. Data from reference (1).

3. RESULTS

3.1. Identification of Sex Steroid and 5α-Reductase mRNAs and/or Protein in Ocular Tissues

To determine whether sex steroid and 5α-reductase mRNAs exist in various ocular sites, tissues were obtained from mice, rats, hamsters, guinea pigs, rabbits and humans and processed for molecular biological procedures. As shown in Table 1, our findings showed that androgen, estrogen and progesterone receptor mRNAs are present in the lacrimal gland, lid, conjunctiva, cornea, iris/ciliary body, lens, retina/uvea, retina/choroid and retinal pigment epithelial (RPE) cells of rats, rabbits and/or humans, and that the expression of these mRNAs, depending upon the specific tissues, is gender- and species-dependent.[1] In addition, our results demonstrated that: (1) androgen receptor mRNA occurs in lacrimal glands of mice, hamsters and guinea pigs;[10] (2) androgen receptor protein exists in the lacrimal gland, conjunctiva, cornea, lens epithelium, retina and RPE cells of rats and/or humans;[7] and (3) the mRNAs for Type 1 and Type 2 5α-reductase are present in the human lacrimal gland, conjunctiva, cornea and RPE cells.[7]

3.2. Regulation of Sex Steroid Receptor mRNAs in the Rat Lacrimal Gland

Sex steroid hormones influence a variety of structural and/or functional attributes of lacrimal tissue,[2] and it appears that the glandular actions of androgen, estrogens and progestins may be modulated by other sex steroids or glucocorticoids.[2] To explore the underlying bases for these endocrine interrelationships, we examined whether sex steroids or corticosteroids control the expression of androgen, estrogen or progesterone receptor mRNAs, as well as androgen receptor protein, in the lacrimal gland. Lacrimal tissues were obtained from ovariectomized or orchiectomized rats (n = 5–9/group/ experiment) following 7 days of treatment with vehicle, testosterone, 17β-estradiol, progesterone or dexamethasone and processed for either immunohistochemistry or molecular biological

Table 2. Influence of sex steroid or glucocorticoid treatment on the level of androgen (AR), estrogen (ER), and progesterone (PR) receptor mRNA and/or protein in the rat lacrimal gland

	AR			
Treatment	mRNA	Protein	ER mRNA	PR mRNA
Testosterone	↓	↑	↓	—
17β-Estradiol	—	—	↓	—
Progesterone	↓	↑	—	—
Dexamethasone	↓	↓	↓	—

Lacrimal glands were obtained from hormone-treated (i.e. 7 days with placebo, testosterone, 17β-estradiol, progesterone or dexamethasone) ovariectomized rats (n = 5-7/group/experiment). Tissues were processed for either immunohistochemistry (to detect the number of AR protein-containing cells; designated "Protein" in the Table) or for the analysis of AR, ER and PR mRNAs by RT-PCR (at exponential phase of amplification), Southern blot hybridization, autoradiography and densitometry. All absorbance data were standardized to the corresponding levels of β-actin mRNA. Symbols mean significantly higher (↑) or lower (↓), or not significantly different from —, the value of placebo-exposed controls. Data are from reference (8).

procedures. As shown in Table 2, our findings with lacrimal glands of ovariectomized rats demonstrated that: (1) administration of testosterone or progesterone, but not 17β-estradiol, enhanced the number of androgen receptor protein-containing cells and suppressed the amount of androgen receptor mRNA. In contrast, dexamethasone exposure attenuated the levels of both androgen receptor protein and mRNA;[8] (2) treatment with testosterone, 17β-estradiol or dexamethasone, but not progesterone, decreased the content of estrogen receptor mRNA;[8] and (3) sex steroid or glucocorticoid administration did not induce any consistent alteration in the amount of progesterone receptor mRNA.[8] For comparison, testosterone's effect on the density of androgen receptor protein-containing cells and the level of androgen receptor mRNA was also observed in lacrimal tissues of orchiectomized rats.[6,8]

4. DISCUSSION

Our findings demonstrate that sex steroid receptor mRNAs and/or proteins, as well as the mRNAs for Types 1 and Type 2 5α-reductase, are present in numerous ocular tissues, suggesting that these sites may represent target organs for androgen, estrogen and progestin action. In addition, our results show that sex steroids and glucocorticoids control the levels of androgen and estrogen receptor mRNAs and/or protein in the rat lacrimal gland. Our ongoing studies are designed to extend these observations and elucidate the precise role of sex steroids in maintaining the health and well-being of the eye.

ACKNOWLEDGMENTS

The authors express their appreciation to Drs. Devinder P. Cheema (Boston, MA), Elizabeth Daher (Boston, MA), Kenneth Kenyon (Boston, MA), Peter A. Rapoza (Boston, MA), Toshimichi Shinohara (Boston, MA), John W. Shore (Boston, MA), Daniel J. Townsend (Boston, MA), Fu-Shin Yu (Boston, MA), Van Luu-The (Quebec, Canada), Kathleen

Dorey, Gail S. Prins, Elizabeth M. Wilson, Geoffrey Greene and Lan Hu and Ms. Laila A. Hanninen (Boston, MA) for their provision of human tissue samples, cells, cDNAs or technical guidance. This research was supported by research grants from NIH (EY05612), Allergan and the Massachusetts Lions Research Fund, and a postdoctoral fellowship award from the Uehara Memorial Foundation.

REFERENCES

1. Wickham LA, Gao J, Toda I, Rocha EM, Sullivan DA. Identification of androgen, estrogen and progesterone receptor mRNAs in rat, rabbit and human ocular tissues. Submitted for publication, 1997.
2. Sullivan DA, Wickham LA, Rocha EM, Silveira LA, Toda I. Influence of gender and sex steroid hormones on the structure and function of the lacrimal gland. *Adv Exp Med Biol* 1997; in press.
3. Sullivan DA, Wickham LA, Krenzer KL, Rocha EM, Toda I. Aqueous tear deficiency in Sjögren's syndrome: Possible causes and potential treatment. In: Pleyer U, Hartmann C and Sterry W, eds. *Oculodermal Diseases - Immunology of Bullous Oculo-Muco-Cutaneous Disorders*. Buren, The Netherlands: Aeolus Press, 1997:95–152.
4. Knepper PA. Method for the prevention of ocular hypertension, treatment of glaucoma and treatment of ocular hypertension. *US Patent* 4,617,299.
5. Prins GS, Birch L, Greene GL. Androgen receptor localization in different cell types of the adult rat prostate. *Endocrr* 1991; 129:3187–3199.
6. Rocha FJ, Wickham LA, Pena JDO, Gao J, Ono M, Lambert RW, Kelleher RS, Sullivan DA. Influence of gender and the endocrine environment on the distribution of androgen receptors in the lacrimal gland. *J Steroid Biochem Mol Biol* 1993; 46:737–749.
7. Rocha EM, Wickham LA, Silveria LA, Krenzer KL, Toda I, Sullivan DA. Identification of androgen receptor protein and 5α-reductase mRNA in human ocular tissues. Submitted for publication, 1997.
8. Wickham LA, Gao J, Toda I, Sullivan DA. Endocrine regulation of androgen receptor protein and sex steroid mRNA levels in the rat lacrimal gland. Submitted for publication, 1997.
9. Chomczynski P, Sacchi N. Single-step method of RNA isolation by acid guanidinium thiocyanate-phenol-chloroform extraction. *Anal. Biochem.* 1987; 162:156–159.
10. Sullivan DA, Edwards JA, Wickham LA, Pena JDO, Gao J, Ono M, Kelleher RS. Identification and endocrine control of sex steroid binding sites in the lacrimal gland. *Curr Eye Res* 1996; 15:279–291.

13

DINUCLEOTIDE REPEAT POLYMORPHISM NEAR THE TEAR LIPOCALIN GENE

Eric Lacazette,[1] Gilles Pitiot,[2] Jacques Mallet,[2] and Anne-Marie Françoise Gachon[1]

[1]Laboratoire de Biochimie Médicale
Faculté de Médecine
Clermont-Ferrand Cedex, France
[2]Laboratoire de Génétique Moléculaire
URM 9923 CNRS, Bâtiment CERVI
Hôpital de la Pitié-Salpétrière, Paris, France

1. INTRODUCTION

Human tear lipocalin, secreted by the main lacrimal gland, belongs to a group of proteins able to bind lipophiles by enclosing them within their structures. We have recently reported the chromosomal location of the gene for human tear lipocalin (LCN1) on the long arm of chromosome 9 by *in situ* hybridization and somatic hybrid analysis.[1,2] These results confirmed the localization of the same tear lipocalin gene by Glasgow et al.,[3] although these workers did not use the gene symbol LCN1. These data together also definitively proved that LCN1 is not assigned to chromosome 8, as erroneously reported by Redl's team.[4] We describe a dinucleotide polymorphism in the 3' region of the tear lipocalin gene, which is known to span 6.2 kb and contains 7 exons.[5]

2. MATERIALS AND METHODS

We screened, using a cDNA LCN1 probe, a copy of the chromosome 9-specific gene library LL09NC01, which was constructed by Dr. J. Allmeman at the Biomedical Sciences Division, Lawrence Livermore National Laboratory, Livermore, CA, under the auspices of the National Gene Library Project sponsored by the US Department of Energy. Positive clones were further analyzed using oligonucleotide probes, and screening with a $(CA)_{10}$ oligonucleotide revealed a microsatellite in the clone P32H3. The Eco R1 region was subcloned in pBluescript® plasmid vector, and the sequence was established, using the standard primers KS and SK. DNA sequences flanking the CA repeat were used to de-

Lacrimal Gland, Tear Film, and Dry Eye Syndromes 2
edited by Sullivan *et al.*, Plenum Press, New York, 1998

sign polymerase chain reaction (PCR) primers. One of the oligonucleotide primers used for each amplification was labeled with a fluorescent product. PCR amplifications were carried out in a total volume of 15 µl containing 8 ng genomic DNA and 0.4 unit *Taq* polymerase (Appligene), with the Perkin-Elmer GeneAmp® PCR System 9600. Reaction conditions consisted of 165 nM of each primer, 10 mM Tris HCl, pH 9, at 25°C, 50 mM KCl, 1.5 mM $MgCl_2$, 0.1% Triton X100, 0.2 mg/ml BSA, and 250 nM of each dNTP. The thermocycler parameters consisted of an initial cycle at 94°C for 3 min followed by 30 cycles at 94°C for 30 sec, 53°C for 1 min, and 70°C for 30 sec. Final elongation step was extended for 10 min at 70°C. The length of the amplified DNA fragments was analyzed by an automated nonradioactive laser fluorescent gel detection (system ABI PRISM™ 377 DNA Sequencer, Perkin Elmer Corp.). 1.5 µl of each sample was mixed with 3.5 µl of loading cocktail consisting of 2 µl formamide, 0.5 µl blue dextran (50 mM EDTA, 50 mg/ml blue dextran), and 1 µl size standard (GS-500 TAMRA, Applied Biosystems). Before loading on a 4% denaturating polyacrylamide gel, the samples were denatured 2 min at 94°C and then chilled on ice. The electrophoresis was performed for 2 h. The results were analyzed with Genotyper 1.1 software (Perkin Elmer Corp.).

3. RESULTS AND DISCUSSION

According to the size of the cosmid insert, the distance between LCN1 and the microsatellite was less than 30 kb on the 3' side of LCN1. The portions flanking the CA repeat were sequenced, and the data have been deposited in the EMBL nucleotide sequence data base (confidential until 30 June 1997). Two sequences were carefully chosen as primers. Co-dominant segregation was observed in three different three-generation families (CEPH families). The frequencies of the relevant alleles were estimated by analyzing 226 chromosomes from 113 unrelated Caucasian individuals; they are given in Table 1. The observed heterozygosity is 0.63.

Recent data on disease loci on chromosome 9 show that XPA (xeroderma pigmentosum) locus is 9q32-q34.[6] In addition, it is known[7] that some of the families with tuberous sclerosis segregate for the TSC1 gene on 9q34. The presence of a new microsatellite marker in the 9q34 region may be useful for linkage analysis and for detecting losses of heterozygosity.

ACKNOWLEDGMENTS

We thank the U384 INSERM research groups (Dir.: B. Dastugue) for free access to their facilities. We are grateful to Dr. Gingrich at the LLNL that gave us access to the

Table 1. Allele frequencies

Allele	Frequency	Number of CA repeats
1	0.03	17
2	0.45	18
3	0.07	19
4	0.41	20
5	4.10^{-3}	21
6	0.04	23

LLO9NC01 library, to Drs. Obermayer and Frischauf for duplicating it at ICRF, and to Dr. Soularue at Genethon for spotting it. We also thank S. Fauré (Généthon II) for providing the CEPH samples, and I. Creveaux and M. Petit for providing the 113 Caucasian samples. This work was supported by a grant from the Ministère de l'Enseignement et de la Recherche (MESR: N° 93Y 0113/1), by University funding (BQR), and by the local committee of La Ligue Nationale contre le Cancer.

REFERENCES

1. Lassagne H, Nguyen VC, Mattei MG, Gachon AMF. Assignment of LCN1 to human chromosome 9 is confirmed. *Cytogenet Cell Genet.* 1995;71:104.
2. Lassagne H, Ressot C, Mattei MG, Gachon AMF. Assignment of the human tear lipocalin gene (LCN1) to 9q34 by in situ hybridization. *Genomics.* 1993;18:160–161.
3. Glasgow BJ, Heinzmann C, Kojis T, Sparkes R, Mohandas T, Bateman JB. Assignment of tear lipocalin gene to human chromosome 9q34–9qter. *Curr Eye Res.* 1993;12:1019–1023.
4. Holzfeind P, Redl B. Structural organization of the gene encoding the human tear prealbumin and synthesis of the recombinant protein in *Escherichia coli. Gene.* 1994;139:177–183.
5. Baumgartner M, Holzfeind P, Redl B. Assignment of the gene for human tear prealbumin (LCN1), a member of the lipocalin superfamily, to chromosome 8q24. *Cytogenet Cell Genet.* 1994;65:101–103.
6. Satokata I, Tanaka K, Miura N, et al. Characterization of a splicing mutation in group A xeroderma pigmentosum. *Proc Natl Acad Sci USA.* 1990;87:9908–9912.
7. Povey S, Burley MW, Attwood J, et al. Two loci for tuberous sclerosis: One on 9q34 and one on 16p13. *Am J Hum Genet.* 1994;58:107–127.

14

STUDIES OF LIGAND BINDING AND CD ANALYSIS WITH Apo- AND Holo-TEAR LIPOCALINS

Ben J. Glasgow, Adil R. Abduragimov, Taleh N. Yusifov, and Oktai K. Gasymov

Departments of Pathology and Ophthalmology
University of California at Los Angeles School of Medicine
Los Angeles, California

1. INTRODUCTION

Tear lipocalins (TL) are novel members of the lipocalin family, a group of proteins that are capable of binding and transporting small hydrophobic molecules.[1,2] Most lipocalins, such as retinol binding protein and pheromaxeine bind specific ligands. However,TL are capable of binding a broad array of lipid molecules.[3] Cholesterol, fatty acids, phospholipids, and gly-colipids are bound to TL in tears. TL are promiscuous and have been shown to bind retinol, synthetic analogs of cholesterol, fatty acids, fatty alcohols, phospholipids, glycolipids, and even environmental contaminants such as dioctyl phthalate.[1,3]

Comparison of the primary structure and crystallography data of other lipocalins suggests that TL form a barrel composed of anti-parallel ß sheets.[4] Circular dichroism (CD) analysis confirms the predominantly ß structure,[5] and electron paramagnetic reso-nance shows that ligands are immobilized with the hydrophobic portion inserted inside the TL cavity and hydrophilic moieties projecting out of the cavity.[3] Preliminary data with synthetic analogs indicate that the binding affinity of ligands for TL relate to the length, hydrophobicity, and steric interactions of the alkyl chain.[3] Very little data exist regarding the relative binding affinity for native ligands. The validity of ligand affinity experiments using apo-TL is predicated on the retention of basic structure, including disulfide bonds and ß structure, after delipidation with organic solvents.

As an initial step toward understanding the relative ligand affinity for TL, we evalu-ated the structural characteristics of apo-TL. We explored the possibility that TL is dena-tured by organic solvents and compared the disulfide motif that is known for holo-TL with apo-TL.[5] We searched for perturbations in secondary structures by CD analysis and com-pared equilibrium binding in assays of native and delipidated TL.

Lacrimal Gland, Tear Film, and Dry Eye Syndromes 2
edited by Sullivan *et al.*, Plenum Press, New York, 1998

2. MATERIALS AND METHODS

2.1. TL and Retinol Preparation

Holo-TL were purified from pooled tear samples by column chromatography as previously described. [3,6] Purification was confirmed by analytical SDS tricine PAGE[3] and by electrospray mass spectrographic analysis.[5] Apo-TLwas prepared by delipidation of holo-TL described previously.[3] Protein concentration was determined by the biuret method.[7]

Disulfide reduction of apo-TL was performed with dithiothreitol.[8] After reduction, protein was separated from excess reducing agent by gel filtration with Sephadex G-25 with 25 mM Tris-HCl, pH 8.4, containing 0.5 mM mercaptoethanol.

Retinol was obtained by reducing all trans-retinal isomer with sodium borohydrate.[9] The concentration of retinol was determined from absorbance at 325 nm using the extinction coefficient of 53000.[10]

2.2. Circular Dichroism Spectroscopy

CD spectra was recorded on a Jasco 600 spectropolarimeter. The path length of cells was 0.2 mm for far-UV CD spectra and 10 mm for near-UV. Protein concentration was 1.2 mg/ml. Spectra were obtained by averaging either 8 or 16 scans for wavelength 190–250 nm and 250–320 nm for far- and near-UV, respectively. Results are reported in m degrees.

2.3. Lipid Binding Assays

Binding of [9,10^{3}H] palmitic acid to apo-TL and native TL was performed with a Lipidex assay.[11,12] For the assay, 150 µl of 10 mM Tris-HCl buffer, pH 8.0, and 50 µl protein solution (25 µM apo- or 10 µM holo-TL) were mixed with 50 µl of varying concentrations of tritiated palmitic acid in a 1.5 ml Eppendorf polypropylene reaction vial. After incubation for 30 min at 37°C, the vials were centrifuged for a few seconds to remove condensation from the wall and then placed on ice. From each vial, 50 µl was pipetted into a scintillation vial to assess the actual fatty acid concentration in the aqueous solution. To the remaining volume, 75 µl of continuously stirred ice-cold Lipidex 1000 suspension (50% v/v suspension in 10 mM potassium phosphate buffer, pH 7.5) was added and mixed three or four times on a vortex mixer during 30 min of incubation at 0°C. Finally the vials were centrifuged (8 min, 10,000 *g*, at 4°C), and 100 µl of supernatant was pipetted into scintillation vials to quantify the amount of bound fatty acid.

Blank values were obtained for each fatty acid dilution by measuring the radioactivity of incubations in which 50 µl Tris-HCl buffer without protein was added. The measured blank values were subtracted from the fatty acid binding data.

2.4. Retinol Binding Experiments for CD Analysis

To 70 nmoles of TL in 25 mM Tris-HCl, pH 8.4 (without and withβ-mercaptoethanol), 2 µl of retinol (70 mM solution in ethanol) was added. Special care was taken to ensure that the addition of the ethanol solution was made directly into the aqueous medium and not onto glass walls. The content was mixed gently, then incubated at 22°C for 30 min. Then solutions were used for spectral analysis.

3. RESULTS

The far- UV CD spectrum demonstrates a minimum at 212 nm and crossover at about 200 nm with a positive CD band below 195 nm, indicative of ß sheet structure in native, apo, and reduced apo-TL. With delipidation, there is decreased optical activity at 190–196 nm and increased optical activity at 206–208 nm (random coil formation) (Fig. 1). Disulfide reduction of the apoprotein results in marked increase of optical activity at 190–196 nm, 216 nm (ß sheet contribution), and 222 nm (α helical contribution) when compared to the native or apoprotein (Fig. 1).

The near- UV CD spectrum provides information about the environment of the aromatic side chains of the protein. The near- UV CD spectra of apo-TL demonstrates reduction of a trough in this range when compared to native TL (Fig. 2).

Retinol was chosen as the ligand to measure CD spectra of binding of TL because free retinol is optically inactive, whereas bound retinol shows optical activity at 320 nm. The addition of retinol to delipidated TL resulted in a negative CD band in a range including 320 nm, indicative of binding (Fig. 3). Also there was a marked diminution of the trough at 280 nm that was greater in apo-TL than in holo-TL, indicating that a conformational change in the protein involves aromatic residues during ligand binding. A definite

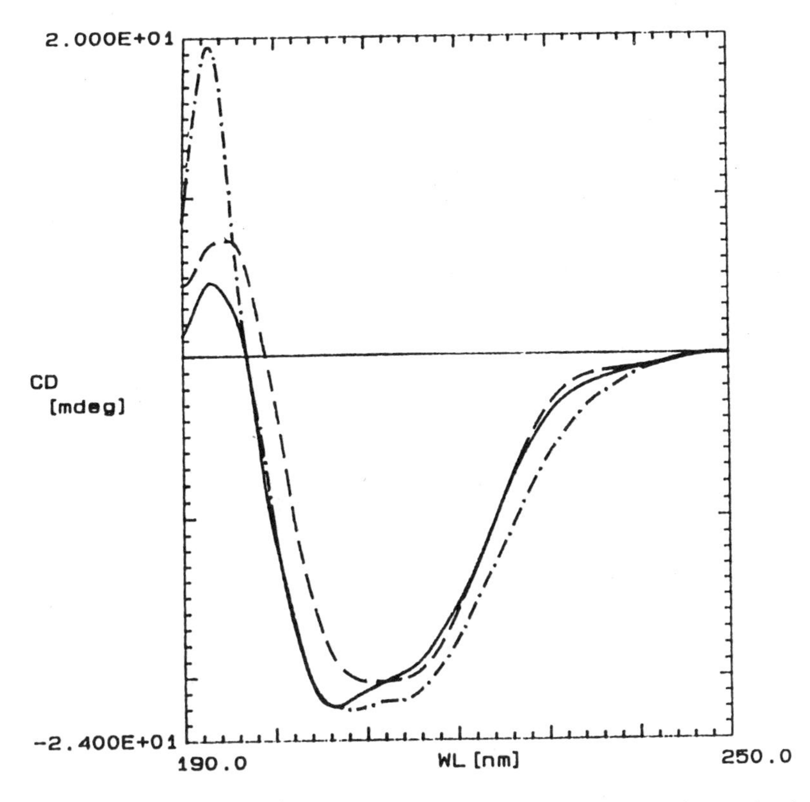

Figure 1. Far- UV spectrum of TL in different states. The apoprotein (solid line) shows decreased optical activity from holo-TL (– – –) at 190–196 nm (decreased secondary structure) and increased activity at 206–208 nm (random coil), but retains the trough at 216–220 nm (β sheet structure). The reduced apoprotein (– • – • –) shows marked increase in optical activity at 190–196 nm (secondary structure) and similar optical activity at 206–208 nm (random coil).

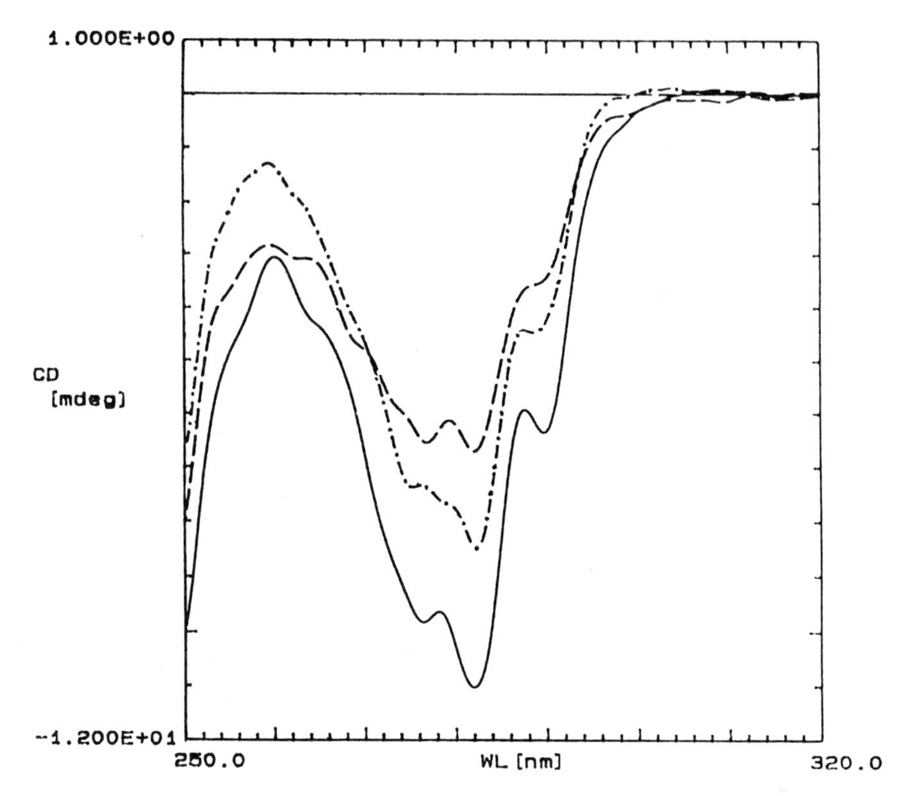

Figure 2. Near- UV spectrum of TL in different states. There is general reduction in the optical activity in both the apo- TL (–•–•–) and reduced apo-TL (– – –) indicating a relaxation in protein structure from the native protein (solid line). The marked reduction in optical activity at 280–290 nm is indicative of conformational changes involving the aromatic amino acid resudues. Delipidated TL has a pronounced loss of optical activity at 260 nm. This indicates a loss of conformational asymmetry involving aromatic residues. A more pronounced diminution of the trough in this region occurred with reduced apoprotein. In addition, there is alteration of optical activity throughout the spectra indicative of relaxation in structure that occurs with delipidation and also with reduction of the disulfide bonds. A subtle increase in optical activity is observed at 260 nm in the reduced apoprotein. In addition, greater alteration in optical activity was observed for the reduced apo-TL compared to apo-TL at 276 nm than at 282 nm (Fig. 2). Optical activity in this area is accounted for by exposed tyrosine residues.

but subtle change occurred in the spectra at 290–292 nm, indicating that the change in conformation involves tryptophan (Fig. 3). With the addition of retinol to the apo-TL, an alteration at 260–276 nm was observed. This alteration is accounted for by reduced exposure of tyrosine residues.

The reduced apoprotein showed marked alteration of the CD spectra at both 280 and 320 nm. The upward CD band at 280 nm indicates a radical change in the aromatic residues with reduction of the disulfide and the addition of retinol. Retinol is bound, but the radical shift in optical activity at 300–340 nm suggests a different conformation in the bound structure.

Scatchard analysis revealed for holo-TL, B_{max}=0.85 moles ligand/mole protein, K_d= 13.5 μM (Fig. 4), and for apo-TL, B_{max}= 0.52 moles ligand/mole protein, K_d=8.3 μM (Fig. 5).

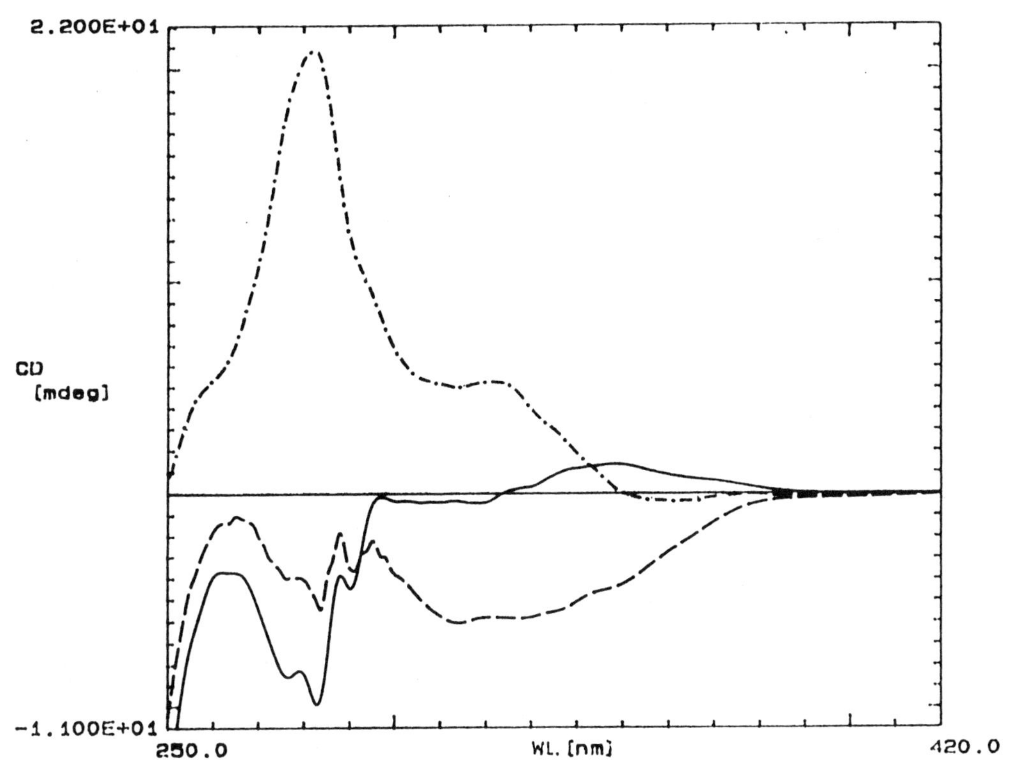

Figure 3. Near-UV spectrum of the different states of TL in the presence of retinol. The upward CD band at 320 nm in holo-TL (solid line) is due to binding of retinol. Apo-TL (–•–•–) shows greater optical activity, indicating much stronger binding. However, the reduced form of apo-TL (–––) shows a marked alteration in optical activity throughout the spectrum. The positive CD band at 280 nm is indicative of the participation of aromatic amino acid residues in these changes.

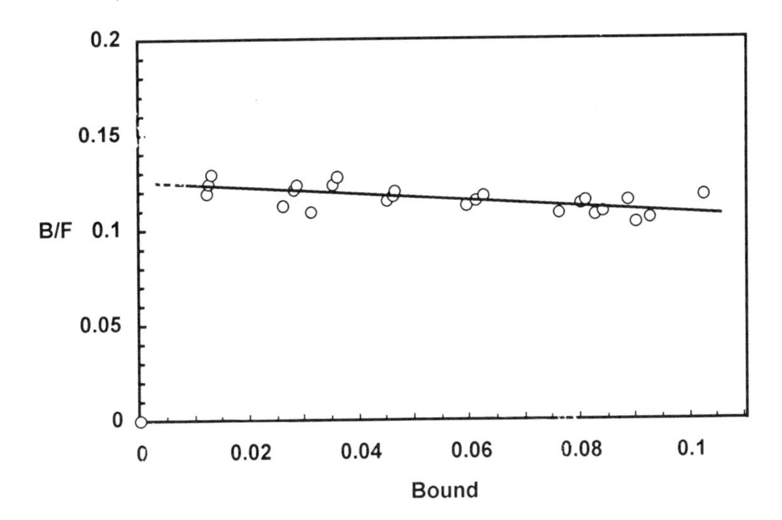

Figure 4. Scatchard analysis of the binding of various concentrations of tritiated palmitic acid with native tear lipocalin. The slope of the curve represents the apparent dissociation constant (K_d). All values for ligand binding are expressed as moles palmitic acid bound per mole of TL.

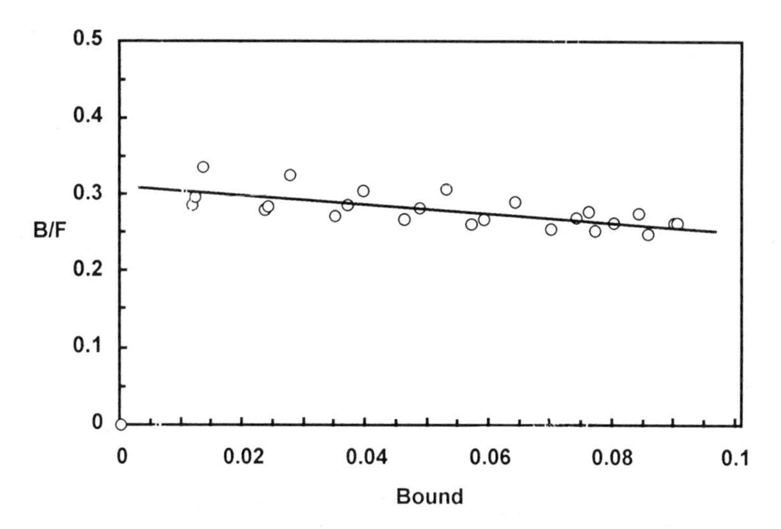

Figure 5. Scatchard analysis of the binding of various concentrations of tritiated palmitic acid with delipidated tear lipocalin. All values for ligand binding are expressed as moles palmitic acid bound per mole of delipidated TL.

4. DISCUSSION

The CD experiments offer insight into protein structural changes that occur with delipidation and disulfide reduction. Despite the harsh conditions of delipidation, overall ß sheet architecture is preserved in the apo and even reduced apo forms of TL. However, some loss of secondary structure and increase in random coil formation were observed. The relative contribution of the α and ß structure is difficult to distinguish in these spectra. One possible explanation for the loss of secondary structure after delipidation is that the protein-lipid complex contributes to stability of secondary structure. This is presumably due to hydrophobic interactions within the cavity of TL. Alternatively, it is possible that the organic solvents used in delipidation resulted in direct protein perturbations and random coil formation.

Reduction of the disulfide bond in the apoprotein resulted in altered secondary structure compared to both the native TL and the non-reduced apoprotein. In this instance, the spectra clearly show increased optical activity at 192 and 222 nm, which when taken together are indicative of increasedα helix formation without a change in random coil formation. This suggests that reduction of the disulfide resulted in a more relaxed configuration of the protein, permitting enhanced interaction of the polypeptide chains already involved inα helical and ß sheet structure.

The near- UV spectrum reveals information about the alteration of the structure that involves aromatic amino acid residues. In TL, conformational changes occur with delipidation and/or reduction that involve aromatic residues. It is apparent from Figure 2 that optical activity is relatively less at 260 nm for apoTL compared to holo-TL or reduced TL indicating fewer exposed Tyrosine residues. Tyrosine residues are buried by delipidation of holo-TL. The reduction of the apoprotein shows relatively more optical activity at 260 nm than can be accounted for by relaxation of structure alone. There is a relatively larger contribution of optical activity by tyrosine residues in the apoprotein at 260 nm. Since exposed tyrosine residues contribute more to optical activity in this region than do buried residues, we conclude that disulfide reduction in combination with delipidation leads to exposure of tyrosine residues that were presumably buried in a hydrophobic pocket.

The retinol binding experiments offer useful information regarding the functional state of the different forms of TL. Delipidated TL binds retinol more strongly than native TL because there is more optical activity. The binding of retinol results in alteration of apo- TL structure that involves the chromophore residues—specifically tyrosine and tryptophan. The marked alterations that occur with reduction of apo-TL indicate marked conformational changes that involve aromatic residues. The markedly positive CD band at 300–340 nm in the delipidated reduced protein with retinol suggests a different mechanism of binding than in the non-reduced form.

These findings have a practical signficance for studies of ligand binding of TL. First, the overall ß structure is resistant to conditions of delipidation and disulfide bond reduction. However, perturbations do occur in the secondary structure with delipidation that involve chromophore residues and alter the mechanism of retinol binding. At least for ligands such as retinol, native protein rather than delipidated may be preferable to study the biological ligand binding affinity. Furthermore, disulfide bonds greatly influence the mechanism of binding of retinol to the apoprotein.

Ligand binding assays were hampered by the lack of saturation at concentrations of tritiated palmitic acid that were soluble in aqueous solution. It was necessary to extrapolate from the data points to the x intercept to determine the maximum ligand bound for both native and apo-TL. These values suggest that the delipidated protein binds only 0.52 moles of ligands as compared to 0.85 moles for the native protein. This raises the possibility of dimer formation as a result of the delipidation process under these conditions.

Changes occur with delipidation and especially with disulfide reduction that may undermine experiments using apo- forms of the protein for functional ligand studies. We suggest that ligand binding assays of TL should be performed with native protein if at all possible.

Despite the aforementioned caveats, our data demonstrate that a number of changes occur in the conformation of TL during ligand binding. Structural and conformational changes are reflected in optical activity that involve aromatic amino acid residues. The data suggest that one or more tyrosine residues shift from a buried site to an exposed site when ligands are removed from the reduced apoprotein. Four of the five tyrosine molecules are situated in close proximity in the linear structure of TL.[1,13–15] It is plausible that tyrosine residues participate in a large hydrophobic ligand binding pocket. Additional experiments will be necessary to further define the dynamic processes that occur during binding and the structural basis for the broad affinity of TL for its ligands.

ACKNOWLEDGMENTS

This work was supported by the Karl Kirschgessner and Stein-Oppenheimer Foundations and by USPHS NIH grant EY11224.

REFERENCES

1. Redl B, Holzfeind P, Lottspeich F. cDNA cloning and sequencing reveals human tear prealbumin to be a member of the lipophilic-ligand carrier protein superfamily. *J Biol Chem*. 1992;267:20282–20287.
2. Delaire A, Lassagne H, Gachon AMF. New members of the lipocalin family in human tear fluid. *Exp Eye Res*. 1992;55:645–647.

3. Glasgow BJ, Abduragimov AR, Farahbakhsh Z, Faull KF, Hubbell WL. Tear lipocalins bind a broad array of lipid ligands. *Curr Eye Res.* 1995;14:363–372.
4. Flower DR. The lipocalin protein family: Structure and function. *Biochem J.* 1996;318:1–14.
5. Glasgow BJ, Abduragimov AR, Yusifov TN, Faull KF, Hubbell WL, Horwitz J. Characterization of the disulfide motif in the major isoform of tear lipocalins. ARVO Abstracts. *Invest Ophthalmol Vis Sci.* 1996;37:S849.
6. Glasgow BJ. Tissue expression of lipocalins in human lacrimal and von Ebner's glands: Colocalization with lysozyme. *Graefes Arch Clin Exp Ophthalmol.* 1995;233:513–522.
7. Bozimowski D, Artiss JD, Zak B. The variable reagent blank: Protein determination as a model. *J Clin Chem Clin Biochem.* 1985;23:683–689.
8. Elman GL. Tissue sulfhydryl group. *Arch Biochem Biophy.* 1959;82:70–77.
9. Heller J, Horwitz J. Conformational changes following interaction between retinol isomers and human retinol-binding protein and between the retinol-binding protein and prealbumin. *J Biol Chem.* 1973;248:6308–6316.
10. Hubbard R, Brown PK, Bounds D. Methodology of vitamin A and visual pigments. *Methods Enzymol.* 1971;18:628–629.
11. Glatz JFC, Veerkamp JH. A radiochemical procedure for the assay of fatty acid binding by proteins. *Anal Biochem.* 1983;132:89–95.
12. Vork MM, Glatz JFC, Surtel DAM, van der Vusse GF. Assay of the binding of fatty acids by proteins: Evaluation of the Lipidex 1000 procedure. *Mol Cell Biochem.* 1990;98:111–117.
13. Lassagne H, Gachon AMF. Cloning of a human lacrimal lipocalin secreted in tears. *Exp Eye Res.* 1993;56:605–609.
14. Glasgow BJ, Heinzmann C, Kojis T, Sparkes RS, Mohandas T, Bateman JB. Assignment of tear lipocalin gene to human chromosome 9q34–9qter. *Curr Eye Res.* 1993;12:1019–1023.
15. Bläker M, Kock K, Ahlers C, Buck F, Schmale H. Molecular cloning of human von Ebner's gland protein, a member of the lipocalin superfamily high expressed in lingual salivary glands. *Biochim Biophys Acta.* 1993; 1172:131–137.

15

SIGNAL TRANSDUCTION PATHWAYS ACTIVATED BY CHOLINERGIC AND α_1-ADRENERGIC AGONISTS IN THE LACRIMAL GLAND

Darlene A. Dartt, Robin R. Hodges, and Driss Zoukhri

Schepens Eye Research Institute
and Department of Ophthalmology
Harvard Medical School
Boston, Massachusetts

1. INTRODUCTION

Nerves are a major pathway to stimulate lacrimal gland protein, electrolyte, and water secretion. They provide a rapid secretory response to protect the ocular surface from environmental insults and challenges by production of the aqueous layer of the tear film. The cellular mechanisms by which nerves stimulate lacrimal gland secretion (signal transduction) have recently been elucidated and will be the focus of the present article.

The lacrimal gland is a tubuloalveolar exocrine gland containing pyramidally shaped acinar cells (the major cell type) joined at the apical end of their lateral membranes to form an acinus (the functional unit) (Fig. 1).

Acinar cells secrete proteins, electrolytes, and water. Regulated proteins are synthesized in the acinar cell endoplasmic reticulum, modified in the Golgi apparatus, and stored in secretory granules that fill the apical region of the cell. Upon stimulation, the secretory granule membrane fuses with the apical membrane, and the regulated secretory proteins are released into the lumen. Electrolytes and water are secreted by activation of ion transport proteins and ion channels located in the apical and basolateral membranes. Activation of these transporters causes movement of ions and water across the apical and lateral membranes into the lumen. The fluid secreted by the acinar cells travels through the duct system, where it is modified by ductal cell secretion of electrolytes, water, and proteins and finally exits onto the surface of the eye.

Acinar cells, because of the presence of apically located tight junctions, form a highly polarized epithelium. Nerves that innervate the gland surround the basolateral

Lacrimal Gland, Tear Film, and Dry Eye Syndromes 2
edited by Sullivan *et al.*, Plenum Press, New York, 1998

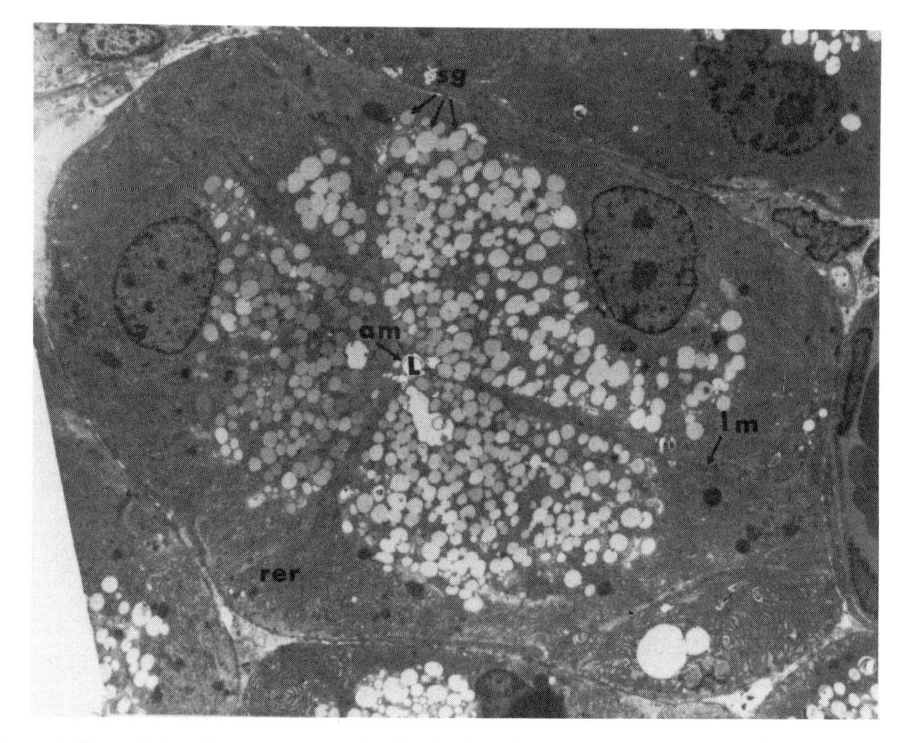

Figure 1. Transmission electron micrograph of lacrimal gland acinus. Acinus consists of acinar cells joined together at the apical membrane (am) to form a lumen (L). Nerves are located around the basal and lateral membranes (bm and lm, respectively). Nerves stimulate secretion of proteins that have been synthesized in the rough endoplasmic reticulum (rer) and stored in secretory granules (sg). Proteins are secreted across the apical membrane into the lumen. Original magnification X4000. (Courtesy of Laila Hanninen and Kenneth Kenyon. From Dartt DA. Tear physiology and biochemistry. In: Albert DM, Jakobiec FA, eds. Philadelphia: Saunders; 1994:1043–1049. *Principles and Practice of Ophthalmology.*

membrane, whereas secretion occurs at the opposite side of the cell at the lumen. A major focus of research has been to determine how the signal to secrete (activation of nerves) is transduced across the cell from the basolateral to the apical side. The signal is transduced by a cascade of intracellular, biochemical processes; each cascade is specific to the type of neurotransmitter and the type of cell being stimulated.

2. FUNCTIONAL INNERVATION

The lacrimal gland is innervated by both parasympathetic and sympathetic nerves, with the former predominating. Parasympathetic nerves release the neurotransmitters acetylcholine (ACh) and vasoactive intestinal peptide (VIP). Sympathetic nerves release norepinephrine and neuropeptide Y. ACh, norepinephrine, and VIP are the major stimuli of lacrimal gland secretion, and each activates a separate signal transduction pathway.[1] ACh activates a muscarinic, cholinergic pathway, and norepinephrine activates an α_1-adrenergic pathway. VIP induces an adenylate cyclase, cAMP-dependent pathway. In most secretory tissues studied, cholinergic and α_1-adrenergic agonists activate the same signal transduction pathway. Thus, when cholinergic and α_1-adrenergic agonists are added simultaneously, the resultant secretory response is the same as that obtained when either agonist is added alone. In contrast, in the lacrimal gland, if these two agonists are added together, secretion is greater than that obtained when each agonist is added alone.[2] Based on this finding, we hypothesized that cholinergic and α_1-adrenergic agonists activate separate, different signal transduction pathways in the lacrimal gland. Subsequent identification of the biochemical processes activated by each agonist has supported this hypothesis.

3. OVERVIEW OF SIGNAL TRANSDUCTION PATHWAYS IN THE LACRIMAL GLAND

3.1. Cholinergic Agonist Activated Pathway

As in most other tissues, cholinergic agonists in the lacrimal gland activate a Ca^{2+}- and protein kinase C (PKC)-dependent pathway (Fig. 2).[3]

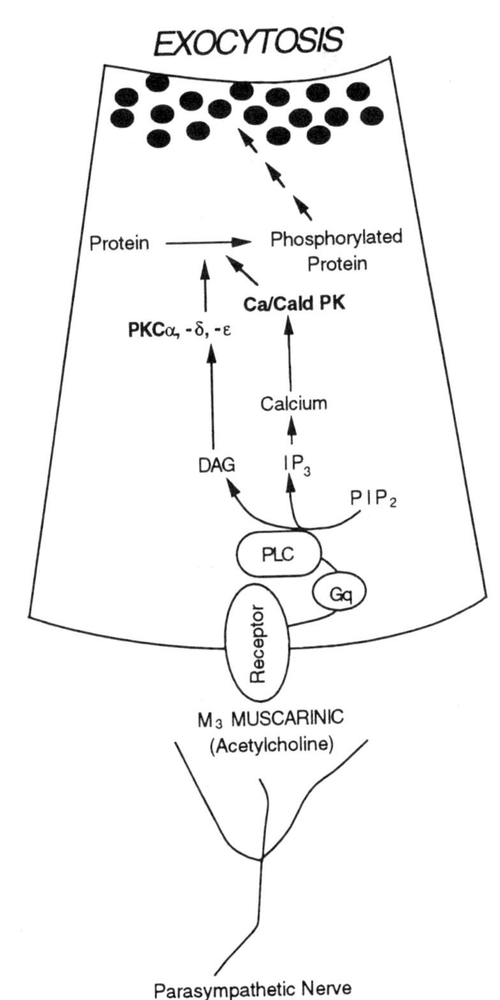

Figure 2. Cholinergic agonist stimulated signal transduction pathway in lacrimal gland acinar cells. DAG, diacylglycerol. G, G protein. PLC, phospholipase C. PIP_2, phosphatidylinositol 4,5-bisphosphate. IP_3, inositol 1,4,5-trisphosphate. Ca/Cald PK, calcium calmodulin-dependent protein kinase.

ACh released by stimulation of parasympathetic nerves activates muscarinic receptors of the M_3 subtype located on the basolateral membranes.[4] These receptors then activate guanine nucleotide-binding proteins (G proteins) of the $G_{\alpha q}$ or $G_{\alpha 11}$ subtype.[5] The activated G protein then stimulates phospholipase C (PLC) to break down phosphatidylinositol 4,5-bisphosphate (PIP_2) into inositol 1,4,5-trisphosphate (IP_3) and diacylglycerol (DAG).[6,7] IP_3 causes release of intracellular Ca^{2+} and diacylglycerol activates PKC to induce protein secretion. An increase in either Ca^{2+} or PKC activity is a potent stimulus of secretion.

3.2. α_1-Adrenergic Agonist Activated Pathway

Norepinephrine released from sympathetic nerves activates α_1-adrenergic receptors in the lacrimal gland (Fig. 3), but the subtype of receptor used ($\alpha_{1A/B,\,-C,}$ or $_{-D}$) is unknown.

Activation of α_1-adrenergic receptors then stimulates a G protein, but its identity is unknown. The activated G protein then induces an effector enzyme, which is also unknown. It is known that the effector enzyme is not PLC or phospholipase D (PLD).[2,8] Even

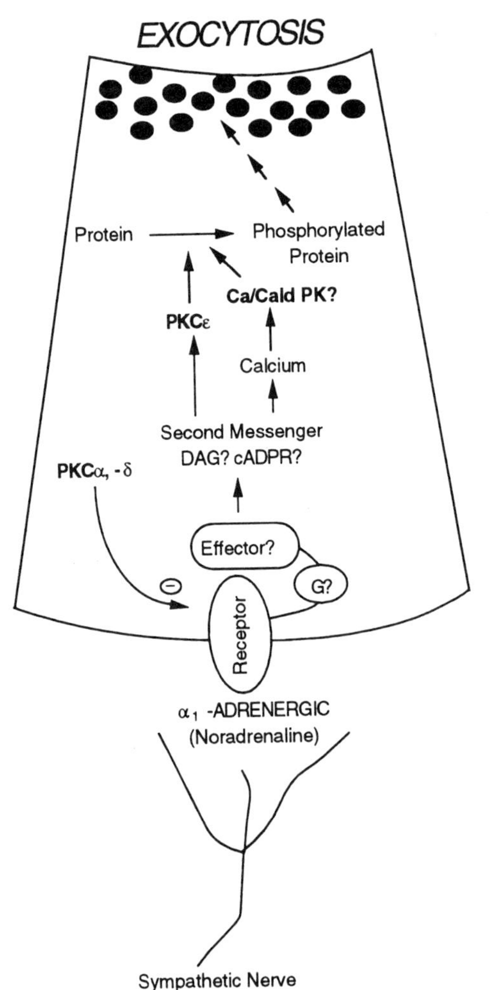

Figure 3. α_1-Adrenergic agonist stimulated signal transduction pathway in lacrimal gland acinar cells. cADPR, cyclic ADPribose. Ca/Cald PK, calcium calmodulin-dependent protein kinase.

though the effector enzyme is unknown, α_1-adrenergic agonists do increase the intracellular Ca^{2+} concentration and do activate PKC. The increase in Ca^{2+} by these agonists is small, significantly smaller than for cholinergic agonists; thus the Ca^{2+}-dependent arm of the $\alpha 1$-adrenergic agonist pathway is a minor stimulus of secretion.[2] This is in contrast to the cholinergic agonist pathway in which Ca^{2+} is a major stimulus of secretion. Activation of PKC by α_1-adrenergic agonists, similar to cholinergic agonists, is a major stimulus of secretion.

4. CA^{2+}-DEPENDENT PORTION OF THE SIGNAL TRANSDUCTION PATHWAYS

4.1. Cholinergic Agonist Activated

IP_3 released by cholinergic stimulation interacts with specific IP_3 receptors on the endoplasmic reticulum. This interaction releases Ca^{2+} thereby increasing the intracellular Ca^{2+} concentration. Depletion of the endoplasmic reticulum Ca^{2+} store causes influx of extracellular Ca^{2+} by an unknown mechanism.[9] For a more detailed discussion of this process, see Putney chapter in this volume. The increase in Ca^{2+} activates protein, electrolyte, and water secretion, either directly by stimulating secretory granule fusion with the apical membrane or indirectly by stimulating Ca^{2+} and calmodulin-dependent protein kinases.[10] These kinases are thought to induce secretion by phosphorylation (and hence activation) of specific protein substrates. These substrates, however, have yet to be identified.

4.2. α_1-Adrenergic Agonist Activated

Dissing's laboratory[11] has proposed that the small α_1-adrenergic agonist increase in Ca^{2+} may perhaps result from activating the production of cyclic ADP ribose. They found that cyclic ADP ribose releases Ca^{2+} by binding to ryanodine receptors, which are located on endoplasmic reticulum membranes, and that ryanodine inhibits the α_1-adrenergic agonist increase in Ca^{2+}. If α_1-adrenergic agonists use the Ca^{2+} store released by ryanodine receptors, this store would be different from that released by IP_3 receptors. Thus, in the lacrimal gland, cholinergic and α_1-adrenergic agonists would use different Ca^{2+} stores, which could partially account for the additivity of cholinergic and α_1-adrenergic stimulated protein secretion.

5. PROTEIN KINASE C DEPENDENT PORTION OF THE SIGNAL TRANSDUCTION PATHWAY

PKC is not just one enzyme, but a family of 11 different isozymes.[12] These isozymes have been divided into three groups based on differences in their structure and activators (Fig. 4).

The classical PKC isozymes, α, βI, βII, and γ, are activated by Ca^{2+} and DAG/phorbol esters. The new PKC isoforms, δ, ε, μ, η, and θ, are activated by DAG/phorbol esters, but not by Ca^{2+}. The atypical PKC isozymes ι/λ and ζ are not activated by Ca^{2+} or DAG/phorbol esters. Differential activation of PKC isoforms is responsible for specificity of action of two different agonists in the same tissue.[14] Evidence that PKC isoforms acti-

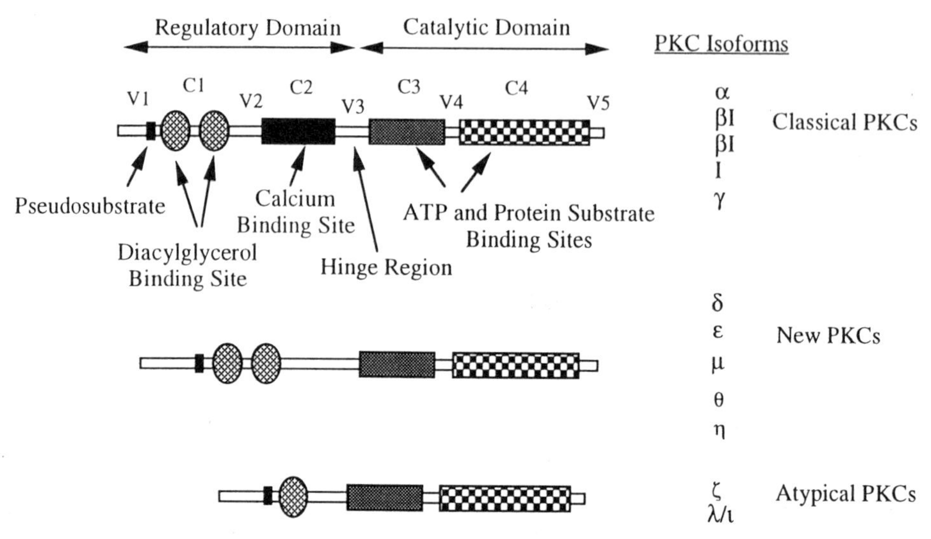

Figure 4. Schematic diagram showing structures of protein kinase C isoforms. C1–4, constant regions. V1–5, variable regions.

vate specific processes include: (i) Different tissues express different isoforms of PKC. (ii) PKC isoforms are differentially localized in a tissue-specific manner. (iii) Stimuli move (translocate) specific isozymes to specific locations within the cell. In vitro, PKC isoforms exhibit limited substrate selectivity. Thus, the subcellular location of a PKC isoform when it is activated is important, as it will activate the substrate that it is near.

Using antibodies specific to PKC isoforms, we found that the lacrimal gland contains five different PKC isoforms, PKC α, -δ, -ε, -μ, and -ι/λ (Zoukhri et al., this volume). Each of these isoforms was found in an isoform-specific location.

5.1. PKC Isozymes Used by Cholinergic Agonists

DAG released by breakdown of PIP_2 activates PKC, which accounts for about 50% of cholinergic-agonist-induced protein secretion.[13] To determine which PKC isozymes are used for this process, we took advantage of the fact that PKC isozymes each have a unique amino acid sequence in the pseudosubstrate domain of the enzyme. The function of the pseudosubstrate domain is to interact with the substrate-binding site of PKC to keep it inactive.[14] When PKC is activated, the pseudosubstrate domain is released and the substrate-binding site can bind to and phosphorylate substrates. We used peptides that correspond to the unique pseudosubstrate regions of PKCα, -δ, and -ε and myristoylated them to make them membrane permeant.[15] These peptides enter the cell, bind to the substrate-binding site of a specific activated PKC isozyme, and inhibit the activity of that isozyme. Inhibition of protein secretion in the presence of a specific PKC pseudosubstrate peptide indicates that that PKC isozyme was involved in secretion. We found that cholinergic agonist-induced secretion was inhibited 70% by myristoylated PKCα pseudosubstrate peptide, 22% by myristoylated PKCδpeptide, and 50% by myristoylated PKCε peptide (see Zoukhri et al., this volume). We concluded that cholinergic agonists primarily use PKCα, use PKCεto a lesser extent, and use PKCδ to a minor extent to stimulate protein secretion (Table 1).

Table 1. Role of protein kinase C isozymes in lacrimal gland secretion

PKC Isozyme	Stimulus	Second messenger	Translocation	Protein secretion
α	Phorbol esters	—	+	+++
	Cholinergic	DAG + Ca^{2+}	–	+++
	α_1-Adrenergic	DAG?, + Ca^{2+}	–	+ (⁻)
δ	Phorbol esters	—	++	–
	Cholinergic	DAG	+	+
	α_1-Adrenergic	?	–	++(⁻)
ε	Phorbol esters	—	+++	++
	Cholinergic	DAG	+++	++
	α_1-Adrenergic	?	–	+++
μ	Phorbol esters	—	+	?
	Cholinergic	DAG	+	?
	α_1-Adrenergic	?	–	?

DAG, diacylglycerol.

Activation of PKC isoforms can also be accompanied by translocation (movement) from the cytoplasm to a specific membrane. Using the antibodies specific to PKC isoforms and Western blotting techniques, we found that cholinergic agonists did not translocate PKCα, did translocate PKCε, and transiently translocated PKCδ. (Table 1).[16]

We concluded that cholinergic agonists use PKCα to stimulate protein secretion, but do not translocate it. Thus PKCα would phosphorylate a substrate found in the same location as PKCα under basal conditions. Cholinergic agonists also use PKCε to stimulate protein secretion, but translocate this isozyme. Thus PKCε would phosphorylate a substrate in the membranes to which PKCε has translocated. Finally, cholinergic agonists use PKCδ to only a limited extent to stimulate protein secretion, and transiently translocate this PKC isozyme. PKCδ could be used by cholinergic agonists for a cellular process other than protein secretion. For example, Zoukhri et al. (this volume) present evidence that PKCδ is involved in cellular Ca^{2+} handling.

5.2. PKC Isozymes Used by α_1-Adrenergic Agonists

The major pathway used by α_1-adrenergic agonists to stimulate protein secretion was previously unknown. α_1-Adrenergic agonists did not activate PLC or PLD and gave only a slight elevation in the intracellular Ca^{2+}.[2,8] We used the myristoylated PKC pseudosubstrate peptides to show that α_1-adrenergic agonists do activate PKC and activate specific isozymes of PKC.[15] We found that myristoylated PKCε pseudosubstrate peptide inhibited protein secretion 86%, suggesting that PKCε plays a major role in α_1-adrenergic agonist activated protein secretion (Table 1). In contrast, myristoylated PKCα and -δ peptides increased protein secretion, suggesting that α_1-adrenergic agonist activation of these isozymes inhibits protein secretion.

We also determined the effect of α_1-adrenergic agonists on translocation of PKC isozymes and found that these agonists did not translocate PKCα, -δ, or -ε (Table 1). We concluded that α_1-adrenergic agonists primarily use PKCε to stimulate protein secretion, but do not translocate it. Thus the α_1-adrenergic agonist-dependent PKCε pool would phosphorylate a substrate found in the same location as PKCε under basal conditions. In contrast, the cholinergic agonist-dependent PKCε pool would phosphorylate substrates in membranes to which PKCε has been moved. We also concluded that α1-adrenergic agonists use PKCα and -δ, but use them to inhibit protein secretion, and do not translocate them. Thus PKCα and -δ would phosphorylate substrates found in the same locations as PKCα and -δ under basal conditions. This further implies that cholinergic and α_1-adrenergic agonists use different pools of PKCα, because cholinergic agonists use PKCα to stimulate secretion, whereas α_1-adrenergic agonists use it to inhibit secretion, and neither agonist translocates PKCα. Cholinergic and α_1-adrenergic receptors could be located in different microdomains and activate the PKCα/substrate complex in that domain. Finally, α_1-adrenergic agonists use PKCδ to inhibit protein secretion, whereas cholinergic agonists use PKCδ to a limited extent to stimulate protein secretion. Because α1-adrenergic agonists do not translocate PKCδ, but cholinergic agonists transiently translocate PKCδ, these two agonists could use the same pool of PKCδ, but activate it in different cellular locations.

6. CONCLUSIONS

Cholinergic and α_1-adrenergic agonists use separate signal transduction pathways to stimulate lacrimal gland protein secretion (Table 1). Cholinergic agonists activate PLC to generate IP_3 and DAG. IP_3 stimulates protein secretion by releasing intracellular Ca^{2+} from IP_3-sensitive stores, whereas DAG stimulates protein secretion by activating PKCα, to a lesser extent PKCε, and to a minor extent PKCδ. During this activation, PKCε is translocated, PKCδ is transiently translocated, and PKCα is not translocated. Because each PKC isoform would stimulate secretion by phosphorylating a substrate to which it is in proximity, differential translocation of PKC isoforms could account for their distinct roles in protein secretion. Both the Ca^{2+}- and the PKC-dependent pathways are equally potent in stimulating protein secretion.

α_1-Adrenergic agonists activate an unknown effector enzyme, but do slightly increase the intracellular Ca^{2+} level and activate PKC. In contrast to cholinergic agonists, α_1-adrenergic agonists may produce cyclic ADP ribose that releases a small amount of Ca^{2+} from ryanodine-sensitive Ca^{2+} stores. Also in contrast to cholinergic agonists, α_1-adrenergic agonists use PKCε to stimulate protein secretion, and PKCα and -δ to inhibit protein secretion. During this activation, none of the PKC isoforms are translocated. Thus PKC isoforms stimulated by α_1-adrenergic agonists would phosphorylate substrates different from those activated by cholinergic agonists. α_1-Adrenergic agonist stimulation differs from cholinergic stimulation in that the Ca^{2+}-dependent pathway is a minor stimulus of secretion, and the PKC-dependent pathway is a major stimulus.

Although cholinergic and α_1-adrenergic agonists both use PKC, they translocate different PKC isozymes and use different PKC isozymes to stimulate protein secretion. Differential use of PKC isoforms, along with differential use of Ca^{2+} stores, account for the additivity of the secretory response to cholinergic and α1-adrenergic agonists in the lacrimal gland.

ACKNOWLEDGMENTS

This work was supported by U.S. Public Health Service grant EY06177 from the National Eye Institute, National Institutes of Health, Bethesda, MD,

REFERENCES

1. Dartt DA. Regulation of inositol phosphates, calcium and protein kinase C in the lacrimal gland. *Prog Ret Eye Res*. 1994; 13:443–478.

2. Hodges RR, Dicker DM, Rose PE, Dartt DA. α_1-Adrenergic and cholinergic agonists use separate signal transduction pathways in lacrimal gland. *Am J Physiol.* 1992; 262:G1087-G1096.
3. Berridge MJ. Inositol trisphosphate and calcium signaling. *Nature* . 1993; 361:315–325.
4. Mauduit P, Jammes H, Rossignol B. M_3 muscarinic acetylcholine receptor coupling to PLC in rat exorbital lacrimal acinar cells. *Am J Physiol.* 1993; 264:C1550-C1560.
5. Meneray MA, Fields TY, Bennett DJ. Involvement of $G_{q/11}$ and G_s in cholinergic stimulation of secretion in rabbit lacrimal gland. ARVO Abstracts. *Invest Ophthalmol Vis Sci.* 1996; 37:S998.
6. Dartt DA, Dicker DM, Ronco LV, Kjeldsen IM, Hodges RR, Murphy SA. Lacrimal gland inositol trisphosphate isomer and inositol tetrakisphosphate production. *Am J Physiol.* 1990; 259:G274-G281.
7. Godfrey PP, Putney JW. Receptor-mediated metabolism of the phosphoinositides and phosphatidic acid in rat lacrimal acinar cells. *Biochem J.* 1984; 218:187–195.
8. Zoukhri D, Dartt DA. Cholinergic activation of phospholipase D in lacrimal gland acini is independent of protein kinase C and calcium. *Am J Physiol.* 1995; 268:C713-C720.
9. Putney JWJ. Capacitative calcium entry revisited. *Cell Calcium.* 1990; 11:611–624.
10. Dartt DA. Signal transduction and control of lacrimal gland protein secretion: a review. *Curr Eye Res.* 1989; 8:619–636.
11. Gromada J, Jorgensen TD, Dissing S. Cyclic ADP-ribose and inositol 1,4,5-trisphosphate mobilize Ca^{2+} from distinct intracellular pools in permeabilized lacrimal acinar cells. *FEBS Lett.* 1995; 360:303–306.
12. Nishizuka Y. The molecular heterogeneity of protein kinase C and its implications for cellular regulation. *Nature.* 1988; 334:661–665.
13. Zoukhri D, Hodges RR, Dicker DM, Dartt DA. Role of protein kinase C in cholinergic stimulation of lacrimal gland protein secretion. *FEBS Lett.* 1994; 351:67–72.
14. Newton AC. Protein kinase C: Structure, function and regulation. *J Biol Chem.* 1995; 270:28495–28498.
15. Zoukhri D, Hodges RR, Sergheraert C, Toker A, Dartt DA. Lacrimal gland PKC isoforms are differentially involved in agonist-induced protein secretion. *Am J Physiol.* 1997; 272:C263-C269.
16. Zoukhri D, Hodges RR, Willert S, Dartt DA. Immunolocalization of lacrimal gland PKC isoforms: Effect of phorbol esters and cholinergic agonists on their cellular distribution. *J Membr Biol.* in press.

16

CALCIUM SIGNALLING IN LACRIMAL ACINAR CELLS

James W. Putney, Jr., Yi Huang, and Gary St. J. Bird

Calcium Regulation Section
Laboratory of Signal Transduction
National Institute of Environmental Health Sciences – NIH
Research Triangle Park, North Carolina

1. INTRODUCTION

The predominant signalling pathway by which hormones and neurotransmitters control the secretion of lacrimal fluids and proteins involves an elevation of cytoplasmic calcium. We have investigated this signalling system in mouse lacrimal acinar cells activated through their muscarinic cholinergic receptors. These receptors initiate signalling by activation of a phospholipase C enzyme that degrades specifically a minor membrane phospholipid, phosphatidylinositol bisphosphate.[1,2] The resulting products are 1,2–diacylglycerol (DG) and inositol 1,4,5–trisphosphate (IP_3); both play important roles in determining the extent and duration of lacrimal calcium signalling.

2. PRINCIPLES OF CALCIUM SIGNALLING IN LACRIMAL CELLS

The time course of intracellular calcium ($[Ca^{2+}]_i$) signalling in a single lacrimal acinar cell is shown in Fig. 1. A single cell was loaded with the fluorescent Ca^{2+} indicator fura–2 and treated with the muscarinic cholinergic agonist methacholine. This produces a rapid rise in $[Ca^{2+}]_i$, followed by a prolonged and sustained elevation in $[Ca^{2+}]_i$ that remains elevated above the basal level as long as the stimulating agonist is present. Fig. 1 also shows the result of an experiment in which the cell is activated in a medium lacking added Ca^{2+} ($[Ca^{2+}] \sim 10\ \mu M$). Methacholine addition still induces a substantial increase in $[Ca^{2+}]_i$, but the response is transient. Re–addition of Ca^{2+} extracellularly restores the sustained phase of the response. This result indicates that $[Ca^{2+}]_i$ signalling in lacrimal acinar cells comprises two phases--an initial phase due to release from intracellular Ca^{2+} stores, and a second phase due to increased entry of Ca^{2+} across the plasma membrane.

Lacrimal Gland, Tear Film, and Dry Eye Syndromes 2
edited by Sullivan *et al.*, Plenum Press, New York, 1998

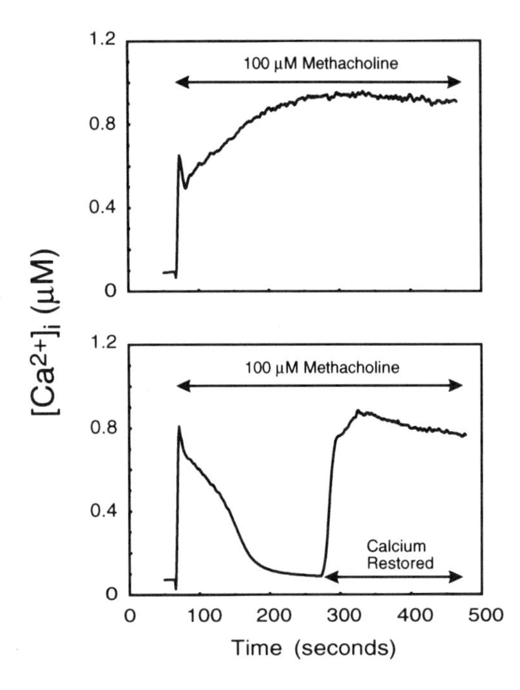

Figure 1. Calcium signalling in a single lacrimal acinar cell. A single lacrimal acinar cell was loaded with the fluorescent Ca^{2+} indicator fura–2, and changes in $[Ca^{2+}]_i$ were assessed by monitoring the ratio of fluorescence intensities with alternating excitation at 340 and 380 nm. (Top) Activation of muscarinic–cholinergic receptors with methacholine causes a rapid rise in $[Ca^{2+}]_i$ and a sustained elevation above baseline. (Bottom) With no Ca^{2+} present extracellularly, the rapid increase in $[Ca^{2+}]_i$ is still seen, but the response rapidly returns to baseline. The sustained response can be restored by restoration of extracellular Ca^{2+}.

The release of intracellular Ca^{2+} results from the action of IP_3, which binds to and activates a receptor/ion channel present on the endoplasmic reticulum. The entry of Ca^{2+} also results from IP_3 action. However, this entry appears to be signalled not by a direct action of IP_3 at the plasma membrane, but rather by an indirect mechanism called capacitative calcium entry.[3,4] The process of capacitative calcium entry is believed to involve a signal for calcium entry that is produced or activated as a result of depletion of intracellular stores. The evidence for this idea has been summarized in a number of recent reviews. Perhaps the simplest means for illustrating this phenomenon is by use of specific inhibitors of intracellular Ca^{2+}–ATPases, such as thapsigargin. When thapsigargin is applied to a lacrimal cell, the endoplasmic reticulum stores are passively depleted by inhibiting the Ca^{2+} pump. As shown in Fig. 2, this leads to activation of Ca^{2+} entry. The figure also shows that addition of methacholine to a thapsigargin–activated cell does not further increase Ca^{2+} entry, indicating that depletion of intracellular stores has activated all of the Ca^{2+} entry normally under control of the more physiological stimulus through the muscarinic receptor.

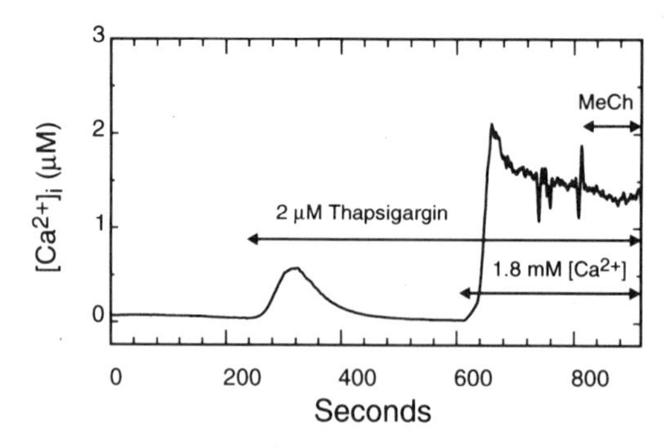

Figure 2. Failure of methacholine to increase Ca^{2+} entry further in a single thapsigargin–activated lacrimal acinar cell. Sustained Ca^{2+} entry was activated in a fura–2–loaded lacrimal acinar cell by 2 μM thapsigargin. Addition of 100 μM methacholine (MeCh) causes no further increase in $[Ca^{2+}]_i$.

3. THE NATURE OF INTRACELLULAR CALCIUM POOLS IN LACRIMAL CELLS

What is the evidence that this mobilization of calcium comes specifically from the endoplasmic reticulum and not, for example, from the mitochondria, as was believed for some time? Perhaps the most straightforward evidence is the finding from many laboratories that in virtually all cell types examined (including lacrimal cells[5,6]), application of the Ca^{2+}–mobilizing signal IP_3 to permeabilized cells releases Ca^{2+} from ATP–dependent Ca^{2+} stores that are insensitive to mitochondrial inhibitors. Rather, these stores are sensitive to the ER calcium pump inhibitor, thapsigargin. In addition, in permeable cells, the thapsigargin–sensitive stores can be further subdivided into those that can be released by the Ca^{2+}–mobilizing ligand (1,4,5)IP_3 and a residual that is insensitive to (1,4,5)IP_3. However, we have examined calcium mobilization by various treatments in single, attached fura–2–loaded mouse lacrimal acinar cells.[7] In contrast to previous work with populations of lacrimal cells in suspensions,[5] the attached cells appeared to be in a much better physiological condition as indicated by (i) lower basal levels of $[Ca^{2+}]_i$; (ii) larger increases in $[Ca^{2+}]_i$ in response to the muscarinic agonist, methacholine, or in response to thapsigargin; and (iii) in unstimulated cells, there was no elevation in $[Ca^{2+}]_i$ in changing from a Ca^{2+}–deficient medium to a Ca^{2+}–containing medium. When these cells were treated, in the absence of extracellular Ca^{2+}, with the phospholipase C–linked agonist methacholine, a large transient rise in $[Ca^{2+}]_i$ was observed. Subsequent treatment with either thapsigargin or the calcium ionophore ionomycin produced no additional rise in $[Ca^{2+}]_i$ (Fig. 3). This finding, in contrast to what was observed in populations of cells, indicates that lacrimal cells maintained in a proper physiological status contain a single, homogeneous intracellular Ca^{2+} pool. This pool is uniformly sensitive to (1,4,5)IP_3 (produced in response to methacholine) or to thapsigargin, and thus no significant mitochondrial pool of Ca^{2+} was detected.

In this same study,[7] when we treated lacrimal cells with methacholine in the presence of extracellular Ca^{2+} for a prolonged period (30 min), and then extracellular Ca^{2+} was withdrawn, a small but significant peak of Ca^{2+} was released by addition of ionomycin. The loading of this pool of Ca^{2+} was blocked in cells previously microinjected with the mitochondrial calcium uptake inhibitor ruthenium red (Fig. 3). Thus, although resting cells contain insignificant quantities of mitochondrial Ca^{2+}, mitochondria clearly accumulate Ca^{2+} during periods of elevated cytoplasmic Ca^{2+} due to cell activation.

4. ROLE OF PROTEIN KINASE C IN LACRIMAL CALCIUM SIGNALLING

When lacrimal cells are activated by submaximal concentrations of methacholine, $[Ca^{2+}]_i$ appears to oscillate. These oscillations are roughly symmetrical fluctuations usually superimposed on a raised basal level of $[Ca^{2+}]_i$. The most significant characteristic of the oscillations is their constant frequency at different agonist concentrations.[8–13] These sinusoidal oscillations are considerably simpler than the baseline spike type of $[Ca^{2+}]_i$ oscillations,[2] and can be explained most simply by a single negative feedback on the $[Ca^{2+}]_i$ signalling mechanism. Our work indicates that for the lacrimal acinar cell activated through its muscarinic cholinergic receptor, the negative feedback responsible for the sinusoidal oscillations is due to protein kinase C.[13] Sinusoidal oscillations in lacrimal aci-

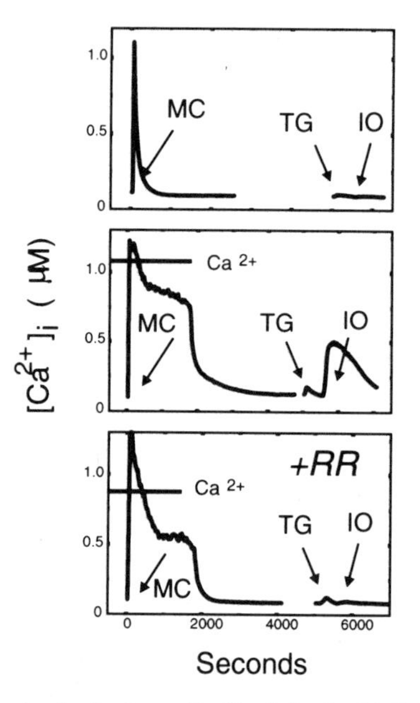

Figure 3. Pools of calcium in mouse lacrimal acinar cells. Single lacrimal cells were incubated in medium lacking added Ca^{2+} (top) or in medium containing 1.8 mM extracellular Ca^{2+} (center and bottom). $[Ca^{2+}]_i$ was measured by fura–2 fluorescence. (Top) Addition of 100 μM methacholine (MC) to a single lacrimal cell caused a transient increase in $[Ca^{2+}]_i$. Subsequent additions of 2 μM thapsigargin (TG) and 10 μM ionomycin (IO; in the presence of 200 μM EGTA) had no significant effect on $[Ca^{2+}]_i$. Thus, under these conditions a single homogeneous pool of stored Ca^{2+} exists entirely sensitive to the Ca^{2+}-mobilizing actions of (1,4,5)IP_3 produced from cholinergic receptor activation. (Center) After 30 min stimulation with 100 μM methacholine, the cell was placed in a nominally Ca^{2+}–free medium, but in the continued presence of MC. 2 μM TG and 10 μM IO (the latter in the presence of 200 μM EGTA) were added as indicated. Addition of ionomycin resulted in a transient, significant increase in $[Ca^{2+}]_i$. This result indicates that with prolonged activation, Ca^{2+} is accumulated into a new, agonist–insensitive pool. (Bottom) A single lacrimal cell was microinjected with ruthenium red (RR; 1 mM pipette solution), and then the procedure described for the experiment was followed. No release was observed with ionomycin. This indicates that the agonist–insensitive pool likely represents the mitochondria. The results shown are illustrative of findings in 5 independent experiments. (Redrawn from data originally presented in ref. 7.)

nar cells are completely blocked by pharmacological activation, inhibition, or down–regulation of protein kinase C. Complete inhibition occurs regardless of the level of agonist activation or the level of $[Ca^{2+}]_i$, indicating that the inhibition does not result simply from alterations in the degree of phospholipase C activation. The site of negative inhibition appears to be on or proximal to phospholipase C, because (i) pharmacological activation of protein kinase C by phorbol esters inhibits the production of (1,4,5)IP_3 in response to muscarinic agonists, (ii) bypassing (1,4,5)IP_3 production by intracellular application of stable analogs of (1,4,5)IP_3 always gives sustained, non–oscillating $[Ca^{2+}]_i$ signals, and (iii) phorbol esters do not inhibit the effects of injected (1,4,5)IP_3 analogs.[13] Activation of phospholipase C increases diacylglycerol, which in turn activates protein kinase C which feeds back and inhibits phospholipase C. This leads to a diminution in diacylglycerol production, diminished protein kinase C activity, and relief of the inhibition of phospholipase C. Continuous cycling of this feedback loop generates oscillations in phospholipase C activity, which, in the absence of some feed-forward input, gradually damp down to a sustained level under tonic control by the opposing forces of receptor activation and protein kinase

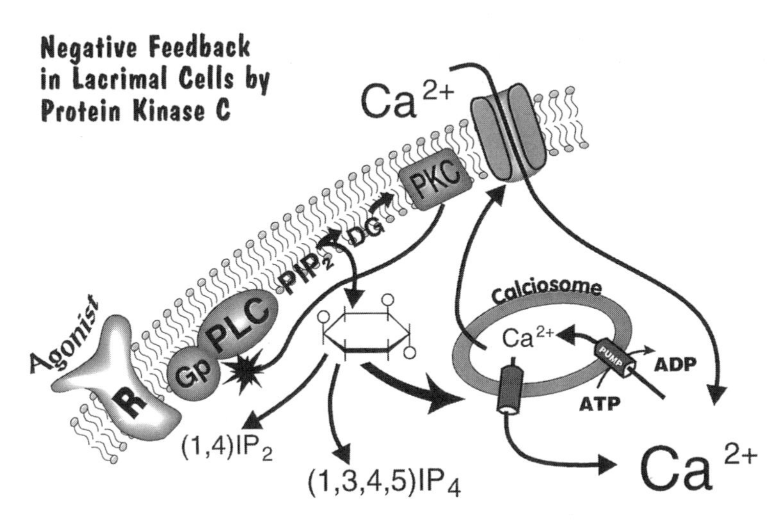

Figure 4. Sinusoidal oscillations result from negative feedback in lacrimal cells by protein kinase C. Phospholipase C (PLC) activation leads to production of $(1,4,5)IP_3$ and mobilization of Ca^{2+}, but also to protein kinase C (PKC) activation. With some delay, this inhibits PLC, causing a fall in $(1,4,5)IP_3$ and attenuation of the $[Ca^{2+}]_i$ signal. The inhibition of PLC also attenuates the PKC signal, leading to a cyclical increase in $[Ca^{2+}]_i$. Continuing cycling of this feedback loop gives rise to sinusoidal oscillations.

C inhibition. It is important to point out that in this particular scheme, $[Ca^{2+}]_i$ is simply a passive follower of the oscillating $(1,4,5)IP_3$ production, and plays no obvious active role in generating or modulating the oscillations. Thus, this particular type of oscillation might more appropriately be called diacylglycerol oscillations or phospholipase C/protein kinase C oscillations (Fig. 4).

REFERENCES

1. Berridge MJ. Inositol trisphosphate and calcium signalling. *Nature.* 1993;361:315–325.
2. Putney JW Jr., Bird GSJ. The inositol phosphate-calcium signalling system in non-excitable cells. *Endocr Rev.* 1993;14:610–631.
3. Putney JW Jr. A model for receptor-regulated calcium entry. *Cell Calcium.* 1986;7:1–12.
4. Putney JW Jr. Capacitative calcium entry revisited. *Cell Calcium.* 1990;11:611–624.
5. Kwan CY, Takemura H, Obie JF, Thastrup O, Putney JW Jr. Effects of methacholine, thapsigargin and La^{3+} on plasmalemmal and intracellular Ca^{2+} transport in lacrimal acinar cells. *Am J Physiol.* 1990;258:C1006-C1015.
6. Bird GSJ, Putney JW, Jr. Effect of inositol 1,3,4,5-tetrakisphosphate on inositol trisphosphate-activated Ca^{2+} signalling in mouse lacrimal acinar cells. *J Biol Chem.* 1996;271:6766–6770.
7. Bird GSJ, Obie JF, Putney JW Jr. Functional homogeneity of the non-mitochondrial Ca^{2+}-pool in intact mouse lacrimal acinar cells. *J Biol Chem.* 1992;267:18382–18386.
8. Jacob R. Calcium oscillations in electrically non-excitable cells. *Biochim Biophys Acta.* 1990;1052:427–438.
9. Gray PTA. Oscillations of free cytosolic calcium evoked by cholinergic and catecholaminergic agonists in rat parotid acinar cells. *J Physiol.* (Lond). 1988;406:35–53.
10. Yule DI, Gallacher DV. Oscillations of cytosolic calcium in single pancreatic acinar cells stimulated by acetylcholine. *FEBS Lett.* 1988;239:358–362.
11. Tsunoda Y, Stuenkel EL, Williams JA. Oscillatory mode of calcium signalling in rat pancreatic acinar cells. *Am J Physiol.* 1990;258:C147-C155.

12. Sage SO, Adams DJ, Van Breemen C. Synchronized oscillations in cytoplasmic free calcium concentration in confluent bradykinin-stimulated bovine pulmonary artery endothelial cell monolayers. *J Biol Chem.* 1989;264:6–9.
13. Bird GSJ, Rossier MF, Obie JF, Putney JW Jr. Sinusoidal oscillations in intracellular calcium due to negative feedback by protein kinase C. *J Biol Chem.* 1993;268:8425–8428.

17

VOLTAGE- AND Ca^{2+}-DEPENDENT CHLORIDE CURRENT ACTIVATED BY HYPOSMOTIC AND HYPEROSMOTIC STRESS IN RABBIT SUPERIOR LACRIMAL ACINAR CELLS

George H. Herok,[1,2] Thomas J. Millar,[1,2] Philip J. Anderton,[1,3] and Donald K. Martin[1,4]

[1]Co-operative Research Centre for Eye Research and Technology
University of New South Wales
Kensington, Australia
[2]Department of Biological Sciences
University of Western Sydney/Nepean, Australia
[3]School of Optometry
University of New South Wales
Kensington, Australia
[4]Department of Biological Sciences
University of Technology, Sydney
New South Wales, Australia

1. INTRODUCTION

Tears provide a protective and lubricating film that covers the anterior surface of the eye. The bulk of tears is a salty serous fluid, containing a variety of proteins associated with bacterial defense. In rabbits, tears are produced by the superior and inferior lacrimal glands. Secretion of the serous fluid is dependent on the distribution and nature of ion channels in lacrimal acinar cells.[1,2] We have identified a voltage- and Ca2+-dependent outwardly rectifying Cl- (Cl_{or}) channel in rabbit superior lacrimal gland acinar cells. The purpose of this study was to evaluate the role of this Cl_{or} channel in secretion by the rabbit lacrimal gland acinar cells.

2. MATERIALS AND METHODS

New Zealand white rabbits of both sexes were killed by an overdose of sodium pentobarbital (45 mg/kg) and the glands were excised. The glands were cut into fine pieces and incubated in a collagenase (200 U/ml) and hyaluronidase (600 U/ml) solution at 37°C for 45 min. The cellular pellet was washed twice with the NaCl-rich bath solution containing bovine serum albumin (2 mg/ml). Whole-cell currents were recorded using an Axopatch 200 A amplifier and CV201A headstage (Axon Instruments). The currents were recorded under non-physiological conditions (140 mM NaCl solution in both the bath and pipette with Cs^+ added to the bath to block K^+ activity). Cell volume changes were observed using a confocal microscope.

3. RESULTS

The resting membrane potential under physiological conditions of these cells was -40 ± 2 mV (n = 17), which was similar to the resting membrane potential of -37 ± 8 mV (n = 50) obtained by Kikkawa[3] using intracellular electrodes. Tetraethyl-ammonium (TEA) reduced the total membrane conductance of the outward currents by 70%, and the slope conductance was reduced from 24 ± 1 nS (n = 12) to 16.2 ± 1.5 nS (n = 8). The nature of this TEA-insensitive component of the outward current was identified and characterized as being due to a Cl_{or} current using ion substitution. Fig. 1 shows that this Cl_{or} current is both voltage- and Ca^{2+}-activated.

Fig. 2 shows the response of lacrimal acinar cells when exposed to hyperosmotic and hyposmotic solutions. Exposure to both increased the activity of the Cl_{or} channel. Cells exposed to a hyposmotic solution (222 mOsm) showed a large increase in the Cl_{or} current at +80 mV by 142 ± 18% (n = 8). However, cells exposed to a hyperosmotic solution (398 mOsm) at +80 mV showed an increase of only 70 ± 11% (n =6) in the Cl_{or} activity.

During exposure to hyposmotic solutions, after a couple of minutes a large transient increase in Cl_{or} current activity occurs, followed by current activity quickly returning to basal levels. Cells exposed to a hyposmotic solution significantly changed in morphology. They appeared to increase in size, and some cells, which initially had healthy-looking round membranes, after a few minutes of exposure to this solution had a shrivelled appearance. With time, the cells started to lose this appearance and, following a bath change, regained their original healthy-looking membranes.

As shown in Fig. 3, the largest volume changes occurred within the first 6 min of exposure. Interestingly, this is the same period of time over which the large transient changes in current activity were recorded, as shown in Fig. 2A.

The morphology of cells in a hyperosmotic solution did not change significantly, although we observed some cells to have shrunk in size. With time, in cells exposed to hy-

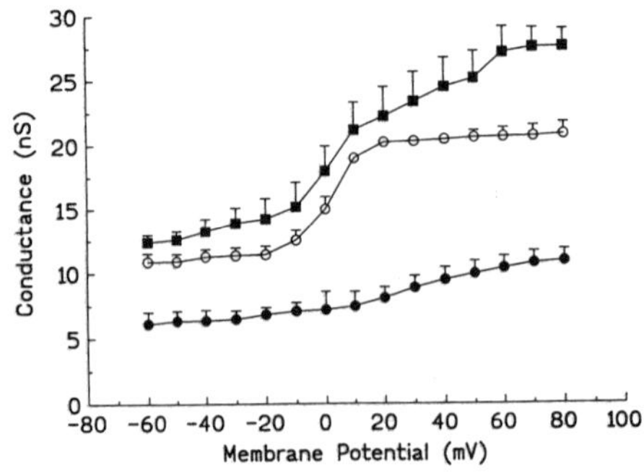

Figure 1. Relation between membrane potential and total membrane conductance for the Cl_{or} channels for free Ca^{2+} concentrations in the pipette solution of 1.0 x 10^{-9} M (●, n =6), 1.5 x 10^{-7} M (O, n =8), and 1.0 x 10^{-6} M (■, n = 6). In each experiment, the pipette solution contained Na^+ ions rather than K^+ ions, and 5 mmol/L Cs^+ was present in the bath solution. The bars represent the SEM when these were larger than the symbols.

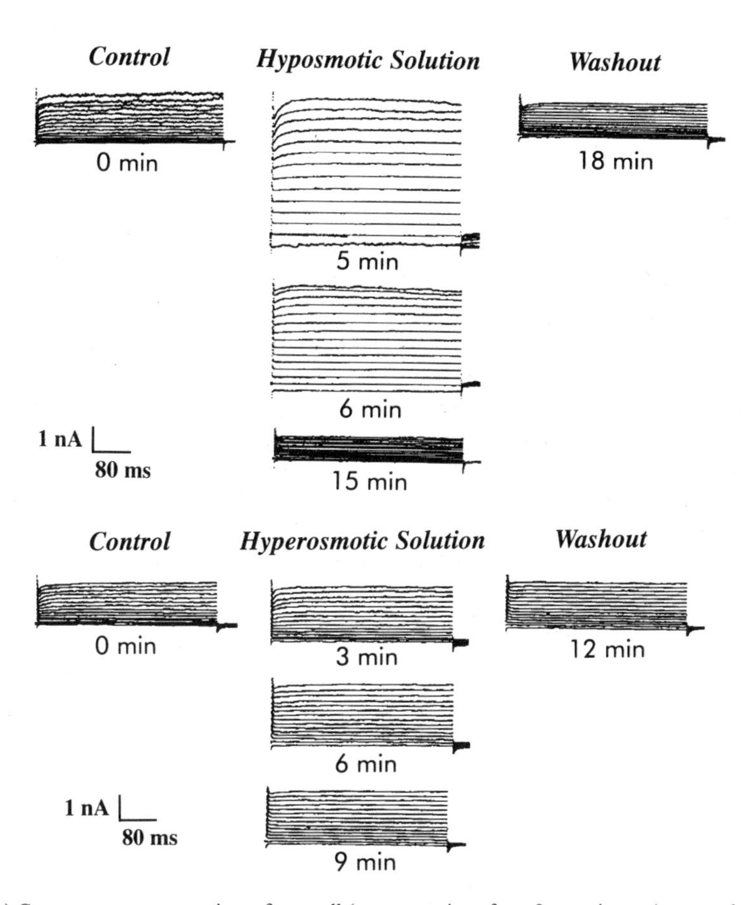

Figure 2. (A) Current response over time of one cell (representative of $n = 8$ experiments) exposed to a hyposmotic solution. (B) Current response over time of one cell (representative of $n = 6$ experiments) exposed to a hyperosmotic solution.

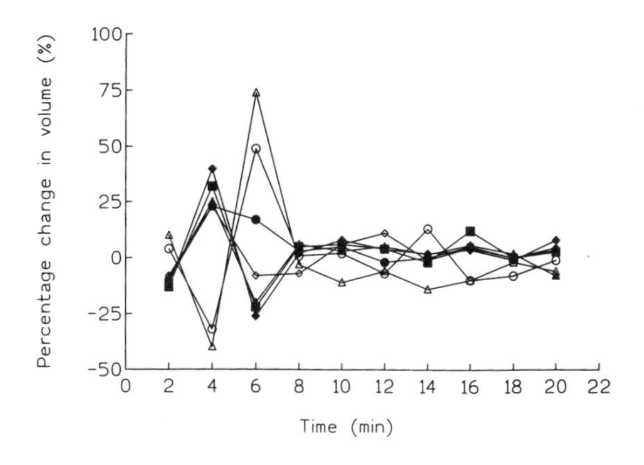

Figure 3. Percentage change in volume over time in each of 7 cells exposed to hyposmotic solutions.

perosmotic solutions, the Cl_{or} current activity increased gradually, peaking, and then gradually decreasing toward basal levels.

4. DISCUSSION

So how can these cell volume changes be explained with respect to current activity of this chloride channel, since in both hypo- and hyperosmotic solutions there is an increase in current activity. Hyperosmotic solutions cause cells to shrink. This would in turn activate this channel, causing an influx of Cl^- ions, followed by the passive movement of water to restore cell volume. However, when cells are exposed to a hyposmotic solution they initially swell and after some time undergo regulatory volume decrease (RVD).[4–7] In many cells, this is observed by an increase in K^+ and Cl^- channel activity, leading to an extrusion of these ions followed by the passive movement of water. In salivary cells, for example, it has been shown that during normal secretion there is a decrease in cell volume.[8,9] In these rabbit lacrimal cells following this RVD, the cells lose volume, which then triggers an increase in inward movement of chloride ions which corresponds to the transient Cl_{or} current activity recorded. This regulatory volume increase response is most likely to balance some of the water loss following RVD.

It is also known that cell volume is reduced during secretion in rat salivary glands. Thus it is possible that Cl_{or} channels may also be involved in the aqueous secretion of tears by the rabbit lacrimal gland even though our data suggest that this channel is associated primarily with volume regulation.

REFERENCES

1. Mircheff AK. Lacrimal fluid and electrolyte secretion: A review. *Curr Eye Res.* 1989;8:607–617.
2. Hille B. In: Hille B, ed. *Ionic Channels of Excitable Membranes.* Sunderland, MA: Sinauer Associates; 1992;218–219.
3. Kikkawa T. Secretory potentials in the lacrimal gland of the rabbit. *Jpn J Ophthalmol.* 1970;14:25–40.
4. Hoffmann EK, Simonsen LO. Membrane mechanisms in volume and pH regulation in vertebrate cells. *Physiol Rev.* 1989;69:315–382.
5. Sarkadi B, Parker JC. Activation of ion transport pathways by changes in cell volume. *Biochim Biophys Acta.* 1991;1071:407–427.
6. McCarthy NA, O'Neil RG. Calcium signaling in cell volume regulation. *Physiol Rev.* 1992;72:1037–1061.
7. Pierce SK, Politis AD. Ca^{2+}- activated cell volume recovery mechanisms. *Annu Rev Physiol.* 1990;52:27–42.
8. Foskett JK, Melvin JE. Activation of salivary secretion: Coupling of cell volume and $[Ca^{2+}]$ in single cells. *Science.* 1989;244:1582–1585.
9. Foskett JK. $[Ca^{2+}]$ modulation of Cl- content controls cell volume in single salivary acinar cells during fluid secretion. *Am J Physiol.* 1990;259:C998-C1004.

18

G PROTEIN COUPLING OF RECEPTOR ACTIVATION TO LACRIMAL SECRETION

Michele A. Meneray and Tammy Y. Fields

Department of Physiology
Louisiana State University Medical Center
New Orleans, Louisiana

1. INTRODUCTION

In lacrimal gland, protein secretion involves transduction of extracellular signals generated by neurotransmitters and neuromodulators of the parasympathetic and sympathetic nerve fibers that innervate the gland. In this gland, as in other exocrine tissues, transmembrane signalling is accomplished by the adenosine 3',5'-cyclic monophosphate (cAMP) and the inositol 1,4,5-trisphosphate (IP_3) pathways. In the cAMP pathway, vasoactive intestinal peptide (VIP) stimulates and met-enkephalin inhibits the regulatory effector enzyme adenylyl cyclase and the consequent alterations in intracellular cAMP that result in stimulation or inhibition of protein release by lacrimal acinar cells.[1,2] Agonists that activate phosphatidylinositol turnover also stimulate release of protein into the tears.[3,4] In this pathway, M_3-muscarinic receptor activation results in an increase in phosphatidylinositol 4,5-bisphosphate (PIP_2)-specific phospholipase C (PLC) activity and the production of IP_3.[5] Within both intracellular pathways, heterotrimeric G proteins couple receptor activation to regulation of the effector enzymes adenylyl cyclase and PLC. We have demonstrated that the stimulatory G protein G_s and inhibitory G proteins of the G_i/G_o family are present in lacrimal gland membranes and that the α subunits of these G proteins are specifically associated with VIP and enkephalin regulation of adenylyl cyclase.[6,7] We have also demonstrated that $G_{q/11}$, known to be coupled to PLC,[8] is present in lacrimal acinar cell membranes.[9] Thus, our previous work has established G protein coupling of receptors to adenylyl cyclase and provided preliminary evidence for coupling of $G_{q/11}$ to PLC in lacrimal gland. In the work presented here, we have extended our studies to provide direct evidence that G proteins couple receptor activation to secretion in intact cells. This was accomplished by the introduction of antisera directed against peptide sequences of the a subunits of G_s, $G_{q/11}$ and members of the G_i/G_o family into cultured lacrimal acinar cells and the measurement of stimulation or inhibition of protein release by cholinergic

Lacrimal Gland, Tear Film, and Dry Eye Syndromes 2
edited by Sullivan *et al.*, Plenum Press, New York, 1998

and peptidergic agonists. Using this methodology, we have demonstrated that G_s links VIP to stimulation of secretion and that G proteins of the G_i/G_o family couple enkephalin to inhibition of protein release. We have also shown that within the PLC pathway, $G_{q/11}$ links cholinergic agonists to enhanced protein release by cultured acini. Finally, our results suggest that G_s may have a role in the regulation of vesicular traffic and exocytosis that is independent of cell-surface receptor activation.

2. METHODS

Three-day primary cultures of lacrimal acini were established from the lacrimal glands of adult male New Zealand rabbits.[10] To determine the G proteins involved in secretion, acini were permeabilized and incubated with polyclonal antibodies directed against C-terminal or internal decapeptide sequences of the α subunits of selected G proteins prior to the measurement of the release of secretory protein. Protein release was measured in response to 100 μM carbachol, a cholinergic agonist that stimulates lacrimal secretion by activation of the IP_3 pathway,[5] 100 nM VIP, a peptidergic agonist that stimulates lacrimal secretion by activation of the cAMP pathway through a GTP-dependent activation of adenylyl cyclase,[1,2] or 100 μM carbachol plus 10 μM D-ala^2-met-enkephalinamide (DALA), an enkephalin analog that inhibits carbachol stimulation of secretion through an unknown mechanism but is known to inhibit adenylyl cyclase.[1,2] In brief, cultured acini were washed with a permeabilization buffer containing 145 mM KCl, 2.0 mM $MgCl_2$, 1.2 mM KH_2PO_4, 10 μM $CaCl_2$, 10 mM glucose, 10 mM HEPES, 0.2% BSA, and 0.1% soybean trypsin inhibitor, pH 7.0.[11] Acini were permeabilized with 0.4 U/ml streptolysin-O (SLO) in the permeabilization buffer for 30 min at 37°C with non-immune rabbit serum or polyclonal antibodies to the α subunits of G_s, $G_{q/11}$, G_i, or G_o. The acini were washed with a balanced salt solution and equilibrated with one wash between two time periods. During a final 20 min period, medium was supplemented with the appropriate agonists or vehicle. The medium was then removed and assayed for total protein.[12]

3. RESULTS AND DISCUSSION

Since G proteins couple receptor activation to the effector enzymes adenylyl cyclase and PLC and since agonists such as VIP or carbachol increase the second messengers cAMP and IP_3 that lead to protein release, it has been presumed that G proteins couple receptor activation to secretion in intact cells. In the work presented here, we tested this hypothesis by measuring the effect of insertion of antibodies to $G_{s\alpha}$ and $G_{q/11\alpha}$ into lacrimal acini on VIP- or carbachol-stimulated secretion. Exposure of acini to anti-$G_{s\alpha}$ antiserum resulted in a significant reduction in the VIP-stimulated release of protein (Fig. 1). Because the C-terminus of $G_{s\alpha}$ is the site of interaction with receptors,[13] these results suggest that the antiserum blocks the VIP-receptor-$G_{s\alpha}$ interaction that would otherwise result in cAMP-dependent protein release. Insertion of anti-$G_{q/11\alpha}$ antiserum had no effect on VIP-stimulated secretion, and the effect of the combination of antisera on secretion was no greater than the effect of anti-$G_{s\alpha}$ alone.

In similar experiments, we tested the effect of the same antisera on cholinergic stimulation of protein release (Fig. 2). In contrast to the lack of effect of anti-$G_{q/11\alpha}$ antiserum on VIP-stimulated secretion, introduction of this antibody into cultured acini re-

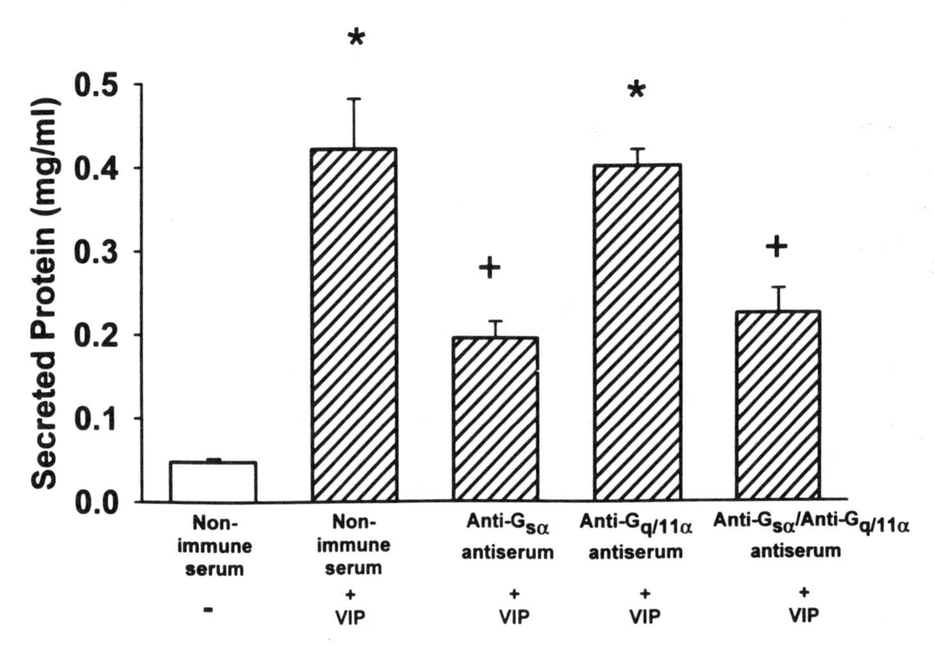

Figure 1. Effect of anti-$G_{s\alpha}$ and anti-$G_{q/11\alpha}$ antiserum alone or in combination on VIP-induced protein release by permeabilized cells. Cultured acini were permeabilized by streptolysin-O (SLO) with simultaneous exposure to non-immune serum or the indicated antibodies. Protein release was measured in the presence of vehicle or 100 nM VIP. *, significantly different from non-immune serum; +, significantly different from non-immune serum and non-immune serum plus agonist.[9]

duced the response to the cholinergic agonist carbachol by 70%. The mechanism by which this antiserum blocked cholinergic stimulation of secretion most likely is inactivation of $G_{q/11\alpha}$ coupling of muscarinic receptors to PLC, as occurs in other exocrine tissues.[8] The reversal of atropine sensitive carbachol-induced secretion by anti-$G_{q/11\alpha}$ antiserum demonstrates that muscarinic activation of secretion involves $G_{q/11\alpha}$. However, confirmation of coupling of the M_3 receptor to PLC through this G protein remains to be tested.

Introduction of anti-$G_{s\alpha}$ antiserum reduced the cholinergic secretory response by 37% (Fig. 2). Simultaneous introduction of both anti-$G_{q/11\alpha}$ and anti-$G_{s\alpha}$ antisera resulted in complete inhibition of the effects of carbachol on protein release by cultured acini. The function of G_s in the cholinergic secretory response most likely is not stimulation of adenylyl cyclase, since alterations in cAMP cannot be detected in response to cholinergic stimulation of lacrimal gland. G_s also is not likely to be involved in PLC activation, since in most cells PLC is coupled to receptor activation by $G_{q/11\alpha}$ and is not known to be coupled to G_s.[14] An alternative function for G_s in cholinergic stimulation of secretion is involvement in vesicular trafficking and exocytotic events.[15,16] If the mechanism by which G_s effects cholinergic stimulation is through vesicular trafficking, it is likely that this G protein has the same function in VIP-induced protein release in addition to linking the receptor to adenylyl cyclase. This hypothesis is currently under investigation in our laboratory.

To determine the G proteins that couple enkephalins to inhibition of lacrimal secretion, we tested the effect of antibodies directed against the a subunit of the inhibitory G proteins G_{i1}, G_{i2}, G_{i3}, and G_o. Enkephalins are present in the nerve fibers that innervate mammalian lacrimal gland,[17] and we have previously shown that the met-enkephalin ana-

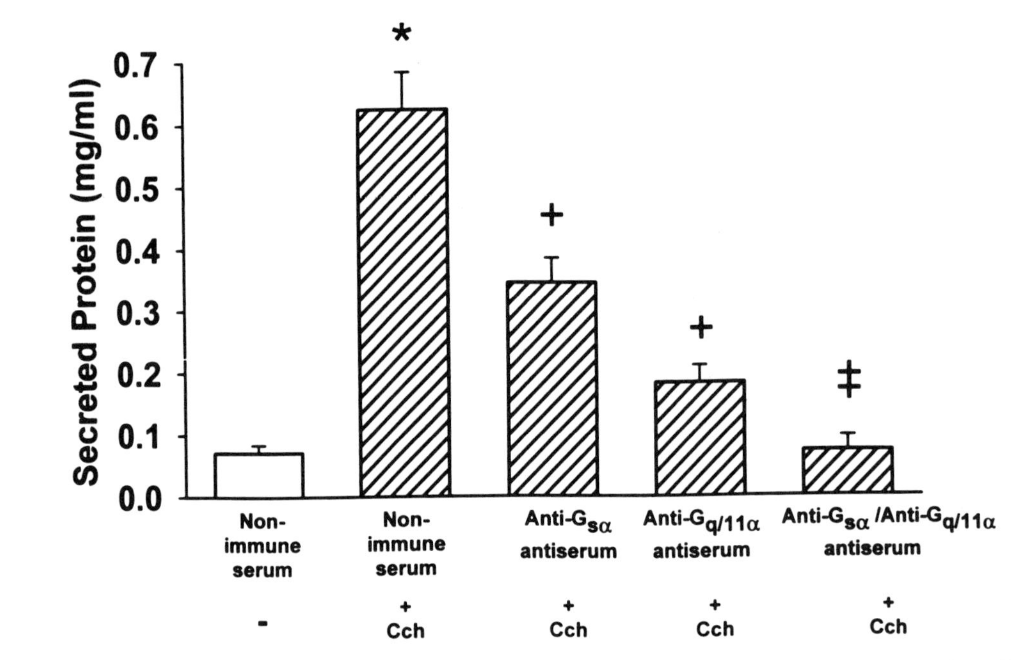

Figure 2. Effect of anti-$G_{s\alpha}$ and anti-$G_{q/11\alpha}$ antiserum alone or in combination on carbachol (Cch)-induced protein release by permeabilized cells. Cultured acini were permeabilized by streptolysin-O (SLO) with simultaneous exposure to non-immune serum or the indicated antibodies. Protein release was measured in the presence of vehicle or 100 μM carbachol. *, significantly different from non-immune serum; +, significantly different from non-immune serum and non-immune serum plus agonist; ‡, significantly different from non-immune serum plus agonist.[9]

log DALA inhibits adenylyl cyclase activity in lacrimal membranes. DALA also inhibits carbachol-stimulated protein release by lacrimal gland slices.[1,2] The intracellular interactions between the cAMP pathway that is inhibited by DALA and the PLC pathway that is stimulated by carbachol have not been determined. Members of the G_i/G_o family, however, are present in lacrimal membranes[9] and are known to couple receptor activation to inhibition of adenylyl cyclase. In the experiments shown in Fig. 3, DALA significantly reduced the secretory response to carbachol by cultured acini. Insertion of anti-$G_{i1\alpha}$ antiserum had no effect on DALA inhibition of secretion. Insertion of antibodies directed against a peptide sequence common to both G_{i1} and G_{i2}, however, significantly reversed inhibition of the carbachol response. Inhibition of secretion by DALA was also reversed with anti-$G_{i3\alpha}$ and anti-$G_{o\alpha}$ antisera. The functional role for G_{i2}, G_{i3}, and G_o, all of which are present in rabbit lacrimal membranes,[6,7] most likely is to couple receptor activation to inhibition of adenylyl cyclase; however, this remains to be tested.

A clear understanding of the intracellular events that lead to inhibition of secretion has particular importance for the diminished release of tear proteins associated with autoimmune disease.[18] Because lymphocytes, a normal component of the lacrimal gland,[19] increase in number in Sjögren's syndrome in particular and can synthesize and release peptides including opioids,[20] inhibition of secretion by communication between lymphocytes and acinar tissues through an increase in inhibitory peptides becomes an important concept.[21] An extension of this concept is that inhibitory peptides present in nerve terminals in the normal lacrimal gland function to prevent excessive stimulation and operate to maintain a homeostatic state, while

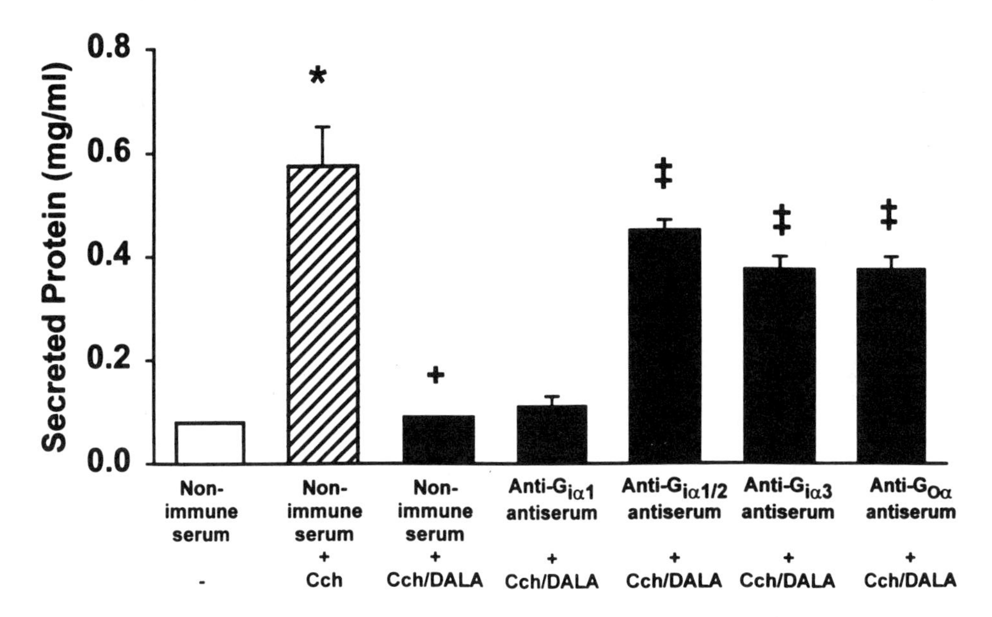

Figure 3. Effect of anti-G_i/G_o antisera on DALA inhibition of carbachol (Cch) stimulated protein release by permeabilized cells. Cultured acini were permeabilized by streptolysin-O (SLO) with simultaneous exposure to non-immune serum or the indicated antibodies. Protein release was measured in the presence of vehicle, 100 μM carbachol or 100 μM carbachol plus 10 μM DALA. *, significantly different from non-immune serum; +, significantly different from Cch; ‡, significantly different from Cch plus DALA.

an infiltrate of lymphocytes may result in abnormally high levels of inhibitors or combination of inhibitors that act synergistically to override neural control.

It is now clear that modulators and their receptors important for adequate lacrimal secretion are linked to specific G proteins that transduce the extracellular signal and that activation of these G proteins regulates the release of protein by secretory acinar cells. In the work presented here, we have demonstrated that VIP and cholinergic stimulation as well as enkephalin inhibition are G protein dependent. From this and previous studies, as well as the work of others, we can state that VIP receptor activation of lacrimal protein release is mediated via G_s coupling to the effector adenylyl cyclase and that $G_{q/11}$ couples M_3 muscarinic receptors to PLC. Enkephalin inhibition of secretion is mediated by G_i/G_o coupling to inhibition of adenylyl cyclase, which by an unknown mechanism blocks cholinergic stimulation of secretion. Finally, we suggest that G_s may have a receptor-second messenger independent role in the regulation of vesicular traffic and exocytosis.

ACKNOWLEDGMENT

This work was supported by NIH grant EY07380.

REFERENCES

1. Cripps MM, Bennett DJ. Peptidergic stimulation and inhibition of lacrimal gland adenylate cyclase. *Invest Ophthalmol Vis Sci.* 1990;31:2145–2150.

2. Cripps MM, Patchen-Moor K. Inhibition of stimulated lacrimal secretion by [D-Ala^2]Met-enkephalinamide. *Am J Physiol.* 1989;257:G151-G156.
3. Putney JW Jr, Van deWalle CM, Leslie BA. Stimulus-secretion coupling in the rat lacrimal gland. *Am J Physiol.* 1978;235:C188-C198.
4. Godfrey PP, Putney JW Jr. Receptor-mediated metabolism of the phosphoinositides and phosphatidic acid in rat lacrimal acinar cells. *Biochem J.* 1984;218:187–195.
5. Mauduit P, Jammes H, Rossignol B. M_3 muscarinic acetylcholine coupling to PLC in rat exorbital lacrimal acinar cells. *Am J Physiol.* 1993;264:C1550-C1560.
6. Meneray MA, Bennett DJ. Identification of GTP-binding proteins in lacrimal gland. *Invest Ophthalmol Vis Sci.* 1995;36:1173–1180.
7. Meneray MA, Fields TY. Identification and characterization of G proteins in the mammalian lacrimal gland. This volume.
8. Sawaki K, Hiramatsu Y, Baum BJ, Ambudkar IS. Involvement of $G_{\alpha q/11}$ in m_3-muscarinic receptor stimulation of phosphatidylinositol 4,5- bisphosphate-specific phospholipase C in rat parotid gland membranes. *Arch Biochem Biophys.* 1993;305:546–550.
9. Meneray MA, Fields TY, Bennett DJ. G_s and $G_{q/11}$ couple VIP and cholinergic stimulation to lacrimal secretion. *Invest Ophthalmol Vis Sci.* 1997;38:1261–1270.
10. Meneray MA, Fields TY, Bromberg BB, Moses RL. Morphology and physiologic responsiveness of cultured rabbit lacrimal acini. *Invest Ophthalmol Vis Sci.* 1994;35:4144–4158.
11. Piiper A, Stryjek-Kaminska D, Stein J, Caspary WF, Zeuzem S. Tyrphostins inhibit secretagogue-induced 1,4,5-IP_3 production and amylase release in pancreatic acini. *Am J Physiol.* 1994;266:G363-G374.
12. Lowry OH, Rosebrough NJ, Farr AL, Randall RJ. Protein measurement with the Folin phenol reagent. *J Biol Chem.* 1951;193:265–275.
13. Masters SB, Sullivan K, Miller R, et al. Carboxyl terminal domain of $G_{s\alpha}$ species coupling of receptors to stimulation of adenylyl cyclase. *Science.* 1988;241:448–451.
14. Exton JH. Phosphoinositide phospholipases and G proteins in hormone action. *Annu Rev Physiol.* 1994;56:349–369.
15. Muller L, Picart R, Barret A, Bockaert J, Homburger V, Tougard C. Identification of multiple subunits of heterotrimeric G proteins on the membrane of secretory granules in rat prolactin anterior pituitary cells. *Mol Cell Neurosci.* 1994;5:556–566.
16. Watson EL, DiJulio D, Kauffman D, Iverson J, Robinovitch MR, Izutsu KT. Evidence for G proteins in rat parotid plasma membranes and secretory granule membranes. *Biochem J* . 1992;285:441–449.
17. Lehtosalo J, Uusitalo H, Mahrberg T, Panula P, Palkama A. Nerve fibers showing immunoreactivities for proenkephalin A-derived peptides in lacrimal glands of the guinea pig. *Graefes Arch Clin Exp Ophthalmol.* 1989;227:455–458.
18. Van Bijsterveld OP, Mackor AJ. Sjögren's syndrome and tear function parameters. *Clin Exp Rheumatol.* 1989;7:151–154.
19. Wieczorek R, Jacobiec FA, Sacks EH, Knowles DM. The immunoarchitecture of the normal human lacrimal gland. *Ophthalmology.* 1988;95:100–109.
20. Blalock JE. Production of peptide hormones and neurotransmitters by the immune system. In: Blalock JE, ed. *Neuroimmunoendocrinology.* Basel: Karger; 1992:1–24.
21. Kaslow HR, Guo Z, Warren DW, Wood RL, Mircheff AK. Autoimmune events and lacrimal acinar: Active participants or passive targets? This volume.

19

MICROTUBULES AND INTRACELLULAR TRAFFIC OF SECRETORY PROTEINS IN RAT EXTRAORBITAL LACRIMAL GLANDS

Philippe Robin, Marie-Noëlle Raymond, and Bernard Rossignol

Laboratoire de Biochimie des Transports Cellulaires
Université Paris XI
91 405 Orsay Cedex, France

1. INTRODUCTION

Proteins destined to be secreted via the regulated pathway are synthesized in the endoplasmic reticulum (ER), transit through the various cisternae of the Golgi apparatus where they become modified, and finally are stored in secretory granules. Under stimulation of the cells, secretory granules fuse with the apical plasma membrane, and their content is released into the extracellular medium. It is now well established that every step of the transport occurring between these different compartments involves vesicular carriers,[1] but the role of the cytoskeleton in the secretory pathway is still under discussion. In many cell types, the rate of the intracellular traffic of proteins is reduced in the presence of microtubule depolymerizing agents, so microtubules were first thought to play a role in the movement of the secretory vesicles.[2] Recent in vitro results have shown that vesicle budding, movement, and fusion do not require the cytoskeleton, so the microtubules are now considered mainly to play a role in the maintenance of a correct location of the organelles rather than in a direct guiding of the secretory vesicles.[3]

Studies in our laboratory showed that, in rat lacrimal glands, the breakdown of the microtubule network evoked a disturbance in the secretory process.[4] In the present work, we specify the level of the secretory pathway at which the microtubules are involved and the role they play in the organization of the Golgi apparatus. We used nocodazole and docetaxel (a taxol analog) respectively to destroy and stabilize microtubules, and brefeldin A (BFA) to disorganize the Golgi complex. We tested the effect of these drugs (or combinations of them) on the regulated secretion of newly synthesized proteins and on the structure of the microtubule network and the Golgi apparatus.

Lacrimal Gland, Tear Film, and Dry Eye Syndromes 2
edited by Sullivan *et al.*, Plenum Press, New York, 1998

2. MATERIALS AND METHODS

Male Sprague-Dawley rats (5 weeks old) were used throughout this study. They were killed by 1 min CO_2 inhalation. Lacrimal glands were dissected at room temperature to eliminate connective tissue and subsequently rinsed twice with a large volume of KRBG buffer (Krebs-Ringer bicarbonate buffer containing 1 g/*l* glucose) at 37°C.

2.1. Protein Secretion

Gland lobules were pulse-labeled for 10 min with ^{3}H-leucine (8 μCi per gland), washed three times in KRBG buffer containing 1 mM leucine to stop the labeling of newly synthesized proteins, and subjected to a chase period. This chase period overlaps the steps of intracellular transit of ^{3}H-leucine-labeled proteins and their apical storage in the secretory granules. In most cases, 10 μM nocodazole or docetaxel (gift of Rhône Poulenc Rorer) was added, respectively, 60 and 120 min before the pulse and conserved until the end of the chase period (75 min). Then gland lobules were washed again three times in KRBG buffer-1 mM leucine. Release of the ^{3}H-labeled proteins was then triggered by addition of either 2 μM carbachol or 10 μM isoproterenol in the presence of 10 μM papaverine.

After 3h 30 min of stimulation, duplicate aliquots of the incubation medium were taken and subjected to two cycles of 20% trichloroacetic acid-0.1% phosphotungstic acid precipitation, and the protein-associated radioactivity was counted. The lobules were dried on paper, homogenized in water with an Ultra Turrax homogenizer, and centrifuged (38,000g, 20 min). Two fractions of the supernatants were subjected to the same treatment as the incubation medium fractions.

Net secretion was expressed as the amount of ^{3}H-labeled proteins released in the incubation medium relative to the total ^{3}H-labeled soluble proteins. Basal secretion value was subtracted from stimulated secretion value.

2.2. Galactose Incorporation in Glycoproteins and Glycoprotein Secretion

Lacrimal gland lobules, incubated in KRBG buffer at 37°C, were labeled for 1 h with ^{14}C-galactose (0.8 μCi per gland). Then lobules were washed three times in KRBG buffer, and 2 μM carbachol was immediately added to trigger protein release. Control experiments were performed in parallel without addition of carbachol. After 3 h stimulation, lobules were removed from the incubation medium, dried on paper, and homogenized in water. When the effects of nocodazole or docetaxel were tested, they were added respectively 1 h and 2 h before the beginning of the labeling and maintained until the end of the stimulation. Homogenates were then centrifuged (at 38,000g, 20 min) and the supernatants, as well as the incubation media, were subjected to TCA precipitation as described for ^{3}H-labeled proteins.

Incorporation of ^{14}C-galactose in glycoproteins is the sum of ^{14}C-labeled proteins released in the incubation medium and ^{14}C-labeled soluble proteins remaining in the tissues. Values were normalized to 1 mg of cellular protein and expressed as percent of control.

Net secretion of glycoproteins was expressed as the amount of ^{14}C-labeled proteins released in the incubation medium relative to the total ^{14}C-labeled soluble proteins. The basal secretion value was subtracted from that of stimulated secretion; results are expressed as percent of control.

2.3. Immunofluorescence Microscopy

After appropriate treatment in KRBG buffer at 37°C, lacrimal lobules were fixed for 30 min in 3% paraformaldehyde in PBS, washed with PBS, and frozen in isopentane cooled by liquid nitrogen. Thin sections (5 μm) were made and incubated with either a monoclonal antibody against α-tubulin (N.356, Amersham) (dilution 1/500) or a monoclonal antibody against the 58 kDa Golgi protein (G-2404, Sigma) (dilution 1/20). The sections were then stained with an FITC-labeled secondary antibody (F-2266, Sigma) (dilution 1/100).

3. RESULTS AND DISCUSSION

3.1. Effect of Nocodazole and Docetaxel on the Microtubule Network

Rat lacrimal glands are constituted of small cells organized in acini. The nuclei are located in the basal region of the cells. Immunofluorescence of tubulin on control tissue reveals a dense microtubular mesh extending throughout the cytoplasm (Fig. 1a). The labeling seems slightly more intense in the apical part of the cell, in the vicinity of the lumen. When the lobules were treated with nocodazole, the microtubules were mainly destroyed (Fig. 1b), and the tubulin staining appeared rather diffuse. Nevertheless, some short fragments of microtubules are still visible in the proximity of the apical membrane. When the lobules were incubated in the presence of docetaxel, the microtubules gathered in bundles at the periphery of the cells (Fig. 1c); the cytoplasm frequently appears empty.

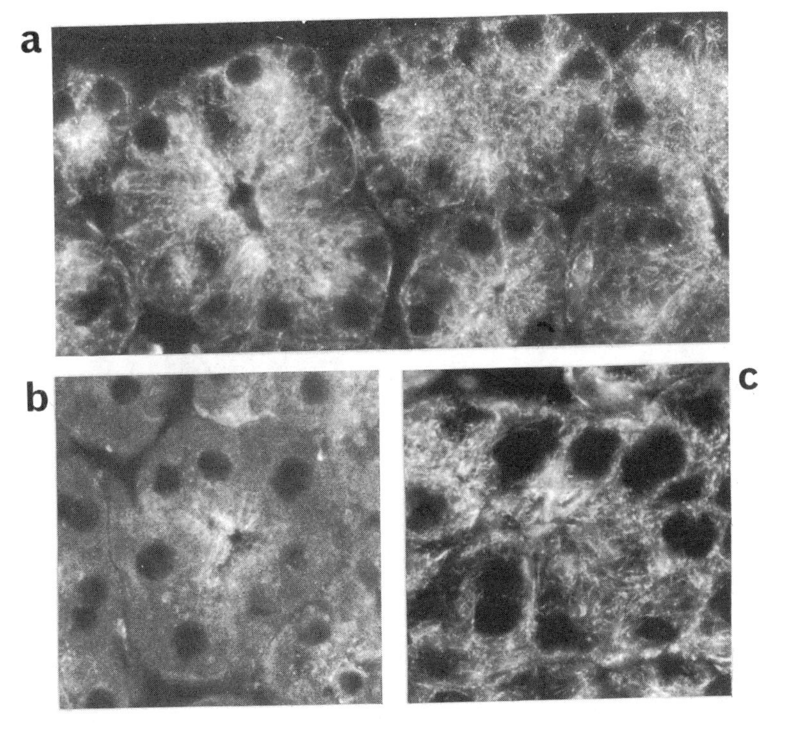

Figure 1. Effect of nocodazole and docetaxel on the microtubule network. Lobules were incubated for 2 h (a) in the absence of drug, (b) in the presence of 10 μM nocodazole, or (c) in the presence of 10 μM docetaxel.

3.2. Effect of Nocodazole and Docetaxel on Protein and Glycoprotein Secretion

We compared the influence of microtubule depolymerization or stabilization on the secretion of newly-synthesized proteins triggered by stimulation of either β-adrenergic or muscarinic receptors using isoproterenol and carbachol respectively. If nocodazole or docetaxel is added at the end of the chase period, when the radiolabeled proteins are still within the secretory granules, protein release is not inhibited by the drug (data not shown). When the lobules are incubated with the drug before the pulse, during the pulse, and all along the chase the release of radiolabeled proteins is affected. Fig. 2a shows that the inhibitory effect of nocodazole reaches 75%, whereas docetaxel inhibition is close to 50%. The inhibition levels are independent of the nature of the secretagogue used, so we can eliminate a possible direct effect of the drug directly on the signal transduction pathways.

To get more information on the transit step affected by the disorganization of the microtubule network, we studied the effect of nocodazole and docetaxel on ^{14}C-galactose incorporation, a trans-Golgi event, and on the release of radiolabeled glycoproteins. Fig. 2b shows that galactose incorporation is slightly reduced by the drugs, whereas the level of inhibition of glycoprotein secretion reaches around 50% and 80% for respectively docetaxel and nocodazole. This indicates that the steps of the secretory pathway occurring after the galactosylation are the most sensitive to microtubule disturbance. However, docetaxel was less effective than nocodazole.

3.3. Effect of Nocodazole and Brefeldin A on the Organization of the Golgi Apparatus

We studied the morphology of the Golgi apparatus using the 58 kDa protein. This marker is a peripheral membrane protein that colocalizes with mannosidase II and does not dissociate from the Golgi complex under BFA treatment.[5] In control lacrimal cells, the

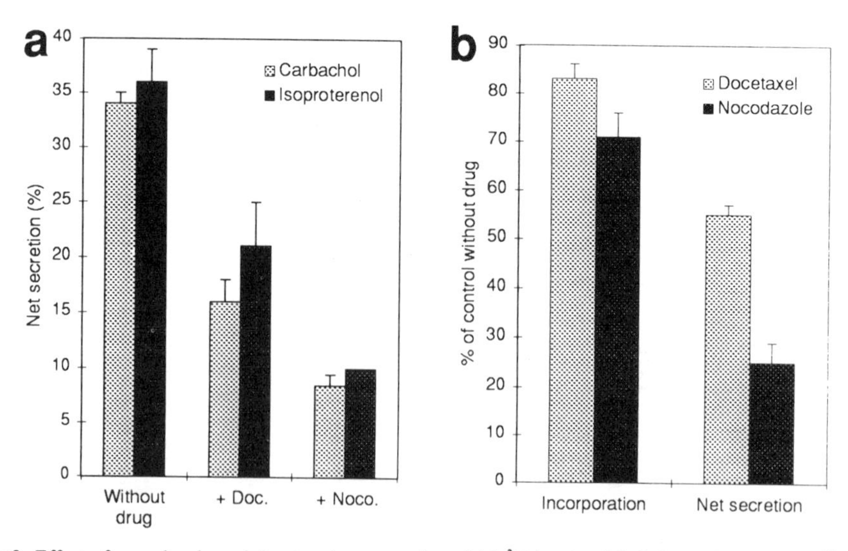

Figure 2. Effect of nocodazole and docetaxel on secretion of (a) ^{3}H-leucine-labeled proteins, and (b) ^{14}C-galactose-labeled glycoproteins. Results are expressed as mean ± SEM (n=3–9).

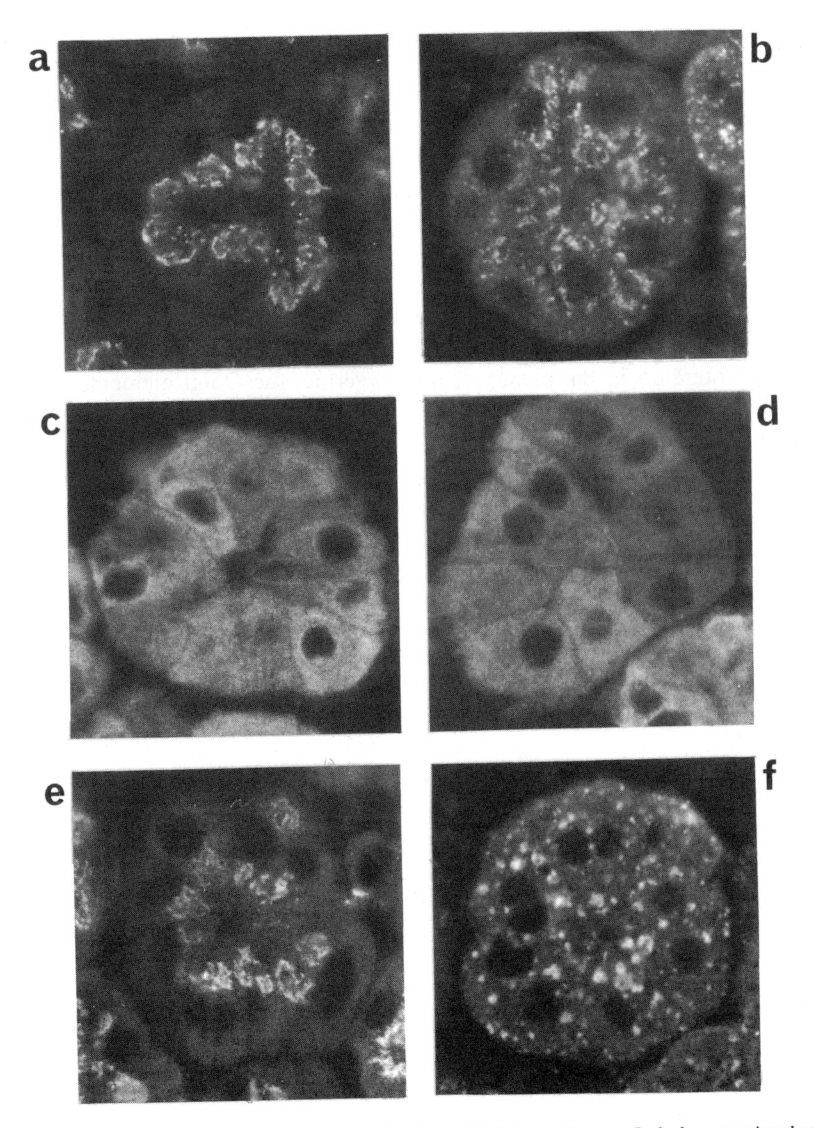

Figure 3. Effect of nocodazole and BFA on the distribution of Golgi membranes. Lobules were incubated for 1 h in the absence (a,b) or the presence (c,d) of 10 μM BFA. BFA-treated lobules were then incubated in the presence of 0.1 μM BFA and 10 μM isoproterenol (e,f) for 1 h. Lobules shown in b,d,f were additionally treated with 10 μM nocodazole.

Golgi staining shows abundant and curve structures localized between the nucleus and the apical membrane (Fig. 3a). When the lobules are treated with docetaxel, the Golgi organization does not seem to be modified (data not shown), whereas after treatment with nocodazole, the Golgi elements are scattered through the cytoplasm (Fig. 3b).

To get more precise information about the role of the microtubule network in the Golgi organization, we performed experiments using nocodazole and BFA. This drug drastically modifies the Golgi structure but does not affect the microtubule network. It has been proposed[6] that under BFA treatment extensive Golgi tubules are formed and extend along microtubules toward the ER. This mechanism may explain the redistribution of the Golgi proteins in the ER after action of the drug. Fig. 3c shows in lacrimal gland the effect of BFA on the Golgi structure. The cells appear completely fluorescent, and sometimes a dense tangled net of curved structures resembling the ER can be seen. This redistribution of the Golgi elements is, in lacrimal cells, independent of the microtubule network: when the lobules were incubated in the presence of nocodazole prior to the addition of BFA, no difference was seen (Fig. 3d).

As we recently showed[7] that an increase in the cAMP level (triggered by addition of isoproterenol) promotes a reorganization of the Golgi apparatus after BFA treatment (Fig. 3e), we studied the effect of nocodazole on that restructuring step. Fig. 3f shows the result of such an experiment. When lobules are submitted to the action of nocodazole and BFA and then to isoproterenol in the presence of papaverine, the Golgi elements appear dispersed in small fluorescent patches in the whole cytoplasm. It seems that in these conditions Golgi elements are reformed, but in the absence of a microtubule network, they are unable to gather in the vicinity of the nucleus.

Now we can draw these conclusions: (a) Microtubules play a key role in the intracellular transport of secretory proteins after the galactosylation step (at the trans-Golgi level) but before the exocytosis step. (b) Microtubules are involved in the dynamics of the Golgi complex. The microtubule network is not required for dispersion of the Golgi elements under BFA action. On the other hand, the reorganization of the Golgi apparatus following removal of BFA has two phases: a microtubule-independent reformation of Golgi elements, and a microtubule-dependent reclustering of the reformed elements in the juxtanuclear area.

REFERENCES

1. Rothman JE, Wieland FT. Protein sorting by transport vesicles. *Science*. 1996; 272:227–234.
2. Burgess TL, Kelly RB. Constitutive and regulated secretion of proteins. *Annu Rev Cell Biol.* 1987;3:243–293.
3. Allan V. Membrane traffic motors. *FEBS Lett.* 1995;369:101–106.
4. Busson-Mabillot S, Chambault-Guerin AM, Ovtracht L, Muller P, Rossignol B. Microtubules and protein secretion in rat lacrimal glands: localization of short term effects of colchicine on the secretory process. *J Cell Biol.* 1982;95:105–117.
5. Donaldson JG, Lippincott-Schwartz J, Bloom GS, Kreis TE, Klausner RD. Dissociation of a 110-kD peripheral membrane protein from the Golgi apparatus is an early event in brefeldin A action. *J Cell Biol.* 1990;111:2295–2306.
6. Cole NB, Lippincott-Schwartz J. Organization of organelles and membrane traffic by microtubules. *Curr Opin Cell Biol.* 1995;7:55–64.
7. Robin P, Rossignol B, Raymond M-N. Recovery of protein secretion after brefeldin A treatment of rat lacrimal glands: Effect of cAMP. *Am J Physiol.* 1996;271:C783–C793.

20

EFFECTS OF NEUROPEPTIDES ON SEROTONIN RELEASE AND PROTEIN AND PEROXIDASE SECRETION IN THE ISOLATED RAT LACRIMAL GLAND

Jaipaul Singh,[1] Keith A. Sharkey,[2] Robert W. Lea,[1] and Ruth M. Williams[1]

[1]Department of Applied Biology
University of Central Lancashire
Preston, England
[2]Department of Medical Physiology
University of Calgary
Alberta, Canada

1. INTRODUCTION

The secretion of tears by the lacrimal gland is regulated by the autonomic nervous system[1] and by non-cholinergic, non-adrenergic (NCNA) nerves that are believed to release a number of biologically active peptides that can function as neurotransmitters of NCNA nerves regulating the secretory processes.[2,3] The neuropeptides have been investigated for their potential role in modulating the immune system. In many organs and tissues from several animal species, a morphological and functional association was found between mast cells and neuropeptide-containing nerves that contain substance P (Sub P) and calcitonin gene-related peptide (CGRP), or both.[4,5] Mast cells have a secretory capacity. They released mediators which include serotonin (5-HT), and histamine which can evoke pronounced biological effects resulting in cellular regulation.[6] It is now apparent that neuropeptides can modulate some immune response directly,[7] and several peptides--in particular, Sub P have been shown to induce mast cell mediator release.[8] It is possible that neuropeptides may function as immunological effectors in inflammatory responses in the lacrimal gland. Moreover, in a previous study we showed that the lacrimal gland is innervated with numerous neuropeptides, and there is a consistent association between peptidergic nerves and mast cells.[9] This study was designed to investigate any functional interaction between mast cells and peptidergic neurotransmitters Sub P and CGRP.

2. METHODS

2.1. Experimental Procedure

Adult Sprague-Dawley rats were killed by a blow to the head followed by cervical dislocation. The lacrimal glands were quickly excised and placed in a modified oxygenated (95% O_2, 5% CO_2) Krebs-Henseleit (KH) solution comprising (mM) NaCl, 118; KCl, 3.7; $CaCl_2$, 2.56; $NaHCO_3$, 25; KH_2PO_4, 2.2; $MgCl_2$, 1.2; and glucose 10, pH 7.4. The connective tissue capsule was removed from each gland which was subsequently sliced into small segments (5–10 mg). Lacrimal tissue aliquots (3–35 mg) were washed (5 times) with the oxygenated KH solution and placed in 5 ml plastic tubes containing either KH solution alone (control) or KH solution containing different concentrations of the test substances. The aliquots were incubated for 20 min at 37°C in a shaking water bath. The following test substances were used: Substance P (Peninsula), CGRP (Peninsula), 5-HT (Sigma), calcium ionophore A23187 (Sigma), and sodium cromoglycate (a gift from Fisons Pharm.). At the end of the 20 min incubation period, effluent samples were analyzed for peroxidase, total protein, or 5-HT. At the end of the experiment, the tissue was blotted and weighed.

2.1.1. Protein Assay. The protein content of the samples was determined spectrophotometrically using a modification of the Bradford assay,[10] as described previously by Bromberg and Welch.[11] Samples of incubation medium were added to the protein reagent and mixed. Optical density was recorded at 595 nm using a Visi spectrophotometer (Pharmacia, UK). Absorbance was measured after 30 min using KH solution as a blank and bovine serum albumin (Sigma) as a standard. All assays were performed in duplicate. Concentrations were determined from standard curves constructed for each series of experiments. Protein secretion was expressed as mg ml^{-1} (100 mg $tissue)^{-1}$.

2.1.2. Peroxidase Assay. Peroxidase activity was measured as a marker of acinar cell secretion using a modification of the method of Herzog and Fahimi.[12] Briefly, 200 μl of the incubation medium was added to 3,3'-diaminobenzidine (5 mM) in sodium phosphate buffer (0.1 M, pH 7.4). Hydrogen peroxide was added and the contents mixed. The optical density was measured immediately at 460 nm using a Visi spectrophotometer. Absorbance was measured over 3 min, and the slope of the measurements was used to determine peroxidase activity using a peroxidase (Sigma) standard and KH solution as a blank. Values are expressed as ng ml^{-1} (100 mg $tissue)^{-1}$.

2.1.3. Seotonin (5-HT) Release. Perchloric acid (100 μl) was added to a sample of the incubation medium (900 μl) to precipitate protein. The mixture was shaken and left for 10 min on ice. After centrifugation (10,000g 1 min), the supernatant was analyzed for 5-HT by high performance liquid chromatography (HPLC) using an electrochemical detection system. Known concentrations of 5-HT were used as standards. Samples were separated on a Beckman Ultrasphere Octadecylsilica column (4.5x50 mm, 5 μm) using a mobile phase consisting of 0.2 M acetic acid, 0.05 M sodium acetate, 100 mg/L EDTA, 100 mg/L SDS, and 25% methanol (v/v) at a flow rate of 0.5 ml min^{-1} at 40°C. Detection was by means of a BAS LC4B amperometric detector with a glassy carbon electrode operating at +0.60 V. The signal was digitized, stored, and quantitated by the external standard method using a Waters Maxima data acquisition system. Results are expressed as pmol 100 mg^{-1} tissue. To assess total 5-HT content, the lacrimal glands were homogenized in perchloric acid and the supernatant analyzed as described above.

2.1.4. Statistical Analysis. Data are presented as mean ± standard error of the mean (SEM) from 5–8 experiments using glands from different animals. Data were compared using ANOVA followed by Newman-Keuls post hoc test, or unpaired Student's t-tests as appropriate. Differences between values were considered significant at $p<0.05$.

3. RESULTS

Basal protein secretion varied between individual glands. The basal protein output in one particular experiment was 1.46 ± 0.25 mg ml^{-1} (100 mg $tissue)^{-1}$ (n=6). Fig. 1 shows amylase output from lacrimal segments in the control and during different experimental conditions. The Ca^{2+} ionophore A23187, Sub P CGRP, or 5-HT evoked marked increases in total protein output compared to basal. All the responses were significantly ($p<0.05$) reduced by pretreatment of the tissues with 10^{-4} M sodium cromoglycate (NaC), except for CGRP.

Basal peroxidase secretion from lacrimal segments was 32.8 ± 4.4 ng ml^{-1} (100 mg $tissue)^{-1}$ (n=24). Fig. 2 shows peroxidase secretion from rat lacrimal segments in the control and following stimulation with 5-HT, CGRP, and Sub P. The effects of Sub P on peroxidase secretion in combination with CGRP or NaC are shown for comparison. 5-HT,

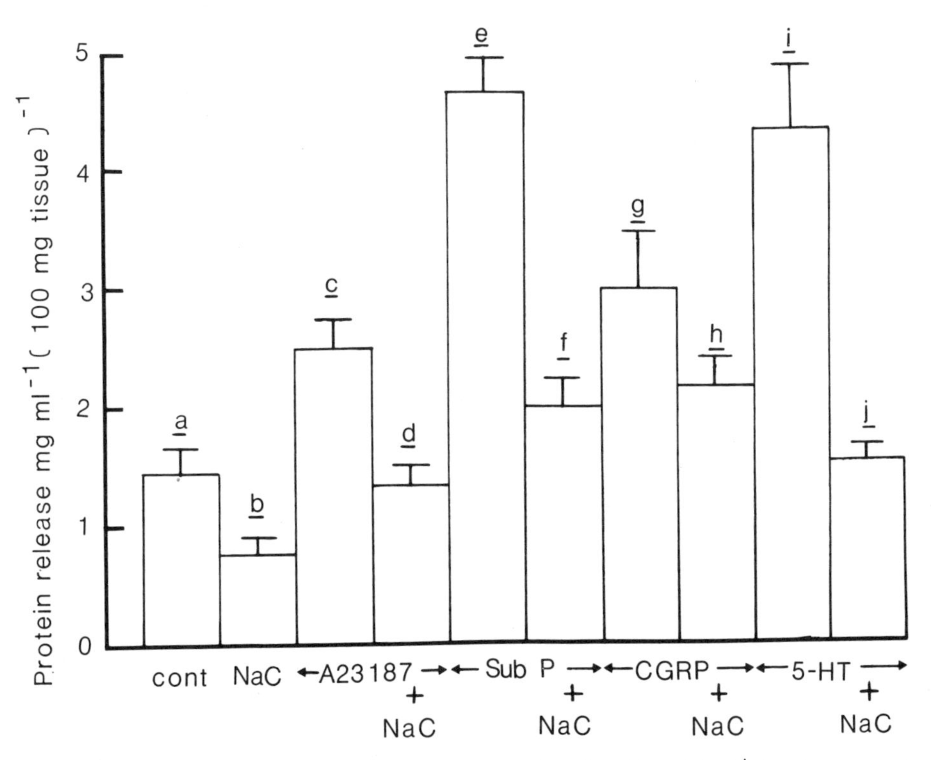

Figure 1. Protein output from rat lacrimal gland segments. (a) Control condition, (b) 10^{-4} M sodium cromoglycate (NaC), (c) 10^{-6} M A23187 alone, (d) 10^{-6} M A23187 plus 10^{-4} M NaC, (e) 10^{-6} M Substance P (Sub P) alone (f) 10^{-6} M Sub P plus 10^{-4} M NaC, (g) 10^{-6} M CGRP alone, (h) 10^{-6} M CGRP plus 10^{-4} M NaC, (i) 10^{-5} M 5-HT alone, (j) 10^{-5} M 5-HT plus 10^{-4} M NaC. Values are mean ± SEM (n=5–8).

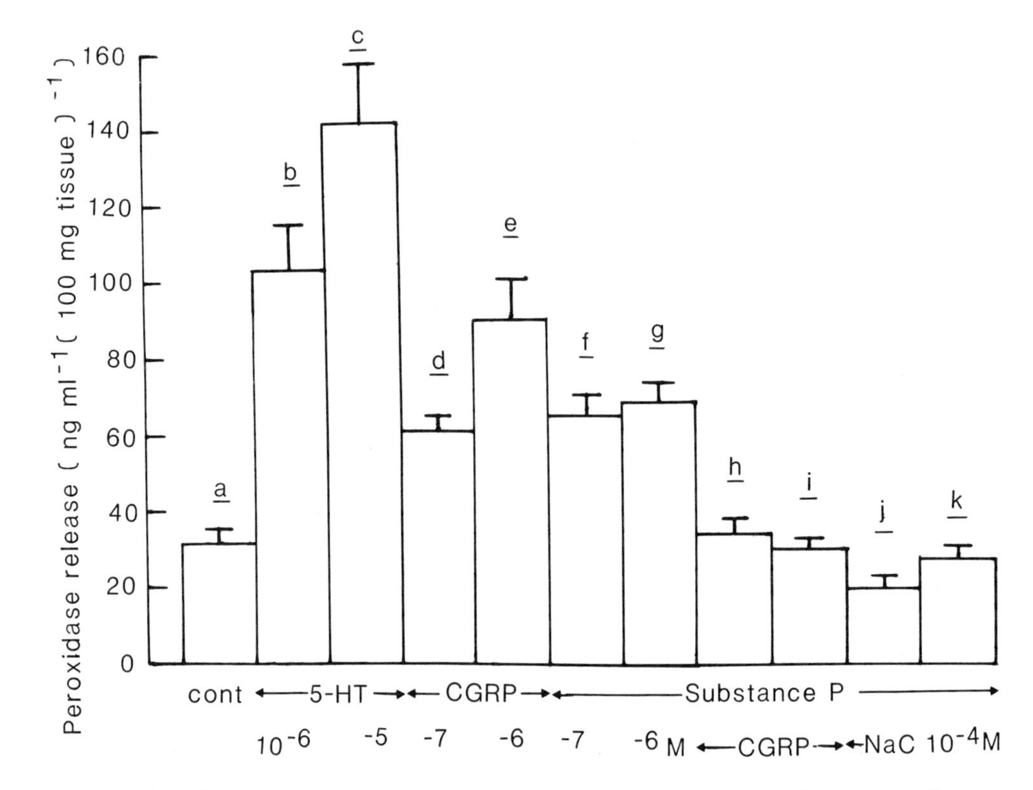

Figure 2. Peroxidase release from rat lacrimal segments in (a) control condition, (b) 10^{-6} M and (c) 10^{-5} M 5-HT, (d) 10^{-7} M and (e) 10^{-6} M CGRP, (f) 10^{-7} M and (g) 10^{-6} M Sub P alone, (h) 10^{-7} M Sub P plus 10^{-7} M CGRP, (i) 10^{-6} M Sub P plus 10^{-6} M CGRP, (j) 10^{-7} M Sub P and 10^{-4} M NaC, (k) 10^{-6} M Sub P plus 10^{-4} M NaC. Values are mean ± SE (n=6–8).

CGRP, and Sub P evoked marked increases in peroxidase secretion. The Sub P-induced secretory responses were attenuated by both CGRP and NaC.

Basal 5-HT release was 4.25 ± 0.4 pmol 100 mg tissue^{-1} (n = 15). Fig. 3 shows 5-HT release from rat lacrimal segments in the basal state and during stimulation with A23187 in the absence and presence of NaC, Sub P, and CGRP. The Ca^{2+} ionophore A23187 and Sub P significantly (p<0.05) increased 5-HT release from lacrimal segments. In contrast, CGRP not only inhibited 5-HT release but also antagonized the stimulatory effect of Sub P. Similarly, NaC attenuated the Sub P-evoked 5-HT release.

4. DISCUSSION

The results of this study have shown that exogenous application of either Sub P, CGRP, 5-HT or A23187 results in large increases in total protein output and peroxidase secretion from rat lacrimal segments. The secretory effects of 5-HT, Sub P, and A23187, but not CGRP, were markedly attenuated by the mast cell stabilizer sodium cromoglycate (NaC). Similarly, Sub P and A23187, but not CGRP, stimulated the release of 5-HT from lacrimal gland segments. These responses were attenuated by NaC. Furthermore, when Sub P was combined with CGRP, there was a significant reduction in 5-HT release com-

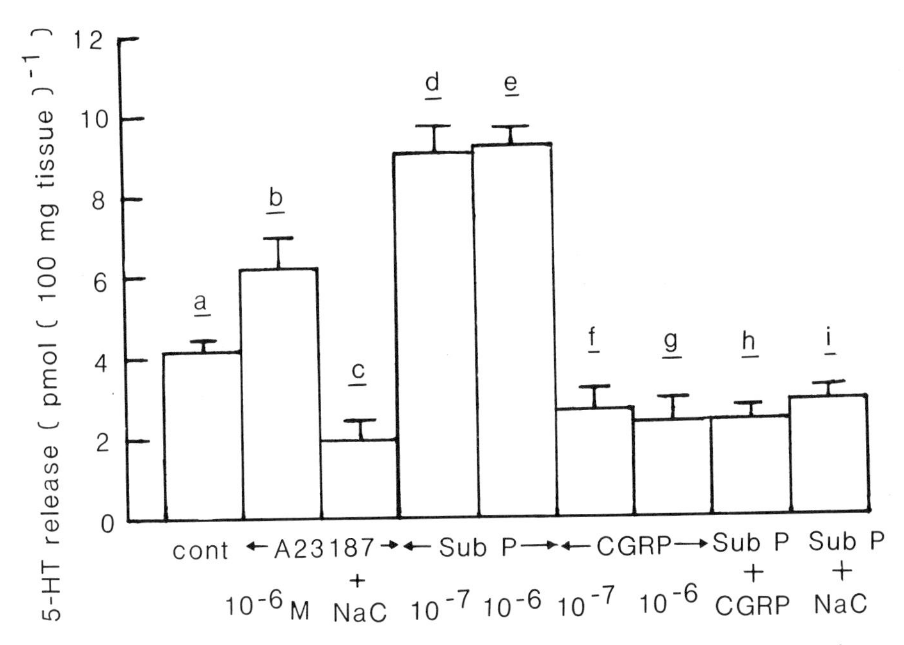

Figure 3. 5-HT release from lacrimal segments in (a) control condition, (b) 10^{-6} M A23187, (c) 10^{-6} M A23187 plus 10^{-4} M NaC, (d) 10^{-7} and (e) 10^{-6} M Sub P, (f) 10^{-7} M and (g) 10^{-6} M CGRP, (h) 10^{-6} M Sub P plus 10^{-6} M CGRP, (i) 10^{-6} M Sub P plus 10^{-4} M NaC. Values are mean ± SEM (*n*=5–8).

pared to the effect of Sub P alone. These findings implicate the involvement of Sub P, CGRP, and 5-HT in the control of lacrimal gland protein secretion. These secretagogues may be acting directly on acinar cells to elicit secretion or indirectly via mast cell. Moreover, the ability of CGRP to antagonize the secretory responses of Sub P indicates that CGRP is either interacting with Sub P at receptor level or preventing degranulation of mast cells. However, this interaction between Sub P and CGRP either at receptor level or during mast cell degranulation is less understood and deserves further study.

The present study has also demonstrated that the neuropeptides such as Sub P and CGRP represent putative neurotransmitters of peptidergic nerves that are distributed in close association in the lacrimal gland.[9,13] It was also found that pretreatment of tissue with the mast cell stabilizer NaC resulted in a significant attenuation in 5-HT release evoked by Sub P. This interesting result suggests that Sub P may primarily stimulate protein and peroxidase secretion indirectly by causing mast cell degranulation, with the releasing mediator(s) exerting a stimulatory effect on the lacrimal acini. Sub P has been shown to be a potent mast cell secretagogue in a variety of tissues.[8] This proposal is supported by the observation that the mast cell mediator 5-HT effectively stimulated lacrimal protein secretion. Taken together, the present results provide corroborative evidence to support an important role of mast cells in the events involved in the secretion of enzymes and proteins from the lacrimal gland. In addition, the finding that Ca^{2+} ionophore A23187 can stimulate 5-HT release and that the response was antagonized by sodium cromoglycate suggests that an influx of Ca^{2+} into mast cells is a crucial event in the release of 5-HT.

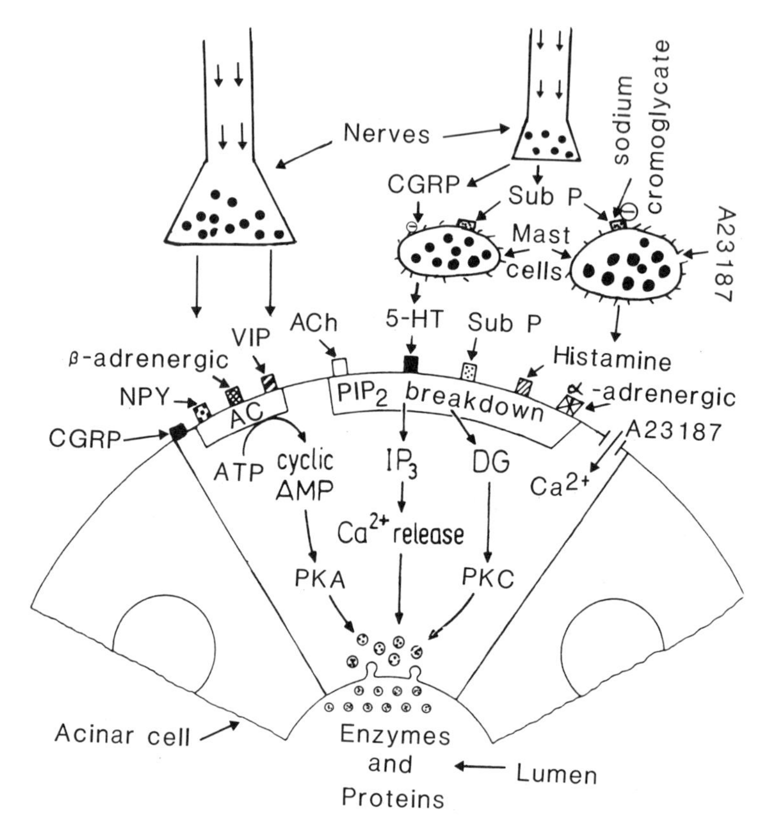

Figure 4. A schematic model showing the involvement of intrinsic nerves, mast cells, and various neurotransmitters, mediators, and drugs during protein and enzyme secretion from rat lacrimal acinar cells. AC, adenylate cyclase. PIP_2, phosphatidyl inositol 1,4-bisphosphate. IP_3, inositol 1,4,5-trisphosphate. DG, diacylglycerol. PKA, protein kinase A. PKC, protein kinase C.

In summary (see Fig. 4 for main conclusions), the results show that activation of intrinsic secretomotor nerves in the lacrimal gland can result in the release of a number of endogenous neurotransmitters including Sub P and CGRP. Some of these neurotransmitters can stimulate acinar cells either directly or indirectly to secrete proteins and enzymes. It is suggested that Sub P may act on the mast cells to bring about degranulation resulting in the release of 5-HT, which in turn stimulates the acinar cells to release proteins. In contrast, CGRP is believed to inhibit degranulation of mast cells resulting in decrease in 5-HT release. These effects of Sub P can be antagonized by either CGRP or sodium cromoglycate. The calcium ionophore can also stimulate 5-HT release and protein secretion, indicating a role for Ca^{2+} in the stimulus secretion coupling mechanism. These secretory responses can also be mimicked by exogenous application of secretagogues.

ACKNOWLEDGMENTS

This work was supported by the British Council, the Wellcome Trust, the MRC of Canada, and the Alberta Heritage Foundation.

REFERENCES

1. Bromberg BB. Autonomic control of protein secretion. *Invest Ophthalmol Vis Sci.* 1981;20:110.
2. Dartt DA. Signal transduction and control of lacrimal gland protein secretion: A review. *Curr Eye Res.* 1989;8:619–636.
3. Dartt DA. Regulation of tear secretion. *Adv Exp Med Biol.* , 1994;350:1–9.
4. Foreman JC. Peptides and neurogenic inflammation. *British Med Bull.* 1987;43:386.
5. Skofitsch S, Donnerer J, Petronnijevic S, Saria A, Lembeck F. Release of histamine by neuropeptides from perfused rat hind quarters. *Naunyn Schmiedebergs Arch Pharmacol.* 1983;322:153.
6. Befus AD, Bienenstock J, Denburg JA. *Mast Cell Differentiation and Heterogenicity.* New York: Raven Press;1986;
7. Shanahan F, Anton P. Neuroendocrine modulation of the immune system: Possible implications for inflammatory bowel disease. *Dig Dis Sci.* 1988;33 (Suppl 3):415.
8. Shanahan F, Denburg, JA, Fox J, Bienenstock J, Befus AD. Mast cell heterogenicity effects of neurogenic peptides on histamine release. *J Immunol.* 1985;135:1331.
9. Williams RM, Singh J, Sharkey KA. Innervation and mast cells of the rat exorbital lacrimal gland: Effect of age. *J Auton Nerv Syst.* 1994;47:95.
10. Bradford MM. A rapid and sensitive method for the quantification of microgram quantities of protein utilizing the principle of protein dye binding. *Anal Biochem.* 1976;72:248.
11. Bromberg BB, Welch MH. Lacrimal protein secretion: Comparison of young and aged rats. *Exp Eye Res.* 1985;40:313.
12. Herzoz V, Fahimi H. A new sensitive colorimetric assay for peroxidase using 3,3'-diaminobenzidine as a hydrogen donor. *Anal Biochem.* 1973;55:554.
13. Matsumoto Y, Tanake T, Ueda S, Kawata M. Immunohistochemical and enzymehistochemical studies of peptidergic, aminergic and cholinergic innervation of the lacrimal gland of the monkey (*Macaca fuscata*). *J Auton Nerv Syst.* 1991;37:207.

21

ANALYSIS OF PHOSPHODIESTERASE ISOENZYMES IN THE OCULAR GLANDS OF THE RABBIT

Thomas J. Millar[1,2] and Harry Koutavas[2]

[1]Department of Biological Sciences
University of Western Sydney
Nepean, New South Wales, Australia, 2747
[2]Co-operative Research Centre for Eye Research and Technology
University of New South Wales
Kensington, New South Wales, Australia

1. INTRODUCTION

Cyclic nucleotides, cAMP and cGMP, are important in regulating cellular function.[1,2] The levels are controlled by the balance between the synthesis, catalyzed by adenylate or guanylate cyclase, and the breakdown, catalyzed by 3',5'-cyclic nucleotide phosphodiesterases (PDE), of which there are seven distinct families.[3–5] These are: PDE I, Ca^{2+}/calmodulin activated and nicardipine inhibited; PDE II, cGMP activated and EHNA inhibited; PDE III, cGMP inhibited and milrinone inhibited; PDE IV, cAMP specific and rolipram inhibited; PDE V, cGMP specific (non-visual) and zaprinast inhibited; PDE VI, cGMP specific (visual) and found only in the retina; and PDE VII, cAMP specific and IBMX and rolipram insensitive.

Cyclic nucleotide levels in cells can be increased by applying phosphodiesterase inhibitors. Since cyclic nucleotides in lacrimal gland acinar cells play a role in tear secretion,[6] determination of the particular phosphodiesterase enzymes present in the lacrimal gland may lead to a target for pharmacological manipulation of tear secretion by using a class-specific phosphodiesterase inhibitor. We examined the superior and inferior lacrimal glands for class-specific PDE activity and compared the distribution of activities with those found in other orbital glands--the Harderian gland, a lipid-secreting gland, and the zygomatic gland, a purely mucus-secreting salivary gland located in the inferior orbit.

2. MATERIALS AND METHODS

Male and female New Zealand albino rabbits (1.5–2.5 kg) were overdosed with sodium pentobarbital injected into an ear vein in accordance with the Australian NH and MRC guidelines. Lacrimal Harderian and zygomatic glands were removed and homogenized in Tris-HCl 50 mM, leupeptin 20 mM, soybean trypsin inhibitor 10 mg mL^{-1}, pepstatin 10 mg mL^{-1}, chymostatin 10 mg mL^{-1}, and (*p-amidinophenyl*) methanesulfonyl fluoride hydrochloride 1 μM and centrifuged (50,000*g*, 60 min). The supernatant was applied to a Mono-Q HR 5/5 anion-exchange column and eluted with a NaCl gradient (0.15-1.0 M). Fractions (1 ml) from the column were analyzed for PDE activity using cAMP as the substrate and monitoring its levels by a reverse-phase HPLC method.[7] The different PDE classes were determined by carrying out incubations for enzyme activity in the presence of class-specific activators or inhibitors. For example, for PDE I, incubation was carried out in the presence of Ca^{2+} and calmodulin. Kinetic studies of PDE activities were also undertaken using standard techniques to determine Km and V values.

3. RESULTS

A typical profile from the reverse-phase HPLC column representing PDE activity is shown in Fig. 1.

Analysis of PDE activity from each of the glands showed that the superior and the inferior lacrimal gland had the same PDE classes, but the pattern of PDE classes in the other glands was different (Table 1).

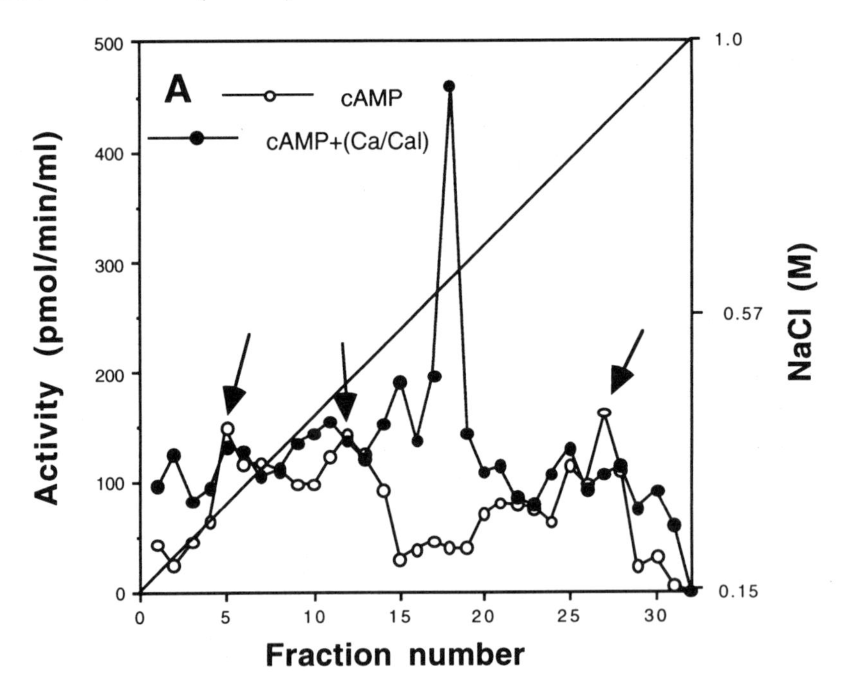

Figure 1. Averaged PDE activity profile for protein fractions obtained from the rabbit lacrimal gland. When the incubation was carried out with cAMP alone, three peaks of activity were observed (arrows). When Ca^{2+}/calmodulin was included in the incubation mixture, a major peak of activity was observed at about fraction 17 (PDE Type I activity).

Table 1. Predominant PDE activity in different ocular glands

Ocular gland	Predominant PDE activity
Lacrimal	PDE I, III, and IV
Harderian	PDE II and IV
Zygomatic	PDE III and IV

The Km and V for each of the different peak PDE activities were measured (Table 2).

PDE type I was found only in the lacrimal gland. We therefore tested the effects of nicardipine, a PDE I specific inhibitor, on protein secretion from cultured cells obtained from lacrimal, Harderian, and zygomatic glands. Cells were isolated based on the method of Hann et al.[8] and grown on 12-well Costar tissue culture plates in DMEM (Dulbecco's modified Eagle, s medium) supplemented with 10% FCS (foetal calf serum). After 8 days in culture, the media was replaced with 1 mL of fresh media containing either 5 mM nicardipine, 5 mM IBMX (3-isobutylmethylxanthine), a non-specific PDE inhibitor, or 5 mM carbachol, a cholinergic agonist known to stimulate lacrimal secretion.[9] At various times, samples of supernatant from the cultures were analyzed using SDS-PAGE. The results are shown in Fig. 2.

Notable is the appearance of a band at 20 kDa in media from cultures treated with nicardipine or IBMX. This band corresponds to lipocalin.[10] It increased in time with incubation and was less pronounced in the supernatant from cultures incubated with IBMX. In cultures incubated with carbachol, only a very weak band was seen in this position. Nicardipine had no effect on secretion from cultures of the Harderian or zygomatic glands.

4. DISCUSSION

Our analysis of PDE activity in the lacrimal, Harderian, and zygomatic glands reveal that each gland has a specific array of PDE classes. PDE type I was predominant in the lacrimal gland, PDE II in the Harderian gland, and PDE III in the zygomatic gland. All glands contained PDE IV. Application of PDE I inhibitor specifically increased the secretion of a 20 kDa protein, possibly lipocalin, with no other apparent effects. IBMX similarly increased the se-

Table 2. Km and V values obtained from different peak PDE activities separated from rabbit lacrimal, Harderian, and zygomatic glands

PDE	Km cAMP (μM)	Km cGMP (μM)	V_{max} cAMP (pmol/min/mL)	V_{max} cGMP (pmol/min/mL)
Lacrimal				
I	25.5	ND	608	ND
III	7.8	28.6	403	153
IV	84.2	ND	172.4	ND
Harderian				
II	6.4	70.2	237	110.2
IV	38.8	ND	113.1	ND
Zygomatic				
III	4.4	46.9	284	348.9
IV	31.8	ND	305.7	ND

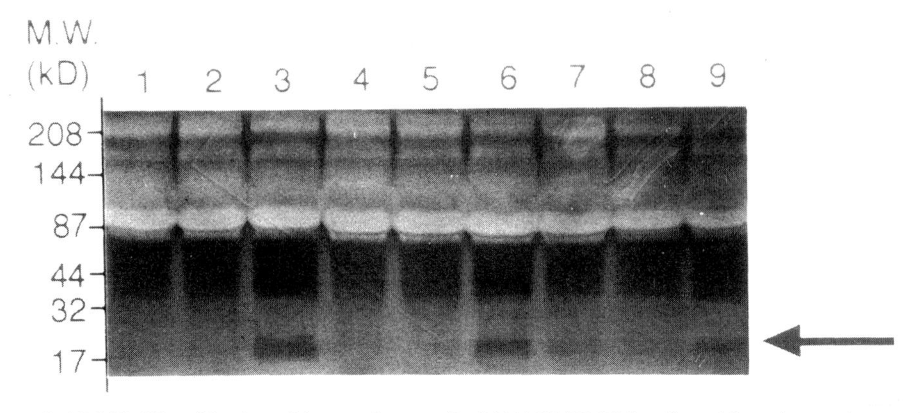

Figure 2. SDS-PAGE purification of tissue culture media (10% FCS/DMEM) collected from tissue culture wells: (1, 4, 7) no pharmacological agents, $t = 0$; with nicardipine, (2) $t = 30$ min, (3) $t = 24$ h; with IBMX, (5) $t = 30$ min, (6) $t = 24$ h; with carbachol, (8) $t = 30$ min, (9) $t = 24$ h. A prominent band at 20 kDa appears after 24 h incubation (arrow). Many bands between 32 and 208 kDa are believed to be associated with protein in FCS.

cretion of this protein but to a lesser degree. Carbachol is known to increase water and electrolyte secretions and hence appears to be operating through a different intracellular pathway from that involved with protein secretion. However, there could be some interaction between these pathways because a small amount of 20 kDa protein was seen in cultures incubated for 24 h with carbachol. The selectivity of nicardipine for increasing only one protein indicates that there may be several intracellular pathways linked to the secretion of particular proteins.[8] This holds great promise pharmacologically because it suggests that the levels of individual proteins may be manipulated by targeting specific second-messenger pathways.

REFERENCES

1. Butcher RW, Sutherland EW. Adenosine 3',5'-phosphate in biological materials. I. *J Biol Chem.* 1962;237:1244–1250.
2. Jahn R, Padel U, Porsch PH, Soling HD. Adrenocorticotropic hormone and α-melanocyte stimulating hormone induced secretion and protein phosphorylation in the rat lacrimal gland by activation of a cAMP-dependent pathway. *Eur J Biochem.* 1982;126;623–629.
3. Beavo JA, Reifsnyder DK. Primary sequence of cyclic nucleotide phosphodiesterase isoenzyme and the design of selective inhibitors. *Trends Pharmacol Sci.* 1990;11:150–155.
4. Thompson WJ, Appleman MM. Multiple cyclic nucleotide phosphodiesterase activities from rat brain. *Biochemistry.* 1970;10:311–316.
5. Michaeli T, Bloom TJ, Martins T, et al. Isolation and characterisation of a previously undetected human cAMP phosphodiesterase by complementation of cAMP phosphodiesterase deficient *Saccharomyces cerevisiae. J Biol Chem.* 1993;268:12925–12932.
6. Gilbard JP, Rossi SR, Heyda KJ, Dartt DA. Stimulation of tear secretion by topical agents that increase cyclic nucleotide levels. *Invest Ophthalmol Vis Sci.* 1990;31:1381–1388.
7. Spoto G, Whitehead E, Ferraro A, Di Terrlizzi P, Turano C, Riva F. A reverse-phase HPLC method for cAMP phosphodiesterase activity. *Anal Biochem.* 1991;196:207–210.
8. Hann LE, Tatro JB, Sullivan DA. Morphology and function of lacrimal gland acinar cells in primary culture. *Invest Ophthalmol Vis Sci.* 1989;30:145–158.
9. Dartt DA. Regulation of inositol phosphates, calcium and protein kinase C in the lacrimal gland. *Prog Retinal Eye Res.* 1994;13:443–478.
10. Fullard RJ, Tucker DL. Changes in human tear protein levels with progressively increasing stimulus. *Invest Ophthalmol Vis Sci.* 1991;32:2290–2301.

22

IMMUNOHISTOCHEMISTRY AND SECRETORY EFFECTS OF LEUCINE ENKEPHALIN IN THE ISOLATED PIG LACRIMAL GLAND

Jaipaul Singh,[1] Peter K. Djali,[1] and Ernest Adeghate[2]

[1]Department of Applied Biology
University of Central Lancashire
Preston, England, United Kingdom
[2]Department of Human Anatomy
United Arab Emirates University
Al-Ain, United Arab Emirates

1. INTRODUCTION

The innervation of the lacrimal gland is provided mainly by the parasympathetic nerves and to a lesser extent by the sympathetic nerves of the autonomic nervous system.[1–3] The sites of nerve termination vary from species to species, but in general they seem to terminate near blood vessels and basement membrane of secretory acini and ducts. Stimulation of parasympathetic nerves resulted in the release of acetylcholine, which in turn activates cholinergic muscarinic receptors. This leads to the metabolism of phosphatidylinositol 1,4- bisphosphate, resulting in cellular calcium mobilization.[2,3] On the other hand, stimulation of adrenergic nerves results in the release of noradrenaline, which activates both α– and β–adrenergic receptors, resulting in the mobilization of Ca^{2+} and the metabolism of adenosine 3,5-cyclic monophosphate, respectively.[2,3] In addition to parasympathetic and sympathetic nerves, there is much evidence for the presence of non-cholinergic, non-adrenergic nerves that contain a number of neuropeptides, some of which can elicit lacrimal protein secretion.[2–4] However, the distribution and secretory effects of enkephalin-related peptides in the lacrimal gland of the pig is less understood despite its close morphological relation to man. This study investigates the localization of leucine enkephalin (LEU-ENK) and its involvement with lacrimal protein secretion.

2. METHODS

All experiments were performed on isolated pig lacrimal glands obtained from a local abattoir. The glands were removed immediately after killing adult male and female pigs, placed in an ice-cold oxygenated Krebs-Henseleit (K-H) solution, and transported to the laboratory within 15–20 min. Segments of lacrimal glands were cut into small pieces and fixed overnight in paraformaldehyde solution containing picric acid. After fixation, the specimens were cut into 40–50 μm thick sections using vibroslice equipment (Chapden Instruments, USA). The sections were incubated in 3% H_2O_2 in methanol, washed three times in PBS, pH 7.4, and incubated for 1 h in normal goat serum. They were later incubated in polyclonal antiserum against LEU-ENK for 24 h at 4°C at a dilution of 1:500. This step was followed by incubation of the sections in biotinylated anti-rabbit immunoglobulin for 1 h at room temperature. After washing in PBS, the specimens were treated for 1 h with avidin-biotin peroxidase complex. Sites of immunoreaction were detected by incubating the sections in 5 mg diaminobenzidine in 10 ml Tris buffer (pH 7.8) containing 120 μl of 0.3% H_2O_2 for 5 min. The sections were washed in PBS, dehydrated in ascending concentrations of ethanol, cleared in xylene, and mounted in De-Pex (Serva, Heidelberg, Germany). Sections incubated with primary antiserum-inactivated excess antigen (100 μg ml^{-1} of diluted antiserum) served as controls.

The lacrimal glands from 20 animals were cut into small segments (5–10 mg), and a total weight of about 250–350 mg was placed into a Perspex flow chamber (2 ml volume) and continuously superfused at a constant rate of 1 ml min^{-1} with a K-H solution comprising (mM): NaCl, 118; KCl, 3.7; $CaCl_2$, 2.56; $NaHCO_3$, 25; KH_2PO_4, 2.2; $MgCl_2$, 1.2; and glucose, 10. The solution was kept at pH 7.4 and 37°C while being continuously gassed with a mixture of 95% O_2, 5% CO_2. Total protein output in effluent samples was measured by an automated on-line colorimetric method.[5] During stimulation, known concentrations of pharmacological agents [LEU-ENK (10^{-12}–10^{-7} M), atropine (10^{-5} M), phentolamine (10^{-5} M), propranolol (10^{-5} M), naloxone (10^{-5} M), and theophylline (10^{-3} M)] were added to the perfusing medium. Bovine serum albumin (BSA) was used as standard. All values were expressed as micrograms of protein per milliliter effluent per 100 mg tissue above basel level. Data are expressed as mean ± standard error of the mean (SEM). Data were compared using Student's t-test, and only values with $p<0.05$ were accepted as significant. All reagents were obtained from Sigma (UK).

3. RESULTS

Fig. 1 demonstrates the presence and distribution of LEU-ENK-immunoreactive neurons in the interlobular region and immunoreactive nerves in the interacinar areas in sections of the pig lacrimal gland. These results are in agreement with a previous study,[6] which demonstrated that LEU-ENK-immunopositive intrinsic nerves are located in the interacinar and interlobular areas, especially around interacinar areas, indicating a physiological role for this neuropeptide in the control of lacrimal function. Thus, it was relevant to ascertain whether LEU-ENK could evoke lacrimal protein secretion.

Mean (± SEM) basal protein secretion in this series of experiments was 16.36 ± 1.24 μg ml^{-1} (100 mg $tissue)^{-1}$ (n=45). Superfusion of isolated pig lacrimal segments with different concentrations (10^{-12}–10^{-7} M) of LEU-ENK resulted in marked increases in total protein output (Fig. 2A). Even at 10^{-12} M, LEU-ENK had a marked secretagogue effect. In an attempt to find out more about the nature of the LEU-ENK-evoked protein secretion, it was relevant to combine it with different pharmacological agents. The results are shown in Fig. 2B. Combining 10^{-5} M atropine with 10^{-9} M LEU-ENK had no significant effect on the response compared to LEU-ENK alone. In contrast, combining phentolamine and propranolol (all 10^{-5} M) with LEU-ENK resulted in a significant ($p<0.05$) reduction in LEU-

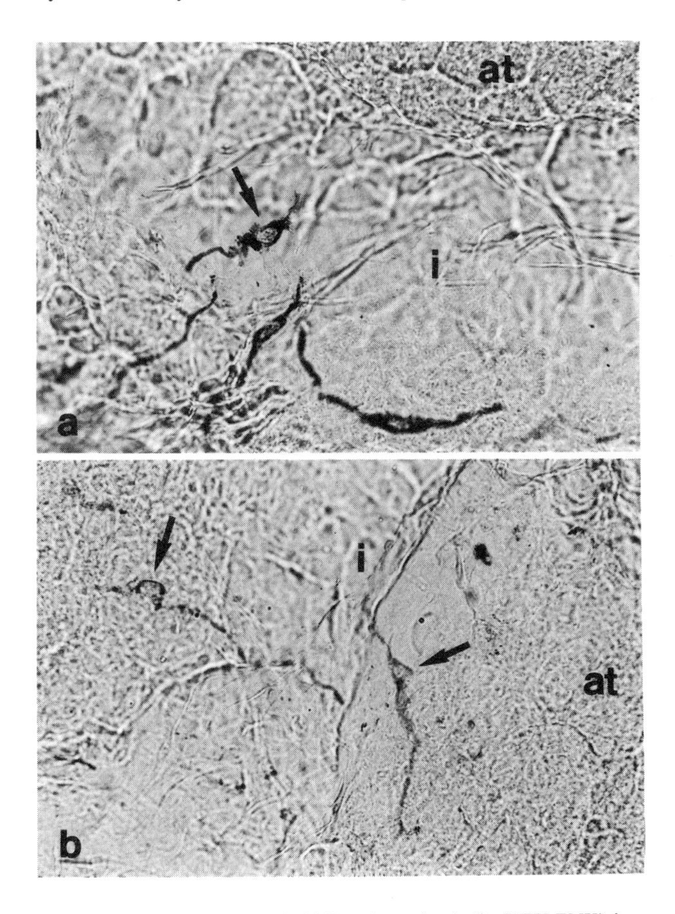

Figure 1. Light micrographs of pig lacrimal gland. (a) Leucine-enkephalin (LEU-ENK)-immunoreactive neurons (arrows) in the interlobular region. (b) LEU-ENK-immunopositive nerves in the interacinar areas at, acinar tissue. i, interlobular space. X312. These micrographs are typical of six such glands.

ENK-evoked protein output. Similarly, naloxone (10^{-5} M) alone had no detectable effect on protein output, but when combined with LEU-ENK, it resulted in a significant ($p<0.05$) decrease in LEU-ENK-induced protein secretion. Superfusion of lacrimal segments with 10^{-3} M theophylline caused a marked increase in protein output. However, when theophylline was combined with LEU-ENK, there was a significant ($p<0.005$) potentiation of the secretory response.

4. DISCUSSION

The results of this study demonstrate that LEU-ENK is well distributed in the nerves of the pig lacrimal gland. These LEU-ENK-immunoreactive nerves were located in the interlobular and interacinar areas, indicating that the neuropeptide may be an important neuromodulator of lacrimal secretion. Similar immunohistochemical studies have identified LEU-ENK-immunopositive nerve fibers near the basal surface of lacrimal acinar cells of a number of species, including the pig,[6] rat,[7] and guinea pig.[8] Exogenous application of

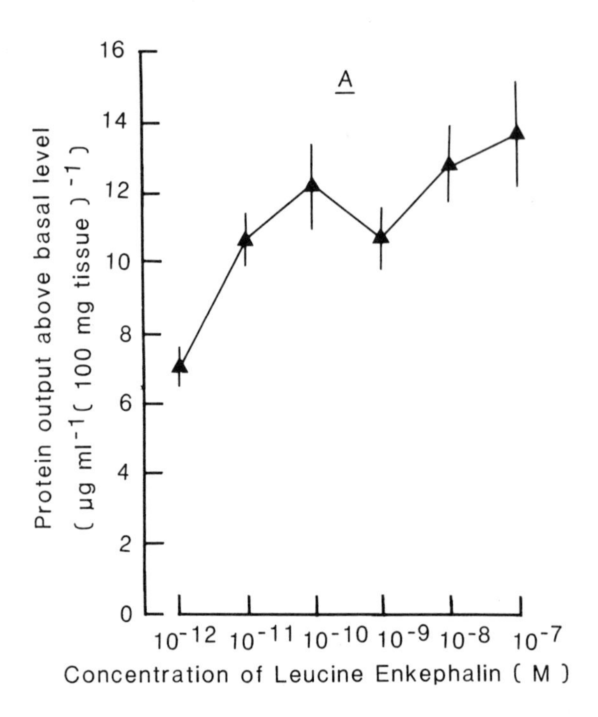

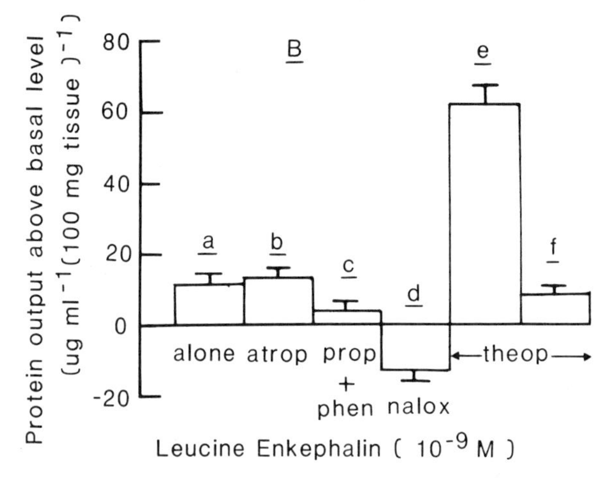

Figure 2. (A) Dose-response effect of leucine enkephalin (LEU-ENK) on total protein output from pig lacrimal gland segments. (B) Effect of 10^{-9} M LEU-ENK in the absence (a) and the presence of (b) 10^{-5} M atropine, (c) 10^{-5} M propranolol (prop) and 10–5M phentolamine (phen), (d) 10^{-5} M naloxone, and (e) 10^{-3} M theophylline (theop). The effect of theop alone is shown for comparison (f). Values are mean ± SEM (n = 3–20).

different concentrations of LEU-ENK resulted in marked increases in total protein output from pig lacrimal gland segment. A clear response was obtained at a physiological concentration of 10^{-12} M, indicating a secretory role for the neuropeptide in the control of pig lacrimal secretion. In contrast to LEU-ENK, the methionine enkephalin analog [D-ala^2] methionine-enkephalinamide has been shown to inhibit lacrimal protein output and adenylate cyclase activity, indicating that the enkephalins can have both stimulating and inhibitory secretory effects in the lacrimal gland.[9] Combining LEU-ENK with the endorphin antagonist naloxone in this study resulted in a significant inhibition of total protein output, whereas the cholinergic muscarinic antagonist atropine had no effect on the LEU-ENK-evoked secretory responses. Surprisingly, in the combined presence of the adrenergic receptor blockers propranolol and phentolamine, but not in the presence of either alone, the LEU-ENK-induced protein output was significantly reduced, but not abolished. These results indicate that LEU-ENK is activating opiate receptors, rather than cholinergic receptors, to elicit protein secretion. However, this effect of LEU-ENK may be associated in part with the adrenergic signal transduction mechanism, since the response is partially blocked by phentolamine and propranolol.

In this study, the phosphodiesterase inhibitor theophylline alone evoked a marked increase in total protain output, somewhat similar in magnitude to the response obtained with 10^{-9} M LEU-ENK alone. However, when theophylline was combined with LEU-ENK, there was a marked potentiation of the secretory response, indicating that LEU-ENK may be acting via a different signal transduction pathway to elicit protein secretion.[2,3] It is possible that LEU-ENK is utilizing cellular calcium to elicit protein secretion. In a previous study, it was shown that LEU-ENK can elicit Ca^{2+} uptake and release of magnesium from pig lacrimal gland.[10] Taken together, these results demonstrated the presence of LEU-ENK in the pig lacrimal gland. The neuropeptide, in turn, can stimulate protein secretion, possibly via the signal transduction pathway involving cellular Ca^{2+} mobilization.

ACKNOWLEDGMENTS

Supported by the Wellcome Trust and the British Council.

REFERENCES

1. Bromberg BB. Autonomic control of lacrimal gland protein secretion. *Invest Ophthalmol Vis Sci.* 1981;20:1110.
2. Dartt DA. Signal transduction and control of lacrimal protein secretion: A review. *Curr Eye Res.* 1989;8:619.
3. Dartt DA. Regulation of tear secretion. *Adv Exp Med Biol.* 1994;350:1.
4. Williams RM, Singh J, Sharkey KA. Innervation and mast cells of rat exorbital gland: Effect of age. *J Auton Nerv Syst.* 1994;47:95.
5. Adeghate E, Singh J, Burrows S, Howarth FC, Donath T. Secretory responses and aminergic and peptidergic innervation of rat lacrimal gland. *Biogenic Amines.* 1994;10:487.
6. Adeghate E, Singh J. Immunohistochemical identification of galanin and leu-enkephalin in porcine lacrimal gland. *Neuropeptides.* 1994;27:285.
7. Walcott B. Leucine-enkephalin-like immunoreactivity and the innervation of the rat exorbital lacrimal gland. ARVO Abstracts. *Invest Ophthalmol Vis Sci.* 1990;31:S44.
8. Lehtosalo J, Uusitalo H, Marberg T, Pannula P, Palkama A. Nerve fibres showing immunoreactivities for pro-enkephalin A-derived peptides in the lacrimal glands of the guinea-pig. *Graefes Arch Clin Exp Ophthalmol.* 1989;227:455.
9. Cripps MM, Patchen-Moor K. Inhibition of stimulated lacrimal secretion by [D-ala^{2}] met-enkephalinamide. *Am J Physiol.* 1989;257:G151A.
10. Djali PK, Singh J, Adeghate E. Immunohistochemical study of leucine enkephalin and its secretory effects in the isolated pig lacrimal gland. *Graefes Arch Clin Exp Ophthalmol.* 1996;234:264.

23

INTERACTION BETWEEN VASOACTIVE INTESTINAL POLYPEPTIDE (VIP) AND NEUROPEPTIDE Y (NPY) IN THE ISOLATED RAT LACRIMAL GLAND

Jaipaul Singh,[1] Ruth M. Williams,[1] Robert W. Lea,[1] and Ernest Adeghate[2]

[1]Department of Applied Biology
University of Central Lancashire
Preston, England, United Kingdom
[2]Department of Human Anatomy
Faculty of Health Sciences
United Arab Emirates University
Al-Ain, United Arab Emirates

1. INTRODUCTION

Lacrimal protein output and fluid secretion are regulated by the autonomic nerves that innervate the acini, duct, and blood vessels. The known putative neurotransmitters released by the autonomic nerves include acetylcholine (ACh), noradrenaline (NA), and vasoactive intestinal polypeptide (VIP).[1–3] Moreover, exogenous applications of ACh, NA, and VIP can elicit protein secretion from the rat lacrimal gland.[2,3] Recently, several studies have shown that neuropeptide Y (NPY) is widely distributed in the body and to co-exist with VIP in nerves.[4,5] However, the presence of NPY-immunoreactive nerves in the lacrimal gland, and more so its ability to stimulate protein secretion and to interact with VIP, has not yet been fully established. This study employs the technique of electrical field stimulation (FS) and exogenous agonists and antagonists to investigate the involvement of VIP and NPY in the control of lacrimal protein secretion. The technique of immunohistochemistry was used to demonstrate the distribution of the two neuropeptides in the lacrimal gland.

2. METHODS

All experiments were performed on the lacrimal gland from adult male and female Sprague-Dawley rats. Immediately after killing the animals, the glands were removed, placed in oxygenated Krebs-Henseleit (K-H) solution, and cut into small segments (5–10 mg). A total weight of about 200–250 mg was placed into a Perspex flow chamber (2 ml volume) and continuously superfused at a constant rate of 1 ml min^{-1} with a K-H solution comprising (mM): NaCl, 118; KCl, 3.7; $CaCl_2$, 2.56; $NaHCO_3$, 25; KH_2PO_4, 2.2; $MgCl_2$, 1.2; and glucose, 10. The solution was kept at pH 7.4 and 37°C while being gassed with a mixture of 95% O2, 5% CO_2. Total protein output in effluent samples was measured by an automated on-line colorimetric method.[5] Following a stabilization period of 40 min, the tissue was stimulated electrically (amplitude, 50 V; pulse duration, 1 msec; frequency, 20 Hz) by silver wire electrodes embedded on both sides of the perfusion chamber. In some experiments the K-H solution contained atropine, phentolamine, or propranolol (all 10^{-5} M) in the absence and presence of either the nerve-blocking drug tetrodotoxin (TTX, 10^{-6} M) or the VIP receptor antagonist ([4-Cl-D-Phe^6-Leu^{17}] VIP, 10^{-6} M). During agonist stimulation, known concentrations of either VIP, NPY, a combination of NPY and VIP, or the VIP receptor antagonist were added to the perfusing medium. Bovine serum albumin (BSA) was used as standard. All values are expressed as micrograms of protein per milliliter effluent per 100 mg tissue above basal level.

Adult Sprague-Dawley rats were perfused transcardially with Zamboni's fixative. The lacrimal glands were removed and immersion fixed for 6 h in phosphate- buffered (pH 7.4) paraformaldehyde (4%)-glutaraldehyde (0.5%) solution containing 0.2% vol/vol picric acid. The specimens were cut into 20–40 μm slices using vibroslice equipment (Chapden Instruments, USA). The sections were later treated with 3% H_2O_2 solution for 5 min. They were then incubated in normal goat serum (NGS) for 1 h after washing in PBS. This was followed by incubation of the sections with either rabbit anti-VIP or anti-NPY (all 1:1000 dilution) sera for 24 h at 4°C. After several washings in PBS, the sections were incubated for 1 h at room temperature in biotinylated anti-rabbit immunoglobulin and later in avidin-biotin-peroxidase complex. Sites of immunoreactions were detected using diaminobenzidine (DAB) solution. In all rinses, PBS solution containing 0.5% Triton was used. After staining, the sections were dried, dehydrated, and mounted on De-Pe-X. Sections incubated with primary antiserum inactivated by the addition of the antigen in excess (100 μg ml^{-1} of diluted antiserum) served as controls. VIP and NPY were obtained from Sigma (UK) and rabbit anti VIP, NPY, biotinylated anti-rabbit immunoglobulin, and avidin-biotin-peroxidase complex from Amersham (UK). NGS was purchased from Humanolto (Godollo, Hungary) and De-Pe-X from Serva (Heidelberg, Germany). Data are expressed as mean ± standard error of the mean (SEM). Data were compared by Student's t-test, and only values with $p<0.05$ were accepted as significant.

3. RESULTS

The mean ± SEM basal protein output from superfused rat segments in this series of experiments was 20.6 ± 3.2 mg ml^{-1} (100 mg $tissue)^{-1}$, $n = 75$. Fig. 1 shows histograms of mean (± SEM) total protein output above basal level. The tissue was pretreated with either TTX, VIP receptor antagonist, or atropine, phentolamine, and propranolol 10 min prior to stimulation. The duration of each stimulatory parameter was 6 min. The results demonstrate that FS of intrinsic secretomotor nerves can cause a marked increase in lacrimal protein output. The response was reduced but not abolished by the combined presence of cholinergic and adrenergic antagonists. In addition, when the VIP receptor antagonist was present, the FS-evoked response was reduced further, and the nerve-blocking drug TTX abolished the FS-induced protein secretion. The results also show that both VIP and NPY

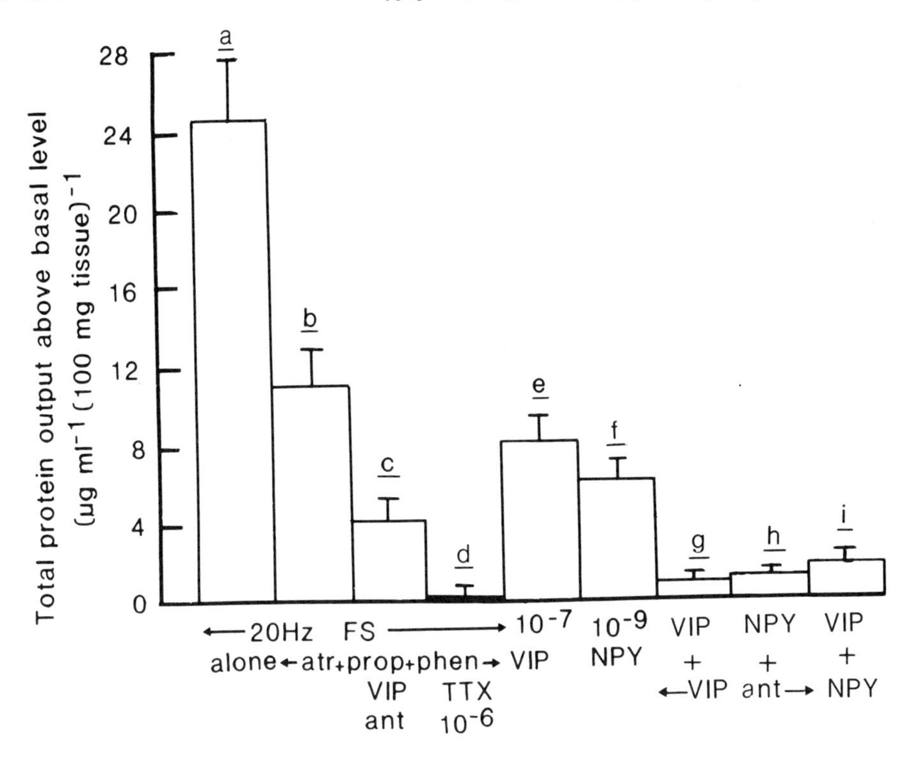

Figure 1. Histograms showing total protein output from lacrimal segments during (a) electrical field stimulation (FS) alone, (b) FS in the combined presence of atropine (atr), phentolamine (phen), and propranolol (prop) (all 10^{-5} M), and in the additional presence of either (c) the VIP receptor antagonist 10^{-6} M, or (d) 10^{-6} M TTX. The effect of exogenous application of either (e) 10^{-7} M VIP or (f) 10^{-9} M NPY, or a combination of either (g) VIP or (h) NPY with the VIP receptor antagonist, or (i) the combined presence of VIP and NPY are shown for comparison. All values are mean ± SEM (n=6–8).

have clear secretagogue effects on the rat lacrimal gland. These responses were blocked by the VIP receptor antagonist. Moreover, the two neuropeptides seemed to antagonize the effect of each other.

Since VIP and NPY can elicit lacrimal protein output, it was relevant to ascertain whether they are distributed in the lacrimal gland. VIP and NPY were identified in the nerves supplying the rat lacrimal gland. VIP-immunoreactive nerve fibers were observed mainly in the wall of blood capillaries of the gland. They followed the course of the blood capillaries found on the basal surfaces of the lacrimal acini from where they branch off to end up as parenchymal nerves in the basolateral wall of the acinar cells (Fig. 2a). The number of VIP-positive nerves was relatively many. NPY-immunoreactive nerves were discernible in the basal surfaces of the lacrimal acini and on the wall of the lacrimal ducts (Fig. 2b). The topographical relationship of these VIP- and NPY-immunoreactive nerves and the lacrimal acini was very close.

4. DISCUSSION

The technique of electrical field stimulation (FS) is a useful physiological tool to investigate the nervous control of secretory responses in exocrine glands.[6] In this study, FS

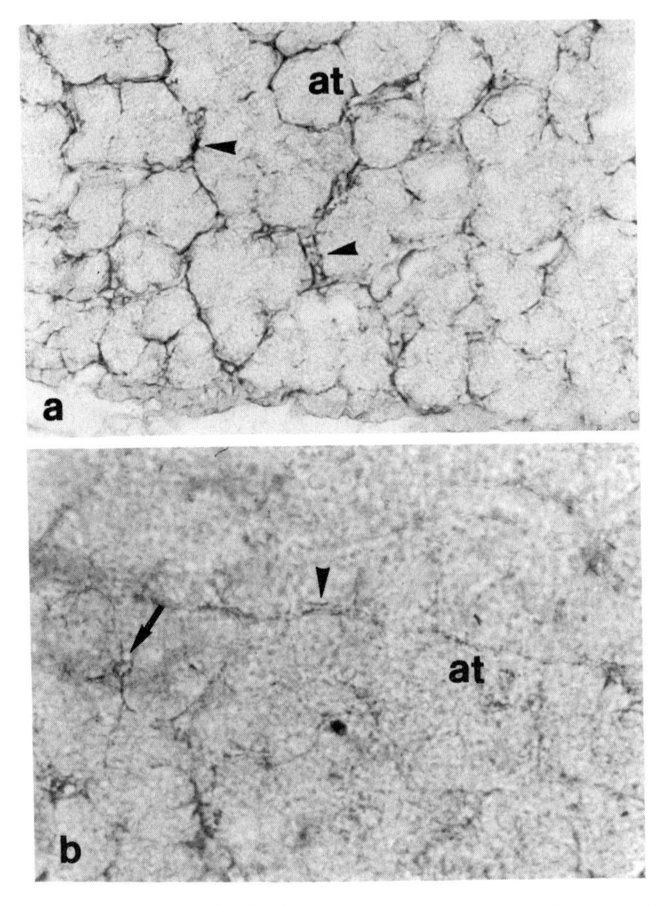

Figure 2. Light micrographs of the rat lacrimal gland. (a) VIP-positive fibers (arrow) in the wall of blood capillary (asterisk) and fine nerve fibers (arrowheads) around the basal surface of the acini (ac). (b) NPY-immunopositive nerves (arrows) around secretory acini (ac). D, lacrimal duct. Bar = 50 μm. The micrographs are typical of five such glands.

of intrinsic secretomotor nerves of lacrimal segments resulted in a large increase in total protein output. The FS-evoked response was reduced but not abolished by the combined presence of cholinergic and adrenergic receptor antagonists, indicating a non-cholinergic, non-adrenergic (NC/NA) component of the secretory response. The FS-induced protein output was further reduced in the presence of the VIP receptor antagonist ([4-Cl-D-Phe^{16}-Leu^{17}]-VIP) and totally abolished by the nerve-blocking drug tetrodotoxin. Taken together, the results indicate that NC/NA nerves are present in the rat lacrimal gland. FS of these nerves caused the release of an endogenous neurotransmitter(s), which in turn stimulated the acinar cells to secrete protein. The ability of the VIP receptor antagonist to further reduce the FS-evoked NC/NA response suggests that VIP may be one of the putative neurotransmitter substances released during FS. This observation is further corroborated by the finding that exogenous application of VIP can result in protein secretion and the response was blocked by the VIP receptor antagonist. Previous studies have also shown that VIP can stimulate lacrimal protein output from the rat lacrimal gland[7,8] and response is mediated via intracellular cyclic AMP.[2,3,8]

This study has also demonstrated that the secretory response to FS was not totally abolished by the combined presence of cholinergic, adrenergic, and VIPergic antagonists, suggesting that in addition to VIP, other NC/NA neurotransmitter(s) or endogenous mediator(s) may be released during FS and can evoke protein secretion. Moreover, exogenous NPY can also stimulate total protein output. It is possible that NPY is released in conjunction with VIP during FS. It has also been observed in this study that when NPY was combined with either VIP or the VIP receptor antagonist, there was a marked attenuation in protein secretion compared to the larger response obtained with either NPY or VIP alone. This observation indicates an interaction between VIP and NPY during the stimulus-secretion coupling mechanism. Ligand-binding studies in our laboratory have shown that the interaction occurs at receptor level in the lacrimal acinar cell (data not shown). Since both VIP and NPY can stimulate protein secretion, it was relevant to ascertain whether the neuropeptides are distributed in the lacrimal gland of the rat. The results have demonstrated the presence of both VIP- and NPY-immunoreactive nerve fibers throughout the lacrimal gland, especially in the basal surface of the lacrimal acini and in the wall of blood capillaries. In conclusion, this study has shown that lacrimal protein output is regulated by NC/NA nerves that are distributed in the lacrimal gland, and that the putative NC/NA neurotransmitters seem to be VIP and NPY.

ACKNOWLEDGMENTS

This work was supported by the Wellcome Trust and the British Council.

REFERENCES

1. Bromberg BB. Autonomic control of lacrimal protein secretion. *Invest Ophthalmol Vis Res*. 1981;20:110.
2. Dartt DA. Signal transduction and control of lacrimal gland proteins: A review. *Curr Eye Res*. 1989;8:619.
3. Dartt DA. Regulation of tear secretion. *Adv Exp Med Biol*. 1994;350:1.
4. Grunditz T, Eknian R, Hankainson R, Sundler F, Uddman R. Neuropeptide Y and vasoactive intestinal polypeptide co-exist in the rat nerve fibres emanating from the thyroid ganglion. *Regul Pept*. 1988;23: 193.
5. Adeghate E, Singh J, Burrows S, Howarth FC, Donath T. Secretory responses and aminergic and peptidergic innervation of rat lacrimal gland. *Biogenic Amines*. 1984;10:487.
6. Pearson GT, Singh J, Petersen OH. Adrenergic nervous control of cyclic AMP-mediated amylase secretion in the rat pancreas. *Am J Physiol*. 1984;246:G563.
7. Hussain M, Singh J. Is VIP the putative non-cholinergic non-adrenergic neurotransmitter controlling protein secretion in the rat lacrimal gland? *Q J Exp Physiol*. 1988;73:135.
8. Dartt DA, Barker AK, Valliant C, Rose P.E. Vasoactive intestinal polypeptide stimulation of protein secretion from rat lacrimal acini. *Am J Physiol*. 1984;247:G502.

24

IDENTIFICATION AND CELLULAR LOCALIZATION OF THE COMPONENTS OF THE VIP SIGNALING PATHWAY IN THE LACRIMAL GLAND

Robin R. Hodges, Driss Zoukhri, Jessica P. Lightman, and Darlene A. Dartt

Schepens Eye Research Institute
and Department of Ophthalmology
Harvard Medical School
Boston, Massachusetts

1. INTRODUCTION

Vasoactive intestinal peptide (VIP) is a 28 amino acid regulatory peptide that is found in parasympathetic nerves of most exocrine glands. VIP binds to its receptor on the basolateral membrane, which is coupled by guanine nucleotide-binding proteins (G proteins) to adenylyl cyclase (AC). Two distinct VIP receptors (types I and II)[1,2] and at least nine types of adenylyl cyclases (I-IX) have been identified.[3] Multiple forms of the three G protein subunits (α, β, and γ) also exist, making for a large number of possible combinations. ACs are large, complex structures that are multiply regulated by the subunits of various G proteins, protein kinase C (PKC), and Ca^{2+}.

VIP stimulates the exchange of GDP for GTP on the α subunit of the G_s protein. After GTP is bound, the $\beta\gamma$ subunits of the complex dissociate, leaving a free α subunit. Activation of AC by $G_{s\alpha}$ generates the second messenger cyclic AMP (cAMP) from ATP. cAMP binds to protein kinase A (PKA), which then phosphorylates specific protein substrates leading to exocytosis.

There is substantial evidence for reciprocal regulation between the cAMP and the phospholipase C signaling pathways. PKC and Ca^{2+}, both part of the phospholipase C pathway, have been shown to activate or inhibit various isoforms of AC.[3] In the pancreas and submandibular gland, cAMP can inhibit agonist-induced Ca^{2+} signals, and generation of cAMP in the liver can produce Ca^{2+} transients or increase inositol trisphosphate (IP_3) mobilization of Ca^{2+}.[4] This can occur either by increasing the amount of Ca^{2+} released or by increasing the sensitivity of the IP_3 receptor to IP_3.

The lacrimal gland is composed of a branched acinar system.[5] Acinar and myoepithelial cells are the main cell types of the secretory unit. In other exocrine glands, it is believed that they regulate the movement of neurohormones from the blood supply to the acini. Because they contain a network of α-smooth muscle actin, it is also believed that they are involved in contraction to eject the secretory product from the acini.

Previously, we have shown that VIP-like immunoreactivity is present throughout the rat lacrimal gland and distributed primarily around the acini.[6] We have also shown that VIP causes a dose-dependent increase in cAMP,[7] activates Ca^{2+}-dependent K^+ channels which play an essential role in fluid secretion,[8] and is a potent stimulator of lacrimal gland protein secretion.[6,9] Meneray and Bennett have shown that function-blocking antibodies against $G_{s\alpha}$ decreased VIP-stimulated cAMP production while antibodies against $G_{o\alpha}$ stimulated cAMP production.[10]

In the present study, we show that all components required for VIP to stimulate protein secretion are present in the lacrimal gland. In addition, we examine the effect of VIP on intracellular $[Ca^{2+}]$ ($[Ca^{2+}]_i$).

2. MATERIALS AND METHODS

2.1. Materials

Male Wistar rats 125–150 g were obtained from Charles River Laboratories (Wilmington, MA). Anti-VIP receptors type I and II polyclonal antibodies and preabsorbed serum controls were generous gifts from Dr. Edward J. Goetzl (Department of Medicine, Microbiology and Immunology, University of California, San Francisco). Anti-adenylyl cyclase polyclonal antibodies were from Santa Cruz (Santa Cruz, CA), the monoclonal antibody against α-smooth muscle actin from Sigma (St. Louis, MO), anti-$G_{s\alpha}$ antibody from Calbiochem (La Jolla, CA), and alkaline phosphatase reagents from Promega (Madison, WI).

2.2. Methods

2.2.1. Immunohistochemistry. Lacrimal glands were fixed in 4% formaldehyde for 4 h at 4°C. After cryopreservation overnight in 30% sucrose at 4°C in PBS, containing in mM: 145 NaCl, 7.3 Na_2HPO_4, and 2.7 NaH_2PO_4, pH 7.2, the tissue was frozen in O.C.T. embedding medium. Cryostat sections (6 μm) were placed on gelatin-coated slides and air dried for 2 h.

Antibody to VIP receptor type I was raised against the amino terminus of the receptor. An antibody to VIP receptor II was raised against the first extracellular loop.[2] The secondary antibody was conjugated to FITC. The anti-mouse secondary antibody conjugated to rhodamine (for α-smooth muscle actin) was used in co-localization experiments. Negative controls were performed with either preabsorbed serum (anti-VIPR antibodies) or preincubation of the antibodies with their specific peptides (anti-AC antibodies). Sections were viewed using a Nikon UFX II microscope equipped for epi-illumination or a Leitz Leica TCS4D confocal scanning microscope equipped with a krypton-argon laser (Heidelberg, Germany).

2.2.2. Western Blots. The lacrimal glands were homogenized with a dounce homogenizer in 10 mM Tris-HCl, pH 7.5, containing 10% sucrose and 0.5 mM PMSF. After homogenization, the samples were centrifuged (8,000 x *g*, 5 min) at 4°C to remove unbroken cells and nuclei. The supernatant was then centrifuged (100,000 x *g*, 1 h) at 4°C. The pellet was solubilized in ice-cold buffer containing in mM: 20 Tris, pH 8.0, 1 DTT, 100 NaCl, 1 EDTA, and 1% sodium cholate for 1 h. Nonsolubilized membranes were removed by 13,000 x *g* centrifugation for 3 min. Proteins were separated by SDS-PAGE on a 12% acrylamide gel. Proteins were transferred to nitrocellulose membrane and blocked overnight at 4°C in 5% dried milk in buffer containing 10 mM Tris-HCl, pH 8.0, 150 mM NaCl, and 0.05% Tween-20. The membranes were incubated for 1 h at room temperature with the primary antibody (1 μg/ml). The secondary antibody was conjugated to alkaline phosphatase. Immunoreactivity was detected by incubation with the alkaline phosphatase reagents 5-bromo-1-chloro-3-indolyl phosphate/nitro blue tetrazolium. Controls were performed by incubating the blots in the absence of primary antibodies. Membranes from rat liver were used as positive controls.

2.2.3. Preparation of Rat Lacrimal Gland Acini. The lacrimal glands were minced and incubated in Krebs-Ringer bicarbonate buffer (in mM: 119 NaCl, 4.8 KCl, 1 $CaCl_2$, 1.2 $MgSO_4$, 1.2 KH_2PO_4, and 25 $NaHCO_3$) supplemented with 10 mM Hepes, 5.5 mM glucose (KRB-Hepes), and 0.5% BSA, pH 7.4, containing collagenase (150 U/ml). Lacrimal gland lobules were subjected to gentle pipetting, ten times at regular time intervals, through tips of decreasing diameter. The preparation was then filtered through nylon mesh (150 μm pore size), and the acini pelleted with 2 min centrifugation at 50 x *g*. The pellet was washed twice by centrifugation (50 x *g*, 2 min) through a 4% BSA solution made in KRB-Hepes buffer. The dispersed acini were allowed to recover for 30 min in 5 ml fresh KRB-Hepes buffer containing 0.5% BSA.

2.2.4. Measurement of $[Ca^{2+}]_i$. Acini were incubated in KRB-Hepes buffer containing 0.5% BSA, 0.5 μM fura-2 tetra-acetoxymethyl ester, 10% Pluronic F127, and 250 μM sulfinpyrazone for 60 min at 22°C. In experiments performed without extracellular Ca^{2+}, fura-2 loaded acini were washed twice and assayed in KRB-Hepes buffer without any added Ca^{2+} and containing 2 mM EGTA. Fluorescence was measured at excitation wavelengths of 340 and 380 nm and emission wavelength of 505 nm. To calculate $[Ca^{2+}]_i$, 5.6 mM EGTA, 7.5 mM Tris-HCl, pH 7.5, and 1% Triton X-100 were added at the end of the reaction to obtain minimum fluorescence. Maximum fluorescence was determined by the addition of 14.5 mM $CaCl_2$. The dissociation constant of 135 nM for fura-2 at 22°C was used to calculate $[Ca^{2+}]_i$.

3. RESULTS

3.1. Location of VIP-Containing Nerves

Because VIP is a potent stimulator of lacrimal gland protein secretion[6,9] and activates Ca^{2+}-dependent K^+ channels, we investigated whether the individual components of the cAMP-dependent signal transduction pathway were present in the lacrimal gland. Using confocal microscopy, we stacked 12 1-μm images to illustrate the extent of VIP innervation of the lacrimal gland acini. Confocal microscopy demonstrates that VIP-containing nerves surround the basolateral membranes of the majority of acini (Fig. 1). A few fine varicose fibers also appear to penetrate the lateral spaces between individual acinar cells in an acinus.

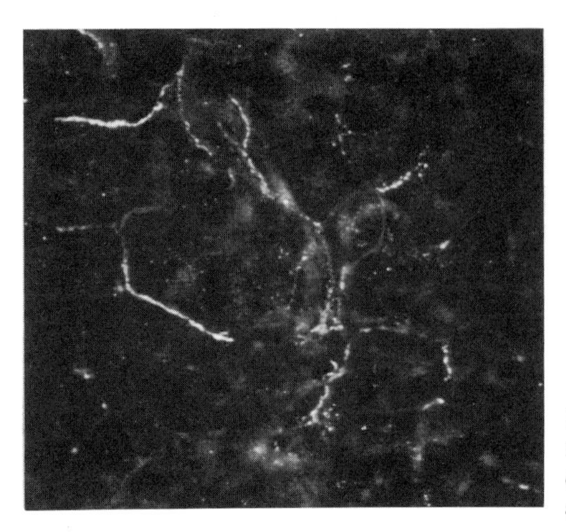

Figure 1. Confocal micrograph of 12 1-μm images stacked together showing VIP-containing nerves. VIP-containing nerves were seen in the basolateral spaces around acinar and myoepithelial cells.

3.2. Location of VIP Receptors

Next, we investigated the identity and localization of VIP receptors. Lacrimal gland sections were incubated with antibodies raised against both types of known VIP receptors. A confocal micrograph representing 12 1-μm sections stacked together shows that VIPR I is present on the basolateral membranes of some acinar cells (Fig. 2A). Immunoreactivity was also occasionally seen around lobes and on the basolateral membranes of ducts (Fig. 2B). In double-labelling experiments using this antibody with an antibody against α-smooth muscle actin, a marker for myoepithelial cells, VIPR I, did not appear to be localized on myoepithelial cells.[7]

An antibody to the first extracellular loop of VIPR II also showed the presence of VIPR II on the basolateral membranes of some acinar cells (Fig. 2C). Immunoreactivity was also seen on the basolateral membranes of ducts and blood vessels and occasionally on myoepithelial cells.[7]

Use of preabsorbed serum from each antibody substantially reduced the fluorescence. These results indicate that both types of VIP receptors are present in the acinar and ductal cells of the lacrimal gland with similar distributions, but only VIPR II was seen on myoepithelial cells.

3.3. Presence of G Proteins

To determine if the G protein, $G_{s\alpha}$, was present in the lacrimal gland, proteins in the membrane fractions were separated by SDS-PAGE and subjected to Western blotting. Membranes from liver were used as positive controls. Two major bands with molecular weights of 47 and 42 kDa corresponding to the large and small forms of $G_{s\alpha}$ were detected in both lacrimal gland and liver (Fig. 3).

3.4. Localization of Adenylyl Cyclases

Four antibodies to five isoforms of adenylyl cyclase: II, III, IV, and V/VI, were used for immunofluorescence studies. The antibody directed against AC II appeared to bind exclusively to the myoepithelial cells and did not appear to label the acini themselves. Double-labelling experiments with the anti-AC II antibody and the antibody raised against

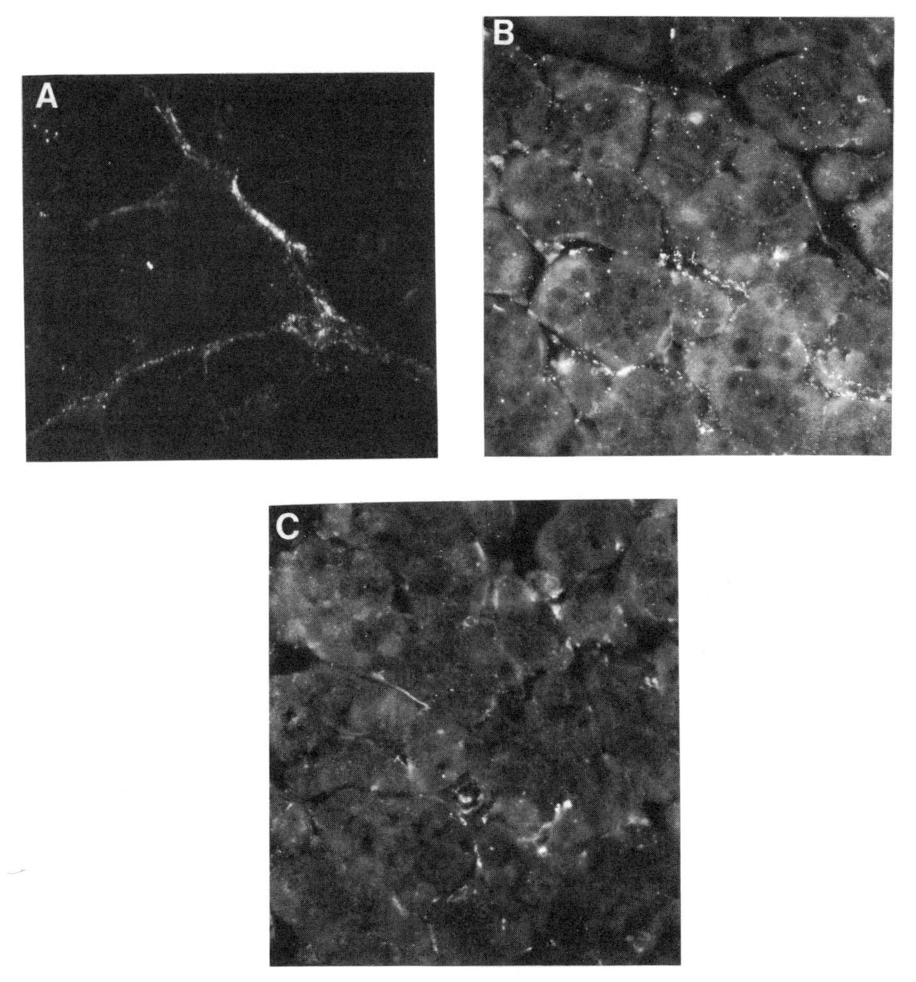

Figure 2. (A) Confocal micrograph of 12 1-μm images stacked together showing localization of VIPR I. (B) Conventional micrograph showing VIPR I. (C) Conventional micrograph showing VIPR II. X 430.

α-smooth muscle actin confirmed this finding. AC II immunoreactivity was present on most myoepithelial cells that surrounded the acini, ducts, and blood vessels.[7] Thus, AC II appears to be on the myoepithelial cells but not on acinar, ductal, or vascular cells. The antibody directed against AC III showed limited binding in the lacrimal gland with an occasional duct, blood vessel, or myoepithelial cell showing binding. Structures in the extracellular space, possibly nerves, were also labelled.[7]

An antibody directed against AC IV showed binding on the intracellular membranes, possibly Golgi apparatus or endoplasmic reticulum, of all acini. It also appeared to be present in ductal or vascular cells.[7] Use of preabsorbed peptide controls reduced fluorescence substantially. No immunoreactivity was seen in the lacrimal gland using an antibody directed against AC V/VI. These results suggest that at least three isoforms of adenylyl cyclase, each with a different distribution, are present in the lacrimal gland.

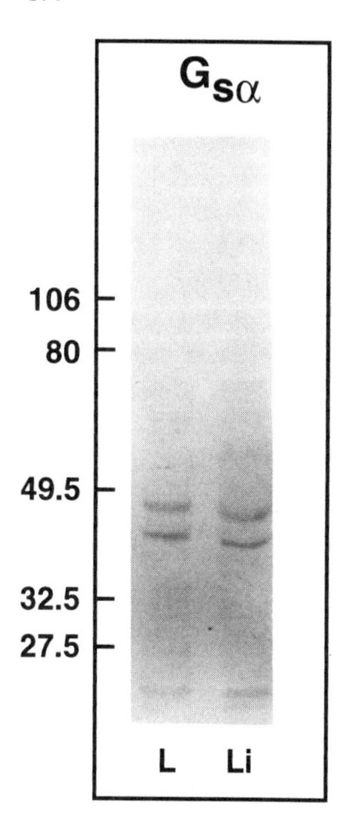

Figure 3. Membranes from lacrimal gland and liver were separated by SDS-PAGE, transferred to nitrocellulose, and probed with anti-$G_{s\alpha}$ antibody. Similiar results were obtained in at least two other experiments. L, lacrimal gland. Li, liver.

3.5. Effect of VIP on $[Ca^{2+}]_i$

In some tissues, cAMP can increase $[Ca^{2+}]_i$.[4] Thus, we tested the effect of VIP on $[Ca^{2+}]_i$ in fura-2 loaded acini. VIP (10^{-8} M) increased the peak levels of $[Ca^{2+}]_i$ over basal by 26 ± 9 nM. The increase in Ca^{2+} was dependent upon the concentration of VIP with a maximum increase at 10^{-8} M.[7] This increase was small, but statistically significant. In comparison, in acini from the same animal, the cholinergic agonist carbachol (10^{-5} M), which is known to activate the phospholipase C signaling pathway and increase inositol trisphosphate and $[Ca^{2+}]_i$, significantly increased $[Ca^{2+}]_i$ over basal by 121 ± 14 nM ($n = 3$).

To determine if the rise in $[Ca^{2+}]_i$ was due to influx of Ca^{2+}, fura-2 loaded acini were assayed in buffer without added Ca^{2+} and with 2 mM EGTA. The change in $[Ca^{2+}]_i$ over basal was decreased 70% at 10^{-8} M (Fig. 4; n=3).

4. DISCUSSION

VIP, a potent stimulator of lacrimal gland peroxidase and newly synthesized protein secretion, was localized in nerves surrounding the basolateral membranes of most acini. This indicates that VIP can either interact directly with the cells to stimulate protein secretion or diffuse to acini that are not in close proximity to the nerves.

VIP receptor types I and II were present on the basolateral membranes of some acini and myoepithelial cells. Not all acini showed immunoreactivity with antibodies to either VIP or VIP receptors, suggesting that only a subpopulation of acini can respond directly to VIP. Lemullois et al. also found a similar heterogeneity among lacrimal gland acini using an antibody against M_3 muscarinic receptor.[11] Because VIP is a potent stimulator of protein secretion, it is possible that the number of secreting acini is greater than the actual number that appear to express the receptor for VIP.

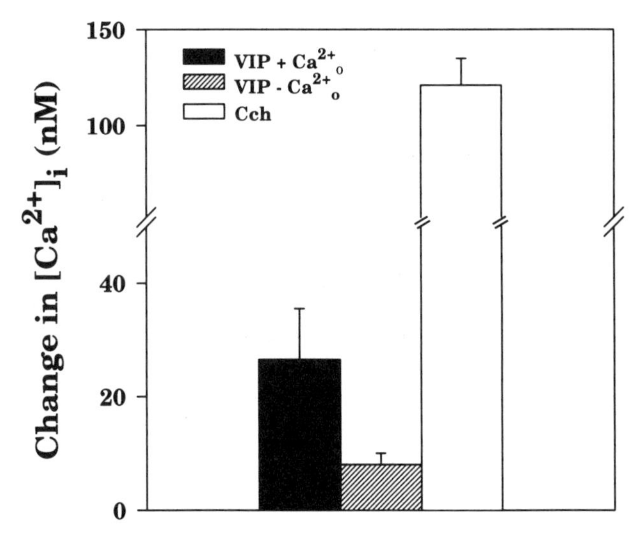

Figure 4. Effect of VIP (10^{-8} M) on intracellular $[Ca^{2+}]$ in the presence or absence of extracellular Ca^{2+} ($[Ca^{2+}]_o$).

Like acini, some myoepithelial cells showed immunoreactivity with the antibody to VIPR II. Lemullois et al. also found M_3 muscarinic receptors on myoepithelial cells.[11] This suggests that VIP could directly stimulate them, possibly to contract.

Use of immunofluorescence microscopy also showed the presence of three isoforms of adenylyl cyclases in the lacrimal gland. AC II was found almost exclusively in myoepithelial cells and seemed to outline the cells. Though the role of AC II on the myoepithelial cells remains to be determined, it is possible that it is involved in contraction to eject the secretory product from the acinar cells. AC III had only limited binding on ducts, blood vessels, and perhaps nerves. Although AC IV was found on acini, its binding appeared to be intracellular, perhaps on Golgi apparatus or endoplasmic reticulum. We have not yet been able to demonstrate the presence of AC V/VI in the lacrimal gland. Interestingly, no AC immunoreactivity was seen with any antibody on the basolateral membrane where it would be expected and where binding was seen with the antibodies against VIP and VIPRs I and II. It is possible that the AC necessary for signal transduction could be recruited from intracellular compartments upon stimulation with VIP. The studies described here were performed on unstimulated sections of lacrimal glands. Recruitment of AC IV from intracellular compartments might explain the short lag period (10 min) seen in VIP-

Table 1. Location of adenylyl cyclase isoforms in the lacrimal gland

Adenylyl cyclase isoforms	Acinar cells	Ductal cells	Myoepithelial cells
AC II	ND	ND	+++
AC III	ND	+	+
AC IV	+++	++	ND
AC V/VI	ND	ND	ND

Amount of fluorescence was graded on a scale of + (weakest) to +++ (strongest). ND, not detected.

stimulated protein secretion (newly synthesized and peroxidase).[6,9] It is also possible that another isoform of AC (I, VII, VIII, XI) might be present in the lacrimal gland and linked to the VIP receptors.

Stimulation of acini with VIP at 10^{-8} M caused a small increase in $[Ca^{2+}]_i$. Previously, we found that maximum peroxidase secretion also occurred at 10^{-8} M VIP.[6] Mauduit et al.[9] previously showed that VIP does not increase IP_3 levels; thus we tested the effect of VIP on Ca^{2+} influx by omitting extracellular Ca^{2+}. Omission of extracellular Ca^{2+} caused the VIP-induced Ca^{2+} response to decrease substantially. This implies that VIP is activating Ca^{2+} influx through plasma membrane channels. The cholinergic agonist carbachol, known to increase IP_3 and thus $[Ca^{2+}]_i$, caused a twofold increase in $[Ca^{2+}]_i$. Because the increase in $[Ca^{2+}]_i$ stimulated by VIP was small compared to that seen with carbachol, it seems unlikely that the increase is due to release of $[Ca^{2+}]_i$ by IP_3. This small increase in $[Ca^{2+}]_i$ stimulated by VIP might satisfy the requirement of Ca^{2+} for exocytosis.[12]

In some cell types, generation of cAMP can increase IP_3-induced mobilization of Ca^{2+} either by increasing the amount of Ca^{2+} released or by increasing the sensitivity of the IP_3 receptor to IP_3.[4] This did not occur in the lacrimal gland as VIP did not change the elevation of $[Ca^{2+}]_i$ in response to carbachol.

We conclude that the lacrimal gland contains all necessary components for the VIP signaling pathway: VIP-containing nerves, VIP receptor types I and II, the α subunit of G_s, and at least three types of AC (II, III, and IV). We also conclude that VIP affects Ca^{2+} influx.

ACKNOWLEDGMENT

Supported by National Eye Institute grant EY06177.

REFERENCES

1. Adamou JE, Aiyar N, Horn SV, Elshourbagy NA. Cloning and functional characterization of the human vasoactive intestinal peptide (VIP)-2 receptor. *Biochem Biophys Res Commun.* 1995;209:385–392.
2. Ichikawa S, Sreedharan SP, Owen RL, Goetzl EJ. Immunochemical localization of type I VIP receptor and NK-1-type substance P receptor in rat lung. *Am J Physiol.* 1995;268:L584-L588.
3. Sunahara RK, Dessauer CW, Gilman AG. Complexity and diversity of mammalian adenylyl cyclases. *Annu Rev Pharmacol Toxicol.* 1996;36:461–480.
4. Toescu EC. Temporal and spatial heterogeneities of Ca^{2+} signaling: Mechanisms and physiological roles. *Am J Physiol.* 1995;269:G173-G185.
5. Murakami M, Nagato T, Tanioka H. A scanning electron microscope study of myoepithelial cells in the intercalated ducts of rat parotid and exorbital lacrimal glands. *Okajimas Folia Anat Jpn.* 1990;67:309–314.
6. Dartt DA, Baker AK, Vaillant C, Rose PE. Vasoactive intestinal polypeptide stimulation of protein secretion from rat lacrimal gland acini. *Am J Physiol.* 1984;247:G502-G509.
7. Hodges RR, Zoukhri D, Sergheraert C, Zieske JD, Dartt DA. Identification of VIP receptor subtypes in the lacrimal gland and their signal transducing components. *Invest Ophthalmol Vis Sci.* 1997;38:610–619.
8. Lechleiter JD, Dartt DA, Brehm P. Vasoactive intestinal peptide activates Ca^{2+}-dependent K^+ channels through a cAMP pathway in mouse lacrimal cells. *Neuron.* 1988;1:227–235.
9. Mauduit P, Herman G, Rossignol B. Newly synthesized protein secretion in rat lacrimal gland: Post-second messenger synergism. *Am J Physiol.* 1987;253:C514-C524.
10. Meneray MA, Bennett DJ. Identification of G proteins in lacrimal gland. *Invest Ophthalmol Vis Sci.* 1995;36:1173–1180.
11. Lemullois M, Mauduit P, Rossignol B. Immunolocalization of myoepithelial cells in isolated acini of rat exorbital lacrimal gland: Cellular distribution of the muscarinic receptors. *Biol Cell.* 1996;86:175–181.
12. Augustine GJ, Burns ME, DeBello WM, Pettit DL, Schweizer FE. Exocytosis: Proteins and perturbations. In: Cho AK, Blaschke TF, Ho IK, Loh HH, eds. Palo Alto: Annual Reviews; 1996;36:659–701.

25

KINESIN ACTIVATION DRIVES THE RETRIEVAL OF SECRETORY MEMBRANES FOLLOWING SECRETION IN RABBIT LACRIMAL ACINAR CELLS

S. F. Hamm-Alvarez,[1,3] S. R. da Costa,[1] M. Sonee,[1] D. W. Warren,[2] and A. K. Mircheff[3]

[1]Department of Pharmaceutical Sciences
University of Southern California School of Pharmacy
Los Angeles, California
[2]Department of Cell and Neurobiology
[3]Department of Physiology and Biophysics
University of Southern California School of Medicine
Los Angeles, California

1. INTRODUCTION

Recent studies in the lacrimal gland have focused on how changes in membrane trafficking may contribute to the development of tear insufficiency and the development of autoimmunity in the lacrimal gland. We have focused on resolving the roles that microtubule-based vesicle transport and the individual motor proteins, kinesin and cytoplasmic dynein, play in facilitating membrane-trafficking events in lacrimal acini.

2. BACKGROUND

Microtubules (MTs) are polymeric assemblies of α- and β-tubulin that constitute a major component of the cytoskeleton.[1] The MT has an intrinsic biochemical polarity, and the distinct ends are referred to as plus- and minus-ends. Organization of these ends within cells can provide polarity. In addition to serving roles in cell shape and migration, interphase MTs provide a network that supports the movement of cellular membrane vesicles driven by MT-based motor proteins. Two classes of cytoplasmic motor proteins are known: kinesins and cytoplasmic dyneins.[1] Each of these motors uses nucleotide hydroly-

Lacrimal Gland, Tear Film, and Dry Eye Syndromes 2
edited by Sullivan *et al.*, Plenum Press, New York, 1998

sis to propel membrane vesicles along the MT path. Kinesin is an MT plus-end-directed motor; cytoplasmic dynein is an MT minus-end-directed motor. The membrane vesicle movements driven by these motor proteins have been implicated in numerous membrane-trafficking events in endocytosis, apical secretion and recycling, and traffic between intracellular membrane compartments.[2]

3. RESULTS AND DISCUSSION

3.1. Kinesin Activity from Stimulated Lacrimal Acini Is Higher Than from Controls

Our initial goal was to determine whether the activity of the kinesin motor was different in control versus stimulated (10 μM carbachol, 30 min, 37°C) lacrimal acini. The lacrimal acinar cells utilized are isolated from rabbit lacrimal gland and cultured as described[3]; they can reconstitute into acinar structures with distinct apical and basal-lateral domains. Kinesin was concentrated by MT-affinity purification of high-speed cell supernatant and released in 10 mM MgGTP. In the presence of MTs and GTP, kinesin adsorbed to the surface of a glass coverslip can promote the movement of MTs along the surface of the glass (MT gliding). Stimulated cells yielded a kinesin MT gliding activity that was 2.2X higher than the activity from control cells.[4] The finding that increased kinesin activity was correlated with stimulation of secretion following carbachol exposure implicated kinesin in regulated secretion.

3.2. Kinesin and the Secreted Protein, β-Hexosaminidase, Are Redistributed to Golgi Compartments from a Putative Secretory Compartment Following Stimulation

To establish a further link between kinesin and secretion, we examined changes in kinesin-membrane association following stimulation of secretion. Membranes from control and stimulated (10 μM carbachol, 30 min, 37°C) acini were separated over sorbitol density gradients. The membrane compartment composition of each fraction was determined by comparing the distributions of the biochemical markers marking each different membrane compartment. In resting cells, kinesin and the secreted protein, β-hexosaminidase, were enriched in the "Hex1" compartment in fraction 10 (Fig. 1). Stimulation caused significant decreases in the relative amounts of kinesin and β-hexosaminidase in the peak fraction (fraction 10) and the appearance of a new peak of kinesin in fraction 8. Additional β-hexosaminidase could be measured in the peak containing the recovered kinesin in fraction 8.

The observation that kinesin and β-hexosaminidase were lost from one membrane compartment and recovered in another suggested that kinesin might actively drive one or more steps in traffic between these compartments. However, this interpretation was complicated by the observation that the separation of membranes by sorbitol density gradient centrifugation was not complete. Significant overlap of kinesin and β-hexosaminidase with Golgi and endoplasmic reticulum membranes was observed in resting and stimulated cells.[4] Membranes were therefore subjected to a second separation by partitioning analysis into a dextran-polyethylene glycol two-phase system to resolve more accurately the individual membrane populations participating in the redistributions. This analysis, which included separations at different pHs, confirmed that kinesin and β-hexosaminidase from Hex1 were redistributed to a Golgi compartment.[4] This finding and the measured increase

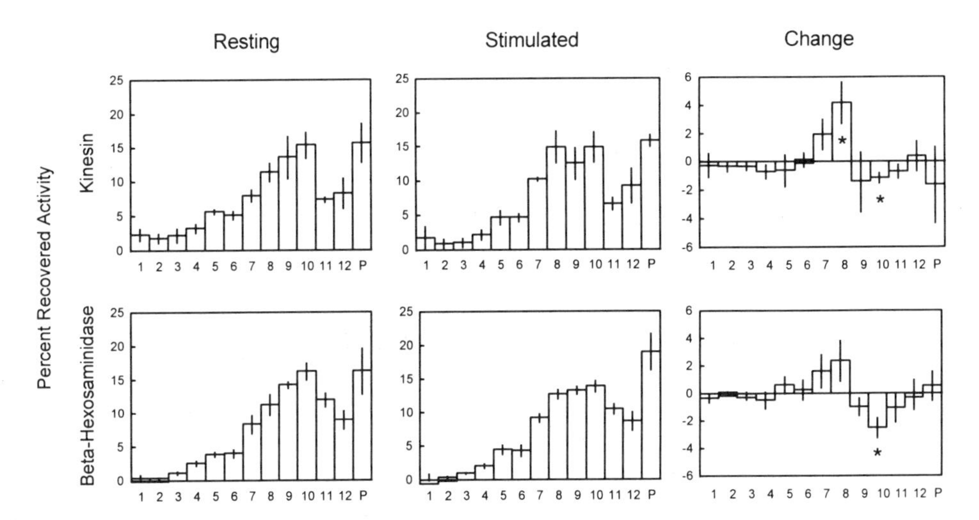

Figure 1. Carbachol-induced changes in density gradient distributions of kinesin and β-hexosaminidase. Acini were harvested, resuspended in culture medium with and without 10 μM carbachol for 30 min, washed, and lysed for fractionation analysis. Low-speed supernatants were applied to sorbitol density gradients. After centrifugation, membranes were sedimented from each density gradient fraction. Enzymatic activities in each fraction were calculated in standard units (nmoles/h). Values presented are percentages of the totals recovered in summed density gradient fractions. For β-hexosaminidase activity, $n = 6$ resting and $n = 6$ stimulated cell preparations. For kinesin signal, $n = 4$ resting and $n = 3$ stimulated cell preparations. Vertical lines at ends of bars represent SEM. *indicates $p \leq 0.05$. (Reproduced with permission from ref. 4.).

in activity of kinesin from stimulated acini are consistent with a role for kinesin in driving either the movement of vesicles from Hex1 to the apical plasma membrane or the retrieval of apical membranes containing adsorbed secretory proteins to the Golgi. However, an alternative explanation is that carbachol treatment altered the MT array. Such effects could result in redistribution of kinesin and β-hexosaminidase to Golgi membranes, through an indirect organizational mechanism not requiring kinesin-driven vesicle transport.

3.3. MT and Microfilament Distribution Were Not Altered during Stimulation of Secretion

To test this hypothesis, we used confocal microscopy to examine whether changes in organization, distribution, or density of MTs or microfilaments was observed following carbachol stimulation (10 μM or 1 mM carbachol, ≤ 30 min, 37°C). No disassembly of MTs or microfilaments was seen with stimulation at either dose at any time (unpublished data). These data were confirmed by measurement of the cytoskeletal and soluble tubulin and actin following exposure to carbachol using quantitative Western blot analysis.

3.4. Secretion of β-Hexosaminidase Is Partially Reduced by MT-Targeted Drugs

Recent preliminary studies have focused on effects of the MT-targeted drug taxol on stimulated secretion of β-hexosaminidase. Although significant stimulation ($p \leq 0.05$) is elicited by both 10 μM and 1 mM carbachol in taxol (10 μM, 60 min, 37°C)- treated cells,

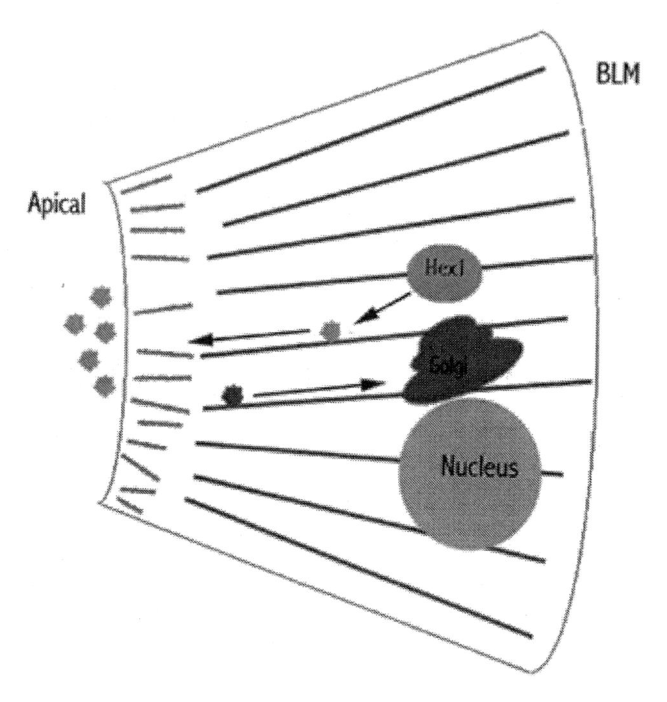

Figure 2. Model for kinesin-driven vesicle transport during stimulated secretion and recycling. (Adapted from ref. 4.)

the magnitude of the stimulation in the presence of 10 μM carbachol was reduced by taxol to approximately 60–70% of controls. These experiments suggest that release of secretory proteins may have an MT-dependent component.

We hypothesize that the MT-based motor, kinesin, is a target protein for stimulation of secretion. Additionally, we propose that kinesin drives the retrieval of secretory membrane from the apical surface, based on studies of MT polarity in other polar epithelia that are proposed to have MT plus-ends organized toward the basal-lateral surface. Since MT polarity is unknown in lacrimal acini and may have MT plus-ends to the basal-lateral surface, it is possible that kinesin drives the movement of secretory membranes to the apical surface. The measurement of kinesin activation and the redistribution of kinesin and secreted protein to Golgi membranes from Hex1 following stimulation are consistent with this hypothesis. We find no evidence for a role for MT disassembly in stimulation of secretion. This working model is summarized in Fig. 2.

REFERENCES

1. Walker RA, Sheetz MP. Cytoplasmic microtubule-associated motors. *Annu Rev Biochem.* 1993;62:429–451.
2. Cole NB, Lippincott-Schwartz J. Organization of organelles and membrane traffic by microtubules. *Curr Opin Cell Biol.* 1995;7:55–64.
3. Rismondo V, Gierow JP, Lambert RW, Golchini K, Feldon SE, Mircheff AK. Rabbit lacrimal acinar cells in primary culture: Morphology and acute responses to cholinergic stimulation. *Invest Ophthalmol Vis Sci.* 1994;35:1176–1183.
4. Hamm-Alvarez SF, da Costa S, Yang T, Wei X-H, Gierow P, Mircheff AK. Cholinergic stimulation of lacrimal acinar cells promotes redistribution of membrane-associated kinesin and the secretory protein, β-hexosaminidase, and increases kinesin motor activity. *Exp Eye Res.* 1997;64:141–156.

26

PROTEIN KINASE C ISOFORMS DIFFERENTIALLY CONTROL LACRIMAL GLAND FUNCTIONS

Driss Zoukhri,[1] Robin R. Hodges,[1] Christian Sergheraert,[2] and Darlene A. Dartt[1]

[1]The Schepens Eye Research Institute and Harvard Medical School
Boston, Massachusetts
[2]Institut Pasteur de Lille
CNRS URA 1309
Lille, France

1. INTRODUCTION

Lacrimal gland protein secretion is primarily under the control of cholinergic muscarinic and α1-adrenergic receptors.[1] Cholinergic agonists are coupled to the activation of phospholipase C (PLC),[2] which leads to the production of two second messenger molecules: inositol 1,4,5-trisphosphate (IP_3) and diacylglycerol (DAG). IP_3 increases the cytoplasmic concentration of calcium ($[Ca^{2+}]_i$) and DAG activates protein kinase C (PKC), two events that are thought to trigger protein secretion.[1] Lacrimal gland α_1-adrenergic receptors are of particular interest. Unlike those in other tissues, they are not coupled to the PLC pathway, although their activation leads to a slight increase in $[Ca^{2+}]_i$.[3] We have also shown that these receptors are not linked to the activation of phospholipase D in lacrimal gland acini.[4] Thus the transduction pathway(s) used by the α_1-adrenergic receptors to trigger lacrimal gland protein secretion remains to be identified.

PKC was originally described as a Ca^{2+}- and phospholipid-dependent protein kinase activated by DAG produced by the receptor-mediated breakdown of phosphoinositides. Molecular cloning and biochemical techniques have shown that PKC is a family of closely related enzymes consisting of at least 11 different isozymes in three categories.[5,6] A first group, termed classical or conventional PKCs (cPKC), including PKCα, -βI, -βII, and -γ isoforms, have a Ca^{2+}- and DAG-dependent kinase activity. A second group, termed new or novel PKCs (nPKC), including PKCε, -δ, -θ, -η, and -μ isoforms, are Ca^{2+}-independent and DAG-stimulated kinases. A third group, termed atypical PKCs (aPKC), including PKCζ, and -ι/λ isoforms, are Ca^{2+}- and DAG-independent kinases. All PKC isoforms, ex-

Lacrimal Gland, Tear Film, and Dry Eye Syndromes 2
edited by Sullivan *et al.*, Plenum Press, New York, 1998

cept PKCμ, have a pseudosubstrate sequence in their N-terminal part which is thought to interact with the catalytic domain to keep the enzyme inactive in resting cells.[7]

In previous studies, we showed that lacrimal gland acini express three isoforms of PKC: PKCα, -δ, and -ε.[8] In the present study, we report the identification of two additional isoforms, PKCμ and -ι/λ. Using immunofluorescence techniques, we show that these isoforms are differentially located. Using Western blotting techniques, we show that lacrimal gland PKC isoforms translocate differentially in response to phorbol esters and cholinergic agonists. Using N-myristoylated pseudosubstrate-derived peptides, we show that PKC isoforms differentially control lacrimal gland protein secretion and cholinergic-induced Ca^{2+} elevation.

2. MATERIALS AND METHODS

2.1. Materials

Phorbol esters were from LC Laboratories (Woburn, MA), collagenase type CLS III from Worthington Biochemical (Freehold, NJ). Polyclonal antibodies to PKCα, -δ, -ε, -μ, and -ι/λ and their corresponding control peptides were obtained from Santa Cruz Biotechnology (Santa Cruz, CA). Other chemicals were from Sigma (St. Louis, MO).

2.2. Methods

2.2.1. Immunohistochemistry. Lacrimal glands were removed from male Wistar rats that had been anesthetized with CO_2 for 1 min and then decapitated. Lacrimal glands were fixed in 4% formaldehyde in PBS for 4 h at 4°C. After cryopreservation overnight in 30% sucrose in PBS, the tissue was frozen in O.C.T. embedding medium. Cryostat sections (6 μm) were incubated with the indicated primary antibody for 1 h; then the secondary antibody (1:50), conjugated to FITC, was applied for 1 h at room temperature. Sections were viewed using a Nikon UFXII microscope equipped for epi-illumination. Negative controls included omission of the primary antibody and preabsorption of the primary antibody with the corresponding immunizing peptide.

2.2.2. Preparation of Lacrimal Gland Acini and Measurement of Protein Secretion and $[Ca^{2+}]_i$. Dispersed acini were isolated by collagenase digestion. Peroxidase secretion, an index of lacrimal gland protein secretion and thus referred to as protein secretion, was measured as described previously.[8,9] For $[Ca^{2+}]_i$ measurement, acini were incubated in KRB-Hepes buffer containing 0.5% bovine serum albumin, 0.5 μM fura-2-tetra-acetoxymethyl ester, 10% Pluronic F127, and 250 μM sulfinpyrazone for 60 min at 22°C. Fluorescence was measured at excitation wavelengths of 340 and 380 nm and emission wavelength of 505 nm as previously described.[3]

2.2.3. Electrophoresis and Immunoblotting. Lacrimal gland acini were incubated for the indicated period with or without agonists. Acini were then homogenized, and soluble and particulate fractions were prepared by high-speed centrifugation. Proteins in both fractions were separated by SDS-PAGE and transferred to nitrocellulose membranes as previously described.[8] Immunmoreactive bands were visualized using the ECL method according to manufacturer's protocol. The films were digitally scanned using BDS-Image and analyzed using National Institutes of Health Image software.

2.2.4. Pseudosubstrate Peptide Synthesis. Myristoylated and acetylated peptides were synthesized by BOC strategy on a MHBA resin (Novabiochem, Meudon, France) using an Applied Biosystems 430A automated synthesizer (Foster City, CA). Myristic acid was coupled to the peptide using DCC and HOBt. Acetylation was carried out using a 20% acetic anhydride solution in dichloromethane. Peptides were purified by reverse-phase HPLC on a Vydac C_4 (30 x 0.9 cm) preparative column. Peptide integrity was monitored by amino acid analysis and mass spectrometry.

3. RESULTS

3.1. Immunolocalization of PKC Isozymes

Using immunofluorescence microscopy, we determined the cellular location of the PKC isoforms present in the lacrimal gland. The results are summarized in Table 1. PKCα immunofluorescence was seen on the basolateral and apical membranes of all acini clearly outlining the individual acinar cells in the acinus. Immunofluorescence was also detected near the apical membrane, suggesting that PKCα might be associated with the membranes of the secretory granules. PKCε immunoreactivity was more cytoplasmic than PKCα, and binding was also seen on the basolateral and apical membranes of acinar cells. PKCε immunoreactivity was also occasionally detected on myoepithelial cells. PKCδ also had a cytoplasmic distribution, and binding was also seen on some myoepithelial cells. PKCι/λ had a cytoplasmic distribution and was present on intracellular membranes, possibly the endoplasmic reticulum or Golgi apparatus. PKCμ localization was similar to that of PKCα in that immunofluorescence is clearly membranous, on the basolateral and apical membranes, outlining the individual acinar cells in an acinus, but was also found in an intracellular filamentous network. These results show that five isoforms of PKC are present in the lacrimal gland, each with a different distribution.

3.2. Phorbol Ester and Cholinergic Agonist-Induced Translocation of PKC Isozymes

Classical and novel, but not atypical, PKC isoforms translocate from the soluble (cytosol) to the particulate (membrane) fraction in response to phorbol esters or other agonists.[5] To determine if lacrimal gland PKC isoforms translocate in response to phorbol esters and cholinergic agonists, we used cell fractionation and SDS-PAGE/Western blotting techniques with antisera specific to PKC isoforms. As summarized in Table 2, phorbol esters stimulated the translocation of PKCα, -δ, -ε, and -μ, but not -ι/λ. Translocation occurred at 1 min and was still measurable at 10 min. In contrast, the cholinergic agonist

Table 1. Immunolocalization of PKC isoforms in the lacrimal gland

PKC isoform	Immunolocalization
α	In basolateral and apical membranes of acinar cells, and in membranes of the secretory granules
δ	Cytosolic distribution in acinar cells, and in myoepithelial cells
ε	In basolateral and apical membranes of acinar cells, and in myoepithelial cells
μ	In basolateral and apical membranes of acinar cells, and in an intracellular filamentous network
ι/λ	Cytosolic distribution in acinar cells near the Golgi area

Table 2. Effect of phorbol esters and cholinergic agonists on the cellular distribution of PKC isoforms in the lacrimal gland

Amount of membrane-bound PKC (% of total)*	α	δ	ε	μ
Control	54	52	45	72
PdBu (10^{-6} M)	85	86	77	90
Carbachol (10^{-3} M)	46	72	73	84

*An increase in % indicates translocation from cytosol to membranes. Data are means of 2-3 separate experiments.

carbachol stimulated the translocation of PKCδ, -ε, and -μ, but, surprisingly, not -α. Similarly to PdBu, carbachol did not translocate the DAG-independent isoform, PKCι/λ. Translocation of PKCδ and -μ was transient (back to control level by 5 min), whereas that of PKCε persisted at 10 min.

These results show that lacrimal gland PKC isoforms translocate differentially in response to phorbol esters and cholinergic agonists.

3.3. Effect of Myristoylated Pseudosubstrate-Derived Peptides on Agonist-Induced Protein Secretion

We have synthesized and N-myristoylated three peptides derived from the pseudosubstrate sequences of PKCα, -δ, and -ε (myr-PKCα[15–28], myr-PKCδ[142–153], and myr-PKCε[149–164]), and tested their effects on agonist-induced protein secretion. As summarized in Table 3, preincubation of lacrimal gland acini for 1 h with the myristoylated peptides altered agonist-induced protein secretion. Myr-PKCα inhibited phorbol ester- and cholinergic agonist-induced protein secretion to a larger extent than did myr-PKCε, whereas myr-PKCδ had only a minor effect. In contrast, α_1-adrenergic-induced protein secretion was inhibited only by myr-PKCε, whereas myr-PKCα and myr-PKCδ had a stimulatory effect. The inhibitory effect of the peptides on cholinergic- and α_1-adrenergic-induced protein secretion was due solely to inhibition of PKC as the peptides did not alter the changes in $[Ca^{2+}]_i$ induced by these agonists (data not shown).

Table 3. Effect of the myristoylated pseudosubstrate-derived peptides of PKCα, -δ, and -ε on agonist-induced lacrimal gland protein secretion

	Percent Inhibition of Peroxidase Secretion		
Agonist	Myr-PKCα	Myr-PKCδ	Myr-PKCε
PdBu (10^{-6} M)	80 ± 17	11 ± 11	46 ± 20
Carbachol (10^{-5} M)	70 ± 16	22 ± 19	51 ± 14
Phenylephrine (10^{-4} M)	0*	0*	86 ± 7

*In fact, myr-PKCα increased phenylephrine-induced protein secretion 1.8-fold and myr-PKCδ increased it 2.5-fold. Acinar cells were preincubated for 60 min in the presence of 10^{-7} M of each myristoylated peptide before addition of the indicated agonist. Data are means ± SEM of 4-6 separate experiments.

Table 4. Effect of the myristoylated pseudosubstrate-derived peptides of PKCα, -δ, and -ε on phorbol ester-induced inhibition of cholinergic-stimulated $[Ca^{2+}]_i$ elevation in the lacrimal gland

Condition	Peak $[Ca^{2+}]_i$ (nM)	Percent of inhibition
Carbachol (Cch, 10^{-5} M)	107 ± 14	–
PMA+Cch	53 ± 5	50%
Myr-PKCα+PMA+Cch	58 ± 7	46%
Myr-PKCδ+PMA+Cch	88 ± 23	17%
Myr-PKCε+PMA+Cch	104 ± 28	3%

Fura-2-loaded acinar cells were preincubated for 60 min with or without 10^{-7} M of each myristoylated peptide. PMA (5×10^{-7} M) was then added or not for another 5 min before addition of carbachol. Data are means ± SEM of 4 separate experiments.

3.4. Effect of Myristoylated Pseudosubstrate-Derived Peptides on Cholinergic-Induced $[Ca^{2+}]_i$ Elevation

It is well documented that cholinergic agonists stimulate the release of $[Ca^{2+}]_i$ and influx of Ca^{2+} in the lacrimal gland. Recent reports showed that preactivation of PKC results in negative feedback on cholinergic-induced Ca^{2+} release in the lacrimal gland, implicating PKC in the mechanism of desensitization of the cholinergic response. The results in Table 4 show that when fura-2-loaded lacrimal gland acini are pretreated with the phorbol ester PMA, carbachol-induced $[Ca^{2+}]_i$ elevation is inhibited by 50%. To determine which isoform of PKC is mediating the inhibitory effect of PMA, we used the myristoylated pseudosubstrate-derived peptides. As summarized in Table 4, preincubation of lacrimal gland acini with myr-PKCε or myr-PKCδ, but not myr-PKCα, prevented the inhibitory effect of PMA.

These results suggest that the Ca^{2+}-independent isoforms PKCδ and -ε negatively modulate cholinergic-induced $[Ca^{2+}]_i$ elevation in the lacrimal gland

4. DISCUSSION

The results of the present study show that the lacrimal gland expresses five isoforms of PKC: a classical isoform, PKCα; three novel isoforms, PKCδ, -ε, and -μ; and one atypical isoform, PKCι/λ. Using immunofluorescence techniques, we showed that, under basal (unstimulated) conditions, these isoforms were targeted to different loci. PKCα and -μ immunoreactivity were associated with the plasma membrane and with structures at the apical side of the acinar cells that resemble secretory granules. The unique location of PKCα and -μ to the apical membrane of the acinar cells suggests a pivotal role for these isoforms in controlling exocytosis in the lacrimal gland. PKCδ, -ε, and -ι/λ immunoreactivity was detected mainly in the cytosol. PKCδ and occasionally -ε immunoreactivity was also localized on myoepithelial cells. The role of myoepithelial cells in the lacrimal gland is not yet understood. However, due to their resemblance to smooth muscle cells, it is suggested that they might contract in response to stimuli to help the acinar cells eject their secretory products into the lumen.

In another series of experiments, using cell fractionation and Western blotting techniques, we showed that all PKC isoforms except PKCι/λ translocated from the soluble to the particulate fraction in response to phorbol esters. Translocation in response to carbachol was subtle and transient for PKCδ and -μ, whereas that for PKCε was more extensive and was still apparent at 10 min. In contrast to PdBu, carbachol did not stimulate the translocation of PKCα at any of the times measured. This finding is intriguing since we found that this isoform plays a major role in cholinergic-induced protein secretion in the lacrimal gland. Indeed, inhibition of PKCα by its myristoylated pseudosubstrate-derived peptide resulted in 70% inhibition of cholinergic-induced protein secretion. This may imply that translocation of PKCα is not necessary for it to stimulate protein secretion.

An important finding with the myristoylated pseudosubstrate-derived peptides is that α_1-adrenergic agonists stimulate PKCε to induce lacrimal gland protein secretion. Indeed, up to now, the second messenger system used by these agonists was unknown in the lacrimal gland. The question that remains now to be answered is how the α_1-adrenergic receptors activate PKCε since they do not stimulate PLC or PLD to produce DAG, the endogenous activator of PKC.

Another finding with the myristoylated pseudosubstrate-derived peptides was that the Ca^{2+}-independent isoforms PKCδ and -ε negatively modulate cholinergic-induced $[Ca^{2+}]_i$ elevation in the lacrimal gland. The site of action of these isoforms might be either the muscarinic receptor itself, protein G, or PLC, to decrease the amount of the Ca^{2+}-mobilizing messenger, IP_3. Another possibility is that PKCδ and/or -ε might inhibit cholinergic-induced entry of Ca^{2+} by phosphorylating the plasma membrane Ca^{2+} channels.

In conclusion, our studies show that lacrimal gland acini express five isoforms of PKC each with a unique location, that these isoforms are differentially translocated by phorbol esters or cholinergic agonists, that they are differentially involved in agonist-induced protein secretion, and that they differentially control cholinergic-induced $[Ca^{2+}]_i$ elevation.

ACKNOWLEDGMENT

Supported by National Eye Institute grant EY06177.

REFERENCES

1. Dartt DA. Regulation of inositol phosphates, calcium and protein kinase C in the lacrimal gland. *Prog Retinal Eye Res.* 1994;13:443–478.
2. Mauduit P, Jammes H, Rossignol B. M_3 muscarinic acetylcholine receptor coupling to PLC in rat exorbital lacrimal acinar cells. *Am J Physiol.* 1993;264:C1550-C1560.
3. Hodges RR, Dicker DM, Rose PE, Dartt DA. α_1-Adrenergic and cholinergic agonists use separate signal transduction pathways in lacrimal gland. *Am J Physiol.* 1992;262:G1087-G1096.
4. Zoukhri D, Dartt DA. Cholinergic activation of phospholipase D in lacrimal gland acini is independent of protein kinase C and calcium. *Am J Physiol.* 1995;268:C713-C720.
5. Newton AC. Protein kinase C: Structure, function, and regulation. *J Biol Chem.* 1995;270:28495–28498.
6. Nishizuka Y. Protein kinase C and lipid signaling for sustained responses. *FASEB J.* 1995;9:484–496.
7. House C, Kemp BE. Protein kinase C contains a pseudosubstrate prototope in its regulatory domain. *Science.* 1987;238:1726–1728.
8. Zoukhri D, Hodges RR, Dicker DM, Dartt DA. Role of protein kinase C in cholinergic stimulation of lacrimal gland protein secretion. *FEBS Lett.* 1994;351:67–72.
9. Dartt DA, Baker AK, Vaillant C, Rose PE. Vasoactive intestinal polypeptide stimulation of protein secretion from rat lacrimal gland acini. *Am J Physiol.* 1984;247:G502–G509.

ROLE OF PROTEIN KINASES IN REGULATION OF APICAL SECRETION AND BASAL-LATERAL MEMBRANE RECYCLING TRAFFIC IN RECONSTITUTED RABBIT LACRIMAL GLAND ACINI

J. Peter Gierow[1] and Austin K. Mircheff[2]

[1]Department of Natural Sciences
University of Kalmar
Kalmar, Sweden, and
Department of Physiology and Biophysics
University of Southern California School of Medicine
Los Angeles, California
[2]Departments of Physiology and Biophysics and of Ophthalmology
University of Southern California School of Medicine
Los Angeles, California

1. INTRODUCTION

Cholinergic stimulation activates at least two different membrane traffic functions in lacrimal gland acinar cells: (i) exocytic release of stored secretory proteins across the apical plasma membranes, which is followed by retrieval and recycling of secretory vesicle constituents, and (ii) recycling traffic between basal-lateral plasma membranes and endomembrane compartments, e.g., membranes involved in translocation of neurotransmitter and neuropeptide receptors and of Na,K-ATPase and other ion transport proteins. The signal transduction and effector mechanisms involved in regulated protein secretion have been studied most extensively in rat extraorbital lacrimal glands.[1–4] Work with this preparation has produced considerable evidence implicating both branches of the phosphoinositide cascade, including Ca^{2+}-mediated activation of Ca^{2+}/calmodulin-dependent protein kinase and diacylglycerol-mediated activation of protein kinase C.[5,6] A role for phospholipase D has also been suggested.[7] In contrast, relatively little is known about regulation of the basal-lateral plasma membrane-endomembrane recycling traffic. Therefore, the goals of the present study were to assess the effects of protein kinase activators and inhibitors

Lacrimal Gland, Tear Film, and Dry Eye Syndromes 2
edited by Sullivan *et al.*, Plenum Press, New York, 1998

on secretion of β-hexosaminidase, a marker for function of the apical secretory pathway,[8] and on endocytosis of Lucifer Yellow, a marker for function of the basal-lateral plasma membrane-endomembrane recycling pathway, in rabbit lacrimal gland acinar cells.[9,10] Accordingly, we employed the phorbol ester, phorbol-12-myristate-13-acetate (PMA); the Ca^{2+}-ionophore, A23187; the protein kinase C inhibitor, calphostin c; and the Ca^{2+}/calmodulin-dependent protein kinase (CaM kinase II) inhibitor, KN62.

2. MATERIALS AND METHODS

2.1. Materials

Female New Zealand White rabbits were obtained from Irish Farms (Norco, CA) and were used in accordance with the Guide for the Care and Use of Laboratory Animals. Cell preparation reagents, culture media, and supplements are described in detail elsewhere.[9–11] Calphostin c, PMA, and A23187 were from Calbiochem (La Jolla, CA), KN62 from Seikagaku (Rockville, MD), and Lucifer Yellow from Molecular Probes (Eugene, OR). Carbamylcholine (carbachol) and 4-methyl-umbelliferyl-*N*-acetyl-β-D-glucosaminide were from Sigma (St. Louis, MO).

2.2. Methods

2.2.1. Lacrimal Gland Acinar Cell Isolation and Primary Culture. Cells were isolated essentially as described elsewhere[9,11] and placed in a standard, serum-free culture medium.[10] The cells were kept in culture for 2 days, during which period they reorganized into acinus-like structures. They were then harvested, washed once with Ham's medium supplemented with bovine serum albumin (5 mg/ml), and washed once with a medium similar to the primary culture medium but lacking laminin, carbachol, and thyroxine, and supplemented with 10 mM HEPES, final pH 7.6. Washed acini were resuspended in the latter medium, diluted 100-fold with Hank's balanced salt solution (without Ca^{2+} and Mg^{2+} but supplemented with 10 mM HEPES, final pH 7.6), and allowed to equilibrate at room temperature.

2.2.2. Fluid Phase Endocytosis and Secretion. Cells were preincubated for 30 min at 37°C in the presence of 0.1 μM calphostin c or 20 μM KN62 or with vehicle alone (1% DMSO), then incubated 20 min at 37°C in the presence of 1 mg/ml Lucifer Yellow under basal conditions or in the presence of 0.1 mM carbachol, 1 μM PMA, or 1 μM A23187. Additional control incubations were performed in parallel at 4°C to account for contributions from damaged cells. At the end of the incubation period, suspensions were diluted 5-fold with ice-cold buffered Hank's solution supplemented with 10 mM HEPES (pH 7.6) and placed on ice. After removal of aliquots for determination of total protein, cells were sedimented by centrifugation at (100g, 5 min) at 4°C. The supernatants were collected and saved for analyses of released protein and β-hexosaminidase catalytic activity. Sedimented cells were washed and analyzed for endocytosed Lucifer Yellow.

2.2.3. Biochemical Determinations. The determination of β-hexosaminidase activity was based on the procedure of Barrett and Heath[12] with 4-methyl-umbelliferyl-*N*-acetyl-β-D-glucosaminide as a substrate. Reactions were started by addition of 100 μl reaction medium at room temperature and terminated by addition of 2 ml ice-cold quench solution.

Fluorescence of the product was determined with excitation at 365 nm and emission at 460 nm. Endocytosed Lucifer Yellow released and total cell protein were measured as described earlier.[9]

3. RESULTS

3.1. Secretion of Protein and β-Hexosaminidase

Secretion of bulk protein in the absence of kinase inhibitors was stimulated more than 2.5-fold over basal values by both carbachol and PMA. Pretreatment with calphostin c completely abolished the effect of PMA. Calphostin c appeared to inhibit carbachol-stimulated protein secretion partially (30%), but this effect was not statistically significant.

Secretion of β-hexosaminidase in the absence of kinase inhibitors was stimulated more than 2-fold by carbachol, PMA, and A23187. PMA and A23187 in combination appeared to have additive effects. Calphostin c completely inhibited the effect of PMA, but only partially (60%) inhibited the carbachol-stimulated secretion of β-hexosaminidase. Pretreatment with the CaM kinase II inhibitor, KN62, completely inhibited the A23187-stimulated β-hexosaminidase secretion, but only partially inhibited PMA- and carbachol-stimulated release.

3.2. Fluid Phase Endocytosis

Carbachol and PMA both stimulated Lucifer Yellow endocytosis roughly 2-fold over basal values in the absence of protein kinase inhibitors. In contrast, the Ca^{2+}-ionophore, A23187, had no effect. The protein kinase C inhibitor, calphostin c, completely inhibited the effect of PMA, but had no effect on carbachol-stimulated endocytosis. A combination of A23187 and calphostin c had, as expected, no effect. KN62, the inhibitor of CaM kinase II, had no effect on PMA-stimulated endocytosis, but decreased carbachol-stimulated uptake to levels not significantly different from basal values.

4. DISCUSSION

These results indicate that both protein kinase C and CaM II kinases play important roles in muscarinic stimulation of regulated apical secretion in the rabbit lacrimal gland acinar cells, and that neither is the sole effector.

Our results also indicate that, even though protein kinase C is capable of eliciting an endocytic response, it is not the sole effector in muscarinic stimulation of basal-lateral membrane-endomembrane recycling traffic in rabbit lacrimal gland acinar cells. Inhibition of carbachol-stimulated Lucifer Yellow endocytosis by KN62 would seem to suggest a role for CaM II kinase in this process, but this conclusion is inconsistent with the lack of effect of A23187. Thus, the possibility arises that KN62 may have non-specifically inhibited some other effector of the basal-lateral membrane-endomembrane traffic.

We have previously shown that protein secretion and fluid phase endocytosis are at least partially independent phenomena. The observation that the two processes exhibited different carbachol dose-response relationships suggested that the endocytic process responsible for uptake of fluid phase markers was not limited to the retrieval of secretory vesicle membranes that had discharged their contents.[8] This conclusion was confirmed by subcellular fractionation analyses that dissected the basal-lateral membrane-endomembrane recycling pathway.[10] The present study furthers these observations by indicating differences in the intracellular signalling pathways that mediate apical protein secretion and basal-lateral plasma membrane-endomembrane recycling.

ACKNOWLEDGMENTS

This work was supported by a University of Kalmar Faculty Research Grant (JPG) and by NIH grant EY05801 (AKM).

REFERENCES

1. Busson-Mabillot S, Chambaut-Guerin A-M, Ovtracht L, Muller P, Rossignol B. Microtubules and protein secretion in rat lacrimal glands: Localization of short-term effects of colchicine on the secretory process. *J Cell Biol.* 1982;95: 105–117.
2. Cripps MM, Bromberg BB, Patchen-Moor K, Welch MH. Adrenocorticotropic hormone stimulation of lacrimal peroxidase secretion. *Exp Eye Res.* 1987;45:673–683.
3. Dartt DA, Baker AK, Vaillant C, Rose PE. Vasoactive intestinal peptide stimulation of protein secretion from rat lacrimal gland acini. *Am J Physiol.* 1984;247:G502-G509.
4. Putney JW, Van de Walle CM, Leslie BA. Stimulus-secretion coupling in the rat lacrimal gland. *Am J Physiol.* 1978;235:C188-C198.
5. Dartt DA, Rose PE, Joshi VM, Donowitz M, Sharp WG. Role of calcium in cholinergic stimulation of lacrimal gland protein secretion. *Curr Eye Res.* 1985;4:475–483.
6. Dartt DA, Ronco LV, Murphy SA, Unser MF. Effect of phorbol esters on rat lacrimal gland protein secretion. *Invest Ophthalmol Vis Sci.* 1988;29:1726–1731.
7. Zoukhri D, Dartt DA. Cholinergic activation of phospholipase D in lacrimal gland acini is independent of protein kinase C and calcium. *Am J Physiol.* 1995;268:C713-C720.
8. Gierow JP, Zeng H, Okamoto CT, Mircheff AK. Regulation of β-hexosaminidase and secretory component secretion in reconstituted rabbit lacrimal gland acini. *Invest Ophthalmol Vis Sci.* 1996;37:S928.
9. Gierow JP, Lambert RW, Mircheff AK. Fluid phase endocytosis by isolated rabbit lacrimal gland acinar cells. *Exp Eye Res.* 1995;60:511–525.
10. Gierow JP, Yang T, Bekmezian A, et al. Endosomal Na,K-ATPase in lacrimal gland acinar cells: Correcting a case of mistaken identity, *Am J Physiol.* 1996;271:C1685-C1698.
11. Rismondo V, Gierow JP, Lambert RW, Golchini K, Feldon SE, Mircheff AK. Functional and ultrastructural characteristics of rabbit lacrimal acinar cells in short-term culture. *Invest Ophthalmol Vis Sci.* 1994;35:1176–1183.
12. Barrett AJ, Heath MF. Lysosomal enzymes. In: Dingle JT, ed. *Lysosomes, a Laboratory Handbook.* Amsterdam: Elsevier; 1977:118–120.

BREFELDIN A DETOXIFICATION IN RAT EXTRAORBITAL LACRIMAL GLANDS

Philippe Robin, Bernard Rossignol, and Marie-Noëlle Raymond

Laboratoire de Biochimie des Transports Cellulaires
Université Paris XI
Orsay Cedex, France

1. INTRODUCTION

Brefeldin A (BFA) has been extensively used, during the last 10 years, as a tool for the study of intracellular trafficking mechanisms. This antibiotic, produced by several fungi, was found to inhibit intracellular protein transit and to disrupt the Golgi apparatus structure. Although it is now well established that BFA acts by preventing the association of the ADP ribosylation factor (ARF) to the Golgi membranes, its molecular target has not yet been isolated and characterized. One major feature of BFA is that its action is rapidly reversible after removal of the drug.[1] Nevertheless, a spontaneous reversion was observed in hepatocyte cultures even in the presence of the drug.[2,3] This suggested that BFA could be metabolized to an inactive form. More recently, Brüning et al.[4] showed that, in Chinese hamster ovary cells, BFA could be converted to glutathione derivatives by glutathione-S-transferase, a well-known enzyme involved in the detoxification of numerous compounds.[5,6] In the general glutathione-based detoxification pathway, the toxic compound is intracellularly converted to a glutathione derivative by glutathione-S-transferase. The glutathione derivative is then extruded from the cells and extracellularly converted to a cysteine derivative by γ-glutamyl transpeptidase.

In the present work, we investigated the detoxification of BFA in lacrimal glands using radioactively labeled BFA and HPLC analysis techniques.

2. MATERIALS AND METHODS

2.1. Preparation of Gland Lobules

Male Sprague-Dawley rats (5 weeks old) were killed by 1 min of carbon dioxide inhalation. Lacrimal lobules were immediately dissected in Krebs-Ringer bicarbonate buffer supplemented with 1 g/l glucose (KRBG buffer) and rinsed twice with a large volume of the same buffer.

2.2. Analysis and Purification of BFA Derivatives

For the analysis of the BFA derivatives, lacrimal lobules were incubated with [^{3}H]BFA (kind gift of Dr. W. Nickel and Dr. F. Wieland, University of Heidelberg, Germany) in KRBG buffer (1 gland for 3 ml) at 37°C. Then the media were removed and diluted 1:1 (v/v), with methanol. The lobules were dried on paper, homogenized in 50% methanol/water (v/v), and centrifuged for 20 min at 38000*g*. The radioactive compounds present in the media and the supernatants were then analyzed by reverse phase HPLC as previously described.[7]

For the purification of the BFA derivatives, at the end of a 4 h incubation with 100 μM [^{3}H]BFA in KRBG buffer (1 gland per ml), methanol [50% final concentration (v/v)] was added to the medium, and the lobules were directly homogenized in the medium. The BFA derivatives were purified by HPLC under the same conditions as for analytical HPLC. The fractions containing the pure BFA derivatives were immediately subjected to evaporation and stored dry.

2.3. Secretion of Newly Synthesized Proteins

The secretion of newly synthesized proteins was determined as described by Robin et al.[8] Gland lobules were pulse-labeled for 10 min with [^{3}H]leucine (8 μCi per gland) and subjected to a 75 min chase period. Then the release of the [^{3}H]labeled proteins was triggered by addition of 2 μM carbachol. BFA or its purified derivatives were added just after the pulse and maintained until the end of the stimulation. At various times of the stimulation, aliquots of the incubation medium were precipitated with 20% trichloroacetic acid-0.1% phosphotungstic acid, and the protein-associated radioactivity was counted. At the end of the stimulation, the lobules were dried on paper, homogenized, and centrifuged for 20 min at 38,000*g*. Fractions of the supernatants were subjected to the same treatment as the incubation medium fractions.

Net secretion was expressed as the amount of [^{3}H]labeled proteins released in the incubation medium relative to the total [^{3}H]labeled soluble proteins. Basal secretion value was subtracted from stimulated secretion value.

3. RESULTS AND DISCUSSION

To test the existence of a BFA degradation system in lacrimal glands, we incubated the glands in the presence of [^{3}H]BFA and homogenized them in their medium. The homogenate was then analyzed by reverse phase HPLC. Fig. 1 shows that the amount of native BFA decreased, whereas two main derivatives, respectively identified by mass spectroscopy as glutathione-BFA (GS-BFA) and cysteine-BFA (Cys-BFA), were formed. These products were purified to homogeneity by HPLC, and their effect on the secretion of [^{3}H]labeled proteins was tested. Fig. 2 shows that neither GS-BFA nor Cys-BFA was effective in blocking the secretion of the newly synthesized proteins.

To get more information about the formation of the two conjugation derivatives of BFA, we followed, in the incubation medium, the kinetics of BFA degradation and of GS-BFA and Cys-BFA formation. Fig. 3 shows the amounts of the three compounds at differ-

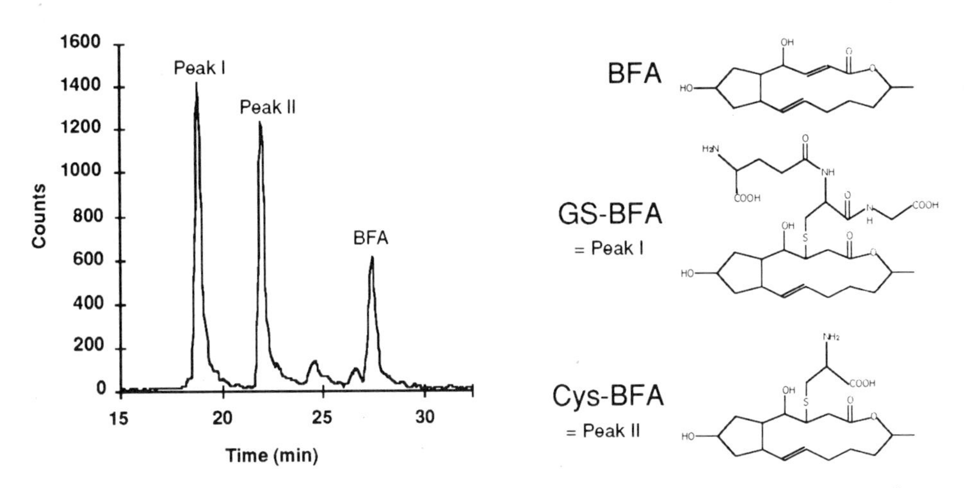

Figure 1. Analysis of the derivatives of BFA. Lacrimal lobules were incubated for 4 h with 100 μM [^{3}H]BFA. Then the lobules were homogenized in their medium, and the radioactive products were analyzed by HPLC.

ent times. The amount of BFA slowly decreases and reaches around 15% of the initial amount after 7 h of incubation; this corresponds to a concentration of 1.5 μM. As, at this concentration, the secretion of radiolabeled proteins is still inhibited,[7] no spontaneous reversion of the BFA effects could be detected in lacrimal gland after 5 or 6 h incubation in the presence of 10 μM BFA.

The amount of GS-BFA in the medium increases during the first 4 h and then decreases, whereas the amount of Cys-BFA continuously increases and reaches about 60% of the total amount of the radiolabeled compounds after 7 h of incubation. At that time, Cys-BFA is the main product of the incubation medium. These results seem to indicate that Cys-BFA is a metabolite of GS-BFA, which fits with the knowledge concerning the metabolism of many glutathione conjugates[5,6]: the glutamyl and the cysteinyl residues of the glutathione derivative are cleaved extracellularly by theγ-glutamyl transpeptidase and

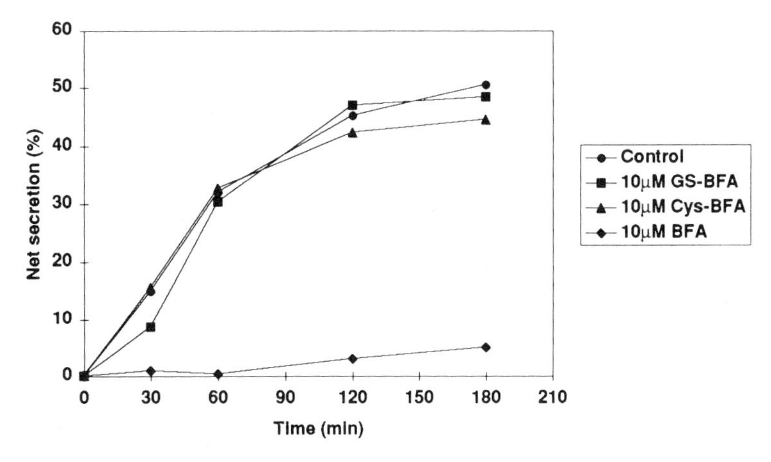

Figure 2. Effect of BFA and its derivatives on protein secretion.

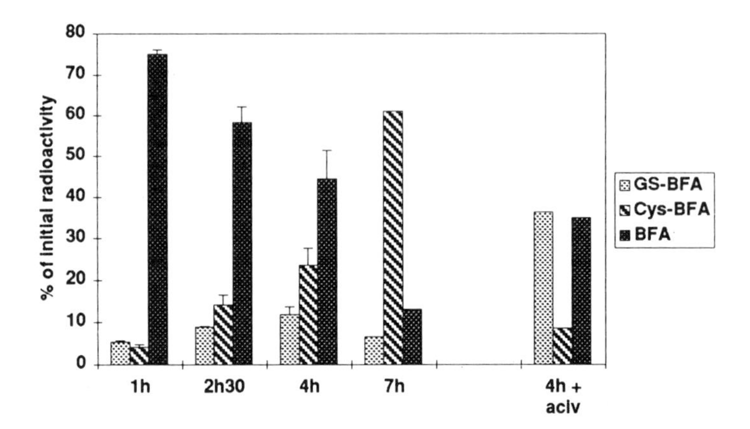

Figure 3. Kinetics of conjugates formation. Lobules were incubated in the presence of 10 μM [^{3}H]BFA, and at the indicated times the media were analyzed by HPLC. One experiment was performed in the presence of 10 μM acivicin (+ aciv). The data are expressed as percent of the radioactivity present at the beginning of the experiment. Values are means ± ranges.

a dipeptidase. To check if such a reaction occurs in lacrimal glands, we incubated the lobules for 4 h in the presence of 10 μM BFA and 10 μM acivicin, a γ-glutamyl transpeptidase inactivator. Fig. 3 shows that, in the presence of acivicin, the amount of Cys-BFA in the medium is lower than in the control, whereas the amount of GS-BFA is higher.

We can conclude from these results that in lacrimal glands BFA is metabolized via a classical pathway involving a conjugation reaction to glutathione and a degradation of the glutathione conjugate formed. We also conclude that this metabolization of BFA by this conjugation system is effectively a detoxification pathway, as the two products formed are inactive towards protein secretion.

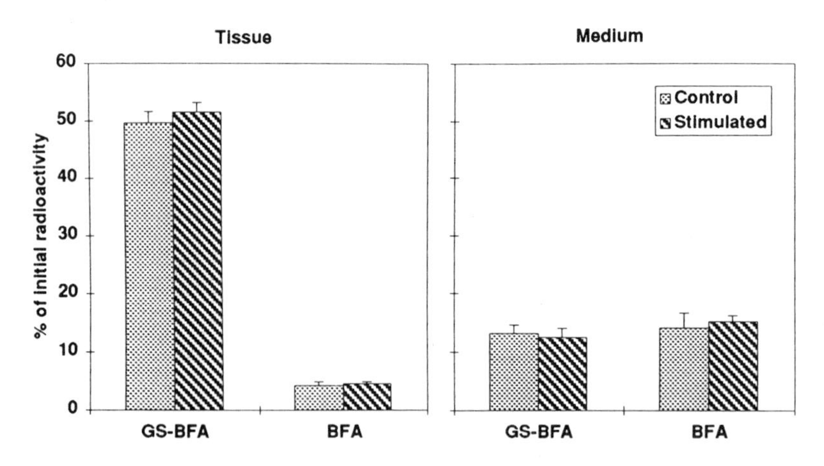

Figure 4. Effect of isoproterenol on BFA conjugation. Lobules were first incubated for 1 h in the presence of 10 μM [^{3}H]BFA. Concentration of BFA was then lowered to 0.1 μM, and lobules were stimulated or not for 1 h with 10 μM isoproterenol in the presence of 10 μM papaverine. The tissues and the media were then analyzed by HPLC. The data are expressed as percent of the radioactivity present at the beginning of the experiment. Values are means ± SE.

Recently we showed that in some particular experimental conditions, the reversion of the BFA effects on lacrimal cells could be promoted by an increase of the cAMP level.[7] Considering our present results, one could imagine that cAMP increases either the rate of BFA conjugation to glutathione or the rate of GS-BFA export, or both. Fig. 4 shows that when the lacrimal cells are stimulated by isoproterenol (which provokes an elevation of the cAMP level), the amounts of BFA and GS-BFA found in the tissue as well as in the medium are not modified. We can thus conclude that cAMP promotes the reversion of BFA effects by a pathway that is independent of the detoxification reactions.

REFERENCES

1. Klausner RD, Donaldson JG, Lippincott-Schwartz J. Brefeldin A: Insights into the control of membrane traffic and organelle structure. *J Cell Biol.* 1992;116: 1071–1080.
2. Misumi Y, Misumi Y, Miki K, Takatsuki A, Tamura G, Ikehara Y. Novel blockade by brefeldin A of intracellular transport of secretory proteins in cultured rat hepatocytes. *J Biol Chem.* 1986;261:11398–11403.
3. Fujiwara T, Oda K, Yokota S, Takatsuki A, Ikehara Y. Brefeldin A causes disassembly of the Golgi complex and accumulation of secretory proteins in the endoplasmic reticulum. *J Biol Chem.* 1988;263:18545–18552.
4. Brüning A, Ishikawa T, Kneusel RE, Matern U, Lottspeich F, Wieland FT. Brefeldin A binds to glutathione S-transferase and is secreted as glutathione and cysteine conjugates by Chinese hamster ovary cells. *J Biol Chem.* 1992;267:7726–7732.
5. Meister A, Anderson ME. Glutathione. *Annu Rev Biochem.* 1983;52:711–760.
6. Sies H, Ketterer B. *Glutathione Conjugation: Mechanisms of Biological Significance.* London: Academic Press; 1988.
7. Robin P, Rossignol B, Raymond M-N. Recovery of protein secretion after brefeldin A treatment of rat lacrimal glands: Effect of cAMP. *Am J Physiol.* 1996;271:C783-C793.
8. Robin P, Rossignol B, Raymond M-N. Effect of microtubule network disturbance by nocodazole and docetaxel (Taxotere) on protein secretion in rat extraorbital lacrimal and parotid glands. *Eur J Cell Biol.* 1995;67:227–237.

29

IDENTIFICATION AND CHARACTERIZATION OF G PROTEINS IN THE MAMMALIAN LACRIMAL GLAND

Michele A. Meneray and D. Jean Bennett

Department of Physiology
Louisiana State University Medical Center
New Orleans, Louisiana

1. INTRODUCTION

The regulation of protein secretion in exocrine tissues such as the lacrimal gland involves coupling of cell surface receptor activation by neurotransmitters and neuropeptides to the generation of intracellular messengers.[1,2] In lacrimal gland, components of the cyclic adenosine 3', 5'- monophosphate (cAMP) pathway include seven transmembrane receptors whose activation leads to stimulation or inhibition of the effector enzyme adenylyl cyclase and alterations in the intracellular levels of cAMP, as well as alterations in the release of proteins into the tears.[2] In virtually all tissues, heterotrimeric G proteins that reversibly bind guanine nucleotides link cell surface receptor activation to adenylyl cyclase and thus are critical components of signal transduction events in the cAMP pathway. In lacrimal gland, experimental evidence for G protein-dependent regulation of adenylyl cyclase includes stimulation of the enzyme by NaF, a direct activator of Gs_{α}, and by guanine 5'-0–3'-thiosphosphate (GTPγS), a non-hydrolyzable analog of GTP.[3,4] Vasoactive intestinal peptide (VIP) stimulation of the enzyme is also dependent upon GTP.[5] In the work presented here, we have used toxin-catalyzed ADP-ribosylation and immunoreaction with antisera directed against peptide sequences of the α subunits of specific G proteins to identify and characterize the G proteins present in the mammalian lacrimal gland and that mediate lacrimal secretion through transduction of extracellular signals and alterations in adenylyl cyclase activity.

2. METHODS

Membrane fractions were prepared from lacrimal tissues of male Sprague-Dawley rats and New Zealand rabbits. Starting materials for the membrane preparations[5] included lacrimal gland slices, freshly isolated acini, and primary acinar cultures.[6] ADP-ribosylation of the membranes was accomplished by addition of preactivated cholera or pertussis toxin to a membrane suspension containing 8 μM NAD, 2 μM $^{[32]}$NAD, 10 mM thymidine, 1 mM ATP, 0.4 mM GTP, 0.6 mM EDTA, 3 mM $MgCl_2$, 5 mM DTT, and 50 mM potassium phosphate buffer (pH 7.4), and incubation for 30 min at 30°C. When the effects of ribosylation on adenylyl cyclase were determined, $^{[32]}$NAD was omitted, and the NAD final concentration was 1 mM. For identification of toxin-sensitive G proteins, ribosylated proteins were separated by SDS-PAGE and detected by autoradiography and laser scanning densitometry. Immunoblotting was accomplished by separation of membrane proteins by SDS-PAGE, transfer to PVDF membranes, reaction with polyclonal antibodies to peptide sequences of the α subunits of selected G proteins (1:1000 dilution), and detection with alkaline-phosphatase-conjugated goat anti-rabbit IgG (1:1500). To determine the effect of anti-G protein antisera on adenylyl cyclase activity, membranes were incubated in the presence of non-immune rabbit serum or antisera for 60 min at 30°C.[7] For both ribosylated and immunoreacted membranes, adenylyl cyclase activity was measured by the generation of cAMP in the presence of 4 mM $MgCl_2$, 0.5 mM ATP, 1 mM DTT, 0.1 mg/ml BSA, 0.1 mM GTP, 0.1 mM EDTA, 0.1 mM IBMX, 10 mM creatine phosphate, and 50 U/ml creatine phosphokinase by a protein kinase binding method.[8]

3. RESULTS AND DISCUSSION

3.1. ADP-Ribosylation

Cholera and pertussis toxin covalently modify α subunits of G proteins by catalyzing the transfer of ADP-ribose to amino acid residues of the G protein. With the incorporation of $^{[32]}$P, this procedure allows the detection of toxin-sensitive G proteins by SDS-PAGE and autoradiography. Cholera toxin treatment of membranes resulted in the ADP-ribosylation of two proteins at 42 and 45 kDa (Fig. 1). Ribosylation in the presence of cholera toxin was both dose (0–100 μg/ml) and time (0–45 min) dependent. Because the α subunit is the site of choleratoxin-catalyzed ribosylation, these results indicated two

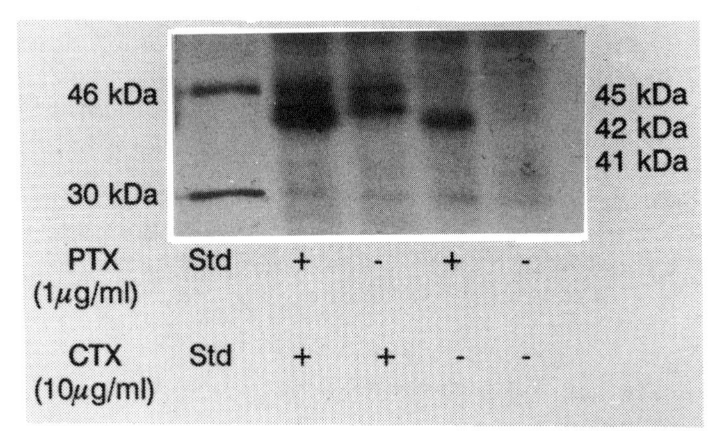

Figure 1. ADP-ribosylation of proteins in lacrimal membranes incubated with 1 μg/ml pertussis toxin (PTX) plus 10 μg/ml cholera toxin (CTX); 10 μg/ml cholera toxin; 1 μg/ml pertussis toxin; or no toxin for 30 min. (From ref. 4.)

forms of the α subunit of choleratoxin-sensitive G proteins in lacrimal tissues. In contrast, pertussis toxin treatment resulted in the ADP-ribosylation of a single band at 41 kDa (Fig. 1). This protein clearly differed in electrophoretic mobility and in its specificity as a substrate for pertussis toxin-mediated ADP-ribosylation from the 42 kDa subunit that was ribosylated by cholera toxin. Like the cholera toxin substrates, the ribosylation of the pertussis toxin-sensitive protein was dose (0–25 μg/ml) and time (0–30 min) dependent. Toxin enhancement of adenylyl cyclase paralleled incorporation of $^{[32]}$P by ribosylation: stimulation of adenylyl cyclase was dose-dependent for cholera or pertussis toxin (Fig. 2). Stimulation of the enzyme by GTPγS was equivalent to pretreatment with the maximum doses of either toxin, and the effects of ribosylation and activation by GTPγS were not additive. Thus, toxin effects on adenylyl cyclase were not separate from the mechanisms regulated by the binding of guanine nucleotide.

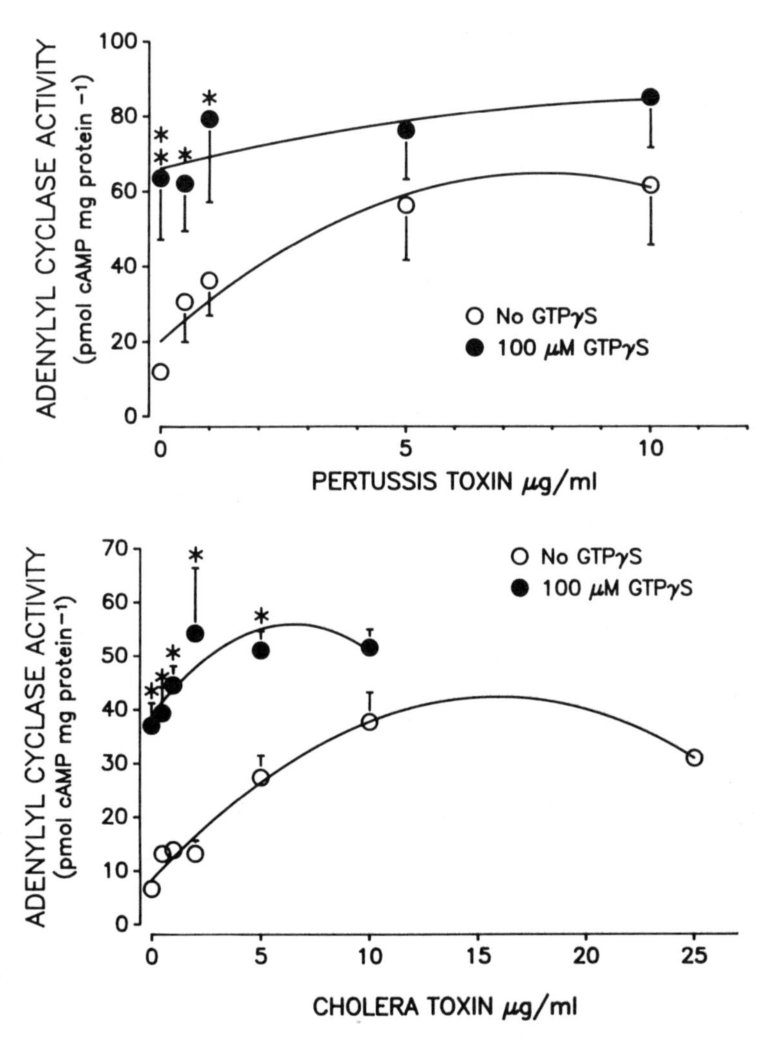

Figure 2. Effect of toxin pretreatment of lacrimal acinar membranes on adenylyl cyclase. Membranes were preincubated with cholera or pertussis toxin and assayed in the absence or presence of 100 μM GTPγS. (From ref. 4.)

3.2. Immunoblotting

Because cholera and pertussis toxin substrates include the α subunits of G_s and G proteins of the G_i/G_o family, we sought to identify the G proteins present in lacrimal gland with specific antibodies raised against peptide sequences of the α subunits of G_s, G_{i1},G_{i2}, G_{i3}, and G_o (Fig. 3). Immunoblotting of rabbit lacrimal membranes prepared from slices, isolated acini, and cultured acini with anti-$G_{s\alpha}$ antiserum detected two immunoreactive bands at 44 and 47 kDa. Bands were not detected with anti-$G_{i\alpha 1}$ antiserum but were with anti-$G_{i\alpha 1/2}$ at 40–41 kDa. Immunoreactivity was also present at 40–41 kDa in blots reacted with anti- $G_{i\alpha 3}$, and the intensity of the reactivity was increased at the same kDa when the blots were reacted with antiserum directed against a peptide sequence common to both $G_{i\alpha 3}$ and $G_{o\alpha}$.

3.3. Effect of Antibodies on Adenylyl Cyclase

With the identity of the toxin-sensitive G proteins established by immunoblotting, evidence for the functional interaction of the G proteins in lacrimal membranes with receptors that regulate adenylyl cyclase was derived by measurement of enzyme activity in membranes preincubated with non-immune serum, anti-$G_{s\alpha}$, or anti- $G_{o\alpha}$ antisera and stimulated with VIP (Fig. 4). Anti-$G_{s\alpha}$ antiserum significantly reduced VIP stimulation of the enzyme. In contrast, VIP-stimulated adenylyl cyclase activity was enhanced in membranes preincubated with anti-$G_{o\alpha}$. Because the C-terminus of the α subunit of G_s is the site of interaction with the receptor,[9] our results demonstrate that G_s couples VIP receptor activation to adenylyl cyclase in lacrimal gland. The enhanced response in membranes pretreated with anti-$G_{o\alpha}$ antiserum suggests that in addition to mediating receptor-dependent inhibition of adenylyl cyclase, inhibitory G proteins may play some role in the integrated response of the gland to stimulation. One possibility is the presence of a pathway in which dissociation of the subunits of G_o occurs at some basal level. This could result in the association of the released βγ subunits with the VIP-activated $G_{s\alpha}$. The resulting heterotrimer ($G_{s\alpha}\beta\gamma$) would not activate adenylyl cyclase. Interestingly, direct activation of the catalytic subunit of adenylyl cyclase with forskolin was also reduced by anti-$G_{s\alpha}$ antiserum. Partial inhibition of forskolin-stimulated activity is consistent with the suggestion that although forskolin activates the enzyme by direct interaction with the catalytic unit, it may also stabilize the interaction between $G_{s\alpha}$ and adenylyl cyclase.[10] Furthermore, in the absence of G protein receptor activation, the $G_{s\alpha}$-adenylyl cyclase complex is thought to be in equilibrium with the uncomplexed form of the enzyme.[11] Thus, a portion of the total activity of forskolin-stimulated adenylyl cyclase might be subject to the effect of blocking interactions of $G_{s\alpha}$ with the enzyme.

It is now clear that G proteins represent a pivotal site in signal transduction pathways. In exocrine tissues, such as the lacrimal gland, coupling of receptors to effector enzymes such as adenylyl cyclase by G proteins is an essential component of effective stimulus-secretion coupling that results in the release of proteins into the tears. We have identified both stimulatory and inhibitory G proteins in the lacrimal gland and have shown that these proteins have multiple roles in the regulation of adenylyl cyclase. Because adenylyl cyclase is important in regulation of secretion by the gland and is coupled to receptors through G proteins, characterization of the G proteins and their functions, both receptor- and non-receptor-dependent, is essential in understanding the intracellular interactions that regulate lacrimal protein secretion.

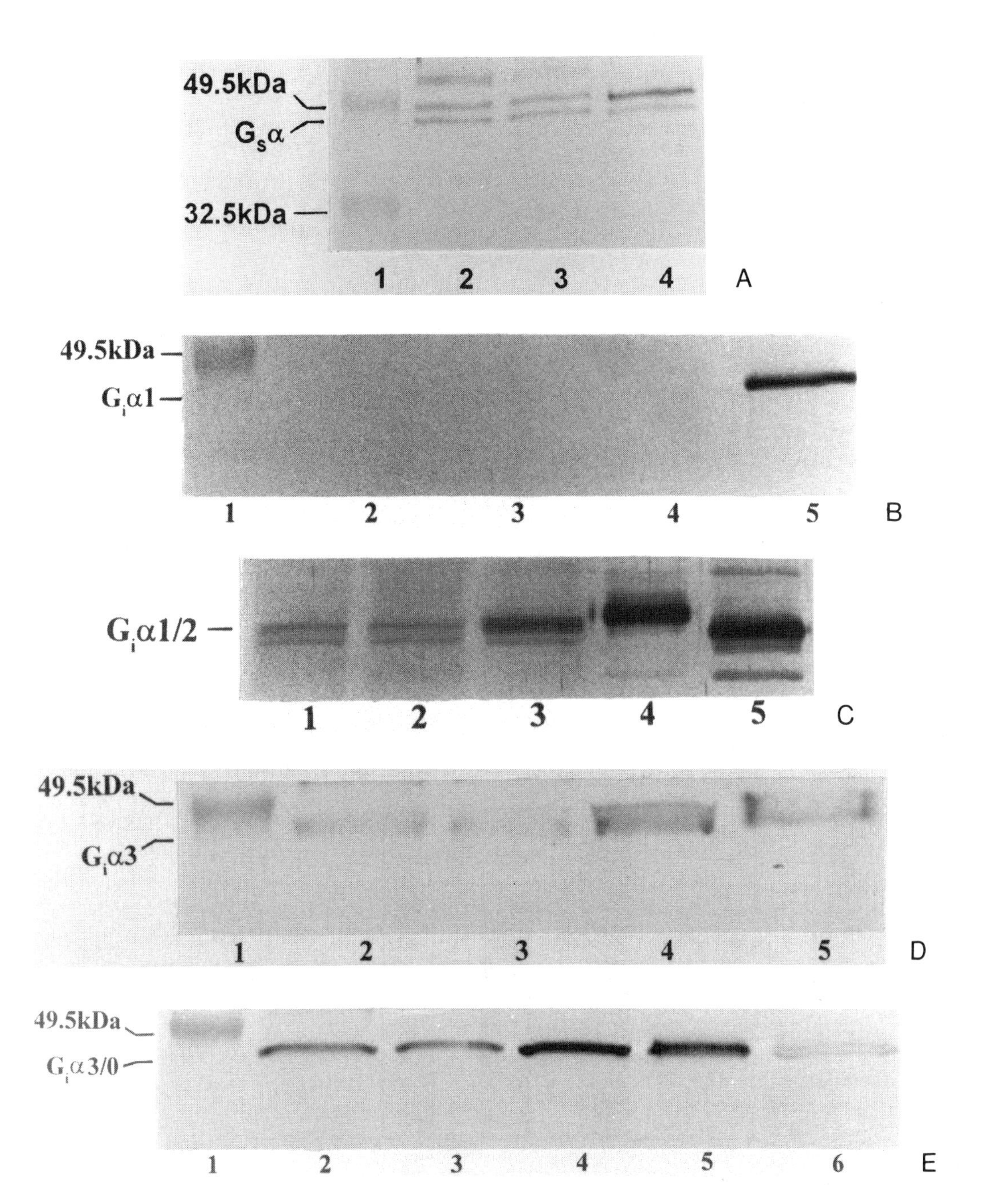

Figure 3. (A) $G_{s\alpha}$ immunoblot of lacrimal membrane proteins. 25 μg protein/lane. 1, Prestained standards. 2, Fragments. 3, Acini. 4, Cultured acini. (B) $G_{i\alpha 1}$ immunoblot of lacrimal membrane proteins. 50 μg protein/lane. 1, Prestained standards. 2, Fragments. 3, Acini. 4, Cultured acini. 5, $G_{i\alpha 1}$ recombinant standard. (C) $G_{i\alpha 1/2}$ immunoblot of lacrimal membrane proteins. 6 M urea gel; 50 μg protein/lane. 1, Fragments. 2, Acini. 3, Cultured acini. 4, $G_{i\alpha 1}$ recombinant standard. 5, $G_{i\alpha 2}$ recombinant standard. (D) $G_{i\alpha 3}$ immunoblot of lacrimal membrane proteins. 25 μg protein/lane. 1, Prestained standards. 2, Fragments. 3, Acini. 4, Cultured acini. 5, $G_{i\alpha 3}$ recombinant standard. (E) $G_{i\alpha 3/o}$ immunoblots of lacrimal membrane proteins. 25 μg protein/lane. 1, Prestained standards. 2, Fragments. 3, Acini. 4, Cultured acini. 5, $G_{i\alpha 3}$ recombinant standard. 6, $G_{o\alpha}$ recombinant standard.

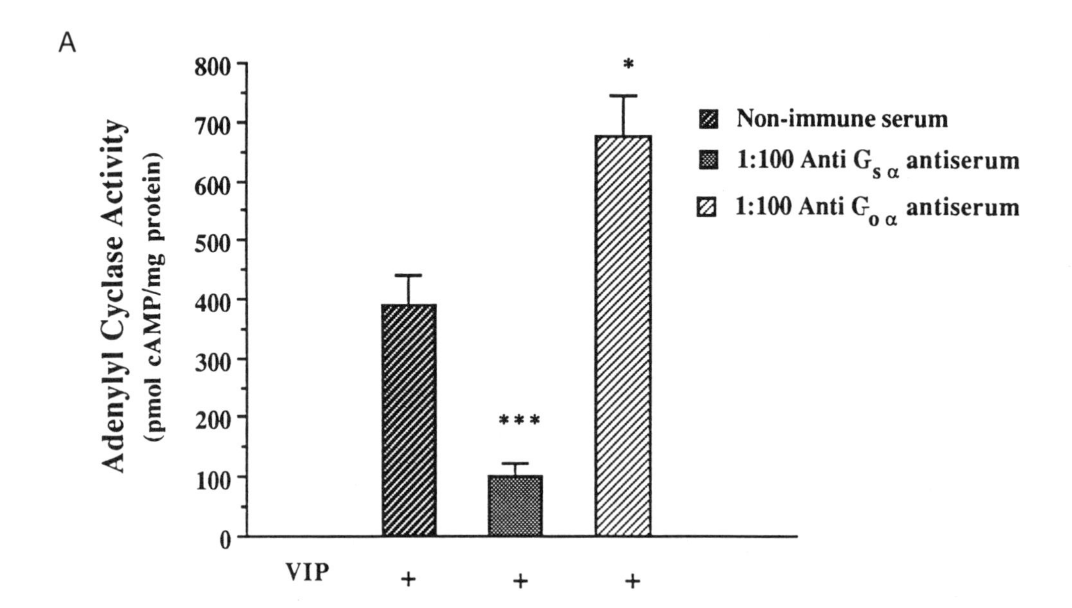

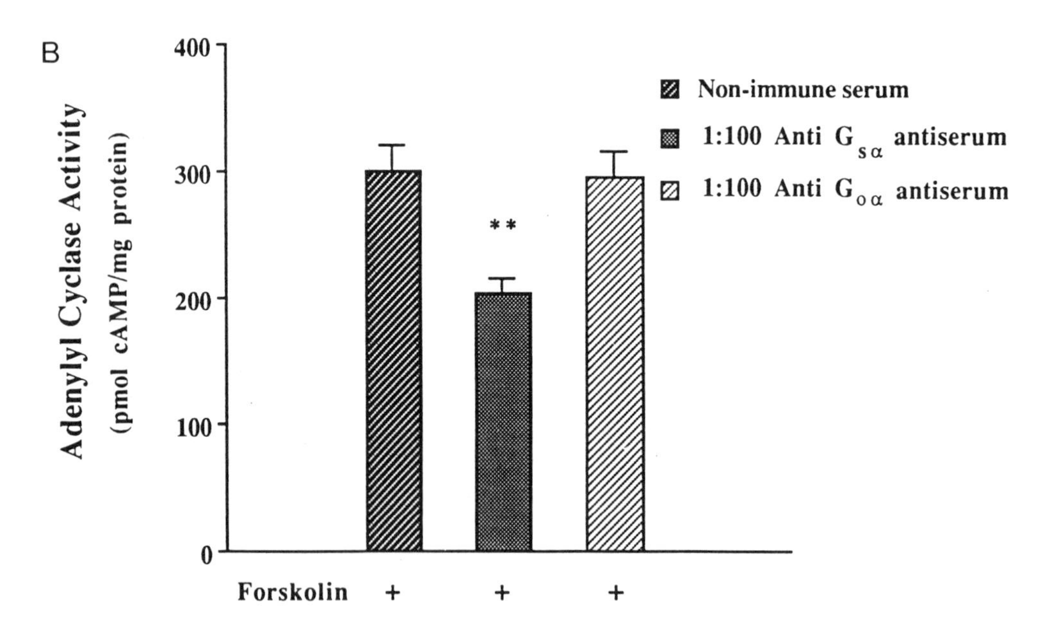

Figure 4. Effect of pretreatment of rabbit lacrimal membranes with antiserum on adenylyl cyclase activity and in the presence of (A) 100 nM vasoactive intestinal peptide (VIP) or (B) 10 μM forskolin. $*p<0.05$, $**p<0.02$, $***p<0.01$. (From ref. 7.)

ACKNOWLEDGMENTS

This work was supported by NIH grant EY07380.

REFERENCES

1. Harper JF. Stimulus-secretion coupling: Second messenger-regulated exocytosis. In: Greengard P, Robison GA, eds. *Advances in Second Messenger and Phosphoprotein Research.* New York: Raven Press;1988:193–231.
2. Dartt DA. Regulation of tear secretion. *Adv Exp Med Biol.* 1994;350:115–119.
3. Mircheff AK, Conteas CN, Lu CC, Santiago G, Gray M, Lipson LG. Basal-lateral and intra-cellular membrane populations of rat exorbital lacrimal gland. *Am J Physiol.* 1983;245:G133.
4. Cripps MM, Bennett DJ. Guanine nucleotide binding proteins in the dual regulation of lacrimal function. *Invest Ophthalmol Vis Sci.* 1992;33:3592–3600.
5. Cripps MM, Bennett DJ. Peptidergic stimulation and inhibition of lacrimal gland adenylate cyclase. *Invest Ophthalmol Vis Sci.*1990;31:2145–2150.
6. Meneray MA, Fields TY, Bromberg BB, Moses RL. Morphology and physiologic responsiveness of cultured rabbit lacrimal acini. *Invest Ophthalmol Vis Sci.* 1994;35:4144–4158.
7. Meneray MA, Bennett DJ. Identification of GTP-binding proteins in lacrimal gland. *Invest Ophthalmol Vis Sci.* 1995;36:1173–1180.
8. Brown BL, Albano JDM, Ekins RP, Sgherzi AM. A simple and sensitive saturation assay method for the measurement of adenosine 3',5'-cyclic monophosphate. *Biochem J.* 1971;121:561–562.
9. Masters SB, Sullivan K, Miller RT, et al. Carboxyl terminal domain of $G_{s\alpha}$ specifies coupling of receptors to stimulation of adenylyl cyclase. *Science.* 1988;241:448–451.
10. Bouhelal R, Guillon G, Homburger V, Bockaert J. Forskolin-induced change of the size of adenylate cyclase. *J Biol Chem.* 1985;260:10901–10904.
11. Morris D, McHugh-Sutkowski E, Moos M, Simonds WF, Spiegel AM, Seamon KB. Immunoprecipitation of adenylate cyclase with an antibody to a carboxyl-terminal peptide from $G_{s\alpha}$. *Biochemistry.* 1990;29:9079–9084.

30

INWARD-RECTIFYING POTASSIUM CHANNELS IN THE RABBIT SUPERIOR LACRIMAL GLAND

George H. Herok,[1,2] Thomas J. Millar,[1,2] Philip J. Anderton,[1,3] and Donald K. Martin[1,4]

[1]Co-operative Research Centre for Eye Research and Technology
University of New South Wales
Kensington, Australia
[2]Department of Biological Sciences
University of Western Sydney/Nepean, Australia
[3]School of Optometry
University of New South Wales
Kensington, Australia
[4]Department of Biological Sciences
University of Technology, Sydney
New South Wales, Australia

1. INTRODUCTION

Inwardly rectifying currents have been described in a wide variety of cells including both excitable (e.g., skeletal and cardiac muscle) and non-excitable (e.g., renal distal tubule and retinal Müller) cells. In these tissues, they have been reported as having several physiological roles. First, the inward rectifier is involved in setting and stabilizing the resting membrane potential. Second, in some cells, e.g., glial cells,[1] its function is to buffer transient increases in extracellular K^+ concentration. Third, it may be involved in recycling K^+ to prevent its accumulation in the cell in response to Na^+-H^+ exchange. In this process, extruding H^+ from the cell increases intracellular Na^+. As the Na^+-K^+ pump extrudes these Na^+ ions, K^+ ions accumulate. These K^+ ions may then leave via the inward rectifier.[2]

This study investigates and characterizes the properties of an inwardly rectifying K^+ (K_{ir}) current in the rabbit superior lacrimal gland and tries to determine its physiological role.

Lacrimal Gland, Tear Film, and Dry Eye Syndromes 2
edited by Sullivan *et al.*, Plenum Press, New York, 1998

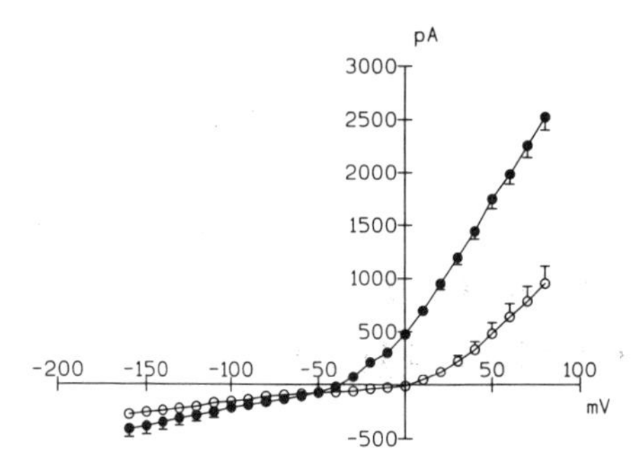

Figure 1. Steady-state whole-cell I-V relations of single rabbit superior lacrimal gland (RSLG) acinar cells before (-●- mean of 17 experiments) and after (-O- mean of 8 experiments) the addition of 10 mM TEA to the extracellular bath solution. In each experiment the recordings were under physiological conditions, ie, the pipette contained a KCl-rich solution and the extracellular bath contained a NaCl-rich solution. The vertical bars represent the SEM when these were larger than the symbols.

2. MATERIALS AND METHODS

New Zealand white rabbits of both sexes were killed by an overdose of sodium pentobarbital (45 mg/kg), and the superior lacrimal glands were excised. They were cut into fine pieces and incubated in a collagenase (200 U/ml) and hyaluronidase (600 U/ml) solution at 37°C for 45 min. The cellular pellet was washed twice with the NaCl-rich bath solution containing bovine serum albumin (2 mg/ml). Whole-cell and single-channel currents were recorded using an Axopatch 200A amplifier and CV201A headstage (Axon Instruments). The currents were recorded under physiological conditions (140 mM NaCl solution in the bath, and 140 mM KCl solution in the pipette). The K^+ pipette concentration was varied in some single-channel experiments.

3. RESULTS

Fig. 1 shows the steady-state whole-cell I-V relations of single rabbit superior lacrimal gland (RSLG) acinar cells before and after the addition of 10 mM TEA (a common K^+ channel blocker) to the extracellular bath solution. This caused a shift of the reversal potential from -40 mV to -1 mV. The total membrane conductances for inward currents was reduced by 18% and for outward currents by 72%, which suggested that TEA-sensitive currents made up the majority of the outward currents and the minority of the inward currents. Increasing the intracellular Ca^{2+} concentration from 150 nmol/L to 1000 nmol/L produced an increase in the slope conductance of the K_{ir} current from 2.7 ± 0.16 nS to 5.2

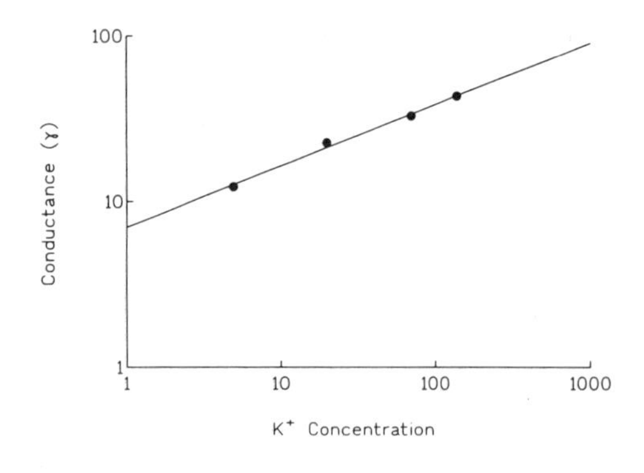

Figure 2. Log-log plot of single-channel conductance of the inward K^+ channel in cell-attached patches as a function of the pipette K^+ concentration. The straight line is the least-squares line of best fit and is given by $\log \gamma = 0.37 \log [K^+]_o + 0.85$ (pS, $r^2 = 0.98$, $p < 0.05$).

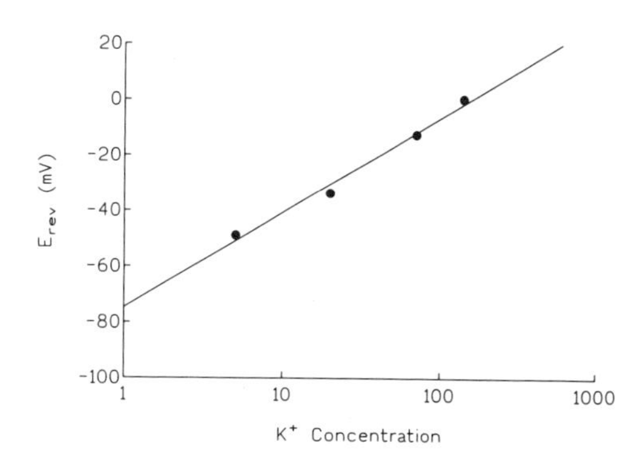

Figure 3. Semi-logarithmic plot of the reversal potential of the inward-rectifying K^+ channel as a function of pipette K^+ concentration, recorded in cell-attached patches. The straight line is the least-squares line of best fit and is given by $E_{rev} = 35.45 \log [K^+] - 77.8$ (mV, $r^2 = 0.99$, $p < 0.05$).

± 0.45 nS (n = 6). The K_{ir} current was activated, with steep voltage dependence, on hyperpolarization, and the conductance is approximately dependent on the square root of the external K^+ concentration $[K]_o$. Internal substitution of K^+ with various cations gave the following permeability sequence: $K^+(1.0) > Rb^+(0.83) > Li^+(0.15)$. The K_{ir} current was inhibited by Ba^{2+} (100 μm), TEA (10 mM), and Cs^+ (5 mM) and was insensitive to 4-AP (5 mM). The single-channel conductance was 43 ± 2.7 pS (*n*= 11), and the relationship between single-channel conductance (γ) and external K^+ concentration $[K]_o$ is given by: $\gamma = 7.04[K]_o^{0.37}$ pS (Fig. 2). The relationship between $[K]_o$ and zero current potential (E_{rev}) is given by: $E_{rev} = 35.45 \log[K]_o - 77.8$ mV (Fig. 3).

4. DISCUSSION

The K_{ir} current we identified in these rabbit lacrimal acini satisfies the criteria for identifying an inward rectifier, ie, the channels tend to open with steep voltage dependence on hyperpolarization, and the conductance is approximately dependent on the square root of the external K^+ concentration. This K_{ir} current has similar dependence on $[K]_o$ as other inward rectifiers observed in exocrine glands[3] and other tissues.[4–6] However, this K_{ir} current is Ca^{2+} activated, whereas the K_{ir} in sheep parotid[3] and rat hepatocytes[4] is Ca^{2+}-insensitive. The K_{ir} current is inhibited by TEA, whereas in the sheep parotid no inhibition is observed with TEA. This inward current contributes the minor component of the total inward current (18%) in these rabbit lacrimal acinar cells. Also, since the resting membrane potential in the rabbit lacrimal gland under physiological conditions was -40 ± 2 mV (*n* = 17), which was similar to the resting membrane potential of -37 ± 8 mV (*n* = 50) obtained by Kikkawa[7] using intracellular electrodes, we can conclude from our data that the physiological role of K_{ir} in the rabbit superior lacrimal acini seems to be involved with stabilizing the resting membrane potential.

REFERENCES

1. Barres BA, Chun LLY, Corey DP. Ion channels in vertebrate glia. *Annu Rev Neurosci.* 1990;13:441–474.
2. Wang W, White S, Geibel J, Giebisch G. A potassium channel in the apical membrane of rabbit thick ascending limb of Henle's loop. *Am J Physiol.* 1990;258:F244-F253.
3. Ishikawa T, Wegman EA, Cook DI. An inwardly rectifying potassium channel in the basolateral membrane of sheep parotid secretory cells. *J Membr Biol.* 1993;131:193–202.
4. Henderson RM, Graf J, Boyer JL. Inward-rectifying potassium channels in rat hepatocytes. *Am J Physiol.* 1989;256:G1028-G1035.

5. Cooper K, Rae JL, Dewey J. Inwardly rectifying potassium current in mammalian lens epithelial cells. *Am J Physiol.* 1991;261:C115-C123.
6. McKinney LC, Gallin EK. Inwardly rectifying whole-cell and single channel K currents in the murine macrophage cell line J774.1. *J Membr Biol.* 1988;103:41–53.
7. Kikkawa T. Secretory potentials in the lacrimal gland of the rabbit. *Jpn J Ophthalmol.* 1970;14:25–40.

ELECTROPHYSIOLOGICAL EVIDENCE FOR REDUCED WATER FLOW FROM LACRIMAL GLAND ACINAR EPITHELIUM OF NZB/NFW F1 MICE

Peter R. Brink, Elizabeth Peterson, Kathrin Banach, and Benjamin Walcott

Department of Physiology and Biophysics
and Department of Neurobiology and Behavior
State University of New York at Stony Brook
Stony Brook, New York

1. INTRODUCTION

1.1. Secretion of Solutes and Solvent

One manifestation of dry eye is a reduction in tear volume. In mammals, much of the tear volume is thought to arise from the acinar secretory and duct cells of the lacrimal gland. The subjacent tissue of the gland contains reticular-like connective tissue with scattered lymphocytes. Plasma cells within the connective tissue produce IgA, which is sequestered intracellularly by acinar cells and processed and finally secreted into the lumen of the gland.[1] Many other secretory products are generated by the acinar cells. Secretagogue-triggered vesicle release is the process used by acinar cells to secrete proteins.

Another important secretory aspect of acinar/duct cells centers on their ability to transport specific solutes such as Cl^- and K^+ via membrane channels and transporters from interstitium to lumen, and, as a consequence, to move water to maintain osmotic equilibrium. Water diffuses through plasma membranes quite rapidly. The permeability of water to phospholipid bilayers is $\sim 1X10^{-4}$ cm/sec. This is three to four orders of magnitude higher than the permeability of monovalent solutes like K^+ to the plasma membrane. In addition, many cell types have water channels (aquaporins) that further facilitate the transport of water. Aquaporins can increase water permeability by a factor of 10X.[2,3] In red blood cell membranes, water permeability is $5X10^{-3}$cm/sec, which is about an order of magnitude greater than the water permeation rate for an equivalent bilayer ($1X10^{-4}$).[2–5] A number of water channels have been identified, and aquaporin 5 (AQP5) has been found in mammalian cornea and lacrimal gland.[5] Even the absence of water channels in NZB/NFW F1 (NZB) or MRL/MR-lpr-lpr (MRL) acinar/duct cells would not be expected

Lacrimal Gland, Tear Film, and Dry Eye Syndromes 2
edited by Sullivan *et al.*, Plenum Press, New York, 1998

to dramatically alter water flow, as the plasma membrane permeability is at least a thousand times greater than the plasma membrane permeability of solutes being transported transepithelially (Na^+, K^+, Cl^-). Thus, if tear volume is reduced in dry eye, a logical starting point is solute transport across the acinar and duct epithelium.

Patients with dry eye have the notable symptom of increased tear osmolarity.[6–8] The increased osmolarity presumably arises as a result of exposure of the tear film to air. The tear film bathing the corneal surface normally is covered by a thin lipid layer, which is secreted by the meibomian glands. Disruption of the lipid layer and direct exposure of the tear fluid to air is one of the calling cards of dry eye.

Another point of interest, as already indicated, is the possibility of decreased water flow in dry eye patients. Data presented by Holly[9] show that sustained or steady-state tear secretion rates in dry eye patients are half those of normal individuals. This strongly suggests a defect in the secretory process of the lacrimal gland acini and/or ducts. In this scenario, reduced flow would be most easily explained as a reduced ability to move solutes to the lumen, resulting in less total solvent flow to maintain osmotic equilibrium. The reduced solute transport across the epithelium could occur because of (i) reduced synthesis and insertion rates of ion/solute channels/transporters, (ii) changes in open probability of channels or binding affinity of transporters, and/or (iii) changes in driving forces for solutes.

The acinar cell is a polar epithelial cell that contains a variety of channel types. Patch clamp methods have revealed cation channels,[10] K^+ channels,[11–13] Cl^- channels,[13,14] and gap junction channels.[15] Many of these channel types are not just necessary for the maintenance of the resting potential of the acinar cell. A number of studies have illustrated the need for K^+ and Cl^- channels, for example, in the regulation of secretagogue-triggered vesicle release.[16,17] The acinar cells of the lacrimal gland also contain Na/K ATPase and Na/H and Cl/HCO_3 antiporters.[18,19] These transporters not only facilitate transepithelial transport but most likely are important in the regulation of cell volume. As already indicated, acinar/duct cells contain aquaporin 5.[5]

1.2. The Antithesis of Dry Eye

The ability of channels to affect water movement is best illustrated with an example from the gastrointestinal epithelium. Epithelial crypt cells of the mucosa of the small intestine contain Cl^- channels on their apical surfaces that under normal electrophysiological conditions generate an outward Cl^- flux, despite the fact that the Cl^- chemical gradient favors inward movement of Cl^-. The reason is that the Cl^- equilibrium potential is usually more positive than the resting potential of the cell. Cytosolic Cl^- is elevated in the crypt cells by basally located cotransporters of cations and Cl^-.[20] Cholera toxin has the ability to increase the apical Cl^- channel activity via cAMP-mediated phosphorylation, which results in increased Cl^- in the lumen. Cl^- absorption from the lumen occurs via passive diffusion but is hampered in rate by extensive tight junctions between all of the epithelial cells of the villi and crypts. The end result of the enhanced activity of the apical Cl^- channel is to increase the number of solutes in the lumen, and solvent will follow or remain. This increased volume is not absorbed, and diarrhea is the usual outcome. If the Cl^- equilibrium potential is more negative than the resting potential of the cell, a Cl^- influx will occur. This last circumstance would result in water resorption.

1.3. A Fluid Reduction Hypothesis

Our approach has been to monitor the electrophysiological properties of two mammalian model systems, the NZB mouse and the MRL mouse, and compare them with control

(SW) mice. We used patch clamp methods to monitor whole-cell currents and single channels.[21] The question simply put is: Are there differences in properties between control and NZB/MRL mouse lacrimal acinar cells that indicate a possible malfunction in transepithelial transport of solutes that might reduce water movement from interstitium to lumen?

2. METHODS

2.1. Cell Isolation

NZB or control (SW) mice were gassed in halothane for 10 min. Under sterile conditions, the lacrimal glands were removed and placed in warmed (37°C) soybean trypsin inhibitor (STI) containing 1% penicillin and streptomycin. With two sharp scalpel blades, the glands were chopped into pieces less than 2 mm in size. The STI was removed and the pieces washed in warmed Hank's buffered saline solution (HBSS). The HBSS was removed, and the gland fragments were placed in sterile warmed HBSS containing 0.7 mg/ml EDTA for 15 min in a 37°C water bath. The EDTA HBSS was pipetted off and the tissue washed with STI. The STI was removed and replaced with collagenase (1 mg/ml), hyaluronidase (1 mg/ml; 5000 units/mg), and DNase (2000 units/ml) made in 5 ml of DMEM (CHDI) that was filtered through a 0.45 μm membrane. The CHDI-bathed cells were kept at 37°C for 25 min. The tissue (CHDI bathed) was collected and centrifuged at 1200 rpm for 5 min. The supernatant was removed and replaced with 5 ml of DMEM that contained carbamyl choline, EGF, ascorbic acid, and insulin. The pellet was resuspended and then pipetted and filtered through a sterile polypropylene mesh (500 μm) to remove larger tissue pieces. The cells were centrifuged again at 1200 rpm for 5 min and resuspended in 2 ml of enhanced DMEM (insulin, etc). 0.5 ml aliquots of cell suspension were plated into 35 mm plates coated with 100 ml of Matragel and allowed to sit for 5 min. Subsequently 2 ml of DMEM was added, and then the plated cells were placed in an incubator (37°C) overnight. Cells were successfully patched 24, 48, and 72 h after isolation.

2.2. Patch Clamp

Isolated acinar cells were perfused with NaCl saline containing 110 mM NaCl, 1 mM $MgCl_2$, 1 mM $CaCl_2$, and 10 mM HEPES, pH 7.2. Electrodes were produced on a P-97 Sutter puller. An Axopatch 200 series patch amplifier was also employed. The pipette solution contained 110 mM KCl, 0.9 mM EGTA, 0.1 mM $CaCl_2$, 0.1 mM $MgCl_2$, 10 mM HEPES, pH 7.05. Whole-cell and detached recording modes were utilized.[21] In the whole-cell configuration, cells were held at 0 mV or at their resting potentials. The step regime used was 20 mV steps starting at +10 mV from the holding potential, followed by a polarity switch of equal magnitude. For recordings in the detached configuration, holding potentials were applied for at least 30 sec and for as long as 10 min.

3. RESULTS

Two examples of whole-cell patch clamp are shown in Fig. 1.

Two features are different for the two examples. First, the outward rectification of the control is much less prevalent in the NZB cell, and the total cell conductance of the

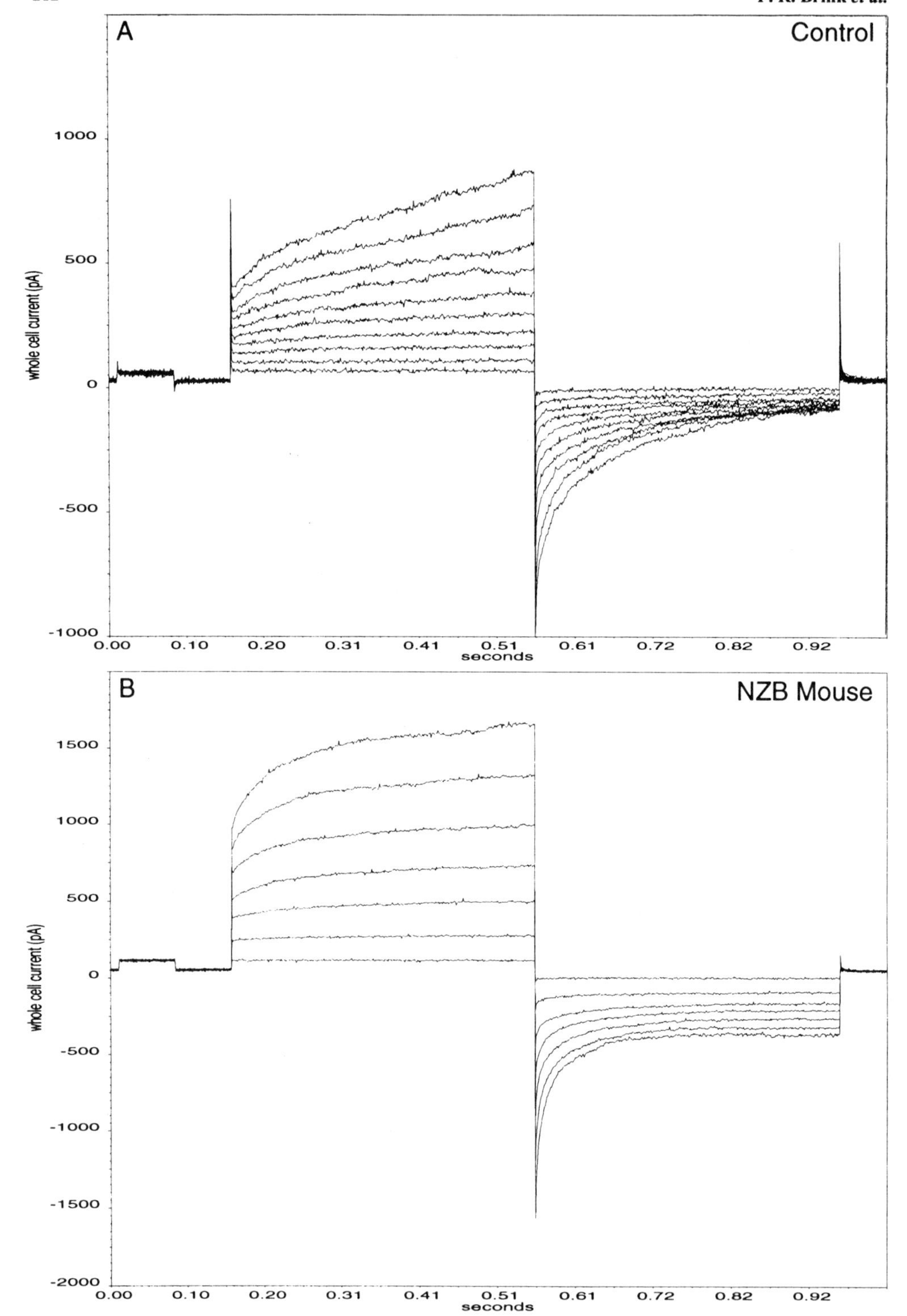

Figure 1. Whole-cell currents from (A) control acinar cell and (B) 5 month NZB mouse acinar cell. Step protocol was to step from +10 to +150 mV followed by a step of equal magnitude of opposite polarity. For the NZB cells, the steps did not exceed ± 130 mV. Holding potential = 0 mV.

Table 1. Resting potentials

	Average (mV)	Range (mV)	No. of experiments
Control (SW)	-28	-48 to -13	5
NZB	-4	-21 to + 8	5

NZB is ~2X larger than the control. The same step protocol was used for both preparations, except for the NZB where the steps did not exceed ± 130 mV. Another feature that can be determined from Fig. 1 is the resting potential of the cells. For the control cell, V_m=-20 mV; for the NZB, -2 mV. Table 1 shows the data taken from 5 control and 5 NZB cells. The NZB cells are depolarized relative to the control cells by 24 mV.

The whole-cell data strongly suggest that there is an increase in the number of active cation and/or anion channels in the membranes of NZB acinar cells relative to controls. Fig. 2 shows an example of a commonly found channel type in the control acinar cells: the maxi-K channel. The data shown here are from a detached (inside-out) patch where the bath is NaCl saline and the pipette solution is 110 mM KCl (see Methods). The slope conductance is 130 pS, and the reversal potential appears to be the K equilibrium potential. In three of six patches made from control mice, maxi-K channels were found. In 1 one of eight patches made from NZB cells, maxi-K channels were found.

In both cell preparations it was possible to find channel activity that was a result of either cation or anion channels, or both. Fig. 3 shows a detached patch (inside-out) taken from an NZB cell. The holding potential was held constant and, as indicated, a Cl^- channel blocker (flufluamic acid) was added to the bath. The application of the drug resulted in reduced channel activity. These channels are most likely cation channels. The I-V relationship shows the reversal potential (Er) to be +5 mV. Thus, this can not be a K^+ channel (Er~-100 mV). The +5 mV translates into a channel that is almost equally permeable to K^+ and Na^+. Previous reports indicate that the unitary conductance of the cation channel is 20–30 pS, which is similar to the 22 pS found here. Previous reports indicate that the Cl^- channel conductance is 10 pS.

The only parameter measured that did not seem to be affected in NZB or MRL mice was junctional membrane conductance. Fig. 4 shows an example of junctional currents generated in response to transjunctional voltage steps from ± 10 to ± 150 mV from an MRL mouse acinar cell pair (1.5 months). It is the same as control or NZB cells. Steady-state junctional conductance in MRL and control mice shows no difference in the voltage dependence or the magnitude of the conductance.

4. DISCUSSION

4.1. Other Fluid Sources

The conjunctiva itself is a potential fluid source, much like the acinar and duct cells, but no secretory proteins would be expected from this source. These cell types might also display electrophysiological behaviors concomitant with reduced transepithelial fluid flow.

Another explanation for reduced tear volume is related to transepithelial solute transport. The focus here is on the concentration of material within the secreted vesicles of

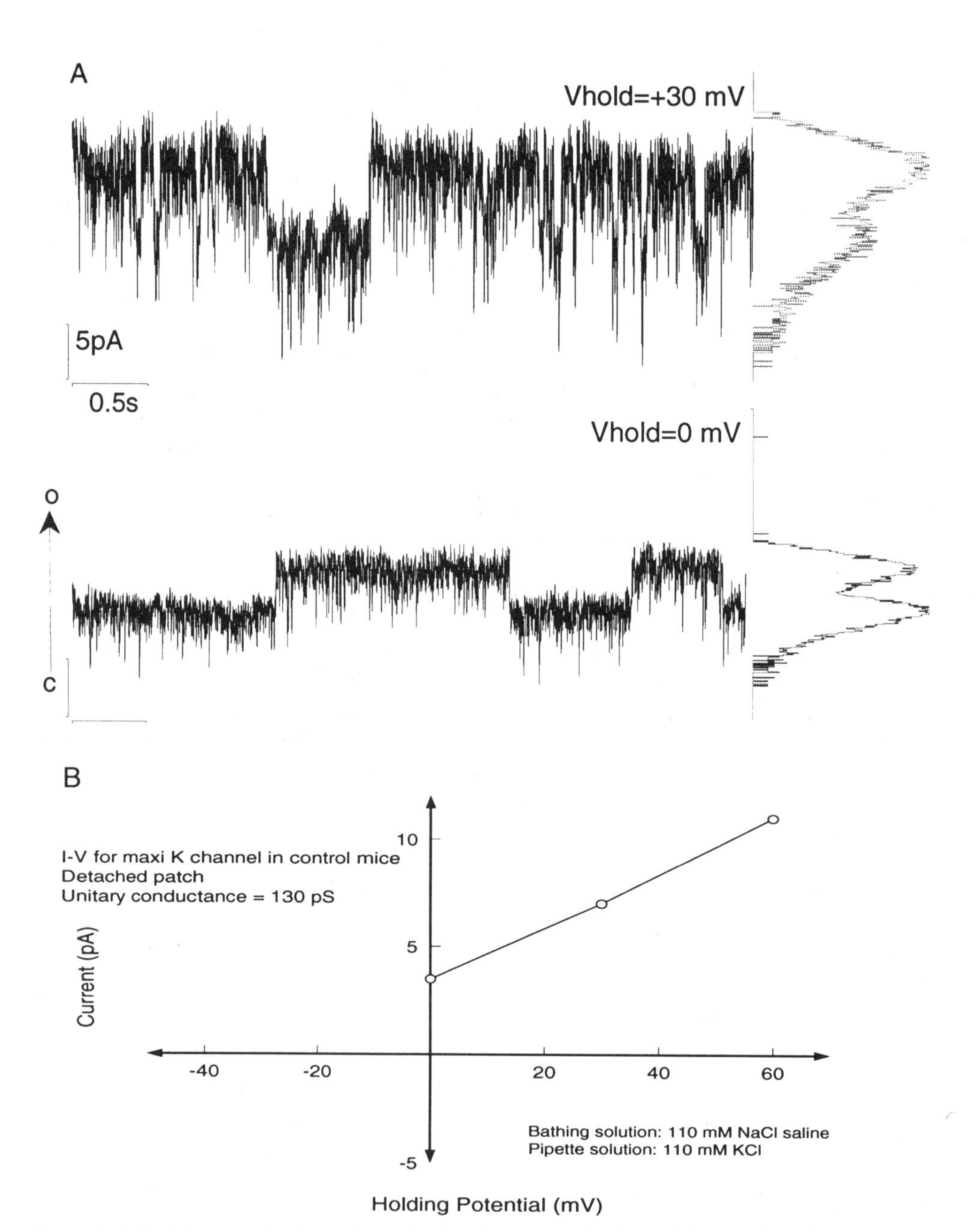

Figure 2. (A) Inside-out patch of a membrane detached from a control acinar cell showing channel activity at two holding potentials (+30 and 0 mV). The bath contained NaCl saline and the pipette KCl solution. (B) I-V plot of data in (A). The slope conductance was 130 pS.

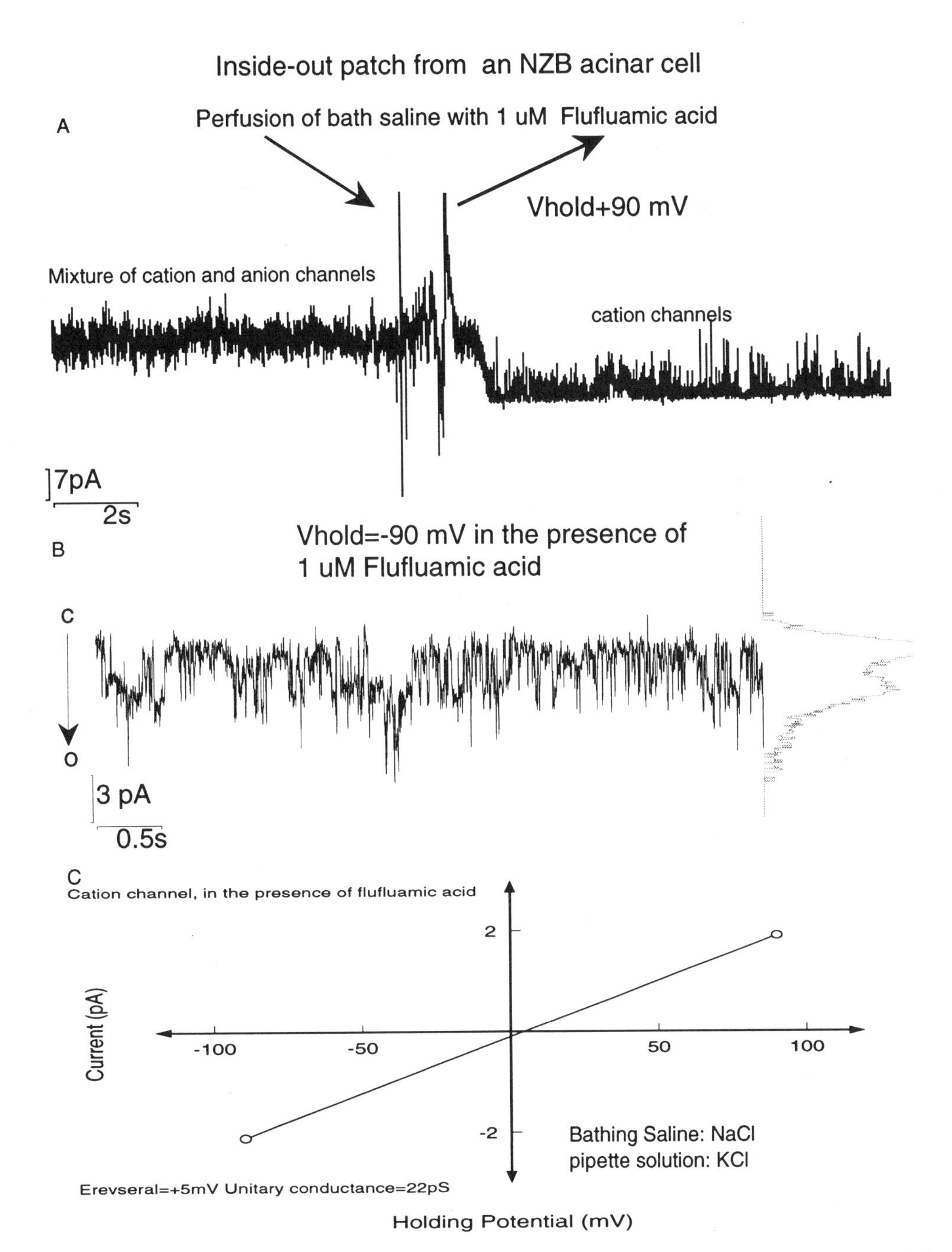

Figure 3. Same conditions as in Fig. 2, but the patch comes from an NZB cell (5 months). (A) The records indicate some Cl^- current present. The magnitude of the current decreases after the application of a Cl^- channel blocker (flufluamic acid, 1 μM). (B) Remaining channel activity. (C) Remaining activity has a unitary conductance of 22 pS and a reversal potential of +5 mV. These data are consistent with previous data reported for cation channels in acinar cells.

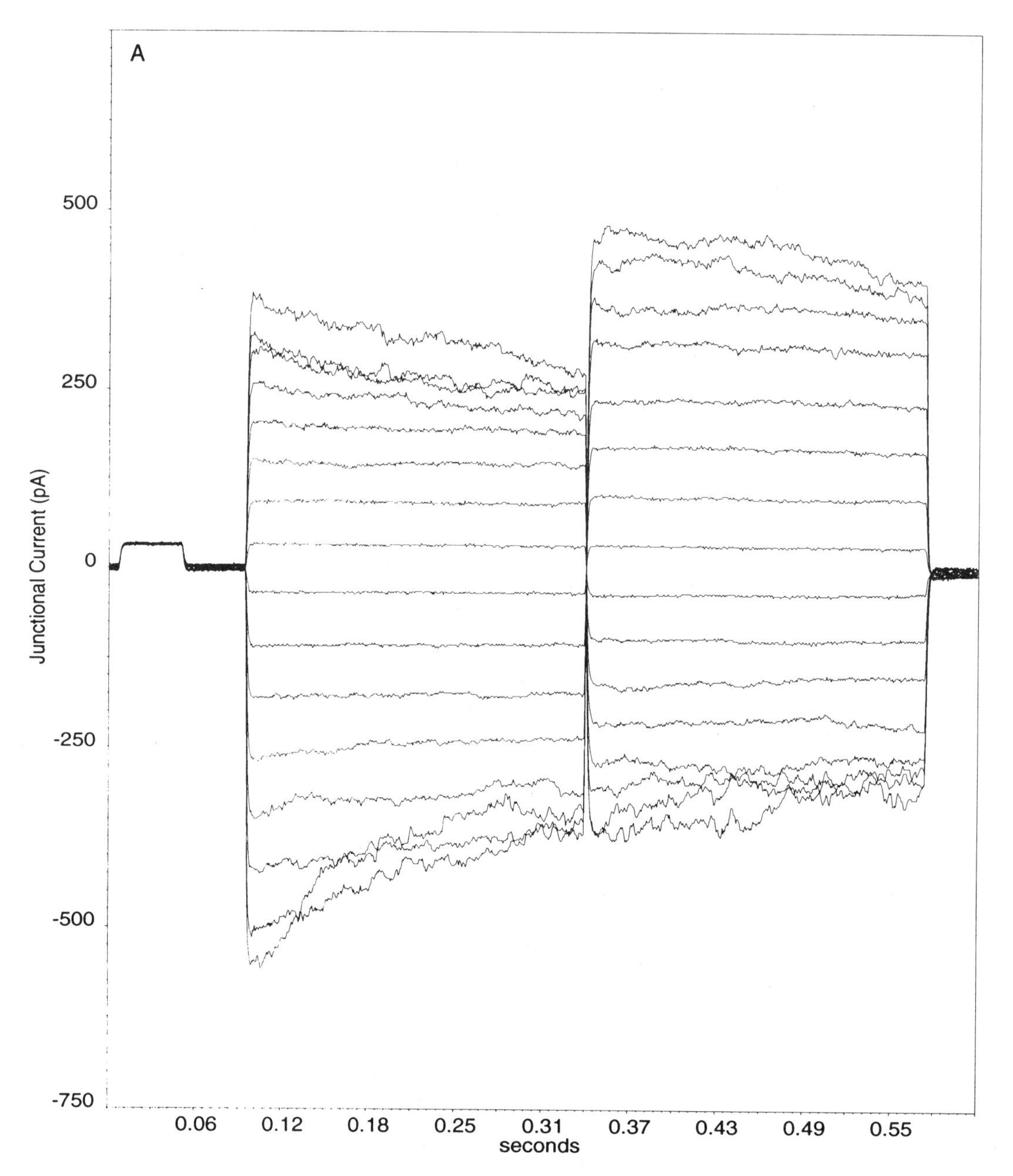

Figure 4. (A) Gap junctional membrane currents in MRL mice generated in response to transjunctional voltage steps varying in magnitude and polarity from ± 10 mV to ± 150 mV. (B) Steady-state conductance from a control mice acinar cell pair, MRL cell pair, and normal rat cell pair. NZB cell pairs behaved the same.

the acinar cells. Under normal conditions, the contents of the vesicles are concentrated relative to the cytosol. The vesicles are hyperosmotic. Thus, release of the vesicles in normal individuals would deliver to the lumen solutes that would cause the movement of water from interstitium to lumen. If acinar cells in dry eye individuals lost the ability to concentrate vesicles, the result would be less water movement into the lumen. One possi-

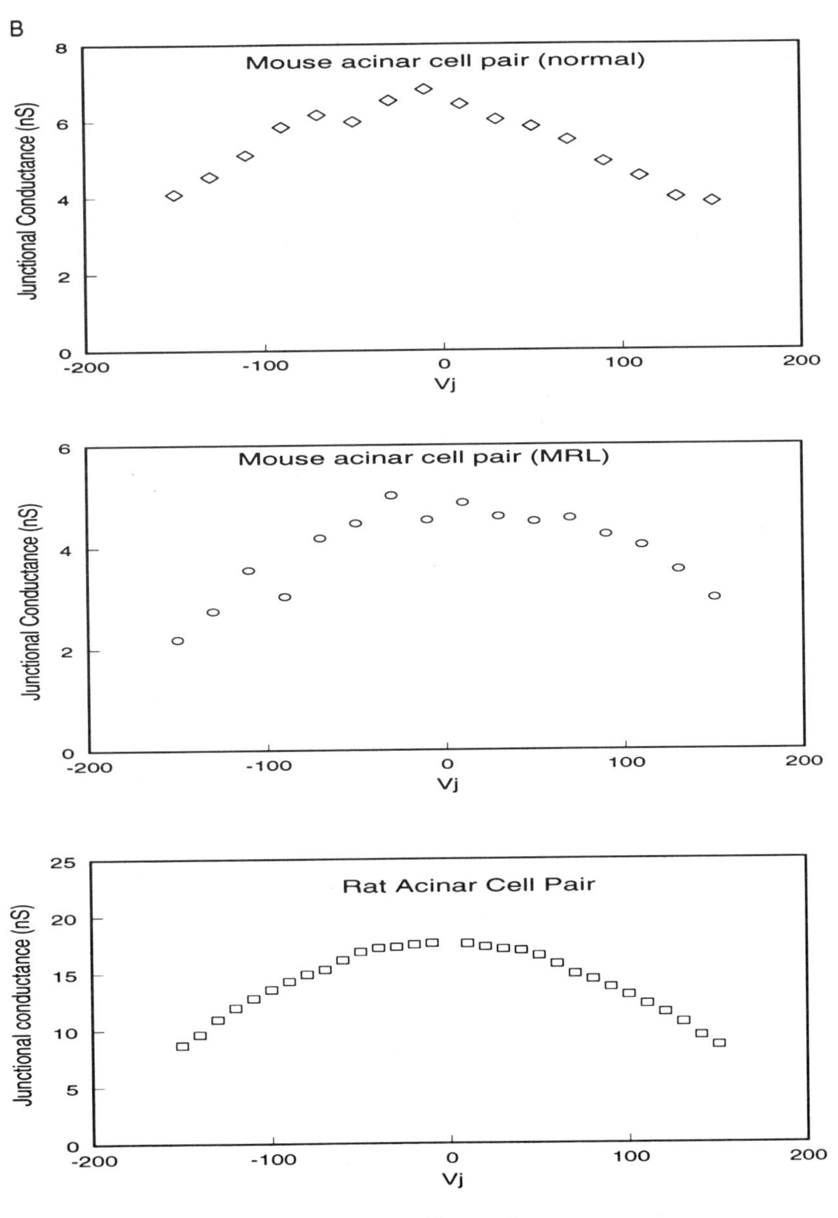

Figure 4. *(Continued).*

ble mechanism could be the conversion of acinar cells vesicle release from secretagogue-like secretion to constitutive secretion. This latter case results in isosmotic vesicles. For this constitutive model, the prediction is that receptor-mediated secretion would be absent or greatly reduced under dry eye conditions. Data presented by Walcott et al.[22] argue against the secretagogue/constitutive conversion hypothesis. The NZB acinar cells do possess a secretagogue secretory component, and there appears to be no elevated baseline (constitutive) secretion relative to control cells.

4.2. An Innervation Explanation

The ultimate source of fluid for the acinar and duct cells is the vasculature underlying the epithelium. Sustained sympathetic input could cause reduced fluid flow due to vasoconstriction, as can be the case in salivary gland. There are no data on sympathetic innervation to the vasculature in lacrimal gland, controls or otherwise; thus this possibility remains unanswered.

4.3. Efflux to Influx Cl^-, K^+, Na^+

The data presented are consistent with reduced fluid flow from interstitium to lumen of the lacrimal gland. The reduced resting potentials of the cells and domination of nonrectifying currents (cation and anion) and the apparent reduction in K^+ activity argue for reduced water movement.

Given the reduced efficiency of outward K^+ current and the elevated cation and anion currents found in NZB, a situation in which little solute is moved from the interstitium or cytosol into the lumen of the gland is plausible, especially if the reversal potentials for the anion and cation channels are similar to the resting potential (no net driving force). If the Cl^- reversal potential is not the same as the resting potential, then for Cl^- to effect an efflux, the intracellular concentration would have to exceed the extracellular Cl^- concentration, an unlikely situation. Recall the membrane potential is depolarized (-4 mV) relative to controls (-28 mV). If the Cl^- reversal potential is more negative than the resting potential, a Cl^- influx ensues, resulting in solute removal from the lumen (assuming an apically placed Cl^- channel). For this latter case, water would flow out of the lumen back into the interstitium.

Cl^- conductance in the mouse acinar cells is known to be sensitive to intracellular calcium levels.[13,14] It is possible that the apparent elevation of whole-cell conductance in NZB cells is due to increases in the total number of active channels, either by increased production of channels or by elevation of the cytosolic free calcium and subsequent increases in open probability, or both.

Na^+ would be expected to diffuse into the lumen via the extracellular space. Tight junctions are the rate-limiting factor in the extracellular path for Na^+ and K^+ and Cl^- as well. Whether the tight junction-mediated path between interstitium and lumen is altered in NZB or MRL cells remains to be determined. Changes in tight junction permeability to solutes could play an important role in water movement in lacrimal gland acinar and duct cells. Although if tight junctions are sufficiently leaky in acini and ducts, the tear fluid would be expected to be isosmotic, and it is hard to envision this case resulting in lower tear volume. Further, normal tears are slightly hypotonic relative to plasma (300 vs 320 mosM),[7,20] implying a relatively tight epithelium.

In the experiments shown here, no ATP was present in the extracellular bath or in the pipette. We chose not to use ATP to test for the presence of a specific ATP-activated cation channel found in lacrimal gland acini.[10] However, this channel type will also have to be investigated with regard to dry eye.

Our results suggest that there are molecular changes within the acinar cells of NZB mice that could affect water transport. In addition, we have indicated a number of other possibilities and predictions for dry eye conditions where tear volume generated by the lacrimal gland is reduced.

ACKNOWLEDGMENT

The work was support by NIH grant EY09406.

REFERENCES

1. Montgomery PC, O'Sullivan NL, Martin LB, Skandera CA, Peppard JV, Pockley AG. Regulation of lacrimal gland immune responses. *Adv Exp Med Biol.* 1994;350:161–168.
2. Verkman, AS, Van Hoek AN, Ma T, et al. Water transport across mammalian cell membranes. *Am J Physiol.* 1996;270:C12-C30.
3. Agre P, Preston GM, Smith BL, et al. Aquaporin CHIP: The archetypal molecular water channel. *Am J Physiol.* 1993;265:F465-F476.
4. Grompert BD. *The Plasma Membrane: Models for Structure and Function.* New York: Academic Press; 1977, pp 42–45.
5. Agre P, Brown D, Nielsen S. Aquaporin water channels: Unanswered questions and unresolved controversies. *Curr Opin Cell Biol.* 1995;7:472–483.
6. Gilbard JP, Farris RL, Santamaria J. Osmolarity of tear microvolumes in keratoconjunctivitis Sicca *Arch Ophthalmol.* 1978;96:677–681.
7. Farris RL. Tear osmolarity- a new gold standard. *Adv Exp Med Biol.* 1994;350:495–504.
8. Gilbard JP, Rossi SR. Changes in tear ion concentration in dry-eye disorders. *Adv Exp Med Biol.* 1994;350:529–532.
9. Holly FJ. Lacrimation kinetics as determined by a Schirmer-type technique. *Adv Exp Med Biol.* 1994;350:543–548.
10. Thorn P, Petersen OH. Nonselective cation channels in exocrine gland cells. In: Siemaen D, Hescheler J, eds. *Nonselective Cation Channels: Pharmacology, Physiology and Biophysics.* Basel: Birkhauser; 1993:185–200.
11. Park KP, Beck JS, Douglas IJ, Brown PD. Ca^{2+} activated K^{+} channels are involved in regulatory volume decrease in acinar cells isolated from the rat lacrimal gland. *J Membr Biol.* 1994;141:193–201.
12. Findlay I. A patch-clamp study of potassium channels and whole cell currents in acinar cells of the mouse lacrimal gland. *J Physiol (Lond).* 1984;350:179–195.
13. Marty A, Tan YP, Trautmann A. Three types of calcium dependent channel in rat lacrimal gland. *J Physiol (Lond).* 1984;357:293–325.
14. Evans MG, Marty A. Calcium dependent chloride currents in isolated cells from rat lacrimal glands. *J Physiol (Lond).* 1986;378: 437–460.
15. Neyton J, Trautmann A. Single channel currents of an intercellular junction. *Nature.* 1985;317:331–335.
16. Petersen OH, Maruyama Y. Calcium activate potassium channels and their role in secretion. *Nature.* 1984;307:693–696.
17. Findlay I, Petersen OH. ACh stmulates a Ca^{2+} dependent Cl^{-} conductance in mouse lacrimal acinar cells. *Pflugers Arch.* 1985;403 328–330.
18. Mircheff AK, Lambert W. Ross, Lambert RW, Maves CA, Gierow JP, Wood RL. *Adv Exp Med Biol.* 1994;350:79–86.
19. Lambert RW, Bradley ME, Mircheff AK. Cl^{-}/HCO_3 antiport in rat lacrimal gland. *Am J Phyisol.* 1988;255:G367-G373.
20. Berne RM, Levy MN. *Physiology.* St. Louis: C.V. Mosby; 1993;327–328.
21. Hamill OP, Marty A, Neher E, Sakaman B, Sigworth FJ. Improved patch clamp techniques for higher resolution current recording for cells and cell-free membrane patches. *Pflugers Arch.* 1981;391:85–101.
22. Walcott B, Carlos N, Patel A, Brink PR. Age-related decrease in innervation density of the lacrimal gland in mouse models of dry eye. This volume.

CELLULAR ORIGIN OF MUCINS OF THE OCULAR SURFACE TEAR FILM

Ilene K. Gipson and Tsutomu Inatomi

Schepens Eye Research Institute, and Department of Ophthalmology
Harvard Medical School
Boston, Massachusetts

1. INTRODUCTION

The surface of the eye is covered by a tear film that has been estimated to be 40 µm thick.[1,2] The film is extraordinarily regular over the cornea where it is the primary refractive surface of the eye. The mucus within the tear film provides the major structural component of the layer, and recent estimates of its thickness, 30 µm, suggest that it contributes most of the thickness of the tear film. The mucus layer has been estimated to be thicker and more viscous at the very apical surface of the epithelial cells, with a diminution of viscosity towards the tear film surface.[3]

2. MUCINS: MOLECULAR CHARACTERISTICS

Mucins are the structural or skeletal molecules of the mucus layer. These highly glycosylated proteins are at least 50% carbohydrate by mass. Saccharides—mono, di, or oligo—are O-linked through serine and threonine to the mucin protein backbone. Heterogeneity in glycosylation of mucins has been reported; heterogeneity is evident between tissues, individuals, and species (for review, see refs. 4–6).

Because of their heavy glycosylation and large size, characterization of mucins at the molecular level has been slow. Recent reports indicate that at least eight or nine mucin genes are expressed in human tissues (for review, see ref. 6). The mucins have been designated MUC1–4, 5AC, 5B, 6–8. Of these, complete cDNA sequence is available for MUC1 (4–7 kb), MUC2 (15.5 kb), and MUC7 (2.3 kb). Sequence data from these three mucins, plus partial sequences from the remaining mucins, show several structural features of this class of glycoproteins. Each has tandem repeats of amino acids in the protein backbone.

Lacrimal Gland, Tear Film, and Dry Eye Syndromes 2
edited by Sullivan *et al.*, Plenum Press, New York, 1998

These tandem repeat units are rich in serine and threonine, but between the individual mucin genes (for review, see ref. 6) the units vary in length, nucleotide sequence, and number of times they are repeated in the protein backbone.

Several categories of mucins can be designated based on their structural features. One of the cloned mucins, MUC1, has a membrane-spanning domain. This is the most studied of the mucins. In addition to its membrane-spanning sequence, MUC1 has a small, cytoplasmic tail that is conserved between species and a long, extracellular domain with a variable number of tandem repeats, each of which has 20 amino acids. The number of tandem repeats is variable due in part to the polymorphic expression pattern of the mucin gene. The extracellular domain of MUC1 has been estimated to extend 200–500 nm into the glycocalyx of cells (see ref. 6 for review). Thus the molecule could extend some 2–5 μm outside the cell membrane into a mucus layer.

Four (and possibly five) of the cloned mucins are secreted, gel-forming mucins. These mucins, MUCs 2, 5AC, 5B, and 6 (and possibly 3) are all very large and have cysteine-rich regions outside their tandem repeat domains that are available for intermolecular disulfide bonding[6] (MUC6 data, N. Toribara, personal communication). The cysteine-rich regions of the gel-forming mucin genes (at least MUCs 2, 5AC, 5B, and 6) share homologies, and there is homology between parts of these regions and the D4, C1, and C-terminal domains of von Willebrand factor.[4,5] This blood coagulation factor, like mucins, forms dimers or polymers through disulfide bonds; such dimerization is required for polymerization and packaging of the factor in cytoplasmic vesicles.[6,7] In MUCs 2 and 6, both 5' and 3' regions have cysteine-rich domains that flank the tandem repeat region(s). These amino and carboxyl-terminal regions may be responsible not only for gel formation but also for packaging of the mucins in secretory vesicles. Experimental evidence that the carboxyl-terminal, cysteine-rich region of mucins is involved in dimerization comes from transfection of COS 7 cells with the carboxyl-terminal, cysteine-rich region of a salivary mucin gene. The transfection yielded secreted proteins that formed dimers.[8]

These large, gel-forming mucins may be the major structural components of mucus layers on all wet-surfaced mucosal epithelia in the body. Tissue-specific patterns of expression of these gel-formers is evident, i.e., trachea/bronchus expresses MUCs 2, 5AC, and 5B; stomach expresses 5AC and 6; colon expresses 2, 3, and 5B, and endocervix expresses 5AC, 5B, and 6.[4,9]

Of the remaining three mucins, MUC7 is the best understood. It is a small, secreted, monomeric, non-gel-forming mucin, first described in the salivary gland.[12] It appears to have limited tissue distribution. Little is known of the structural and functional features of MUCs 4 and 8. MUC4 appears to have wide tissue distribution and, unlike other mucins, is expressed not only in simple epithelium but also in stratified, squamous, non-keratinizing epithelium, i.e., ectocervix and trachea.[4,9] Since only tandem repeat sequence is available for MUC4, it is not clear which category this mucin falls into, secreted or transmembrane.

3. EXPRESSION OF MUCIN GENES BY OCULAR SURFACE EPITHELIA

It is generally assumed that the goblet cells of the conjunctiva provide the major mucins that assemble to form the mucus layer of the tear film. Greiner et al.[10] and Dilly[11] provided histochemical data that led to their suggestion that the stratified epithelium of the conjunctiva is a second source of mucins for the tear film. Other cellular sources of mucins have been proposed. Based on studies of sialic acid content or presence of acidic glycoproteins, several authors suggested that the lacrimal gland may secrete mucins for the tear film.[12–14] By analogy, one could also suggest that accessory lacrimal glands may be a source of mucins for the ocular surface.

Recent data from our laboratory demonstrate that at least three mucin genes are expressed by the ocular surface epithelia.[15,16] Using Northern blot analysis and in situ hybridization, we have shown that both corneal and conjunctival epithelium express the transmembrane mucin MUC1. Message for the mucin gene is present within cells of all layers of the epithelium. Curiously, antibodies to the extracellular domain of MUC1 bind only to the apical cells of the corneal epithelium. In conjunctiva, basal as well as apical cells bind the MUC1 antibodies.

The function of MUC1 is not entirely clear, although it has been proposed that it is a molecule that functions as a disadhesive. Several lines of evidence support this hypothesis. Cells transfected with the MUC1 gene and expressing the mucin at high levels were less adhesive to one another.[17] Similarly, in Muc1 null mice, blastocysts can adhere to the uteri of null mice anytime during the estrus cycle; normal mice in which blastocyst adherence occurs only at a specific time in the estrus cycle, just after ovulation, appear to downregulate Muc1 at the time of implantation.[18] Taken together, these two studies suggest that MUC1 prevents adhesion of cells to one another. The presence of MUC1 on the apical, luminal surfaces of epithelia may thus prevent adherence of other cells, foreign debris, or pathogens. On the ocular surface, MUC1 may prevent adhesion of the lid epithelium or inflammatory cells to the corneal surface as well as prevent adhesion of pathogens or other debris to the surface of epithelia. Similarly, it may prevent tight adherence of the mucus coat of the tear film to the ocular surface. One can envision a loose association of the mucus of the tear film for the movement of the layer which may be required to remove surface debris.

A major gel-forming mucin on the surface of the eye appears to be MUC5AC.[16] (We define a major mucin as one whose mRNA is detectable by Northern blot analysis and in situ hybridization.) Northern blot analysis of RNA from human conjunctiva demonstrated binding of a cDNA probe to MUC5AC mRNA in the polydispersed pattern characteristic of the gel-forming mucins. In situ hybridization on ocular surface epithelia, using this probe, demonstrated the message in conjunctival goblet cells. Fluorescence in situ hybridization studies show that the message is detected primarily in the region of the goblet cell between the nucleus and the apical mucin packet-filled region of the cell (Fig. 1A,B). Since none of the other gel-forming mucins were detected in ocular surface epithelia (see below), this mucin may be the major structural mucin of the tear film mucus.

We could not detect by Northern blot analysis any of the other gel-forming mucins in ocular surface epithelial RNA. At the 1995 ARVO meeting, several reports were made that MUC2 was detectable by Northern and Western blot analyses.[19,20] We too detected the mucin by PCR, but upon repeated study could not find MUC2 message that was detectable by Northern blot analysis or in situ hybridization either in human or rat conjunctiva (for complete description, see ref. 19). We conclude from these studies that MUC2 is not a major mucin of the ocular surface.

There have been no previous reports on the presence of the gel-forming mucin MUC5B in ocular surface epithelia. Using a cDNA probe to the tandem repeat of the mucin, MUC5B was not detected by Northern blot analysis of conjunctival RNA (Fig. 2).

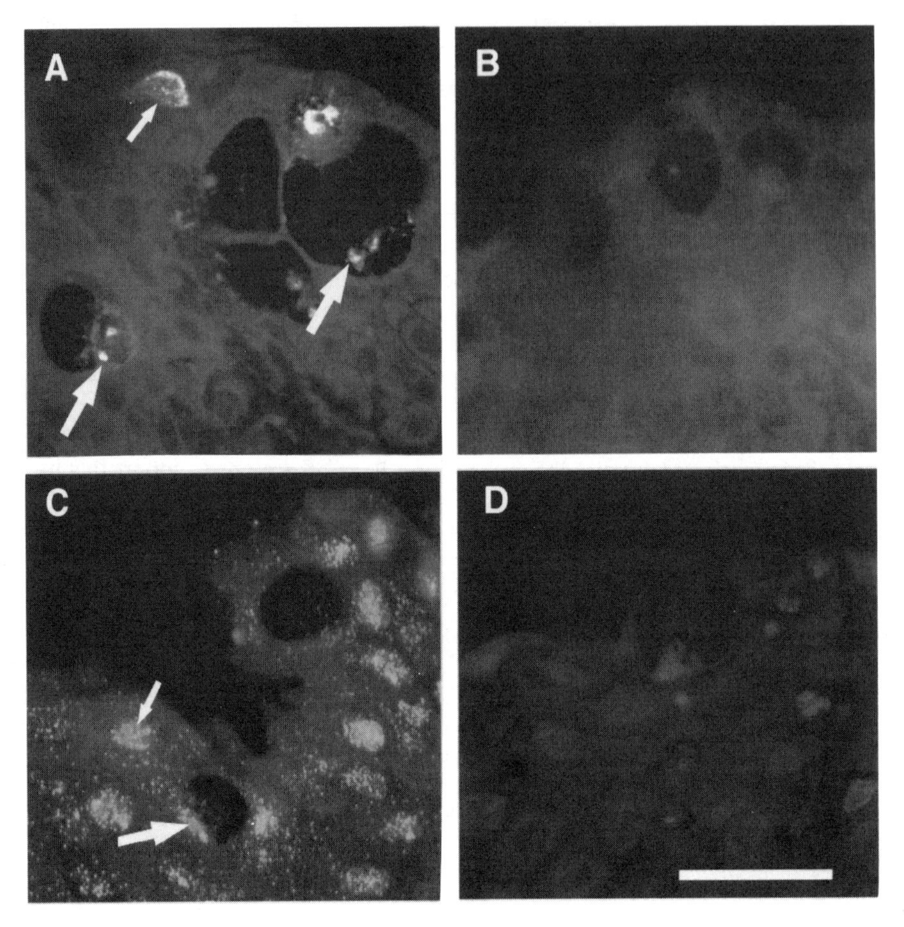

Figure 1. Fluorescence in situ hybridization showing localization of MUC5AC (A) and MUC4 (C) mRNA in sections of conjunctival epithelium. (B) and (D) show respective sense control incubations. MUC5AC message (A) is present only in goblet cells and is localized between the nuclei of goblet cells and the mucin packet region (large arrows). One labeled cell has no apparent mucin packet region (small arrow). This could be a result of the plane of the section or may indicate an undifferentiated cell prior to development of the goblet cell phenotype. MUC4 message (C) appears to be present in both stratified epithelial cells (small arrow) and goblet cells (large arrow). Message appears especially dense around the nucleus. Sense controls for both MUC5AC (B) and MUC4 (D) show no label. Bar = 20 μm.

Thus, like MUCs 2, 3, and 6, MUC5B does not appear to be a major mucin of the ocular surface epithelia tested to date.

The relatively uncharacterized mucin MUC4 is expressed by the conjunctival epithelium.[16] By Northern blot analysis, message for the mucin is detectable as a smear in blots of conjunctival RNA. Fluorescence in situ hybridization demonstrates that message is primarily in the stratified cells of the epithelium, but one can detect some goblet cell labeling as well. Curiously, the label appears to be concentrated around the nucleus of cells; RNase digestion of sections eliminated binding of the MUC4 oligonucleotide probe to the

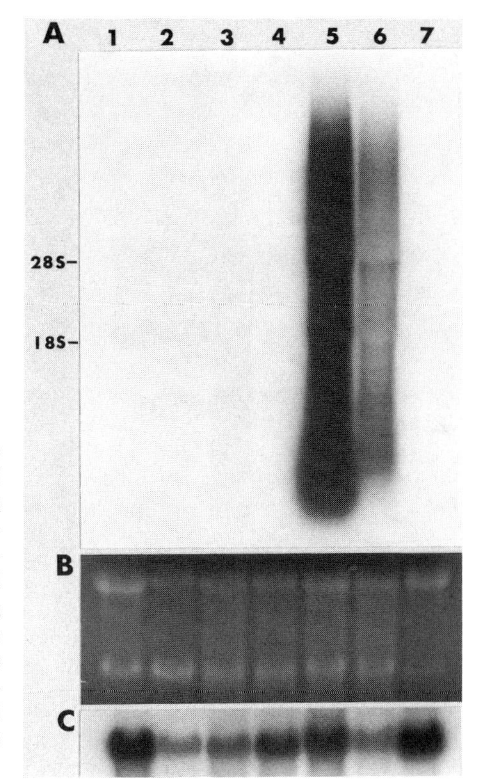

Figure 2. Northern blot analysis for MUC5B mRNA in corneal and conjunctival epithelia. Ten micrograms of total RNA from various human tissues was loaded in each lane and probed for MUC5B using a cDNA probe and techniques described previously (A).[11] MUC5B mRNA expression was not detected in either cultured human corneal epithelial or conjunctival RNA. Ethidium bromide staining (B) and β-actin (C) show the integrity of RNA of samples shown in (A). Molecular weights of 28S and 18S ribosome RNA are indicated. Lane 1: cultured human corneal epithelium. Lane 2: conjunctiva. Lane 3: stomach. Lane 4: small intestine. Lane 5: trachea. Lane 6: salivary gland. Lane 7: human umbilical vein endothelial cell.

conjunctival sections, indicating that the probe is binding mRNA (Fig. 1C,D). The function of MUC4 at the ocular surface is unclear, and since little sequence data are available for this mucin, it has yet to be determined if it is a secreted, membrane-spanning, gel-forming, or soluble mucin. Likely it is not a soluble mucin since it appears to have a large message size and it exhibits a polydispersed pattern on Northern blot analyses.[16]

We report here that the lacrimal gland does not appear to be expressing any of the secretory mucins cloned to date. We screened lacrimal gland RNA by Northern blot analysis, with probes to MUCs 2, 3, 4, 5AC, 5B, 6, and 7 (for methods, see ref. 19). We could not detect message for any of these mucins (Fig. 3). These data do not rule out the presence of other soluble, uncloned mucins similar to MUC7. Studies of secretory products of the lacrimal gland as well as the accessory glands may yield new mucin genes.

4. SUMMARY

In summary, we have demonstrated that the ocular surface epithelia express at least three mucin genes. We suggest that the gel-forming mucin MUC5AC is a major mucin forming the mucus gel of the tear film. We further suggest that MUC1 facilitates the spread of the MUC5-containing mucus on the ocular surface and, along with the mucus gel, prevents cell and debris adhesion to the ocular surface. The function of MUC4 at the ocular surface remains to be elucidated.

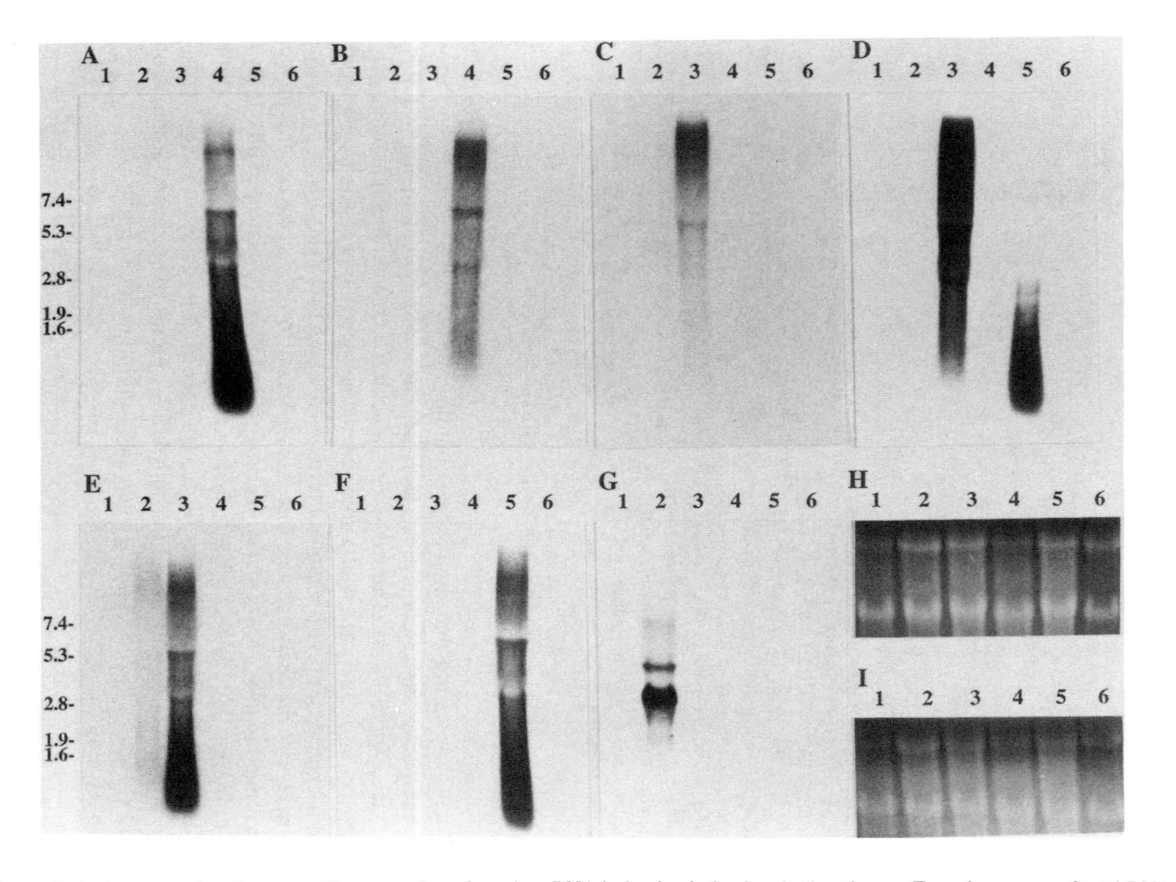

Figure 3. Northern blot analysis demonstrating tissue-specific expression of mucin mRNA in lacrimal gland and other tissues. Ten micrograms of total RNA from human lacrimal gland and various tissues was loaded per lane. Membranes were hybridized with probes for either MUC2 (A), MUC3 (B), MUC4 (C), MUC5AC (D), MUC5B (E), MUC6 (F), or MUC7 (G). Integrity of RNA is shown by ethidium bromide staining of the two blots used for A-D and E-G, and are shown in H and I, respectively. Techniques and probes have been described previously.[11,19] No binding was observed to the lacrimal gland RNA. Ethidium bromide staining of the two membranes used for assays for specific mucins showed integrity of RNA. Molecular weights of marker are indicated. Lane 1: lacrimal gland. Lane 2: salivary gland. Lane 3: trachea. Lane 4: small intestine. Lane 5: stomach. Lane 6: salivary gland.

REFERENCES

1. Prydal JI, Campbell FW. Study of precorneal tear film thickness and structure by interferometry and confocal microscopy. *Invest Ophthalmol Vis Sci.* 1992;33:1996–2005.
2. Prydal JI, Artal P, Woon H, Campbell FW. Study of human precorneal tear film thickness and structure using laser interferometry. *Invest Ophthalmol Vis Sci.* 1992;33:2006–2011.
3. Dilly PN. Structure and function of the tear film. *Adv Exp Med Biol.* 1994;350:239–247.
4. Carlstedt I, Sheehan JK, Corfield AP, Gallagher JT. Mucous glycoproteins: A gel of a problem. *Essays Biochem.* 1985;20:40–76.
5. Strous GJ, Dekker J. Mucin-type glycoproteins. *Crit Rev Biochem Mol Biol.* 1992;27:57–92.
6. Gendler SJ, Spicer AP. Epithelial mucin genes. *Annu Rev Physiol.* 1995;57:607–634.
7. Keates AC, Nunes DP, Afdhal NH, Troxler RF, Offner GD. Molecular cloning of a major human gallbladder mucin. Complete carboxyl-terminal sequence and genomic organization of the D4 domain of MUC5B. *Biochem J.* in press.
8. Sadler JE. von Willebrand Factor. *J Biol Chem.* 1991;266:22777–22780.
9. Shelton-Inloes BB, Titani K, Sadler JE. cDNA sequences for human von Willibrand factor reveal five types of repeated domains and five possible protein sequence polymorphisms. *Biochemistry.* 1986;25:3164–3171.
10. Perez-Vilar JA, Eckhardt E, Hill RL. Porcine submaxillary mucin forms disulfide-bonded dimers between its carboxyl-terminal domains. *J Biol Chem.* 1996;271:9845–9850.
11. Gipson IK, Ho SB, Spurr-Michaud SJ, et al. Mucin genes expressed by human female reproductive tract epithelia. *Biol Reprod.* 1997;56:999–1011.
12. Bobek LA, Tsai H, Biesbrock AR, Levine, MJ. Molecular cloning sequence, and specificity of expression of gene encoding the low molecular weight human salivary mucin (MUC7). *J Biol Chem.* 1993;268:20563–20569.
13. Greiner JV, Weidman TA, Korb DR, Allansmith MR. Histochemical analysis of secretory vesicles in non goblet conjunctival epithelial cells. *Acta Ophthalmol.* 1985;63:89–92.
14. Dilly PN. On the nature and the role of the subsurface vesicles in the outer epithelial cells of the conjunctiva. *Br J Ophthalmol.* 1985;69:477–481.
15. Jensen OA, Falbe-Hansen I, Jacobsen T, Michelsen A. Mucosubstances of the acini of the human lacrimal gland (orbital part). *Act Ophthalmol.* 1969;47:605–619.
16. Kreuger J, Sokoloff N, Botelho SY. Sialic acid in rabbit lacrimal gland fluid. *Invest Ophthalmol Vis Sci.* 1976;15:479–481.
17. Allen M, P Wright, Reid L. The human lacrimal gland. A histochemical and organ culture study of the secretory cells. *Arch Ophthalmol.* 1972;88:493–497.
18. Inatomi T, Spurr-Michaud S, Tisdale AS, Gipson IK. Human corneal and conjunctival epithelia express MUC1 mucin. *Invest Ophthalmol Vis Sci.* 1995;36:1818–1827.
19. Inatomi T, Spurr-Michaud S, Tisdale AS, Zhan Q, Feldman ST, Gipson IK. Expression of secretory mucin genes by human conjunctival epithelia. *Invest Ophthalmol Vis Sci.* 1996;37:1684–1692.
20. Ligtenberg MJL, Buijs F, Yos HL, Hilkens J. Suppression of cellular aggregation of high levels of episialin. *Cancer Res.* 1992;52:2318–2324.
21. Braga VM, Gendler SJ. Modulation of Muc-1 mucin expression in the mouse uterus during the estrus cycle, early pregnancy and placentation. *J Cell Sci.* 1993;105:397–405.
22. Bolis S, Devine P, Morris CA. Detection and characterization of human ocular surface mucins. ARVO Abstracts. *Invest Ophthalmol Vis Sci.* 1995;36:S421.
23. Jumblatt MM, Geohegan TE, Jumblatt JE. Mucin gene expression in human conjunctiva. ARVO Abstracts. *Invest Ophthalmol Vis Sci.* 1995;36:S997.

33

SOLUBLE MUCIN AND THE PHYSICAL PROPERTIES OF TEARS

John M. Tiffany, Jyotin C. Pandit, and Anthony J. Bron

Nuffield Laboratory of Ophthalmology
University of Oxford, United Kingdom

1. INTRODUCTION

Human tears have a non-Newtonian viscosity,[1] meaning that the coefficient of viscosity varies with the value of shear rate at which it is measured. Human tears display shear-thinning, so high-shear viscosity is low, approaching that of water at very high shear rates, but high at very low shear rates. This variation is advantageous in minimizing the drag during blinking but resisting gravitational drainage in the open eye. Tears also have a low surface tension with a mean value of 43–44 mN/m in normal subjects[2] compared to about 72 mN/m for water. The standard belief for many years has been that both these properties depend upon the presence of goblet-cell (secreted) mucin dissolved in the aqueous tears, in equilibrium with the gel layer of mucus coating the conjunctival and corneal surfaces. We might suspect that unstimulated tears would contain a higher concentration of soluble mucin than stimulated tears, since they are in contact with the conjunctiva for longer than the more rapidly flowing stimulated tears. Model mucus solutions closely resembled human tears in their rheological[3] and surfactant[2,4] behavior. We have investigated these properties in relation to the macromolecular content of stimulated and unstimulated tears.

2. METHODS

2.1. Materials

Tear protein analogues (lysozyme, lactoferrin, and secretory IgA) and bovine submaxillary mucin obtained from Sigma (UK) were used as models for tear components. Samples of purified rabbit conjunctival mucin,[5] and human conjunctival mucin (from conjunctival swabs, solubilized in 6 M guanidine hydrochloride) were also used. Both un-

Lacrimal Gland, Tear Film, and Dry Eye Syndromes 2
edited by Sullivan *et al.*, Plenum Press, New York, 1998

stimulated (UT) and stimulated (ST) human tears were collected at the slit lamp in 5 μl or 20 μl glass capillary tubes and immediately transferred to plastic Eppendorf capsules. Samples of UT were taken from several subjects and pooled until a sufficient volume for rheology was reached; enough ST could usually be collected from a single subject. Both UT and ST samples were centrifuged before use to remove debris and any streamers of gelatinous mucus that might interfere with centering of the rheology cup.

2.2. Methods

The viscosity of UT and ST samples, and of the tear protein analogues at typical tear concentrations, was measured in the Contraves Low-Shear 30 rheometer at room temperature.[1] Surface tension was measured by the microcapillary technique of Tiffany et al.[2] on UT and ST as well as on model tear proteins and mucins, under various conditions of concentration and temperature, and after storage at 4°C in sealed glass capillary tubes. The macromolecular content of tears and mucin preparations was investigated by SDS-PAGE and agarose gel slab electrophoresis, and by HPLC on a TSK-3000 size-exclusion column. Agarose gels were vacuum-blotted onto nitrocellulose membranes and the blots subjected to lectin-ABC probing with wheat germ agglutinin (WGA).

3. RESULTS

3.1. Viscosity

Fig. 1 shows the coefficient of viscosity for ST and UT as a function of shear rate. Pronounced shear-thinning is apparent in both cases, and the two curves coincide within the limits of error of the method. This is inconsistent with the suggestion that UT should contain more soluble mucin than ST, because the aqueous tear fluid is in longer contact with the conjunctival surfaces. Our results suggest either that UT and ST contain the same level of mucin, or that their mucin content (if any) does not influence viscosity.

The viscosities of lysozyme, lactoferrin, and sIgA solutions at concentrations typical of tears (3.3, 1.8, and 0.5 mg/ml, respectively[6]), and combinations of two or three of these

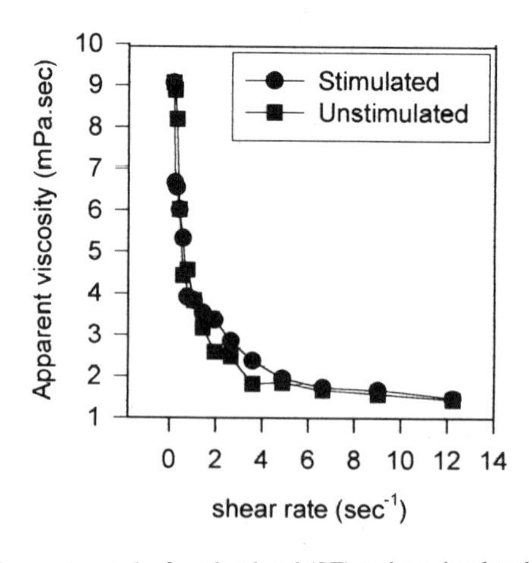

Figure 1. Viscosity/shear rate graphs for stimulated (ST) and unstimulated (UT) human tears.

Table 1. Surface tension of human tears and model solutions

Solution	Concentration (mg/ml)	Surface tension (mN/m)
Unstimulated tears	—	41
Stimulated tears	—	46.5
Rabbit ocular mucin	5	68.5
	10	54
Tear protein mixture	Lysozyme 1.3 Lactoferrin 1.8 sIgA 0.5	67

proteins, were measured under the same conditions. All the viscosities of pure protein solutions were low (1.3 mPa.sec or less) and showed no shear-thinning (not shown). Thus they would seem to make little contribution to overall tear viscosity, although further combinations of these proteins with mucin and/or lipocalin, or other known tear components, are not ruled out.

3.2. Surface Tension

The surface tensions of both ST and UT fell within the normal range of 40–46 mN/m reported by Tiffany et al.,[2] although that of ST was the higher. The model tear proteins, at concentrations typical of tears, gave higher surface tensions, whether singly or in combinations. Rabbit ocular mucin gave 54 mN/m, even at 10 mg/ml suggesting that previous reports that the surface tension of mucin solutions resembled that of tears may have depended on protein contaminants (Table 1).

Tiffany et al.[2] also reported that cold storage of tear samples resulted in higher surface tensions, and suggested that this might be due to loss of mucin from solution by adsorption on the interior walls of the capillary tube. Our tear storage experiments tested this over 7 days of storage (Fig. 2). UT, ST, and rabbit ocular mucin (10 mg/ml) all showed a rise in surface tension over 1–2 days, but UT and ST gave a plateau at 46 and 54 mN/m, respectively, whereas the mucin solution lost all surface activity and showed the same surface tension as water (72 mN/m). Although this appears to agree with the original sugges-

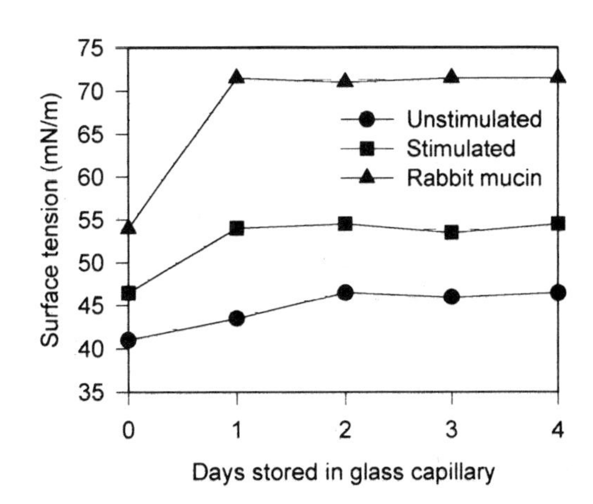

Figure 2. Change in surface tension with storage time in sealed glass capillary tubes of stimulated (ST) and unstimulated (UT) human tears and purified rabbit ocular mucin (10 mg/ml).

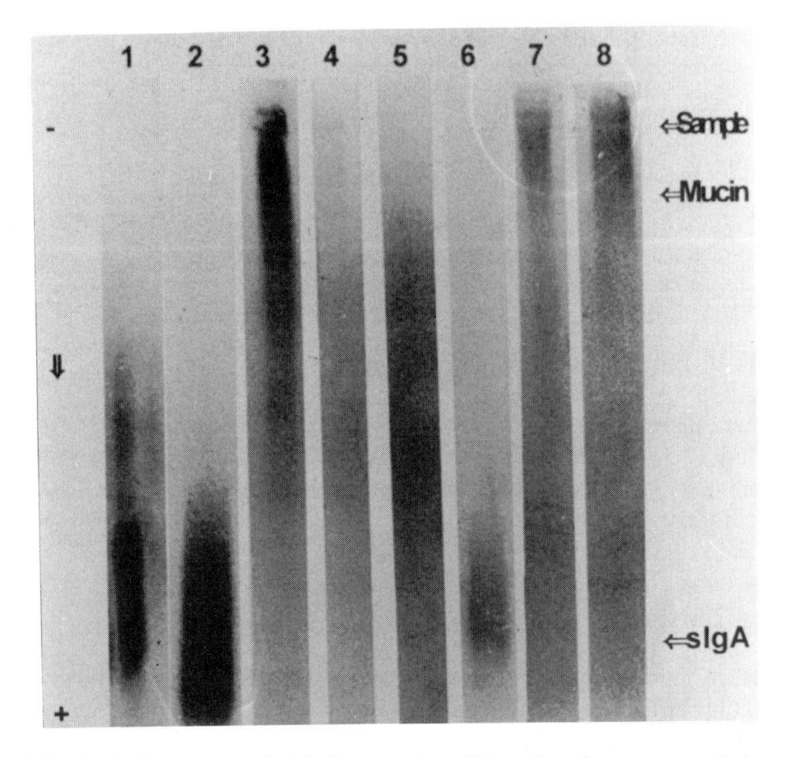

Figure 3. ABC-stained wheat germ agglutinin blots on nitrocellulose sheet from agarose gel electrophoresis of tears, mucins, and sIgA. Lanes contain: 1, unstimulated tears (UT); 2, stimulated tears (ST); 3–5, rabbit ocular mucin at 1.25, 0.125, and 0.0125 mg/ml, respectively; 6, human sIgA; 7 and 8, polymeric and reduced human conjunctival mucin.

tion of wall adsorption, it also suggests that mucin is in no way responsible for tear surface tension.

3.3. Agarose Gel Electrophoresis

Attempts to quantify mucin directly by lectin-blotting were unsuccessful because there is no readily available lectin that binds exclusively to mucin, and numerous tear glycoproteins are also detected by this means. Hence WGA was used as a broad-spectrum lectin detecting many species of glycoproteins.

Fig. 3 shows the WGA-probed blots from agarose electrophoresis gels. In lanes 1 and 2, UT and ST, respectively, show large spots with about the same migration as secretory IgA (lane 6), but distinctly different from the high-MW spots seen with polymeric or reduced human conjunctival mucin (lanes 7 and 8, respectively). Rabbit ocular mucin was run at three different concentrations (lanes 3–5). Mucin at 1.25 mg/ml in lane 3 shows a large spot at very high MW similar to that given by human mucin, and a trace of this spot can also be seen at 0.125 mg/ml rabbit mucin (lane 4), but not at 0.0125 mg/ml (lane 5). Although epithelial-type mucins may be present, their M_rs are in the same range as sIgA (less than 400 kDa). We conclude that any mucin of conjunctival type in UT or ST must be present at less than 0.125 mg/ml. Note that under these running conditions the major tear proteins have all run off the bottom of the gel.

4. DISCUSSION

The concept that tears contain soluble mucin was a natural one, since aqueous tears and gel-form mucin are intimately mixed together during blinking and eye movements. Mucin strongly influences viscosity[3] and produces shear-thinning solutions with viscosity/shear rate graphs virtually identical in form to those for tears; mucus solutions have been found to be of similar surface activity to human tears and much more so than model proteins in solution.[2,4] We previously estimated that tear viscosity and surface tension could both be achieved by 5 mg/ml mucin. The agarose gel results show that this level of polymeric conjunctival mucin cannot be present. We also previously found that solutions of hyaluronan (rheologically similar to polymeric mucin) do not show the shear-thinning characteristic of both mucins and hyaluronan at M_rs less than 1–1.5 MDa (not shown). Hence, even if epithelial mucins were present in tears at concentrations on the order of 5 mg/ml, their M_r would be too low to produce the required viscosity.

The finding that there is little difference between ST and UT, plus the agarose gel electrophoresis results, clearly indicate that the polymeric secreted mucin from goblet cells is essentially absent from fluid aqueous tears. Although epithelial mucins have been demonstrated on corneal and conjunctival cell surfaces,[7,8] their concentration in tears is unknown; these mucins typically are 200–400 kDa and hence probably too short to show pronounced shear-thinning in solution except at high concentrations. Hyaluronan is also detectable in tears[9] but only at low levels. Tear viscosity may well depend on complexes of several tear macromolecules, but these results suggest that combinations of the major proteins alone will be insufficient.

The surface tension of tears similarly appears not to depend on polymeric mucin content, although the epithelial mucins may be as effective as proteins, but have not yet been tested. Some material is lost during cold storage, presumably by adsorption onto the capillary tube walls, but this is more marked with mucin than with tears. The surface tensions of UT and ST before storage lie within the normal human range.[2]

Mucin itself is much less surface active than previously thought. This may be because we have used a highly purified sample, whereas previous reports by ourselves and others used commercial samples which generally contain residual protein contaminants.

The tear components responsible for the observed physical and physiological properties of aqueous tear fluid remain unknown. We are still far from understanding how tears work.

REFERENCES

1. Tiffany JM. The viscosity of human tears. *Int Ophthalmol.* 1991;15:371–376.
2. Tiffany JM, Winter N, Bliss G. Tear film stability and tear surface tension. *Curr Eye Res.* 1989;8:507–515.
3. Kaura R, Tiffany JM. The role of mucous glycoproteins in the tear film. In: Holly FJ, ed. *The Preocular Tear Film in Health, Disease, and Contact Lens Wear.* Lubbock, TX: Dry Eye Institute; 1986:728–732.
4. Holly FJ. Formation and stability of the tear film. *Int Ophthalmol Clin.* 1973; 13:73–96.
5. Kaura R, Tiffany JM. Preliminary characterization of the mucous glycoprotein component of rabbit ocular mucus. *Biochem Soc Trans.* 1984;12:651.
6. Fullard RJ, Snyder C. Protein levels in nonstimulated and stimulated tears of normal human subjects. *Invest Ophthalmol Vis Sci.* 1990;31:1119–1126.
7. Bogart B, Sack RA, Beaton A, Lew G, Kim HC. sIgA, glycoproteins and soluble mucin in reflex and closed eye tears: Does the epithelium shed its membrane-bound mucin? ARVO Abstracts. *Invest Ophthalmol Vis Sci.* 1994;35:S1560.

8. Gipson IK, Spurr-Michaud SJ, Tisdale AS, Kublin C, Cintron C, Keutmann H. Stratified squamous epithelia produce mucin-like glycoproteins. *Tissue Cell*. 1995;27:397–404.
9. Frescura M, Berry M, Corfield A, Carrington S, Easty DL. Evidence of hyaluronan in human tears and secretions of conjunctival cultures. *Biochem Soc Trans*. 1994;22:228S.

34

CHARACTERIZATION AND ORIGIN OF MAJOR HIGH-MOLECULAR-WEIGHT TEAR SIALOGLYCOPROTEINS

Robert A. Sack,[1] Bruce Bogart,[2] Sonal Sathe,[1] Ann Beaton,[1] and George Lew[2]

[1]SUNY State College of Optometry
New York, New York
[2]New York University Medical Center
Department of Cellular Biology
New York, New York

.1. INTRODUCTION

The purpose of this study was to characterize the nature and origin of the major high molecular weight soluble sialoglycoproteins in the tear fluid in open and closed eye environments.

2. METHODS AND RESULTS

Tear samples, 1 ml reflex (R) and 2–4 μl overnight closed eye (C), were recovered by capillary tube and centrifuged. Resultant R pellets were characterized and shown to consist of debris and desquamated epithelium (Fig. 1). The much larger C pellets contained debris, very large numbers of polymorphonuclear leukocytes (PMNLs), and a smaller amount of epithelium. R pellets were extracted in acidic PBS. This extract and R and C supernatants were separated isotonically on an SW 4000 size exclusion HPLC column. This resulted in the recovery of a crude glycoprotein fraction that eluted from the column well before sIgA (385 kDa). This HPLC fraction and the original fluids were characterized on SDS PAGE under reducing conditions. The gels were probed for glycoproteins and sialoglycoproteins (SGs) using, respectively, periodate oxidation followed by silver (Fig. 1) or Alcian blue staining.

Samples were blot transferred and immunoprobed before and after deglycosylation using a panel of Abs for mucins and carbohydrate epitopes. The principal SGs in tears and in the R pellet extracts reacted specifically and highly selectively with a monoclonal anti-

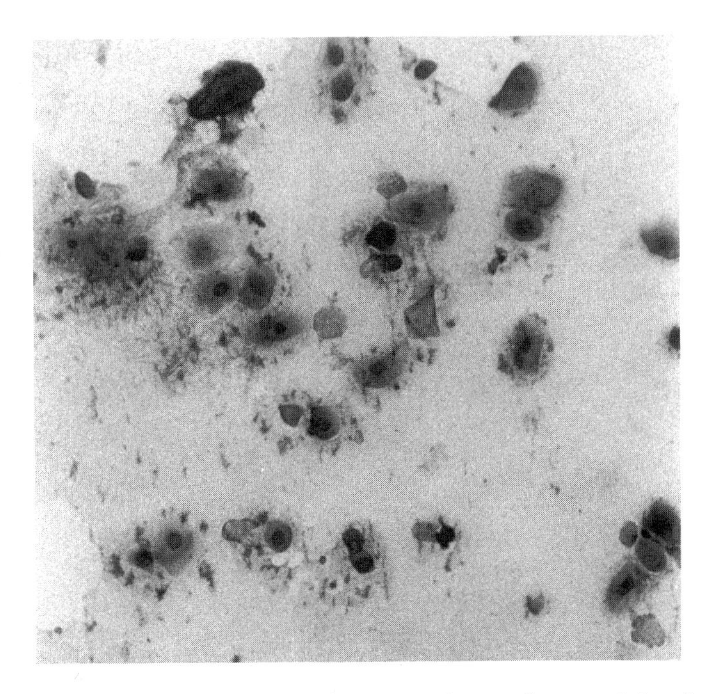

Figure 1. Cells recovered in pellet from typical reflex tear samples, revealing a population of epithelial cells.

body (MAb) to salivary mucin with an affinity for a sialyl Lewis[a] epitope. This MAb was used for immunoaffinity purification of the tear SGs. The major SG in R fluid was also separated by SDS PAGE, blot transferred, and preliminary amino acid analysis data obtained. Immunofluorescence microscopy and extract analysis were also carried out on human tissue. The principal high-molecular-weight, non-reducible glycoproteins in R fluid consisted of a non-reducible SG of ~450–500 kDa, a larger asialoglycoprotein, and a small amount of higher-molecular-weight SGs (Fig. 2).

The major SG exhibited considerable intersubject variability in terms of molecular weight and heterogeneity. In tears from some individuals, the major SG migrated as a single band of variable size. In other instances, it migrated as two partially resolved entities. This SG, which represents <1% of the protein in R fluid, represented as much as >85% of the protein in R pellet saline wash, a finding suggestive of an epithelial surface origin. Analysis of reflex tears collected over a period of time showed a progressive decrease in the concentration of SGs with further sampling. This finding excludes an inducible lacrimal gland secretion as a major source for tear SGs. C fluid was associated with a marked increase in SG concentration and a shift to higher-molecular-weight SGs. All major SGs exhibited a common pattern of antigenicity reacting specifically with the MAb for salivary mucin specific for the sialyl Lewis[a] epitope. SGs indistinguishable in terms of size and antigenicity from those in tear fluid were also found in extracts from human conjunctival and corneal epithelium but not from stroma (Fig. 3).

Immunofluorescence microscopy of conjunctiva revealed a localization restricted to the epithelial plasma membrane. This increased in intensity from the basal-to-apical direction. No staining was seen in goblet cells. These SGs exhibit many properties in common with a MUC1 entity. However, no cross-reactivity was observed between two MUC1

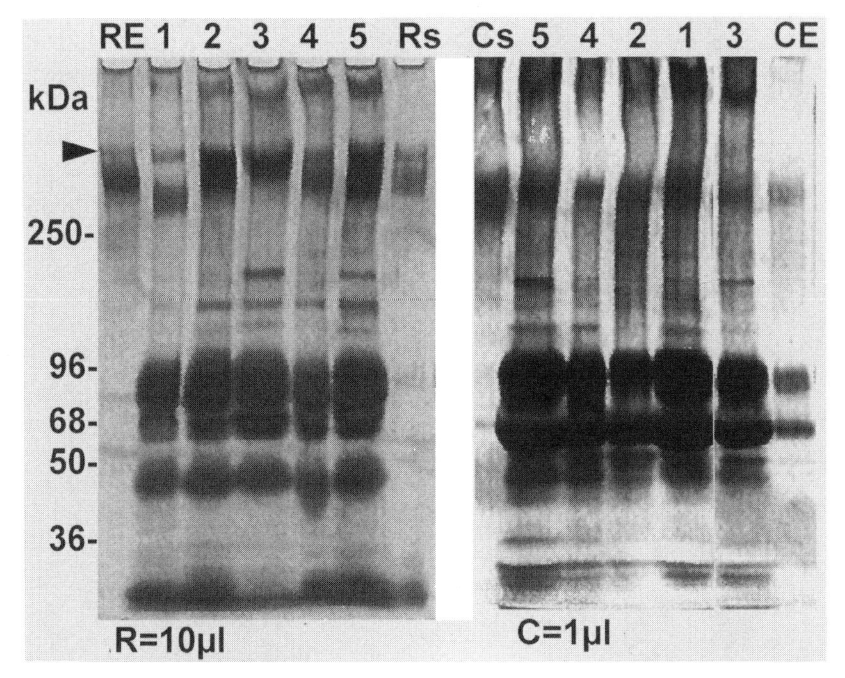

Figure 2. Reflex (10 μl) and closed eye (1 μl) tear samples, donors 1–5, and R pellet extracts (RE) and HPLC fraction (Rs) and the corresponding C pellet extracts (CE) and C HPLC fractions (Cs) were separated on 2–15% SDS PAGE (reduced) and periodate silver stained. Arrowhead points to the asialoglycoprotein, which stains black with periodate silver staining and does not stain with Alcian blue. Two other major high-molecular-weight glycoproteins (>250 kDa) stain brown with periodate silver and stain with Alcian blue and are therefore considered SGs. Note that on a 4% gel with stacking gel, a third high-molecular-weight SG can be seen at the interface.

standards and the MAb for the Lewis[a] epitope that intensely stained the tear SGs. Moreover, no reactivity was seen between the partially deglycosylated purified SGs and a panel of MAbs specific for epitopes to the core protein of MUC1. Tentative amino acid compositional data on the 450–500 kDa is also suggestive of a unique SG.

3. DISCUSSION

From this data we suggest that the principal SGs in tears are likely to be of epithelial origin with different pools tapped in open- and closed-eye conditions. The SG in R-type tear secretions appear to be derived from the epithelial plasma membrane where it is at least in part non-covalently bound.

There are two possible explanations for our findings: (i) The absence of immunoreactivity could be due to loss or blockage of core protein epitopes resulting from extensive glycosylation or transcriptional or proteolytic processing. In support of this hypothesis, others have shown by probing human conjunctiva for the MUC1 core protein that staining is restricted primarily to the basal epithelial cell layer and is greatly reduced on the apical epithelial surface.[1,2] This would suggest a loss or dramatic change in the availability of the MUC1 core protein. (ii) Alternatively, these SGs might represent a unique entity with an epithelial plasma membrane localization. Definitive identification awaits core protein sequencing data.

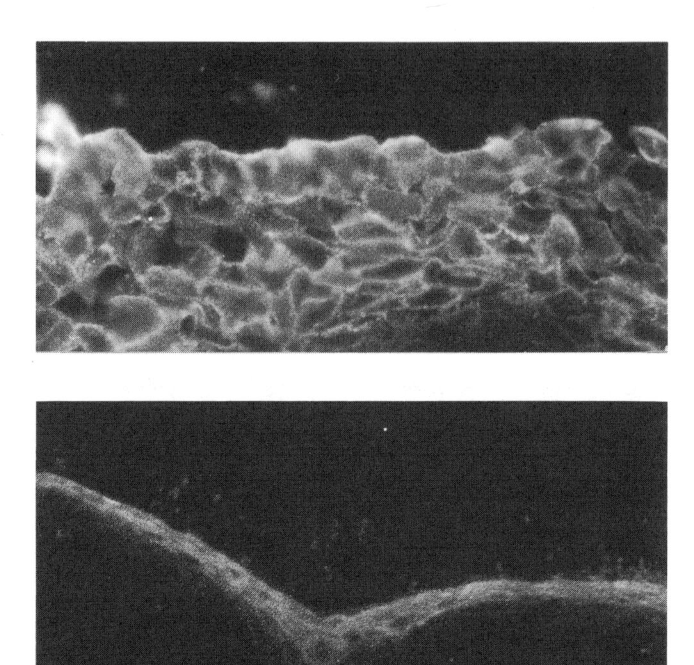

Figure 3. Immunofluorescence of human conjunctiva probed with MAb for sialyl Lewis[a] epitope. Note that staining was localized to the epithelium, with the intensity of staining increasing in the basal to apical direction.

These findings show that the principal soluble non-reducible SGs in tears, regardless of identity, are likely to be of external ocular epithelial origin. Further details of this and related work will be presented elsewhere.

REFERENCES

1. Jumblatt MM, Georghegan TE, Jumblatt JE. Mucin gene expression in human conjunctiva. *Invest Ophthalmol Vis Sci.* 1995;36(suppl.):S997.
2. Inatomi T, Spurr-Michaud S, Tisdale AS, Gipson IK. Human corneal and conjunctival epithelia express MUC1 mucin. *Invest Ophthalmol Vis Sci.* 1995;36:1818–1827.

35

DETECTION AND QUANTIFICATION OF CONJUNCTIVAL MUCINS

James E. Jumblatt and Marcia M. Jumblatt

Department of Ophthalmology and Visual Sciences
University of Louisville School of Medicine
Louisville, Kentucky

1. INTRODUCTION

Mucins are a family of high-molecular-weight glycoproteins secreted by the goblet and non-goblet epithelial cells of mucosal tissues. Molecular studies of human respiratory, intestinal, and mammary mucins have identified eight distinct mucins (MUC1-MUC8), which are expressed in distinct tissue-specific patterns.[1] Mucins characteristically contain distinct serine- and threonine-rich tandem repeat domains that are heavily O-glycosylated. O-glycosylation of the repeat region of mucin apoproteins begins with the addition of *N*-acetylgalactosamine (GalNAc) to serine and threonine residues. Straight and branched sugar side chains are subsequently elaborated, resulting in extremely high-molecular-weight (>10^6 Da) glycoproteins containing 50–85% carbohydrate by weight.[2]

Although mucin glycoproteins are major constituents of the pre-ocular tear film, little is known of the regulation of their synthesis or secretion. Morphological evidence suggests that mucin is secreted by rat conjunctival goblet cells in response to topical application of VIP, serotonin, or adrenergic agonists,[3,4] and molecular studies have recently identified several specific mucin transcripts in human conjunctival tissues.[5] Although cholinomimetic drugs are also reported to stimulate conjunctival mucin secretion,[6] little is known about the underlying receptors or cellular mechanisms that mediate mucin discharge.

In the present study, we utilized *Helix pomatia* agglutinin (HPA) to detect and quantify mucins secreted from isolated rabbit conjunctival tissue. HPA is a lectin that binds selectively to GalNAc, a major saccharide component of most mammalian mucins. A sensitive enzyme-linked lectin assay (ELLA), based on binding of HPA-horseradish peroxidase (HPA-HRP) conjugates to mucins absorbed to microtiter plates, was employed to investigate the role of calcium in muscarinic receptor-mediated mucin secretion.

Lacrimal Gland, Tear Film, and Dry Eye Syndromes 2
edited by Sullivan *et al.*, Plenum Press, New York, 1998

2. METHODS

2.1. Animals and Tissues

New Zealand white rabbits (2.5 kg) were euthanized by injection of pentobarbital into the marginal ear vein. The nictitans were excised and stored in ice-cold oxygenated Krebs-Ringer solution until use in the secretion studies. Samples of rabbit orbital lacrimal gland were also collected and processed for gel electrophoresis.

2.2. Assay of Mucin Secretion

Fresh rabbit nictitans were placed in polypropylene vials containing 3 ml Krebs-Ringer buffer (composition in mM: NaCl 118, KCl 4.8, $CaCl_2$ 1.3, KH_2PO_4 1.2, NaHCO3 25, MgSO4 2.0, dextrose 10, pH 7.4). The vials were incubated at 32°C in a Dubnoff shaking incubator in an atmosphere of 95% O_2/5% CO_2. The external medium was replaced with fresh medium at 30 min intervals. Tissues received three successive 30 min incubations (S1-S3). Secretagogues were present during the final incubation (S3). Aliquots of the incubation media were assayed for mucin using a modified ELLA (see below), and the ratio of mucin released during S3 and S2 (S3/S2 ratio) was calculated for each tissue sample to provide an internally normalized index of mucin secretion that is independent of tissue mass. To evaluate the effects of secretagogues or other treatments, the S1/S2 ratios of treated tissues were compared to those obtained from untreated controls. Results are expressed as % of control values. Statistical evaluation of drug responses was performed by Student's t-test (unpaired).

Mucin content of incubation medium was measured using an ELLA based on the binding of *Helix pomatia* agglutinin-horseradish peroxidase conjugates (HPA-HRP; Sigma) to mucins absorbed to 96-well microtiter plates. 50 μl aliquots of incubation media were placed in microtiter wells and allowed to dry overnight in a vacuum desiccator. The wells were rinsed three times with 200 μl of ice-cold PBS and blocked with 100 μl 1% bovine serum albumin in PBS for 60 min at 4°C. Subsequently, the wells were again rinsed three times with PBS and incubated for 2 h at 22°C with 100 μl PBS containing 12.5 μg/ml HPA-HRP. The lectin solution was aspirated and each well rinsed three times with 100 μl PBS. 200 μl of substrate solution (0.01 M sodium phosphate, pH 6.0; 0.001 M

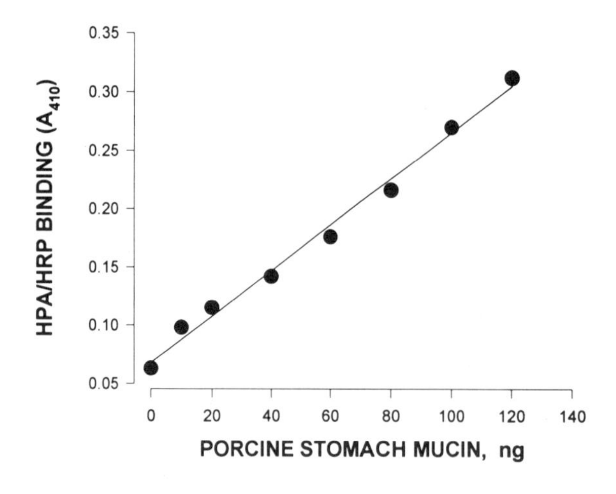

Figure 1. Enzyme-linked lectin assay (ELLA). Porcine stomach mucin (0–120 ng) was absorbed onto a 96-well plate, blocked, and labeled with HPA-HRP (12.5 mg/ml). Enzyme activity was detected with *o*-dianisidine/H_2O_2 and read at 410 nm. ELLA response was linear over this concentration range. Points on the graph represent the means of triplicate samples.

EDTA, 0.02% *o*-dianisidine, 0.003% H_2O_2) was then added to each well and incubated for 15 min at 22°C. Absorbances were read at 410 nm using a microplate reader (Cambridge Technologies). Mucin samples were assayed in triplicate and compared to a standard curve of porcine stomach mucin. As shown in Fig. 1, the standard curve was linear in the range of 1–120 ng mucin/well.

2.3. Chromatographic Analysis of Secreted Mucins

For metabolic labeling of conjunctival mucins, tissues were incubated for 4 h at 37°C in Eagle's Minimal Essential Medium (MEM) containing 5% fetal bovine serum and 2.5 μCi/ml ^{3}H-glucosamine. Subsequently, tissues were rinsed repeatedly in non-radioactive MEM, transferred to Krebs-Ringer buffer, and utilized for secretion studies as described above. Samples of the incubation media were centrifuged (14,000*g*, 10 min) and the supernatants analyzed by gel filtration on a Sepharose CL-4B (45 x 1.8 cm) column in 0.1 M Tris buffer, pH 8.0. Aliquots of column eluent were analyzed for radioactivity by liquid scintillation spectroscopy and for mucin by ELLA.

2.3.1. Gel Electrophoresis. SDS polyacrylamide gel electrophoresis (SDS-PAGE) of tissue extracts was performed under reducing conditions.[7] Homogenized samples of nictitans or lacrimal gland were diluted 1:10 in sample buffer, as was a stock solution of 0.1 mg/ml porcine stomach mucin. Samples were heated to 100°C for 10 min, and 20 μl of each sample and of pre-stained molecular-weight standards (Biorad) were loaded onto a 4% polyacrylamide gel. Electrophoresis was carried out in a minigel apparatus (Biorad) at 100 V for 1 h. Electrophoretically separated proteins were transferred to nitrocellulose. The membrane was then blocked with 1% BSA in PBS for 1 h, incubated with HPA-HRP (10 μg/ml in blocking buffer) for 1 h, rinsed with PBS, and reacted with diamino-benzidine and hydrogen peroxide.

2.3.2. Histology. Rabbit nictitans were fixed in 10% neutral buffered formalin for 1h and embedded in paraffin. Sections (10 μm) were cut and placed on glass slides. For HPA-HRP labeling, sections were deparaffinized, rehydrated, and washed with PBS. Sections were incubated with HPA-HRP (5 μg/ml in PBS) for 1 h, washed in PBS, and developed using DAB as a chromogen. To assess non-specific lectin binding, GalNAc (20 ng/ml) was included during the lectin incubation. For PAS staining, rehydrated sections were incubated in 0.5% periodic acid for 5 min, rinsed in deionized water, and developed in Schiff reagent (Fisher Scientific) for 15 min. For morphological studies of ionophore-induced mucin secretion, conjunctival tissues were exposed to 10 μM ionomycin for 30 min prior to fixation, embedding, sectioning, and PAS staining. Slides were coverslipped, then viewed and photographed using a Nikon photomicroscope.

3. RESULTS

As shown in Fig. 2a, PAS stains both goblet cells and the proteoglycan-rich connective tissue of the conjunctival stroma. HPA-HRP strongly labels goblet cells of the conjunctiva and lightly labels the cytoplasm of non-goblet conjunctival epithelial cells (Fig. 2b). A layer of HPA reactive material, presumably tear film mucin, is localized at the apical surface of the conjunctiva, and the connective tissue of the stroma is unlabeled. HPA-HRP labeling is abolished by the addition of excess GalNAc to the lectin incubation medium (Fig. 2c).

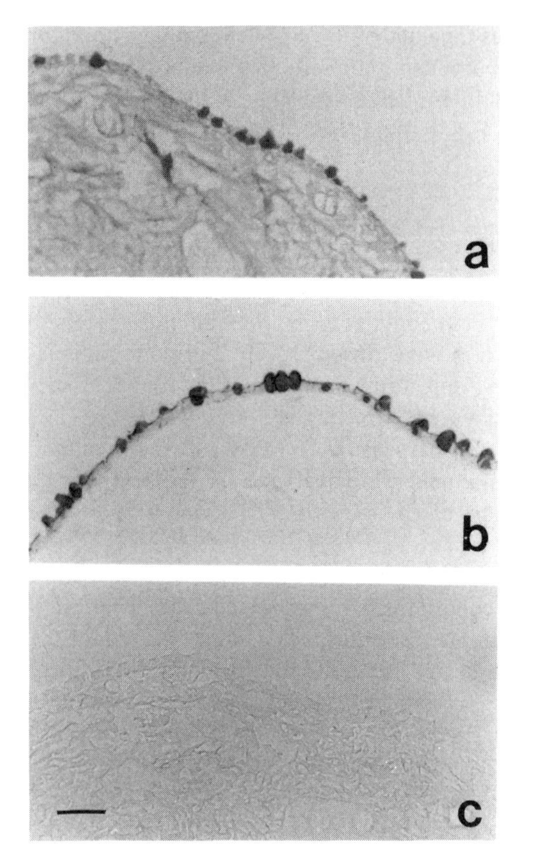

Figure 2. Histological detection of conjunctival mucins. Sections of fixed rabbit conjunctiva were either stained with PAS (a) or labeled with HPA-HRP (b, c). PAS stains goblet cells and the glycoproteins of the conjunctival stroma. HPA-HRP 2b) labels goblet cells and, to a lesser extent, non-goblet epithelial cells, but does not label stromal components. The addition of excess GalNAc completely abolished HPA-HRP labeling (c). Bar = 100 mm.

Rabbit conjunctival and lacrimal gland extracts and porcine stomach mucin were analyzed by SDS-PAGE on 4% polyacrylamide gels. Following transfer to nitrocellulose, mucin glycoproteins were labeled with HPA-HRP. As shown in Fig. 3, polydisperse bands of high-molecular-weight material are evident near the top of the gel in both conjunctival and porcine stomach mucin samples, but are not present in lacrimal gland extracts. A lower-molecular-weight (approx. 180 kDa), HPA-positive band is seen in the conjunctival extract and may represent an under-glycosylated, intracellular form of mucin.

The calcium ionophores A23187 and ionomycin were tested for their ability to stimulate mucin secretion in rabbit conjunctiva. At a concentration of 10 μM, both ionophores triggered a three- to four-fold increase in mucin release (Fig. 4), consistent with the established role of calcium in goblet cell exocytosis.[8] This response was partially blocked by removal of extracellular calcium and further reduced by the inclusion of the calcium chelator BAPTA, suggesting that release of calcium from intracellular stores may also contribute to ionophore-stimulated mucin secretion.

The nature of the material secreted in response to 10 μM A23187 was determined by pre-labeling conjunctival tissues with ^{3}H-glucosamine followed by chromatography of released mucins on Sepharose Cl-4B columns (Fig. 5). HPA-HRP bound exclusively to high-molecular-weight material (MW>10^6Da) eluted in the void volume of the column, whereas ^{3}H-glucosamine-labeled material was present in both included and excluded column fractions. Material eluting in the void volume was concentrated and analyzed further

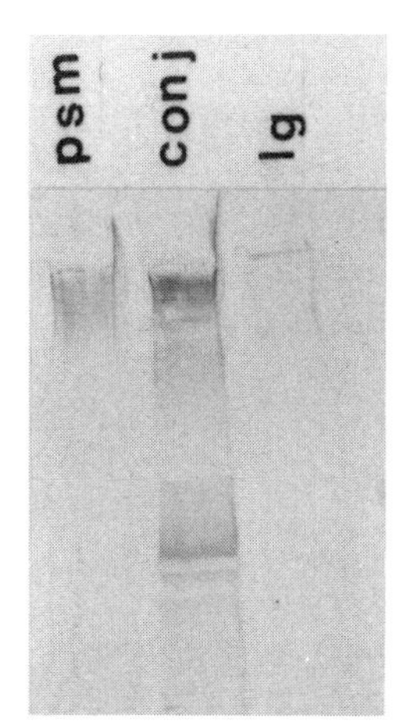

Figure 3. SDS-PAGE analysis of conjunctival mucin. Porcine stomach mucin (psm), conjunctival tissue (conj), and lacrimal gland tissue (lg) were electrophoresed on 4% gels, blotted onto nitrocellulose, and examined by HPA-HRP labeling. High-molecular-weight (>300 kD), HPA-positive mucin is present in the porcine stomach mucin and conjunctival samples, but not in the lacrimal gland extract. An additional conjunctival band at about 180 kDa may represent an intracellular mucin.

by isopycnic centrifugation in CsCl (not shown). The gradient yielded a peak of ^{3}H-labeled, HPA-binding material with a buoyant density of 1.40–1.48 g/ml, consistent with reported values for purified mucin glycoproteins.[2]

The response of conjunctival goblet cells to ionomycin was further studied by histologic evaluation of PAS-positive goblet cells in stimulated and control nictitans. Rabbit nictitans were incubated for 30 min in 10 μM ionomycin, fixed, sectioned, and stained with PAS. As seen in Fig. 6b, ionomycin treatment resulted in a substantial depletion of

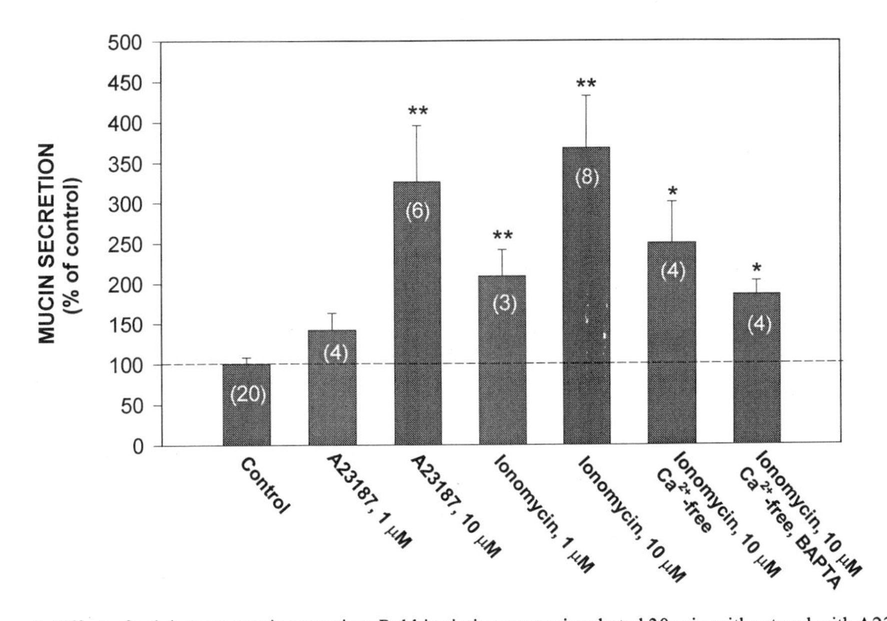

Figure 4. Effect of calcium on mucin secretion. Rabbit nictitans were incubated 30 min without and with A23187, ionomycin, and BAPTA, as described. The calcium ionophores A23187 and ionomycin enhance mucin secretion as determined by ELLA. Enhancement was reduced in calcium-free medium and in the presence of BAPTA. Bars represent the means ± SEM of 3 to 20 replicates. **, indicates $p<0.05$ compared to control. *, $p<0.01$ compared to control.

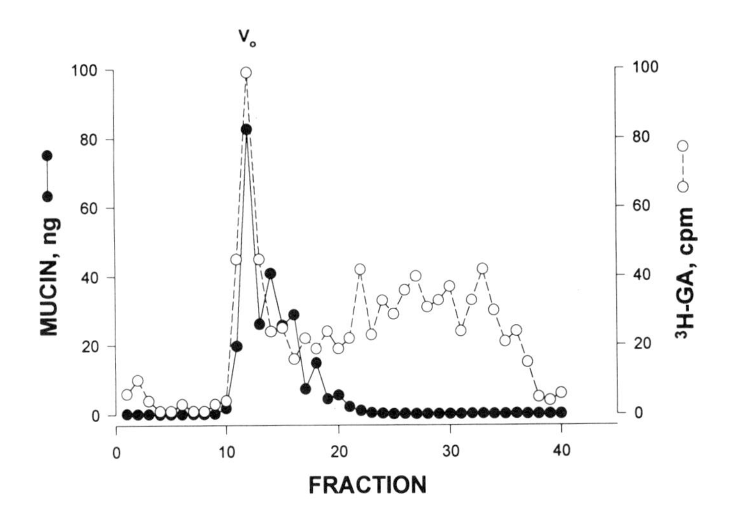

Figure 5. Sepharose CL-4B chromatography of secreted mucin. Rabbit nictitans were pre-labeled with ^{3}H-glucosamine, stimulated with A23187, and secreted products were analyzed by Sepharose CL-4B chromatography. ELLA detected mucin void volume only (Σ), while ^{3}H-glucosamine labeled glycoconjungates were present in both the void and included volumes.

PAS-positive conjunctival goblet cells as compared to the control tissue shown in Fig. 6a. This finding is consistent with the observed extracellular release of HPA-HRP-positive, mucin-like glycoconjugates following ionophore stimulation (see Fig. 4).

Mucin exocytosis in response to the cholinergic agonist carbamylcholine was studied in isolated rabbit nictitans. As shown in Fig. 7, addition of carbamylcholine (10^{-7} M) resulted in an 80% increase of mucin release above control. This release was blocked by the muscarinic antagonist atropine (10^{-6} M), showing the response to be muscarinic receptor-mediated. Carbamylcholine-induced mucin secretion was only partially inhibited by removal of extracellular calcium.

4. SUMMARY AND CONCLUSIONS

These results demonstrate that the lectin conjugate HPA-HRP is a selective probe for rabbit conjunctival mucins, as indicated by histological labeling of conjunctival goblet cells and binding to high-molecular-weight, mucin-like glycoconjugates identified by Sepharose CL-4B chromatography and SDS-PAGE. Based on this selectivity, we have developed a sensitive, quantitative assay (ELLA) for analysis of mucins secreted from isolated conjunctival tissues. Compared to previously described lectin-based assays,[9] the present assay is relatively simple to perform and has a wide dynamic range (~1–200 ng mucin). Our preliminary findings suggest that this assay is also suitable for quantifying human conjunctival and tear film mucins (Jumblatt MM and Jumblatt JE, unpublished data).

The inhibitory effects of Ca^{2+} removal on mucin secretion in response to calcium ionophores or carbamylcholine demonstrate that mucin release is a calcium-dependent process, involving Ca^{2+} from both external and intracellular sources. Furthermore, carbamylcholine-induced mucin secretion in this tissue is mediated by muscarinic acetyl-

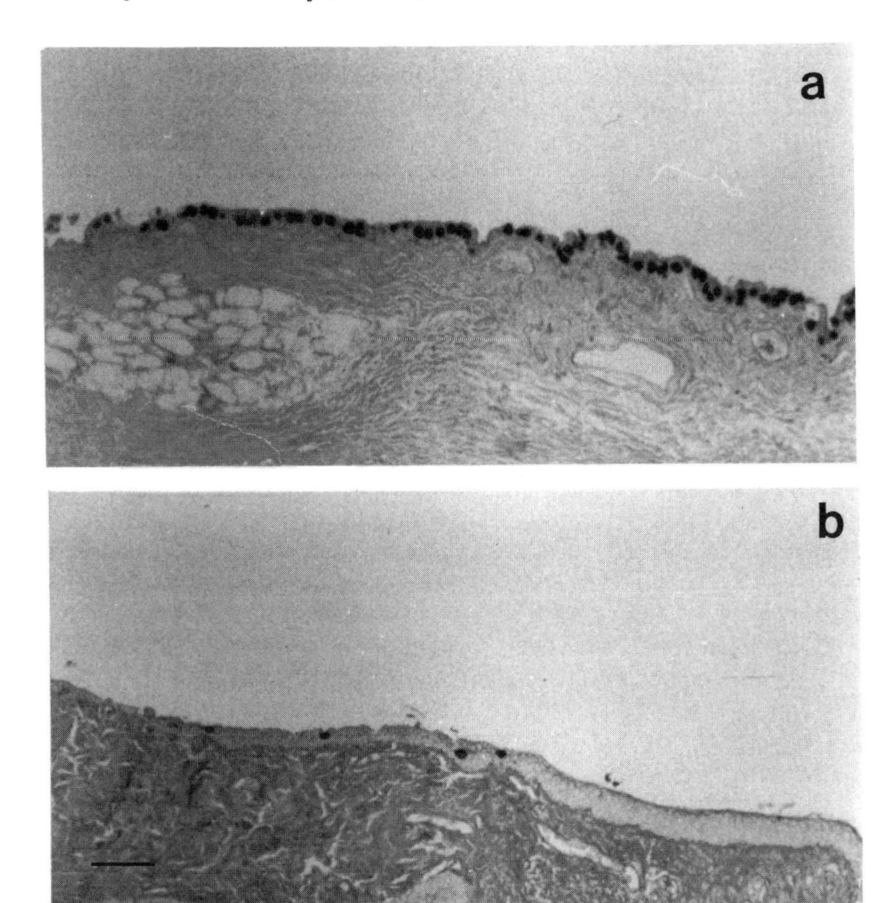

Figure 6. Ionomycin-induced goblet cell exocytosis. Nictitans were incubated without (6a) or with (6b) ionomycin (10 mM) for 30 min, fixed, sectioned, and stained with PAS. Ionomycin treatment resulted in extensive exocytosis as reflected by the greatly reduced number of darkly stained goblet cells in (b) Bar = 100 mM.

choline receptors, in agreement with in vivo studies showing muscarinic stimulation of goblet cell exocytosis in the rat conjunctiva (Kessler and Dartt 1993). Secretion in response to calcium ionophores was greater than that seen in response to carbamylcholine, possibly reflecting differences in the magnitude of intracellular calcium elevation induced by these agents. Alternatively, carbamylcholine may trigger only a subset of secretory goblet cells. In any case, the ELLA developed for this study should prove useful as a screening device for evaluating potential secretagogues to enhance conjunctival mucin secretion in human dry-eye diseases.

ACKNOWLEDGMENTS

We thank Rita Hackmiller for her excellent technical assistance. This work was supported by a research grant from the National Institutes of Health EY10736, an unrestricted grant from Research to Prevent Blindness, and the Kentucky Lions Eye Foundation.

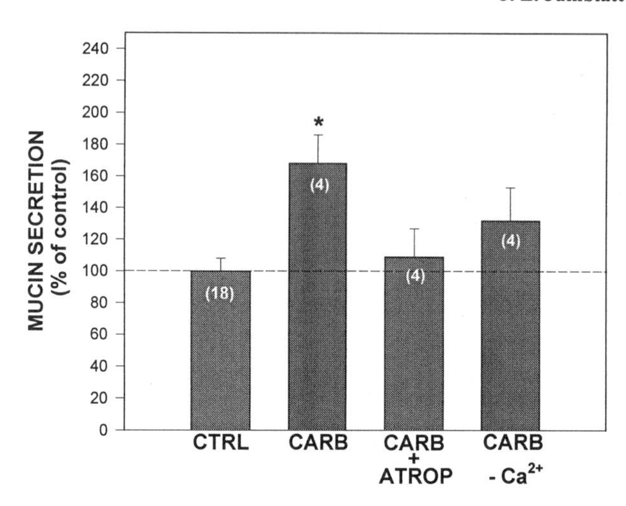

Figure 7. Receptor-mediated stimulation of mucin secretion. Rabbit nictitans were incubated in the absence or presence of carbachol (10^{-7} M), and mucin secretion was determined by ELLA. Carbachol enhanced mucin secretion, and this enhancement was blocked by atropine. Carbachol-induced mucin secretion was diminished in Ca^{2++}-free medium. Bars represent the means of quadruplicate samples. *, $p<0.05$ as compared to control.

REFERENCES

1. Gendler SJ, Spicer AP. Epithelial mucin genes. *Annu Rev Physiol.* 1995;57:607–634.
2. Strous GJ, Dekker J. Mucin-type glycoproteins. *Crit Rev Biochem Mol Biol.* 1992;27:57–92.
3. Kessler TL, Dartt DA. Neural stimulation of conjunctival goblet cell mucous secretion in rats. *Adv Exp Med Biol.* 1994;350:393–398.
4. Dartt DA, Kessler TL, Chung EH, Zieske JD. Vasoactive intestinal peptide-stimulated glycoconjugate secretion from conjunctival goblet cells. *Exp Eye Res.* 1996;63:27–34.
5. Inatomi T, Spurr-Michaud S, Tisdale AS, Zhan Q, Feldman ST, Gipson IK. Expression of secretory mucin genes by human conjunctival epithelia. *Invest Ophthalmol Vis Sci.* 1996;37:1684–1692.
6. Kessler TL, Dartt DA. Effect of adrenergic and cholinergic agonists and neural stimulation on conjunctival goblet cell mucous secretion in rats. ARVO Abstracts. *Invest Ophthalmol Vis Sci.* 1993;34:S822.
7. Laemmli UK. Cleavage of structural proteins during the assembly of the head of bacteriophage T4. *Nature.* 1970;227:680–685.
8. Verdugo P. Goblet cells, secretion and mucogenesis. *Annu Rev Physiol.* 1990;52:157–176.
9. Dwyer TM, Szebeni A, Diveki K, Farley JM. Transient cholinergic glycoconjugate secretion from swine tracheal submucosal gland cells. *Am J Physiol.* 1992;262:L418-L426.

36

MUCOUS CONTRIBUTION TO RAT TEAR-FILM THICKNESS MEASURED WITH A MICROELECTRODE TECHNIQUE

Philip Anderton[1,2] and Sophia Tragoulias[1]

[1]Co-operative Research Centre for Eye Research and Technology
[2]School of Optometry
University of New South Wales
Sydney, Australia

1. INTRODUCTION

The biochemistry and microstructure of the intermediate aqueous component of the tear film has been a subject of considerable interest over the past few years. Classical models of the tear film[1,2] represent this layer as an aqueous fluid phase, 4–7 μm thick, bound to the hydrophobic corneal epithelium by a very thin layer of mucus.

More recent evidence suggests levels of mucus in aqueous tears that may be higher than originally proposed. Electron microscopy shows the mucus layer extending at least 1 μm into the aqueous layer.[3] Aqueous tears resist movement under gravitational forces to a far greater degree than would be predicted from in vitro measurements of tear fluid viscosity.[4] Optical measures using laser interferometry[5] and confocal microscopy[6] indicate a tear-film thickness that exceeds the original estimate of 4–7 μm by a factor of three or four times.

This evidence suggests an alternative model of tear-film structure in which the aqueous layer is replaced by a layer of aqueous/mucus gel. This structure would impart substance to the tear film, especially immediately adjacent to the surface of the corneal epithelium.

The validity of the optical measurements may be questioned since, in optical terms, the posterior boundary between tears and epithelium is relatively poorly defined. Therefore, the question of tear-film thickness was pursued in the present study with an entirely different technique, one not subject to the limitations imposed by the indefinite optical character of the posterior boundary of the tear film. The results confirm optical measures and demonstrate that a mucolytic agent, *N*-acetyl-cysteine, causes a significant reduction in tear-film thickness. This provides further support for models of the tear film that incorporate a relatively thick "aqueous" layer, supported by an elaborate underlying mucous gel.

Lacrimal Gland, Tear Film, and Dry Eye Syndromes 2
edited by Sullivan *et al.*, Plenum Press, New York, 1998

2. METHODS

Three Wistar rats, 250–300 g, of either sex were used. All procedures complied with the Australian National Health and Medical Research guidelines for the use of animals in research and were approved by the University of NSW Animal Care and Ethics Committee.

Animals were anesthetized with an intraperitoneal injection of ketamine (Ketapex, 80 mg/kg) and xylazine (Rompun, 10 mg/kg). Eyelids were held closed by small clamps or adhesive tape until the start of experiments. The animal was mounted in a non-traumatic head-holder (Narishege Scientific Instruments, SH-8) firmly fixed to a rat stereotaxic frame (Narishege, SJ-2), which supported the electrode holder of a hydraulic micromanipulator (Narishege, MO-10). The assembly was configured so that, on opening the eye, the electrode tip was held 1–2 mm from the tear film, and approximately normal to its surface. The animal was electrically earthed with a thin, subcutaneous silver/silver chloride wire.

Electrodes were pulled on a Brown-Flaming electrode puller to have impedances of 100–200 megohms when filled with 5 M potassium acetate solution. Acetate solution was used to minimize problems with tip crystallization and dehydration. Such electrodes have tip diameters ~ 0.1 μm and are commonly used in intracellular recordings.

Fig. 1 shows a schematic diagram of the apparatus. Electrodes were connected to the positive input terminal of a high-performance custom DC amplifier with an option for current injection into the electrode. The indifferent input was connected to ground and a Faraday cage that enclosed the preparation. The amplitude of the voltage waveform produced by a small square-wave current injection was monitored continuously and used as an index of electrode resistance.

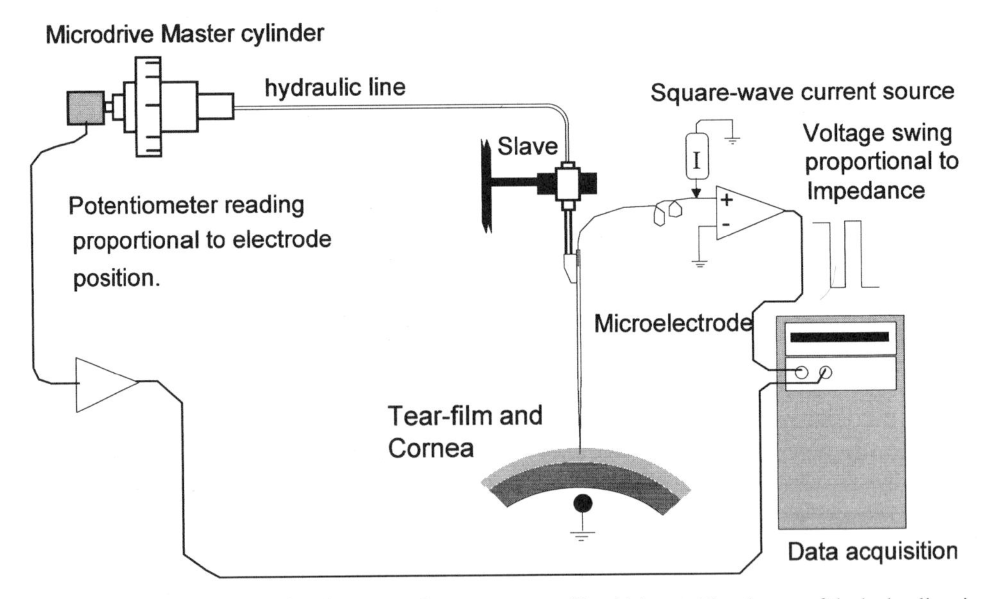

Figure 1. Schematic diagram of equipment used to measure tear-film thickness. The plunger of the hydraulic microdrive slave cylinder carried the electrode holder and electrode. It was rigidly mounted to a stereotaxic head-holder.

A 10-turn potentiometer was fitted to the shaft of the micrometer screw of the micro-drive master cylinder, and a constant voltage supply was attached to its end terminals. The DC voltage at the center-tap terminal was thus directly proportional to electrode displacement.

Electrode position and impedance signals were recorded continuously during experiments by a computer-based Data Acquisition device (AMLAB International, Sydney).

Electrode impedance was infinite with the electrode in air. The electrode was moved to a position just above the tear film with the course control. Data acquisition streaming was then enabled, and with further movement under fine control, contact with the eye was indicated by an abrupt reduction of impedance to the actual electrode value (point A in Fig. 2), usually 100–200 megohms. A profile of impedance against distance was obtained as the electrode was moved slowly toward the corneal epithelium.

Small respiratory movements, circulatory pulse, and eye movements contributed to noise recorded with each profile. If these were excessive, the anesthetic dose was increased. These effects were generally transient and did not confound average impedance measures. Contact with the corneal epithelium was indicated at the position where tip impedance began to rise consistently with further movement of the electrode (point B in Fig. 2). This was due either to an increase in pressure of the tip onto the epithelial surface, or to progressive movement of the tip into the epithelial tissue.

Tear-film thickness was measured as the total range of electrode movement for which there was no consistent change in impedance, and this was measured from position/impedance plots such as that shown in Fig. 2.

3. RESULTS

3.1. Effect of *N*-Acetyl-Cysteine on Tear-Film Thickness

Tear-film thickness was measured after the eye had received a droplet of either saline (control) or 20% *N*-acetyl-cysteine solution in saline (test). The eye was closed and

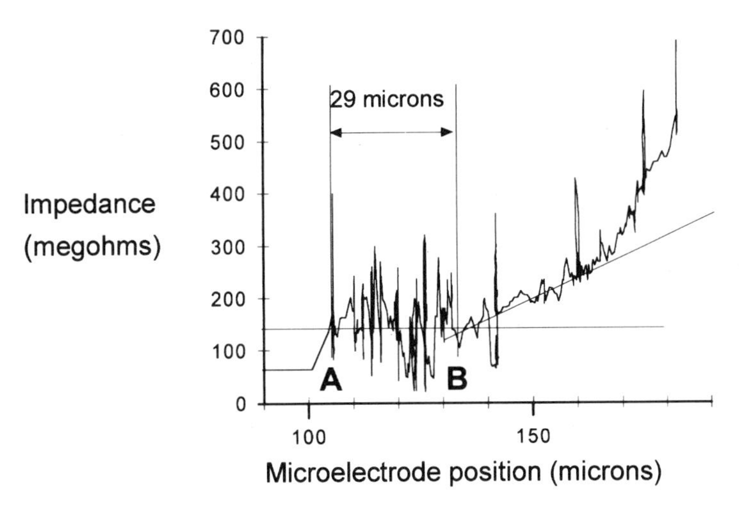

Figure 2. A typical trace of the change in microelectrode impedance with movement through the tear film. Contact with the tearfilm is made at A; contact with the corneal epithelium, at B. Noise is due to circulatory, respiratory, and other small eye movements.

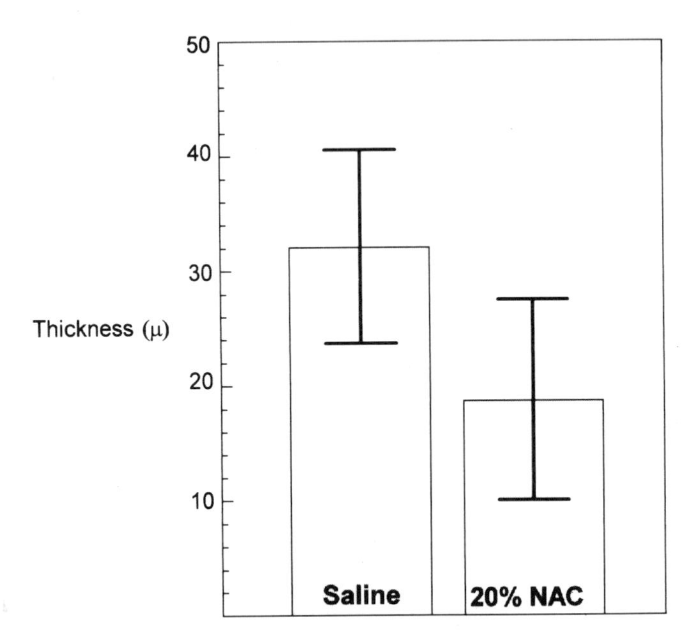

Figure 3. Effects of saline and 20% *N*-acetyl-cysteine in saline on the thickness of the rat tear film.

the contents massaged to distribute the solutions throughout the conjunctival cul-de-sac. Measurements were made as soon as the eye was opened to control for drying effects: they were repeated three times on each eye. Fig. 3 shows a summary diagram comparing the results of nine repetitions of test and control treatments on three different eyes where noise was limited and electrode performance was optimum. Error bars indicate standard deviations. On eyes subjected to control treatments, the mean tear-film thickness was 32 μm. Under test conditions, in the presence of *N*-acetyl-cysteine, mean thickness was 19 μm. The difference is significant at the $p < 0.005$ level as estimated by the two-sample t-test, assuming unequal sample variances.

3.2. Effect of *N*-Acetyl-Cysteine on Tear Proteins

N-acetyl-cysteine degrades the structure of mucins by breaking disulfide bonds. *N*-acetyl-cysteine may also affect other discrete proteins in the tear film. For this reason we examined the effect of *N*-acetyl-cysteine on the protein composition of tears. Tears were collected from the lateral canthus of the eyes of anesthetized rats using 10 μL glass capillary tubes. Average sample volumes were 8–10 μL. A 2 μL aliquot was mixed with an equal volume of control (saline) or test (10% *N*-acetyl-cysteine in saline) solution. Electrophoresis of these samples in the presence of sodium dodecylsulfate was carried out by modification of the Laemmli[7] method on 3–15% polyacrylamide gradient gels.

Fig. 4 shows parallel electrophoresis runs of test and control samples treated with silver stain. Protein profiles are similar except at positions A, B, and C. The normal protein density at A is reduced by *N*-acetyl-cysteine. The band at B is entirely removed. At C, the broad band of protein density in the control sample has been replaced by two thinner bands, and the overall density is reduced. The exact nature of these elements in the tear protein profiles is the subject of further investigation. However, it is clear that *N*-acetyl-

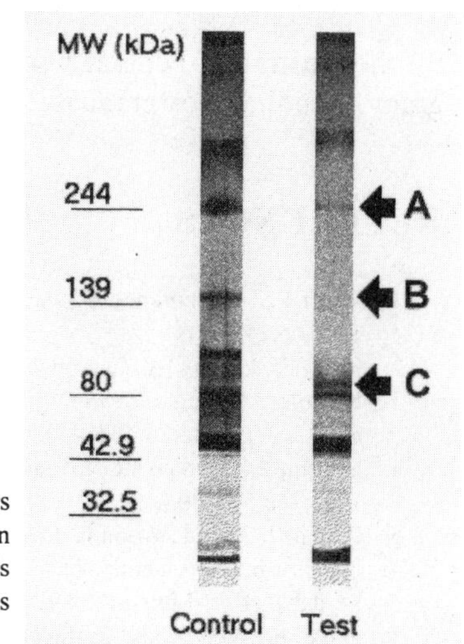

Figure 4. Silver stain of SDS-PAGE (3–15%) protein profile of rat tears before (control) and after (test) treatment with *N*-acetyl-cysteine. Known molecular weight standards are indicated at the left. Comparison of gels at A, B, and C indicates that *N*-acetyl-cysteine affects tear-film proteins as well as mucus.

cysteine has markedly altered the nature of this profile. These findings have been confirmed in three different experiments.

4. DISCUSSION

These electromechanical measurements confirm recent estimates of rat tear-film thickness made with optical techniques, providing further evidence that the tear film is considerably thicker than the classical value of 4–7 μm.

The utility of the technique is subject to the accuracy with which it measures the positions of the anterior and posterior mechanical boundaries of the tear film. It is likely that noise included in most recordings originated in small eye movements originating in circulatory pulse, respiration, and slow eye movements. These effects were minimized by ensuring that the animal was deeply anesthetized during each procedure.

It is clear from our gels that *N*-acetyl-cysteine has significant effects on the lower-molecular-weight protein components of tears, in addition to its mucolytic action. This is a significant factor that must be considered whenever *N*-acetyl-cysteine is used to examine effects of mucolysis on biofilm structure. It is possible that some protein components may play a secondary role in the maintenance of a normal gel structure in these films. If so, the effects of the mucolytic may reflect not only mucolysis, but also changes in the nature and concentration of these secondary proteins.

Depressant anesthetics reduce basal tear production in experimental animals.[8,9] The anesthetic used in these studies was a dissociative anesthetic, and tear volumes were not depressed, even in deep anesthesia. It is possible that the changes described may be related to an effect of anesthetics, and this will be checked in future investigations.

ACKNOWLEDGMENTS

The authors acknowledge the contributions of Dr. Carol Morris and Dr. Thomas Millar, with whom they have had many valuable discussions. Technical assistance was provided by Lucila Flores and Katherine Hollis-Watts. This research was partly supported by the Australian Federal Government through the Co-operative Research Centres Program/Australian Postgraduate Award - Industry Scheme.

REFERENCES

1. Wolff E. Mucocutaneous junction of lid-margin and distribution of tear fluid. *Trans Opthal Soc. U.K.* 1946;66:291–308.
2. Holly FJ, Lemp MA. Tear physiology and dry eyes. *Surv Ophthalmol.* 1977;22:69–87.
3. Nichols BA. Demonstration of the mucous layer of the tear film by electron microscopy. *Invest Ophthalmol Vis Sci.* 1985;26:464–473.
4. Hodson S, Earlam R. Of an extracellular matrix in human pre-corneal tear film. *J Theor Biol.* 1994;168:395–398.
5. Prydal JI, Artal P, Woon H, Campbell FW. Study of human precorneal tear film thickness and structure using laser interferomentry. *Invest Ophthalmol Vis Sci.* 1992;33:2006–2011.
6. Prydal JI, Kerr Muir MG, Dilly PN. Comparison of tear film thickness in three species determined by the glass fibre method and confocal microscopy. *Eye.* 1993;7:472–475.
7. Laemmli UK. Cleavage of structural proteins during assembly of the head of bacteriophage T4. *Nature.* 1970;227:680–685.
8. Chrai SS, Patton TF, Mehta A, Robinson JR. Lacrimal and instilled fluid dynamics in rabbit eyes. *J Pharm Sci.* 1973;62:1112–1121.
9. Cross DA, Krupin T. Implications of the effects of anaesthesia on basal tear production. *Anesth Analg.* 1977;56:35–37.

STRUCTURAL ANALYSIS OF SECRETED OCULAR MUCINS IN CANINE DRY EYE

Stephen D. Carrington,[1] Sally J. Hicks,[1] Anthony P. Corfield,[2]
Renee L. Kaswan,[3] Nicki Packer,[4] Shirley Bolis,[5] and Carol A. Morris[5]

[1]Department of Anatomy
School of Veterinary Science
University of Bristol, England, United Kingdom
[2]Department of Medicine Laboratories
Bristol Royal Infirmary
Bristol, England, United Kingdom
[3]College of Veterinary Medicine
University of Georgia
Athens, Georgia
[4]Macquarie University Centre for Analytical Biochemistry
Macquarie University
Sydney, Australia
[5]Cooperative Research Centre for Eye Research and Technology
University of New South Wales
Sydney, Australia

1. INTRODUCTION

The tear film is vital for the normal function of the ocular surface, influencing transparency, optical quality, and defense against the external environment. A functional understanding of the tear film depends on a knowledge of its components and their interactions. Recent data indicate the thickness of the trilaminar tear film to be 35–40 µm, the majority (~30 µm) contributed by a mucous gel situated adjacent to the epithelial surface.[1–3] This gel contains mainly secreted mucins, which interact with a variety of other components, including a number of highly glycosylated components of the ocular surface glycocalyx.[4–7]

Ocular mucins were initially believed to be entirely derived from conjunctival goblet cells. More recently, an alternative synthetic pathway has been defined in the stratified epithelial cells of the ocular surface.[1,8–12] Mucins are large glycoconjugates with molecular weights in the range 5×10^5 - 4×10^7 Da. Native secreted (synthetically complete and undegraded) mucins share a common structural organization, displaying subunits linked end-to-end through disulfide bridges.[13–15]

Sugar residues contributing up to 85% of dry weight are mainly O-linked to characteristic tandem repeating sequences of the core polypeptide backbone. Each serine and threonine residue in the mucin polypeptide backbone is a potential linkage site for *N*-acetylgalactosamine, which is the first sugar in all O-linked oligosaccharide chains. The oligosaccharide side chains typically contain galactose, fucose, *N*-acetyl-hexosamines, and sialic acids, but no uronic acids and little mannose.[13–15] Chains vary in length and branching, with charges ranging from neutral to highly negative, depending on the presence of sialic acid and sulfate.[16,17]

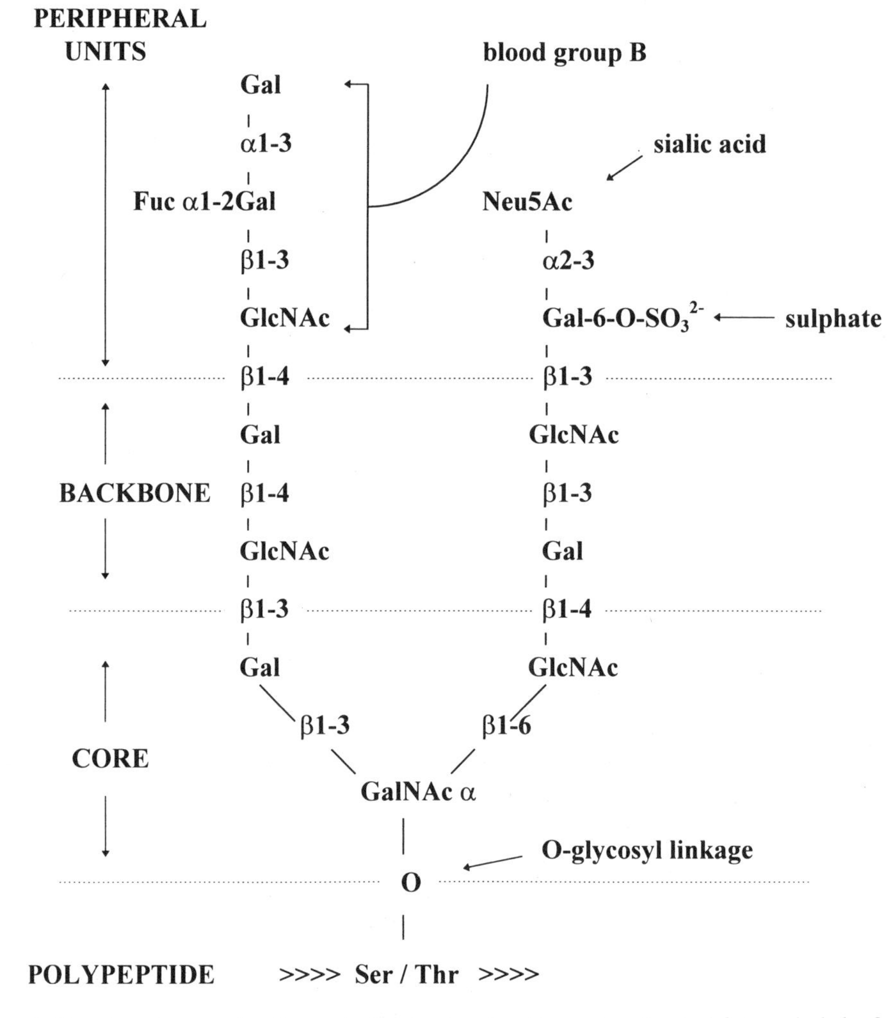

Figure 1. Hypothetical composite oligosaccharide structure found in mucins. Principal features include: O-glycosidic linkage of *N*-acetyl-D-galactosamine to serine and threonine residues on the polypeptide chain; CORE region, with core type 2 as an example; BACKBONE of poly-*N*-acetyl-lactosamine type; PERIPHERAL groups such as sialic acid, sulfate, and a blood group antigen. Many combinations based on these arrangements are possible.

Three regions can be identified in the oligosaccharides: core, backbone, and periphery (Fig. 1). Variation exists in each region. Thus, for any mucin molecule, hundreds of O-linked oligosaccharide structures may occur. Mucin glycosylation is tissue-specific and related to the functional requirements at each mucosal site.[16–18] Thus, it is important to identify these structures to correlate them with tissue-specific function and with pathological processes. A detailed consideration of the metabolism and molecular biology of ocular mucins can be found elsewhere.[6,19]

The accumulation of abnormal mucus in keratoconjunctivitis sicca (KCS) contributes to the pain and discomfort of filamentary keratitis.[20] Altered biosynthetic processes and/or mucin interactions are implicated in the accumulation of this material. This may occur through changes in tear film osmolarity, the presence of abnormal non-mucin components in the tear film, an increase in the secretion of gel-forming components of the tear film, changes in the proportions or aggregation of such gel-forming components, and/or changes in the core protein structure of mucins. Altered glycosylation of the core protein of mucins may also be implicated. The current study examines this final possibility. Detailed studies of the glycosylation patterns of ocular mucins are hampered by the ethical and practical difficulties of collecting large samples of mucus from affected human subjects. In an attempt to define an experimental model in which these constraints are overcome, we have recently been developing the dog as a model for KCS in humans.

KCS may be induced by nutritional, environmental, toxic, and other factors including autoimmune disease such as rheumatoid arthritis.[21–24] KCS is common in dogs, the West Highland white terrier and the cocker spaniel being particularly prone to a condition like Sjögren's syndrome in man.[23,24] In common with the human condition, female dogs are more commonly affected than males.[25] A copious secretion of viscous mucus is characteristic of the canine condition.[23,26] Dogs affected with KCS accumulate up to ~ 0.5 ml of this material, which can aspirate from such eyes. This situation is ideal for biochemical studies on ocular mucins, since it is possible to isolate milligram quantities of purified mucins by pooling material collected from relatively few animals. It is also possible to collect significant quantities of mucus, for the purposes of comparison, from the conjunctival sac of normal dogs: typically up to 0.1 ml can be collected per eye.

The aim of this study was to examine patterns of oligosaccharides isolated from purified ocular mucin fractions. These were purified from pooled samples of mucus aspirated from the eyes of normal and KCS dogs.

2. MATERIALS AND METHODS

2.1. Collection of Mucus Samples

Ocular mucus was collected from the ocular surface of three dogs with KCS (2 female, 1 male) over a 23-day period and of 28 normal dogs (female).

2.2. Purification of Secreted Ocular Mucins

Based on our previous investigations,[7] we have established a routine method for purifying secreted mucins. Briefly, samples are collected using low-level suction from a water-powered vacuum pump into 4 M GuHCl in PBS containing a cocktail of protease inhibitors[27] and dispersed by stirring overnight at room temperature. Samples are adjusted to 1.40–1.41 g/ml with solid CsCl and fractionated using a single-step density gradient

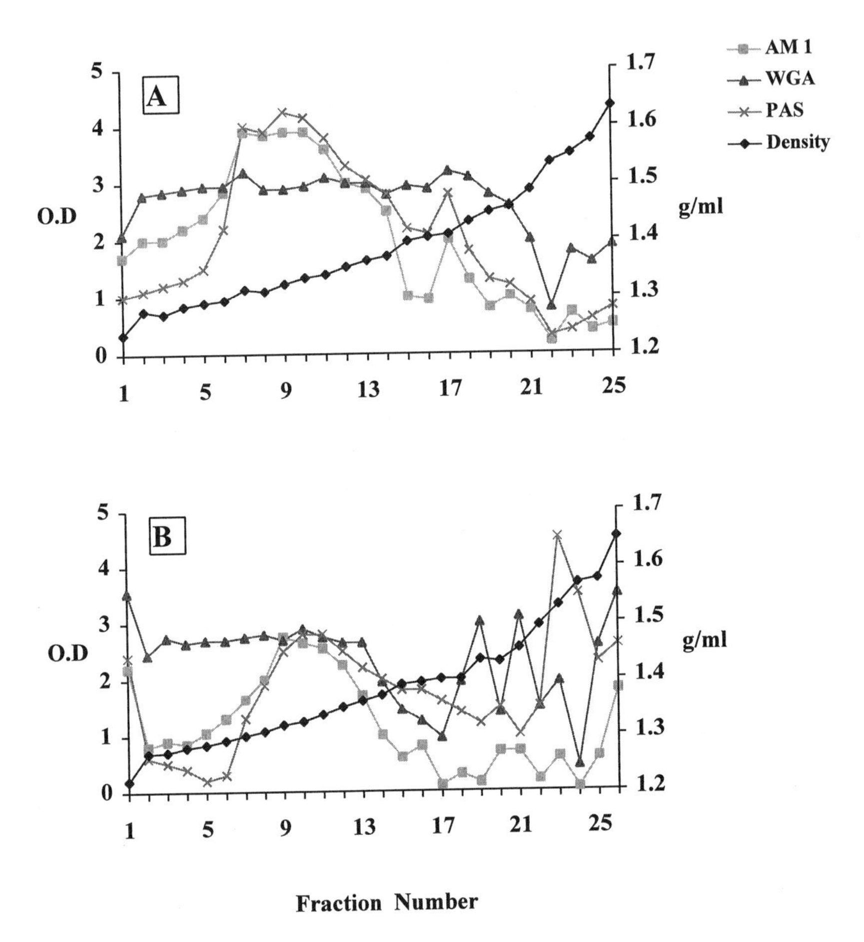

Figure 2. Density gradient profiles of ocular mucus (starting density ~1.4 g/ml) from (A) normal and (B) KCS dogs. Fractions are stained with periodic acid-Schiff (PAS), anti-mucin antibodies (AM1), and wheat germ agglutinin (WGA).

centrifugation method[28] (Fig. 2; Table 1). Three high-molecular-weight secreted mucin fractions are purified by gel filtration on Sepharose CL-4B or CL-2B (Fig. 3) and desalting on Sephadex G10, followed by lyophilization.

2.3. Analysis of Monosaccharide and Amino Acid Content

100 μg samples of purified mucins were analyzed for monosaccharide and amino acid content using standard methods.[29,30]

2.4. β-Elimination of O-linked Oligosaccharide Side Chains

An alkali-catalyzed β-elimination reaction was carried out in the presence of sodium boro[^{3}H]hydride on 1 mg samples of purified mucins.[31] Released oligosaccharides were size-fractionated on Biogel P4 (200–400 mesh; 90 x 1 cm column) in 0.1 M pyridinium acetate, pH 5.0, and elution profiles were determined by counting incorporated radioactivity.

Table 1. Characteristics of mucin fractions

Mucin Fraction	Density Range (g/ml)	Staining Characteristics
1	< 1.30	WGA + AM1 –
2	1.30–1.36	WGA + AM1 +
3	1.36–1.47	WGA + AM1 –

WGA, wheat germ agglutinin. AM1, anti-mucin antibodies.

2.5. Thin-Layer Chromatography

Fractions representing the major peaks were desalted and subjected to thin-layer chromatography.[32] Certain fractions were chromatographed before and after digestion with sialidase (neuraminidase Type X, from *Clostridium perfringens*).

3. RESULTS

3.1. Monosaccharide and Amino Acid Content

The monosaccharide and amino acid analyses of fractions derived from normal ocular mucus are indicated in Table 2. O-Linked oligosaccharides are linked through a single *N*-acetylgalactosamine (GalNAc). Therefore, the molar ratio of this sugar has been adjusted to unity. The fractions all show the signature of purified mucins: an absence of mannose, a high molar ratio of GalNAc to other sugars, and a content of serine+threonine+proline > 30%. Fraction 1 shows similar results. However, its low buoyant density (<1.30 g/ml) suggests that it may contain mucin that is lipid complexed. The molar ratio of Neu5Ac is <0.5 for all three fractions, indicating a mixture of neutral and sialylated oligosaccharides. Overall, the high molar ratio of GalNAc to other sugars indicates that the majority of oligosaccharides are short.

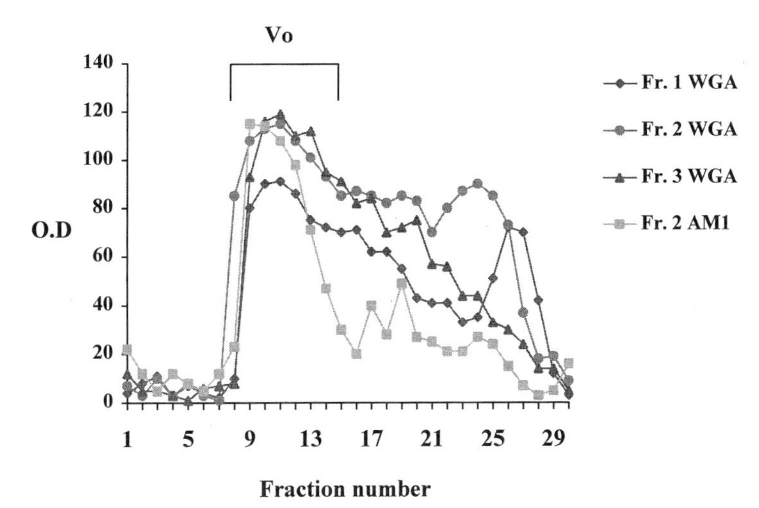

Figure 3. Typical gel filtration profile of canine secreted ocular mucin (Fraction 2) on Sepharose CL-4B. High molecular weight 'excluded' material containing mucin (V_0 range = Fractions 8–15) and lower molecular weight 'included' material are present.

Table 2. Monosaccharide and amino acid content of normal mucin fractions

			A. Monosaccharide content (molar ratio)			
Fraction	Neu5Ac	Fuc	GlcNAc	Gal	Man	GalNAc
1	0.31	1.33	0.46	1.08	0.00	1.00
2	0.26	1.92	0.36	1.07	0.00	1.00
3	0.46	1.74	0.64	1.18	0.00	1.00

Neu5Ac, sialic acid. Fuc, fucose. GlcNAc, *N*-acetylglucosamine. Gal, galactose. Man, mannose.

	B. Amino acid content (% of total protein)			
Fraction	Serine	Threonine	Proline	Total %
1	12.55	10.05	9.60	32.20
2	13.55	10.85	9.60	34.00
3	12.45	11.05	11.95	35.45

Table 3 summarizes the results for the equivalent fractions derived from KCS ocular mucins. Fractions 2 and 3 have comparable monosaccharide profiles which are consistent with a mucin identity. Fraction 1, however, contains a high molar ratio of mannose, combined with a low content of serine+threonine+proline (<22%). This indicates the presence of *N*-linked oligosaccharides that may be derived from inflammatory exudates. In Fractions 2 and 3, the molar ratios of sugars related to GalNAc indicate an increase in the GlcNAc and Gal content of 80–200%, with the fucose largely unchanged. The proportion of Neu5Ac is markedly increased to ~360% (Fraction 2) and~240% (Fraction 3) of its level in the equivalent normal fractions. The amino acid ratios are similar in Fractions 2 and 3 from normal and KCS material.

3.2. β-Elimination of O-linked Oligosaccharide Side Chains

Analysis on Biogel P4 of released oligosaccharides from normal fractions showed a 'signature' of four main peaks in all three fractions (Fig. 4A). The majority of oligosaccharides were very short, most being in the mono/trisaccharide range. The largest oligosaccharides present were indicated as being penta- to tetrasaccharides. The proportions of the four peaks varied among the fractions.

Table 3. Monosaccharide and amino acid content of KCS mucin fractions

			A. Monosaccharide content (molar ratio)			
Fraction	Neu5Ac	Fuc	GlcNAc	Gal	Man	GalNAc
1	0.12	2.13	2.43	3.35	3.22	1.00
2	0.93	1.72	0.86	1.98	0.13	1.00
3	1.10	1.81	2.03	2.42	0.00	1.00

See Table 2A footnote for abbreviations.

	B. Amino acid content (% of total protein)			
Fraction	Serine	Threonine	Proline	Total %
1	7.75	5.60	7.85	21.20
2	15.75	11.05	10.30	37.10
3	11.70	11.55	10.80	34.05

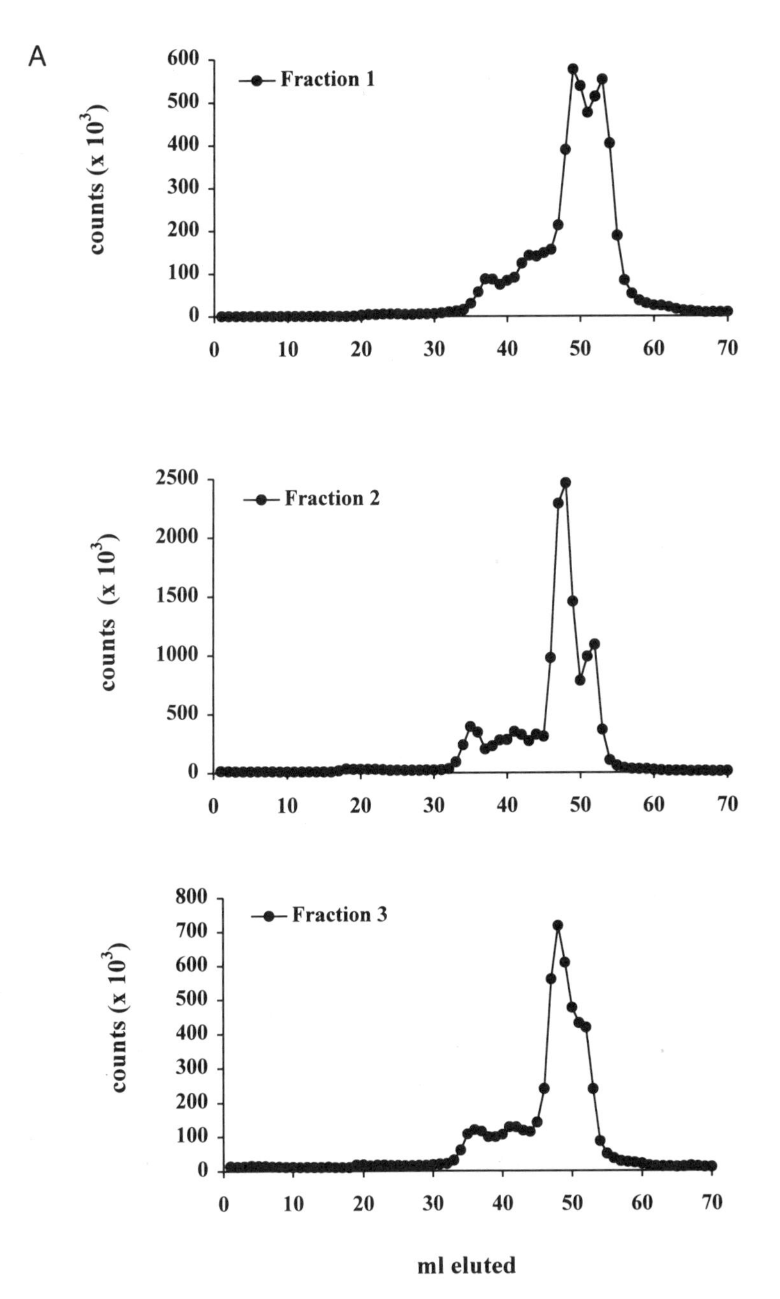

Figure 4. Size-fractionation on Biogel P4 of β-eliminated oligosaccharides from (A) normal and (B) KCS mucin fractions. The major peaks are short chains, eluting in similar fashion to sialyl *N*-acetylgalactosaminitol and *N*-acetylgalactosaminitol, which elute in the range 46–48 ml and 51–53 ml respectively (not shown).

Biogel P4 profiles of KCS oligosaccharides (Fig. 4B) each showed the presence of four peaks which appeared equivalent to the 'signature' seen for the normal samples, but the proportions of the peaks differed from the normal samples. The most marked change was an increase in the size of the second peak, which was attributable to larger oligosaccharides (shown by asterisks).

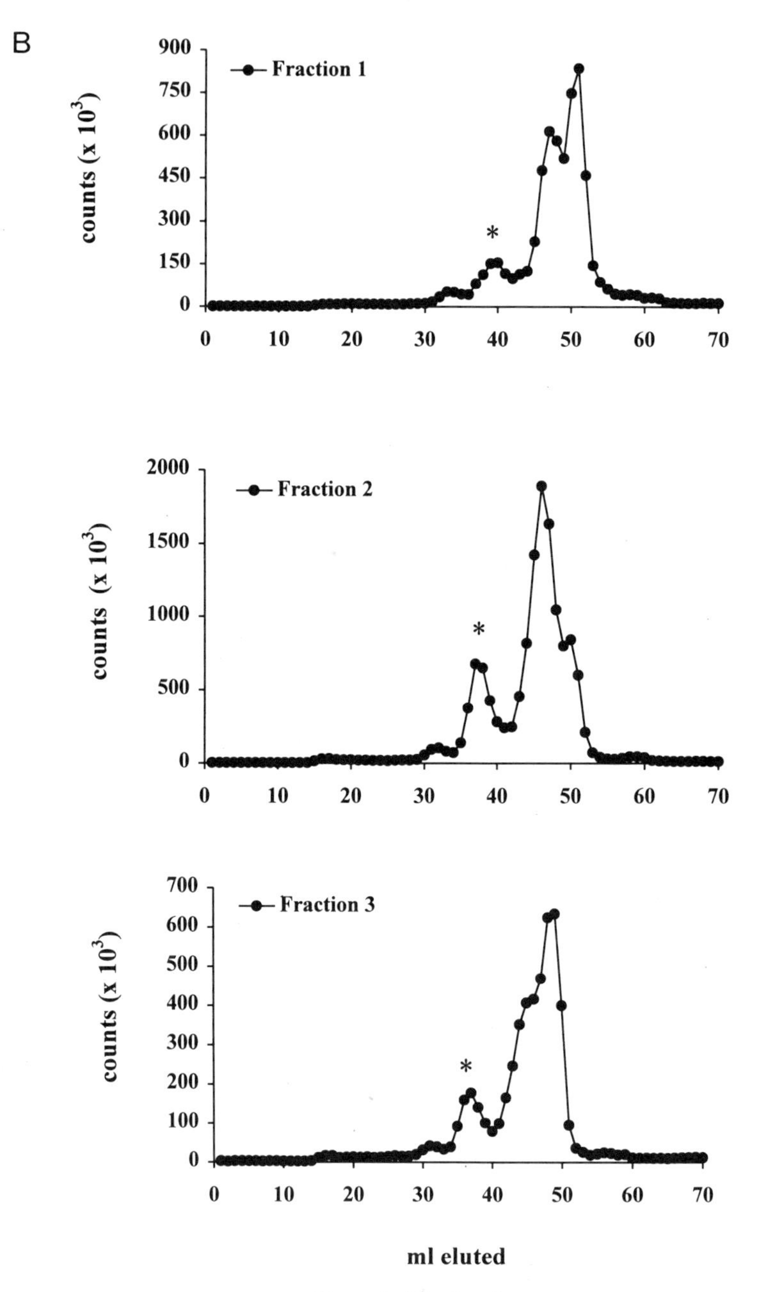

Figure 4. *(Continued)*.

3.3. Thin-Layer Chromatography

This was used to assess the number of different structures represented within each of the four 'signature' peaks on Biogel P4 profiles. Comparable patterns were observed for all fractions in both normal and KCS dogs. Representative results from Fraction 2 are illustrated in Fig. 5. The presence of more slowly migrating bands in the earlier peaks from

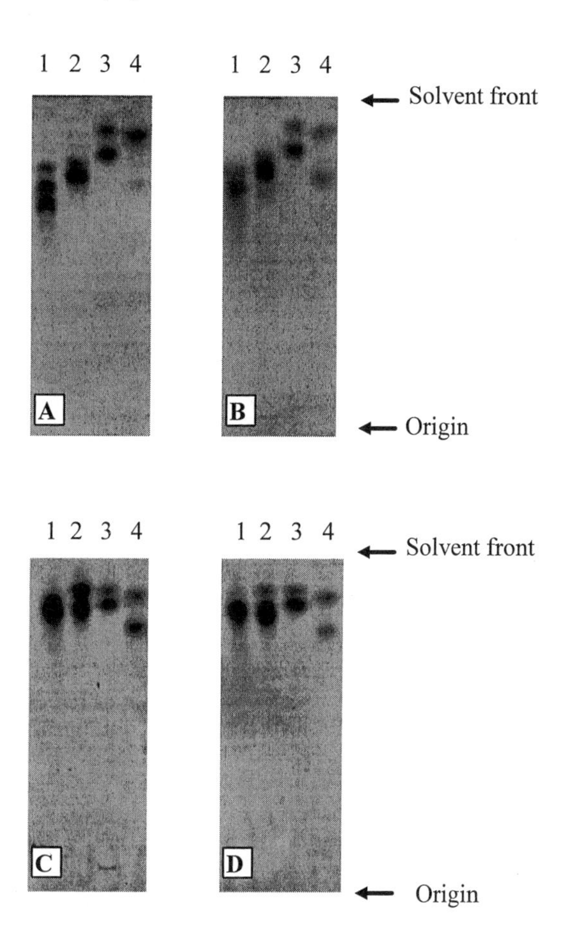

Figure 5. Thin-layer chromatography of oligosaccharides released from mucin Fraction 2 (1.30–1.36 g/ml) for (A) normal and (B) KCS samples. Lanes 1, 2, 3, and 4 represent 'signature peaks' from Biogel P4 profiles (see Fig. 4). (C) Normal, after sialidase digestion. (D) KCS, after sialidase digestion. Each lane contains 50,000 cpm.

the Biogel profiles is consistent with the presence of larger oligosaccharides. In all cases each peak is represented by at least two bands, some appearing to be partially superimposed (e.g., Fig. 5A, Lane 2). Sialidase digestion typically causes a simplification of the oligosaccharide banding pattern. This is represented as a convergence of Rf values across the plate, and a reduction in the number of bands in Lane 1. (Compare Fig. 5A & C, and Fig. 5B & D.) Lanes 1 and 2 (representing Peaks 1 and 2) showed a marked shift in Rf after sialidase digestion, confirming that a high proportion of this material is sialylated. In contrast, the Rf of material in Lanes 3 and 4 (representing Peaks 3 and 4) was not appreciably altered.

4. DISCUSSION

This study has shown that the pathological accumulation of mucus observed in KCS is parallelled by an increase in the sialylation of ocular mucins. This change in glycosylation may contribute to an alteration of the viscoelasticity properties of the gel layer of the tear film, affecting the elimination of mucus from the ocular surface. The negative charge of sialic acid contributes to water binding by mucin molecules.[33] This property may be a physiological adaptation in response to desiccation of the ocular surface in KCS.

Our results indicate that the oligosaccharide side chains of canine ocular mucins are short, probably less than five sugars. A similar pattern of glycosylation has been found in bovine and ovine salivary mucins[34,35] and is associated with malignancy in man,[36] but is otherwise uncommon. Nevertheless, analyses of ocular mucins from humans[37] and cattle (Morris, Bolis, and Packer, unpublished data) show similar results. This suggests that this characteristic may relate to the local physiological requirements of the normal tear film.

Analysis of oligosaccharides on Biogel P4 has shown that normal and KCS ocular mucins have a ‘signature’ comprising four comparable peaks in the monosaccharide-to-pentasaccharide range. Analysis by thin-layer chromatography indicates that comparable peaks from normal and KCS fractions contain structures with similar Rf values, and that each peak contains more than one component. Future work will resolve individual oligosaccharide structures using high-resolution techniques, such as Dionex HPLC, and glycomapping using specific glycosidases.

ACKNOWLEDGMENT

This work was supported by the Wellcome Trust Project Grant 037530/Z/93/Z/1.5/WRE/JL.

REFERENCES

1. Dilly PN. Structure and function of the tear film. *Adv Exp Med Biol.* 1994;350:239–247.
2. Prydal JI, Artal P, Woon H, Campbell FW. Study of the human precorneal tear film thickness and structure using laser interferometry. *Invest Ophthalmol Vis Sci.* 1992;33:2006–2011.
3. Prydal JI, Kerr Muir MG, Dilly PN. Comparison of tear thickness in three species determined by the glass fibre method and confocal microscopy. *Eye.* 1993;7:472–475.
4. Watanabe H, Fabricant M, Tisdale AS, Spurr-Michaud SJ, Lindberg K, Gipson IK. Human corneal and conjunctival epithelia produce a mucin-like glycoprotein for the apical surface. *Invest Ophthalmol Vis Sci.* 1995;36:337–344.
5. Watanabe H, Tisdale AS, Gipson IK. Eyelid opening induces expression of a glycocalyx glycoprotein of rat ocular surface epithelium. *Invest Ophthalmol Vis Sci.* 1993;34:327–338.
6. Corfield AP, Carrington SD, Hicks SJ, Berry M, Ellingham RB. Ocular mucins: Purification, metabolism and functions. In *Progress in Retinal and Eye Research.* Oxford: Elsevier Science;Vol 16 in press.
7. Hicks SJ, Carrington SD, Kaswan RL, Adam S, Bara J, Corfield AP. Demonstration of discrete secreted and membrane-bound ocular mucins in the dog. *Exp Eye Res.*, in press.
8. Dilly PN. On the nature and the role of the subsurface vesicles in the outer epithelial cells of the conjunctiva. *Br J Ophthalmol.* 1985;69:477–481.
9. Gipson IK, Yankauckas M, Spurr-Michaud SJ, Tisdale AS, Rinehart W. Characteristics of a glycoprotein in the ocular surface glycocalyx. *Invest Ophthalmol Vis Sci.* 1992;33:218–227.
10. Gipson IK, Spurr-Michaud SJ, Tisdale AS, Kublin C, Cintron C, Keutmann H. Stratified squamous epithelia produce mucin-like glycoproteins. *Tissue Cell.* 1995;27:397–404.
11. Greiner JV, Weidman TA, Korb MR, Allansmith MR. Histochemical analysis of secretory vesicles in non-goblet conjunctival epithelial cells. *Acta Ophthalmol.* 1985;63:89–92.
12. Inatomi T, Spurr-Michaud S, Tisdale A, Gipson IK. Human corneal and conjunctival epithelia express MUC1 mucin. *Invest Ophthalmol Vis Sci.* 1995;36:1818–1827.
13. Carlstedt I, Sheehan JK, Corfield AP, Gallagher JT. Mucous glycoproteins: A gel of a problem. *Essays Biochem.* 1985;20:40–76.
14. Forstner JF, Forstner GG. Gastrointestinal mucus. In: Johnson LR, ed. *Physiology of the Gastrointestinal Tract.* New York: Raven Press; 1994:1245–1283.
15. Strous GJ, Dekker J. Mucin-type glycoproteins. *Crit Rev Biochem Mol Biol.* 1992;27:57–92.
16. Montreuil J, Vliegenthart JFG, Schachter H. (eds.) New Comprehensive Biochemistry. Vol 29a *Glycoproteins.* Amsterdam: Elsevier Science BV; 1995.

17. Roussel P, Lamblin G, Lhermitte M., et al. The complexity of mucins. *Biochimie.* 1988;70:1471–1482.
18. Corfield AP, Myerscough N, Gough M, Brockhausen I, Schauer R, Paraskeva C. Glycosylation patterns of mucins in colonic disease. *Biochem Soc Trans.* 1995;23:840–845.
19. Gipson IK, Inatomi TI. Mucin genes expressed by the ocular surface epithelium. In: *Progress in Retinal and Eye Research.* Oxford: Elsevier Science, 1977;16:81–98.
20. Zaidman GW, Geeraets R, Paylor RR, Ferry AP. The histopathology of filamentary keratitis. *Arch Ophthalmol.* 1985;103:1178–1181.
21. Bron AJ, Tiffany JM, Kaura R, Mengher L. Disorders of the tear film lipids and mucous glycoproteins. In: Easty DL, Smolin G, eds. *External Eye Disease.* Butterworth, London. 1985:63–105.
22. McGill J. The tear film in health and disease. In: Easty DL, Smolin, G. eds.. *External Eye Disease.* Butterworth; 1985:106–132.
23. Sansom J, Barnett KC. Keratoconjunctivitis sicca in the dog: A review of two hundred cases. *J Small Anim Pract.* 1985;26:121–131.
24. Carrington SD, Bedford PGC, Guillon J-P, Woodward EG. Polarised light biomicroscopic observations on the pre-corneal tear film. II. Keratoconjunctivitis sicca in the dog. *J Small Anim Pract.* 1987;28:671–679.
25. Barnett KC. Keratoconjunctivitis sicca - sex incidence. *J Small Anim Pract.* 1988;29:531–534.
26. Kaswan RL, Bounous D, Hirsh SG. Diagnosis and management of keratoconjunctivitis sicca. *Vet Med.* 1995;90:539–560.
27. Corfield AP, Paraskeva C. Secreted mucus glycoproteins in cell and organ culture. In: Hounsell E, ed. *Glycoprotein Analysis in Biomedicine* . Vol 14. Totowa: Humana Press; 1993:211–232.
28. Carlstedt I, Lindgren H, Sheehan JK, Ulmsten U, Wingerup L. Isolation and characterisation of human cervical-mucus glycoproteins. *Biochem J.* 1983;211:13–22.
29. Cooper CA, Packer NH, Redmond JW. The elimination of O-linked glycans from glycoproteins under non-reducing conditions. *Glycoconj J.* 1994;11:163–167.
30. Yan JX, Wilkins MR, Ou K, et al. Large scale amino acid analysis for proteome studies. *J Chromatogr A.* 1996;736:291–302.
31. Carlson DM. Structures and immunochemical properties of oligosaccharides isolated from pig submaxillary mucins. *J Biol Chem.* 1968;243:616–626.
32. Veh RW, Michalski JC, Corfield AP, Sander-Wewer M, Gies D, Schauer R. New chromatographic system for the rapid analysis and preparation of colostrum sialyloligosaccharides. *J Chromatogr.* 1981;212:313–322.
33. Tam PY, Verdugo P. Control of mucus hydration as a Donnan equilibrium process. *Nature.* 1981; 292:340–342.
34. Bertolini M, Pigman W. The existence of oligosaccharides in bovine and ovine submaxillary mucins. *Carbohydr Res.* 1970;14:53–63.
35. Corfield AP, do Amaral Corfield C, Veh RW, Wagner SA, Clamp JR, Schauer R. Characterisation of the major and minor mucus glycoproteins from bovine submandibular gland. *Glycoconj J.* 1991;8:330–339.
36. Singhal A, Hakomori S-I. Molecular changes in carbohydrate antigens associated with cancer. *Bioessays.* 1990;12:223–230.
37. Berry M, Ellingham RB, Corfield AP. Polydispersity of human ocular mucins. Invest Ophthalmol Vis Sci., Vol 37 No. 13 pp. 2559–2571, in press.

38

CORNEAL EPITHELIAL TIGHT JUNCTIONS AND THE LOCALIZATION OF SURFACE MUCIN

Henry F. Edelhauser, David E. Rudnick, and Ramzy G. Azar

Emory Eye Center
Emory University
Atlanta, Georgia

1. INTRODUCTION

Studies have shown that the corneal epithelial surface can be protected with artificial tears that have an ionic composition similar to natural tears.[1] More recent studies have reported that the corneal epithelial mucin layer is also maintained to protect the epithelial tight junctions with artificial tears that contain bicarbonate.[2] Although the mechanism of this effect is not yet known, bicarbonate is the physiological buffer in tears.[2–5] The maintenance of a slightly alkaline environment is more compatible with the corneal epithelium than neutral or acidic conditions. Bicarbonate also appears to be important in maintaining the normal electrophysiological parameters of the corneal epithelium. Preliminary data from our laboratory has suggested that an artificial tear solution with bicarbonate will protect and maintain the corneal epithelial potential and resistance of rabbit corneas mounted in modified Ussing chambers. The role of bicarbonate in corneal epithelial homeostasis is not clear; however, Conroy et al.[6] have reported that carbonic anhydrase is present in the epithelium, and recently Candia[7] has shown that there is a net flux of bicarbonate from the cornea to the tears. Rismondo et al.[8] have shown in rabbit tears (HCO^-_3 = 27.5 mM/L) that the concentration of bicarbonate is higher than that obtained from the lacrimal gland (HCO^-_3 = 14.7 mM/L). Thus, tear bicarbonate comes from two sources: lacrimal gland fluid and the corneal epithelium.

The mucin layer is important in maintaining the outer corneal epithelial ultrastructure, and when the outer cell layer becomes damaged, the cells below express the mucin to cover their outer surface.[3]

The purpose of this study is to compare the electrical resistance of the rabbit corneal epithelial tight junction to the localization of surface mucin following exposure to artificial tears buffered with and without bicarbonate and following exposure of the corneal epithelium to benzalkonium chloride (BAC).

Lacrimal Gland, Tear Film, and Dry Eye Syndromes 2
edited by Sullivan *et al.*, Plenum Press, New York, 1998

Table I. Composition of AQA, physiological tears, and human tears (mM/L)

	AQA 5*	Phys Tear≈	Phys Tear≈ w/o HCO^-_3	Human Tears**
Na^+	144	112.9	130	128.7
K^+	4.9	17.4	17.4	17
Cl^+	103.4	132.6	148.7	141.3
Ca^{++}	1.4	0.36	.32	0.32
Mg^{++}	0.6	0.31	.31	0.35
HPO^-_4	0.6	—	—	—
Glucose	26	—	—	—
SO^-_4	7.5	—	—	—
HCO^-_3	20	11.9	—	12.4
Gluconate	2.8	—	—	—
Zn^{++}	—	11μM	11μM	11μM
HEPES	25	—	—	—
Dextran 70	—	0.1%	0.1%	—
HPMC	—	0.3%	0.3%	—
pH	7.8	7.7	7.4	7.7±0.1
mOsm	305	290	296	302

*Klyce et al. IO 12:127-139, 1973.
**Rismondo et al. CLAO J 15:222-229, 1989.
≈Prepared by Alcon Laboratories, Inc., Fort Worth, Texas.

2. METHODS

New Zealand White rabbits were anesthetized intramuscularly with xylazine HCl (20 mg/mL) and ketamine HCl (100 mg/mL) and then euthanized with an overdose of sodium pentobarbital (324 mg/mL) injected intravenously. All interactions with rabbits for these experiments conformed to the Guide for the Care and Use of Laboratory Animals, the ARVO Resolution on the Use of Animals in Research, and the Declaration of Helsinki. The corneas were enucleated and mounted in a modified Ussing chamber according to the procedure of Klyce et al.[9] Inserts were designed to match the curvature of the rabbit cornea. The cornea was gently clamped in the chamber and bathed in a Ringer's solution (AQA 5) similar in ionic concentration to both human and rabbit tears[8] on both the endothelial and epithelial sides of the membrane. The solution was maintained at 37°C with a pH of 7.4 under 95% air/5% CO_2 aeration. Table I lists the composition of AQA, Physiological Tears with and without bicarbonate and human tears.

The bioelectric measurements of the cornea were evaluated using a DVC-1000 Voltage/Current Clamp (World Precision Instruments). Transepithelial potential difference (V_{te}) and short circuit current (I_{sc}) were measured and recorded on a chart recorder (Kipp & Zonen), and resistance (R) was calculated using Ohm's law.

After the mounting procedure, V_{te} of the corneas was allowed to stabilize for about 60 min. At this point, I_{sc} was measured, and then the test solution was added to the epithelial bath. For experiments studying the effects of bicarbonate in artificial tears, solutions with or without bicarbonate were placed on the epithelium for 30 min, followed by an AQA 5 solution wash and a 30 min recovery period. For experiments with BAC, concentrations ranging from 0.01% to 0.0001% were added to the corneas. Corneas were exposed to BAC for 15 or 30 min. Afterward, the corneas from all experiments were removed from the chamber and fixed in an 0.5% cetylpyrimidium chloride–aldehyde fixative to evaluate the mucin layer.[10]

3. RESULTS

The effects of bicarbonate on V_{te} and R are shown in Fig. 1. In corneas exposed to tear formulations containing bicarbonate, the recovery of both V_{te} and R was greater than in corneas exposed to formulations without bicarbonate. The final (mean ± SEM) V_{te} of the bicarbonate-exposed corneas was 28.8 ± 2.0 mV compared to 20.8 ± 5.6 mV in corneas exposed to bicarbonate-free solution. Similarly, R recovered to 6.84 ± 1.0 ½cm^2 of

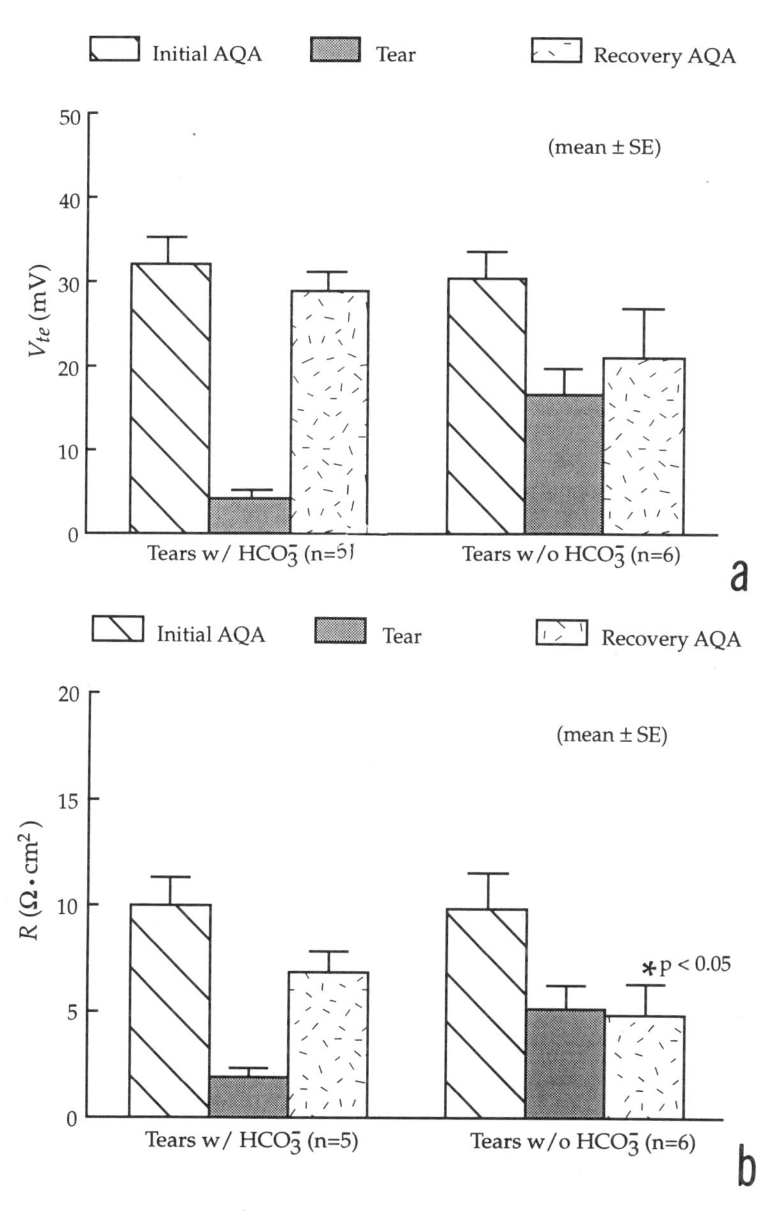

Figure 1. Effects of bicarbonate containing artificial tears versus a tear formulation without bicarbonate. (a) Comparison of V_{te}. (b) Comparison of R.

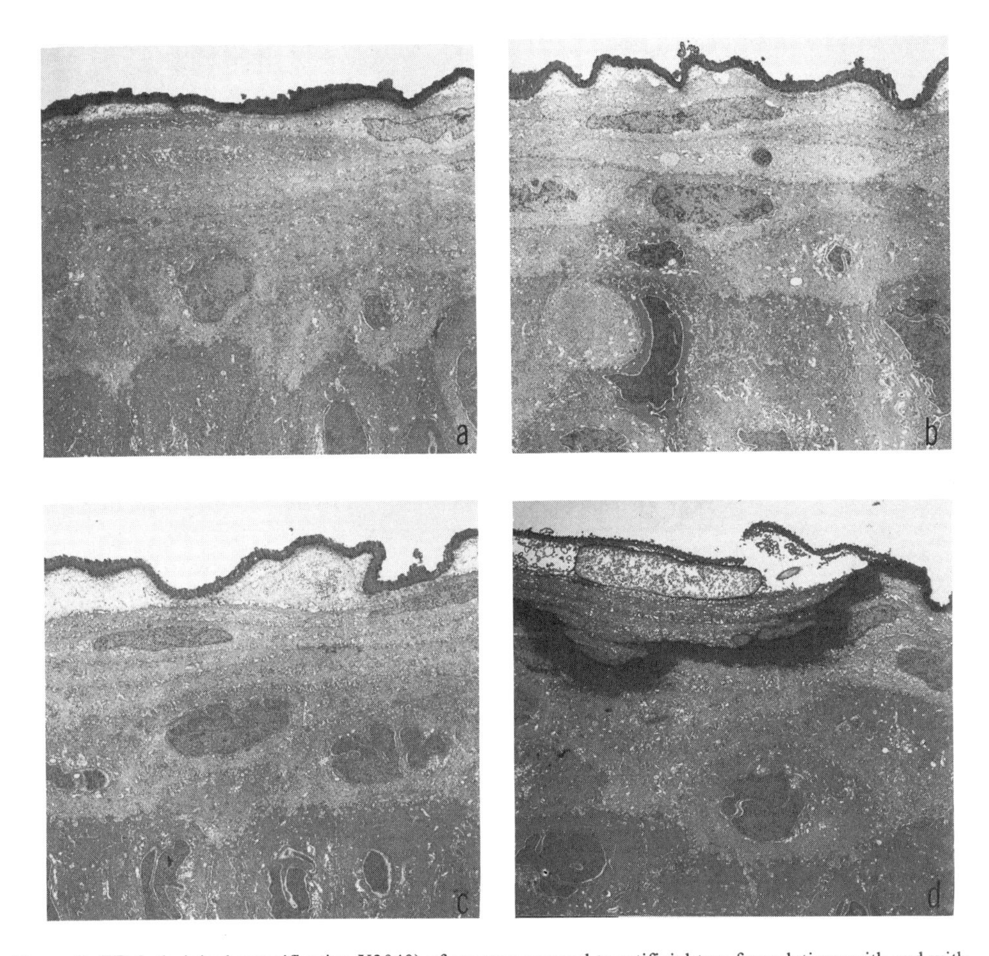

Figure 2. TEMs (original magnification X3040) of corneas exposed to artificial tear formulations with and without bicarbonate. (a) Control. (b) Artificial tear solution containing bicarbonate. (c and d) Artificial tear solution without bicarbonate.

the original R for bicarbonate-exposed corneas compared to 4.9 ± 1.5 ½·cm^2 for bicarbonate-free exposure ($p < 0.05$). The initial values should be considered controls since the corneas had not been exposed to any test solutions at that time. Therefore, the recovery values can also be viewed as change from control conditions. I_{sc} was also measured during the experiment, but did not show any noticeable differences between the evaluated solutions and controls.

In Fig. 2, TEMs of corneas are shown after exposure to artificial tears with and without bicarbonate. The mucin layer was preserved in the fixation process and can be seen by a dark, thick layer covering the cell layers. Under a control situation (Fig. 2a), the mucin layer covers the outermost cells of the epithelium, signifying that the tight junctions between the cells are still intact. This is corroborated by the bioelectric measurements, since breakdown of the tight junctions corresponds to loss of corneal V_{te} and R. When corneas were exposed to a bicarbonate-containing tear solution (Fig. 2b), the mucin layer is maintained intact at the outermost epithelium. However, upon exposure to a bicarbonate-free tear solution (Fig. 2c), the mucin layer is localized on the surface epithelial cells, and the surface cells may be markedly edematous. In some cases the mucin can be found below the superficial epithelium, localizing at the second and third cell layers of the epithelium (Fig. 2d). This indicates a breakdown in the tight junctions of the superficial epithelium. The electron micrographs suggest that the tight junction formation may occur between the cells immediately below the intact mucin layer.

When the corneas are exposed to BAC, 0.01% and 0.001%, there is a loss of epithelial resistance in all cases. The TEMs in Fig. 3 show the effects of BAC exposure to the epithelial mucin layer as compared with a control (Fig. 3a). Below each micrograph is a schematic representation of the mucin layer as shown in the corresponding photo. After a 15 min exposure to a 0.01% solution of BAC, the corneal mucin layer was seen several levels below the surface cells. Also, because of the loss of tight junctions between the cells, the surface cells would slough off the tissue (Fig. 3b). After 15 min of 0.001% BAC, the mucin layer of exposed corneas was usually observed above the second or third cell layer, indicating some loss of tight junctions but less damage than at the higher concentrations of BAC. Micrographs (not shown) of high concentrations of BAC (0.01% or greater) exposed for a prolonged period (³30 min) showed almost complete loss of epithelial squamous cells, at times exposing the basal cell layer.

4. DISCUSSION

The results of this study show that the outer corneal epithelial cell junctions are sensitive to bicarbonate and BAC. Any loss in tight-junction integrity leads to a decrease in epithelial resistance and the secretion of mucin by the corneal epithelial wing and basal cells. Since this is an isolated corneal preparation, no conjunctival cells are present. Greiner et al.[11] and Dilly[12] have previously suggested that the conjunctival epithelial cells secrete mucin, and Gipson et al.[13] have shown immunohistochemically that the corneal epithelial cells also produce and secrete a glycoprotein onto their surface. Our data provide further evidence that when the outer corneal epithelial cells are damaged (via a lack of bicarbonate or with BAC), the epithelial resistance is lost, and the underlying cells differentiate into superficial cells and secrete a layer of mucin (Figs. 2d, 3b, 3c).

These results confirm our previous study[3] that bicarbonate may not be required for maintenance of the thickness of the epithelial mucin layer. The ocular mucin gel[14,15] may, however, as in the gut, form a matrix that maintains the bicarbonate-rich environment es-

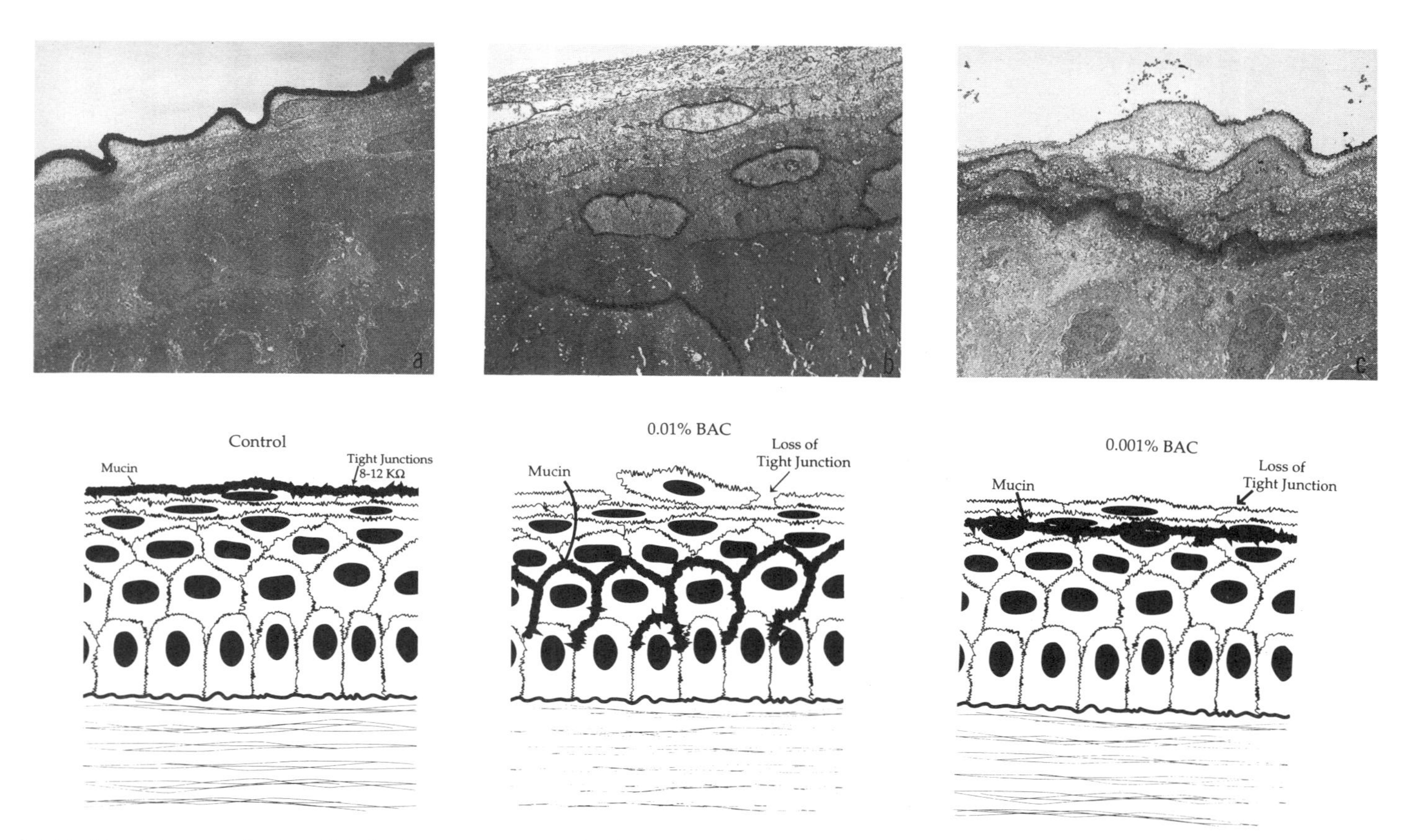

Figure 3. (a) TEM (original magnification X3040) of a control cornea. Notice the pronounced mucin layer covering the superficial epithelium. (b) TEM of a cornea exposed to a solution containing 0.01% BAC. Several layers of the epithelium have slough off, and the mucin layer is exhibited several layers below the surface epithelium. (c) TEM of a cornea exposed to a solution containing 0.001% BAC. Though the superficial cornea has shown a loss of tight junctions, the mucin layer is maintained at the second and third cell layers.

sential for normal corneal epithelial structure and function. Thus, a decrease in tear bicarbonate and BAC preservative toxicity can lead to significant changes in the corneal epithelial tight junctions and electrical properties.

ACKNOWLEDGMENTS

Supported in part by departmental core grant NIH P30EY06360, Alcon Laboratories, Inc., and Research to Prevent Blindness.

REFERENCES

1. Lopez Bernal D, Ubels JL. Quantitative evaluation of the corneal epithelial barrier: Effect of artificial tears and preservatives. *Curr Eye Res.* 1991;10:645–656.
2. Lopez Bernal D, Ubels JL. Artificial tear composition and promotion of recovery of the damaged corneal epithelium. *Cornea.* 1993;12:115–120.
3. Ubels JL, McCartney MD, Lantz WK, Beaird J, Dayalan A, Edelhauser HF. Effects of preservative-free artificial tear solutions on corneal epithelial structure and function. *Arch Ophthalmol.* 1995;113:371–378.
4. Keller N, Morre D, Carper D, Longwall A. Increased corneal permeability induced by the dual effects of transient tear acidification and exposure to benzalkonium chloride. *Exp Eye Res.* 1980;30:203–213.
5. Raber I, Breslin CW. Toleration of artificial tears: The effects of pH. *Can J Ophthalmol.* 1978;13:247–249.
6. Conroy CW, Buck RH, Maren TH. The microchemical detection of carbonic anhydrase in corneal epithelia. *Exp Eye Res.* 1992;55:637–640.
7. Candia OA. A novel system to measure labeled CO_2 and HCO_3 fluxes across epithelia: Corneal epithelium as model tissue. *Exp Eye Res.* 1996;63:137–149.
8. Rismondo V, Osgood TB, Leering P, Hattenhauer MG, Ubels JL, Edelhauser HF. Electrolyte composition of lacrimal gland fluid and tears of normal and vitamin A-deficient rabbits. *CLAO J.* 1989;15:222–229.
9. Klyce SD, Neufeld AH, Zadunaisky JA. The activation of chloride transport by epinephrine and dibutyryl cyclic-AMP in the cornea of the rabbit. *Invest Ophthalmol.* 1973;12:127–138.
10. Nichols BA, Chiappino ML, Dawson CR. Demonstration of the mucous layer on the tear film by electron microscopy. *Invest Ophthalmol Vis Sci.* 1985;26:464–473.
11. Greiner JV, Weidman TA, Korb DR, Allansmith MR. Histochemical analysis of secretory vesicles in non-goblet conjunctival epithelial cells. *Acta Ophthalmol.* 1985;63:89–92.
12. Dilly PN. On the nature and the role of the subsurface vesicles in the outer epithelial cells of the conjunctiva. *Br J Ophthalmol.* 1985;69:477–481.
13. Gipson IK, Yankauckas M, Spurr-Michaud SJ, Tisdale AS, Rinehart W. Characteristics of a glycoprotein in the ocular surface glycocalyx. *Invest Ophthalmol Vis Sci.* 1992;33:218–227.
14. Allen A, Gardner A. Gastric mucus and bicarbonate secretion and their possible role in mucosal production. *Gut.* 1980;21:249–262.
15. Neatia MF, Forstner JF. Gastrointestinal mucus. In: Johnson LR, ed. *Physiology of the Gastrointestinal Tract.* New York: Raven Press; 1987:975–1010.

BREAKUP AND DEWETTING OF THE CORNEAL MUCUS LAYER

An Update

Ashutosh Sharma

Department of Chemical Engineering
Indian Institute of Technology at Kanpur
Kanpur, India

1. INTRODUCTION

The normal precorneal tear film usually remains intact between consecutive blinks, but holes begin to appear and grow at random spots in about 10–60 sec when blinking is prevented. Although the exact mechanism of the tear film breakup has eluded our understanding, it is certain that the breakup is secondary to the nonwettability of the corneal surface. We had earlier proposed a mechanism based on the possibility of the rupture and dewetting of the precorneal mucus layer due to the long-range van der Waals forces.[1–4] The tear breakup was thought to be triggered by the "hydrophobicity," or nonwettability, of the underlying corneal epithelial surface devoid of its mucus covering.

The last ten years have seen significant advances both in the understanding of thin film stability and in characterization of the physico-chemical properties of the cornea-tear film system. Direct microscopic observations have shown that even extremely viscous (viscosities as high as 10,000 times water viscosity) polymer films as thick as 0.2 μm indeed dewet their substrates by formation of holes within a few minutes.[5–11] The rapid breakup of relatively thick viscous films by the long-range van der Waals forces is in part due to a strong slippage (lack of adherence) of the entangled polymers and gels on their substrates.[6,12–14] While these observations support the aforementioned mechanism of the tear film breakup, the following facts should also be considered in a critical reappraisal and update of the mechanism. (i) The normal corneal epithelial surface *sans* mucus is now known to be as hydrophilic and wettable by aqueous tears as mucus itself.[15–18] (ii) The precorneal mucus layer now appears to be much thicker than what was thought to be the case previously.[19,20] (iii) In the original model[1–4] of the mucus layer breakup, the exact numerical value of the Hamaker constant for the epithelium–mucus–aqueous tears system was

Lacrimal Gland, Tear Film, and Dry Eye Syndromes 2
edited by Sullivan *et al.*, Plenum Press, New York, 1998

not known. The Hamaker constant governs the strength of the van der Waals forces (wettability), which cause the dewetting. (iv) Finally, in the original proposal, the role of slippage was not addressed.

The purpose of this paper is to update and restate the mechanism of Sharma and Ruckenstein[1–4] by addressing the above issues.

2. DEWETTING OF THIN FILMS: EXPERIMENTAL EVIDENCE AND IMPLICATIONS

There was a lack of direct controlled experiments on dewetting of thin films by the long-range intermolecular interactions at the time we theoretically proposed the possibility of the precorneal mucus layer breakup in 1985. Since then, a large number of experimental studies have confirmed this phenomenon of spontaneous dewetting of thin viscous films by the formation of growing holes.[5–11] The experiments are usually done with thin films of extremely viscous oils or polymer melts to slow down the kinetics of rupture, thereby facilitating time-resolved observations. Initially uniform thin films are deposited on smooth surfaces (usually variously modified silicon wafers) by spin coating. Details can be found elsewhere,[5–11] but a typical picture of dewetting observed in these experiments is summarized in Figs. 1 and 2 for molten polystyrene films on silicon wafers.[11] Micrograph of Fig. 1 shows the formation of circular holes in an initially uniform polystyrene film of 40 nm thickness. These holes expand and coalesce to eventually form a polygonal network structure, as shown in Fig. 2. Dark threads represent the polymer, and

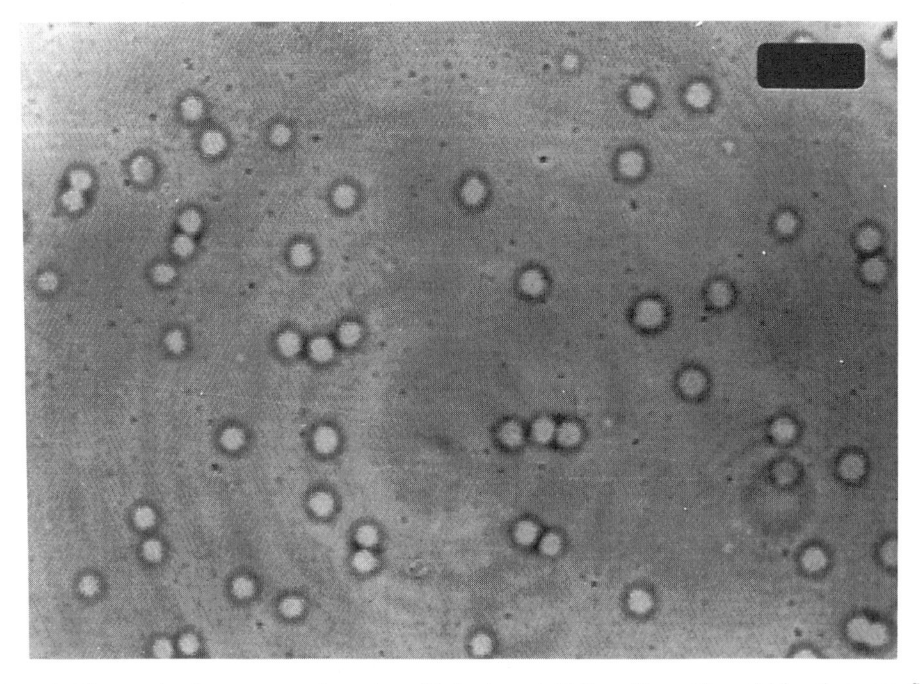

Figure 1. Micrograph of spontaneous formation of holes in an initially uniform 40-nm thick polystyrene film (MW, 28 K) on a silicon wafer. Bar = 100 μm.

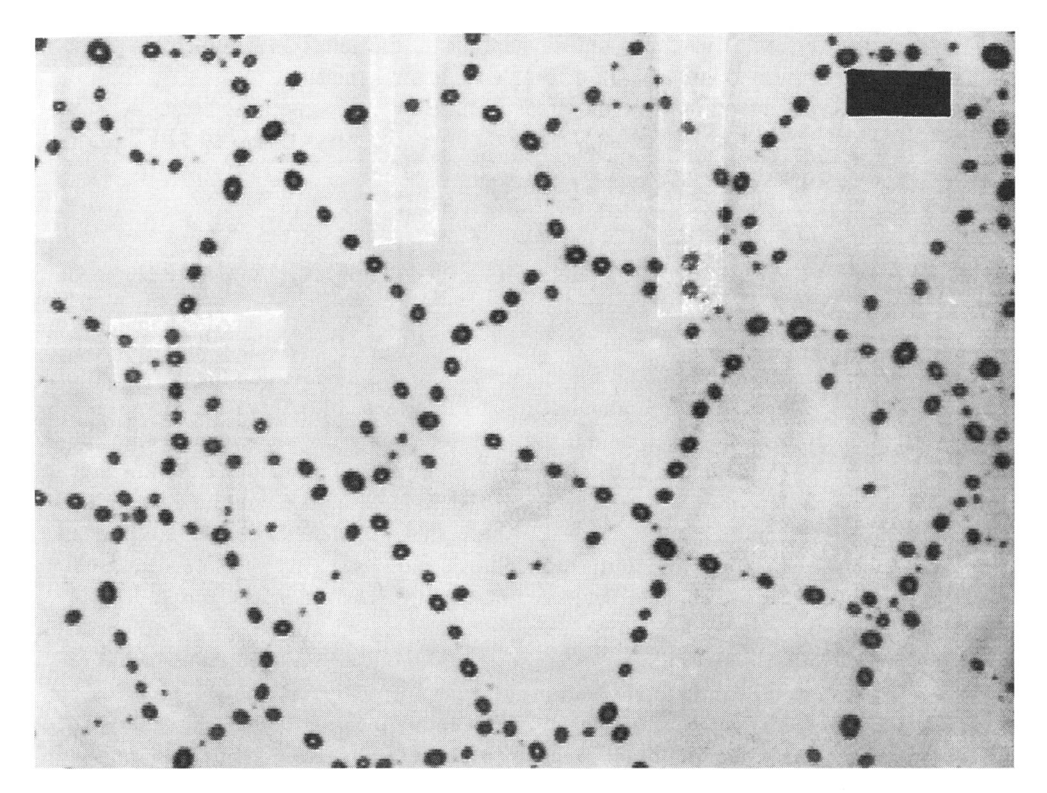

Figure 2. Micrograph of final polygonal network pattern in a 10-nm thick polystyrene film (MW, 660 K) on a silicon wafer. Repeated growth and coalescence of holes shown in Fig. 1 lead eventually to this pattern.

the lighter areas are the bare substrate devoid of the polymer. This picture has a striking similarity to the mucus network visible in filter impressions of conjunctival surface.[21]

The two most important observations that have emerged from the experiments[5–11] are the following. (i) Even extremely viscous films (viscosities some 10,000 times the water viscosity) of high surface tension and thickness (<0.2 μm) can spontaneously dewet within a few minutes. Mucus viscosity is certainly lower, and the mucus--aqueous interfacial tension is close to zero. Both of these factors should further help a more rapid breakup of relatively thick mucus layers. Further, thicker films also dewet when their thickness is locally reduced to within a few tenths of a micrometer by the presence of defects, trapped particles, and foam.[6–8, 22] This factor also appears to be important in the possibility of rupture of the mucus layer with trapped contaminants and desquamating cells. (ii) The process of film rupture and its kinetics are controlled almost entirely by the long-range Lifshitz--van der Waals (LW) forces; the shorter-range forces (e.g., acid–base, AB, interactions) play only a minor role limited to the very last stages of breakup.[11,23–25] The dominant role of long-range forces was nicely demonstrated by experiments[5,11] on rupture of polystyrene films on silicon wafers that were variously modified by thin surface coatings of different wettabilities. The number density of holes was controlled solely by the long-range interactions with the silicon wafer, regardless of the coating properties that determine the shorter-range forces.[11] A detailed theory has been reported elsewhere.[11,23–27] Here it is sufficient to note that the long- and short-range surface interactions engender the apolar (LW) and polar (AB) components, respectively, of the surface and interfacial tensions. A thin film becomes unstable and dewets due to the long-range LW interactions whenever the *apolar* component of its spreading coefficient is negative.[11, 23–25]

In what follows, an estimate of the breakup time for the precorneal mucus layer is obtained.

3. BREAKUP TIME FOR THE PRECORNEAL MUCUS LAYER: THEORY

Our earlier hydrodynamic model for the breakup time was based on the condition of "no-slip" or "perfect-stick" at the epithelial--mucus interface. In reality, the condition of "no-slip" at a solid-liquid interface is realized only for simple low-molecular-weight liquids that adhere well to the underlying substrate. It is now well known that the high-molecular-weight entangled polymers and gels resist shearing (relative motion), leading to elasticity and slippage on their substrates,[6,12–14] unless they are chemically grafted (bonded or anchored) to the substrate. In addition, mucus should slip easily on the normal epithelial surface, because normal uncontaminated mucus (in water) cannot adhere to the corneal surface due to the extreme hydrophilicity of mucus and glycocalyx carrying normal cells.[16–18] On the contrary, hydrophilic surfaces (eg, red blood cells) in aqueous media experience a net repulsion leading to "hydration-pressure."[26,27] This issue has been discussed in detail elsewhere,[17] and will also be presented in another paper at this symposium.[18] The lack of mucus adherence leading to its enhanced mobility should also greatly aid the dual roles of mucus in epithelial cleansing and lubrication.[17,18]

The hydrodynamics of thin films with strong slippage is similar to that of the foam films that display very rapid rupture. Without going into complex mathematical details, the analogy holds due to the fact that the mid-plane of a foam film experiences zero shear, which is also the case at the nominal solid-liquid interface of a strongly slipping (nonadherent) liquid. For such cases, an estimate for the time of rupture T (in sec) is given by the following equation.[28,29]

$$T = 8\,\mu\,H^3 / \mathrm{A} \qquad (1)$$

where μ is viscosity (in Pa.s) of mucus, H is thickness (in m) of the mucus layer, and A is an effective Hamaker constant (in J) that measures the strength of the long-range apolar van der Waals forces responsible for the film breakup.[4] Interestingly, in contrast to our earlier result[1–4] for non-slipping liquids, the breakup time for strongly slipping liquids is independent of the surface tension of the film. In any case, the mucus–aqueous interfacial tension is close to zero, and therefore viscosity remains the sole stabilizing factor that resists rupture. Based on Equation 1, thinner films rupture much faster than thicker films.

The effective Hamaker constant for the epithelium–mucus–aqueous tears system is related to the apolar Lifshitz--van der Waals component of the spreading coefficient, S.[17,23–27]

$$\mathrm{A} = -\,12\,\pi\,d^2\,\mathrm{S} \qquad (2)$$

where d = 0.158 nm is the minimum "cut-off" intermolecular distance,[26,27] and S is the spreading coefficient.

Further, the apolar (LW) component of the spreading coefficient is related to the apolar (LW) components of surface tensions of the epithelium (γ_e), mucus (γ_m), and water (γ_w).[23–27]

$$S = 2 [\gamma_e^{1/2} - \gamma_m^{1/2}] [\gamma_m^{1/2} - \gamma_w^{1/2}] \quad (3)$$

The apolar components of various surface tensions as determined from the contact angle goniometry can be found elsewhere,[16,17] and will also be reported in another paper in this volume.[18] Briefly, $\gamma_w = 21.8$, $\gamma_m = 40$, and $\gamma_e = 28$ (all mJ/m^2). These surface properties, together with Equations 1 and 2, predict $S = - 3$ mJ/m^2, and $A = 3 \times 10^{-21}$J. This magnitude for the effective Hamaker constant is entirely realistic, and it is characteristic of a wide variety of polymeric and biological surfaces in aqueous media.[26,27] A negative value of the apolar component of the spreading coefficient in this case implies that the long-range LW forces favor dewetting of the mucus layer on the normal corneal epithelium in the presence of aqueous tears. Equation 3 is obtained[23–27] from the definition of the spreading coefficient, $S = \gamma_{ew} - \gamma_{mw} - \gamma_{em}$, where γ_{ew}, γ_{mw}, and γ_{em} refer to the apolar components of interfacial tensions at the epithelium–water, mucus–water, and epithelium–mucus interfaces.

Although the exact value of mucus viscosity is not known, normal ocular mucus is known to be a highly hydrated fluid like "sloppy" gel. Equation 1 shows that even a highly viscous (μ = 100 mPa.s; 100 times the water viscosity) mucus layer as thick as 0.3 μm should break up within about 20 sec. For lower viscosity (10 mPa.s), even 0.6 μm thick mucus films would rupture within this time and begin to form a network-like structure consisting of mucus ridges and exposed patches of the epithelium. If average thickness of the precorneal mucus is higher than a few tenths of a micrometer, as some measurements seem to indicate,[19,20] then rupture by this mechanism within the normal range of the BUT is possible only at some localized sites of reduced thickness, which may be due to trapped contaminants and cellular debris. Rupture assisted by submicron-size dust particles is also commonly observed in laboratory experiments on dewetting of thin films.[6–8,22]

The proposed mechanism is applicable only to the *rupture* of the *normal* mucus layer *without any change in its structure*. An excessive contamination of mucus by the apolar and weakly polar entities (eg, cellular debris, lipids) would lead to a *structural* collapse of the mucus gel due to the loss of its polarity and hydrophilicity.[17,18] This should lead to mucus aggregation, clumping, and tenacious attachment to the epithelium. This mechanism, which appears to be more germane in dry eyes, is discussed elsewhere in this volume.[18]

In view of the above considerations, an updated version of a mechanism of the tear film breakup secondary to the rupture of the mucus layer is now stated.

4. A MECHANISM OF TEAR FILM BREAKUP

The precorneal mucus is replenished and resurfaced during a blink due to the shear created by a rapidly moving eyelid. The long-range van der Waals forces cause the rupture of the precorneal mucus layer at the thinnest spots from where the excess mucus retracts to form islands of mucus and a network structure with mucus strands adhering to the thicker pools of mucus. At this stage, the aqueous tears come into direct contact with the exposed patches of the corneal epithelium. The tear film breakup can be initiated only if some of these epithelial sites are nonwettable by the aqueous tears.

Earlier it was thought that the epithelial surface without its mucus covering is hydrophobic and nonwettable by tears,[30] but modern techniques have shown the normal superficial cells with associated glycocalyx to be almost as hydrophilic and as wettable by the aqueous media as is the corneal mucus.[15–18] Interpretations of the epithelial surface properties and its wettability are, however, subtle issues. For one, the corneal surface is a heterogeneous mosaic of cells with varying ages, morphologies, degrees of desquamation and differentiation, and hence, wettability. However, contact angle measurements using millimeter-size drops can determine only the *average* epithelial wettability; they cannot be used to probe sites of cellular dimensions. Thus, due to predominance of mature glycocalyx-bearing cells, the epithelial surface on the macroscopic scales may well appear to be completely wettable, even when a small population of mildly nonwettable cells is also present. Theoretical considerations[31,32] suggest that the tear film breakup can be initiated even by an extremely small, mildly nonwettable corneal site of dimensions comparable to the size of a single superficial squamous cell (~ 20–40 μm in diameter). Once a small hole is formed in the tear film, it would continue to enlarge due to an immediate contamination of the surrounding corneal surface at the three-phase contact line by the superficial lipids. There are two distinct possible mechanisms for the generation of microscopic nonwettable sites on the surface of a normal epithelium in the absence of the mucus layer.

i. The ongoing desquamation of degenerating squamous cells creates small "craters," where the less differentiated, deeper-layer cells lacking in a mature network of extracellular glycocalyx, microvilli, and microplicae are exposed. The deeper-layer cells, especially those uncovered prematurely, are likely to be less wettable,[17] as evidenced also by an increased binding of *Pseudomonas* to the damaged[33] and deeper-layer cells.[34] Thus, some of the sites containing degenerating and newly uncovered cells are likely candidates for initiating the tear breakup secondary to the rupture of the mucus layer.
ii. Based on the surface properties of the epithelium and mucus,[16,17] we have shown elsewhere[17,18] that apolar and weakly polar "hydrophobic" contaminants should be readily absorbed in mucus, but they cannot adhere to the underlying epithelium in the presence of normal hydrophilic mucus. However, a direct attachment of the cellular debris, lipids, and other hydrophobic contaminants to the epithelium becomes energetically favorable in the absence of the corneal mucus covering.[17,18] Such adsorption can mask the polar nature of the glycocalyx and make the bare epithelial patch nonwettable, thus acting as a "trigger" or "nucleus" for the tear film breakup.

In summary, the recent experimental and theoretical advances in the understanding of thin films support our original hypothesis[1–4] of the rupture of the precorneal mucus layer within the observed range of the BUTs in normal eyes. The tear film breakup can then be triggered on exposed epithelial sites of partial wettability, even if these sites are of microscopic dimensions. Finally, in this update, two possible mechanisms—cell desquamation and contamination—are suggested for the generation of microscopic, nonwettable epithelial sites.

REFERENCES

1. Sharma A, Ruckenstein E. Mechanism of tear film breakup and formation of dry spots on cornea. *J Colloid Interface Sci.* 1985;106:12–27.

2. Sharma A, Ruckenstein E. Mechanism of tear film breakup and its implications for contact lens tolerance. *Am J Optom Physiol Opt*. 1985;62:246–253.
3. Sharma A, Ruckenstein E. The role of lipid abnormalities, aqueous and mucus deficiencies in the tear film breakup and implications for tear substitutes and contact lens tolerance. *J Colloid Interface Sci*. 1986;111:8–21.
4. Ruckenstein E, Sharma A. A surface chemical explanation of tear film breakup and its implications. In: Holly FJ, Lamberts DW, MacKeen DL, eds. *The Preocular Tear Film in Health, Disease, and Contact Lens Wear*. Lubbock, TX: Dry Eye Institute; 1986:697–727.
5. Reiter G. Dewetting of thin polymer films. *Phys Rev Lett*. 1992;68:75–78; Unstable thin polymer films. *Langmuir*. 1993;9:13441351.
6. Redon C, Brzoska JB, Brochard-Wyart F. Dewetting and slippage of microscopic polymer films. *Macromolecules*. 1994;27:468–471.
7. Yerushalmi-Rozen R, Klein J, Fetters LJ. Supression of rupture in thin nonwetting liquid films. *Science*. 1994;263:793–795.
8. Yerushalmi-Rozen R, Klein J. Stabilization of nonwetting thin liquid films on a solid substrate by polymeric additives. *Langmuir*. 1995;11:2806–2814.
9. Faldi A, Composto RJ, Winey KI. Unstable polymer layers. *Langmuir*. 1995;11:4855–4861.
10. Guerra JM, Srinivasarao M, Stein RS. Photon tunneling microscopy of polymeric surfaces. *Science*. 1993;262:1395–1400.
11. Sharma A, Reiter G. Instability of thin polymer films on coated substrates. *J Colloid Interface Sci*. 1996;178:383–399.
12. de Gennes PG. *CR Acad Sci*. 1979;228B:219–224.
13. Brochard-Wyart F, de Gennes PG, Hervert H, Redon C. Wetting and slippage of polymer melts on semi-ideal surfaces. *Langmuir*. 1994;10:1566.-1572.
14. Sharma A, Khanna R. *Macromolecules*. Nonlinear stability of microscopic polymer films with slippage. 1996;29:6959–6961.
15. Tiffany JM. Measurements of wettability of the corneal epithelium. I particle attachment method. *Acta Ophthalmol*. 1990;68:175–181; Measurements of wettability of the corneal epithelium. II contact angle method. *Acta Ophthalmol*. 1990;68:182–187.
16. Sharma A. Surface properties of damaged and normal corneal epithelia. *J Dispersion Sci Tech*. 1992;13:459–478.
17. Sharma A. Energetics of corneal epithelial cell-ocular mucus-tear film interactions. *Biophys Chem*. 1993;47:87–99.
18. Sharma A. This volume.
19. Nichols BA, Chiappino ML, Dawson. Demonstration of the mucous layer of the tear film by electron microscopy. *Invest Ophthalmol Vis Sci*. 1985;26:464–473.
20. Prydal JI, Artal P, Woon H, Campbell FW. Study of human precorneal tear film thickness and structure by interferometry. *Invest Ophthalmol Vis Sci*. 1992;33:2006–2011.
21. Adams AD. In: Holly FJ, Lamberts DW, MacKeen DL, eds. *The Preocular Tear Film in Health, Disease, and Contact Lens Wear*. Lubbock, TX: Dry Eye Institute: 1986:304–311.
22. Khesgi HS, Scriven LE. Dewetting: nucleation and growth of dry regions. *Chem Eng Sci*. 1991;46:519–525.
23. Sharma A. Relationship of thin film stability and morphology to macroscopic parameters of wetting in the apolar and polar systems. *Langmuir*. 1993;9:861–869.
24. Sharma A, Jameel AT. Nonlinear stability, rupture and morphological phase separation of thin fluid films. *J Colloid Interface Sci*. 1993;161:190–208; Stability of thin polar films on nonwettable substrates. *J Chem Soc Faraday Trans*. 1994;90:625–628.
25. Khanna R, Jameel AT, Sharma A. Stability and breakup of thin polar films on coated substrates. *Ind Eng Chem Res*. 1996;35:3081–3092.
26. van Oss CJ, Chaudhury MK, Good RJ. Interfacial Lifshitz-van der Waals and polar interactions in macroscopic systems. *Chem Rev*. 1988;88:927–941.
27. van Oss CJ. Acid-base interfacial interactions in aqueous media. *Colloids Surf A*. 1993;78:1–49.
28. Ruckenstein E, Jain RK. Spontaneous rupture of thin liquid films. *J Chem Soc Faraday Trans II*. 1974;70:132–141.
29. Sharma A, Kishore CS, Salaniwal S, Ruckenstein E. Nonlinear stability and rupture of ultrathin free films. *Phys Fluids A*. 1995;9:1832–1840.
30. Holly FJ, Lemp MA. Wettability and wetting of corneal epithelium. *Exp Eye Res*. 1971;11:239–250; Formation and rupture of the tear film. *Exp Eye Res*. 1973;15:515–525.

31. Sharma A, Coles WH. Physico-chemical factors in tear film breakup. ARVO Abstracts. *Invest Ophthalmol Vis Sci.* 1990;31:S552.
32. Sharma A. Perturbation analysis of surface dewetting by formation of holes. *J Colloid Interface Sci.* 1993;156:96–103.
33. Klatz SA, Au YK, Misra RP. A partial thickness epithelial defect increases the adherence of *Pseudomonas aeruginosa* to the cornea. *Invest Ophthalmol Vis Sci.* 1989;30:1069–1074.
34. Sokol JL, Masur SK, Asbell PA, Wolosin JM. Layer-by-layer desquamation of corneal epithelium and maturation of tear-facing membranes. *Invest Ophthalmol Vis Sci.* 1990;31:294–304.

40

THE MEIBOMIAN GLANDS AND TEAR FILM LIPIDS

Structure, Function, and Control

Anthony J. Bron and John M. Tiffany

Nuffield Laboratory of Department of Ophthalmology
University of Oxford
Oxford, United Kingdom

1. INTRODUCTION

In the last few decades, interest in the meibomian glands has grown with the recognition of various forms of posterior blepharitis and of the contribution of meibomian gland disease (MGD) to the evaporative form of dry eye.[1] This paper reviews current knowledge of the structure and function of the meibomian glands.

2. MORPHOLOGY

The meibomian glands are tubulo-acinar holocrine glands embedded in the tarsal plates (Fig. 1); the glands secrete into a ductal system.[2,3]

The terminal duct is lined by modified, keratinized, squamous epithelium.[4] Secretion involves central acinar degeneration, the transfer of released lipid into the acinar lumina and their associated ductules, and the further movement of lipid into the central collecting ducts of the glands. Freeze fracture studies show acinar morphology to vary with stage of differentiation, smooth endoplasmic reticulum and Golgi morphology becoming increasingly prominent with lipid droplet formation. The lamellar, onion-skin organization of the droplets suggests that membranes are incorporated into the droplets.[5]

Meibomian oil, liquid at lid temperature, is delivered from the ducts onto the lid margin where it forms a reservoir on the cutaneous side of the mucocutaneous junction. Delivery occurs via a single row of meibomian orifices, present on both the upper and lower lids (30–40 in the upper lid, 20–25 in the lower lid). When the lids are closed, the independent reservoirs of each lid are pooled on the lid skin between their coapted mar-

Lacrimal Gland, Tear Film, and Dry Eye Syndromes 2
edited by Sullivan *et al.*, Plenum Press, New York, 1998

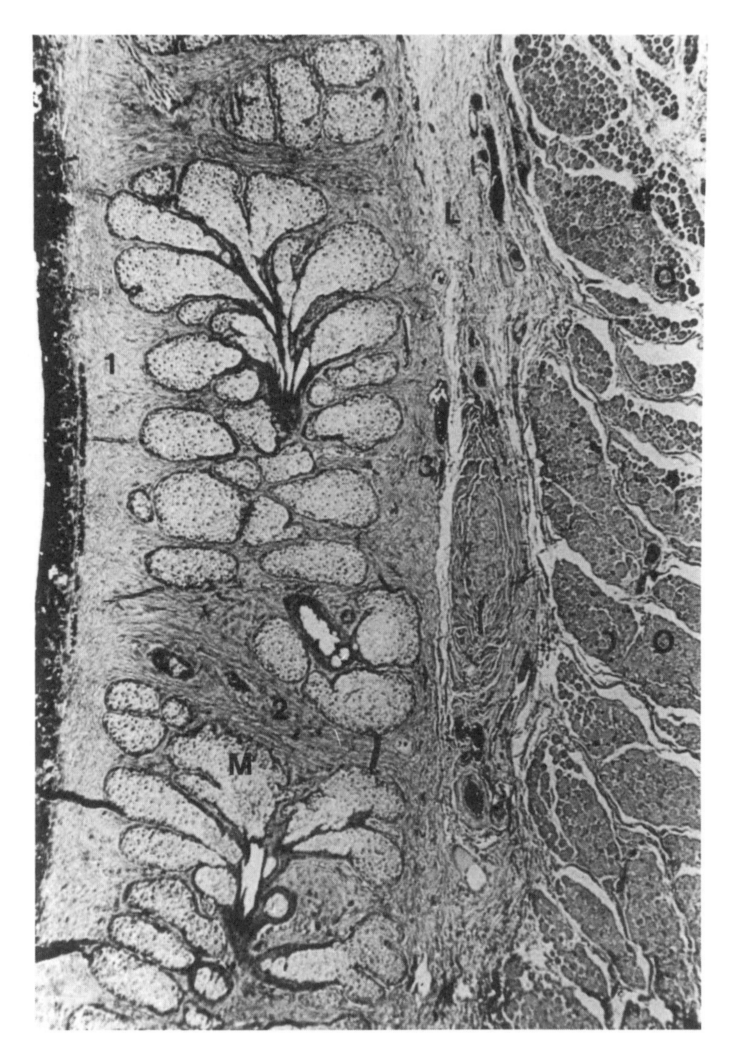

Figure 1. Vertical lid section to show the meibomian gland within the tarsal plate. The orbicularis muscle bundles and the insertion of the levator aponeurosis are seen to the left. (Tripathi RC, Tripathi BJ. (1984) Anatomy of the human eye, orbit and adnexa in the Eye (ed H Davson) Academic Press London pp 40–157.)

gins. In places, the row of orifices may be duplicated.[6] About 7% of orifices lie behind the mucocutaneous junction and deliver their oil beneath the tear meniscus.[7]

3. THE ROLE OF MEIBOMIAN SECRETIONS

Meibomian oil forms the most anterior layer of the preocular tear film and is reconstituted in its entirety with each blink in the time taken to fully open the lids (about 176 msec[8]). As the lids open, oil is drawn from the pooled lid reservoir into the preocular film. The oil is restored to the reservoir with lid closure, presumably spreading forwards onto the lid skin in tidal fashion as the lid margins come together. In this way, there is inevitable mixing of meibomian oil with the sebum of the marginal lid skin.

The chief role of meibomian oil is to retard evaporation from the preocular surface, to reduce the free energy of the anterior surface of the tear film, and to safeguard these functions by forming a barrier to contamination by lid margin sebum. As demonstrated by McDonald,[9] skin sebum has a higher spreading pressure than meibomian oil and therefore has the potential to disrupt the integrity of the preocular oil film. (In his simple experiment, McDonald showed that touching the preocular tear film with a sebum-coated rabbit whisker caused rupture of the native tear oil layer.) Other functions proposed for meibomian secretions include an antimicrobial action due to the presence of fatty acids, complexing with mucin, and, possibly, the production of sexually attractive pheromones.[10]

4. INNERVATION

Human[11] and primate studies indicate that the meibomian glands are richly innervated. The demonstration of both cholinergic and VIPergic fibers by Chung et al.[12] suggests that at least part of this parasympathetic innervation is via the pterygopalatine pathway (7th cranial nerve). Perra et al.[13] found a dense innervation by acetylcholinesterase-positive fibers in human meibomian glands, some of which was closely associated with blood vessels. Although there is substantial sympathetic [tyrosine hydroxylase(TH)-positive] innervation to the vessels in this region as well as a cholinergic and VIPergic innervation, the sparseness of TH innervation to the acini suggests that not all of the moderate NPYergic innervation is sympathetic, and may run via a different pathway. The sparse sensory innervation to the gland by substance P- and CGRP (calcitonin gene related protein)-positive fibers[12,14] could play a role in glandular inflammation. Studies by Kirsch et al.[15] showed varicose axon terminals associated with acinar basal laminae, but never internal to them, and the presence of nitrergic fibers innervating the gland. A number of neuropeptides were coexpressed, and there were differences between the fibers innervating acini and the fibers innervating vessels.

5. HORMONAL CONTROL

Androgens such as testosterone stimulate sebaceous secretion; estrogens and antiandrogens (e.g., ciproterone acetate) reduce it.[16–20] It is thus of interest that meibomian gland disease is associated with seborrheic eczema, which in turn is influenced by androgen levels.

Perra[21] has shown the presence of androgen-metabolizing enzymes in the meibomian gland, and the acini express androgen receptors,[22,23] which suggests that androgens could play a part in regulating secretion. In keeping with this, we have found a sex difference in the resting (or 'casual') level of oil on the lid margin, with a lower level in women, between puberty and the menopause, than in men.[24] The lacrimal glands, too, are known to express androgen receptors,[25] and there are strong arguments to suggest that androgens perform an important supportive role in the maintenance of secretion by lacrimal and salivary glands and that loss of circulating androgens may be a precipitating factor in the occurrence of Sjögren's syndrome dry eye and possibly non-Sjögren dry eye.[26,27] Experimentally, androgens can reverse the inflammatory events in the lacrimal gland in a mouse model for Sjögren syndrome.[26] Clinically, obstructive meibomian gland disease (MGD) may accompany aqueous deficient dry eye, and Shimazaki et al.[28] have reported an increased prevalence of MGD in patients with Sjögren's syndrome. It has been speculated that the association of these two disorders may imply that lowered levels of circulating androgen contribute to both conditions.[29]

6. TECHNIQUES FOR ANALYSIS OF MEIBOMIAN SECRETION

Studies of the chemical composition of meibomian oil have been based on the collection of lipid after expression of the glands, by applying pressure through the tarsal plate with varying degrees of firmness.[30,31]

A variety of methods can be applied to determine both the types or classes of lipid molecules and their more detailed structure. Thus, either thin-layer chromatography (TLC)[32–34] or high-performance liquid chromatography (HPLC)[35] will separate the secretion into classes such as triacylglycerides or free fatty acids, although the steryl esters and wax esters are less easily separated. After recovery of these fractions, the lipids can be hydrolyzed and separated by gas-liquid chromatography (GLC) to define the individual fatty acids and alcohols on the basis of chain length. Further analysis using GLC linked to a mass-spectrometric detector (GC-MS) can give more precise identification of chain branching, sterol identity, and many other details.

7. COMPOSITION

The composition of secreted meibomian oil may differ from that expressed from the glands for the purposes of analysis.[30,36] The chief lipids in expressed secretions are the non-polar wax and sterol esters, which make up about 59% of the total.[31] The remainder includes an important contribution by polar lipids (phospholipids, 15%) and also the variable presence of small amounts of free fatty acids (2%), sterol esters (2%), free sterols (1.6%) and neutral fats (Table 1).

Shine and McCulley have made the interesting observation that the secretions of normal subjects may be either cholesterol ester rich (N[CP]) or poor (N[CA]), with the implication that the former may provide a better substrate for the esterases of commensal lid microorganisms and therefore be important in the evolution of anterior blepharitis.[37,38] Triglyceride fatty acid abnormalities have been reported in patients with MGD, including meibomian keratoconjunctivitis.[39]

Much is known about the fatty acid and fatty alcohol components of the secretions; their branched and unsaturated features confer a low melting point on meibomian oil.[40,42] It is likely that at least some of the reported components of expressed meibomian lipid, including squalene, a sebum-associated sterol precursor,[43] derive from the contamination of

Table 1. Class composition of lipids in human meibomian oil

Class	% by Weight
Hydrocarbons	7.5
Wax esters	32.3
Sterol esters	27.3
Diesters	7.7
Triacyl glycerides	3.7
Free sterols	1.6
Free fatty acids	2.0
Polar lipids	14.8

Data from Nicolaides.[33] Calculation includes some exogenous hydrocarbon.

Table 2. Physical properties of human meibomian lipid

Refractive index[46]	1.48
Viscosity at 35°C[41]	9.7 – 19.5 Pa sec
Melting point range	
Human meibomian lipid	19.5– 32.9°C
Human skin lipid[46]	15-17 – 33-36°C

collected oil with the sebum of the lid margin; some too may derive from organelles and membranes expressed from the holocrine meibomian glands.

About 15% of the meibomian lipid is phospholipid. By its polar nature it interacts with the aqueous phase of the tear film and initiates the reconstitution of the preocular lipid layer with each blink.[44] Its principal components are phosphatidyl ethanolamine and phosphatidyl choline.[45]

Table 2 shows that meibomian oil, which is a lipid mixture, melts over a range of temperatures (19–32°C).[46] These are within the temperature range of the lid margin. Meibomian oil is less fluid than sebum at physiological temperature.[41]

8. SECRETION AND EXCRETION OF LID OIL

For the purposes of description, secretion is taken here to imply the transfer of lipid from the glandular acini to the ductular system. Delivery implies transfer from the ducts to the lid margin reservoir, and excretion is the loss of lipid from the marginal reservoir and the preocular film.

It may be assumed that in normal subjects blinking spontaneously, there is a steady state between secretion, delivery, and excretion. Little is written about excretion, but the most likely major route is onto the skin and lashes of the lid margin, fueled by the movement of lipid as the lid margins meet with each blink. It is likely that the diffusion of lipid across the aqueous phase of the tear film, as postulated by Holly,[47] is a lesser route, but it is a route that, given the complexing of lipid with mucus that occurs,[48] allows the lipid to leave the tear film via the puncta with the mucus thread. Since the puncta are located in mucous membrane, behind the mucocutaneous junction, they cannot directly drain lipid from the oil reservoir, which lies on the cutaneous side of the junction. The possibility that local skin disease (eg, eczema) could increase the excretion of meibomian oil in a clinically important way has been suggested by H. Kaufman (personal communication).

9. MECHANISMS

9.1. Neurohumoral Regulation

The potential for neural and hormonal regulation of meibomian secretion has already been mentioned. We know from the work of Buschke and Fränkel[49] that physostigmine will stimulate the release of meibomian oil in the rabbit. Chung et al.[12] drew attention to the ability of alpha adrenergic agents to release sebaceous secretions from the innervated preputial glands. The influence of retinoids in the normal regulation of oil secretion and delivery is difficult to assess, in view of the reports of obstructive changes that develop on direct exposure to these agents,[50,51] but a physiological role for retinol should be considered.

9.2. Mechanical Factors

A role for blinking in the delivery of lid oil was proposed by Wolff,[52] Linton et al.[53] and McDonald and Brubaker.[54] Later, Josephson[55] observed 'jetting' of oil onto the human preocular film following the blink. Oil globules may be seen floating in the meniscus of the preocular lipid film.[56] Linton et al.[53] suggested that the muscle of Riolan, disposed at the lid margin around the termination of the terminal ductules of the gland, might perform a regulatory function, while Dryja[57] suggested that certain structural features of the glandular acini might impart a valvular action. Tiffany[42] discussed the forces generated by the lids during lid closure. Studies in our laboratory using the technique of meibometry[24,58,59] support a role for lid action in the delivery of lipid to the lid margin, on the basis of the following observations (Figs. 2–4). When meibomian oil is completely removed from the upper and lower lids with hexane, lid oil fails to re-accumulate over a period of 3 min, in the absence of blinking.

When the same experiment is repeated in a group of patients undergoing cataract surgery under general anesthesia, oil accumulates readily over a longer period (25–120 min; mean, 46 min).

The inference from these studies is that in the short term, blink action aids the delivery of meibomian oil onto the lid margin, while in the longer term it is the continuous secretion of oil by the meibomian acini that brings about delivery.

In a further study of the diurnal variation of casual oil levels on the lid margin, levels were highest within the first hour after waking.[58]

We have inferred that during sleep, in the absence of blinking, there is an accumulation of meibomian oil within the acini and ducts of the meibomian glands; on waking, it is assumed that the accumulated excess is discharged. This gives some support to the advice given to patients with obstructive MGD to apply heat and massage to the lids in the morning, shortly after waking.

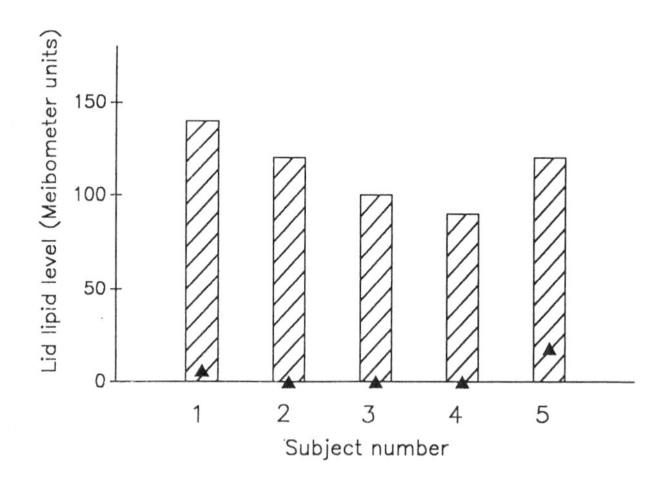

Figure 2. Recovery of lid margin lipid after hexane cleaning of the upper and lower lid margins, following 3 min with the eyes remaining open in five normal subjects. Shaded columns indicate the casual level for each subject; closed triangles indicate the level at 3 min. There is negligible or undetectable delivery in 3 min. (Courtesy of C. Chew.)

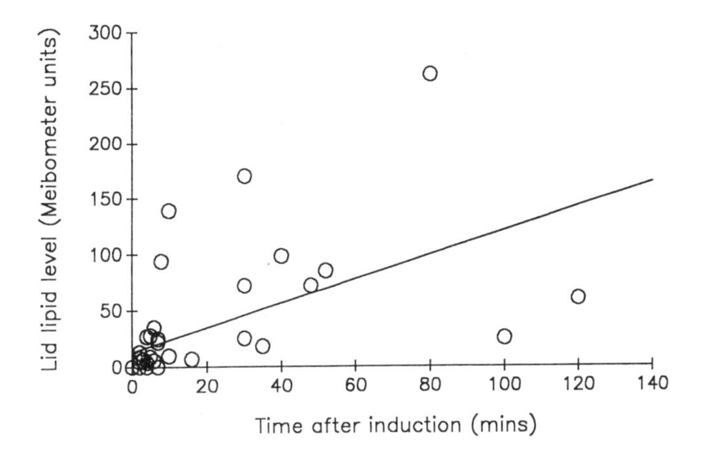

Figure 3. Lipid recovery on the lids at the beginning and end of surgery in 13 subjects after hexane cleaning of the lids. There is a significant rise in lipid over time, indicating that meibomian oil secretion and delivery continue, despite the absence of blinking. (Courtesy of C. Chew.)

10. HYPOTHESIS FOR THE SECRETION AND DELIVERY OF MEIBOMIAN OIL

The above observations have led us to formulate a general concept of meibomian gland secretion (Figs. 5 and 6).

The holocrine process within the meibomian glands, modulated by neurohumoral factors, gives rise to the secretion of meibomian oil into the acinar lumina and into an elastic ductal reservoir. Delivery of oil onto the lid margins occurs when a critical pressure within the terminal ducts is exceeded. At night, when the lids are closed and there is no blink action, the ductal reservoir is expanded by the force of secretion until this critical

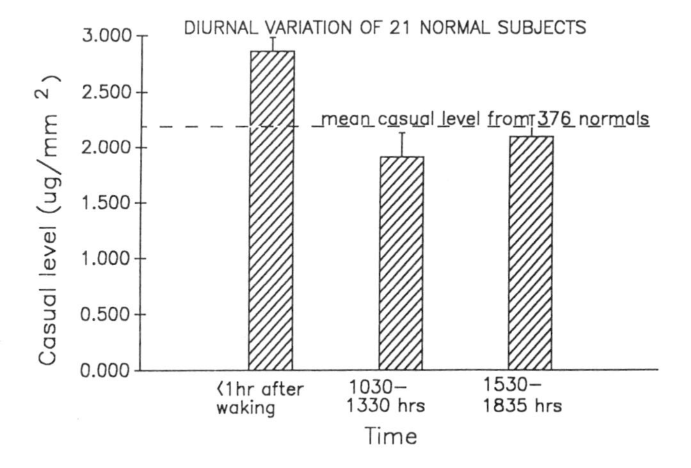

Figure 4. Diurnal variation of casual meibomian lipid levels at the lid margins. The histograms illustrate the mean lipid levels (μg of lipid per mm^2) in a group of 21 adults, measured at different times. The dashed line indicates the mean level of casual lipid measured at various times of day.[24] The early morning casual level within 1 h of waking is significantly higher than at other times of day. (Courtesy of C. Chew.)

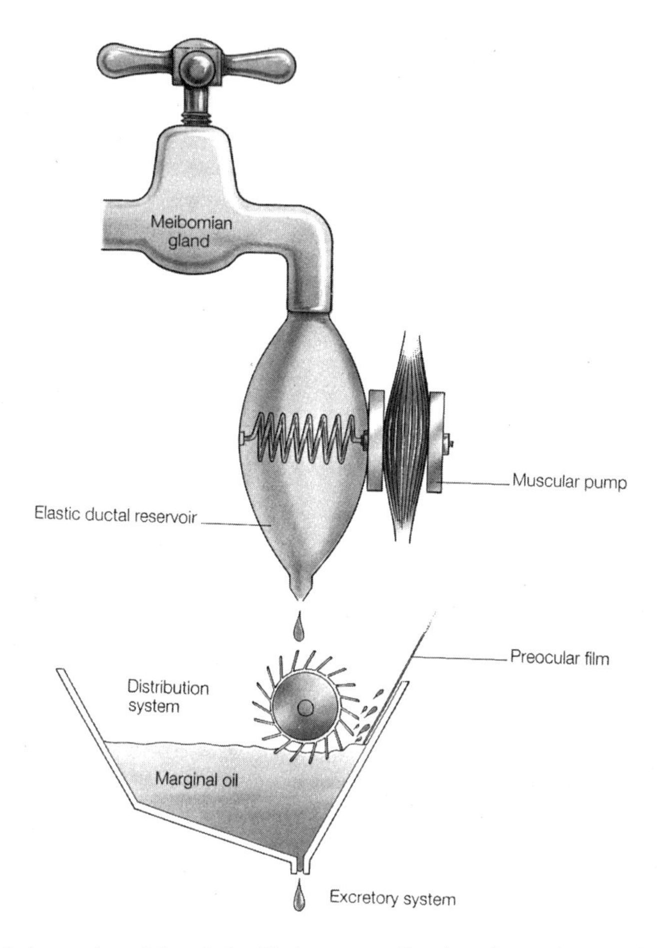

Figure 5. Schematic impression of the relationship between meibomian oil secretion and delivery onto the lid margin. (From Bron,[3] with permission.)

pressure is exceeded and delivery resumes. Thus, delivery of oil during sleep is entirely on the basis of secretory pressure. On waking, and with the return of blinking, the relative excess of oil stored overnight in the ducts and acini is discharged onto the lid margins, so that for a limited period of time the casual level on the lid margins is increased. This temporarily increases the excretory rate, while at the same time the delivery rate falls to a normal, waking level. In time, the casual level is restored to the daytime steady state.

During the day and with the eyes open and blinking, blink action is assumed to apply a contractile force to the duct and acini, which, in a proportion of glands, raises the ductal pressure above the critical pressure required for delivery. Thus, with each blink, these glands might release a 'jet' of oil onto the tear film. Their ducts, being less full, would then be less likely to discharge during immediate subsequent blinks until they had been sufficiently refilled by the secretory process. This concept of intermittent delivery would fit in with Norn's observation that only 44% of glands appear to be stainable with a fat stain at any one time.[7,60]

In keeping with the above observations, rapid and forceful blinking, as in response to an ocular foreign body, results in an increase in thickness of the precorneal lid oil,[61]

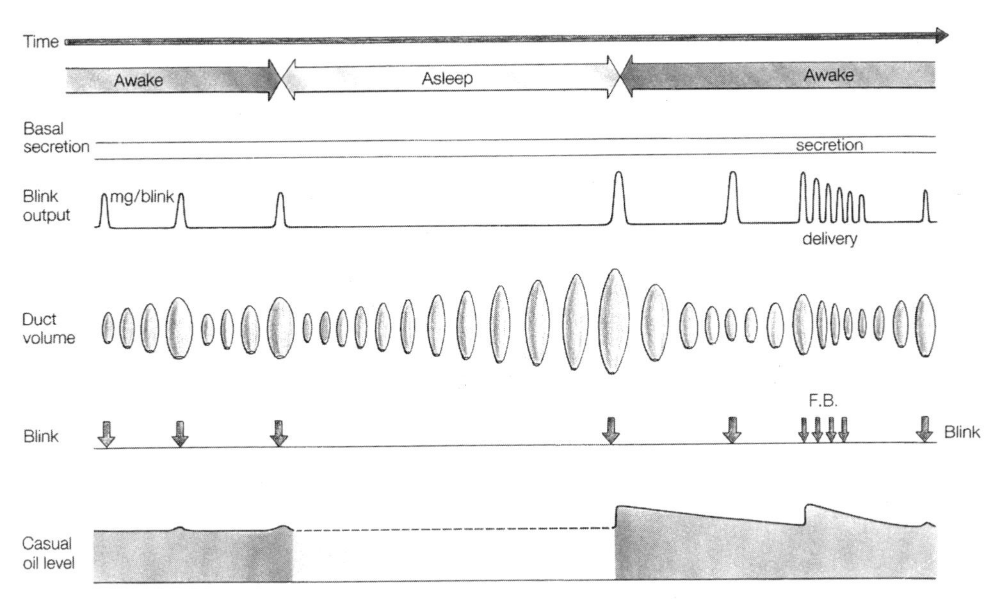

Figure 6. Proposed sequence of events involved in meibomian oil secretion. The events shown are: time; the waking and sleeping states; basal secretion, depicted here as showing no fluctuations with time; blink output of oil - it is assumed that with each blink, some but not all ducts will deliver an aliquot of oil onto the lid margin. After sleep, because of ductal accumulation, the output per blink will be increased, for a short period. With the first blink test a, the output per blink will (-10 blinks at 1 second intervals) gradually fall over 10 sec although the total output will exceed the normal over that time. Duct volume is envisaged to increase infinitesimally in each duct between blinks, but more significantly during sleep or when there is otherwise prolonged lid inaction. On waking, there is a milking-down of the duct to its normal size. The timing of blinks is shown by arrows; FB = fast blink. The casual oil level is depicted at different times – a tiny boost is given with each blink, which subsides with loss, chiefly to the lid skin. No information is available as to the status of the casual oil during sleep or resting lid closure. On waking, the excess oil within the dilated ducts is discharged in the early blinks, and increases the height of the casual level; a similar effect would be achieved by a spell of rapid blinking. (From Bron,[3] with permission.)

which could perform some protective function. Conversely, Franck[62] has noted that the symptoms of "office eye" coincide with the period of oil layer thinning that occurs in the afternoon. It may be assumed that in the steady state, with an individual blinking normally, secretion, delivery, and excretion would be in balance.

11. DELIVERY RATE OF MEIBOMIAN OIL

It would be of interest to measure delivery rate of meibomian oil in normal subjects and in patients with MGD. We have taken some preliminary steps in this direction. In our early studies in anesthetized subjects, we estimate that the rate of accumulation of lid oil during general anesthesia was in the region of 400 μg/h. This value was derived by calculating the amount of lipid that had accumulated on the central, lower lid margin in a fixed amount of time, and extrapolating this to the total length of the upper and lower lids, assuming secretory behavior to be uniform along the total lid length. The value must be regarded as a minimum for the secretory rate, since no correction was made for continued excretion during the assessment. This value is equivalent to 333 μl/h, assuming an average lipid density of 900 μg/μl. If it is assumed that there is a total of about 50 glands in the two lids combined, this would represent an average release of about 6.7 μl/h per gland.

In attempting to transfer this approach to the normal subject in the waking state, we have argued that the normal, spontaneously blinking subject is in a steady state with respect to oil delivery, and that the accumulation of lipid on the lid margin after hexane cleaning should provide a similar index of secretory function to that found in the anesthetized patient. Our preliminary studies with this approach, however, show a wide variation in response which at this stage indicates only that after total removal of lipid from the lid margin, delivery rapidly restores levels towards the resting casual level. The difficulty in performing this delivery test in normals suggests that there will be additional problems of interpretation in patients with obstructive MGD, since in these patients, although the obstructive changes are often diffuse, they are often distributed patchily, or focally, along the upper and lower lid margin.[38,63]

12. QUALITATIVE AND QUANTITATIVE METHODS OF ASSESSING MEIBOMIAN GLAND FUNCTION

In recent years a number of methods have become available to assess the functions of the meibomian glands. Changes in oil gland morphology encountered in obstructive MGD have been classified and a grading system applied.[64] The volume and viscosity of expressed secretions have been graded and the values applied to the assessment of oil gland disease in dry eye.[65] Mathers has used the technique of infrared meibography to demonstrate oil gland dropout with age and in the presence of MGD.[66]

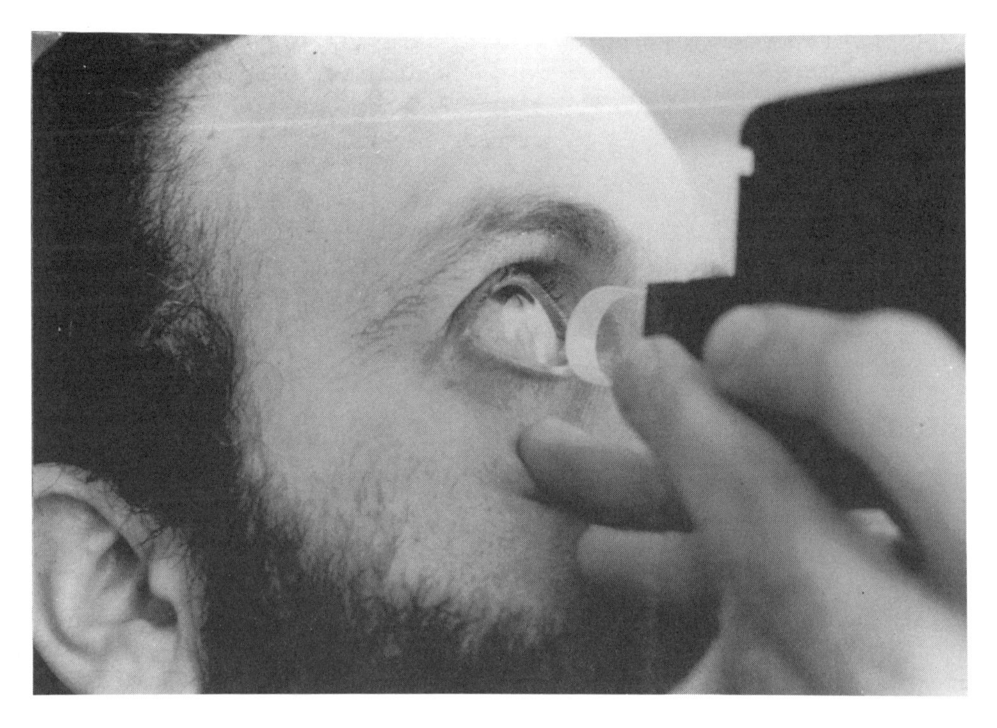

Figure 7. Application of the Meibometer loop to the lid margin.

The technique of meibometry has been developed to assess the size of the lid reservoir of oil and to determine an index of oil delivery. In this technique, the steady state, resting (or casual) level of lid oil is measured by blotting oil from the lid margin with a loop of plastic tape (Fig. 7).

The amount of oil removed is measured densitometrically. Oil is taken up from the central third of the lower lid; from the value obtained, the amount present in the total lid reservoir can be calculated, knowing the percent of lipid picked up and making assumptions of lipid density and lid length. A value of about 312 μg has been calculated for the adult male, which can be compared with values for the total expressible lid lipid of 1.11 mg/lid[67] and 1.27 mg/lid.[36,42]

The casual oil level in the lower lid reservoir rises with age (Fig. 8) and is lower in women during the reproductive period, from about the age of puberty to the menopause, suggesting a potential hormonal influence.[24]

This rise of casual oil level with age is unexplained. Norn[68] noted that lid oil layer thickness did not fall with age, despite reduced expressibility of glands, and suggested that the excretion of meibomian oil is reduced with age; this could certainly explain an age-related rise of oil level.

Various interference techniques have been used to measure oil thickness in the preocular tear film, giving values ranging from 32 to 80 nm; the upper values, such as those of Yokoi et al.,[69] are probably the most representative (Table 3).

Using this approach and adopting an average value of 900 μg/μl as the density of the meibomian oil, we calculated a value of 9 μg as the amount of oil in an average preocular film.[59] It is apparent that with a pooled lid margin reservoir in the region of 300 μg, there is an ample amount on the lid margin to resurface the preocular film with each blink. In the case of a wide palpebral aperture, such as is encountered in endocrine exophthalmos, there are greater demands on the lipid reservoir, which could lead to a thinner preocular lipid layer. By the same token, a smaller aperture, as occurs in relative downgaze or when screwing up the eyes protectively, will make fewer demands on the lipid reservoir and may result in a thicker oil film. Tsubota and Nakamori[74] have pointed out the potential benefits of a smaller palpebral aperture in patients with dry eye, in reducing the surface area available for evaporation. On this basis, they suggest that VDU workers with dry eye should place the VDU terminal below the horizontal. It is apparent that where oil gland deficiency is present in addition to aqueous deficiency, a wide aperture or reduced blink frequency could compound the dry eye problem.

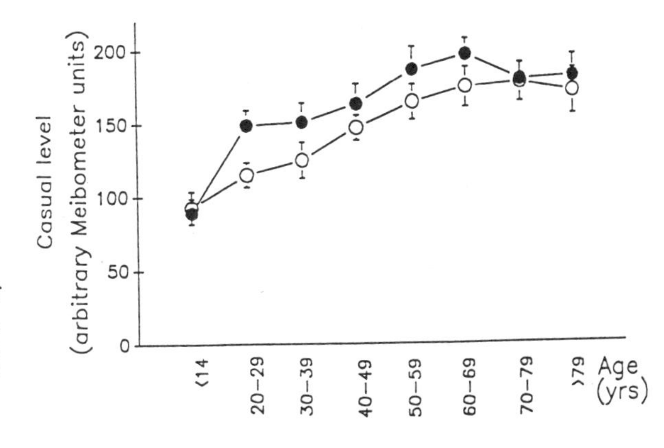

Figure 8. Lid margin casual levels of lipid measured by meibometry in males (solid circles) and females (open circles) of different ages. (From Chew et al.,[59] with permission.)

Table 3. Precorneal lipid layer thickness: values by various methods

Method	Thickness (nm)
Interference microscopy[70]	102 ± 3
Interference microscopy (corrected value)[71]	68
Specular reflectometry[72]	13-70
Specular reflectometry[69]	70-80
Photometric reflectometry[73]	32-46

[70]Norn 1979; [71]Norn 1992; [72]Guillon 1982; [69]Yokoi 1996; [73]Olsen 1985

Preliminary studies have suggested a positive trend between the amount in the lipid reservoir and that in the tear film, although the relationship is not significant.[75] Although it would be of clinical value to know the relationship between tear oil thickness and evaporation rate, this information is at present unavailable. It is therefore not possible to infer yet, on the basis of interferometry, that a given preocular tear oil thickness will give rise to excessive evaporation and, incidentally, that a given degree of oil gland disease will be responsible for an evaporative (as opposed to an aqueous-deficient) dry eye.

Early studies established the role of the preocular oil film in the control of evaporation from the surface of the eye.[76,77] Mishima and Maurice[77] demonstrated that closure of the meibomian oil gland orifices led to increased evaporation from the ocular surface, and this was later confirmed by Gilbard et al.,[78] who demonstrated, also in the rabbit, that meibomian gland closure led to a rise in tear osmolality.

A number of studies of evaporation from the ocular surface have been carried out by various workers, who have assessed the functional role of the tear lipid in the normal eye and the effects of obstructive oil gland disease. Tsubota[82,84] reports a value for evaporation that is about 7.8% of the tear flow rate in normals, and this rate is increased 1.5 to three times in the presence of MGD disease (Table 4).

13. SUMMARY

Meibomian gland disease—and, in particular, obstructive meibomian gland disease—makes an important contribution to ocular surface disease, in the form of mei-

Table 4. Tear evaporation in human subjects: values from various studies

Reference	Normal subjects (μm/min/cm^{-2})	Dry eye (μm/min/cm^{-2})
Rolando et al. (1983)[75]	24.4	52.2
Hamano et al. (1990)[76]	73.4	–
Tomlinson et al. (1991)[77]	33.4-133.6	–
Tsubota & Yamada (1992)[78] 40% relative humidity	42.5	25.9
Mathers et al. (1993)[79] 30% relative humidity	88.2	285.6

Data converted using the notional eye area 2.2 cm^2.

bomian keratoconjunctivitis.[85] With improved methods for the study of meibomian oil composition and function, we are moving closer to the possibility of distinguishing the contribution of meibomian deficiency, as opposed to inflammatory events, to this disorder. More importantly, where aqueous tear deficiency and meibomian gland disease coincide in patients with dry eye, we are closer to the possibility of distinguishing their relative contributions to the dry eye state. This has implications for future therapies.

REFERENCES

1. Lemp MA. Report of the National Eye Institute/Industry Workshop on clinical trials in dry eyes. *CLAO J.* 1995;21:221–231.
2. Obata H, Horiuchi H, Miyata K, Tsuru T, Machinami R. [Histopathological study of the meibomian glands in 72 autopsy cases.] *Nippon Ganka Gakkai Zasshi.* 1994;98:765–771.
3. *Wolff's Anatomy of the Eye and Orbit.* Bron AJ, Tripathi RC, Tripathi B. 8th ed. London: Chapman and Hall; 1997.
4. Jester JV, Nicolaides N, Smith RE. Meibomian gland dysfunction. I. Keratin protein expression in normal human and rabbit meibomian glands. *Invest Ophthalmol Vis Sci.* 1989;30:927–935.
5. Sirigu P, Shen RL, Pinto da Silva P. Human meibomian glands: The ultrastructure of acinar cells as viewed by thin section and freeze-fracture transmission electron microscopies. *Invest Ophthalmol Vis Sci.* 1992;33:2284–2292.
6. Hykin PG, Bron AJ. Age-related morphological changes in lid margin and meibomian gland anatomy. *Cornea.* 1992;11:334–342.
7. Norn M. Meibomian orifices and Marx's line studied by triple staining. *Acta Ophthalmol.* 1985;63:698–700.
8. Doane MG. Interaction of eyelids and tears in corneal wetting and the dynamics of the normal human eyeblink. *Am J Ophthalmol.* 1980;89:507–516.
9. McDonald JE. Surface phenomena of tear films. *Trans Am Ophthalmol Soc.* 1968;66:905–909.
10. Mancini MA, Majumdar D, Chatterjee B, Roy AK. α_{2u}-Globulin in modified sebaceous glands with pheromonal functions: Localization of the protein and its mRNA in preputial, meibomian, and perianal glands. *J Histochem Cytochem.* 1989;37:149–157.
11. Montagna W, Parakkal PF. *The Structure and Function of Skin.* 3rd ed. New York: Academic Press; 1974, pp. 280–331.
12. Chung CW, Tigges M, Stone RA. Peptidergic innervation of the primate meibomian gland. *Invest Ophthalmol Vis Sci.* 1996;37:238–245.
13. Perra MT, Serra A, Sirigu P, Turno F. Histochemical demonstration of acetylcholinesterase activity in human meibomian glands. *Eur J Histochem.* 1996;40:39–44.
14. Seifert P, Spitznas M. Immunocytochemical and ultrastructural evaluation of the distribution of nervous tissue and neuropeptides in the meibomian gland. *Graefes Arch Clin Exp Ophthalmol.* 1996;234:648–656.
15. Kirsch W, Horneber M, Tamm ER. Characterization of meibomian gland innvervation in the cynomolgus monkey (*Macaca fascicularis*). *Anat Embryol (Berl).* 1996;193:365–375.
16. Neumann F, Elger W. The effect of a new anti-androgenic steroid, 6-chloro-17-hydroxy-1,2-methylene pregna-4,6-diene-3,20-dione acetate (cyproterone acetate) on the sebaceous glands of mice. *J Invest Dermatol.* 1966;46:561–564.
17. Strauss JS, Pochi PE. Assay of anti-androgens in man by the sebaceous gland response. *Br J Dermatol.* (Suppl 60) 1970;82:33–42.
18. Shuster S, Thody AJ. The control and measurement of sebum secretion. *J Invest Dermatol.* 1974;62:172–190.
19. Zaun H. Zur Hormoneller Beeinflussung der Talgsekretion. *Fette Seifen Anstrichmittel.* 1979;81:130–133.
20. Thody AJ, Shuster S. Control and function of sebaceous glands. *Physiol Rev.* 1989;2:383–416.
21. Perra MT, Lantini MS, Serra A, Cossu M, De Martini G, Sirigu P. Human meibomian glands: A histochemical study for androgen metabolic enzymes. *Invest Ophthalmol Vis Sci.* 1990;31:771–773.
22. Rocha EM, Toda I, Krenzer KL, Sullivan DA. Identification of androgen receptor protein in ocular tissues of the rat. ARVO Abstracts. *Invest Ophthalmol Vis Sci.* 1996;37:S849.
23. Sullivan DA, Wickham LA, Toda I, Gao J. Identification of androgen, estrogen and progesterone receptor mRNAs in rat, rabbit and human ocular tissues. ARVO Abstracts. *Invest Ophthalmol Vis Sci.* 1995;35:S651.

24. Chew CKS, Jansweijer C, Tiffany JM, Dikstein S, Bron AJ. An instrument for quantifying meibomian lipid on the lid margin: The Meibometer. *Curr Eye Res.* 1993;12:247–254.
25. Rocha FJ, Wickham LA, Pena JDO, et al. Influence of gender and the endocrine environment on the distribution of androgen receptors in the lacrimal gland. *J Steroid Biochem Mol Biol.* 1993;46:737–749.
26. Sullivan DA, Wickham LA, Krenzer KL, Rocha EM, Toda I. Aqueous tear deficiency in Sjögren's syndrome: Possible causes and potential treatment. In: Pleyer U, Hartmann C, Sterry W, eds. *Oculodermal Diseases - Immunology of Bullous Oculo-Muco-Cutaneous Disorders.* Buren, The Netherlands: Aeolus Press, 1997:95–152.
27. Mircheff AK, Warren DW, Wood RL. Hormonal support of lacrimal function, primary lacrimal deficiency, autoimmunity, and peripheral tolerance in the lacrimal gland. *Ocul Immunol Inflamm.* 1996;4:1–28.
28. Shimazaki J, Tsubota K. Meibomian gland dysfunction in patients with Sjögren's syndrome. American Academy of Ophthalmology, Annual Meeting, October 27–31, Chicago, 1996;145.
29. Mamalis N, Harrison D, Hiura G, et al. Are "dry eyes" a sign of testosterone deficiency in women? (Abstract.) 10th International Congress of Endocrinology, 1996.
30. Nicolaides N. Skin lipids. II. Lipid class composition of samples from various species and anatomical sites. *J Am Oil Chemists Soc.* 1965;42:691–702.
31. Tiffany JM. Lipid films in water conservation of biological systems. *Cell Biochem Funct.* 1995;13:177–180.
32. Tiffany JM. Physiological functions of the meibomian glands. *Prog Retinol Eye Res.* 1995;14:47–74.
33. Nicolaides N. Recent findings on the chemical composition of the lipids of steer and human meibomian glands. In: Holly FJ, ed. *The Preocular Tear Film: In Health, Disease and Contact Lens Wear.* Lubbock, TX: Dry Eye Institute; 1986:570–596.
34. Osgood JK, Dougherty JM, McCulley JP. The role of wax and sterol esters of meibomian secretions in chronic blepharitis. *Invest Ophthalmol Vis Sci.* 1989;30:1958–1961.
35. Shine WE, McCulley JP. Meibomian gland secretion: Sterol ester differences associated with *Acanthamoeba* keratitis. ARVO Abstracts. *Invest Ophthalmol Vis Sci.* 1993;34:S854.
36. Nicolaides N, Kaitaranta JK, Rawdah TN, Macy JI, Boswell FM III, Smith RE. Meibomian gland studies: Comparison of steer and human lipids. *Invest Ophthalmol Vis Sci.* 1981;20:522–536.
37. Shine WE, McCulley JP. The role of cholesterol in chronic blepharitis. *Invest Ophthalmol Vis Sci.* 1991;32:2272–2280.
38. Bron AJ, Tiffany J M. Evolution of lid margin changes in blepharitis. World Congress on the Cornea 1996, in press.
39. Shine WE, McCulley J P. Meibomian gland secretion polar lipids associated with chronic blepharitis disease groups. ARVO Abstracts. *Invest Ophthalmol Vis Sci.* 1996;37:S849.
40. Andrews JS. The meibomian secretion. *Int Ophthalmol Clin.* 1973;13:23–28.
41. Butcher EO, Coonin A. The physical propates of human seban. *J Invest Dermatol.* 1949;12:249–254..
42. Tiffany JM. The lipid secretion of the meibomian glands. *Adv Lipid Res.* 1987;22:1–62.
43. Tiffany JM. Individual variations in human meibomian lipid composition. *Exp Eye Res.* 1978;27:289–300.
44. Holly FJ, Lemp MA. Tear physiology and dry eyes. *Surv Ophthalmol.* 1977;22:69–87.
45. Greiner JV, Glonek T, Korb DR, Booth R, Leahy CG. Phospholipids in meibomian gland secretion. *Ophthalmic Res.* 1996;28:44–49.
46. Tiffany JM, Marsden RG. The influence of composition on physical properties of meibomian secretion. In: Holly FJ, ed. *The Preocular Tear Film in Health, Disease and Contact Lens Wear.* Lubbock, TX: Dry Eye Institute; 1986;597–608.
47. Holly FJ. Formation and rupture of the tear film. *Exp Eye Res.* 1973;15:515–525.
48. Moore JC, Tiffany JM. Human ocular mucus: Chemical studies. *Exp Eye Res.* 1981;33:203–212.
49. Buschke A, Fränkel A. Ueber die Funktion der Talgdrüsen und deren Beziehung zum Fettstoffwechsel. *Berl klin Wochenschr.* 1905;24:318–322.
50. Fraunfelder FT, LaBraico JM, Meyer SM. Adverse ocular reactions possibly associated with isotretinoin. *Am J Ophthalmol.* 1985;100:534–537.
51. Lambert RW, Smith RE. Effects of 13-*cis*-retinoic acid on the hamster meibomian gland. *J Invest Dermatol.* 1989;92:321–325.
52. Wolff E. The muco-cutaneous junction of the lid margin and the distribution of the tear fluid. *Trans Ophthalmol Soc UK.* 1946;66:291–308.
53. Linton RG, Curnow DH, Riley WJ. The meibomian glands: An investigation into the secretion and some aspects of the physiology. *Br J Ophthalmol.* 1961;45:718–723.
54. McDonald JE, Brubaker S. Meniscus-induced thinning of tear films. *Am J Ophthalmol.* 1971;72:139–146.
55. Josephson JE. Appearance of the preocular tear film lipid layer. *Am J Optom Physiol Optics.* 1983;60:883–887.

56. Wolff E. *Anatomy of the Eye and Orbit.* 3rd ed. London: Lewis; 1948;187–193.
57. Dryja TP. A possible function for value-like structures in meibomian gland ducts. *Invest Ophthalmol Vis Sci.* 1986;27(Supp):196.
58. Chew CKS. Quantifying meibomian lipid on the human lid margin. M.Sc. Thesis, University of Oxford, Oxford, 1993.
59. Chew CKS, Hykin PG, Jansweijer C, Dikstein S, Tiffany JM, Bron AJ. The casual level of meibomian lipids in humans. *Curr Eye Res.* 1993;12:255–259.
60. Norn MS. Eye ointments: application, elimination, and positive action. *Ophthalmic Digest.* 1977;39:1–6.
61. Korb DR, Baron DF, Herman JP, et al. Tear film lipid layer thickness as a function of blinking. *Cornea.* 1994;13:354–359.
62. Franck C. Fatty layer of the precorneal film in the "office eye syndrome." *Acta Ophthalmol.* 1991;69:737–743.
63. McCulley JP, Dougherty JM, Deneau DG. Classification of chronic blepharitis. *Ophthalmology.* 1982;89:1173–1180.
64. Bron AJ, Benjamin L, Snibson GR. Meibomian gland disease: Classification and grading of lid changes. *Eye.* 1991;5:395–411.
65. Mathers WD, Shields WJ, Sachdev MS, Petroll WM, Jester JV. Meibomian gland morphology and tear osmolarity changes with Acutane therapy. *Cornea.* 1991;10:286–290.
66. Robin JB, Jester JV, Nobe J, Nicolaides N, Smith RE. In vivo transillumination biomicroscopy and photography of meibomian gland dysfunction: A clinical study. *Ophthalmology.* 1985;92:1423–1426.
67. Cory CC, Hinks W, Burton JL, Shuster S. Meibomian gland secretion in the red eyes of rosacea. *Br J Dermatol.* 1973;89:25–27.
68. Norn M. Expressibility of meibomian secretion: Relation to age, lipid precorneal film, scales, foam, hair and pigmentation. *Acta Ophthalmol.* 1987;65:137–142.
69. Yokoi N, Takehisa K, Kinoshita S. Correlation of tear lipid layer interference patterns with the diagnosis and severity of dry eye. *Am J Ophthalmol.* 1996;122:818–824.
70. Norn MS. Semiquantitative interferences study of fatty layer of precorneal film. *Acta Ophthalmol.* 1979;58:766–774.
71. Norn MS. Lipid tests: tear film interference, in Lemp MA, Marquardt R, (eds): *The Dry Eye: A Comprehensive Guide.* Berlin, Springer-Verlag, 1992; pp 160–163.
72. Guillon JP. Tear film photography and contact lens wear. *J Brit Contact Lens Assoc.* 1982;5:84–87.
73. Olsen T. Reflectometry of the precorneal film. *Acta Ophthalmol.* 1985;63:432–438.
74. Tsubota K, Nakamori K. Dry eyes and video display terminals. *N Engl J Med.* 1993;328:584.
75. Tiffany JM, Chew CKS, Bron AJ, Quinlan M. Availability of meibomian oil and thickness of the oil layer on the precorneal tear film. ARVO Abstracts. *Invest Ophthalmol Vis Sci.* 1993;34:S821.
76. Iwata S, Lemp MA, Holly FJ, Dohlman CH. Evaporation rate of water from the precorneal tear film and cornea in the rabbit. *Invest Ophthalmol.* 1969;8:613–619.
77. Mishima S, Maurice DM. The oily layer of tear film and evaporation from the corneal surface. *Exp Eye Res.* 1961;1:39–45.
78. Gilbard JP, Rossi SR, Heyda KG. Tear film and ocular surface changes after closure of the meibomian gland orifices in the rabbit. *Ophthalmology.* 1989;96:1180–1186.
79. Rolando M, Refojo MF, Kenyon KR. Increased tear evaporation in eyes with keratoconjunctivitis sicca. *Arch Ophthalmol.* 1983;101:557–558.
80. Hamano H, Hori M, Mitsunaga S. Application of the evaporimeter to the ophthalmic field. *J Jpn Contact Lens Soc.* 1990;22:101–107.
81. Tomlinson A, Trees GR, Occhipinti JR. Tear production and evaporation in the normal eye. *Ophthalmol Physiol Optics.* 1991;11:44–47.
82. Tsubota K, Yamada M. Tear evaporation from the ocular surface. *Invest Ophthalmol Vis Sci.* 1992;33:2942–2950.
83. Mathers WD, Binarao G, Petroll M. Ocular water evaporation and the dry eye: A new measuring device. *Cornea.* 1993;12:335–340.
84. Tsubota K. New approaches to dry-eye therapy. *Int Ophthalmol Clin.* 1994;34:115–128.
85. McCulley JP, Sciallis GF. Meibomian keratoconjunctivitis. *Am J Ophthalmol.* 1977;84:788–793.

41

TEAR FILM INTERFEROMETRY AS A DIAGNOSTIC TOOL FOR EVALUATING NORMAL AND DRY-EYE TEAR FILM

Marshall G. Doane and M. Estella Lee

Schepens Eye Research Institute
Harvard Medical School
Boston, Massachusetts

1. INTRODUCTION

The technique and method of thin film interferometry is being increasingly employed in the examination of the lipid layer of the in vivo tear film in normal and dry-eye individuals, with and without contact lenses present.[1–10] This technique permits determination of the thickness of the superficial lipid layer that floats upon the normal tear film and of the fluid layer that covers the anterior surface of contact lenses. In general, these thickness estimates have been based upon the general hue (color) of the reflected interference patterns. For reasons discussed later, the determination of overall tear film thickness (i.e., from air boundary to corneal surface) is not easily accomplished, as can be verified by the work of Prydal et al.[11,12]

Basic optical theory of reflectance and thin film interferometry can precisely indicate the hue and saturation that will be seen as a function of the thickness of the transparent layer causing the interference phenomena. However, in the real world, other factors must also be considered. The observed intensity, hue, and saturation of a specific spectral wavelength that is observed as a reflected ray from a transparent thin film is a function of (i) the relative refractive indices of the media involved, (ii) the phase difference of the rays reflected from the refractive index boundaries, (iii) the spectral distribution of the illumination, (iv) the spectral sensitivity of the detector, and (v) the adaptive condition of the human eye observing the colors. Thus, in order for the color, or hue, of the reflected light to be accurately correlated with thin film thickness, the conditions of observation must be exactly specified. A further important factor that must be known for precise thickness determinations is the angle of incidence of the incoming rays of illumination that give rise to the interference colors.

Lacrimal Gland, Tear Film, and Dry Eye Syndromes 2
edited by Sullivan *et al.*, Plenum Press, New York, 1998

2. OPTICAL THIN FILM INTERFERENCE

Interference between two rays of light requires a consistent phase relationship between them. This can easily be accomplished by using either a single light source and its optical image or two different images of the same source. The latter is the method used in thin film interferometry, where a single ray emerging from the light source is reflected from two different surfaces. Interference phenomena must be observed by *specular reflection* from the in vivo tear film. In the instance of a normal tear film, there is a thin oily (lipid) layer that "floats" on the surface of a much thicker aqueous solution, and the observed interference phenomenon of such a tear film is usually associated with only the anterior lipid layer. The two interfering rays are the ray reflected from the front surface of the lipid layer (air/lipid boundary) and the ray reflected from the lower surface of this layer (lipid/aqueous boundary). The two reflected rays originate from the *same* incident ray, so these rays satisfy the requirement of a consistent phase relationship. The situation is quite different when examining the tear film on the anterior surface of a contact lens, but that issue will not be dealt with here.

Specular reflectance is a special condition where the reflecting object acts as a mirror. That is, the angle that the incoming beam makes with respect to the surface equals the angle of observation on the other side of the normal, or perpendicular, to the surface. This is usually stated as "the angle of incidence equals the angle of reflection." Any light seen at other angles is much dimmer and is due to light scattering from the surface. Such light does not maintain its uniform phase relationships with the incoming beam and does not contribute to the interference bands that are seen.

The derivation of the conditions for the generation of thin-film interference bands, or *fringes*, are presented in most basic textbooks in optics. The underlying principle is that each incident ray of light reflected by the thin film will be split into two parts, one that is reflected from the upper boundary of the thin film (usually the air-film interface), and the other that penetrates the thin film and is reflected from the lower boundary (at the film-substrate interface). If these two parts of the original incident ray are brought to a common focus, the intensity of the combined reflected ray is dependent upon both the intensity of the light reflected from the two boundaries and the relative phase relationship between these two rays. The intensity of the reflectance *per se* is determined by the relative refractive index differences; the phase relationship is dependent upon the optical path difference (OPD) between them. Consider these two factors a bit more closely.

2.1. Refractive Index Differences

The fraction of the incident intensity of a light ray that is specularly reflected from a surface (where the surface is acting as a mirror; ie, angle of incidence = angle of reflection) is dependent upon the difference between the refractive indices of the media involved. The greater the differences between the refractive indices of the two sides of the interface or the larger the angle of incidence of the incoming ray (that is, the angle between the incoming ray and the perpendicular to the surface of the interface), the higher will be the proportion of the incident ray's intensity that is reflected. The effects of incident angle do not become significant, however, until such angles become larger than about 30 degrees. If n_0 is the refractive index of the medium anterior to the reflecting surface, and n_f is the refractive index of the medium of the thin film,

Equation 1 holds:

$$\mathbf{r} = \left(\frac{(\mathbf{n}_f - \mathbf{n}_0)}{\mathbf{n}_f + \mathbf{n}_0)} \right)^2 \tag{1}$$

where **r** is the reflectance of the incident ray, given as the ratio of intensity of incident light to intensity of reflected light. Let us examine the case for a ray of incident light traveling in air and reflected from the surface of a lipid film on the aqueous tear layer. In this case, $\mathbf{n}_0$ will be essentially 1.00 (for air), and $\mathbf{n}_f$ will be about 1.45 (estimated for lipid). The reflectance can then be calculated as:

$$\mathbf{r} = \left(\frac{(1.45 - 1.00)}{(1.45 + 1.00)} \right)^2 = 3.4\% \tag{1a}$$

The reflectance from the *bottom* of the lipid layer (i.e., at the lipid/aqueous solution boundary) will be much less, due to the small difference in refractive indices:

$$\mathbf{r} = \left(\frac{(1.34 - 1.45)}{(1.34 + 1.45)} \right)^2 = 0.16\% \tag{1b}$$

Thus, we see that the surface reflectance of the ray will be 3.4% of the intensity of the incident ray, and the reflectance from the *lower* boundary of the lipid film is only 0.16% of the 96.6% of the incident intensity reaching this boundary. This illustrates why the interference colors are of much less intensity than the incident light and of relatively low contrast. Multiple reflections between the two boundaries will be only a small fraction of the 0.16% value, so these reflections from the thin lipid film (or a thin aqueous film on a contact lens, for that matter) will not be an observable factor in most instances.

Consideration of refractive index differences will explain why it is so difficult to obtain interference patterns that represent the *entire* tear film thickness, that is, from the air/lipid boundary to the tear film/corneal epithelium boundary. The corneal epithelium has a rather thick layer of adsorbed mucin on its surface, and it appears that this glycoprotein merely becomes more highly hydrated as it "merges" with the aqueous tear layer. Thus, there is virtually no sharp refractive index boundary between the aqueous tear layer, the mucin, and the epithelial surface. This means that there is very little reflectance, and makes it extremely difficult to obtain optical information from the "bottom" of the tear film.

2.2. Optical Path Difference (OPD)

We can represent the general situation of a ray of light reflected from the upper and lower boundaries of a thin, transparent film as shown in Fig. 1.

Let the thickness of the film layer be **t,** and **Ø** and **Ø'** the angles of incidence and refraction of the ray SA. According to Snell's law for refraction at interfaces of different refractive index, $\mathbf{n}_0 \sin \mathbf{Ø} = \mathbf{n}_f \sin \mathbf{Ø'}$. Since the entering ray traverses the tear layer twice, the OPD between the two reflected rays can be shown by Equation 2.

$$\text{OPD} = 2\mathbf{t}\, \mathbf{n}_f \cos \mathbf{Ø'} \tag{2}$$

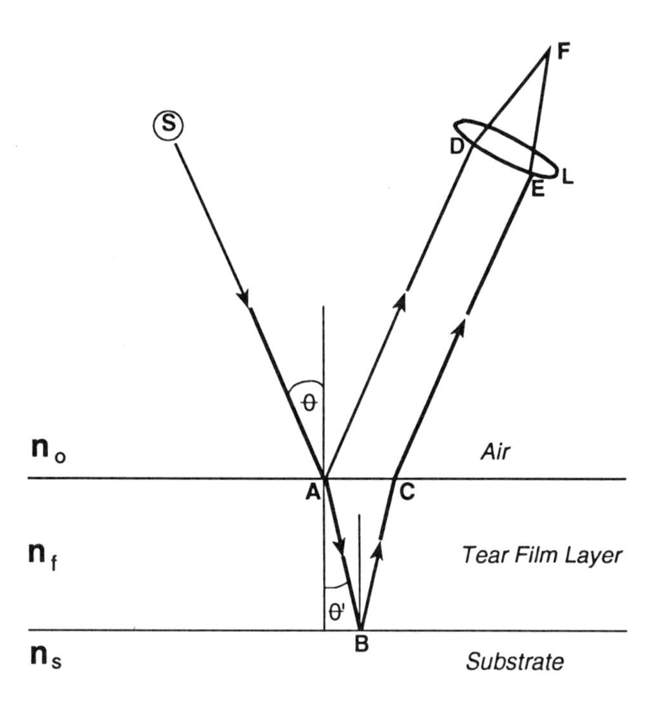

Figure 1. Schematic representation of thin film interference.

If this path difference is an integral multiple of the wavelength of the light, the intensity of the combined rays brought to a common focus will be at a maximum, a condition of *constructive* interference. If the path difference is an odd multiple of half-wavelengths, it will be maximally diminished, and there will be *destructive* interference. Note the presence of the factor cos **Ø'** in Equation 2. The OPD is determined by a multiplier that is the cosine of the angle of the refracted ray. At normal incidence, **Ø'** is zero and cos **Ø'** is unity. At any other angle between normal and the plane of the surface, the factor cos **Ø'** is less than unity, and the OPD will be multiplied by this value. Thus, we see that not only is the OPD dependent upon the thickness of the layer, *but also varies with the angle of incidence of the illumination*. So, the hue, or color, of the resulting interference pattern will also vary with the angle of the incoming light to the surface of the cornea. The practical approach is to make the incident light always normal to the corneal surface over the entire area of specular reflectance. This is accomplished by making the incident light a convergent beam focused at the center of curvature of the cornea, and that is exactly the procedure I have always used. In this manner, there is a direct, unvarying correlation between the observed color and the thickness of the thin film, since the angle factor is always unity.

The simplest situation occurs when considering monochromatic light with a given single wavelength of λ. It can be shown that when the OPD is an *even* multiple of half-wavelengths, there is *constructive* interference; that is, a maximum in the reflected intensity of light occurs. If the path difference is an *odd* multiple of half-wavelengths, the intensity of the reflected light will be at a minimum, and there will be *destructive* interference. Still other factors need to be considered, such as the change of phase by 180 degrees of light reflected from a medium that has a higher refractive index than the medium of the incident light, but that is beyond the scope of this presentation.

3. INTERFERENCE-DERIVED SPECTRAL COLORS

If we now consider the case of the reflectivity of a thin, transparent film on a substrate, we can define the reflectivity of the upper, or first, boundary of the thin film as **r'** and that of the lower boundary of the film (ie, with the substrate) as **r"**. It can be shown (as derived in any basic optics text) that when the refractive index boundaries *and* the path length differences are taken into account, the reflectivity of the thin film as a whole, **R**, is given by Equation 3.

$$\mathbf{R} = \frac{\mathbf{r}'^2 + \mathbf{r}''^2 + 2\mathbf{r}'\mathbf{r}''\cos\Delta}{1 + \mathbf{r}'^2\mathbf{r}''^2 + 2\mathbf{r}'\mathbf{r}''\cos\Delta} \tag{3}$$

where Δ is the phase difference between the two reflected rays. If we utilize the fact that $\cos\Delta = 1 - 2\sin^2\frac{\Delta}{2}$, we obtain

$$\mathbf{R} = \frac{(\mathbf{r}' + \mathbf{r}'')^2 - 4\mathbf{r}'\mathbf{r}''\sin^2 1/2\Delta}{(1 + \mathbf{r}'\mathbf{r}'')^2 - 4\mathbf{r}'\mathbf{r}''\sin^2 1/2\Delta} \tag{3a}$$

Now, substituting the equations for **r'** and **r"**, and rearranging:

$$\mathbf{R} = \frac{\mathbf{n}_{f^2}(\mathbf{n}_s - \mathbf{n}_0)^2 - (\mathbf{n}_{s^2} - \mathbf{n}_{f^2})(\mathbf{n}_{f^2} - \mathbf{n}_{0^2})\sin^2\frac{1}{2}\Delta}{\mathbf{n}_{f^2}(\mathbf{n}_s + \mathbf{n}_0)^2 - (\mathbf{n}_{s^2} - \mathbf{n}_{f^2})(\mathbf{n}_{f^2} - \mathbf{n}_{0^2})\sin^2\frac{1}{2}\Delta} \tag{4}$$

This equation accounts for both the refractive index differences between the initial media, the film media, and the substrate media as well as the thickness of the thin film relative to the wavelength of the incident light.

Fig. 2 shows the variation in reflected light intensity as a function of lipid layer thickness for three widely spaced wavelengths of visible light, 400, 550, and 700 nm. In pure colors, these would be deep blue, green, and deep red, respectively. This represents the case where the lipid layer is floating upon a substrate of aqueous solution. Note that for very thin lipid thicknesses, below where the optical thickness of the film $\mathbf{n}_f\mathbf{t} = 0.05$ µm, the curves of the dependence of the reflectivity for the three wavelengths are quite close together, so no distinct colors will be observed. As the film thickness gradually increases, the spectral differences in the reflectance become further separated, and distinct dispersion of the spectral colors will appear. The color we see will be the result of the *combination* of the relative intensities of *all* the wavelengths at a specific film thickness, not just these three.

For monochromatic light, Equation 4 gives us all the information required for calculating the thickness change represented by adjacent fringes in the interference pattern, the "contour interval." The estimation of film thickness by hue, or color, is much more complicated.

4. ESTIMATION OF FILM THICKNESS BY INTERFERENCE COLORS

Although a given wavelength of light is associated with a specific color, the perception of color by the human eye is not nearly as precise. Not only does a "pure" single

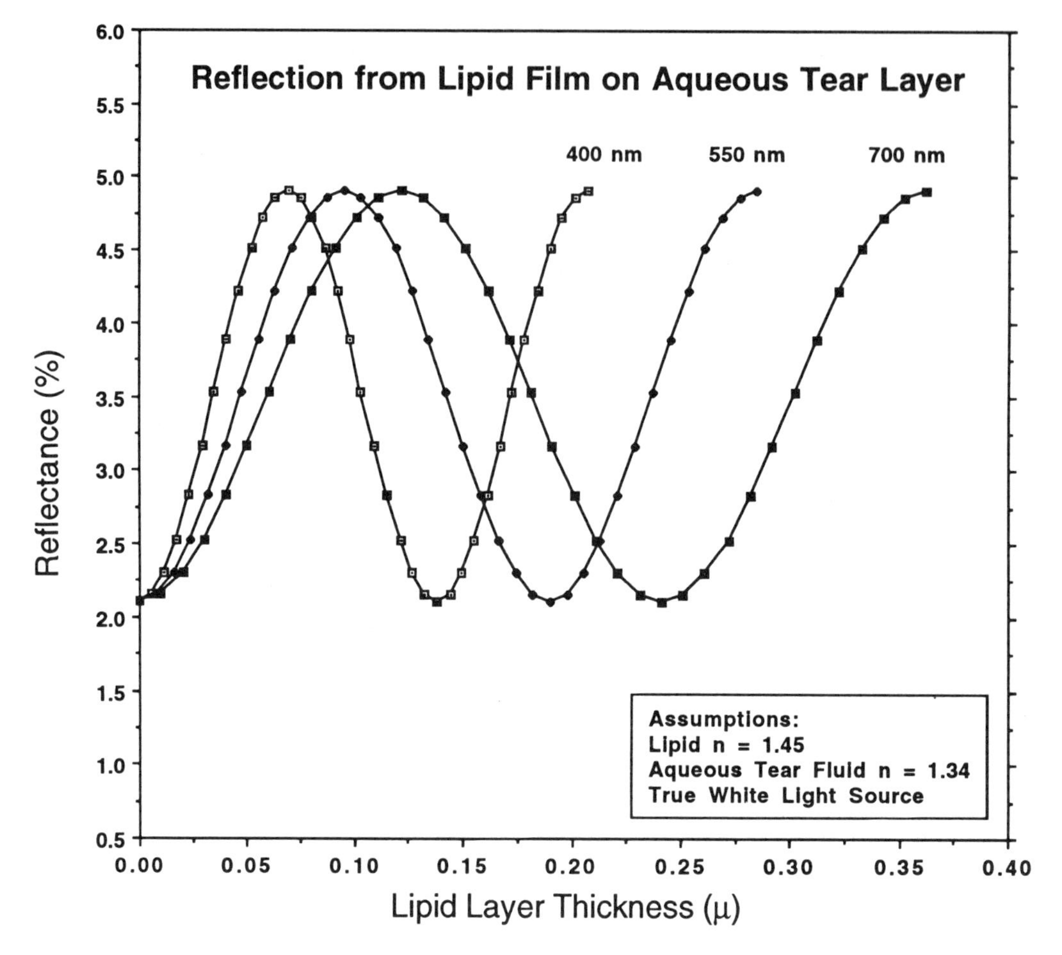

Figure 2. Lipid film on aqueous substrate. Calculated reflectances for three wavelengths.

wavelength of light give rise to the perception of a given color, but in most instances there are many combinations of "pure" wavelengths that give rise to the perception of the same color, a process called color metamerism. This is a consequence of the fact that there are only three different color receptors, or cones, in the human retina. Indeed, color photography, color television, and color printing all depend upon this fact, that a given color perception can be generated by many different combinations of different wavelengths. Thus, a yellow sensation can be created by passing yellow light about 600 nm wavelength into the eye. Exactly the same yellow perception can be created by passing into the eye a mixture of quite different "pure" colors, lights that by themselves would have appeared green and red, respectively. It is not possible to distinguish between the color represented by a single wavelength of light and an appropriate mixture of quite different colors.

The color observed in an interference pattern is also affected by the spectral distribution of the source illumination, since the color we see is dependent upon the relative intensities of the colors of the visible spectrum. A tungsten-halogen lamp has a much different spectral distribution than sunlight, fluorescent light, or an ordinary incandescent lamp. Still another factor in the color we perceive is the adaptive state of the eye and the brightness and color of the adjacent area or field of view.

There is still another factor in the matter of the color that is seen—the spectral sensitivity of the detector being used to observe the colored interference fringes. The detector may be the human retina (as it usually is as the final detector) or the CCD array of a video camera or the color film in a photographic camera.

The major point here is that the color we perceive in the image of a color interference pattern is, of course, dependent upon the thickness of the thin film, but also is strongly dependent upon many other factors, ranging from the angle of incidence of the source illumination upon the eye to the color distribution of that light and the color sensitivity function of the detector. If the goal is to quantitate the thickness of the lipid layer on the surface of the tear film, the most precise and simplest way to do this is to use monochromatic light of known wavelength. The result is not nearly as spectacular in its images, lacking the vivid interference colors obtained when using white light, but it is much more meaningful in its ability to quantitate film thickness and thickness distribution patterns.

REFERENCES

1. Doane MG. An instrument for in vivo tear film interferometry. *Optom Vis Sci.* 1989;66:383–388.
2. Doane MG, Lee ME. Compositional and structural changes in the in-vivo dry-eye tear film. ARVO Abstracts. *Invest Ophthalmol Vis Sci.* 1992;33:S950.
3. Guillon JP. Tear film structure and contact lenses. In: Holly FJ, ed. *The Preocular Tear Film in Health, Disease, and Contact Lens Wear.* Lubbock, TX: Dry Eye Institute; 1986:914–935.
4. Hamano H, Hori M, Kawabe H, et al. Clinical applications of bio differential interference microscope. *Contact Intraoc Lens Med J.* 1980;6:229–235.
5. Hamano H, Hori M. Change of surface patterns of precorneal tear film due to secretion of meibomian gland. *Folia Ophthalmol Japonica.* 1980;31:353–355.
6. Josephson JE. Observations of the specular reflection from the anterior surface of the preocular tear film. In: Holly FJ, ed. *The Preocular Tear Film in Health, Disease, and Contact Lens Wear.* Lubbock, TX: Dry Eye Institute; 1986:554–563.
7. Josephson JE. Appearance of the preocular tear film lipid layer. *Am J Optom Physiol Opt.* 1983;60:883-.
8. Knoll H, Walter H. Pre-lens tear film specular microscopy. *Int Contact Lens Clin.* 1985;12:30–37.
9. McDonald JE. Surface phenomena of the tear film. *Am J Ophthalmol.* 1969;67:56–64.
10. Norn MS. Semiquantitative interference study of fatty layer of precorneal tear film. *Acta Ophthalmol.* 1979;57:766–774.
11. Prydal JI, Campbell FW. Study of precorneal tear film thickness and structure by interferometry and confocal microscopy. *Invest Ophthalmol Vis Sci.* 1992; 33:1996–2005.
12. Prydal JI, Artal P, Woon H, Campbell FW. Study of human precorneal tear film thickness and structure using laser interferometry. *Invest Ophthalmol Vis Sci.* 1992;33:2006–2011.

HUMAN AND RABBIT LIPID LAYER AND INTERFERENCE PATTERN OBSERVATIONS

Donald R. Korb,[1] Jack V. Greiner,[2,3] Thomas Glonek,[4] Amy Whalen,[1] Stacey L. Hearn,[2] Jan E. Esway,[1] and Charles D. Leahy[2]

[1]100 Boylston Street
Boston, Massachusetts
[2]Schepens Eye Research Institute, and
[3]Department of Ophthalmology
Harvard Medical School
Boston, Massachusetts
[4]MR Laboratory
Midwestern University
Chicago, Illinois

1. INTRODUCTION

The lipid layer is extremely important to tear film stability[1] as it is believed to prevent excess evaporation from the ocular surface. The meibomian glands of the upper and lower eyelids are considered to be the principal sites of source lipids destined for the tear film.[2,3] These meibomian secretions in human[4–8] and rabbit[9–11] are composed of triglycerides, neutral lipids, and polar lipids. The phospholipid component of the latter group has also been analyzed.[11]

It is believed that the blinking motion of the eyelids spreads these meibomian lipids over the tear film. The effects of blinking on the thickness and overt physical appearance of the human lipid layer have been studied by observing the interference patterns of the tear film lipid layer.[12,13] Given the use of the rabbit model for studies of tear film dynamics and stability, the present study examines the physical features of the rabbit lipid layer immediately following blinking and during the interblink period, and compares them to those of the human.

2. METHODS

Human subjects (*n* = 32) age 21–30 years and rabbits (*n* = 20) weighing 3–4 kg were positioned in a custom-built system[12] designed to photographically record birefringence of

Lacrimal Gland, Tear Film, and Dry Eye Syndromes 2
edited by Sullivan *et al.*, Plenum Press, New York, 1998

the tear film lipid.[2,12] Human subjects and rabbits had no evidence of ophthalmologic disease other than refractive error. Qualitative analysis of the tear film lipid layer was based on observations of color interference effects in zones of specular reflection.[12–22] The magnification of the color interference patterns as viewed in real time on a monitor could be varied from 6x-40x for accurate analysis. Trueness of color representation on the video monitor was controlled by calibrating the instrument with Eastman Kodak color reference standards (Wratten filters) prior to analysis.

The light source was adjusted to attain an area of observation limited to approximately 2.5 mm vertically and 5.0 mm horizontally. In the case of humans, a fixation target was used to elevate the eye 5–10 degrees above the primary position of gaze. This allowed consistent illumination of the desired area of the tear film lipid layer. Though less cooperative, rabbits remained relatively still for the period of observation, and in general their position of gaze remained constant. Measurements were taken both with and without light general anesthesia with intramuscular injection of acepromazine (1–2 mg/kg), and no difference in tear film lipid layer thickness was observed.

Baseline lipid layer thickness values were established for each human subject by observing the lipid layer interference pattern for 60 sec. For rabbits, the observation period was extended to as long as 30 min. These time periods allowed examination of the lipid layer between multiple blinks. Ambient relative humidity in the examination room was 38–42%.

The thickness values assigned to specific observed colors were derived from previous observations of interference colors from the lipid layer of the tear film.[12,13] To further standardize the procedure, observation of the tear film lipid layer was directed to a specific portion of the cornea. This area corresponded to the region approximately 1 mm above the lower meniscus to slightly below the inferior pupillary margin, averaging 2.5 mm in height by 5 mm in width. The thickness assigned to the lipid layer corresponded to that represented by the predominant color or absence of color if that was the dominant feature within this defined area. To compensate for the subjective nature of color observation, video tape recordings were independently graded by more than one observer.

Blink frequency of human subjects was measured over 5 min on two separate occasions while subjects were seated in an examining room. Human subjects were engaged in conversation and unaware that blink rate measurements were being recorded by two observers. Blink frequency of the rabbits was also recorded by two observers simultaneously. The rabbits were observed in their usual housing environment, precluding any deviation from naturally occurring, spontaneous blinking patterns caused by unusual stimuli. Rabbits were observed at a distance of >1 m to avoid a threatened response that may alter the blink rate. Using a stopwatch, a minimum of 25 recordings were taken for each animal on three separate days.

3. RESULTS

The tear film lipid layer thickness varied from <60 nm to 180 nm (average 78.59 nm) in humans and appeared to be >180 nm in rabbits. Of further interest was the character of the lipid layer. In humans, the lipid layer color interference patterns had a dynamic nature most pronounced upon, and immediately following, upward movement of the eyelid, revealing a variety of patterns and wave formations. In contrast, the lipid layer in rabbits was nearly stationary with only minimal movement after a blink and practically no complex patterns and minimal wave formations. Unlike the human subjects, with lipid layers exhibiting many colors and variable wave formations across the tear film, the rabbit lipid layer was generally white and colorless, rarely exhibiting colors even during the period immediately following blinking. When present, the colors were restricted to a small surface area and remained for several minutes. The average interblink period was 7.01 sec (8.56 blinks/min) in humans and 313.2 sec (0.19 blinks/min) in rabbits.

4. DISCUSSION

The lipid layer of the rabbit tear film appears thicker and more stable than that of the human. It is perhaps these attributes that allow the rabbit to enjoy a significantly longer interblink period than the human, since a thicker lipid layer would theoretically delay the dryness signal which is commonly believed to trigger blinking. The apparent increased thickness and stability of the rabbit lipid layer relative to the human may be related to a variety of factors including species differences in meibom composition,[4–11] a smaller area of exposure of the interpalpebral fissure in rabbit,[23] and the potential contribution of lipids from the rabbit Harderian glands.[24–28] This inherent stability of the rabbit lipid layer is further evidenced by its ability to resist major perturbations following meibomian gland occlusion.[29]

The physical appearance of the rabbit lipid layer and its behavior following blinking are noticeably different from that of the human. Such obvious differences raise questions about the suitability of the rabbit as a model for studying tear film stability and dynamics.

REFERENCES

1. Mishima S, Maurice DM. The oily layer of the tear film and evaporation from the corneal surface. *Exp Eye Res.* 1961;1:39–45.
2. Korb DR, Greiner JV. Increase in tear film lipid layer thickness following treatment of meibomian gland dysfunction. *Adv Exp Med Biol.* 1994;350:293–298.
3. Mishima S, Maurice DM. The effect of normal evaporation on the eye. *Exp Eye Res.* 1961;1:46–52.
4. Tiffany JM. Individual variations in human meibomian lipid composition. *Exp Eye Res.* 1978;27:289–300.
5. Nicolaides N, Kaitaranta JK, Rawdah TN, Macy JI, Boswell FM, Smith RE. Meibomian gland studies: Comparison of steer and human lipids. *Invest Ophthalmol Vis Sci.* 1981;20:522–536.
6. Nicolaides N, Ruth EC. Unusual fatty acids in the lipids of steer and human meibomian gland excreta. *Curr Eye Res.* 1982;2:93–98.
7. Nicolaides N, Santos EC, Papadakis K. Double-bond patterns of fatty acids and alcohols in steer and human meibomian gland lipids. *Lipids.* 1984;19:264–277.
8. Nicolaides N, Santos EC. The di- and triesters of the lipids of steer and human meibomian glands. Lipids. 1985;20:454–467.
9. Tiffany JM. The meibomian lipids of the rabbit. I. Overall composition. *Exp Eye Res.* 1979;29:195–202.
10. Tiffany JM, Marsden RG. The meibomian lipids of the rabbit. II. Detailed composition of the principal esters. *Exp Eye Res.* 1982;34:601–608.
11. Greiner KV. Glonek T, Korb DR, Booth R, Leahy CD. Phospholipids in meibomian gland secretion. *Ophthalmic Res.* 1996;28:44–49.
12. Korb DR, Baron DF, Herman JP, et al. Tear film lipid layer thickness as a function of blinking. *Cornea.* 1994;13:354–359.
13. Korb DR, Greiner JV, Glonek T, Esbah R, Finnemore VM, Whalen AC. Effect of periocular humidity on the tear film lipid layer. *Cornea.* 1996;15:129–134.
14. McDonald JE. Surface phenomenon of the tear film. *Arch Ophthalmol.* 1969;67:56–64.
15. Norn MS. Semiquantitative interference study of fatty layers of precorneal film. *Acta Ophthalmol.* 1979;57:766–774.
16. Hamano H, Hori M, Kawabe H, et al. Biodifferential interference microscope observations on anterior segment of eye. *J Jpn Contact Lens Soc.* 1979;21:229–231.

17. Hamano H, Hori M, Kawabe H, et al. Clinical applications of biodifferential interference microscope. *Contact Intraocular Lens Med J.* 1980;6:229–235.
18. Hamano H, Mitsunga S. Clinical examinations and research on tear. In: Tanaka K, Anan N, Mikami M, et al., eds. Menicon Toyo's 30th Anniversary Special Compilation of Research Reports. Tokyo: Toyo Contact Lens Co.; 1982: Chap 2.
19. Guillon J-P. Tear film photography and contact lens wear. *J Br Contact Lens Assoc.* 1982;5:84–87.
20. Josephson JE. Appearance of the preocular tear film lipid layer. *Am J Optom Physiol Optics.* 1983;60:883–887.
21. Kilp H, Schmid E, Kirchner L, Zipf-Pohl A. Tear film observation by reflecting microscopy and differential interference contrast microscopy. In: Holly FJ, ed. *The Preocular Tear Film in Health, Disease, and Contact Lens Wear.* Lubbock, TX: Dry Eye Institute; 1986:564–569.
22. Maurice DM. The effect of the low blink rate in rabbit on topical drug penetration. *J Ocul Pharmacol Ther.* 1995;11:297–304.
23. Greiner JV, Glonek T, Korb DR, et al. Volume of the human and rabbit meibomian gland system. This volume.
24. Prince JH, ed. *The Rabbit in Eye Research.* Springfield, IL: Charles C Thomas; 1964. p 50.
25. Cogan DG, Fink R, Donaldson DD. X-ray irradiation of orbital glands of rabbit. *Radiology.* 1955;64:731–737.
26. Paule WJ. *The Comparative Histochemistry of the Harderian Gland.* Ohio State University, 1957.
27. Schneir ES, Hayes ER. The histochemistry of the Harderian gland of the rabbit. *Natl Cancer Inst J.* 1951;12:257.
28. Walter A. Ueber die HautderÅsen mit Lipoidsekretion bei Nagern. *Bietr path Anat. Path.* 1924;73:142–167.
29. Greiner JV, Glonek T, Korb DR, et al. Effect of meibomian gland occlusion on tear film lipid layer thickness. This volume.

43

ABNORMAL LIPID LAYERS

Observation, Differential Diagnosis, and Classification

Jean-Pierre Guillon

Scope Research Ltd. and Dry Eye Symptoms Clinic
Institute of Optometry
London, England, United Kingdom

1. INTRODUCTION

The preocular tear film stability is governed by its structure. The superficial lipid layer provides a barrier to limit evaporation. From basic surface physics, it is also believed that the lipid layer plays a very important part in the formation and stabilization of the full film following the eye opening. The appearance of abnormal ocular lipid layers in humans has been rarely studied. The purpose of this paper is to describe this appearance to the clinician, to differentiate between various types, and to propose a classification based on appearance.

There is now increasing evidence that abnormal lipid layer appearance is linked to increased evaporation and decreased stability.[1] This will, in turn, produce symptoms of dryness that have an evaporative cause and must be differentiated from symptoms induced by decreased lacrimal production or created by surfacing abnormalities.

2. TEAR FILM EXAMINATION

By normal slit lamp examination, the tear film is transparent, and only gross surface irregularities like clumps or debris can be seen. The superficial structure of the tear film is visible by specular reflection, and the Keeler TEARSCOPE Plus provides 360° specular reflection using the white light emission of a cold cathode. The instrument produces a white background against which the tear film structures are visible. The observations and diagnosis are based on the intensity and visibility of the structure reflected, on the regularity of the layer, on the presence of interference colors, and on the integrity or the appearance of a break in the lipid layer or a full break in the tear film.

The design and use of the TEARSCOPE Plus are described elsewhere in this volume. This paper is limited to the observation of abnormal lipid layers, their structure, and differential diagnosis. Unfortunately, full-color pictures can not be reproduced here.

Lacrimal Gland, Tear Film, and Dry Eye Syndromes 2
edited by Sullivan *et al.*, Plenum Press, New York, 1998

3. NORMAL LIPID LAYER APPEARANCE

The normal lipid layer has been classified in increasing order of thickness and described by its appearance (Fig. 1, Table 1).

4. ABNORMAL LIPID LAYERS

The presence of abnormal lipid layers may have various causes, which include the following: meibomian over-secretion, abnormal secretion, abnormal spreading, lipid break-up, natural contamination, effect of cosmetics, effect of eyedrop instillation, and effect of contact lenses.

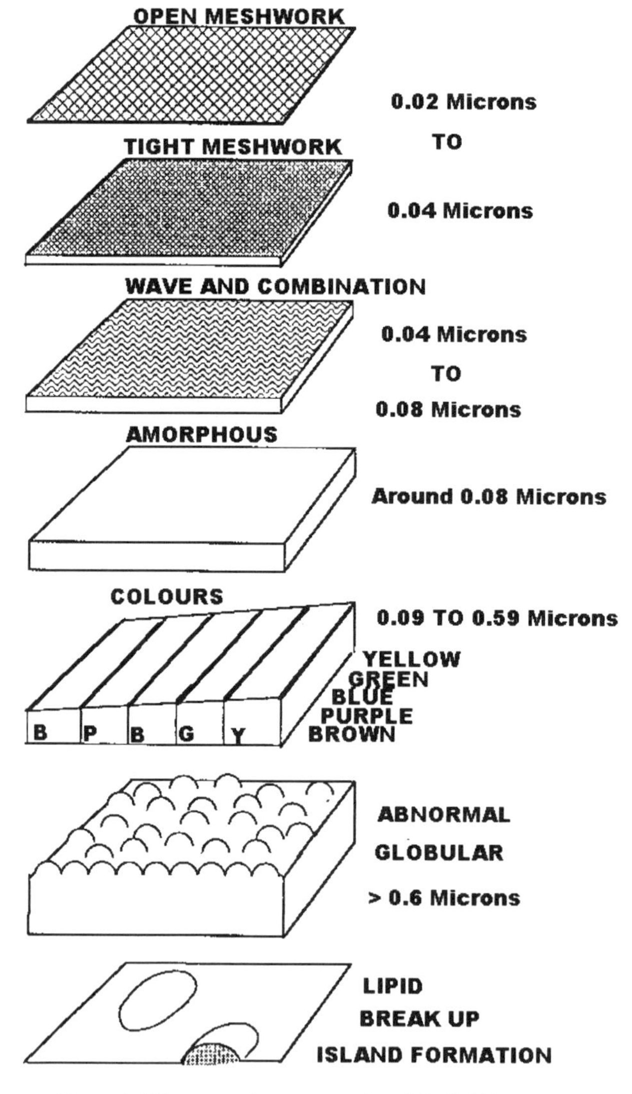

Figure 1. Diagrammatic representation of the lipid layer.

Table 1. Classification of normal lipid layer

Pattern	Description	Observation
Open meshwork	Faint mesh-like	Low visibility
Tight meshwork	Darker mesh	—
Meshwork & wave	Mixed	—
Wave	Flow-like	—
Wave & amorphous	Mixed	—
Amorphous	Even surface	Good visibility
Wave & colors	Mixed	Irregular spreading
Amorphous & colors	Mixed	Even background with colors
Colors	Interference 1st order	Brown & blue

Meibomian over-secretion may be present during infection, inflammation, as part of reflex tearing, or following lid manipulation. The secretion appears as a layer of large thickness variation where interference colors of the 2nd, 3rd, and 4th orders of interference are visible within a small area.

Abnormal secretion will occur in cases of under-secretion in meibomian gland dysfunction or blockage. A lipid layer will be present, but so thin that it will hardly be visible at low magnification and only be noticed by its movement following the blink under high-magnification specular microscopy. This appearance has been linked to a reduced tear film stability and a short break-up time.[1]

In *blepharitis*, the secretion may have an abnormal composition and/or may be contaminated by abnormal secretions from the other glands or from debris and detritus from the lid edge surface. The secretion will appear irregular in surface and will aggregate in globules and not spread evenly. The worse the condition, the greater are the size and number of globules. In extreme cases, the globular structure is so packed that it appears as an irregular, silvery reflecting structure in specular illumination.

Abnormal spreading is also a cause of abnormal appearance of the lipid layer. In cases of incomplete blinking, the lipid layer is not reformed properly and may be thinner in the upper region. Variable thickness will become apparent, and localized variation may occur.

Poor mixing of newly secreted lipid will appear as areas of different thickness, taking the form of surface plaques or streaks or of oily lenses. These secretions will take longer to come to a stop following the blink and will require a few blinks to start mixing with the background lipids. This may suggest lipids of different surface activity or of different classes. These surface irregularities may affect the evaporation rate and the underlying aqueous phase.

Break-up of the lipid layer is obvious when viewed with the TEARSCOPE Plus, as it appears as a grey area separated by zones of highly reflective lipid cover. The metallic grey reflection seen in the break-up area is due to the destructive interference condition occurring during a reflection of the incident light at a surface of higher refractive index and undergoing a phase change of 90°. When a break-up of the lipid layer is visible, evaporation from the unprotected aqueous phase will be very high with reduced tear film stability,[1] ocular surface irritation, and corneal epithelium staining if reflex tearing does not occur.

Break-up may occur in different situations and is most obvious in cases of surface contamination. In blepharitis, the abnormal nature of the secretion can induce a superficial break-up when it is mixed with skin lipids that find their way across the lid margin and onto the tear film surface. In those cases, aqueous droplets resting at the lid edge between the lashes and the tear prism are visible.

During *reflex tearing*, an abnormal quantity of aqueous fluid may destabilize a thin lipid layer and temporarily break the superficial lipid structure. Here again, contact may occur with contaminants present at the surface of the lashes or the skin, when the tear film overflows. Any contaminants settling in the area between the lashes and the tear meniscus will interfere with the chemical barrier that prevents skin lipid from invading the tear film, and any breach in this barrier will create a bridge for continuous invasion of surface-active products to destabilize the tear film. When contamination is present, some areas of the lid edge will become wettable and will be covered by droplets of water. This is the main sign confirming the presence of contamination.

Abnormal lipid layers will also be affected by mucous strands present in the tear film, and the appearance of foam at the edge corners of the lid or along the lid margins will be further evidence of contamination.

5. EFFECT OF COSMETICS

One of the most common causes of abnormal appearance of the lipid layer is cosmetics. They can affect the layer in various degrees, from simple evidence of their presence at the surface of the film to a full, constant destruction of the layer induced by every blink when the offending product is coating the lid margin and the lid edges. In the latter case, the product induces massive evaporation and can reach the ocular surface and induce arcuate corneal epithelium staining to fluorescein similar to that found in blepharitis with equivalent dry eye symptoms.

The main problem with cosmetics is that they are designed to stick to and penetrate the skin surface and alter its wetting properties. Further, their effect will depend on the skill with which they have been applied. The worst product offenders are make-up removers, which are oily and are used for removing heavily applied make-up. The nature of their oil induces, by their surface effect, a temporary destruction of the spreading capabilities of the lipid layer. When these products are coating the corner of the lids, they continuously affect the lipid layer with every blink, and their effect may be visible the day after their use (more than 20 h later). In those cases, dry eye symptoms will be present, and only observation with the TEARSCOPE Plus permits the differential diagnosis of the problem.

Eyeliners applied directly to the rim of the lid will interfere with the proper function of the meibomian glands by blocking their orifices and contaminating their secretion. Further, their removal will require contact with the lid edge and possible further disturbance to the tear film.

Mascara can flake or break up and increase the debris level on the superficial lipid layer.

Eye shadow can distribute its particles on the tear film surface and increase debris formation.

Moisturizers are commonly used by people suffering from marginal dry eye symptoms as they commonly have dry or combination skin. When present in the tear film, these products appear as oily floaters or oily plaques. Their effect is worse when they have been applied to the lid corners, as repeated contact with the superficial tear film increases the disturbance.

6. EFFECT OF ARTIFICIAL TEAR FORMULATION

It has been shown that the simple instillation of one drop of non-preserved saline can affect the structure of the superficial lipid layer and cause it to break up temporarily. Only repeated blinking can reform the lipid layer to its normal appearance. Any instillation will have an effect on the lipid surface; ointments and gels will disturb the whole tear film, including the lipid layer. These effects will be long lasting. The use of ointment will produce an irregular superficial surface where interference colors seen in globular formation are formed within the ointment itself. By nature, ointments are not a good structure to support a superficial lipid layer, and the laycr will be missing in most cases. Observation of any ointment use by wide-field specular reflection is mandatory to judge the effectiveness of that treatment in dry eye cases.

7. ABNORMAL LIPID LAYER IN CONTACT LENS WEAR

With soft contact lenses, the lipid layer in always thinner than that found in the preocular tear film. Commonly an incomplete layer covering a thinning aqueous phase is visible. The layer is prone to invasion by surface contaminants, which will migrate to the contact lens surface and produce discrete non-wetting areas. The poor integrity of the lipid layer is one cause of increased evaporation and reduced break-up time in soft lens wear. With rigid lenses, the lipid layer is most commonly absent or will provide only partial coverage over the aqueous phase. Commonly the lipid layer is limited to isolated islands floating at the surface of the aqueous phase and sticking to the lens surface after evaporation and drainage of the aqueous phase.

8. CONCLUSION

The superficial lipid layer can be classified based on its appearance when viewed in wide-field specular reflection with the TEARSCOPE Plus. The classification is also in order of increasing thickness. The abnormal lipid layers can also be classified by their appearance. Further studies are necessary to link appearance to composition and surface properties.

REFERENCE

1. Craig JP, Tomlinson A. Importance of the lipid layer in human tear film stability and evaporation. Optometry and Vision Science. 1997;74(1):8–13.

44

ASSOCIATION OF TEAR LIPID LAYER INTERFERENCE PATTERNS WITH SUPERFICIAL PUNCTATE KERATOPATHY

Aoi Komuro, Norihiko Yokoi, Yoko Takehisa, and Shigeru Kinoshita

Department of Ophthalmology
Kyoto Prefectural University of Medicine
Kyoto, Japan

1. INTRODUCTION

It is sometimes difficult to differentiate dry eye from other ocular surface diseases exhibiting severe superficial punctate epithelial keratopathy. Observation of tear lipid layer interference patterns has been found useful in differentiating dry eyes from non-dry eyes.[1–3] In this study, we investigated whether dry eye and other ocular surface diseases exhibiting severe superficial punctate keratopathy can be differentiated by observing tear lipid layer interference patterns.

2. MATERIALS AND METHODS

The study included 15 eyes of 15 ocular surface disease patients without dry eye (non-dry eye group) and 20 eyes of 20 dry eye patients (dry eye group). All eyes had moderate to severe superficial punctate keratopathy and corneal fluorescein staining scores exceeded A2D2, according to the classification of Miyata et al.[4] In this classification, "A" stands for area, and "D" for density of corneal epithelial lesions; values range from 0 to 3, depending on the severity of damage. In the Schirmer I test, all dry eye patients showed less than 5 mm per 5 min, and all non-dry eye patients more than 5 mm per 5 min. Ocular surface diseases included 12 cases of drug-induced corneal epithelial damage, one of atopic keratopathy, and two of phlyctenular keratitis. Normal eyes served as controls.

Specular images of precorneal tear film at the central cornea (2 mm diameter area) were observed in all subjects, using a newly developed specular reflection video-recording system.[3] Each image was recorded on videotape for about 30 sec under spontaneous blinking. Then a representative image from each recording was printed out using a video

Lacrimal Gland, Tear Film, and Dry Eye Syndromes 2
edited by Sullivan *et al.*, Plenum Press, New York, 1998

printer, and classified into one of five grades based on the judgment of three ophthalmologists:

- Grade 1: grayish color, uniform distribution.
- Grade 2: grayish color, nonuniform distribution.
- Grade 3: a few colors, nonuniform distribution.
- Grade 4: many colors, nonuniform distribution.
- Grade 5: corneal surface partially exposed.

3. RESULTS AND DISCUSSION

The grade distribution in each group, as % of total, was as follows (Fig. 1): in the dry eye group, 0, 0, 20%, 30%, and 50%, for Grades 1, 2, 3, 4, and 5, respectively; in the non-dry eye group, 0, 67%, 13%, 0, and 20%, respectively. The difference in grade distribution between the two groups was significant (p=0.015, Wilcoxon test). Grade 2 was seen only in the non-dry eye group.

The present findings in dry eye correlate well with our previous results,[3] in which dry eye with relatively severe corneal involvement either had thicker precorneal lipid layer (Grade 3 or 4) or lacked lipid (Grade 5). As discussed previously,[3] the aqueous volume of the tear film may contribute to this mechanism. Briefly, the interpalpebral space increases in the closed eye in dry eye patients because of decreased aqueous volume. That space may be occupied by excessive lipids from the meibomian gland, resulting in a thickening of the lipid layer. Unlike dry eye grades 1 through 4, in Grade 5 dry eye, aqueous volume is too small for the lipid to expand on, resulting in a lack of lipid interference.

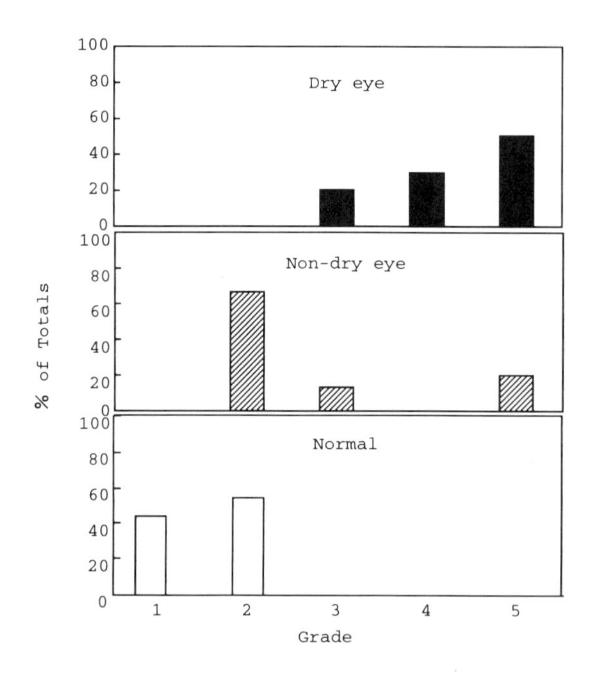

Figure 1. Grade distribution in dry eye and non-dry eye ocular surface disease cases and in normal eyes (normal eye data are from ref. 3).

Tear film instability related to severe superficial epithelial damage also contributes to this proposed mechanism, as supported by the fact that even some non-dry eye cases showed Grade 5. Given the fact that the non-dry eye group showed a normal range in the Schirmer I test, it is understandable why this group often showed normal lipid interference (Grade 2) rather than colored interference.

Tear lipid layer interference pattern observation may become a useful, non-invasive, and simple method of differentiating dry eye from other ocular surface diseases exhibiting severe superficial punctate keratopathy.

ACKNOWLEDGMENTS

Supported in part by a research grant from the Kyoto Foundation for the Promotion of Medical Science, and the Intramural Research Fund of Kyoto Prefectural University of Medicine.

REFERENCES

1. Norn MS. Semiquantitative interference study of fatty layer of precorneal film. *Acta Ophthalmol.* 1979;57:766–774.
2. Doane MG. Abnormalities of the structure of the superficial lipid layer on the in vivo dry-eye tear film. *Adv Exp Med Biol.* 1994;350:489–493.
3. Yokoi N, Takehisa Y, Kinoshita S. Correlation of tear lipid layer interference patterns with the diagnosis and severity of dry eye. *Am J Ophthalmol.* 1996;122:818–824.
4. Miyata K, Sawa M, Nishida T, Mishima H, Miyamoto Y, Otori T. Classification of severity of superficial punctate keratitis. *Rinsho Ganka (Jpn J Clin Ophthalmol).* 1994;48:183–188.

MEIBOMIAN SECRETIONS IN CHRONIC BLEPHARITIS

James P. McCulley and Ward E. Shine

Department of Ophthalmology
The University of Texas Southwestern Medical Center at Dallas
Dallas, Texas

1. INTRODUCTION

Secretions of the eyelid meibomian glands form the outer layer of the ocular tear film. Two conditions are necessary for this lipid material (meibum) to be effective: first, it must be secreted in appropriate amounts, and, second, it must form an effective lipid layer over the hydrophilic aqueous-mucin layer of the tear film. The many functions of this lipid layer have been enumerated by others.[1] The focus of this discussion will be the importance of meibum lipid composition and the ramifications of lipid abnormalities, as well as the relationship of ocular microflora to these differences. Thus, signs associated with chronic blepharitis, such as secretion consistency, inflammation, microbial infections, and dry eye syndrome (previously defined more narrowly only as keratoconjunctivitis sicca), will be discussed in terms of the specific lipid composition of meibomian gland secretions.

2. THE TEAR FILM LIPID LAYER AND ASSOCIATED AQUEOUS PHASE

In order to explore the conditions necessary to produce an effective tear film lipid layer, one must discuss not only some of its important functions but also the environment in which it functions. In particular one must appreciate that the tear film aqueous phase, adjacent to the lipid layer, may affect its stability. For example, characteristics (stability, elasticity, and solubilities in the aqueous layer) of lipid layers formed from fatty acids or monoglycerides are greatly affected by the type and concentration of sugar dissolved in the aqueous layer.[2,3] Some electrolytes (eg, sodium, calcium, chloride) present in the aqueous phase and its pH also affect lipid layers.[4] This was particularly true for lipid layers formed from PE (phosphatidylethanolamine) and oleic acid.[5] With the actual tear film,

Lacrimal Gland, Tear Film, and Dry Eye Syndromes 2
edited by Sullivan *et al.*, Plenum Press, New York, 1998

sugars are replaced by mucins that seem to be spread throughout the tear film,[6] just as the sugars are. These proteoglycans have viscoelastic properties that are affected by hydration, sodium and calcium, and pH[7]; which of these factors, if any, have important effects on the structure of water[8] and its association with the lipid layer is an open question. Finally we would like to point out that two constituents of tears, bicarbonate and lactate,[1] bind significant amounts of calcium.[9] However, it is the buffering effect of bicarbonate[9] that is more important, acting to maintain the open-eye aqueous layer at pH 7.8.[10]

The lipid layer over this aqueous-mucin layer must also have certain characteristics. We therefore propose the following as a model for an effective human tear film lipid layer. Adjacent to the aqueous-mucin phase are the polar phospholipids such as PE, PC (phosphatidylcholine), SM (sphingomyelin), and others,[11] and various ceramides and cerebrosides. (Sulfatides may be present only in those normals lacking cholesterol esters.) It is important that only saturated fatty acids are esterified to these polar lipids; a large fraction of these fatty acids must be hydroxylated. Small amounts of triglycerides, up to 3%, may also be incorporated into this polar monolayer,[12] and here unsaturation is important.[13] Hydrocarbons most likely intercalate into the polar lipid phase and act as a "glue" to strengthen the structure (especially important when the eye is open), thus increasing the tear film break-up time. The longer-chain hydrocarbons, such as C28 (28 carbon chain length) may also decrease the water vapor transmission rate[14] of the lipid layer. It has also been noted that C16 hydrocarbons behave very differently from the shorter C12 hydrocarbons,[15,16] and therefore only the longer hydrocarbons should be incorporated. If wax esters intercalate into the polar lipid phase, the alcohol portion may bridge into the nonpolar lipid phase. When cholesterol esters are present, small amounts may intercalate in a like manner.[17] Small amounts of free fatty acids18 may also be quite important, since their presence can promote the formation of a gel phase (e.g., SM and palmitic acid) and swelling of the lipid-water gel phase, especially in the presence of monoglycerides.[19] However, the total lipid layer consists not only of the polar lipids but also of much larger amounts of nonpolar lipids. The majority of these nonpolar lipids (wax esters, cholesterol esters, triglycerides, and hydrocarbons) form the bulk of the tear film lipid layer. Free fatty acids may also be important in this nonpolar part of the tear film lipid layer. This model more closely approaches that proposed with the mucin distributed throughout the aqueous layer and the polar lipids adjacent to this layer[20] than a model with separate mucin and aqueous layers;[21] in a recent report, photomicrographs also suggest a continuous aqueous-mucin layer and a two phase lipid layer.[22] As suggested by others,1,6 we believe that the tear film (lipid and aqueous-mucin layers including the glycocalyx) acts as a thixotropic system that is essential for the proper fluidization and restructuring of the tear film, especially the lipid phase, during a blink.

3. THE IMPORTANCE OF THE LIPID LAYER IN CHRONIC BLEPHARITIS

Although human meibomian gland lipids have been well characterized,[23] investigations concerning the relationships between lipid composition and disease signs have not been extensive. For a number of years we have been investigating how changes in meibomian gland lipids and in ocular microbial populations are related to chronic blepharitis. Among our earlier observations was that 35% of the chronic blepharitis patients had associated keratoconjunctivitis sicca, but only a few of these cases could have been caused by stagnation of meibomian glands.[24] It is now known that there are at least two causes of dry eye,[25] one involving an insufficient lipid layer due to meibomian gland dropout.[26] Attempts to understand this type of dry eye have involved gross manipulation of the lipids. Thus, the lipid layer thickness can be increased by manually expressing lipid.[27] Alternatively, the thickness of the lipid layer can be increased by increasing the extraocular humidity,[28] but any beneficial effect of this treatment on dry eye has yet to be demonstrated. None of these investigations directly addresses the possibility that specific lipid changes themselves affect susceptibility to dry eye. Thus, we have been investigating signs and symptoms associated with chronic blepharitis and their dependence on specific lipid and microbial populations.

Based on clinical signs, we initially developed a classification system for chronic blepharitis.[29] Our subsequent research dealt with the question: which disease signs are related to specific changes in lipids or microorganisms? Based on this work, answers to these questions are becoming apparent. General information necessary to understand our model has been presented above. However, we are constantly reevaluating this model and our chronic blepharitis classification system as we develop new lipid and microbial data from our investigations.

Our lipid analyses have determined the following important characteristics of meibum. (i) Analysis of polar lipid fatty acids indicates the presence of phospholipids, ceramides, cerebrosides, and possibly sulfatides.[30] (ii) These lipids are composed of both normal and hydroxylated fatty acids, C12-C18 in length, which are saturated only in normal individuals.[31] (iii) Similarly, triglycerides are 80% C12-C19 in length. However, in contrast to the polar lipids, the triglycerides are 50% unsaturated.[32] (iv) Wax ester fatty acids are also 80% C12-C19 and are 30% unsaturated except when cholesterol esters are absent, in which case the wax esters are saturated.[33] (v) Although we have observed hydrocarbons as long as C33, we have never detected any hydrocarbons shorter than C16 and only saturated ones (dominant range C22-C28). (vi) The fatty acid moiety of cholesterol esters is also very long (dominant range C20-C27) and mainly saturated. (vii) We have determined that the three dominant free fatty acids in meibum are palmitic, stearic, and oleic acids.[34]

In our model we believe that all lipids, including free fatty acids shorter than C20, are first and foremost associated with a thin polar lipid phase adjacent to the aqueous-mucin layer. These polar lipids are dominated by PC, SM, PE, cerebrosides, and free fatty acids. Less polar (ceramides, mono- and diglycerides) or nonpolar (triglycerides and possibly wax esters) lipids are associated with this polar lipid phase only to the point of saturation. Other nonpolar lipids (cholesterol esters and hydrocarbons, and the remainder of the triglycerides and wax esters, mono- and diglycerides), and also long-chain (> C20) fatty acids form a thicker nonpolar phase overlaying the polar lipid phase. These phases are weakly associated with each other through hydrophobic bonds accomplished with or without bridging by the hydrocarbons and possibly by a portion of the triglycerides and wax esters. Only a very small fraction of the cholesterol esters would be associated with the polar lipid phase. Instead we suggest that most of the cholesterol and wax esters are associated with each other in the nonpolar phase. We suggest that the polar lipid phase should be considered a structural support phase upon which the main barrier (nonpolar) phase is formed.[35]

4. LIPID DIFFERENCES ASSOCIATED WITH CHRONIC BLEPHARITIS DISEASE SIGNS

Investigation of the meibomian gland lipid secretions resulted in the observation that there are two types of normal populations.[36] They are distinguished by the presence (NCP) or absence (NCA) of cholesterol esters and by the presence of unsaturated fatty acids and alcohols only in the normal population with cholesterol esters (NCP).[33,36] All secretions from individuals with chronic blepharitis disease signs also contain cholesterol esters and associated unsaturated fatty acids and alcohols. An important observation was that both mono- and diunsaturated fatty acids (C18 and C20) are present; triunsaturated fatty acids (C18) are present only in the triglycerides. We have sorted these chronic blepharitis patients into six groups: STAPH, staphylococcal blepharitis; SBBL, seborrheic blepharitis alone; MIX, mixed seborrheic and staphylococcal blepharitis; MBSB, mixed seborrheic blepharitis and meibomian gland seborrhea; 2MEIB, seborrheic blepharitis with secondary meibomianitis; and MKC, meibomian keratoconjunctivitis or primary meibomianitis. We believe that the additional unsaturation (which is almost entirely in the wax esters) when cholesterol esters are present is a direct biochemical response to the presence of the cholesterol esters (which are almost entirely saturated, however). Our view is that this unsaturation is necessary to maintain adequate lipid fluidity and thixotropy. Secretions from the MKC group, however, are distinguished from the other groups by lower unsaturation in fatty acids and alcohols of the wax and cholesterol esters.[33,36] Some MKC patients also have lower fatty acid unsaturation in triglycerides.[32] Thus, in the MKC group, paste-like meibum is at least partly the result of inadequate unsaturation. Although not statistically significant, it may also be an important observation[34] that in the free fatty acids, the monounsaturated fatty acid, oleic acid, was lowest in the MKC group (with paste-like meibum) and highest in the MBSB group (with very fluid meibum). In fact, we suggest that it is increased lipid layer hydration[18,19] resulting from this increased level of free oleic acid (or associated mono- or diglycerides if formed from triglycerides) that is partly responsible for the fluid meibum in the MBSB group. Associated foaming in this (MBSB) group[29] may be related, and could suggest involvement of micelles containing both free fatty acids and monoglycerides.

In the polar lipids, in contrast to the nonpolar lipids, additional unsaturation is limited entirely to chronic blepharitis patients; both normal groups' polar lipids are saturated. Patients are distinguished by meibum that contains sphingolipid fatty acid unsaturation;[31] however, both types of normals contain saturated sphingolipid fatty acids. This sphingolipid unsaturation in patients may act to weaken the cohesiveness of the polar lipid phase. In addition, lipids from one disease group, MKC, also contain unsaturated glycero-lipid (phospholipid) fatty acids,[31] which may further affect the uniformity of the polar lipid phase. In fact, we have analyzed the sphingolipid fatty acids extensively and also find large differences in hydroxylated fatty acids. Thus, in the NCA normal group (without cholesterol esters or wax and cholesterol ester unsaturation), virtually all of the sphingolipid fatty acids are hydroxylated. On the other hand, all other groups (with cholesterol esters and unsaturated wax and cholesterol ester fatty acids) contain about 70% hydroxylation in the sphingolipid fatty acids. Again, the one exception is the MKC group, which contains less than 20% of the sphingolipid hydroxy fatty acids present in the normal (NCP) group.[31] This hydroxylation may aid in maintaining the structure (eg, through hydrogen bonding[37]) of the polar lipid phase; lower hydroxylation in the MKC group may help to explain the lower tear film stability in this group. However, an indirect effect of α-hydroxylation is suggested by the differences in the disease groups MBSB (with a highly fluid meibum), 2MEIB (partial glandular involvement), and MKC (with pasty meibum). Both 2MEIB and especially MKC have low levels of α-hydroxy fatty acids; in contrast, MBSB has high levels.

We believe that the additional unsaturation present in meibum that contains cholesterol esters has importance beyond simple biophysical parameters. These additional mono- and diunsaturated fatty acids may predispose individuals to the inflammation associated with chronic blepharitis. It is now recognized that the chemoattractant (chemokinesis or chemotaxis[38]) 4-hydroxy-2-nonenal (HNE) can readily be formed from linoleic acid, both chemically and by lipoxygenases.[39–41] Furthermore, the recruited neutrophils are also stimulated to produce reactive oxygen species, not only by linoleic acid, but also by the monounsaturated fatty acid, oleic acid.[42] Thus, individuals may be predisposed to blepharitis by the presence of the additional unsaturated fatty acids in meibum with cholesterol esters. Those individuals who remain normal do so because of lower lipase activity (to hydrolyze wax esters or triglycerides), since free fatty acids are necessary ultimately to form HNE. However, if active oxygen is also lower, formation of HNE will be decreasing since active oxygen species are also necessary for HNE formation. Additionally, even if HNE is produced, it can be eliminated by reaction with glutathione or similar biological compounds.[39] It is important to note that even though rabbit secretions also contain cholesterol esters, unsaturation is minimized by substitution of iso- and especially anteiso-fatty acids.[36]

Other chronic blepharitis disease signs can be associated with changes in certain polar lipids. For one specific polar lipid (tentatively identified as a phospholipid), the level in meibum correlates with both the MKC and the MBSB groups. Thus, this lipid, which appears somewhat less polar than PE, is higher in that from the MKC group ($p<0.02$) and lower in meibum from the MBSB group when compared with the other groups. Specific cerebrosides and an unknown lipid of medium polarity appear important for determining other disease groups. Thus, we believe that many of the chronic blepharitis disease signs, other than blepharitis, are associated with specific changes in the polar lipids. The exceptions are non-fluidity in the MKC group, where decreased unsaturation in nonpolar lipids is also important, and possibly increased fluidity in the MBSB group, as already discussed.

5. TREATMENT OF CHRONIC BLEPHARITIS SIGNS

We believe that the effects of specific treatments for tear film lipid layer conditions have been evaluated too narrowly. For example, dry eye conditions due not to lacrimal gland insufficiency but to lipid layer abnormalities have been investigated from the point of view of the total lipid present. Although studies of associated gland dropout[26] are important and the increase in lipid layer thickness with increased extraocular humidity[27] is interesting, understanding is lacking. It has been clearly demonstrated that the tear film lipid layer from dry eye patients behaves dramatically differently from that of normals.[43] What may actually be happening in the case of excess humidity is that the tear film and lipid layer pH is decreasing due to carbon dioxide retention. Then, because of fatty acids or other ionic molecules present in the lipid layer, the pH decrease may result in an expansion of the nonpolar lipid phase.[19] Alternatively, the increase in bicarbonate may result in increased binding of calcium ions[4] with the same result. Our research suggests, however, that very specific lipid changes may be necessary for lipid layer type dry eye to develop. This necessary precondition seems not to be related to gland dropout or tear film lipid layer hydration. We have observed that in our chronic blepharitis patients with dry eye signs (diagnosis KCS), the amount of the polar lipid PE present in meibum is one-third (30%) the level in other patients or in normals (t test $p < 0.01$). The level of sphingomyelin is also lower in patients diagnosed with KCS, but it appears less significant (t test $p < 0.05$). These results suggest that very specific lipid abnormalities are associated with individual disease signs—in this example, lipid layer associated KCS. It should be noted that unlike PC, PE is able to hydrogen bond intermolecularly.[44] Together with the association with KCS, this suggests an important role for PE in the polar lipid phase.

Systemic tetracycline has been found useful in treating some signs associated with chronic blepharitis, especially for MKC. It is convenient to speculate that the only effect of tetracycline is to increase the viscoelastic properties of tear film mucin,[45] which apparently occurs through a direct interaction with the mucin.[46] However, a recent report suggests that there are unusual changes in the MKC meibomian gland lipids that result in a lipid layer that is inhomogeneous and immobile,[47] in contrast to normals (homogeneous and mobile). Thus, tetracyclines have other more direct effects that, in light of the previous discussion, may be more important. For example, tetracycline can bind to certain phospholipids, thus decreasing their polarity.[48] We further suggest that tetracyclines dissolved in nonpolar lipid phases may increase their fluidity. In addition, certain tetracyclines very effectively decrease the level of active oxygen species;[49] some of these are involved in HNE synthesis and the inflammatory process.[39] Finally, our own research has demonstrated that tetracycline inhibits extracellular lipase production at levels that do not inhibit *Staphylococcus* species growth.[50] Besides decreasing free fatty acids, this inhibition could also decrease available polyunsaturated fatty acid precursors necessary for HNE formation.

Related research has suggested that the presence of meibum cholesterol esters may also affect microbial populations. Our premise is that if esterases released free cholesterol from cholesterol esters, it might have an effect on certain microorganisms. We determined that cholesterol stimulated the growth of *Staphylococcus aureus* more than did the other two sterols tested.[51] We also observed that meibum from some patients with *Acanthamoeba* infection contained unusually high levels of both cholesterol esters and polyunsaturated fatty acids,[52] again suggesting an association with the presence of cholesterol esters.

We have not observed a strong association between microflora and disease signs. Our extensive microbial investigations have produced only a few significant associations, namely *Staphylococcus aureus*.[53] We also have noted that higher levels of coagulase-negative *Staphylococcus* species are associated with posterior hordeola, but there is no clear correlation with particular disease groups.

In summary, we have observed a number of chronic blepharitis signs associated with the meibomian gland secretion lipid composition. Some of these are of general importance, occurring in both one group of normals and all blepharitis patients, such as increased wax ester unsaturation in the presence of cholesterol esters. We suggest that this increased unsaturation may be directly related to blepharitis through the production of 4-hydroxy-2-nonenal (HNE) from unsaturated fatty acids under certain necessary conditions. On the other hand, sphingolipid unsaturation is specific to chronic blepharitis patients and phospholipid unsaturation is specific to one group (MKC). Our data suggest that in the disease groups with pasty or very fluid meibum, the degree of unsaturation and possibly unsaturated free fatty acids or mono/diglycerides may be relevant. Furthermore, the level of sphingolipid fatty acid hydroxylation appears important. We also suggest that the presence of cholesterol esters themselves may predispose individuals to microbial disease as well. Whether the effect is from cholesterol alone (e.g., stimulation of microbial growth) or from an effect on the structure and characteristics[54] of the lipid layer has not been determined. In terms of specific signs, we find a close negative correlation between the level of phosphatidylethanolamine and KCS. Also, we find important differences in specific polar lipids and their fatty acids in meibum from individual disease groups. Finally, treatment of disease signs has been difficult. However, new ideas concerning treatments including reasons for the beneficial effect of tetracyclines have been presented. Future treatments may also target meibomian gland Golgi apparatus/endoplasmic reticulum[55] and peroxisomes[56] where many of the meibum lipids are synthesized and modified.

REFERENCES

1. Berman ER. *Biochemistry of the Eye*. New York: Plenum Press; 1991:63–88.
2. Rodriguez-Patino JM, de la Fuente-Feria J, Gomez-Herrera C. Fatty acid films spread on aqueous solutions of compounds containing alcohol radicals: Structure and stability. *J Colloid Interface Sci.* 1992;148:223–230.
3. Rodriguez-Patino JM, Dominguez MR, de la Fuente-Feria J. The effect of sugars on monostearin monolayers. *J Colloid Interface Sci.* 1993;157:343–354.
4. Webb MS, Tilcock CPS, Green BR. Salt-mediated interactions between vesicles of the thylakoid lipid digalactosyldiacyglycerol. *Biochim Biophys Acta.* 1988;938:323–333.
5. Collins D, Conner J, Ting-Beall H-P, Huang L. Proton and divalent cations induce synergistic but mechanistically different destabilizations of pH-sensitive liposomes composed of dioleoyl phosphatidylethanolamine and oleic acid. *Chem Phys Lipids.* 1990;55:339–349.
6. Prydal JI, Artal P, Woon H, Campbell FW. Study of human precorneal tear film thickness and structure using laser interferometry. *Invest Ophthalmol Vis Sci.* 1992;33:2006–2011.
7. Crowther RS, Marriott C, James SL. Cation induced changes in the rheological properties of purified mucus glycoprotein gels. *Biorheology.* 1984;21:253–263.
8. Wiggins PM. Role of water in some biological processes. *Microbiol Rev.* 1990;54:432–449.
9. Hallas J. The association between calcium and acetoacetate, 3-hydroxybutyrate, pyruvate and lactate as determined by potentiometry. *Scand J Clin Invest.* 1987;47:581–585.
10. Chen FS, Maurice DM. The pH in the precorneal tear film and under a contact lens measured with a fluorescent probe. *Exp Eye Res.* 1990;50:251–259.
11. Greiner JV, Glonek T, Korb DR, Booth R, Leahy CD. Phospholipids in meibomian gland secretion. *Ophthalmic Res.* 1996;28:44–49.
12. Hamilton JA. Interactions of triglycerides with phospholipids: Incorporation into the bilayer structure and formation of emulsions. *Biochemistry.* 1989;28:2514–2520.
13. Lee Y-C, Zheng YO, Taraschi TF, Janes N. Hydrophobic alkyl headgroups strongly promote membrane curvature and violate the headgroup volume correlation due to "headgroup" insertion. *Biochemistry.* 1996;35:3677–3684.
14. Martin-Polo M, Voilley A, Bond G, et al. Hydrophobic films and their efficiency against moisture transfer. 2. Influence of the physical state. *J Agric Food Chem.* 1992;40:413–418.
15. Thoma M, Mohwald H. Phospholipid monolayers at hydrocarbon/water interfaces. *J Colloid Interface Sci.* 1994;162:340–349.
16. Brezesinski G, Thoma M, Struth B, Mohwald H. Structural changes of monolayers at the air/water interface contacted with *n*-alkanes. *J Phys Chem.* 1996;100:3126–3130.
17. Smaby JM, Brockman HL. Regulation of cholesteryl oleate and triolein miscibility in monolayers and bilayers. *J Biol Chem.* 1987;262:8206–8212.
18. Minami H, Nylander T, Carlsson A, Larsson K. Incorporation of proteins in sphingomyelin-water gel phases. *Chem Phys Lipids.* 1996;79:65–70.
19. Larsson K, Krog N. Structural properties of the lipid-water gel phase. *Chem Phys Lipids.* 1973;10:177–180.
20. Brauninger GE, Dinesh OS, Kaufman HE. Direct physical demonstration of oily layer on tear film surface. *Am J Ophthalmol.* 1972;73:132–134.
21. Tiffany JM, Bron AJ. Role of tears in maintaining corneal integrity. *Trans Ophthalmol Soc.* U.K. 1978;98:335–338.
22. Chen H-B, Yamabayashi S, Ou B, Tanaka Y, Ohno S, Tsukahara S. Structure and composition of rat precorneal tear film. *Invest Ophthalmol Vis Sci.* 1997;38:381–387.
23. Nicolaides N, Kaitaranta JK, Rawdah TN, et al. Meibomian gland studies: Comparison of steer and human lipids. *Invest Ophthalmol Vis Sci.* 1981;20:522–536.
24. McCulley JP, Sciallis GF. Meibomian keratoconjunctivitis. *Am J Ophthalmol.* 1977;84:788–793.
25. Mathers WD, Daley TE. Tear flow and evaporation in patients with and without dry eye. *Ophthalmology.* 1996;103:664–669.
26. Mathers WD. Ocular evaporation in meibomian gland dysfunction and dry eye. *Ophthalmology.* 1993;100:347–351.
27. Korb DR, Baron DF, Herman JP, et al. ear film lipid layer thickness as a function of blinking. *Cornea.* 1994;13:354–359.
28. Korb DR, Greiner JV, Glonek T, et al. Effect of periocular humidity on the tear film lipid layer. *Cornea.* 1996;15:129–134.

29. McCulley JP, Dougherty JM, Deneau DG. Classification of chronic blepharitis. *Ophthalmology*. 1982;89:1173–1180.
30. Shine WE, McCulley JP. Human meibomian secretion polar lipids associated with chronic blepharitis disease groups. ARVO Abstracts. *Invest Ophthalmol Vis Sci*. 1995;36:S156.
31. Shine WE, McCulley JP. Meibomian gland secretion polar lipids associated with chronic blepharitis. ARVO Abstracts. *Invest Ophthalmol Vis Sci*. 1996;37:S849.
32. Shine WE, McCulley JP. Meibomian gland triglyceride fatty acid differences in chronic blepharitis patients. *Cornea*. 1996;15:340–346.
33. Shine WE, McCulley JP. Role of wax ester fatty alcohols in chronic blepharitis. *Invest Ophthalmol Vis Sci*. 1993;34:3515–3521.
34. Dougherty JM, McCulley JP. Analysis of the free fatty acid component of meibomian secretions in chronic blepharitis. *Invest Ophthalmol Vis Sci*. 1986;27:52–56.
35. Donhowe IG, Fennema O. The effect of relative humidity gradient on water vapor permeance of lipid and lipid-hydrocolloid bilayer films. *J Am Oil Chem Soc*. 1992;69:1081–1087.
36. Shine WE, McCulley JP. The role of cholesterol in chronic blepharitis. *Invest Ophthalmol Vis Sci*. 1991;32:2272–2280.
37. Boggs JM, Koshy KM, Rangaraj G. Effect of fatty acid chain length, fatty acid hydroxylation, and various cations on phase behavior of synthetic cerebroside sulfate. *Chem Phys Lipids*. 1984;36:65–89.
38. Di Mauro C, Cavalli G, Curzio M, et al. Evidences of 4-hydroxynonenal involvement in modulation of phagocyte activities. *Int J Tissue React*. 1995;17:61–72.
39. Esterbauer H, Schaur RJ, Zollner H. Chemistry and biochemistry of 4-hydroxynonenal, malonaldehyde and related aldehydes. *Free Radic Biol Med*. 1991;11:81–128.
40. Gardner HW, Hamberg M. Oxygenation of (3Z)-nonenal to (2E)-4-hydroxy-2-nonenal in the broad bean (*Vicia faba* l). *J Biol Chem*. 1993;268:6971–6977.
41. Baer AN, Costello PB, Green FA. Stereospecificity of the products of the fatty acid oxygenases derived from psoriatic scales. *J Lipid Res*. 1991;32:341–347.
42. Li Y, Ferrante A, Poulos A, Harvey DP. Neutrophil oxygen radical generation: Synergistic responses to tumor necrosis factor and mono/polyunsaturated fatty acids. *J Clin Invest*. 1996;97:1605–1609.
43. Doane MG. Abnormalities of the structure of the superficial lipid layer on the in vivo dry-eye tear film. *Adv Exp Med Biol*. 1994;350:489–493.
44. Brockman H. Dipole potential of lipid membranes. *Chem Phys Lipids*. 1994;73:57–79.
45. Marriott C, Kellaway IW. The effect of tetracyclines on the viscoelastic properties of bronchial mucus. *Biorheology*. 1975;12:391–395.
46. Brown DT, Marriott C, Beeson MF. A rheological study of mucus-antibiotic interactions. In: Chantler EN, Elder JB, Elstein M, eds. *Mucus in Health and Disease. II*. New York: Plenum; 1977:85–88.
47. Kaecher T, Honig D, Mobius D, Welt R. Morphologie dis meibom-lipidfilms. *Ophthalmologe*. 1995;92:12–16.
48. Argast M, Beck CF. Tetracycline diffusion through phospholipid bilayers and binding to phospholipids. *Antimicrob Agents Chemother*. 1984;26:263–265.
49. Miyachi Y, Yoshioka A, Imamura S, Niwa Y. Effect of antibiotics on the generation of reactive oxygen species. *J Invest Dermatol*. 1986;86:449–453.
50. Dougherty JM, McCulley JP, Silvany RE, Meyer DR. The role of tetracycline in chronic blepharitis: Inhibition of lipase production in staphylococci. *Invest Ophthalmol Vis Sci*. 1991;32:2970–2975.
51. Shine WE, Silvany R, McCulley JP. Relation of cholesterol-stimulated *Staphylococcus aureus* growth to chronic blepharitis. *Invest Ophthalmol Vis Sci*. 1993;34:2291–2296.
52. Shine WE, McCulley JP. Meibomian gland secretion sterol ester differences associated with Acanthamoeba keratitis. ARVO Abstracts. *Invest Ophthalmol Vis Sci*. 1993;34:S854.
53. Dougherty JM, McCulley JP. Comparative bacteriology of chronic blepharitis. *Br J Ophthalmol*. 1984;68:524–528.
54. Fennema G, Kester JJ. Resistance of lipid films to transmission of water vapor and oxygen. *Adv Exp Med Biol*. 1991;302:703–719.
55. Sirigu P, Shen R, da Silva PP. Human meibomian glands: The ultrastructure of acinar cells as viewed by thin section and freeze-fracture transmission electron microscopies. *Invest Ophthalmol Vis Sci*. 1992;33:2284–2292.
56. Gorgas K, Volkl A. Peroxisomes in sebaceous glands. IV. Aggregates of tubular peroxisomes in the mouse meibomian gland. *Histochem J*. 1984;16:1079–1098.

46

ANDROGEN REGULATION OF THE MEIBOMIAN GLAND

David A. Sullivan,[1,2] Eduardo M. Rocha,[1,2] M. David Ullman,[5] Kathleen L. Krenzer,[1,2] Jianping Gao,[1,2] Ikuko Toda,[1,2] M. Reza Dana,[1,2,3] Dorothy Bazzinotti,[4] Lilia Aikawa da Silveira,[1,2] and L. Alexandra Wickham[1,2]

[1]Schepens Eye Research Institute
[2]Department of Ophthalmology at Harvard Medical School
[3]Department of Ophthalmology
Brigham & Women's Hospital
[4]New England College of Optometry
Boston, Massachusetts
[5]VA Hospital
Bedford, Massachusetts

1. INTRODUCTION

Androgens are known to control the development, differentiation and lipid production of sebaceous glands throughout the body.[1] Given that the meibomian gland is a large sebaceous gland,[1,2] we hypothesize that androgens may regulate meibomian gland function, enhance the quality and quantity of lipids produced by this tissue and stimulate the formation of the tear film's lipid layer. In addition, we hypothesize that androgen deficiency (e.g. due to menopause, aging, Sjögren's syndrome, anti-androgen medications, inherent insensitivity) may lead to meibomian gland dysfunction and consequent 'evaporative' dry eye.

If our hypotheses are correct, and androgens do regulate meibomian tissue in a manner analogous to that of skin sebaceous glands, then meibomian glands should contain androgen receptor mRNA, androgen receptor protein and 5α-reductase. This enzyme, which has two distinct isozymes (Types 1 and 2) encoded by different genes, converts testosterone and dehydroepiandrosterone (native and sulfated forms) into 5α-dihydrotestosterone, a very potent metabolite.[3] Androgen action in many sebaceous glands is increased by, or dependent upon, the presence of 5α-reductase.[1,3] In addition, if our hypotheses are correct: (1) androgens should be able to control the synthesis and/or secretion of meibomian gland lipids; and (2) androgen deficiency should be linked to meibomian gland dysfunction and functional dry eye.

The purpose of the following studies was to test these hypotheses.

Lacrimal Gland, Tear Film, and Dry Eye Syndromes 2
edited by Sullivan *et al.*, Plenum Press, New York, 1998

2. METHODS

2.1. Animals and Surgical Procedures

Young adult Sprague-Dawley rats and New Zealand White rabbits were obtained from Zivic-Miller Laboratories (Allison Park, PA) and Pine Acre Rabbitry (Brattleboro, VT), respectively, and maintained in constant temperature rooms with light/dark intervals of 12 hours length. When indicated, rabbits were orchiectomized by surgeons at Pine Acre Rabbitry, allowed to recover for 14 days, and then treated for 14 days with systemic placebo, systemic 19-nortestosterone, topical placebo or topical 19-nortestosterone. The systemic exposure was achieved through the implantation of subcutaneous, slow release pellets (Innovative Research of America, Sarasota, FL) into the subscapular region. These pellets, which contained either vehicle or 19-nortestosterone (200 mg), were designed to release physiological amounts of androgen over a period of 21 days. The topical exposure involved the application of a drop of either vehicle or 19-nortestosterone (0.1%, Allergan, Inc., Irvine, CA) to both eyes of rabbits twice a day. All experiments with rats and rabbits adhered to the The Association for Research in Vision and Ophthalmology Resolution on the Use of Animals in Research.

2.2. Clinical Procedures

Clinical studies were conducted with patients taking anti-androgen medications (for prostatic indications), as well as with age-matched and younger controls, as previously described.[4] In brief, after giving their consent, subjects were asked to complete a "Quality of Life" survey (Allergan) and then underwent an ophthalmic exam of the anterior segment, including specific tests of dry eye and a thorough analysis of the lid margins, according to reported protocols.[5,6] Meibomian gland secretions, after expression from the lid margins, were also collected for lipid analyses.[4]

2.3. General Procedures

Meibomian glands, lids and prostatic tissue were obtained following animal sacrifice or during surgical procedures on men and women. LNCaP epithelial cells (human prostate carcinoma) were purchased from American Type Culture Collection (ATCC; Rockville, MD) and cultured according to ATCC recommendations. Samples were processed for molecular biological, immunohistochemical or lipid analytical techniques. Statistical analysis of the data was conducted with the unpaired, two-tailed Student's t test or the Mann Whitney U test.

2.4. Molecular Biological Procedures

Reverse transcription polymerase chain reactions (RT-PCR) were used to identify androgen receptor and Types 1 and 2 5α-reductase mRNAs experimental samples, as previously described.[7,8] These procedures included the: (1) isolation of total RNA from tissues and cells by utilizing TRI-Reagent (Molecular Research Center, Inc., Cincinnati, OH) or modified acid guanidinium-thiocyanate-phenol-chloroform extraction methods;[9] (2) determination of RNA concentration and integrity by spectrophotometry and gel electrophoresis, respectively; (3) preparation of cDNAs from RNA samples (5 µg) by using AMV or MMLV reverse transcriptase, oligo dT and either the First-Strand cDNA Synthesis kit from Invitrogen (San Diego, CA) or the Advantage RT-for-PCR Kit from Clontech Laboratories Inc. (Palo Alto, CA); (4) amplification of cDNAs by employing a Perkin Elmer Cetus GeneAmp PCR System 9600 (Perkin Elmer, Norwalk, CT), Taq DNA polymerase (Gibco/BRL), dNTPs (Perkin-Elmer), PCR buffers and 5' and 3' primers corresponding to rabbit/human androgen receptor, Types 1 and 2 5α-reductase and β-actin cDNA; (5) electrophoresis of PCR products and a 100 bp DNA molecular weight ladder (Gibco) on 1.5% agarose gels and verification of anticipated fragment sizes with ethidium bromide staining; and (6) transfer of amplified cDNA products to GeneScreen nylon membranes (Dupont/NEN, Boston, MA) and hybridization with specific ^{32}P-labeled probes. Primers and oligonucleotide probes were designed as reported,[7,8] and were synthesized by either National Biosciences, Inc. (Plymouth, MN) or Gibco/BRL Custom Primers, Inc. (Palo Alto, CA). cDNA probes for the androgen receptor and β-actin were generously provided by Drs. Elizabeth Wilson (Chapel Hill, NC) and Lan Hu (Boston, MA), respectively. The number of PCR cycles, the amplification conditions and the variety of positive (e.g. RNA from LNCaP cells and prosate) and negative controls have been described in detail.[7,8] The identity of Types 1 and 2 5α-reductase PCR products were confirmed by DNA sequence analysis.[8]

2.5. Histological Procedures

The presence of androgen receptor protein in meibomian glands of rats and humans was assessed by using immunohistochemical techniques, as previously reported.[8] In brief, tissues were embedded in OCT compound, cut into 6 μm sections, fixed sequentially in acetone, then 4% paraformaldehyde (Sigma Chemical Company, St. Louis, MO), and blocked with a 2% 'normal' goat serum (Vector Laboratories, Burlingame, CA). Sections were then exposed to purified rabbit polyclonal antibody to the androgen receptor (gift from Dr. Gail S. Prins, Chicago, IL), incubated serially with avidin D, biotin and biotinylated goat anti-rabbit IgG (Vector), treated with Vectastain Elite ABC reagent (Vector), developed with a solution containing 3-amino-9-ethylcarbazole (Sigma), postfixed in 2% paraformaldehyde and preserved in Crystal Mount (Biomeda, Foster City, CA). The specificity of the staining reaction was confirmed by first antibody competition experiments with defined androgen receptor peptides (from Dr. Prins), as described.[8]

2.6. Lipid Analytical Procedures

The methods for the analysis of lipids and fatty acids from rabbit meibomian gland extracts and human meibomian gland secretions by gas chromatography/mass spectrometry and high pressure liquid chromatography/mass spectrometry have been reported in detail.[4]

3. RESULTS

3.1. Identification of Androgen Receptor mRNA, Androgen Receptor Protein, and 5α-Reductase mRNA in Meibomian Glands

To determine whether the meibomian gland is an androgen target organ, tissues were obtained from rats, rabbits and/or humans and analyzed for the presence of androgen receptor mRNA, androgen receptor protein and 5α-reductase mRNA. Our results showed that: (1) meibomian glands from male and female rabbits and humans contain androgen receptor mRNA;[7] (2) Meibomian glands of rats and humans contain androgen receptor protein within the nuclei of their acinar epithelial cells;[8] and (3) human meibomian glands contain the mRNA for both Types 1 and 2 5α-reductase.[8]

3.2. Androgen Influence on the Lipid Profile within the Rabbit Meibomian Gland

To examine whether androgens influence the lipid profile within the rabbit meibomian gland, orchiectomized animals (n = 5/treatment group) were treated with systemic or topical vehicle or 19-nortestosterone for 14 days, as outlined in the Methods section, and meibomian glands were then processed for lipid evaluation. For comparative purposes, lipid content was also characterized in meibomian tissues of intact, untreated male controls. Our findings demonstrated that androgens exert a significant impact on the lipid pattern of the rabbit meibomian gland.[4] Orchiectomy was associated with a loss of long-chain fatty acids (FA) in the diglyceride and/or triglyceride fraction of meibomian gland lipids, whereas the topical or systemic administration of 19-nortestosterone, but not placebo compounds, restored the FA pattern to that of intact animals.[4] Androgen treatment of orchiectomized rabbits also induced the appearance of specific α-hydroxy FA and aliphatic alcohols, and increased the percentage of long-chain FA, in the total lipid fraction of meibomian glands.[4]

3.3. Effect of Androgen Deficiency on Human Meibomian Gland Function

To determine whether androgen deficiency might be linked to meibomian gland dysfunction, ophthalmic exams were performed, and meibomian gland secretions were collected, from urological patients (n = 15) taking anti-androgen medications, as well as from age-matched (n = 6) and younger (n = 4) controls. Our results, which have been reported in detail,[4] showed that androgen deficiency appears to be associated with meibomian gland dysfunction, an altered lipid profile in meibomian gland secretions, a decreased tear film break up time and functional dry eye. In brief, our clinical and biochemical findings demonstrated that patients, as compared to controls, had: (1) a higher frequency of light sensitivity, painful eyes and blurred vision;[4] (2) a decrease in the tear film break up time, a higher frequency of orifice metaplasia, a poorer quality of meibomian gland secretions and severe meibomian gland disease;[4] and (3) an attenuation in the amounts of cholesterol esters and wax esters, relative to those of cholesterol, as well as a decreased expression of specific molecular species in the diglyceride fraction of meibomian gland secretions.[4]

4. DISCUSSION

Our results show that the meibomian gland is an androgen target organ, that androgens modulate lipid production within this tissue, and that androgen deficiency may possibly cause meibomian gland disease. Given these findings, we suggest that topical androgen administration may potentially serve as a safe and effective therapy for meibomian gland dysfunction and its associated 'evaporative' dry eye syndromes in androgen-deficient patients.

ACKNOWLEDGMENTS

The authors express their appreciation to Natasha Boguslavsky (Boston, MA), Barbara Butler, RN (Boston, MA), Helen DeCosta (Boston, MA), Elizabeth Daher, MD (Boston, MA), Barbara Evans (Waltham, MA), Jim Evans, PhD (Waltham, MA), Laila A. Hanninen (Boston, MA), John LaMothe (Boston, MA), Lorie Lepley (Boston, MA), Van Luu-The (Quebec, Canada), Nancy Moran, RN (Boston, MA), Stephen C. Pflugfelder, MD (Miami, FL), Jerome P. Richie, MD (Boston, MA), Julie Rosado (Boston, MA), Martin Rosado (Boston, MA), Michael M. Rowe, PhD (Irvine, CA), John W. Shore, MD (Boston, MA), Daniel J. Townsend, MD (Boston, MA), Robert Ventura (Bedford, MA), Fu-Shin Yu, PhD (Boston, MA) and Drs. Hu, Prins and Wilson for their provision of human tissue samples, molecular probes, technical help or clinical assistance. This research was supported by research grants from Allergan , NIH (EY05612) and the Massachusetts Lions Research Fund.

REFERENCES

1. Thody AJ, Shuster S. Control and function of sebaceous glands. *Physiol Rev* 1989; 69:383–416.
2. Driver PJ, Lemp MA. Meibomian gland dysfunction. *Surv Ophthalmol* 1996; 40:343–367.
3. Luu-The V, Sugimoto Y, Puy L, Labrie Y, Solache IL, Singh M, Labrie F. Characterization, expression, and immunohistochemical localization of 5α-reductase in human skin. *J Invest Dermatol* 1994; 102:221–226.
4. Sullivan DA, Krenzer KL, Ullman MD, Wickham LA, Toda I, Bazzinotti D, Dana MR. Androgen control of the meibomian gland. Submitted for publication, 1997.
5. Lemp MA. Report of the National Eye Institute/Industry Workshop on clinical trials in dry eyes. *CLAO J* 1995; 21:221–232.
6. Bron AJ, Benjamin L, Snibson GR. Meibomian gland disease. Classification and grading of lid changes. *Eye* 1991; 5:395–411.
7. Wickham LA, Gao J, Toda I, Rocha EM, Sullivan DA. Identification of androgen, estrogen and progesterone receptor mRNAs in rat, rabbit and human ocular tissues. Submitted for publication, 1997.
8. Rocha EM, Wickham LA, Silveria LA, Krenzer KL, Toda I, Sullivan DA. Identification of androgen receptor protein and 5α-reductase mRNA in human ocular tissues. Submitted for publication, 1997.
9. Chomczynski P, Sacchi N. Single-step method of RNA isolation by acid guanidinium thiocyanate-phenol-chloroform extraction. *Anal. Biochem.* 1987; 162:156–159.

DELIVERY OF MEIBOMIAN OIL USING THE CLINICAL MEIBOMETER®

John M. Tiffany[1], Anthony J. Bron,[1] Federico Mossa,[1] and Shabtay Dikstein[2]

[1]Nuffield Laboratory of Ophthalmology
University of Oxford
Oxford, United Kingdom
[2]School of Pharmacy
Hebrew University
Jerusalem, Israel

1. INTRODUCTION

Tear film studies in recent years have emphasized the importance of the oily secretion of the meibomian glands in reducing evaporation from the open eye and in promoting stability of the precorneal film. The thickness of the spread oil film is readily measured, but little information exists on the amount of oil available for the film, or on its quantity or manner of delivery from the glands.

Earlier work[1–3] showed the feasibility of measuring meibomian gland output by a blotting method from the lower eyelid margin, similar to and derived from a commercially available device (the Courage & Khazaka Sebumeter®) for skin surface oil. Difficulties in blotting technique and calibration made this unsuitable for routine clinical use. We have developed a newer version, the Clinical Meibometer®, and used it to examine oil delivery and casual lid margin level. This study reports the casual oil levels found in a small number of normal subjects, and an outline method for estimating the rate of delivery from the glands to the lid margin.

2. METHODS

2.1. Blotting Technique

A preformed loop of meibometry tape is placed in the reading head of the Meibometer to establish the zero setting. The loop is then held in the clip holder with a predetermined length exposed (Fig. 1). The subject's lower lid is gently everted (avoiding any

Lacrimal Gland, Tear Film, and Dry Eye Syndromes 2
edited by Sullivan *et al.*, Plenum Press, New York, 1998

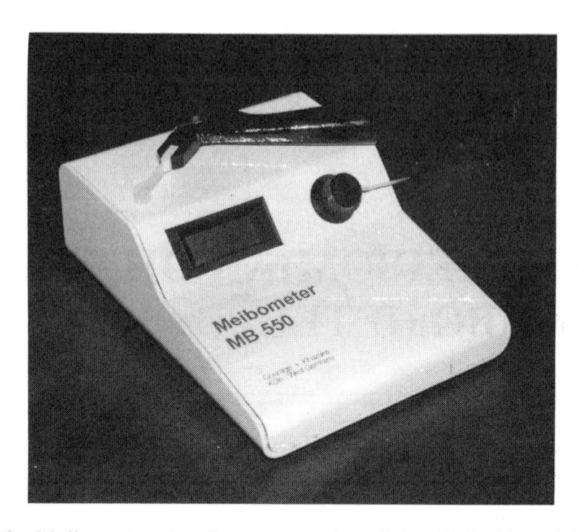

Figure 1. The parts of the Meibometer: the photometer unit and the clip holder with preformed loop of tape.

stretching which might squeeze out oil) with upward-directed gaze, and the loop is pressed onto the central one-third of the lid margin with enough force to cause the straight parts of the loop to buckle (Fig. 2), and held for 10 sec. A crescent-shaped blot of oil is seen across the tape (Fig. 3). The loop is again placed in the reading head of the Meibometer, and its handle rotated from the 1 o'clock to the 5 o'clock position while pressing the "Read" button. In this way the blot is scanned across the reading window of the photosensor. The digital display shows the maximum reading from the densest part of the blot (Fig. 4).

The readout is given in arbitrary instrument units, and is normally in the range 50–300 units. We have carried out a lengthy absolute calibration procedure (details not given here) but are developing calibration tapes that will give a predetermined instrument reading (100 or 200 units). If it is required to convert instrument units into lid-margin oil density in $\mu g/mm^2$, one must know the percentage pick-up of oil by the tape; this has been determined to be close to 30% for the human eyelid, leading to the simple relationship:

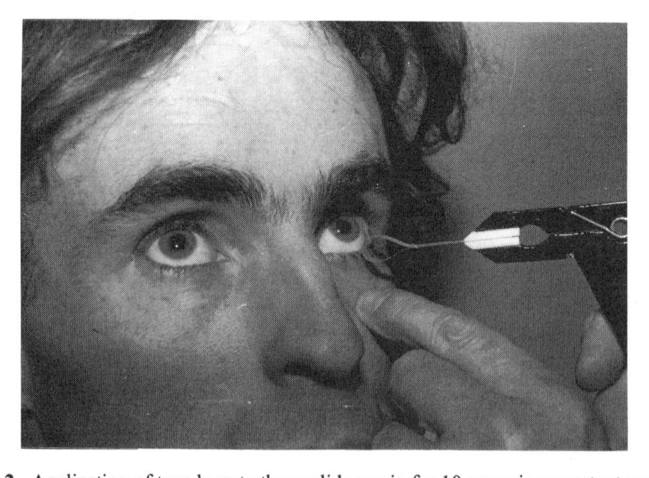

Figure 2. Application of tape loop to the eyelid margin for 10 sec using constant pressure.

Figure 3. Appearance of the oil blot on the tape.

$$\text{Oil loading } (\mu g/mm^2) = 0.0142 \text{ x (instrument units)}$$

2.2. Subjects

Thirty-three normal adult subjects (8 male, 25 female; ages 20–55) were selected from patients and staff in Oxford Eye Hospital. None was suffering from any anterior segment disease, using eyedrops or taking any drug that could influence meibomian gland function, or wearing eye cosmetics or contact lenses.

2.3. Procedures

Casual lid-margin loadings of oil were determined by blotting for all 33 subjects. The oil was removed by wiping the lid margins with a cotton applicator soaked in hexane. Some lids were blotted again to ensure that the post-cleaning reading was zero. Subjects

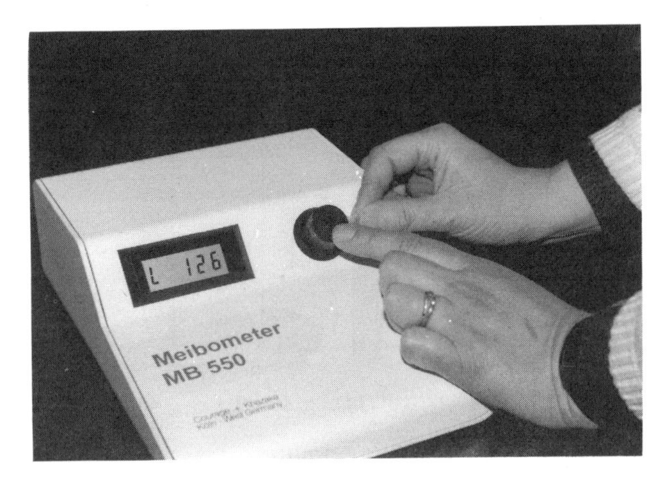

Figure 4. A loop of tape inserted into the photodetector head of the Meibometer for reading. The tape is rotated so as to scan the blot across the photodetector.

then made ten deliberate but light complete blinks in 10 sec, and the casual level was measured again.

For ten subjects (5 male, 5 female), recovery with normal unforced blinking was investigated. Topical anesthetic (1 drop 0.4% benoxinate) was applied so that no abnormal pattern of blinking would be induced by the cleaning procedure. Then after measuring the casual level, the lid margins were cleaned. A second blot was taken after 15 min of recovery at normal rates of unforced blinking.

3. RESULTS

3.1. Variation of Casual Level with Age

Fig. 5 shows the distribution of values of casual level (in instrument units) for male and female subjects grouped by age decades. Although this is a small group compared to the 421 subjects reported by Chew et al.,[3] and does not cover as wide an age range (here, 20–55; Chew et al., 1–94), the main characteristics are similar. The mean level rises slightly with age in women, but the number of male subjects is too low to draw such a conclusion; the difference between the 20–29 and the 30–39 groups for men is not significant.

These results indicate that the clinical version of the Meibometer agrees in general terms with the original experimental version described by Chew et al.[2]

3.2. Recovery after Cleaning

Fig. 6 shows the results, expressed as a percentage of the original casual level, of the two recovery methods:

10 forced blinks: The mean reading ± SEM after 10 blinks in 10 sec was 191 ± 9.3 units, relative to a mean casual level before lid cleaning of 250 ± 7.7 units. This corresponds to a 10-blink recovery of 76.4 ± 3.7% of the original casual level.

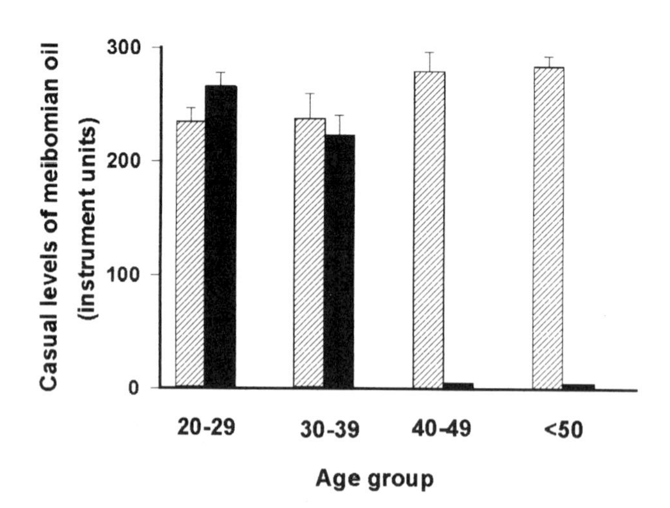

Figure 5. Casual level of lid-margin oil as a function of age and sex for 33 normal subjects. Shaded bars, women; black bars, men.

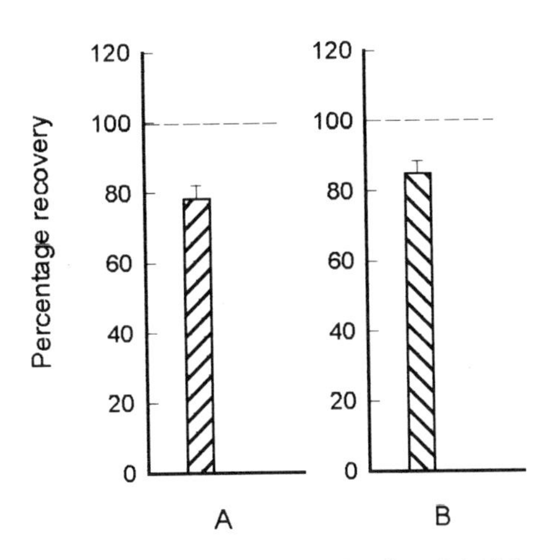

Figure 6. Recovery as percentage of original casual level resulting from (A) 10 forced blinks in 10 sec (n = 33) and (B) 15 min of unforced blinking (n = 10).

15-min unforced blinking: For 10 subjects only, the mean reading ± SEM after 15 min was 248 ± 12.6 units relative to 276 ± 12.9 units before cleaning, corresponding to a recovery of 89.9 ± 4.6% in 15 min.

4. DISCUSSION

Using the instrumental calibration and a value for the percentage pick-up of oil by the tape (ideally re-determined by the user for local clinical populations), the original loading of oil on the lid margin at the site of blotting can be calculated. For most purposes, however, it is quite satisfactory to report results in instrumental units. The range of casual levels recorded for the 33 subjects used here was 165–337 units.

The percentage pick-up, P%, or fractional pick-up, p, is found by blotting the lower lid a first time (reading A), then with a fresh tape immediately blotting the same area again, without any intervening blink (reading B). If the first blot picks up p, then $(1 - p)$ is left, and the second blot picks up $p(1 - p)$. Hence, B/A = $(1 - p)$, or p = (1 - A/B), and $P = 100p$.

The results of casual levels in a selected group of normal subjects are consistent with our earlier studies,[3] although in such a small group we cannot fully distinguish the variations due to age and sex. However, the ease of operation of the instrument would make collection of such data much quicker than before. At present, we are also accumulating readings on dry eye patients with the new instrument, and correlating these with other measurements such as the associated thickness of the oil layer on the preocular tear film.

Our understanding of the operation of the meibomian gland is that the central duct acts as an elastic reservoir which under conditions of quiet unforced blinking is almost completely full.[4] Repeated forced blinks, which inevitably generate higher pressures within the lid than the light, unforced blinks in the second method, will deplete this reservoir, so the apparent recovery of casual lid-margin oil after ten such blinks is probably misleading. In the 15-min study of recovery, we used anesthetic to minimize extra blinking as a result of the cleaning procedure, and consider that this represents the near-equilibrium state of the gland where delivery (expulsion) is matched by synthesis of oil. Other recovery times are being studied.

The Meibometer® in its revised form is simple to use in the clinical environment and gives a reliable and repeatable reading in instrumental units of the amount of oil picked up under standard conditions from the lower lid margin. We are developing a system of strips of preformed loops, so that for each patient a fresh loop will be torn from the strip and mounted in the clip holder. The total time for zeroing the instrument, blotting the lid, and reading the blot can be as little as 2 min. We foresee considerable use of the technique in assessment of meibomian gland dysfunction, as another indicator for the diagnosis of dry eye, and in environmental studies similar to those of Franck in the workplace in relation to the "office eye syndrome."[5,6]

ACKNOWLEDGMENTS

We thank Elaine Grande and Beatrix Nagyova for help in preparing the poster.

REFERENCES

1. Chew CKS, Tiffany JM, Dikstein S, Bron AJ. Lipid levels on the lid margins of patients with meibomian gland dysfunction. *Invest Ophthalmol Vis Sci.* 1992;33:950.
2. Chew CKS, Hykin PG, Jansweijer C, Dikstein S, Tiffany JM, Bron AJ. The casual level of meibomian lipids in humans. *Curr Eye Res.* 1993;12:255–259.
3. Chew CKS, Jansweijer C, Tiffany JM, Dikstein S, Bron AJ. An instrument for quantifying meibomian lipid on the lid margin: the Meibometer. *Curr Eye Res.* 1993;12:247–254.
4. Tiffany JM, Chew CKS, Bron AJ. Delivery of meibomian oils to the eyelid margin. *Exp Eye Res.* 1994;59(Suppl. 1):S111.
5. Franck C. Fatty layer of the precorneal film in the "office eye syndrome." *Acta Ophthalmol.* 1991;69:737–743.
6. Franck C, Bach E, Skov P. Prevalence of objective eye manifestations in people working in office buildings with different prevalences of the sick building syndrome compared with the general population. *Int Arch Occup Environ Health.* 1993;65:65–69.

48

VOLUME OF THE HUMAN AND RABBIT MEIBOMIAN GLAND SYSTEM

Jack V. Greiner,[1,2] Thomas Glonek,[3] Donald R. Korb,[4] Amy C. Whalen,[4] Eric Hebert,[4] Stacey L. Hearn,[1] Jan E. Esway,[4] and Charles D. Leahy[1]

[1]Schepens Eye Research Institute, and
[2]Department of Ophthalmology
Harvard Medical School
Boston, Massachusetts
[3]MR Laboratory
Midwestern University
Chicago, Illinois
[4]100 Boylston Street
Boston, Massachusetts

1. INTRODUCTION

Meibomian gland secretion (meibom) in human[1–5] and rabbit[6,7] comprises triglycerides, neutral lipids, and polar lipids. Phospholipids have been further identified and quantified.[8] Collectively, these lipids are believed to form the lipid layer of the tear film, which can be observed in the interpalpebral fissure by interferometry.[9–11]

The amount of lipid on the tear film surface appears related to the number of meibomian glands[12–14] and their functioning.[9] With meibomian gland dysfunction, a thinner lipid layer is observed.[9] Lipid layer thickness can be increased with increased or improved blinking[10] or following treatment for meibomian gland dysfunction.[9]

Based on measurements of interference patterns seen in specular reflection, it appears that the thickness of the human tear film lipid layer is less than that of the rabbit.[15] The purpose of the present study was to determine the volume, and thus the capacity, of the meibomian gland system in humans and rabbits, and to relate these findings to other parameters associated with the stability and thickness of the tear film lipid layer.

Lacrimal Gland, Tear Film, and Dry Eye Syndromes 2
edited by Sullivan *et al.*, Plenum Press, New York, 1998

2. METHODS

2.1. Species

For the human subjects, the potential age differences[16] were controlled by examining eyes (n = 40) from young adults (n = 20; age 20–35 years). Eyes (n = 40) of New Zealand albino rabbits (n = 20; 3–4 kg) were used.

2.2. Determination of the Area of the Interpalpebral Fissure

In both species the area of the interpalpebral fissure was determined by photographing eyes in the normal anatomical position. Each photograph included a millimeter ruler. Projecting 35 mm transparencies, tracings of the interpalpebral fissures were made and then cut out with hands covered in latex powderless gloves to avoid contamination of the tracing paper with cutaneous oils. Additionally, using the millimeter ruler in the photograph, cutouts were made measuring the equivalent of 1 mm square. The cutouts of the interpalpebral fissure and of the squares were weighed. A size-weight proportion was made to determine the average exposed ocular surface area. This calculation yields an approximation of the interpalpebral fissure; the exact area cannot be calculated using this method because it does not take into consideration the convexity of the ocular surface.

2.3. Meibomian Gland Counts

Meibomian glands of humans were counted in both the upper and lower eyelids using a biomicroscope. In addition, the upper and lower eyelids were everted and positioned perpendicular to a camera equipped with a 50 mm lens, thus permitting 1:1 image size. Photographs were made of the tarsal plate; included in each photograph was a millimeter ruler. Photographic transparencies (35 mm) were projected to verify glandular counts.

Following euthanasia of rabbits, the upper and lower eyelids were removed completely. With the eyelids positioned with the conjunctival surface upward, meibomian glands were counted using an operating microscope. To verify meibomian gland counts, eyelids were positioned and photographed as described above.

2.4. Determination of Meibomian Gland Volume

The volume of human and rabbit meibomian glands was determined using photographic transparencies of the tarsal plates that permitted observation of the meibomian gland contents. The millimeter ruler adjacent to the tarsal plate in each photograph permitted determination of the length and width of the meibomian gland. Assuming the meibomian glands to have a cylindrical morphology, glandular volume could be calculated by assuming the width of each gland to represent the diameter of the cylinder. The eyelids were arbitrarily divided equally into nasal, central, and temporal zones. The average volume of at least three representative glands from each of the three lid zones was calculated for each sample. In humans, where gland length was at times indistinct, these measurements were recorded as approximations based on length of adjacent glands, as adjacent glands were nearly always similar throughout all eyelid zones. The most notable changes in length occurred in the far temporal zone. The total glandular volume was estimated by determining the average gland volume for each of the three zones and then multiplying by the number of glands in each zone. Total meibomian gland volume of each tarsal plate was then determined by adding the volumes for all three zones.

A second method to determine meibomian gland volume in rabbit eyelids utilized a Zeiss operating microscope to examine eyelids dissected *en toto*, which permitted measurement of each individual gland. The width of each gland was measured 1–2 mm deep to the glandular orifice. The length of each gland was measured and included only that portion of the gland containing visible secretion. Again, assuming a cylindrical morphology, glandular volume was calculated. This same method was also utilized in humans using a Haag-Streit slit lamp to examine eyelids in vivo. By using this method, the total glandular volume in each eyelid was calculated by adding together the individual meibomian gland volume calculations for each section of the tarsal plate.

3. RESULTS

The area of the interpalpebral fissure was 1.80 cm^2 in humans and 1.44 cm^2 in rabbits. The number of meibomian glands in the upper lids of human and rabbit was similar; likewise, the lower lids of these species were comparable in their meibomian gland counts (Table 1). Glandular length in the upper and lower eyelids was determined for nasal, central, and temporal zones; these data, along with the average length of all glands in each lid, are presented in Table 1, as are the corresponding gland widths. The average length of the human meibomian glands in the upper eyelid is more than double that of the rabbit upper eyelid. In contrast, the average length of the human meibomian glands in the lower eyelid is similar to that of the rabbit lower eyelid. Corresponding widths of the upper eye-

Table 1. Meibomian gland data

Parameter	Area of measurement	Human	Rabbit
Number of glands[a]	Total: Upper lid	31.60 ± 4.33	32.30 ± 3.43
	Total: Lower lid	25.28 ± 3.10	27.15 ± 2.35
Length of Glands[b] (mm)	Total: Upper lid	4.81 ± 1.18	2.09 ± 0.54
	Upper lid: Nasal	3.49 ± 1.31	2.14 ± 0.63
	Upper lid: Central	5.44 ± 0.59	2.37 ± 0.37
	Upper lid: Temporal	3.70 ± 0.82	1.77 ± 0.41
	Total Lower lid	1.92 ± 0.41	1.78 ± 0.39
	Lower lid: Nasal	1.85 ± 0.44	1.90 ± 0.40
	Lower lid: Central	2.06 ± 0.33	1.68 ± 0.30
	Lower lid: Temporal	1.71 ± 0.43	1.77 ± 0.43
Width of glands[c] (mm)	Total: Upper lid	0.46 ± 0.12	0.43 ± 0.14
	Upper lid: Nasal	0.49 ± 0.08	0.46 ± 0.16
	Upper lid: Central	0.44 ± 0.13	0.43 ± 0.12
	Upper lid: Temporal	0.52 ± 0.12	0.40 ± 0.14
	Total: Lower lid	0.58 ± 0.12	0.45 ± 0.14
	Lower lid: Nasal	0.61 ± 0.10	0.45 ± 0.15
	Lower lid: Central	0.58 ± 0.11	0.49 ± 0.14
	Lower lid: Temporal	0.56 ± 0.14	0.42 ± 0.12

[a]Number of glands = number of meibomian gland ostia observed along eyelid margin.
[b]Length of glands = portion of gland observed to contain secretion.
[c]Width of glands = width of widest portion of gland as measured 1-2 mm beneath surface of eyelid margin.

Table 2. Volume of meibomian glands

Area of measurement	Human (μl)	Rabbit (μl)
Total: Upper lid	26.13 ± 6.31	10.52 ± 2.35
Upper lid: Nasal	0.68 ± 0.27	0.40 ± 0.31
Upper lid: Central	0.96 ± 0.60	0.36 ± 0.20
Upper lid: Temporal	0.85 ± 0.48	0.25 ± 0.18
Total: Lower lid	13.40 ± 2.21	8.32 ± 3.23
Lower lid: Nasal	0.54 ± 0.28	0.38 ± 0.30
Lower lid: Central	0.59 ± 0.25	0.35 ± 0.20
Lower lid: Temporal	0.38 ± 0.18	0.27 ± 0.16

lid are essentially the same; the widths of the human meibomian glands in the lower lid, however, are somewhat larger than in the rabbit. In both the human and the rabbit upper eyelid, glands in central regions were longer than those in temporal or nasal regions. Regional length variation was not observed in the lower eyelid. Gland width appeared fairly constant in both the upper and lower eyelids.

The total meibomian gland volume in the human upper eyelid was more than twice that calculated for the rabbit (Table 2). The total meibomian gland volume in the human lower eyelid was approximately 1.5 times that of the rabbit. The total meibomian gland volume (both lids) in the human (39.53 mm^3) is twice that of the rabbit (18.84 mm^3). As expected, there is a much greater meibomian gland volume in the human upper eyelid than in the lower eyelid. In the rabbit, however, the meibomian gland volume in the upper eyelid is approximately the same as in the lower eyelid.

4. CONCLUSIONS

The ratio of the total glandular volume to the normally exposed ocular surface (interpalpebral fissure) is 1.7 times greater in the human than in the rabbit. Assuming the meibomian gland origin of tear film lipids, it might be expected that this greater ratio would result in a significantly thicker tear film lipid layer. Preliminary data do not support this expectation and suggests that the rabbit tear film lipid layer thickness exceeds that of the human.[15] The apparent discrepancy revealed in this study may be a reflection of species differences in secretion rate, physical and chemical properties, and source[17–21] of lipids, which may result in a thicker lipid layer in the rabbit. Although little is known about the secretion rate or physical properties of the meibomian gland lipids, differences in the chemical composition of the lipid in human[1–5] and rabbit[6,7,22] have been reported.

We report a range of interspecies differences in the meibomian glands. For example, the large difference in meibomian gland volume between the upper and lower eyelids in human is in contrast to the similar volume in the upper and lower eyelids of the rabbit. Species differences in meibomian gland volume, lipid chemistry, tear film lipid layer thickness, source(s) of lipids, and blink rate urge caution when extrapolating rabbit tear film data to the human.

REFERENCES

1. Tiffany JM. Individual variations in human meibomian lipid composition. *Exp Eye Res*. 1978;27:289–300.

2. Nicolaides N, Kataranta JKJ, Rawdah TN, Macy JI, Boswell FM III, Smith RE. Meibomian gland studies: Composition of steer and human lipids. *Invest Ophthalmol Vis Sci.* 1981;20:522–536.
3. Nicolaides N, Ruth EC. Unusual fatty acids in the lipids of steer and human meibomian gland excreta. *Curr Eye Res.* 1982;2:93–98.
4. Nicolaides N, Santos EC, Papadakis K. Double-bond patterns of fatty acids and alcohols in steer and human meibomian gland lipids. *Lipids* 1984;19:264–277.
5. Nicolaides N, Santos EC. The di- and triesters of the lipids of steer and human meibomian glands. *Lipids.* 1985;20:454–467.
6. Tiffany JM. The meibomian lipids of the rabbit. I. Overall composition. *Exp Eye Res.* 1979;29:195–202.
7. Tiffany JM, Marsden RG. The meibomian lipids of the rabbit. II. Detailed composition of the principal esters. *Exp Eye Res.* 1982;34:601–608.
8. Greiner JV, Glonek T, Korb DR, Booth R, Leahy CD. Phospholipid profile of meibomian gland secretion: A phosphorus-31 nuclear magnetic resonance spectroscopic study. *Ophthalmic Res.* 1996;28:44–49.
9. Korb DR, Greiner JV. Increase in tear film lipid layer thickness following treatment of meibomian gland dysfunction. *Adv Exp Med Biol.* 1994;350:293–298.
10. Korb DR, Baron DF, Herman JP, et al. Tear film lipid layer thickness as a function of blinking. *Cornea.* 1994;13:354–359.
11. Korb DR, Greiner JV, Glonek T, Esbah R, Finnemore VM, Whalen AC. Effect of periocular humidity on the tear film lipid layer. *Cornea.* 1996;15:129–134.
12. Ekins MB, Waring GO. Absent meibomian glands and decreased corneal sensation in hydrotic ectodermal dysplasia. *J Pediatr Ophthalmol Strabismus.* 1981;18:44–47.
13. Mondino BJ, Bath PE, Foos BV. Absent meibomian glands in the ectodactyl, ectodermal dysplasia and cleft lip-palate syndrome. *Am J Ophthalmol.* 1984; 97:496–500.
14. Bron AJ, Mengher GO. Congenital deficiency of meibomian glands. *Br J Ophthalmol.* 1987;71:312–314.
15. Korb DR, Greiner JV, Glonek T, et al. Human and rabbit lipid layer and interference pattern observations. This volume.
16. Hykin PG, Bron AJ. Age-related morphological changes in lid margin and meibomian gland anatomy. *Cornea.* 1992;11:334–342.
17. Prince JH, ed. *The Rabbit in Eye Research.* Springfield, IL: Charles C. Thomas;1964, p 50.
18. Cogan DG, Fink R, Donaldson DD. X-ray irradiation of orbital glands of rabbit. *Radiology.* 1955;64:731–737.
19. Paule WJ. *The Comparative Histochemistry of the Harderian Gland.* Ohio State University, 1957.
20. Schneir ES, Hayes ER. The histochemistry of the Harderian gland of the rabbit. *Natl Cancer Inst J.* 1951;12:257.
21. Walter A. Ueber die HautderÅsen mit Lipoidsekretion bei Nagern. Bietr. path. Anat. Path. 1924;73:142–167.
22. Greiner JV, Glonek T, Korb DR, Leahy CD. Meibomian gland phospholipids. *Curr Eye Res.* 1996;15:371–375.

EFFECT OF MEIBOMIAN GLAND OCCLUSION ON TEAR FILM LIPID LAYER THICKNESS

Jack V. Greiner,[1,2] Thomas Glonek,[3] Donald R. Korb,[4] Stacey L. Hearn,[1] Amy C. Whalen,[4] Jan E. Esway,[4] and Charles D. Leahy[1]

[1]Schepens Eye Research Institute and
[2]Department of Ophthalmology
Harvard Medical School
Boston, Massachusetts
[3]MR Laboratory
Midwestern University
Chicago, Illinois
[4]100 Boylston Street
Boston, Massachusetts

1. INTRODUCTION

The function of the meibomian glands is to provide the oil that forms a layer covering the aqueous component of the tear film, thereby preventing excess water evaporation. Should these glands become compromised, an adequate lipid layer may not form over the aqueous portion of the tear film, resulting in increased evaporation and the development of a dry eye condition. This scenario can be experimentally induced by occluding the meibomian gland orifices in the rabbit via cauterization, thereby denying the meibomian oils access to the tear film. Such an experimental paradigm has permitted studies on the role of the oil layer in tear film osmolarity[1] and evaporation.[2,3] In the latter studies, cauterization resulted in rapid evaporation of the tear film and the conclusion that the meibomian glands were solely responsible for generating the oily layer of the tear film.

The superficial oily layer of the tear film can be observed using specular reflection. This can be most easily accomplished using the slit lamp, although such a technique permits visualization of only a relatively small region of the tear film. Using a custom-built instrument, however, it is possible to observe the oily layer of the tear film over a large corneal surface area. This instrument has allowed studies of the effects of blinking,[4] humidity,[5] and meibomian gland expression[6] on the thickness of the human tear film lipid layer. The present study applies this technique for observing the tear film to the condition resulting from meibomian gland occlusion in the rabbit.

Lacrimal Gland, Tear Film, and Dry Eye Syndromes 2
edited by Sullivan *et al.*, Plenum Press, New York, 1998

2. METHODS

New Zealand white rabbits ($n = 6$; 3–4 kg), of either sex, were administered general anesthesia using intramuscular ketamine (100 mg/kg) and xylazine (10 mg/kg). The meibomian gland orifices in one eye were individually closed by applying an ACCU Temp disposable cautery. Heat was applied to the meibomian gland orifice for ~1 sec; closure was confirmed by slit lamp observation at high magnification. The contralateral eye served as a paired control. Tear film lipid layer thickness (LLT) was measured at baseline prior to cauterization and again after a recovery period, using the custom-designed tear film apparatus and methods previously described.[4] Re-cauterization was performed, and the tear film LLT measured again after a recovery period.

After nearly 4 weeks, cyanoacrylate glue was applied to the eyelid margin of experimental eyes to ensure the absolute closure of the meibomian gland orifices, and the tear film LLT was remeasured. Animals were sacrificed and eyelids treated *in situ* with 2.5% glutaraldehyde fixative solution. Eyelids were dissected and samples were dehydrated in graded ethyl alcohols, embedded in plastic, sectioned, and stained for light microscopic analysis to confirm meibomian gland obstruction.

3. RESULTS

Baseline LLT in rabbits prior to closure of the meibomian gland orifices averaged >180 nm in both eyes. The rabbit tear film and lipid layer were static and did not move unless blinking or significant movement of the nictitating membrane occurred. The rabbit color interference patterns tended to exhibit patchy, mottled structures rather than horizontal wave patterns. After cauterization, eschar was present at the cautery site for at least 3 days postoperatively. LLT measurements were recorded at postoperative day 5. LLT in control eyes was unchanged; measurements in treated eyes averaged Å180 nm. Since measurements in control and treated eyes were not markedly different, re-cauterization was performed; results were unchanged. Subsequent application of cyanoacrylate glue to the eyelid margins of experimental eyes did not affect LLT measurements for the duration of the experiment (Table 1).

4. CONCLUSIONS

LLT of the rabbit tear film was only minimally reduced after cauterization and subsequent gluing of the meibomian gland orifices. In contrast, obstruction and inspissation of the human meibomian glands result in reduction of LLT.[6]

There are a number of possible explanations for the differences observed between rabbit and human systems when the meibomian glands are compromised. First, ocular surface structures in the rabbit may serve as alternative and/or additional sources of oil for the lipid layer of the tear film. A possible candidate for this role may be Harder's gland in the rabbit, which has been shown to contain neutral

fats.[7–11] Another potential source of oil in both human and rabbit species is the oil-producing sebaceous glands in the caruncle. Although the contribution of these secretions to the tear film is relatively minor, possible quantitative and qualitative differences between human and rabbit sebaceous secretions may result in species differences in lipid layer formation and stability. The sebaceous oils of the caruncle have not been charac-

Table 1. Tear film lipid layer thickness after occlusion of rabbit meibomian gland orifices

Date	Procedure	Control Eye (n=6)	Treated Eye (n=6)
Day 1	Baseline	>180 nm	≥180 nm
Day 3	Initial Cauterization	Unremarkable	Eschar and excess mucus
Day 8	TFA analysis	>180 nm	≈180 nm
Day 12	Re-cauterization	Unremarkable	Eschar and excess mucus
Day 14	TFA analysis	>180 nm	90-180 nm*
Day 27	Cyanoacrylate gluing	Unremarkable	Excess mucus
Day 29	TFA analysis	>180 nm	≈180 nm
Day 58	TFA analysis	>180 nm	≈180 nm

TFA analysis, tear film apparatus analysis.[4]
*4 rabbits with lipid layer thickness (LLT) of 180 nm; 2 rabbits with LLT of 90 nm.

terized, but species differences between human[12–16] and rabbit[17–19] meibomian lipids have been reported.

It is also important to consider whether the cauterization procedure itself disrupts the mucocutaneous junction, which may normally serve to anchor the lipid layer of the tear film.[20] The disruption of this normal lid boundary may result in a loss of integrity of the tear film lipid layer membrane[17] in the periphery of the interpalpebral fissure. Such a situation may permit the sebaceous glands associated with the surrounding lid hairs to spread sebaceous oils forward onto the tear film surface. The rabbit lid is densely populated with hair follicles that may contribute to sebaceous oil delivery, in contrast to the human lid which is not as well endowed with hair follicles in this region.

Differences in rabbit and human responses to meibomian gland obstruction may also reflect species differences, qualitative or quantitative, in the nature of the aqueous layer that underlies the lipid layer of the tear film. For example, the rabbit lacrimal gland may contribute surfactant-type materials (eg, high concentrations of glycolipid and/or glycoproteins) to the tear film that mimic the water-conserving effect normally attributed to the lipid layer in humans. This issue has not been addressed experimentally.

The results obtained with the rabbit model in this study were not anticipated and suggest that the tear film lipid layer of the rabbit is inherently different from its human counterpart. This hypothesis is further supported by observations of normal rabbit tear films during the course of this investigation and their comparison with previous studies on the human tear film.[4–6] The color interference patterns of the human are dynamic and wavelike, exhibiting movement and maximum color intensity immediately following a blink.[4] Color subsequently fades during the interblink period and requires further blink motion to renew its intensity. The color interference patterns vary from predominant to subtle in both color intensity and area occupied. In marked contrast, the normal rabbit tear film is static and demonstrates a homogeneous or a patchy and mottled appearance, rather than horizontal wave patterns.[20] Additionally, given the significantly longer interblink period in the rabbit (one blink per 5 min) vs. the human (one blink per 7 sec),[20] the rabbit tear film and lipid layer appear more static in nature, since less overall lid movement results in less movement of the lipid layer itself. As such, with occlusion there was increased appearance of color interference patterns although the tear film lipid layer appeared to remain thick at ~180 nm (Table 1).

In summary, although there is a great deal of support for the notion that the meibomian glands contribute significantly to the tear film lipid layer in humans,[2,6,21–25] the situation in rabbits is less clear, based on the results of the present study. Alternative sources of oil in the rabbit may play a role in that species' tear film structure. The present results, and the fact that the blinking mechanism (rate) in rabbit is very different from human,[20] underscore the point that the rabbit is a poorly understood model for studying human tear film structure and dynamics.

REFERENCES

1. Gilbard JP, Rossi SR, Heyda KG. Tear film and ocular surface changes after closure of the meibomian gland orifices in the rabbit. *Ophthalmology* 1989;96:1180–1186.
2. Mishima S, Maurice DM. The oily layer of the tear film and evaporation from the corneal surface. *Exp Eye Res.* 1961;1:39–45.
3. Mishima S, Maurice DM. The effect of normal evaporation on the eye. *Exp Eye Res.* 1961;1:46–52.
4. Korb DR, Baron DF, Herman JP, et al. Tear film lipid layer thickness as a function of blinking. *Cornea* 1994;13:354–359.
5. Korb DR, Greiner JV, Glonek T, Esbah R, Finnemore VM, Whalen AC. Effect of periocular humidity on the tear film lipid layer. *Cornea.* 1996;15:129–134.
6. Korb DR, Greiner JV. Increase in tear film lipid layer thickness following treatment of meibomian gland dysfunction. *Adv Exp Med Biol.* 1994;350:293–298.
7. Prince JH, ed. *The Rabbit in Eye Research.* Springfield, IL: Charles C Thomas; 1964, p 50.
8. Cogan DG, Fink R, Donaldson DD. X-ray irradiation of orbital glands of rabbit. *Radiology.* 1955;64:731–737.
9. Paule WJ. *The Comparative Histochemistry of the Harderian Gland.* Columbus, Ohio State University, 1957.
10. Schneir ES, Hayes ER. The histochemistry of the Harderian gland of the rabbit. *Natl Cancer Inst J.* 1951;12:257.
11. Walter A. Ueber die HautderÅsen mit Lipoidsekretion bei Nagern. Bietr. path. Anat. Path.1924;73:142–167.
12. Nicolaides N, Kataranta JKJ, Rawdah TN, Macy JI, Boswell FM III, Smith RE. Meibomian gland studies: Composition of steer and human lipids. *Invest Ophthalmol Vis Sci.* 1981;20:522–536.
13. Nicolaides N, Ruth EC. Unusual fatty acids in the lipids of steer and human meibomian gland excreta. *Curr Eye Res.* 1982;2:93–98.
14. Nicolaides N, Santos EC, Papadakis K. Double-bond patterns of fatty acids and alcohols in steer and human meibomian gland lipids. *Lipids* 1984;19:264–277.
15. Nicolaides N, Santos EC. The di- and triesters of the lipids of steer and human meibomian glands. *Lipids.* 1985;20:454–467.
16. Tiffany JM. Individual variations in human meibomian lipid composition. *Exp Eye Res.* 1978;27:289–300.
17. Greiner JV, Glonek T, Korb DR, Booth R, Leahy CD. Phospholipids in meibomian gland secretion. *Ophthalmic Res.* 1996;28:44–49.
18. Tiffany JM. The meibomian lipids of the rabbit. I. Overall composition. *Exp Eye Res.* 1979;29:195–202.
19. Tiffany JM, Marsden RG. The meibomian lipids of the rabbit. II. Detailed composition of the principal esters. *Exp Eye Res.* 1982;34:601–608.
20. Korb DR, Greiner JV, Glonek T, et al. Human and rabbit lipid layer and interference pattern observations. This volume.
21. Wolff E. Mucocutaneous junction of lid margin and distribution of tear fluid. *Trans Ophthalmol Soc UK.* 1946;66:291–308.
22. Bron AJ, Mengher GO. Congenital deficiency of meibomian glands. *Br J Ophthalmol.* 1987;71:312–314.
23. Brown SI, Dervichian DG. The oils of the meibomian gland. *Arch Ophthalmol.* 1969;82:537–549.
24. Holly FJ. Formation and stability of the tear film. *Int Ophthalmol Clin.* 1973;13(1):73–96.
25. Holly FJ, Lemp MA. Tear physiology and dry eyes. *Surv Ophthalmol.* 1977;22:69–87.

50

MEIBOMIAN GLAND LIPIDS, EVAPORATION, AND TEAR FILM STABILITY

William D. Mathers and James A. Lane

University of Iowa Hospitals and Clinics
Department of Ophthalmology
Cornea Service, Iowa City, Iowa

1. INTRODUCTION

The tear film is a relatively complex structure of lipids, proteins, and mucins riding on the hydrophobic surface of the epithelium. It is a dynamic structure and not very stable even in normals. The ever-changing interaction between these three basic elements requires reconstitution every few seconds to maintain the health of the epithelium. To understand the tear film as a dynamic process, we need first to consider individually a number of elements that are its constituents.

The largest component of the tear film is the aqueous product of the lacrimal glands, the output of which is usually referred to as the tear production. We cannot actually measure production directly without stimulating the eye in some way. Schirmer's test without anesthetic measures the absorption by filter paper of the majority of the lacrimal gland production, and the volume can be directly measured from the length of the wetted strip. This test is, however, actually measuring the reflex capacity of the system following the variable degree of stimulation induced by the paper strip on the conjunctiva. Although this is usually considered a very important test, it is also widely recognized as being at least moderately imprecise.[1–3]

A more precise measure of tear production is obtained by using a fluorescein dilution method. This test measures the tear flow into the ocular surface vs. loss, assuming a steady-state tear volume and a steady-state tear production.[4–6]

We measured the steady-state tear production in an unstimulated eye using a 0.5 μl drop of 0.5% fluorescein applied to the bulbar conjunctiva with an accurately calibrated plastic micropipette.[7] Great care was taken to avoid stimulating the bulbar conjunctiva with the fluorescein application. This test was performed in a relatively low light and a wind-free environment. The level of fluorescein on the corneal surface was measured initially at 1 min, then every 3 min for 19 min using a Fluoromaster fluorophotometer (Palo

Alto, CA). From the measurement of the tear volume, one can regress the log of the concentration to time zero, and through a simple formula calculate the tear volume and the tear flow in microliters per minute.[5]

In some patients, reflex tearing produced a more rapid decrease in concentration immediately following instillation, usually lasting 4–5 min. This was easily detected by comparing the slope of the regression in the first 5 min with later measurements. If the initial regression was steeper by more than [0.06], then the initial slope was used to calculate tear volume and the remaining data points to calculate tear flow and tear turnover.

We examined a large number of normal individuals, with no obvious pathology by standard ocular examination, no complaints of dry eyes, and no history of contact lens wear. We obtained a set of data points that describe normal tear volume, tear flow, and Schirmer's tests in these normals for each decade of life.[8]

To gain a more complete picture of tear film stability, we also measured tear osmolarity using a sample of tears collected from the tear lake with a drawn capillary tube. Typically, 100 nl was removed, as atraumatically as possible, and tested with a freeze point depression osmometer.[9–11]

In addition, we evaluated meibomian gland function, another important key to understanding tear film stability.[12,13] Meibomian gland function is difficult to analyze, and cannot be evaluated with a single test. We started with an assessment of the volume and appearance of the meibomian gland secretions. Digital expression of meibomian glands of the lower lid produces, after a few seconds of digital pressure, a volume of lipid extruded from each meibomian orifice. The diameter of the extruded lipid dome was measured in millimeters, and the viscosity, or thickness, was estimated using a scale of 1 to 4: 1 = clear, 2 = slightly viscous, 3 = viscous and partly opaque, and 4 = very thick and opaque.

Evaporation, first clearly described by Rolando and Refojo, was another measurement that provided information regarding tear film stability. It was relatively easy to measure and did not require expensive equipment.[14,15] We constructed a latex mold that conformed to the natural contour of the bony orbital rim. The mold isolated a space over the ocular surface enclosing approximately 14 cm^3 in volume and was fitted with humidity and temperature sensors. After filling with dry air (less than 5% relative humidity), the inflow and outflow vents were clamped and the humidity allowed to rise as evaporation proceeded from the ocular surface and from the skin around the eyelids. From the slope of the rise of the humidity, one calculated the relative rate of evaporation from all of these surfaces. We then compared the eye-open rate with the eye-closed rate. The difference in evaporative rate was assumed to be coming only from the ocular surface of exposed cornea and conjunctiva. The evaporative rate was recorded at the arbitrary relative humidity of 30%, and reported as grams per centimeter squared per second. Our results have been confirmed by Tsubota and Yamada using a very similar device of their creation.[16]

Data from these tests collected on 72 normal individuals of different ages gave us a picture of how the tear film changes with age. There appear to be changes with age in all categories except tear film decay.[17] The changes in evaporation, tear flow, and Schirmer's test are particularly striking (Table 1). Some previous investigators have also found changes with age.[18,19] Our results explain why contact lenses are more tolerated in teenagers and young adults and more problematic in the older population. We need more data to explore the differences between men and women and to determine the effects of menopause and other alterations in hormonal status which may influence tear production and possibly other aspects of tear film function.

We then applied the same battery of examinations to 156 patients with ocular surface disease—blepharitis, dry eye, or other disease states, and compared them with the 72

Table 1. Mean tear film values of normals by age decade

Age (yr)	No.	Tear Osmo.	Tear Vol.	Tear Flow	Tear Decay	Tear Evap.	Drop Out	Lipid Visc. High	Lipid Visc. Low	Lipid Visc. Avg.	Lipid Vol. High	Lipid Vol. Low	Lipid Vol. Avg.	Schirm Test
10-20	12	297.58	5.38	0.40	0.08	11.27	0.00	1.00	1.00	1.00	0.78	0.58	0.68	26.25
21-30	11	295.90	3.42	0.21	0.06	12.64	0.05	1.10	1.00	1.05	0.70	0.59	0.64	23.70
31-40	9	295.11	2.78	0.24	0.10	14.20	0.11	1.00	1.00	1.00	0.78	0.64	0.71	23.22
41-50	8	306.13	2.68	0.15	0.06	13.95	0.69	1.56	1.00	1.20	0.76	0.53	0.65	12.38
51-60	12	305.17	1.81	0.10	0.06	13.37	2.42	1.54	1.08	1.27	0.63	0.45	0.54	10.92
61-70	10	310.09	1.73	0.12	0.07	19.50	2.36	1.64	1.09	1.34	0.71	0.48	0.61	12.00
70 plus	10	311.20	1.15	0.08	0.08	21.49	1.85	1.70	1.10	1.25	0.64	0.45	0.54	7.30
Test for linear trend: p-value		<0.0001	<0.0001	<0.0001	0.9160	0.0016	0.0257	<0.0001	0.0922	0.0006	0.2291	0.0209	0.0756	0.0001

normals.[17] One difficulty in studying blepharitis and dry eye patients is that there is no standard way to define these entities. Patients were entered with an original diagnosis based on standard clinical definitions of blepharitis and dry eye, which were therefore relatively vague. After we performed the tests, we constructed a second, physiologic diagnosis based on the following definitions:

Dry eye: Schirmer test without anesthetic of <10 mm of wetting, or tear flow of ≤0.10 μl/min.

Obstructive meibomian gland dysfunction: Gland drop-out greater than one gland per lid, and/or a lipid volume of <0.5 mm.

Seborrheic meibomian gland dysfunction: Lipid volume of ≥ 0.8 mm.

Allergic status: Thickening of the conjunctiva with a papillary response and symptoms consistent with ocular allergy.

Normals: No ocular symptoms or obvious eyelid disease by standard slit lamp examination; meibomian gland volume of 0.5–0.7 mm; no contact lens use.

Rosacea: Presence of dilated blood vessels on the face, and a hyperemic flush over the malar skin and nose.

Infection: Presence of pus at the base of the cilia, or a very inflamed, tender lid margin with a serous discharge.

We have in this manner created a classification system for blepharitis and dry eye based primarily on the appearance of the expressed meibomian gland, meibomian gland morphology, evaporation, and tear production assessed with Schirmer's test and fluorescein dilution (Table 2).

With this data, we examined the various parameters that might have affected tear film stability. It is our thesis that tear film stability actually refers primarily to evaporation rate; a stable tear film is one in which a minimum amount of tears evaporates. We postulate that this evaporation rate is determined mostly by the interaction between the lipid layer on the surface and the protein constituents of the tear film (primarily lipocalin), mucin which coats the surface of the epithelial cells, and the aqueous component of lacrimal secretions. The lipid layer is thus the key to this problem.

Since there are obvious differences in the appearance of meibomian secretions in individuals with various types of blepharitis and meibomian gland dysfunction, the changes in appearance may be due in part to differences in the composition of the lipids. The composition of the meibomian gland lipids is complex, however. This has been studied by a number of investigators over the last 20 years, and a considerable amount of detail has emerged regarding the makeup of the different types of lipids in meibomian secretions.[20,21] These lipids can be divided into many categories. In a pilot study, we analyzed individually the secretions of 22 patients and found seven groups of lipids: cholesterol esters, waxy esters, short-chain waxy esters, di-esters, triglycerides, free fatty acids, and choles-

Table 2. Presenting diagnosis vs. physiologic diagnosis

Presumptive Diagnosis		Final Diagnosis	
Blepharitis	42	Meibomian dysfunction	36
Blepharitis/dry eye	37	Meibomian dysfunction/dry eye	73
Allergic disease	4	Allergic disease	10
Dry eye	73	Dry eye	37
Normal	72	Normal	72
Total	228	Total	228

Table 3. Lipid class composition of meibum (weight %)

	CE	WE	scWE	DE	TG	FFA	CH
Mean	39.4	45.2	5.9	2.3	3.1	2.8	1.2
SD	3.1	3.4	1.1	0.8	2.2	1.3	0.5

(n=22)
CE, cholesterol esters. WE, waxy esters. scWE, short-chain wax esters. DE, di-esters. TG, triglycerides. FFA, free fatty acids. CH, cholesterol.

terol. Our analytical methods were thin layer chromatography (TLC) and gas chromatography with mass spectrometry (GC MS). The approximate percentages are depicted in Table 3.

It is relatively simple to divide lipids into large categories and analyze their approximate compositions; it is much more complex to identify differences between groups of patients. Dr. McCulley's work with Ward Shine in this area has progressed over many years of painstaking effort.[22–25] They have analyzed lipid samples from individual patients, grouped according to their own clinical classification system, and have identified many differences between groups, some of which are listed in Table 4. One major finding was that normals can be divided into those with and without cholesterol esters, whereas all the blepharitis groups had cholesterol.[22] The exhaustive analysis of the lipid types and chain structures in these diagnostic subgroups is very complex, as many small differences were found. It is, however, difficult to relate these changes to a theory of some disease process.

At least part of the difficulty may have been the study design. The investigators attempted to use lipid analysis to verify their clinical diagnosis and classification system, which is not based on measurable physiologic parameters. It might be better to apply the reverse logic, and use the data on lipid analysis to construct a classification system that could be applied clinically (Table 4).

The most interesting group to examine here regarding stability is the meibomian keratoconjunctivitis (MKC) group This group is, I believe, the one most closely associated with our definition of obstructive meibomian gland dysfunction (MGD), which we have identified as having an unstable tear film, increased evaporation, gland drop-out, viscous lipid, and low lipid volume.

Table 4. Differences is lipid composition of blepharitis subgroups

Meibomian keratoconjunctivitis (MKC)	
Increased:	Sphingolipid N-acyl fatty acid
	n 17 n 20 isomer triglycerides
	Ceramides lacking hydroxyl fatty acids
	Iso branched C 22:0 free fatty acids
	Cholesterol
Decreased:	Cerebroside sphingolipid
	n 18 oleic acid
	n 24:1 triglycerides
Seborrheic blepharitis group	
Increased:	n 14:1 triglycerides
Decreased:	Anti-iso branched C 23:0 free fatty acids
Staph blepharitis	
Increased:	Glycerol lipid phosphatidylcholine

Table 5. Thin layer chromatography data

Group	Chol. est.	Wax est.	Shrt wax est.	Trig.	Free FA	Chol.
Normal	39.14	45.26	6.05	3.14	2.96	1.3
	2.71	2.7	1.09	2.47	1.43	0.56
Seb. MGD	42.7	41.53	6.4	3.5	2.5	1.17
	1.57	2.99	0.89	1.84	1.15	0.12
Obst. MGD	38.73	48.8	4.7	2.23	2.17	0.83
	4.31	4.33	1.01	0.93	0.95	0.31

Chol. Est., cholesterol esters. Wax Est., waxy esters. Shrt Wax Est., short-chain waxy esters. Trig., triglycerides. Free FA, free fatty acids. Chol., cholesterol. Seb. MGD, seborrheic meibomian gland dysfunction. Obst. MGD, obstructive meibomian gland dysfunction.

We performed a TLC analysis of a small group of blepharitis patients and compared the normals with our definition of seborrheic MGD and obstructive MGD. Both normals and seborrheics have low evaporation rates and therefore a stable tear film. We found the seborrheic patients had an increase in cholesterol esters, whereas the obstructive MGD patients had an increase in waxy esters and short-chain waxy esters and a decrease in cholesterol and tryglycerides. These results are based on only a few patients and should be expanded considerably (Table 5).

In the next phase of our analysis, we will attempt to correlate differences in the lipid composition with evaporation rate. We will be looking for an association between increased lipid viscosity, gland drop-out, and increased evaporation rate. It is unlikely we will identify a single factor or a particular lipid component responsible for increased evaporation, because control of evaporation is likely to be an interactive process. There may, however, be one or more components associated with low or high evaporation, because some normals and seborrheics have strikingly low evaporation rates. Obviously if a particular component can be correlated with this effect, it might improve our understanding of how the barrier function of the lipid controls evaporation.

Cholesterol ester monolayer experiments may be helpful in shedding some light on this question. There is evidence in rabbits, which have a higher level of cholesterol esters, that monolayers of lipids are more stable with branched-chain fatty acids and a low percent of unsaturated fatty acids. In humans, the stability of the monolayer is increased when the cholesterol esters are *cis*-unsaturated rather than saturated or *trans*-fatty acids. Meibum contains 4% triglyceride, and this level is associated with increased stability of cholesterol ester monolayers at normal body temperature. It is not clear whether this has any bearing on the human disease process.[22]

The lipid layer does not function alone; it acts in concert with the protein and mucin components of the tear film. It is likely that the protein previously referred to as tear specific albumin, and now as a lipocalin, may be important in this process.[26–28] The lipocalins are used to transport lipids such as vitamin A in other organ systems. The hollow shape of these proteins is particularly suited to this task. It is reasonable to assume that lipocalins may bind various meibomian lipids and allow them to form a stable layer over the aqueous portion of the tear film. Lipocalins represent the largest single protein component in the tear film and their role is under active investigation at the present time.

Another means of investigation that may shed some light on meibomian gland lipid function and tear film stability is visualization of the lipid surface interference patterns using confocal microscopy. Interference patterns have been identified in the tear film for many years. Investigations by Ehlers, McDonald, Hamano, and others have described the interference colors generated by the thickness of the tear film lipid layer.[29–31] These colors are generated by destructive interference of white light and reveal information concerning the thickness of the layer. From these data we know that, in a normal individual, the thickness of the lipid layer is probably less than 80 nm. However, with partial lid closure, in seborrheics or other increased lipid states, the thickness probably is several times greater.

We have adapted our confocal microscope to use a 10X dry objective to focus on the tear film and observe the interference patterns in normal and diseased states.[32] Our confocal microscope uses a mercury arc lamp, which is nearly monochromatic, and a black and white camera. Therefore, interference colors are not seen. However, a sharp demarcation is obtained from the constructive and destructive interference and one can estimate the thickness of the lipid layer from these data. Further information is obtained by deliberately altering the thickness of the lipid layer with meibomian gland expression.

For these examinations, we devised a set of descriptive terms to identify various aspects of the appearance of this lipid layer: debris, pattern variability, linearity, dry spots, and thickness. We used a nonparametric scale of 1 to 10 for this study.

Debris measured the amount of particulate matter seen in the tear film (1= none or minimum, 10=more than 10 particles).

Pattern variability measured the degree to which the tear film changed over the first 1 min of observation (non-variable=1, highly variable=10).

Linearity was that property of the lipid surface appearance whereby the interference pattern produced a linear configuration resembling oil on water (highly linear=1, non-linear=10).

Dry spots appeared as localized thinning of the tear film which created a valley (none=1, many=10).

Thickness measured the relative amount of dark area, with darker indicating thicker lipid (minimum dark area=1, maximum=10).

We examined 20 subjects identified as normal and 33 patients with a diagnosis of either dry eye or some form of blepharitis, or both. Our results indicate that several physiologic properties of the tear film, such as osmolarity, lipid viscosity, and volume, correlated with the appearance of the tear film (Table 6).

We found the appearance of the tear film was different for seborrheic MGD, obstructive MGD, and dry eye compared to normals. As expected, the lipid layer was thinner in obstructive MGD, which has a low amount of lipid, and thickest in seborrheic MGD. The linearity of the interference pattern was greatest in seborrheics, who have the highest lipid volume, and least in obstructive meibomian gland dysfunction, with the lowest amount of lipid. The lack of correlations between evaporation and lipid thickness or other visible parameters is notable. We speculate that this analysis is confounded by the normals who have a thin lipid layer and a low evaporation rate whereas obstructive MGD also has

Table 6. Correlation coefficients for physiologic measurements of the tear film and values for the confocal appearance variables

Property	Osmo.	Evap.	Tear vol.	Tear flow	Dec.	Schirm	Dropout	Visc. hi	Visc. avg.	Vol. hi	Vol. avg.
Debris	—	−0.21	0.21	0.21	—	0.25	−0.28	—	—	—	—
Variability	—	—	—	—	−0.31	—	—	—	—	0.26	0.22
Linearity	0.41	—	—	−0.23	—	−0.3	—	0.32	—	−0.45	−0.45
Dry spot	—	−0.28	—	—	0.26	—	—	0.26	—	—	—
Thickness	—	—	—	—	—	—	—	−0.2	−0.21	—	—

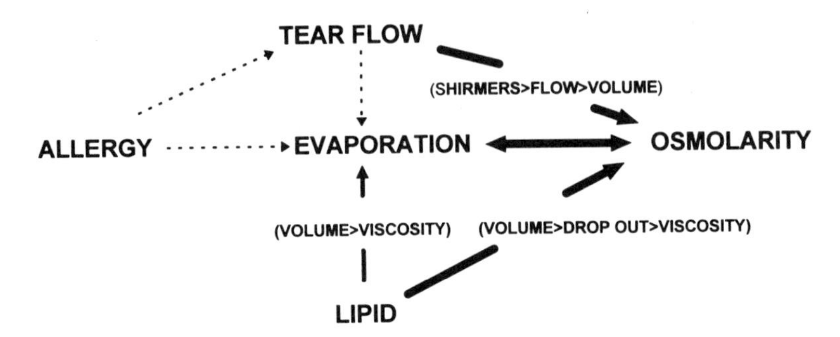

Figure 1. Parameter correlations for all patients.

a thin layer but increased evaporation. It is apparent that thickness alone does not determine the effectiveness of the lipid layer as a barrier to water.

Using our data on 228 patients, both normals and diseased, we created a model of tear film function that allocates tear osmolarity as the final end point of tear film homeostasis and evaporation as the center of this process (Fig. 1).[17] To construct this model, we obtained correlation coefficients between each of these variables which help to define their relationship (Table 7). For example, lipid volume is correlated with evaporation and with osmolarity, and tear flow is correlated with osmolarity but not with evaporation.

These relationships generally hold true regardless of which group one analyzes (Fig. 2). However, some correlations become stronger when subsets are analyzed individually (Table 8). For example, MGD patients have a higher correlation between lipid variables and evaporation or osmolarity compared with dry eye patients without lipid abnormalities. Although we found that the volume of lipid correlates with evaporation such that an increased volume correlated with a decreased evaporation rate, this certainly did not fully explain the process or account for all the factors that control evaporation.

As stated, normals frequently had very thin lipid layers, as seen by confocal microscopy, with very low evaporation rates. Thus it appears that a thin lipid layer can form a very stable tear film and also have a high degree of barrier function. We postulate that the interaction between lipocalin and lipid is highly effective in the normal and may be perturbed in the diseased state. A simple increase in the volume of the lipid will only partially compensate for this ineffective interaction.

Water conservation scientists have long sought an effective barrier to retard evaporation in reservoirs. The most effective compounds so far have been n-alkoxy ethanols

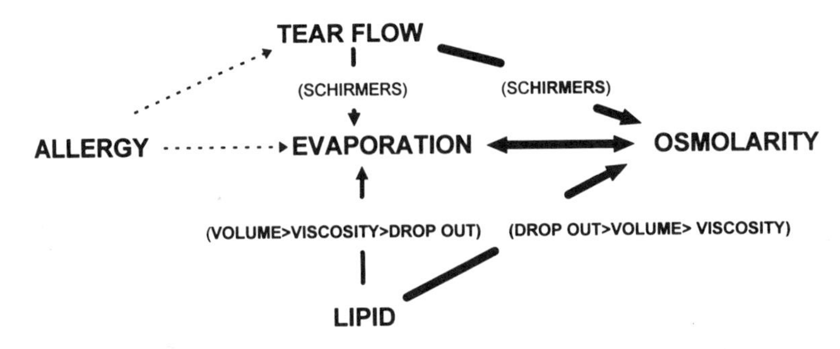

Figure 2. Parameter correlations for meibomian gland dysfunction patients.

Table 7. Correlation results for all patients ($n = 228$)

	Tear Osmo.	Tear Vol.	Tear Flow	Tear Decay	Tear Evap.	Dropout	Visc. Hi	Visc. Lo	Visc. Avg	Vol. Hi	Vol. Lo	Vol. Avg	Schirm
Mean	310	2.59	0.16	0.07	20.4	2.22	1.76	1.21	1.56	0.64	0.45	0.55	11.7
Corr. with Osmo.	n/a	–0.3	–0.32	–0.12	0.36	0.39	0.39	0.31	0.32	-0.3	-0.4	-0.4	-0.4
Corr. with Evap.	0.36	–0.08	–0.09	–0.03	n/a	0.18	0.22	0.26	0.22	–0.2	–0.3	–0.3	–0.1
Age adjusted:													
Osmo	n/a	–0.16	–0.19	–0.11	0.32	0.33	0.32	0.29	0.26	–0.28	–0.4	–0.35	–0.34
Evap	0.32	–0.02	–0.03	–0.02	n/a	0.15	0.2	0.25	0.2	–0.23	–0.28	–0.26	–0.06

Table 8. Mean values and correlations for meibomian gland dysfunction patients

	Tear Osmo.	Tear Vol.	Tear Flow	Tear Dec.	Tear Evap	Dropout	Visc. Hi	Visc. Lo	Visc. Avg.	Vol. Hi	Vol. Lo	Vol. Avg.	Schirm
All MGD pts.	313	2.38	0.14	0.07	22	3.7	2.1	1.3	1.8	0.6	0.4	0.5	9.8
MGD obst. and DE	315	2.08	0.1	0.06	27	4.7	2.7	1.6	2.3	0.4	0.2	0.3	8.9
MGD seb. and DE	311	2.03	0.1	0.08	12	0.4	1.5	1.09	1.3	0.9	0.6	0.7	5.7
MGD rosacea and DE	317	4.18	0.2	0.07	12	10.1	2.4	1.4	2.1	0.4	0.2	0.3	4.8
MGD seb. + obst. and DE	309	1.85	0.08	0.04	19	2.3	1.7	1.2	1.6	0.9	0.5	0.7	7.4
All MGD pts.													
Correlation with Osmo.	n/a	−0.15	−0.17	−0.11	0.3	0.4	0.26	0.26	0.19	−0.3	0	−0.3	−0.3
Correlation with Evap.	0.3	−0.08	−0.01	−0.01	n/a	0.2	0.15	0.2	0.14	−0.3	0	−0.3	0.4

which have been demonstrated to decrease evaporation 60 to 70%.[33] This effectiveness pales in comparison to the 95% reduction typically achieved by the lipid of the tear film. Without an intact lipid barrier, a pure water surface will evaporate at a rate of 170 X 10^{-7}g/cm^2/sec at a temperature of 34°C and a relative humidity of 30%, whereas the eye of a young human subject has an evaporation of 10 X 7^{-7} g/cm^2/sec.[34] We have attempted to modify and increase the stability of the tear film by the addition of a number of lipid compounds. Each lipid that we tried increased the evaporation rate, demonstrating that the tear film stability is probably not dependent upon meibomian lipid acting as a simple barrier. How this instability occurs is still a mystery.

ACKNOWLEDGMENTS

Supported in part by a grant from RPB Inc. and NEI grant EY10151.

REFERENCES

1. Schirmer O. Studien zur Physiologie und Pathologie der praden Absoderung Tränen abfur. *Graefes Arch Clin Exp Ophthalmol.* 1903;56:197–291.
2. Farris RL, Gilbard JP, Stuchell RN, Mandell ID. Diagnostic tests in keratoconjunctivitis sicca. *CLAO J.* 1983;9:23–28.
3. Clinch TE, Benedetto DA, Felberg NT, Laibson PR. Schirmer's test: A closer look. *Arch Ophthalmol.* 1983;101:1383–1386.
4. Maurice DM. The use of fluorescein in ophthalmological research. *Invest Ophthalmol.* 1967;6:464–477.
5. Brubaker RF, Maurice DM, McLaren JW. Fluorometry of the anterior segment. In: Master B, ed. *Noninvasive Techniques in Ophthalmology*. New York: Springer-Verlag: 1990: 248–280.
6. Jordan A, Baum J. Basic tear flow: Does it exist? *Ophthalmology*. 1980;87:920–930.
7. Mathers WD, Binarao G, Petroll WM. Ocular evaporation and the dry eye. *Cornea.* 1993;12:335–340.
8. Mathers WD, Lane JA, Zimmerman MB. Tear film changes with normal ageing. *Cornea.* 1996;15:229–235.
9. Gilbard JP, Farris RL, Santamaria J. Osmolarity of tear microvolumes in keratoconjunctivitis sicca. *Arch Ophthalmol.* 1978;96:677–681.
10. Nelson DS, Wright JC. Tear film osmolarity determination: An evaluation of potential errors in measurement. *Curr Eye Res.* 1986;5:677–681.
11. Farris RL. Tear osmolarity variation in the dry eye. *Trans Am Ophthalmol Soc.* 1986;34:250–268.
12. Mathers WD, Daley TE, Verdick R. Video imaging of the meibomian gland. *Arch Ophthalmol.* 1994;112:448–449.
13. Mathers WD. Ocular evaporation in meibomian gland dysfunction and dry eye. *Ophthalmology.* 1993;100:347–351.
14. Rolando M, Refojo MF. Tear evaporimeter for measuring water evaporation rate from the tear film under controlled conditions in humans. *Exp Eye Res.* 1983;36:25–33.
15. Mishima S, Maurice DM. The oily layer of the tear film and evaporation from the corneal surface. *Exp Eye Res.* 1961;1:39–45.
16. Tsubota K, Yamada M. Tear evaporation from the ocular surface. *Invest Ophthalmol Vis Sci.* 1992;33:2942–2950.
17. Mathers WD, Lane JA, Zimmerman MB. Model for ocular tear film function. *Cornea.* 1996;15:110–119.
18. Seal DV. The effect of aging and disease on tear constituents. *Trans Ophthalmol Soc UK.* 1985;104:355–361.
19. McGill J, Liakos G, Seal DV, Goulding N. Tear film changes in healthy and dry eye conditions. *Trans Ophthalmol Soc UK.* 1983;103:313–317.
20. Tiffany JM. The lipid secretion of the meibomian glands. In: Paoletti R, Kritchevsky D, eds. *Advances in Lipid Research.* San Diego: Academic Press; 1987:1–62.
21. Dougherty JM, Osgood JK, McCulley JP. The role of wax and sterol ester fatty acids in chronic blepharitis. *Invest Ophthalmol Vis Sci.* 1991;32:1932–1937.

22. Shine WE, McCulley JP. The role of cholesterol in chronic blepharitis. *Invest Ophthalmol Vis Sci.* 1991;32:2272–2280.
23. Shine WE, McCulley JP. Role of wax ester fatty alcohols in chronic blepharitis. *Invest Ophthalmol Vis Sci.* 1993;34:3515–3521.
24. Shine WE, McCulley JP. Human meibomian secretion polar lipids associated with chronic blepharitis disease groups. ARVO Abstracts. *Invest Ophthalmol Vis Sci.* 1995;36:S156.
25. Shine WE, McCulley JP. Meibomian gland secretion polar lipids associated with chronic blepharitis disease groups. ARVO Abstracts. *Invest Ophthalmol Vis Sci.* 1996;37:S849.
26. Fullard RJ, Kissner HE. Purification of the isoforms of tear specific prealbumin. *Curr Eye Res.* 1991;10:613–628.
27. Redl B, Holzfeind P, Lottspeich F. CDNA cloning and sequencing reveals human tear prealbumin to be a member of the lipophilic-ligand carrier protein superfamily. *J Biol Chem.* 1992;267:20282–20287.
28. Glasgow BJ, Heinzmann C, Kojis T, Sparkes RS, Mohandas T, Bateman JB. Assignment of tear lipocalin to human chromosome 9q34–9qter. *Curr Eye Res.* 1993;11:1019–1023.
29. Ehlers N. The precorneal tear film: Biomicroscopical, histological and chemical investigations. *Acta Ophthalmol.* 1965;81:1–134.
30. McDonald JE. Surface phenomena of tear films. *Am J Ophthalmol.* 1969;67:56–64.
31. Hamano J, Hori M, Kawabe H, Umeno M, Mitsunaga S. Bio-differential interference microscopic observations on anterior segment of eye. *Jpn J Contact Lens Soc.* 1979; 21:229–238.
32. Mathers WD, Daley TE. In vivo observation of the human tear film by tandem scanning confocal microscopy. *Scanning.* 1994;16:316–319.
33. Katti SS, Kulkarni SB, Gharpurey MK, Biswas AB. Control of water evaporation by monomolecular films. *J Sci Industr Res.* 1962;21D:434–437.
34. Hisatake K, Tanaka S, Aizawa Y. Evaporation rate of water in a vessel. *J Appl Physics.* 1993;73:7395–7401.

51

SURFACE–CHEMICAL PATHWAYS OF THE TEAR FILM BREAKUP

Does Corneal Mucus Have a Role?

Ashutosh Sharma

Department of Chemical Engineering
Indian Institute of Technology at Kanpur
Kanpur, India

1. INTRODUCTION

Numerous experimental[1–6] and theoretical[3–8] studies have conclusively demonstrated that relatively thick (< 100 μm) fluid films on *nonwettable* substrates dewet spontaneously by the nucleation and growth of initial defects or dry spots. As expected, the thinner films on the less wettable substrates are more unstable. Interestingly, the theory also shows that the tear film breakup can be initiated by even an extremely small, mildly nonwettable corneal site of dimensions comparable to the size of a single superficial squamous cell.[9]

Over two decades ago, the seminal work[10,11] of Holly and Lemp suggested that the conversion of a completely wettable corneal surface to a partially wettable surface during the interblink period may engender the precorneal tear film breakup. While this view is in accord with the modern understanding of the stability of thick films, there are several unanswered questions about the specific processes that actually trigger corneal surface nonwettability in normal and dry eyes.

Based on measurements of the *apolar* surface properties obtained with the critical surface tension method of Zisman, it has been thought that the corneal epithelium is hydrophobic and nonwettable by the aqueous tears, but that a hydrophilic mucus coating renders it wettable.[10–12] However, it has been argued that artifacts in the preparation of the corneal surface, use of apolar diagnostic liquids, and the methods of interpretation (e.g., Zisman's method) may have been responsible for these conclusions.[13–17] Certainly, applications of modern techniques of measurement and interpretation of surface properties have repeatedly shown normal epithelial cells with associated glycocalyx to be almost as hydrophilic and wettable by aqueous media as by corneal mucus.[14–17] The conclusion is not surprising, because the superficial corneal epithelial cells, like any number of hydrophilic macromolecules and cells with an extracellular coat of glycosylated glycocalyx,[18–21] display a strong electron-donor type of polarity that engenders hydrogen bonding with water molecules and thereby makes the surface wettable.[16,17] Thus, the question whether the precorneal mucus has a direct role in tear film stability needs to be reexamined in view of the known apolar and polar surface properties of the corneal epithelium, mucus, and tear fluid. This issue has a direct bearing on the understanding of the surface–chemical mechanisms of the tear film breakup.

The purpose of this paper is to build a holistic view of the surface–chemical interactions among the corneal epithelium, mucus, tear film, and tear film contaminants/foreign bodies (eg, lipids, cell debris, bacteria). Based on this approach, the previously offered mechanisms of corneal wetting are critically assessed, and new possibilities for understanding normal and abnormal tear film breakup are suggested.

2. MATERIALS AND METHODS

2.1. Theory of Apolar and Polar Surface Properties

It is now well known that both apolar (Lifshitz–van der Waals; LW) and polar (acid–base; AB) interactions determine the overall surface properties such as surface and interfacial tensions, wettability, and adhesion. However, for macromolecules and biological surfaces in aqueous media, the polar acid–base interactions, including hydrogen bonding and hydration pressure, are the dominant interactions.[16–22] In what follows, we briefly outline the meaning of apolar and polar surface properties, and a methodology for their measurements. Detailed derivations and applications of the theory to a variety of macromolecules and biological surfaces may be found in extensive reviews.[19–21] The polar acid–base properties of the cornea can be found elsewhere.[16,17]

The total surface and interfacial tensions are the sum of their respective apolar (LW) and polar (AB) components. All of these can be found from three fundamental parameters that completely characterize a material: the apolar component of surface tension (γ^{LW}), an electron-donor (or proton-acceptor) parameter (γ^-), and a conjugate electron-acceptor (proton-donor) parameter (γ^+).[19–21] The total surface tension of a material *i*, and interfacial tension between materials *i* and *j* are obtained from the following relations, respectively.[19–21]

$$\gamma_i = \gamma_i^{LW} + \gamma_i^{AB} = \gamma_i^{LW} + 2\sqrt{\gamma_i^+\gamma_i^-} \tag{1}$$

$$\gamma_{ij} = \gamma_{ij}^{LW} + \gamma_{ij}^{AB} = \left(\sqrt{\gamma_i^{LW}} - \sqrt{\gamma_j^{LW}}\right)^2 + 2\left(\sqrt{\gamma_i^+\gamma_i^-} + \sqrt{\gamma_j^+\gamma_j^-} - \sqrt{\gamma_i^+\gamma_j^-} - \sqrt{\gamma_i^-\gamma_j^+}\right) \tag{2}$$

The equilibrium contact angle, θ, of a liquid drop (L) on a solid surface (S) is given by the Young equation, which can also be written in terms of apolar and polar parameters.[19–21]

$$(1+\cos\theta)\gamma_L = 2\left[\sqrt{\gamma_S^{LW}\gamma_L^{LW}} + \sqrt{\gamma_S^+\gamma_L^-} + \sqrt{\gamma_S^-\gamma_L^+}\right] \tag{3}$$

The three unknown surface tension parameters (γS^{LW}, γS^{+}, γS^{-}) for a surface can thus be found from Equation 3 by measuring advancing equilibrium contact angles of at least three different well-defined probe liquids, of which at least two must be polar (e.g., water, glycerol).[16–22]

Another useful thermodynamic concept is the free energy of adhesion, which is the difference (per unit area) between the interfacial tensions (energies) after and before the attachment (adhesion) of two surfaces. The free energy of adhesion between surfaces 1 and 2 in a medium 3 (e.g., water) is also given by the sum of its apolar (LW) and polar (AB) components:[16–22]

$$\begin{aligned}\Delta G = \Delta G^{LW} + \Delta G^{AB} &= \gamma_{12} - \gamma_{13} - \gamma_{23} \\ &= 2\left(\sqrt{\gamma_1^{LW}} - \sqrt{\gamma_3^{LW}}\right)\left(\sqrt{\gamma_3^{LW}} - \sqrt{\gamma_2^{LW}}\right) \\ &+2\left[\sqrt{\gamma_3^{+}}\left(\sqrt{\gamma_1^{-}} + \sqrt{\gamma_2^{-}} - \sqrt{\gamma_3^{-}}\right) - \sqrt{\gamma_1^{+}\gamma_2^{-}}\right. \\ &\left. + \sqrt{\gamma_3^{-}}\left(\sqrt{\gamma_1^{+}} + \sqrt{\gamma_2^{+}} - \sqrt{\gamma_3^{+}}\right) - \sqrt{\gamma_1^{-}\gamma_2^{+}}\right] \qquad (4)\end{aligned}$$

Negative values of the free energy signify attraction between the surfaces leading to adhesion, whereas positive free energy denotes repulsion, or a lack of attachment. More negative values indicate stronger adhesion. For biological surfaces in water, the apolar (LW) component of the free energy is almost always negative, so the LW interactions always favor adhesion. However, the polar (AB) component can readily become positive and dominant for *polar hydrophilic* surfaces. The hydrophilic repulsion or hydration pressure thus engendered is in fact responsible for almost all of the physico-chemical properties of aqueous macromolecular solutions[18–22] and for the stability (lack of attachment) of glycocalyx-carrying cells (eg, red blood cells).[18] In contrast, *apolar or weakly polar* "hydrophobic" surfaces in aqueous media experience a "hydrophobic attraction," because the polar component of the free energy also becomes negative.[19–21]

The above theory of apolar and polar interactions in aqueous media has been widely applied for characterization of surface properties, wetting, and adhesion for a variety of polymeric and biological surfaces including gels and cells.[16–22] Zisman's method cannot be used for measurements of polar surface properties and interactions in aqueous media.

2.2. Experiments

Briefly, surfaces of successively deeper corneal structures were exposed by processing corneas of adult New Zealand albino rabbits in four different ways: Group A corneas were gently rinsed with cold saline, which leaves the superficial mucus intact. Group B corneas were irrigated with 20% *N*-acetylcysteine solution for 30–45 min, which exposes the squamous epithelial cells. Group C corneas were dried by Kim wipes after the *N*-acetyl treatment, so a slight dehydration and damage of the cell surface is anticipated. Group D corneas were cleaned and dried by cotton swabs with or without an additional *N*-acetyl pretreatment, which should dehydrate and extensively damage the superficial cells and partially expose the wing cells. Details may be found elsewhere.[16–17]

Equilibrium contact angles of small drops of three well-defined probe liquids—diiodomethane, physiologic saline, and glycerol—were measured rapidly on the corneal surfaces of groups A to D by a goniometer. A new corneal site was chosen for each measurement. Variations in the contact angle were also recorded with room temperature drying of the corneas for a period of 45 min. Contact angles of the probe liquids thus determined are reported elsewhere.[16]

3. RESULTS AND DISCUSSION

The apolar and polar surface properties of different corneal layers (groups A to D) are summarized in Table 1. The apolar component of the surface tension did not change significantly during drying, but the polar properties did. The apolar component of surface tension for the corneal surface (27–29 mJ/m^2) and mucus (about 40 mJ/m^2) agree with the respective critical surface tensions obtained by Zisman's method.[10–12] However, all of the corneal surfaces and mucus (group A) also display significant electron-donor type of polarity (reflected in γS^- values), which declines with drying, damage, and progression to the deeper corneal layers from groups A to D. With progressive drying and damage, the hydrophilicity of the corneal surface (as measured by the negative free energy of binding to water, ΔG_{SW}) also declines and, consequently, the interfacial tension against water increases.[16,17]

Our visual observations showed that surfaces of mucus-covered corneas (group A) and of corneas with intact cells (group B) remained smooth and glossy even after 30–45 min of exposure to air, but damaged corneas (group D) lost their luster. This observation correlates well with the decreased hydrophilicity (water-retaining capacity) of damaged corneas (Table 1).

A variety of biological and synthetic polymers such as fibrinogen, fibronectin, dextran, and ribosomal RNA also display about the same magnitudes of electron-donor type polarity and of hydrophilicity.[18–22] In fact, the apolar and polar surface properties of the human erythrocyte ($\gamma S^{LW} = 26.5$, $\gamma S^- = 52.1$, $\gamma S^+ = 4.8$; in mJ/m^2)[18] are remarkably similar to the properties of glycocalyx-bearing corneal epithelial cells of group B (Table 1).

Due to their polar properties, the corneal mucus (group A) and epithelial cells with glycocalyx (group B) are about equally hydrophilic, and both should be completely wetted by a normal tear film of surface tension <~50 mJ/m^2. This result would appear to suggest a minimal role for mucus in corneal wetting, but it must be remembered that the contact an-

Table 1. Values (in mJ/m^2) of the acid–base parameters (γ^+, γ^-) and the apolar (γ^{LW}) and polar (γ^{AB}) components of surface tensions for corneal surfaces, as determined from contact angle measurements

Group		γ_S^{LW}	γ_S^+	γ_S^-	γ_S^{AB}	γ_S	ΔG_{SW}
A	Hydrated	40	0.4	64	9	49	−146
	Dry		0.6	57	11	51	−143
B	Hydrated	27	3.2	56	27	54	−143
	Dry		3.4	53	27	54	−141
C	Hydrated	27	5.7	49	33	60	−144
	Dry		6.6	40	33	59	−138
D	Hydrated	29	4.7	33	25	53	−127
	Dry		4.9	18	19	47	−115

See text for descriptions of Groups -D.
Hydrated, initial contact angles on hydrated surfaces. Dry, after 45 min corneal drying at room temperature. ΔG_{sw}' negative free energy of binding to water.

gle measurements can only probe the surface properties over large, millimeter-size areas, whereas, the tear film breakup can be initiated even by micron-size nonwettable sites,[8,9] which may form due to cell desquamation and local epithelial contamination. The surfaces of degenerating, desquamating, and deeper layer cells are likely to be a lot less polar and more nonwettable compared even to the group D corneal surface. Further, in epithelial surface abnormalities (keratinization, glycocalyx deficiency, erosions), the surface may become nonwettable due to the loss of surface polarity, and a normal mucus layer would then indeed be needed for wetting. In what follows, the role of mucus in *maintenance* of the corneal wettability is explored based on the surface properties.es (Table 1).

Table 2 displays the total free energy of adhesion of mucus, as well as its apolar and polar components, to a variety of corneal surfaces in water. These values are calculated based on the surface properties (Table 1) and Equation 4. (For water as medium, $\gamma^{+} = \gamma^{-} = 25.5$, $\gamma^{LW} = 21.8$, $\gamma = 72.8$; all in mJ/m^2.) Positive free energies signify repulsion or the lack of adhesion of surfaces or macromolecules; the negative values indicate attraction leading to adhesion. While the apolar (LW) interactions favor adhesion, the polar repulsion engendered by water-retaining capacities (hydrophilicity) of mucus and cell surfaces prevents adhesion of mucus to itself, and to normal and moderately damaged and dehydrated cells. These results explain as to why uncontaminated mucus forms a highly expanded and hydrated fluid like "sloppy" gel, which cannot adhere tightly to the epithelium, even at the sites of recent cell desquamation or transient micro-erosions. These qualities of mucus assist in corneal hydration, cleaning, lubrication, and spreading of mucus.[17] Anchoring of mucus to the epithelium would transmit the shear produced by the rapidly moving lids during blinking to the epithelium, causing damage to fragile cells. Further, adhesion of mucus to the epithelial surface would prevent its movement and rejuvenation in the vicinity of the epithelium-mucus interface, leading to accumulation of contaminants (eg, degenerated cells). The situation with the precorneal mucus therefore seems analogous to that of the respiratory tract, where the mucous gel floats on top of ciliated epithelium and periciliary fluid. More detailed calculations[17] of the surface forces reveal that the precorneal mucus layer should be separated from the cell glycocalyx at an equilibrium distance in excess of about 5 nm. The intervening space where mucus meets with the cell glycocalix is therefore largely filled with an electrolyte–water mix.

An intimate binding of mucus to the epithelium, however, becomes possible if either the cell or mucus loses almost all of their polar properties (see Table 2), which can occur due to a combination of factors such as dehydration, damage, contamination, and epithelial abnormalities. Further, the free energy of adhesion of contaminated apolar or weakly polar mucus to itself in water also becomes negative,[17] which should lead to col-

Table 2. Free energies of adhesion (in mJ/m^2) of mucus to itself and to various types of corneal surfaces in water

Surfaces involved in adhesion (surface 1-water-surface 2)	ΔG^{LW}	ΔG^{AB}	ΔG (total)
Mucus–mucus (group A)	−5.5	+43.0	+37.5
Hydrated mucus (group A) – hydrated normal cell (group B)	−1.9	+40.9	+39.0
Dehydrated mucus (group A) – dehydrated normal cell (group B)	−1.9	+35.3	+33.4
Hydrated mucus (group A) – damaged cell (group D)	−2.2	+22.6	+20.4
Dehydrated mucus (group A) – damaged cell (group D)	−2.2	+8.4	+6.2
Hydrated mucus–severely damaged/abnormal apolar epithelium	−2.2	−15.6	−17.8
Contaminated apolar mucus–normal cell (group B)	−1.9	−8.0	−9.9

lapse of hydrophilic gel into relatively dehydrated, hydrophobic aggregates and clumps. A strong attachment of these "hydrophobic" mucous strings and clumps to the apolar epithelium is a likely precursor for the formation of the corneal filaments.[17]

Although the apolar lipids of the superficial oily layer have been thought to be the most important source of mucus contamination,[10–12] a major source of hydrophobic contaminants is the epithelium itself which sheds a few hundreds of degenerated squamous cells every minute.[23,24] The role of mucus in the maintenance of the corneal wettability by the removal of nonwettable contaminants can be understood by an inspection of Table 3, which displays the free energies of adhesion of the apolar (or weakly polar) contaminants to mucus, and to the epithelium. In Table 3, the surface tension of the apolar contaminants (particles) is assumed to be 30 mJ/m^2, which is characteristic of the apolar lipids and cell membranes.[10,11,17] However, none of the qualitative conclusions arrived at in Table 3 are dependent on this assumption, since the AB (polar) component of the free energy is independent of this detail. The apolar or weakly polar contaminants (eg, cell debris, lipids, gram- negative bacteria, air-borne colloidal particles) are readily adsorbed and absorbed by mucus, but interestingly, they cannot absorb on the epithelium *in the presence of mucus* (Table 3). Thus, a normal mucus layer not only forms an efficient surface–chemical *trap* for the capture and removal of hydrophobic contaminants, but also a surface–chemical *barrier* preventing the epithelial contamination. However, in the absence of the mucus layer, direct attachment and accumulation of the contaminants on the epithelium becomes energetically favorable (Table 3), which can mask the hydrophilic nature of the cell glycocalyx.

The surface energetics of attachment of foreign bodies, cells, and bacteria to various types of the corneal surfaces are discussed in the greater detail elsewhere.[17] In particular, binding of bacteria is encouraged by the loss of polar properties of the corneal surface, which may be due to mucus deficiency, epithelial damage, ocular surface abnormalities, or tear film insults (e.g., hyperosmolarity, preservatives).

The above results from the surface energetics of the epithelium—tear film system suggest several surface–chemical pathways of the corneal surface nonwettability, and tear film breakup in normal and dry eyes.

Since the normal, uncontaminated epithelium is as hydrophilic and wettable as the mucus layer, the role of mucus is not in masking the epithelial "hydrophobicity", but in maintaining the corneal wettability during the BUT by preventing the epithelial contamination, and by a continuous trapping and removal of nonwettable contaminants. The epithelial cleansing, hydration, and lubrication are made possible by the *polar* properties of uncontaminated mucus, which prevent its adhesion to itself and to the normal epithelium, thereby allowing its hydration and free movement.

Table 3. Free energies of adhesion (in mJ/m^2) of apolar or weakly polar contaminants to mucus, and to the corneal epithelium with or without mucus

Surfaces adhering in medium 3 (surface 1 – medium 3 – surface 2)	ΔG^{LW}	ΔG^{AB}	ΔG (total)
1. Apolar particle–water–hydrated mucus (group A)	−2.7	−15.6	−18.3
2. Apolar particle–water–dehydrated mucus (group A)	−2.7	−18.5	−21.2
3. Apolar particle–mucus–cell (group B, hydrated)	−1.9	+18.5	+16.6
4. Apolar particle–mucus–cell (group D, dehydrated)	−1.7	+21.4	+19.7
5. Apolar particle–water–cell (group B, hydrated)	−0.9	−8.1	−9.0
6. Apolar particle–water–cell (group D, dehydrated)	−1.1	−37.0	−38.1

Let us first consider the more important case of dry eyes secondary to, or concurrent with, the epithelial surface abnormalities. For example, in keratoconjunctivitis sicca, the following abnormalities are often present:[25–28] increased exfoliation of superficial cells followed by a premature uncovering of less differentiated deeper cells, abnormalities of glycocalyx synthesis or its extracellular attachment, degeneration of cells, micro-erosions, and keratinization of conjunctival cells. All of these abnormalities should promote greater adhesion and immobilization of mucus secondary to the loss of epithelial polarity. Further, an increased cell loss concurrent with reduced mucus turnover (mobility) would also encourage greater contamination of mucus itself, leading to the loss of its polar properties. Such a change enhances the strength of adhesion and encourages a partial collapse of hydrated mucus gel into less hydrophilic aggregates. In view of this, corneal hydration and lubrication are also likely to be affected, and the shear transmitted during blinking to the adherent mucus may lead to further epithelial damage. Thus, even in the event of sufficient mucus production, epithelial abnormalities can initiate a vicious cycle leading progressively to accumulation of contaminated, dehydrated, and adherent mucus. The tear film breakup would be immediate on such a surface. Mucus deficiency can only accelerate the chain of events leading to the tear film breakup. Thus, depending on the severity of the epithelial disorder, a coexisting deficiency in the production or spreading of mucus may also be required for the completion of the vicious cycle during the interblink period. In the proposed mechanism, the role of aqueous tear deficiency is limited to aiding the contamination, dehydration, and epithelial damage, eg, by increasing the osmolarity and shear. In conclusion, in the case of corneal surface abnormalities leading to the loss of epithelial polarity (hydrophilicity), a coating of *functional* mucus is indeed required for tear film stabilization. However, the surface–chemical events noted above can conspire to make this an impossibility.

In view of the above mechanism, the effects of preservatives (especially of the cationic variety) and oils on the tear film breakup[29] can also be explained in terms of their binding with mucus to reduce its polar properties, hydrophilicity, and mobility. These would be further compromised by the epithelial damage and increased cell loss.[30] Due to a small turnover of mucus and the epithelial involvement, effects on the tear film stability should persist for some time even after the offending substance has been washed out from the aqueous tears.

What may happen in the case of decreased thickness and turnover of the corneal mucus due to deficiencies in its production, spreading, or removal? It is possible that a severe mucus deficiency alone can initiate the vicious cycle consisting of increased mucus contamination and adherence, decreased turnover and mobility, clumping of mucus, cell damage by dehydration and shear, and finally, loss of wettability accompanied by a more rapid tear film breakup.

All of the above surface–chemical pathways and their interactions leading to the tear film breakup in dry eyes are summarized in Fig. 1.

There is another surface–chemical possibility for the tear film breakup in the case of mucus deficiency, namely, breakup of the mucus layer itself leading to exposure of the epithelial patches to the aqueous tears.[31,32] The epithelial surface without its mucus coating attracts nonwettable contaminants (see Table 3). The tear film breakup may be triggered either by epithelial contamination, or even by micron–size nonwettable sites of degenerating and desquamating cells.[9] The mechanics of the mucus layer breakup was suggested by Sharma and Ruckenstein earlier.[31,32] An updated version of this mechanism and its implications for the tear film breakup will be presented in detail elsewhere in this volume.

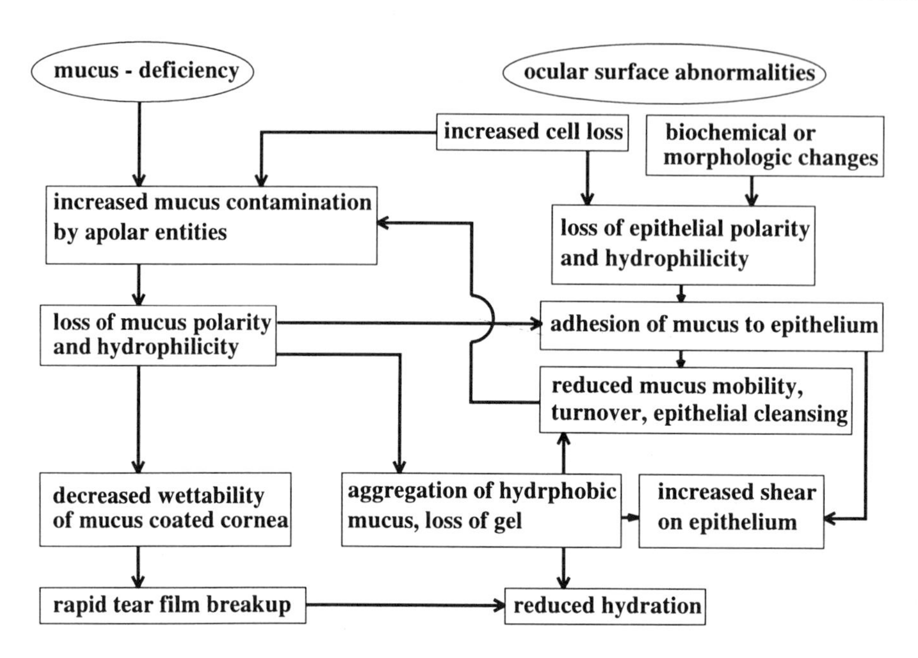

Figure 1. Summary of the proposed surface–chemical pathways leading to the tear film breakup.

The earliest and the most quoted surface-chemical mechanism of Holly and Lemp[10–12] proposed the diffusion and adsorption of the apolar lipids of the superficial oily layer to the mucus–aqueous interface, leading to its nonwettability and the tear film breakup. Our surface–chemical results (Table 3) indeed support the feasibility of adhesion of apolar lipids to mucus. However, the question as to whether the low solubility/diffusivity lipids can diffuse across the tear film and adsorb in sufficient quantities within the normal range of the BUTs needs further exploration. This mechanism may be most germane for the tear breakup in normal eyes with long BUTs. A monolayer coverage of the mucus–aqueous interface by lipids is sufficient for nonwettability, so it is difficult to imagine an immediate role for mucus thickness and turnover in this mechanism. Since mucus contamination in this model begins at the mucus–aqueous interface, coverage of the interface with lipids of aqueous origin should take roughly the same time regardless of the mucus layer thickness. Further, mucus turnover and resurfacing during blinks should be most efficient in the vicinity of the mucus–aqueous interface, so a moderate decline in mucus turnover is not likely to inhibit the cleansing of the contaminated zone. Finally, the tear film breakup engendered by lipid contamination leaves no significant role for the ocular surface abnormalities, which is in contrast to the surface–chemical picture proposed in this paper.

Perhaps it is an oversimplification to assume a single mechanism for the corneal surface nonwettability and the tear film breakup in the normal eyes, as well as in a variety of dry eye disorders. The overall functioning of a complex multicomponent biological system, such as the tear film–corneal surface, depends not only on the structure of its components, but most crucially on the *interactions* among them. We have tried to evolve such a view in this paper, which may suggest some new avenues for the design and interpretation of future experiments.

ACKNOWLEDGMENTS

This work was made possible by fruitful discussions with C. J. van Oss and J.M. Tiffany, that kept alive the author's interest in the problem of corneal wetting. This work was partially supported by a grant from the All India Council for Technical Education.

REFERENCES

1. Paddy JF. Cohesive properties of thin films on liquids adhering to a solid surface. *Spec Disc Faraday Soc.* 1970;1: 64–74.
2. Doughman DJ, Holly FJ, Dohlman CH. Paper presented at the ARVO Meeting, 1971.
3. Taylor GI, Michael, DH. On making holes in a sheet of fluid. *J Fluid Mech* 1973;58:625–639.
4. Khesgi HS, Scriven LE. Dewetting: Nucleation and growth of dry regions. *Chem Eng Sci* . 1991;46:5: 519–525.
5. Redon C, Brochard-Wyart F, Rondelez F. Dynamics of dewetting. *Phys Rev Lett* 1991;66:715–718.
6. Sykes C, Andrieu C, Detappe V, Deniau S. Critical radius of holes in liquid coating. *J Phys III (France)* 1994;4:775–781.
7. de Gennes P G. Wetting: Statics and dynamics. *Rev Mod Phys.* 1985;57:827–863.
8. Sharma A. Perturbation analysis of surface dewetting by formation of holes. *J Colloid Interface Sci* . 1993;156:96–103.
9. Sharma A, Coles WH. Physico-chemical factors in tear film breakup. *Invest Ophthalmol Vis Sci.* 1990;31S: 552.
10. Holly FJ, Lemp MA. Wettability and wetting of corneal epithelium. *Exp Eye Res.* 1971;11: 239–250, 1971.
11. Holly FJ, Lemp MA. Formation and rupture of tear film. *Exp Eye Res.* 1973;15:515–525.
12. Holly FJ. Wettability and bioadhesion in ophthalmology. In: Schrader ME, Loeb G, eds. *Modern Approaches to Wettability: Theory and Applications.* Plenum Press; 1992:213–248.
13. Cope C, Dilly PN, Kaura R, Tiffany JM. Wettability of the corneal surface: A reappraisal.*Curr Eye Res.* 1986;5:777–785.
14. Tiffany JM. Measurements of wettability of the corneal epithelium: I particle attachment method. *Acta Ophthalmol.* 1990;68:175–181.
15. Tiffany JM. Measurements of wettability of the corneal epithelium: II contact angle method.*Acta Ophthalmol.* 1990;68:182–187.
16. Sharma A. Surface properties of damaged and normal corneal epithelia. *J Dispersion Sci Tech.* 1992;13:459–478.
17. Sharma A. Energetics of corneal epithleial cell-ocular mucus-tear film interactions.*Biophys Chem.* 1993;47:87–99.
18. van Oss CJ. Energetics of cell-cell and cell-biopolymer interactions. *Cell Biophys.* 1989;14:1–16.
19. van Oss CJ, Chaudhury MK, Good RJ. Monopolar surfaces. *Adv Colloid Interface Sci.* 1987;28:35–65.
20. van Oss CJ, Chaudhury MK, Good RJ. Interfacial Lifshitz-van der Waals and polar interactions in macroscopic systems. *Chem Rev.* 1988;88:927–941; and references therein.
21. van Oss CJ. Acid-bse interfacial interactions in aqueous media. *Colloids Surf A.* 1993;78:1–49; and references therein.
22. Bhattacharjee S, Sharma A, Bhattacharya PK. Surface interactions in osmotic pressure controlled flux decline during ultrafiltration. *Langmuir* . 1994;10:4710–4720.
23. O' Leary DJ, Wilson G. Tear side regulation of desquamation in the rabbit corneal epithelium. *Clin Exp Optom.* 1961;69:22–26.
24. Sharma A, Coles WH. Kinetics of corneal epithelial maintenance and graft loss: A population balance model. *Invest Ophthalmol Vis Sci.* 1989;30:1962–1970.
25. Dilly PN. Corneal of the epithelium to the stability of the tear film. *Trans Ophthalmol Soc UK.* 1985;104:381–389.
26. Liotet S, van Bijsterveld OP, Kogbe O, and Laroche L. A new hypothesis on tear film stability. *Ophthalmologica (Basel).* 1987;195:119–124.
27. Lemp MA, Gold JB, Wong S, Mahmood M, Guimaraes R. An in vivo study of corneal surface morphologic features in patients with keratoconjunctivitis sicca. *Am J Ophthalmol.* 1981;98: 426–428.

28. Bron AJ, and Mengher LS. The ocular surface in keratoconjunctivitis sicca. *Eye* 1989;3: 428–437.
29. Norn MS, Opauszki A. Effects of ophthalmic vehicles on the stability of the precorneal tear film. *Acta Ophthalmol.* 1977;55:23–34.
30. Burstein NL. Corneal cytotoxicity of topically applied drugs vehicles and preservatives. *Surv Ophthalmol.* 1980;25:15–30.
31. Sharma A, Ruckenstein E. The role of lipid abnormalities, aqueous and mucus deficiencies in the tear film breakup and implications for tear substitutes and contact lens tolerance. *J Colloid Interface Sci.* 1986;111:8–21.
32. Ruckenstein E, and Sharma A. A surface chemical explanation of tear film breakup and its implications. In: Holly FJ, Lamberts DW, Mackeen DL, eds. Dry Eye Institute; Lubbock, TX: *The Preocular Tear Film: In Health, Disease and Contact Lens Wear.* 1986:697–727.

52

THE BIOPHYSICAL ROLE IN TEAR REGULATION

Alan Tomlinson,[1] Jennifer P. Craig,[1] and Gerald E. Lowther[2]

[1]Department of Vision Sciences
Glasgow Caledonian University
Glasgow, Scotland, United Kingdom
[2]School of Optometry
Indiana University
Bloomington, Indiana

1. INTRODUCTION

The mechanism that determines and regulates tear production in the human eye has not been established. One suggested mechanism involves mediation through changes in the physical parameters of tears. Von Bahr[1] was the first to propose that tear osmolality was a function of tear production and elimination. This view was elaborated by Mishima[2] who suggested that the regulation of tear production was controlled by changes in tear tonicity, increases in tear tonicity leading to an increase in tear production in the normal eye, to protect the cornea from harmful changes in osmotic balance.[3,4] In the instance in which low tear evaporation or an impaired drainage system occurred, this would lead to the accumulation of tears in the eye, low tonicity, and the consequent inhibition of tear production. Conversely, if high evaporation occurred in the presence of a patent drainage system this would lead to reduced tear volume, an increase in tonicity, and a stimulus to production.

Clinical evidence exists to support a relationship between production and elimination of tears from the eye. Norn[5] and Mishima[2] showed that a diminished outflow of tears is followed by a decrease in secretion. Dalgleish[6] found that idiopathic obstruction to outflow in older patients was rarely accompanied by epiphora. Allen[7] found the almost complete lack of lacrimation in individuals with congenital absence of lacrimal puncta. Francois and Neetens[8] reported a few subjective complaints of epiphora in older patients with no connection between the lacrimal sac and nose or in patients with lymphosarcoma of the lacrimal sac. This led them to suggest the possibility of receptors in the wall of the lacrimal sac playing a part in an autoregulatory mechanism controlling the production and

Lacrimal Gland, Tear Film, and Dry Eye Syndromes 2
edited by Sullivan *et al.*, Plenum Press, New York, 1998

drainage of lacrimal fluid from the eye. Thus, a decrease in flow through the sac would initiate a decrease in tear production.

Previous investigations have attempted to show a relation between the elimination of tears from the eye (by evaporation) and the tear turnover rates in the normal eye, where these turnover rates were measured either by fluorimetry[9] or the Schirmer test.[10] Additional studies have been carried out relating evaporation and tear tonicity.[11] These studies failed to show the anticipated associations. The results may have been attributable to the small sample size, significant inter-subject variability in tear physiology, or the cross-sectional nature of the study design.

A recent study[12] investigated simultaneously the relation among tear tonicity, production (measured by gamma scintigraphy), evaporation, and drainage in normal subjects. Again, direct correlations between tear parameters were not established, but the results suggested that, similar to other systems in the body, tear production may be under homeostatic control. In such a homeostatic system no direct correlations would be anticipated among tear production, osmolality, and evaporation. Only in the event of tear osmolality above or below critical threshold values would changes in production be triggered. A feedback loop would control the physiology of the lacrimal system in such a way as to maintain tear parameters within narrow limits rather than at constant levels.[9–11]

The present study used a longitudinal study design in which the effect of altering one tear parameter was followed prospectively for a number of days. To investigate the regulatory responses of the normal lacrimal system, it was thought to be of interest to intervene and alter one tear parameter, and evaluate the reaction of others. One way of achieving this would be to create, in normal subjects, an artificial obstruction to tear outflow via the drainage system, which would mimic the clinical situation of pathology.[2,5–8] Non-dissolvable plugs were applied to the upper and lower puncta of subjects' eyes, and the effects on evaporation, turnover rate, and osmolality of the tears were monitored. Previous clinical research on the use of punctal plugs has indicated that although intermittent epiphora is common in the majority of patients on initial plugging,[13] this diminished after the first few days. This is anecdotal evidence for a change in tear production consequent upon the procedure. In addition to providing information on the regulatory mechanisms of tears, the present experiment would provide other important information on the effect of the clinical procedure of punctal occlusion on tear physiology.

2. METHODS

2.1. Subjects

Fourteen subjects (7 males, 7 females) aged 18–30 years were recruited from the staff and students of Glasgow Caledonian University. All subjects were free of ocular and general pathology, were taking no systemic medications, and were not wearing contact lenses. All subjects signed informed consent forms for the experiment, which was approved by the Ethics Committee of the University.

2.2. Procedure

Evaporation rate, osmolality, and tear turnover rate were measured. The subjects also filled out a questionnaire concerning tearing and comfort with the plugs. The protocol called for recordings to be made at the following intervals:

Baseline	2 days before punctal occlusion
Day 1	2 h after punctal occlusion
Day 4	3 days after punctal occlusion
Day 7	6 days after punctal occlusion (immediately preceding plug removal)
Day 15	1 week after plug removal.

In 11 of the 14 subjects, this was the pattern of measurements. Three subjects still reported excessive tearing at day 7, and the decision was made to leave the plugs in these subjects for a further 7 days. Therefore, in these subjects the readings immediately preceding removal were taken at day 15 and (after removal) at day 22.

2.3. Conditions of Measurement

Measurements on each subject were made consecutively in the same room within a 40 min period. The mean temperature and relative humidity throughout were 22.87±1.66°C and 42.02±3.00%.

Measurements were made in the same order for each subject, beginning with the least invasive (evaporimetry) and concluding with the most invasive (tear turnover rate).

3. TECHNIQUES

3.1. Tear Evaporation Rate

This was assessed with the modified Servomed Evaporimeter, which measured relative humidity and temperature at two points above the evaporative surface. The evaporation rate was determined from the vapor pressure gradient between these two points. The method was originally described by Trees and Tomlinson[14] and recently modified by Craig and Tomlinson.[15]

3.2. Nanoliter Osmometry

Tear osmolality was measured with the Clifton Technical Physics freezing point depression nanoliter osmometer by the procedure described previously.[16]

3.3. Tear Turnover Rate

Turnover rate was determined from the rate of decay of fluorescence of 1 μl of 1% fluorescein instilled onto the lower conjunctiva. Tear film fluorescence was measured with an automated scanning fluorophotometer (Fluorotron Master, Coherent Radiation) equipped with the anterior segment adaptor.[17] Four measurements were taken of baseline fluorescence prior to instillation of fluorescein and instillation at 1 min intervals for 20 min following the basal tear turnover rate [(Tt_0), defined as the percentage decrease of fluorescein concentration in tears in 1 min)] was determined from the formula:[17,18]

$$T_t(t_o) = 100 \cdot \frac{[C_t(t_o) - C_t(t_o + 1)]}{C_t(t_o)} (\% \cdot \min^{-1}) \tag{1}$$

where $C_t(t)$ is the concentration of fluorescein in tears at time t in min (ie, measured fluorescence minus corneal autofluorescence). Assuming a monophasic decay of tear film fluorescence with a decay constant k_t (min^{-1}):

$$C_t(t) = C_t(0)\, e^{-kt.t}\ (ng/ml) \quad (2)$$

It follows that

$$T_t = 100\,(1 - e^{-kt.1})\ (\%/min) \quad (3)$$

Assuming that the tear film fluorescence is proportional to the fluorescein concentration in the tear film, kt is equal to the decay constant obtained by fitting an exponential regression to the data.

The method of calculating tear turnover rate in this experiment was similar to that adopted in the European concerted action on ocular fluorimetry.[18] The Fluorotron employed was equipped with the software for data processing developed by the European group ("ANT-SEGMENT tear").[18] Basal tear turnover rate was determined by ignoring the first 5 min of readings following fluorescein instillation, consistent with the accepted practice of other workers in the field.[17,18]

3.4. Punctal Plugs

The non-dissolvable plugs were developed by Eagle Vision of Tennessee. These are known as Tapered Shaft Punctum Plugs in the United States and as Minim punctal plugs in the United Kingdom. All four lacrimal puncta were occluded. In the majority of cases, the "small" and "medium" plugs were employed, but the largest plug that could be inserted was used.

3.5. Subjective Responses to Punctal Plugging

Subjects were asked to assess subjectively the tearing (onto the facial skin) experienced in the period after punctal occlusion together with any sensation of discomfort. Subjects were issued a "diary" in which they recorded daily their subjective responses to excess tearing and discomfort on visual analog scales.[13]

4. RESULTS

The data obtained for evaporation rate, osmolality, and tear turnover rate were analyzed by time of measurement--baseline, day 1, day 4, etc. The values obtained for tear osmolality and turnover rate were similar at baseline and following plug removal ($t = -1.68$, $p=0.12$, and $t = 0.62$, $p = 0.55$, respectively. Therefore, the effects of punctal occlusion on these aspects of tear physiology were not evident 7 days after plug removal. Evaporation did show a reduction after plug removal ($t = 2.95$, $p = 0.01$).

The effect of punctal occlusion on tear physiology was determined from the values obtained at days 1, 4, and 7. An analysis of variance using a balanced design (Minitab for Windows) indicated a significant decrease in evaporation and tear turnover rate from days 1 through 7 with punctal occlusion ($F = 7.72$, $p = 0.002$; $F = 5.51$, $p = 0.01$, respectively) (Figs. 1 and 2). Tear osmolality showed a slight, but not significant, trend towards an in-

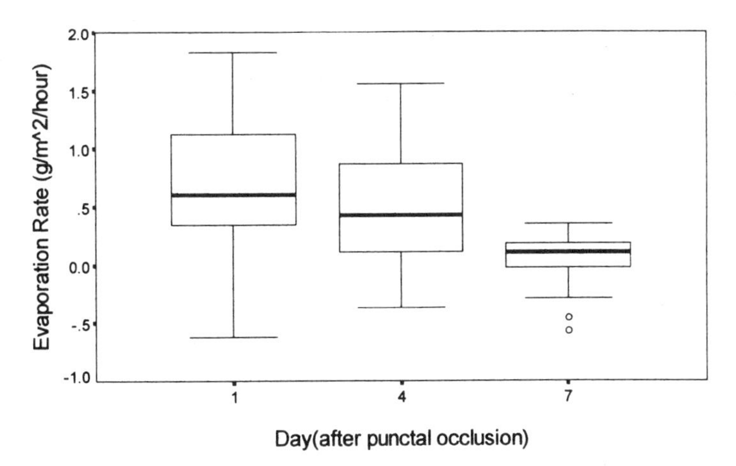

Figure 1. Evaporation rate after punctal occlusion. The median value for all subjects is shown by the thick line. Other values represent the 25, 75 percentiles and the minimum and maximum. (Outliers are seen as open circles.)

crease from a lower value obtained immediately following plug insertion to a higher value 7 days later ($F = 2.96, p = 0.07$) (Fig. 3).

The observations made from these measurements of tear parameters are supported by the subjective data included in the diary kept by the subjects. Subjective reports of epiphora, or spillage of tears over the lid margin, were highest immediately after plug insertion (Table 1). The record of epiphora showed a decrease through the period of punctal occlusion. This trend was significant by a one-factor, repeated-measures ANOVA ($F = 29.96, p = 0.0001$). The subjective reports of discomfort with the plugs showed some increase at plug insertion (Table 2). The discomfort noted on initial insertion decreased with time ($F = 7.48, p = 0.0001$). The initial discomfort noted with the plugs dropped to a lower level at days 4 and 7, a level similar to that after plug removal. In general, complaints of discomfort were more related to epiphora than to mechanical irritation from the plugs.

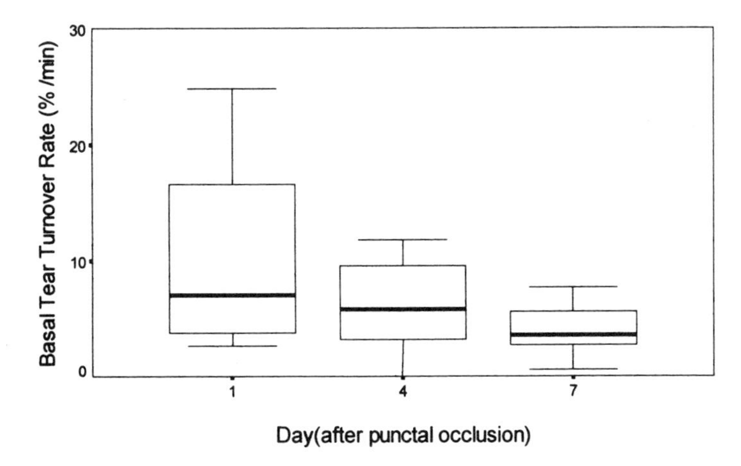

Figure 2. Basal tear turnover rate after punctal occlusion; values are as in fig 1. (Outliers are indicated with an asterisk.)

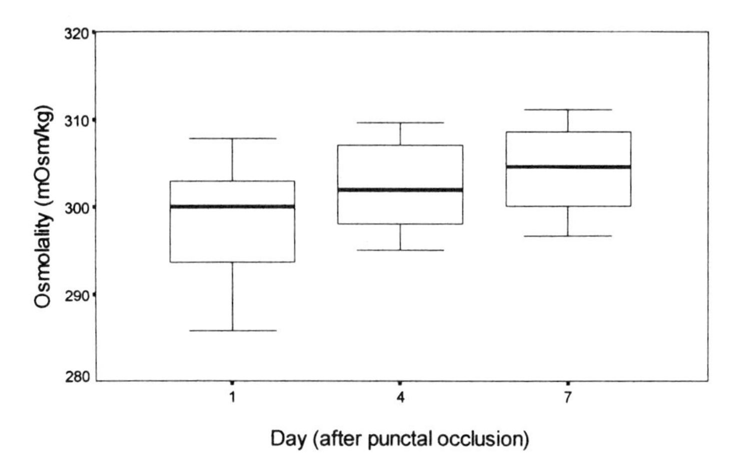

Figure 3. Osmolality after punctual occlusion. Values are as in Fig. 1.

To determine if inter-relationships existed between the tear parameters measured in this experiment, correlations were derived between parameters at days 1, 4, and 7 following plug insertion. Generally no significant correlations were found. Of the nine comparisons made, only osmolality and basal tear turnover correlated on the day of occlusion ($r^2 = 0.40, p = 0.016$).

5. DISCUSSION

The results from this study provide information about the impact of punctal occlusion on tear physiology, suggesting advantages of the procedure for dry eye patients. These advantages are subject to the caveat that extrapolation of results from patients with normal tear physiology may not be applicable to those with dry eye. The insertion of plugs into both puncta of subjects caused an increase in fluid volume in the eye, probably as a result of some stimulus to reflex production as well as enhanced retention of fluid in the eye. Obviously, these are desired effects in of dry eye. If tear turnover rate subsequently decreases in the same way in dry eyes as found for normals in this study, a consequent de-

Table 1. Subjective impression of epiphora after punctal occlusion

Day	No. of subjects	Mean level (%)	Standard deviation
1	14	80.17	14.57
4	14	44.84	20.32
7	14	33.34	18.26
15	11	0.80	1.07
15	3*	32.57	15.82
22	3*	0.53	0.92

Plugs were *in situ* on days 1, 4, and 7 in all subjects and through day 15 for 3 subjects*. All observations were made on visual analog scales.

Table 2. Subjective impression of discomfort with punctal plugs

Day	No. of subjects	Mean level (%)	Standard deviation
1	14	38.30	27.16
4	14	16.32	19.88
7	14	13.72	19.89
15	11	0.73	0.99
15	3*	17.2	12.82
22	3*	0.53	0.92

Plugs were *in situ* on days 1, 4, and 7 for all subjects and through day 15 on 3 subjects*. All observations were made on visual analog scales.

crease in fluid volume over time could offset the initial beneficial effects. However, if punctal occlusion is combined with the use of artificial tears, a dry eye patient will benefit even in the situation where changes in tear physiology take place.

Evaporation was relatively high in the normal subjects after punctal occlusion--a similar finding to that reported by Tsubota and Yamada.[19] This was probably a result of enlargement of the envelope of tear fluid to be contained by the lipid layer. The efficacy of the lipid in inhibiting evaporation was likely to have decreased by this enlargement. In dry eye cases in which there is an aqueous deficiency and decreased tear fluid in the eye, the effect of punctal occlusion would be beneficial, providing the lipid film is normal. However, in those cases where the dry eye is due to lipid deficiency, ie, in cases of wet dry eye,[20] punctal occlusion may exacerbate the problem. The increased volume of tear fluid in the eye after plugging would result in the deficient lipid layer experiencing even greater stress and lead to an increase in tear evaporation rate. The significant reduction in tear osmolality reported in dry eye patients after punctal occlusion is not confirmed in this study.[21] The osmolality was found to decrease after plugs were inserted, but not at a statistically significant level.

A concern with previous studies of the effect of punctal occlusion using non-dissolvable intra-cannicular plugs[13] or collagen inserts[22] is the possibility that complete occlusion did not take place. This is less likely with the plugs employed in this study, which are raised with a visible dome section, sitting above the lid margin. It is still possible that some drainage took place around the plugs, although the largest plug that could be inserted into the punctum was used in each case. Certainly the presence of the plugs could be verified, but total occlusion could not, as a tracer visible beneath the skin (eg, a radioactive tracer used in previous experiments[12]) was not employed in this experiment. However, the considerable increase in fluid volume in the eye on occlusion and the subjective complaint of initial tearing onto the facial skin suggest that a significant amount of obstruction to drainage was achieved. In addition, the amount of blockage of the drainage channel obtained with the plugs was probably constant throughout the six days of occlusion (no plugs were dislodged), so observations of the effect of plugging on tear physiology are valid.

The model suggested by von Bahr[1] and Mishima[2] for the regulation of tear production through changes in osmolality would predict that an increase in osmolality would result in an increased production, or, conversely, a decrease in osmolality would result in a reduction in production rate. Previous attempts to show these relationships failed to show a linear correlation between osmolality and evaporation[11] or between evaporation and tear turnover rate.[9] The absence of a close linkage in a later experiment[12] relating evaporation, osmolality, and tear turnover rate (by gamma scintigraphy) has led to the suggestion that tear production is controlled by a yet-unknown, homeostatic mechanism. The results of the present experiment lend support to a homeostatic regulatory mechanism for tear production, but not necessarily mediated through changes in tear osmolality. The intervention in the normal regulation of tear production in the study subjects was achieved by altering the drainage facility from the eye by the use of punctal occlusion. Tear evaporation rate was higher on initial plugging, but was not large enough to compensate for the absence of drainage facility in the occluded subjects.

Plugging the puncta produced increased tear production, which may have been in part due to a reflex stimulus from awareness of the plugs (Table 1). The presence of reflex tearing can be seen from the higher level of turnover and the large scatter of the day 1 data, although the median level of turnover remained within the normal range of basal values (Fig. 2). As the subjects became more accustomed to the plugs over the six days, the reflex tearing reduced, as seen by the reduced scatter of data at day 7. This reflex component of tear production during the experimental period overlays changes taking place in the basal level of tear production during punctal occlusion. However, a significant decrease in basal turnover is evidenced by the reduction in values during this period (Fig. 2) and by the decrease in subjects observations of epiphora after two to three days of punctal occlusion (Table 1).

The method of determining tear production in this experiment is indirect, measuring a change (rate of decay) in fluorescence of a fluorophore instilled onto the conjunctiva. This fluorescence decay is dependent on the change in concentration of fluorescein in the tears and is affected by several factors. This rate will be reduced with increases in evaporation of tears and increased by episodes of epiphora and increased production of new tears. At day 1, the evaporation rate is higher than at day 7, mediating against the decrease of turnover seen in the experiment. Epiphora is reportedly more frequent at day 1 and absent in most subjects by day 7 and could explain the overall results for turnover rate at the two measurement points. However, as the capacity of the eye to retain fluid is unchanged between days 1 and 7, the reduction in epiphora at day 7 is likely to be due to a decrease in the actual production of tears. Therefore, we feel the weight of evidence in this experiment is in favor of a change in actual production of tears over the seven days following punctal plugging. The decrease in tear turnover was accompanied by a significant reduction in evaporation rate during the period of occlusion. Osmolality showed a slight increase in values from day 1 to day 7 but this was not significant. Therefore the change in tear turnover in this period may not have been a consequence of changes in osmolality and cannot be seen as direct support for the model of tear regulation hypothesized by von Bahr[1] and Mishima.[2]

If this change in tear production rate is triggered by a regulatory mechanism, at what point would this mechanism become effective? Previous work has failed to show a direct correlation between osmolality and production[12] or between osmolality and tear elimination (via evaporation).[11] However, if the linkage was not a direct one but one affected by homeostasis mediated by osmolality or some other trigger stimulus, the effect would be produced only if the change in the trigger exceeded a normal range of values. In an experiment such as the one described here in which evaporation, osmolality, and tear turnover rate were measured consecutively (within a relatively short period of time), but not concurrently, the identification of trigger events may be difficult. But as all the changes in physical tear parameters observed in this experiment were long term and took place over several days (Figs. 1–3), the effect of consecutive measurements over a short space of time was probably not important. Scatter plots of individual parameters generally suggest no relationships. The results, then, of these correlations and the failure to find direct relationships between any of the parameters evaluated during punctal occlusion are consistent with those observed in previous experiments[9,11,12] and lend support to a looser feedback/homeostatic mechanism, rather than one with a direct (precise) linkage between physical tear parameters.

If a homeostatic mechanism downregulated tear production during the six days of punctal occlusion in this experiment, what is the most likely site and trigger for it? François and Neetens[8] have argued for a detection system monitoring tear flow through the lacrimal sac. It is true that the condition of punctal occlusion will reduce flow through the sac and may be said to provide evidence supporting this theory. However, we find it illogical in physiological terms that a regulatory mechanism would be designed to reduce tear production further in the event of a reduced tear flow being monitored in the drainage system. Additionally, this would be an inappropriate site for such a governor, as it would not monitor the adequacy of tears at the point where they are required, on the ocular surface. Instead, we favor a homeostatic system that is regulated, at the ocular surface, by a detector that monitors either volume in the eye or concentration of a macromolecule in the tear fluid.

ACKNOWLEDGMENTS

The punctal plugs used in this study were donated by CIBA Vision Ophthalmics Inc. and Eagle Vision.

REFERENCES

1. von Bahr G, Könnte der Flüssigkeitsabgang durch die corne von physioogischer bedeutung sein. *Acta Ophthalmol.* 1941;19:125–134.
2. Mishima S. Some physiological aspects of the precorneal tear film. *Arch Ophthalmol.* 1965;73:233–241.
3. Balik J. The lacrimal fluid in keratoconjunctivitis sicca: A quantitative and qualitative investigation. *Am J Ophthalmol.* 1952;35:773–782.
4. Gilbard JP, Rossi SR, Gray KL, Hanninen LA, Kenyon KR. Tear film osmolarity and ocular surface disease in two rabbit models for keratoconjunctivitis sicca. *Invest Ophthalmol Vis Sci.* 1988;29:374–378.
5. Norn MS. Tear secretion in diseased eyes. *Acta Ophthalmol.* 1966;44:25–32.
6. Dalgleish R. Incidence of idiopathic acquired obstruction in the lacrimal drainage apparatus. *Br J Ophthalmol.* 1964;48:373–376.
7. Allen J. Congenital absence of the lacrimal punctum. *J Pediatr Ophthalmol.* 1968;5:176–178.
8. François J, Neetens A. Tear flow in man. *Am J Ophthalmol.* 1973;76:351–358.
9. Tomlinson A, Trees GR, Occhipinti JR. Tear production and evaporation in the normal eye. *Ophthalmic Physiol Opt.* 1991;11:44–47.
10. Cedarstaff TH, Tomlinson A. Human tear volume, quality and evaporation: A comparison of Schirmer, tear break-up time and resistance hygrometry techniques. *Ophthalmic Physiol Opt.* 1983;3:239–245.
11. Tomlinson A, Simmons PA. Regulation of tear production: Relation between tear evaporation and osmolarity in normals. ARVO abstracts. *Invest Ophthalmol Vis Sci.* 1992;33:S950.
12. Craig JP. Tear physiology in normal and dry eyes. PhD Thesis, Glasgow Caledonian University. 1995.
13. Slusser TG. Effects of lacrimal drainage occlusion with non-dissolvable intracanalicular plugs on hydrogel contact lens related dry eye. MS Thesis, University of Alabama at Birmingham, 1994.
14. Trees GR, Tomlinson A. Effect of artificial tear solutions and saline on tear film evaporation. *Optom Vis Sci.* 1990;67:886–890.
15. Craig JP, Tomlinson A. Importance of the lipid layer in human tear film stability and evaporation. *Optom Vis Sci.* 1997;74:8–13.

16. Craig JP, Tomlinson A. Effect of age on tear osmolality. *Optom Vis Sci.* 1995;72:713–717.
17. Kok JC, Boets EPM, van Best JA, Kijlstra A. Fluorophotometric assessment of tear turnover under trial contact lenses. *Cornea.* 1992;11:515–517.
18. van Best JA, del Castillo JMB, Coulangen LM. Measurement of basal tear turnover using a standardised protocol. *Graefes Arch Clin Exp Ophthalmol.* 1995;233:1–7.
19. Tsubota K, Yamada M. Tear evaporation from the ocular surface. *Invest Ophthalmol Vis Sci.* 1992;33:2942–2950.
20. Rolando M, Refojo MF. Tear evaporimeter for measuring water evaporation from the tear film under controlled conditions in humans. *Exp Eye Res.*. 1983;36:25–33.
21. Gilbard JP, Rossi SR, Azar DT, Heyda KG. Effect of punctal occlusion by Freeman silicone plug insertion on tear osmolarity in dry eye disorders. *CLAO J.* 1989;15:216–218.
22. Lowther GE, Semes L. Effect of absorbable intracanalicular collagen implants in hydrogel contact lens patients with drying symptoms. *ICLC* 1995;22:238–243.

53

LONGITUDINAL ANALYSIS OF PRECORNEAL TEAR FILM RUPTURE PATTERNS

Etty Bitton and John V. Lovasik

École d'optométrie
Université de Montréal
Montréal, Québec, Canada

1. INTRODUCTION

The precorneal tear film is a complex structure that has attracted much attention over the years, especially in relation to dry eyes and contact lens (CL) wear. A healthy and stable tear film is essential for comfortable vision and successful CL wear.[1] With an estimated 40% of the North American population suffering from dry eyes[2] and some 60 million CL wearers worldwide,[3] the need for simple, accurate, and cost-effective tests of tear stability has spawned much research both in the ophthalmic industry and in clinical practice.

A thorough case history and an assessment of the production, distribution, elimination, and stability of the precorneal tear film (PCTF) are essential parts of a CL and dry eye patient work-up. Several clinical tests are available for evaluating the PCTF, but many of these remain qualitative and highly dependent upon the skill and experience of the observer. The two principal challenges that confront clinicians in the area of tear physiology relate to universally accepted definitions of "dry eye" and "CL failure" and to standardization of tear tests such as the Tear Break-Up Time (TBUT), to allow valid comparisons of results across studies.[4]

Among the numerous tear tests available,[1] tear film rupture has been studied widely and used clinically to assess tear film stability.[5,6] Tear film rupture has been traditionally observed by coloring the PCTF with fluorescein, an inert ophthalmic dye, and observing the break-up of the dye on the surface of the cornea. The interval between a blink and the first spontaneous tear rupture has been termed the Tear Break-Up Time (TBUT).[5–7] Although the TBUT has had many important clinical applications, it has relatively poor specificity and sensitivity and demonstrates substantial variability within and across subjects.[8] Nonetheless, its ease of use and low cost have made it a mainstream diagnostic technique for over 30 years.[9,10]

Lacrimal Gland, Tear Film, and Dry Eye Syndromes 2
edited by Sullivan *et al.*, Plenum Press, New York, 1998

The TBUT remains a static measurement representing a singular temporal event and does not offer any qualitative or quantitative information about the form of the ruptured tear area. Since the distribution of tears across the cornea is a multifactorial dynamic process, it may be more appropriate to study tear film phenomena longitudinally for a better understanding of their physical characteristics. With the ever-increasing sophistication in digital imaging technology and faster and more specialized digital image capture cards, many rapid biological events can now be imaged and analyzed in 'real time.' In the area of tear physiology, the tear rupture event can be imaged and studied as it changes over time as opposed to a single 'snapshot' that reflects the TBUT.[11] With this in mind, the objective of the present study was to apply on-line digital imaging technology to study the fluorescein-colored PCTF with respect to the time of a first rupture in the tear layer (ie, the TBUT) and subsequent evolution of the pattern of tear rupture areas.

2. MATERIALS AND METHODS

Ten healthy subjects (5 women and 5 men), ages 21 to 44 years (mean ± SD, 27.1 ± 7.5 years) volunteered for the present study. The subjects had best corrected visual acuities of 6/6 (20/20) or better in each eye, did not wear contact lenses, were not taking any medications, and had no complaints of dry eye. The tear film was highlighted with fluorescein (Barnes-Hind 0.6 mg Ful-Glo® sterile strips) premoistened with unpreserved saline (Allergan Lens Plus, aerosol) and instilled in the inferior temporal conjunctival cul-de-sac. The precise amount of fluorescein instilled was not quantified since previous studies found that the exact quantity of fluorescein did not affect TBUT measurements.[4] The fluorescein-labelled PCTF was highlighted using a cobalt filter (transmission peak 485 nm) and a barrier filter (transmission peak 525–530 nm) positioned in front of the slit lamp objectives to further improve tear visibility. All images were recorded in the reduced ambient lighting typically used in routine biomicroscopy.

Immediately prior to recording, each subject was positioned comfortably in the head/chin rest of a Zeiss 75 SLT*FL biomicroscope, asked to look straight ahead, and then blink two or three times after the fluorescein instillation to ensure uniform distribution of the fluorescein across the corneal surface. A full illumination beam[4,12] positioned 40–50 degrees temporally and a total magnification that filled the computer monitor screen with the cornea were set prior to recording. A complete recording included a complete blink and the interval thereafter, wherein the subject resisted blinking as long as comfortable. The eyelids were not manipulated in any way throughout the test, as this reduces the TBUT.[4,6] Three such blink records were made for each eye, starting with the right eye, all in a single test session. A typical recording session lasted 20 - 40 min. The test room was not susceptible to large changes in temperature, humidity or ventilation, thus minimizing any environmental contaminants that might affect the tear film dynamics. If fluorescein was excessively diluted through lacrimation, additional dye was instilled for subsequent recordings.

The fluorescein-colored tear film was imaged with a high-resolution (480 H lines), low light sensitivity (0.5 lux) Zeiss HVS 1470, RGB camera and digitized and compressed with an Adobe Premiere board housed in a MacIntosh IIci computer. Full color digitization occurred at a rate of 30 frames per sec (fps). These data were relayed to an external optical disk storage system (Topline) with a 2 gigabyte capacity. The National Institutes of Health (NIH) *Image* software was used for image analysis. This software has an 8 bit resolution capacity; hence, prior to analysis, stored images were transformed from 24 bit to 8 bit resolution using the Adobe Photoshop image editing software.

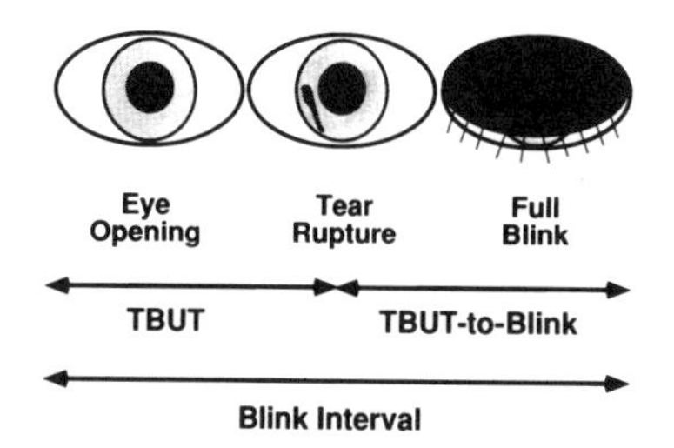

Figure 1. Blink interval terminology illustrating the TBUT, TBUT-to-Blink, and Complete Blink Intervals.

To constrain the volume of data for analysis to manageable levels, a temporal resolution of 0.5 sec was selected. The frame containing the fully developed rupture area was isolated through a frame-by-frame visual search procedure and then 'thresholded' to allow quantification of various parameters of this ruptured area as well as all preceding ones at 0.5 sec intervals. The same threshold criterion was applied to every 15th frame in a single recording sequence. Only the first spontaneously ruptured area was followed for longitudinal changes.

For each tear rupture sequence, the following attributes were measured: the sector of the cornea in which the rupture pattern was found using the corneal coordinate system proposed by the Australian Cornea and Contact Lens Research Unit (CCLRU);[13] the rupture pattern morphology; the timing of the rupture (ie, the TBUT); the rupture evolution rate; the duration of the rupture (i.e., interval between the TBUT and the blink); and finally the complete blink interval (Fig. 1).

3. RESULTS

Three distinct rupture patterns of the PCTF were identified in the present study which totaled 60 separate recordings (3 per eye in 10 subjects). The rupture morphologies detected were characterized as either Streaks, Dots, or Pools.[14,15] A rupture pattern was identified as a Streak if it had a linear shape upon its appearance and subsequent development. The orientation of the Streak could be along any meridian. A pattern with a circular-like morphology was categorized as a Dot. A disturbance of the tear film exhibiting neither a circular nor a linear shape was categorized as a Pool rupture pattern (Fig. 2).

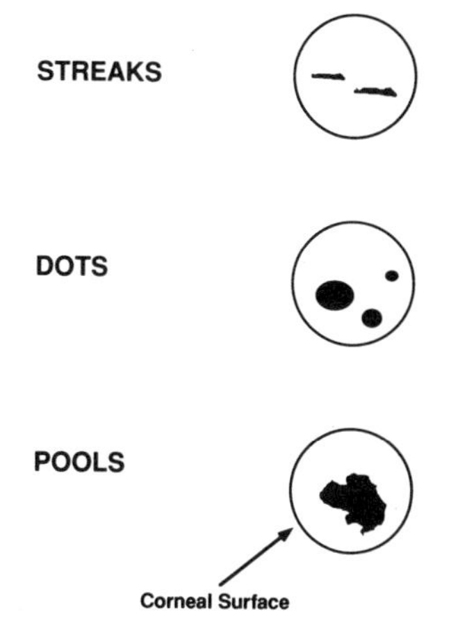

Figure 2. Tear rupture pattern morphology.

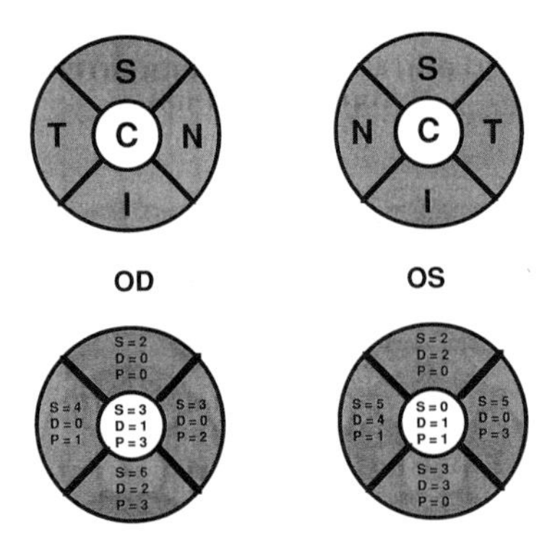

Figure 3. Corneal sector allocation: S, superior. I, inferior. N, nasal. T, temporal. C, central. Frequency distribution of each rupture pattern type over the reference corneal sectors: S, Streaks. D, Dots. P, Pools.

The number and sectorial distribution of each rupture pattern for right and left eyes are illustrated in Fig. 3. A total of 33 Streaks, 14 Pools, and 13 Dot-like rupture patterns were found in the recordings. Regardless of the rupture category, the superior corneal areas of either eye had the lowest frequency of disruption of the tear film, whereas the corneal band across the 3 o'clock to 9 o'clock position was most densely populated by ruptures falling across all three categories.

Streak-like rupture patterns exhibited the shortest TBUT (mean ± SD, 3.4 ± 0.55 sec) and were significantly different (t-test, $p = 0.002$) from the Dots, which revealed the longest TBUT (10.3 ± 1.73 sec) (Fig. 4).

The rate at which the area of rupture increased over time was also compared across the rupture patterns. Time zero was considered to be the video frame where the first thresholded rupture area was detected. The size of the thresholded area was expressed as a percentage of the Region of Interest (ROI). Subsequent changes in the size of the rupture area were tracked in a similar fashion over time. The averaged size of the rupture area by rupture pattern was then plotted as a function of time based on a 0.5 sec resolution (i.e., every 15th frame). The change in the size of the rupture area over time for Streaks, Dots,

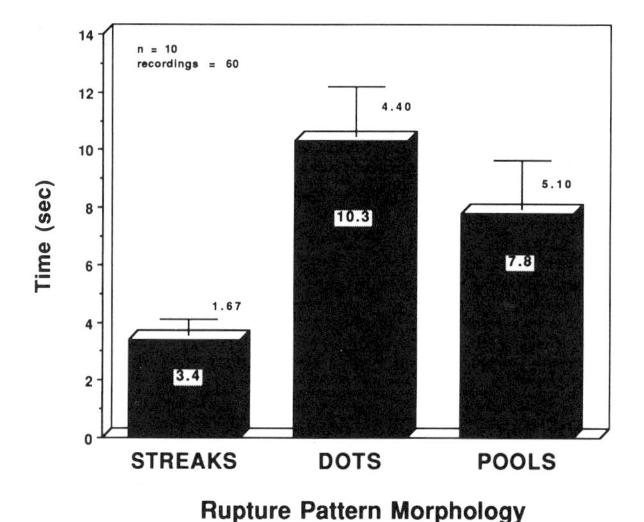

Figure 4. Tear breakup time across rupture pattern type.

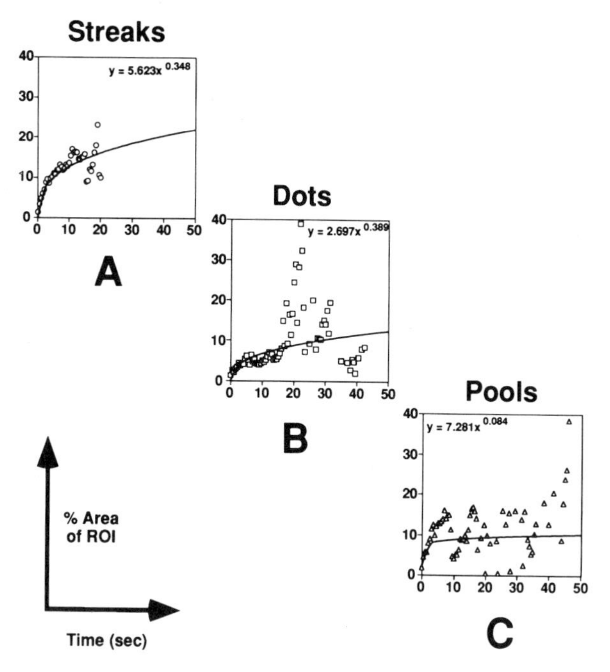

Figure 5. Tear rupture pattern evolution across rupture pattern type.

and Pools is shown in Fig. 5. To simplify analyses, the linear data were transformed into a natural log format, as shown by the equations below. The slope of the linear regression line through each data set gives the average rate of rupture per pattern. The regression formulas defining each function are as follows: y (Streaks) = $5.623x^{0.348}$, y (Dots) = $2.697x^{0.389}$ and y (Pools) = $7.281x^{0.084}$.

Linear equation:	$y = a + bx$
Transformation to natural log:	$\ln y = \ln a + b \ln x$
Power function equation:	$y = a\, x^b$

The persistence of the rupture pattern (i.e., the TBUT-to-Blink time) and the Blink Interval were measured and compared across tear rupture patterns. The Streak-like rupture pattern revealed the shortest timing of these intervals; the Dot-like morphology, the longest (Figs. 6 and 7). These intervals as well as the TBUT revealed significant differences for the Streak-like and Dot-like rupture patterns (Table 1).

4. DISCUSSION

The mechanisms underlying the normal spontaneous rupture of the tear film are yet to be fully elucidated. The physical and biochemical factors causing the tear film to rupture into either Streaks, Pools, or Dot-like patterns within and across individuals remain unknown. If lipid contamination of the mucin layer[16–18] is one of the causal mechanisms for tear film rupture, the rapidly evolving area for the Streak pattern (Fig. 4) found in the

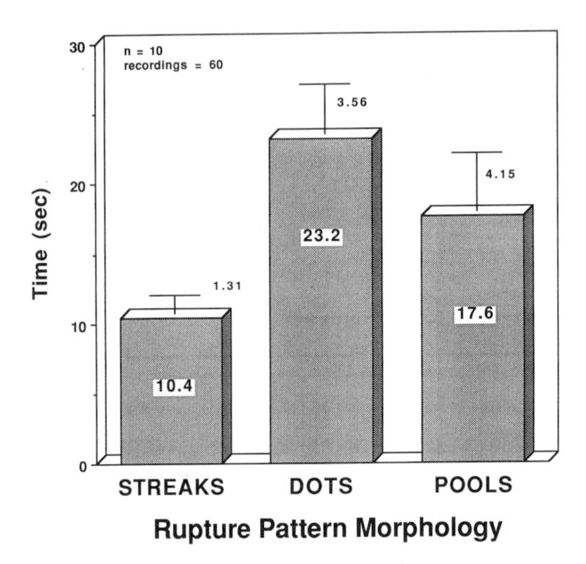

Figure 6. Tear rupture persistence (TBUT-to-Blink) interval across rupture pattern type.

present study may be indicative of subjects whose mucin is more susceptible to lipid contamination, and hence they may be poorer candidates for contact lens wear. On the other end of the spectrum, subjects with Dot-like rupture patterns would appear to have a greater resistance to mucin contamination by lipid and hence demonstrate a longer-lasting tear film layer seen here as a more slowly evolving rupture pattern.

Very few studies have correlated the shape and location of tear film rupture patterns in systematic fashion. Until recently, the absence of suitable technology made such studies very difficult if not impossible to perform on a quantitative basis. Early studies on corneal

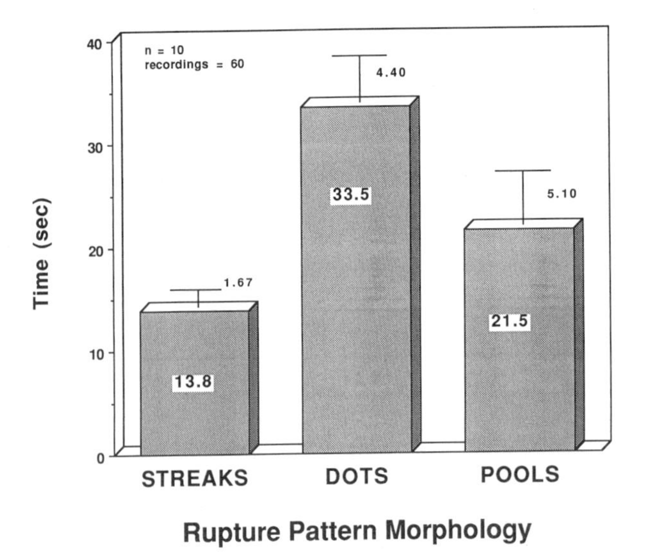

Figure 7. Blink interval across rupture pattern type.

Table 1. Statistical analyses of the components of the Blink Interval across pattern type

Interval	t-Test (separate variance)	*p*
TBUT		
Streaks vs. Dots	3.79	0.002
Dots vs. Pools	1.05	0.306
Streaks vs. Pools	2.51	0.023
TBUT-to-Blink		
Streaks vs. Dots	3.38	0.004
Dots vs. Pools	1.04	0.31
Streaks vs. Pools	1.65	0.119
Blink Interval		
Streaks vs. Dots	4.20	0.001
Dots vs. Pools	1.79	0.086
Streaks vs. Pools	1.44	0.17

desiccation described corneal dry spots as having a linear or dot-like shape that occurred rather randomly and predominantly in the inferior and temporal regions of the cornea.[12,19,20] The results of the present study support some earlier findings. A ratio of approximately 3:1 has been calculated for ruptures in the inferior vs. superior cornea.[12,20] Our findings of about 2.8/1 are in close agreement with those earlier studies. An increased exposure of the infero-temporal region of the cornea to air currents has been proposed as a possible explanation for the tendency for this area to desiccate before other areas of the cornea.[20]

Streak-like rupture patterns revealed the shortest TBUT and the Dot-like pattern the longest. Traditionally the TBUT has been used clinically, in combination with other tear tests, to identify candidates exhibiting unstable tear films, albeit with considerable variability in measurements.[1,7] Rupture patterns may become a useful clinical tool for identification of poor contact lens candidates. With three recordings per eye, the pattern showed at least 66% repeatability, making it less variable than the static TBUT measurement. Tear rupture patterns may be a better index of tear film stability, since they provide information on the dynamic aspects of tear rupture and morphology. Further clinical studies will provide additional information on the stability and reliability of tear film rupture patterns as an index of tear film stability.

Our study investigated the influence of rupture pattern type upon initiating a blink reflex. Most patients reported that they could no longer keep their eyes open due to a stinging or irritating 'dryness' from the slit lamp light source. Streak-like rupture patterns were noted to have the shortest time interval to the closing blink (mean, 10.31 sec), while Dot-like patterns exhibited the longest TBUT-to-Blink interval (mean, 23.24 sec). These latter findings provided additional support for the theory that rupture patterns and their evaluation rates can be used as indicators of PCTF stability. A Streak-like pattern ruptures more quickly (i.e., shorter TBUT) and evolves more rapidly, exposing more of the corneal surface and destabilizing the PCTF more quickly. The irritation response results in a faster reflex blink to redistribute the PCTF more quickly and hence cause a shorter TBUT-to-Blink interval. Conversely, the Dot-like rupture patterns take longer to rupture (ie, longer TBUT), evolve more slowly, and disturb the PCTF less, thereby providing a more stable PCTF as witnessed by a longer TBUT-to-Blink interval. Interestingly, the blink interval *per se* could also distinguish between Streak and Dot-like rupture patterns. This could be extremely useful in simplifying the screening of poor contact lens candidates since it could be performed without the necessity of fluorescein.

Our results demonstrate that Streak-like rupture patterns are significantly different from Dot-like patterns. Rupture patterns exhibiting Pool-like morphology are not statistically different from the other two rupture pattern populations. There may be several explanations for this: (i) An error of classification, ie, a rupture pattern was identified as a Pool when in fact it was really a Streak or Dot. (ii) Pool-like rupture patterns are a separate class of rupture patterns exhibiting partial characteristics of Streak and Dot-like patterns. (iii) Small sample size. Further studies on tear rupture patterns are required to determine morphological subgroups more precisely.

5. CONCLUSIONS

When combined with other tests of tear film integrity, the conventional TBUT provides a simple, quantifiable, and clinically usable index of tear film stability. However, since tears are in continuous motion due to chemical, physical, and mechanical factors, an evaluation of the *evolution* of a rupture in the tear film may provide an improved physiological index of corneal wettability. In the clinical setting, patients exhibiting a short TBUT and/or a rapidly expanding tear rupture area coupled with a short TBUT-to-Blink time (as is characteristic of Streak-like rupture patterns) may be identified more reliably as having an unstable tear film and may be more prone to dry eye or contact lens-related problems.

Additional investigations are needed to assess tear film rupture patterns in patients wearing various types of contact lenses and in patients presenting with dry eye symptoms. Further studies of tear film phenomena are needed to improve our understanding of the fundamental corneal wetting properties of the PCTF. The recording and analysis system developed for the present study offers a new, potentially productive approach for addressing some of the issues raised by our present findings.

ACKNOWLEDGMENTS

We thank all our subjects for their participation in this study. We acknowledge Dr. Jacques Gresset for his help with statistical analyses. We also gratefully acknowledge grant support from the Canadian Optometric Education Trust Fund (COETF) to E. Bitton and J.V. Lovasik, and a Natural Sciences and Engineering Research Council of Canada (NSERC) research grant (no. OGP0116910) to J.V. Lovasik.

REFERENCES

1. Caffery BE, Josephson JE. The preocular tear film in contact lens wear. *Pract Optom.* 1992;3:88–95.
2. Forst G. The precorneal tear film and "dry eyes." *International Contact Lens Clinic.* 1992;19:136–140.
3. Holden B. International association of contact lens educators expands globally. *International Contact Lens Clinic.* 1992;19:171–184.
4. Cho P, Brown B. Review of the tear break-up time and a closer look at the tear break-up time of Hong Kong Chinese. *Optom Vis Sci* . 1993;70:30–38.
5. Norn MS. Desiccation of the precorneal film I. Corneal wetting time. *Acta Ophthalmol.* 1969;47:865–880.
6. Lemp MA, Hamill JR. Factors affecting tear film breakup in normal eyes. *Arch Ophthalmol.* 1973; 89:103–105.

7. Lupelli L. A review of lacrimal function tests in relation to contact lens practice. *Contact Lens J.* 1986;14:4–17.
8. Cho P. Stability of the precorneal tear film: A review. *Clin Exp Optom.* 1991;74:19–25.
9. Norn MS. Vital staining of cornea and conjunctiva. *Acta Ophthalmol.* 1962;40:389–401.
10. Norn MS. Vital staining in practice. *Acta Ophthalmol.* 1964;42:1046–1053.
11. Bitton E, Lovasik JV. Modelling the tear film rupture pattern using dynamic digital imaging techniques. *Can J Optom.* 1994;56:94–98.
12. Cho P, Brown B, Chan I, Conway R, Yap M. Reliability of the tear break-up time technique of assessing tear stability and the locations of the tear break-up in Hong Kong Chinese. *Optom Vis Sci.* 1992;69:879–885.
13. Terry RL, Schnider CM, Holden BA, et al. *cclru* standards for success of daily and extended wear contact lenses. *Optom Vis Sci.* 1993;70:234–243.
14. Bitton E, Lovasik JV. Digital morphometry of tear rupture patterns. *Optom Vis Sci.* 1993;70:64.
15. Bitton E, Lovasik JV. Influence of tear film rupture patterns on the blink interval. *Invest Ophthalmol Vis Sci.* 1994;35:1576.
16. Holly FJ. Formation and rupture of the tear film. *Exp Eye Res.* 1973;15:515–525.
17. Holly FJ. Tear film physiology. *Am J Optom Physiol Opt.* 1980;57:252–257.
18. Holly FJ. Tear film formation and rupture: An update. In: Holly F, ed. *The Preocular Tear Film in Health, Disease and Contact Lens Wear.* Lubbock, TX: Dry Eye Institute; 1986:634–645.
19. Marx E. De la sensibilité et du desséchement de la cornée. *Ann 'Oc.* 1921;150:774–789.
20. Rengstorff RH. The precorneal tear film: Breakup time and location in normal subjects. *Am J Optom Physiol Optics.* 1974;51:765–769.

54

THE ROLE OF TEAR PROTEINS IN TEAR FILM STABILITY IN THE DRY EYE PATIENT AND IN THE RABBIT

Ronald D. Schoenwald,[1] Sangeeta Vidvauns,[1] Dale Eric Wurster,[1] and Charles F. Barfknecht[2]

[1]Division of Pharmaceutics, and
[2]Division of Medicinal and Natural Products Chemistry
College of Pharmacy
The University of Iowa
Iowa City, Iowa

1. INTRODUCTION

Sigma receptors have been identified in the lacrimal glands of rabbits; stimulation of these receptors results in the modulation of protein secretion from acinar cells.[1–4] When acinar cells were incubated with various sigma ligands, it was established that increases and decreases in protein release could be used to identify agonists and antagonists.[4] Following the application of a newly designed agonist, *N,N*-dimethyl-2-phenylethylamine HCl (AF2975), to the rabbit eye, a statistically significant increase in total protein was observed, when compared to either baseline values or the fellow eye.[3] Tears were collected, and protein fractions were separated into various fractions with the use of size-exclusion high-pressure liquid chromatography (SE-HPLC).[4] In particular, a 23 min protein fraction (about 16–18 kDa) was found to increase by 150% and 90% at 10 and 60 min, respectively following the topical application of AF2975 (50 μl of 0.15%) to the rabbit eye. Other protein peaks also showed an increase, but much less than the 23 min peak. Following desalting and concentrating, it was possible to separate the 23 min protein peak from rabbit tears into five isoforms using isoelectric chromatofocusing and a protein standard with a pI of 4.6. The isoforms were acidic with pIs higher than 4.6; however, the 23 min protein peak has not been specifically identified, which would require determination of its amino acid sequence. The function of the small-molecular-weight 23 min protein fraction is not known, but has been contrasted to the human tear fraction commonly referred to as tear-specific prealbumin (TSP) or as proteins migrating faster than albumin (PMFA).[5–7]

Lacrimal Gland, Tear Film, and Dry Eye Syndromes 2
edited by Sullivan *et al.*, Plenum Press, New York, 1998

In human tears, the key proteins that predominate in tears but are not present in serum are lysozyme, lactoferrin, secretory IgA, and tear lipocalin (TL). Sixty proteins have been identified in human tears, but they account for approximately 65–75% of the total tear proteins.[6,8] With the exception of TL, these proteins are important in providing antimicrobial activity to tears. Gachon et al.[6] and Fullard[9] found that TL is a highly acidic protein that electrophoretically could be separated into six isoforms in the pH range of 4.6–5.4. Although the specific function of TL has not been established, its amino acid sequence has been identified and found to match closely the lipocalin family of proteins.[10] Glasgow et al.[11] established that TL binds to a host of endogenous fatty ligands, namely stearic acid, palmitic acid, lauric acid, and cholesterol. In particular, it is conceivable that TL could bind fatty acid ligands to the outer surface of the tear film and thus contribute significantly to its stability.

In this report, both rabbit tears and the tears of dry eye patients were used to test whether tear protein extracts are important to the stability of the tear film. Surface tension and in vitro breakup time (BUT), the latter a newly designed in vitro procedure, were used to measure the surface properties of protein extracts.

2. MATERIALS AND METHODS

2.1. Tear Collection from Dry Eye Patients

A preliminary study of AF2975 in dry eye patients was sponsored by Angelini Pharmaceuticals, Inc. and conducted at The University of Iowa Hospitals & Clinics, Department of Ophthalmology. Experiments separate from the study protocol were conducted with protein extracted from Schirmer strips. Schirmer strips were obtained from 14 dry eye volunteers, 12 females and 2 males, between the ages of 26 and 68, who were without significant systemic or ocular disease, except for dry eye syndrome. Fourteen dry eye volunteers placed one or two drops of AF2975 0.15% in one eye and a vehicle in the other eye four times a day for 21 days at 8 AM, 12 noon, 4 PM, and 8 PM.

After day 7 and again after day 21, Schirmer test strips were inserted in the eyes, collected after 5 min, measured for wetting, and placed dry under refrigeration

(-20°C). The time that the strips were removed from the eye after the last dose was recorded. After all of the strips had been collected (1–2 h after dosing), they were extracted and assayed for total protein and TL. Tear extracts were then measured for surface tension as well as in vitro tear breakup time. The dosing protocol, double masked and randomized, was also unknown during the period of analysis.

2.2. Extraction of Tear Proteins

Tear proteins were extracted from Schirmer strips with 600 μl of 0.9 % w/v NaCl in pH 6.24 phosphate buffer (0.066 M). The protein extract was used to analyze for total protein, measurement of TL, surface tension, and in vitro BUT.

2.3. Measurement of Total Tear Protein

The protein content of each Schirmer strip was measured using a modified Coomassie reagent according to a procedure previously described.[4] Bovine serum albumin (BSA, Sigma, St. Louis, MO) was used as a standard. A volume of 0.1 ml of the extract was reacted with 2.5 ml of modified Coomassie reagent in a stoppered quartz cell (10 mm; VWR Scientific, San Francisco, CA). After 1 min, the absorbance was measured at 595 nm using a UV spectrophotometer (UV-Vis Recording Spectrophotometer, Shimadzu, Kyoto, Japan). The average protein secretion of each eye at the end of week 1 and week 3 was calculated and statistically compared with the control eye using the Student's t-test.

2.4. Measurement of Tear Lipocalin (TL)

A volume of 150 μl of tear protein extract was injected on a size exclusion (SE-HPLC) column (Spherogel® TSK 3000 SW, 10 μ, 7.5 mm X 30 cm, Beckman, Fullerton, CA). The HPLC system consisted of a solvent delivery pump (Model LC-600, Shimadzu), a variable-wavelength UV detector (Model SPD-6A UV/VIS spectrophotometric detector, Shimadzu), an auto injector (Model SIL-9A, Shimadzu) and a chart recorder (Model CR601 integrator, Shimadzu). The mobile phase consisted of 0.5 M NaCl/0.1 M phosphate buffer (pH 5) pumped at a rate of 0.5 ml/min. The detection (retention time, 22–23 min) was carried out at 280 nm. The average peak area for the treated eye was statistically compared (t-test) with the control eye at the end of week 1 and week 3 for TL.

2.5. Measurement of Surface Tension of Tear Protein Extracts

A horizontal capillary method for measuring the surface tension of small fluid volumes was developed by Ferguson and Kennedy[12] and applied to tear samples by Tiffany et al.[13] Surface tension was calculated by knowing the exact pressure necessary to flatten the meniscus at the end of a capillary so that the fluid was exactly planar with the open end. Fig. 1 shows a diagram of the apparatus used for the measurements, which were made in 10 μl capillary tubes (Fisher, Fair Lawn, NJ). Pressure to exactly equalize the meniscus was generated by pushing gently on the barrel of the syringe. The pressure was measured with an electronic micromanometer (Alnor® ECO series, model 530, Alnor, Skokie, IL). A volume of ~2 μl was required to conduct the measurements. Therefore, the use of tears ob-

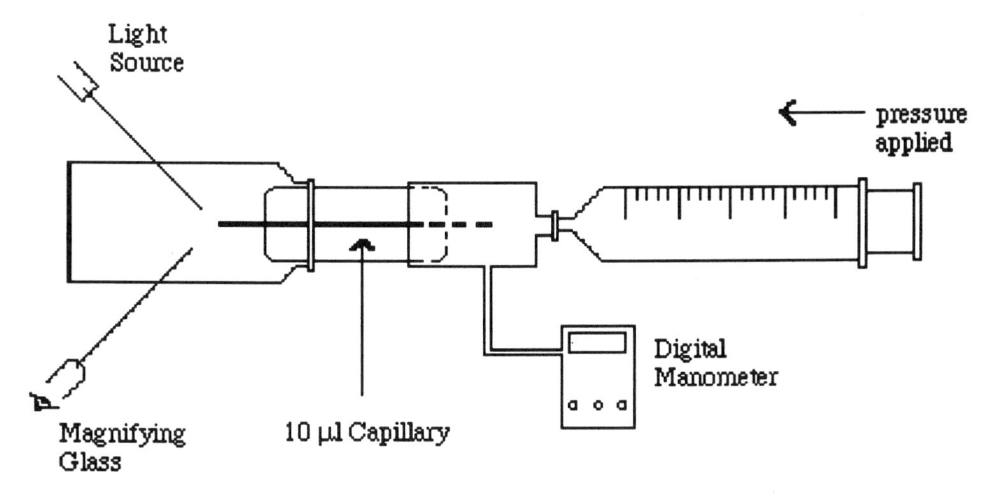

Figure 1. Measurement technique for determining surface tension of protein extracts of rabbit tears using the horizontal capillary method.

tained directly from dry eye patients was not possible, since their tear volume was insufficient. Measurements were carried out at room temperature.

2.6. Measurement of in Vitro Breakup Time

A technique was developed for measuring film integrity in vitro, which would be similar in principle (but not structure) to the in vivo BUT. A small loop with a 4 mm diameter was constructed from 0.1 mm diameter platinum wire. A cylindrical well constructed to accept the wire loop was filled with 40 μl of tear protein extract. The loop was placed on the bottom of the well, which allowed fluid to overflow the edge of the wire loop. Measurements were made in an enclosed humidity chamber (RH ~100%) to prevent the evaporation of water from the film. The stability of the film was determined in the horizontal position and measured by the length of time from loading to loss of the film. Fig. 2 shows a schematic representation of the experiment. The average in vitro BUT for extracts from the treated eye was statistically compared (t-test) to the control eye for week 1 and week 3.

2.7. Drug Administration and Collection of Tears

New Zealand white rabbits of either sex, 1.8–2.2 kg, were restrained in specially designed holders for the collection of tears following stimulation of tear proteins by AF2975 (50 μl of 0.15%) or inhibition of tear proteins using 0.15% haloperidol. Drugs were administered to one eye and vehicle to the other eye. The vehicle consisted of an isotonic phosphate buffer (pH 7.4) with 1.5% w/v of hydroxypropyl methylcellulose (Methocel®, E50LV Premium grade, Dow, Midland, MI) added as a viscolyzer (20 centipoise).

Tears were collected at 0 (just before dosing), 10, and 60 min post-dosing. In a previous study, AF2975 and haloperidol were classified pharmacologically as sigma agonist and antagonist, respectively, and therefore could be used to alter protein content in tears.[5] AF2975 has been shown to stimulate protein secretion following the incubation of acinar cells obtained from rabbit lacrimal glands and to increase protein secretion in rabbit tears when administered topically.[2,3] A detailed procedure for dosing and for collecting tears has been described previously.[4] Proparacaine (25 μl of 0.5% w/v) was instilled 8 min be-

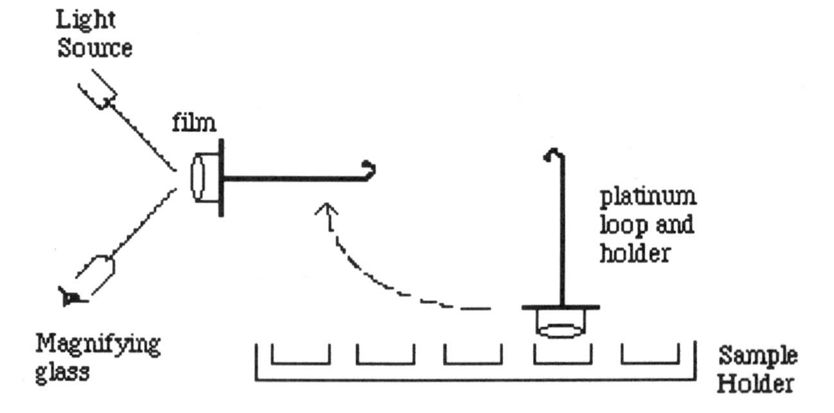

Figure 2. Measurement technique for determining in vitro BUT of films obtained from protein extracts of tears using a platinum ring at ~100% relative humidity.

fore inserting the Schirmer strip at 0, 10, and 60 min after dosing the drug or vehicle. After an additional 5 min, the strips were removed, dried, and stored at -20°C for up to 1 week before further analysis.

2.8. Extraction and Separation of Rabbit Tear Proteins, and Measurement of Surface Tension and in Vitro BUT

Tear proteins were extracted from Schirmer strips using 1 ml of 0.9% w/v NaCl in pH 6.24 phosphate buffer (0.066 M). The Schirmer strips from either the vehicle- or drug-treated eyes were mixed vigorously with solution for 30 sec and measured for total protein content as described above. Total protein was measured from a volume of 100 μl of the 1 ml extract. A volume of 3–5 μl of the extract was used in the measurement of surface tension, 40 μl in the measurement of in vitro BUT, and the remainder to make multiple measurements of surface tension and in vitro BUT. Using the same group of rabbits, treatments were repeated at a 1–2 week interval and Schirmer strips used to collect tears for separation into the various protein fractions. Surface tension and in vitro BUT determinations were also measured for each protein fraction. A volume of 1 ml of extract was used to separate the proteins into the various fractions by size exclusion (SE-HPLC). In a separate experiment, an extraction efficiency of 97% was estimated from the use of bovine serum albumin 0.2% (Sigma), which had been taken up by the Schirmer strips, extracted by the same procedure and then assayed using the modified Coomassie reagent.

A volume of 1 ml of extract was injected onto a size exclusion (SE-HPLC) column as described for TL above. The detection of each protein peak was measured at 280 nm, and yielded three major and two minor peaks with retention times of 13, 18, 21, 24, and 30 min.

Each protein fraction was separately collected and concentrated (Amicon stirred system, Series 800, Amicon, Beverly, MA), dissolved in 1 ml of extraction solution, and then treated identically in order to compare tear samples following either vehicle or drug treatment. Each protein fraction was measured for surface tension and in vitro BUT as described above.

3. RESULTS FROM DRY EYE PATIENTS

A summary of the results for the patient volunteers after instillation of AF2975 (weeks 1 and 3) is presented in Table 1. A more complete summary for each individual patient has been published previously.[14] Based on Schirmer strip results, no significant effect on tear flow was observed. However, statistical increases for total protein and TL for the treated eye, compared to the control eye, were measured and are presented in Table 1. In addition, extracts of total protein were measured for surface tension and in vitro BUT and yielded a statistically significant decrease in surface tension and an increase in in vitro BUT. The improvement in surface properties of the tear extracts of the dry eye patients as a result of administering AF2975 correlated with the increase in total protein and TL for each patient.[14]

The results in Table 1 suggest that the measured responses for the treated eyes for total protein, TL, and in vitro BUT are slightly better for week 3 than week 1 and that the drug's effect may be increasing upon continuous dosing. However, the control eye is also increasing by a small percentage, which may be due to a contralateral effect or to randomness (the slight increases are not statistically significant).

Table 1. Statistical summary of results for dry eye patients after instillation of 0.15% AF 2975 (n=14)

Analysis	Treated eye	Vehicle eye	% Change	p Value
Week 1:				
Total protein (μg/ml)	136.1 ± 54.1	93.9 ± 30.0	44.9%	0.008
Tear lipocalin (HPLC area)	151,341 ± 126,120	109,604 ± 123,088	38.1%	0.298
Surface tension (dynes/cm)	52.6 ± 3.00	56.0 ± 2.04	6.07%	0.014
In vitro BUT (sec)	24.4 ± 1.83	22.5 ± 1.83	8.44%	0.050
Week 3:				
Total protein (μg/ml)	149.8 ± 59.7	116.3 ± 45.1	28.8%	0.021
Tear lipocalin (HPLC area)	156,060 ± 177,899	85,226 ± 110,222	83.1%	0.062
Surface tension (dynes/cm)	52.5 ± 2.26	56.3 ± 1.87	6.75%	0.002
In vitro BUT (sec)	25.3 ± 1.37	22.5 ± 2.04	12.4%	0.002

Previous work has indicated a correlation between tear protein content and tear film stability.[15–17] However, there are no direct measurements of tear film stability. BUT, or its more recent modification, non-invasive breakup time (NIBUT), are used as clinical assessments of tear stability along with other indicators for the detection of dry eye. BUT, or NIBUT, is an in vivo assessment of the relative instability of the tear film, which is also the basis of our in vitro BUT measurements.

As shown by Tiffany et al.,[13] and using the same method described here for measuring the surface tension of tears, dry eye patients were found to have higher surface tension measurements than normal patients, 49.6 ± 2.2 vs. 43.6 ± 2.7 mN/m, respectively. Additional experiments by Tiffany et al.,[13] using standard solutions of serum proteins and mucus from pig stomach, showed that both mucus and proteins substantially influence surface tension. However, these investigators did not measure proteins known to exist in tears nor did they extract proteins from either normal or dry eye patients. In the present study, tear extracts containing higher levels of total protein and particularly TL were observed to lower surface tension.

4. RESULTS FROM RABBITS

Although TL has not been identified in rabbit tears, a similar approach was undertaken with rabbit tears, with additional experiments conducted to determine if a sigma antagonist, haloperidol, would decrease protein content in tears. The surface properties of individual protein fractions were also measured to determine the extent to which individual fractions contribute to the surface properties of tears.

4.1. Effect of AF2975 and Haloperidol on Total Protein Stimulation

Following the instillation of a 0.15% solution of AF2975 to a separate group of rabbits (n=23 per sampling time), the increase in total protein was higher at 10 than at 60 min; both measurements were statistically significant compared to the vehicle-treated eye. The increases at 10 and 60 min, compared to the vehicle-treated eye, were 63.5 and 33%, respectively.

Table 2. Surface tension measured by the horizontal capillary method for protein fractions of tears after instillation of AF2975 or haloperidol

SE-HPLC Fractions	10 min post-dosing (dynes/cm)			60 min post-dosing (dynes/cm)		
(R.T.,min)*	Treated	Vehicle	*p* Value	Treated	Vehicle	*p* Value
AF2975:						
13.4	61.99 (0.61)	61.49 (0.23)	0.775	62.25 (0.75)	61.58 (0.31)	0.356
18.1	61.03 (1.71)	61.01 (1.54)	0.977	61.04 (1.92)	61.22 (1.57)	0.877
20.6	59.83 (1.61)	60.47 (1.15)	0.330	60.15 (1.58)	60.23 (1.45)	0.929
23.6	45.97 (1.23)	58.48 (0.26)	0.002	54.08 (0.88)	59.30 (0.46)	0.007
29.8	58.47 (0.77)	59.55 (1.02)	0.187	57.22 (2.00)	59.57 (0.66)	0.120
Haloperidol:						
13.4	62.53 (1.69)	61.39 (1.13)	0.132	62.09 (1.67)	61.54 (1.12)	0.132
18.1	61.59 (1.10)	61.54 (1.13)	0.855	61.58 (1.15)	61.48 (1.19)	0.728
20.6	61.16 (0.55)	61.63 (0.72)	0.372	61.14 (0.84)	61.60 (0.94)	0.505
23.6	66.34 (0.53)	59.45 (0.51)	0.001	62.05 (0.63)	59.41 (0.64)	0.005
29.8	61.05 (0.86)	59.95 (0.46)	0.187	61.03 (0.92)	59.89 (0.27)	0.120

*Retention time in minutes for protein fractions separated by size-exclusion high pressure liquid chromatography.
Values are average of 3 eyes ± standard deviation determined at room temperature.

Table 2 lists the results for the surface tension measurements for tear extracts removed at 10 min and 60 min following instillation of AF2975. The results for the drug-treated eye show a significant ($p<0.001$) decrease of 24 and 14% at 10 and 60 min, respectively. Haloperidol, a sigma ligand[18] that decreases protein secretion when incubated with lacrimocytes,[4] produced a statistically significant increase in surface tension ($p<0.001$) of 13 and 5% for total protein extract following instillation in normal rabbit eyes at 10 and 60 min.

AF2975 increased in vitro BUT by 72 and 22% when compared to vehicle-treated eyes ($p<0.001$), whereas haloperidol decreased in vitro BUT by 35 and 20% ($p<0.001$) at 10 and 60 min, respectively. The magnitude of the change for each drug depends on the dose-effect response, the corresponding protein release from acini cells, and the physicochemical relationship between protein concentration and the phenomena measured--surface tension and in vitro BUT. Based upon a reduction in surface tension measurements, the direction of the percentage change for in vitro BUT was expected.

4.2. Surface Tension and in Vitro BUT Measured for Individual Protein Peaks Following Instillation of AF2975 or Haloperidol

The purity of each protein peak was determined over 200–370 nm using a diode array detector (model SPD-M6A, Shimadzu) and found to be over 98% for the 20 and 23

Table 3. In vitro breakup time measured with a platinum wire loop for protein fractions of tears after instillation of AF2975 or haloperidol

SE-HPLC Fractions (R.T., min)*	10 min Post-dosing			60 minutes Post-dosing		
	Treated	Vehicle	*p* Value	Treated	Vehicle	*p* Value
AF2975:						
13.4	23.66 (2.37)	22.52 (0.42)	0.544	22.36 (1.69)	22.49 (0.33)	0.918
18.1	20.85 (0.60)	20.82 (0.38)	0.940	20.92 (0.57)	20.84 (0.53)	0.855
20.6	22.97 (2.32)	23.46 (1.20)	0.707	22.80 (2.23)	23.50 (1.27)	0.615
23.6	37.29 (2.41)	21.51 (0.45)	0.005	27.59 (0.62)	21.42 (0.61)	0.002
29.8	24.28 (1.84)	22.72 (0.89)	0.300	23.16 (0.80)	22.72 (0.62)	0.400
Haloperidol:						
13.4	20.34† (1.67)	21.71 (1.18)	0.060	20.82 (1.67)	21.63 (1.15)	0.079
18.1	21.04 (1.05)	20.39 (0.71)	0.096	21.03 (1.00)	20.37 (0.62)	0.102
20.6	22.00 (1.09)	21.68 (0.99)	0.583	22.02 (1.23)	21.68 (1.06)	0.624
23.6	18.26 (0.47)	21.76 (0.59)	0.016	19.71 (0.56)	21.85 (0.42)	0.046
29.8	21.85 (0.83)	22.32 (0.55)	0.374	21.82 (0.75)	22.35 (0.43)	0.158

*Retention time in minutes for protein fractions separated by size-exclusion high pressure liquid chromatography.

†Values are average of 3 eyes ± standard deviation. Units are seconds and represent length of time film remains intact on platinum ring at 100% humidity.

min peaks; the minor peaks at 18 and 21 min varied between 80 and 95% purity, indicating interference from some other UV-sensitive components. Although no terminal amino acid determinations were made for the 23 min peak, it was subjected to capillary electrophoresis analysis. Using isoelectric focusing and a protein standard with a pI of 4.6, it was possible to separate the 23 min protein peak into five or six isoforms with pI values above 4.6. It is therefore conceivable that the 23 min peak may be an example of a "rabbit lipocalin" protein.

Table 3 gives the results of surface tension measurements of extracts of individual protein fractions following the administration of AF2975 or haloperidol. Of the three major peaks, only the 23 min peak showed a statistically significant effect on surface tension ($p<0.005$). Compared to the vehicle-treated eye, the drug-treated eye showed 23 and 8% decreases in surface tension for the 23 min peak at 10 and 60 min, respectively, after dosing AF2975. The 30 min peak showed a 5% decrease at 60 min after dosing, but this was not statistically significant ($p=0.120$). When haloperidol was administered, increases in surface tension of 14 and 7% were measured for TL at 10 and 60 min after dosing ($p<0.005$), respectively. No significant change was observed for the other protein peaks.

In vitro BUT also showed statistically significant changes for the 23 min peak after dosing either AF2975 or haloperidol ($p<0.01$), but no statistically significant changes for the other protein fractions (Table 4). For AF2975, there were increases in in vitro BUT of

Table 4. Total protein, surface tension, and in vitro BUT after instillation of AF2975 or haloperidol

	Pre dose measurement			10 Min post-dosing			60 Min post dosing		
Measurement	Treated	Vehicle	*p* Value	Treated	Vehicle	*p* Value	Treated	Vehicle	*p* Value
AF2975 (50 μl of 0.15%):									
Total protein*	127.1 (±42.8)	130.7 (±37.5)	–	188.9 (±49.1)	136.9 (±37.4)	0.0001	160.3 (±46.4)	131.9 (±37.6)	0.015
Surface tension†	57.8 (±0.95)	55.4 (±0.72)	–	43.9 (±1.25)	55.5 (±0.72)	0.0001	49.8 (±1.06)	55.2 (±0.55)	0.0001
in vitro BUT†	21.2 (±0.29)	20.8 (±1.52)	–	36.5 (±1.47)	20.6 (±1.26)	0.001	25.9 (±0.43)	20.9 (±1.10)	0.001
Haloperidol (50 μl of 0.15%):									
Surface tension†	56.5 (±0.71)	55.8 (±0.41)	–	65.4 (±0.64)	55.8 (±0.51)	0.0001	60.9 (±0.69)	55.8 (±0.62)	0.0001
in vitro BUT†	21.6 (±0.64)	21.2 (±0.72)	–	13.2 (±1.08)	21.2 (±0.81)	0.001	16.3 (±0.83)	21.3 (±0.94)	0.001

*n=23; †n=6. Values are average ± standard deviation.

70 and 26%, whereas haloperidol showed decreases of 18 and 12% at 10 and 60 min after dosing, respectively.

5. CONCLUSIONS

The data in Tables 2–4 suggest that the 23 min peak, but not other protein fractions, is involved in reducing the surface tension of tears and also in decreasing film disruption. In rabbit tears, the 23 min protein fraction may be important in maintaining a stable tear film across the corneal surface by virtue of its relatively high tear concentration and other properties yet to be determined. Although the 23 min protein fraction in rabbit tears has not been identified, it may prove to be analogous to the 23 min protein fraction in human tears, which has been identified as tear lipocalin.[5–7]

For human tears, it is conceivable that TL, when bound to the fatty acid components of tears, prevents their precipitation (e.g., strands of mucus are present in dry eye patients). Therefore, if a TL deficiency occurs, the lipid components could migrate out of the eye by either precipitation or rapid drainage, leading to a disruption of the tear film. It is also conceivable that TL could simultaneously interact with both the tear film and the corneal surface (or mucin). Thus, TL may be responsible for promoting a stable tear film.

It is evident from our findings that protein contributes to improved wetting of the aqueous/mucin interface or the epithelium/mucus interface, since both surfaces have been hypothesized as the origin of the tear film.[19,20]

REFERENCES

1. Schoenwald RD, Barfknecht CF, Xia E, Newton RE. The presence of σ-receptors in the lacrimal gland. *J Ocular Pharmacol.* 1993;9:125–139.
2. Shirolkar S, Schoenwald RD, Barfknecht CF, et al. Lacrimal secretion stimulants: Sigma receptors and drug implications. *J Ocular Pharmacol.* 1983;9:211–227.
3. Shirolkar S, Schoenwald RD, Barfknecht CF, Yang Y-S, Vidvauns S. Determination of tolerance to tear protein release following topical application of *n,n*-dimethyl-2-phenylethylamine HCl. *J Ocular Pharmacol.* 1995;11:41–47.

4. Schoenwald RD, Barfknecht CF, Shirolkar S, Xia E. The effects of sigma ligands on protein release from lacrimal acinar cells: A potential agonist/antagonist assay. *Life Sci.* 1995;56:1275–1285.
5. Gachon AMF, Kpamegan G, Kantelip B, Dastugue B. Relationship between lacrimal gland, isolated cells (lacrimocytes) and tears: biochemical and histological studies in the rabbit eye. *Curr Eye Res.* 1986;5:647–654.
6. Gachon AM, Lambin P, Dastugue B. Human tears: electrophoretic characteristics of specific proteins. *Ophthalmic Res.* 1980;12:277–285.
7. Delair A, Lassagne H, Gachon AMF. New members of the lipocalin family in human tear fluid. *Exp Eye Res.* 1992;55:645–647.
8. Berman ER. Tears, introduction and general description. In*: Biochemistry of the Eye*. New York: Plenum Press, 1980: 63–88.
9. Fullard RJ, Kissner DM. Purification of the isoforms of tear specific prealbumin. *Curr Eye Res.* 1991;10:613–628.
10. Redl B, Holzfeind P, Lottspeicht F. cDNA cloning and sequencing reveals human tear prealbumin to be a member of the lipophilic-ligand carrier protein superfamily. *J Biol Chem.* 1992;267:20282–20287.
11. Glasgow BJ, Abduragimov AR, Farahbakhsh ZT, Faull KF, Hubbell WL. Tear lipocalins bind a broad array of lipid ligands. *Curr Eye Res.* 1995;14:363–372.
12. Ferguson A, Kennedy SJ. Notes on surface-tension measurement. *Phys Soc.* 1932;24:511–520.
13. Tiffany JM, Winter N, Bliss G. Tear film stability and tear surface tension. *Curr Eye Res.* 1989;8:507–515.
14. Schoenwald RD, Vidvauns S, Wurster DE, Barfknecht CF. Tear film stability of protein extracts from dry eye patients administered a sigma agonist. *J Ocul Pharmacol Ther.* 1997;13:151–161.
15. Synder C, Fullard RJ. Clinical profiles of non dry eye patients and correlations with tear protein levels. *Int Ophthalmol.* 1991;15:383–389.
16. Fullard RJ, Kaswan RM, Bounous DI, Hirsh SG. Tear protein profiles vs. clinical characteristics of untreated and cyclosporine-treated canine KCS. *J Am Optom Assoc.* 1995;66:397–404.
17. Kissner DM, Fullard RJ. Reduced tear lipocalin levels in KCS patients entering a multicenter cyclosporine clinical trial. *Invest Ophthalmol Vis Sci.* 1996;37:S906.
18. Su T-P, Junien J-L. Sigma receptors in the central nervous system and the periphery. In: Itzhak Y, ed. *Sigma Receptors: Neuroscience Perspectives.* San Diego, Academic Press; 1994:21–44.
19. Lemp MA. Basic principle and classification of dry eye disorder. In: Lemp MA, Marquardt R, eds. *The Dry Eye: A Comprehensive Guide.* Berlin, Springer-Verlag; 1992:1090.
20. Tiffany JM. Composition and biophysical properties of the tear film: Knowledge and uncertainty. *Adv Exp Med Biol.* 1994;350:231–238.

RELATIONSHIP BETWEEN PRE-OCULAR TEAR FILM STRUCTURE AND STABILITY

Michel Guillon,[1,2] Cecile Maissa,[1,3] and Elaine Styles[1]

[1]Contact Lens Research Consultants
London, England
[2]Laboratoire de Biochimie des Transports Cellulaires
Université de Paris-Sud
Orsay, France
[3]Biomaterials Research Unit
Aston University
Birmingham, England

1. INTRODUCTION

The pre-ocular tear film (POTF) has several functions that can be grouped under three general headings: optical, protective, and lubricative.[1] The tear film provides a smooth optical surface to the strongest refractive element of the ocular system by compensating for the micro-irregularities of the corneal epithelium or the contact lens surface.[2] The tear film also plays an essential role in the maintenance of ocular integrity, by removing foreign bodies from the front surface of the eye, thus supplying antimicrobial[3] and mechanical[4] protection to the corneal epithelium. Any anomaly of the POTF may therefore have significant deleterious effects. Such anomalies have a multitude of causes, from excessive contamination to abnormal structure or insufficient tear volume.

To identify any relationship between tear film stability and structure, we investigated a population of myopic subjects attending our research clinic for contact lens fitting. The hypothesis tested was that the characteristics of the POTF control its stability.

2. CLINICAL DATA

The study was carried out in 239 consecutive myopic patients (119 male, 120 female), aged 30.2 ± 6.8 (18–55) years attending Contact Lens Research Consultants for soft contact lens fitting. The clinical data that form the basis of the current analysis has been reported in detail.[5] The subjects, non-wearers and daily soft contact lens wearers, were examined prior to contact lens wear on the day of their visit.

The key population POTF characteristics were:

- The visible pattern formed by the different classes of lipids mixing within the lipid layer was most commonly a meshwork pattern.
- The lipid layer was most commonly free of contamination by debris. When contaminated, the severity was usually minimal (spot). The presence of floating lipid plaque was very rare.
- The type of break observed was usually a localized line or spot break. Overall, surface breaks, indicative of a rapidly destabilizing tear film, were very rare.
- Estimation of tear volume by measurement of the tear prism height revealed (i) a distribution skewed towards very large tear volume for a few rare cases and (ii) a good correlation between the two eyes of the same subject.
- The stability of the tear film revealed large intersubject variability. For 60% of cases, the break-up time (BUT) was 5–15 sec, with 10% of cases < 5 sec and 12% > 45 sec.

3. STATISTICAL ANALYSIS

3.1. Multiple Linear Regression

After re-coding by creating dummy variables for all non-continuous variables, data entry was carried out by backward elimination with $p<0.1$ acceptance criteria. Five predictors (lipid mixing pattern, lipid contamination, break-up type, break-up position, and tear prism height) of the dependent variable (median NIBUT) were assessed. The results revealed a significant predictability (F=73.105; $p<0.001$). The model chosen (Fig. 1) explained 63% of the variability. The height of the tear prism and the position of the initial break were the main factors in predicting the non-invasive break-up time (NIBUT).

MEDIAN NIBUT (SEC)

=

0.7

+12.90	**Tear Prism Height (mm)**		
+35.1	**"if Break Position**	**=**	**None"**
+5.1	**"if Break Position**	**=**	**Horizontal zones"**
+2.5	**"if Break Position**	**=**	**Vertical zones"**
+7.4	**"if Break Type**	**=**	**Blink"**
+5.3	**"if Break Type**	**=**	**Spots/Line"**
+2.3	**"if Lipid pattern**	**=**	**Wave"**
+6.9	**"if Lipid pattern**	**=**	**Amorphous"**
+3.1	**"if Lipid pattern**	**=**	**Colours"**
+8.9	**"if Lipid pattern**	**=**	**Other"**

Figure 1.

3.2. Discriminant Analysis

The five predictors were re-coded by creating dummy variables for all non-continuous variables. The dependent continuous variable (median NIBUT) was classified into three groups: group 1, lowest 25% values; group 2, mid 50% values; group 3, highest 25% values. Data entry was by backward elimination with 0.001 acceptance criteria. The results obtained produced a territorial map (Fig. 2) that overall predicted correctly 59% of group membership.

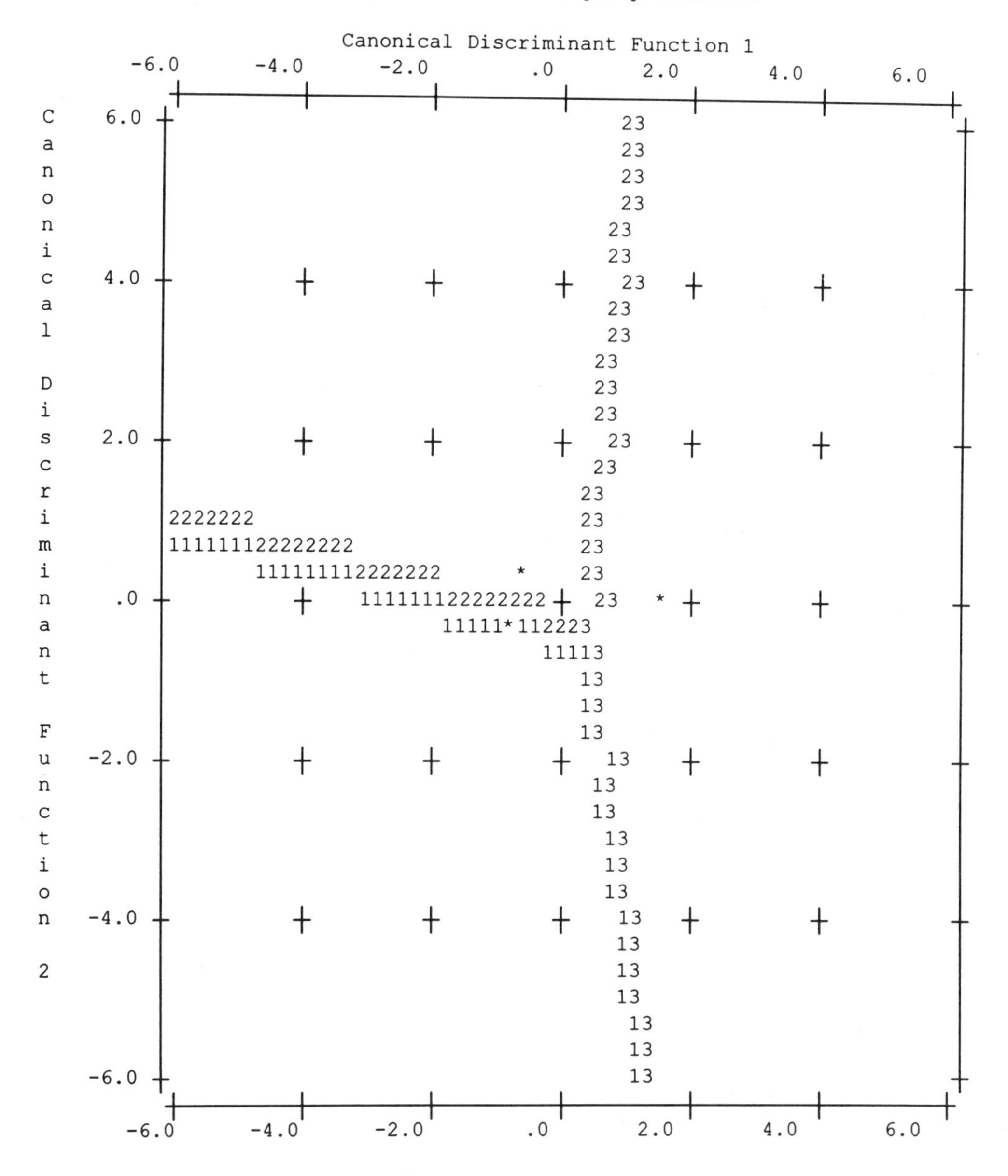

Figure 2.

3.3. Chi-square Automatic Interaction Detector (CHAID)[6]

All continuous variables were classified into three groups, defined as in the previous section. Using the same set of five predictors as for MLR and discriminant analysis, CHAID carried automated chi-square analysis, giving preference in all cases to the most discriminant parameter. The results obtained (Fig. 3) determined 10 final outcomes as to the relative influence/interaction of the various factors. The model gave a better representation of the clinical situation whereby specific but very different tear film characteristics may lead to the absence of a very stable tear film (long break-up time). For example, both a blink-induced tear break-up, with a low tear prism height (low tear volume), and a peripherally positioned tear break-up with a thick, irregular lipid layer and overall area type of break, were non-compatible with a long break-up time.

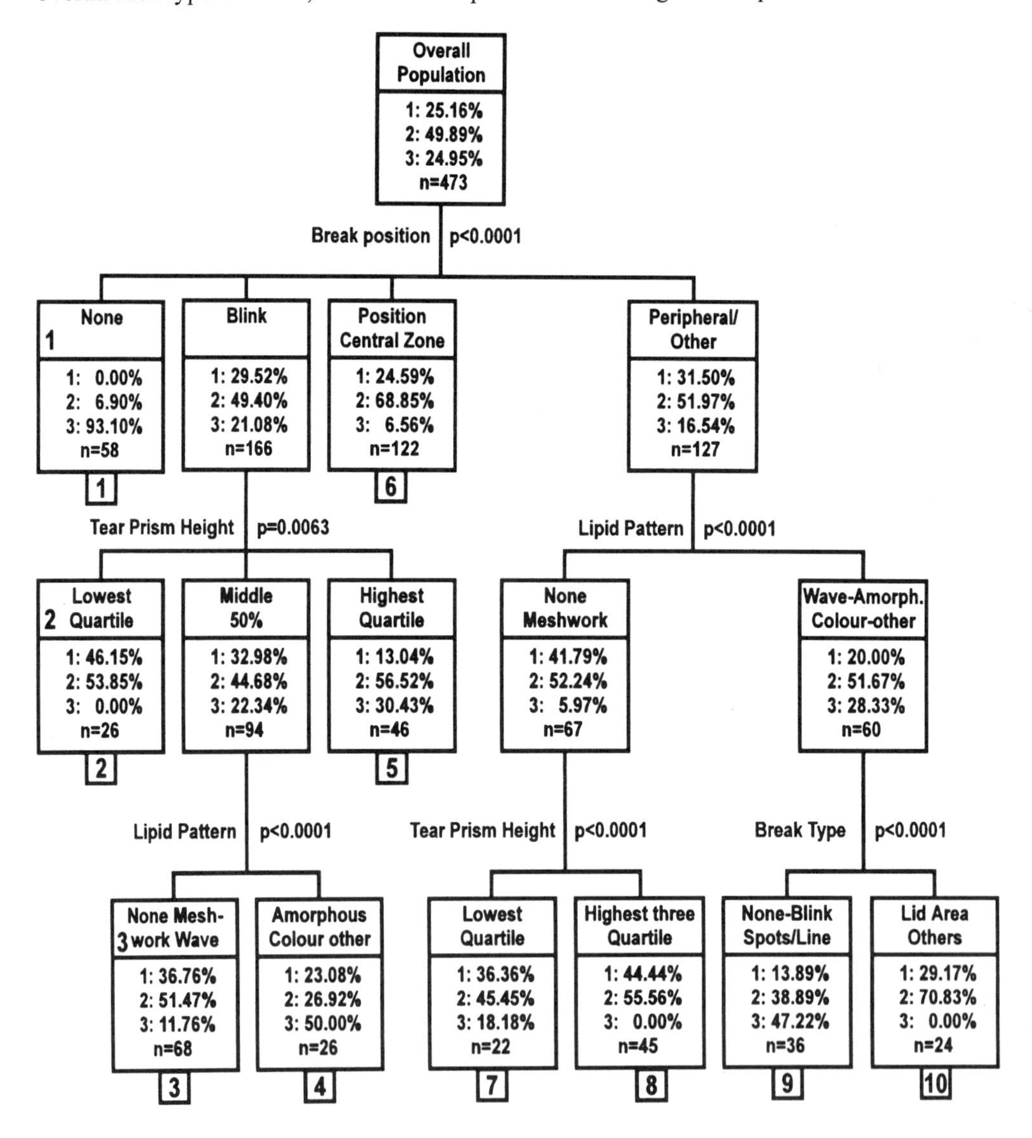

Figure 3.

4. CONCLUSION

Our results showed that the identification of factors influencing POTF stability positively or negatively was possible by CHAID analysis. Break position, tear prism height, and lipid patterns were the key factors. CHAID analysis produced different complementary models that account for the enhanced influence of the extreme characteristics of any predictor.

REFERENCES

1. Maurice D. The Charles Prentice Award Lecture 1989. The physiology of tears. *Optom Vis Sci.* 1990;67:391–399.
2. Guillon JP, Guillon M. The role of tears in contact lens performance and its measurement. In: Ruben M, Guillon M, eds. *Contact Lens Practice.* London: Chapman & Hall; 1994:452–483.
3. Fullard RJ, Snyder C. Protein levels in non-stimulated and stimulated tears of normal subjects. *Invest Ophthalmol Vis Sci.* 1990;31:1119–1129.
4. Lemp MA, Holly FJ, Iwata S, Dohlman CH. The precorneal tear film. I. Factors in spreading and maintaining a continuous tear film over the corneal surface. *Arch Ophthalmol.* 1970;83:89–94.
5. Guillon M, Styles E, Guillon JP, Maissa C. Pre-ocular tear film characteristics for non-wearers and soft contact lens wearers. *Optom Vis Sci.* submitted for publication.
6. Magidson J. Chi-square analysis of a scalable dependent variable. Proceedings of 1992 Annual Meeting of the American Statistical Association, Education Statistics Section.

56

ASSOCIATION OF PRECORNEAL AND PRECONJUNCTIVAL TEAR FILM

Yoko Takehisa, Norihiko Yokoi, Aoi Komuro, and Shigeru Kinoshita

Department of Ophthalmology
Kyoto Prefectural University of Medicine
Kyoto, Japan

1. INTRODUCTION

In a previous study, we showed that interference patterns from the tear film lipid layer at the central cornea in normal and dry eyes could be classified into five grades: normal eyes were Grade 1 or 2, and dry eyes were Grade 2, 3, 4, or 5.[1] In addition, we found a significant correlation between these grades and abnormalities in other dry eye examinations, such as rose bengal staining and tear film breakup time, and suggested that Grade 5 is the severest form of dry eye.

As tear problems affect the conjunctival as well as the corneal epithelium, it is thought that the conjunctival epithelium and preconjunctival tear film are also involved in dry eye. Grades used for precorneal tear film interference can also be applied to the evaluation of conjunctival epithelial damage and preconjunctival tear film. Therefore, in the present study, we examined preconjunctival tear film and conjunctival epithelial damage in dry eye and compared the results with grades of precorneal lipid layer interference patterns.

2. SUBJECTS AND METHODS

Subjects comprised 49 eyes of 27 dry eye patients (all females) aged 26–82 years (57.7±12.0, mean ± SD). Dry eye was diagnosed on the same standard as described in our previous report—one or more abnormal result from both tear and ocular surface examinations.[1] Tear examinations were the Schirmer I test, the cotton thread test, and fluorescein breakup time measurement. For the ocular surface examinations, fluorescein staining for corneal epithelium and rose bengal staining for corneal and conjunctival epithelium were performed.

Lacrimal Gland, Tear Film, and Dry Eye Syndromes 2
edited by Sullivan *et al.*, Plenum Press, New York, 1998

The specular reflection from the tear film lipid layer (2 mm in diameter) at the central corneal portion was recorded on a video system. These patterns were classified as either Grade 5 or the other grades (2, 3, or 4), because this study was especially concerned about the amount of tears on the corneal epithelial surface; Grade 5 tears have less surface on which lipid is overlaid compared to the other grades.

Specular images from the tear film lipid layer on the interpalpebral temporal bulbar conjunctiva (4.0 mm from the limbus) were also observed and classified into two grades, A or B. Grade A eyes showed smooth lipid layer movement with blinking, indicating a reasonable amount of water covering the conjunctival surface. Grade B eyes had no lipid layer movement with blinking, indicating that the amount of tears was insufficient to cover the conjunctival surface.

For the conjunctival epithelial evaluation, rose bengal (1.0%, 2 μl) staining was scored from 0 to 3 points at the nasal and temporal bulbar conjunctiva, in accordance with the method of van Bijsterveld.[2] The scores of Grade A and B eyes were then compared.

3. RESULTS AND DISCUSSION

Specular images from the precorneal lipid layer were classified as Grade 5 in 10 eyes and Grade 2, 3, or 4 in 39 eyes. Regarding the images from the preconjunctival lipid layer, Grade 5 eyes were all classified as Grade B (Fig. 1), whereas 38 of 39 eyes in Grade 2, 3, or 4 met the criteria for Grade A (Fig. 2). This result implied that eyes classified as Grade 5 from the precorneal lipid layer interference patterns had less tear volume on the temporal bulbar conjunctival surface than eyes in other grades.

Figure 1. Representative specular image from tear film lipid layer on the interpalpebral temporal bulbar conjunctiva in Grade B. These cases showed no lipid layer movement with blinking, which indicates not enough tears present to cover the conjunctival surface. The bulbar conjunctival surface is exposed (arrows).

Figure 2. Representative specular image from tear film lipid layer on the interpalpebral temporal bulbar conjunctiva in Grade A. These cases showed smooth movement of the lipid layer with blinking. Lipid layer interference patterns on the bulbar conjunctiva are seen (arrows).

From conjunctival epithelial staining by rose bengal, Grade 5 eyes had significantly higher scores than the other grades ($p<0.001$; Fig. 3). This result reasonably allows us to conclude that the abnormal conjunctival epithelial staining by rose bengal was induced by the decrease in preconjunctival tear film. The conjunctival epithelial damage determined by rose bengal may involve mucin layer maldistribution on the bulbar conjunctiva.[3]

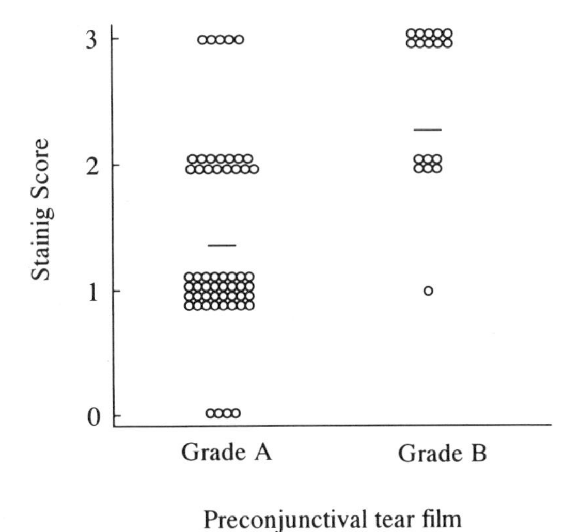

Figure 3. Rose bengal staining in Grade A and B eyes. There was a significant difference in the staining scores (Wilcoxon test, $p<0.001$). Bar = mean.

From this study, it is evident that Grade 5 dry eye that was classified non-invasively by the precorneal lipid layer interference patterns is the severest form of dry eye. This is also true when we consider the tear film abnormalities on the conjunctiva. Moreover, it should be emphasized that observation of precorneal tear lipid layer interference also provides information about the tear and epithelial condition of the conjunctiva. Therefore, it is important to note that classification of dry eye as Grade 5 may require intense therapeutic intervention.

ACKNOWLEDGMENTS

Supported in part by a research grant from the Kyoto Foundation for the Promotion of Medical Science, and the intramural research fund of Kyoto Prefectural University of Medicine.

REFERENCES

1. Yokoi N, Takehisa Y, Kinoshita S. Correlation of tear lipid layer interference patterns with the diagnosis and severity of dry eye. *Am J Ophthalmol.* 1996;122:818–824.
2. van Bijsterveld OP. Diagnostic tests in the sicca syndrome. *Arch Ophthalmol.* 1969;82:10–14.
3. Feenstra RPG, Tseng SCG. Comparison of fluorescein and rose bengal staining. *Ophthalmology.* 1992;99:605–617.

57

AGE AND GENDER EFFECTS ON THE NORMAL TEAR FILM

Jennifer P. Craig and Alan Tomlinson

Department of Vision Sciences
Glasgow Caledonian University
Glasgow, Scotland

1. INTRODUCTION

The incidence of dry eye is believed to increase in the older population, and probably more so in females, but previous studies have failed to provide conclusive evidence. On the contrary, symptomatic evaluation has shown a decrease in dry eye symptoms with advancing age,[1] possibly attributable to the corresponding decrease in corneal sensitivity[2] or avoidance of provocative stimuli. Alternatively, it is possible that the older normal eye is fully functional, without an age-related change in physiology, and that the dry eye observed in some older individuals is due to pathological intervention. The purpose of this series of studies was to investigate the effect of age and gender on tear physiology.

2. MATERIALS AND METHODS

The left eyes of 145 normal subjects were examined; these comprised 66 females and 79 males with a mean age of 35.8 ± 17.5 years and range of 13 to 75 years. The groups examined for each tear parameter are given in Table 1. None of the subjects had dry eye or anterior segment pathology or contact lens wear for at least three months. The absence of dry eye was defined as the absence of symptoms on a standard dry eye questionnaire.[1] Subjects were receiving no topical or systemic medication.

Tear evaporation was measured by vapor pressure evaporimetry,[3] and osmolality by freezing-point depression nanoliter osmometry.[4] The lipid layer structure and non-invasive break-up time (NIBUT) of all eyes were evaluated interferometrically with the Keeler Tearscope™.[5] Tear thinning time (TTT) was measured with the HIRCAL grid.[6] Tear production was determined by gamma scintigraphy from the rate of disappearance of Tc^{99m}-labelled DTPA from the conjunctival sac.[7] Reflex tear production was measured from the readings from the first 5 min, and basal tearing from 5 to 15 min, after instillation of the radioactive tracer.

Lacrimal Gland, Tear Film, and Dry Eye Syndromes 2
edited by Sullivan *et al.*, Plenum Press, New York, 1998

Table 1. Tear parameters by gender and age

Parameter	No. of subjects All	Males	Females	Age (yr) mean ± SD	Mean values ± SD All	Males	Females
Evaporation rate	122	69	53	36.8 ± 18.2	1.07 ± 1.23	1.018 ± 1.171	1.145 ± 1.71
Osmolality	127	70	57	37.2 ± 17.7	303.1	306.4 ± 12.1	301.3 ± 10.2
Lipid structure	145	79	66	35.8 ± 17.5	F	F	M(c)
NIBUT	125	64	61	37.8 ± 17.9	29.4 ± 23.4	31.3 ± 25.4	23.8 ± 22.1
TTT	83	45	38	33.8 ± 17.0	23.5 ± 20.6	25.3 ± 19.3	21.4 ± 22.1
Basal tear production	41	22	19	43.0 ± 19.0	0.56 ± 0.32	0.49 ± 0.24	0.63 ± 0.38
Reflex tear production	41	22	19	43.0 ± 19.0	3.33 ± 1.95	2.94 ± 1.79	3.79 ± 2.07

Parameter	Male/Female difference (p value)	<41/≥41 diff (p value) All	Males	Females	Linear correlation with Age (p value) All	Males	Females
Evaporation rate	0.472	0.117	0.885	0.055	0.093	0.560	0.103
Osmolality	0.012	0.096	0.861	0.005	0.203	0.406	0.002
Lipid structure	0.051	0.424	0.844	0.338	–	–	–
NIBUT	0.027	0.938	0.992	0.871	0.128	0.167	0.951
TTT	0.138	0.407	0.084	0.940	0.492	0.128	0.153
Basal tear production	0.151	0.949	0.095	0.184	0.861	0.130	0.452
Reflex tear production	0.146	0.302	0.206	0.710	0.294	0.290	0.818

Units for individual parameters: $g/m^2/h$ for evaporation rate, mOsm/kg for osmolality, sec for stability measurements, and μl/min for production rates.
NIBUT, non-invasive break-up time. TTT, tear thinning time.

The Student's t-test was used to identify significant differences between the mean values for each parameter for males and females. Subjects were divided by age into two groups, < and ≥41 years, and the parameter values compared. Modal lipid layer appearances are recorded for lipid layer data, and the Mann-Whitney U test was used to establish differences between mean values for gender and for age. The various tear parameters were linearly regressed on age (where appropriate) and the significance of the slope of the regression line from one with a gradient of zero is given by the p values.

3. RESULTS

Tear evaporation rate, osmolality, and production data were normally distributed and thus suitable for parametric statistical analysis. Both measures of tear stability were positively skewed. After natural log transformation, these data were normally distributed and subsequently analyzed with parametric tests. Lipid layer appearance was divided into six categories. All subjects had a visible lipid layer. The distribution of the lipid layer patterns observed is shown in Table 2. The lipid layer results were not continuous and, consequently, were analyzed by non-parametric tests.

No significant gender or age effects were found for tear evaporation, lipid layer characteristics, and tear production. Tear stability was constant with age, but females had

Table 2. Distribution of lipid layer patterns

Lipid layer pattern	Abbrev.	Count	Frequency (%)
Absent	Abs	0	0.0
Open meshwork	M(o)	29	20.0
Closed meshwork	M(c)	38	26.2
Flow	F	39	26.9
Amorphous	A	27	18.6
Normal colored fringe	CF(n)	11	7.6
Others	Oth	1	0.7

"Others" indicates abnormal and/or contaminated patterns.

a lower mean tear stability than males. (The difference was significant for the NIBUT data, but not for the TTT data.) All values of tear osmolality were within physiologically normal limits. For all subjects, there was no effect of age on tear osmolality, but when the group was divided by gender, osmolality increased significantly in females but not in males. This was due to the low osmolality in younger females (<41 years) shown in Table 3. No significant difference was found between the osmolalities of younger males, older males, and older females ($p > 0.05$ in all cases).

4. DISCUSSION

Overall, the results showed no significant change in tear physiology with age and support the hypothesis that there is no inherent predisposition to dry eye in older normal individuals.

In agreement with most published work,[8,9] tear evaporation was found to remain constant with age. It has been demonstrated in previous work that no significant differences in tear evaporation exist where there is confluent lipid cover[3] (as found in 99.3% of subjects in the current study). Given the relative constancy of the lipid layer throughout life,[10] a change in tear evaporation with age would not be expected. Recently, however, Mathers et al. found a significant increase in tear evaporation with age.[11] The reason for this difference is unknown.

Consistent with earlier work by the authors,[12] tear osmolality was found to be significantly lower in females than in males, although results for all subjects fell within normal limits. The lower osmolality in females was due primarily to values in younger females (<41 years). This may be explained by the finding of a higher basal tear flow rate in young females than in males,[13] although this observation was not confirmed by the data from the limited sample in the current study. Similar gender differences in osmolality have been reported in the literature,[14,15] but most authors have offered little explanation,

Table 3. Mean tear osmolalities by age and gender

	Osmolality (mOsm/kg)	
Age (yr)	Males	Females
<41	306.6 ± 14.3	298.7 ± 11.4
≥41	306.0 ± 6.9	305.5 ± 6.0

and in most cases have treated the differences as artifacts. Age did not affect osmolality when all subjects were considered, but, when separated by gender, a significant relationship was found to exist only for females. Mathers et al. found a significant increase in osmolality with age,[11] but unfortunately their data was not separated by gender. Since the ratio of females to males in that study was 51:21, it is possible that the significant effect they observed was due to the preponderance of females.

The significant gender-related difference in NIBUT found in this study has been reported previously,[16,17] although several workers have failed to show a statistically significant difference.[18,19] However, in all cases, females exhibited lower mean tear stabilities than males. This was the case with the TTT data from the current study and may be a factor in the greater propensity for dry eye problems in females.

The flow pattern, the modal lipid layer appearance in the males in this study, has been shown in previous work to be slightly more stable than the closed meshwork marmoreal pattern, the modal pattern exhibited by the females.[3] The slight difference in mean tear stability between the males and females appears to be consistent with this slight (but non-significant) difference in modal lipid pattern. The absence of a change in the lipid pattern with age for all subjects in the current study is consistent with previous work.[10,11]

Most authors have shown no significant change in basal tear flow with age.[20,21] The results of this study are consistent with this majority. No reduction in reflex tear flow was observed with age, contrary to previous reports.[22,23] Unfortunately, the technique in the current study was not ideal for determining reflex tearing, since there was considerable intersubject variation in the effect of drop instillation on reflex tearing, and this probably accounts for the contradiction.

In conclusion, we found that the biophysical properties of tears remain within physiological limits, and fairly constant, throughout life, in the normal eye. This suggests that dry eye occurs by pathological intervention affecting one or more of the tear parameters.

REFERENCES

1. McMonnies C, Ho A. Patient history in screening for dry eye conditions. *J Am Optom Assoc.* 1987;58:296–301.
2. Millodot M. The influence of age on the sensitivity of the cornea. *Invest Ophthalmol Vis Sci.* 1977;16:240–241.
3. Craig JP, Tomlinson A. Importance of the lipid layer in human tear film stability and evaporation. *Optom Vis Sci.*, in press.
4. Craig JP, Simmons PA, Patel S, Tomlinson A. Refractive index and osmolality of human tears. *Optom Vis Sci.* 1995;72:718–724.
5. Guillon JP, Guillon M. Tear film examination of the contact lens patient. *Contax.* 1988;May:14–18.
6. Hirji N, Patel S, Callander M. Human tear film pre-rupture phase time (PR-RPT) - a non-invasive technique for evaluating the pre-corneal tear film using a novel keratometer mire. *Ophthalmic Physiol Opt.* 1989;9:139–142.
7. Craig JP. Tear physiology in the normal and dry eye. PhD Thesis. Glasgow Caledonian University. 1995;42–45.
8. Rolando M, Refojo MF. Tear evaporimeter for measuring water evaporation from the tear film under controlled conditions in humans. *Exp Eye Res.* 1983;36:25–33.
9. Tomlinson A, Giesbrecht C. The ageing tear film. *J Br Contact Lens Assoc.* 1993;16:67–69.
10. Norn MS. Semiquantitative interference study of the fatty layer of the precorneal tear film. *Acta Ophthalmol.* 1979;57:766–774.
11. Mathers WD, Lane JA, Zimmerman MB. Tear film changes associated with normal aging. *Cornea.* 1996;15:229–334.
12. Craig JP, Tomlinson A. Effect of age on tear osmolality. *Optom Vis Sci.* 1995;72:713–717.
13. Henderson JW, Prough WA. Influence of age and sex on flow of tears. *Arch Ophthalmol.* 1950;43:224–231.

14. Farris RL, Stuchell RN, Mandel ID. Tear osmolarity variation in the dry eye. *Trans Am Ophthalmol Soc.* 1986;84:250–268.
15. Terry JE, Hill RM. Human tear osmotic pressure. Diurnal variations and the closed eye. *Arch Ophthalmol.* 1978;96:120–122.
16. Norn MS. Desiccation of the precorneal tear film. I. Corneal wetting time. *Acta Ophthalmol.* 1969;47:865–880.
17. Cho P, Yap M. Age, gender, and tear break-up time. *Optom Vis Sci.* 1993;70:828–831.
18. Patel S, Farrell JC. Age-related changes in precorneal tear film stability. *Optom Vis Sci.* 1989;66:175–178.
19. Chopra SK, George S, Daniel R. Tear film break up time (BUT) in non-contact lens wearers and contact lens wearers in normal Indian population. *Indian J Ophthalmol.* 1985;33;213–216.
20. Puffer MJ, Neault RW, Brubaker RF. Basal precorneal tear turnover in the human eye. *Am J Ophthalmol.* 1980;89:369–376.
21. Kuppens EVMJ, Stolwijk TR, de Keizer RJW, van Best JA. Basal tear turnover and topical timolol in glaucoma patients and healthy controls by fluorophotometry. *Invest Ophthalmol Vis Sci.* 1992;33:3442–3448.
22. Norn MS. Tear secretion in normal eyes. *Acta Ophthalmol.* 1965;43:567–573.
23. Hamano T, Mitsunaga S, Kotani S, et al. Tear volume in relation to contact lens wear and age. *CLAO J.* 1990;16:57–61.

58

THE KINETICS OF LID MOTION AND ITS EFFECTS ON THE TEAR FILM

A. Berke and S. Mueller

School of Optometry (HFAK)
Cologne, Germany

1. INTRODUCTION

Lid motion is essential for the maintenance of an intact tear film on the ocular surface. Because the lid motion lasts only 250 ms, high-speed cameras are necessary for analyzing the details. We have developed a mathematical formalism that describes lid motion. Our theoretical results show good agreement with the experimental results published by Doane[1] in 1980 and our own high-speed measurements.[2] The only input data needed are the width of lid fissure, the time of lid closure, and the duration of a blink, which can be measured with a video camera.

2. MATHEMATICAL DESCRIPTION OF LID MOTION

2.1. Lid Closure

The tear film is influenced mainly by the vertical motion of the upper lid. Knowledge of the distance-time-relation and the velocity-time-relation of the upper lid motion is necessary for estimating the forces acting on the tear film.

Because the boundary conditions of lid motion must be fulfilled, the closing and opening of the lid must be described appropriately. The distance-vs.-time relation of lid closure can be described by:

$$x(t) = x_0\, t^2 \exp(-at^2) \tag{1}$$

Here x(t) is the vertical space coordinate of the upper lid at time t. The constant x_0 is dependent on the amplitude and duration of lid closure. The constant a is dependent on the duration of lid closure only. The product $x_0 \exp(-at^2)$ is a kind of acceleration. The term $\exp(-at^2)$ describes the decrease of acceleration during lid closure. Equation 1 is chosen in accordance with the well-known space-time law of classical mechanics.

Two boundaries must be fulfilled. First, the velocity of the upper lid must vanish at the end of lid closure. Second, the space coordinates of the lid margin at the end of lid closure is approximately the width L of the lid fissure.

The velocity of lid motion is given by the first derivative of Equation 1.

$$v(t) = 2\, x_0\, t\, (1 - at^2)\, \exp(-at^2) \tag{2}$$

The constants a and x_0 are to be determined. The first boundary condition is:

$$v(t = t_0) = 0$$

so the constant a is given by:

$$a = 1/t^2_0 \tag{3}$$

where t_0 is the duration of lid closure. The constant x_0 is given by the condition:

$$x(t = t_0) = L = x_0\, t^2_0\, \exp(-at^2_0)$$

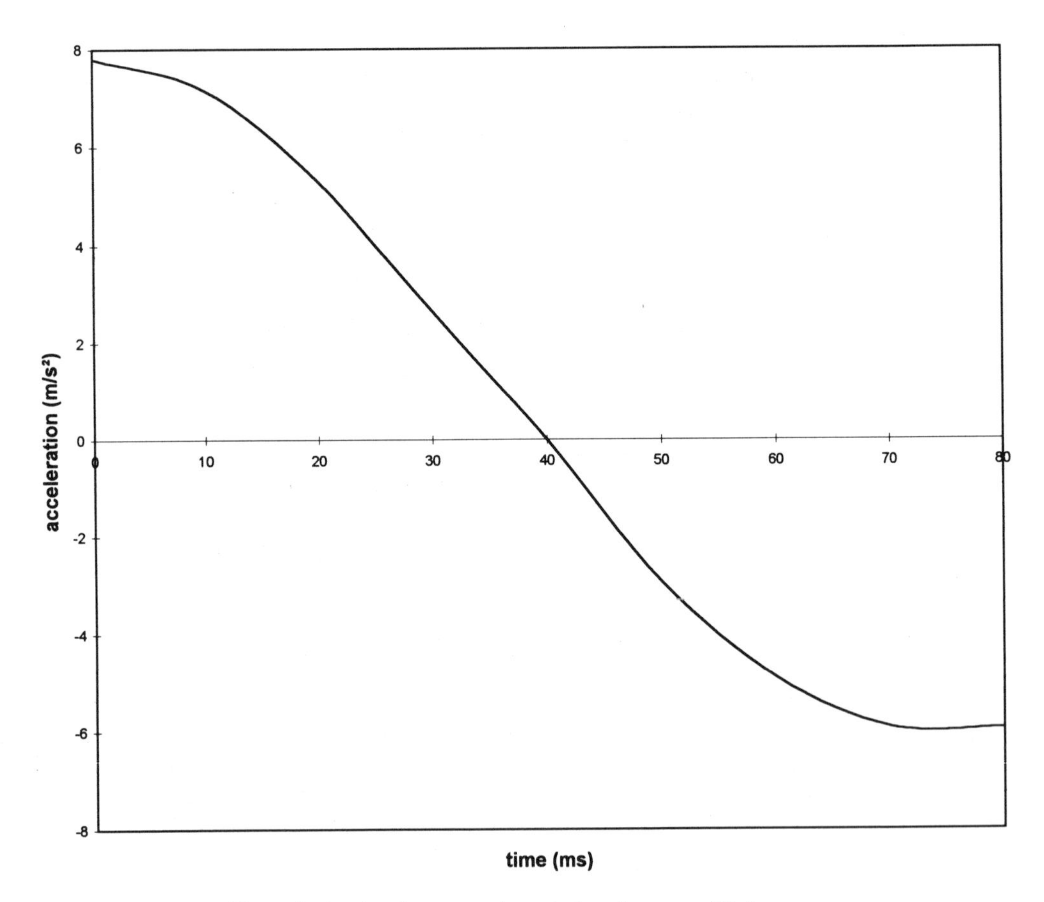

Figure 1. Acceleration-versus-time relation of upper eyelid closure.

and, with $a = 1/t^2_0$, we obtain:

$$x_0 = L / [t^2_0 \exp(-1)] \tag{4}$$

The acceleration of the upper lid, which is essential for estimating the pushing forces acting on the superficial lipid layer, is given by the second derivative of Equation 1.

$$a(t) = 2\, x_0 \,(1 - 5\, at^2 + 2\, a^2t^4) \exp(-at^2) \tag{5}$$

During the first half of lid closure, the lid's acceleration is positive, thus increasing the velocity of lid motion. Maximum velocity is reached at about half time of lid closure. At this point, acceleration is 0. Then acceleration is negative, so that the lid's velocity decreases until zero at the end of lid closure.

2.2. Lid Opening

The opening of the lids can be expressed by the equation:

$$x(t) = L - x_1\, (t - t_0)^2 \exp[-b\,(t - t_0)^2] \tag{6}$$

The velocity of lid motion is given by the first derivative of Equation 6.

$$v(t) = 2\, x_1\, (t - t_0)\, [1 - b(t - t_0)^2] \exp[-b(t - t_0)^2] \tag{7}$$

The constants x_1 and b are determined by the following boundary conditions: first, at the beginning of lid opening, the velocity must be 0, and second, at the end of lid motion at the time ($t = t_e$), the displacement coordinates $x(t = t_e)$ must vanish. So we obtain:

$$b = 1 / (t_e - t_0)^2 \tag{8}$$

$$x_1 = L / [(t_e - t_0)^2 \exp(-1)] \tag{9}$$

The complete lid motion can be described with Equations 1, 2, 6, and 7, and knowledge of the width of lid fissure and duration of lid closure and opening. These are variables that can easily be measured with a video camera, so that high-speed photography of lid motion is not necessary.

The theoretical results are in good agreement with the results of our own high-speed measurements[2] and with the results obtained by Doane.[1] A comparison of Doane's results of the maximum velocities of lid motion with our theoretical results (using $t_0 = 82.1$ ms; $t_e = 257.9$ ms; $L = 10$ mm) is shown in Table 1. Further comparisons of experimental and theoretical data are shown in Figs. 2–5.

Table 1. Maximum velocities of upper eyelid motion: Comparison of experimental[1] and theoretical[2] results

Factor	Doane, 1980[1]	Berke, Mueller, 1996[2]
Maximum closing velocity	18.7 ± 1.7 cm/s	19.4 cm/s
Maximum opening velocity	9.7 ± 0.7 cm/s	9.1 cm/s

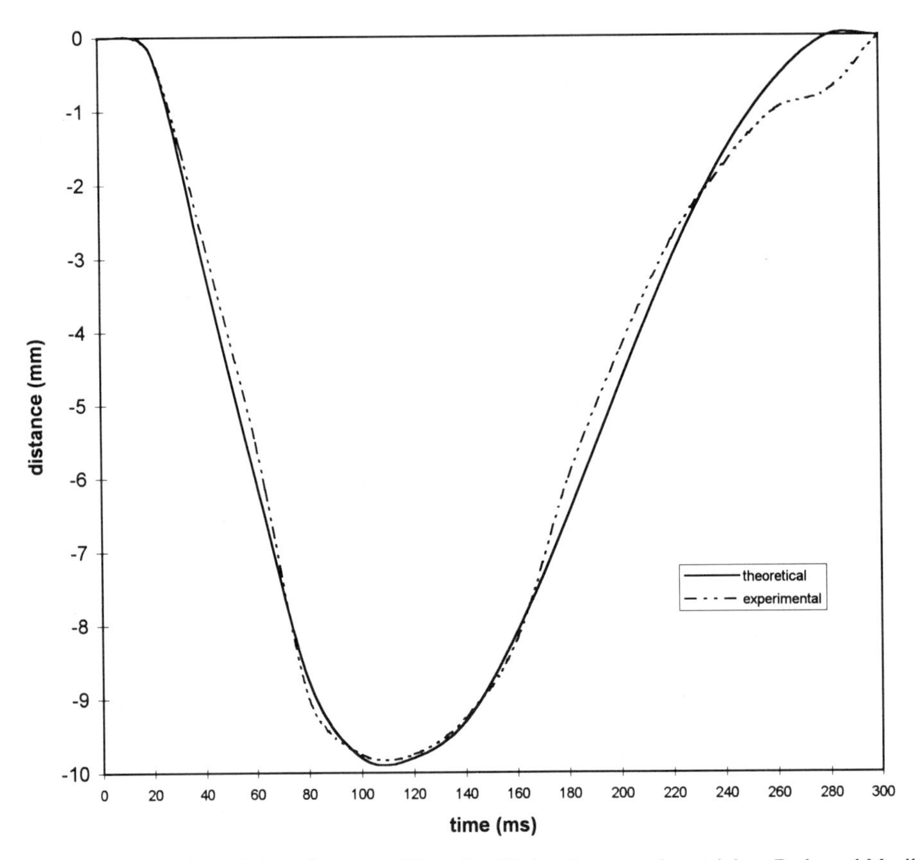

Figure 2. Distance-vs.-time relation of upper eyelid motion. Broken line, experimental data, Berke and Mueller[2]; continuous line, theoretical data.

3. FORCES ACTING ON THE TEAR FILM

3.1. Shearing Forces

Assuming a Newtonian fluid, the sliding motion of the upper lid across the ocular surface causes a stationary laminar flow in the tear film, the Couette flow. This is an exact solution of the Navier Stokes equation. The force of this flow is given by:

$$F = \eta A \, v / d \tag{10}$$

where η is the viscosity of the tear film, A is the area of the ocular surface under the upper lid, v is the velocity of the lid motion, and d is the thickness of the tear film. Assuming a maximum velocity of 18.7 cm/s (see Table 1), an area of 2 cm^2, a viscosity of 10^{-2} g/cm/s, and a thickness of the watery phase of 6 μm, the maximum shearing forces during lid closure are about 6 mN.

As can be seen from Equation 10, the quantity of tear film (given by the thickness d) is essential. A decrease of the thickness by the factor of 2 results in an increase of the

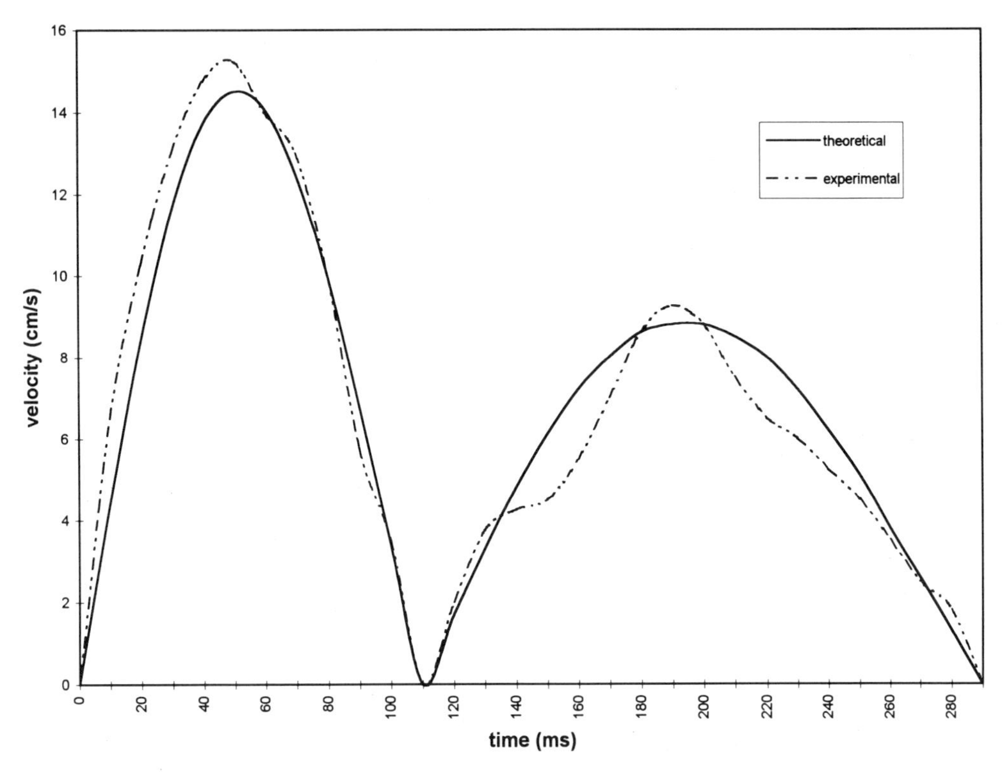

Figure 3. Velocity-vs.-time relation of upper eyelid motion. Broken line, experimental data, Berke and Mueller[2]; continuous line, theoretical data.

shearing forces by the factor of 2. A prolonged duration of the blink, and thus a reduced velocity of lid motion, results in decreased shearing forces.

3.2. Pushing Forces

During lid closure, the upper lid pushes the superficial lipid layer downward to the lower lid margin, while the watery phase remains nearly stable. The lipid layer is compressed between the eyelids while its film pressure is enhanced. The upper lid motion leads to an energy transfer to the superficial lipid layer. This energy is given by:

$$E = \int_0^{x_{max}} F\,dx = \int_0^{t_0/2} F\,v\,dt \tag{11}$$

or, by combining Equations 2 and 5:

$$E = 4m\,x_o^{\,2} \int_0^{t_0/2} \left(t - 6at^3 + 7a^2t^5 - 2a^3t^7\right) \exp(-2at^2)\,dt \tag{12}$$

where m is the mass of the upper lid. The integration limit $t_0/2$ is chosen by the fact that after half of lid closure, the acceleration has a negative sign, resulting in a deceleration of

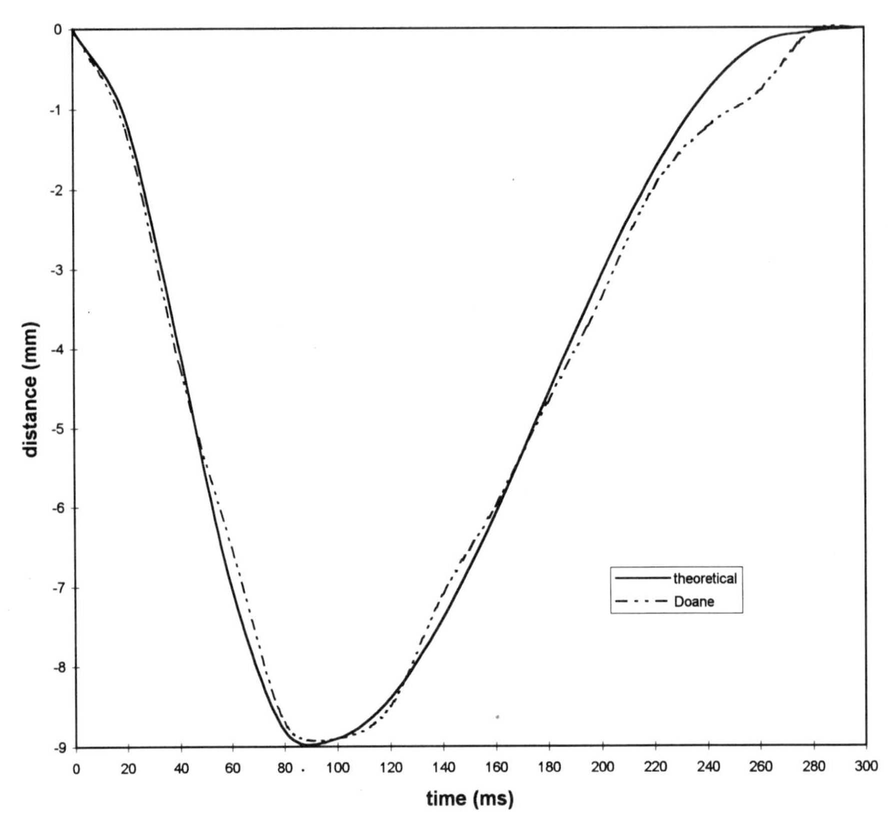

Figure 4. Distance-vs.-time relation of upper eyelid motion. Broken line, experimental data, Doane[1]; continuous line, theoretical data.

the upper lid. There is no more energy transfer to the superficial lipid layer. The results of the numerical integration of Equation 12 are shown in Fig. 6. The energy transfer to the superficial lipid layer is about 10 to 40 μJ depending on the duration of lid closure.

As Doane[1] noted, most of the blinks are incomplete. The energy transfer during an incomplete lid closure is less than during a complete lid closure. Assuming a maximum distance of lid closure of 7.5 mm, the energy transferred to the lipid layer is about 10 to 20 μJ.

The compression of the superficial lipid layer enhances its film pressure. The film pressure π is in the following relationship:

$$dE = - \pi \, dA \tag{13}$$

where dE is the energy uptake of the lipid film, and dA is the decrease of the area of the lipid film. An energy of 10 to 15 μJ transferred to the lipid layer and a decrease of area of 2 cm^2 result in a film pressure of 50 to 75 mN/m.

ACKNOWLEDGMENT

This work is supported by the Zentralverband der Augenoptiker (ZVA), Germany.

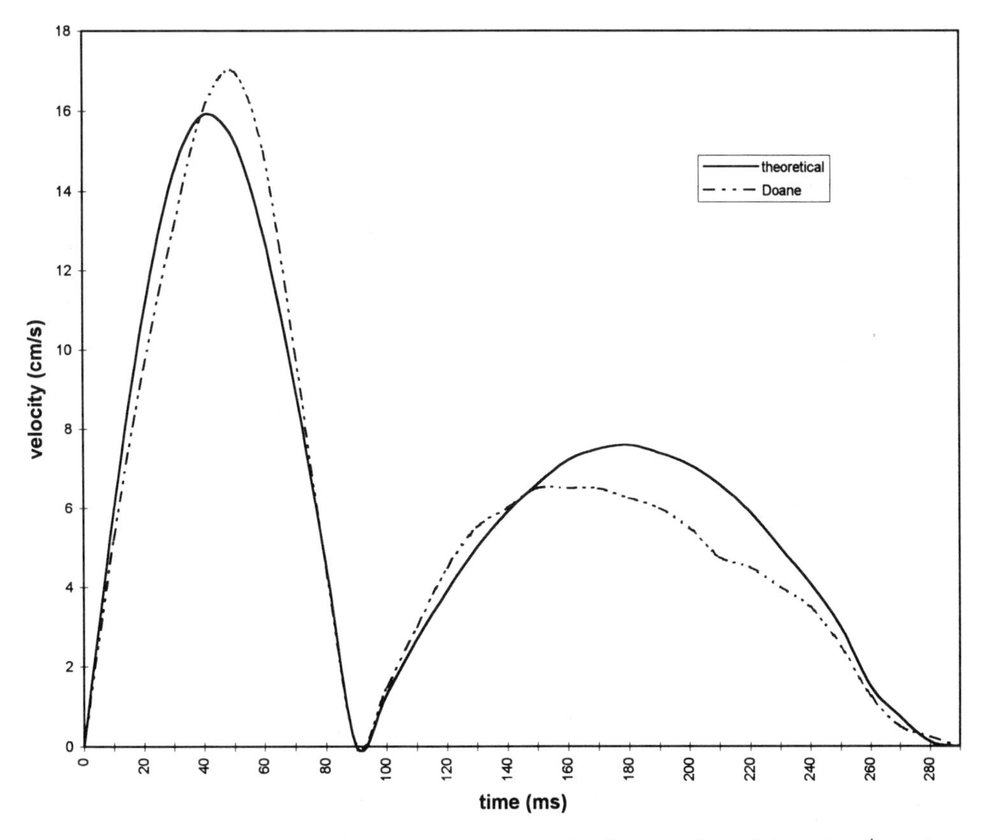

Figure 5. Velocity-vs.-time relation of upper eyelid motion. Broken line, experimental data, Doane[1]; continuous line, theoretical data.

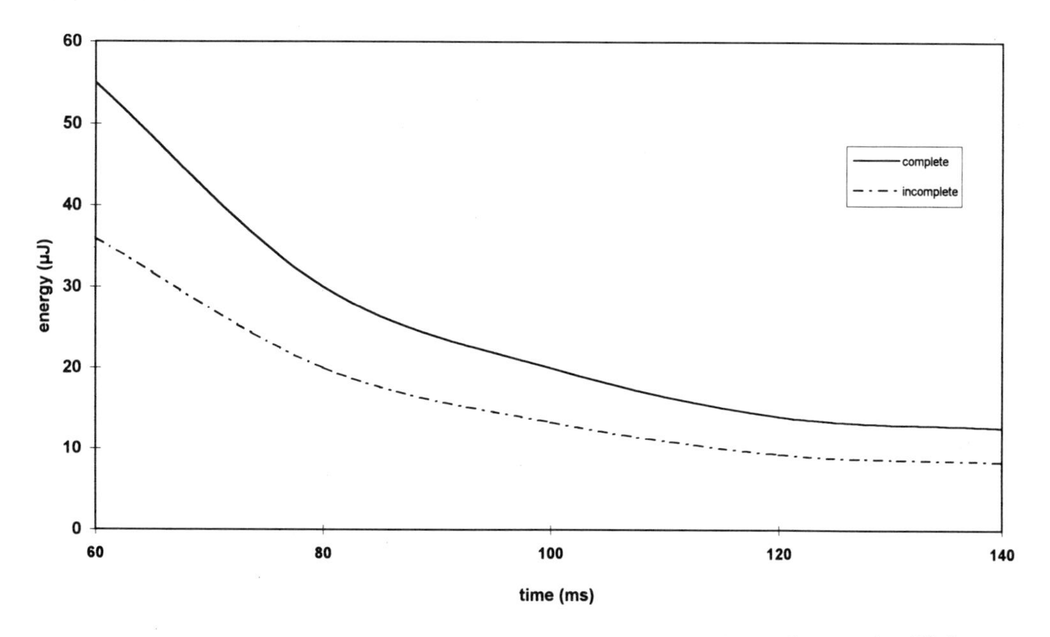

Figure 6. Energy transfer to the superficial lipid layer during lid closure. Continuous line, complete lid closure; broken line, incomplete lid closure.

REFERENCES

1. Doane MG. Interaction of eyelids and tears in corneal wetting and the dynamics of the normal human eye-blink. *Am J Ophthalmol.* 1980;89:507–516.
2. Berke A, Mueller S. Einfluss des Lidschlages auf die Kontaktlinse und die zugrundeliegenden Kräfte. *die Kontaktlinse.* 1/1996:17–26.

59

HYDRODYNAMICS OF MENISCUS-INDUCED THINNING OF THE TEAR FILM

Ashutosh Sharma,[1] Sanjay Tiwari,[1] Rajesh Khanna,[1] and John M. Tiffany[2]

[1]Department of Chemical Engineering
Indian Institute of Technology at Kanpur
Kanpur, India
[2]Nuffield Laboratory of Ophthalmology
University of Oxford
Oxford, England, United Kingdom

1. INTRODUCTION

A concave liquid meniscus is always formed rather rapidly when the surface of a liquid film meets a solid surface that displays partial wetting, that is, the equilibrium contact angle is less than 90°.[1] The same phenomenon is commonly witnessed in the climbing of liquids even against gravity in narrow capillaries and around wettable surfaces placed in a pool of liquid. Similar lacrimal menisci are observed around foreign surfaces (e.g., contact lenses) placed in the tear film, and also along the upper and lower eyelids.[2] After eyelid opening, the border between a lacrimal meniscus and the tear film thins due to Laplace pressure or a "capillary-suction" engendered by the concave meniscus. Continued local thinning adjacent to the meniscus results in the appearance of a "black line" when a fluorescein-stained tear film is viewed under blue light.[2]

After the seminal work of McDonald and Brubaker,[2] there has been little further progress in understanding meniscus-induced local thinning of the tear film. The basic physics and hydrodynamics of local thinning adjacent to a meniscus are now well understood in surface science, especially in the context of foam films.[3–7] However, these ideas have not diffused to "lacrimology," which may be due partly to the mathematical complexity of surface science models.

The objective here is to present the mechanism, hydrodynamics, and kinetics of the meniscus-induced local thinning of the tear film leading to "black lines." The roles of meniscus curvature, tear film thickness, surface tension, and the superficial lipid layer in film thinning are also addressed. Since a quantitative analysis of tear film hydrodynamics requires mathematical modeling, we first summarize the basic physics of film thinning lest it gets obscured by the mathematical details.

Lacrimal Gland, Tear Film, and Dry Eye Syndromes 2
edited by Sullivan *et al.*, Plenum Press, New York, 1998

2. MECHANISM OF MENISCUS-INDUCED THINNING

Movement of the upper eyelid during a blink establishes a thin (about 10 μm) aqueous tear film on the corneal surface, and the excess tears form the upper and lower lacrimal menisci. Fig. 1A depicts an idealized, schematic, cross-sectional view of the tear film complete with one of its menisci (upper or lower) immediately after eyelid opening. The actual shape of the lacrimal menisci immediately after opening of the eyelids is not known. The mean equivalent circular radius of curvature of the meniscus usually ranges from 0.1 to 1 mm; the values in normal eyes (0.545 ± 0.259 mm) being significantly higher than in dry eyes (0.314 ± 0.16 mm).[8] Since the radius of curvature of the cornea (and its tear film) is much larger than the meniscus radius, the tear film surface may be considered essentially flat in the hydrodynamic model. Further, the radius of curvature of the meniscus in the plane perpendicular to the plane of Fig. 1 is also very large: the "black line" appears straight when viewed over a few millimeters. Fortunately, the gravitational force is also unimportant for analysis of the meniscus-tear film system, since the ratio of surface tension to gravity forces is large for small menisci and thin films.[7] This approximation makes the analysis independent of the orientation of the tear film surface with respect to the earth's surface.

The pressures on the two sides of a curved liquid surface are not equal; the magnitude of the pressure difference across a curved surface is given by the Young-Laplace equation of capillarity:[1]

$$\Delta P = \gamma / R \tag{1}$$

where γ is surface tension, and R denotes the local in-plane radius of curvature of the liquid surface. Pressure at the liquid side is higher than atmospheric pressure at places where the surface is convex (when viewed from the liquid side). Conversely, lower liquid pressures prevail at concavely curved regions of the liquid. Thus, a concave lacrimal meniscus

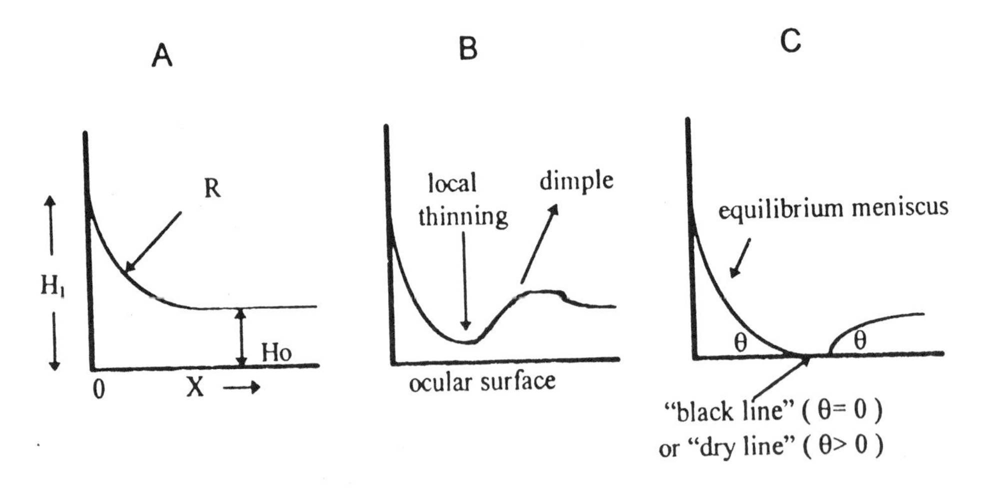

Figure 1. Schematic view of various stages in the meniscus-induced thinning of the tear film leading to the "black line." (A) Immediately after eyelid opening. (B) Local thinning induced by Laplace pressure. (C) A "black line" eventually forms if the ocular surface is completely wettable, but a "dry line" would form instead on a partially wettable surface. (Tear film thickness is greatly exaggerated here for clarity.)

(Fig. 1) has a lower (subatmospheric) pressure than does the adjoining tear film, where pressure is essentially atmospheric due to small curvatures. The lacrimal meniscus therefore acts as a vacuum cleaner that drives the tear fluid away from the film and into the meniscus, which causes a local thinning at the film-meniscus boundary (Fig. 1B). The rate of thinning is very fast initially, but slows down enormously after a few seconds, due to increased viscous resistance for flow at the thinnest spot, and also because of continuously changing curvatures (pressure differences). As shown later in this paper, a small, narrow "dimple" also forms on the film side (Fig. 1B) due to an interplay of all these factors, which is commonly observed during thinning of foam films.[4–7] If blinking is prevented, local thinning continues at ever-declining rates until the minimum thickness becomes smaller than about 0.1 μm. At this time, wettability of the substrate determines whether a true "dry line" can form. The normal ocular surface is completely wettable by the tear fluid,[9,10] so a "black line" of submicrometer thickness should remain stable.

We will now present a quantitative analysis of the meniscus-induced thinning based on computer-assisted simulations of the hydrodynamic equations (Navier-Stokes equations), which are widely used for quantitative descriptions of flows in thin films.

3. HYDRODYNAMICS OF MENISCUS-INDUCED THINNING: THIN FILM EQUATION

The following "thin film equation" is widely used for quantitative analyses of film thinning in a variety of settings, including the foam[5–7] and wetting[11–14] films.

$$3\, C\, \mu\, H_t + \gamma\, [\, H^3\, \{\, H_{xx}\, (1 + H_x^{\,2})^{-3/2}\, \}_x\,]_x = 0 \qquad (2)$$

In this differential equation for film thickness (H), subscripts are used as a shorthand notation for differentiation: $H_t=\partial H/\partial t$, $H_{xx} = \partial^2 H/\partial x^2$, etc. Time is denoted by t, x is the distance along the ocular surface measured from the meniscus corner (Fig. 1), μ is viscosity of tears (about 1 mPa.s at high shear),[15] and γ is surface tension of the tear film (35 to 55 mN/m^2).[16] The constant C=1 for a "free" tear surface of zero shear stress, whereas C=4 when the tear film surface is immobilized due to the lipid layer and adsorption of polar lipids and proteins at the lipid-aqueous interface. Adsorption at the tear surface resists thinning due to the Marangoni flow, and therefore increases the value of the constant C.[11–13,17] The first term in Equation 2 represents the viscous resistance to film thinning, and the second term represents the surface tension force that engenders thinning due to differences in the local curvature.

Four boundary conditions and an initial condition are needed for numerical solutions of Equation 2. At x=0, the meniscus width (H_1) along the lids and the meniscus curvature (R) were specified. Far away from the meniscus region and close to the corneal center, symmetry conditions ($H_x = H_{xxx} = 0$) were imposed. The last three conditions also ensure conservation of the tear volume in the meniscus-film system.[5–7,13] The constant meniscus-width condition implies that the meniscus does not spill beyond the line of gland orifices where the lid surface becomes nonwettable. The initial condition immediately after opening of the eyelids (at t = 0) was chosen to be a circular meniscus attached smoothly to a uniform tear film of constant thickness H_0 (Figs. 1 and 2). An appropriate nondimensional version of Equation 2 was solved with the help of standard finite-difference techniques on a HP-9000 supermini computer. Numerical solutions predicted the shape of the meniscus-

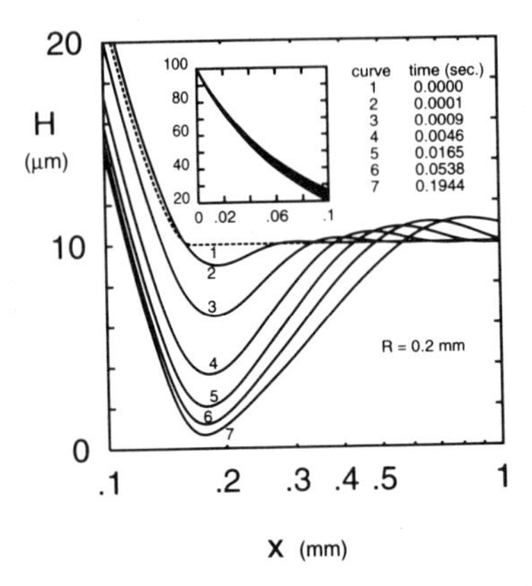

Figure 2. Initial stages of local thinning of the tear film at the meniscus-tear film border. The initial thinning is rapid enough to reduce the minimum film thickness to submicrometer levels within 1 sec. Parameters for the simulation: meniscus radius (R) = 0.2 mm, tear film thickness (H_0) = 10 μm, meniscus width (H_1) = 0.1 mm, viscosity = 1.3×10^{-3} Pa.s, surface tension = 35 mN/m, C = 1. Although the zero shear viscosity of tears is substantially larger,[15] lower viscosities close to that of water are likely during rapid thinning (high shear).

tear film system at different instants of time, from which the rate of thinning at the thinnest spot was also obtained for many different nondimensional combinations of viscosity, surface tension, radius of curvature and width of the meniscus, and tear film thickness. Further mathematical details may be obtained from the first-named author.

4. RESULTS AND DISCUSSION

The evolution of the surface in the vicinity of the lacrimal meniscus-tear film boundary is illustrated in Fig. 2 for a typical set of parameter values. The meniscus-tear film border thins very rapidly with the appearance of a small "dimple" towards the tear film side. However, the tear film remains largely undisturbed beyond a few millimeters away from the meniscus (x >2 mm). A portion of the meniscus close to the lid also remains largely undisturbed (see inset in Fig. 2). The lower bound for the surface tension (35 mN/m)[16] used in Fig. 2 gives the minimum possible rate of thinning. For higher surface tension, the rate of thinning is slightly higher, and the minimum thickness is slightly lower (see Equations 3 and 4 below). For example, the minimum thickness is about 10% lower than in Fig. 2 for a normal tear film (γ= 43.6 mN/m).

Figs. 3 and 4 (log-log plots) show the variation of the minimum thickness (H_{min}) with time. The initial decline is very rapid; a submicrometer-sized spot emerges within 1

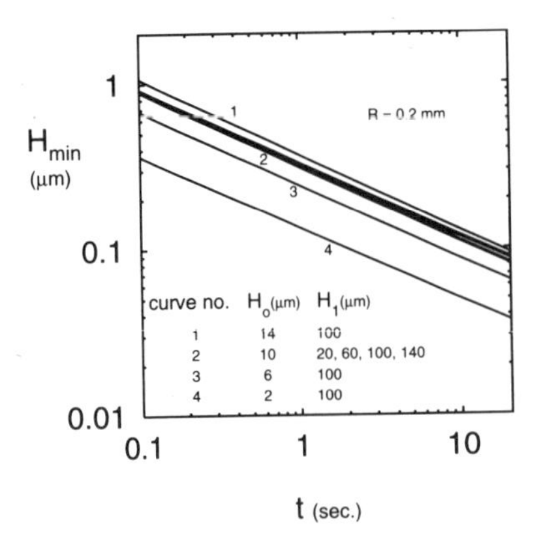

Figure 3. Effect of tear film thickness (H_0) on the minimum thickness at the meniscus-tear film border. The rate of thinning is faster for thinner films. Meniscus width (H_1) has less influence on film thinning (curve 2). All other parameters are the same as in Fig. 2.

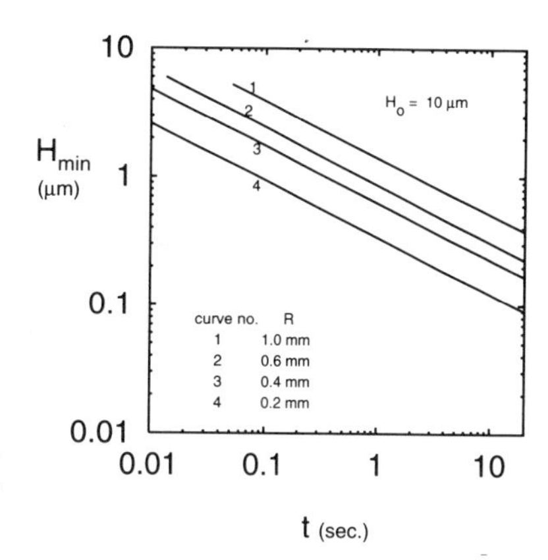

Figure 4. Effect of meniscus radius (R) on the minimum thickness at the meniscus-tear film border. The rate of thinning is faster for smaller radii. All other parameters are the same as in Fig. 2.

sec. Fig. 3 shows a more rapid thinning of initially thinner tear films (smaller H_0). The meniscus-width (H_1) does not play an important role, as evidenced by curve 2 of Fig. 3, which shows very little variation in H_{min} even when H_1 ranges from 20 to 140 μm. Fig. 4 shows the influence of meniscus radius for a fixed tear film thickness (10 μm). The smaller menisci engender faster thinning. Finally, Fig. 5 displays the computed values of the minimum thickness after 10 and 30 sec for a wide spectrum of the meniscus radius and the tear film thickness. For example, a 10 μm thick tear film attached to a 0.4 mm radius meniscus thins locally to about 0.2 μm in the first 10 sec after a blink. Note that simulations in Figs. 2–5 are reported for a "free" film (C=1). As discussed previously, the superficial lipid layer retards film thinning (C=4). For a tangentially immobilized tear surface, H_{min} shown in the figures should be multiplied by a factor of 1.86, as shown by the correlation given below.

Results of our nondimensional simulations could be correlated by the following nondimensional correlation for minimum film thickness.

$$(H_{min} / R) = 1.757\ (H_0 / R)^{0.568}\ (\mu\ R\ C / \gamma\tau)^{0.448} \tag{3}$$

The time taken to attain a prescribed minimum thickness can also be calculated from the above relation.

$$t = 3.52\ (R / H_{min})^{2.232}\ (H_0 / R)^{1.268}\ (\mu\ R\ C / \gamma) \tag{4}$$

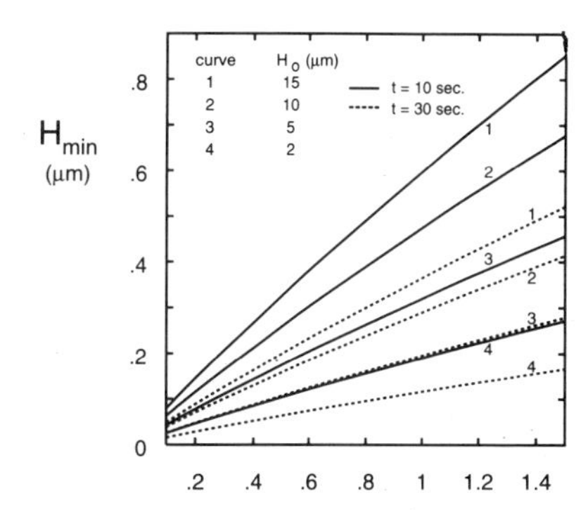

Figure 5. Minimum thickness at the meniscus-tear film border after 10 sec (solid curves) and 30 sec (broken curves). Minimum thickness declines for thinner tear films, and for more "thirsty" menisci of smaller radii. All physical properties are the same as in Fig. 2.

Several important conclusions are apparent from the simulations, results of which are summarized by the correlations, Equations 3 and 4. Equation 4 shows a strong dependence of the time of thinning on the meniscus radius, $t \propto R^2$, which is a well-known result derived by Reynolds over a century ago,[3] and it has been verified for the local meniscus-induced thinning of foam films.[4–7] Clearly, the time for the appearance of a "black line" declines for thinner tear films and for more "thirsty," smaller menisci. Thus, the time of appearance of a "black line" should be a good indicator of aqueous tear deficiency. For example, a "normal" tear film (R=0.4 mm, H_0 =10 μm, γ=45 mN/m, C=4) would locally thin to about 0.2 μm within 25 sec. In a hypothetical "dry" eye (R=0.2 mm, H_0 =4 μm), the same thickness would be reached in 2 sec! Although the actual thickness at the onset of a "black line" in the lacrimal fluorescein test is not known, it is likely to be of the order of a few tenths of a micrometer. In principle, a complete lack or emulsification of the lipid layer, combined possibly with a deficiency of protein adsorption, would also accelerate the formation of the "black line," since the parameter C can decline. This, however, appears to be an extreme scenario, since only a small amount of adsorption is sufficient for the tangential immobility of liquid surfaces. Finally, although the time for the appearance of the "black line" should also depend on tear surface tension and viscosity, the dependence is weaker, and the range of variation of these quantities is usually small.[15,16] For example, surface tensions of normal (43.6 ± 2.7 mN/m) and dry (49.6 ± 2.2 mN/m) eyes would cause little variation in the time for the appearance of the "black line."

In summary, the results obtained here quantify the mechanism and effects of aqueous tear deficiency, surface tension, viscosity, and lipid/protein adsorption in the meniscus-induced local thinning of the tear film around the menisci of eyelids and contact lenses.

REFERENCES

1. Adamson AW. *Physical Chemistry of Surfaces*, 5th ed. New York: John Wiley; 1990:4–52, 544–550.
2. McDonald JE, Brubaker S. Meniscus-induced thinning of tear films. *Am J Ophthalmol.* 1971;72:139–145.
3. Reynolds O. On the theory of lubrication and its applications to Mr. Beauchamp Tower's experiments. *Philos Trans R Soc London.* 1886;177:157–165.
4. Frankel SP, Mysels KJ. On the "dimpling" during the approach of two interfaces. *J Phys Chem.* 1962;66:190–201.
5. Malhotra AK, Wasan DT. Effect of film size on drainage of foam and emulsion films. *AIChE J.* 1987;33:1533–1541.
6. Lin CY, Slattery J. Thinning of a liquid film as a drop or bubble coalesces at a fluid-fluid interface. *AIChE J.* 1982;28:787–794.
7. Joye JL, Miller CA, Hirasaki GJ. Dimple formation and behavior during axisymmetric foam film drainage. *Langmuir.* 1992;8:3083–3092.
8. Mainstone JC, Bruce AS, Golding TR. Tear meniscus measurement in the diagnosis of dry eye. *Curr Eye Res.* 1996;15:653.
9. Tiffany JM. Measurement of wettability of the corneal epithelium. *Acta Ophthalmol.* 1990;68:175–187.
10. Sharma A. Energetics of corneal epithelial cell-ocular mucus-tear film interactions. *Biophys Chem.* 1993;47:87–99.
11. Dimitrov DS. Dynamic interaction between approaching surfaces of biological interest. *Prog Surface Sci.* 1983;14:295–424.
12. Sharma A, Ruckenstein E. An analytical nonlinear theory of thin film rupture and its application to wetting films. *J Colloid Interface Sci.* 1986;113:456–479.
13. Teletzke GF, Davis SH, Scriven LE. How liquids spread on solids. *Chem Eng Commun.* 1987;55:41–81.
14. Sharma A, Jameel AT. Nonlinear stability, rupture and morphological phase separation of thin fluid films on the apolar and polar substrates. *J Colloid Interface Sci.* 1993;161:190–208.
15. Tiffany JM. The viscosity of human tears. *Int Ophthalmol.* 1991;15:371–376.

16. Tiffany JM, Winter N, Bliss G. Tear film stability and tear surface tension. *Curr Eye Res.* 1989;8:507–515.
17. Ruckenstein E, Sharma A. A surface chemical explanation of tear breakup and its implications. In: Holly FJ, Lamberts DW, MacKeen DL, eds. *The Preocular Tear Film: In Health, Disease and Contact Lens Wear.* Lubbock, TX: Dry Eye Institute; 1986;697–727.

NOTE ADDED IN PROOF

Numerical solutions of the thin film equation (equation 2) for the meniscus-induced thinning of the tear film have also been presented in a recent paper by Wong, Fatt and Radke (Wong H, Fatt I, Radke CJ. Deposition and thinning of the human tear film. *J Colloid Interface Sci.* 1996;184:44–51). In this study, the initial tear-meniscus is modeled as a parabolic surface, rather than a circular-arc, as in our model. Although the parabolic approximation cannot be used to study the effects of the meniscus-width on the film thinning, the other results of this model are remarkably similar to those presented here, e.g., equations 3 and 4. An excellent match between the two models proves that the assumptions regarding the initial meniscus shape have a minimal influence on the mechanism and kinetics of the meniscus-induced tear film thinning.

COMPUTER-ASSISTED CALCULATION OF EXPOSED AREA OF THE HUMAN EYE

John M. Tiffany,[1] Bryan S. Todd,[2] and Mark R. Baker[3]

[1]Nuffield Laboratory of Ophthalmology
[2]Computing Laboratory, and
[3]Visual Sciences Unit
University of Oxford
Oxford, United Kingdom

1. INTRODUCTION

Besides nourishing and protecting the corneal epithelium, the preocular film forms the principal refractive element of the visual system. Oil from the meibomian glands spreads on the outer surface of the aqueous film and helps to stabilize it. Another function of this oily layer is to control evaporation, which would otherwise be considerable from a thin aqueous layer at about 35°C and exposed directly to the surrounding air.

A number of studies on the rate of evaporation from the human tear film have been reported, for both normal and dry eyes.[1–7] In some cases the total evaporation per eye is reported, but in others an attempt has been made to calculate the evaporation rate per unit area of exposed tear film. A simple method of estimating the exposed area, showing it to be linearly related to the interpalpebral height, has been proposed,[3] but the method of calculation is not given. These authors also suggested that the corneal curvature could be taken as representative of the whole exposed surface, whereas our preliminary manual calculations suggested that the contribution of the sclera was sufficiently large for its curvature to be included as a separate parameter.

We have written a computer program to calculate exposed ocular areas from digitized photographic images, assuming that both cornea and sclera are spherical surfaces, and used our results to test the linear area formula of Rolando and Refojo[3] and to investigate the contribution of the sclera to total area.

2. METHODS

2.1. Ocular Images

Series of photographs were taken with a long-focus lens of single eyes of volunteers. The fellow eye was patched so that the uncovered eye looked directly at the camera lens. In addition to level gaze, the camera position was adjusted to produce photographs with the ocular axis elevated or depressed up to 12.5° above or below the horizontal. For scaling purposes, a grid of 5-mm squares was placed in front of the photographed eye; photographic parallax errors due to object distance make this appear too large relative to the eye, so a grid reduced to approximately 95% of actual size was used, determined by experiment. Color prints of the photographs were scanned using a Logitech Scanman handheld color scanner and the files stored on computer in uncompressed Targa (TGA) format (Fig. 1). A number of line drawings of eyes were also produced from these images, varying the vertical or lateral positioning of the cornea. In addition, defined areas were painted on a ping-pong ball and photographed.

2.2. Computed Areas

A computer program ("Ocularea") was written in Turbo Pascal 6.0 for use on digitized images of eyes. The program calculates the exposed areas of cornea, sclera, and caruncle from mouse-controlled on-screen outlines. The term "caruncle" is used here to describe that part within the continuous lid margins that has the curvature of neither the cornea nor the sclera; it includes both the anatomical caruncle and the plica or loose conjunctiva at the nasal scleral boundary. Several assumptions were made: (i) the limbus is circular, (ii) the cornea is a spherical cap of constant radius, (iii) the sclera is spherical and of constant radius, (iv) the "caruncle" can be considered to be a flat surface (averaging positive and negative curvatures of its constituent parts), and (v) the center of the limbus lies over the geometric center of the sclera. The program is designed to run on a mouse-controlled PC of level 386 or above, and can operate on images sized 190 x 320 pixels or above.

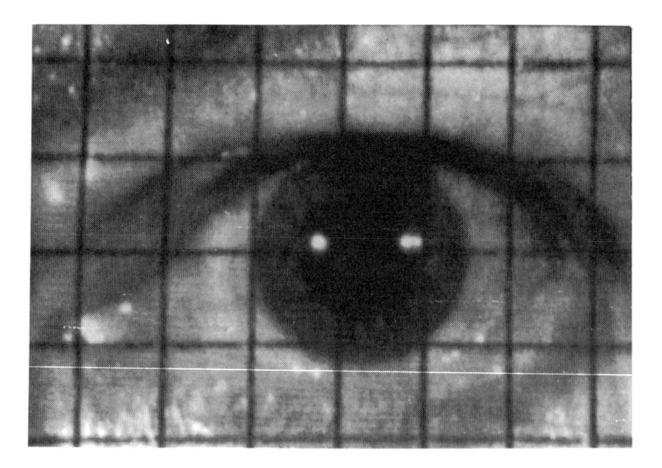

Figure 1. Image of eye with superimposed 5 mm grid.

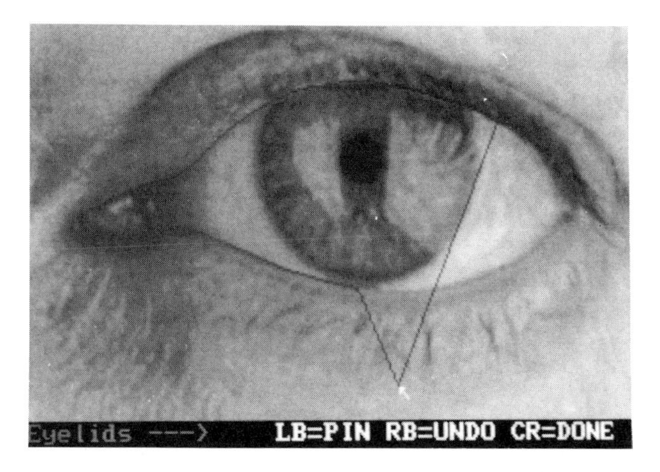

Figure 2. Image of eye with first stage of outline marking (eyelid margin) showing the simulation of "rubber band" stretching around "pins" placed by mouse clicks. Similar outlines are marked for the corneal limbus and the caruncular area (incliding plica).

The first stage of area calculation is the outlining of the three areas. A line is drawn by mouse from a starting point on the lid margin, stretched like a rubber band over points selected by mouse clicks, until the best-fit outline of the whole interpalpebral aperture is obtained (Fig. 2).

Next, the outline of the limbus is marked in the same way; where part of the cornea is obscured by the lids, no estimate of the concealed image position is required. Lastly, the caruncle is outlined in the same way. Here some subjective choice must be made of the correct outline at the plica.

Scaling of the image before calculation is addressed in the second stage. Pairs of parallel lines are positioned by mouse to outline a rectangle of defined size. This can be altered in the software, but in all cases mentioned here it was 10 mm high by 20 mm wide, corresponding to grid lines of the superimposed grid pattern included in the photographs. After this, the program displays the fully scaled set of outlines for various parts of the image, including a best-fit circular outline for the corneal limbus, prior to calculating areas (Fig. 3).

The results of the area calculations are displayed (Fig. 4). These include the 2-D areas of exposed cornea, sclera, and caruncle; the circular radius of the limbus; tables of the 3-D areas of cornea and sclera, calculated for a series of spherical radii from 7.5–8.2 mm (cornea) and 11.0–18.0 mm (sclera). In this case, the 3-D area of the caruncle is, of course, the same as the 2-D area. The total 2-D and 3-D exposed areas, the palpebral height, and the ratio of corneal to scleral 3-D areas were calculated for all the eye images studied.

3. RESULTS

In an initial series of trials using subjects with normal and with widened interpalpebral height, the program demonstrated an inter-observer repeatability error of less than 3% in measuring the exposed surface area of subjects' eyes. Calculation by Ocularea of a de-

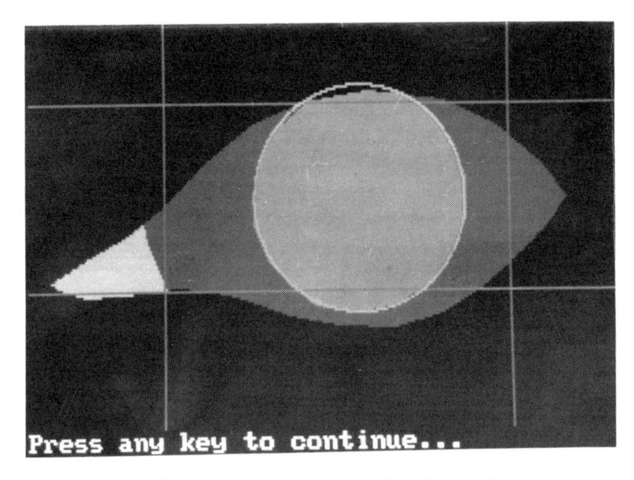

Figure 3. Screen image after outlining areas and setting scaling lines (20 mm horizontal x 10 mm vertical).

fined segmental area painted on a ping-pong ball, with appropriate scaling, indicated an error no larger than 3.5%; since the segment extended to the equator at each side of the ball, these small lateral triangles were tilted at almost 90° to the photographic plane and therefore had a high correction for tilt. On images of real eyes this would be less marked (except possibly in exophthalmic patients), so the overall error would be less than this value.

The corneal curvature can, if required, be measured in vivo by a reflection method (e.g., the Javal-Schiøtz keratometer), but the scleral curvature is less easily measured as it gives much poorer reflections. The Ocularea therefore calculates scleral area for a range of curvatures so that the error introduced by variation of curvature can be assessed. This

"TRIALE17"
Radius of limbus = 0.606 cm

2-D area of cornea = 1.154886 sq cm
2-D area of conjunctiva = 1.006542 sq cm
2-D area of caruncle = 0.144218 sq cm

3-D area of cornea		3-D area of conjunctiva	
Radius (cm)	Area (sq cm)	Radius (cm)	Area (sq cm)
0.7500	1.456477	1.1000	?
0.7550	1.449721	1.1500	?
0.7600	1.443240	1.2000	?
0.7650	1.437016	1.2500	1.490469
0.7700	1.431035	1.3000	1.386441
0.7750	1.425281	1.3500	1.331118
0.7800	1.419740	1.4000	1.290872
0.7850	1.414401	1.4500	1.259470
0.7900	1.409253	1.5000	1.234012
0.7950	1.404284	1.5500	1.212846
0.8000	1.399486	1.6000	1.194924
0.8050	1.394849	1.6500	1.179534
0.8100	1.390366	1.7000	1.166167
0.8150	1.386028	1.7500	1.154449
0.8200	1.381828	1.8000	1.144095

Figure 4. Screen display of results, showing 2-dimensional areas of visible sclera, cornea, and caruncle. The mean circular radius of the limbus is also calculated. 3-Dimensional areas of cornea and sclera are calculated for a wide range of spherical radii.

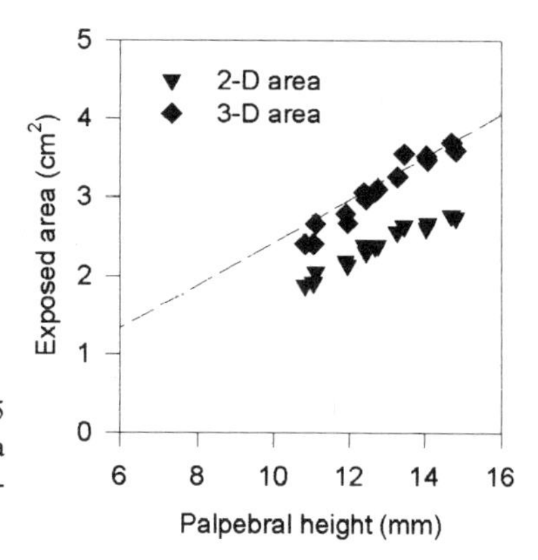

Figure 5. 3-Dimensional calculated areas (for scleral radius = 12.5 mm and corneal radius = 8.0 mm) of eye images using Ocularea, as a function of interpalpebral height. The 2-dimensional areas and the relationship found by Rolando and Refojo[3] are also shown.

was found to be typically 3–8% of the scleral area (about 1.7–4.4% of total exposed ocular area) for an error of 0.5 mm in scleral radius around the mean radius of about 12.5 mm.[8]

The range of palpebral heights measured is considerably less than that of Rolando and Refojo[3]; we have considered only those seen in our very limited series of subjects (10–15 mm). Nevertheless, these seem to fall near the range expected from the Rolando formula (Fig. 5). Our own results suggest a formula: (area) = 0.325 (height) - 1.073, rather than the original: (area) = 0.28 (height) - 0.44.

4. DISCUSSION

The primary reason for development of the Ocularea program was to allow use of the 2-dimensional image plane projection to estimate the area of an exposed ocular surface that is curved in the third dimension.

We conclude that the exposed area of the human eye can be calculated with a good degree of accuracy from the interpalpebral height, as indicated by Rolando and Refojo, despite the dubious explanation given in the primary publication.[3] The slight difference in the linear relationships is possibly due to differences in the mean values taken for corneal and scleral curvature. The images clearly show, however, that the scleral curvature, and not only the corneal curvature, is important in measuring total area. The proportion of area represented by cornea varies with palpebral height, from about 50% at 11 mm to about 35% at 15 mm (Fig. 6).

More interestingly, there appears to be a constant relationship between the 2-dimensional and the 3-dimensional area of the image, indicating that 3-dimensional areas could be calculated by simple counting (pixel counts or mechanical planimeter). Fig. 7 shows a slight dependence of this ratio on palpebral height, although the correlation coefficient is poor ($r^2 = 0.43$). The mean value of 1.294 ± 0.029 (SD) may be preferable.

Menozzi et al.[9] suggest that there is a preferrred, or most comfortable, direction of gaze, at 12.3° below the horizontal, and recommend that users of visual-display or computer screens should ensure that their principal direction is at this angle. It is therefore relevant to estimation of evaporation rates in this position to know the exposed area.

It has not yet been established whether there is any difference in evaporation rate between cornea and sclera. The aqueous film over the cornea would seem to be uniform in structure, and therefore presumably uniform in evaporation rate; the same cannot be said for the sclera, where the surface is obviously rougher. Measurements of evaporation as a

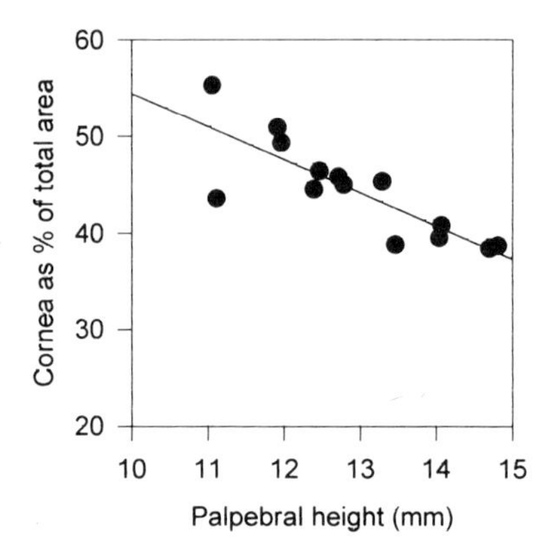

Figure 6. Percentage of total area represented by cornea, as a function of interpalpebral height.

function of elevation or depression of gaze (i.e., of variation of the corneal/scleral area ratio) may resolve this question.

We conclude that the Rolando and Refojo formula relating interpalpebral height to ocular area is a good measure of exposed area, but that the dominance of the corneal area is much less than suggested. It is relatively difficult to calculate the exposed area from a 2-dimensional image; the relative dominance one must give to cornea, sclera, caruncle, and the extreme temporal component of sclera can materially affect the assessed area. We have tried to form a consistent approach to the area of the caruncle, which obviously has convex and concave elements, and have concluded that this is best expressed as a planar area. It is possible that the extreme parts of the sclera, at the temporal canthus where the angular compensation for surface tilt is greatest, could also be best represented by a plane, but we have not yet explored all these avenues.

The program requires a minimum of computer hardware, and can be run on a standard PC with VGA monitor and mouse. It can be easily modified to accommodate 3-D models of the anatomy of the eye other than the simple spherical cap model we have initially considered. The program is a useful tool for investigating relationships between ocular parameters, such as the dependence of exposed surface area of the eye on both interpalpebral height and elevation or depression of gaze. It is clear that area measurement presents many difficulties, and we are further exploring possible sources of error.

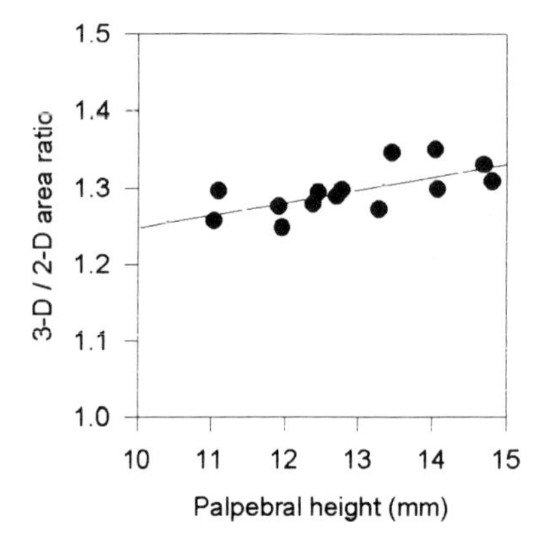

Figure 7. Ratio of 3-D and 2-D areas as a function of interpalpebral height. The regression line is (ratio) = 1.103 + 0.015 (height).

REFERENCES

1. Hamano H, Hori M, Mitsunaga S. Measurement of evaporation rate of water from the precorneal tear film and contact lenses. *Contacto.* 1981;25:7–14.
2. Mathers WD. Ocular evaporation in meibomian gland dysfunction and dry eye. *Ophthalmology.* 1993;100:347–351.
3. Rolando M, Refojo MF. Tear evaporimeter for measuring water evaporation rate from the tear film under controlled conditions in humans. *Exp Eye Res.* 1983;36:25–33.
4. Tomlinson A, Trees GR, Occhipinti JR. Tear production and evaporation in the normal eye. *Ophthalmic Physiol Opt.* 1991;11:44–47.
5. Tsubota K, Yamada M. Tear evaporation from the ocular surface. *Invest Ophthalmol Vis Sci.* 1992;33:2942–2950.
6. Mathers WD, Binarao G, Petroll M. Ocular water evaporation and the dry eye: A new measuring device. *Cornea.* 1993;12:335–340.
7. Mathers WD, Daley TE. Tear flow and evaporation in patients with and without dry eye. *Ophthalmology.* 1996;103:664–669.
8. Duke-Elder S, Wybar KC. *The Anatomy of the Visual System. In: Duke-Elder S, ed. Vol. 2, System of Ophthalmology.* London: Henry Kimpton; 1961:80–81.
9. Menozzi M, von Buol A, Krueger H, Miège C. Direction of gaze and comfort: Discovering the relation for the ergonomic optimization of visual tasks. *Ophthalmic Physiol Opt.* 1994;14:393–399.

61

CYTOKINES

An Overview

James T. Rosenbaum, Beatriz Brito, Young Boc Han, Jongmoon Park, and Stephen R. Planck

Oregon Health Sciences University
Casey Eye Institute
Portland, Oregon

1. DEFINING INFLAMMATION

Inflammation has been aptly described as a wound that does not heal. The definition recognizes that the factors that contribute to the normal repair process also control the disease process of inflammation. Cytokines are the protein signals used by the cells in the immune system. The cytokines include at least 18 interleukins, the tumor necrosis factors, the interferons, colony-stimulating factors, chemokines, and a variety of growth factors including transforming growth factor β and the family of fibroblast growth factors. Differentiating normal control of cytokine function from the dysregulation that characterizes inflammation is a challenge. This review will consider a few basic principles of cytokine function, describe the research to date that specifically relates cytokines to dry eyes, and then consider how genetically altered animals may help resolve some of the paradoxes relating to an understanding of cytokines.

2. BASIC PRINCIPLES OF CYTOKINE FUNCTION

An appreciation of cytokines requires a knowledge of several basic principles. First, cytokines function within a network. This interwoven communication system has multiple implications. In some cases, it may mean that interrupting the function of a single cytokine such as interleukin-1 (IL-1) or tumor necrosis factor α (TNFα) could result in 'downstream' events such that multiple additional cytokines are impacted as well.

Second, cytokine function is frequently redundant. The functional activities of IL-1α, IL-1 β, and TNFα are virtually indistinguishable. Consequently, the inhibition of a single cytokine may be obscured by other cytokines fulfilling the role of the inhibited mediator.

Lacrimal Gland, Tear Film, and Dry Eye Syndromes 2
edited by Sullivan *et al.*, Plenum Press, New York, 1998

Third, cytokine function is pleiotropic. The role of IL-1 in the hypothalamus is primarily the control of thermoregulation. The role of IL-1 is quite different in a vascular bed or when the target is the liver or a lymphocyte. Fourth, cytokines have receptor-mediated effects that are local. Cytokines typically affect adjacent cells or even the cell of origin. This is in contrast to a hormone, that is designed to function at a distance by virtue of being secreted into the blood. Unlike endocrine communication, cytokines work as autocrine, paracrine, juxtacrine or intracrine mediators.

3. STUDIES ON CYTOKINES IN SJÖGREN'S SYNDROME

A number of studies have attempted to relate cytokines directly to Sjögren's syndrome. The majority of these studies relate more directly to minor salivary glands which are more accessible for study than lacrimal glands. Several groups have used RT-PCR or immunohistology to detect the expression of cytokines or cytokine transcripts in affected glands.[1–11] IL-1, TNF, and γ interferon are among those cytokines which have frequently been detected in diseased tissue. Both murine models as well as human samples have been utilized in these characterizations. At this symposium, reports by Lambert, Barton et al., Fukagawa et al., Rocha et al., Robinson et al., Schechter et al., Jones et al., Smith et al., and P. Fox et al. further describe the presence or function of cytokines in diseased exocrine tissue. It needs to be emphasized, however, that the mere presence of a cytokine does not ensure that it is playing a pathologic role. Increased levels of albumin are invariably present at a site of inflammation without directly contributing to the inflammation. A critical assessment requires distinguishing a bystander from a direct contributor to a pathologic process. In addition, since cytokines play beneficial as well as harmful roles, the increased expression of cytokines might indicate a repair rather than a destructive process. New technologies allow the creation of genetically altered animals. Mice that fail to express the cytokine, transforming growth factor beta, develop an exocrinopathy that is an excellent model for Sjögren's syndrome.[12]

4. INHIBITION OF CYTOKINES

Since cytokines function to control inflammation, an obvious hypothesis is that the inhibition of cytokines could reduce inflammation. Cytokines could be blocked by synthesis inhibitors such as corticosteroids, by inhibitors of intracellular signalling such as MAP kinase inhibitors, by antibodies to either the ligand or its receptor, by antisense technology directed at either the ligand or its receptor, by soluble receptors that would act as a 'sponge' prohibiting the binding of the cytokine to its intended target, or by natural inhibitors such as the interleukin-1 receptor antagonist or interleukin-10, which inhibits the synthesis of several other cytokines.

5. CYTOKINE INHIBITION AND CLINICAL DISEASE

The difficulty and potential of cytokine inhibition are illustrated by the experience with the use of cytokine blockade in the treatment of rheumatoid arthritis. Early clinical trials described some success in treating this disease by the inhibition of TNF via either soluble receptors[13] or neutralizing antibodies[14] or by the blockade of interleukin-1 with the interleukin-1 receptor antagonist.[15] The long-term efficacy of this approach and the toxicities are still incompletely known. In addition, in some diseases such as sepsis in which cytokines are obvious candidates to explain the pathophysiology, the inhibition of the cytokine, specifically TNF, has actually worsened the disease outcome.[16] To date, the addition of cytokines to treat autoimmune disease has enjoyed as much success as the inhibition of cytokines. Currently, β_2 interferon is being used to treat multiple sclerosis;[17] α interferon may have a role in treating vasculitis secondary to hepatitis C;[18] and γ interferon probably has some limited efficacy in the treatment of scleroderma.[19] These observations emphasize the delicate balance in the cytokine network and show how cytokines are neither all good nor all bad.

6. CYTOKINES IN ENDOTOXIN-INDUCED UVEITIS

Our own laboratory's experience in unraveling the role of cytokines in anterior uveitis illustrates many of the principles outlined here. Endotoxin-induced uveitis (EIU) serves as an instructive paradigm. The subcutaneous injection of endotoxin in most strains of rats results in a cellular infiltrate in the anterior uveal tract within 24 h of the injection. Endotoxin is a potent inducer of cytokines. Our laboratory and others have demonstrated that the injection of endotoxin at a remote site in the rat induces the expression of transcripts for many cytokines in the iris within 1h of the injection. Although cytokines are cleared rapidly, cytokines can be detected in the aqueous humor as well. Despite the obvious hypothesis that EIU is cytokine dependent, the inhibition of either IL-1 or TNF has not proven effective in endotoxin-dependent models of inflammation in rabbits or rats.[20] In fact, several groups have found that the inhibition of TNF exacerbates EIU.[21,22]

Several hypotheses could account for the failure of cytokine inhibition to ameliorate EIU. The cytokine may, in fact, have a beneficial effect. The cytokine inhibitor may fail to have adequate biodistribution in the critical milieu. The redundancy of the system may negate any benefit from the inhibition of a single mediator.

7. INFLAMMATION IN MICE WITH GENE DELETIONS

Gene deletion, or 'knockout' mice, offer an alternative to pharmacologic inhibition. In this approach the inhibition is absolute. Properly designed deletion should create a rodent that completely fails to express the targeted transcript. Pharmacologic inhibition often fails, because the drug never 'gets to where the money is, ie., the inhibitor never adequately achieves the biodistribution necessary to block the intended target. Gene deletion solves this problem. Disadvantages of gene deletion include the potential that the chronic absence of a cytokine may force compensation for the loss such that the role of the cytokine is diminished. For example, a mouse might thrive in the absence of a receptor for interleukin-1 if it has been able to upregulate the expression of TNF.

Interferon γ has been studied with regard to its role in EIU.[23] IFN-γ expression is induced by endotoxin.[24] In theory, its effects should contribute to inflammation. For example, by promoting the synthesis of oxygen radicals, IFN should contribute to endothelial damage and vascular permeability. Indeed, blocking IFN-γ with antibodies helps to protect animals from the lethal effects of endotoxin.[25] Surprisingly, the inhibition of IFN-γ has been reported to worsen EIU.[23]

We have tested the importance of IFN-γ in EIU with mice that fail to synthesize IFN-γ.[26] These mice were the generous gift of David Hinrichs, Oregon Health Sciences

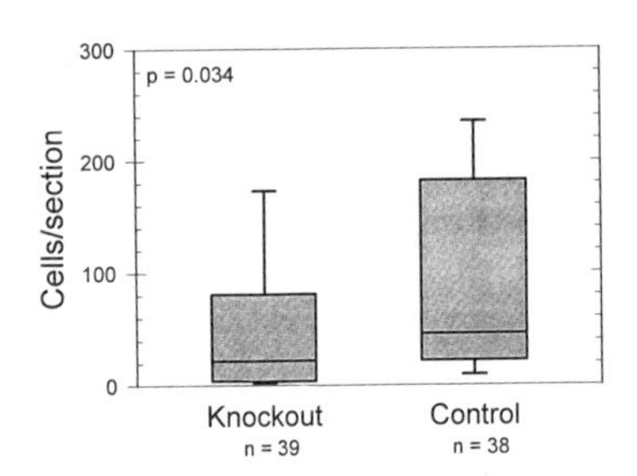

Figure 1. Number of leukocytes in representative cross-sections of eyes 24 h after intravitreal endotoxin injection. The boxes extend from the 25th to the 75th percentile. The horizontal line within the box shows the median; the vertical bars indicate the 10th and 90th percentiles. The cellular infiltrate is significantly lower in knockout mice that do not express IFN-γ (p=0.034 by Mann-Whitney rank sum test).

University. The mice are overtly healthy and breed well, although their susceptibility to certain infections may be increased.

IFN-γ knockout mice and BALB/c congenic controls were injected bilaterally with *Escherichia coli* endotoxin intravitreally at a dose of 200 μg. Twenty-four hours later, animals were euthanized, and the degree of inflammation was judged by histology and direct counting of leukocytes in a cross-section of the eye. The evaluator was masked to the experimental protocol. As shown in Fig. 1, γ-interferon-deficient mice show a statistically significant reduction in the cellular infiltrate in the uveal tract in response to an endotoxin injection. However, the reduction in inflammation is mild such that interferon-deficient mice still mount a considerable inflammatory response.

RT-PCR was used to study cytokine message expression in the iris and ciliary body after endotoxin injection. As shown in Fig. 2, cytokine transcripts such as IL-1, IL-6, and TNF do not appear to be expressed disproportionately in the IFN knockout animals.

Similar studies to clarify the role of IL-1, TNF, IL-6, IL-8, IL-10, and MIP-1α (macrophage inflammatory peptide) are underway in our laboratory.

8. CONCLUSIONS

Thus, ctyokines have been implicated in many forms of inflammation including Sjögren's syndrome. However, the presence of a cytokine is not by itself sufficient evidence to prove that the cytokine is playing a pathologic role. Furthermore, inhibiting the cytokine is a logistic challenge. And the cytokine itself should not be considered completely evil or solely beneficial. While future therapies of Sjögren's syndrome might ultimately involve a thermostatic fine tuning of the cytokine response, this laudable goal still requires much additional clarification of the pathogenesis of Sjögren's syndrome and the contribution of cytokines.

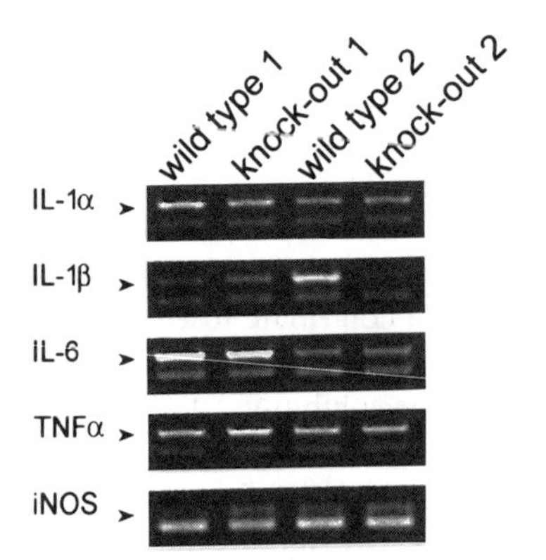

Figure 2. RT-PCR detection of cytokine mRNA expressed in the iris of IFN-γ knockout mice or congenic controls. Arrowheads point to the cytokine transcript; the unlabelled transcript is the housekeeping mRNA for GAPDH (glyceraldehyde 3-phosphate dehydrogenase). The knockout mice do not show a consistent inability to express IL-1α, IL-1β, IL-6, TNFα, or inducible nitric oxide synthase (iNOS).

ACKNOWLEDGMENTS

The authors are grateful for the technical assistance of Xiao Na Huang, Leslie O'Rourke, and Mary Williams. This research was supported by the National Eye Institute (EY06484, EY06477, EY10572) and Research to Prevent Blindness, New York City.

REFERENCES

1. Oxholm P, Daniels TE, Bendtzen K. Cytokine expression in labial salivary glands from patients with primary Sjögren's syndrome. *Autoimmunity.* 1992;12:185–191.
2. Saito I, Terauchi K, Shimuta M, Nishiimura S, Yoshino K, Takeuchi T, et al. Expression of cell adhesion molecules in the salivary and lacrimal glands of Sjögren's syndrome. *J Clin Lab Anal.* 1993;7:180–187.
3. Ohyama Y, Nakamura S, et al. Cytokine messenger RNA expression in the labial salivary glands of patients with Sjögren's syndrome. *Arthritis Rheum.* 1996;39:1376–1384.
4. Takahashi M, Mimura Y, Hamano H, Haneji N, Yanagi K, Hayashi Y. Mechanisms of the development of autoimmune dacryodenitis in the mouse model for primary Sjögren's syndrome. *Cell Immunol.* 1996;170:54–62.
5. Ogawa N, Dang H, Lazaridis K, McGuff HS, Aufdemorte TB, Talal N. Analysis of transforming growth factor β and other cytokines in autoimmune exocrinopathy (Sjögren's syndrome). *J Interferon Cytokine Res.* 1995;15:759–767.
6. DeVita S, Dolcetti R, Ferraccioli G, et al. Local cytokine expression in the progression toward B cell malignancy in Sjögren's syndrome. *J Rheumatol.* 1995;22:1674–1680.
7. Boumba D, Skopouli FN, Moutsopoulos HM. Cytokine mRNA expression in the labial salivary gland tissues from patients with primary Sjögren's syndrome. *Br J Rheumatol.* 1995;34:326–333.
8. Cauli A, Yanni G, Pitzalis C, Challacombe S, Panayi GS. Cytokine and adhesion molecule expression in the minor salivary glands of patients with Sjögren's syndrome and chronic sialoadenitis. *Ann Rheum Dis.* 1995;54:209–215.
9. Jones DT, Monroy D, Ji Z, Atherton SS, Pflugfelder SC. Sjögren's syndrome: Cytokine and Epstein-Barr viral gene expression with the conjunctival epithelium. *Invest Ophthalmol Vis Sci.* 1994;35:3493–3504.
10. Fox RI, Kang HI, Ando D, Abrams J, Pisa E. Cytokine mRNA expression in salivary gland biopsies of Sjögren's syndrome. *J Immunol.* 1994;152:5532–5539.
11. Hamano H, Saito I, Haneji N, Mitsuhashi Y, Miyasaka N, Hayashi Y. Expression of cytokine genes during development of autoimmune sialadenitis in MRL/lpr mice. *Eur J Immunol.* 1993;23:2387–2391.
12. Dang H, Geiser AG, Letterio JJ, et al. SLE-like autoantibodies and Sjögren's syndrome-like lymphoproliferation in TGF-β knockout mice. *J Immunol.* 1995;155:3205–3212.
13. Hasler F, van de Putte L, Baudin M, et al. Chronic TNF neutralization (up to 1 year) by Lenercept (TNF R55-IgG1, Ro 45–2081) in patients with rheumatoid arthritis: Results of an open-label extension of a double-blind single-dose phase I study. *Arthritis Rheum.* 1996;39:S243. (Abstract)
14. Elliot M, Maini R, Feldman M, et al. Repeated therapy with monoclonal antibody to tumour necrosis factor (cA2) in patients with rheumatoid arthritis. *Lancet.* 1994;344:1125–1127.
15. Bresnihan B, Lookabaugh J, Witt K, Musikic P. Treatment with recombinant human interleukin-1 receptor antagonist (rhIL-1ra) in rheumatoid arthritis (RA): Results of a randomized double-blind, placebo-controlled multicenter trial. *Arthritis Rheum.* 1996;39:S73. (Abstract)
16. Fisher CJ Jr, Agosti JM, Opal SM, et al. Treatment of septic shock with the tumor necrosis factor receptor:Fc fusion protein. The Soluble TNF Receptor Sepsis Study Group. *N Engl J Med.* 1996;334:1697–1702.
17. Weinstock-Guttman B, Ransohoff RM, Kinkel RP, Rudick RA. The interferons: Biological effects, mechanisms of action, and use in multiple sclerosis. *Ann Neurol.* 1995;37:7–15.
18. Misiani R, Bellavita P, Fenili D, et al. Interferon α-2a therapy in cryoglobulinemia associated with hepatitis C virus. *N Engl J Med.* 1994;330:751–756.
19. Pope J. Treatment of systemic sclerosis. *Curr Opin Rheum.* 1993;5:792–801.
20. Rosenbaum JT, Boney RS. Activity of an interleukin-1 receptor antagonist in rabbit models of uveitis. *Arch Ophthalmol.* 1992;110:547–549.
21. Kasner L, Chan C-C, Whitcup SM, Gery I. The paradoxical effect of tumor necrosis factor α (TNFα) in endotoxin-induced uveitis. *Invest Ophthalmol Vis Sci.* 1993;34:2911–2917.

22. Rosenbaum JT, Boney RS. Failure to inhibit endotoxin-induced uveitis with antibodies that neutralize tumor necrosis factor. *Reg Immunol.* 1993;5:299–303.
23. Kogiso M, Tanouchi Y, Mimura Y, Nagasawa H, Himeno K. Endotoxin-induced uveitis in mice. 1. Induction of uveitis and role of T lymphocytes. *Jpn J Ophthalmol.* 1992;36:281–290.
24. Cockfield SM, Ramassar V, Halloran PF. Regulation of IFN-γ and tumor necrosis factor-α expression in vivo. *J Immunol.* 1993; 150:342–352.
25. Heremans H, Van Damme J, Dillen C, Dijkmans R, Billiau A. Interferon γ, a mediator of lethal lipopolysaccharide-induced Shwartzman-like shock reactions in mice. *J Exp Med.* 1990; 171:1853–1869.
26. Dalton DK, Pitts-Meek S, Keshav S, Figari IS, Bradley A, Stewart TA. Multiple defects of immune cell function in mice with disrupted interferon-γ genes. *Science.* 1993;259:1739–1745.

GENDER- AND ANDROGEN-RELATED IMPACT ON THE EXPRESSION OF PROTO-ONCOGENES AND APOPTOTIC FACTORS IN LACRIMAL AND SALIVARY GLANDS OF MOUSE MODELS OF SJÖGREN'S SYNDROME

Ikuko Toda, L. Alexandra Wickham, Eduardo M. Rocha,
Lilia Aikawa da Silveira, and David A. Sullivan

Schepens Eye Research Institute and Department of Ophthalmology
Harvard Medical School
Boston, Massachusetts

1. INTRODUCTION

Sjögren's syndrome, a complex autoimmune disorder that occurs almost exclusively in females, is one of the leading causes of aqueous tear deficiency throughout the world.[1] This disease is associated with an extensive lymphocyte accumulation in the lacrimal gland, a destruction and/or dysfunction of epithelial cells, a significant decline in tear secretion and consequent dry eye.[2,3] The precise etiology of Sjögren's syndrome is unknown, but the progression of this disease has been linked to the action of sex steroids,[3–5] as well as to the inappropriate expression of cytokines, proto-oncogenes and other apoptotic factors related to programmed cell death.[3,6–13]

Recently, our research has shown that androgen treatment dramatically suppresses the inflammation in lacrimal tissues of the female MRL/Mp-lpr/lpr (MRL/lpr) and NZB/NZW F1 [F1] mouse models of Sjögren's syndrome.[3] To explain this hormone effect, we have hypothesized that: (1) fundamental, gender-related differences exist between lacrimal glands of males and females that may promote the inflammatory process and that may be due, at least in part, to the influence of sex steroids; and (2) the anti-inflammatory action of androgens is a unique, tissue-specific effect, that is mediated through a hormone interaction with epithelial cell receptors, which then elicit an altered expression and/or activity of cytokines, proto-oncogenes and apoptotic factors in the lacrimal gland.[3] In support of this hypothesis, we have found that significant, gender-associated differences exist in the the levels of cytokine mRNAs in lacrimal tissues of MRL/lpr mice,[14] that epithelial cells are the target cells for androgen action in the lacrimal gland,[15] and that androgen administration modulates the content of anti- and pro-inflammatory cytokine mRNA and/or protein in female MRL/lpr lacrimal tissue.[14,16] However, whether apoptotic factors play a role in the gender-related incidence of, or the androgen effects on, lacrimal gland inflammation is unclear.

Therefore, to further test our hypothesis, the present investigation was designed to: (1) determine whether gender-associated differences occur in the expression of proto-oncogenes and apoptotic factor mRNAs in lacrimal glands of autoimmune mice, and if so, to assess whether such differences may be linked to inflammation or are also present in tissues of 'normal' mice; and (2) examine whether the anti-inflammatory action of androgens involves an alteration in the levels of proto-oncogene and apoptotic factor mRNAs in the lacrimal gland. For comparative purposes, we also evaluated submandibular glands in these studies and explored whether the mRNAs for various apoptotic factors are present in human ocular tissues.

2. METHODS

2.1. Animals, Hormone Treatment, and Human Tissue Collection

Adult, male and/or female MRL/Mp-lpr/lpr (MRL/lpr), NZB/NZW F1 (F1), non-obese diabetic (NOD), C3H/lpr, C3H/HeJ, BALB/c, BALB/b and C3H/HeN mice were obtained from The Jackson Laboratory (Bar Harbor, Maine) or Taconic Laboratories (Germantown, NY). Animals were housed in constant temperature rooms with fixed light/dark intervals of 12 hours length. After the onset of disease or at designated ages, age-matched mice were either sacrificed by CO_2 inhalation or administered subcutaneous implants of placebo- or testosterone (10 mg)-containing pellets in the subscapular area. These pellets were purchased from Innovative Research of America (Sarasota, FL) and were constructed to release physiological (i.e. for a male) amounts of vehicle or hormone over a 21 day period. Following animal sacrifice, lacrimal (exorbital) and submandibular glands were removed and processed for histological, molecular biological or acinar epithelial cell isolation procedures. All studies with mice adhered to the resolution of The Association for Research in Vision and Ophthalmology on the use of animals in research. Human ocular tissues were obtained during the course of ophthalmic surgical procedures.

2.2. Histological, Cell Isolation, and Statistical Procedures

Lacrimal and submandibular glands were fixed in 4% paraformaldehyde or Bouin's solution, embedded in paraffin, sectioned (2/tissue) and stained with H&E, as previously described.[17] Sections were evaluated for the number of immune foci, the total area of lymphocyte infiltration and the percentage of tissue inflammation by microscopy and computer-assisted image analysis, according to reported methods.[17] Acinar epithelial cells were isolated from lacrimal glands by using previously outlined techniques.[18] Statistical evaluation of the data between two groups was conducted by using either Student's unpaired, two-tailed t test or the Mann Whitney U test. Statistical comparisons between the means of multiple groups (i.e. lacrimal gland weight, total RNA content) were performed by analysis of variance and Fisher's PLSD with a significance level of 95%.

2.3. Molecular Biological Procedures

The levels of proto-oncogene and apoptotic factor mRNAs in exocrine glands and cells were determined by RT-PCR, Southern blot hybridization, autoradiography and densitometry, as previously described in detail.[18,19] In brief, total RNA was isolated from lacrimal glands (≥ 2 tissues/sample), submandibular glands (≥ 3 tissues/ sample) and acinar epithelial cells by using either TRI reagent (Molecular Research Center, Cincinnati, OH) or another modified acid guanidinium-thiocyanate-phenol-chloroform extraction procedure.[20] The resulting RNA preparations were analyzed by spectrophotometry to measure their concentration and examined on agarose gels to verify their integrity. cDNAs were transcribed from total RNA samples (5 μg) by using AMV or MMLV reverse transcriptase, oligo dT priming and either the First-Strand cDNA Synthesis kit from Invitrogen (San Diego, CA) or the Advantage RT-for-PCR Kit from Clontech (Palo Alto, CA). PCR amplification of the cDNAs was conducted by utilizing a Perkin Elmer Cetus GeneAmp PCR System 9600 (Perkin Elmer, Norwalk, CT), Taq DNA polymerase (Gibco/BRL), dNTPs, PCR buffer (Invitrogen) and 0.4 μM of each 5' and 3' primers corresponding to mouse Fas antigen, Fas ligand, bcl-2, Bax, c-myb, c-myc, p53 and β-actin mRNAs. Primers and oligonucleotide probes were synthesized by National Biosciences, Inc. (Plymouth, MN) or obtained from Clontech. The primers for Fas antigen analysis were designed to detect 4 different transcripts, including mRNA sequences situated before the insertion site of the early transposable element (ETn) in the second intron of the Fas antigen gene, as well as sequences that either spanned or occured beyond the ETn location. The sense and antisense primer sequences, as well as the various PCR conditions and sample controls, have been reported.[18,19] The number of PCR cycles selected for each primer set was determined through extensive experimentation, and chosen so as to fall within the exponential phase of the amplification.[18,19] After amplification, PCR products and DNA molecular weight ladders (1/gel; Gibco) were electrophoresed on agarose gels, and then stained with ethidium bromide, in order to confirm the anticipated fragment sizes. Amplified cDNA products were transfered to GeneScreen nylon membranes (Dupont/NEN, Boston, MA), fixed by UV cross-linking and incubated with specific ^{32}P-labeled probes, that were phosphorylated with γ-^{32}P-dATP (Dupont/NEN) by an end-labelling method with T4 Polynucleotide Kinase (New England Biolabs, Beverly, MA) or with α^{32}P-dCTP (NEN/Dupont) by using the Random Primer Extension Labeling System (NEN/Dupont) or the Random Primers DNA Labeling System (Gibco/BRL). Following an overnight hybridization, Southern blots were processed for autoradiography, densitometry and image analysis, as previously reported.[18,19] Absorbance data were standardized to the corresponding level of β-actin mRNA and comparative results are presented in terms of the sample mRNA/β-actin mRNA ratio.

3. RESULTS

3.1. Influence of Gender on the Extent of Lymphocyte Inflammation in Exocrine Tissues of Mouse Models of Sjögren's Syndrome

A number of murine strains, including MRL/lpr,[21] F1,[22] C3H/lpr[23] and NOD,[24] have been proposed as models for human Sjögren's syndrome, and in particular, for the lacrimal and salivary gland inflammation that occurs in this autoimmune disease. However, whether all of these strains, as in humans, show significant, gender-related differences (i.e. female greater than male) in the magnitude of lymphocyte infiltration in exocrine tissues is unclear. Such information is important, not only to validate the 'model' status of these strains, but also to permit evaluation of the possible role of proto-oncogenes and apoptotic factors in the gender-associated incidence of glandular inflammation. Therefore, to address this issue, lacrimal and submandibular glands were obtained from MRL/lpr, F1, C3H/lpr and NOD mice (n = 5–10/group) after the onset of autoimmune disease and processed for histological analysis. Our results demonstrated that striking gender-, strain- and tissue-related differences exist in the extent of autoimmune exocrinopathy. Thus: (1) the number of lymphoid foci, the magnitude of lymphocytic accumulation and the percentage inflammation were significantly greater in lacrimal glands of female, as compared to male, MRL/lpr and C3H/lpr mice.[17] In contrast, at the ages tested, neither of these strains demonstrated gender-associated differences in the severity of salivary gland inflammation;[17] (2) all immune parameters were significantly higher in submandibular, but not lacrimal, tissues of female, relative to those of male, F1 mice;[17] and (3) the degree of inflammation in NOD mice showed a tissue-specific pattern: disease expression was far worse in lacrimal glands of males, whereas immune pathology was far greater in salivary tissues in females.[17]

3.2. Effect of Gender on the Expression of Apoptotic Factor mRNAs in Murine Lacrimal and Submandibular Tissues

To determine whether gender-related differences exist in the expression of proto-oncogene and apoptotic factor mRNAs in lacrimal and submandibular glands of autoimmune mice, exocrine tissues were obtained from MRL/lpr, F1 and NOD mice (n = 2–6 glands/sample; n = 5–6 samples/gender) after the onset of disease and processed for the analysis of Fas antigen, Fas ligand, bcl-2, Bax, c-myb, c-myc and p53 mRNAs by RT-PCR, Southern blot hybridization and densitometry. As shown in Table 1, our findings demonstrated that: (1) the amounts of bcl-2, c-myb, c-myc and/or p53 mRNAs are higher, and the content of partial or complete Fas antigen mRNA is lower, in lacrimal glands of female, as compared to male, MRL/lpr and F1 mice;[18,19] (2) NOD lacrimal tissues display a similar, gender-associated profile in Fas antigen and c-myc mRNA levels, but a strikingly different pattern (i.e. male greater than female) for Fas ligand, bcl-2, c-myb and p53 mRNAs;[18] and (3) the apoptotic factor mRNA expression in the submandibular gland of MRL/lpr mice is unlike that of lacrimal tissue, and shows few gender-related variations.[19]

Table 1. Influence of gender on the expression of proto-oncogene and apoptotic factor mRNAs in the lacrimal and submandibular glands of autoimmune and normal mice

mRNA	Lacrimal gland					Submandibular gland
	MRL/lpr	F1	NOD	C3H/HeJ	BALB/c	MRL/lpr
Fas antigen	M*	M	M	M	M	–
Fas ligand	–	–	M	–	–	–
bcl-2	F	F	M	F	F	–
Bax	–	–	–	–	–	–
c-myb	F	–	M	–	–	–
c-myc	F	–	F	–	–	F
p53	F	F	M	–	–	F
β-Actin	–	–	–	–	–	–

RNA was isolated from exocrine glands of MRL/lpr, F1, NOD, C3H/HeJ and BALB/c mice and processed for RT-PCR and Southern blot hybridization to compare mRNA levels of proto-oncogenes and apoptotic factors. The letters "M" and "F" refer to male and female, respectively, and indicate which gender had the higher mRNA content. The symbol "–" means that no gender-related difference was found in mRNA content. The truncated (exons 1-2), but typically not the more complete (i.e. exons 1-3, spanning the ETn insertion site), form of Fas antigen mRNA was detected. Data in this table are from the following references (18,19).

To assess whether the gender-associated differences in proto-oncogene and apoptotic factor mRNA levels may be linked to inflammation, or possibly represent fundamental differences between glands of males and females, lacrimal tissues were obtained from normal adult C3H/HeJ and BALB/c mice (n = 2–6 glands/sample; n = 5–6 samples/gender) and processed for mRNA analysis. Our findings showed that gender-related variations occurred in the content of both Fas antigen (i.e. male > female) and bcl-2 (i.e. female > male) mRNA, in a pattern identical to that of MRL/lpr and F1 mice.[18,19] For comparison, additional studies demonstrated that Fas antigen mRNA amounts were higher in lacrimal glands of male, relative to female, BALB/b mice.[18]

Of interest, our preliminary research has also shown that: (1) Fas antigen, Fas ligand, bcl-2, Bax, c-myb, c-myc and p53 mRNAs are all expressed in acinar epithelial cells of lacrimal tissues from female C3H/HeN mice;[18] and (2) human ocular tissues, including the lacrimal gland, meibomian gland, bulbar conjunctiva, cornea and/or retinal pigment epithelium, contain Fas antigen, Fas ligand, bcl-2 and Bax mRNAs.[25] However, the expression of these factors in human autoimmune disease remains to be determined.

3.3. Impact of Testosterone Administration on the Expression of Proto-Oncogene and Apoptotic Factor mRNAs in Lacrimal Glands of Autoimmune Mice

To examine whether the anti-inflammatory action of androgens involves an alteration in the levels of proto-oncogene and apoptotic factor mRNAs in the lacrimal gland, female MRL/lpr, F1 and C3H/HeJ mice (7–15 mice/group) were treated with vehicle of testosterone for 21 days and lacrimal tissues (n = 2–6 glands/sample; n = 4–6 samples/group) were then processed for mRNA measurement. Our results showed that testosterone treatment induced an increase in Bax, but a decrease in bcl-2, mRNA content in lacrimal glands of MRL/lpr, F1 and C3H/HeJ mice.[18,19]

4. DISCUSSION

Our results demonstrate that marked gender-, strain- and tissue-dependent differences exist in the magnitude of lymphocyte infiltration and the expression of proto-oncogene and apoptotic factor mRNAs in lacrimal and submandibular glands of autoimmune mice. In addition, our findings show that androgens regulate the expression of Bax and bcl-2 mRNA in lacrimal tissue. How these apoptotic factor differences and hormone effects relate to the possible pathogenesis or progression of autoimmune sequelae in the lacrimal gland, as well as to the possible maturation, function and longevity of epithelial cells and lymphocytes, are subjects of continuing investigation in our laboratory.

ACKNOWLEDGMENTS

The authors express their appreciation to Dr. Shigekazu Nagata, Lan Hu, Fu-shin Yu (Boston, MA) and Tsutomu Inatomi (Boston, MA) for their provision of molecular reagents, scientific guidance and/or technical assistance. This research was supported by grants from NIH (EY05612) and the Massachusetts Lions' Research Fund, and Postdoctoral Fellowship awards from the Uehara Memorial Foundation and the Sjogren's Syndrome Foundation.

REFERENCES

1. Homma M, Sugai S, Tojo T, Miyasaka N, Akizuki M, eds. *Sjögren's Syndrome. State of the Art.* Amsterdam: Kugler Press, 1994.
2. Lemp MA. Basic principles and classification of dry eye disorders. In: Lemp MA, Marquardt R, eds. *The Dry Eye.* Berlin: Springer-Verlag, 1992:101–131.
3. Sullivan DA, Wickham LA, Krenzer KL, Rocha EM, Toda I. Aqueous tear deficiency in Sjögren's syndrome: Possible causes and potential treatment. In: Pleyer U, Hartmann C and Sterry W, eds. *Oculodermal Diseases - Immunology of Bullous Oculo-Muco-Cutaneous Disorders.* Buren, The Netherlands: Aeolus Press, 1997:95–152.
4. Fox RI, Saito I. Sjögren's syndrome: immunologic and neuroendocrine mechanisms. *Adv Exp Med Biol* 1994; 350:609–621.
5. Ahmed SA, Talal N. Sex hormones and the immune system-part 2. Animal data. *Bailliere's Clin Rheum* 1990; 4:13–31.
6. Watanabe-Fukunaga R, Brannan CI, Copeland NG, Jenkins NA, Nagata S. Lymphoproliferation disorder in mice explained by defects in Fas antigen that mediates apoptosis. *Nature* 1992; 356:314–317.
7. Chu JL, Drappa J, Parnassa A, Elkon K. The defect in Fas mRNA expression in MRL/lpr mice is associated with insertion of the retrotransposon, ETn. *J Exp Med* 1993; 178:723–730.
8. Mountz JD. Steinberg AD, Klinman DM, Smith HR, Mushinski JF. Autoimunity and increased c-myb transcription. *Science* 1984; 226:1087–1089.
9. Kitajima T, Furukawa F, Kanauchi H, Imamura S, Ogawa K, Sugiyama T. Histological detection of c-myb and c-myc proto-oncogene expression in infiltrating cells in cutaneous lupus erythematosus-like lesions of MRL/l mice by in situ hybridization. *Clin Immunol Immunopathol* 1992; 62:119–123.
10. Sugai S, Saito I, Masaki Y, Takeshita S, Shimizu S, Tachibana J, Miyasaka N. Rearrangement of the rheumatoid factor-related germline gene Vg and bcl-2 expression in lymphoproliferative disorders in patients with Sjogren's Syndrome. *Clin Immunol Immunopathol* 1994; 72:181–186.
11. Takahashi T, Tanaka M, Brannan CI, Jenkins NA, Copeland NG, Suda T, Nagata S. Generalized lymphoproliferative disease in mice, caused by point mutation in the Fas ligand. *Cell* 1994; 76:969.
12. Sibbitt WL. Oncogenes, growth factors and autoimmune diseases. *Antican Res* 1991; 11:97–113.
13. Firestein GS. Cytokines in autoimmune diseases. *Concepts Immunopathol* 1992; 8:129–160.
14. Rocha EM, Toda I, Wickham, LA, Silveira LA, Sullivan DA. Effect of gender, androgens and cyclophosphamide on cytokine mRNA levels in lacrimal tissues of mouse models of Sjögren's syndrome. Submitted for publication, 1997.
15. Ono M, Rocha FJ, Sullivan DA. Immunocytochemical location and hormonal control of androgen receptors in lacrimal tissues of the female MRL/Mp-lpr/lpr mouse model of Sjögren's syndrome. *Exp. Eye Res.* 1995; 61:659–666.
16. Rocha EM, Wickham LA, Huang Z, Toda I, Gao J, Silveira LA, Sullivan DA. Presence and testosterone influence on the levels of anti- and pro-inflammatory cytokines in lacrimal tissues of a mouse model of Sjögren's syndrome. *Adv Exp Med Biol* 1997; in press.
17. Toda I, Rocha EM, Siveira LA, Wickham LA, Sullivan DA. Gender-related difference in the extent of lymphocyte infiltration in lacrimal and salivary glands of mouse models of Sjögren's syndrome. Submitted for publication, 1997.
18. Toda I, Wickham LA, Sullivan DA. Gender and androgen-related influence on the expression of proto-oncogene and apoptotic factor mRNAs in lacrimal glands of autoimmune and non-autoimmune mice. Submitted for publication, 1997.
19. Toda I, Wickham LA, Sullivan DA. Influence of gender and androgen treatment on the mRNA expression of proto-oncogenes and apoptotic factors in lacrimal and salivary tissues of the MRL/lpr mouse model of Sjögren's syndrome. Submitted for publication, 1997.
20. Chomczynski P, Sacchi N. Single-step method of RNA isolation by acid guanidinium thiocyanate-phenol-chloroform extraction. *Anal. Biochem.* 1987; 162:156–159.
21. Hoffman RW, Alspaugh MA, Waggie KS, Durham JB, Walker SE. Sjogren's syndrome in MRL/l and MRL/n mice. *Arthritis Rheum.* 1984; 27:157–165.
22. Kessler HS. A laboratory model for Sjogren's syndrome. *Am J Pathol* 1968; 52:671–678.
23. Johnson BC, Morton JI, Trune DR. Lacrimal and salivary gland inflammation in the C3H/lpr autoimmune strain mouse: a potential mode for Sjögren's syndrome. *Otolaryngol Head Neck Surg* 1992; 106:394.
24. Moore PA, Bounous DI, Kaswan RL, Humphreys-Beher MG. Histologic examination of the NOD-mouse lacrimal glands, a potential model for idiopathic autoimmune dacryoadenitis in Sjogren's syndrome. *Lab Anim Sci* 1996; 46:125–128.
25. Toda I, Wickham LA, Rocha EM, Silveira LA, Sullivan DA. Identification of proto-oncogene and apoptotic factor mRNAs in human ocular surface tissues. Submitted for publication, 1997.

63

APOPTOSIS IN THE LACRIMAL GLAND AND CONJUNCTIVA OF DRY EYE DOGS

Jianping Gao, Tammy A. Gelber-Schwalb, John V. Addeo, and Michael E. Stern

Department of Biological Science
Allergan, Inc.
Irvine, California

1. INTRODUCTION

Keratoconjunctivitis sicca (KCS), or dry eye syndrome, is characterized by the development of ocular surface damage. KCS can occur as an individual phenomenon, or in association (as a secondary phenomenon) with various types of systemic autoimmune disorders. Sjögren's syndrome (SS), with the signature symptoms of dry eye and dry mouth, has the strongest association with KCS. The most prominent characteristic of SS is a progressive, follicular, lymphoid infiltration found in biopsy specimens of the lacrimal and salivary glands. These infiltrates consist of primarily CD4+ T cells and B cells demonstrated by immunohistological staining. Functional studies have shown that this type of impairment of the lacrimal gland results in decreased tear secretion and altered tear protein composition.[1]

Although the exact etiology of dry eye syndromes is still unknown, it is believed to be multifactorial and to involve multiple systemic deficiencies including genetic, neural, endocrine, and immunological systems.[2,3] Recently, however, the factors initiating the accumulation of these autoreactive lymphocytes by suppressing their pre-programmed cell death are suggested to be important as well. Apoptosis, or programmed cell death, is an active process of a gene-directed physiological event that occurs in most natural regulatory systems. It differs from necrosis in the lack of effects on contiguous cells and in the absence of inflammatory cells.[4] Dysregulation of this physiological mechanism for cell death has been implicated in a variety of diseases. In particular, apoptosis appears to play a critical role in autoimmunity.[5] A study conducted in the autoimmune mouse model MRL/lpr indicated that the progression of this syndrome may, in part, be related to the suppression of lymphocytic apoptosis in the salivary gland. This disorder of lymphoproliferation may be explained by defects in fas antigen, a cell surface protein that induces apoptosis.[6] In vitro, IFN-γ-induced acinar epithelial cell apoptosis was demonstrated in

Lacrimal Gland, Tear Film, and Dry Eye Syndromes 2
edited by Sullivan *et al.*, Plenum Press, New York, 1998

cultured human salivary gland, suggesting a potential involvement of the epithelial cells in SS disease mechanism.[7] Additionally, dysfunction of apoptosis has been postulated to lead to inappropriate longevity of autoreactive B cells in systemic lupus erythematosus (SLE) patients. Increased mRNA levels of bcl-2, a proto-oncogene inhibitory of apoptosis, have been reported in peripheral blood mononuclear cells from 19 of 24 SLE patients in comparison to normal healthy controls.[8]

To gain a better understanding of the relationship between apoptosis and the autoimmune process, we have established a colony of dogs with chronic KCS to evaluate apoptosis, a potential mechanism of dry eye. The effect of topical cyclosporin A (CsA), an immunosuppressive drug, on the level of apoptosis in the target tissues of KCS dogs has also been demonstrated. We propose that the lymphocytic accumulation of the lacrimal gland and conjunctiva, which results in destruction of acinar and conjunctival epithelial cells, is, in part, due to abnormal levels of apoptosis. We hypothesize that, in the lacrimal gland and conjunctiva of KCS dogs: (i) the accumulation of lymphocytes is due, in part, to a lack of apoptosis; (ii) the lacrimal acinar and conjunctival epithelial cells have an elevated apoptosis compared with normal subjects; (iii) topical CsA may facilitate apoptosis in lymphocytes and suppress lacrimal acinar and conjunctiva epithelial apoptosis, and (iv) various mediators such as p53, fas, and fas ligand may be involved in the initiation of this apoptosis process.

2. MATERIALS AND METHODS

2.1. Animal Selection, Treatment, and Tissue Collection

Spontaneously dry eye dogs were diagnosed by a board-certified veterinary ophthalmologist and housed in environmentally controlled rooms. All animal-related procedures comply with the applicable sections of the ARVO resolution on the use of animals in research. The group of dogs comprised one Pekinese, three cocker spaniels, four beagles, and five pugs including 3 male and 13 female dogs. They were randomly separated into two groups. One group of 10 dogs with 19 treated eyes (one KCS dog had monocular dry eye) was treated with 0.2% cyclosporin A (Allergan); the other group of 3 dogs with 6 tested eyes was treated with vehicle. Additionally, 4 normal beagles were utilized as the baseline control. Animals were dosed with one drop in each eye of the designated compounds, twice a day at 8 AM and 4 PM. The lubricant eye drops, Refresh Plus™ (Allergan), were applied at 12 PM between treatments. The treatment period was 12 weeks.

For surgical biopsy procedures, all dogs were pretreated with atropine (1.0 ml/kg) subcutaneously, then anesthetized by intravenous injection of a short-duration anesthetic mixture, ketamine and valium (1:1 v/v, 1 ml/kg), and topical proparacaine. Biopsy specimens were taken from the nictitans membrane lacrimal gland (NMG) and the lateral limbal conjunctiva of the eye predosing and at week 12 (post dosing). Biopsies were also obtained from 4 normal dogs as control.

2.2. Apoptosis Evaluation

Apoptosis was first evaluated by terminal deoxynucleotidyl transferases (TdT)-mediated dUTP-digoxigenin nick end labeling (TUNEL), using the modified procedure of the ApopTag™ Kit (Oncor), as described previously.[9] Paraffin-embedded biopsies (6 μm) from the NMG and conjunctiva were deparaffinized, dehydrated, and rehydrated followed by 30 min digestion with 20 μg/ml proteinase K (Sigma) at 37°C. After quenching endogenous peroxidase, the samples were incubated with TdT in the presence of 11-digoxigenin dUTP at 37°C for 90 min. The reaction was terminated by the stop/wash buffer. The samples were then blocked by 2% BSA at room temperature (RT) for 10 min followed by 30 min incubation with anti-digoxigenin-peroxidase at RT. The samples were then visualized by reacting with 2.24 mM diaminobenzidine (DAB) and counterstained with 1.0% (w:v), pH 4.0, methyl green at RT for 10 min. After washing with double distilled water, the tissues were dehydrated in butanol, rinsed in xylene, and mounted in Permount. The samples were finally evaluated under the light microscope (Leica) in which the apoptotic cells were labeled brown, whereas normal cells were light green.

Apoptosis was further confirmed by the DNA fragmentation assay modified from Prigent's method.[10] NMG and conjunctival biopsies were dissected from the normal control and pre- and post-CsA-treated dogs. Tissues from the same group were combined (n=2 for normal dog, n=4 for KCS dog) and homogenized in 600 μl of lysing buffer (10 mM Tris, pH 8.0; 0.5 mM EDTA, pH 8.0; 75 mM NaCl; 0.2% SDS, and 150 μg/ml proteinase K) (Sigma), and incubated at 60°C for 2 h. The lysates were centrifuged at 14,000 X *g* for 20 min. The supernatant was recovered for DNA precipitation by 1 volume isopropanol with 0.2 M NaCl. The DNA pellet was washed in 75% ethanol, dissolved in TE, and finally separated on 1.5% agarose gel.

2.3. Antibodies and Immunohistochemistry

A polyclonal antibody against recombinant bacterial expressed p53 amino terminal region (aa 37–45) (Oncogene Sciences) was utilized to demonstrate p53 protein expression in the dog tissues. Paraffin-embedded sections (6 μm) mounted on ProbeOn Plus glass slides (Fisher) were deparaffinized in xylene followed by serial dehydration and rehydration steps based on standard histology precedures. Before staining, sections were heated for 10 min in two changes of DAKO Target Retrieval Solution in a microwave. Tissue sections were then treated with 3% hydrogen peroxide for 10 min and preblocked by DAKO Protein Block Serum-Free solution for 15 min at RT. After washing in PBS, sections were first incubated with p53 primary antibody (dilutions: 1:500) overnight at 4°C, then with secondary antibody (biotinylated rabbit anti-sheep IgG; Oncogene Science) for 60 min at RT. Samples were then allowed to react with streptavidin conjugated to horseradish peroxidase (DAKO). Color development was achieved by incubation for 10 to 15 min at RT with a chromogen solution (AEC: 3% 3-amino-9-ethylcarbazole).

3. RESULTS

3.1. Apoptosis Evaluation

Apoptosis in the NMG and conjunctiva of the normal and KCS dogs, as well as pre- and post-CsA-treated KCS dogs, were evaluated and compared using the TUNEL assay which identifies 3'hydroxyl ends (DNA fragments) as an early step in apoptosis. Apoptotic signals in the normal control tissues were found to be limited. There were very few positively stained acinar and conjunctival epithelial cells dispersed throughout whole sections. A small number of labeled lymphocytes were traffic through and detected in the interstitium. In contrast, the acinar and conjunctival epithelial cells of KCS dogs exhibited strong positivity. About 60–70% of lacrimal acinar cells and more than 70% of conjunctival epithelial cells were labeled, whereas much of the accumulated lymphocytes were virtually uniformly negative (more than 60%). Additionally, strong staining was found in ductal cells in the KCS dog lacrimal gland. However, after 12 weeks of 0.2% topical CsA treatment, epithelial cell apoptosis in the lacrimal and conjunctival tissues were markedly decreased. The ratio of positive vs. negative cells was significantly reduced in comparison to that of pre-CsA-treated KCS dogs. In some cases (8 eyes from 6 dry eye dogs), the apoptosis level was almost back to the normal baseline. Additionally, the apoptosis level in the lymphocytic infiltrate increased in the post-CsA-treated lacrimal gland and conjunctiva. Among these infiltrates, more than 80% of total lymphocytes underwent apoptosis in comparison to that in the pre-CsA samples, where less than 40% of infiltrates were positive by TUNEL staining. These results indicated that apoptosis was involved in the dysfunction of lacrimal gland acinar cells and conjunctival epithelial cells in the dogs with dry eye syndromes.

3.2. DNA Fragmentation Analysis Using Agarose Gel Electrophoresis

To confirm the TUNEL results, the profile of nuclear DNA structure was analyzed. One of the most distinct features of apoptosis is DNA laddering, which takes part at the fairly early stage of the programmed cell death, and has a unique 180–200 bp increment resulting from specific endonuclease cleavage between the internucleosomal units. The genomic DNA was extracted from the lacrimal gland and conjunctiva of the normal as well as the KCS dogs. DNA fragments were resolved on an agarose gel. The typical pattern of DNA ladder was detected in the lacrimal samples only from pre-CsA-treated KCS dogs, but not in the normal control and post-treatment groups.

3.3. Protein Expression of Tumor Suppresser p53

The distribution and the amount of the tumor-suppresser protein p53 were evaluated in the lacrimal gland of KCS dogs before and after 12 weeks of topical CsA administration. In the KCS dog prior to treatment, p53 protein located exclusively in the nucleus of acinar cells, but not in ductal cells. Acinar nucleus of the pre-CsA-treated KCS dog lacrimal gland was heavily loaded with reddish-brown signals representing p53 protein immunoreactivity. However, after CsA administration, the signal of positively stained acinar cells for p53 protein was diminished. For control purposes, the level of p53 protein was also measured for the normal dog lacrimal tissue. Both the acinar and ductal epithelial cells exhibited very low levels of p53 protein in the normal dog lacrimal gland.

4. DISCUSSION

So far, other than lubricants, there is no effective therapeutic drug available for KCS or dry eye. Besides insufficient knowledge of the cause of the disease, lack of an appropriate animal model is another major problem. Currently, most animal models for primary/secondary Sjögren's syndrome or systemic lupus erythematosus have been developed in mice.[11–14] A problem with the mouse model, however, is that the eyes are small, which makes topical treatment difficult. Studies advocating topical administration of cyclosporine using dry eye dogs have been reported to be successful.[15–17] Canine KCS was considered to be an immune-related disorder, with similarities to Sjögren's syndrome and human aqueous tear deficiency.[15–17] Therefore, we established the Spontaneously Dry

Eye Dog model in house, to study the mechanisms of chronic KCS or Sjögren's syndrome as well as the effect of topical cyclosporine treatment. The lacrimal gland and conjunctiva of these dogs contain extensive lymphocytic infiltration believed responsible for this syndrome. We hypothesize that this accumulation of lymphocytes is due, in part, to a lack of apoptosis. Meanwhile, an elevated apoptosis in lacrimal acinar cells and conjunctival epithelial cells may also play a role in the altered lacrimal gland and conjunctiva function. The results of this study have confirmed our hypothesis and demonstrated that apoptosis does play an important role in the process of dry eye syndromes.

Two methods were employed to evaluate apoptosis. The first was TUNEL assay. In the normal tissues, the number of apoptotic epithelial cells was extremely small. Light staining was occasionally found in the myoepithelial cells in the normal lacrimal gland or in the superficial epithelial cells in the normal conjunctiva. These phenomena are reasonable and predictable, since the normal acinar and conjunctival epithelial cells represent a stable population, have a long turn-over time, and should have little apoptotic activity. Interestingly, the dispersed lymphocytes in the interstitium of lacrimal lobules were found to be positive. Unlike the parotid gland, the lacrimal gland does not possess its own lymph nodes. The scattered lymphocytes normally traffic through the gland, en route to the regional lymph nodes where they eventually undergo apoptosis.[18,19] Our results suggest that lymphocytic apoptosis may be initiated prior to reaching the final destination. Although the normal conjunctiva consists of lymphocytes and plasma cells as well as lymphoid aggregates,[20] we have found abnormally large amounts of lymphocytes in the conjunctival stroma of KCS dogs in comparison to that in the normal control. Most of these lymphocytes, however, were not stained by the TUNEL assay, implying that the infiltrating cells were still functional. We also assessed the level of apoptosis in the CsA-treated tissues. The number of positive epithelial cells in both the lacrimal gland and conjunctiva was significantly reduced compared with the value from the pre-CsA-treated group. In contrast, not only the total number of infiltrated lymphocytes was dramatically diminished in the lacrimal gland and conjunctiva of the post-CsA-treatment group, but the large portion of remaining lymphocytes exhibited raised apoptotic activity. In summary, the TUNEL assay results suggest that: (i) the distinct, specific staining and the lack of randomly destroyed chromatin confirm the presence of apoptotic cell death, and (ii) the elevated level of apoptosis in the lacrimal acinar and conjunctival epithelial cells of KCS dogs may be responsible for their altered cellular function and tissue atrophy. It has recently been reported that the regression of the rabbit lacrimal gland after castration appeared to be due to apoptosis.[21] Also, data in the rat parotid gland illustrates that the duct obstruction-induced glandular atrophy is mediated via apoptosis.[22]

To confirm the TUNEL assay results, we further utilized gel electrophoresis to analyze the nuclear chromatin structure. Gel electrophoresis of DNA extracted from the lacrimal and conjunctival biopsies of KCS dogs without CsA treatment revealed a typical apoptotic DNA laddering pattern, which has the appropriate 180–200 bp increment. In contrast, the DNA from normal specimens remained intact. However, we have not been able to demonstrate DNA ladders from either lacrimal or conjunctival tissues post-CsA treatment. It might be due to the insufficient amount of "apoptotic DNA," since most cellular DNA comes from lacrimal acinar or conjunctival epithelial cells, in which the level of apoptosis was greatly reduced following topical CsA administration. This raises another question --- Where does the fragmented DNA come from? Does it originate from dying epithelial cells, or infiltrated lymphocytes, or both? Although this is not conclusive from the current experiment, the electrophoresis data confirm that apoptosis is involved in lacrimal gland as well as conjunctival destruction. Another possibility for the failure of observing small DNA fragments is that certain types of cells simply do not produce ladders with small oligonucleosomal units. It has been reported that formation of oligonucleosomal fragments requires the presence of both Mg^{2+} and Ca^{2+}, but will not occur in the absence of either of the two divalent cations.[23]

Currently, there is much discussion on the regulation of apoptosis, specifically on the intracellular molecular signaling directing stimulation or suppression of apoptosis. Various apoptotic inducers as well as inhibitors have been identified. The critical role of p53 is believed to be involved in defective apoptosis in the development of tumor cells.[24] The importance of p53 association with autoimmunity has been indicated recently. The knockout mouse that lost the p53 function did not undergo radiation-induced apoptosis.[25] T cells in patients with breast cancer were able to recognize the wild-type p53 protein and then highly proliferate in response to it.[26] These data were thought to be relevant to the immune processes leading towards autoimmunity. In this study, we demonstrated that canines with KCS expressed a high level of p53 protein, specifically in the nucleus of lacrimal acinar cells, whereas the normal control dogs exhibited a limited p53 expression, which was consistent with previous findings.[27,28] The p53 level was diminished following 12 weeks of CsA treatment. This result demonstrated the positive correlation between levels of apoptosis and p53 protein expression. Further investigation is necessary to clarify which initiators may cause the accumulation of p53 protein in the lacrimal tissue of the KCS dog, and whether p53 protein is one of the co-factors for the initiation of lacrimal autoimmunity. Particularly, is it possible that p53, as a self protein, binds to a foreign antigen such as a viral antigen following viral infection of the lacrimal gland, and triggers an immune response? This proposed mechanism is supported by evidence that high-titer autonuclear antibodies specific for p53 protein can be induced in mice injected with purified complexes of murine p53 and SV40 large T antigen, but not in mice treated with either protein alone.[29] Given the hypothesis that dry eye or KCS is commonly associated with viral infection on the ocular surface or directly in the lacrimal gland, the p53-mediated apoptosis of host cells is likely under the influence of viruses. The viral manipulation of host cell apoptosis was reported to be frequently via the interactions with cellular proteins such as p53.[24] The involvement of other apoptotic mediators are currently under investigation. Preliminary data have shown the presence of fas and fas ligand protein in the lacrimal gland of dry eye dogs (data not shown).

Cyclosporine is a non-cytotoxic immunosuppressive drug. It has been known for many years, due to its effective application to prevent rejection after organ transplantation[30] and to treat a variety of autoimmune diseases.[31,32] The proposed mechanism of its immunosuppressive action is that the binding of cyclosporine to a specific nuclear protein, cyclophilin, is required for the initiation of the inhibition of T cell activities, which subsequently prevents the production of inflammatory cytokines such as IL-2 and IFN-γ, and eventually interrupts the autoimmune response.[33] The lymphocytic infiltrates in SS consist primarily of CD4+ T cells and B cells.[1] Since T cells are specifically affected by cyclosporine, it is reasonable to propose that cyclosporine therapy would be effective for immune-related causes of dry eye syndromes. It has been demonstrated that cyclosporine was effective in suppressing the autoimmune disease including the ocular and lacrimal gland lesion, in MRL/lpr mice,[34] as well as in KCS dogs.[35–37] CsA also exhibited anti-inflammatory activity on animals and humans with chronic ocular surface diseases.[15,38,39] Corneal ulcer and/or conjunctival hyperplasia are commonly seen in both the human and dogs with KCS, demonstrating the clear signs of the inflammatory reaction. There is evidence of the expression of the inflammatory cytokine IL-6 in the conjunctiva of patients with SS, suggesting that the ocular surface likely contributes to the pathogenesis of dry eye.[40] In this paper, we demonstrated the differential effects of CsA on the level of apoptosis in the epithelial cells and lymphocytic infiltrates in the lacrimal gland and conjunctiva of dry eye dogs. The apoptotic level and p53 expression in the lacrimal acinar cells was significantly reduced by CsA treatment, suggesting that the inhibitory impact of CsA on the acinar epithelial apoptosis appeared to be mediated through a p53-dependent mechanism. It is likely that CsA can differentially alter the gene expression of certain apoptosis mediators, such as p53, fas, and fas ligand in the epithelial cells of target tissues and in the infiltrated lymphocytes, to reestablish their normal apoptotic balance.

REFERENCES

1. Pflugfelder SC, Wilhelmus KR, Osato MS, et al. The autoimmune nature of aqueous tear deficiency. *Ophthalmology*. 1986;93:1513–1517.
2. Fox RL. Epidemiology, pathogenesis, animal models, and treatment of Sjögren's syndrome. *Curr Opin Rheumatol*. 1994;6:501–508.
3. Sullivan DA, Wickham LA, Krenzer KL, et al. Aqueous tear deficiency in Sjögren's syndrome: possible causes and potential treatment. In: Pleyer U, Hartmann C, Sterry W, eds. *Oculodermal Diseases: Immunology of Bullous Oculo-Muco-Cutaneous Disorders*. Buren, The Netherlands: Aeolus Press, 1997:95–152.
4. Wyllie AH, Kerr JFR, Currie AR. Cell death: the significance of apoptosis. *Int Rev Cytol*. 1980;86:251–306.
5. Duke RC, Ojaius DM, Young JD. Cell suicide in health and disease. *Sci Am*. 1996;275:80–87.
6. Watanabe-Fukunaga R, Brannan CI, Copeland NG. Lymphoproliferation disorder in mice explained by defects in fas antigen that mediates apoptosis. *Nature*. 1992;356:314–317.
7. Wu AJ, Chen ZJ, Tsokos M, et al. Interferon-gamma induced cell death in a cultured human salivary gland cell line. *J Cell Physiol*. 1996;167:297–304.
8. Rose LM, Latchman DA, Isenberg DA. Bcl-2 and fas, molecules which influence apoptosis. A possible role in systemic lupus erythematosus. *Autoimmmunity*. 1994;17:271–278.
9. Gao J, Gelbert-Schwalb TA, Addeo JV, Stern MS. Apoptosis in the rabbit cornea after photorefractive keratectomy. *Cornea*. 1997;16:200–208.
10. Prigent P, Blanpied C, Aten J, Hirsch F. A safe and rapid method for analyzing apoptosis-induced fragmentation of DNA extracted from tissues or cultured cells. *J Immunol Methods*. 1993;160:139–140.
11. Fox RI, Pisa P, Pisa EK, Kang HI. Lymphoproliferative disease in SCID mice reconstituted with human Sjögren's syndrome lymphocytes. *J Clin Lab Anal*. 1993;7:46–56.
12. Haneji N, Hamano H, Yanagi K, Hayashi Y. A new animal model for primary Sjögren's syndrome in NFS/sld mice.. *J Immunol*. 1994;153:2769–2777.
13. Johnson BC, Morton JI, Trune DR. Lacrimal and salivary gland inflammation in the C3H/Ipr autoimmune strain mouse: a potential mode for Sjögren's syndrome. *Otolaryngol Head Neck Surg*. 1992;106:394–399.
14. Sullivan DA. Androgen-induced suppression of autoimmune disease in lacrimal glands of mouse models of Sjögren's syndrome. *Adv Exp Med Biol*. 1994;350:683–690.
15. Kaswan RL. Keratoconjunctivis sicca: immunological evaluation of 62 canine cases. *Am J Vet Res*. 1985;46:376–383.
16. Kaswan RL, Salisbury MA, Ward DA. Spontaneoud canine keratoconjunctivitis sicca. A useful model for human keratoconjunctivitis sicca: Treatment with cyclosporine eye drops. *Arch Ophthalmol*. 1989;107:1210–1216.
17. Kaswan RL. Characteristics of a canine model of KCS: effective treatment with topical cyclosporine. *Adv Exp Med Biol*. 1994;350:583–594.
18. Allansmith M. Immunology of the tears. *Int Ophthalmol Clin*. 1973;13:47–72.
19. Wieczorek R, Jacobiec FA, Sacks EH, Knowles DM. The immunoarchitecture of the normal human lacrimal gland. *Ophthalmology*. 1988;95:100–109.
20. Fine BS, Yanoff M. *Ocular Histology*. Harper & Row, 1979;308–309.
21. Azzarolo AM, Olsen E, Huang ZM, et al. Rapid onset of cell death in lacrimal glands after ovaritectomy. *Invest Ophthalmol Vis Sci*. 1996;37:S856.
22. Walker NI, Gobe GC. Cell death and cell proliferation during atrophy of the rat parotid gland induced by duct obstruction. *J Pathol*. 1987;153:333–344.
23. Sun X-M, Cohen GM. Mg^{2+} - dependent cleavage of DNA into kilobase pair fragments is responsible for the initial degradation of DNA in apoptosis. *J Biol Chem*. 1994;269:14857–14860.

24. Hale AJ, Smith CA, Sutherland LC, et al. Apoptosis: molecular regulation of cell death. *Eur J Biochem.* 1996;236:1–26.
25. Fox RL. In: Lieberman R, ed. Apoptosis, cancer and p53 tumour suppressor gene. *Cancer Metastasis Rev.* 1995;14:149–161.
26. Tilkin AF, Lubin R, Soussi T, et al. Primary proliferative T cell response to wild-type p53 protein in patients with breast cancer. *Eur J Immunol.* 1995;25:1765–1769.
27. Strasser A, Harris AW, Jacks T, Cory S. DNA damage can induce apoptosis in proliferating lymphoid cells via p-53 independent mechanisms inhibitable by bcl-2. *Cell.* 1994;79:329–339.
28. Fritsche M, Haessler C, Brandner G. Induction of nuclear accumulation of the tumor-suppressor protein p53 by DNA damaging agents. *Oncogene.* 1993;8:307–318.
29. Dong X, Hamilton KJ, Satoh M, et al. Initiation of autoimmunity to the p53 tumor suppressor protein by complexes of p53 and SV40 large T antigen. *J Exp Med.* 1994;179:1243–1252.
30. Starzl TE. New approaches in the use of Cyclosporine: with particular reference to the liver. *Transplant Proc.* 1988;20:356–360.
31. Gunn HC, Ryffel B. Successful treatment of autoimmunity in (NAB X NZW) F1 mice with cyclosporin and (Nva)-Cyclosporin. II. reduction of glomerulonephritis. *Clin Exp Immunol.* 1986;64:234–242.
32. Liu SH, Zhou D-H, Gottsch JD, et al. Treatment of experimental autoimmune dacryoadenitis with Cyclosporin A.. *Clin Immunol Immunopathol.* 1993;67:78–83.
33. Hess AD. Mechanisms of action of cyclosporine: Considerations for the treatment of autoimmune diseases. *Clin Immunol Immunopathol.* 1993;68:220–228.
34. Jabs DA, Lee B, Burek CL, et al. Cyclosporine therapy suppresses ocular and lacrimal gland diseases in MRL/MP-lpr/lpr Mice. *Invest Ophthalmol Vis Sci.* 1996;37:377–383.
35. Stern MS, Gelber TA, Gao J, et al. The effects of topical cyclosporine A (CsA) on dry eye dogs (KCS). ARVO Abstracts. *Invest Ophthalmol Vis Sci.* 1996;37:S1026.
36. Morgan RV, Abrams KL. Topical administration of cyclosporine for treatment of keratoconjunctivitis sicca in dogs. *J Am Vet Med Assoc.* 1991;199:1043–1046.
37. Olivero DK, Davidson MG, English RV, et al. Clinical evaluation of 1% cyclosporine for topical treatment of keratoconjunctivitis sicca in dogs. *J Am Vet Med Assoc.* 1991;199:1039–1042.
38. Zierhut M, Thiel HJ, Weidle EG, et al. Topical treatment of severe corneal ulcers with cyclosporine A. *Graefes Arch Clin Exp Ophthalmol.* 1989;227:30–35.
39. Liegner JT, Yee RW, Wild JH. Topical cyclosporine therapy for ulcerative keratitis associated with rheumatoid arthritis. *Am J Ophthalmol.* 1990;109:610–612.
40. Jones DT, Monroy D, Ji Z, et al. Sjögren's syndrome; cytokine and Epstein-Barr virus gene expression within the conjunctival epithelium. *Invest Ophthalmol Vis Sci.* 1994;35:3493–3503.

CYTOKINES AND TEAR FUNCTION IN OCULAR SURFACE DISEASE

Keith Barton,[1,2] Alexandra Nava,[1] Dagoberto C. Monroy,[1] and Stephen C. Pflugfelder[1]

[1]Corneal and External Disease Service
Bascom Palmer Eye Institute
University of Miami School of Medicine
Miami, Florida
[2]Moorfields Eye Hospital
London, England, United Kingdom

1. INTRODUCTION

It is likely that rates of tear production and turnover are important in determining the cytokine environment of ocular surface tissues, by supplying cytokines in tear fluid and by clearing those produced at the ocular surface. Epidermal growth factor (EGF), which is released into tear fluid by the lacrimal gland,[1–4] influences healing of corneal epithelial and conjunctival wounds.[5–10] These concentrations vary inversely with reflex tear secretion,[3,4] but under normal conditions in vivo, the relationships between tear concentrations of EGF and other variables such as age, gender, tear production, and clearance have not been examined. This information potentially could contribute to a better understanding of the role of EGF in maintenance of ocular surface integrity under normal circumstances, so in the first part (*Tear EGF Concentrations in Normals*) of a two-part study, we examined these influences in a large group of normal subjects.[11]

In the second part (*Tear Cytokines in Rosacea*), we sought to compare the differences in tear fluid concentrations of two fundamental pro-inflammatory cytokines, IL-1α and TNF-α, and of EGF, between normal controls and ocular rosacea patients, and to determine if these differences are related to alterations in aqueous tear production or turnover.[12] Rosacea is a chronic, progressive condition of unknown etiology affecting both the facial skin and the eye,[13] characterized by recurrent episodes of facial erythema, papules, pustules, telangiectasia, and rhinophyma. Many symptoms of ocular rosacea are similar to those experienced by patients with aqueous tear deficiency, though tear production is usually normal in the former.[14] Similarly, the ocular signs are often those seen in isolated meibomian gland disease.[15] The cause of the irritation has not been established, but instability of the tear film[16] and elevated tear film osmolarity associated with increased evaporation secondary to deficiency in the lipid component of tears[17] are possible mechanisms. The presence of external ocular inflammatory signs and improvement with topically applied corticosteroids suggest that rosacea may be an appropriate condition in which to examine the interaction of inflammatory cytokines with tear production and clearance.

2. METHODS

2.1. Selection of Subjects

2.1.1. Tear EGF Concentration in Normals Study. In the first part of the study, 68 volunteers with no symptoms of ocular irritation, history of external eye disease or ocular surgery in the last year, contact lens wear, or ocular medication usage other than preserved artificial tears were recruited.

2.1.2. Tear Cytokines in Rosacea Study. For the second part, three groups were included: (i) 14 adult patients with ocular irritation and facial signs of rosacea, specifically excluding those using systemic immunosuppressives, topical retinoid therapy, or topical medications other than non-preserved artificial tear preparations in addition to the above exclusion criteria; (ii) normal controls, frequency-matched for age, with no symptoms of ocular irritation or biomicroscopic features of ocular surface disease; (iii) 15 "ideal" control subjects who had no symptoms of ocular irritation or biomicroscopic signs of ocular surface disease. In addition, these subjects had no evidence of meibomian gland disease, unlike the age-matched group (group ii), in whom clinical evidence of meibomian gland disease was the norm.

2.2. Clinical Examination

Slit-lamp biomicroscopy was used to assess inferior tear meniscus, debris and mucus strands in the preocular tear film, lid margin vascular injection (brush marks) and irregularity of the posterior lid margin, bulbar and palpebral conjunctival hyperemia, corneal punctate epithelial erosions, corneal adherent mucus, and filamentary keratitis. The meibomian glands were examined by slit-lamp biomicroscopy and graded for the presence of orifice metaplasia, expressibility of meibum, and meibomian gland acinar dropout, as described previously.[18]

Fluorescein tear break-up time (TBUT), fluorescein and rose bengal staining on slit-lamp biomicroscopy, and the Schirmer test without anesthesia were performed as previously described.[18] No subject was included if the corneal fluorescein or rose-bengal staining score was greater than 2 out of 12, as previously described.[14]

The tear clearance rate (TCR) was determined in a manner similar to that of Xu et al.,[19] except that the Schirmer strip was inserted into the unanesthetized conjunctival sac 15 min after the instillation of 5 μl of 2% fluorescein. The TCR was derived by comparing the dilution of fluorescein on the strip with a nomogram.[11] The logarithm of the tear clearance rate (LN[TCR]) was used because the TCR values are defined according to doubling dilutions, and are therefore logarithmic in nature.

2.3. Laboratory Investigations

Tear collection was performed atraumatically in a dimly lit room, prior to other tests, to minimize reflex tearing. In the first part of the study, 5 μl was collected, using a polyester wick,[11] from one eye which was chosen randomly. For the second part, 10 μl was taken from each eye and pooled. Saturated rods were placed in a microliter pipette tip, stored in a 500 μl polypropylene Eppendorf tube, numbered, and kept at -84°C for 10 to 30 days prior to ELISA.[12] In nine randomly chosen subjects, tears were collected at 9 AM and again at 3 PM to look for diurnal fluctuation in tear EGF concentration.

ELISA assays were performed using a commercial kit (R&D Systems, Minneapolis, MN). Prior to each analysis, tears were extracted from the rods by centrifugation,[12] diluted in ELISA buffer (supplied by manufacturer), transferred to a microtiter plate, and the assay performed according to the manufacturer's instructions. Cytokine concentrations were determined from a standard curve, as previously described.[20]

2.4. Statistical Analyses

Multiple linear regression analysis was used to identify relevant factors in the first part of the study, with age as the independent variable. Logistic regression was used in the second part, with subject group as the independent variable. The influence of tear function parameters, Schirmer I, TCR, LN(TCR), TFI, and EGF was examined. Student's t-test was used to compare means of relevant factors and to look for gender diferences in EGF concentrations and tear function parameters.

3. RESULTS

3.1. Tear EGF Concentration in Normals

In the first part of the study, an asymptomatic control group of 68 subjects (33 males and 35 females) of median age 48 years (range 21–88) was recruited. When the TCR results were examined for normality of distribution, it was found that the most appropriate measure of tear clearance was the natural logarithm of the TCR, LN(TCR).

EGF levels were obtained from tear samples in 65 cases (Table 1).[11] No significant difference was noted in EGF samples from tears collected in the afternoon, as compared with the morning, so diurnal fluctuation was not believed to be a significant confounding variable. Tears from male subjects contained significantly higher concentrations of EGF than those from females ($p = 0.043$). No gender difference was observed in Schirmer I or LN(TCR). A significant gender difference in TBUT was noted ($p = 0.01$), and TBUT was found to correlate with Schirmer I value ($r = 0.29$, $p = 0.02$), but not with LN(TCR) or other parameters.

Age was found to have a negative influence on the tear clearance rate LN(TCR) but not on the Schirmer I value or tear EGF levels (Table 2).

Table 1. Schirmer, LN(TCR), and EGF concentrations in normals (mean ± SEM)

Parameter	Males	Females	All
Schirmer I (mm)	18 ± 2	18 ± 2	18 ± 1
LN(TCR)	3.68 ± 0.16	3.66 ± 0.16	3.67 ± 0.11
Tear fluid EGF concentration (ng/ml)	3.38 ± 0.33*	2.39 ± 0.35*	2.86 ± 0.25

*$p = 0.043$.

Table 2. Relationships between Schirmer, LN(TCR), EGF concentration, and age in normals

Parameters	Correlation coefficient (r)	Significance (*p*)
Schirmer I vs. LN(TCR)	0.36	0.002
Age vs. Schirmer I	−0.06	0.61
Age vs. LN(TCR)	−0.29	0.015
Age vs. EGF	−0.12	0.33
EGF vs. Schirmer I	−0.28	0.026
EGF vs. LN(TCR)	−0.23	0.067

The inter- and intra-assay repeatability of the EGF ELISA was also examined. A single aliquoted sample measured three separate times on each of three consecutive days gave a standard deviation of 0.05 ng/ml, compared with 1.99 ng/ml for the study group as a whole. Variance component analysis revealed that the day-to-day component accounted for 78% of the variation, with 22% unexplained.

3.2. Tear Cytokines in Rosacea

Fourteen patients (8 males and 6 females) with signs of facial rosacea and symptoms of ocular irritation were recruited, age 71 ± 4 years (mean ± SEM). Twelve controls (5 males and 7 females) frequency-matched for age (73 ± 3 years) without symptoms, and 15 *ideal* controls (7 males and 8 females) (age 36 ± 2 years) were included. The rosacea patients complained of a variety of symptoms including dryness, burning, foreign body sensation, blurring, and mucus discharge. Photophobia was a less common complaint. By definition, the control groups were symptom-free.

Schirmer I wetting was not significantly different in rosacea patients (14 ± 2 mm, mean ± SEM) from age-matched controls (20 ± 2), but was significantly lower than in ideal controls (22 ± 2; $p = 0.013$). LN(TCR) was significantly higher in both age-matched controls (3.7 ± 0.21; $p = 0.048$) and ideal controls (4.1 ± 0.05; $p = 0.002$) than in rosacea patients (3.0 ± 0.28), indicating significantly reduced tear turnover in rosacea.

Epidermal growth factor was detected in the tears in all subject groups. Tear concentrations were not significantly higher in rosacea patients (3.7 ± 1.0 ng/ml; mean ± SEM) than in age-matched controls (2.1 ± 0.36), but were significantly higher than in ideal controls (1.2 ± 0.3; $p = 0.027$). EGF concentrations were not correlated with LN(TCR) or Schirmer I score in this part of the study.

IL-1α was detected in the tears of all subject groups. Concentrations were significantly higher (45.4 ± 4.6 pg/ml, mean ± SEM) in rosacea patients than in either age-matched (22.6 ± 5.0; $p = 0.003$) or ideal (17.1 ± 3.4; $p < 0.001$) controls. The relationship between time of day at which the tear collection was performed and the IL-1α concentration observed was examined by linear regression. No significant diurnal fluctuation was noted in tear IL-1α concentration.

The relationship of IL-1α concentration to tear function was examined in the group as a whole. Although an inverse correlation was observed between Schirmer I wetting ($r = -0.39$, $p = 0.012$), this was lower than the correlation with LN(TCR) ($r = -0.58$, $p < 0.0001$). There were not sufficient numbers in the individual subject groups to examine this relationship in each group.

No TNF-α was detected in the tears of any subject, implying a tear concentration of less than 10 pg/ml. The positive control sample was positive in all cases. The two control groups were compared for differences in mean Schirmer I score, LN(TCR), tear EGF, and IL-1α concentrations. None of these achieved significance.

4. CONCLUSIONS

4.1. Tear EGF Concentration in Normals

Radioimmunoassay (RIA) has been demonstrated to correlate well with the biological activity of EGF in tears,[2] and levels found using this and other immunodetection techniques[2–4] are similar to those reported in our study.[11] The low variability we observed using the ELISA, together with the availability of a commercial kit, should facilitate comparison of EGF concentrations in future studies. The values reported in our study appear to fall within a relatively constant range, as suggested previously,[3] but the lack of a significant change with age appears to be a novel finding (Table 2).

The gender differences in EGF levels contrast with the findings of others,[4] although in that study the higher levels in females were not significant. It is possible that the lower concentrations of EGF that we found in the tears of females are due partly to a lower threshold for reflex tearing. The inverse correlation of Schirmer I score with EGF levels supports previous work that reported that EGF concentrations decline with increased tear fluid flow.[4] Those with greater Schirmer I values probably have more reflex tearing, but tear collection time did not correlate with Schirmer test results or EGF levels in our study, and therefore reflex tearing is unlikely to account for the gender differences we observed.

Reduced androgenic influences on the lacrimal gland are also likely to be important. In fetal development, EGF mRNA expression in the developing male mouse genital tract is directly dependent on the release of testosterone.[21] Similarly, testosterone appears to stimulate prostatic cancer cells by increasing concentrations of mitogenic cytokines.[22,23] Androgens are profoundly stimulatory to lacrimal gland secretory function,[24–26] acinar volume,[27] and weight, with an inhibitory effect on autoimmune inflammation.[28]

The failure of the Schirmer I value to decline with age is surprising, and contrasts with previous findings,[29–31] although it is supported by Xu.[32] These discrepancies are probably attributable to the acknowledged variability of this test, although in both this study and that by Xu[32] subject numbers were large. The significant decline in tear clearance with age and the correlation between Schirmer I and tear clearance would have been predicted, but has not been found previously.[32]

Diagnostic tests for keratoconjunctivitis sicca, such as the Schirmer test and conjunctival rose-bengal staining score, have often been criticized for their low specificity and high variability. The tear function index (TFI) has been described previously by Xu et al.[19] This is a similar but less elaborate test for assessing tear production and turnover of tears on the ocular surface than the fluorescein clearance test.[33] The TFI was devised to increase the predictive accuracy of the Schirmer test in the diagnosis of dry eye conditions by combining the length of Schirmer wetting with the TCR. Xu demonstrated that the TCR correlated well with tear turnover as defined by Jordan.[29,32] However, we believe that the actual effect of combining these two factors is to exaggerate the effect of the Schirmer test, rather than to emphasize tear clearance, as suggested by Xu. Schirmer I wetting gives a rough estimate of tear production[34] and LN(TCR) of turnover (i.e., a function of production, drainage, and evaporation).[32] The TFI combines a measure of production (Schirmer) with a measure of production and drainage (TCR), thus exaggerating the importance of tear production and reducing the overall influence of drainage in the analysis. The slightly greater correlation between TFI and EGF than the individual components are almost certainly due to exaggeration of the correlation with production. On the other hand, when the Schirmer score is divided by LN(TCR), no correlation is found with EGF, as one index of production is divided by one index combining production and drainage, potentially cancelling out a significant proportion of the effect of production, so that this value may be related to drainage. This conclusion is supported by the finding that, for the study as a whole, the correlation coefficient (r) for TFI vs. Schirmer was 0.94, whereas that for TFI vs. LN(TCR) was 0.59. We believe that the Schirmer I test and the LN(TCR) are more valuable when considered individually.

4.2. Tear Cytokines in Rosacea

In this study we report low but detectable levels of IL-1α in the tear fluid of ideal normal asymptomatic subjects. There was no significant difference in tear fluid IL-1α concentrations between control groups, but levels were significantly higher in patients with ocular rosacea than in either control group. We were unable to detect TNF-α in any subject, suggesting that, if present at all in tear fluid, its concentrations are below 10 pg/ml.

The mode of release of IL-1α in healthy cells is uncertain, but its release is induced by cell injury. Necrosis results in release and processing of IL-1α, whereas IL-1β is released only in the unprocessed pro-form. In contrast, during apoptosis, mature forms of IL-1α and IL-1β are released.[35] Soluble IL-1 can be detected in the plasma and other body fluids, and its concentration increases during inflammation.[36,37] IL-1 is produced by many cell types, including monocytes, macrophages, fibroblasts, and stratified epithelium including the corneal epithelium.[38–40] In addition to its reported biological effects such as activation of type II helper T cells (TH2), increase in matrix metalloproteinase production by corneal stromal keratocytes, and induction of systemic responses such as fever and anorexia,[36,40,41] IL-1α also has an effect on corneal epithelial wound healing in tissue culture that is additive to that of EGF.[8]

There are two possible causes of elevated IL-1 concentrations in the tear fluid of rosacea patients: increased release by epithelial or inflammatory cells on the ocular surface, and reduced tear drainage. Although we did not examine the former, our observations of the latter included the finding that LN(TCR) was significantly reduced in rosacea patients when compared with both control groups, indicating a significant reduction in tear turnover in this condition. On the other hand, tear production (Schirmer I score) was not reduced in comparison with age-matched controls. Furthermore, the strong inverse correlation between tear fluid IL-1 concentrations and LN(TCR), as well as the inverse correlation between IL-1 concentration and Schirmer, suggest that overall, reduced tear turnover may be the main reason for the elevated IL-1 level in rosacea patients. In addition, the absence of TNF-α, which is not normally present at the ocular surface, supports the suspicion that the elevation of tear IL-1 levels in rosacea is largely due to IL-1 normally produced at the ocular surface but which is concentrated in rosacea by delayed tear turnover.

Reduced tear turnover may therefore be an important risk factor for chronic symptomatic ocular surface disease in rosacea. Elevated IL-1 in the tear fluid may also aggravate eyelid margin inflammation and edema of the lacrimal drainage system that could result in further reduction in tear turnover. Furthermore, it is possible that some or all of the clinical manifestations of ocular rosacea may be mediated by abnormally elevated tear fluid concentrations of IL-1, or other, as yet unidentified inflammatory factors. This theory is supported by the therapeutic effectiveness of topical corticosteroid therapy in patients with ocular rosacea.[42]

Meibomian gland disease is also observed in many asymptomatic older patients.[43] The onset of symptoms of irritation in these patients could be related to the development of delayed tear turnover with secondary increase in tear fluid IL-1α.

Tear EGF concentrations were significantly greater in rosacea patients than in ideal normal controls, and they were 76% higher in rosacea patients than in the age-matched control group, although this difference was not statistically significant. The association between EGF and Schirmer I score, noted in the first part of the study, may explain the greater EGF concentrations in the rosacea group because they had lower Schirmer I scores. Perhaps the correlation between aqueous tear production and tear EGF concentration was not as great in this study as in our previous study, because of the smaller sample size and greater standard deviation for tear EGF concentration in the rosacea group. Another possible but unproven cause of the increased tear EGF concentration in rosacea patients is increased production by the conjunctival epithelium in response to chronic ocular surface disease.

5. SUMMARY

In summary, tear EGF levels correlate most strongly with tear production in normals, and it is likely that some form of homeostatic mechanism exists to provide a constant supply to the ocular surface. Commercial ELISA kits appear to measure EGF in tears with good consistency and may be useful in the future to improve comparability of data from different studies. In addition, in ocular rosacea, which mimics keratoconjunctivitis sicca in a number of respects, there is a differential increase in the level of the inflammatory cytokine IL-1α in the tear fluid. Much of this elevation appears to be the result of reduced tear turnover, which may form an important positive feedback mechanism encouraging tear stagnation and the perpetuation of ocular surface inflammation.

ACKNOWLEDGMENTS

The data presented in this paper have been reproduced courtesy of *Cornea* (Reference #11) and *Ophthalmology* (Reference #12).

REFERENCES

1. Yoshino K, Monroy DC, Pflugfelder SC. Cholinergic stimulation of lactoferrin and EGF secretion by the human lacrimal gland. *Cornea.* 1996;15:617–621.
2. Ohashi Y, Motokura M, Kinoshita Y, et al. Presence of epidermal growth factor in human tears. *Invest Ophthalmol Vis Sci.* 1989;30:1879–1882.
3. van Setten G, Viinikka L, Tervo T, Pesonen K, Tarkkanen A, Perheentupa J. Epidermal growth factor is a constant component of normal tears. *Graefes Arch Clin Exp Ophthalmol.* 1989;227:184–187.
4. van Setten G. Epidermal growth factor in human tear fluid: Increased release but decreased concentrations during reflex tearing. *Curr Eye Res.* 1990;9:79–83.
5. Watanabe K, Nakagawa S, Nishida T. Stimulatory effects of fibronectin and EGF on migration of corneal epithelial cells. *Invest Ophthalmol Vis Sci.* 1987;28:205–211.

6. Mishima H, Nakamura M, Murakami J, Nishida T, Otori T. Transforming growth factor-beta modulates effects of epidermal growth factor on corneal epithelial cells. *Curr Eye Res.* 1992;11:691–696.
7. Grant MB, Khaw PT, Schultz GS, Adams JL, Shimizu RW. Effects of epidermal growth factor, fibroblast growth factor, and transforming growth factor-beta on corneal cell chemotaxis. *Invest Ophthalmol Vis Sci.* 1992;33:3292–3301.
8. Boisjoly HM, Laplante C, Bernatchez SF, et al. Effects of EGF, IL-1 and their combination on in vitro corneal epithelial wound closure and cell chemotaxis. *Exp Eye Res.* 1993;57:293–300.
9. Nishida T, Nakamura M, Murakami J, Mishima H, Otori T. Epidermal growth factor stimulates corneal epithelial cell attachment to fibronectin through a fibronectin receptor system. *Invest Ophthalmol Vis Sci.* 1992;33:2464–2469.
10. Wilson SE, He YG, Weng J, Zieske JD, Jester JV, Schultz GS. Effect of epidermal growth factor, hepatocyte growth factor, and keratinocyte growth factor, on proliferation, motility and differentiation of human corneal epithelial cells. *Exp Eye Res.* 1994;59:665–678.
11. Nava A, Barton K, Monroy DC, Pflugfelder SC. The effects of age, gender and fluid dynamics on the concentration of tear film epidermal growth factor. *Cornea.* 1997;16:430–438.
12. Barton K, Nava A, Monroy DC, Pflugfelder SC. Inflammatory cytokines in the tears of patients with ocular rosacea. *Ophthalmology.* 1997;104:In press.
13. Jenkins MS, Brown SI, Lempert SL, Weinberg RJ. Ocular rosacea. *Am J Ophthalmol.* 1979;88:619–622.
14. Pflugfelder SC, Tseng SCG, Yoshino K, Monroy DC, Felix C, Reis BL. Correlation of goblet cell density and mucosal epithelial membrane mucin expression with rose bengal staining in patients with ocular irriation. *Ophthalmology.* 1997;104:222–235.
15. Shimazaki J, Sakata M, Tsubota K. Ocular surface changes and discomfort in patients with meibomian gland dysfunction. *Arch Ophthalmol.* 1995;113:1266–1270.
16. McCulley JP, Sciallis GF. Meibomian keratoconjunctivitis. *Am J Ophthalmol.* 1977;84:788–792.
17. Mathers WD. Ocular evaporation and meibomian gland dysfunction in dry eye. *Ophthalmology.* 1993;100:347–351.
18. Lemp MA. Report of the National Eye Institute/industry workshop on clinical trials in dry eyes. *CLAO J.* 1995;21:221–232.
19. Xu K, Yagi Y, Toda I, Tsubota K. Tear function index: A new measure of dry eye. *Arch Ophthalmol.* 1995;113:84–88.
20. Gupta A, Monroy DC, Zhonghua J, Yoshino K, Huang A, Pflugfelder SC. Transforming growth factor beta-1 and beta-2 in human tear fluid. *Curr Eye Res.* 1996;15:605–614.
21. Gupta C, Singh M. Stimulation of epidermal growth factor gene expression during the fetal mouse reproductive tract differentiation: Role of androgen and its receptor. *Endocrinology.* 1996;137:705–711.
22. Motta M, Dondi D, Moretti RM, et al. Role of growth factors, steroid and peptide hormones in the regulation of human prostatic tumor growth. [Review]. *J Steroid Biochem Mol Biol.* 1996;56:107–111.
23. Yang Y, Chisholm GD, Habib FK. Epidermal growth factor and transforming growth factor alpha concentrations in BPH and cancer of the prostate: Their relationships with tissue androgen levels. *Br J Cancer.* 1993;67:152–155.
24. Sullivan DA, Bloch KJ, Allansmith MR. Hormonal influence on the secretory immune system of the eye: Androgen regulation of secretory component levels in rat tears. *J Immunol.* 1984;132:1130–1135.
25. Sullivan DA, Allansmith MR. Hormonal influence on the secretory immune system of the eye: Endocrine interactions in the control of IgA and secretory component levels in tears of rats. *Immunology.* 1987;60:337–343.
26. Gao J, Lambert RW, Wickham LA, Banting G, Sullivan DA. Androgen control of secretory component mRNA levels in the rat lacrimal gland. *J Steroid Biochem Mol Biol.* 1995;52:239–249.
27. Cornell-Bell AH, Sullivan DA, Allansmith MR. Gender-related differences in the morphology of the lacrimal gland. *Invest Ophthalmol Vis Sci.* 1985;26:1170–1175.
28. Sato EH, Sullivan DA. Comparative influence of steroid hormones and immunosuppressive agents on autoimmune expression in lacrimal glands of a female mouse model of Sjögren's syndrome. *Invest Ophthalmol Vis Sci.* 1994;35:2632–2642.
29. Jordan A, Baum J. Basic tear flow: Does it exist? *Ophthalmology.* 1980;87:920–930.
30. McGill JI, Liakos G, Seal DV. Normal tear protein profiles and age-related changes. *Br J Ophthalmol.* 1984;68:316–320.
31. Mathers WD, Lane JA, Zimmerman MB. Tear film changes associated with normal aging. *Cornea.* 1996;15:229–234.
32. Xu K, Tsubota K. Correlation of tear clearance rate and fluorophotometric assessment of tear turnover. *Br J Ophthalmol.* 1995;79:1042–1045.

33. Pflugfelder SC, Tseng SCG, Pepose JS, Fletcher MA, Klimas NG, Feuer W. Epstein-Barr viral infection and immunologic dysfunction in patients with aqueous tear deficiency. *Ophthalmology.* 1990;97:313–323.
34. Mishima S, Gasset A, Klyce SD, Jr., Baum JL. Determination of tear volume and tear flow. *Invest Ophthalmol. Vis Sci* 1966;5:264–276.
35. Hogquist KA, Nett MA, Unanue ER, Chaplin DD. Interleukin-1 is processed and released during apoptosis. *Proc Natl Acad Sci USA.* 1991;88:8485–8489.
36. Dinarello CA, Wolff SM. The role of interleukin-1 in disease. *N Engl J Med.* 1993;328:106–113.
37. Dinarello CA. Interleukin-1 and its biologically related cytokines. *Adv Immunol.* 1989;44:153–205.
38. Wilson SE, He YG, Lloyd SA. EGF, EGF receptor, basic FGF, TGF beta-1, and IL-1 alpha mRNA in human corneal epithelial cells and stromal fibroblasts. *Invest Ophthalmol Vis Sci.* 1992;33:1756–1765.
39. Wilson SE, Schultz GS, Chegini N, Weng J, He Y. Epidermal growth factor, transforming growth factor alpha, transforming growth factor beta, acidic fibroblast growth factor, basic fibroblast growth factor, and interleukin-1 proteins in the cornea. *Exp Eye Res.* 1994;59:63–70.
40. Schmidt JA, Tocci MJ. Interleukin-1. In: Sporn MD, Roberts AB, eds. *Peptide Growth Factors and Their Receptors.* New York: Springer-Verlag; 1991:473–521.
41. Girard MT, Matsubara M, Fini ML. Transforming growth factor beta and interleukin-1 modulate metalloproteinase expression by corneal stromal cells. *Invest Ophthalmol Vis Sci.* 1991;32:2441–2454.
42. Dinarello CA. Interleukin-1 and interleukin-1 antagonism. *Blood.* 1991;77:1627–1652.
43. Hykin PG, Bron AJ. Age-related morphological changes in lid margin and meibomian gland anatomy. *Cornea.* 1992;11:334–342.

65

CHEMOKINE PRODUCTION IN CONJUNCTIVAL EPITHELIAL CELLS

Kazumi Fukagawa,[1,2] Kazuo Tsubota,[2,3] Shigeto Simmura,[2,3] Hirohisa Saito,[1] Hiroshi Tachimoto,[1] Akira Akasawa,[1] and Yoshihisa Oguchi[2]

[1]Department of Allergy and Immunology
National Children's Medical Research Center
Tokyo, Japan
[2]Department of Ophthalmology
Keio University
Tokyo, Japan
[3]Department of Ophthalmology
Tokyo Dental College
Chiba, Japan

1. ROLE OF CHEMOKINES IN INFLAMMATION

1.1. Selective Recruitment of Inflammatory Cells

The inflammatory cells that accumulate in inflammatory sites differ depending upon the cause of the inflammation. Bacterial infection or the precipitation of IgG immunocomplexes leads to the accumulation of neutrophils; viral infection or delayed hypersensitivitiy, to the accumulation of monocytes and lymphocytes; and parasite infection and the late-phase reaction of allergic inflammation, to the accumulation of eosinophils and basophils.

What is the mechanism of this selective recruitment of inflammatory cells from blood vessels to the inflammatory sites? Adhesion molecules, cytokines, and chemokines[1] are thought to play important roles in this process (Fig. 1). In this section we focuson the function of chemokines.

1.2. Subfamilies of Chemokines

Chemotactic factors such as FMLP (N-formyl-methionyl-leucyl-phenylalanine), C5a, and LTB4 (leukotriene B4) induce chemotaxis of inflammatory cells. After interleukin IL-8 was found to be a cytokine with chemotactic activity,[2] many chemokines (cy-

Lacrimal Gland, Tear Film, and Dry Eye Syndromes 2
edited by Sullivan *et al.*, Plenum Press, New York, 1998

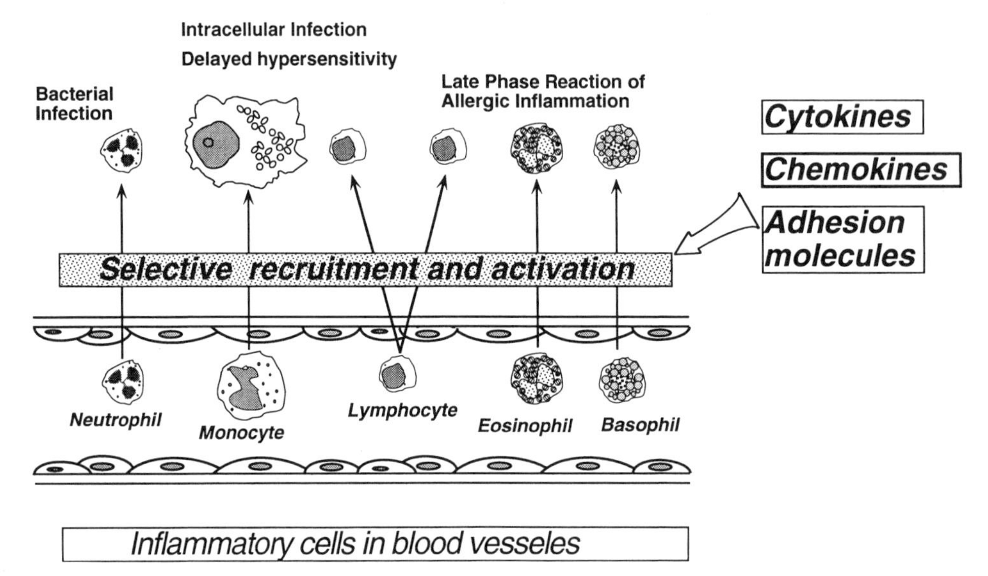

Figure 1. Selective recruitment of inflammatory cells from blood vessels into the inflammatory sites.

tokines with chemotactic activities) were found and investigated.[3] These chemokines have structural similarities: four cysteine residues, two disulfide bonds, and 20–50% homology in their predicted amino acid sequences. Based on the position of these conserved cysteines, the chemokine family is divided into two subfamilies (Fig. 2).[4] In the C-X-C, or α, subfamily, the first two cysteines are separated by another amino acid residue. In the C-C, or β, subfamily, the first two cysteines are adjacent. The genes for human C-X-C chemokines have been found on chromosome 4q; the genes for C-C chemokines are mapped on chromosome 17q.[4,5]

1.3. IL-8 and RANTES

IL-8 is one of the well-studied C-X-C chemokines (Table 1).[4,6] It is produced by macrophages, mast cells, fibroblasts, keratinocytes, endothelial cells, and epithelial cells. IL-8 is both a potent chemoattractant for neutrophils and also an activator of neutrophils. RANTES[7] (regulated upon activation, normal T cell expressed and secreted) is one of the C-C chemokines; this subfamily also includes MCP (monocyte chemotactic protein)[1,2,3] and MIP (macrophage inflammatory protein)-1a and 1b. RANTES is produced by macrophages, T lymphocytes, platelets, fibroblasts, mesangial cells, endothelial cells, and epithelial cells. RANTES is a chemoattractant for T cells,[8] eosinophils,[9,10] and basophils,[11,12] but it has no chemotactic activity for neutrophils.

1.4. Chemokines in the Recruitment of Inflammatory Cells

Each chemokine is produced from a variety of cells, and each chemokine has a variety of targets; moreover, some of them share common receptors.[13] Chemokines also have

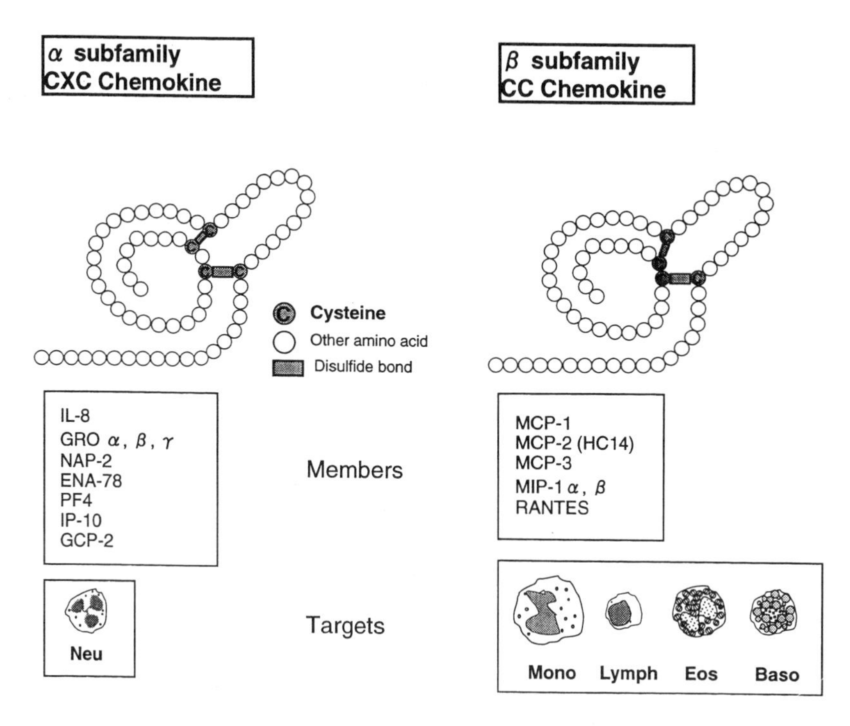

Figure 2. Structure and members of C-X-C and C-C chemokine families.

selectivities towards the targets of chemotactic activities. Although exact physiological and pathological features are not yet clearly defined, simplified generalities can be made. C-X-C chemokines are chemoattractants for neutrophils; C-C chemokines are chemoattractants for monocytes, T lymphocytes, eosinophils, and basophils (Fig. 2). The variety and selectivity of chemokine production and targets may play an important role in the accumulation of inflammatory cells.

Table 1. Characteristics of chemokines

Subfmily	Chemokine	Produced by	
CXC(α)	IL-8	Mϕ/mono, mast	fibro, endo, epi, keratino, tumor
	GRO (α,β,γ)	Mϕ/mono	fibro, endo, epi, tumor
	IP10	Mϕ/mono, T, plt	fibro, endo, keratino, tumor
CC(β)	MCP1	Mϕ/mono	fibro, endo, keratino, tumor
	MCP3	Mϕ/mono	tumor
	MIP1α	Mϕ/mono, T, B, baso, mast	fibro, tumor
	MIP1β	Mϕ/mono, T, B,	fibro
	RANTES	Mϕ/mono, T, plt	fibro, endo, epi, mesangial

IL-8 (Interleukin-8), GRO (Growth Regulated), IP10 (Interferon-gamma inducible Protein 10), MCP (Monocyte Chemotactic Protein), MIP (Macrophage Inflammatory Protein), RANTES (Regulated upon Activation, Normal, T cell Expressed and Secreted), Mϕ: macrophage, mono: monocyte, T: T lymphocyte, B:B lymphocyte, baso: basophil, mast: mast cell, plt: platelet, fibro: fibroblast, endo: endothelial cell, epi: epithelial cell, keratino: keratinocyte, mesangial: mesangial cell.

2. ROLE OF EPITHELIAL CELLS IN ALLERGIC REACTION

2.1. Intercellular Communication in Allergic Inflammation

Since the vigorous investigations of cytokines, chemokines, and adhesion molecules in the 1980's, intercellular communication has been studied as the mechanism of many physiological phenomena. The mechanisms of allergic inflammation, such as differentiation and activation of inflammatory cells, production of IgE antibodies, and accumulation of inflammatory cells, are also thought to be maintained by intercellular communications with cytokines, chemokines, and adhesion molecules (Fig. 3).

Several cellular candidates supply cytokines and regulate allergic inflammation. Helper T cells type 2[14–16] (Th2) produce several allergic cytokines such as IL-4, which induce IgE production from plasma cells, and IL-5, which induce differentiation and activation of eosinophils. Mast cells have been reported[17–19] to produce IL-4, IL-5, TNF (tumor necrosis factor)-α, GM-CSF (granulocyte-macrophage colony stimulating factor), IL-8, MIP-1a, and MCP-3, and mast cells are the trigger of type 1 allergic reaction. Moreover, structural cells have also been reported to accumulate and maintain allergic reaction by producing cytokines and chemokines.[20] In this section, we focus on the tissue structural cells, especially epithelial cells, as the supplier of cytokines and chemokines in allergic inflammation.

2.2. Tissue Structural Cells in Allergic Inflammation

Tissue structural cells were once thought of as the targets of inflammatory cells in the allergic reaction, but they have been shown to play a role in the allergic cytokine net-

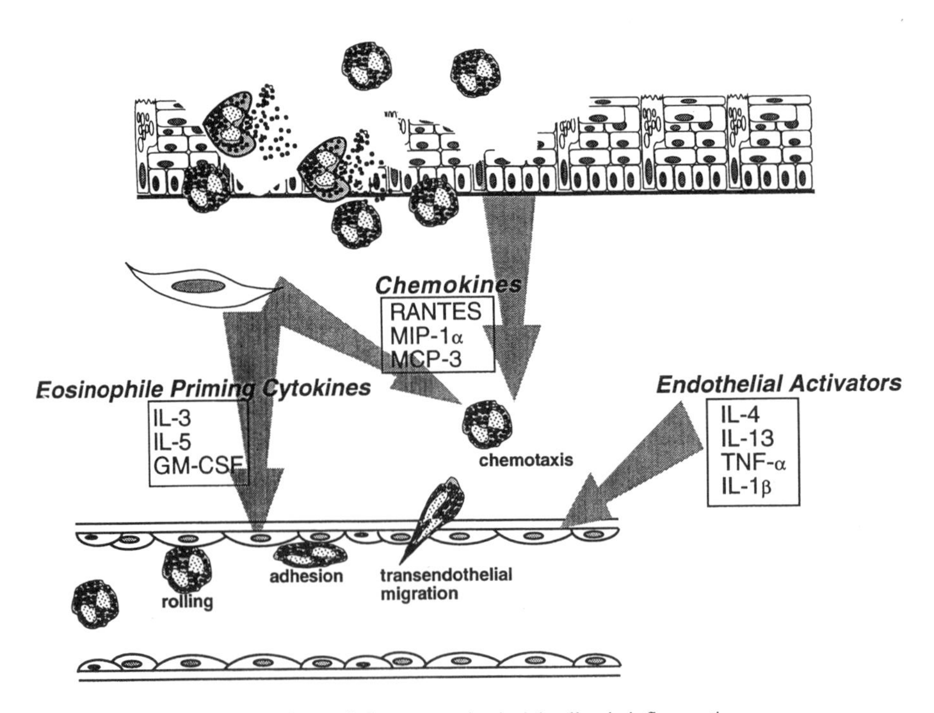

Figure 3. Intercellular communications in allergic inflammation.

work[20] through the production of cytokines, chemokines, and adhesion molecules. Fibroblasts produce SCF (stem cell factor, c-kit ligand), TNF-α and IL-1β. Endothelial cells express adhesion molecules, by which inflammatory cells adhere and migrate from the blood vessels, and also produce IL-8 and GM-CSF. The epithelial cells of mucous membrane, such as bronchial epithelial cells, also produce chemical mediators, such as PAF (platelet activating factor) and PGs (prostaglandins), cytokines,[21] such as GM-CSF and IL-6, and chemokines, such as RANTES,[22] IL-8,[21,23] and MCP-1. Although epithelial cells have barrier functions against foreign substances, they may be involved in activating or maintaining allergic inflammation.

2.3. Role of Chemokines from Epithelial Cells in Allergic Inflammation

The mechanisms of inflammatory cell recruitment from blood vessels are well studied.[24–26] While the endothelial activating cytokines induce the expression of adhesion molecules on endothelial cells, the priming cytokines activate inflammatory cells. After transendothelial migration, inflammatory cells migrate into subepithelial tissue, and then finally migrate into the epithelium and undergo secretion. Chemokines from the epithelial cells are believed to play an important role[22, 27–30] in this recruitment process into the epithelium and secretion (Fig. 4), which may cause corneal tissue damage in the presence of ocular allergy.

3. CHEMOKINE PRODUCTION IN CONJUNCTIVAL EPITHELIAL CELLS

Vision-threatening eye diseases, such as vernal keratoconjunctivitis (VKC)[31] and atopic keratoconjunctivitis (AKC),[32,33] are reported to have eosinophils, basophils, type 2

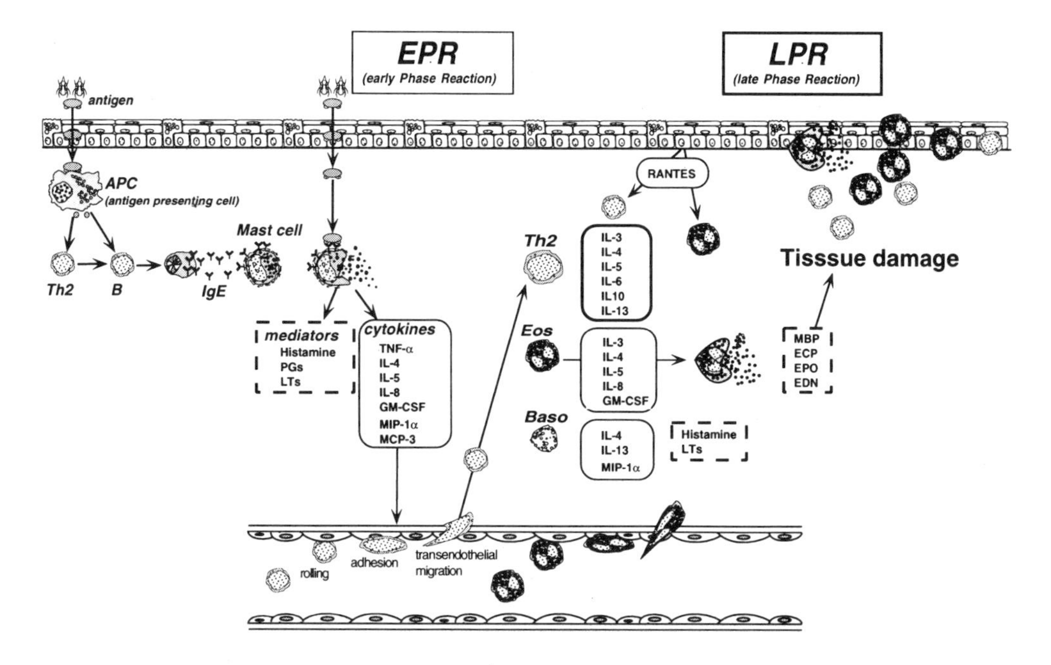

Figure 4. Participation of chemokines from epithelial cells in eosinophil recruitment.

helper T lymphocytes (Th2), mast cells, and neutrophils existing in the hypertrophic conjunctival tissue in affected patients. Eosinophils are known to release mediators such as major basic protein (MBP)[34] and eosinophil cationic protein (ECP),[35] which cause tissue damage. Neutrophils are also known to cause tissue damage by producing several mediators, such as cationic protein, and reactive oxygen species (ROS). Eosinophils and neutrophils that infiltrate from conjunctival tissue may play a role in damaging the avascular corneal tissue.

We investigated the expression and production of RANTES and IL-8 from human conjunctival epithelial cells in vivo and in vitro.[27] Conjunctival epithelium from a patient with AKC stained positively for both RANTES and IL-8 (data not shown). We have reported that TNF-α induces de novo production of RANTES from a human conjunctival epithelial cell line; Wong-Kilbourne derived human conjunctiva (WK-hC), and IFN-γ synergistically increased the TNF-α-dependent production of RANTES from the cell line (Fig. 5).[28] We found that TNF-α also induced de novo production of IL-8 from WK-hC, but there was no synergistic increase with IFN-γ (data not shown).[30] IL-4, IL-5, and IL-6 did not induce de novo production of RANTES or IL-8, and these cytokines did not affect the TNF-α-dependent production of RANTES or IL-8 in this cell line (data not shown).[30] RANTES and IL-8 production in WK-hC were both suppressed by dexamethasone, while cyclosporin A suppressed only IL-8 production, and genistein suppressed only RANTES production (Fig. 6).[28,30]

Taken together, human conjunctival epithelial cells were capable of producing IL-8 as well as RANTES, and the mechanisms of producing these chemokines may be different. Dexamethasone may suppress the inflammation of the ocular surface through suppressing chemokine production from conjunctival epithelial cells. These findings may be helpful in understanding the mechanisms of chemokine production in human conjunctival epithelium and the treatment of allergic ocular diseases.

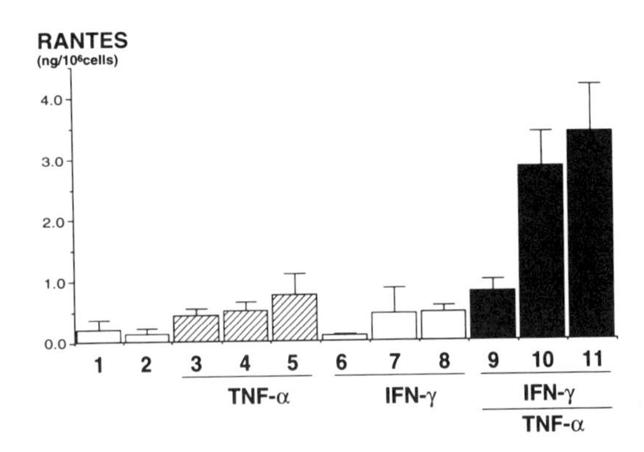

Figure 5. Induction of RANTES production by TNF-α, IFN-γ, and the combination of the two in the conjunctival epithelial cell line WK-hC. RANTES secretion as detected in cell supernatants by ELISA after 24 h incubation with PMA 100 ng/ml + A23187 10^{-7} M (bar 2) (n=3; p=NS); TNF-α (1, 10, 100 ng/ml) (bars 3, 4, and 5, respectively) (n=5; *p<0.05 compared with control release and with release induced by TNF-α (10, 100 ng/ml)); IFN-γ (1, 10, 100 ng/ml) (bars 6, 7, and 8, respectively) (n=5 for IFN-γ 100 ng/ml; p=NS). Synergistic induction of RANTES secretion as detected in the supernatants of cells treated with the combinations of TNF-α (10 ng/ml) and IFN-g (1, 10, 100 ng/ml) (bars 9, 10, and 11, respectively) (n=5; **p<0.05 compared with the control release induced by TNF-α (10 ng/ml) alone and with the release induced by IFN-γ (10, 100 ng/ml) and TNF-α (10 ng/ml)). Each bar and extension represents the mean and SD.

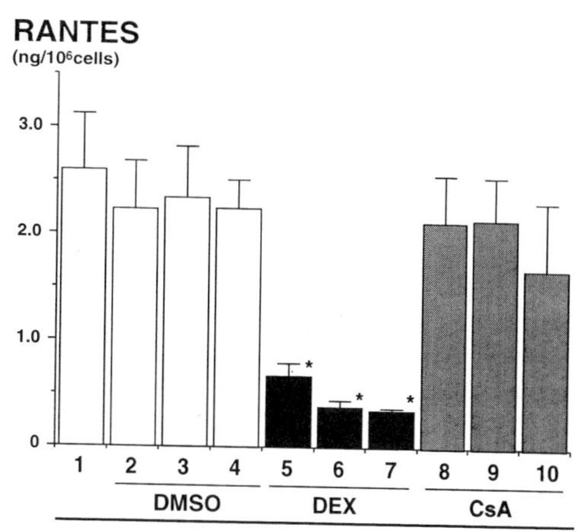

Figure 6. Inhibition of RANTES production in the conjunctival epithelial cell line WK-hC. RANTES secretion as detected in the supernatants of cells pre-exposed to either DMSO diluent (10^5,10^4, and 10^3 times diluted) (bars 2, 3, and 4, respectively), dexamethasone (10^{-8}, 10^{-7}, and 10^{-6} M) (bars 5, 6, and 7, respectively) (n=3; *p<0.05 compared with control release pre-exposed to DMSO diluent) or cyclosporin A (10^{-7}, 10^{-6}, and 10^{-5} M) (bars 8, 9, and 10, respectively) (n=3; p=NS), and then treated with the combinations of TNF-α (10 ng/ml) and IFN-g (100 ng/ml) for an additional 24 h. Each bar and extension represents the mean and SD.

REFERENCES

1. Baggiolini M et al. CC chemokines in allergic inflammation. *Immunol Today*. 1994;15:127–133.
2. Baggiolini M et al. Neutrophil-activating peptide-1/interleukin 8, a novel cytokine that activates neutrophils. *J Clin Invest*. 1989;84:1045–1049.
3. Kunkel S, Strieter RM, Lindley IJ, Westwick J. Chemokines: new ligands, receptors and activities. *Immunol-Today*. 1995;16:559–61.
4. Baggiolini M et al. Interleukin-8 and related chemotactic cytokines - CXC and CC chemokines. *Adv Immunol*. 1994;55:97–179.
5. Schall TJ. Biology of the RANTES/sis cytokine family. *Cytokine*. 1991;3:165–183.
6. Krieger M et al. Activation of human basophils through the IL-8 receptor. *J Immunol*. 1992;149:2662–2667.
7. Schall TJ et al. A human T cell-specific molecule is a member of a new gene family. *J Immunol*. 1988;141:1018–1025.
8. Schall TJ et al. Selective attraction of monocytes and T lymphocytes of the memory phenotype by the cytokine RANTES. *Nature*. 1990;34:669–671.
9. Kameyoshi Y et al. Cytokine RANTES released by thrombin-stimulated platelets is a potent attractant for human eosinophils. *J Exp Med*. 1992;176:587–592.
10. Rot A et al. RANTES and macrophage inflammatory protein 1α induce the migration and activation of normal human eosinophil granulocytes. *J Exp Med*. 1992;176:1489–1495.
11. Kuna P et al. RANTES, a monocytes and T lymphocyte chemotactic cytokine releases histamine from human basophils. *J Immunol*. 1992;149:636–642.
12. Bischoff SC et al. RANTES and related chemokines activate human basophil granulocytes through G protein-coupled receptors. *Eur J Immunol*. 1993;23:761–767.
13. Baruch BA et al. Monocyte chemotactic protein-3 (MCP3) interacts with multiple leukocyte receptors. C-C CKR1, a receptor for macrophage inflammatory protein-1 alpha/RANTES, is also a functional receptor for MCP3. *J Biol Chem*. 1995;270:22123–22128.
14. Corrigan C et al. T cells and eosinophils in the pathogenesis of asthma. *Immunol Today*. 1992;13:501–507.
15. Mosmann TR et al. Two types of murine helper T cell clone. *J Immunol*. 1986;136:2348–2357.
16. Maggie E et al. Accumulation of Th2-like helper T cells in the conjunctiva of patients with vernal conjunctivitis. *J Immunol*. 1991;146:1169–1174.
17. Saito H et al. Characterization of cultured human mast cells. *Int Arch Allergy Immunol*. 1995;107:63–65.
18. Ebisawa M et al. Human mast cells produce TNF-α, MIP1-α, and IL-5 following activation through FceR I. *J Allergy Clin Immunol*. 1995;95:219 (abstr).
19. Plaut M et al. Mast cell lines produce lymphokines in response to cross-linkage of FceR I or to calcium ionophores. *Nature*. 1989;346:64–67.

20. Denburg JA et al. Structural cell-derived cytokines in allergic inflammation. *Int Arch Allergy Appl Immunol.* 1991;94:127–132.
21. Cromwell O et al. Expression and generation of interleukin-8, IL-6 and granulocyte-macrophage colony-stimulating factor by bronchial epithelial cells and enhancement by IL-1β and tumor necrosis factor-α. *Immunology.* 1992;77:330–337.
22. Stellato C et al. Expression of chemokine RANTES by a human bronchial epithelial cell line. *J Immunol.* 1995;155:410–418.
23. Nakamura H, Yoshimura K, Jaffe HA, Crystal RG. Interleukin-8 gene expression in human bronchial epithelial cells. *J Biol Chem.* 1991;266:19611–19617.
24. Ebisawa M et al. Eosinophil transendothelial migration induced by cytokines. I. Role of endothelial and eosinophil adhesion molecules in IL-1β-induced transendothelial migration. *J Immunol.* 1992;149:4021–4028.
25. Springer TA. Adhesion receptors in the immune system. *Nature.* 1990;346:425–434.
26. Rensick MB et al. Mechanisms of eosinophil recruitment. *Am J Respir Cell Mol Biol.* 1993;8:349–355.
27. Fukagawa K et al. RANTES production in cultured human conjunctival epithelial cell line. *Cornea.* submitted.
28. Fukagawa K. Chemokine produced by conjunctival epithelial cells. *J Eye.* 1996;13:1201–1207.
29. Fukagawa K. Allergic ocular disease. *J Eye.* 1996;13:853–861.
30. Fukagawa K et al. Different regulations of RANTES and IL-8 production from cultured human conjunctival epithelial cell line. *J Immunol.*, submitted.
31. Tsubota K, Takamura E, Hasegawa T, Kobayashi T. Detection by brush cytology of mast cells and eosinophils in allergic and vernal conjunctivitis. *Cornea.* 1991;10:525–31.
32. Foster CS et al. Immunopathology of atopic keratoconjunctivitis. *Ophthalmology.* 1991; 98:1190–1196.
33. Foster CS. The pathophysiology of ocular allergy: Current thinking. *Allergy.* 1995;50(21 Suppl):6–9, 34–38.
34. Trocme SD et al. Eosinophil granule major basic protein deposition in corneal ulcers associated with vernal keratoconjunctivitis. *Am J Ophthalmol.* 1993;115:640–643.
35. Leonardi A et al. Eosinophil cationic protein in tears of normal subjects and patients affected by vernal keratoconjunctivitis. *Allergy.* 1995;50:610–613.

66

MOLECULAR BASIS AND ROLE OF DIFFERENTIAL CYTOKINE PRODUCTION IN T HELPER CELL SUBSETS IN IMMUNOLOGIC DISEASE

Andrea Keane-Myers, Vincenzo Casolaro, and Santa Jeremy Ono

Schepens Eye Research Institute
Department of Ophthalmology and Committee on Immunology
Harvard Medical School
Boston, Massachusetts

1. INTRODUCTION

T cells mediate a wide range of immunologic functions including the capacity to help B cells develop into antibody-producing cells, the capacity to increase the microbicidal action of monocytes, and the mobilization of the immune response.[1] These effects depend on the expression of specific cell surface molecules known as T cell receptors, which recognize antigen in the presence of major histocompatability molecules on the surface of antigen presenting cells.[2]

Many of the functions of activated T cells are elaborated by the production of an array of proteins known as cytokines or lymphokines.[3] In the past decade, the development of cloning techniques for murine and human T lymphocytes has helped to clarify the role of cytokines in the immune response. Using such techniques, Mosmann and colleagues found that repeated stimulation of T helper lymphocytes in vitro with given antigens resulted in the development of a restricted and stereotyped pattern of lymphokine production.[4–8] This production pattern falls into two phenotypes. Th1 cells produce interleukin-2 (IL-2) and interferon-gamma (IFN-γ), but not IL-4, IL-5, IL-6, or IL-10, and are responsible for both humoral and cell-mediated immune responses including macrophage activation and delayed-type hypersensitivity.[8] Conversely, Th2 cells produce IL-4, IL-5, IL-6, and IL-10, but not IL-2 or IFN-γ, and provide optimal help for humoral immune responses, as well as assisting mucosal immunity through production of mast cell and eosinophil growth and differentiation.[8] Therefore, the functional capabilities of the murine Th1 and Th2 CD4+ subsets correlate well with the types of cytokines they secrete.

Lacrimal Gland, Tear Film, and Dry Eye Syndromes 2
edited by Sullivan *et al.*, Plenum Press, New York, 1998

Cytokines present during the initiation of a CD4+ T cell response can determine the subsequent development of a particular Th cell phenotype. Thus, IL-12, produced by cells of the monocyte lineage, is a critical factor driving the development of Th1 cells from antigen-specific naive CD4+ T cells, [9–11] primarily through its ability to enhance IFN-γ production by Th1 clones. IFN-γ at relatively low concentrations inhibits proliferation of murine Th2 clones exposed to IL-2 or IL-4, and through this effect favors the development of Th1 cells.[12–17] In contrast, Th2 cells develop when naive T cells are stimulated in vitro in the presence of IL-4, which prevents the production of Th1 cells by downmodulating the production of IL-12 by monocytes.[9,11,18,19]

Commitment of Th1 or Th2 populations developing during either an immune response to a pathogen or an inappropriate immune response to an allergen or autoantigen may determine the difference between health and chronic disease.[20,21] Mouse and human Th1 cells preferentially develop during infections with intracellular bacteria, protozoa, and viruses, whereas Th2 cells tend to dominate during helminthic parasite infections and in response to allergens.[20,21] Whether a specific individual will respond to a particular pathogen in an appropriate Th1 or Th2 specific manner largely depends upon that individual's genetic makeup. The availability of inbred strains of mice, including a wide variety of congenic, recombinant, and mutant strains, has provided ideal tools for studying the genetic basis of the immune response in disease.[1,22] In this article, we will examine the role of divergent T cell subsets in two disease models. Susceptibility to Lyme disease, caused by infection with the spirochete *Borrelia burgdorferi*, has been associated with a strong Th1 response, with disease resistance correlating with a strong Th2 response.[26] Conversely, in allergic disease, a strong Th2 response correlates with disease formation, and Th1 responses are associated with disease protection.

Lyme disease is a chronic, multisystem illness caused by infection with the bacteria, *Borrelia burgdorferi*. Among the pathologic manifestations of the disease are arthritis,[23] carditis,[24] and neurologic disorders.[25] The establishment of a mouse model of Lyme disease has provided a useful experimental system for investigating aspects of immunity to this pathogen.[25–28] A critical role for CD4+ T cells in protection was suggested by the finding that mice depleted in vivo of CD4+ T cells were significantly more susceptible to *Borrelia* infection than undepleted mice.[26] The likelihood that cytokines were mediating these events was supported by our finding that outcomes in mice depleted in vivo of the type 2 cytokine IL-4 were similar to those in mice depleted of CD4+ cells (ie, greater susceptibility to disease), whereas outcomes in mice depleted of the type 1 cytokine IFN-γ had increased disease resistance. [26] Other researchers using a footpad model of *Borrelia* infection have observed similar protective effects of IL-4 and disease-promoting effects of IFN-γ.[20] Finally, results from studies using recombinant IL-4 to treat susceptible mice early in disease showed IL-4 to be sufficient for disease protection, as these animals had significant reductions in joint swelling and in the numbers of spirochetes recovered from their joints and skin when compared with sham-treated mice.[27]

Like Lyme disease, asthma is also a chronic and complex disorder that appears to be at least partially mediated by T helper responses. However, the protective and disease-inducing phenotypes in asthma are the exact opposite of what is observed in *Borrelia* infections. Hence, a number of studies have correlated the pathoneumonic airway inflammation in asthma with infiltration by T lymphocytes bearing the T helper 2 phenotype,[29,30] and by infiltration of Th2 cytokine-dependent eosinophils and mast cells.[31–35] That Th2 cytokines are important in disease formation is emphasized by experiments in which susceptible mice were depleted in vivo of the type 2 cytokines IL-4 and IL-5. These Th2-depleted animals had a much better disease outcome (ie, no asthma) than what was observed in the sham-treated animals. In addition, if these susceptible mice were treated with either recombinant IFN-γ (Keane-Myers A, and Wills-Karp M, unpublished data) or recombinant IL-12 (both associated with the production of a Th1 response),[36] they were protected from disease. Conversely, resistant mice depleted of IL-12 were rendered susceptible to disease, again suggesting that a Th1 response is important in protection (Keane-Myers A, Wills-Karp M, unpublished data).

In spite of the clear role that Th subclasses play in disease manifestations, molecular distinctions between the Th1 and Th2 subsets of CD4+ T helper cells remain incompletely defined. However, recent data from a number of laboratories now implicate a number of nuclear factors in Th differentiation. Moreover, analysis of these factors and the *cis*-elements with which they interact has provided potential clues to genetic predisposition for immunologic diseases. Even if these factors turn out not to be true lineage commitment factors, their abilities to skew the Th response may form the foundation for new approaches toward the treatment or management of these diseases. We will briefly outline the newly implicated factors.

Much of the information has come from the Harvard laboratories of Grusby and Glimcher and from Kishimoto in Japan. In the decade since Mossman made the seminal discovery that mammals have two major types of T helper cells, data from these and several other laboratories have provided clear evidence that the balance between Th1 and Th2 cells is intimately linked to disease pathogenesis. As noted previously, this balance is "tipped" in favor of Th2 cells in allergic disease. Thus, Glimcher's discovery that the c-maf transcription factor is expressed only in Th2 cells, produces a Th2 specific footprint in the IL-4 proximal promoter, and activates IL-4 transcription in non-Th2 cells has major implications for those investigating the genetics of atopy.[37] Most recently, this laboratory has isolated a novel factor, NIP45, which interacts with the Rel homology domain of NF-ATp and which synergistically activates IL-4 gene transcription with c-maf in B lymphoma cells.[28] Several laboratories are in the process of determining whether c-maf and NIP45 are atopy genes, and are investigating the cell surface interactions and intracellular signals that impact on c-maf expression and lineage commitment down the Th2 pathway. In particular, the roles that the B7–1 and B7–2 molecules as well as the cytokines IL-4, -12, -13, TNF-α, and IFN-γ play in this commitment process are the focus of intense investigation.[28]

A third factor that also appears to be a component of Th2 but not Th1 enhanceosomes is the Elf-related factor CP2 (Casolaro V, Ono SJ, unpublished data). We have found that the protein is a Th2 transactivator and Th1 repressor, and that the protein requires multiple promoter bound factors on the human IL-4 promoter for recruitment into the preinitiation complex. Additional evidence that the factor is biologically relevant comes from our finding that a dominant negative CP2, lacking a C-terminal dimerization domain, is able to selectively extinguish IL-4 gene expression in T lymphocytes. We are currently developing studies in transgenic mice to assess more carefully the importance of CP2 in Th lineage commitment and disease pathogenesis.

Clues to the signal transduction events that control the early stages of Th lineage commitment (prior to c-maf expression) have come from several groups. Using knockout mice, the laboratories of Grusby and Kishimoto have demonstrated that members of the Stat family of proteins are early and essential factors for the initial differentiation process. Stat4 is required for the development of Th1 cells, and Stat6 is required for the development of Th2 cells.[38] Moreover, naive T cells from the Stat4 knockout mice are unresponsive to IL-12. The genes encoding these proteins are therefore good candidates for atopy genes. Certainly, other factors proximal to c-maf (and distinct from the Stats) may also represent atopy genes. Since IL-4 synthesis and binding to the IL-4 receptor on the naive

T cell is an obligatory step in Th2 differentiation, factors that might drive c-maf independent IL-4 gene expression in T cells and possibly mast cells and basophils also deserve careful analysis. The identification and initial characterization of a mast cell specific enhancer in the mouse IL-4 gene by the Brown group is therefore of great interest.[39] Finally, it is also important to keep in mind that the Th commitment process may be significantly different in humans and mice, and that genes operating at this level may be distinct in the two species. For example, our group and those of Lichtman and Abbas have independently implicated members of the Rel family of transcription factors in the commitment process.[40,41] In addition, allelic polymorphisms in the PubB element of the IL-4 promoter appear to be critical in determining the strength of that promoter in particular individuals.[42]

In our experiments on the human IL-4 promoter in the Jurkat cell line, and in Lichtenstein's studies on human cytokine production from purified basophils, there are clear distinctions in the signal transduction pathways activating Th1 and Th2 type cytokines. In the human IL-4 promoter, the P-element is sufficiently different from that of the mouse such that it constitutes a high affinity Rel binding site and a comparatively weak NFAT site. Thus, the calcineurin-dependent activation of IL-4 promoter activity is actually inhibited by PMA and by overexpression of RelA. Recent data also provide new support that this mechanism may participate in murine Th differentiation.[43] Serfling and coworkers have demonstrated that HMG I(Y) interferes with NF-AT binding to the murine Pu-bB site, and that Th2 clones frequently have low levels of HMG I(Y). In addition, HMG I(Y) interaction with Pu-bB would facilitate the interaction of Rel factors with the site, thus silencing IL-4 gene transcription in Th1 cells. Taken together, these results, from multiple independent laboratories, indicate that the equilibrium between NF-AT and Rel proteins, influenced by HMG I(Y), may play an important role in Th cell differentiation.

It has been argued that a role for Rel proteins in Th lineage commitment is hard to reconcile, in view of the probable universal nuclear translocation of these proteins in response to TCR cross-linking, but this does not take into account the possibility of an unknown third signal that might modify the nature of the Rel response. Sen has shown that the different Rel proteins are differentially sequestered in the cytoplasm, and that the I-kBα and β inhibitory molecules may differentially release the different Rel family members in response to extracellular stimuli.[44] For example, NF-κB and RelB are induced by phorbol ester stimulation alone, whereas c-Rel requires both phorbol ester and calcium ionophore. Thus, we would argue that it is essential to remain sensitive to potential differences in Th commitment pathways in human and mouse T cells, and to continue to focus on other external signals that might play a major role in Th differentiation.

In summary, the research carrried out in several laboratories is shedding light on molecular mechanisms controlling Th cell differentiation. Since the balance between Th1 and Th2 cells plays a decisive role in disease pathogenesis, it is hoped that the identification of lineage commitment factors will provide novel targets for the therapy of these diseases. Our own laboratory is focusing on the biology and genetics of allergic diseases of the eye,[45] and we hope that our molecular approach will turn out to be effective in identifying new targets and in designing new rational drugs.

REFERENCES

1. Fitch FW, Lancki DW, Gajewski TF. T-cell-mediated immune regulation. In: Paul WE, ed. *Fundamental Immunology.* New York: Raven Press, Ltd., 1993;733.
2. Grusby MJ, Glimcher LH. Immune responses in MHC class II-deficient mice. *Ann Rev Immunol.* 1995;13:417–35.

3. Paul WE. *Fundamental Immunology*. New York: Raven Press, Ltd., 1993. Paul W, ed.
4. Cherwinski HM, Schumacher JH, Brown KD, Mosmann TR. Two types of mouse T-helper clone. III. Further differences in lymphokine synthesis between Th1 and Th2 clones revealed by RNA hybridization, functionally monospecific bioassays, and monoclonal antibodies. *J Exp Med.* 1987;166:1229–1244.
5. Coffman RL, Seymour BW, Lebman DA, et al. The role of T helper cells in mouse B cell differentiation and isotype regulation. *Immunol Rev.* 1988;102:5–28.
6. Mosmann TR, Cherwinski H, Bond MW, Giedlin MA, Coffman RL. Two types of murine T cell clones-definition according to profiles of lymphokine activities and secreted proteins. *J Immunol.* 1986;136:2348–57.
7. Mosmann TR, Coffman RL. Two types of mouse helper T cell clone: implications for immune regulation. *Immunol Today* 1987;8:223–227.
8. Street NE, Mosmann T. Functional diversity of T lymphocytes due to secretion of different cytokine patterns. *FASEB J.* 1991;5:171–177.
9. Trinchieri G. Interleukin-12: a proinflammatory cytokine with immunoregultory functions that bridge innate resistance and antigen-specific adaptive immunity. *Ann Rev Immunol.* 1995;13:251–76.
10. D'Andrea A, Rengaraju M, Valiante NM, et al. Production of natural killer cell stimulatory factor (interleukin 12) by peripheral blood mononuclear cells. *J Exp Med.* 1992;176:1387–13.
11. Scott P. Il-12: Initiation Cytokine for Cell-Mediated Immunity. *Science* 1993;260: 496–497.
12. Gallin JI, Farber JM, Holland SM, Nutman T. Interferon-γ in the Management of Infectious Diseases. *Ann Intern Med.* 1995;123:216–224.
13. Rousset F, Robert J, Andary M, et al. Shifts in Interleukin-4 and Interferon-γ production by T cells of patients with elevated serum IgE levels and the modulatory effects of these lymphokines on spontaneous IgE synthesis. *J Allergy Clin Immunol*, 1991;87:58–69.
14. Keane-Myers A, Nickell SP. Role of IL-4 and IFN-γ in Modulation of Immunity to Borrelia burgdorferi in Mice. *J Immunol.* 1995;155:2020–2028.
15. Finkleman FD, Katona IM, Mossman T, Coffman RL. IFN-γ regulates the isotypes of Ig secreted during in vivo humoral responses. *J Immunol.* 1988;140:1022.
16. Bradley LM, Dalton DK, Croft M. A direct role for IFN-γ in regulation of Th1 cell development. *J Immunol* 1996;157:1350–1358.
17. Gajewski TF, Joyce J, Fitch FW. Anti-proliferative effect of IFN-γ in immune regulation. III. Differential selection of Th1 and Th2 murine helper T lymphocyte clones using recombinant IL-2 and recombinant IFN-γ. *J Immunol.* 1989;143:15–22.
18. Gazzinelli RT, Wysocka M, Hayashi S, et al. Parasite-Induced IL-12 Stimulates Early IFN-γ Synthesis and Resistance During Acute Infection with Toxoplasma gondii. *J Immunol.* 1994;153:2533–2543.
19. Meyaard L, Hovenkamp E, Otto SA, Miedema F. IL-12-Induced IL-10 production by Human T Cells As a Negative Feedback for IL-12-Induced Immune Responses. *J Immunol.* 1996;156:2776–2782.
20. Matyniak J, Reiner SL. T helper phenotype and genetic susceptibility in experimental Lyme disease. *J Exp Med.* 1995;181:1251.
21. Romagnani S. Lymphokine production by human T cells in disease states. *Ann Rev Immunol.* 1994;12:227.
22. Cheers C, McKenzie IFC. Resistance and susceptibility of mice to bacterial infection; genetics of listeriosis. *Inf Immun.* 1978;19:755–762.
23. Steere AC, Schoen RT, Taylor E. The clinical evolution of Lyme arthritis. *Ann Intern Med.* 1987;107:725.
24. Zimme G, Schaible UE, Kramer MD, Mall G, Museteanu C, Simon MM. Lyme carditis in immunodeficient mice during experimental infection of Borrelia burgdorferi. *Virch.Arch.Pathol Anat Histopathol.* 1990;417:129.
25. Luft BJ, Steinman CR, Neimark HC, et al. Invasion of the central nervous system by Borrelia burgdorferi in acute disseminated infection. *J Am Med A* ssoc. 1992;267:1364.
26. Keane-Myers A, Nickell SP. T Cell Subset-Dependent Modulation of Immunity to Borrelia burgdorferi in Mice. *J Immunol.* 1995;154:1770–1776.
27. Keane-Myers A, Maliszewski CR, Finkleman FD, Nickell SP. Recombinant IL-4 treatment augments resistance to Borrelia burgdorferi infections in both normal susceptible and antibody deficient susceptible mice. *J Immunology* 1996;156:2488–2494.
28. Keane-Myers A, Gause WC, Linsley PS, Chen S-J, Wills-Karp M. B7-CD28/CTLA-4 Costimulatory pathways are required for the development of Th2-mediated allergenic airway responses to inhaled antigens. *J Immunol.* 1997;158:2042–2049.
29. Gavett SH, Chen X, Finkleman F, Wills-Karp M. Depletion of murine CD4+ T lyphocytes prevents antigen-induced airway hyperreactivity and pulmonary eosinophilia. *Am J Respir.Cell Mol Biol.* 1994;10:587–593.

30. Robinson DS, Hamid Q, Ying S, et al. Predominant TH2-Like Bronchoalveolar T-Lymphocyte Population in Atopic Asthma. *N Eng J Med.* 1992;326:298–304.
31. Lukacs NW, Strieter RM, Chensue SW, Kunkel SL. Interleukin-4-dependent Pulmonary Eosinophil Infiltration in a Murine Model of Asthma. *Am J Respir Cell Mol Biol.* 1994;10:526–532.
32. Corry DB, Folkesson HG, Warnock ML, et al. Interleukin 4, but not Interleukin 5 or Eosinophils is Required in a Murine Model of Acute Airway Hyperreactivity. *J Exp Med.* 1996;183:109–117.
33. Foster PS, Hogan SP, Ramsay AJ, Matthaei KI, Young IG. Interleukin 5 Deficiency Abolishes Eosinophila, Airways Hyperreactivity, and Lung Damage in a Mouse Model of Asthma. *J Exp Med.* 1996;183:195–201.
34. Bradding P, Feather IH, Howarth PH, et al. Interleukin 4 is Localized to and Released by Human Mast Cells. *J Ex. Med*. 1992;176:1381–1386.
35. De Monchy JG, Kauffman HF, Venge P, et al. Bronchoalveolar Eosinophilia During Allergen-induced Late Asthmatic Reactions. *Am Rev Respir Dis*. 1985;131:373–376.
36. Gavett SH, O'Hearn DJ, Li X, Huang S-K, Finkleman FD, Wills-Karp M. Interleukin 12 Inhibits Antigen-Induced Airway Hyperresponsiveness, Inflammation, and TH2 Cytokne Expression in Mice. *J Exp Med.* 1995; 182: 1527–1536.
37. Ho IC, Hodge MR, Rooney JW, Glimcher LH. The proto-oncogene c-maf is responsible for tissue-specific expression of interleukin-4. *Cell*. 1996;85:973–983.
38. Kaplan MH, Schindler U, Smiley ST, Grusby MJ. Stat6 is required for mediating responses to IL-4 and for development of Th2 cells. *Immunity*. 1996;4:313–319.
39. Henkel G, Brown MA. PU.1 and GATA: Components of a mast cell-specific interleukin-4 intronic enhancer. *Proc Natl Acad Sci USA*. 1994; 91:7737–7741.
40. Lederer JA, Liou JS, Kim S, Rice N, Lichtman AH. Regulation of NF-kappa B activation in T helper 1 and T helper 2 cells. *J Immunol.* 1996; 156:56–63.
41. Casolaro V, Georas S, Song Z, al. e. Inhibition of NFATp-dependent transcription by NFkB: Implications for differential gene expression in T cell subsets. *Proc Natl Acad Sci USA*. 1995.
42. Song Z, Casalaro V, Chen C, Georas SN, Manos D, Ono SJ. Polymorphic Nucleotides Within the Human IL-4 Promoter that Mediate Overexpression of the Gene. *J Immnuol.* 1996;156:424–429.
43. Klein-Hessling S, Schneider G, Heinfling A, Chuvpilo S, Serfling E. HMG I(Y) interferes with the DNA binding of NF-AT factors and the induction of the interleukin-4 promoter in T cells. *Proc Natl Acad Sci USA* . 1996;93:15311–15316.
44. Venkataraman L, Wang W, Sen R. Differential regulation of c-rel translocation in activated B and T cells. *J Immunol.* 1996;157:1149–1155.
45. Casolaro V, Georas SN, Song Z, Ono SJ. Biology and Genetics of Atopic disease. *Current Opinions in Immunology*. 1996;8:796–803.

PRESENCE AND TESTOSTERONE INFLUENCE ON THE LEVELS OF ANTI- AND PRO-INFLAMMATORY CYTOKINES IN LACRIMAL TISSUES OF A MOUSE MODEL OF SJÖGREN'S SYNDROME

Eduardo M. Rocha, L. Alexandra Wickham, Zhiyan Huang, Ikuko Toda, Jianping Gao, Lilia Aikawa da Silveira, and David A. Sullivan

Schepens Eye Research Institute and Department of Ophthalmology
Harvard Medical School
Boston, Massachusetts

1. INTRODUCTION

During the past several years our laboratory has shown that testosterone treatment suppresses the inflammation in, and enhances the functional activity of, lacrimal glands in female mouse models (MRL/Mp-lpr/lpr [MRL/lpr] and NZB/NZW F1 [F1]) of Sjögren's syndrome.[1–5] This hormone action seems to be a unique, tissue-specific effect, that may be reproduced by therapy with a variety of androgen analogues, but not by the administration of danazol, estradiol, dexamethasone, cyclosporine A, or an experimental, non-androgenic steroid.[3,4] The precise mechanism(s) underlying this androgen influence is unclear, but we hypothesize that this hormone effect is mediated through an interaction with receptors in epithelial cell nuclei, which then cause an altered expression of proto-oncogenes, apoptotic factors and cytokines in lacrimal tissue, thereby leading to the contraction of immunopathological lesions and an improvement in glandular function.[6] In support of this hypothesis, we have found that epithelial cells are the target cells for androgen activity,[7] and that androgen exposure elicits significant changes in the levels of apoptotic factor mRNAs in the lacrimal glands of autoimmune mice.[8,9]

To extend these findings, and to further advance our understanding of the mechanism(s) underlying the androgen-induced reduction of lacrimal tissue inflammation, we examined in the present study whether: (1) the mRNAs for anti-inflammatory and pro-inflammatory cytokines may be detected in lacrimal glands of autoimmune, as well as non-autoimmune, mice; and (2) testosterone treatment alters the levels of cytokine mRNA and/or protein in lacrimal tissues of female MRL/lpr mice.

Lacrimal Gland, Tear Film, and Dry Eye Syndromes 2
edited by Sullivan *et al.*, Plenum Press, New York, 1998

2. METHODS

2.1. Animals and Hormone Treatment

Adult, male and female MRL/lpr, F1, C3H/lpr, BALB/c, C57BL/6 and C3H/HeJ mice were purchased from The Jackson Laboratory (Bar Harbor, Maine) or Taconic Laboratories (Germantown, NY) and maintained in constant temperature rooms with fixed light/dark periods of 12 hours length. After the onset of disease (i.e. autoimmune mice), or at post-pubertal ages (i.e. non-autoimmune mice), age-matched mice were either sacrificed by CO_2 inhalation or treated for 21 days with subcutaneous implants of placebo-, testosterone (10 mg)- or cyclophsophamide (10 mg)-containing pellets (Innovative Research of America, Sarasota, FL). These pellets, which were placed in the subscapular region, were designed for the slow, but continuous, release of vehicle or compound over the experimental interval. Following animal sacrifice, exorbital lacrimal glands, submandibular glands, spleens and/or peritoneal cells were removed and processed for molecular biological, cell culture or protein extraction procedures. All studies adhered to the resolution of The Association for Research in Vision and Ophthalmology on the use of animals in research.

2.2. Cell Cultures and General Procedures

To serve as positive controls for cytokine mRNA analysis, RNA was isolated from the following preparations: (1) peritoneal exudate cells (for granulocyte monocyte-colony stimulating factor; GM-CSF), which were obtained after thioglycollate (DIFCO Laboratories, Detroit, MI; 20 ng/ml, 1.5 ml/mouse) injection into the peritoneal cavity of C57BL/6 mice, and which were harvested after a 2 hour incubation in DMEM and 10% fetal bovine serum (Hyclone, Logan, Utah), according to published methods;[10] (2) P388D1 mouse monocyte/macrophage cells (ATCC, Rockville, MD; for IL-1α, IL-6, IL-7, TGF-β and TNF-α), which were cultured for 48 hours in the presence of lipopolysaccaride (LPS; DIFCO; 40 μg/ml), according to a reported protocol;[11] (3) C57BL/6 mouse spleen cells (for IL-2, IL-3, IL-4, IL-5 and TNF-β), which were isolated and then incubated (2.5×10^6 cells/ml) for 48 hours at 37°C with phorbol 12-myristate 13-acetate (PMA; 10 ng/ml; Sigma Chemical Co., St. Louis, MO) and ionomycin (500 ng/ml; Calbiochem-Behring, La Jolla, CA), according to published procedures;[11] and (4) splenic tissues from MRL/lpr and BALB/c mice (for IL-1β, IL-2R, IL-6, IL-7, TGF-β, TNF-α, TNF-β and IFN-γ). Protein levels in lacrimal gland extracts were determined by the Hartree method,[12] using bovine serum albumin (Calbiochem-Behring) as the standard. Statistical analysis of the data between two groups was performed by utilizing either Student's unpaired, two-tailed t test or the Mann Whitney U test. Statistical comparisons between the means of multiple groups was conducted by analysis of variance and Fisher's PLSD with a significance level of 95%.

2.3. Molecular Biological Procedures

To detect and measure cytokine mRNAs in experimental samples, specific reverse transcription polymerase chain reactions (RT-PCR) were used. Total RNA was isolated from lacrimal (≥ 2 glands/sample), submandibular (≥ 2 glands/sample), and splenic tissues, as well as cells, by utilizing modified acid guanidinium-thiocyanate-phenol-chloroform extraction procedures.[13] The resulting RNA preparations were analyzed by spectrophotometry at 260 nm to determine their concentration and evaluated on 1.2% agarose (Gibco/BRL, Grand Island, NY) gels to verify RNA integrity. cDNAs were transcribed from total RNA samples (5 μg) by utilizing AMV or MMLV reverse transcriptase, oligo dT or random priming and either the First-Strand cDNA Synthesis kit from Invitrogen (San Diego, CA) or the Advantage RT-for-PCR Kit from Clontech Laboratories Inc. (Palo Alto, CA), according to modifications of the manufacturer's protocol. PCR amplification (final reaction volume = 25 or 50 μl) of the cDNAs was conducted with a Perkin Elmer Cetus GeneAmp PCR System 9600 (Perkin Elmer, Norwalk, CT) by using Taq DNA polymerase (Gibco/BRL, Grand Island, NY), dNTPs (Perkin-Elmer), PCR buffer (Invitrogen, Perkin-Elmer or Gibco/BRL) and 0.4 μM of each 5' and 3' primer corresponding to mouse IL-1α, IL-1β, IL-2, IL-2 receptor (IL-2R), IL-3, IL-4, IL-5, IL-6, IL-7, TGF-β, TNF-α, TNF-β, IFN-γ, GM-CSF, β-actin and glyceraldehyde-3-phosphate dehydrogenase. Primers and oligonucleotide probes were obtained from Clontech or synthesized by National Biosciences, Inc. (Plymouth, MN) or Gibco/BRL Custom Primers, Inc. (Palo Alto, CA). The number of PCR cycles chosen for each primer set was determined experimentally and, for comparative measurements, occurred in the exponential phase of the polymerase chain reaction. Additional controls in the RT-PCR procedures included primer kit standards, reaction mixtures without cDNA, and DEPC-treated water without other reaction components. Positive and negative control cDNAs were run in parallel, but separate, tubes in all PCR procedures. Following amplification, the PCR products were analyzed on 1.5% agarose gels, containing a 100 bp DNA molecular weight ladder (Gibco) and stained with ethidium bromide, in order to verify the anticipated fragment sizes. When indicated, the amplified cDNA products were transfered to GeneScreen nylon membranes (Dupont/NEN, Boston, MA) by positive pressure, fixed by UV cross-linking and incubated with specific ^{32}P-labeled probes that had been phosphorylated by random priming (Random Primer Labeling System, Dupont/NEN, Boston, MA) or end-labelling (T4 Polynucleotide Kinase, New England BioLabs, Beverly, MA) methods. Oligonucleotide probes were designed by using published cytokine sequences and Oligo Version 4 Primer Analysis Software (National Biosciences). Following an overnight hybridization in an oscillating water bath, the Southern blots were washed and processed for autoradiography by utilizing X-OMAT AR film (Kodak, Rochester, NY), with or without intensifying screens, at -80°C. To quantitate mRNA levels, band densities in autoradiographs and agarose gel internegatives were analyzed with a computer-assisted image analysis system, as previously described.[14] In the hormone-related studies, absorbance results were standardized to the corresponding level of β-actin mRNA and data are reported in terms of the sample mRNA/β-actin mRNA ratio.

2.4. Measurement of TGF-β1 Protein in Lacrimal Gland Extracts

To determine the levels of TGF-β1 protein in MRL/lpr mouse lacrimal glands, tissue proteins were extracted after glandular RNA isolation by exposing the supernatants to isopropanol, precipitating the protein by centrifugation and washing the pellets with guanadine HCl/ethanol. Additional protein was recovered from the isopropanol-associated supernatants by the use of 1 M acetic acid. Protein samples were acid-activated with HCl, neutralized with NaOH and then analyzed with Quantikine ELISA kits from R&D Systems, Inc. (Minneapolis, MN), according to the manufacturer's instructions. Colorimetric absorbance was determined at 450 nm, with automatic subtraction of the background absorbance at 450 nm and the reference absorbance at 630 nm, by use of a Dynatech MR700 Microplate Reader. To measure TGF-β1 levels in samples, standard curves were included in each assay and analyzed by log logit transformation. No evidence of interference by experimental samples was found in this assay.

3. RESULTS

3.1. Identification of Cytokine mRNAs in Murine Lacrimal Glands

To determine whether the mRNAs for anti-inflammatory and pro-inflammatory cytokines may be detected in lacrimal glands of autoimmune and non-autoimmune mice, tissues were obtained from MRl/lpr, F1, C3H/lpr, BALB/c and C3H/HeJ mice and processed for RT-PCR, agarose gel electrophoresis and, for certain cytokines, Southern blot hybridization. For comparison, we also evaluated submandibular gland RNA from MRL/lpr mice. As shown in Table 1, the mRNAs for IL-1α, IL-1β, IL-2 receptor, IL-6, IL-7, TGF-β and TNF-α were detected consistently in lacrimal and salivary tissues of autoimmune female MRL/lpr mice, and many of these cytokine mRNAs were also present in lacrimal glands of other strains. In contrast: (1) bands for TNF-β and GM-CSF were either very faint or absent in various samples; (2) no IL-2, IL-3 or IL-5 mRNAs were evident in exocrine tissues; and (3) IFN-γ was only detected in the salivary gland. The results for IL-4 were in-

Table 1. Cytokine mRNAs in exocrine glands of autoimmune and non-autoimmune mice

Factor	Lacrimal gland					Submandibular gland
	MRL/lpr	F1	C3H/lpr	BALB/c	C3H/HeJ	MRL/lpr
IL-1α	+			+		+
IL-1β	+	+	+	+	+	+
IL-2	–			–		–
IL-2R	+			+		+
IL-3	–					–
IL-4	–*					–*
IL-5	–					–
IL-6	+			–		+
IL-7	+			+		+
TGF-β	+	+	+	+	+	+
TNF-α	+	+	+	+	+	+
TNF-β	±			–		±
GM-CSF	±			–		–
IFN-γ	–			–		+
G3PDH	+			+		+
β-Actin	+	+	+	+	+	+

RNA was isolated from exocrine glands of MRL/lpr, F1, C3H/lpr and C3H/HeJ mice ($n \geq 6$ samples; $n \geq 2$ glands/sample), as well as lacrimal tissues of BALB/c mice ($n = 2$ samples; $n = 9$-10 glands/sample), reverse transcribed and then amplifed by PCR. Amplified PCR products were identified by agarose gel electrophoresis and/or Southern blot hybridization (IL-1β, IL-6, TGF-β, TNF-α and β-actin in MRL/lpr lacrimal tissue). For positive controls in these assays, cDNAs were prepared from RNA from mouse peritoneal exudate cells, LPS-stimulated P388D1 mouse monocyte/macrophage cells, PMA- and ionomycin-stimulated mouse spleen cells and spleens from MRL/lpr and BALB/c mice. Additional controls were as outlined in the Methods. Symbols refer to detectable (+), very faint (±) and not detectable (–) under the conditions employed. The asterisk (*) associated with the IL-4 results indicates that no bands were detected in any sample, including the positive controls.

conclusive, given that no amplified PCR products were observed in either the experimental samples or positive controls.

3.2. Influence of Testosterone Treatment on the Levels of IL-1β, TNF-α and TGF-β mRNAs in Lacrimal Glands of MRL/lpr Mice

To determine whether androgen administration alters the levels of pro- and/or anti-inflammatory cytokine mRNAs in lacrimal tissues of female MRL/lpr mice, animals (n = 11–13 mice/group) were treated after the onset of disease with placebo- or testosterone (10 mg)-containing compounds for 21 days. Lacrimal glands (n = 4–6 glands/sample; n = 5 samples/group) were then obtained and processed for the analysis of IL-1β, TNF-α and TGF-β mRNAs by RT-PCR, Southern blot hybridization and densitometry. All data were standardized to the corresponding levels of β-actin mRNA. Our results showed that androgen administration caused a significant ($p < 0.05$) decline in the levels of IL-1β and TNF-α mRNA, but had no effect on the amount of TGF-β or β-actin mRNA, in lacrimal tissue, as compared to mRNA amounts in placebo-treated controls.[14]

3.3. Effect of Testosterone Therapy on the Level of TGF-β1 Pprotein in Lacrimal Tissues of MRL/lpr Mice

Both testosterone and the immunosuppressive compound, cyclophosphamide, are known to reduce inflammation in lacrimal tissues of MRL/lpr mice.[3] To assess whether these effects are paralleled by an increase in the amount of the anti-inflammatory protein, TGF-β1, adult female MRL/lpr mice (n = 10–11 mice/group) were administered placebo-, testosterone (10 mg)- or cyclophosphamide (10 mg)-containing pellets for 3 weeks, and lacrimal glands (n = 3–6 glands/sample; n = 5 samples/group) were then removed, homogenized, acid extracted and processed for protein measurement. As shown in Figure 1, testosterone, but not cyclophosphamide, treatment stimulated a significant ($p < 0.05$) rise in the total amount of TGF-β1 protein in lacrimal glands, as compared to levels in tissues of placebo controls. This hormone action was also associated with a significant ($p < 0.05$) increase in the amount of total protein per gland (Figure 1), as well as lacrimal gland weight (placebo = 10.7 ± 0.3 mg/gland; testosterone = 20.1 ± 0.6 mg/gland), relative to placebo-exposed controls.

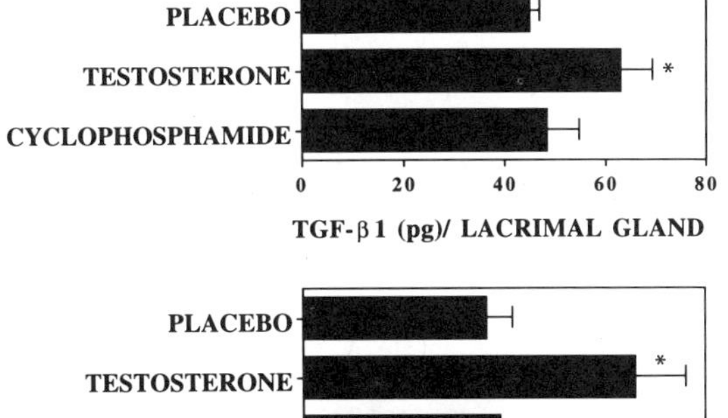

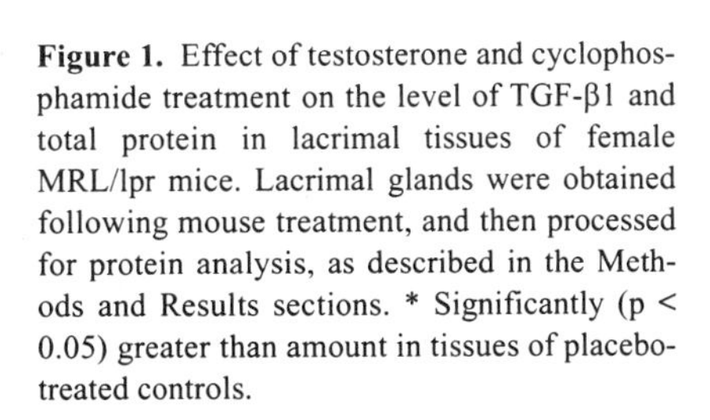

Figure 1. Effect of testosterone and cyclophosphamide treatment on the level of TGF-β1 and total protein in lacrimal tissues of female MRL/lpr mice. Lacrimal glands were obtained following mouse treatment, and then processed for protein analysis, as described in the Methods and Results sections. * Significantly ($p < 0.05$) greater than amount in tissues of placebo-treated controls.

4. DISCUSSION

The present findings demonstrate that a variety of cytokine mRNAs exist in lacrimal glands of both autoimmune and non-autoimmune mice, and that testosterone treatment causes a marked attenuation in the mRNA amount of the pro-inflammatory cytokines IL-1β and TNF-α in female MRL/lpr lacrimal tissue. In addition, our results show that androgen exposure increased the total level of the anti-inflammatory TGF-β1 protein, but had no effect on the amount of TGF-β mRNA, in MRL/lpr lacrimal glands. This apparent discrepancy might be explained by several considerations: (1) androgen treatment may have increased TGF-β mRNA content in acinar epithelial cells, but decreased TGF-β mRNA levels in the lymphoid compartment (i.e. through immunosuppression), thereby resulting in similar mRNA amounts per 5 μg RNA in placebo- and hormone-exposed tissues. In support of this possibility, TGF-β mRNA has been identified in rat acinar epithelial cells,[15] and androgens are known to enhance TGF-β mRNA amounts in 'normal' rat lacrimal glands[16] and reduce inflammation in MRL/lpr lacrimal tissues;[1–4] (2) the total level of TGF-β mRNA may actually have risen in lacrimal glands of female MRL/lpr mice following hormone therapy, given that androgens induce a significant increase in the total weight and RNA content[8] of these tissues, and cytokine mRNA data in the current investigation were based upon the relative content in a fixed 5 μg amount of RNA; and (3) the half-life of TGF-β mRNA in the lacrimal gland may be relatively short,[17] whereas that of the corresponding protein may be longer. In contrast to the androgen action, cyclophosphamide administration had no impact on TGF-β1 protein levels on MRL/lpr lacrimal tissues, indicating that the anti-inflammatory mechanism of this compound is different than that of testosterone.

Overall, the observed hormone effects on IL-1β and TNF-α mRNAs, as well as on TGF-β1 protein, could provide one possible explanation for the androgen-related suppression of autoimmune disease in lacrimal glands of MRL/lpr mice. In the future we hope to determine the underlying mechanisms involved in this hormonal control of cytokine expression. Such analysis may help lead to the development of unique, therapeutic strategies to safely and effectively treat lacrimal gland inflammation in Sjögren's syndrome.

ACKNOWLEDGMENTS

The authors express their appreciation to Dr. Lan Hu (Boston, MA) for her provision of β-actin cDNA. This research was supported by grants from NIH [EY05612], Allergan, the Massachusetts Lions Research Fund and the Sjogren's Syndrome Foundation.

REFERENCES

1. Ariga H, Edwards J, Sullivan DA. Androgen control of autoimmune expression in lacrimal glands of MRL/Mp-lpr/lpr mice. *Clin Immunol Immunopath.* 1989;53:499–508.
2. Vendramini AC, Soo CH, Sullivan DA. Testosterone-induced suppression of autoimmune disease in lacrimal tissue of a mouse model (NZB/NZW F1) of Sjögren's Syndrome. *Invest Ophthalmol Vis Sci.* 1991;32:3002–3006.
3. Sato EH, Sullivan DA. Comparative influence of steroid hormones and immunosuppressive agents on autoimmune expression in lacrimal glands of a female mouse model of Sjögren's syndrome. *Invest Ophthalmol Vis Sci.* 1994;35:2632–2642.
4. Rocha FJ, Sato EH, Sullivan BD, Sullivan DA. Effect of androgen analogue treatment, and androgen withdrawal, on lacrimal gland inflammation in a mouse model (MRL/Mp-lpr/lpr) of Sjögren's syndrome. *Reg Immunol.* 1994; 6:270–277.

5. Sullivan DA, Edwards J. Androgen stimulation of lacrimal gland function in mouse models of Sjögren's syndrome. *J. Ster. Biochem. Mol. Biol.* 1997; 60:237–245.
6. Sullivan DA, Wickham LA, Krenzer KL, Rocha EM, Toda I. Aqueous tear deficiency in Sjögren's syndrome: Possible causes and potential treatment. In: Pleyer U, Hartmann C and Sterry W, eds. *Oculodermal Diseases - Immunology of Bullous Oculo-Muco-Cutaneous Disorders*. Buren, The Netherlands: Aeolus Press, 1997:95–152.
7. Ono M, Rocha FJ, Sullivan DA. Immunocytochemical location and hormonal control of androgen receptors in lacrimal tissues of the female MRL/Mp-lpr/lpr mouse model of Sjögren's syndrome. *Exp. Eye Res.* 1995; 61:659–666.
8. Toda I, Wickham LA, Sullivan DA. Influence of gender and androgen treatment on the mRNA expression of proto-oncogenes and apoptotic factors in lacrimal and salivary tissues of the MRL/lpr mouse model of Sjögren's syndrome. Submitted for publication, 1997.
9. Toda I, Wickham LA, Sullivan DA. Gender and androgen-related influence on the expression of proto-oncogene and apoptotic factor mRNAs in lacrimal glands of autoimmune and non-autoimmune mice. Submitted for publication, 1997.
10. Thorens B, Mermod JZ, Vassalli P. Phagocytosis and inflammatory stimuli induce GM-CSF mRNA in macrophages through posttranscriptional regulation. *Cell* 1987; 48:671–679.
11. Gutierrez-Ramos JC, Olsson C, Palacios R. Interleukin (IL1 to IL7) gene expression in fetal liver and bone marrow stromal clones: cytokine-mediated positive and negative regulation. *Exp Hematol* 1992; 20:986–990.
12. Hartree EF. Determination of protein: a modification of the Lowry method that gives a linear photometric response. *Analyt Biochem.* 1972;48:422–427.
13. Chomczynski P, Sacchi N. Single-step method of RNA isolation by acid guanidinium thiocyanate-phenol-chloroform extraction. *Anal. Biochem.* 1987; 162:156–159.
14. Rocha EM, Toda I, Wickham, LA, Silveira LA, Sullivan DA. Effect of gender, androgens and cyclophosphamide on cytokine mRNA levels in lacrimal tissues of mouse models of Sjögren's syndrome. Submitted for publication, 1997.
15. Ono M, Huang Z, Wickham LA, Gao J, Sullivan, DA. Analysis of androgen receptors and cytokines in lacrimal glands of a mouse model of Sjögren's syndrome. *Invest Ophthalmol Vis Sci Suppl* 1994; 35:1793.
16. Huang Z, Gao J, Wickham LA, Sullivan, DA. Influence of gender and androgen treatment on TGF-β1 mRNA levels in the rat lacrimal gland. *Invest Ophthalmol Vis Sci Suppl* 1995; 35:S991.
17. Thompson HW, Beuerman RW, Cook J, Underwood LW, Nguyen DH. Transcription of message for tumor necrosis factor-alpha by lacrimal gland is regulated by corneal wounding. *Adv Exp Med Biol* 1994; 350:211–217.

68

INFILTRATING LYMPHOCYTE POPULATIONS AND CYTOKINE PRODUCTION IN THE SALIVARY AND LACRIMAL GLANDS OF AUTOIMMUNE NOD MICE

Christopher P. Robinson,[1,2] Janet Cornelius,[1] Denise I. Bounous,[3]
Hideo Yamamoto,[2] Michael G. Humphreys-Beher,[2] and Ammon B. Peck[1]

[1]Department of Pathology and Laboratory Medicine
[2]Department of Oral Biology
University of Florida
Gainesville, Florida
[3]Department of Pathology
College of Veterinary Medicine
University of Georgia
Athens, Georgia

1. INTRODUCTION

It is only in the last five years that significant interest has developed in detailing the autoimmune destruction of the salivary and lacrimal tissues in the non-obese diabetic (NOD) mouse. Pioneering studies have demonstrated that the lymphocytic infiltration of the salivary and lacrimal glands correlates with a functional decline in saliva flow and tear production, respectively, that is not attributable to the loss of blood glucose regulation in NOD mice.[1,2] Together, these studies indicate that the NOD mouse represents the first-described animal model for the spontaneous, autoimmune-induced loss of both tear production and saliva flow, and is, therefore, emerging as an excellent animal model for the study of human Sjögren's syndrome.

Sjogren's syndrome is a human autoimmune disease characterized by dry eye and dry mouth syndromes due to the destruction of exocrine tissue.[3] Clinically patients develop chronic lymphocytic infiltration of the salivary and lacrimal glands as well as autoantibodies to ribonuclear proteins. Diagnosis of Sjögren's syndrome often relies on the detection of lymphocytic infiltration in labial lip biopsies excised from patients.[3]

Lacrimal Gland, Tear Film, and Dry Eye Syndromes 2
edited by Sullivan *et al.*, Plenum Press, New York, 1998

In this study we have characterized the infiltrating lymphocyte repertoire and cytokine mRNA profile of the exocrine glands of NOD mice over the time course of immunopathogenesis. This has allowed for the qualitative comparison of lymphocyte populations infiltrating the pancreas, submandibular gland (SMG), and lacrimal gland (LAC) tissues of NOD mice, as well as providing a detailed account of the anti-exocrine autoimmune response similar to that seen in human Sjögren's syndrome.

2. MATERIAL AND METHODS

Female NOD mice from Jackson Laboratories (Bar Harbor, ME) were maintained in the animal facility at the University of Florida (Gainesville, FL). Mice were sacrificed in groups of five at 2 week intervals from 8 through 20 weeks. The following tissues were removed: spleen, pancreas, LAC, parotid (PAR), and SMG. A small piece was cut from each tissue and placed in 10% buffered formalin for histologic sectioning and staining. Each tissue was processed as a pool from the five mice per group. A small aliquot of each pool was removed for RNA isolation; all remaining tissue was prepared for flow cytometric analysis.

Single-cell suspensions of splenic leukocytes were obtained as controls. Red blood cells were lysed with 0.84% ammonium chloride. After washing, the remaining leukocytes were aliquoted at 1 x 10^6/tube and washed with FACS buffer (PBS with 0.1% NaN_3) and 0.5% BSA prior to antibody staining.

Aliquots of all cell populations were resuspended in 100 μl FACS buffer and stained with antibody at 1 μg/10^6 cells. Cells were stained first with anti-CD3 for 40 min, washed with FACS buffer, and stained with the appropriate second antibody. After a final wash, cells were suspended in FACS buffer for analysis. Flow cytometric analysis was performed using a FACScan flow cytometer (Becton Dickinson, Mountain View, CA) equipped with a 15 milliwatt, 488 nm air-cooled argon-ion laser and using LYSYS™ II software. Pooled tissues were minced in PBS and placed in lysis buffer; then mRNA was isolated using a Micro-FastTrack™ Kit (Invitrogen, San Diego, CA). Equal quantities of reverse transcribed cDNA from each sample (1 ng per reaction) were amplified by PCR for 40 cycles at 60°C annealing (1 min) and 72°C elongation (2 min). Specific PCR products were identified using the Genius™ system of nonradioactive DNA labeling and detection (Boehringer Mannheim) according to the manufacturer's protocols.

3. RESULTS

Each tissue was stained with hematoxylin and eosin (H&E) and examined for leukocytic infiltration. At 8 weeks of age, the SMG of two of the five mice showed small focal areas of infiltration; the LAC and PAR remained normal. By 12 weeks, the LAC showed infiltration; however, these were fewer and smaller lymphocytic foci than in the SMG, which by 14 weeks were heavily infiltrated in all mice.

FACS gates set on the spleen lymphocyte population were used to select the infiltrating populations of the other tissue samples. Within these gated populations, the percentage of $CD3^+$ cells tended to remain relatively constant over the time course of the study at 48% for the spleen, 41% for the islet, 37% for the LAC and 39% for the PAR (Table 1).[4] In the SMG, there was an increase in $CD3^+$ cells from 41% at 8 weeks to 70% at 20 weeks. The $CD4^+$ population in the SMG approximated the spleen (68%); the LAC

Table 1. Lymphocyte cell populations in lacrimal and submandibular glands in NOD mice

	10 Weeks		20 Weeks	
Cell Phenotype	LAC	SMG	LAC	SMG
CD3+/CD4+	33%	70%	56%	67%
CD3+/CD8+	6%	13%	15%	18%
CD3-/B220+	4%	15%	31%	17%
CD3+/CD45RBhi	7%	14%	7%	17%
CD3+/CD45RBlo	17%	19%	19%	28%
CD3+/Vβ8+	–	–	14%	23%
CD3+/Vβ6+	–	–	10%	15%
CD3+/Vβ3+	–	–	5%	2%

LAC, lacrimal gland. SMG, submandibular gland.

$CD4^+$ population was about half that of the spleen at 8 weeks and increased over time to the level of the spleen. Generally, fewer than 10% of the $CD3^+$ population were $CD8^+$ in these glands at 8 weeks, which was lower than in the spleen, which averaged 27%. The percentage of $CD8^+$ cells increased in all three glands as the disease progressed.

As indicated by the spleen population, the ratio of $CD4^+$ to $CD8^+$ cells was relatively constant in the periphery at about 2.5:1. Both the LAC and PAR had higher ratios early, which decreased to spleen levels at 12 weeks (Table 1).[4] The $CD3^-B220^+$ B-cell population in the SMG increased from ~1% at 8 weeks to ~15% at 10 weeks and maintained that level. In the LAC, $CD3^-B220^+$ cells was ~1% at 8 weeks and increased to ~33%. The presence of a $CD3^+B220^+$ cell population was detected in both of these tissues and did not exceed 10% in the SMG or 17% in the LAC. The $CD3^+B220^+$ phenotype has been reported as a double negative population in MRL/lpr mice and as a lymphokine activated killer phenotype (LAK cell)[5] in other studies.[6] In SMG at 8 weeks, 3.1% of the $CD3^+$ cells were $CD45RB^{hi+}$ (naive T-cell) and from 10 weeks on increased to ~16%. The percentage of $CD3^+CD45RB^{lo+}$ (memory T- cells) population was 15% at 8 weeks, increased to 27% at 12 weeks and remained at that level. The $CD3^-CD45RB^{hi+}$ population in the SMG remained constant at ~25% from 10 - 20 weeks. In the LAC, the $CD45RB^{hi+}$ population (both $CD3^+$ and $CD3^-$) increased to maximum of ~55% at 12 weeks.

TCR Vβ usage was analyzed by using selected monoclonal antibodies (Table 1).[4] In the Vβs tested, there was no appreciable time-related variation in response. Both Vβ6 and Vβ8 were significantly increased over background as represented by Vβ3, and, though less significant, Vβ9 and Vβ17 were also increased. These increases were much more dramatic in the SMG than in the LAC.

We further investigated temporal mRNA expression of a large number of pro-inflammatory and effector cytokines. Cytokine transcripts were detected at both an earlier age and greater intensity in the LAC than in the SMG of similar age groups (Table 2). mRNA transcripts for IL-4 were absent from all sample groups. In the LAC tissue, IL-1β, IL-5, IL-6, IL-7, IL-10, TNFα, and CD28 mRNA transcripts were detected as early as 8 weeks. By 10 weeks of age, IFNγ and IL-2R were first detected, and levels of TNFα increased dramatically and maintained a high level of transcription through 20 weeks. No IL-2 was detected in the LAC early, with a burst in IL-2 production detected between 12 and 14 weeks of age. Similarly, iNOS was first detected after 12k. Levels of IFN, TNFα, iNOS, IL-2R, and CD28 increased with age, while IL-5 declined. Interleukins-6 and -10 yielded variable results in each age group.

Table 2. RT-PCR determination of mRNA cytokine levels

Gland/Age (wk)	IL-1β	IL-2	IL-4	IL-5	IL-6	IL-7	IL-10	IL-12	IFNγ	TNF-α	iNOS
LAC/10	++	–	–	+++	+++	++	++	–	+	++	–
LAC/20	+++	+	–	++	++	++	+	+	+++	++	++
SMG/10	+	+	–	–	–	–	–	+	–	–	–
SMG/20	++	+	–	–	++	++	++	+	+++	++	++

The cytokine mRNA expression in the SMG was characterized by little or no expression before 12 weeks. The earliest detectable transcripts in the SMG include IL-1β, IL-5, IL-7, TNFα, CD28 and IL-2R. The most interesting picture, however, appears between 12 and 14 weeks, as levels of IL-1β, IL-6, IL-7, TNFα and IFNγ expression increased dramatically. Transcription of these cytokines was maintained at high levels, while expression of IL-2, IL-10, and iNOS appeared between 14 and 16 weeks. These results indicate that considerable changes in the cytokine profiles occur between 12 and 16 weeks of age in the SMG of these mice, despite the fact that lymphocytic infiltration was present in the tissue at an earlier age.

4. DISCUSSION

This study offers the first description of the temporal progression of the autoimmune attack against both the salivary and lacrimal glands of NOD mice. The unique feature of both the NOD mouse and humans during the pathogenesis of Sjögren's syndrome is the loss of secretory function, which results in the clinical presentation of xerostomia and xerophthalmia.[1,3]

Many recent studies have focused on the potential roles of cytokines as cytotoxic effectors of exocrine tissue destruction. To date, there is substantial evidence that IL-1, IFNγ, TNFα, and nitric oxide may be key mediators in the pathogenic process of islet cell destruction in both humans and NOD mice.[7–10] Of particular interest, production of high levels of IFNγ by islet infiltrating cells has been repeatedly demonstrated in NOD mice and appears tightly linked to β cell destruction.[11] Furthermore, IFNγ has been noted to induce cell death of both islet and salivary gland cells *in vitro*, and may represent a non-specific mediator of exocrine cell destruction *in vivo*.[12] However, a caveat to this argument is the detection of high levels of IFNγ and TNFα in the salivary glands of MRL/lpr mice, which do not lose secretory function.[13]

Our data demonstrate numerous similarities in both lymphocyte phenotypes (CD4/CD8 ratios, Vβ TCR restriction, and B220 populations) and cytokine expression (increased IL-1β, IL-2, IL-10, TNFα, and IFNγ) between the NOD mouse and other animal models for autoimmune sialoadenitis and Sjögren's syndrome.[3,13,14] What, then, distinguishes the loss of exocrine function in humans and NOD mice as opposed to other disease models? Although the initiating agent for Sjögren's syndrome remains unknown, many investigators believe that extrinsic factors (ie, viral agent) may be responsible for the breakdown of tolerance in immunologically susceptible individuals. In contrast, however, NOD mice contain the necessary intrinsic elements for the breakdown of salivary gland tolerance in their genetic background. Using the NOD*scid* mouse, our group has recently described multiple salivary gland abnormalities of NOD*scid* mice that do not appear to be immunologically related.[15] These alterations are detectable starting at 8–10 weeks of age, and include morphological abnormalities, aberrant gene expression, and increased proteolytic activity. It is clear that any number of immunological changes can result in aggregation of lymphocytes in exocrine tissues, such as the lpr/gld mutation of MRL mice, graft vs. host models,[16] TGF-β knockouts,[17] and bcl-2 overexpression.[18] However, only with the appearance of developmental defects or extrinsic destruction of the exocrine tissue do we see the development of secretory dysfunction through loss of immune regulation.

REFERENCES

1. Humphreys-Beher MG, Hu Y, Nakagawa Y, Wang PL, Purushotham KR. Utilization of the non-obese diabetic mouse as an animal model for the study of secondary Sjögren's syndrome. *Adv Exp Med Biol.* 1994;350:631–636.
2. Hu Y, Nakagawa Y, Purushotham KR, Humphreys-Beher MG. Functional changes in the salivary glands of autoimmune NOD mice. *Am J Physiol.* 1992;263:E607-E614.
3. Fox P, Speight P. Current concepts in autoimmune exocrinopathy: immunologic mechanisms in the salivary pathology of Sjögren's Syndrome. *Crit Rev Oral Biol Med.* 1996;7:144–158.
4. Robinson CP, Cornelius J, Bounous DI, Yamamoto H, Humphreys-Beher MG, Peck AB. Characterization of the changing lymphocyte populations and cytokine expression in the exocrine tissues of autoimmune NOD mice. *Autoimmunity* 1997; in press.
5. Ballas Z, Rasmussen W. Lymphokine activated killer (LAK) cells. *Cell Immunol.* 1991;134:296–313.
6. Zhou T, Bluethmann H, Eldridge J, Bery K, Mountz J. Origin of CD4-CD8-B220+ T cells in MRL-lpr/lpr mice. Clues from a T cell receptor beta transgenic mouse. *J Immunol.* 1993;150:3651–3657.
7. Rabinovitch A, Saurez-Pinzon W, Sorenson O, Bleakley R, Power R. IFN-gamma gene expression in pancreatic islet-infiltrating mononuclear cells correlates with autoimmune diabetes. *J Immunol.* 1995;154:4874–4882.
8. Rabinovitch A, Saurez-Pinzon W, Sorenson O, Bleackley R. Inducible nitric oxide synthase in pancreatic islets of NOD mice: identification of iNOS expressing cells and relationships to cytokines expressed in these islets. *Endocrinolology.* 1996;137:2093–2099.
9. Yang X, McDevitt H. Role of TNF-alpha in the development of autoimmunity and the pathogenesis of insulin dependent diabetes in NOD mice. *Circ Shock.* 1994;43:995–1004.
10. Welsh M, Welsh N, Bendtzen K. Comparison of mRNA contents of IL-1β and nitric oxide synthase in pancreatic islets isolated from male and female NOD mice. *Diabetologia.* 1995;38:153–160.
11. Sarvetnick N, Shizuru J, Liggitt D. Loss of pancreatic tolerance induced by Beta cell expression of IFN gamma. *Nature.* 1993;346:844–847.
12. Wu A, Chen ZJ, Tsokos M. Interferon gamma induced cell death in a cultured human salivary gland cell line. *J Cell Physiol.* 1996;167:297–304.
13. Hamono H, Saito I, Haneji N. Expression of cytokine genes during development of sialadenitis in MRL/Ipr mice. *Eur J Immunol.* 1993;23:2387–2391.
14. Skarstein K, Holmdahl R, Johannessen AC, Jonsson R. Characterization of the T cell receptor repertoire and anti-Ro/SS-A autoantibodies in relation to sialoadenitis in NOD mice. *Immunol.* 1993;83:497–501.
15. Robinson C, Yamamoto H, Peck A, Humphreys-Beher M. Genetically programmed development of salivary gland abnormalities in autoimmune NOD mice. *Clin Immunol Immunopathol.* 1996;79:50–59.
16. Cutler L, Rozenski D, Coolens J. Experimental autoallergic sialadenitis in the LEW rat. *Cell Immunol.* 1991;135:335–345.
17. Dang H, Geiser A, Letterio J. SLE-like autoantibodies and Sjögren's Syndrome-like lymphoproliferation in TGF-B knockout mice. *J Immunol.* 1995;155:3205–3212.
18. Strasser A, Whittingham S, Vaux D. Enforced bcl-2 expression in B-lymphoid cells prolongs antibody responses and elicit autoimmune disease. *Proc Natl Acad Sci USA.* 1993;88:8661–8665.

DO CYTOKINES HAVE A ROLE IN THE REGULATION OF LACRIMAL GLAND ACINAR CELL ION TRANSPORT AND PROTEIN SECRETION?

R. William Lambert

Division of Ophthalmology
Department of Surgery
Southern Illinois University School of Medicine
Springfield, Illinois

1. INTRODUCTION

The role of cytokines in the regulation of immune-related events is well documented. An emerging area of research is the study of cytokine influence beyond these classic immunoregulatory boundaries to the basic physiology of a given tissue. Thus, it has become evident that cytokines elicit phenotypic changes in a variety of non-immune cells, including epithelial cells. In fact, cytokines exert control over a number of epithelial functions, including transmembrane flux of substrate. For example, IFN-γ decreases carbachol-stimulated chloride secretion by T84 cells,[1] and IL-1α contributes to the regulation of renal amino acid transport.[2]

The most common cause of dry eye conditions is insufficient aqueous tear production by the lacrimal glands.[3] In several instances (e.g., Sjögren's syndrome), the underlying cause of lacrimal gland hypofunction is unknown, but is associated with the presence of focal lymphocytic infiltrates adjacent to otherwise normal-appearing secretory epithelial (acinar) cells.[4] This observation raises the possibility that cytokines secreted by the invading immunocytes contribute to the regulation of lacrimal gland physiology in general, and aqueous tear production in particular.

Previous research has demonstrated that cytokines influence the regulation of secretory component production by lacrimal acinar cells.[5] Pilot studies in this laboratory have indicated that certain cytokines may be involved in the regulation of acinar cell Na^+/H^+ exchange. The present study was designed to explore further the potential for cytokine involvement in the regulation of lacrimal acinar cell physiology. Studies initially focused on

Lacrimal Gland, Tear Film, and Dry Eye Syndromes 2
edited by Sullivan *et al.*, Plenum Press, New York, 1998

the Na^+/H^+ exchanger[6] due to its central role in the lacrimal secretory apparatus. Most recently, the potential for cytokine regulation of protein (β-hexosaminidase) secretion has been addressed.

2. MATERIALS AND METHODS

Studies were performed with lacrimal glands from female New Zealand White rabbits, 3–4 kg. Acinar cells were isolated from gland fragments as described previously[7] with slight modifications. Fragments were dissociated using a series of incubations with digestive enzymes (400 U/ml collagenase, 700 U/ml hyaluronidase, and 10 U/ml DNase I in DMEM) and chelation of divalent cations (0.76 mg/ml EDTA in HBSS) at 37°C in a shaking (120 cycles/min) water bath. The resulting digest was filtered sequentially through 500 μm and 20 μm nylon mesh, then centrifuged (50*g*, 5 min). Cells were layered over a Ficoll step gradient (2%, 3%, and 4% in DMEM) and centrifuged (50*g*, 5 min). Yield (two glands yield approximately 2 x 10^8 cells) and viability (~85%) were determined by hemocytometry and trypan blue exclusion rates, respectively. Cells were either immediately processed for 'acute' treatment studies or introduced to culture to allow 'chronic' treatment with cytokines.

Cells were cultured (37°C, 5% CO_2) in 100 mm^2 dishes (typically 1 x 10^7 cells/10 ml media/dish) using a medium consisting of a 1:1 mixture of DMEM and Ham's F-12 Nutrient Mixture, supplemented with penicillin (100 U/ml), streptomycin (0.1 mg/ml), glutamine (2 mM), sodium butyrate (2 mM), linoleic acid (0.084 μg/ml), insulin (5 μg/ml), transferrin (5 μg/ml), and selenium (5 ng/ml).[8] Composition of BSS was: 130 mM NaCl, 6.2 mM KCl, 1.6 mM $MgCl_2$, 1.0 mM $CaCl_2$, 10 mM Tris, 10 mM HEPES, 10 mM glucose, and 0.5% BSA, pH 7.4.

Cytokines (IL-1α [2 ng/ml], IL-1β [10^3 U/ml], IL-6 [10^3 U/ml], and IFN-γ [10^3 U/ml]) (Genzyme) were chosen by matching those identified in lacrimal or salivary glands with those shown to regulate substrate transport in other systems.

To initiate acute treatment, freshly isolated cells were divided among 15 ml tubes, washed 2X with BSS, and resuspended in BSS with or without carbachol (10^{-5} M) and appropriate cytokine or vehicle (BSS + 0.1% BSA).

To initiate chronic treatment, freshly isolated cells were plated at 1 x 10^7 cells/dish (5 dishes/group) in culture media containing appropriate cytokine (or vehicle). Culture continued for at least 24 h before cells were harvested. At termination of culture, gross morphology and cell viability were assessed.

2.1. Functional Assays

2.1.1. β-Hexosaminidase Assay. Freshly isolated or cultured acinar cells were transferred to 15 ml polypropylene tubes, rinsed twice with BSS, and resuspended in 1 ml BSS with or without carbachol (10^{-5} M) and incubated at 4°C or 37°C for 20 min. Reactions were stopped with the addition of 3 ml ice-cold BSS. Cells were pelleted before an aliquot of supernatant was removed from each tube and saved for assay.[9]

2.1.2. Total Protein Assay. A 100 μl aliquot of incubating cells was removed from each tube during the 20 min reaction period and saved for protein determination. These samples were treated with 0.5% SDS (final) prior to assay.[10]

2.1.3. pH_i Measurements/BCECF Fluorescence. Cells were loaded with BCECF (2',7'-bis(carboxyethyl)-5(6)-carboxyfluorescein, Molecular Probes) by incubation with its membrane-permeant acetoxymethyl ester, BCECF-AM (final concentrations 3 μM BCECF-AM and 0.01% DMSO in BSS) for 20 min at 37°C. Following loading, cells were washed three times with BSS, suspended in BSS, and retained at 23°C in a gently shaking water bath until use. BCECF fluorescence was monitored by alternating excitation wavelengths between 440 and 490 nm at 3 sec intervals and recording fluorescence at 530 nm. Fluorescence was calibrated by the K^+-nigericin technique. Unknown pH_i values were determined from a linear regression line obtained by the least-squares fit of calibration data.

3. RESULTS

3.1. Acute Treatment

Previous work[6] in this laboratory demonstrated that acute (i.e., 20 min) treatment of either freshly prepared or cultured acinar cells with cytokines (IL-1α, IFN-γ, or IL-6) did not affect either the carbachol-induced rise in pH_i or amiloride-sensitive recovery from an acid load (i.e., Na^+/H^+ exchange activity) relative to control. To expand these studies, freshly isolated acinar cells were treated to determine whether acute exposure of acinar cells to IFN-γ' IL-1α, or IL-6 resulted in a change in either basal or carbachol-stimulated β-hexosaminidase secretion. In three experiments with 5 replicates/group, none of the cytokines tested consistently altered either basal or carbachol-stimulated β-hexosaminidase secretion (data not shown).

3.2. Chronic Treatment

Since many cytokine-induced effects require at least several hours to become apparent, it was of interest to determine whether longer ('chronic') treatment of acinar cells with cytokines would result in changes in the secretory capacity of acinar cells. To perform these studies, acinar cells were isolated and placed into primary culture for 24 h in the presence or absence of cytokines, as described in Methods. Following culture, cells were harvested and tested for changes in Na^+/H^+ exchange activity or basal and carbachol-stimulated β-hexosaminidase secretion. Chronic treatment with IL-1α or IFN-γ, but not IL-6, increased the initial rate of amiloride-sensitive recovery from an intracellular acid load, suggesting an increase in Na^+/H^+ exchange (not shown). As concerns protein secretion (Fig. 1), chronic treatment with either IL-6 or IFN-γ did not significantly alter either basal or carbachol-stimulated β-hexosaminidase secretion.

Acinar cells were also tested for their response to chronic exposure to IL-1α and IL-1β. Fig. 2 depicts results from one of three representative experiments. Compared to carbachol-stimulated secretion without cytokine treatment (1.86-fold increase over basal), carbachol induced a 1.54-fold and 1.67-fold increase in protein secretion in cells treated with IL-1α and IL-1β, respectively. However, when cells were treated with a combination of IL-1α and IL-1β, the ability of carbachol to stimulate β-hexosaminidase secretion was decreased by approximately 80%, to a 1.19-fold increase. Similar results were obtained in all three experiments such that carbachol stimulated an increase in β-hexosaminidase secretion as follows (mean ± SD): control (1.82 ± 0.17-fold); IL-1α-treated (1.77 ± 0.21-fold); IL-1β-treated (1.63 ± 0.17-fold); and IL-1α- and IL-1β-treated (1.17 ± 0.08-fold).

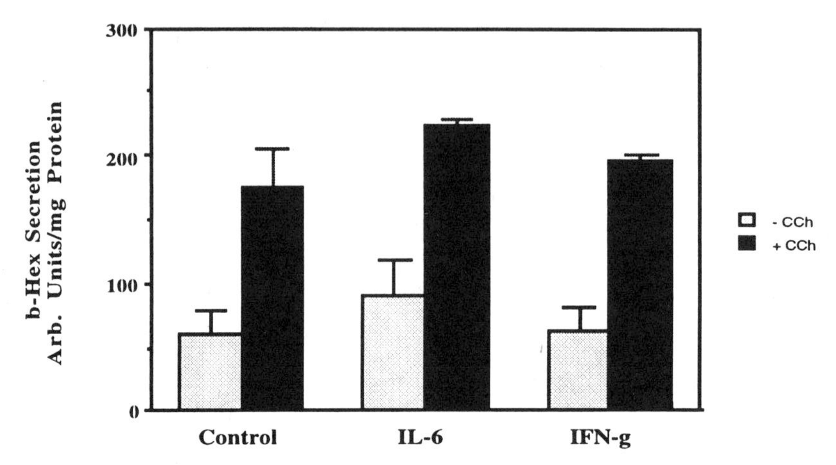

Figure 1. Data from a representative experiment testing whether chronic (i.e., 24 h) treatment of acinar cells with IL-6 or IFN-γ changed either basal or carbachol-stimulated β-hexosaminidase secretion. Each cytokine was tested in three different experiments as described in Methods, with 5 replicates/group/experiment. Values are mean ± SD. There was no significant difference in either basal or carbachol-stimulated protein secretion between the treatment groups.

4. SUMMARY AND CONCLUSION

The present study explored the potential involvement of cytokines (IL-1α, IL-1β, IL-6, and IFN-γ) in the regulation of basic physiological processes carried out by lacrimal acinar cells. Overall, evidence gathered supports the hypothesis that cytokines may be involved in the regulation of lacrimal secretory processes.

When combined with earlier studies,[6] these data suggest that acute treatment (i.e., 20 min) of acinar cells with cytokines does not significantly impact either basal or carbachol-stimulated ion transport or protein secretion. However, chronic treatment of cultured aci-

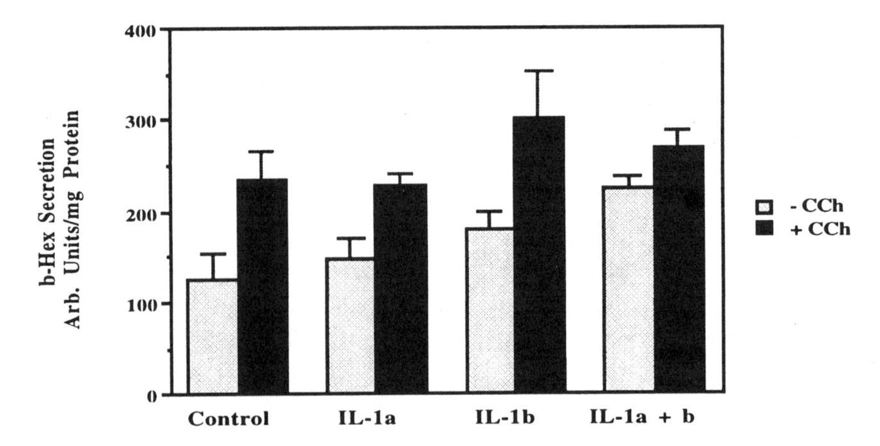

Figure 2. Data from one of three experiments testing whether chronic treatment with IL-1α, IL-1β, or a combination of the two cytokines altered either basal or carbachol-stimulated protein secretion. Data are expressed in relative units of β-hexosaminidase secretion/mg total protein and represent the mean ± SD of 5 replicates/group.

nar cells with cytokines does appear, in some instances, to influence these cell functions. For example, 24 h treatment of acinar cells with IL-6 or IFN-γ did not alter basal or carbachol-stimulated protein (β-hexosaminidase) secretion, whereas a combination of IL-1α and IL-1β decreased carbachol-stimulated β-hexosaminidase secretion by 80%. The mechanism behind this effect is unknown, but it appears to stem from an increase in the basal secretion rather than a decrease in the stimulated secretion. Future studies will attempt to identify the mechanism behind the actions of IL-1α and IL-1β, in addition to testing other pertinent cytokines (e.g., TNF-α, TGF-β).

The study of cytokine involvement in the regulation of lacrimal acinar cell physiology has received relatively little attention. Thus, substantial gaps remain concerning the identity of cytokines, their source(s) and interplay, receptor distribution, second messenger involvement, and ultimate influence over the secretion of aqueous tears.

ACKNOWLEDGMENT

This research was supported by NIH grant EY10563.

REFERENCES

1. Holmgren J, Fryklund J, Larsson H. Gamma-interferon-mediated down-regulation of electrolyte secretion by intestinal epithelial cells: A local immune response. *Scand J Immunol.* 1989;30:499–503.
2. Kohan DE, Schreiner GF. Interleukin 1 modulation of renal epithelial glucose and amino acid transport. *Am J Physiol.* 1988;254:F879-F886.
3. Smolin G, Friedlaender MH, eds. *Dry Eye.* Boston: Little, Brown; 1994.
4. Fox RI, Kang HI. Pathogenesis of Sjögren's syndrome. *Rheum Dis Clin North Am.* 1992;18:517–538.
5. Kelleher RS, Hann LE, Edwards JA, Sullivan DA. Endocrine, neural, and immune control of secretory component output by lacrimal gland acinar cells. *J Immunol.* 1991;146:3405–3412.
6. Lambert RW. Cytokine regulation of Na/H exchange in lacrimal acinar cells of the female rabbit. ARVO Abstracts. *Invest Ophthalmol Vis Sci.* 1995;36:S992.
7. Hann LE, Kelleher RS, Sullivan DA. Influence of culture conditions on the androgen control of secretory component production by acinar cells from the rat lacrimal gland. *Invest Ophthalmol Vis Sci.* 1991;32:2610–2621.
8. Rismondo V, Gierow JP, Lambert RW, Golchini K, Feldon SE, Mircheff AK. Rabbit lacrimal acinar cells in primary culture: Morphology and acute responses to cholinergic stimulation. *Invest Ophthalmol Vis Sci.* 1994;35:1176–1183.
9. Dingle JT, ed. *Lysosomes, a Laboratory Handbook.* Amsterdam: Elsevier; 1977.
10. Lowry OH, Rosebrough NJ, Farr AL, Randall RJ. Protein measurement with the Folin phenol reagent. *J Biol Chem.* 1951;193:265.

THE RAT EXORBITAL LACRIMAL GLAND AS A SITE OF SYNTHESIS OF EGF-LIKE GROWTH FACTORS

Hervé Marechal,[1] Hélène Jammes,[2] Bernard Rossignol,[1] and Philippe Mauduit[1]

[1]Laboratoire de Biochimie des Transports Cellulaires
Centre National de la Recherche Scientifique
Unité de Recherche Associée 1116
Université Paris-Sud, Orsay, France
[2]Unité d'Endocrinologie Moléculaire
Institut National de la Recherche Agronomique
Jouy en Josas, France

1. INTRODUCTION

The lacrimal gland produces the complex aqueous portion of tears that contains many components including electrolytes and proteins.[1,2] The pH, electrolyte concentration, and protein composition of lacrimal fluids are crucial to maintain the health of the ocular surface. The proteins that lacrimal glands synthesize and secrete are very specific, but only a few of them have been characterized. These proteins are thought to be involved in the bacteriostatic action of tears. It has been shown that human[3,4] and mouse[5] tears contain epidermal growth factor (EGF) or an EGF-like immunoreactivity. The EGF precursor mRNA has been detected by Northern blot in mouse[6] and by RT-PCR in human[7] lacrimal glands. Moreover, EGF-like immunoreactivities have been detected in mouse,[6] rabbit,[8] and rat[9,10] lacrimal tissues, but the immunoreactive proteins have not been further characterized. The EGF receptor mRNA has also been detected, by RT-PCR, in the human lacrimal gland,[11] and by RT-PCR and Northern blot in the rat exorbital lacrimal gland.[12] The EGF receptor protein has been identified and shown to be activated by EGF in the rat tissue.[12]

Since the work of Savage and Cohen[13] showing the stimulating effect of EGF on corneal epithelial cell proliferation, the role of EGF in corneal wound healing has been studied extensively. The results indicated that EGF and other growth factors are involved in the stimulation of re-epithelialization processes as well as keratinocyte and corneal en-

Lacrimal Gland, Tear Film, and Dry Eye Syndromes 2
edited by Sullivan *et al.*, Plenum Press, New York, 1998

dothelium proliferation.[14,15] In light of these results, it was suggested that the lacrimal gland may be important in the process of corneal wound healing by producing growth factors secreted in tears.[15] The presence of EGF and/or EGF-related molecules in lacrimal glands of both humans and rodents further suggested that this may be an important property of this tissue. Furthermore, the presence of EGF receptors at least in the rat exorbital lacrimal gland[12] further suggested that EGF produced in the lacrimal gland could also serve a paracrine and/or autocrine function. However, de novo synthesis and secretion of EGF and/or EGF-like molecules need to be more directly demonstrated. The EGF-containing molecules present in the tissue (taking into account that EGF was potentially synthesized as a high-molecular-weight, membrane-associated precursor) also remains to be clearly identified in order to analyze correctly the regulation of the EGF secretory process. So, in view of the above hypothesis, and because the rat lacrimal gland tissue is very easy to obtain compared to its human counterpart, we decided to look for the presence of EGF or EGF-related molecules at both nucleic acid and protein levels in the rat exorbital lacrimal gland. This led us to develop specific cDNA and antibodies that were used to demonstrate the presence of the EGF precursor in this tissue. In the future, these tools would allow us to study the mechanism(s) of regulation of EGF secretion in tears.

2. MATERIALS AND METHODS

2.1. Preparation of Animals and Tissues

Adult male albino Sprague-Dawley rats from IFFA CREDO, France were used throughout this study. Rats were killed by carbon dioxide inhalation. Tissues were removed rapidly and washed extensively to eliminate blood contamination. Then they were fragmented into small pieces and either used directly or frozen and stored at -80°C.

2.2. RNA Extraction

Total RNA isolated from brain, heart, liver, kidney, parotid glands, submaxillary glands, whole lacrimal glands, and lacrimal acinar cells was used for RT-PCR amplification or to purify the poly(A)$^+$ RNA (mRNA) used for Northern blot analysis as previously described.[12]

2.3. RT-PCR

The prepared RNA (5 mg) was first denatured in the presence of 100 pM of random primers and incubated for reverse transcription (RT) as described previously.[12] The PCR reaction, using specific primers derived from the rat EGF, transforming growth factor alpha (TGFa), and heparin binding-EGF (HB-EGF) sequences, was used to amplify the cDNA produced by the RT reaction. For each couple of primers, 30 cycles of amplification were performed, and amplified cDNA was analyzed by agarose gel electrophoresis as previously described.[12]

2.4. Northern Blot

Poly(A)$^+$ RNA (20 mg) from different rat tissues was analyzed by Northern blot to detect the presence of the EGF transcript. The EGF probe used was the 617 base-pair (bp) cDNA fragment amplified from submaxillary gland RNA. Probe labelling and Northern blot analysis were performed as previously described.[12]

2.5. Radioreceptor Assay and Radioimmunoassay

To quantify EGF-like molecules in submaxillary and lacrimal glands as well as in kidney, 1 g of tissue was homogenized at 4°C in 8 volumes of 10 mM phosphate - 250 mM sucrose buffer (pH 7.4). The homogenate was centrifuged (40,000g, 20 min). The supernatant (cytosol) was used as a source of soluble growth factors, and the pellet was solubilized in 1% triton X100 containing buffer and used as a source of membrane-associated growth factor. Radioreceptor assay (RRA) was performed using either ^{125}I-mouse EGF (mEGF) or ^{125}I-rat EGF (rEGF) as radioligands and rat liver membrane fraction as a source of EGF receptors. Bound and free radioligands were separated by rapid filtration through glass fiber filter as described previously.[16] EGF-like activity was measured by comparing the competing activity of an aliquot of the soluble fraction from the different tissues with authentic mEGF or rEGF. Radioimmunoassay (RIA) was performed using ^{125}I-rEGF as radioligand and a polyclonal rabbit anti-rEGF antibody developed in our laboratory as antibody. Immunoreactive rEGF (irEGF) in both cytosolic (submaxillary gland, lacrimal gland, and kidney) and detergent solubilized (lacrimal gland and kidney) membrane fractions were quantified both before and after trypsin hydrolysis as described previously.[17] After overnight incubation at 4°C, antigen-antibody complexes were precipitated in the presence of 10% polyethyleneglycol (PEG). irEGF in the different fractions was estimated by comparison with a competition curve obtained with authentic rEGF.

3. RESULTS AND DISCUSSION

Fig. 1 shows the RT-PCR amplification of the total RNA fractions from different rat tissues. Sequences of the specific sense and antisense primers for the amplification of EGF, TGFa, and HB-EGF mRNA sequences were deduced from previously published sequences (Genbank data). Sizes of the predicted amplification products were 617, 779, and 647 bp, respectively, for EGF, TGFa, and HB-EGF cDNA. As could be anticipated from previous reports, the strongest amplification signal for EGF was observed in the submaxillary gland and kidney. However, both preparations of lacrimal gland, that is, RNA prepared from either whole gland or acinar cells, were strongly positive. Parotid gland and liver showed lower amplification signals, and brain and heart were negative. Amplification of TGFa mRNA was observed in similar amounts in all tissues tested except heart. The strongest HB-EGF mRNA amplifications were observed in lacrimal cells, kidney, brain, and heart; other tissues were only faintly positive. The specificity and identity of the different amplification products were tested by restriction mapping analysis and by Southern blotting using a mouse cDNA probe for EGF (data not shown).

The results concerning the presence of EGF and TGFa mRNA in the rat lacrimal gland tissue agree with those previously reported in the literature. EGF mRNA has been demonstrated in human[7] and mouse[6] tissues and TGFa mRNA in rat tissue.[18] To our knowledge, this work reports for the first time the presence of HB-EGF mRNA in a lacrimal gland, but we cannot explain the amplification difference observed between the whole gland and the acinar cell RNA preparations.

The 617 bp EGF cDNA amplified from rat submaxillary gland was purified and used as a probe for a Northern blot analysis. As shown in Fig. 2, a high amount of 5 kb

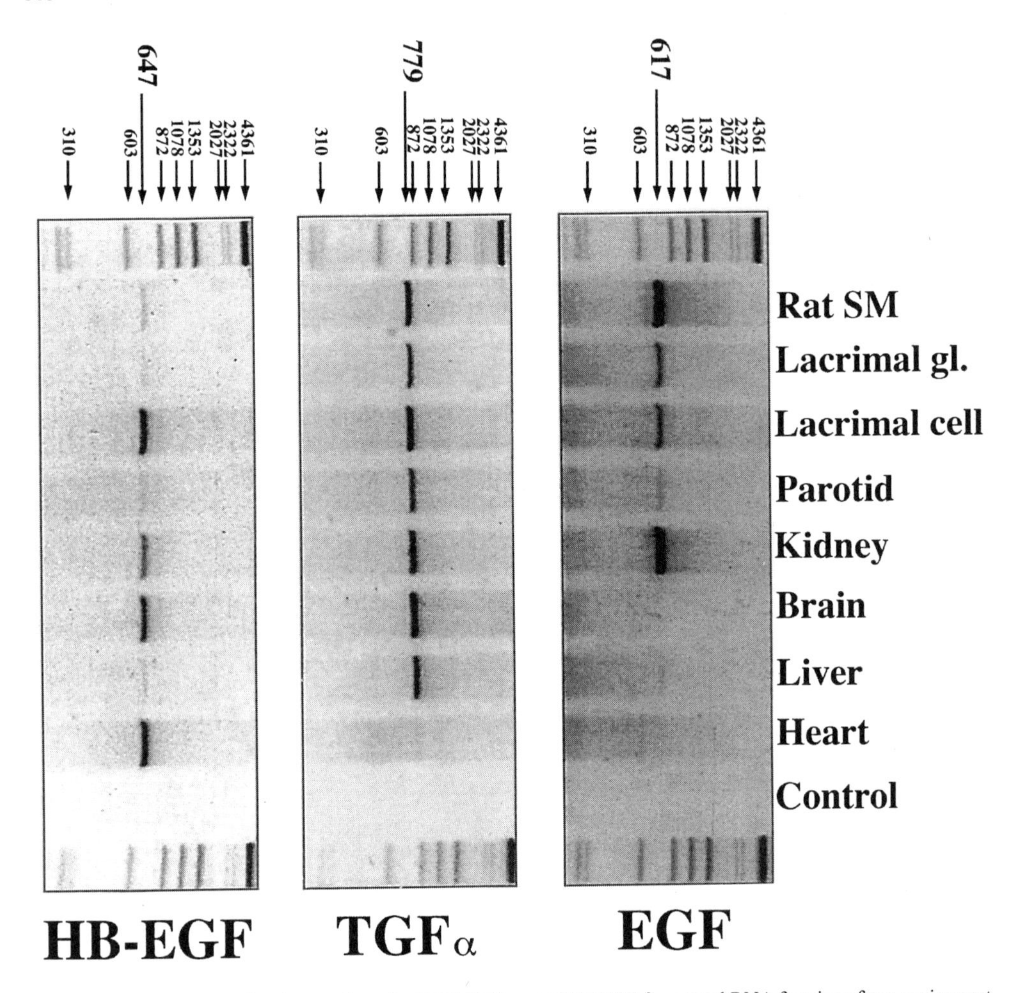

Figure 1. RT-PCR amplification products for EGF, TGFa, and HB-EGF from total RNA fractions from various rat tissues. SM, submaxillary gland. Lacrimal gl., whole lacrimal gland. Control incubations were performed in the absence of any RNA.

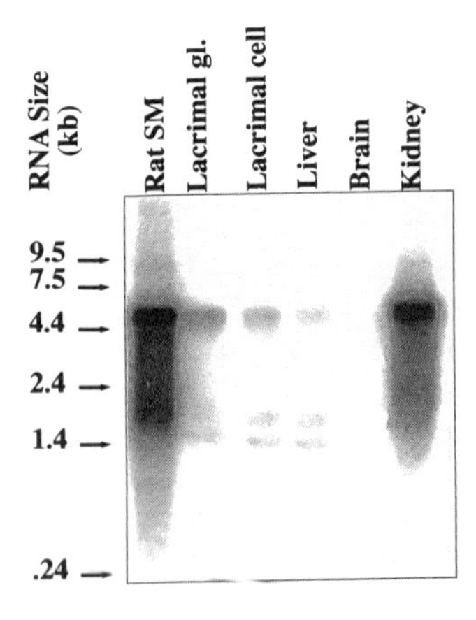

Figure 2. Northern blot analysis of the EGF precursor transcript in the polyA$^+$ RNA fraction from different rat tissues. SM, submaxillary gland. Lacrimal gl., whole lacrimal gland.

transcript was observed in both submaxillary gland and kidney. This transcript was also observed in highly significant but lower amount in mRNA preparations from lacrimal gland and acinar cells. Liver was slightly positive; brain was negative. The size of the observed transcript is sufficient to encode the full-length EGF precursor (preproEGF, or ppEGF) (open reading frame 3402 pb for a known sequence of at least 4801 pb).[19] Thus, the results of the Northern blot analysis taken together with those of the RT-PCR experiments indicate that the rat submaxillary gland and kidney, like mouse tissues, express the highest amount of preproEGF mRNA, and that the lacrimal gland contains relatively high amounts of this transcript.

The next experiments were designed to identify EGF or EGF-containing protein (ppEGF) in the lacrimal gland tissue. RRA was first used to test for the presence of soluble EGF receptor (EGFR) binding proteins. Seven-week-old rats were used for this study. As shown in Table 1, the submaxillary gland, as could be predicted, contains the highest amount of activity (5700 pg of rEGF equivalent per mg of tissue). Kidney and lacrimal gland contain similar amounts of rEGF equivalent but at least 15 times less than submaxillary gland. RRA did not discriminate among all the EGFR binding proteins. Thus the amount of activity measured in this way could not be solely attributed to EGF. We thus developed a specific antibody directed against rEGF and used it for RIA experiments. This antibody was specific for rEGF and shows 100 times less affinity for mEGF (80% identity with rEGF) and did not bind human EGF. The irEGF was measured in the soluble fraction from all three tissues. As shown in Table 1, in the absence of trypsin, the amount of irEGF in submaxillary gland was nearly identical to the amount of rEGF equivalent determined by RRA. However, irEGF was undetectable in the lacrimal gland and very low in kidney. This indicates that the EGFR binding activity measured in these fractions might be due to some other growth factor of the EGF family (TGFa, for example, in the lacrimal gland[18] or HB-EGF, as suggested from the RT-PCR experiments).

Since trypsin hydrolysis was proposed as a convenient way to release soluble EGF from its high-molecular-weight precursor,[17] we used this approach to evaluate the presence of ppEGF in both lacrimal gland and kidney. Trypsin treatment did not affect the amount of irEGF in lacrimal gland cytosol, whereas it greatly increased the irEGF in the kidney fraction (X8) (Table 1). This result suggests that kidney cytosol contains some sol-

Table 1. Quantification by radioreceptor assay (RRA) and radioimmunoassay (RIA) of EGF-like activities in cytosolic and membrane fractions from rat tissues

		RIA	
Tissue	RRA (pg/mg)	−Trypsin (pg/mg)	+Trypsin (pg/mg)
Submaxillary gland	5700	5200	–
Lacrimal gland			
Cytosol	200	ND	ND
Membrane	–	1.6	25.7
Kidney			
Cytosol	340	2.9	23
Membrane	–	9.5	110

Results are expressed as μg equivalent rEGF (RRA) or immunoreactive rEGF (RIA) per mg wet weight. –, experiment can not be performed. ND, not detectable.

uble high-molecular-weight EGF-containing molecules that are not efficiently recognized by the antibody until they are subjected to trypsin hydrolysis. In contrast to cytosol, the solubilized membrane fraction from both lacrimal gland and kidney appeared to contain an irEGF activity that is greatly increased upon trypsin treatment. Thus, our results are in very good agreement with the literature concerning the presence of a high-molecular-weight and membrane-associated EGF precursor in rat kidney.[17] We also demonstrate for the first time that this precursor is also present in the membrane fraction from rat lacrimal gland. Taken together with the absence of detectable soluble irEGF in this tissue, we suggest that tears EGF may originate, at least partly, from the proteolytic (regulated ?) cleavage of this membrane-associated precursor.

REFERENCES

1. Alexander JH, van Lennep EW, Young JA. Water and electrolyte secretion by exorbital gland of the rat studied by micropuncture and catheterization techniques. *Pflugers Arch.* 1972;337:229–309.
2. Bothelo SY. Tears and the lacrimal gland. *Sci Am.* 1964;311:78–86.
3. Ohashi Y, Motokura M, Kinoshita Y, et al. Presence of epidermal growth factor in human tears. *Invest Ophthalmol Vis Sci.* 1989;30:1879–1882.
4. Van Setten GB, Vinikka L, Tervo T, Pesonen K, Tarkkanen A, Perheentupa J. Epidermal growth factor is a constant component of normal human tear fluid. *Graefes Arch Clin Exp Ophthalmol.* 1989;227:184–187.
5. Tsutsumi O, Tsutsumi A, Oka T. Epidermal growth factor-like, corneal wound healing substance in mouse tears. *J Clin Invest.* 1988;81:1067–1071.
6. Kasayama S, Ohba Y, Oka T. Expression of epidermal growth factor gene in mouse lachrymal gland: Comparison with that in the submandibular gland and kidney. *J Mol Endocrinol.* 1990;4:31–36.
7. Wilson SE, Lloyd SA. Epidermal growth factor messenger RNA production in human lacrimal gland. *Cornea.* 1991;10:519–524.
8. Steineman TL, Thompson HW, Maroney KM, et al. Changes in epithelial epidermal growth factor receptor and lacrimal gland EGF concentration after corneal wounding. ARVO Abstracts. *Invest Ophthalmol Vis Sci.* 1990;31:S55.
9. Van Setten GB, Tervo K, Virtanen I, Tarkkanen A, Tervo T. Immunohistochemical demonstration of epidermal growth factor in the lacrimal and submandibular glands of rats. *Acta Ophthalmol.* 1990;68:477–480.
10. Watanabe H, Ohashi Y, Kinoshita S, Manabe R, Ohshiden K. Distribution of epidermal growth factor in rat ocular and periocular tissues. *Graefes Arch Clin Exp Ophthalmol.* 1993;231:228–232.
11. Wilson SE, Lloyd SA. Epidermal growth factor and its receptor, basic fibroblast growth factor, transforming growth factor beta 1, and interleukin-1 alpha messenger RNA production in human corneal endothelial cells. *Invest Ophthalmol Vis Sci.* 1991;32:2747–2756.
12. Marechal H, Jammes H, Rossignol B, Mauduit P. EGF receptor mRNA and protein in rat lacrimal acinar cells: Evidence of its EGF-dependent phosphotyrosilation. *Am J Physiol.* 1996;270:C1164-C1174.
13. Savage CR, Cohen S. Proliferation of corneal epithelium induced by epidermal growth factor. *Exp Eye Res.* 1973;15:361–366.
14. Burstein NL. Review: Growth factor effects on corneal wound healing. *J Ocul Pharmacol.* 1987;3:263–277.
15. Wilson SE. Lacrimal gland epidermal growth factor production and the ocular surface. *Am J Ophthalmol.* 1991:111:763–765.
16. Mauduit P, Jammes H, Rossignol B. M_3 muscarinic acetylcholine receptor coupling to PLC in rat exorbital lacrimal acinar cells. *Am J Physiol.* 1993;264:C1550-C1560.
17. Schaudies RP, Nonclercq D, Nelson L, et al. Endogenous EGF as a potential renotrophic factor in ischemia-induced acute renal failure. *Am J Physiol.* 1993;265:F425-F434.
18. Van Setten GB, Macauley S, Humphrey-Beher M, Chegini N, Schultz G. Detection of transforming growth factor-a mRNA and protein in rat lacrimal glands and characterization of transforming growth factor-a in human tears. *Invest Ophthalmol Vis Sci.* 1996;37:166–173.
19. Saggi SJ, Safirstein R, Price PM. Cloning and sequencing of the rat preproepidermal growth factor cDNA: Comparison with mouse and human sequences. *DNA Cell Biol.* 1992;11:481–487.

71

THE DISTRIBUTION OF FGF-2 AND TGF-β WITHIN THE LACRIMAL GLAND OF RABBITS

Joel Schechter, Dwight W. Warren, and Richard L. Wood

Department of Cell and Neurobiology
University of Southern California School of Medicine
Los Angeles, California

1. INTRODUCTION

Growth factors have diverse functions and are distributed throughout the body. A number have been detected in the lacrimal gland, eg, EGF,[1,2] FGF-2,[3] TGF-α,[4] TGF-β,[5] and HGF.[6] In many tissues, FGF-2 and TGF-β have an inverse functional relationship. FGF-2, released in response to tissue damage, activates proteases that function in degradation of extracellular matrix molecules. TGF-β, released during later stages of tissue repair, inhibits proteases, thereby stabilizing extracellular matrix molecules. Although both FGF-2 and TGF-β are present within the lacrimal gland, their sites of synthesis and localization have not been documented. FGF-2 and TGF-β also have been shown to exert inverse functional effects on the synthesis and release of PRL, an immunomodulating hormone that we have localized within the lacrimal gland.[7] The present study was undertaken to determine the sites of localization of FGF-2 and TGF-β within the lacrimal gland, as a basis for elucidating their functional roles in the lacrimal gland's response to corneal injury or other initiators of lacrimal gland inflammation.

2. MATERIALS AND METHODS

2.1. Light Microscopic Immunohistochemistry

New Zealand White rabbits, male and female, juvenile and adult, were narcotized with a mixture of ketamine and xylazine and given an overdose of sodium pentobarbital for euthanasia. Inferior lacrimal glands were removed, and tissue fragments were placed directly in OCT (Miles, Elkhart, IN) and rapidly frozen with liquid nitrogen. Cryostat sections were cut at 6–8 μm in thickness. Sections were affixed to glass slides and fixed briefly with acetone or 4% formaldehyde; non-specific IgG binding was blocked with 5% BSA in 0.05 M Tris buffer, pH 7.4. The primary antibodies were monoclonals: a-hFGF-2 (Sigma) and a-hTGF-ß (Chemicon). Endogenous peroxidase activity was blocked by treatment with 0.2% H_2O_2. Sections were exposed to second antibody (goat anti-mouse IgG) conjugated with biotin. Slides were then reacted with ABC reagent (Vector, Burlingame, CA) and diaminobenzidine, and mounted with Immumount (Shandon/Lipshaw, Pittsburgh, PA). Negative controls consisted of exposure to non-immune mouse IgG, to second antibody alone, or to ABC reagent alone. Sections of rabbit anterior pituitary gland served as positive controls.

2.2. Corneal Adenovirus Inoculation

Viral inoculation was done by injection of 10 µl aliquots of live virus into five corneal sites, followed by scratching the corneal surface at each injection site and exposure to 50.0 µl of additional virus-containing media.[8]

2.3. Collection of Lacrimal Fluid

We have devised a method for collection of lacrimal fluid[9] by cannulating the lacrimal duct with small-bore polyethylene centrifuge tubing.

2.4. Polyacrylamide Gel Electrophoresis (SDS-Page) and Immunoblotting

Frozen, unfixed lacrimal tissues (intact or isolated lacrimal cells) were homogenized and then centrifuged at low speed to remove large fragments. The suspension was subjected to SDS-PAGE. Lacrimal fluid was used directly. The gels were stained with Coomassie blue under reducing and non-reducing conditions, or the proteins transferred to nitrocellulose paper, blocked with 5% BSA, and exposed overnight to antibodies. Immunobinding was visualized with peroxidase-labeled secondary antibodies and enhanced chemiluminescence (ECL). Lanes loaded with pituitary gland and the appropriate factor of interest were used as positive controls. Negative controls consisted of absence of primary antibody.

3. RESULTS

Sections immunostained for FGF-2 revealed dramatic immunopositivity in a basally located population of cells in both ducts and acinar epithelium, in some interstitial cells, and in basement membrane-associated extracellular matrix at the periphery of acini and ducts. Matching sections immunostained for TGF-β revealed staining of blood vessels and scattered cells within the interstitium, but no evidence of immunopositivity within either acinar or duct epithelium. Western blots from lacrimal gland specimens—homogenates of lacrimal gland membrane fractions of males and females, isolated and cultured acinar cells, and secretory fluid collected from lacrimal ducts of females—demonstrated that under reducing conditions each of these samples had a broad protein band centered at 7.5 kDa that was strongly reactive for FGF-2, plus minor bands at 18 and 24 kDa. Reactivity for TGF-β was located in protein bands running at 10–12 kDa in homogenates from intact glands (males and females), but no reactivity was detected in samples from isolated lacrimal cells or lacrimal fluid.

Immunopositivity for FGF-2 changed after corneal inoculation with adenovirus (21 days). Immunopositivity for FGF-2 was dramatically decreased in basal cells of the smaller ducts and acini and from basement membranes. Twenty-one days after corneal inoculation with adenovirus, immunostaining for TGF-β was comparable to that described in normal rabbits, ie, scattered immunopositive cells within the interstitium. The intensity of staining for FGF-2 was not changed in Western blotting after viral challenge, indicating a similar overall content of FGF-2 despite the difference in distribution detected by immunocytochemistry. There was a modest decrease in intensity of the TGF-β protein bands after viral challenge.

4. DISCUSSION

Our study has demonstrated that FGF-2 is localized in consistent sites within the lacrimal glands of normal male and female rabbits, and is also present in lacrimal gland secretory fluid. Basal cells immunopositive for FGF-2 within both duct and acinar epithelia were frequently elongate, suggesting the possibility that they could represent myoepithelial cells. However, our ultrastructural studies of rabbit lacrimal gland have failed to detect myoepithelial cells in this species. Immunopositivity in interstitial cells appears likely to be macrophages, a cell known to be a source of FGF-2 in other tissues.[10] It is also now well established that FGF-2 frequently is complexed in association with basement membrane molecules.[11–13] Complexed in this location, it is generally considered to be inactive with respect to its role as a protease activator, angiogenic factor, or stimulus of wound repair or inflammatory response. FGF-2 is reported to have a molecular weight of 16.5–24 kDa or higher. The broad band at 7.5 kDa under reducing conditions presumably represents fragments of the FGF-2 molecule. The monomer form of TGF-β runs at a molecular weight of 12.5 kDa and the dimer form at 25 kDa.

Adenovirus injection into the cornea elicits an inflammatory response in the lacrimal gland.[14] Our data, both immunohistochemical and Western blots, prompt us to conclude that with respect to the lacrimal gland's response to this corneal insult, ducts and acini are releasing FGF-2 basally and possibly apically. FGF-2 would then be acting as a trophic hormone within the tear fluid (apical release) for corneal repair, and at the local, *in situ* level (basal release) where it likely plays a multi-functional role—as an angiogenic factor, a protease acivator, a stimulus for PRL release, etc.

REFERENCES

1. Ohashi Y, Motokura M, Kinoshita Y, et al. Presence of epidermal growth factor in tears. *Invest Ophthalmol Vis Sci.* 1989;30:1879–1882.
2. van Setten GB, McCauley S, Humphreys-Beher M, Chegini N, Schultz G. Detection of transforming growth factor-α mRNA and protein in rat lacrimal glands and characterization of transforming growth factor-α in human tears. *Invest Ophthalmol Vis Sci.* 1996;37:166–173.
3. Wilson SE, Lloyd SA, Kennedy RH. Basic fibroblast growth factor (FGFb) and epidermal growth factor (EGF) messenger RNA production in human lacrimal gland. *Invest Ophthalmol Vis Sci.* 1991;32:2816–2820.
4. van Setten GB, Tervo T, Virtanen I, Trakkanen A, Tervo TT. Immunohisto-chemical demonstration of epidermal growth factor in the lacrimal and submandibular glands of rats. *Acta Ophthalmol.* 1990;68:477–480.
5. Pflugfelder SC. Capturing the essence of tears - dry eye. New York: Science Writers Eye Research; 1995:32–33.

6. Li Q, Weng J, Mohan RJ, et al. Hepatocyte growth factor (HGF) and HGF receptor in lacrimal gland, tears, and cornea. *Invest Ophthalmol Vis Sci.* 1996;37:727–739.
7. Wood RL, Park KH, Gierow JP, Mircheff AK. Immunogold localization in acinar cells of lacrimal gland. *Adv Exp Med Biol.* 1994;350:75–77.
8. Trousdale MD, Nobrega R, Wood RL, Stevenson D, Klein D, McDonnell PJ. Studies of adenovirus-induced eye disease in the rabbit. *Invest Ophthalmol Vis Sci.* 1995;36:2740–2748.
9. Takashima Y, Takagi H, Reinach PS, Mircheff AK, Warren DW, Yoshimura N. Detection of endothelin-1 in rat tears and its localization in ocular glands. *Curr Eye Res.* 1996;15:768–773.
10. Baird A, Mormede P, Ying SY, et al. A nonmitogenic pituitary function of fibroblast growth factor: Regulation of thyrotropin and prolactin secretion. *Proc Natl Acad Sci USA.* 1985;82:5545–5549.
11. Baird A, Ling N. Fibroblast growth factors are present in the extracellular matrix produced by endothelial cells *in vitro:* Implications for a role of heparinase-like enzymes in the neovascular response. *Biochem Biophys Res Commun.* 1987;142:428–435.
13. Folkman J, Klagsbrun M, Sasse J, Wadzinski M, Ingber D, Vlodavsky I. A heparin-binding angiogenic protein, basic fibroblast growth factor, is stored within basement membrane. *Am J Pathol.* 1988;130:393–400.
14. Wood RL, Trousdale MD, Stevenson D, Azzarolo AM, Mircheff AK. Adenovirus infection of cornea causes histopathologic changes in the lacrimal gland. *Curr Eye Res.* 1997;16:459–466.

TEAR FLUID CHANGES AFTER PHOTOREFRACTIVE KERATECTOMY

Minna Vesaluoma and Timo Tervo

Department of Ophthalmology
Helsinki University Central Hospital
Helsinki, Finland

1. INTRODUCTION

Correction of refractive errors by excimer laser photorefractive keratectomy (PRK) is based on remodelling of the corneal curvature by highly controlled photoablation of the anterior corneal stroma.[1] Tear fluid cytokines or other modulators produced by inflammatory cells, nerve cells, corneal epithelial or stromal cells, or extracellular matrix are likely to regulate the healing of the photoablated area.[2–9] As the healing of the wound determines the refractive result, it is most important to study the mechanisms that control the healing response.

2. CORNEAL HEALING

2.1. Epithelial Healing and Plasmin Activity

The epithelialization of the wounded area is initiated by migration of epithelial cells.[10] The migrating epithelial cells of the leading edge probably utilize the plasminogen-plasmin system for directional proteolysis.[11] PRK induces increased release of plasmin (μIU/min) in tear fluid (Fig. 1), and the release is elevated as long as there is an epithelial defect (2–3 days).[2] Plasmin activity per volume of tears (IU/l), however, decreases during the first postoperative days because of reflex tearing (Figs. 2 and 3). This probably prevents excessive proteolysis at the wound surface.

2.2. Stromal Healing, Fibronectin, and Tenascin

PRK induces keratocyte degeneration in the superficial stroma, presumably due to apoptosis,[12] followed by migration of the remaining keratocytes to the site of the injury.

Lacrimal Gland, Tear Film, and Dry Eye Syndromes 2
edited by Sullivan *et al.*, Plenum Press, New York, 1998

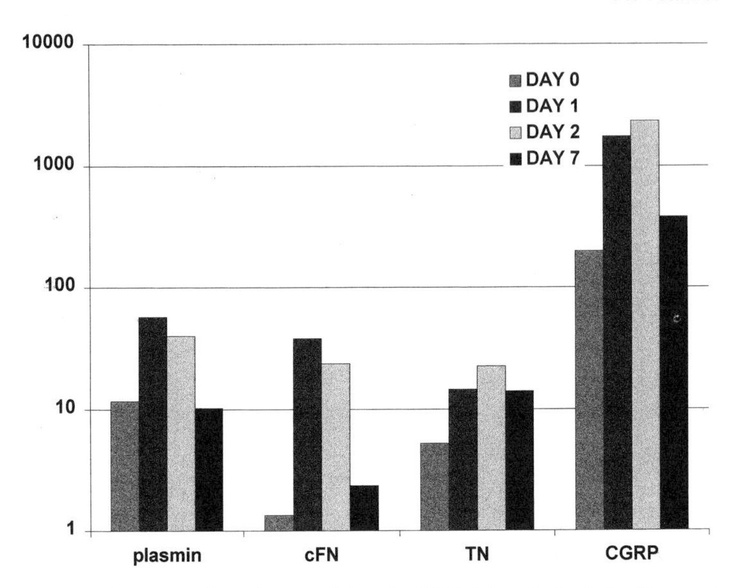

Figure 1. Mean pre- and postoperative releases of plasmin (μIU/min) and of cellular fibronectin (cFN), tenascin (TN), and calcitonin gene-related peptide (CGRP) (ng/min).

Then keratocytes turn into activated myofibroblast-like contractile cells and start producing new extracellular matrix components, such as collagen, hyaluronic acid, proteoglycans, tenascin, and cellular fibronectin.[13–16] Fibronectin serves as an adherent surface to the migrating epithelial cells during the early stages of healing. Tenascin, another extracellular matrix protein, is presumed to be involved in cell proliferation and migration, or in modulation of interactions between cells and fibronectin matrix.[17] The synthesis of both fibronectin and tenascin is enhanced during the first 3 months after wounding and

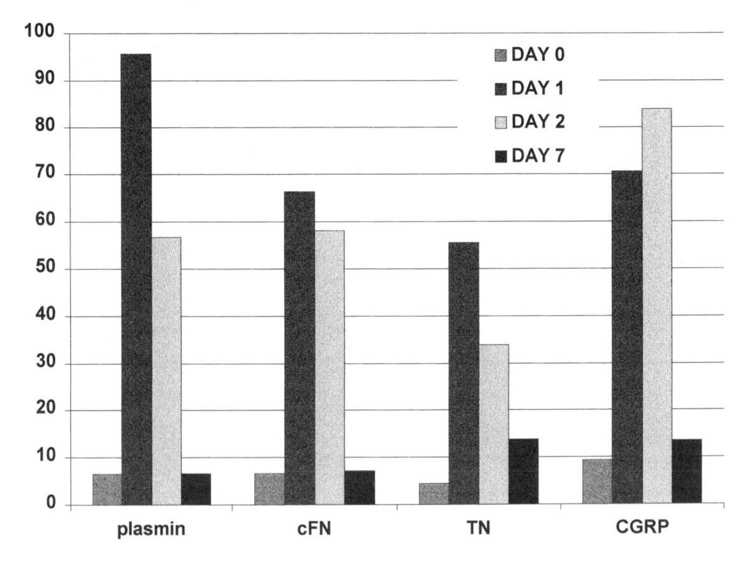

Figure 2. Mean pre- and postoperative flows in the collection capillary (μl/min). cFN, cellular fibronectin. TN, tenascin. CGRP, calcitonin gene-related peptide.

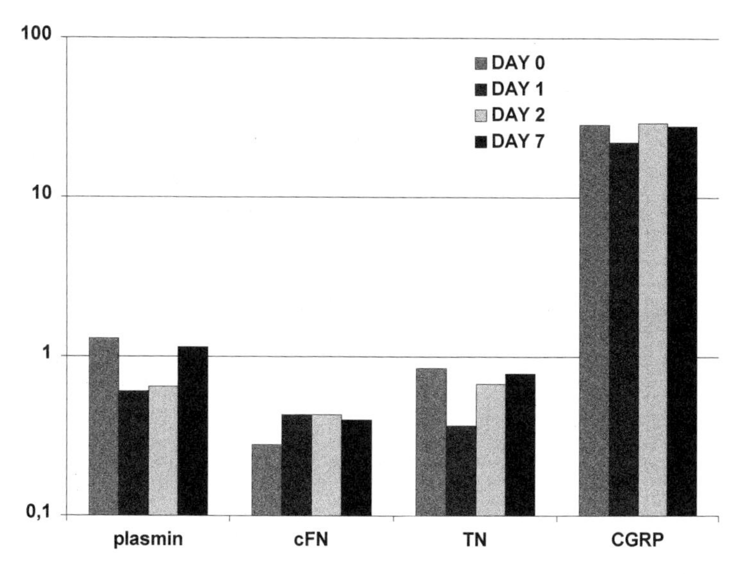

Figure 3. Mean pre- and postoperative activities of plasmin (IU/l) and concentrations of cellular fibronectin (cFN), tenascin (TN), and calcitonin gene-related peptide (CGRP) (μg/ml).

then declines slowly.[13] Following PRK, the release of both cellular fibronectin and tenascin in tears is highest when the epithelial defect is present (Fig. 1).[3,4]

2.3. Corneal Nerves and Calcitonin Gene-Related Peptide (CGRP)

PRK ablates epithelial and anterior stromal nerves.[18] Sensitivity returns in 3–4 weeks, but morphologically the epithelial innervation is regenerated in about 3 months. However, the subepithelial nerve plexus and the anterior stromal nerves do not seem to regenerate completely. Release of CGRP in tears is enhanced after PRK, but once the corneal surface is reepithelialized, CGRP returns to the preoperative level (Fig. 1).[5] Despite the marked hypersecretion of tears, CGRP concentration does not decrease after PRK, suggesting a concomitant increase in CGRP release by corneal sensory nerves and/or lacrimal glands (Figs. 2 and 3).

3. GROWTH FACTORS

3.1. Tear Fluid Growth Factors following PRK

Human corneal epithelial and stromal cells as well as the lacrimal gland synthesize a number of growth factors or their receptors.[19,20] Several cytokines have been shown to be present in tear fluid.[21–26] The release of HGF, TGF-β1, vasoendothelial growth factor (VEGF), platelet-derived growth factor-BB (PDGF-BB), and TNF-α in human tear fluid increases significantly following an excimer laser-induced corneal wound (Fig. 4).[6–9] The release of each cytokine returns to the preoperative level by the seventh postoperative day, by which time the epithelial defect has healed. PDGF-BB could be measured in preoperative tears of only 17% of the patients, whereas other growth factors are constantly present. PDGF-BB concentration, however, increases significantly on the second postoperative

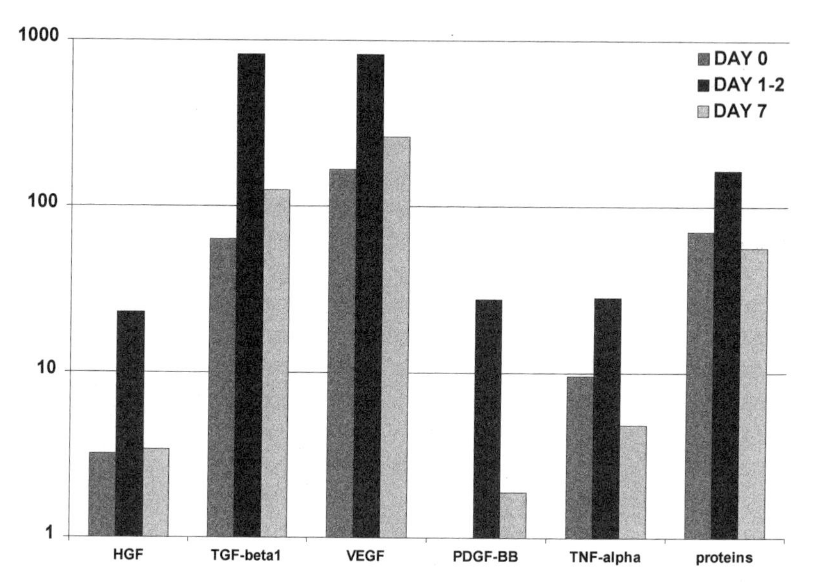

Figure 4. Mean pre- and postoperative releases of HGF, TGF-ß1, VEGF, PDGF-BB, TNF-α (pg/min) and of total proteins (μg/min).

day, and all patients show measurable concentrations. The concentrations of TGF-β1 and TNF-α, on the other hand, remain on a constant level, while those of HGF and VEGF decrease during the first two postoperative days, possibly due to hypersecretion of tears (Figs. 5 and 6). The VEGF concentration returns to the preoperative level by the seventh postoperative day, but the HGF level is still lower than preoperatively.

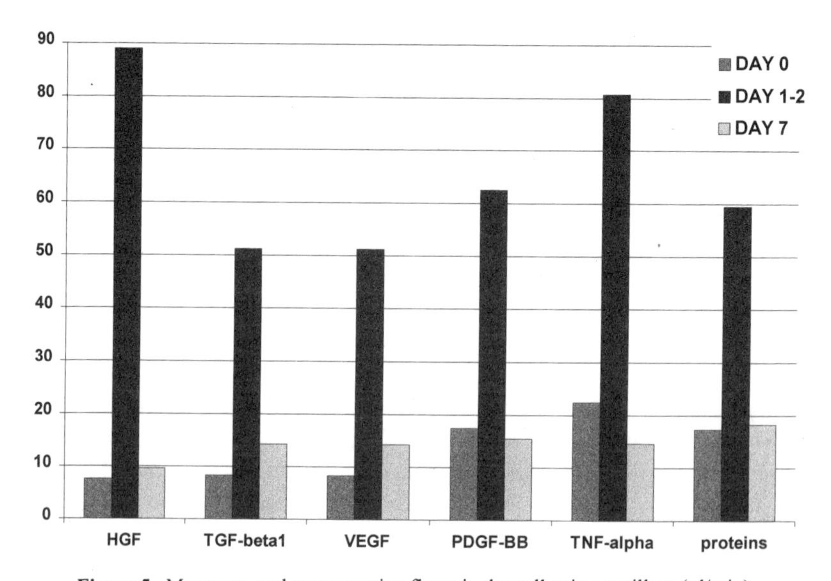

Figure 5. Mean pre- and postoperative flows in the collection capillary (μl/min).

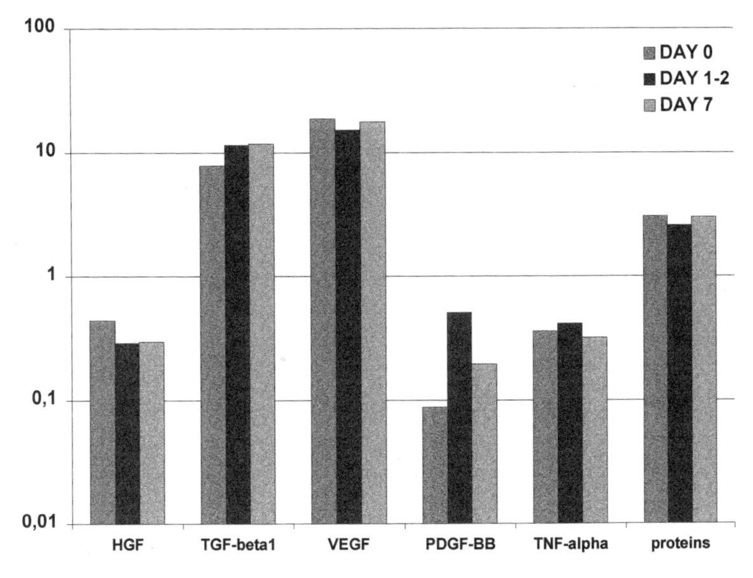

Figure 6. Mean pre- and postoperative concentrations of HGF, TGF-β1, VEGF, PDGF-BB, TNF-α (ng/ml) and of total proteins (mg/ml).

3.2. Potential Functions of Growth Factors in Corneal Wound Healing

HGF is a classical paracrine mediator produced by fibroblasts to modulate the functions of epithelial cells. It stimulates proliferation and motility, but inhibits differentiation of corneal epithelial cells.[27] TGF-ß, on the contrary, inhibits the proliferation of corneal epithelium, but enhances the formation of stromal extracellular matrix.[28,29] Furthermore, it is a very potent chemoattractant for human corneal epithelial cells and stromal fibroblasts.[30] On the other hand, it inhibits extracellular matrix degradation.[31] It also appears to stimulate myofibroblast transformation of stromal keratocytes.[32] VEGF mRNA has been found in hyperproliferative healing cutaneous epithelium, suggesting a role in wound healing.[33] The presence of VEGF in normal tears, as well as the presence of mRNA for VEGF and its receptor FLT-1 in cultured keratocytes, suggests a regulatory role in the stroma.[7,34] Li and Tseng suggest that PDGF-BB is produced by corneal epithelial cells to modulate the action of stromal fibroblasts.[20] It might, however, facilitate epithelial hyperplasia and migration during wound healing through induction of HGF.[35] PDGF stimulates healing of rabbit corneal scrape wound *in vivo*.[36] The potential stromal events stimulated by PDGF-BB during corneal wound healing include, eg, fibroblast chemotaxis, fibroblast proliferation, and extracellular matrix synthesis. TNF-α has been shown to stimulate fibronectin-induced migration of corneal epithelial cells and fibroblast proliferation.[37,38] It also inhibits extracellular matrix protein production and stimulates collagenase or prostaglandin E2 secretion.[39,40]

ACKNOWLEDGMENT

The financial support by Suomen Silmäsäätiö is gratefully acknowledged.

REFERENCES

1. Trokel SL, Srinivasan R, Braren B. Excimer laser surgery of the cornea. *Am J Ophthalmol.* 1983;96:710–715.
2. Tervo T, Virtanen T, Honkanen N, Härkönen M, Tarkkanen A. Tear fluid plasmin activity after excimer laser photorefractive keratectomy. *Invest Ophthalmol Vis Sci.* 1994;35:3045–3050.
3. Virtanen T, Ylätupa S, Mertaniemi P, Partanen P, Tuunanen T, Tervo T. Tear fluid cellular fibronectin levels after photorefractive keratectomy. *J Refract Surg.* 1995;11:106–112.
4. Vesaluoma M, Ylätupa S, Mertaniemi P, Tervo K, Partanen P, Tervo T. Increased release of tenascin in tear fluid after photorefractive keratectomy. *Graefes Arch Clin Exp Ophthalmol.* 1995;233:479–483.
5. Mertaniemi P, Ylätupa S, Partanen P, Tervo T. Increased release of immunoreactive calcitonin gene-related peptide (CGRP) in tears after excimer laser keratectomy. *Exp Eye Res.* 1995;60:659–665.
6. Tervo T, Vesaluoma M, Bennett GL, et al. Tear hepatocyte growth factor (HGF) production increases markedly after excimer laser surface ablation. *Exp Eye Res.*, 1997, in press.
7. Vesaluoma M, Teppo A-M, Grönhagen-Riska C, Tervo T. Release of TGF-β1 and VEGF in tear fluid following photorefractive keratectomy. *Curr Eye Res.* 1997;16:19–25.
8. Vesaluoma M, Teppo A-M, Grönhagen-Riska C, Tervo T. Increased release of tumor necrosis factor-α in human tear fluid after excimer laser-induced corneal wound. *Br J Ophthalmol.* 1997;81:145–149.
9. Vesaluoma M, Teppo A-M, Grönhagen-Riska C, Tervo T. Platelet-derived growth factor-BB in tear fluid - a potential modulator of corneal wound healing. *Curr Eye Res.* 1997, in press.
10. Dua HS, Gomes JAP, Singh A. Corneal epithelial wound healing. *Br J Ophthalmol.* 1994;78:401–408.
11. Tervo T, van Setten G-B. Aprotinin for inhibition of plasmin on the ocular surface: Principles and clinical observations. In: Beuerman RW, Crosson CE, Kaufman HE, eds. *Healing Processes of the Cornea.* Houston: Gulf Publishing; 1989:151–160.
12. Wilson SE, He Y-G, Weng J, et al. Epithelial injury induces keratocyte apoptosis: Hypothesized role for interleukin-1 system in the modulation of corneal organization and wound healing. *Exp Eye Res.* 1996;62:325–337.
13. Latvala T, Tervo K, Tervo T. Expression of cellular fibronectin and tenascin in the rabbit cornea after excimer laser PRK: Twelve months results. *Br J Ophthalmol.* 1995;79:65–69.
14. Hanna KD, Pouliquen Y, Waring GO III, et al. Corneal stromal wound healing in rabbits after 193-nm excimer laser surface ablation. *Arch Ophthalmol.* 1989;107:895–901.
15. SundarRaj N, Geiss MJ III, Fantes F, et al. Healing of excimer laser ablated monkey corneas. *Arch Ophthalmol.* 1990;108:1604–1610.
16. Fitzsimmons TD, Fagerholm P, Härfstrand A, Schenholm M. Hyaluronic acid in the rabbit cornea after excimer laser superficial keratectomy. *Invest Ophthalmol Vis Sci.* 1992;33:3011–3016.
17. Erickson HP, Bourdon MA. Tenascin: An extracellular matrix protein prominent in specialized embryonic tissues and tumors. *Annu Rev Cell Biol.* 1989;5:71–92.
18. Tervo K, Latvala T, Tervo T. Recovery of corneal innervation following photorefractive keratoablation. *Arch Ophthalmol.* 1994;112:1466–1470.
19. Gipson IK, Inatomi T. Extracellular matrix and growth factors in corneal wound healing. *Curr Opin Ophthalmol.* 1995;6(4):3–10.
20. Li D-Q, Tseng SCG. Three patterns of cytokine expression potentially involved in epithelial-fibroblast interactions of human ocular surface. *J Cell Physiol.* 1995;163:61–79.
21. Ohashi Y, Motokura M, Kinoshita Y, et al. Presence of epidermal growth factor in human tears. *Invest Ophthalmol Vis Sci.* 1989;30:1879–1882.
22. van Setten GB, Viinikka L, Tervo T, Pesonen K, Tarkkanen A, Perheentupa J. Epidermal growth factor is a constant component of normal human tear fluid. *Graefes Arch Clin Exp Ophthalmol.* 1989;227:184–187.
23. van Setten GB, Schultz G. Transforming growth factor alpha is a constant component of human tear fluid. *Graefes Arch Clin Exp Ophthalmol.* 1994;232:523–526.
24. van Setten GB. Basic fibroblast growth factor in human tear fluid: Detection of another growth factor. *Graefes Arch Clin Exp Ophthalmol.* 1996;234:275–277.
25. Kokawa N, Sotozono C, Nishida K, Kinoshita S. High total TGF-β2 levels in normal human tears. *Curr Eye Res.* 1996;5:341–343.
26. Gupta A, Monroy D, Ji Z, Yoshino K, Huang A, Pflugfelder SC. Trans-forming growth factor beta-1 and beta-2 in human tear fluid. *Curr Eye Res.* 1996;15:605–614.
27. Wilson SE, He YG, Weng J, Zieske JD, Jester JV, Schultz GS. Effect of epidermal growth factor, hepatocyte growth factor, and keratinocyte growth factor on proliferation, motility and differentiation of human corneal epithelial cells. *Exp Eye Res.* 1994;59:665–678.

28. Kruse FE, Tseng SCG. Transformierender Wachstumsfaktor beta 1 und 2 hemmen die Proliferation von Limbus- und Hornhautepithel. *Ophthalmologe.* 1994;91:617–623.
29. Ohji M, SundarRaj N, Thoft RA. Transforming growth factor-β stimulates collagen and fibronectin synthesis by human corneal stromal fibroblasts. *Curr Eye Res.* 1993;12:703–709.
30. Schultz G, Chegini N, Grant M, Khaw P, MacKay S. Effect of growth factors on corneal wound healing. *Acta Ophthalmol.* (Suppl) 1992;70:60–66.
31. Girard MT, Matsubara M, Fini ME. Transforming growth factor-beta and interleukin-1 modulate metalloproteinase expression by corneal stromal cells. *Invest Ophthalmol Vis Sci.* 1991;32:2441–2454.
32. Jester JV, Barry PA, Cavanagh HD, Petroll WM. Role of TGF-β1 in myofibroblast transformation and corneal wound healing. (Abstract.) Ocular Cell & Molecular Biology Symposium II, p. 7, San Diego, CA, Aug 20–22, 1995.
33. Brown LF, Yeo K-T, Berse B, et al. Expression of vascular permeability factor (vascular endothelial growth factor) by epidermal keratinocytes during wound healing. *J Exp Med.* 1992;176:1375–1379.
34. Bednarz J, Weich HA, Rodokanaki-von Schrenck A, Engelmann K. Expression of genes encoding growth factors and growth factor receptors in differentiated and dedifferentiated human corneal endothelial cells. *Cornea.* 1995;14:372–381.
35. Li D-Q, Tseng SCG. Differential regulation of cytokine and receptor transcript expression in human corneal and limbal fibroblasts by epidermal growth factor, transforming growth factor-α, platelet-derived growth factor B, and interleukin-1β. *Invest Ophthalmol Vis Sci.* 1996;37:2068–2080.
36. McCartney MD, Thomas DM, Mahendroo PP. An electron microscopic and nuclear magnetic resonance spectroscopic evaluation of rabbit corneal epithelial wound healing. *Invest Ophthalmol Vis Sci.* 1992;33:2917–2925.
37. Wang X, Kamiyama K, Iguchi I, Kita M, Imanishi J. Enhancement of fibronectin-induced migration of corneal epithelial cells by cytokines. *Invest Ophthalmol Vis Sci.* 1994;35:4001–4007.
38. Sugarman BJ, Aggarwal BB, Hass PE, Figari IS, Palladino MA, Shepard HM. Recombinant tumor necrosis factor-α: Effects on proliferation of normal and transformed cells in vitro. *Science.* 1985;230:943–945.
39. Mauviel A, Daireaux M, Redini P, Galera G, Loyau G, Piyol J. Tumor necrosis factor inhibits collagen and fibronectin synthesis in human dermal fibroblasts. *FEBS Lett.* 1988;236:47–52.
40. Dayer JM, Beutler B, Cerami A. Cachectin/tumor necrosis factor stimulates collagenase and prostaglandin E2 production by human synovial cells and dermal fibroblasts. *J Exp Med.* 1985;162:2153–2168.

73

TIME COURSE OF APOPTOSIS IN LACRIMAL GLAND AFTER RABBIT OVARIECTOMY

Ana Maria Azzarolo,[1] Richard L. Wood,[1] Austin K. Mircheff,[2,3] Eric Olsen,[1] Zuo Ming Huang,[1] Ramin Zolfagari,[1] and Dwight W. Warren[1,2,3]

[1]Department of Cell and Neurobiology
[2]Department of Physiology and Biophysics
[3]Department of Ophthalmology
University of Southern California
School of Medicine
Los Angeles, California

1. INTRODUCTION

Lacrimal insufficiency is the most frequent cause of dry eye conditions. Dry eye greatly decreases the quality of life through symptoms ranging from discomfort and contact lens intolerance to persistent pain and burning sensation and potential vision-threatening corneal damage. The incidence of dry eye, regardless of origin [primary lacrimal deficiency (PLD) or Sjögren s syndrome] is gender-related, affecting women much more frequently than men. PLD occurs especially frequently after menopause and during pregnancy, lactation, and oral contraceptive use, suggesting that hormonal factors play an important role in regulating lacrimal gland function. The greater incidence of Sjögren's syndrome in women indicates that hormonal factors also influence the incidence and progression of autoimmune phenomena that affect the lacrimal glands. This concept has been recently reviewed.[1]

In addition to the incidence of dry eye, several morphological, biochemical, and functional characteristics of the lacrimal gland have been shown to be sexually dimorphic, including parameters related to the ability of the gland to secrete electrolytes and water. Considerable evidence supports the thesis that androgens play a central role in maintaining many of these gender-related differences. We have shown that lacrimal glands rapidly regress when rats are hypophysectomized and when rabbits are ovariectomized, and that dihydrotestosterone (DHT) prevents this regression by exerting a general trophic effect. (For a review, see Warren et al., this volume.)

These observations led us to postulate that decreases in available androgen levels, which generally occur during menopause and states of elevated estrogen levels, such as pregnancy, lactation, and use of oral contraceptives, could trigger glandular regression and, therefore, lacrimal insufficiency. The mechanism by which the lacrimal glands regress after removal of androgens, as well as how androgens prevent this regression, is not clear yet. Recently, we have compiled data showing evidence of nuclear DNA fragmentation in sexually mature female rabbit lacrimal glands at 2, 4, and 6 h after ovariectomy that differed significantly from sham-operated animals ($p < 0.05$).[2] These results were obtained by using the TUNEL method (APOPtag^{TM} kit) and by detection of DNA laddering in agarose gel electrophoresis. DNA fragmentation was prevented when DHT (4 mg/kg) was administered at the time of the ovariectomy. These observations suggest that lacrimal regression after ovariectomy involves programmed cell death that can be prevented by the administration of androgens.[2]

Cell death has been classified as either necrotic or apoptotic. This distinction was originally made on the basis of morphology as observed by electron microscopy.[3] Apoptosis is an active, energy-dependent process of programmed cellular self-destruction, whereas necrosis is a more random, degenerative process where the cells lose capacity for volume homeostasis and undergo uncontrolled swelling.[4] Apoptosis is also marked by specific morphological changes, including condensation of chromatin, reduction in nuclear size, shrinkage of total cell volume, compaction of cytoplasmic organelles, and dilation of Endoplasmic Reticulum, with the mitochondria remaining morphologically normal. This is followed by budding and separation of both nucleus and cytoplasm into multiple, small, membrane-bound apoptotic bodies of various sizes.[5] Apoptotic bodies are then recognized and rapidly phagocytosed by neighboring cells and by macrophages transiently recruited to the area. Thus, apoptosis does not usually result in a mass of disorganized cell debris such as seen in necrosis following injury or toxicity. Nuclear chromatin is cleaved at the linker regions between nucleosomes, resulting in different-sized DNA fragments yielding a ladder banding pattern on neutral agarose gel electrophoresis.[5]

Although apoptosis is a critical mechanism of embryonic development and essential in the normal functioning of the immune system, it is also known to be promoted or inhibited in certain differentiated mammalian tissues by hormone withdrawal or stimulation.[3] This observation has been well documented in many organs, eg, prostatic cells deprived of testosterone, adrenal cortical cells deprived of adrenocorticotropic hormone, and endometrial cells deprived of progesterone.[6] Therefore, we hypothesized that withdrawal of androgenic support triggers lacrimal gland atrophy and that this atrophy could be induced through the process of apoptosis.

Our preliminary observations, however, did not clarify the cells involved or the nature of cell death observed in the lacrimal gland after ovariectomy. Neither did they explain the lacrimal regression observed previously at 10 days after ovariectomy. Therefore, the purposes of this study were to determine the nature, time course, and kind of cells involved in the cell death observed previously in the lacrimal gland of ovariectomized rabbits.

2. METHODS

We used sexually mature female New Zealand rabbits. Rabbits were ovariectomized and killed after 1, 2, 4, 6, and 24 h, and 3, 6, and 9 days. Additional groups of ovariectomized rabbits were treated with 4 mg/kg/day of DHTand killed after 4 h and 6 days. At each time period, sham-operated rabbits were used as controls. Lacrimal glands were removed and processed for morphological analysis and detection of DNA fragmentation using the in situ apoptosis detection kit Apoptag^{TM} (Oncor, Gaithersburg, MD). The labeled nuclei were quantified using a densitometer and Methamorph software.

3. RESULTS

The time course of appearance of DNA fragmentation exhibited two peaks, one at 4–6 h after ovariectomy, the other at 6 days after ovariectomy. Nuclear DNA fragmentation values at 24 h and 3 and 9 days were similar to control values. Ovariectomy-induced DNA degradation occurred primarily in interstitial cells and to a lesser extent in acinar cells. The continuous presence of DHT prevented ovariectomy-induced DNA degradation. When viewed by light microscopy, most of the interstitial cells undergoing apoptosis were plasma cells. The interstitial area of the ovariectomized animals appeared to be larger and with an increased number of macrophages compared to the control group. In addition, some of the nuclei of interstitial and acinar cells appeared more condensed. The acinar size appeared smaller, and the nuclei of acinar cells seemed closer together, suggesting a loss of cell mass. There were massive regions of cellular degeneration and cellular debris in the ovariectomized rabbits; these regions were identified as acinar cells. Acinar size appeared to be increased after DHT treatment, compared to the ovariectomized group. In addition, some necrosis and mitotic figures in acinar cells were characteristic of all the rabbits treated with DHT.

4. DISCUSSION

Both apoptosis and necrosis seem to be involved in the regression of lacrimal glands after ovariectomy. Apoptosis affects principally interstitial cells, mainly plasma cells, whereas necrosis seems to be the primary mode of cell death in acinar cells.

The mechanism by which DHT prevents lacrimal gland regression is not clear. One possibility is that androgens exert a direct effect on acinar cells. A second possibility is that androgens exert an indirect effect on acinar cells through a factor produced by the interstitial cells. Thus, androgens could promote the synthesis and release of growth factors or cytokines in interstitial cells, which could cause acinar cells to grow and divide. The causes for the appearance of mitotic figures after DHT treatment is also not clear. One explanation could be that DHT promotes polyploidy, a phenomenon that has been described in the exorbital lacrimal gland in developing rats.[7] Another possibility could be that DHT increases the number of nuclei, to render binucleate acinar cells or increase the number of acinar cells as a means of replacing the damaged cells. Regardless of the mechanism of action by which DHT prevents lacrimal gland regression, our results support the hypothesis that decreased available androgen levels may account for much of the lacrimal insufficiency observed in post-menopausal women and other states in which androgen levels are low.

Insufficient androgen levels could also be a factor in the initiation of Sjögren's syndrome. Thus, apoptotic and/or necrotic cell fragments such as those observed after ovariectomy in rabbits may also be the source of autoantigens that are processed and presented by MHC II positive acinar cells or by professional antigen-presenting cells, initiating an autoimmune reaction. Lacrimal gland acinar cells have already been shown to be able to express Class II major histocompatibility complex molecules (MHC II).[8] Thus, withdrawal of hormonal support may trigger a sequence of events that transforms the acinar cell into an effective autoantigen-presenting cell, one capable of processing and presenting autoantigens in a fashion that leads to $CD4^+$ T-cell activation. A role of apoptosis in autoimmunity has already been suggested in patients with systemic lupus erythematosus. Apoptotic skin keratinocytes in these patients were found to contain surface blebs expressing autoantigens such as DNA, Ro, La, and snRPNs.[9]

If this working hypothesis proves to be correct, it will provide a strong rationale for using androgen supplementation in hormone replacement therapy as a strategy to prevent the initiation and expansion of local autoimmunity that contributes to the development of Sjögren's disease and to restimulate glands that are already degenerated.

ACKNOWLEDGMENTS

The authors thank Barbara Platler for her expert technical assistance. This work was supported by NIH grants EY09405, EY05801, and EY10550.

REFERENCES

1. Azzarolo AM, Mircheff AK, Kaswan RL, et al. Androgen support of lacrimal gland function. *Endocrine*, 1997;6:39–45.
2. Azzarolo AM, Olsen E, Huang ZM, et al. Rapid onset of cell death in lacrimal gland after ovariectomy. ARVO Abstracts. *Invest Ophthalmol Vis Sci.* 1996;37:S856.
3. Kerr JFR, Wyllie AH, Currie AR. Apoptosis: A basic biological phenomenon with wide ranging implications in tissue kinetics. *Br J Cancer.* 1972;26:239–257.
4. Wyllie AH. Cell death. *Int Rev Cytol.* 1987;17:755–785.
5. Arends MJ, Morris RG, Wyllie AH. Apoptosis: The role of the endonuclease. *Am J Pathol.* 1990;136:593–608.
6. Koury MJ, Bondurant MC. Prevention of programmed death in hematopoietic progenitor cells by hematopoietic growth factors. *News in Physiological Sciences.* 1993;8:170–174.
7. Paulini K, Mohr W, Beneke G, Kulka R. Age- and sex-dependent changes in glandular cells: III. Cytomorphometric and autoradiographic investigations on the glandula lacrimalis, glandula infraorbitalis and glandula orbitalis externa of the rat. *Gerontologia.* 1972;18:147–156.
8. Mircheff AK, Gierow JP, Wood RL. Traffic of Class II MHC molecules in rabbit lacrimal gland acinar cells. *Invest Ophthalmol Vis Sci.* 1994;35:3943–3951.
9. Casciola-Rosen LA, Anhalt G, Rosen A. Autoantigens targeted in systemic lupus erythematosus are clustered in two populations of surface blebs on apoptotic keratinocytes. *J Exp Med.* 1994;179:1317–1330.

74

CO-EXPRESSION OF BCL-2 AND CD44S IN BASAL LAYERS OF HUMAN OCULAR SURFACE EPITHELIA

Andrew J. W. Huang,[1] Cheng-Hui Li,[1] You-Wei Chen,[2] and Lilly Y. W. Bourguignon[2]

[1]Bascom Palmer Eye Institute
[2]Department of Cell Biology and Anatomy
University of Miami School of Medicine
Miami, Florida

1. INTRODUCTION

The regulating mechanism of life and death of cells is a closely coordinated process of adaptation and survival. Selective and controlled cell death is a dominant force in renewing the tissue composition of complex organisms. Elimination of unwanted cells occurs during the normal course of tissue differentiation and also when cells are damaged by various noxious forces.[1] The two major mechanisms of cell death are apoptosis (programmed cell death) and necrosis.[2] Apoptosis is an active process in which a central biochemical "program" is activated causing nuclear fragmentation, formation of a rigid apoptotic envelope, and shrinkage of the cells into small fragments that are then phagocytosed, often by surrounding parenchymal cells not usually involved in phagocytosis. In necrosis, cells are passive targets of extensive membrane damage, cell swelling, colloidal osmotic lysis, and release of cellular contents. In the development and differentiation of multicellular organisms, apoptosis is the preferred mechanism for eliminating unwanted cell population. In actively cycling cells, proliferation and apoptosis are two opposing pathways for regulating cellular activities. Dysregulation of apoptosis is an important factor in cancer, leading to proliferation of cells, with a mutation in genes favoring proliferation or blocking apoptosis.[3] The cell cycle, which includes cellular proliferation and differentiation pathways, is regulated by multiple genes, some of which are oncogenes.[4,5] Oncogenes have been divided into two categories: genes that promote cellular proliferation and growth, and genes that regulate apoptosis. Although altered function of most common oncogenes may frequently lead to increased cell proliferation, the bcl-2 proto-oncogene encodes an inner mitochondrial membrane protein that blocks programmed cell

Lacrimal Gland, Tear Film, and Dry Eye Syndromes 2
edited by Sullivan *et al.*, Plenum Press, New York, 1998

death.[6] It has also been shown in vitro that gene product of bcl-2 inhibits programmed cell death without interfering with cell proliferation.[6] All hematopoietic lineages derived from a renewing stem cell display bcl-2. A limited number of non-lymphoid tissues that demonstrate bcl-2 can be grouped as (i) glandular epithelium in which hormones or growth factors regulate hyperplasia and evolution, (ii) complex differentiating epithelium in skin and intestine characterized by long-lived stem cells, and (iii) long-lived postmitotic cells such as neurons. Within these tissues that demonstrate apoptotic cell turnover, bcl-2 is often topographically restricted to long-lived or proliferating cell zones.[7] In normal mature human skin, bcl-2 expression is limited to cells within the basal layers of the epidermis. It is possible that bcl-2 positive cells may serve to maintain the stem-cell pool from which the proliferating stem cells continuously renew themselves and avoid postmitotic differentiation with eventual senescence and death.[8] Bcl-2 may thus function as an antidote to apoptosis, and confer longevity to progenitor and effector cells in these tissues. CD44s is the receptor for hyaluronic acid and is crucial for modulating epithelial adhesion and migration. Engagement of CD44 adhesion molecule receptors of T-lymphocytes, by either hyaluronic acid or anti-CD44 monoclonal antibodies, can inhibit DNA fragmentation and apoptosis induced by dexamethasone or anti-CD3 antibodies.[9] Since ocular surface epithelia are self-renewing and under the control of epithelial migration and differentiation during wound healing, we investigated the expressions of bcl-2 and CD44s in normal human ocular surface epithelia, and their implications.

2. MATERIALS AND METHODS

Human ocular surface epithelia were obtained from normal cadaver eyes. Reverse transcription-polymerase chain reaction (RT-PCR) and Southern hybridization were performed to study bcl-2 gene expression using primers (MBR 1 and 3) and an internal oligoprobe (MBR 2) specific for the normal allele of bcl-2 genes. The PCR primer oligonucleotides (TransPrimers® set) were obtained from Oncogene Science (Uniondale, NY). The 5' and 3' primers were used with the GeneAmp® DNA Amplification Reagent Kit from Perkin-Elmer in the Perkin-Elmer Cetus DNA Thermal Cycler, to detect the normal bcl-2 genes. The c-DNA products purified from RT-PCR were digoxigenin-labeled (DIG DNA labeling kit, Boehringer Mannheim) and used as probes for Northern hybridization. Both Southern and Northern blots were visualized with chemiluminescence.

To study the co-expression of bcl-2 oncoprotein and CD44s in ocular surface epithelia, dual-labeling immunohistochemistry with confocal microscopy was performed on frozen tissue sections with monoclonal mouse anti bcl-2 (DAKO, Carpinteria, CA) and monoclonal rat anti CD44 (CMB-Tech, Miami, FL) antibodies. Texas red-conjugated goat anti-mouse IgG (Jackson ImmunoResearch, West Grove, PA) and FITC-conjugated goat anti-rat IgG (Sigma, St. Louis, MO) were used as secondary antibodies, respectively. All experiments were performed at room temperature. Confocal microscopy was performed with a laser scanning microscope (Multiprobe 2001, Molecular Dynamics, Sunnyvale, CA), equipped with an argon-krypton laser. Acquisition and analysis of the fluorescent images were performed using software provided by Molecular Dynamics for the Silicon Graphics workstation. Immunoperoxidase staining with biotinylated goat anti-mouse IgG and Streptavidin-conjugated horseradish peroxidase (Sigma) and DAB as substrate (Vector Kit) was used to study the expression of bcl-2 in normal and pathological ocular surface tissues.

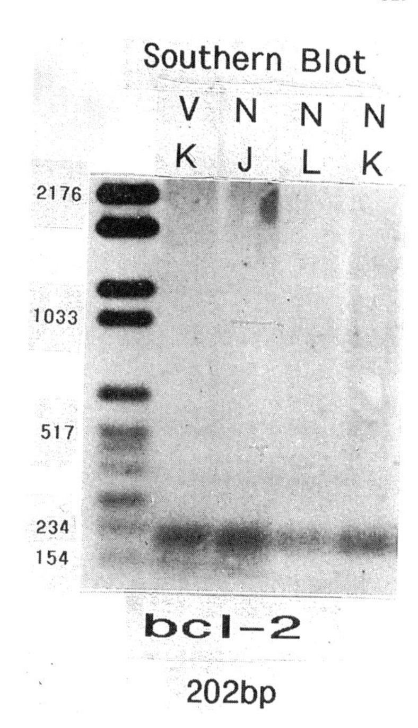

Figure 1. Southern hybridization. A 202 base band specific for bcl-2 was noted in vascularized cornea (VK), normal conjunctiva (NJ), normal limbus (NL), and normal cornea (NK). Expression of bcl-2 gene was more evident in conjunctival derivatives, such as normal conjunctiva and vascularized cornea.

3. RESULTS

By RT-PCR, m-RNA of bcl-2 was detected in all ocular surface epithelial tissues (including cornea, limbus, and conjunctiva; Fig. 1) with a preferential expression in the conjunctival epithelium. A positive band for bcl-2 was visible at the 202 base, as specified by the primers used in RT-PCR. The band was present in all three tissues since the bcl-2 gene was present in all normal cells. The specificity of RT-PCR was confirmed by Southern hybridization using MBR 2 as an internal oligo-nucleotide probe (Fig. 1). Northern hybridization using labeled c-DNA as a probe further confirmed the presence of bcl-2 genes in all three epithelial tissues (data not shown).

Immunostaining of tissue sections revealed the preferential expression of bcl-2 protein in the basal layers of stratified epithelia of conjunctiva and limbus (Fig. 2). The staining was markedly attenuated in the perilimbal cornea and absent in the central cornea. No staining of bcl-2 was noted in the basal corneal epithelium in contact with Bowman's layer (Fig. 2). Positive staining was also noted in the goblet cells of palpebral and forniceal conjunctiva. Co-localization of bcl-2 and CD44s was confirmed by dual-labeling confocal microscopy. Similar to the findings for bcl-2, CD44s was expressed in the basal epithelia of conjunctiva and limbus but not in that of the entire cornea (data not shown).

4. DISCUSSION

Consistent with previous reports that the bcl-2 protein was expressed in the basal layers of several stratified or pseudostratified epithelia (esophagus, palate, trachea, and conjunctiva),[7,8] our results confirm that bcl-2 is also expressed in the basal epithelial layer of limbus and conjunctiva but not in cornea. Since bcl-2 oncogenes are present in all nor-

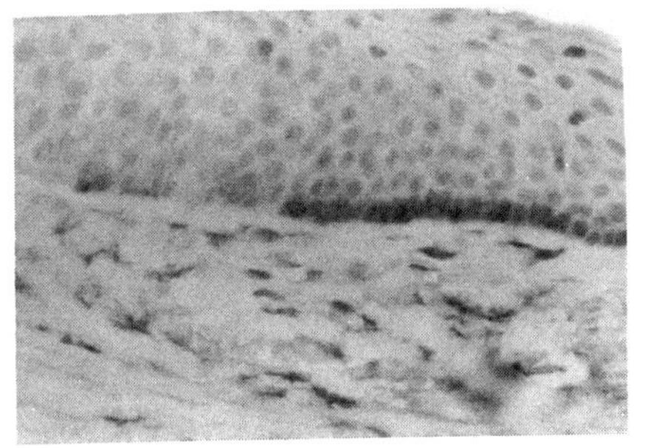

Figure 2. Immunoperoxidase staining of human limbus and peripheral cornea. Expression of bcl-2 protein was noted in the limbal basal epithelium and the adjacent conjunctival basal epithelium. Arrows indicate the termination of Bowman's layer. No evident bcl-2 was noted in the basal epithelia of central and perilimbal cornea. (Upper panel, 20X; lower panel, 40X.)

mal cells, it is not surprising that we detected bcl-2 genes in all three ocular surface epithelial tissues. Research has indicated that the stem cells of cornea reside in the limbal basal layer.[10] Our findings further confirm this notion and are therefore consistent with previous findings that bcl-2 is often topographically restricted to long-lived or proliferating stem cell zones such as those in glandular epithelium, skin, and intestine.[7]

The topographical distribution of bcl-2 protein in the conjunctival epithelium and goblet cells noted by us is very similar to what has been observed in the gastrointestinal system. Bcl-2 was confined to the lower half of the crypts in small and large intestines, with the stem cell of these tissues being in the base of the crypts.[7] The staining patterns in the intestines appeared diffusely cytoplasmic in enterocytes as well as goblet cells. The location of conjunctival stem cells remains unclear. Our finding of preferential expression of bcl-2 protein in the forniceal goblet cells and conjunctival basal epithelium may implicate the location of conjunctival fornix, similar to that of basal intestinal crypts, as the stem cell zone.

Confocal microscopy with dual labeling indicates the restrictive co-expression of bcl-2 and CD44s in the limbal and conjunctival basal layers, presumably the stem cell-rich zones. Since corneal basal epithelium comprises mainly transient amplifying cells (proliferating cells derived from stem cells) or more differentiated post-mitotic cells, it is not surprising that there were no evident stainings of bcl-2 or CD44s detected in the corneal basal layer. It has been reported that loss of cell adhesion to extracellular matrix or ground substance was paralleled by a substantial reduction of the multifunctional cell surface adhesion molecule CD44.[11] It is conceivable that decreased CD44s expression in corneal basal epithelium may lead to loss of cell adhesion, thereby compromising the vital stroma-basal epithelium dialogues and resulting in upregulation of apoptosis in the cornea. Even though binding to CD44 by hyaluronic acid can inhibit apoptosis,[9] it remains yet to be studied whether the anti-apoptotic effect exerted through CD44 is related to the parallel expression of bcl-2 protein or the proliferative potential of the basal epithelium.

In conclusion, preferential expression of bcl-2 and CD44s in the basal layers of conjunctival and limbal epithelium implicates the proliferative potentials and self-renewing nature of these cells. This finding suggests that bcl-2 and CD44s positive basal epithelia may represent the stem cells of ocular surface epithelia. Absence of bcl-2 and CD44s expression in normal cornea suggests that epithelial cells of normal cornea are more differentiated and susceptible to the regulation of apoptosis than those of limbus and conjunctiva.

REFERENCES

1. Arends MJ, Wyllie AH. Apoptosis. *Int Rev Exp Pathol.* 1991;32:223–254.
2. Cohen JJ. Apoptosis. *Immunol Today.* 1993;14:126–130.
3. Williams GT. Programmed cell death. *Cell.* 1991;65:1097–1098.
4. Yunis JJ. The chromosomal basis of human neoplasia. *Science.* 1983;221:227–236.
5. Bishop JM. The molecular genetics of cancer. *Science.* 1987;235:305–311.
6. Hockenbery DM, Nunez G, Milliman C, Schreiber RD, Korsmeyer SJ. Bcl-2 is an inner mitochondrial membrane protein that blocks programmed cell death. *Nature.* 1990;348:334–336.
7. Hockenbery DM, Zutler M, Hickey W, Nahm M, Korsmeyer SJ. Bcl-2 protein is topographically restricted in tissue characterized by apoptotic cell death. *Proc Natl Acad Sci USA.* 1991;88:6961–6965.
8. LeBrun DP, Wamke RA, Cleary ML. Expression of bcl-2 in fetal tissues suggests a role in morphogenesis. *Am J Pathol.* 1993;142:743–753.
9. Ayroldi E, Cannarile L, Migliorati G, Bartoli A, Niocoletti I. Riccardi C. CD44 (Pgp-1) inhibits CD3 and dexamethasone-induced apoptosis. *Blood.* 1995;86:2672–2678.
10. Schermer A, Galvin S, Sun TT. Differentiation-related expression of a major 64K corneal keratin in vivo and in culture suggests limbal location of epithelial stem cells. *J Cell Biol.* 1986;103:49–62.
11. Gunthert AR, Strater J, von Reyher V, et al. Early detachment of colon carcinoma cells during CD95 (APO-1/Fas)-mediated apoptosis. I. De-adhesion from hyaluronate by shedding of CD44. *J Cell Biol.* 1996;134:1089–1096.

75

ALTERATIONS OF OCULAR SURFACE GENE EXPRESSION IN SJÖGREN'S SYNDROME

David T. Jones, Dagoberto Monroy, Zhongua Ji, and Stephen C. Pflugfelder

Ocular Surface and Tear Center
Department of Ophthalmology
Bascom Palmer Eye Institute
University of Miami School of Medicine
Miami, Florida

1. INTRODUCTION

Sjögren's syndrome (SS), an idiopathic autoimmune disorder primarily affecting the lacrimal and salivary glands, leads to characteristic ocular surface changes within the conjunctival epithelium including lymphocytic infiltration, squamous cell metaplasia, epithelial hyperplasia, and a reduction in goblet cell number. We have previously reported that molecular markers of the inflammatory cascade, including two cell surface markers [class II MHC antigen (HLA-DR) and intercellular adhesion molecule-1 (ICAM-1)] and the inflammatory cytokine, interleukin-6 (IL-6), are expressed at elevated levels in SS conjunctival epithelium.[1] As a result of these studies, we have proposed that in addition to mechanical surface abrasion secondary to aqueous tear deficiency, local inflammatory processes likely contribute to the ocular surface disease associated with SS.

To further characterize the proliferative and differentiative status of the conjunctival epithelium of SS patients, we compared the S-phase DNA synthetic activity and the expression of the differentiation marker, MUC-1, in the conjunctival epithelium of SS and non-dry eye control patients. In addition, we utilized semi-quantitative RT-PCR to evaluate the expression of different inflammatory cytokines within the conjunctival epithelium of SS and non-dry eye patients.

2. MATERIALS AND METHODS

Clinical samples were collected from SS and non-dry eye control patients, after obtaining fully informed consent under University of Miami Institutional Review Board guidelines and according to the tenets of the Declaration of Helsinki. Diagnostic criteria for SS were as described.[1,2]

To evaluate epithelial proliferative capacity, conjunctival epithelial cells isolated by direct biopsy were incubated for 1 h with bromodeoxyuridine (BrDU) followed by fixation and immunofluorescent staining with FITC-conjugated BrDU specific antisera as described by the manufacturer (Boehringer-Mannheim, Indianapolis, IN). The proliferative capacity was estimated by averaging the total number of BrDU-labeled cells counted per three high-power fields (20X magnification) in eight SS and eight control specimens. Messenger RNA isolated from impression cytology specimens of temporal bulbar conjunctival epithelium was amplified by heterologous semi-quantitative RT-PCR using cyclophilin as an internal control.[3] RT-PCR primers for TNF-α, IL-1α, IL-1β, IL-6, IL-8, TGF-β1, and TGF-β2 were obtained from Clontech (Palo Alto, CA). Amplified RT-PCR products were quantified by densitometry after gel electrophoresis. Conjunctival epithelial expression of MUC-1 was analyzed by immunofluorescence utilizing the specific antiserum, HMFG-1, directed to the conserved VNTR repeat of MUC-1.[4] Specimens were analyzed before and after exposure to neuraminidase enzymatic digestion of terminal sialic acid residues.[5]

3. RESULTS

The conjunctival epithelial proliferative capacity was estimated by comparing the average number of S-phase cells of SS patients and non-dry eye controls. The SS patients displayed an increased number of S-phase cells per high-power field (13.3) compared with non-dry eye controls (1.1). Moreover, in the control epithelium the S-phase cells were confined largely to the basal layers, whereas in the SS epithelium the S-phase cells were distributed throughout all layers.

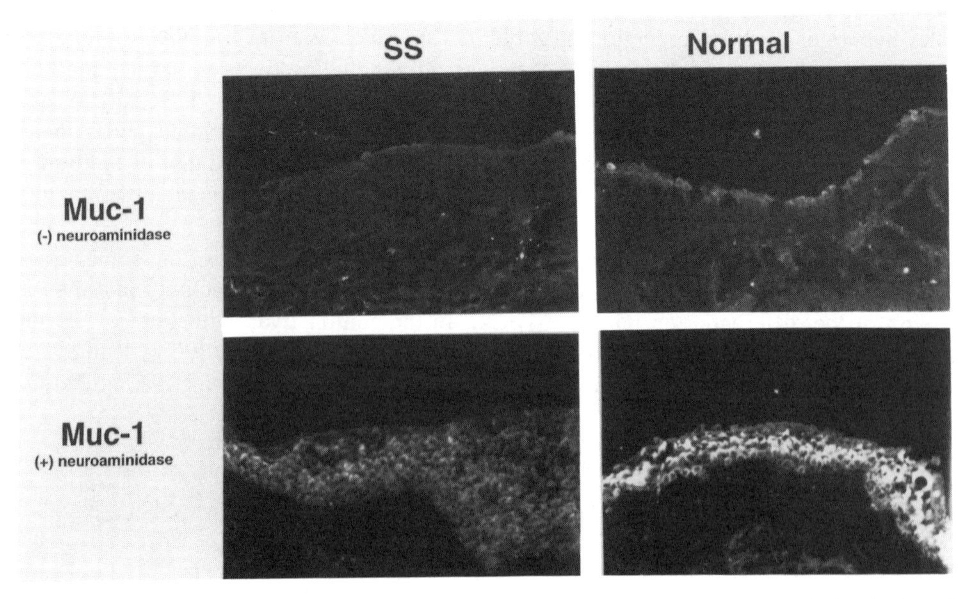

Figure 1. Immunofluorescent staining of Sjögren's syndrome (SS) and non-dry eye conjunctival epithelium for MUC-1. Tissues were analyzed before [(-) neuraminidase treatment] and after [(+) neuraminidase treatment] treatment with neuraminidase.

Immunofluorescent staining of conjunctival epithelium for MUC-1 revealed that the mature MUC-1 antigen [(-) neuraminidase treatment] is expressed weakly along the surface of normal epithelium and at a diminished level in the SS conjunctiva (Fig. 1). In contrast, treatment with neuraminidase, an enzyme that selectively removes terminal sialic acid residues, exposed a MUC-1 epitope that was localized predominantly in the basal and suprabasal cell layers of both normal and SS epithelium [(+) neuraminidase treatment].

The levels of mRNA expression of several different cytokines determined by semi-quantitative RT-PCR were compared between Sjögren's syndrome (n=3) and non-dry eye control patients (n=3) (Fig. 2). We found that TNF-α, IL-1α and β, IL-6, IL-8, and TGF-β1 were all expressed at significantly elevated levels in SS compared with non-dry eye controls. TGF-β2 also appeared to be expressed at elevated levels in SS conjunctival epithelium, but the difference was not statistically significant in this small population.

4. SUMMARY

We have demonstrated that conjunctival epithelium of SS patients displays increased numbers of S-phase cells compared with non-dry eye controls. Moreover, in SS patients, these S-phase cells are distributed throughout all strata of the epithelium. The ex-

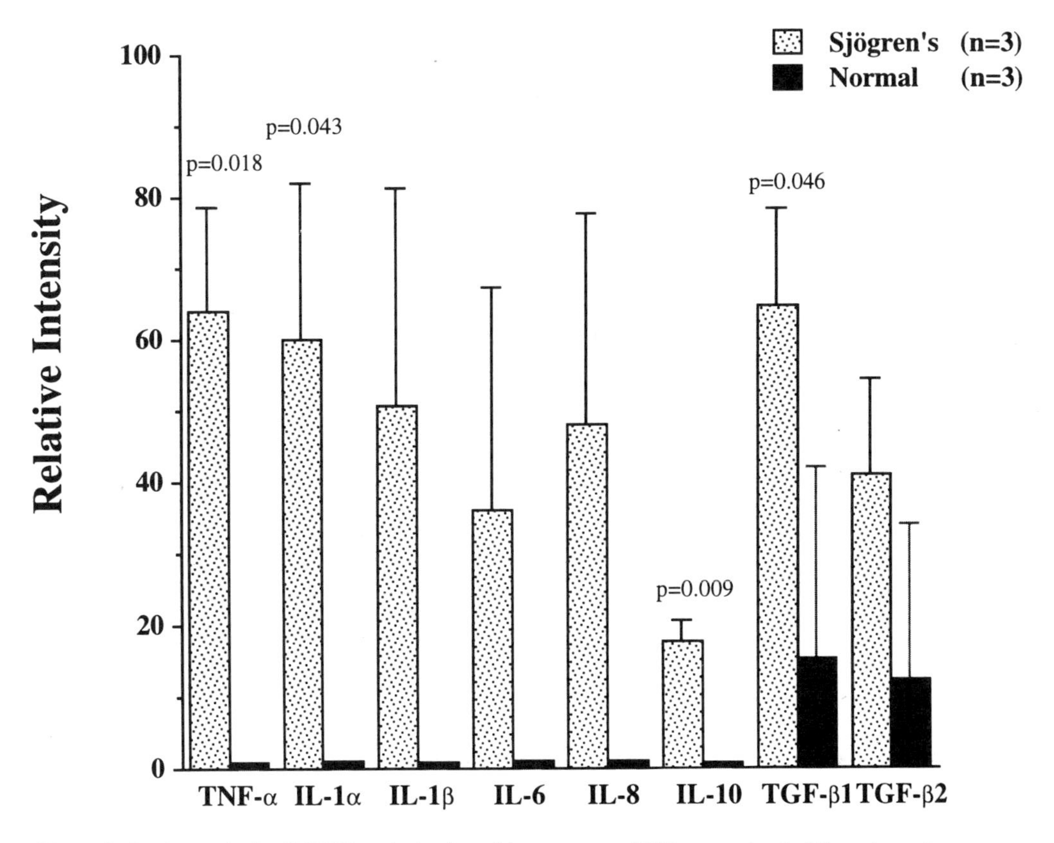

Figure 2. Semi-quantitative RT-PCR analysis of cytokine messenger RNA expression in Sjögren's syndrome vs. non-dry eye controls. The p values were determined by Student's paired t-test.

pression of MUC-1, a cell surface marker indicative of terminally differentiated epithelium, is localized to the conjunctival epithelial surface in SS and control patients. However, MUC-1 surface immunoreactivity appears to be reduced in SS epithelium, suggesting disruption of normal epithelial differentiation. A MUC-1 epitope exposed by pretreatment with neuraminidase is expressed in the basal and suprabasal layers of both patient populations. This antigen likely represents nascent, partially processed MUC-1[6] and may serve as a marker of the preterminally differentiated epithelial phenotype. Messenger RNA encoding several different inflammatory cytokines, including TNF-α, IL-1α and β, IL-6, IL-8, IL-10, and TGF-β1, is expressed at elevated levels within the conjunctival epithelium of SS patients compared with non-dry eye controls.

Based on these observations, we have formulated a model to explain the ocular surface pathology of Sjögren's syndrome. We hypothesize that mechanical abrasion secondary to aqueous tear deficiency creates an inflammatory environment where conjunctival epithelial cells and lymphocytes are stimulated to produce and secrete various cytokines (i.e., IL-1, TNF-α, IL-6, IL-8, etc.) into the tear film. Elevated cytokine levels within the tear film, perhaps combined with reduced concentrations of essential lacrimal gland-derived factors (ie, EGF, retinol), create an environment in which terminal differentiation of the ocular surface epithelium is impaired. As a consequence, the epithelium becomes hyperplastic, displaying increased mitotic activity, and loses the ability to express mature protective surface molecules including the membrane-bound mucin, MUC-1. This would imply that anti-inflammatory medications (ie, corticosteroids or cyclosporine) that suppress the inflammatory component of this cascade may ameliorate the ocular surface disease and discomfort experienced by SS patients.

ACKNOWLEDGMENT

This research was supported in part by Research to Prevent Blindness, Inc.

REFERENCES

1. Jones DT, Monroy D, Ji Z, Atherton S, Pflugfelder SC. Sjögren's syndrome: Cytokine and Epstein-Barr viral gene expression within the conjunctival epithelium. *Invest Ophthalmol Vis Sci.* 1994;35:3493–3504.
2. Fox RI, Robinson CA, Curd JG, et al. Sjögren's Syndrome. Proposed criteria for classification. *Arthritis Rheum.* 1986;29:577–585.
3. Siebert PD, Larrick JW. PCR MIMICS: Competitive DNA fragments for use as internal standards in quantitative PCR. *BioTechniques.* 1993;14:244–249.
4. Burchell J, Gendler S, Taylor-Papadimitriou J, Girling A, Lewis A, Mills R. Development and characterisation of breast cancer reactive monoclonal antibodies directed to the core protein of the human milk mucin. *Cancer Res.* 1987;47:5476–5482.
5. Hey NA, Graham RA, Seif MW, Alpin JD. The polymorphic epithelial mucin, MUC-1, in human endometrium is regulated with maximal expression in the implantation phase. *J Clin Endocrinol Metab.* 1994;78:337–342.
6. Patton S, Gendler SJ, Spicer AP. The epithelial mucin, MUC1, of milk, mammary gland and other tissues. *Biochim Biophys Acta.* 1995;1241:407–424.

GROWTH FACTOR INTERACTION WITH CONTACT LENSES: EVIDENCE FOR AND POSSIBLE EFFECTS OF STORAGE OF BASIC FGF IN CONTACT LENSES

Preliminary Results

G.-B. van Setten[1] and N. Zengin[1,2]

[1]St. Eriks Eye Clinic
Lab of DOHF
Karolinska Institutet
Stockholm, Sweden
[2]Selcuk üniversitesi Tip Fakültesi Göz Hastaliklari Anablim Dali
42080 Konya, Turkey

1. INTRODUCTION

Growth factors are part of the cascade triggering neovascularization.[1,2] Such neovascularization occurs in the cornea (CNV), although usually not very aggressively, as a possible side effect of contact lens (CL) wear.[3] Among the factors involved in this CNV process are fibroblast growth factors such as basic fibroblast growth factor (bFGF). The presence of various vasculogenic growth factors, such as epidermal growth factor (EGF) and transforming growth factor alpha (TGF-α), has been reported in active form in the tear fluid of CL wearers.[4,5] The present study investigated the effect of the presence of CL on the concentration of fibroblast growth factors in the surrounding medium by determining the concentration of bFGF in the presence of unworn and worn CLs.

2. MATERIALS AND METHODS

2.1. Materials

A total of 15 soft contact lenses (diameter 14.5 mm) were used. A standard solution of recombinant human bFGF was prepared at a concentration of about 3.4 pg/ml in a pilot study

Lacrimal Gland, Tear Film, and Dry Eye Syndromes 2
edited by Sullivan *et al.*, Plenum Press, New York, 1998

and of 4 pg/ml in a second study. Three of the contact lenses were never worn and served as controls. Twelve contact lenses were worn by patients from a few days to a few weeks.

2.2. Methods

After use, the contact lenses were incubated in the standard recombinant human bFGF solution for 2 h at room temperature. Then, a sample was taken from each incubation vial (400 μl) and analyzed in duplicate for the concentration of bFGF in the medium.

Determination of bFGF concentrations in duplicate was performed by the use of a high-sensitivity ELISA. The details of this assay have been reported.[6] Briefly, monoclonal antibodies specific for bFGF were bound to microtiter plates. After incubation for 2 h on an orbital shaker, an enzyme-linked polyclonal bFGF specific antibody was added to the wells. After additional incubation for 2 h, a substrate solution (in this assay, NADPH) was added, followed by incubation for 1 h at room temperature. Addition of amplifier solution was followed by additional incubation for 30 min at room temperature. Then the reaction was stopped with 2 N sulfuric acid. Optical density of the wells was determined using a spectrophotometer set to 490 nm with a correction wavelength of 690 nm. The standard curve was established for standard dilutions of recombinant hbFGF dilutions in the range of 0.0–36 pg/ml, in duplicate determinations. Linear regression was 0.997. The inter- and intra-assay precision deviation is, according to the manufacturer, less than 1 pg/ml; the sensitivity is 0.28 pg/ml, and no cross-reactivity with other known cytokines is known. All incubations were run on an orbital shaker at room temperature.

3. RESULTS

We found that the concentration of bFGF decreased in the medium after the incubation with contact lenses, but this was true only for worn CLs (Table 1). In the pilot study, the decrease was 18.5%; in the second study, with a larger number of CLs, this figure was 6%. As the study is ongoing (now about 40 used CLs), we currently observe a decrease of about 8% of the bFGF concentration in the medium.

4. DISCUSSION

It appears that used soft contact lenses have an increased ability to adsorb or absorb bFGF. This study does not give any idea whether this material is located within the CLs or

Table 1. bFGF concentration in the presence of new and worn contact lenses

Contact lens type	bFGF (pg/ml)	SD
Pilot Study:		
Control solution	3.36	0.09
New CLs	3.36	0.22
Worn CLs	2.82	0.26
Second Study:		
Control solution	4.18	0.08
New CLs	4.18	0.07
Used CLs	3.94	0.26

on their surface. However, the CL material itself does not undergo any other remarkable changes during short-time use other than a rather rapid coating of the surface. The resulting biofilm not only attracts bacteria,[7] but also suggests that the amount of bFGF shown to have been withdrawn from the surrounding medium could be located on the lens surface. This is in accordance with unpublished studies of the authors. This observation might help us reach a better understanding of CNV as a possible CL wear-associated side effect. It has been suggested that biological biomaterials could be used to carry and deliver growth factors to stimulate wound healing.[8] Although our study did not investigate CLs as carriers, its findings allow us to consider the biofilm-coated CL as a foreign body on the surface of the eye with a certain reservoir of angiogenic growth factors on its surface. The pathophysiological effects of the bound growth factors could be triggered by other extracellular matrix proteins, such as fibronectin, bound on the lens surface. Due to the observation that the biological activities of bFGF are regulated by a complex array of biochemical interactions with extracellular matrix proteins, such as laminin and fibronectin,[9] the simultaneous presence of both fibronectin and bFGF on the lens surface and in the surrounding medium, i.e., the tear film,[6,10] deserves attention. As especially bFGF is highly angiogenic in corneal tissue, any increased availability of this growth factor to surface structures of the eye could trigger the onset of CNV. This could be facilitated because CL wear may lead to minor surface damage,[11] resulting in repair mechanisms in which, in turn, ECM proteins are naturally involved.[12] Various parameters of the CL such as diameter apparently do influence the ability of the CL to absorb or adsorb bFGF from the surrounding medium (unpublished results by the authors). Further investigation of these variables will help us to increase the safety of CL wear and to understand better the interactions between the ocular surface, contact lenses, and the tear film.

ACKNOWLEDGMENTS

The author is very thankful for the financial support of the Karolinska Institutes Forskningsfonder, the Research Fund of the Visually Disabled (Synfrämjandets Forskningsfond), and the Carmen and Bertil Régniers Fund for investigation of ocular diseases, Stockholm, Sweden.

REFERENCES

1. Folkman J, Klagsbrun M. Angiogenic factors. *Science.* 1987;235:442–447.
2. Gospodarowicz D, Neufeld G, Schweigerer L. Molecular and biological characterization of fibroblast growth factor, an angiogenic factor which also controls the proliferation and differentiation of mesoderm and neuroectoderm derived cells. *Cell Differ.* 1986;19:1–17.
3. Arentsen JJ. Corneal neovascularization in contact lens wearers. *Int Ophthalmol Clin.* 1986;26:15–23.
4. van Setten GB, Tervo T, Andersson R, Perheentupa J,Tarkkanen A. Effects of contact lens wear on the concentrations of plasmin and EGF in human tear fluid. *Ophthalmic Res.* 1990;22:233–240.
5. van Setten G-B, Schultz GS. Transforming growth factor alpha is a constant component of human tear fluid. *Graefes Arch Clin Exp Ophthalmol.* 1994;232:523–526.
6. van Setten GB. Basic fibroblast growth factor in human tear fluid: Detection of another growth factor. *Graefes Arch Clin Exp Ophthalmol.* 1996;234:275–277.
7. Stapleton F, Dart J. Pseudomonas keratitis associated with biofilm formation on a disposable soft contact lens. *Br J Ophthalmol.* 1995;79:864–890.
8. Roy F, DeBlois C, Doillon CJ. Extracellular matrix analogs as carriers for growth factors: In vitro fibroblast behavior. *J Biomed Mater Res.* 1993;27:389–397.

9. Feige JJ, Bradley JD, Fryburg K, et al. Differential effects of heparin, fibronectin, and laminin on the phosphorylation of basic fibroblast growth factor by protein kinase C and the catalytic subunit of protein kinase A. *J Cell Biol.* 1989;109:3105–3114.
10. Jensen OL, Gluud BS, Eriksen HO. Fibronectin in tears following surgical trauma to the eye. *Acta Ophthalmol.* 1985;63:346–350.
11. Lohman LE. Corneal epithelial response to contact lens wear. *CLAO J.* 1986;12:153–156.
12. Tervo K, van Setten GB, Beuerman RW, Virtanen I, Tarkkanen A, Tervo T. Expression of tenascin and cellular fibronectin in rabbit cornea after anterior keratectomy. *Invest Ophthalmol Vis Sci.* 1991;32:2912–2918.

77

NEURAL, ENDOCRINE, AND IMMUNE SYSTEM INTERACTIONS

Relevance for Health and Disease

Jan A. Moynihan,[1,2] Barbara Kruszewska,[1] Gary J. Brenner,[2] and Nicholas Cohen[1,2]

[1]Department of Psychiatry
[2]Department of Microbiology and Immunology
University of Rochester Medical Center
The Center for Psychoneuroimmunology Research
Rochester, New York

1. INTRODUCTION

The integration of the central and peripheral nervous systems and the immune system has become a focal point in the understanding of human health and disease. The nervous system communicates with the immune system via the hypothalamo-pituitary-adrenal (HPA) axis and the sympathetic nervous system (SNS). The immune system, particularly via proinflammatory cytokines such as IL-1, can, in turn, regulate CNS activity. That the nervous system can modulate immunity has been inferred from several lines of evidence.

i. The work of Ader, Cohen, and their colleagues,[1] as well as that of others, documents the classical behavioral conditioning of immunity.
ii. Physical or psychological stressors have significant consequences for immune responses and/or disease outcome in humans and in animal models.[2,3] Although in some studies, stressor-associated immunomodulation correlates with altered levels of glucocorticoids[4–7] or opioids,[8,9] such altered levels are not observed in all studies of effects of a stressor on an immune response.[10–13] At least 17 neuroendocrine peptides and transmitters have immunomodulatory potential and can be altered during a stress response.[10,14] Thus, multiple neurochemical pathways are involved in alteration of immune function.
iii. In addition to neuroendocrine effects on immunity, a significant body of literature documents that the SNS plays a role in immune regulation. The studies of the Feltens and colleagues have shown that there is noradrenergic (NA) innerva-

Lacrimal Gland, Tear Film, and Dry Eye Syndromes 2
edited by Sullivan *et al.*, Plenum Press, New York, 1998

tion of primary and secondary lymphoid organs.[15,16] Further, hematopoietic cells express adrenergic receptors on their surfaces.[17–19] Thus, lymphoid cells are exposed to and have the machinery to respond to norepinephrine (NE), the principal neurotransmitter of the SNS. In addition, adrenergic agonism or antagonism has immunoregulatory effects both in vivo and in vitro.[20,21]

2. T HELPER CELLS AND CYTOKINE PRODUCTION

Our recent work and the work of others have focused on stress or other psychosocial factors, the SNS, and T cell cytokines and effector functions. T lymphocytes express either the cell surface antigen CD8 (a marker for T cytotoxic cells) or CD4 (which delineates T helper cells). Recent advances in immunology include the identification of two subsets of mature $CD4^+$ T helper (Th) cells: Th1 and Th2, found in rodents and humans, differentiated on the basis of the cytokines they produce following antigenic stimulation.[22–24] Differentiation into distinct Th1 or Th2 cells appears to be a property of mature effector T cells. Th1 cells typically produce interleukin (IL)-2 and interferon (IFN)-γ, which are important for the generation of cell-mediated immunity; IFN-γ is also important for production of IgG2a Th2 cells produce cytokines such as IL-4, IL-5, IL-6, and IL-10, and these cytokines generally promote humoral immune responses and isotype switching (particularly production of IgG1, IgA, and IgE). Production of these cytokines is, however, not limited to Th1 and Th2 cells. Preferential skewing toward a Th1- or Th2-dominant response is important for the generation of an adaptive immune response to antigens (including parasites, as well as a number of viruses). A number of factors influence the activation of these two subsets, including the nature of the antigen, antigen concentration, route of inoculation, and host factors such as the strain of the experimental subject.[22,23,25–28] IL-12 is currently considered to be a key cytokine for cellular immune regulation,[24] is required for optimal production of IFN-γ,[23,29,30] and may be the basis for the Th1/Th2 strain differences described above.[31,32] Thus, there is an established cytokine network that biases the overall immune effector functions activated following antigen stimulation. Our laboratory has shown that the neurochemical state of the host can also be a factor in production of these cytokines and effector functions.[33,34]

3. PSYCHOSOCIAL FACTORS AND THE SNS CAN MODULATE TUMOR METASTASES

Although it is quite clear that external stimuli that evoke neural changes can indeed alter disease via changes in immunity,[2,3] it may not always be the case that changes in the susceptibility to or progression of disease are a direct result of changes in immune function. For example, we have studied the role of both psychosocial factors and the SNS in the growth of the alveolar carcinoma line 1, a tumor derived from a BALB/c mouse that is sensitive to natural killer (NK) cell lysis. The number of lung metastases occurring following the injection of line 1 tumor cells is significantly increased in mice that were handled daily for 2 weeks prior to tumor cell injection (Table 1). Although the number of metastases observed varies between experiments, it is noteworthy that differences between groups are always observed. This effect of handling can be mimicked by chemical ablation of the SNS using the neurotoxin 6-hydroxydopamine (6-OHDA). Injection of 6-OHDA results in loss of peripheral sympathetic NE from innervated tissues, including primary and secondary lymphoid organs.

Table 1. Pulmonary metastases following injection of 10^5 line 1 tumor cells in mice

	Treatment			
	Handled	Unhandled	6-OHDA	Vehicle control
Expt. 1	15.0±3.1*	3.2±0.9	—	—
Expt. 2	149.0±11.0*	106.0±15.0	—	—
Expt. 3	5.4±1.3*	2.0±0.8	—	—
Expt. 4	15.0±5.0*	5.0±2.5	—	—
Expt. 5	—	—	25.2±4.9*	12.8±2.3
Expt. 6	—	—	27.6±6.0*	13.5±3.0

Data are expressed as mean±SEM of 8-12 mice/group/experiment. BALB/c mice were handled for 2 min daily for 2 weeks prior to injection of tumor cells (handled); maintained undisturbed in their cages (unhandled); injected with 125 mg/kg of 6-OHDA 2 days prior to tumor cell injection (6-OHDA); or injected with ascorbate vehicle (vehicle control). Asterisks denote a difference compared to the relevant control group, $p<0.05$.
From ref. 34.

Although numbers of pulmonary metastases are increased following manipulation of BALB/c mice, NK cell function is unaffected by either handling or denervation, as determined in vitro against the YAC-1 lymphoma cell line or in vivo by clearance of ^{51}Cr-labeled line 1 tumor cells. Further, denervation is not associated with changes in the ability of BALB/c mice to be primed with irradiated line 1 cells. That is, following injection of irradiated line 1 cells, both denervated and vehicle control mice have similar decreases in numbers of metastases following injection of live tumor cells. These data suggest that the observed increase in numbers of pulmonary metastases following either a psychosocial manipulation or denervation may not be the result of altered specific immune or innate defense effector functions. It is possible that sympathectomy affects non-immunological processes involved in tumor progression, such as extravasation, development of blood supply to a growing tumor, or a number of other physiological measures.

4. PSYCHOSOCIAL AND NEUROENDOCRINE FACTORS AND THE SNS CAN MODULATE AUTOIMMUNE DISEASE

The best-studied animal model of autoimmune disease in relation to neuroendocrine regulation is experimental autoimmune encephalomyelitis (EAE), a demyelinating, paralytic disease that is widely used as a model for multiple sclerosis. MBP-specific Th1 cells can induce the disease, whereas Th2 cells specific for the same peptide-MHC complexes can not.[35] During the peak of the disease, IL-2, IFN-γ, and tumor necrosis factor (TNF)-α (Th1 cytokines) are present in CNS tissues, but IL-4 and IL-10 are not.[36] During the recovery phase, IFN-γ messenger RNA (mRNA) is absent, and IL-10 mRNA expression becomes dominant.[37] In addition, spontaneous recovery from the disease is associated with production of MBP-specific antibody.[38]

In the early 1960s, Levine et al.[39] and later others[40,41] reported that multiple, prolonged sessions of physical restraint beginning prior to and following the induction of EAE decreased the incidence and clinical severity of disease in male and female rats. Further, restraint could prevent relapses of disease that had been observed in female Lewis rats.[42] Daily sessions of electric shock also decreased the incidence and severity of histological lesions in EAE-susceptible DA rats when shock was administered after, but not before, injection of guinea pig spinal cord.[43] Most of these studies focused on disease onset, severity, and recovery as outcome measures, and did not include immunological outcomes. However, other data[40,41,44,45] suggest that stressful experiences affect EAE in susceptible strains of rats via direct effects on the immune system. The effects of stressful experiences on EAE appear to be mediated in part by glucocorticoids and/or opioids.[46,47] In addition, Chelmicka-Schorr et al.[45,48] observed that EAE severity is augmented in sympathectomized Lewis rats.

There is good evidence that the onset of and recovery from EAE are orchestrated by Th1 and Th2 cytokine production. There is substantial evidence that stressful life experiences, acting via glucocorticoids or the sympathetics, can decrease the clinical severity of the disease, as well as histopathological lesions and production of at least one Th1-derived cytokine, IL-2.[40] Future studies should determine if, and how, stressful life experiences influence the Th1/Th2 balance in autoimmune disease.

5. STRESS CAN ALTER HERPES SIMPLEX VIRUS SPECIFIC IMMUNITY AND VIRAL CLEARANCE

Although the increased number of line 1 lung metastases associated with handling or chemical sympathectomy does not appear to be immunologically mediated, there are clear instances in which stress-induced changes in disease pathogenesis are correlated directly with immune function. For example, the virus-specific immune responses of C57Bl/6 female mice following primary infection in the footpad with herpes simplex virus-type 1 (HSV-1) can be altered by either restraint stress[49,50] or mild electric footshock stress.[51] Using the footshock model, we[51] have observed that stressor administration results in suppression of spleen and popliteal lymph node cell number in infected mice, in vitro cytotoxic T cell responses to virally infected cells, IL-2 production following stimulation with virally infected cells in vitro, and humoral immune responses. In addition, significantly more infectious virus is recovered from the footpads of stressed mice than from the footpads of either apparatus control or home cage mice (44.5±1.9, 9.1±3.5, and 1.0±0 plaque-forming units x 10^5, respectively). Thus, impaired antiviral responses are associated with impaired clearance of virus from the site of infection.

6. SYMPATHECTOMY-INDUCED CHANGES IN CYTOKINE PRODUCTION AND IMMUNE EFFECTOR FUNCTION

Our laboratory has shown that following a single intraperitoneal injection of 6-OHDA, splenic NE levels in mice are reduced by approximately 90%. This denervation is reversible, and 20 days following injection, NE levels return to approximately 50% of control values.[52] We have utilized chemical sympathectomy to assess the immunoregulatory contribution of noradrenergic innervation of primary and secondary lymphoid tissue to antibody responses and antigen-specific cytokine production. C57Bl/6 or BALB/c mice are denervated by a single injection of 6-OHDA 2 days prior to immunization with 100 μg KLH; these two strains of mice were originally chosen because of differences in their cytokine and immune effector responses. In general, C57Bl/6 mice produce less antibody and Th2 cytokines compared to BALB/c mice, and C57Bl/6 mice respond to antigen with a bias toward cell-mediated immune responses, whereas BALB/c mice are better antibody producers.[22,23,25]

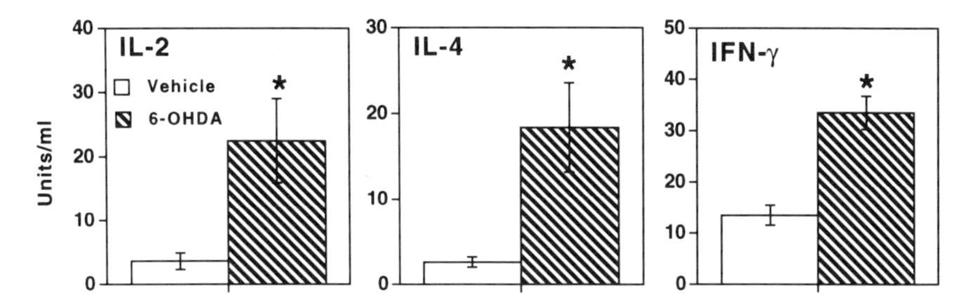

Figure 1. KLH-stimulated cytokine production in vitro by splenocytes from C57Bl/6 mice. Spleen cells were cultured with 160 μg/ml KLH for 24 h (IL-2), 48 h (IL-4), or 72 h (IFN-γ). Data are expressed as the mean±SEM for 10–12 mice/group. Asterisks indicate significant differences ($p<0.05$) compared to vehicle controls.

We have observed that denervation is associated with an increase in KLH-stimulated production of the Th1-derived cytokines IL-2 and IFN-γ and the Th2-derived cytokine IL-4 by spleen cells from C57Bl/6 mice (Fig. 1). Spleen cells from denervated BALB/c mice produce increased antigen-induced IL-2 and IL-4, but no change in IFN-γ, compared to vehicle control mice (Fig. 2). Interestingly, these cytokine patterns are associated with differential antibody responses. Denervated C57Bl/6 mice make more KLH-specific IgM, IgG, IgG1, and IgG2a (IgA was not examined) (Fig. 3). In contrast, denervated BALB/c do not produce enhanced KLH-specific antibody titers compared to vehicle control mice (Fig. 4).

While these intriguing results illustrate that cytokine and antibody responses can be modulated in a strain-dependent fashion by either sympathetically derived NE or other colocalized peptides, how the observed pattern of cytokines might affect antibody responses is less clear. The increased IgM and IgG1 antibody titers in denervated C57Bl/6 mice may be due to an increased Th2 cytokine response (as evidenced by increased IL-4), and enhanced IgG2a titers can be induced by IFN-γ.[23] The lack of increased IgM and IgG1 antibody production by denervated BALB/c mice, despite the high levels of IL-4 production, remains puzzling. Clearly, increased levels of IL-4 and increased antibody titers do not have a one-to-one relationship. Other immunological and neurochemical factors can bias the ultimate level of humoral response in BALB/c mice. For example, we have observed that following exposure to odors of stress conspecific mice, KLH-immunized BALB/c mice produce significantly increased levels of serum IgM and IgG antibody to KLH. Cor-

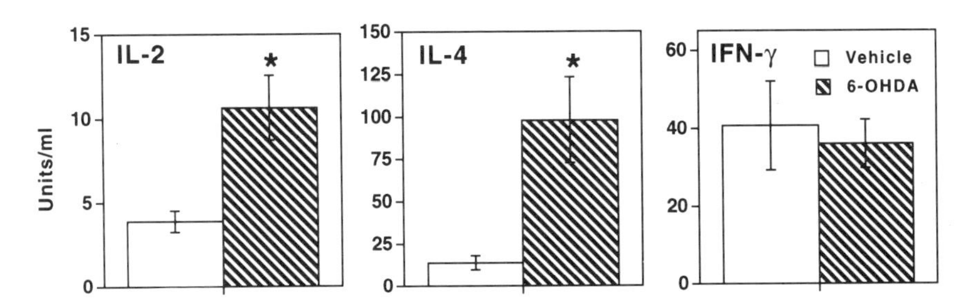

Figure 2. KLH-stimulated cytokine production in vitro by splenocytes from BALB/c mice. Spleen cells were cultured with 160 μg/ml KLH for 24 h (IL-2), 48 h (IL-4), or 72 h (IFN-γ). Data are expressed as the mean±SEM for 10–12 mice/group. Asterisks indicate significant differences ($p<0.05$) compared to vehicle controls.

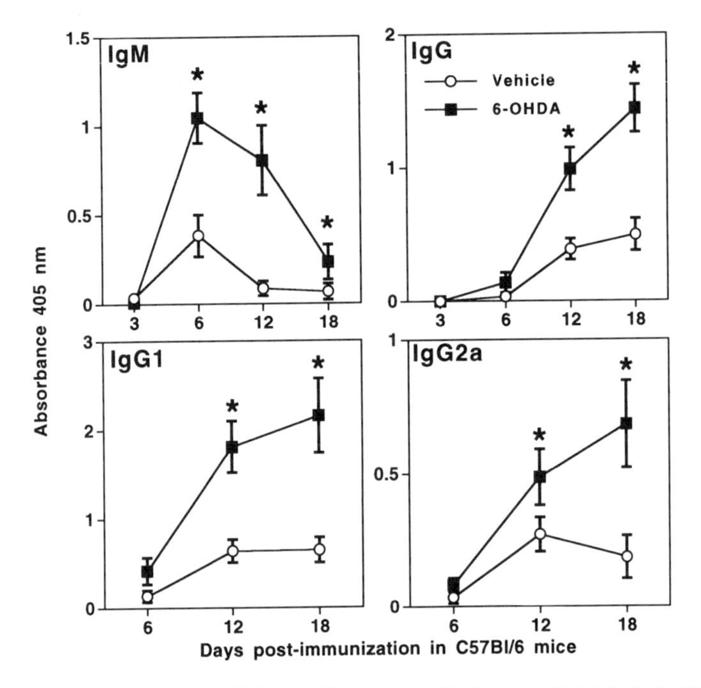

Figure 3. Effect of sympathectomy on KLH-specific serum antibody titers (IgM, IgG, IgG1, and IgG2a) in C57Bl/6 mice as determined by ELISA. Results are expressed as the mean±SEM for 10–12 mice/group. Asterisks indicate significant differences ($p<0.05$) compared to vehicle controls.

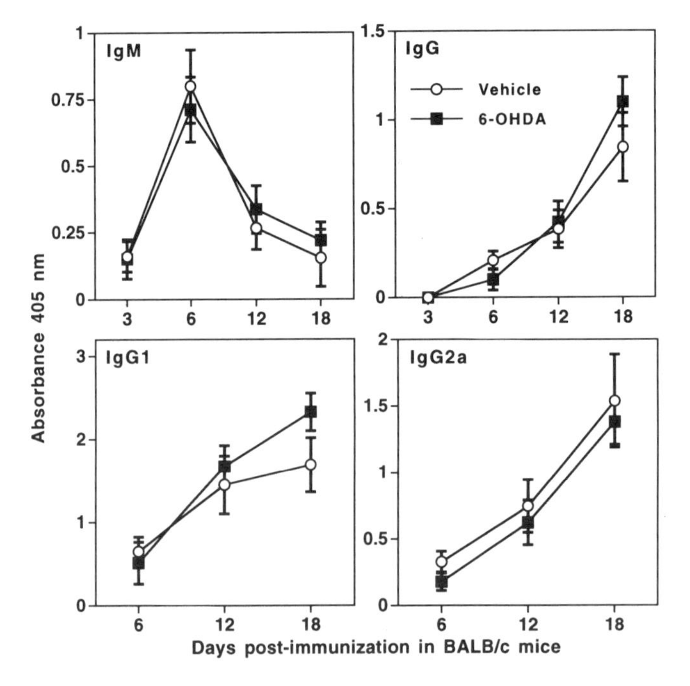

Figure 4. Effect of sympathectomy on KLH-specific serum antibody titers (IgM, IgG, IgG1, and IgG2a) in BALB/c mice as determined by ELISA. Results are expressed as the mean±SEM for 10–12 mice/group.

related with this increase in serum antibody response is an increase in antigen-specific IL-4 production by spleen cells, and either decreased or unchanged production of IL-2.[33,34]

7. OCULAR MUCOSAL IMMUNITY AND THE ROLE OF THE CENTRAL NERVOUS SYSTEM

The mucosal immune system is critical for host defense at the ocular surface, and secretory IgA, the principal effector of secretory immunity in response to ocular pathogens, can be regulated by Th2 cytokines.[53] Ocular immune privilege (anterior chamber-associated immune deviation) results from immune responses characterized by $CD8^+$ T cells and $CD4^+$ Th2 responses. Thus, factors that modulate Th2 cytokines will have importance for the host response to antigenic challenge in the eye. The work of Sullivan and colleagues has demonstrated that hormones and neurotransmitters, such as testosterone and vasoactive intestinal peptide (VIP), can affect production of secretory component or IgA.[54,55] Although the eye and, in particular, the lacrimal gland are richly innervated by sympathetic and parasympathetic fibers,[56,57] Sullivan et al. did not observe a role for these nerves in the regulation of total secretory IgA expression.[58] Perhaps, however, different results would be obtained if changes in antigen-specific IgA and cytokine production by ocular lymphocyte populations, rather than total IgA and IgA-expressing cells, were examined. An investigation of the effects of hormones/transmitters on cytokines and effector functions in response to virus infections (e.g., herpes viruses) that are proposed to play a role in autoimmune lacrimal tissue destruction and Sjögren's syndrome, could be particularly enlightening.

ACKNOWLEDGMENTS

This research was supported by grants MH45681 and MH42076 from the USPHS.

REFERENCES

1. Ader R, Cohen N. Psychoneuroimmunology: Conditioning and stress. *Annu Rev Psychol.* 1993;44:53–85.
2. Moynihan JA, Ader R. Psychoneuroimmunology: Animal models of disease. *Psychosom Med.*, 1997;58:546–558.
3. Sheridan JF, Dobbs C, Brown D, Zwilling B. Psychoneuroimmunology: Stress effects on pathogenesis and immunity during infection. *Clin Microbiol Rev.* 1994;7:200–212.
4. Keller SE. Weiss JM, Schleifer SJ, Miller NE, Stein M. Stress-induced immunosuppression of immunity in adrenalectomized rats. *Science.* 1983;221:1301–1304.
5. Okimura T, Ogawa M, Yamauchi T, Sasaki Y. Stress and immune responses. IV. Adrenal involvement of antibody responses in adrenalectomized rats. *Jpn J Pharmacol.* 1986;41:237–245.
6. Coe CL, Rosenberg LT, Fisher M, Levine S. Psychological factors capable of preventing the inhibition of antibody responses in separated infant monkeys. *Child Dev.* 1987;58:1420–1430.
7. Blecha F, Kelley KW, Satterlee DG. Adrenal involvement in the expression of delayed-type hypersensitivity to DNFB in mice. *Proc Soc Exp Biol Med.* 1982;169:247–252.
8. Shavit Y. Stress-induced immunomodulation in animals: Opiates and endogenous opioid peptides. In: Ader R, Felten DL, Cohen N, eds. *Psychoneuroimmunology*, 2nd ed. New York: Academic Press; 1991:789–806.
9. Shavit Y, Lewis JW, Terman GW, Gale RP, Liebeskind JC. Opioid peptides mediate the suppressive effects of stress on natural killer cell activity. *Science.* 1984;223:188–190.
10. Moynihan JA, Cohen N. Stress and immunity. In: Schneiderman N, McCabe P, Baum A, eds. *Stress and Disease Processes: Perspectives in Behavioral* Medicine. Hillsdale, New Jersey: Lawrence Erlbaum; 1992:27–53.

11. Ben-Eliyahu S, Yirmiya R, Shavit Y, Liebeskind JC. Stress-induced suppression of natural killer cell cytotoxicity in the rat: A naltrexone-insensitive paradigm. *Behav Neurosci.* 1990;104:235–238.
12. Cunnick JE, Lysle DT, Armfield A, Rabin BS. Shock-induced modulation of lymphocyte responsiveness and natural killer cell activity: Differential mechanisms of induction. *Brain Behav Immun.* 1988;2:102–113.
13. Rabin BS, Cunnick JE, Lysle DT. Stress-induced alteration of immune function. *Neuro Endocrin Immunol.* 1990;3:116–124.
14. Khansari DN, Murgo AJ, Faith RE. Effects of stress on the immune system. *Immunol Today.* 1990;11:170–175.
15. Felten DL, Felten SY, Bellinger DL, et al. Noradrenergic sympathetic neural interactions with the immune system: Structure and function. *Immunol Rev.* 1987;100:225–260.
16. Felten SY, Felten DL, Bellinger DL, et al. Noradrenergic sympathetic innervation of lymphoid organs. *Prog Allergy.* 1988;43:14–36.
17. Fuchs BA, Albright JW, Albright JF. β-Adrenergic receptors on murine lymphocytes: Density varies with cell maturity and lymphocyte subtype and is decreased after antigen administration. *Cell Immunol.* 1988;114:231–245.
18. Fuchs BS, Campbell K, Munson AE. Norepinephrine and serotonin content of the murine spleen: Its relationship to lymphocyte β-adrenergic receptor density and the humoral immune response in vivo and in vitro. *Cell Immunol.* 1988;117:339–351.
19. Bishopric NH, Cohen HJ, Lefkowitz RJ. Beta adrenergic receptors in lymphocyte subpopulations. *J Allergy Clin Immunol.* 1980;65:29–33.
20. Sanders VM, Powell-Oliver FE. β2-Adrenoceptor stimulation increases the number of antigen-specific precursor B lymphocytes that differentiate into IgM-secreting cells without affecting burst size. *J Immunol.* 1992;148:1822–1828.
21. Sanders VM, Munson AE. Beta adrenoceptor mediation of the enhancing effect of norepinephrine on the murine primary antibody response in vitro. J *Pharmacol Exp Ther.* 1984;230:183–192.
22. Scott P. Selective differentiation of $CD4^+$ T helper cell subsets. *Curr Opin Immunol.* 1993;5:391–397.
23. Paul WE, Seder RA. Lymphocyte responses and cytokines. *Cell.* 1994;76:241–251.
24. Lamont AG, Adorini L. IL-12: A key cytokine in immune regulation. *Immunol Today.* 1996;17:214–217.
25. Golding B, Zaitseva M, Golding H. The potential for recruiting immune responses toward type 1 or type 2 T cell help. *Am J Trop Med Hyg.* 1994;50 Suppl.:33–40.
26. Bretscher PA, Wei G, Menjon JN, Bielefeldt-Ohmann H. Establishment of stable, cell-mediated immunity that makes 'susceptible' mice resistant to Leishmania major. *Science.* 1994;257:539–542.
27. Brenner GJ, Cohen N, Moynihan JA. Similar immune responses to subcutaneous footpad infection with herpes simplex virus-1 in sensitive (BALB/c) and resistant (C57Bl/6) strains of mice. *Cell Immunol.* 1994;157:510–524.
28. Belosevic M, Finbloom DS, van der Meade P, Slayter MV, Nacy CA. Administration of monoclonal anti-IFN-γ antibodies in vivo abrogates natural resistance of C3H/HeN mice to infection with Leishmania major. *J Immunol.* 1989;143:266–274.
29. Scott P. IL-12: Initiation cytokine for cell-mediated immunity. *Science.* 1993;260:496–497.
30. Trinchieri G. Interleukin 12 and its role in the generation of Th1 cells. *Immunol Today.* 1993;14:335–338.
31. Guler ML, Gorham JD, Hsieh C, et al. Genetic susceptibility to Leishmania: IL-12 responsiveness in Th1 cell development. *Science.* 1996;271:984–987.
32. Noben-Trauth N, Kropf P, Muller I. Susceptibility to Leishmania major infection in interleukin-4-deficient mice. *Science.* 1996;271:987–990.
33. Moynihan JA, Karp JD, Cohen N, Cocke R. Alterations in interleukin-4 and antibody production following pheromone exposure: Role of glucocorticoids. *J Neuroimmunol.* 1994;54:51–58.
34. Moynihan JA, Brenner GJ, Cocke R, et al. Stress-induced modulation of immune function in mice. In: Glaser R, Kiecolt-Glaser JK, eds. *Handbook of Human Stress and Immunity.* New York: Academic Press, 1994:1–22.
35. Liblau RS, Singer SM, McDevitt HO. Th1 and Th2 $CD4^+$ T cells in the pathogenesis of organ-specific autoimmune diseases. *Immunol Today.* 1995;16:34–38.
36. Khoury SJ, Hancock WW, Weiner HL. Oral tolerance to myelin basic protein and natural recovery from experimental autoimmune encephalomyelitis are associated with downregulation of inflammatory cytokines and differential upregulation of transforming growth factor β, interleukin 4, and prostaglandin E expression in the brain. *J Exp Med.* 1992;176:1355–1364.
37. Kennedy MK, Torrance DS, Picha KS, Mohler KM. Analysis of cytokine mRNA in the central nervous system of mice with experimental autoimmune encephalomyelitis reveals that IL-10 mRNA expression correlates with recovery. *J Immunol.* 1992;149:2496–2505.

38. MacPhee IAM, Day MJ, Mason DW. The role of serum factors in the suppression of experimental autoimmune encephalomyelitis: Evidence for immunoregulation by antibody to the encephalitogenic peptide. *Immunology*. 1992;70:527–534.
39. Levine S, Strebel R, Wenk EJ, Harman PJ. Suppression of experimental allergic encephalomyelitis by stress. *Proc Soc Exp Biol Med*. 1962;109:294–298.
40. Griffin AC, Lo WD, Wolny AC, Whitacre CC. Suppression of experimental allergic encephalomyelitis by restraint stress: Sex differences. *J Neuroimmunol*. 1993;44:103–116.
41. Kuroda Y, Mori T, Hori T. Restraint stress suppresses experimental allergic encepthalomyelitis. *Brain Res Bull*. 1994;34:15–17.
42. Levine S, Saltzman A. Nonspecific stress prevents relapse in experimental allergic encephalomyelitis. *Brain Behav Immun*. 1987;1:336–341.
43. Bukilica M, Djordjevic S, Maric I, Dimitrijevic M, Markovic BM, Jankovic BD. Stress-induced suppression of experimental allergic encephalomyelitis in the rat. *Int J Neurosci*. 1991;59:167–175.
44. McCombe PA, deJersey J, Pender MP. Inflammatory cells, microglia, and MHC class II antigen-positive cells in the spinal cord of Lewis rats with acute and chronic relapsing experimental allergic encephalomyelitis. *J Neuroimmunol*. 1992;149:2496–2505.
45. Chelmicka-Schorr E, Kwasniewski MN, Wollmann RL. Sympathectomy augments adoptively transferred experimental allergic encephalomyelitis. *J Neuroimmunol*. 1992;37:99–103.
46. Mason DM, MacPhee IAM, Antoni F. The role of the neuroendocrine system in determining genetic susceptibility to experimental allergic encephalomyelitis in the rat. *Immunology*. 1992;70:1–5.
47. Jankovic BD, Maric D. Enkephalins and autoimmunity: Differential effect of methionine-enkephalin on experimental allergic encephalomyelitis in Wistar and Lewis rats. *J Neurosci Res*. 1987;18:88–94.
48. Chelmicka-Schorr E, Checinski M, Arnason BGW. Chemical sympathectomy augments the severity of experimental allergic encephalomyelitis. *J Neuroimmunol*. 1988;17:347–350.
49. Doutlik S, Kutinova L, Benda R, et al. Some immunological characteristics of subjects suffering from frequent herpes simplex virus recrudescences. *Acta Virol*. 1989;33:435–445.
50. Bonneau RH, Sheridan JF, Feng N, Glaser R. Stress-induced suppression of herpes simplex virus (HSV)-specific cytotoxic T lymphocyte and natural killer cell activity and enhancement of pathogenesis following local HSV infection. *Brain Behav Immun*. 1991;5:170–192.
51. Kusnecov AV, Grota LJ, Schmidt SG, et al. Alterations in herpes virus immunity and clearance following stressor administration in mice. *J Neuroimmunol*. 1992;38:129–138.
52. Kruszewska B, Felten SY, Moynihan JA. Sympathectomy increases Th1 andTh2 cytokine production and antibody production in BALB/c and C57Bl/6 mice. *J Immunol*. 1995;155:4613–4620.
53. Zierhut M, Bieber T, Brocker EB, Forrester JV, Foster CS, Streilein JW. Immunology of the skin and eye. *Immunol Today*. 1996;17:448–450.
54. Kelleher RS, Hann LE, Edwards JA, Sullivan DA. Endocrine, neural, and immune control of secretory component output by lacrimal gland acinar cells. *J Immunol*. 1991;146:3405–3412.
55. Sullivan DA. Ocular mucosal immunity. In: Sullivan DA, ed. *Handbook of Mucosal Immunity*. New York: Academic Press; 1994:569–597.
56. Hosoya Y, Matsushita M, Sugiura Y. A direct hypothalamic projection to the superior salivatory nucleus neurons in the rat: A study using anterograde autoradiographic and retrograde HRP methods. *Brain Res*. 1983;266:329–333.
57. Walcott B, Cameron RH, Brink PR. The anatomy and innervation of lacrimal glands. *Adv Exp Med Biol*. 1994;350:11–18.
58. Sullivan DA, Hann LE, Soo CH, Yee L, Edwards JA, Allansmith MR. Neural-immune interrelationship: Effect of optic, sympathetic, temporofacial, or sensory denervation on the secretory immune system of the lacrimal gland. *Reg Immunol*. 1991;3:204–212.

78

INDUCTIVE SITES FOR RAT TEAR IgA ANTIBODY RESPONSES

Deanne M. Ridley Lathers, Randall F. Gill, Nancy L. O'Sullivan, and Paul C. Montgomery

Department of Immunology and Microbiology
Wayne State University Medical School
Detroit, Michigan

1. INTRODUCTION

The mucosal immune system is an important mediator of acquired immunity at the ocular surface.[1,2] In mammals, the lacrimal gland serves as the predominant source of tear IgA[3,4] and functions as an effector site for mucosal antibody responses.[1,2] Evidence for the linkage of lacrimal glands to the mucosal network has come from cell trafficking[5,6] as well as antibody expression data.[7–9] In a number of species, oral or gastrointestinal antigen stimulation has resulted in tear IgA antibody induction.[7–12] In addition, topical application of particulate antigen[7–9] or microparticle-encapsulated soluble antigen[13] to the conjunctiva (ocular topical, OT) has been documented to be an effective means of eliciting tear IgA antibody responses. Furthermore, evidence indicates that nasal-associated lymphoid tissue (NALT) is a major inductive site in the mucosal immune network,[14,15] and NALT appears to serve as an inductive site for tear IgA antibody responses.[16]

Although pathways leading to tear IgA antibody induction following gastrointestinal (GI) immunization are defined, little information is available on the sequence of events following OT or IN (intranasal, NALT) stimulation. It is not known if tear IgA responses result from conjunctival antigen processing, antigen accessing NALT via the nasolacrimal canal, antigen drainage to gastrointestinal-associated lymphoid tissue (GALT), or a combination of these pathways. For conjunctival processing, antigen would be taken up by the conjunctiva (CONJ) and possibly drain to the superior cervical lymph nodes (sCLN). For NALT processing, antigen would be taken up by microfold (M) cells, appear in NALT, and possibly drain to the posterior cervical lymph nodes (pCLN). For GALT processing, antigen would be taken up by M cells and appear in the Peyer's patches (PP). Although the site of B cell triggering is not known for conjunctival or NALT processing, B cells are known to be triggered within the PP of GALT. Subsequently, antigen reactive IgA com-

Lacrimal Gland, Tear Film, and Dry Eye Syndromes 2
edited by Sullivan *et al.*, Plenum Press, New York, 1998

mitted B cells from any of the inductive pathways would gain access to circulation, traffic to major lacrimal glands, and differentiate into IgA-producing cells.

The present study was designed to determine the primary mucosal inductive pathway for ocular mucosal IgA antibody responses. The rat model was used to determine the capacity of fluorescent microspheres to access various tissues postulated to function in the inductive process and to assess the ability of antigen-containing microparticles administered by the OT or IN routes to elicit tear IgA antibody responses.

2. MATERIALS AND METHODS

Ten to 12-week-old female Fischer 344 rats (126–140 g) (Harlan Sprague-Dawley, Indianapolis, IN) were housed under conventional conditions, and given water and laboratory chow (Purina) *ad libitum*. Animal care and treatment were in compliance with NIH guidelines.

Fluorescent microspheres (1.0 μm diameter, 2.5% Fluoresbrite™ Carboxylate Microspheres, BB, Polysciences) were diluted in saline, and administered by the IN, OT, or GI route to anesthetized rats. IN-administered microspheres were diluted 1:2 in saline and 20 μl applied to each nostril. OT-administered microspheres (10 μl per eye, undiluted) were delivered in 5 μl increments to each conjunctival surface following blotting of excess tears. Rats receiving microspheres orally were intubated with 50 μl of microspheres diluted in 450 μl of saline. Rats were sacrificed 10 min, 30 min, 6 h, and 18 h after the initial administration of the microspheres. CONJ, sCLN, pCLN, and PP were surgically removed. NALT was removed using the method of Carr et al.[16] All tissues were embedded in OCT compound (Miles Scientific, Naperville, IL) and frozen in liquid nitrogen. Cryostat sections containing CONJ, sCLN, pCLN, and NALT were cut at 8 μm, placed on microscope slides, and mounted using Gel Mount (Polysciences, Warrington, PA). Cryostat sections containing PP were cut at 12 μm to maintain better tissue integrity. Sections were examined for the presence of fluorescent microspheres using a Nikon Optiphot microscope (Columbus, OH) and scored by the following criteria: 0=0–5; 1=5–20; 2=20–100; 3=100–250; 4=>250 microspheres per mm^2.

DNP-BSA was prepared as detailed previously.[13] Poly(lactide-co-glycolide) (PLG) microparticles (MP) containing DNP-BSA were prepared using the oil-in-water solvent evaporation method described by Jeffery et al.[17] The PLG-MP (1.7 μm mean diameter) were washed, lyophilized, and their protein concentration determined using a BCA assay (Sigma, St. Louis, MO).[13]

The PLG-MP dosage, based on the encapsulated protein concentration, was 0.5 mg total protein delivered over 3 days. Groups of six anesthetized rats received a primary immunization of DNP-BSA-PLG-MP on days 0, 1, and 2. PLG-MP were reconstituted in sterile PBS (pH 7.2) and administered IN (20 μl per nostril) or OT (10 μl per conjunctival surface). Samples were collected for antibody quantitation on days 11 and 35. A secondary immunization of 0.5 mg of total protein was administered on days 37, 38, and 39. Samples were collected on days 4, 7, 14, 21, 28, and 42 post secondary immunization. Tertiary immunizations (0.5 mg of total protein) were administered on days 83, 84, and 85. Samples were again collected on days 7, 14, 21, 28, and 42 post tertiary immunization.

Rats were anesthetized and tear and serum samples collected. Serum was collected only on alternate samplings, pooled by group, and frozen until assay. Tears were collected using 5 μl calibrated glass microcapillary tubes (Fisher, Pittsburgh, PA). Tear samples were diluted 1:5 in PBS-azide-Tween 20 (PBS-A-T20), pooled by group, and frozen until assay. A solid-phase radioimmunoassay (RIA) was used to quantitate IgA in the pooled tears and IgG in the pooled serum.[4] A standard curve was generated using IgA and IgG DNP-specific standards from Biosource (Camarillo, CA). Dilutions of samples were added to wells of DNP-BSA-coated Falcon PVC 96-well U-shaped plates and incubated overnight at 4°C. Plates were washed with PBS-A-T20, and ^{125}I (Amersham, Arlington Heights, IL)-labeled rabbit anti-rat IgA or IgG (Serotec, Raleigh, NC) was added to each well. Plates were incubated for 5 h at room temperature and unbound antibody removed by washing with PBS-A-T20. The plates were dried for 1 h, 50 μl of scintillant was added to each well, and the wells were counted using a Wallac 1450 Microbeta Plus counter (Gaithersburg, MD). Standard curves were generated based on the CPM obtained from the IgA and IgG standards and the sample antibody concentrations determined from the standard curves using the forecast function (Excel, Microsoft Software, Redmond, WA).

3. RESULTS

To determine the ability of antigen to access tissues with postulated roles in the induction of ocular mucosal IgA antibody responses, the uptake and distribution of fluorescent microspheres were studied at various time points following IN, OT, or GI delivery. Table 1 shows the results of this study. At all time points (10 min, 30 min, 6 h, and 18 h), IN-delivered microspheres were found in NALT. They were also found in pCLN at 6 and 18 h and in PP at 6 h post delivery. OT-delivered microspheres were found in CONJ at all time points, in NALT at 30 min, 6 h, and

18 h and in sCLN at 6 and 18 h post delivery. OT-delivered microspheres were also found in PP at 6 and 18 h. As expected, GI-delivered microspheres were not found in

Table 1. Tissue distribution of fluorescent microspheres at various times following intranasal, ocular, or oral administration

	10 min	30 min	6 h	18 h
Intranasal administration (IN)				
CONJ	0	0	0	0
NALT	3-4	2	2-3	2-3
pCLN	0	0	0-1	2
PP	0	0	1-2	0
Ocular administration (OT)				
CONJ	4	3-4	4	2-3
NALT	0	2-3	1	1-2
sCLN	0	0	1-2	1-2
PP	0	0	1-2	1
Oral administration (GI)				
CONJ	0	0	0	0
NALT	0	0	0	0
CLN	0	0	0	0
PP	0	1-2	2-3	0

Tissue sections were scored as follows: 0=0-5; 1=5-20; 2=20-100; 3=100-250; 4= >250 microspheres per mm^{2}. Scores represent data obtained from a minimum of 10 tissue sections.

CONJ, conjunctiva. NALT, nasal-associated lymphoid tissue. CLN, cervical lymph nodes. PP, Peyer's patches. p, posterior. s, superior.

CONJ, NALT, or pCLN and sCLN at any time point, but were found in PP 30 min and 6 h post delivery. In all cases the numbers of microspheres were greatest in tissues at the delivery site. For IN and OT delivery, microsphere uptake occurred at the respective delivery sites (NALT and CONJ) at the earliest time point (10 min). For GI delivery, the PP uptake was not evident until 30 min.

Antibody expression following exposure to PLG-MP-encapsulated DNP-BSA was studied; the data are presented in Fig. 1. Fig. 1A shows tear IgA DNP-BSA antibody responses obtained after primary, secondary, and tertiary IN or OT immunization. IN antigen delivery elicited the highest tear IgA responses after each immunization. Interestingly, IgA antibody levels were most vigorous after secondary immunization. OT immunization did not elicit significant tear IgA antibody responses until after tertiary immunization, and the response was both delayed and significantly less vigorous (with the exception of the day 42 time point) than those noted following the tertiary IN delivery. As expected, the control (unimmunized) group did not display tear IgA antibody responses. Fig. 1B shows

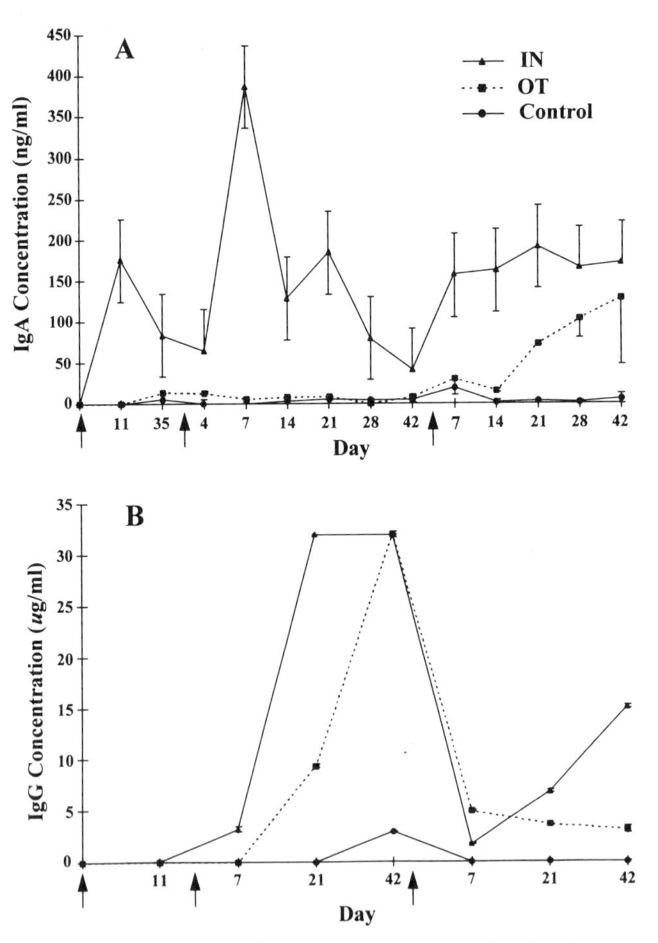

Figure 1. DNP-BSA antibody responses following primary, secondary, and tertiary immunization (arrows) with PLG-MP containing DNP-BSA. Data represent the mean ± SD of (A) tear IgA or (B) serum IgG antibody concentrations. Rats were immunized by the intranasal (IN) or ocular topical (OT) routes. Control rats were unimmunized.

serum IgG DNP-BSA antibody responses following primary, secondary, and tertiary IN or OT immunizations. Neither group showed serum IgG responses following primary immunization. However, both IN and OT delivery resulted in significant secondary serum IgG responses, with IN delivery resulting in more vigorous responses at earlier time points. Following tertiary IN immunization, the serum IgG responses increased, although not to the levels obtained following secondary immunization. IgG responses did not increase for the OT group following tertiary immunization. With the exception of one time point, IgG serum antibodies were not detected in the control (unimmunized) group.

4. DISCUSSION

The fluorescent microsphere data show that uptake is most rapid and pronounced in tissues at the targeted delivery sites. IN administration resulted in NALT uptake by 10 min and retention for the 18 h period studied. Appearance of microspheres by 6 h in pCLN, which drain NALT, reflects transport resulting from either excess microspheres in NALT or an inability to process completely the carboxylate particles. PP uptake by 6 h most likely reflects particle drainage into the gastrointestinal tract following IN delivery. OT administration resulted in CONJ uptake by 10 min and retention for the 18 h period studied. By 6 h, microspheres were detected in sCLN, which drain the CONJ and reflects transport resulting from excess or incomplete processing of microspheres. Interestingly, OT-delivered microspheres also access NALT by 30 min, and this appears to result from nasolacrimal canal passage. These latter data are consistent with earlier studies documenting microsphere NALT access and uptake following OT administration.[16] PP uptake by 6 h after OT delivery appears to result from drainage of a portion of the microspheres into the gastrointestinal tract subsequent to nasolacrimal canal passage. Support for this latter explanation comes from studies showing that radiolabeled soluble antigen can reach the stomach of rats 1–2 h after topical application to the ocular surface.[18] Importantly, oral administration of microspheres resulted in uptake only by PP in GALT. The earlier appearance of PP uptake (30 min vs. 6 h) following oral delivery appears to be associated with efficiencies of the intubation method.

The antibody induction data indicate that IN immunization results in an earlier and more vigorous tear IgA antibody response than does OT immunization. These findings are consistent with a previous report that compared tear IgA antibody responses in rats immunized with the particulate antigen, dinitrophenylated *Streptococcus pneumoniae.*[16] Secondary tear IgA antibody responses were greatest in animals receiving IN (as compared with OT and GI) immunization. These latter data, coupled with our findings, suggest that IN immunization is the most effective method to elicit tear IgA antibody responses. With respect to serum IgG antibodies, both IN and OT immunization elicit strong secondary responses with the maximal IN antibody levels preceding those obtained following OT delivery. Interestingly, following tertiary immunization, IN IgG antibody levels begin to rise again while the OT levels do not. This latter observation has been noted previously in rats receiving OT immunization with PLG-MP-encapsulated DNP-BSA.[13] Although the explanations for the enhanced efficiency of the IN route to elicit antibody responses require further investigation, it appears that the potential to process antigen at multiple sites (CONJ and NALT) in OT animals does not contribute to enhanced antibody responses.

In conclusion, these findings show that delivery-site specific antigen uptake can occur in the CONJ and NALT as well as GALT and, taken together with the antibody response data, indicate that NALT functions as a primary inductive site for ocular mucosal IgA antibody responses.

ACKNOWLEDGMENTS

This work was supported by NIH grants EY05133 and EY04068.

REFERENCES

1. Sullivan DA. Ocular mucosal immunity. In: Ogra PL, Lamm ME, McGhee JR, Mestecky J, Strober W, Bienstock J, eds. *Handbook of Mucosal Immunology*. New York: Academic Press; 1994:569–597.
2. Montgomery PC, Whittum-Hudson J. Mucosal immunity in the ocular system. In: Kiyono H, Ogra PL, McGhee JR, eds. *Mucosal Vaccines*. New York: Academic Press;1996:403–423.
3. Sullivan DA, Allansmith MR. Source of IgA in tears of rats. *Immunology*. 1984;53:791–799.
4. Peppard JV, Montgomery PC. Studies on the origin and composition of IgA in rat tears. *Immunology*. 1987;62:193–198.
5. Montgomery PC, Ayyildiz A, Lemaitre-Coelho IM, Vaerman JP, Rockey JH. Induction and expression of antibodies in secretions: The ocular immune system. *Ann NY Acad Sci*. 1983;409:428–440.
6. Montgomery PC, Skandera CA, Majumdar AS. Evidence for migration of IgA bearing lymphocytes between peripheral mucosal sites. *Coll Protides Biol Fluids*. 1985;32:43–46.
7. Montgomery PC, Majumdar AS, Skandera CA, Rockey JH. The effect of immunization route and sequence of stimulation on the induction of IgA antibodies in tears. *Curr Eye Res*. 1984;3:861–865.
8. Montgomery PC, Rockey JH, Majumdar AS, Lemaitre-Coelho IM, Vaerman JP, Ayyildiz A. Parameters influencing the expression of IgA antibodies in tears. *Invest Ophthalmol Vis Sci*. 1984;25:369–373.
9. Peppard JV, Mann RV, Montgomery PC. Antibody production in rats following ocular-topical or gastrointestinal immunization: Kinetics of local and systemic antibody production. *Curr Eye Res*. 1988;7:471–481.
10. Mestecky J, McGhee JR, Arnold RR, Michalek SM, Prince SJ, Babb JL. Selective induction of an immune response in human external secretions by ingestion of a bacterial antigen. *J Clin Invest*. 1978;61:731–737.
11. Gregory RL, Filler SJ. Protective secretory immunoglobulin A antibodies in humans following oral immunization with *Streptococcus mutans*. *Infect Immun*. 1987;55:2409–2415.
12. Bergman KC, Waldman RH. Stimulation of secretory antibody following oral administration of antigen. *Rev Infect Dis*. 1988;10:939–950.
13. Rafferty DE, Elfaki MG, Montgomery PC. Preparation and characterization of a biodegradable microparticle antigen/cytokine delivery system. *Vaccine*. 1996;14:532–538.
14. Spit BE, Hendriksen EGJ, Bruijnjes JP, Kuper C.F. Nasal lymphoid tissue in the rat. *Cell Tiss Res*. 1989;255:193–198.
15. Kuper CF, Koornstra PJ, Hameleers DMH, et al. The role of nasopharyngeal lymphoid tissue. *Immunol Today*. 1992;13:219–224.
16. Carr RM, Lolachi CM, Albaran RG, Ridley DM, Montgomery PC, O'Sullivan NL. Nasal-associated lymphoid tissue is an inductive site for rat tear IgA antibody responses. Immunol Invest. 1996;25:387–396.
17. Jeffery H, Davis SS, O'Hagan DT. The preparation and characterization of poly(lactide-co-glycolide) microparticles. II. The entrapment of a model protein using a (water-in-oil)-in-water emulsion solvent evaporation technique. *Pharmacol Res*. 1993;10:362–368.
18. Sullivan DA, Yee L, Conner AS, Hann LE, Olivier M, Allansmith MR. Influence of ocular surface antigen on the postnatal accumulation of immunoglobulin-containing cells in the rat lacrimal gland. *Immunology*. 1990;71:573–580.

79

ANATOMY OF MAMMALIAN CONJUNCTIVAL LYMPHOEPITHELIUM

James Chodosh,[1] Robert E. Nordquist,[1] and Ronald C. Kennedy[2]

[1]Department of Ophthalmology
Molecular Pathogenesis of Eye Infection Research Center
Dean A. McGee Eye Institute
[2]Departments of Microbiology and Immunology
University of Oklahoma Health Sciences Center
Oklahoma City, Oklahoma

1. INTRODUCTION

The eye's external surfaces (conjunctiva and cornea) encounter both airborne and contact pathogens.[1] The eyelids, an intermittent barrier, periodically wipe the eye's surface free of debris and spread the preocular tear film. Aqueous tears from the main and accessory lacrimal glands bathe the ocular surface with lysozyme, lactoferrin, and other factors of innate defense, and deliver monospecific secretory immunoglobulin A (sIgA). Goblet cell-derived and intrinsic mucins mix with the aqueous tears and obfuscate attachment of pathogenic organisms. Meibomian gland-derived lipids lubricate opposing ocular surfaces, and indirectly bolster innate ocular surface defense by delay of aqueous tear evaporation. The conjunctiva and cornea are anatomically proximate, are awash in the same tear film, and encounter the same microbes. Yet, the constitutive immune armaments of the conjunctiva and cornea are dramatically different (Table 1).

The cornea and its preocular tear film form the major refractive interface. Because corneal inflammation and antecedent scar formation reduce vision, the cornea's fine function is best served by a reduced immunologic responsiveness. Considered an immune-privileged site due to the high success rate of corneal transplantation, the cornea is avascular and without lymphatics, lacks resident lymphoid cells or Langerhans cells centrally,[2] expresses Fas ligand on its surface epithelium,[3] and demonstrates reduced delayed-hypersensitivity responses.[4] Inflammation of the infected central cornea follows corneal stromal fibroblast chemokine synthesis.[5] The cornea is susceptible to conjunctival dysfunction, because much of the preocular tear film derives from resident conjunctival cells or intraconjunctival glands. Therefore, inflammation and scarring of the conjunctiva may lead indirectly to reduced corneal clarity and blurred vision.

Lacrimal Gland, Tear Film, and Dry Eye Syndromes 2
edited by Sullivan *et al.*, Plenum Press, New York, 1998

Table 1. Comparison of corneal and conjunctival immunology

	Cornea	Conjunctiva
Blood vessels/lymphatics	No	Yes
Resident lymphoid cells	No	Yes
Langerhans cells[2]	Peripheral cornea only	Yes
Fas ligand[3]	Yes	(Not known)
Delayed-type hypersensitivity[4]	Reduced	Normal
Allograft success	High probability with only minimal need for local immunosuppression	Requires systemic immunosuppression for graft survival

The conjunctiva is a loose and redundant tissue that houses many of the glands of tear production, and along with the preocular tear film allows the eye free movement with a minimum of friction. In contrast to the cornea, the conjunctiva is very vascular, contains lymphatic channels, resident lymphoid cells, and Langerhans cells, and demonstrates classical hypersensitivity responses. Although the immunology of the cornea is under intense investigation,[4] this report will focus on known aspects of the immunology of the conjunctiva, where research efforts have been less concentrated.

2. CELLULAR IMMUNE ANATOMY OF HUMAN CONJUNCTIVA

Human conjunctiva contains lymphoid cells in isolation, within the substantia propria and epithelium, and in aggregate, within the substantia propria[6] (Table 2). A number of studies have examined the immune cell types within conjunctiva. Differences in methodology, site and extent of conjunctival biopsies, and age of human subjects tested make such studies difficult to compare. We will attempt to describe and review these studies in light of our own recent observations.

Early investigations[7] demonstrated by light and transmission electron microscopy the presence of plasma cells within human limbal conjunctiva. By light microscopy, Allansmith and coworkers[8] evaluated the plasma cell content of whole conjunctiva (six deceased donors), inferior fornix (nine subjects), superior tarsal conjunctiva (seven subjects), and limbus (two subjects). Lacrimal gland and accessory lacrimal gland-associated plasma cells outnumbered plasma cells within non-lacrimal gland-associated substantia propria by 3:2. In a subsequent study, Allansmith and coworkers[9] evaluated the inflammatory cell content of clinically uninflamed human conjunctiva biopsied from the upper tarsus (15 normal subjects) and lower forniceal conjunctiva (ten normal subjects). Lymphocytes accounted for the majority of inflammatory cells and were seen in both epi-

Table 2. Lymphocyte subpopulations within human conjunctiva

Author	Intraepithelial	Substantia propria	Follicles
Allansmith et al.[8,9]	LC	LC > PC	Present
Bhan et al.[10]	CD8	PC > CD4 = CD8	nd
Sachs et al.[11]	CD8	CD4 = CD8	B cell, CD4
Dua et al.[12]	CD8	CD4 = CD8	B cell, nd

LC, lymphocytes, not otherwise specified. PC, plasma cells. nd, otherwise not described. CD8, CD8+ suppressor/cytotoxic T lymphocytes. CD4, CD4+ helper T lymphocytes. B cell, B lymphocytes.

thelium and substantia propria. All samples contained plasma cells and mast cells in the substantia propria, but not in the epithelium. One-half of the forniceal conjunctival biopsies and one-third of the upper tarsal samples contained conjunctival follicles. In both studies, determinations of inflammatory cell phenotype were based on light microscopy alone.

Investigations by Bhan and colleagues,[10] using biopsies of clinically normal inferior bulbar conjunctiva from three patients, characterized T-cell subsets with a panel of lymphocyte- specific monoclonal antibodies and an immunoperoxidase technique. Suppressor/cytotoxic (CD8+) T cells predominated in normal conjunctival epithelium, while helper (CD4+) and suppressor/cytotoxic T cells populated the substantia propria in about equal numbers. It was of interest that the most predominant inflammatory cells in the substantia propria were plasma cells. With a similar immunohistochemical technique, Sachs and coworkers[11] characterized lymphocyte subpopulations in 2 mm diameter biopsy specimens from epibulbar, tarsal, and forniceal conjunctiva (20 patients undergoing cataract or oculoplastic surgery). The authors found mostly CD8+ suppressor/cytotoxic T cells in the epithelium, and a roughly equal distribution of CD8+ and CD4+ T cells in the substantia propria. Using the murine monoclonal antibody OKT10 that stains a cluster of differentiation antigens expressed by both plasma cells and activated T cells, the authors identified no plasma cells within substantia propria except for those within the accessory lacrimal glands. B cells rarely were seen except in conjunctival follicles.

Dua and coworkers[12] examined the immunophenotype of bulbar conjunctival lymphocytes and, in agreement with the above studies, found predominantly CD8+ T cells within the epithelium. Many of the CD8+ cells also expressed the Human Mucosal Lymphocyte antigen (HML-1). CD4+ and CD8+ phenotypes populated the substantia propria in roughly equal numbers. Within conjunctival follicles, CD4+ T cells predominated over CD8+ T cells by 2:1. Follicles also contained CD22+ B lymphocytes.

As with plasma cells in conjunctiva, a controversy exists regarding the expression of secretory component by conjunctival epithelial cells. Lai, Fat, and coworkers[13] showed evidence of secretory component production by cultured human conjunctiva. However, the technique employed, namely immunoelectrophoresis of culture supernatants, did not exclude lacrimal accessory glands within the cultured tissue. Allansmith and Gillette[14] examined whole conjunctiva (seven deceased donors) and biopsy specimens (four autopsy eyes and three normal patients) by immunofluorescence for human secretory component. Secretory component synthesis was observed within lacrimal gland acini, but not within conjunctiva. In this latter study, it was not reported whether conjunctival follicles were contained within the studied specimens. Tagawa and coworkers[15] biopsied five patients with follicular conjunctivitis, and found secretory component expression in the conjunctival epithelium over activated follicles. The debates over plasma cells and secretory component expression in conjunctiva warrant additional studies.

3. THE CONJUNCTIVAL LYMPHOID FOLLICLE: AN ORGANIZED MUCOSA-ASSOCIATED LYMPHOID TISSUE?

Foreign antigens and microbes bathe our mucosal surfaces. Gastrointestinal and respiratory mucosae possess accumulations of lymphoid cells organized as primary or secondary lymphoid follicles in isolation and in aggregate; these follicles sample the mucosal lumen for polyvalent antigens and activate production of protective secretory antibodies. Sites of antigen sampling and immune induction have been termed organized mucosa-associated lymphoid tissue (O-MALT), and include lymphoid nodules within the adenoids, tonsils, bronchi, and intestines.[16] Organized mucosa-associated lymphoid tissue, first described in mammalian ileal Peyer's patches,[17] is defined by the distinctive ultrastructure of mucosal epithelium over Peyer's patches. In O-MALT, epithelial goblet cells are replaced by ultrastructurally distinct antigen-absorptive cells, termed M cells (or microfold cells) for the appearance of their apical cell membranes. Within Peyer's patch lymphoepithelium, enterocytes surround and isolate each M cell, so that M cells are not seen immediately adjacent to one another.

Peyer's patch M cells are uniquely suited for antigen uptake and subsequent transit. In comparison to adjacent enterocytes, the M cell's limited glycocalyx and shorter microvilli present a reduced barrier to lumen pathogens. The apical microfolds allow a greater surface area for pathogen adherence and subsequent endocytosis. A less organized terminal web and sparser cytoplasm permit rapid transit of endocytosed pathogens. A pocket of CD4+ T helper cells, B lymphocytes, and macrophages invaginates the M cell's basolateral surface. Only a thin rim of apical cytoplasm separates the intestinal lumen from the mucosal immune system. Transport of antigen from lumen to lymphocyte is rapid.[16]

Antigen-primed IgA-committed B lymphoblasts from M cell pockets proliferate within local lymphoid follicle germinal centers, migrate to adjacent as well as distant mucosal surfaces, differentiate into plasma cells, and secrete dimeric and polymeric IgA. Basolateral expression of secretory component by mucosal epithelial cells allows transport of monospecific sIgA into mucosal secretions.

M cells are most proximal on the afferent limb of the mucosal immune system, but present a double-edged sword in mucosal defense.[16] Pathogenic bacteria such as *Salmonella typhi* and *Shigella flexneri* first breach the intestinal barrier via M cell uptake; gastrointestinal epithelial ulcerations associated with these infections begin over lymphoid follicles. Reovirus, poliovirus, human immunodeficiency virus, and other pathogenic human viruses can initiate infection via M cells. Therefore, despite their central role in initiation of the mucosal immune response, M cells also present a weak link in mucosal barrier function.

Conjunctival lymphocyte aggregates are recognized clinically as lymphoid follicles.[6] Follicular hyperplasia follows infection of the conjunctiva with *Chlamydia*, viruses, and hyperpathogenic bacteria such as *Neisseria gonorrhoeae*, and is associated with palpable enlargement of ipsilateral preauricular lymph nodes, suggesting immune activation. Yet, despite the recognized importance of lymphoid hyperplasia as a clinical sign, the biologic function of conjunctival follicles is not known. Axelrod and Chandler[18] suggested that conjunctival lymphoid follicles represent an O-MALT and are part of the common mucosal immune system. We will now review existing anatomic studies of conjunctival lymphoid follicles.

3.1. Avian

Fix and Arp[19] first described multiple nodules of lymphocytes beneath the inferior palpebral conjunctiva and fornix in normal and *Bordetella avium*-infected turkeys. Absolute numbers of follicles and the prevalence of germinal centers within follicles increased with bird age and conjunctival infection. The epithelium over lymphoid follicles lacked goblet cells and by transmission electron microscopy possessed short, sparse microvilli. The authors postulated local antigen presentation within avian conjunctival lymphoid tissue, and equated avian conjunctival lymphoepithelium with M cells of Peyer's patches. Subsequently, these investigators described a similar ultrastructural anatomy in chickens,[20] and demonstrated selective uptake of carbon and iron oxide by the conjunctival lymphoid tissue of chickens[21] and turkeys.[22]

3.2. Rodent

McMaster and coworkers[23] reported sparse numbers of lymphocytes, but no lymphoid follicles, within the conjunctiva of NIH-Swiss mice and Fisher rats. Similarly, Setzer and colleagues[24] demonstrated rare lymphocytes and macrophages within Sprague-Dawley rat conjunctiva, but no plasma cells or lymphoid aggregates. A role for rodent conjunctiva in adaptive immune defense does not appear plausible based on these observations.[25]

Several laboratories have attempted to stimulate mucosal immunity[26,27] or tolerance[28] in rodents with a soluble antigenic determinant delivered in an eyedrop. Implicit in these efforts is the assumption that conjunctival lymphoid tissue is the site of immune activation or tolerance induction. Such studies do not account for the lack of organized lymphoid tissue in rodent conjunctiva. In addition, because the tears drain into the nasopharynx and from there into the bronchus and gut, topically applied antigens[29–33] may not reliably test the contribution of conjunctiva to systemic mucosal immune responses.

3.3. Rabbit

Axelrod and Chandler[18] reported the presence of lymphoid follicles within palpebral rabbit conjunctiva, and suggested the designation conjunctiva-associated lymphoid tissue (CALT). Despite an absence of follicle germinal centers, mitotic figures were common in follicle lymphocytes. Nearby lymphatics contained many lymphoid cells. By immunofluorescence for rabbit immunoglobulin heavy chains, no plasma cells could be identified within the follicles. These investigators also showed scanning electron microscopic evidence for longer microvilli on the lymphoepithelium overlying rabbit CALT than on the epithelium not associated with lymphoid follicles,[34] a puzzling finding in light of the short, irregular microvilli on classical M cells. The authors suggested a function for rabbit conjunctival lymphoepithelium in antigen uptake. Their proposed local mucosal immune arc extended from the conjunctiva to regional lymph nodes and thoracic duct, into the bloodstream, and back to the lacrimal gland, the site of sIgA secretion into the tears. Other investigators demonstrated by immunofluorescence in rabbit lymphoid follicles a high proportion of IgA-committed B cells, but not plasma cells, and suggested that conjunctiva acts as a site of lymphocyte activation for dissemination to distant mucosal sites.[35] Yet, Franklin and coworkers[36] identified plasma cells within rabbit conjunctival substantia propria, and along with Liu and colleagues,[37] showed secretory component expression in rabbit conjunctival epithelium, findings consistent with immune effector function.

3.4. Guinea Pig

McMaster and coworkers[23] demonstrated dramatic numbers of lymphocytes in guinea pig conjunctiva, even in those animals raised in a germ-free environment. Guinea pig conjunctival lymphocytes tended to gather in masses of lymphoid follicles. Dwyer and colleagues[38] studied conjunctiva from fetal, newborn, and adult female albino Dunkin-Hartley guinea pigs. They found lymphoid aggregates in all eyes studied, and suggested that lymphoid follicles were a normal component of guinea pig conjunctiva. Lymphoid nodules were also described within the forniceal conjunctiva of adult pigmented guinea pigs.[39] By transmission electron microscopy, all the conjunctival epithelium over lymphoid nodules had short, irregular microvilli compared to conjunctival epithelium not associated with lymphoid tissue. In a follow-up study, the author suggested that guinea pig conjunctival lymphoepithelial microvilli lacked core actin filaments.[40] These data seem to suggest that the entire epithelium overlying conjunctival lymphoid follicles is specialized to sample antigens on the conjunctival surface. Such a model remains unproven, but is similar to that proposed for bronchus-associated lymphoid tissue.[41]

Evidence regarding specific uptake of antigen by mammalian conjunctival lymphoepithelium would lend support to the concept of CALT. In this light, investigators reported that uptake of horseradish peroxidase by guinea pig lymphoepithelium was no greater than by non-lymphoepithelium.[42] In contrast, other investigators reported that guinea pig conjunctival lymphoepithelium phagocytized horseradish peroxidase faster and more efficiently than did non-lymphoepithelium.[43] It remains unclear why these latter two studies, performed with similar methods, yielded such contrasting results.

3.5. Nonhuman Primate

Ruskell[44] described lymphoid tissue within the conjunctiva of macaque monkeys (*Macaca fascicularis*). The upper and lower conjunctival fornices contained lymphoid follicles, and most follicles contained germinal centers. By transmission electron microscopy, the follicles' lymphoid cells included lymphocytes, reticulocytes, macrophages, dendritic cells, and plasma cells. Follicles possessed an overlying lymphoepithelium similar to that previously described in guinea pig conjunctiva with small, irregular microvilli. Unlike Peyer's patch lymphoepithelium, conjunctival lymphoepithelial ultrastructure appeared uniform and lacked clear distinctions between putative M cell and adjacent non-M cell.

3.6. Human

In 1944, Osterlind[45] used gallocyanine staining of freshly excised cadaver conjunctiva along with light microscopic confirmation to demarcate and count human conjunctival lymphoid follicles in 92 human subjects, newborn to 81 years old. Conjunctival follicles, absent at birth, developed rapidly to a peak of over 30 follicles on average by age 10, and then slowly declined with age.

Wotherspoon and coworkers[46] described lymphoid follicles with germinal centers in 31% of conjunctivae from 88 human autopsy subjects. The authors concluded that O-MALT was not a normal finding in human conjunctiva, but did not specify if serial sections were reviewed.

Biopsied conjunctiva from five patients with follicular conjunctivitis was examined by immunohistochemistry to characterize the immunophenotype within different areas of each follicle.[15] The authors concluded that conjunctival follicle germinal centers contained B lymphocytes positive for surface IgM and IgD expression, and T lymphocytes (T helper > T suppressor/cytotoxic cells). Parafollicular areas contained mostly T helper cells. Subepithelial areas contained plasma cells positive for IgA, and follicular epithelium expressed secretory component, suggesting that secretory component expression by conjunctival epithelium might be regulated by underlying lymphoid tissue.

4. SUMMARY

Ocular surface immune mechanisms are subservient to the fine function of the eye. A clear cornea with a smooth, well-lubricated facade is prerequisite to lucid vision. Hence, corneal inflammation and post-inflammatory scarring are intolerable, and the cornea contains a minimum of lymphoid elements. Although conjunctival dysfunction and consequent tear film deficiency can malign the corneal surface, conjunctival inflammation is tolerated to a considerable degree. In contrast to the human cornea, human conjunctiva contains an abundance of lymphoid tissue.

Certain aspects of human conjunctival immunology elicit little debate. Langerhans cells are abundant in conjunctival epithelium. Isolated CD8+ suppressor/cytotoxic T cells predominate in conjunctival epithelium, while T cells in the substantia propria distribute equally between CD4+T helper cells and CD8+ cells. Yet the presence of plasma cells in human conjunctiva, the expression of secretory component by human conjunctival epithelium, and the function of human conjunctival lymphoid follicles are in dispute.[47] Confusion may derive in part from the use of inappropriate animal models; rodent conjunctiva does not appear to be a worthy facsimile for human conjunctiva. Discrepancies between different human studies likely result from variance in subject age, biopsy site and extent, histologic or histochemical technique, and perhaps the degree of inflammation present at the time of biopsy.[45] Careful immunohistochemical and in situ molecular assays on well-defined loci within the conjunctiva of comparable human subjects may resolve such questions in the future.

Organized mucosa-associated lymphoid tissue is rigorously defined as mucosal lymphoid follicles with an ultrastructurally distinct overlying lymphoepithelium.[16] Based on available evidence, the epithelium overlying mammalian conjunctival lymphoid follicles does not contain distinct M cells. Whether zonal differences in morphology reflect real differences in the capacity to sample tear film antigens for presentation to the mucosal immune system remains to be established.

ACKNOWLEDGMENTS

This work was supported in part by Public Health Service grant EY00357 from the National Institutes of Health and a Research to Prevent Blindness Career Development Award to Dr. Chodosh.

REFERENCES

1. Smolin G. The defence mechanisms of the outer eye. *Trans Ophthalmol Soc UK.* 1985;104:363–366.
2. Rodrigues MM, Rowden G, Hackett J, Bakos I. Langerhans cells in the normal conjunctiva and peripheral cornea of selected species. *Invest Ophthalmol Vis Sci.* 1981;21:759–765.
3. Griffith TS, Brunner T, Fletcher SM, Green DR, Ferguson TA. Fas ligand-induced apoptosis as a mechanism of immune privilege. *Science.* 1995;270:1189–1192.
4. Streilein JW. Ocular immune privilege and the Faustian dilemma. *Invest Ophthalmol Vis Sci.* 1996;37:1940–1950.
5. Oakes JE, Monteiro CA, Cubitt CL, Lausch RN. Induction of interleukin-8 gene expression is associated with herpes simplex virus infection of human corneal keratocytes but not human corneal epithelial cells. *J Virol.* 1993;67:4777–4784.
6. Duke-Elder S. *Diseases of the Outer Eye*, part 1. *System of Ophthalmology*, Vol VIII. St. Louis: Mosby; 1965:56–58.
7. Quigley HA, Kenyon KR. Russell bodies and plasma cells in human conjunctiva. *Am J Ophthalmol.* 1973;76:957–966.
8. Allansmith MR, Kajiyama G, Abelson MB, Simon MA. Plasma cell content of main and accessory lacrimal glands and conjunctiva. *Am J Ophthalmol.* 1976;82:819–826.
9. Allansmith MR, Greiner JV, Baird RS. Number of inflammatory cells in the normal conjunctiva. *Am J Ophthalmol.* 1978;86:250–259.
10. Bhan AK, Fujikawa LS, Foster CS. T-cell subsets and Langerhans cells in normal and diseased conjunctiva. *Am J Ophthalmol.* 1982;94:205–212.

11. Sachs EH, Wieczorek R, Jakobiec FA, Knowles DM. Lymphocyte subpopulations in the normal human conjunctiva: A monoclonal antibody study. *Ophthalmology*. 1986;93:1276–1283.
12. Dua HS, Gomes JAP, Jindal VK, et al. Mucosa specific lymphocytes in the human conjunctiva, corneoscleral limbus and lacrimal gland. *Curr Eye Res*. 1994;13:87–93.
13. Lai A, Fat RFM, Suurmond D, van Furth R. In vitro synthesis of immunoglobulins, secretory component and complement in normal and pathological skin and the adjacent mucous membranes. *Clin Exp Immunol*. 1973;14:377–395.
14. Allansmith MR, Gillette TE. Secretory component in human ocular tissues. *Am J Ophthalmol*. 1980;89:353–361.
15. Tagawa Y, Saito M, Kosaka T, Matsuda H. Lymphocyte subsets of human conjunctival follicles. In: Secchi AG, Fregona IA, eds. *Modern Trends in Immunology and Immunopathology of the Eye*. Milan: Masson; 1986:344–345.
16. Neutra MR, Frey A, Kraehenbuhl J-P. Epithelial M cells: Gateways for mucosal infection and immunization. *Cell*. 1996;86:345–348.
17. Owen RL, Jones AL. Epithelial cell specialization within human Peyer's patches: An ultrastructural study of intestinal lymphoid follicles. *Gastroenterology*. 1974;66:189–203.
18. Axelrod AJ, Chandler JW. Morphologic characteristics of conjunctival lymphoid tissue in the rabbit. In: Silverstein AM, O'Connor GR, eds. *Immunology and Immunopathology of the Eye*. New York: Masson, 1979:292–301.
19. Fix AS, Arp LH. Conjunctiva-associated lymphoid tissue (CALT) in normal and *Bordetella avium*-infected turkeys. *Vet Pathol*. 1989:26:222–230.
20. Fix AS, Arp LH. Morphologic characterization of conjunctiva-associated lymphoid tissue in chickens. *Am J Vet Res*. 1991;52:1852–1859.
21. Fix AS, Arp LH. Quantification of particle uptake by conjunctiva-associated lymphoid tissue (CALT) in chickens. *Avian Dis*. 1991;35:174–179.
22. Fix AS, Arp LH. Particle uptake by conjunctiva-associated lymphoid tissue (CALT) in turkeys. *Avian Dis*. 1991;35:100–106.
23. McMaster PRB, Aronson SB, Bedford MJ. Mechanisms of the host responses in the eye. IV. The anterior eye in germ-free animals. *Arch Ophthalmol*. 1967;77:392–399.
24. Setzer PY, Nichols BA, Dawson CR. Unusual structure of rat conjunctival epithelium: Light and electron microscopy. *Invest Ophthalmol Vis Sci*. 1987;27:531–537.
25. Gudmundsson OG, Sullivan DA, Bloch KJ, Allansmith MR. The ocular secretory immune system of the rat. *Exp Eye Res*. 1985;40:231–238.
26. Montgomery PC, Rockey JH, Majumdar AS, Lemaitre-Coelho IM, Vaerman J-P, Ayyildiz A. Parameters influencing the expression of IgA antibodies in tears. *Invest Ophthalmol Vis Sci*. 1984;25:369–373.
27. Pu Z, Pierce NF, Silverstein AM, Prendergast RA. Conjunctival immunity: Compared effects of ocular or intestinal immunization in rats. *Invest Ophthalmol Vis Sci*. 1983;24:1411–1412.
28. Dua HS, Donoso LA, Laibson PR. Conjunctival instillation of retinal antigens induces tolerance: Does it invoke mucosal tolerance via conjunctiva associated lymphoid tissues (CALT)? *Ocul Immunol Inflamm*. 1994;2:29–36.
29. Hall JM, Pribnow JF. Topical ocular immunization of rabbits. *Invest Ophthalmol Vis Sci*. 1981;21:753–756.
30. Hall JM, Pribnow JF. Ocular immunization of guinea pigs. *Curr Eye Res*. 1987;6:817–824.
31. Hall JM, Pribnow JF. IgG and IgA antibody in tears of rabbits immunized by topical application of ovalbumin. *Invest Ophthalmol Vis Sci*. 1989;30:138–144.
32. MacDonald AB, McComb D, Howard L. Immune response of owl monkeys to topical vaccination with irradiated *Chlamydia trachomatis*. *J Infect Dis*. 1984;149:439–442.
33. Taylor HR, Pierce NF, Pu Z, Schachter J, Silverstein AM, Prendergast RA. Secretory immune cellular traffic between the gut and the eye. In: O'Connor GR, Chandler JW, eds. *Advances in Immunology and Immunopathology of the Eye* New York: Masson; 1985:208–211.
34. Chandler JW, Axelrod AJ. Conjunctiva-associated lymphoid tissue: A probable component of the mucosa-associated lymphoid system. In: O'Connor GR, ed. *Immunologic Diseases of the Mucous Membranes: Pathology, Diagnosis, and Treatment*. New York: Masson, 1980:63–70.
35. Franklin RM, Remus LE. Conjunctival-associated lymphoid tissue: Evidence for a role in the secretory immune system. *Invest Ophthalmol Vis Sci*. 1984;25:181–187.
36. Franklin RM, Prendergast RA, Silverstein AM. Secretory immune system of rabbit ocular adnexa. *Invest Ophthalmol Vis Sci*. 1979;18:1093–1096.
37. Liu SH, Tagawa Y, Prendergast RA, Franklin RM, Silverstein AM. Secretory component of IgA: A marker for differentiation of ocular epithelium. *Invest Ophthalmol Vis Sci*. 1981;20:100–109.

38. Dwycr RStC, Darouger S, Monnickendam MA. Unusual features in the conjunctiva and cornea of the normal guinea pig: Clinical and histological studies. *Br J Ophthalmol.* 1983;67:737–741.
39. Latkovic S. The ultrastructure of the normal conjunctival epithelium of the guinea pig. III. The bulbar zone, the zone of the fornix and the supranodular zone. *Acta Ophthalmol.* 1979;57:305–320.
40. Latkovic S. Ultrastructure of M cells in the conjunctival epithelium of the guinea pig. *Curr Eye Res.* 1989;8:751–755.
41. Pankow W, von Wichert P. M cell in the immune system of the lung. *Respiration.* 1988;54:209–219.
42. Stock EL, Sobut RA, Roth SI. The uptake of horseradish peroxidase by the conjunctival epithelium of the guinea-pig. *Exp Eye Res.* 1987;45:327–337.
43. Shoji J, Inada N, Kasai H, Ishii Y, Kitano S. Immunoresponses to the external antigen in conjunctival-associated lymphoid tissue. *Acta Soc Ophthalmol Jpn.* 1992;96:432–439.
44. Ruskell GL. Organization and cytology of lymphoid tissue in the cynomolgus monkey conjunctiva. *Anat Rec.* 1995;243:153–164.
45. Osterlind G. An investigation into the presence of lymphatic tissue in the human conjunctiva, and its biologic and clinical importance. *Acta Ophthalmol.* 1944;23:1–79.
46. Wotherspoon AC, Hardman-Lea S, Isaacson PG. Mucosa-associated lymphoid tissue (MALT) in the human conjunctiva. *J Pathol.* 1994;174:33–37.
47. Sullivan DA. Ocular mucosal immunity. In Ogra PL, Mestecky J, Lamm ME, et al., eds. *Handbook of Mucosal Immunology.* San Diego: Academic Press; 1994:569–597.

BINDING OF A CYTOPATHIC OR AN INVASIVE STRAIN OF *P. AERUGINOSA* TO CYTOSKELETAL, BASEMENT MEMBRANE, OR MATRIX PROTEINS OF WOUNDED CORNEA IS SIMILAR AND DOES NOT RELY ON INTERACTION WITH ACTIN FILAMENTS

Linda D. Hazlett and Sharon Masinick

Department of Anatomy/Cell Biology
Wayne State University School of Medicine
Detroit, Michigan

1. INTRODUCTION

Pseudomonas aeruginosa is a typically opportunistic bacterium. Although it is widely distributed in human biology and produces numerous virulence-associated factors, *P. aeruginosa* does not infect human tissues unless the normal host defense mechanisms have been impaired. The bacteria does not exhibit tissue specificity, shown by many human pathogens; it infects tissues of diverse origins but only ones that have been damaged in some way. Why the affinity for damaged tissues by this organism remains unanswered. In this regard, several studies have examined *P. aeruginosa* adherence to unmasked extracellular matrix (ECM) following tissue damage in both human[1,2] and experimental[3] systems. Tissue damage may result in exposure of the basement membrane, which consists mainly of collagen (Type IV) and glycoproteins such as fibronectin (FN), laminin (LN), and heparan sulfate proteoglycan. *P. aeruginosa* has been shown to bind to soluble FN, but not to polymerized cellular or plasma FN,[4] and to laminin.[5] But simply being able to adhere or even being able to enter a host cell is not sufficient, and pathogenicity usually exhibits a multifactoral phenotype.[6] Pili are the major adhesin by which *P. aeruginosa* interacts with mammalian host tissue,[7] but studies have shown that pilus-deficient mutants of *P. aeruginosa* bind to respiratory mucus[8] and to corneal epithelium[9] as well as to other non-epithelial cell types.[10] Binding to laminin, for example, may in-

Lacrimal Gland, Tear Film, and Dry Eye Syndromes 2
edited by Sullivan *et al.*, Plenum Press, New York, 1998

volve a non-pilus adhesin from the bacterial outer membrane.[5] It also has been shown that bacterial elastase and alkaline protease cleave soluble laminin,[11] providing for tissue invasion and hemorrhagic tissue necrosis, as well as exposure of lipase-sensitive receptors, allowing enhanced bacterial binding to the corneal epithelium.[12] Type IV collagen also is degraded by bacterial elastase[13] and may contribute to extensive basement membrane degradation. In addition, involvement of the epithelial intracellular cytoskeleton in pathogenesis has been shown for human[14] and mouse[15] corneal tissue. In both, cytokeratins (10 nm filaments) have been shown to provide binding sites for *P. aeruginosa* when the eye is injured, although no involvement of other cytoskeletal proteins such as microfilaments or microtubules were tested. These warrant attention, since many pathogenic organisms are capable of altering the host cytoskeleton architecture as manifested by a dramatic rearrangement of microtubule[16] and microfilament[17] proteins. For many bacterial, parasitic, and viral pathogens, the ability to bind to basement membrane/ECM glycoproteins, including FN, LN, and vitronectin, is considered an important component of their pathogenesis.[18] No study has addressed possible differences in tissue binding in a damaged epithelium between pseudomonal cytopathic vs. invasive bacterial strains in cornea. Therefore, the purpose of the current study was to compare in vitro binding of two strains of *P. aeruginosa*, one identified as cytopathic and non-invasive (ATCC 19660), and the other as cytopathic and invasive (PAO1),[19] in an injured cornea model to determine if these strains demonstrated differences in bacterial affinity for cytoskeletal components such as intermediate (10 nm) filaments (IFs), microfilaments, and microtubules. In addition, bacterial adherence to ECM and/or basement membrane components including FN, LN, and Type IV collagen also was examined. An in vitro eye organ culture model employing binding inhibition assays with antibodies specific for each of the above components was used.

2. MATERIALS AND METHODS

2.1. Mice

Female Swiss (ICR) mice (Harlan Sprague-Dawley, Indianapolis, IN), or male and female $\beta_{2m}^{-/-}$ mice, a transgenic knock-out mouse homozygous for the null allele of the β2 microglobulin gene, or C57Bl/6J x129/J x $\beta_2M^{-/-}$ mice bred within our mouse colony were used. Mice were housed in clear plastic cages, fed laboratory rodent chow and acidified water *ad libitum*. Experiments were done humanely and according to the ARVO Resolution on the Use of Animals in Research.

2.2. Bacterial Cell Cultures

Stock cultures of *P. aeruginosa*, American Type Culture Collection (ATCC) 19660 or PAO1 (kindness of Dr. S. Lory) were used for inoculation of 25 ml of broth medium containing 5% peptone and 0.25% Trypticase Soy broth (Sigma, St. Louis, MO).

Cultures were grown on a rotary shaker (150 rpm) at 37°C for 18 h to an OD_{540} of 1.6, centrifuged (6000*g*, 10 min) at 15°C, washed with sterile saline (0.85%, pH 7.2), and resuspended in saline to a concentration of 2.0×10^{10} colony forming units (cfu) per ml, using a standard curve relating viable counts to optical density at 540 nm. An inoculum of 5 μl containing 5.0×10^8 cfu was used in each experiment.

2.3. Organ Culture

Detailed methods for organ culture and wounding of mouse cornea have been described.[9] Briefly, mice were anesthetized with isofluorane (Aerrane Anaquest, Madison, WI), killed by cervical dislocation, and each central cornea incised (1 mm X 3) with a sterile 25 gauge needle. Under sterile conditions, eyes were enucleated and placed with corneas superiorly oriented into sterilized, prepared culture wells [each with 2–3 ml of Minimal Essential Medium (MEM) with Earle's salts, L-glutamine, and non-essential amino acids without sodium bicarbonate, pH 7.3 (Gibco)] at 37°C.

2.4. Scanning EM Adherence Assay

A 5 µl bacterial cell suspension containing 5.0 x 10^8 cfu *P. aeruginosa* ATCC 19660 or strain PAO1 was topically delivered onto the surface of each eye in culture with a calibrated micropipette (Cole-Parmer, Chicago, IL) fitted with a sterile, disposable tip. Eyes were put in a water-jacketed CO_2 incubator (American Scientific Products, McGaw Park, IL) at 37°C and 5% CO_2 for 60 min. Then the eyes were vigorously rinsed in PBS, fixed, dehydrated, and critical point dried for scanning EM, as described.[9]

2.5. Quantitation of Adherence

Adherent bacteria were quantitated at 60 min after ocular application using well-standardized scanning EM procedures.[9] Counts were done with a square cut to measure 80 mm^2 at 6000X. The data were expressed as the mean number of bacteria per counted field (n=15). An unpaired, two-group Student's t-test with a two-tailed hypothesis was used to test for significance ($p\ ^{2}\ 0.05$).

2.6. Antibody Treatment

The following panel of polyclonal cytoskeleton-reactive antibodies were used: anti-tubulin (Miles-Yeda), anti-actin (kindness of Dr. M. Ireland), J7 (kindness of Dr. M. Kurpakus) (which reacts with a 55 kDa cytokeratin), and a monoclonal anti-IF antibody (Pruss, ATCC, TIB 131), which recognizes all classes of IFs.[20] Basement membrane/ECM reactive antibodies used included a polyclonal anti-collagen Type IV (Collaborative), polyclonal anti-LN (Collaborative), and polyclonal anti-FN (Cappel). The antibodies were tested in binding inhibition assays for their ability to inhibit binding of *P. aeruginosa* 19660 (cytopathic) vs. the invasive strain PAO1. Antibodies were diluted in PBS at concentrations of 375, 188, and 94 µg/ml, and 5 µl was applied to the wounded surface of the mouse cornea just before application of 5.0 x 10^8 cfu bacteria. Adherence assays were done as described above.

3. RESULTS

3.1. Effect of Antibodies on the Binding of *P. aeruginosa* to Mouse Corneal Epithelium

Two of the antibodies against cytoskeletal components, anti-tubulin and Pruss (anti-IFs), both at 375 µg/ml concentration, significantly reduced binding of strain 19660 to mouse cornea when compared to PBS-treated control eyes ($p^2$0.04 and $p^2$0.0001 respec-

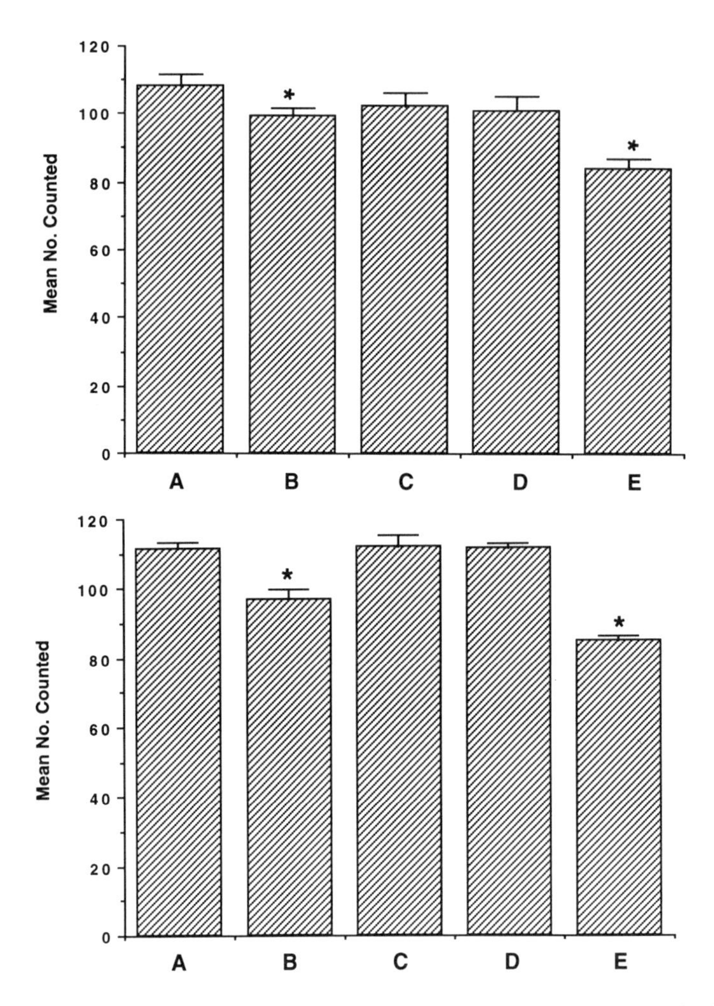

Figures 1 and 2. Inhibition of bacterial binding. Only anti-tubulin (B) and Pruss (E) antibody significantly (*) reduced bacterial binding (concentration=375 μg/ml) compared with the mean number of bacteria bound in the PBS-treated control group (A) for both strains 19660 (Fig. 1) and PAO1 (Fig. 2). Anti-actin (C) and J7 (D) antibody at similar concentration had no significant effect on binding for either bacterial strain.

tively) (Fig. 1). At the same concentration, similar reduction in binding of PAO1 was seen for both anti-tubulin ($p^{2}0.0006$) and Pruss ($p^{2}0.001$) antibodies when compared to binding in PBS-treated control eyes (Fig. 2). At decreased concentrations of antibody (188 and 94 μg/ml), only Pruss significantly reduced the number of bound bacteria of strain 19660 ($p^{2}0.05$ and $p^{2}0.05$ respectively) (Fig. 3) and of PAO1 ($p^{2}0.0004$ and $p^{2}0.0009$, respectively) (Fig. 4). All three of the basement membrane/ECM antibodies, including anti-collagen Type IV, anti-LN, and anti-FN, inhibited binding of *P. aeruginosa* strain 19660 at a concentration of 375 μg/ml ($p^{2}0.0001$, $p^{2}0.0002$, and $p^{2}0.0001$ respectively) (Fig. 5) and of PAO1 ($p^{2}0.003$, $p^{2}0.0008$, and $p^{2}0.0001$) (Fig. 6) when compared to the PBS control group. When lower concentrations of these antibodies were tested, treatment only with anti-LN antibody (188 μg/ml) reduced the number of bound bacteria for strain 19660 significantly ($p^{2}0.0016$) (Fig. 7). In contrast, for PAO1, significant reduction in the number

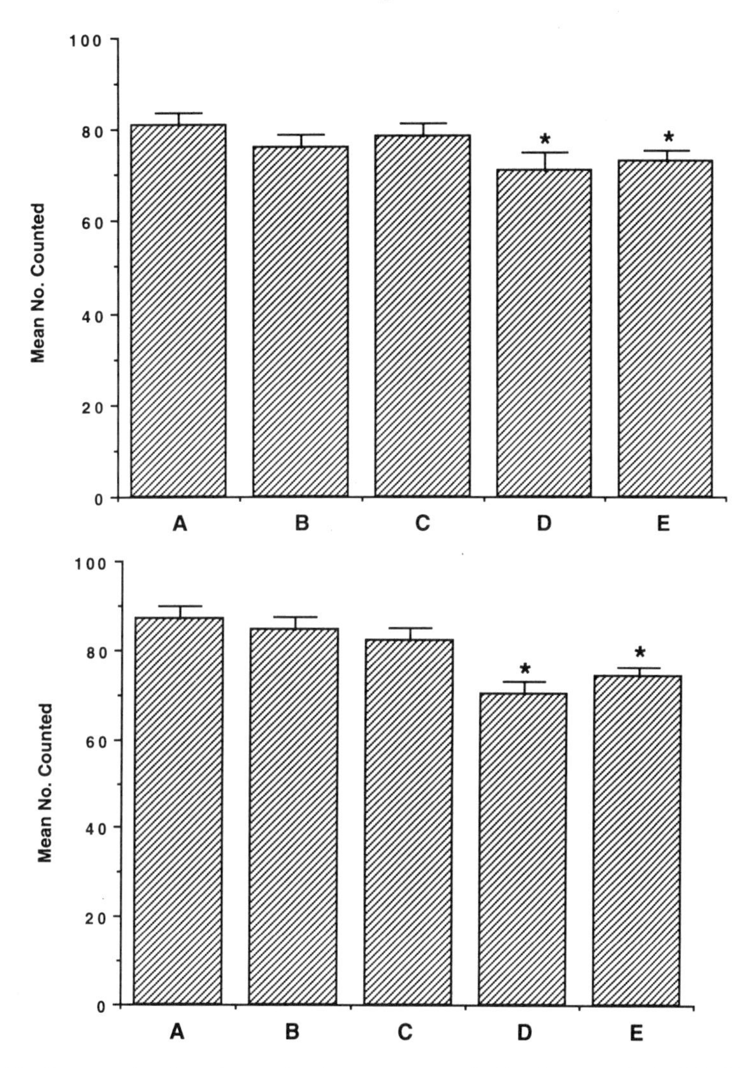

Figures 3 and 4. Concentration dependency of anti-cytoskeletal antibodies. Pruss antibody [188 (D) and 94 (E) μg/ml concentrations] significantly (*) decreased the mean number of bacteria bound for strain 19660 (Fig. 3) and strain PAO1 (Fig. 4) when compared with the respective PBS-treated control group (A). No significant effect was seen with anti-tubulin at either 188 μg/ml concentration (B) or 94 μg/ml concentration (C) for either bacterial strain.

of bacteria bound to cornea was seen following treatment with anti-collagen Type IV (188 μg/ml; $p^{2}0.05$), anti-LN (at both 188 and 94 μg/ml concentration; $p^{2}0.0001$ and $p^{2}0.0096$ respectively), and anti-FN antibody (188 μg/ml; $p^{2}0.026$) when compared to the PBS controls (Fig. 8).

4. DISCUSSION

In the normal body, mucosal-associated structures are covered by a protective barrier of plasma membrane-enclosed epithelial cells. Thus, intracellular components, such as

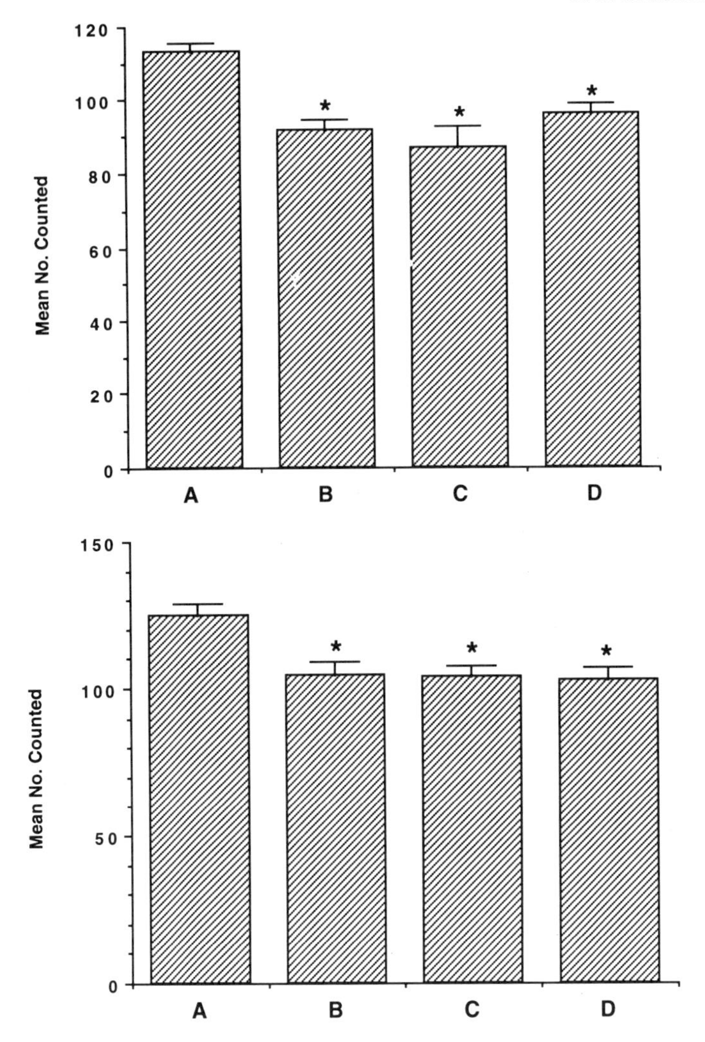

Figures 5 and 6. Inhibition of bacterial binding by anti-ECM/basement membrane antibodies. At 375 μg/ml, anti-collagen Type IV (B), anti-LN (C), and anti-FN (D) antibodies significantly (*) decreased the mean number of bacteria bound for strain 19660 (Fig. 5) and strain PAO1 (Fig. 6) when compared with the respective PBS-treated (A) control groups.

cytoskeletal proteins and ECM/basement membranes, are not readily available for bacterial adherence. Tissue damage resulting from various forms of trauma may expose the intracellular epithelial compartment as well as basement membrane/ECM, permitting bacterial colonization and infection. Many opportunistic pathogens express surface components that can interact with basement membrane/ECM molecules[4,5] as well as with the cytoskeletal[14,15] compartment of a cell. Binding of both cytopathic and invasive *Pseudomonas* strains to these cellular components was strikingly similar when tested in binding inhibition assays employing cytoskeletal, ECM/basement membrane specific antibodies. Admittedly, inhibition of binding of the invasive (PAO1) strain was more significant at lower concentrations of the antibodies than for the cytopathic strain (19660), particularly with antibodies against ECM/basement membrane. This difference may reflect decreased

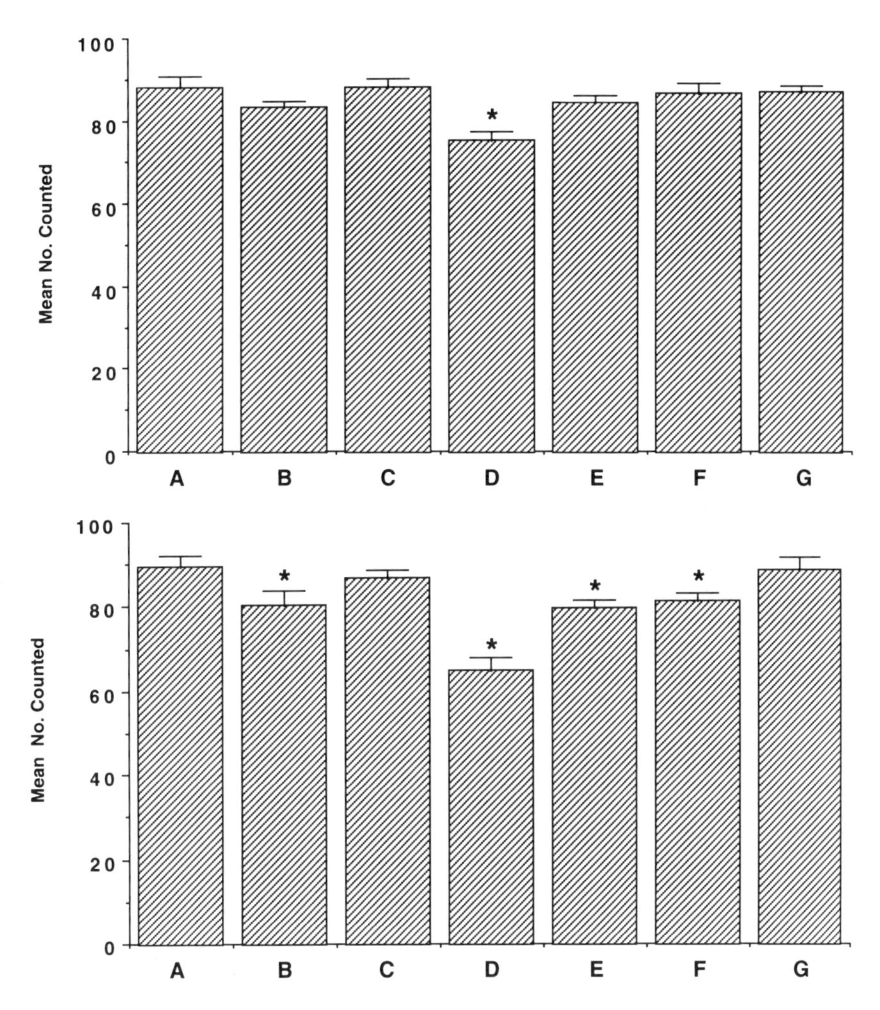

Figures 7 and 8. Concentration dependency of anti-ECM/basement membrane antibodies. At reduced concentrations, only anti-LN (D, 188 μg/ml concentration) significantly (*) decreased the mean number of bacterial strain 19660 bound when compared to the PBS-treated control (A, Fig. 7). Anti-collagen Type IV at 188 (B) or 94 (C) μg/ml, anti-LN at 94 μg/ml (E), and anti-FN at 188 (F) or 94 μg/ml (G) had no inhibitory effect. For strain PAO1 (Fig. 8), anti-collagen Type IV (B, 188 μg/ml), anti-LN (D, 188, and E, 94 μg/ml), and anti-FN (F, 188 μg/ml) all decreased the mean number of bound bacteria when compared to the PBS-treated control (A). No significant effect on binding was seen for anti-collagen Type IV (C, 94 μg/ml) or anti-FN (G, 94 μg/ml).

protease production in vitro by PAO1 when compared with 19660,[9] which could result in less proteolysis and exposure of fewer binding sites by this strain. These data suggest that invasive and cytopathic bacterial strains have affinity for cytoskeletal and ECM/basement membranes, which is of biological significance in the pathogenesis of *P. aeruginosa* ocular infection.

ACKNOWLEDGMENTS

This work was supported by NIH grants EY02986 and EY04068.

REFERENCES

1. Baltimore R, Christie C, Smith G. Immunohistopathological localization of *Pseudomonas aeruginosa* in lungs from patients with cystic fibrosis. *Am Rev Respir Dis.* 1989;140:1650–1661.
2. Philippon S, Streckert HJ, Morgenroth K. *In vitro* study of the bronchial mucosa during *Pseudomonas aeruginosa* infection. *Virchows Arch A Pathol Anat Histopathol.* 1993;423:39–43.
3. Hazlett L, Barrett R, Klettner C, Berk RS, Singh A. *Pseudomonas aeruginosa*: A probe to detail change with age in corneal membrane receptors. *Ophthalmic Res.* 1991;23:141–146.
4. Woods DE, Strauss DC, Johnson WG, Bass JA. Role of fibronectin in the prevention of adherence of *Pseudomonas aeruginosa* to buccal cells. *J Infect Dis.* 1981;143:784–790.
5. Plotowski MC, Tournier J-M, Puchelle E. *Pseudomonas aeruginosa* strains possess specific adhesins for laminin. *Infect Immun.* 1996;64:600–605.
6. Bliska JB, Galan JE, Falkow S. Signal transduction in the mammalian cell during bacterial attachment and entry. *Cell.* 1993;73:903–920.
7. Irwin RT, Doig P, Lee KK, et al. Characterization of the *Pseudomonas aeruginosa* pilus adhesin: Confirmation that the pilin structural protein subunit contains a human epithelial cell-binding domain. *Infect Immun.* 1989;57:3720–3726.
8. Ramphal R, Koo L, Ishimoto S, Totten PA, Lara JC, Lory S. Adhesion of *Pseudomonas aeruginosa* pilin-deficient mutants to mucin. *Infect Immun.* 1991;59:1307–1311.
9. Hazlett LD, Moon MM, Singh A, Berk RS. Analysis of adherence, piliation, protease production and ocular infectivity of several *P. aeruginosa* strains. *Curr Eye Res.* 1991;10:351–362.
10. Plotkowski MC, Saliba AM, Pereira SHM, Cervante MP, Bajolet-Laudinat O. *Pseudomonas aeruginosa* selective adherence to and entry into human endothelial cells. *Infect Immun.* 1994;62:5456–5463.
11. Heck LW, Morihara K, Abrahamson DR. Degradation of soluble laminin and depletion of tissue-associated basement membrane laminin by *Pseudomonas aeruginosa* elastase and alkaline protease. *Infect Immun.* 1986;54:149–153.
12. Gupta SK, Masinick SA, Hobden JA, Berk RS, Hazlett LD. Bacterial proteases and adherence of *P. aeruginosa* to mouse cornea. *Exp Eye Res.* 1996;62:641–650.
13. Bejarano PA, Langeveld JPM, Hudson BG, Noelken ME. Degradation of basement membranes by *Pseudomonas aeruginosa* elastase. *Infect Immun.* 1989;57:3783–3787.
14. Wu X, Kurpakus MK, Hazlett LD. Some *P. aeruginosa* pilus-binding proteins of human corneal epithelium are cytokeratins. *Curr Eye Res.* 1996;15:782–791.
15. Hobden JA, Gupta SK, Masinick SA, Wu X, Kernacki KA, Hazlett LD. Topically applied antibodies against host corneal receptors prevent perforation in experimental *Pseudomonas aeruginosa* keratitis. *Immunol Cell Biol.* 1996;74:258–264.
16. Giron JA, Lange M, Baseman JB. Adherence, fibronectin binding, and induction of cytoskeleton reorganization in cultured human cells by *Mycoplasma penetrans. Infect Immun.* 1996;64:197–208.
17. Rosenshine I, Donnenberg MS, Kaper JB, Finlay BB. Signal transduction between enteropathogenic *Escherichia coli* (EPEC) and epithelial cells: EPEC induces tyrosine phosphorylation of host cell proteins to initiate cytoskeletal rearrangement and bacterial uptake. *EMBO J.* 1992;11:3551–3560.
18. Hook M, Switalski L. Interaction of pathogenic microorganisms with fibronectin. In: Mosher DF, ed. *Biology of Extracellular Matrix: A Series. Fibronectin.* London: Academic Press; 1989.
19. Fleiszig SM, Zaidi T, Preston MJ, Grout M, Evans DJ, Pier GB. Relationship between cytotoxicity and corneal epithelial cell invasion by clinical isolates of *Pseudomonas aeruginosa. Infect Immun.* 1996;64:2288–2294.
20. Pruss RM, Nirsky R, Ruff MC, Thorpe R, Dowding AJ, Anderton BH. All classes of intermediate filaments share a common antigenic determinant defined by a monoclonal antibody. *Cell.* 1981;27:419–429.

SECRETORY IgA RESPONSES ON THE HUMAN OCULAR SURFACE

Aize Kijlstra

Department of Ophthalmo-Immunology
The Netherlands Ophthalmic Research Institute
and the Department of Ophthalmology
University of Amsterdam
Amsterdam, The Netherlands

1. INTRODUCTION

The eye is in continuous contact with infectious microorganisms inhabiting its environment. Furthermore, the external ocular surface is also prone to infection after reactivation of viruses residing in the nerves that innervate the cornea and conjunctiva. To cope with this heavy infectious load, the ocular surface is endowed with powerful defense mechanisms. In view of the different requirements of the visual system among different species and the vast differences of the microenvironment to which the eye of different animals is exposed, it is not surprising that the composition of the proteins involved in the ocular defense is markedly different for various species. These differences are reflected mainly in differences in the defense mechanisms present on the ocular surface. Most species have a specific immune defense mechanism that is characterized by the presence of secretory IgA in the tear film. Secretory immunoglobulin A (s-IgA) is the predominant antibody in tears[1] as well as in other external secretions. IgA in external secretions is present mainly in a dimeric form containing the polypeptides J (joining) chain and a glycoprotein called secretory component, and has a molecular weight of 385 kDa. The s-IgA molecule is a product of two different cell types: IgA with J chain is produced by plasma cells, and secretory component by epithelial cells. External secretions have been thought to contain about equal amounts of the IgA1 and IgA2 subtypes.[2]

IgA in tears is produced by interstitial plasma cells in the main and accessory lacrimal gland tissue.[3–6] Intracellular transport of IgA by acinar epithelial cells and secretion to the lacrimal lumen requires dimerization and combination of IgA by J-chain and subsequent coupling to secretory component. In the lacrimal glands, secretory component is synthesized by acinar epithelial cells,[5–7] and it acts as a membrane receptor for IgA,[8] taking up interstitial IgA, after which the complex is endocytosed and subsequently secreted into the lacrimal ducts.

Lacrimal Gland, Tear Film, and Dry Eye Syndromes 2
edited by Sullivan *et al.*, Plenum Press, New York, 1998

Apart from main and accessory lacrimal IgA secretion, there may be additional sources for this protein in tears, including the conjunctiva. This has been investigated in both animals and humans, and to date the reports are still controversial. IgA- containing plasma cells have been detected in rabbit conjunctival stroma along with secretory component in conjunctival epithelial cells.[9] In the human conjunctiva, numerous IgA-containing plasma cells have also been observed,[4] but conjunctival epithelial cells appeared to lack expression of secretory component.[7] In the rat conjunctiva, neither IgA-containing plasma cells nor secretory component could be detected.[6]

The function of s-IgA has been considered to take place mainly on the ocular surface. It has been hypothesized that IgA may also play a role in the vicinity of its site of synthesis. IgA synthesized in the interstitial space of the lacrimal gland could pick up infectious agents, and, via uptake by the polymeric IgA receptor on acinar epithelial cells, these microorganisms could be eliminated from the gland. Another site of action could be within the cytoplasm of the acinar epithelial cells themselves. Numerous viruses have been implicated in lacrimal gland disease, including the Epstein-Barr virus, HIV, and the hepatitis C virus. As yet, the hypothetical role of local s-IgA in the immune defense against these viruses has not been addressed; the additional functions of s-IgA as listed in Table 1 have been derived from other mucosal sites.

Levels of s-IgA are highly dependent on the flow rate of tears. Under conditions of a flow rate of 1 µl/min, the level of s-IgA can be in the milligram range, whereas an increase in flow rate up to 100 µl/min results in a rapid drop of s-IgA level to approximately 100–400 µg/ml.[10]

During sleep, the level of s-IgA shows a marked increase, and, upon awakening, levels as high as 10 mg/ml have been observed.[11] These observations indicate either a different stimulus response of s-IgA as compared to other lacrimal gland proteins such as lactoferrin and lysozyme or a possible release of s-IgA from other sites, such as the conjunctiva.

2. IgA LEVELS IN VARIOUS HUMAN OCULAR CONDITIONS

2.1. Secretory IgA Deficiency

Selective IgA deficiency is the most common immune deficiency in humans; its prevalence is approximately 1/500. Little is known about the occurrence of ocular infections in individuals with IgA deficiency. Approximately half of such individuals show enhanced synthesis of s-IgM at mucosal surfaces, and this immunoglobulin may compensate for the absence of s-IgA at these sites.[12–15]

2.2. Graves' Ophthalmopathy

Lacrimal gland involvement has been suggested to play a role in Graves' ophthalmopathy. In certain cases the gland may be palpable, and in other cases an enlargement of

Table 1. Role of secretory IgA in ocular mucosal immunity

1.	Binding of foreign substances on the ocular surface, thereby preventing attachment or entry
2.	Intra-epithelial neutralization of pathogens
3.	Neutralization of foreign material within the interstitial space of the lacrimal gland or conjunctiva

the gland has been noted on orbital tomography. Using HPLC analysis of tear proteins, we observed a marked increase in the tear s-IgA levels in 25% of cases with Graves' ophthalmopathy.[16] Similar findings had been noted earlier by Berta et al. who used SDS-polyacrylamide electrophoretic analysis.[17]

2.3. Contact Lens Wearers

Contact lens wear is associated with an increased risk for microbial keratitis. The development of microbial keratitis in a contact lens wearer depends on multiple factors. The first factor of importance is contamination of the contact lens with bacteria, whereby the inoculum often arises from a contaminated contact lens case or a direct ocular binding of the bacteria from the environment or the fingers of the patient.[18–20] The second factor includes the integrity of the corneal epithelium. Many studies have demonstrated that binding of *P. aeruginosa* to the corneal epithelium depends on the condition of these cells, whereby corneal hypoxia and epithelial defects can promote binding of the microorganisms to the cells. A third factor that has not received much attention until now involves the role of mucosal immunity in contact lens-associated bacterial keratitis. A number of reports have appeared that addressed the effect of contact lens wear on the total IgA levels in the tear film. In a prospective study, we showed that contact lens wear is associated with an initial decrease in the tear IgA levels during the first months of lens wear.[21] In a number of individuals, the IgA levels return to normal values again after 1 year, but more than half of the persons investigated had an IgA level in their tear fluid that was below the values before contact lens wear. This finding might explain the higher risk of corneal infections in contact lens wearers, although other factors may also be involved. Earlier studies by Vinding et al.[22] showed a more pronounced decrease in secretory IgA levels when comparing a group of contact lens wearers with a control group. In contrast to the observation that IgA levels are decreased in the tear fluid of contact lens wearers, Mannucci et al.[23] reported a more than twofold increase after extended wear of contact lenses, whereas individuals using daily-wear lenses did not differ from controls.

As mentioned above, most studies dealing with the s-IgA response on the ocular surface of contact lens wearers dealt with total IgA and not with the specificity of the local IgA response. Since *P aeruginosa* is the organism most frequently isolated from corneal ulcers in contact lens wearers, we addressed the question whether the s-IgA response against this bacterium was altered during lens wear.[24] As a positive control group, we tested tears obtained from patients with cystic fibrosis, since the respiratory tract of these individuals is often colonized by *P. aeruginosa*. Our first observation was that 10–20% of both contact lens wearers and healthy controls did not have a detectable response against *Pseudomonas* in their tears, whereas all the cystic fibrosis patients showed a very strong s-IgA response (Table 2). When comparing the mean anti-*P. aeruginosa* IgA response, a significantly lower value was observed in extended-wear contact lens users as compared to non-contact lens wearers. Three conclusions were drawn from this study. One is that many individuals in the population lack a detectable IgA response against *P. aeruginosa* in their tears and may thus be susceptible to *P. aeruginosa* keratitis when coming into contact with this organism under circumstances whereby the physiological condition of their cornea is compromised. A second conclusion is that extended lens wear not only results in a susceptible "damaged" corneal surface but is also associated with a weakened IgA response against this bacterium, which in combination may explain the higher incidence of microbial keratitis in this group of lens wearers. The third conclusion pertains to the observation that cystic fibrosis patients show a high response against the bacterium on their ocular surface, which indicates that a proper antigenic stimulus can re-

Table 2. Levels of *Pseudomonas aeruginosa* IgA antibodies in tears of contact lens wearers, cystic fibrosis patients, and healthy controls

Patient group (*n*)	IgA anti-pseudomonas (U/ml) (mean ± SD)
Healthy controls (23)	82 ± 15
Daily-wear hard lenses (23)	67 ± 11
Daily-wear soft lenses (22)	52 ± 9
Extended-wear lenses (17)	38 ± 6*
Cystic fibrosis (5)	7474 ± 10760*

*Significantly different from controls.

sult in a strong specific local immune response and opens venues to embark on vaccination programs to prevent microbial keratitis in contact lens wearers. In view of the large number of contact lens wearers worldwide and its increasing popularity, such a program may be cost effective.

Besides microbial keratitis, the conjunctiva may also be affected in contact lens wearers. The main disorder seen is giant papillary conjunctivitis (GPC), which is associated with deposit formation on the lenses. In a study of the tear protein profiles in patients with this disorder, it was noted that the IgA levels were decreased approximately 25% as compared to controls.[25]

2.4. Protein Malnutrition

Children with protein malnutrition show an increased incidence of mucosal infections. In 1976 a study was published showing that children with a marked protein-calorie malnutrition had a significantly lower IgA concentration in their tears.[26] These findings support a role for s-IgA in the local defense mechanisms of human mucosal surfaces.

2.5. Sjögren's Syndrome

Sjögren's syndrome is an autoimmune disease of the lacrimal and salivary glands. Destruction of the lacrimal gland results in a marked decrease in the proteins originating from the gland, such as lactoferrin and lysozyme. In a study we performed to assess the IgA levels in tears of Sjögren patients, no difference was observed between patients and controls.[27] The fact that the specific immune response was not affected may explain earlier observations showing that the incidence of ocular infections is not increased in this patient group.[28]

2.6. Viral Antibodies

Little is known about the viral specificity of tear immunoglobulins in humans. Coyle et al., analyzed the presence of IgA antibodies against seven different viruses in the tears of normal human subjects.[29] Approximately 60–70% of individuals had detectable antibodies against HSV-1 and EBV, whereas 20–30% had antibodies against mumps, rubella, and VZV. Only 5% had antibodies against CMV. Earlier, Fox et al., had already reported the presence of herpes simplex virus specific IgA antibodies in tears and showed that analysis of these antibodies was helpful in the diagnosis of active infection.[30]

Other viruses for which a specific IgA response was investigated in the tear film include the influenza virus and HIV. Oral immunization of volunteers with an inactivated influenza vaccine resulted in a marked rise of specific IgA antibodies in tears that reached a maximum 5–7 weeks following immunization.[31] The detection of IgA antibodies against HIV in tears of children born to seropositive mothers proved to be a specific and reliable test to diagnose HIV infection in babies before the age of 15 months.[32]

2.7. *Chlamydia* Antibodies

Chlamydia trachomatis is an important cause of conjunctival disease and, depending on the serotype, will cause an inclusion conjunctivitis (serotype D to K) or trachoma (serotypes A, B, and C). The latter disease is one of the leading causes of blindness in the world. Prevention of trachoma involves measures to improve hygiene, but enhancement of the local secretory IgA response might also be employed to combat this disease. Various authors have shown the presence of IgA antibodies against *Chlamydia* in the tears of chlamydial conjunctivitis patients and have suggested that the presence of these antibodies can be detected before overt disease is present.[33–35]

2.8. Bacterial Antibodies

Analysis of the human secretory IgA response against bacteria has been focused on those microorganisms known to colonize mucosal surfaces. Different groups have shown that oral ingestion of *Streptococcus mutans* resulted in a selective appearance of specific IgA antibodies in tears and saliva but not in serum.[36,37] A second challenge resulted in even higher specific antibody levels. Comparison of the IgA response in tears and saliva against bacteria (*Staphylococcus epidermidis*, *Streptococcus mutans*) representative for each site (eye, oral cavity) revealed interesting results.[38] IgA responses were detected against both bacteria in both sites, but higher IgA levels against *Staphylococcus* were found in tears and higher levels against *Streptococcus mutans* in saliva, suggesting a role for local antigenic stimulation.

3. FUTURE PROSPECTS

The items discussed above show that the secretory immune system of the ocular mucosa can be stimulated by the microorganisms that have been implicated in causing external ocular disease such as *Pseudomonas aeruginosa*, *Chlamydia trachomatis, Staphylococcus epidermidis,* herpes simplex type 1, and Epstein-Barr virus. Most studies indicate that the presence of secretory IgA against a certain microorganism is shortlived, but that reinfection with the same organism leads to a rapid reappearance of specific antibodies in the tearfilm. This suggests that lasting protection against, for instance, *Pseudomonas aeruginosa* in contact lens wearers or *Chlamydia trachomatis* in third-world countries would necessitate a continuous stimulation of the local immune response. Whether such strategies are feasible economically is not yet clear. Furthermore, the fact that mucosal administration of certain antigens may lead to development of tolerance of these antigens is also a matter of concern.

REFERENCES

1. McClellan BH, Whitney CR, Newman LP, Allansmith MR. Immunoglobulins in tears. *Am J Ophthalmol.* 1973;76:89.

2. Mestecky J, McGhee JR. Immunoglobulin A (IgA): Molecular and cellular interactions involved in IgA biosynthesis and immune response. *Adv Immunol.* 1987;40:153.
3. Janssen PT, van Bijsterveld OP. Origin and biosynthesis of human tear proteins. *Invest Ophthalmol Vis Sci.* 1987;24:623.
4. Allansmith MR, Kajiyama GA, Abelson MB, Simon MA. Plasma cell content of main and accessory lacrimal glands and conjunctiva. *Am J Ophthalmol.* 1976;82:819.
5. Gillette TE, Allansmith MR, Greiner JV, Janusz MA. Histologic and immunohistologic comparision of main and accessory lacrimal tissue. *Am J Ophthalmol.* 1980;89:724.
6. Gudmundsson OG, Sullivan DA, Bloch KJ, Allansmith MR. The ocular secretory immune system of the rat. *Exp Eye Res.* 1985;40:231.
7. Allansmith MR, Gillette TE. Secretory component in human ocular tissues. *Am J Ophthalmol.* 1980;89:353.
8. Ahnen DJ, Brown WR, Kloppel TM. Secretory component: The polymeric immunoglobulin receptor. *Gastroenterology.* 1985;89:667.
9. Franklin RM, Prendergast RA, Silverstein AM. Secretory immune system of rabbit ocular adnexa. *Invest Ophthalmol Vis Sci.* 1978;18:1093.
10. Fullard RJ, Tucker D. Tear protein composition and the effects of stimulus. *Adv Exp Med Biol.* 1994;350:309.
11. Sack RA, Tan KO, Tan A. Diurnal tear cycle: Evidence for a nocturnal inflammatory constitutive tear fluid. *Invest Ophthalmol Vis Sci.* 1992;33:626.
12. Arnold RR, Prince SJ, Mestecky J, Linch D, Linch M, McGhee JR. Secretory immunity and immunodeficiency. *Adv Exp Med Biol.* 1978;107:401.
13. Brandtzaeg P, Fjellanger I, Gjeraldsen ST. Immunoglobulin M: Local synthesis and selective secretion in patients with immunoglobulin A deficiency. Science. 1968;160:789.
14. Coelho IM, Pereira MT, Virella G, Thompson RA. Salivary immunoglobulins in a patient with IgA deficiency. *Clin Exp Immunol.* 1974;18:685.
15. Kuizenga A, Stolwijk TR, van Agtmaal EJ, van Haeringen NJ, Kijlstra A. Detection of secretory IgM in tears of IgA deficient individuals. *Curr Eye Res.* 1990;9:997.
16. Khalil HA, de Keizer RJW, Kijlstra A. Analysis of tear proteins in Graves' ophthalmopathy by high performance liquid chromatography. *Am J Ophthalmol.* 1988;106:186.
17. Berta A, Kalman K, Patvaros I, Kelenhegy C. Changes in tear protein composition in Graves' ophthalmopathy. *Orbit.* 1986;5:97.
18. Schein OD, Glynn RJ, Poggio EC, Seddon JM, Kenyon KR, Microbial Keratitis Study Group. The relative risk of ulcerative keratitis among users of daily-wear and extended-wear soft contact lenses: A case-control study. *N Engl J Med.* 1989;321:773.
19. Dart KG, Stapleton F, Minassian D. Contact lenses and other risk factors in microbial keratitis. *Lancet.* 1991;338:650.
20. Donzis PB, Mondino BJ, Weissman BA, Bruckner DA. Microbial contamination of contact lens care systems. *Am J Ophthalmol.* 1987;104:325.
21. Kijlstra A, Polak BCP, Luyendijk L. Transient decrease of secretory-IgA in tears during rigid gas-permeable contact lens wear. *Curr Eye Res.* 1992;11:123–126.
22. Vinding T, Eriksen JS, Nielsen NV. The concentration of lysozyme and secretory IgA in tears from healthy persons with and without contact lens use. *Acta Ophthalmol.* 1987;65:23.
23. Mannucci LL, Pozzau M., Fregona I, Secchi AG. The effect of extended wear contact lenses on tear immunoglobulins. *CLAO J.* 1984;10:163.
24. Cheng KH, Spanjaard L, Rutten H, Dankert J, Polak BCP, Kijlstra A. Immunoglobulin A antibodies against *Pseudomonas aeruginosa* in the tear fluid of contact lens wearers. *Invest Ophthalmol Vis Sci.* 1996;37:2081.
25. Suttorp MSA, Luyendijk L, Kok JHC, Kijlstra A. HPLC analysis of tear proteins in giant papillary conjunctivitis. *Doc Ophthalmol.* 1989;72:235.
26. Reddy V, Raghuramulu N, Bhaskaram C. Secretory IgA in protein-calorie malnutrition. *Arch Dis Child.* 1976;51:871.
27. Wehmeyer A, Das PK, Swaak T, Gebhart W, Kijlstra A. Sjögren syndrome: Comparative studies in local ocular and serum immunoglobulin concentrations with special reference to secretory IgA. *Int Ophthalmol.* 1991;15:147.
28. van Bijsterveld OP. Local antibacterial defense in the sicca syndrome. I. In: Holly FJ, ed. *The Preocular Tear Film in Health, Disease and Contact Lens Wear.* Lubbock, TX: Dry Eye Instititute; 1986:857.
29. Coyle PK, Sibony PA. Viral antibodies in normal tears. *Invest Ophthalmol Vis Sci.* 1988;29:1552.

30. Fox PD, Khaw PT, McBride BW, McGill JI, Ward KA. Tear and serum antibody levels in ocular herpetic infection: Diagnostic precision of secretory IgA. *Br J Ophthalmol.* 1986;70:584.
31. Bergmann KC, Waldman RH, Tischner H, Pohl WD. Antibody in tears, saliva and nasal secretions following oral immunization of humans with inactivated influenza virus vaccine. *Int Arch Allergy Appl Immunol.* 1986;80:107.
32. Renom G, Bouquety JC, Lanckriet C, Georges AJ, Siopathis MR, Martin PM. Detection of anti-HIV IgA in tears of children born to seropositive mothers is highly specific. *Res Virol.* 1990;141:557.
33. Aoki K, Sato C, Hashimoto N, Chiba S, Numazaki K. Clinical and etiological studies of chlamydial conjunctivitis in Sapporo, Japan. *Jpn J Ophthalmol.* 1988;32:444.
34. Herrmann B, Stenberg K, Mardh PA. Immune response in chlamydial conjunctivitis among neonates and adults with special reference to tear IgA. *APMIS.* 1991;99:69.
35. Buisman NJ, Ossewaarde JM, Rieffe M, van Loon AM, Stilma JS. Chlamydia keratoconjunctivitis: Determination of *Chlamydia trachomatis* specific secretory immunoglobulin A in tears by enzyme immunoassay. *Graefes Arch Clin Exp Ophthalmol.* 1992;230:411.
36. McGhee JR, Mestecky J, Arnold RR, Michalek SM, Prince SJ, Babb JL. Induction of secretory antibodies in humans following ingestion of *Streptococcus mutans. Adv Exp Med Biol.* 1978;107:177.
37. Gregory RL, Filler SJ. Protective secretory immunoglobulin A antibodies in humans following oral immunization with *Streptococcus mutans. Infect Immun.* 1987;55:2409.
38. Gregory RL, Allansmith MR. Naturally occurring IgA antibodies to ocular and oral microorganisms in tears, saliva and colostrum: Evidence for a common mucosal immune system and local immune response. *Exp Eye Res.* 1986;43:739.

A METHOD TO STUDY INDUCTION OF AUTOIMMUNITY IN VITRO: CO-CULTURE OF LACRIMAL CELLS AND AUTOLOGOUS IMMUNE SYSTEM CELLS

Harvey R. Kaslow,[1] Zhijun Guo,[1] Dwight W. Warren,[2] Richard L. Wood,[2] and Austin K. Mircheff[1]

[1]Department of Physiology and Biophysics
[2]Department of Cell and Neurobiology
University of Southern California School of Medicine
Los Angeles, California

1. LACRIMAL INFLAMMATION AND AUTOIMMUNE DISEASE: A CAUSE OF LACRIMAL INSUFFICIENCY

Keratoconjunctivitis sicca, or dry eye, is the most frequently encountered form of ocular morbidity in developed countries.[1] In the United States alone, 2 million people, 90% of them women, are estimated to suffer from dry eye associated with Sjögrens syndrome, an autoimmune disease. Perhaps another 8 million suffer from dry eye resulting either from unknown etiologies, termed "primary lacrimal deficiency" (PLD), or from the side effects of drug and/or radiation therapies.[2] In addition to Sjögrens syndrome, immune system disorders such as sarcoidosis, graft-vs.-host disease,[3] and diffuse infiltrative lymphocytosis syndrome (DILS)[4] can cause dry eye.

The lacrimal insufficiencies arising from immune system disorders are among the most severe encountered.[5] The various immune disorders affecting lacrimal function have distinguishing characteristics. Sarcoidosis is characterized by granulomatous infiltrates with a core of macrophages and monocytes and a periphery of T and B lymphocytes.[6] DILS infiltrates consist primarily of $CD8^+$ T cells.[4] The infiltrates in graft-vs.-host disease[3] and Sjögren's syndrome[7] consist primarily of $CD4^+$ T cells and IgM- and IgG-B cells. Sjögren's syndrome is often, but not always, accompanied by circulating antibodies to Golgi-associated proteins[8] and the ribonucleoprotein autoantigens Ro (SSA) and La (SSB).[9,10] Thus, in some cases, inflammatory infiltrates may produce cytotoxic T cells and factors that destroy acinar cells.[11,12] These observations have led to considerable study of the regulation of lacrimal gland immune responses.[13–24]

The overwhelming evidence that autoimmune disease contributes to decreased lacrimal function has prompted hypothetical roles for lacrimal cell types in this process. Cell-mediated immune responses can be initiated when lymphoid-derived antigen-presenting cells (APCs) take up potential antigens, and process them into small peptides that are presented to $CD4^+$ T cells via Class II major histocompatibility complex molecules (MHC II). Along with antigen presentation, APCs display and secrete co-stimulatory signals that can lead to either stimulation or tolerance. Because of this crucial role, lymphoid-derived APCs are termed *professional* APCs. Humoral immune responses also depend on dual signals—for example, antigen-specific signals generated by recognition of antigen by surface B cell immunoglobulins, and co-stimulatory signals from antigen-activated helper $CD4^+$ cells.

The observation that acinar cells can express MHC II suggests that the role of other lacrimal gland cells, such as acinar cells, may go beyond simply releasing or displaying antigen to immune system cells. There is an ongoing debate whether presentation of autoantigens via MHC II on non-lymphoid cells can cause autoimmune responses.[25,26] The numbers of MHC II-positive epithelial cells in human, rabbit, and mouse lacrimal glands increase with increasing lymphocytic infiltration. When rabbit lacrimal cells are placed in primary culture, MHC II is expressed, and cholinergic agonists increase this expression. This MHC II appears to participate in a rapid recycling traffic involving endocytic compartments that contain both the La-cross-reactive protein and cathepsin B. Thus, lacrimal acinar cells have components used by professional APCs to stimulate immune responses.[27,28] These components, combined with display of antigens recognized by immunoglobulins on B cells, may cause acinar cells to participate in stimulative triadic relationships with $CD4^+$ and B cells (Mircheff *et al.*, this volume).

2. AN *IN VITRO* METHOD TO STUDY AUTOIMMUNITY: CO-CULTURE OF AUTOLOGOUS LACRIMAL CELLS AND IMMUNE SYSTEM CELLS

A properly designed cell culture system can help identify mechanisms relevant to *in vivo* functions. We have sought to develop such a co-culture system of lacrimal and immune system cells for the study of lacrimal autoimmune events, particularly the induction of such events. Previous work had established that when cultured in conventional serum-supplemented media, lacrimal acinar cells lose specialized morphological features and functions, and fibroblast-like cells overgrow the culture. However, a serum-free, hormone-supplemented medium, termed PCM, maintains rabbit lacrimal acinar cells with reasonably normal morphology[29] and secretory function.[30–32] Although little or no MHC I or II is detected on freshly prepared acinar cells, after a few hours of culture in PCM, MHC II is clearly evident.[27]

Based on these results, initial experiments involving the co-culture of lacrimal gland cells (LGC) and splenic lymphocytes (SPL) in PCM were promising: co-culturing LGC with either spleen cells (SPC) or lymphocytes purified from spleen cells (SPL) stimulated proliferation, measured using [^{3}H]thymidine incorporation and cell filtration. Roughly equivalent incorporation was obtained using lacrimal gland and immune system cells from either the same or different outbred rabbits, indicating stimulation was not due to allogeneic rejection phenomena as occurs in a typical mixed lymphocyte reaction (MLR) (data not shown). Importantly, irradiation of the lacrimal cells prior to initiation of co-cul-

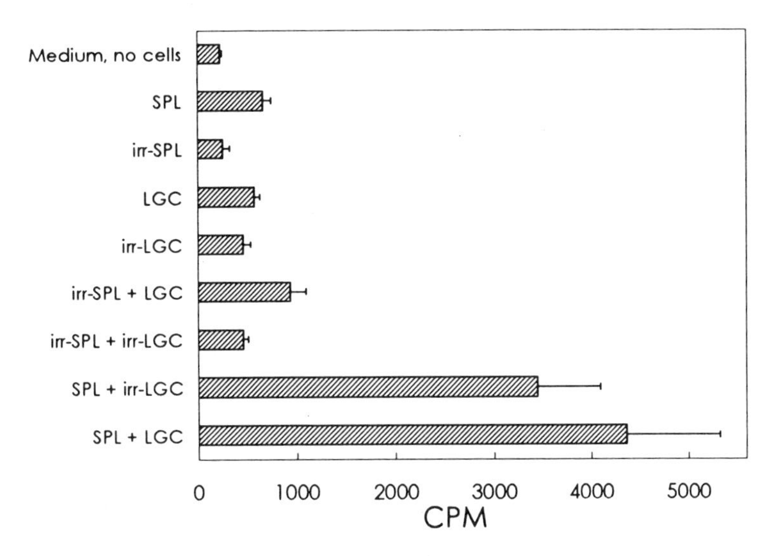

Figure 1. Irradiation of splenic lymphocytes (SPL), but not lacrimal gland cells (LGC), blocks proliferation. Control or irradiated (irr) SPL and LGC were mono-cultured for two days, and then co-cultured. After 5 days, [^{3}H]thymidine was added, and the next day ^{3}H incorporation into the cells determined. Values are means of triplicate determinations ± SD.

ture had little effect, whereas irradiation of lymphocytes obliterated incorporation (Fig. 1). Thus, proliferation of lymphocytes was the likely cause of the ^{3}H incorporation. The time course of proliferation in this *autologous* system (Fig. 2) was the same as that typically seen for lymphocyte proliferation in conventional *heterologous* MLRs.[33]

In initial experiments, as shown in Fig. 2, stimulation was erratic, as reflected by the large standard deviations in this figure. In fact, in some experiments, no stimulation of proliferation was observed (data not shown). We thus hypothesized that the culture system involving serum-free PCM was not optimal. To test this hypothesis, we characterized the effects of various media on proliferation rates in a traditional (heterologous) MLR, using rabbit spleen cells. MLR proliferation in PCM was found to be virtually non-existent, whereas proliferation was sustained in either DME or RPMI when supplemented with 10% fetal bovine serum (FBS) and 50 μM 2-mercaptoethanol (data not shown).

These results prompted us to speculate that 2-mercaptoethanol and FBS aided lymphocyte proliferation, and, if present in the autologous LGC-SPL mixed cell reaction, would reduce the erratic nature of our results. We further speculated that once stabilized after 2 days of culture in PCM [which is based on 50% Hams F-12 and 50% DME (low glucose formula)], lacrimal acinar cells could be transferred into 50% PCM plus 50% DME supplemented with 100 μM 2-mercaptoethanol and 20% FBS, and retain the acinar phenotype. This speculation was tested by visual comparison of cells grown in PCM to cells grown in either supplemented DME or RPMI. Lymphocytes grown in supplemented DME retained higher viability longer than cells grown in supplemented RPMI (data not shown), and therefore DME was chosen for further studies. We found that after culture for 2 days in 100% PCM, the lacrimal preparation retained acinar morphology up to 5–7 days after being transferred into 50% PCM and 50% DME, with the DME supplemented to yield final concentrations in the mixture of 10% serum, 15 mM NaHEPES, pH 7.3, 50 μM 2-mercaptoethanol, 2 mM glutamine, and non-essential amino acids (data not shown). In

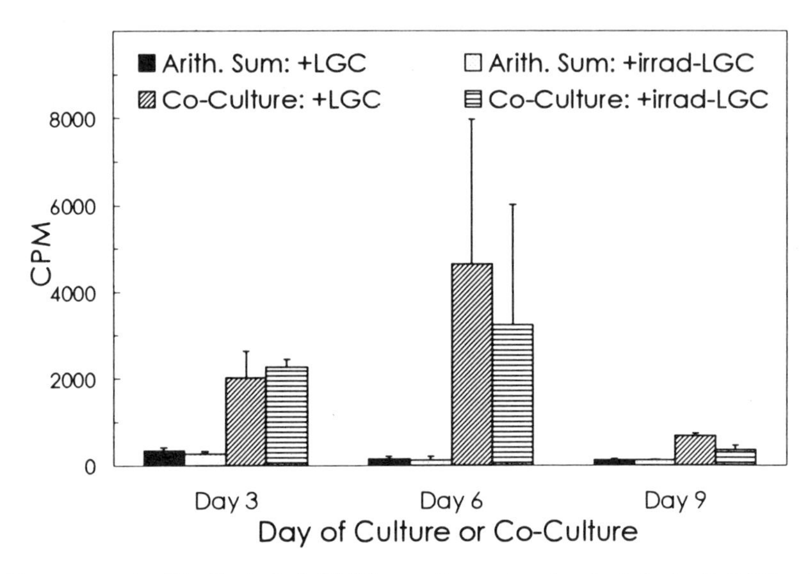

Figure 2. Time course of proliferation in the LGC-SPL co-culture system. Lacrimal gland cells (LGC) and splenic lymphocytes (SPL) were cultured alone for 2 days: LGC in PCM, and SPL in supplemented DME. The SPL were washed and transferred into PCM. The LGC were either not treated, or exposed to 1000 rad (irrad-LGC). The two cell types were then cultured either alone or together. After 2, 5, or 8 days, 1 μCi of [^{3}H]thymidine was added, and 24 h later the cells were harvested, and their ^{3}H content determined. "Arith. Sum" refers to the sum of the CPM incorporated into SPL and either LGC or irrad-LGC cultured alone. Data are means of triplicate determinations; error bars indicate SD. Culture was in a 96-well plate; 6 x 10^5 SPL and/or 6 x 10^5 LGC or irrad-LGC per well.

subsequent work using the PCM/DME culture conditions, we obtained consistent stimulation of proliferation, and much smaller variance (compare Figs. 1 and 2).

The procedure used in experiments such as that in Fig. 1 for preparing LGC was based on described methods.[27,29] After removal from the anesthetized or sacrificed rabbit, the lacrimal gland was cut into small pieces in Ham's F-12 medium supplemented (s-Ham's) with penicillin (100 U/ml), streptomycin (0.1 mg/ml), glutamine (2 mM), linoleic acid (0.084 mg/L), soybean trypsin inhibitor (0.05 mg/ml), BSA (0.5%), butyrate (2 mM), and HEPES (10 mM, pH 7.6) using a pair of scalpels. The pieces were then washed in Hank's medium without Ca^{2+} and Mg^{2+}, and supplemented with EDTA (0.76 mg/ml) and HEPES (10 mM, pH 7.6) (termed s-Hank's), and then incubated in the s-Hank's medium for 15 min (37°C, shaking water-bath). The pieces were again washed, this time in s-Ham's medium, twice, and then incubated in s-Ham's to which were added collagenase (200 U/ml), DNase (type II, 40 U/ml), and hyaluronidase (500 U/ml) (CDH-Ham's) for 15 min (37°C, shaking water-bath). To remove the CDH-Ham's medium, the resulting cell suspension was centrifuged at 700 rpm for 5 min and resuspended in s-Hank's, incubated for 15 min (37°C, shaking water-bath), and centrifuged again. The cells were resuspended in CDH-Ham's and incubated for 25 min. The cells were then centrifuged and resuspended in s-Ham's twice and filtered through a 500 μm mesh and then a 25 μm mesh. The cell suspension was then centrifuged in a step gradient of 5 ml 2%, 5 ml 3%, and 5 ml 4% Ficoll in s-Ham's (20*g*, 5 min). The cells were then collected from the 5% Ficoll fraction, transferred into PCM, and washed twice by centrifugation, and then cultured at a density of 2–3 x 10^6/ml at 37°C and 5% CO_2. The cells were then cultured in 96-well plates with 100 μl of the cell suspension in a well. PCM = 50% Ham's F-12 + 50% DME (low glucose) supplemented with penicillin (100 U/ml), streptomycin (0.1 mg/ml), glutamine (4 mM), hydrocortisone (5 nM), transferrin (5 μg/ml), insulin (5 μg/ml), butyrate (2 mM), linoleic acid (0.084 mg/L), carbachol (1 μM), and laminin (5 mg/L).

SPL were prepared using standard techniques.[33] After the rabbit was sacrificed, the spleen was removed, and teased with a needle in DME (high-glucose formula) supplemented (s-DME) with FBS (10%), penicillin (100 U/ml), streptomycin (0.1 mg/ml), glutamine (2 mM), HEPES (10 mM, pH 7.3), and 2-mercaptoethanol (50 μM). The cell suspension was further dispersed by pushing it four or five times through a 19G needle using a syringe. The cell suspension was placed in a conical tube for 1 min to let large tissue pieces settle, and the supernatant filtered through a 100 μm mesh. The spleen cells were then washed with s-DME twice by centrifugation and resuspension with s-DME (200*g*, 4°C, 10 min). The spleen cell suspension (SPC) was then subjected to centrifugation (15 min, 700*g*, room temperature) by layering 5 ml of the cell suspension containing 0.5–1.0 x 10^8 cells/ml over 5 ml Histopaqe (Sigma, 1,083d). The lymphocytes were recovered from the interface, and cultured in s-DME at a density of 1 x 10^7/ml at 37°C, 5% CO_2.

After 2 days of culture in their respective media, the lymphocytes were harvested from their culture dish, and collected by centrifugation (200*g*, 10 min). The pellet was resuspended in DME (high-glucose) supplemented with 20% FBS and 100 μM 2-mercaptoethanol to achieve a cell density of 3 x 10^6/ml. 100 μl of the SPL was then added to the wells containing the 100 μl of the LGC. All culture took place in incubators set at 37°C, 5% CO_2, and 100% humidity.

3. SUMMARY AND CONCLUSION

Co-culturing autologous lacrimal gland cells and immune system cells can lead to spleen cell proliferation with a time course similar to that for proliferation in a typical heterologous MLR. Although these results are consistent with the hypothesis that lacrimal acinar cells are a source of antigen, and may or may not serve in part as an APC, future studies of this preparation are required to test these hypotheses.

We are unaware of reports demonstrating that co-culturing control epithelial tissue and autologous splenic lymphocytes from apparently healthy animals leads to lymphocytic proliferation. Our results suggest that the appropriate co-culture of tissues and immune cells from healthy animals, perhaps such as detailed above, should help identify mechanisms contributing to the induction of autoimmune disease. Knowledge regarding such mechanisms should help efforts to prevent such disease, and perhaps reverse it.

ACKNOWLEDGMENTS

The authors thank Barbara Platler for outstanding technical help, and the National Institutes of Health for financial support from NIH grants EY05801, EY09405, and EY10550.

REFERENCES

1. National Institutes of Health. Publication No. 84–2472. *Vision Research. A National Plan*. Report of the corneal diseases panel. *NIH Publications* 1984;2part 2:45–60.
2. Lemp ME. Report of the National Eye Institute/Industry Workshop on Dry Eye. *CLAO J.* 1995;21:221–232.

3. Gratwhol AA, Moutsopoulos HM, Chused TM, et al. Sjögren-type syndrome after allogeneic bone marrow transplantation. *Ann Intern Med.* 1977;87:703–706.
4. Itescu S, Brancata LJ, Buxbaum J, et al. A diffuse infiltrative CD8 lymphocytosis syndrome in normal immunodeficient virus (HIV) infection: A host immune response associated with HLA-DR5. *Ann Intern Med.* 1990;122:3–11.
5. Tsubota K, Toda I, Yagi I, Ogawa Y, Ono M, Yoshino K. Three different types of dry eye. *Cornea.* 1994;13:202–209.
6. Weinreb RN, Tessler H. Laboratory diagnosis of ophthalmic sarcoidosis. *Surv Ophthalmol.* 1984;28:653–664.
7. Pflugfelder SC, Wilhelmus KR, Osato MS, Matoba AY, Font RL. The autoimmune nature of aqueous tear deficiency. *Ophthalmology.* 1986;93:1513–1517.
8. Kooy J, Toh BH, Gleeson PA. Heterogeneity of human anti-Golgi antibodies. *Immunol Cell Biol.* 1994;72:123–127.
9. Reveille JD, Arnett FC. The immunogenetics of Sjögren's syndrome. *Rheum Dis Clin North Am.* 1992;18:539–550.
10. Chan EKL, Andrade LEC. Antinuclear antibodies in Sjögren's syndrome. *Rheum Dis Clin North Am.* 1992;18:551–570.
11. Tsubota K, Saito I, Miyasaka N. Gramzyme A and perforin expressed in the lacrimal glands of patients with Sjögren's syndrome. *Am J Ophthalmol.* 1994;117:120–121.
12. Fox RI, Kang I. Pathogenesis of Sjögrens syndrome. *Rheum Dis Clin North Am.* 1992;18:517–538.
13. O'Sullivan NL, Skandera CA, Chin YH, Montgomery PC. In vitro adhesive interactions between rat lymphocytes and lacrimal gland acinar cells: Phenotype of adherent lymphocytes and involvement of adhesion molecules. *J Immunol.* 1994;152:1684–1692.
14. Pockley AG, Montgomery PC. Identification of lacrimal gland associated immunomodulatory activities having differential effects on T and B cell proliferative responses. *Reg Immunol.* 1990–91;3:198–203.
15. Sullivan DA, Sato E. Potential therapeutic approach for the hormonal treatment of lacrimal gland dysfunction in Sjögren's syndrome. *Clin Immunol Immunopathol.* 1992;64:9–16.
16. Liu SH, Prendergast RA, Siverstein AM. Experimental autoimmune dacryoadenitis. I. Lacrimal gland disease in the rat. *Invest Ophthalmol Vis Sci.* 1987;28:270–275.
17. Liu SH, Zhou DH. Experimental autoimmune dacryoadenitis: Purification and characterization of a lacrimal gland antigen. *Invest Ophthalmol Vis Sci.* 1992;33:2029–2036.
18. Liu SH, Zhou DH, Franklin RM. Lacrimal gland-derived lymphocyte proliferation potentiating factor. *Invest Ophthalmol Vis Sci.* 1993;34:650–675.
19. Stepkowski SM, Li T, Franklin RM. Interleukin-2-like molecule produced by acinar cells in lacrimal gland dysfunction in Sjögren's syndrome. ARVO Abstracts. *Invest Ophthalmol Vis Sci.* 1993;34:S1486.
20. Mircheff AK, Warren DW, Wood RL, Tortoriello PJ, Kaswan RL. Prolactin localization, binding, and effects on peroxidase release in rat exorbital lacrimal gland. *Invest Ophthalmol Vis Sci.* 1992;33:642–650.
21. McMurray R, Keisler D, Kanuckel K, Izui S, Walker SE. Prolactin influences autoimmune disease activity in the female b/W mouse. *J Immunol.* 1991;147:3780–3787.
22. Berczi I. Prolactin, pregnancy and autoimmune disease. *J Rheumatol.* 1993;20:1095–1100.
23. Jabs DA, Prendergast RA. *Murine models of Sjögren's syndrome. Adv Exp Med Biol.* 1994;350:623–630.
24. Jabs DA, Burns WH, Prendergast RA. Paradoxic effect of anti-CD4 therapy on lacrimal gland disease in MRL/Mp-*lpr/lpr* mice. *Invest Ophthalmol Vis Sci.* 1996;37:246–250.
25. Bottazzo GF, Pujol-Borrell R, Hanafusa T, Feldmann M. Role of aberrant HLA-DR expression and antigen presentation in induction of endocrine autoimmunity. *Lancet.* 1983;2:1115–1119.
26. Todd I, Bottazzo GF. On the issue of inappropriate HLA Class II expression on endocrine cells: An answer to a sceptic. *J Autoimmun.* 1995;1995:312–322.
27. Mircheff AK, Gierow JP, Lambert RW, Lee LM, Akashi RH, Hofman FM. Class II antigen expression by lacrimal epithelial cells: An updated working hypothesis for antigen presentation by epithelial cells. *Invest Ophthalmol Vis Sci.* 1991;32:2302–2310.
28. Mircheff AK, Gierow JP, Yang T, Wood RL. Autoimmunity of the lacrimal gland. *Int Ophthalmol Clin.* 1994;34:1–18.
29. Rismondo V, Gierow JP, Lambert RW, Golchini K, Feldon SE, Mircheff AK. Rabbit lacrimal acinar cells in primary culture: Morphology and acute responses to cholinergic stimulation. *Invest Ophthalmol Vis Sci.* 1994;35:1176–1183.
30. Gierow JP, Bekmesian A, Rafisolyman S, et al. Secretion of β-hexosaminidase by rabbit lacrimal gland and reconstituted acini. *Invest Ophthalmol Vis Sci.* 1995;36:S990.
31. Meneray MA, Fields TY, Bromberg BB, Moses RL. Morphology and physiologic responsiveness of cultured rabbit lacrimal acini. *Invest Ophthalmol Vis Sci.* 1994;35:4144–4158.

32. Zeng H, Yang T, Zhang J, Wood RL, Mircheff AK, Okamoto CT. Carbachol mobilizes PIgR from lacrimal acinar cell trans-Golgi network and stimulates secretory component secretion. ARVO Abstracts. *Invest Ophthalmol Vis Sci.* 1996;37:S855.
33. Coligan JE, Krusibecek AM, Margulies DH, Shevach EM, Strober W, eds. *Current Protocols in Immunology*. New York: John Wiley; 1991.

RAT LACRIMAL GLANDS CONTAIN ACTIVATED AND RESTING MATURE T CELLS, RECENT THYMIC EMIGRANTS, AND POSSIBLY EXTRATHYMIC POPULATIONS

Nancy L. O'Sullivan, Cheryl A. Skandera, and Paul C. Montgomery

Department of Immunology and Microbiology
Wayne State University Medical School
Detroit, Michigan

1. INTRODUCTION

Lacrimal glands (LG) possess both transient and stationary lymphocyte populations, distinct from those in peripheral lymphoid sites, which regulate and effect humoral and cell-mediated immune responses contributing to the protection of ocular surfaces.[1,2] Further, lacrimal glands are sites of lymphocytic infiltration in inflammatory disorders such as Sjögren's syndrome. Earlier work has established that rat lacrimal gland lymphocytes display a distinct phenotypic profile compared to nonmucosal lymphoid tissues.[3,4] Of particular interest is the presence of lymphocytes expressing Thy-1, which, in the rat, is found on immature B and T lymphocytes in bone marrow and thymus[5,6] as well as on their recently released progeny in the periphery. The Thy-1 molecule is lost with increasing maturity.[7,8] In the case of T cells, developmental stages, including immature recent thymic emigrants (RTE), mature common peripheral T cells (CPT), and activated T cells that have recently encountered antigen, can be identified on the basis of size and expression of Thy-1 and CD45RC.[7,8]

T cell receptor (TCR) $\gamma\delta$ T cells are a major lymphocyte population in mouse mucosal tissues with 45–65% of the intestinal intraepithelial lymphocyte (IEL) T cells expressing this isotype.[9] Although humans have a preponderance of TCR$\alpha\beta$ T cells in skin and mucosa, TCR$\gamma\delta$ T cells are still more numerous in these tissues than in peripheral lymphoid organs.[10] Similarly, in rats, only 5–10% of rat IEL express TCR$\gamma\delta$, but other nonmucosal lymphoid populations have even smaller percentages (1.5% of lymph node T cells and 1.9% of thymocytes).[11,12] Although 5–15% of mouse salivary gland T cells are TCR$\gamma\delta^+$,[9] expression in human, mouse, or rat lacrimal glands has not been reported.

Lacrimal Gland, Tear Film, and Dry Eye Syndromes 2
edited by Sullivan *et al.*, Plenum Press, New York, 1998

T cell populations of extrathymic origin have been identified in mice. These are the only T cells present in athymic nude mice and are also found in small percentages in normal animals. The extrathymic pathway of T cell differentiation becomes predominant under conditions of autoimmunity,[13,14] malignancy,[15] bacterial stimulation,[16] and aging.[17] These T cell populations can be distinguished from the regular, intrathymic T cells by their characteristic lymphoblastic morphology, high degree of granularity, and spontaneous cell proliferation. They constitutively express high levels of LFA-1 and IL-2Rβ and, in a population first identified in the liver, have TCRαβ or CD3 of lower density. Analogous populations have been identified in the rat.[18]

The purpose of this investigation was to phenotypically characterize lacrimal gland lymphocytes with properties of extrathymic and thymus-derived T cells and compare them with populations found in spleen, liver, and thymus.

2. MATERIALS AND METHODS

Rat (F344, Harlan Sprague/Dawley, Indianapolis, IN) lymphocytes were obtained from LG by collagenase digestion or from spleen (SPL), liver (LIV), or thymus (THY) by mechanical disruption. Lacrimal, SPL, and THY mononuclear cells (MNC) were enriched by Percoll (75%, 55%, 40%) and LIV MNC by Lympho-Paque density gradient centrifugation, then stained for two- or three-color FACS analysis using anti-CD3 (R73), -TCRαβ (G4.18), -TCRγδ (V65), -LFA-1 (WT.1), -Thy-1 (OX7), and -CD45RC (OX22) mAbs. Appropriate isotype control mAbs were included in each experiment. Flow cytometric analysis was done using a FACScan and PCLYSYS software (BDIS, San Jose, CA). Lymphocytes were identified by their forward *vs.* side light scatter (FS *vs.* SS) profiles, and regions were established to define small and large lymphocytes. T cells were identified by anti-CD3 staining, then regions were established for electronically gating total $CD3^+$ cells as well as subpopulations with high and low CD3 staining intensity ($CD3^{high}$ and $CD3^{low}$).

3. RESULTS AND DISCUSSION

In the rat, thymocytes (THY) and recent thymic emigrants (RTE) display a distinct profile of lymphocyte surface markers ($Thy\text{-}1^+CD45RC^-$).[8,19,20] RTE differentiate into $Thy\text{-}1^-CD45RC^+$ naive common peripheral T cells (CPT) within about a week of leaving the thymus.[8,19] Upon antigen encounter, CD45RC is downregulated, and the $Thy\text{-}1^-CD45RC^-$ population represents 'primed' T cells poised at the effector stage.[20–24] Lymphocyte populations were stained for CD3, CD45RC, and Thy-1, and expression was determined by FACS analysis. All thymocytes, both $CD3^{low}$ and $CD3^{high}$, were $Thy\text{-}1^+CD45RC^-$ (Fig. 1A, middle tier). The majority of thymocytes expressed very high levels of Thy-1, although

Figure 1. FACS analysis of CD3/CD45RC/Thy-1 triple-stained (A) thymic, (B) splenic, and (C) lacrimal gland lymphocytes. The upper histograms show CD3 expression. The vertical bars indicate the regions used for gating $CD3^{low}$ and $CD3^{high}$ T cells, and the numbers are percentages of cells in these regions. The central contour plot(s) show expression of CD45RC and Thy-1 by $CD3^{low}$ (thymus) and $CD3^{high}$ (all tissues) gated T cells, with numbers representing percentages of T cells in each quadrant. The numbers in parentheses are percentages of T cells with Thy-1 intensity as great as the majority of thymocytes. The lower light scatter profiles show the size distribution of T cells gated from the quadrants indicated by the arrows.

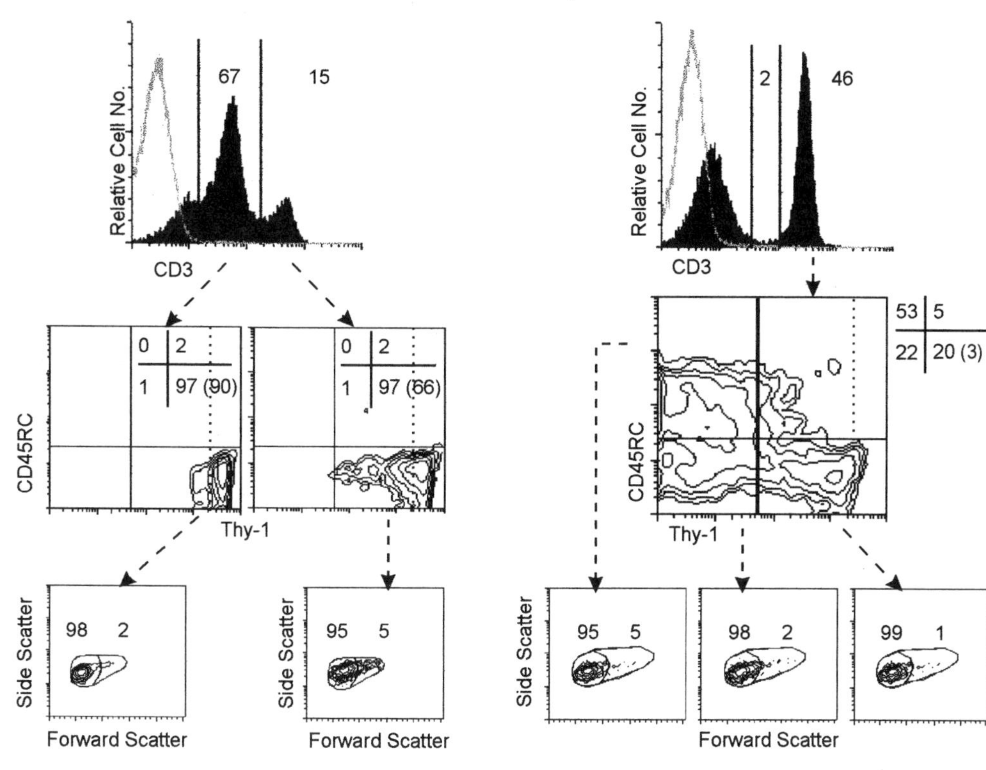
A. Thymus
Relative Cell No.
67
15
CD3
CD45RC
0
2
1
97 (90)
0
2
1
97 (66)
Thy-1
Side Scatter
98
2
Forward Scatter
Side Scatter
95
5
Forward Scatter
B. Spleen
Relative Cell No.
2
46
CD3
53
5
22
20 (3)
CD45RC
Thy-1
Side Scatter
95
5
98
2
99
1
Forward Scatter

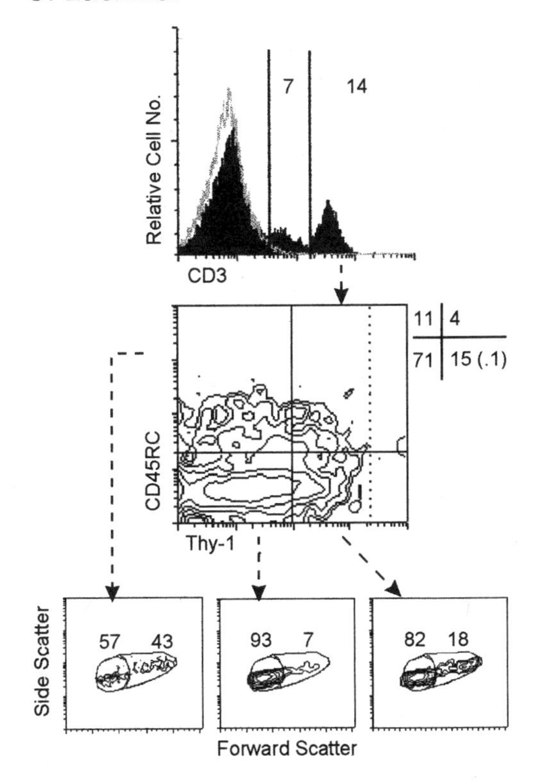
C. Lacrimal
Relative Cell No.
7
14
CD3
11
4
71
15 (.1)
CD45RC
Thy-1
Side Scatter
57
43
93
7
82
18
Forward Scatter

about a third of the $CD3^{high}$ (more mature) had lower Thy-1 levels. Since the majority of peripheral T cells of thymic origin are $CD3^{high}$,[8] expression of CD45RC and Thy-1 by $CD3^{high}$-gated T cells (Fig. 1, top tier) was determined. 20% of SPL T cells expressed the RTE phenotype (Fig. 1B, lower left quadrant). Further, 3% of SPL T cells showed a Thy-1 staining intensity equal to the brightest population of THY, indicating the most recent emigrants. The LG RTE subpopulation was slightly smaller than those present in SPL (Fig. 1C, 15%), but only 0.1% had high levels of the Thy-1 molecule, and overall, the Thy-1 staining intensity was lower than seen in SPL [mean fluorescence intensity (MFI): LG, 310; SPL, 946]. Activated mature T cells may re-express the $Thy\text{-}1^{+}CD45RC^{-}$ phenotype, but these can be distinguished by their larger size and expression of activation markers such as IL-2 receptor.[25] To determine the level of activation, size of T cells expressing the RTE phenotype was examined by FS *vs.* SS, and 99% of SPL $Thy\text{-}1^{+}CD45RC^{-}$ T cells were found to be small lymphocytes (Fig. 1B, lower tier). Nearly 20% of LG T cells with the RTE phenotype were large and are likely to represent activated lymphocytes. T cells coexpressing Thy-1 and CD45RC were considered transitional between the RTE and CPT phenotypes and were found in low percentages in SPL and LG (4–5%). $Thy\text{-}1^{-}CD45^{+}$ T cells have been identified as CPT, which are small, resting lymphocytes and which include both naive, antigen-inexperienced T cells as well as memory T cells, which remain Thy-1 negative but re-express CD45RC.[19] This population is found in the upper left quadrant of each of the contour plots (Fig. 1, middle tier). Fewer T cells with the CPT phenotype were present in LG (11%) compared with SPL (53%); none of the thymocytes expressed this phenotype. The numbers of true CPT are likely to be even lower in LG, since nearly half (43%) of this subpopulation was found to be large cells by light scatter analysis. $Thy\text{-}1^{-}CD45RC^{-}$ T cells were the most prevalent subpopulation in LG (71%), but not in SPL (22%). This population represents those T cells that have recently encountered antigen, resulting in modulation of the CD45RC form of the leukocyte common antigen.

From the above data, it appears that RTE are present as a minor population in LG (8% of total T cells) compared with SPL (19% of T cells). Further, compared with SPL, LG contain decreased numbers of cells with the CPT phenotype, which includes more mature, but still antigen-inexperienced, T cells, as well as long-term memory T cells. Naive T cells have distinct recirculation patterns compared to antigen-experienced memory T cells. Memory T cells selectively recirculate from blood to peripheral tissues (skin, mucosal sites, etc.), then to lymph nodes (*via* afferent lymph). In contrast, naive T cells (CPT) selectively traffic from blood to lymph node (*via* high endothelial venules), bypassing the peripheral tissues.[26] The present data are consistent with this model, since percentages of lymphocytes with the CPT phenotype were lower in LG than in SPL. Antigen-activated cells do not recirculate, but traffic instead to effector tissue sites where terminal differentiation to mature effector or helper cells occurs.[26] Consistent with being an effector site, LG is enriched in $Thy\text{-}1^{-}CD45RC^{-}$ T cells compared with SPL.

Since T cell receptor (TCR) $\gamma\delta$ expressing T cells appear in greater numbers in mucosal effector sites, TCR isotype distribution was examined in LG as well as in SPL and THY populations. Table 1 presents data on the expression of $TCR\alpha\beta$ and $TCR\gamma\delta$ by T ($CD3^{+}$) and $Thy\text{-}1^{+}$ T cells. Although the majority of live-gated $CD3^{+}$ lymphocytes in each tissue was $TCR\alpha\beta^{+}$, LG was significantly ($p < 0.05$) enriched for $TCR\gamma\delta^{+}$ T cells (6.8%) compared with SPL (3.4%) or THY (<1%). Variability in Thy-1 staining intensity, ranging from dull to bright (evident in Fig. 1), precluded live-gating on a discrete $Thy\text{-}1^{+}$ population. Therefore, light scatter-gated lymphocytes were analyzed for dual expression of Thy-1 and $TCR\alpha\beta$ or $TCR\gamma\delta$. The percentages of $Thy\text{-}1^{+}$ T cells were estimated by considering the sum of $TCR\alpha\beta$ plus $TCR\gamma\delta$ to equal the total percentage of T cells. The percentages of

Table 1. Expression of TCRαβ and TCRγδ by T cells and Thy-1$^+$ T cells

	T Cells[a]		Thy-1$^+$ T Cells[b]	
Lymphocyte source	TCRαβ	TCRγδ	TCRαβ	TCRγδ
Lacrimal gland	88.5 ± 1.6	6.8 ± 0.4	100.0 ± 0.0	0.0 ± 0.0
Spleen	97.3 ± 0.4	3.4 ± 0.1	98.5 ± 0.5	1.5 ± 0.5
Thymus	99.5 ± 0.1	0.1 ± 0.1	99.9 ± 1.7	0.1 ± 0.0

[a]Dual fluorescence staining. Percentages of live-gated CD3$^+$ (mAb R73) cells that expressed either TCRαβ (mAb G4.18) or TCRγδ (mAb V65) were determined from 3 or 4 separate experiments for each tissue.

[b]Dual fluorescence staining. Percentages of double-positive Thy-1 (mAb OX-7) and TCRαβ or TCRγδ expressing cells were determined by quadrant analysis of light scatter-gated lymphocytes. The percentages of Thy-1$^+$ T cells that were TCRαβ$^+$ (or TCRγδ$^+$) were calculated using the formula [% Thy-1$^+$TCRαβ$^+$ (or TCRγδ$^+$) ÷ (% Thy-1$^+$TCRαβ$^+$ + Thy-1$^+$TCRγδ$^+$)] X 100, and represent 3 or 4 separate experiments for each value.

Thy-1$^+$ T cells expressing TCRαβ or TCRγδ were calculated, and TCRγδ expression by Thy-1$^+$ T cells in SPL was found to be less than half (1.5% vs. 3.4%) compared to the bulk CD3$^+$ T cell population. Surprisingly, TCRγδ was not expressed by LG Thy-1$^+$ T cells, indicating either a more mature T cell subpopulation or perhaps a distinct population that never expresses Thy-1.

Since it appeared that LG was not the recipient of substantial numbers of newly formed RTE or recirculating CPT compared to SPL (Fig. 1B) or lymph nodes (data not shown), experiments were done to determine whether LG possessed T cell populations similar to the extrathymically differentiated lineages defined in the mouse[13–17] and rat.[18] MNC were isolated from LG and SPL as well as from liver, a tissue with a large proportion of extrathymic T cells.[14–18] Forward *vs.* side light scatter (FS *vs.* SS) analysis was used to estimate the size and granularity of the MNC (Fig. 2, upper tier). 38% of LG were larger (mean FS channel, 610) and more granular (mean SS channel, 81) than the small lymphocyte population (FS, 348; SS, 45). Spleen MNC were smaller and less granular than LG MNC, with only 4% falling in the large cell region (FS, 543; SS, 60) and 96% in the small lymphocyte population (FS, 307; SS, 32). The liver profile was similar to LG, with 83% of the MNC included in the small cell region (FS, 334; SS, 62) and 17% in the large cell region (FS, 503). LIV MNC were the most granular of the populations (SS, 109). Single-color analysis of CD3 expression (central histograms) demonstrated a heterogeneity of staining. 10% of LG MNC (36% of CD3$^+$ cells) and 7% of LIV MNC (18% of CD3$^+$ cells) expressed a lower density of the CD3 molecule. Percentages of CD3^{low} MNC were much lower in SPL (2%). In each of the three tissues, light scatter analysis (Fig. 2, middle tier) showed that the CD3^{low} subpopulation had greater percentages of large/granular cells (LG, 46; SPL, 13; LIV, 36) compared to the CD3^{high} subpopulation (LG, 15; SPL, 4; LIV, 7). Expression levels of LFA-1 and CD3 were determined by two-color staining (Fig. 2, lower tier). These markers were used since extrathymic T cells (and NK cells) in mice are known to constitutively express the highest levels of LFA-1. CD3^{low} LFA-1^{high} expression was seen on 4% of both LG and LIV and on less than 1% of SPL MNC. This was calculated to be 14%, 2%, and 10% of LG, SPL, and LIV CD3^{low} T cells. CD3^{high} LFA-1^{high} staining occurred on 15% of LG (MFI, 1406), 10% of SPL (MFI, 537), and 6% of LIV (MFI, 1475) MNC. The percentages of CD3^{high} LFA-1^{high} T cells were 54% for LG, 20% for SPL, and 15% for LIV. It may be that these T cells are analogous to the mouse CD3^{high} extrathymic intestinal IEL population.[27] Further experiments using thymectomized

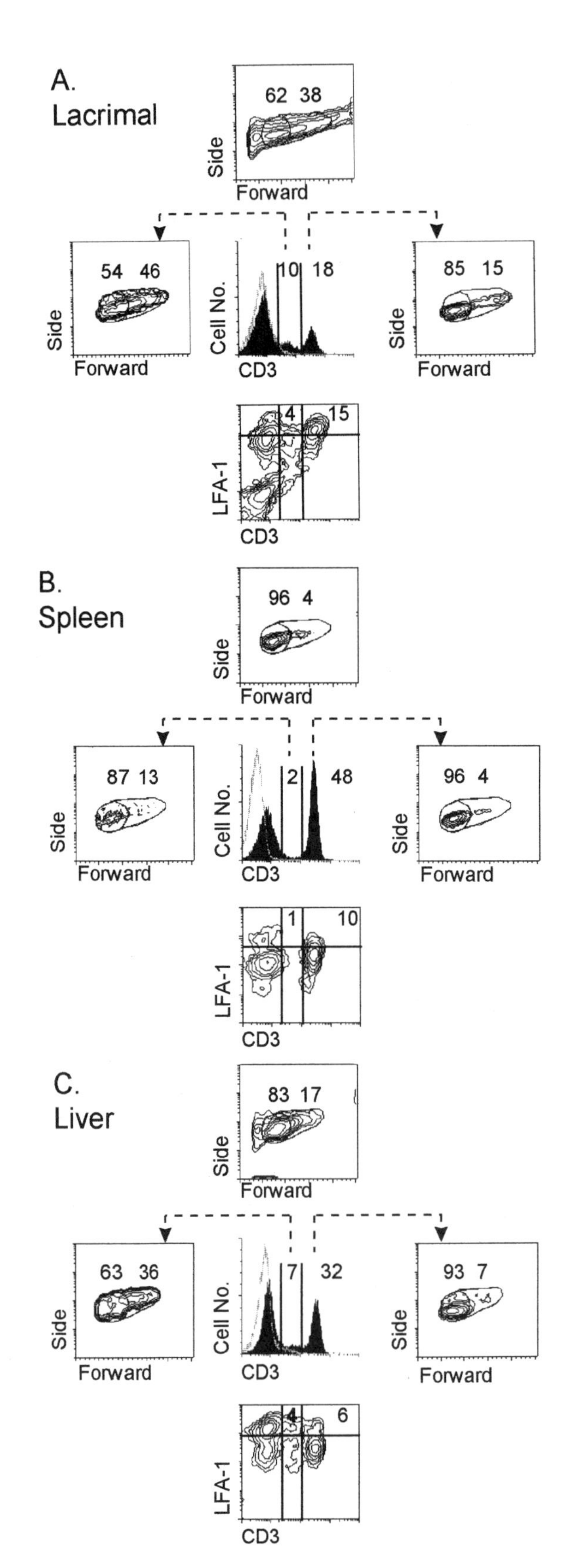

Figure 2. FACS analysis of CD3/LFA-1 dual-stained (A) lacrimal, (B) splenic, and (C) liver mononuclear cells. The upper tier of each panel shows the forward (linear) *vs.* side (log) light scatter profiles of ungated cells and the regions used for gating the lymphoid population. Numbers represent the percentages of gated lymphocytes that were small or large. The central histograms depict CD3 expression, and the numbers are percentages of $CD3^{low}$ and $CD3^{high}$. The left and right contour plots depict the size distribution of $CD3^{low}$ or $CD3^{high}$ gated lymphocytes, respectively. The bottom tiers depict LFA-1 expression by $CD3^{-}$, $CD3^{low}$, and $CD3^{high}$ lymphocyte populations. The numbers denote percentages of $CD3^{low}$ and $CD3^{high}$ cells that are LFA-1^{high}.

or athymic nude rats and other surface markers are needed to demonstrate conclusively the presence of extrathymic T cell populations, but it appears that rat lacrimal glands do contain lymphocytes that resemble extrathymic populations found in mice.

4. CONCLUSIONS

Lacrimal glands of normal rats contain: (i) small percentages of immature recent thymic emigrants, mature common peripheral T cells, and sizable numbers of recently antigen-activated T cells, (ii) a small population of $TCR\gamma\delta^+$ T cells (all Thy-1^-), and (iii) possibly, extrathymic T cell populations. These phenotypic characteristics should be useful for investigating T cell development in lacrimal glands of rats under various conditions of immune stimulation.

ACKNOWLEDGMENT

Supported by NIH grant EY05133.

REFERENCES

1. McGhee JR, Mestecky J, Dertzbaugh MT, Eldridge JH, Hirasawa M, and Kiyono H. The mucosal immune system: from fundamental concepts to vaccine development. *Vaccine*. 1992;10:75–88.
2. Sullivan DA. Ocular mucosal immunity. In: Ogra PL, Mestecky J, Lamm ME, Strober W, McGhee JR, and Bienenstock J, eds. *Handbook of Mucosal Immunology*. San Diego: Academic Press; 1994:569–597.
3. Montgomery PC, Peppard JV, and Skandera CA. A comparison of lymphocyte subset distribution in rat lacrimal glands with cells from tissues of mucosal and non-mucosal origin. *Curr Eye Res*. 1990;9:85–93.
4. Montgomery PC, Skandera CA, and O'Sullivan NL. Phenotypic profiles of lymphocyte populations isolated from rat major salivary glands. *Oral Microbiol Immunol*. 1996;11:248–253.
5. Ritter MA, Gordon LK, and Goldschneider I. Distribution and identity of Thy-1 bearing cells during ontogeny in rat hemopoietic and lymphoid tissue. *J Immunol*. 1987;121:2463–2471.
6. Crawford JM, and Goldschneider I. Thy-1 antigen and B lymphocyte differentiation in the rat. *J Immunol*. 1993;124:969–976.
7. Hosseinzadeh H, and Goldschneider I. Recent thymic emigrants in the rat express a unique antigenic phenotype and undergo post-thymic maturation in peripheral lymphoid tissues. *J Immunol*. 1993;150:1670–1679.
8. Kroese FGM, de Boer NK, de Boer T, Nieuwenhuis P, Kantor AB, and Deenen GJ. Identification and kinetics of two recently bone marrow-derived B cell populations in peripheral lymphoid tissues. *Cell Immunol*. 1985;162:185–193.
9. Kiyono H, and McGhee JR. T helper cells for mucosal immune responses. In; Ogra PL, Mestecky J, Lamm ME, Strober W, McGhee JR, and Bienenstock J, eds. *Handbook of Mucosal Immunology*. San Diego: Academic Press; 1994:263–274.
10. Kaufman SHE. γ/δ and other unconventional T lymphocytes: what do they see and what do they do? *Proc Nat Acad Sci USA*. 1996;93:2272–2279.
11. Kühnlein P, Park J-H, Herrmann T, Elbe A, and Hünig T. Identification and characterization of rat γ/δ T lymphocytes in peripheral lymphoid organs, small intestine, and skin with a monoclonal antibody to a constant determinant of the γ/δ T cell receptor. *J Immunol*. 1994;153:979–986.
12. Kühnlein P, Vicente A, Varas A, Hünig T, and Zapata A. γ/δ T cells in fetal, neonatal, and adult rat lymphoid organs. *Develop Immunol*. 1995;4:181–188.
13. Ohteki T, Seki S, Abo T, and Kumagai K. Liver is a possible site for the proliferation of abnormal $CD3^+4^-8^-$ double-negative lymphocytes in autoimmune MRL-*lpr/lpr* mice. *J Exp Med*. 1990;172:7–12.
14. Seki S, Abo T, Ohteki T, Sugiura K, and Kumagai K. Unusual $\alpha\beta$-T cells expanded in autoimmune *lpr* mice are probably a counterpart of normal T cells in the liver. *J Immunol*. 1991;147:1214–1221.

15. Seki S, Abo T, Masuda T, et al. Identification of activated T cell receptor γδ lymphocytes in the liver of tumor-bearing hosts. *J Clin Invest.* 1990;86:409–415.
16. Abo T, Ohteki T, Seki S, et al. The appearance of T cells bearing self-reactive T cell receptor in the livers of mice injected with bacteria. *J Exp Med.* 1991;174:417–424.
17. Ohteki T, Okuyama R, Seki S, Abo T, et al. Age-dependent increase of extrathymic T cells in the liver and their appearance in the periphery of older mice. *J Immunol.* 1992;149:1562–1570.
18. Hanawa H, Tsuchida M, Matsumoto Y, Watanabe H, Abo T, et al. Characterization of T cells infiltrating the heart in rats with experimental autoimmune myocarditis. *J Immunol.* 1993;150:5682–5694.
19. Kampinga J, Groen H, Klatter F, et al. Post-thymic T cell development in rats: an update. *Biochem Soc Trans.* 1992;20:191–197.
20. Yang C-P, and Bell EB. Functional maturation of recent thymic emigrants in the periphery: development of alloreactivity correlates with the cyclic expression of CD45RC isoforms. *Eur J Immunol.* 1992;22:2261–2269.
21. Sparshott SM, Bell EB, and Sarawar SR. CD45R CD4 T cell subset-reconstituted nude rats: subset-dependent survival of recipients and bi-directional isoform switching. *Eur J Immunol.* 1991;21:993–1000.
22. Sarawar SR, Sparshott SM, Stutton P, Yang C-P, Hutchinson IV, and Bell EB. Rapid re-expression of CD45RC on rat CD4 T cells in vitro correlates with a change in function. Eur *J Immunol.* 1994;3:103–107.
23. Sparshott SM, and Bell EB. Membrane CD45R isoform exchange on CD4 T cells is rapid, frequent and dynamic in vivo. *Eur J Immunol.* 1994;24:2573–2578.
24. Bell EB, Yang C-P, Sarawar SR, and Sparshott SM. The cyclic expression of CD45R isoforms on CD4 T cells. *Transplant.* 1992;20:198–202.
25. Paterson DJ, Jefferies WA, Green JR, Brandon MR, Corthesy P, Puklavec M, and Williams AF. Antigens of activated rat T lymphocytes including a molecule of 50,000 Mr detected only on CD4 positive T blasts. *Mol Immunol.* 1987;24:1281–1290.
26. Mackay CR, Marston WL, and Dudler L. Naive and memory T cells show distinct pathways of lymphocyte recirculation. *J Exp Med.* 1990;171:801–817.
27. Takimoto H, Nakamura T, Takeuchi M, Sumi Y, Tanaka T, Nomoto K, and Yoshikai Y. Age-associated increase in number of CD4+CD8+ intestinal intraepithelial lymphocytes in rats. *Eur J Immunol.* 1992;22:159–164.

IMMUNOHISTOCHEMICAL EXAMINATION OF LACRIMAL GLAND TISSUE FROM PATIENTS WITH OCULAR SARCOIDOSIS

Janine A. Smith,[1] Chi-Chao Chan,[1] Charles E. Egwuagu,[1] and Scott M. Whitcup[2]

[1]Laboratory of Immunology
[2]Clinical Branch
National Eye Institute
National Institutes of Health
Bethesda, Maryland

1. INTRODUCTION

Sarcoidosis is a systemic, non-malignant disorder characterized by the accumulation of T lymphocytes and mononuclear phagocytes forming noncaseating granulomas in involved organs. Organ systems commonly involved include the lymph nodes, skin, joint, exocrine glands, liver, and central nervous system. Ophthalmic involvement occurs in 27–38% of patients with sarcoidosis, and 26% of patients who develop chronic systemic sarcoidosis will have ocular involvement at some time during the course of the disease. Although the most common ocular manifestation of sarcoidosis is bilateral, granulomatous anterior uveitis, clinical involvement of the lacrimal gland occurs in 7–26% of patients with sarcoidosis.[1–3] Despite this fact, the rate of positive lacrimal gland biopsy in suspected sarcoidosis is relatively low (5–25%).[4]

Although the lacrimal gland is accessible, the biopsy results are often equivocal. The incisional nature of the biopsy specimen necessarily provides a very small portion of the palpebral lobe of the gland which may not contain the non-caseating granuloma required for a tissue diagnosis. By using immunohistochemical techniques with monoclonal antibodies directed against potential immune effector molecules, more information may be gained about the nature of the local inflammatory response, and the diagnostic yield from lacrimal gland biopsy may be improved. The purpose of this study is to characterize the inflammatory cell infiltrate within lacrimal gland tissue of patients with ocular sarcoidosis utilizing immunohistochemical techniques to detect the localization of immunocompetent cells and to examine their functional characteristics.

Lacrimal Gland, Tear Film, and Dry Eye Syndromes 2
edited by Sullivan *et al.*, Plenum Press, New York, 1998

2. MATERIAL AND METHODS

Incisional lacrimal gland biopsy specimens were obtained from 14 patients with ocular manifestations of sarcoidosis who were followed in the uveitis clinic of the National Eye Institute/NIH. The diagnosis of sarcoidosis had been previously established in all patients by microscopic examination of a biopsy specimen. All patients received topical anesthesia with 0.5% proparacaine hydrochloride drops and local anesthesia with subconjunctival injection of 2% lidocaine with epinephrine 10 mg/ml. A small biopsy specimen was taken from the palpebral lobe of the lacrimal gland and divided. For light microscopy, one portion of the specimen was fixed in 10% buffered formalin and embedded in glycol methacrylate; 3 mm sections were stained with hematoxylin-eosin. Another portion was embedded in optimal cutting temperature (OCT) compound and snap frozen. Serial 4–6 mm cryostat sections of frozen tissue were placed on gelatinized slides. After acetone fixation, slides were washed and absorbed by 10% normal animal serum and TRIS buffer.

2.1. Immunohistochemical Analysis

A panel of primary monoclonal antibodies specific for human pan T lymphocyte (CD3), T helper/inducer (CD4), T suppressor/cytotoxic (CD8), B lymphocytes (CD22), bone marrow derived monocytes (CD11c), major histocompatibility complex class II antigens (HLA-DR and HLA-DQ), adhesion molecules (ICAM-1, LFA-1α, and LFA-1β), cytokines (TNFα, TNFβ, IL-2, and IL-4), and control IgG were utilized. A secondary biotin-labelled anti-mouse IgG antibody was used, and the avidin-biotin-peroxidase (ABC) technique was employed using 3,3'-diaminobenzidine (DAB) substrate. Slides were counterstained with methyl green, dehydrated, and cleared as in routine processing.[5]

Staining intensity of the lacrimal gland biopsy specimens for each immunohistochemical marker was then graded by a single experienced examiner using an established scale from 0 to 4.[6]

2.2. Immunohistochemistry Grading Scale

Scoring of the staining intensity was as follows:

None	0
Rare	0.25
Trace	0.50
Mild	1.00
Moderate	2.00
Marked	3.00
Complete	4.00

3. RESULTS

3.1. Immunohistochemistry

The median score of staining intensity was < 1 (mild) for CD8, CD22, CD25, TNFβ, IL-2, and IL-4, indicating minimal CD8+ T cells and B cells within the inflammatory infiltrate (Figure 1).

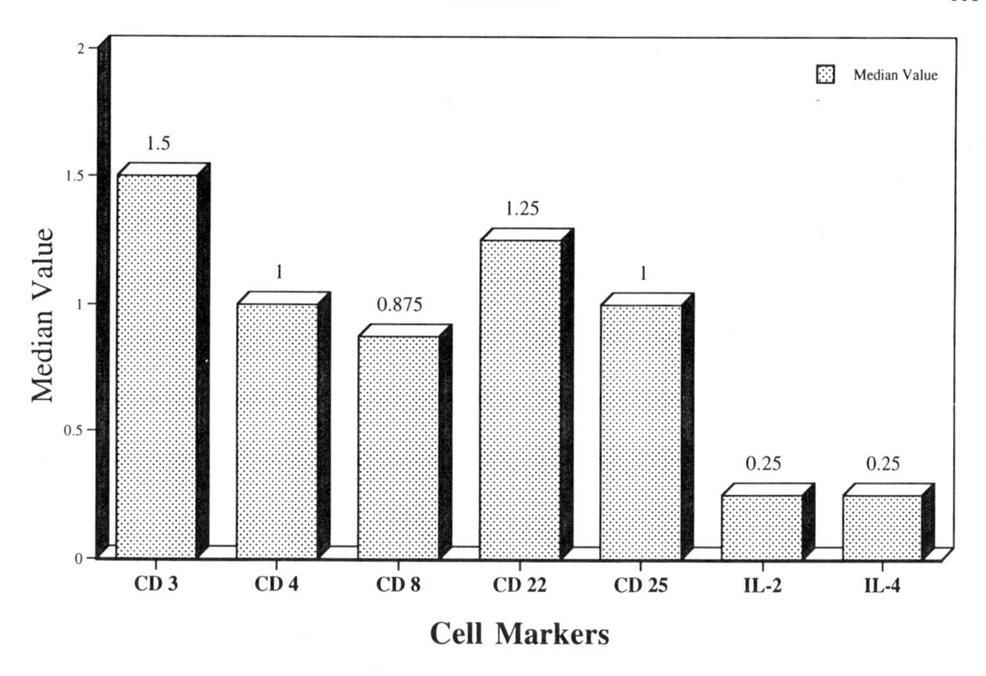

Figure 1. Median scores of immunohistochemical staining for cell markers and cytokines.

The median score was ≥ 1 but < 2 (moderate) for CD3, CD4, ICAM-1, LFA-1α, LFA-1β, TNFα, and CD11c, showing significant amounts of infiltrating CD4+ T cells, activated T cells, and cells of monocyte lineage with concomitant local expression of adhesion molecules (Figures 1 & 2).

The median score was ≥ 2 (marked) for major histocompatibility complex antigens HLA-DR and HLA-DQ (Figure 2). MHC class II molecules were detected within lacrimal gland acinar structures, as well as scattered throughout the interstitium.

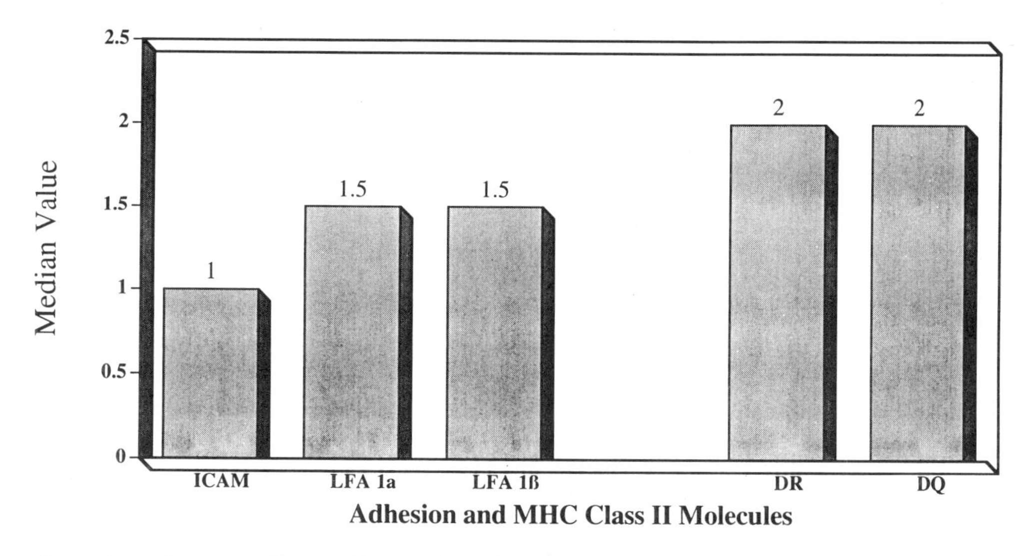

Figure 2. Median scores of immunohistochemical staining for adhesion molecules and MHC Class II molecules.

3.2. Histopathology

Histopathologic examination of lacrimal gland biopsy specimens revealed only one sample with classic non-caseating granuloma formation. Most specimens demonstrated monocytic or lymphoplasmocytic inflitrates. Epithelioid cells were detected in slightly more than one-third of the samples. These results are summarized as follows:

- Monocytic inflitrate 6/14
- Lymphoplasmocytic infiltrate 5/14
- Normal lacrimal gland 3/14
- Plasma cells 6/14
- Epithelioid cells 4/14
- Non-caseating granuloma 1/14

4. DISCUSSION

Although few lacrimal gland biopsy specimens obtained from our patients with ocular sarcoidosis showed the classic histopathologic finding of non-caseating granuloma formation considered diagnostic of sarcoidosis, our immunohisto-chemical studies revealed a consistent preponderance of infiltrating activated CD4+ T cells and macrophages in the samples. These two cell types appear to be the prime initiators of inflammation in several cell-mediated immune reactions, including sarcoidosis. The localized increase in CD4+ T cells and relative paucity of CD8+ T cells in the lacrimal glands of our patients mirrors reports of an elevated CD4/CD8 ratio in lung tissue of patients with pulmonary sarcoidosis, supporting the notion of T cell redistribution within organ systems affected by sarcoidosis.[7]

Demonstration of enhanced expression of HLA-DR and HLA-DQ in these samples suggests that in ocular sarcoidosis the lacrimal glandular epithelium may actively participate in the propagation of the inflammatory reaction with attendant localized T cell activation and chronic antigenic stimulation. Further, the upregulation of ICAM-1 and LFA-1 supports the concept of compartmentalization and T cell redistribution as important in the pathogenesis of ocular manifestations of sarcoidosis as has been demonstrated in the lung.

REFERENCES

1. Obenauf CD, Shaw HE, Sydnor CF, et al. Sarcoidosis and its ophthalmic manifestations. *Am J Ophthalmol.* 1978;86:648–655.
2. Jabs DA, Johns CJ. Ocular involvement in chronic sarcoidosis. *Am J Ophthalmol.* 1986;102:297–301.
3. James DG, Neville E, Langley DA. Ocular sarcoidosis. *Trans Ophthalmol Soc UK.* 1976;96:133–139.
4. Weinreb RN. Diagnosing sarcoidosis by a transconjunctival biopsy of the lacrimal gland. *Am J Ophthalmol.* 1984;97:573–576.
5. Hsu SM, Raine L, Fanger H. The use of avidin-biotin-peroxidase complex (ABC) in immunoperoxidase technique: A comparison between ABC and unlabeled antibody (PAP) procedure. *J Histochem Cytochem.* 1981;29:577–580.
6. Chan CC, Wetzig RP, Palestine AG, Kuwabara T, Nussenblatt RB. Immunohistopathology of ocular sarcoidosis. *Arch Ophthalmol.* 1987;105:1398–1402.
7. Hunninghake GW, Crystal RG. Pulmonary Sarcoidosis: A disorder mediated by excess helper T-lymphocyte activity at sites of disease activity. *N Engl J Med.* 1981;305:429–434.

IMMUNOGLOBULIN LEVELS IN THE TEARS OF PATIENTS WITH CORNEAL GRAFTS AND TRANSPLANT REJECTION

András Berta and Zsolt Lampé

Department of Ophthalmology
University Medical School
Debrecen, Hungary

1. INTRODUCTION

Graft rejection is defined as a permanent loss of the original function of the transplanted tissue or organ, caused by immunological reactions. In case of corneal transplantation, this is not a literal rejection but edema formation, which is followed by invasion of the transplant by host leukocytes, vascularization of the otherwise avascular cornea, and scar formation. In case of transplanted corneas, the permenant loss of transparency due to scarring constitutes the change that makes possible the use of the term transplant rejection.

The diagnostic criteria for corneal transplant rejection are as follows. (i) The process starts three weeks or more in a technically successful and clear graft. (ii) The inflammatory process is limited primarily to the graft. (iii) The process starts at the graft margin nearest to the most proximal blood vessels. (iv) There is movement of inflammatory cells inward from the origin at the host-graft junction to involve the entire graft. (v) The typical pattern of an endothelial or stromal rejection band appears in 25–30% of cases. (vi) Frequently there is an increase in intraocular pressure.[1]

Corneal transplant rejection can be early or late, partial or total. It can affect one or more layers of the cornea (epithelial, stromal, endothelial rejection). In clinical practice we differentiate two types of early corneal rejection. Early Type I rejection is characterized by diffuse edema of the graft, tyndallization of the aqueous humor, and keratic precipitates (precipitates are on edema in the posterior layer of the cornea). Early Type II rejection has characteristic features of a typical endothelial or epithelial rejection line, and sectoral stroma edema that progresses from the limbus towards the center of the corneal graft.[2]

Immunological reactions following corneal transplantation are widely studied. Besides the large variety of animal experiments, in human studies there are three different approaches: histological and immunohistological evaluation of corneal buttons removed during repeated transplantations (rekeratoplasties), serological tests on blood samples, and examination of local immunological reactions (conjunctival cytology, tear tests, aqueous humor). We have used the third approach for several years. The present paper is part of an extensive work examining the local immunological reactions in and around the trasnplanted cornea.[3]

The aim of this study was to investigate the changes in immunoglobulin levels in the tears of patients with corneal grafts and transplant rejection as well as the effects of the administered medications on tear immunoglobulin concentrations, to find out whether tear immunoglobulin determinations could be useful in the postoperative care of corneal transplant patients.

2. PATIENTS AND METHODS

Fourteen patients (six males and eight females) were involved in this study. The mean age (SD) of the patients was 54.3 (±11.7) years. All underwent keratoplasty (corneal transplantation) at the Department of Ophthalmology, University Medical School of Debrecen, Hungary. Four patients suffered from pseudophakic bullous keratopathy, three from keratoconus, three from corneal dystrophy (one Haab-Dimmer, one Gronouw I, one Vogt), three from vascularized leukoma following chemical injury, and one rekeratoplasty (repeated corneal transplantation due to graft rejection). All patients received subconjunctival Dexamethasone injection at the end of the operation and Gentamycin and Bethamethason containing combined eyedrops five times daily during the entire follow-up period. Patients were informed about the purpose and minimal risks of the study, and informed consent was obtained. Clinical and laboratory data were recorded and evaluated anonymously. In all respects we followed the human study regulations of the University Medical School of Debrecen, Hungary.

Tears were collected with capillary tubes before and periodically after surgery, in the morning right before the administration of the first Gentamicin and Bethamethasone containing combined eyedrops. Tear samples were centrifuged immediately and stored at -20°C until analysis. The sediment was subjected to cytological analysis. Data were evaluated retrospectively to determine changes in tear immunoglobulin levels in the postoperative period, changes caused by local steroid therapy, and early signs preceding the appearance of clinical signs of transplant rejection.

IgG, IgA, and IgM levels were determined by laser nephelometry in the tears of patients with corneal graft and transplant rejection. Normal values were determined on 34 normal tear samples (obtained from normal volunteers) using Hyland low-concentration reagents and Hyland immunoglobulin standards, and compared with the values obtained by radial immunodiffusion testing that had been used earlier in our laboratory. The determined normal values (SD) were: IgG, 75.02 (±25.11) mg/L; IgA, 150.12 (±50.07) mg/L; and IgM, 24.90 (±24.77) mg/L.

3. RESULTS

The great variability in immunoglobulin levels in the tears of patients with corneal grafts did not allow statistical evaluation. Interesting tendencies, however, could be observed, including the gradual decrease in immunoglobulin levels in the early postoperative

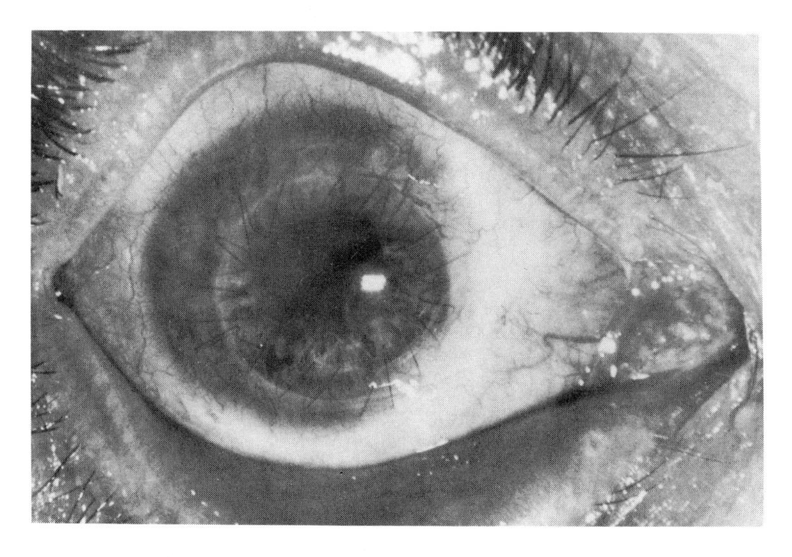

Figure 1. The eye of a patient with Vogt corneal dystrophy 12 days after corneal transplantation.

period and following subconjunctival steroid injection. Very important is the observation that transplant rejection was accompanied by an increase in tear immunoglobulin levels and that this sign may precede the appearance of clinical signs of transplant rejection.

Increased tear levels of IgG, IgA, and IgM were detected in the early postoperative period and gradually disappeared when the signs of irritation subsided. Subconjunctival Dexamethasone injection caused a decrease in the tear levels of all three immunoglobulins in most cases. An increase in IgG, IgA, and IgM levels was also observed at the time and somewhat before the onset of the clinical signs of transplant rejection.

In the patient with Vogt corneal dystrophy, conjunctival hyperemia gradually disappeared in the first 6 days. Excessive tearing was observed, however, during the entire examination period (21 days). The patient received subconjunctival Dexamethasone on the 12th day. Fig. 1 shows the eye 12 days after corneal transplantation.

Fig. 2 shows the IgG, IgA, and IgM levels in the tears of the patient with Vogt corneal dystrophy 3, 6, 9, 12, 15, 18, and 21 days after transplantation. IgG and IgM levels remained within the normal range, showing a gradual decrease. A temporary marked increase in IgG, which was accompanied by a slight decrease in IgA, followed the subconjunctival Dexamethasone injection. IgA levels were unexpectedly low, probably due to excessive tearing.

In one patient with keratoconus, conjunctival hyperemia and tearing disappeared in the first 3 days, and the eye was quiet during the entire 21 days. The patient received subconjunctival Dexamethasone on the 12th day. Fig. 3 shows the eye 12 days after corneal transplantation.

Fig. 4 shows the IgG, IgA, and IgM levels in the tears of a patient (Fig. 3) with keratoconus 3, 6, 9, 12, 15, 18, and 21 days after transplantation. IgG level showed a gradual decrease in the first 12 days, followed by an increase in IgG and a slight decrease in IgA following the subconjunctival Dexamethasone injection. IgA levels were slightly below the normal range, and IgM levels remained normal.

In the patient with corneal leukoma due to lime injury, conjunctival hyperemia and tearing gradually decreased in the first 20 days and then became more pronounced to-

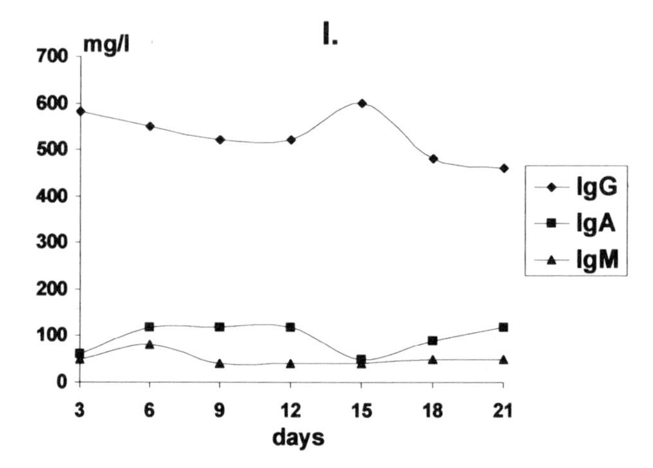

Figure 2. IgG, IgA, and IgM levels in the tears of the patient with Vogt corneal dystrophy after corneal transplantation.

gether with the appearance of transplant edema on the 21st day. The patient received subconjunctival Dexamethasone on the 12th day. Fig. 5 shows the eye 12 days after corneal transplantation.

Fig. 6 shows the IgG, IgA, and IgM levels in the tears of the patient with corneal leukoma due to lime injury 3, 6, 9, 12, 15, 18, and 21 days after transplantation. All three immunoglobulins were above the normal range in the early postoperative period (3 days) and decreased by day 6. The administration of subconjunctival Dexamethasone was followed by a marked decrease in IgG and IgA levels. The appearance of clinical signs of transplant rejection was accompanied by a pronounced increase in IgG and IgA levels and a slight increase in IgM level.

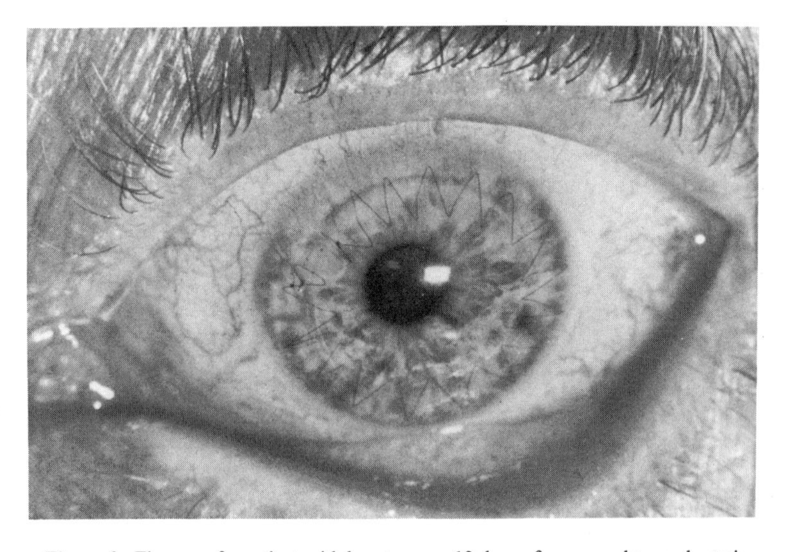

Figure 3. The eye of a patient with keratoconus 12 days after corneal transplantation.

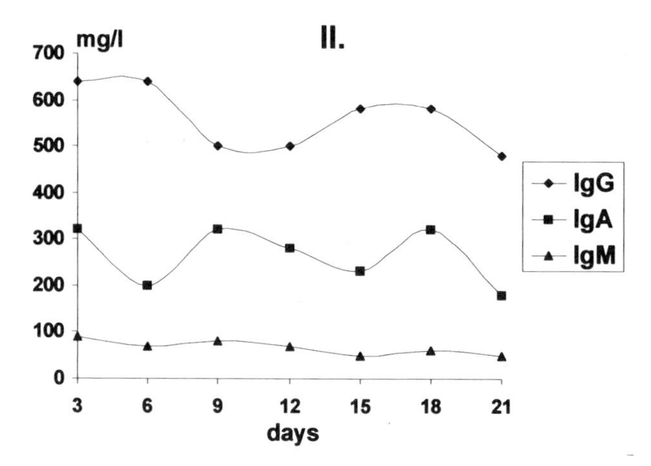

Figure 4. IgG, IgA, and IgM levels in the tears of a patient with keratoconus after corneal transplantation.

4. DISCUSSION

Chandler et al.[4] reported a patient with corneal graft rejection who had a normal level of C4 fraction and no detectable C3 and C1 complement in the tear fluid. We have studied the immunoglobulin and complement levels in a series of experiments in the tears of patients with corneal graft rejection, and detected elevated C4 complement level in the tears of a patient who did not receive systemic immunosuppressive therapy.[5] Both immunoglobulin (IgG, IgA, IgM) and C4 level elevations in tears were found to precede the onset of the clinical symptoms of corneal graft rejection, and were suggested to be used to predict rejection.[6] In an unpublished study, however, we demonstrated that systemic or immunosuppressive treatment suppresses the elevation of immunoglobulins and comple-

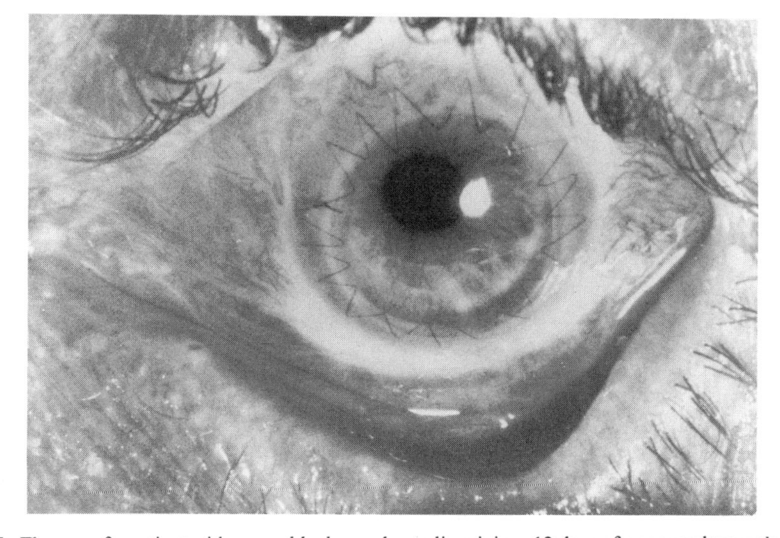

Figure 5. The eye of a patient with corneal leukoma due to lime injury 12 days after corneal transplantation.

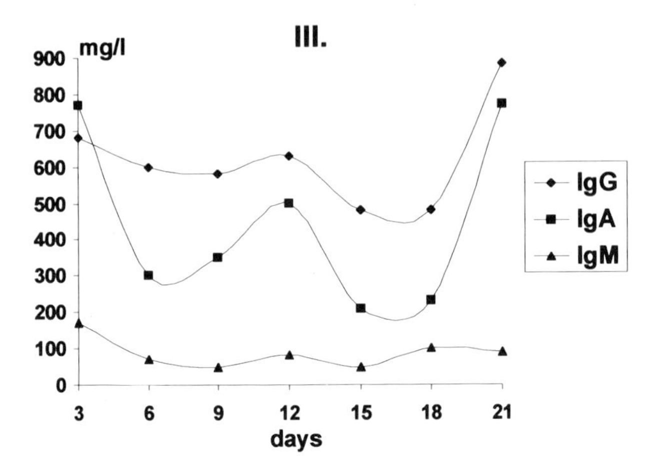

Figure 6. IgG, IgA, and IgM levels in the tears of the patient with corneal leukoma due to lime injury after corneal transplantation.

ment in tears, even in cases in which the treatment is not effective in suppressing the clinical symptoms of corneal graft rejection.

The levels of various immunoglobulins in the tears of patients with corneal graft increase considerably. These changes, however, are only partly the result of local immunological reactions. In freshly operated eyes, because of increased permeability, immunoglobulins, as well as other plasma proteins, readily penetrate the conjunctival vessel walls and cause an elevation in the tear levels. This change in permeability is most apparent for IgG, as IgG is present in the blood in much higher levels than in the tear fluid. Transudation is less effective for IgM, because the higher molecular weight of IgM rarely allows it to penetrate the vessel walls. The IgA level in the plasma is about one tenth that of the tear concentration; therefore, permeability changes would cause a decrease (diluting effect) and never an increase in the tear IgA levels. That is why we consider IgG elevations in the tears as nonspecific signs that can result from various factors (irritation, postoperative inflammation, etc.). Elevations in IgM and IgA levels seem to be more specific for transplant rejection, but other immunological reactions (inflammation caused by infection) may also contribute to local immunoglobulin production.

Corneal grafts are in a privileged situation compared to other transplanted tissues. In most cases the corneal button is transplanted into an avascular host bed. Thus, the tissue does not come in direct contact with the immune system of the recipient. In other cases, however, keratoplasty is performed in more or less vascularized recipient corneas. The degree of vascularization has a decisive influence on the outcome of keratoplasty and on the preoperative prognosis.

However, the corneal graft may remain clear in highly vascularized leukomas, and it may become cloudy despite an avascular recipient bed, as a result of immunological reactions. The exact time of the onset of transplant rejection varies. That is why there is a great need for examination methods suitable for detecting early signs of corneal graft rejection. Such a method would allow the start of immunosuppressive treatment in optimal time, suppressing the rejection reaction before its full development. In addition, the unnecessary administration of immunosuppressive or cytostatic drugs could be avoided. We feel that tear immunoglobulin determinations may become useful tools in the care of patients with corneal transplant rejection.

REFERENCES

1. Polack FM. The corneal graft reaction: An immunological, pathological and clinical perspective. In: *Immunology of the Eye*. Workshop I. Washington, DC. Information Retrieval; 1981:207–219.
2. Alberth B. *Keratoplastik*. Stuttgart: Enke Verlag, 1961;125–146.
3. Berta A. Allergic and Immunological Diseases of the Eye. In: Berta A. *Enzymology of the tears*. Boca Raton: CRC Press; 1992:233–242.
4. Chandler JW, Leder R, Kaufman HE, Caldwell JR. Quantitative determinations of complement components and immunoglobulins in tears and aqueous humor. *Invest Ophthalmol*. 1974;13:151–157.
5. Berta A. Significance of tear protein tests in the prediction of corneal graft rejection. (in Hungarian) *Szemészet*. 1985;122:172–177.
6. Zajácz M, Berta A. Proteolytische Enzymaktivitat und chirurgische Eingriffe bei Ulcus Mooren. *Klin Monatsbl Augenheilkd*. 1985;187:401–402.

86

TEAR FLUID INFLUENCE ON THE OCULAR SURFACE

Stephen C. Pflugfelder

Ocular Surface and Tear Center
Bascom Palmer Eye Institute
Department of Ophthalmology
University of Miami School of Medicine
Miami, Florida

1. INTRODUCTION

A healthy ocular surface is essential for maintaining a comfortable eye and clear vision. Despite many advances in the field of dry eye and ocular surface disease, the role of the tear fluid in maintaining a healthy ocular surface has not been defined. It is unknown whether tear fluid constituents influence the growth and differentiation of the ocular surface epithelium, and if they do, whether this is a primary or redundant function. Furthermore, it is not clear whether the concentration or activity of biologically active tear factors changes in dry eye conditions.

Basic and clinical research provides clues regarding the biological function of the tear fluid on the ocular surface. This research has taught us that the lacrimal gland secretes factors with potential biological activity into the tear fluid, that the concentration of tear fluid cytokines such as epidermal growth factor (EGF) is regulated by sensorineural stimulation and the level of aqueous tear production, and that tear fluid has potent antiproliferative activity. The potential role of the tear fluid on the ocular surface can also be inferred from knowledge that conjunctival squamous metaplasia develops in aqueous tear deficiency, that the ocular surface disease of aqueous tear deficiency is greatest in the exposure zone that is subjected to microtrauma, and that reduced aqueous tear production and/or turnover results in ocular surface inflammation. We will review the research findings that provide indirect information regarding the influence of the tear fluid on the ocular surface.

2. THE LACRIMAL GLAND SECRETES FACTORS WITH POTENTIAL BIOLOGICAL ACTIVITY

An ever-increasing number of factors secreted by the lacrimal gland into the tear fluid are being identified. Although their exact biological function on the ocular surface has not been identified, the reported activity of many of these factors in other systems suggests that lacrimal gland-secreted tear factors can be classified into at least two major functional groups: antimicrobial defense (e.g., lactoferrin, lysozyme, and IgA) and growth regulation (e.g., EGF, TGF-α, vitamin A, and TGF-β). Yoshino and associates[1] have recently reported that cholinergic neurotransmitters stimulate secretion of individual biologically active factors such as lactoferrin and EGF, and that this secretion can be blocked by the muscarinic cholinergic blocker, atropine. Dartt[2] has found that cholinergic neurotransmitters released from parasympathetic nerves provide the major short-term stimulatory regulation of fluid and protein secretion by lacrimal gland acinar cells . These findings suggest that disruption of the cholinergic innervation of the lacrimal gland may result in decreased secretion of biologically relevant tear factors that maintain the health of the ocular surface.

3. TEAR FLUID EGF CONCENTRATION IS REGULATED BY SENSORINEURAL STIMULATION AND LEVEL OF TEAR SECRETION

Van Setten[3] and Ohashi and associates[4] have reported that tear fluid EGF concentrations decrease following induction of reflex tearing. A kinetic study of EGF concentration in tear fluid before and after induction of the nasal-lacrimal reflex by Jones and associates[5] showed a 50–88% decrease in tear fluid EGF concentration that was observed immediately after nasal mucosal stimulation and is sustained for up to 10 min. Nava and associates[6] recently evaluated EGF concentrations in minimally stimulated tear fluid samples obtained from 68 normal subjects (33 male, 35 female) aged 21–88 years and assessed the affects of age, gender, tear production (Schirmer 1), and tear clearance rate (TCR) on EGF concentration. They found that 95% of the samples had tear EGF concentrations of 1–9 ng/ml. There was a slight, but not statistically significant, decline in tear EGF concentration with age. Men had significantly higher tear EGF concentrations than women (mean ± SEM: males 3.38 ± 0.33, females 2.39 ± 0.35). There was inverse correlation of tear EGF concentration with Schirmer 1 test scores ($r = -0.28$, $p = 0.026$) and with the Tear Function Index (TFI)[7] ($r = -0.28$, $p = 0.02$). These findings indicate that elaborate homeostatic mechanisms exist to maintain tear fluid EGF concentration within a defined range that may be optimal for the ocular surface epithelium. They also suggest that disruption of these homeostatic mechanisms (e.g., lacrimal gland disease and loss of reflex tearing) in dry eye conditions may lead to reduced tear EGF concentration that could be a risk factor for ocular surface disease.

4. TEAR FLUID HAS POTENT ANTIPROLIFERATIVE ACTIVITY

Gupta and associates[8] have recently reported that human tear fluid completely inhibited growth of the TGF-β se ns itive mink lung epithelial cell line (CCL-64). The majority of this antiproliferative activity could be neutralized by recombinant TGF-β1 latency associated protein (LAP), suggesting that TGF-β is responsible for the antiproliferative activity of tear fluid. Approximately 50% of TGF-β in the tear fluid was present in a latent form. These findings suggest that tear fluid may maintain the low state of epithelial proliferation on the ocular surface and that diseases that reduce the concentration or bioactivity of TGF-β in the tear fluid could upset the balance of epithelial proliferation on the ocular surface. Indeed, the increased conjunctival epithelial proliferative index observed in the conjunctival epithelium of patients with aqueous tear deficiency (discussed below) could be related to reduced levels of TGF-β activity in the tear fluid.

5. CONJUNCTIVAL SQUAMOUS METAPLASIA DEVELOPS IN AQUEOUS TEAR DEFICIENCY

Aqueous tear deficiency as a result of lacrimal gland secretory dysfunction is correlated with the development of ocular surface disease known as keratoconjunctivitis sicca (KCS). The clinical features of KCS are characterized by superficial punctae keratopathy, intense fluorescein and rose bengal staining of the cornea and conjunctiva, mucus debris in the tear film, and filamentary keratitis.[9] These clinical findings are well correlated with findings of impression cytology that show pathologic squamous metaplasia of the conjunctival epithelium (Fig. 1).[10] Close examination of the pathological changes of conjunctival squamous metaplasia provides clues regarding the function of the tear fluid. These pathologic findings include: (i) increased epithelial stratification and proliferative index, and (ii) abnormal differ-

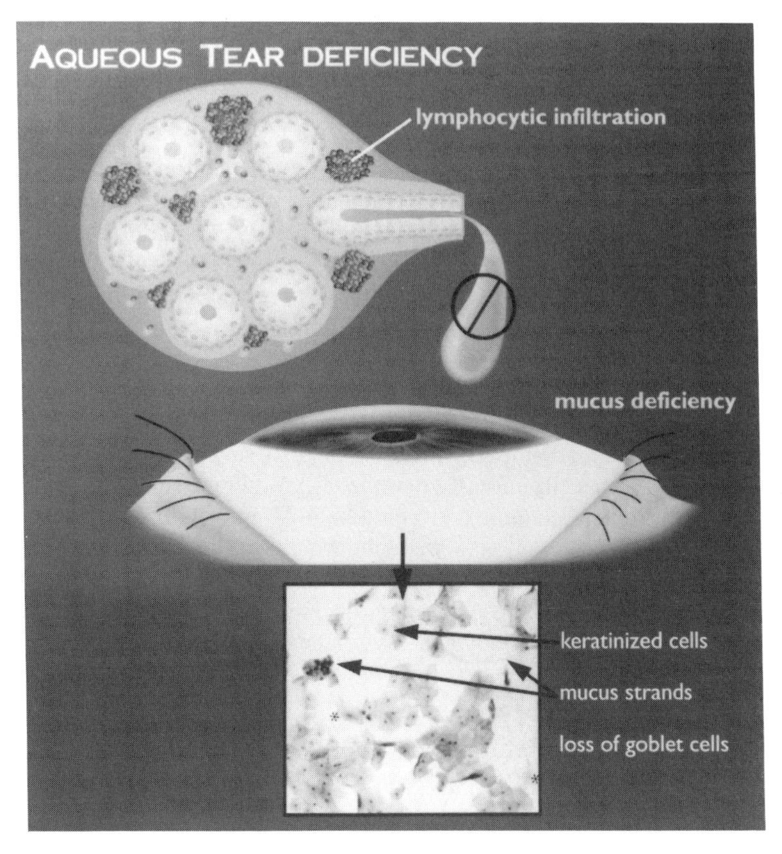

Figure 1. Lymphocytic infiltration of the lacrimal gland in Sjögren's syndrome is associated with aqueous tear deficiency. This results in pathologic changes of the the conjunctival epithelium termed squamous metaplasia. The cytologic features of this condition are loss of mucus-producing goblet cells, enlarged cytoplasmic/nuclear ratio and excessive keratinization of conjunctival squamous cells, and abnormal mucus strands spanning these abnormal squames.

entiation with maintenance of the basal phenotype, reduced cell membrane mucus expression, and reduced goblet cell densities.[11] The severity of these pathologic changes is inversely correlated with aqueous tear production (by Schirmer 1 test) and with loss of nasal-lacrimal reflex tearing .[11] These findings suggest that normal tear fluid inhibits conjunctival epithelial proliferation, promotes terminal differentiation, stimulates epithelial membrane mucus production, and promotes goblet cell differentiation.

6. THE OCULAR SURFACE DISEASE OF AQUEOUS TEAR DEFICIENCY IS GREATEST IN THE EXPOSURE ZONE

It is now well recognized that bulbar conjunctival rose bengal and fluorescein staining scores are greater in the exposed portion of the ocular surface than in the non-exposed zone that is covered with the eyelids. Furthermore, goblet cell densities are lower in the exposed bulbar conjunctiva compared to the non-exposed areas in patients with aqueous tear deficiency, as well as in normal controls.[10] These findings suggest that microtrauma of the exposed bulbar conjunctiva from eyelid shear forces that occur with blinking and environmental stresses may be an important risk factor in the development of KCS in patients with aqueous tear deficiency. Our laboratory has recently performed experiments to determine whether the conjunctival response to microtrauma bears any similarity to the pathological findings of squamous metaplasia. The superficial one or two cell layers of the conjunctival epithelium in rabbits were stripped with a nitrocellulose membrane. This microtrauma induced a transient proliferative response and altered pattern of expression of differentiation markers in the conjunctival epithelium.[12] Epithelial proliferation was induced within 1 h post-trauma, and the number of epithelial cells in the S-phase of the cell cycle remained significantly higher than baseline for 3 days. The number of cell layers in the conjunctival epithelium returned to normal (or was even increased) by day 3; however, an abnormal pattern of expression of epithelial differentiation markers was observed for the first week following microtrauma. Specifically, a marker for basal epithelial cells, cytokeratin 14, was expressed throughout the entire epithelium, and a reduced number of cells staining with an antibody specific for goblet cell mucus was noted. The proliferative rate and pattern of differentiation markers in the conjunctival epithelium returned to baseline within 14 days. Chronically increased epithelial proliferative index and maintenance of a basal phenotype in the conjunctival epithelium of eyes with squamous metaplasia suggest that tear fluid may normally modulate the wound-healing response to conjunctival microtrauma. Perhaps reduced concentrations of biologically active tear constituents in aqueous tear deficiency perpetuates the wound-healing response resulting from physiological trauma such as eyelid blinking and environmental stresses.

7. AQUEOUS TEAR DEFICIENCY IS ASSOCIATED WITH OCULAR SURFACE INFLAMMATION

Inflammation is another important pathologic feature that has been associated with conjunctival squamous metaplasia in the eyes of Sjögren's syndrome-associated aqueous tear deficiency. The conjunctival epithelial cells in most of these eyes shows aberrant expression of the immune activation markers, HLA-DR and ICAM-1.[13] Abnormal T-cell infiltration has also been observed in the conjunctival epithelium and stroma.[10,14]

Furthermore, we have reported that there are significantly increased levels of mRNAs encoding a number of inflammatory cytokines including TNF-α, IL1-α, IL-6, and IL-8 in the conjunctival epithelium of these patients compared to normal controls.[15]

Aberrant expression of HLA-DR antigens by the conjunctival epithelium has also been found to occur in other tear dysfunction states including non-Sjögren syndrome-associated aqueous tear deficiency and ocular rosacea.[11] Patients with ocular rosacea in our study showed levels of conjunctival epithelial HLA-DR positivity equal to that found in specimens obtained from patients with Sjögren's syndrome aqueous tear deficiency; however, Schirmer 1 test scores in these rosacea patients were no different than controls and significantly less than the Sjögren's syndrome group. In contrast, the TCR was similar in the ocular rosacea and Sjögren's syndrome patients, both groups having significantly lower tear turnover than normal controls. Recently reported studies suggest that reduced tear turnover may be a common denominator shared by a number of distinct tear film disorders and ocular surface diseases in-

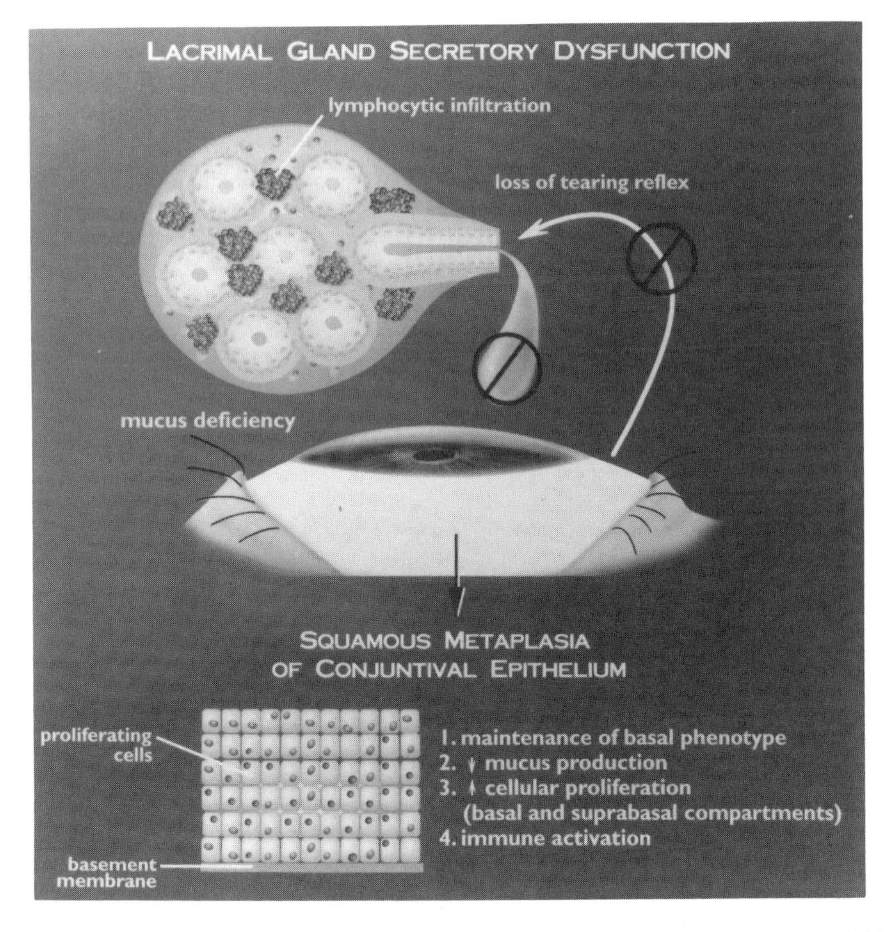

Figure 2. The most severe forms of aqueous tear deficiency, such as Sjögren's syndrome, are accompanied by loss of the ability to reflex tear. This results in pathologic squamous metaplasia of the conjunctival epithelium. The conjunctival epithelium in this condition expresses basal epithelial markers, such as cytokeratin 14, throughout its entire thickness, and shows increased proliferative index (in both the basal and suprabasal compartments), abnormal differentation with reduced epithelial membrane mucus expression by the ocular surface cells, and evidence of immune activation.

cluding aqueous tear deficiency, meibomian gland disease, blink abnormalities, and the closed eye.[7,16,17] We hypothesize that reduced tear turnover may be correlated with increased levels of inflammatory factors in the tear fluid that are capable of inducing epithelial HLA-DR antigen expression. In a recent study, Barton and associates[18] evaluated tear fluid interleukin 1 alpha (IL-1α) concentrations in ideal normal controls, age-matched controls, and ocular rosacea patients. They found that tear fluid IL-1α concentrations were significantly greater in the rosacea patients than in the other two groups. Tear IL-1α concentration was inversely correlated with Schirmer 1 scores ($r = -0.39$, $p = 0.012$) and with TCR ($r = -0.58$, $p = 0.001$). These findings suggest that tear fluid may have anti-inflammatory properties, and that inadequate turnover of tear fluid on the ocular surface leads to increased concentrations of inflammatory factors in the tear fluid, which may cause ocular irritation and have adverse consequences for the ocular surface.

8. CONCLUSIONS

Taken together, these data suggest that tear fluid contains biologically active constituents such as EGF and TGF-β, which are secreted by the lacrimal gland in response to neural stimulation and which may influence the ocular surface by inhibiting conjunctival epithelial proliferation, promoting differentiation, stimulating mucus production, and reducing inflammation. Disruption of this system may lead to conjunctival squamous metaplasia (Fig. 2). Further research is needed to explore these hypotheses and to establish which tear fluid factors influence the health of the ocular surface.

REFERENCES

1. Yoshino K, Monroy D, Pflugfelder SC. Cholinergic stimulation of lactoferrin and epidermal growth factor secretion by the human lacrimal gland. *Cornea.* 1996;15:617–621.
2. Dartt DA. Regulation of tear secretion. *Adv Exp Med Biol.* 1994;350:1–9.
3. Van Setten GB. Epidermal growth factor in human tear fluid: Increased release but decreased concentration during reflex tearing. *Curr Eye Res.* 1990; 9:79–83.
4. Ohashi Y, Motokura M, Kinoshita Y, et al. Presence of epidermal growth factor in human tears. *Invest Ophthalmol Vis Sci.* 1989;30:1879–1882.
5. Jones DT, Monroy D, Pflugfelder SC. Collection of tear fluid proteins: Comparison of polyester wicks with glass capillary tubes. *Cornea*, in press.
6. Nava A, Barton K, Monroy D, Pflugfelder SC. The effects of age, gender and fluid dynamics on the concentration of tear film epidermal growth factor. *Cornea*, in press.
7. Xu K, Yagi Y, Toda I, Tsubota K. Tear function index: A new measure of dry eye. *Arch Ophthalmol.* 1995;113:84–88.
8. Gupta A, Monroy D, Ji Z, Yoshino K, Huang A, Pflugfelder SC. Transforming growth factor beta-1 and beta-2 in human tear fluid. *Curr Eye Res.* 1996;15:605–614.
9. Murillo-Lopez F, Pflugfelder SC. Dry eye. In: Krachmer J, Mannis M, Holland E, eds. *The Cornea.* St Louis: Mosby, 1996: pp 663–686.
10. Pflugfelder SC, Huang AJW, Feuer W, Chuchovski PT, Pereira I, Tseng SCG. Conjunctival cytologic features of primary Sjögren's syndrome. *Ophthalmology.* 1990; 97:985–991.
11. Pflugfelder SC, Tseng SCG, Yoshino K, Monroy D, Felix C, Reis B. Correlation of goblet cell density and mucosal epithelial membrane mucin expression with rose bengal staining in patients with ocular irritation. *Ophthalmology*, 1997;104:223–235.
12. Monroy D, Nava A, Pflugfelder SC. Wound healing response to corneal and conjunctival microtrauma. ARVO Abstracts. *Invest Ophthalmol Vis Sci.* 1996;37:S352.

13. Jones DT, Monroy D, Ji Z, Atherton SS, Pflugfelder SC. Sjögren's syndrome: Cytokine and Epstein-Barr virus gene expression within the conjunctival epithelium. *Invest Ophthalmol Vis Sci.* 1994; 35:3493–3504.
14. Raphael M, Bellefgih S, Piette JC, Le Hoang P, Debre P, Chomette G. Conjunctival biopsy in Sjögren's syndrome: Correlations between histologic and immunohistochemical features. *Histopathology.* 1988;13:191–202.
15. Pflugfelder SC, Ji Z, Naqui R. Immune cytokine RNA expression in normal and Sjögren's syndrome conjunctiva. ARVO Abstracts. *Invest Ophthalmol Vis Sci.* 1996;37:S358.
16. Nelson D, Williams P. Tear turnover and corneal epithelial permeability. ARVO Abstracts. *Invest Ophthalmol Vis Sci.* 1996;37:5906.
17. Prabhasawat P, Tseng SCG. The pathogenic role of delayed tear clearance. AAO Abstracts. *Ophthalmology.* 1996;103:170.
18. Barton K, Monroy D, Nava A, Pflugfelder SC. Inflammatory cytokines in the tear fluid of patients with ocular rosacea. ARVO Abstracts. *Invest Ophthalmol Vis Sci.* 1996;37:S646.

EFFECTS OF LACRIMAL GLAND REMOVAL ON SQUIRREL MONKEY CORNEA

Dmitri Y. Maitchouk, Roger W. Beuerman, Ray J. Varnell, and Lia Pedroza-Schmidt

LSU Eye Center
Louisiana State University Medical Center
New Orleans, Louisiana

1. INTRODUCTION

Lacrimal tissue is responsible for providing the aqueous component of tears, as well as proteins that protect the surface from bacterial attack.[1] Although the discussion of basal tear flow often considers the primary issue to be tear volume, the inclusion of specific tear constituents and their secretory source is equally important. The role of the main and accessory lacrimal glands in the production of basal tear flow and in dry eye conditions has been the subject of considerable research.[1–4] However, this subject has not been completely explored, especially in relation to various types of ocular pathology.[5] The aim of the present study was to determine if accessory lacrimal glands are able to provide the ocular surface with sufficient tears without a secretory contribution from the main lacrimal gland.[6,7] In attempting to determine the relative importance of these two secretory tissues, we took advantage of an unusual anatomical formation found in the squirrel monkey: to provide room for the globe, the main lacrimal gland has migrated to a temporal location outside the bony orbit.

We asked two questions: (i) How much of the aqueous component of the tears does the eye require, and how much of the requirement can be provided by the accessory lacrimal glands? (ii) Are the proteins secreted by the main lacrimal gland similar to those produced by the accessory lacrimal gland? The easiest means to answer these questions was to remove the main lacrimal gland, thereby making the accessory glands solely responsible for ocular surface wetting.

2. MATERIALS AND METHODS

Six squirrel monkeys (1–1.5 kg; two male, four female) were used. Slit-lamp screening examinations with fluorescein were performed prior to surgery to eliminate eyes with corneal complications or existing corneal pathology. Following administration of anesthesia (0.8 ml ketamine IM), the right main lacrimal gland was removed through an 8-mm temporal skin incision. The wound was closed with 6–0 Ethilon sutures. Healing was uneventful in all animals. The left eyes served as controls.

Slit-lamp examinations were performed on operated and control eyes once a week for 10 weeks after surgery. The tear film was stained by wetting a fluorescein strip with saline and touching it to the upper conjunctiva.

Schirmer tear tests were performed before surgery and every 2 weeks after surgery. The wetted portion of the Schirmer tear test paper strips (IOLAB Pharmaceuticals, Claremont, CA) was measured after 5 min. For measuring stimulated reflex tears, the cornea and conjunctiva were gently rubbed with a cotton tip, and the Schirmer test was performed as before.

For protein analysis, tears were collected before surgery and weekly after surgery using 10 μl, fire-polished micropipettes. The lower lid was pulled down gently, and the tip of the capillary was allowed to touch the tear pool in the lower fornix. Care was taken not to touch the conjunctiva or cornea, to avoid reflex tearing. Collected tears (1–2 μl aliquots) were frozen at -70°C. For analysis, tear samples were mixed with a matrix solution containing α-cyano-4-hydroxycinnamic acid (Aldrich Chemical, Milwaukee, WI) and air dried. Mass spectra were obtained using a Voyager Biospectrometry Workstation Matrix-Assisted Laser Desorption Ionization (MALDI) Time of Flight Mass Spectrometer (PreSeptive BioSystems, Framingham, MA). In this spectrometer, a phase transition is effected; proteins introduced as solids are converted into intact, ionized molecules in the gas phase. Subsequently, in the mass analyzer, the mass-to-charge ratios of the protein ions are determined. The principle of laser mass spectrometry is based on methodology whereby intact protein ions are generated in large numbers by laser photon bombardment of protein-containing samples. The production of intact ionized protein molecules occurs when proteins dilutely embedded in a solid matrix are bombarded with intense, short-duration bursts of ultraviolet laser light. The solid matrix consists of low-molecular-weight organic molecules that strongly absorb in the ultraviolet region. The molecular mass range used in tear analysis was 0–22,000 Da.

Light microscopy and transmission electron microscopy were performed to investigate anatomical changes in these tissues.

3. RESULTS

The slit-lamp examinations with fluorescein revealed no corneal abnormalities on the operated side in five of the six squirrel monkeys. In one older female monkey, a small abrasion (about 1.5x2 mm) developed on day 3. The lesion was barely noticeable on slit-lamp examination, but was clearly demarcated after the cornea was stained with fluorescein. The epithelial defect healed on day 10, leaving an anterior stromal haze equivalent to 1 (mild haze not affecting refraction) on a 5-point scale.

Schirmer test results showed that, on the operated side, the average amount of both basic and reflex tears decreased markedly immediately after surgery, then increased gradually over the 10-week course of the study, although never to the levels seen in the unoperated eyes (Tables 1 and 2). Tear production in the operated eye of the older female monkey with the abrasion was similar to the average tear production in the operated eyes of the other monkeys.

Table 1. Average amount of tears after surgery (Schirmer test)

Weeks	Control eyes (mm)	Operated eyes (mm)
0–1	6	1
2–4	6	2
5–10	6	3

Table 2. Average amount of reflex tears after surgery (Schirmer test)

Weeks	Control eyes (mm)	Operated eyes (mm)
0–1	17	2
2–4	17	3
5–10	17	6

MALDI analysis showed generally similar patterns of peaks in the tears from control eyes and from eyes in which the main lacrimal gland had been removed (Fig. 1).

As shown in Fig. 2, the excised main lacrimal gland had the usual acinar structure. By contrast, the accessory glands of Wolfring were present as a collection of separate cellular units, each with an opening to the tears (Fig. 3). The accessory glands of Krause are not prominent in the squirrel monkey.

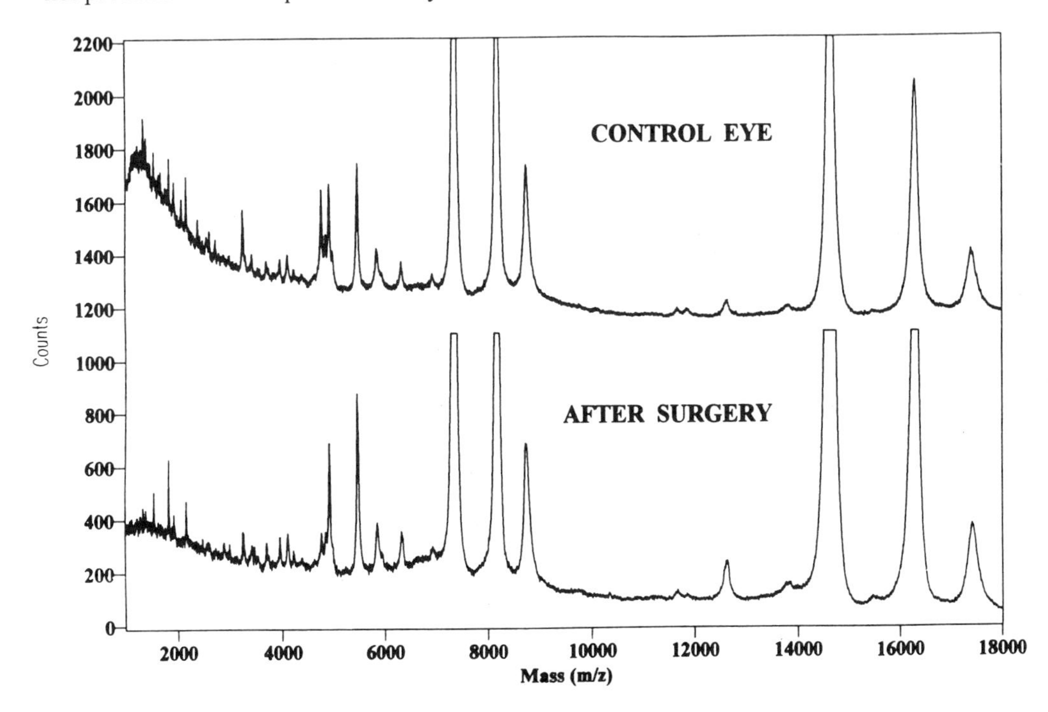

Figure 1. MALDI (mass spectrometry) analysis of tears from operated and control eyes on day 2 after surgery (removal of the main lacrimal gland).

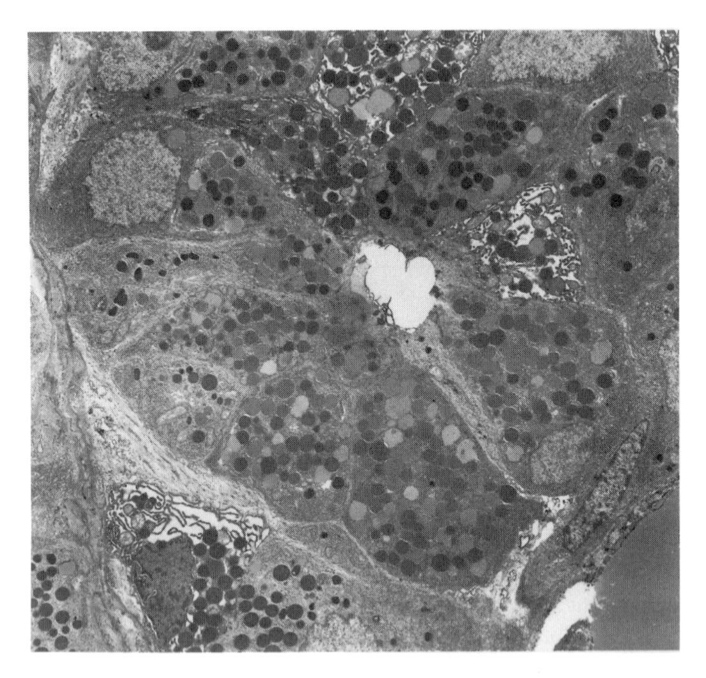

Figure 2. Transmission electron microscopy of the excised main lacrimal gland, showing the usual acinar structure. Original magnification X5,800.

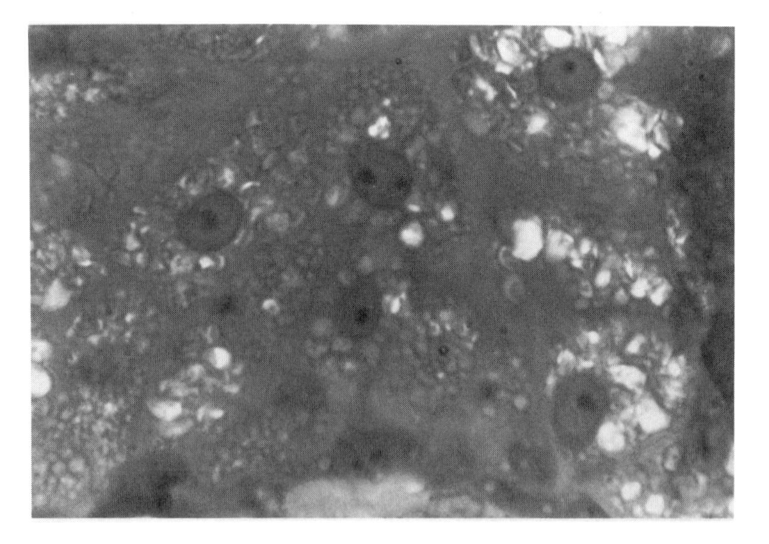

Figure 3. Light microscopic view of accessory lacrimal glands. In this eye, which had undergone removal of the main lacrimal gland, the collections of cells were found in the upper fornix. Original magnification X60.

4. DISCUSSION

Total surgical removal of the main lacrimal gland was used in this study so that there would be no doubt regarding the source of tears. Slit-lamp examinations with instillation of fluorescein and Schirmer testing are used in ophthalmology for the diagnosis of conditions related to dry eye.[6,8] The fact that the only monkey that developed corneal pathology was an older female suggests the interaction between age and endocrine status.[9,10] This finding was interesting as Sjögren's syndrome is more frequent in the older female population.[1,11] Nevertheless, the average amount of tears in the operated eye of this monkey was similar to the average amounts from the operated eyes of the other monkeys, so the thickness of the aqueous layer may not have been a primary factor in the development of the epithelial defect.[12,13] Although basal tear production was decreased, sufficient tears were produced to prevent dry eye. The increase in tear flow observed over time after surgery could be caused by increased secretory activity and/or proliferation of the accessory lacrimal glands.[4] The decrease in stimulated tear production was probably the result of the small size of the accessory lacrimal glands, which could not provide the normal amount of fluid in a short period of time; this change could also be caused by different modes of innervation for the two types of glands.[1,14]

MALDI (mass spectrometry) was found to be valuable for tear analysis, particularly in comparison with gel filtration chromatography and immunoelectrophoresis, as it provides a detailed and precise molecular spectrum and the sensitivity is in the range of 10^{-13} to 10^{-15} moles.[15–17] In addition, our interest included not only previously known proteins, but also unidentified proteins that may be present in tears. The limitation of this technique is difficulty with quantitation, which is inherent to the instrument. We were more concerned with proteins in the molecular weight range of 1,000–22,000 Da, as this weight range contains growth factors that may be important in the development of dry eye.[18,19] The numbers of peaks in the spectra from the control eyes were very similar to the numbers in the spectra from the operated eyes. This means that the proteins secreted into the tears by the main lacrimal gland and the accessory lacrimal glands are similar, and, furthermore, that the accessory lacrimal glands continue to produce these proteins in eyes from which the main lacrimal gland has been removed. As no overt pathology was observed in the eyes without main lacrimal glands, it would appear that accessory glands are sufficient for basal tear flow, and that loss of the main lacrimal gland does not lead to dry eye. Moreover, these findings suggest that other factors may be necessary for dry eye, possibly including inflammation, decreased accessory gland function, and meibomian gland dysfunction.

ACKNOWLEDGMENTS

This work was supported by U.S. Public Health Service grants EY04074 and EY02377 from the National Eye Institute, National Institutes of Health, and by the Department of the Army, Cooperative Agreement DAMD17–93-V-3013. (This does not necessarily reflect the position or the policy of the government, and no official endorsement should be inferred.)

REFERENCES

1. Farris RL. Abnormalities of the tears and treatment of dry eyes. In: Kaufman HE, Barron BA, McDonald MB, Waltman SR, eds. *The Cornea*. New York: Churchill Livingstone; 1988:139–160.

2. Van Haeringen NJ. Clinical biochemistry of tears. *Surv Ophthalmol.* 1981;26:84–96.
3. Holly JF, Lemp MA. Tear physiology and dry eye. *Surv Ophthalmol.* 1977;22:69–87.
4. Seifert P, Spitznas M, Koch F, Cusumano A. Light and electron microscopic morphology of accessory lacrimal glands. *Adv Exp Med Biol.* 1994;350:19–23.
5. Vigneswaran N, Wilk CM, Heese A, Hornstein OP, Naumann GO. Immunohistochemical characterization of epithelial cells in human lacrimal glands. I. Normal major and accessory lacrimal glands. *Graefes Arch Clin Exp Ophthalmol.* 1990;228:58–64.
6. Whitcher JP. Clinical diagnosis of the dry eye. *Int Ophthalmol Clin.* 1987;27(1):7–24.
7. Mackie IA, Seal DV. The questionably dry eye. *Br J Ophthalmol.* 1981;65:2–9.
8. Mackor AJ, van Bijsterveld OP. Tear function parameters in keratoconjunctivitis sicca with and without the association of Sjögren's syndrome. *Ophthalmologica.* 1988;196:169–174.
9. Sullivan DA, Sato EH. Potential therapeutic approach for the hormonal treatment of lacrimal gland dysfunction in Sjögren's syndrome. *Clin Immunol Immunopathol.* 1992;64:9–16.
10. McGill JI, Liakos GM, Goulding N. Normal tear protein profiles and age-related changes. *Br J Ophthalmol.* 1984;68:316–320.
11. Newell FW. *Ophthalmology*. St. Louis: Mosby; 1986:42–80.
12. Lemp MA. General measures in the management of the dry eye. *Int Ophthalmol Clin.* 1987;27(1):36–43.
13. Vanley GT, Leopold IH, Gregg TH. Interpretation of tear film breakup. *Arch Ophthalmol.* 1977;95:445–448.
14. Seifert P, Spitznas M. Demonstration of nerve fibers in human accessory lacrimal glands. *Graefes Arch Clin Exp Ophthalmol.* 1994;232:107–114.
15. Maitchouk DY, Varnell RJ, Beuerman RW, Yan J. Analyzing proteins in tears from patients with dry eye syndrome, with contact lenses, or undergoing photorefractive keratectomy. ARVO Abstracts. *Invest Ophthalmol Vis Sci.* 1996;32:S76.
16. Yoshino K, Tseng SC, Pflugfelder SC. Substrate modulation of morphology, growth, and tear protein production by cultured human lacrimal gland epithelial cells. *Exp Cell Res.* 1995;220:138–151.
17. Karas M, Bahr U, Giebmann U. Matrix-assisted laser desorption ionization mass spectrometry. *Mass Spectrom Rev.* 1992;10:335–357.
18. van Setten GB, Schultz GS, Macauley S. Growth factors in human tear fluid and in lacrimal glands. *Adv Exp Med Biol.* 1994;350:315–319.
19. van Setten GB, Schultz GS. Transforming growth factor-alpha is a constant component of human tear fluid. *Graefes Arch Clin Exp Ophthalmol.* 1994;232:523–526.

LACRIMAL GLAND GROWTH FACTORS AND RECEPTORS: LACRIMAL FIBROBLASTIC CELLS ARE A SOURCE OF TEAR HGF

S. E. Wilson,[1] Q. Li,[2] R. R. Mohan,[1] T. Tervo,[3] M. Vesaluoma,[3] G. L. Bennett,[4] R. Schwall,[4] K. Tabor,[4] J. Kim,[4] S. Hargrave,[2] and K. H. Cuevas[2]

[1]Eye Institute, The Cleveland Clinic Foundation, Cleveland, Ohio
[2]UT Southwestern Medical Center at Dallas, Dallas, Texas
[3]Helsinki University Eye Hospital, Helsinki, Finland
[4]Genentech, Inc., South San Francisco, California

1. INTRODUCTION

The complex aqueous portion of the tear film includes many components that are produced in the lacrimal gland. These tear components are thought to be essential for maintaining a healthy ocular surface and contributing to the corneal optical quality necessary for optimal visual function. Numerous investigations have explored the production of growth factors, cytokines, and receptors in human lacrimal tissue and probed for evidence that these factors function in maintenance of the ocular surface and in corneal wound healing.[1] Studies demonstrating production of growth factors and receptors in the lacrimal gland will stimulate progress towards the more difficult goal of determining their functions in the lacrimal tissue and at the ocular surface. This article will review our current understanding of lacrimal gland and tear growth factors with emphasis on the hepatocyte growth factor (HGF) system.

2. LACRIMAL GLAND GROWTH FACTOR PRODUCTION

There has been increasing support for the hypothesis that the lacrimal gland actively participates in the maintenance of the ocular surface and in corneal wound healing by producing growth factors that interact with receptors on the surface of corneal and conjunctival epithelial cells.[1] For example, studies using immunohistologic methods have shown that epidermal growth factor (EGF), or an EGF-like protein, is present in the tears of mice[2] and hu-

Lacrimal Gland, Tear Film, and Dry Eye Syndromes 2
edited by Sullivan *et al.*, Plenum Press, New York, 1998

mans.[3,4] These studies suggested the lacrimal gland as a source of the EGF present in tears. Subsequent studies determined that mouse,[2] rabbit (Wilson SE, Cook SD, Clarke AF, Thompson H, Beuerman RW, unpublished data), and human[5] primary lacrimal gland tissue contain mRNA coding for EGF precursor protein. EGF-like protein was identified by immunohistologic methods within the lumen of the acini, the cells of the tubular ducts, and at low levels in the acinar cells of the lacrimal gland in the rat.[6] Western blotting methods have also been used to demonstrate conclusively EGF protein in rabbit lacrimal tissue.[7]

Evidence for EGF production in the lacrimal gland and secretion into tears has suggested a potential regulatory role for lacrimal gland-derived EGF in the maintenance of the ocular surface, control of corneal wound healing, and diseases of the ocular surface.[1,8] This relationship between lacrimal gland and tear EGF has, however, become complicated by the observation that corneal epithelium, both ex vivo and in culture, produces EGF mRNA[9,10] and protein.[11] Corneal epithelial cell-produced EGF is likely to have autocrine and/or paracrine effects on the epithelial cells since they are also known to produce EGF receptors.[10] Therefore, a portion of the EGF found in tears collected from the ocular surface could be derived from the ocular surface epithelium itself.

It is likely that many of the growth factors expressed in lacrimal gland tissue have local autocrine or paracrine effects on lacrimal gland cells. It is difficult to prove that these factors are produced by the lacrimal gland to modulate the cells of the ocular surface rather than being washed into the lacrimal fluid after modulating effects within lacrimal tissue. For example, the hypothesis that lacrimal gland-produced EGF has autocrine or paracrine effects in the lacrimal gland is supported by the detection of mRNAs coding for EGF receptor in lacrimal tissue.[12] Little is known, however, regarding which lacrimal cells express EGF receptors or the functions controlled by EGF or any other growth factors in lacrimal tissue.

Recent studies have provided insights into the expression of HGF and HGF receptor in lacrimal gland and cornea and the possible role of this growth factor in maintenance of the ocular surface. HGF is a heparin-binding growth factor that serves as a paracrine mediator of stromal-epithelial interactions in tissues where epithelial and fibroblastic cells conjoin in the formation of a tissue. Thus, HGF is commonly produced by fibroblastic cells within a tissue to regulate functions of epithelial cells expressing HGF receptor. For example, keratocyte cells in the cornea have been shown to express HGF.[13,14] HGF regulates proliferation, motility, and differentiation of corneal epithelial cells expressing HGF receptor.[15] Recent studies,[14] however, have also localized HGF protein expression to fibroblast cells in the interlobular and interalveolar connective tissue of the rabbit lacrimal gland.[14] HGF receptor is expressed in acinar and ductal epithelial cells of lacrimal tissue, suggesting HGF produced in lacrimal gland connective tissue modulates lacrimal epithelial cell functions. Nothing is known about specific functions that are modulated in the lacrimal epithelial cells by HGF.

HGF is also detectable within the lumen of the lacrimal ducts and tears.[14,16] Thus, HGF secreted by the interalveolar connective tissue cells not only binds HGF receptor expressed by lacrimal epithelial cells, but also passes into the lacrimal ducts to become a component of tears.[16] HGF protein is detectable only at the tear film-apical epithelial junction, not in the deeper layers of the epithelium, in the unwounded cornea.[14] This indicates that the lacrimal gland is probably a major source of HGF modulating homeostatic functions of the ocular surface epithelium, since HGF produced by keratocyte cells in the unwounded cornea cannot be detected associated with the basal and more superficial epithelial cells beneath the surface using immunohistochemistry. The accessory lacrimal glands could also contribute HGF to tears, but this has not been thoroughly investigated.

Once the cornea is wounded, there is a change in HGF expression within the cornea-lacrimal gland-tear axis. The best-characterized model is the corneal epithelial scrape wound in the rabbit.[14] HGF production is markedly upregulated within keratocytes during the first few days after injury, and HGF receptor-bound HGF is localized to the entire thickness of the healing corneal epithelium. In the human photorefractive keratectomy model, although the instantaneous concentration of HGF in the tears decreases after wounding, tear flow is markedly increased, and, therefore, total HGF exposed to the ocular surface in the tears is increased more than 1000% the first day after corneal epithelial injury.[16] Since HGF is a heparin-binding growth factor that associates with cellular surface glycosaminoglycans, there is likely to be increased total HGF available for binding to the cognate HGF receptors expressed by the epithelial cells. Increased tear HGF may be derived from both the activated keratocytes beneath the corneal epithelial wound and upregulated expression by the lacrimal gland, although upregulation of lacrimal gland HGF production after corneal wounding has not been conclusively demonstrated. Increased expression of HGF by the lacrimal gland after corneal epithelial wounding is a topic of continuing investigation.

Thus, our studies suggest that HGF secreted by lacrimal gland fibroblastic cells participates in ocular surface homeostasis. After corneal wounding, HGF produced by both keratocyte cells and the lacrimal gland probably contribute to corneal epithelial healing.

3. POTENTIAL ROLE FOR HGF IN THE PATHOGENESIS OF OCULAR SURFACE DISEASE

We have previously hypothesized that a deficiency of lacrimal gland-derived growth factors could have a role in the pathogenesis of keratoconjunctivitis sicca.[1,14] Some studies have demonstrated that ocular surface rose bengal staining associated with keratoconjunctivitis sicca is attributable to deficient preocular tear film protection of corneal epithelial cells.[17,18] We have shown that HGF (and EGF) inhibits the expression of keratin K3, a marker of differentiation of corneal epithelial cells.[15] Our working hypothesis is that altered surface properties of corneal epithelial cells associated with atypical differentiation could contribute to diminished tear film adherence to the surface of individual cells and, therefore, that diminished HGF secretion by diseased lacrimal tissue could have a role in the pathophysiology of keratoconjunctivitis sicca. The lacrimal gland is commonly infiltrated by inflammatory cells in Sjögren's syndrome and keratoconjunctivitis sicca. Infiltrating inflammatory cells could injure interalveolar fibroblast cells that secrete HGF. If this hypothesis is correct, then exogenous HGF or synthetic HGF receptor agonists could have therapeutic effects on ocular surface abnormalities associated with keratoconjunctivitis sicca. HGF would have convenient drug delivery characteristics because of its heparin-binding properties, and therapeutic effect could be seen with a reasonable frequency of topical administration. We hope in the near future to pursue this hypothesis in clinical trials where the efficacy of exogenous HGF can be tested in severe keratoconjunctivitis sicca.

ACKNOWLEDGMENTS

Supported in part by US Public Health Service grant EY10056 from the National Eye Institute, National Institutes of Health, Bethesda, MD, and The Cleveland Clinic Foundation.

REFERENCES

1. Wilson SE. Lacrimal gland epidermal growth factor production and the ocular surface. *Am J Ophthalmol.* 1991;111:763–765.
2. Tsutsumi O, Tsutsumi A, Oka T. Epidermal growth factor-like, corneal wound healing substance in mouse tears. *J Clin Invest.* 1988; 81:1067–1071.
3. Ohashi Y, Motokura M, Kinoshita Y, et al. Presence of epidermal growth factor in human tears. Invest Ophthalmol Vis Sci. 1989;30:1879–1882.
4. van Setten GB, Viinikka L, Tervo T, et al. Epidermal growth factor is a constant component of normal human tear fluid. *Graefes Arch Clin Exp Ophthalmol.* 1989;227:184–187.
5. Wilson SE, Lloyd SA, Kennedy RH. Epidermal growth factor messenger RNA production in human lacrimal gland. *Cornea.* 1991;10:519–524.
6. van Setten GB, Tervo K, Virtanen I, et al. Immunohistochemical demonstration of epidermal growth factor in the lacrimal and submandibular glands of rats. *Acta Ophthalmol.* 1990;68:477–480.
7. Steinemann TL, Thompson HW, Maroney KM, et al. Changes in epithelial epidermal growth factor receptor and lacrimal gland EGF concentration after corneal wounding. ARVO Abstracts. *Invest Ophthalmol Vis Sci.* 1990;31:S55.
8. van Setten GB, Tervo T, Viinikka L, Pesonen K. Ocular disease leads to decreased concentrations of epidermal growth factor in the tear fluid. *Curr Eye Res.* 1991;10:523–527.
9. Wilson SE, He Y-G, Lloyd SA. EGF, basic FGF, and TGF beta-1 messenger RNA production in rabbit corneal epithelial cells. *Invest Ophthalmol Vis Sci.* 1992;33:1987–1995.
10. Wilson SE, Lloyd SA, Kennedy RH. Epidermal growth factor and its receptor, basic fibroblast growth factor, transforming growth factor beta-1, and interleukin-1 alpha messenger RNA production in human corneal endothelial cells. *Invest Ophthalmol Vis Sci.* 1991;32:2747–2753.
11. Wilson SE, Schultz GS, Chegini N, Weng J, He Y-G. Epidermal growth factor, transforming growth factor alpha, transforming growth factor beta, acidic fibroblast growth factor, basic fibroblast growth factor, and interleukin-1 proteins in the cornea. *Exp Eye Res.* 1994;59:63–72.
12. Wilson SE, Lloyd SA, Kennedy RH. Basic fibroblast growth factor (FGFb) and epidermal growth factor (EGF) receptor messenger RNA production in human lacrimal gland. *Invest Ophthalmol Vis Sci.* 1991;32:2816–2820.
13. Wilson SE, Walker JW, Chwang EL, He Y-G. Hepatocyte growth factor (HGF), keratinocyte growth factor (KGF), their receptors, FGF receptor-2, and the cells of the cornea. *Invest Ophthalmol Vis Sci.* 1993;34:2544–2561.
14. Li Q, Weng J, Mohan R, et al. Hepatocyte growth factor (HGF) and HGF receptor protein in lacrimal gland, tears, and cornea. *Invest Ophthalmol Vis Sci.* 1996;37:727–739.
15. Wilson SE, He Y-G, Weng J, Zeiske JD, Jester JV, Schultz GS. Effect of epidermal growth factor, hepatocyte growth factor, and keratinocyte growth factor, on proliferation, motility, and differentiation of human corneal epithelial cells. *Exp Eye Res.* 1994;59:665–678.
16. Tervo T, Vesaluoma M, Bennett GL, et al. Tear hepatocyte growth factor (HGF) availability increases markedly after excimer laser surface ablation. *Exp Eye Res.*, in press.
17. Feenstra RP, Tseng SC. What is actually stained by rose bengal? *Arch Ophthalmol.* 1992;110:984–993.
18. Feenstra RP, Tseng SC. Comparison of fluorescein and rose bengal staining. *Ophthalmology.* 1992;99:605–617.

CLUSTERIN MAY BE ESSENTIAL FOR MAINTAINING OCULAR SURFACE EPITHELIUM AS A NON-KERATINIZING EPITHELIUM

Kohji Nishida, Satoshi Kawasaki, and Shigeru Kinoshita

Department of Ophthalmology
Kyoto Prefectural University of Medicine
Kyoto, Japan

1. INTRODUCTION

Clusterin, a 70–80-kDa glycoprotein, is present in numerous physiological fluids such as plasma, urine, breast milk, cerebrospinal fluid, and semen.[1–3] cDNA sequencing of human clusterin was reported in 1989.[4] Although it has been given various names, including clusterin, apolipoprotein J, SP-40, 40, etc.,[5] clusterin is now the internationally accepted name for this protein. It has potential amphipathic helical domains that allow it to bind to hydrophobic molecules, and potential heparin binding domains that are important for interaction with cell membranes and extracellular matrices.[6] Although several functions of clusterin have been proposed, such as lipid transport,[1] sperm maturation,[7] regulation of the complement cascade,[2] programmed cell death,[5] and membrane recycling,[5,8] the most likely is cell membrane protection, which was reported by Aronow et al.[8]

Regarding the localization of clusterin in the human eye, there have been a few reports that showed that it is present in the retina, vitreous humor, and aqueous humor.[9–11] Since clusterin was the most abundant gene transcripts in the corneal epithelium based on our previous cDNA sequencing analysis[12] (Table 1), it may play an important role in this tissue. The present study was designed to determine the localization of clusterin in the corneal and conjunctival epithelia in normal and diseased states, and we propose a new concept for clusterin function.

2. MATERIALS AND METHODS

2.1. Human Samples

With proper informed consent, in accordance with the Helsinki Declaration regarding research involving human subjects, and upon approval by the Institutional Review Board

Table 1. A gene expression profile of human corneal epithelium

GS	lb	CO	Hp	HL	Gr	Mo	Lg	Cm	Kc	Acc#	Definition	EST
02186		19								X14723	apolipoprotein J	>20
08025		16										0
01919		12				4	3	2	4	J02763	calcyclin	7
02374		11					1			D45370	Human adipose mRNA	0
00504		10	1	4	1		1	1	2	M14328	alpha enolase	10
08067		8								X05421	keratin 3	0
00163		7	1			1	1	2		M17886	acidic ribosomal phosphoprotein P1	>20
00418		7	4	16	1	8				D14531	ribosomal protein homologous to rat L9	>20
00336		6	1	2	1	2		1	1	Z12962	ribosomal protein homologous to yeast L41	>20
00339		6	4							J03934	NAD(P)H:menadione oxidoreductase	0
01141		6			3							8
01808		6				5	1	1	4	S54005	thymosin beta-10	>20
00001		5	1							M77349	Transforming growth factor-beta induced gene product (BIGH3)	7
00177		5	2	2		1				L19739	metallopanstimulin (MPS1)	>20
00293		5	6	9	1	6	3	1	2	u14968	ribosomal protein L27	>20
00437		5	4	4	2	2	1			X57959	ribosomal protein L7	>20
00444		5	6	9	1	3	1		6	M81757	ribosomal protein S19	10
00533		5		1					3	M97164	ferritin heavy chain	>20
00795		5		1	1	2				X79238	ribosomal protein L30	18
01706		5				4	2			M80563	CAPL protein	1
03128		5							4	X15480	anionic glutathione-S-transferase	13
03867		5						1	2	M81457	calpactin 1 light chain	8
04147		5						1				3
05108		5								Z23090	27 kDa heat shock protein	>20

Abundant GS (occurring 5 times or more) in 1057 chromosomally coded GS of corneal epithelium are listed in descending order of occurrence (CO), along with GS number (GS), occurrence of each GS in previously analyzed 745 from HepG2 cells (Hp), 866 clones from HL60 cells (HL), 1087 from granulocytoid cells (Gr), 898 from monocytoid cells (Mo), 878 from lung (Lg), 925 from colonic mucosa (Cm), and 822 from keratinocytes (Kc). Corresponding GenBank entries (Re 90) are listed by their accession number (Acc#) and gene name (Definition). The number of libraries in which they were detected is indicated in the "lb" column by the shading pattern: more than 4, filled area; 3, dark shading; 2, light hatching; and 1, blank. The number of entries in dbST, except for GS we had submitted, regarded as identical to each GS, is shown in the column EST.

(IRB) on Human Experiments of the Research Committee of the Kyoto Prefectural University of Medicine, normal human corneal specimens were obtained from autopsied eyes. Normal human tissues from conjunctiva, epidermis, vagina, esophagus, stomach, small intestine, and colon were obtained during surgery. Also, diseased conjunctival epithelial tissue from a patient with Stevens-Johnson syndrome (56-year-old female) was obtained during surgery.

2.2. Reverse Transcriptase Polymerase Chain Reaction (RT-PCR)

2.2.1. RNA Extraction. Epithelial cells were collected from cornea, conjunctiva,[13] epidermis,[14] vagina, and esophagus, without contamination by underlying cells in the substantia propria. Poly A+ RNA was extracted from these cell specimens using a Quick-Prep mRNA Purification kit (Pharmacia-LKB, Uppsala, Sweden).

2.2.2. Oligonucleotide Primers. Oligonucleotide primers were synthesized by an oligonucleotide synthesizer (Gene Assembler Plus; Pharmacia-LKB), with selection based on the specific nucleotide sequence of the expected amplified clusterin cDNA. An oligonucleotide probe was also synthesized, with selection based on the specific nucleotide sequence of the expected amplified clusterin cDNA. The sequences are given below.

Clusterin primers were:

5'-primer:	GCTTCCCACACTTCTGACTCG	(selected from clusterin gene exon 8)
3'-primer:	GGGCCTGGTTACTTGGTGACG	(selected from clusterin gene exon 9)
Clusterin probe was:	AGCACCGGGAGGAGTGAGATGTGGATGTTG	

The length of the amplified DNA sequence was 293 bp. β-Actin primer sets were purchased (CLONTECH, Palo Alto, CA) as control to confirm cDNA isolation.

2.2.3. Reverse Transcription and PCR. Complementary DNA (cDNA) generation and PCR procedure were performed as previously described.[15] First strand cDNA was heated to 95°C for 10 min, cooled on ice for 5 min, and used as a template for PCR. PCR amplification was performed using a DNA thermal cycler (Perkin-Elmer Cetus, Norwalk, CT) for 30 cycles. The cycle profile consisted of 30 sec at 94°C for denaturation, 1 min at 60°C for annealing, and 2 min at 72°C for primer extension.

2.2.4. Analysis of PCR Product. A 10 μl aliquot of the reaction mixture was then electrophoresed on a 1.5% SeaKem agarose gel (FMC Bio Products, Rockland, ME) containing ethidium bromide, to evaluate amplification and fragment size. The PCR products, separated by 1.5% agarose gel electrophoresis, were then transferred to a nylon filter (Hybond-N+; Amersham, Buckinghamshire, UK) by the capillary transfer method using 0.4 N NaOH. Southern blot hybridization was performed using ^{32}P-labeled oligonucleotide probes specific to clusterin. Hybridization was then carried out at 42°C in a hybridization buffer. Then the filter was washed three times with 2X SSC and 0.1% sodium dodecyl sulfate (SDS) at room temperature for 15 min.

2.3. Immunohistochemical Procedure

For immunohistochemical detection of the clusterin protein, tissue specimens, including cornea, conjunctiva, vagina, esophagus, epidermis, stomach, small intestine, and colon, and conjunctival tissue from a Stevens-Johnson syndrome eye, were snap frozen, sectioned into 7 μm thick slices, and then fixed in absolute ethanol at 4°C for 10 min. The sections were stained by the indirect immunoperoxidase (ABC) method. The primary antibody was a commercially available monoclonal antibody raised against clusterin (Quidel, San-Diego, CA). This antibody, previously characterized as a G7 antibody by Murphy et al.,[16] was diluted to yield an IgG concentration of 15 μg/ml. Control slides were incubated with normal mouse IgG (DAKO, Kyoto, Japan). Finally, the sections were exposed to 3,3'-diaminobenzidine for development.

3. RESULTS

3.1. RT-PCR Analysis

Specific amplification of the clusterin mRNA transcripts was detected in ethidium bromide-stained gel for corneal, conjunctival, vaginal, and esophageal epithelial cells, but

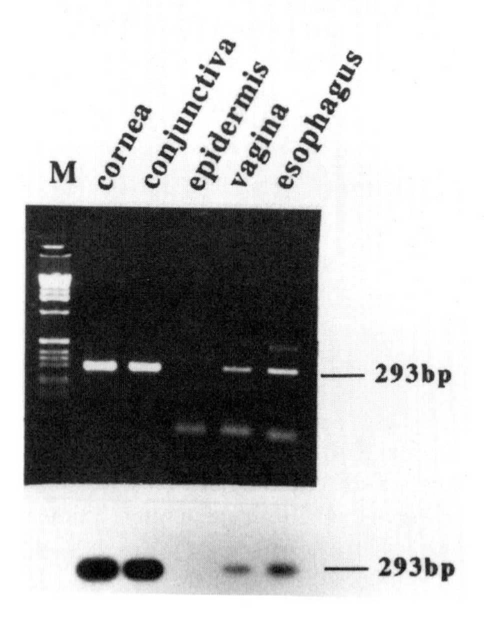

Figure 1. Ethidium bromide-stained gel (top) and Southern blot analysis (bottom) of RT-PCR products for clusterin mRNA transcripts in in vivo human epithelial cells from cornea, conjunctiva, epidermis, vagina, and esophagus. The expected length of the amplified DNA is indicated at the right (293 bp). The lanes shown represent: M, a 1 kilobase-pair DNA ladder.

not for epidermal keratinocytes (Fig. 1). Southern blot analysis showed the same results as the ethidium bromide-stained gel observation. All the samples expressed mRNA transcripts for β-actin, confirming the integrity of the cDNAs and their ability to serve as templates for amplification (data not shown).

3.2. Immunohistochemical Analysis

In the normal stratified squamous epithelia of the human cornea, conjunctiva, vagina, and esophagus, the staining patterns were similar. In the corneal epithelium, the superficial cell layers and upper wing cell layers were intensely stained, whereas the underlying layers were not discernibly stained (Fig. 2A and B). In the conjunctival epithelium, as in the corneal epithelium, positive staining was seen in the upper half of the epithelial regions but not in the control sections (Fig. 2C and D). In the vaginal and esophageal epithelium, positive staining was restricted to the superficial cell layer (Fig. 3B and C). In the normal mucosal epithelia of the stomach, small intestine, and colon, numerous mucin-secreting cells were stained (Fig. 3D, E, and F). In contrast, the epidermis showed no discernible staining in any of the epithelial regions (Fig. 3A). In the diseased conjunctival epithelium with squamous metaplasia (keratinization) obtained from the Stevens-Johnson syndrome eye, staining was markedly reduced, as compared to that in normal conjunctival epithelium (Fig. 2E and F). The control sections exhibited no discernible specific immunoreactive labeling over the entire region in any specimens.

4. DISCUSSION

This study clearly demonstrates that clusterin is synthesized by normal ocular surface epithelial cells and is present mainly in the apical cells of the normal ocular surface epithelium, and that its expression is reduced in the abnormal conjunctival epithelial cells of Stevens-Johnson syndrome. These findings imply that clusterin may have an important role in ocular surface homeostasis at the tear-ocular surface interface.

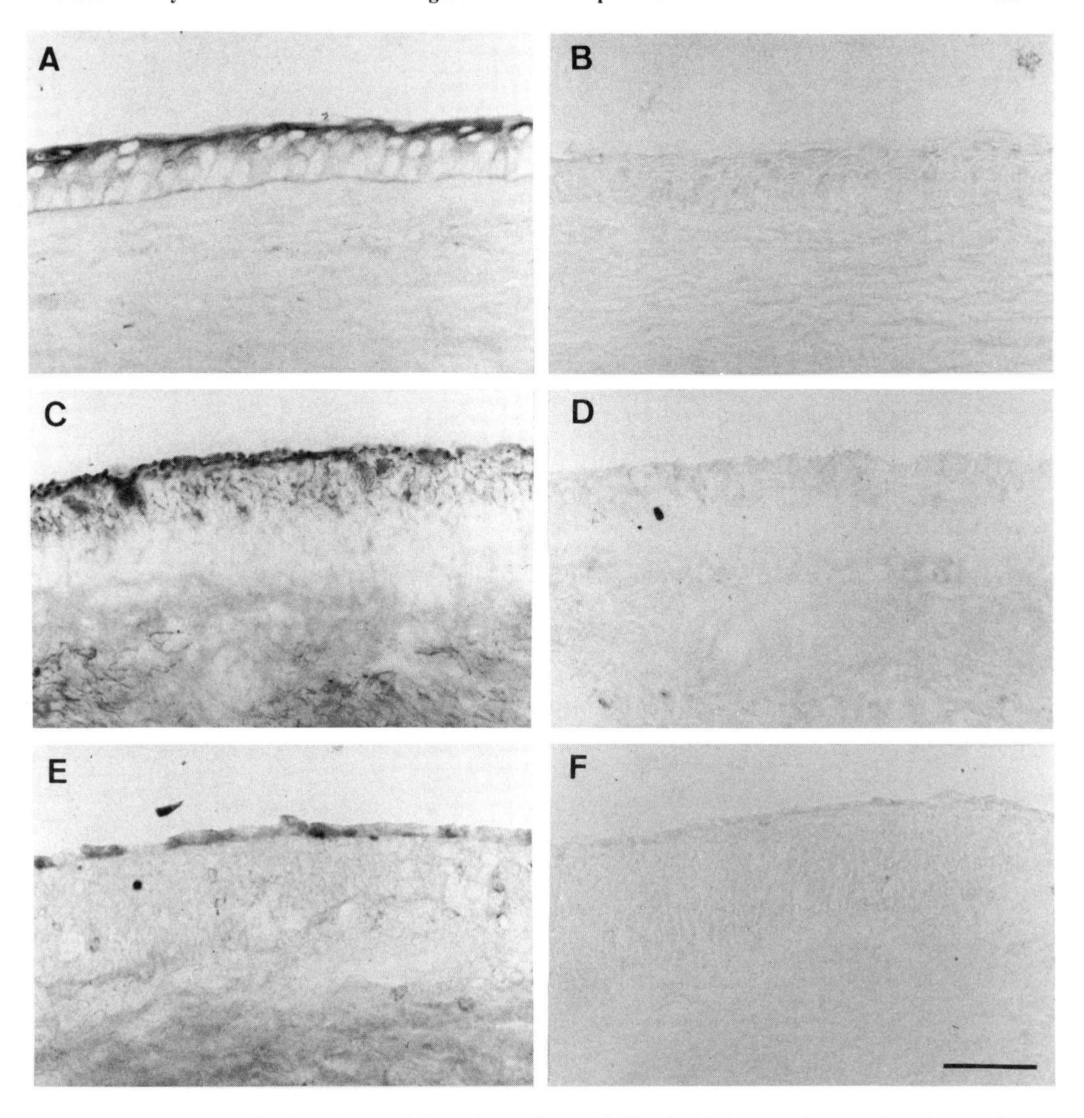

Figure 2. Immunolocalization of clusterin in ocular surface epithelia of adult humans. In corneal sections, staining for clusterin was detected in apical layers of the epithelium (A), whereas control corneal sections incubated with normal mouse serum IgG showed no discernible specific labeling over the entire region (B). In conjunctival tissue sections, staining was detected in the upper half of the epithelial regions (C), whereas control conjunctival sections showed no discernible specific labeling over the entire region (D). In abnormal conjunctival tissue obtained from a patient with Stevens-Johnson syndrome, staining in the epithelium was markedly reduced (E), as compared to normal conjunctiva. Control section showed no discernible specific labeling (F). Bar=50 μm.

The function of clusterin in the ocular surface remains obscure. Its expression was not seen in the epidermis and was markedly reduced in abnormal conjunctival epithelium with squamous metaplasia (keratinization) in Stevens-Johnson syndrome. These facts led us to consider the interesting hypothesis that clusterin is essential for maintaining ocular surface epithelium as a non-keratinizing epithelium. Also, clusterin expression was detected in all of the other mucosal epithelia tested. Therefore, it is possible that clusterin is generally expressed in the mucosal epithelium, and that clusterin may play an important

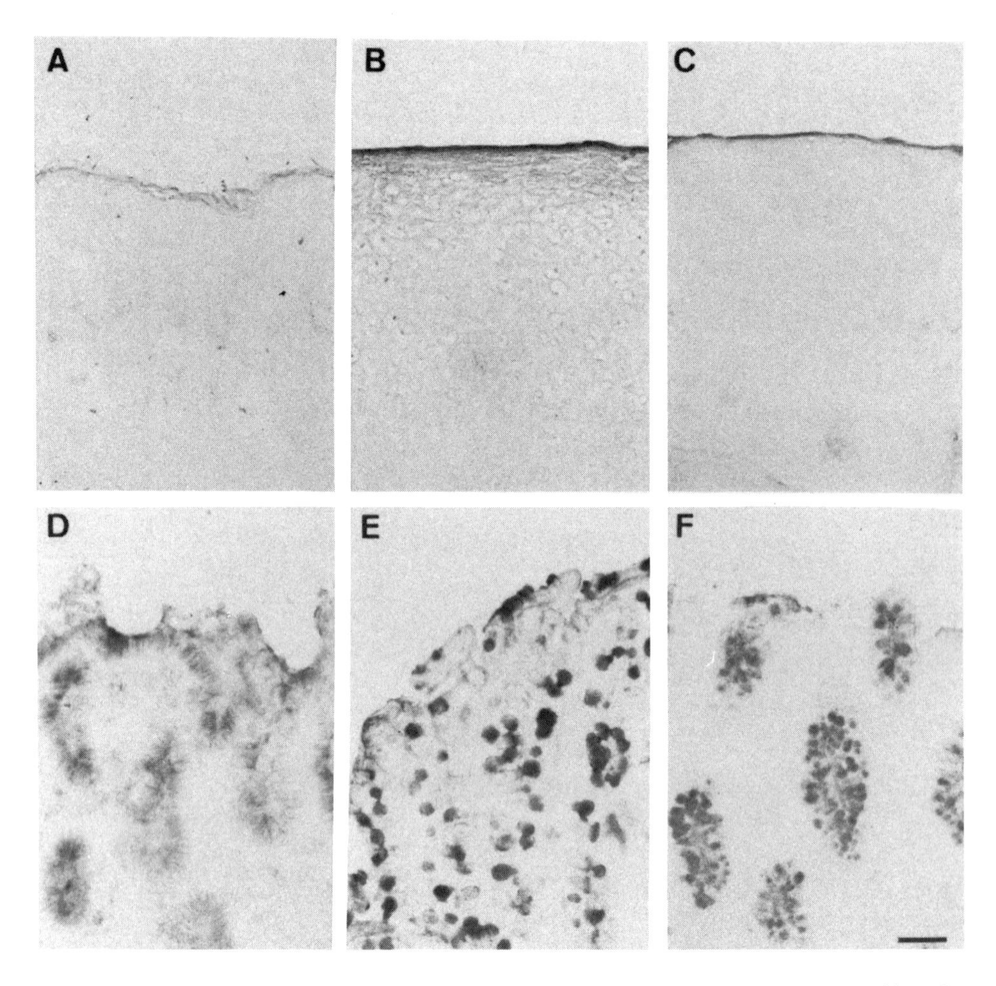

Figure 3. Immunolocalization of clusterin in several human epithelial tissues of adult humans. In epidermal sections, staining for clusterin was not detected in any epithelial regions (A). In sections of vaginal (B) and esophageal (C) epithelium, positive staining was observed in the superficial cell layer. In sections of stomach (D), small intestine (E), and colon (F) epithelium, numerous mucin-secreting cells were stained. Bar=50 μm.

role in maintaining the mucosal epithelium as a non-keratinizing epithelium. To test this hypothesis, further studies are needed, e.g., investigation of clusterin expression in various ocular surface diseases, observation of ocular surface epithelium in which clusterin gene expression is molecular-biologically prevented. Alternatively, as previously suggested regarding other epithelial tissues,[5,8] clusterin may function in ocular surface epithelial cells to provide surface membrane protection against complement factors[7] and against membrane fragments shed by desquamating cells, both of which are found in tears.[17]

REFERENCES

1. de Silva HV, Stuart WD, Duvic CR, et al. A 70-kDa apolipoprotein designated apoJ is a marker for subclasses of human plasma high density lipoproteins. *J Biol Chem*. 1990;265:13240–13247.

2. Jenne DE, Lowin B, Peitsch MC, Böttcher A, Schmitz G, Tschopp J. Clusterin (complement lysis inhibitor) forms a high density lipoprotein complex with apolipoprotein A-I in human plasma. *J Biol Chem.* 1991:266;11030–11036.
3. Jenne DE, Tschopp J. Molecular structure and functional characterization of a human complement cytolysis inhibitor found in blood and seminal plasma: Identity to sulfated glycoprotein 2, a constituent of rat testis fluid. *Proc Natl Acad Sci USA.* 1989:86;7123–7127.
4. Kirszbaum L, Sharpe JA, Murphy B, et al. Molecular cloning and characterization of the novel, human complement-associated protein, SP-40,40: A link between the complement and reproductive systems. *EMBO J.* 1989;8:711–718.
5. Jordan-Starck TC, Witte DP, Aronow BJ, Harmony JAK. Apolipoprotein J: A membrane policeman? *Curr Opin Lipidol.* 1992;3:75–85.
6. de Silva HV, Harmony JAK, Stuart WD, Gil CM, Robbins J. Apolipoprotein J: Structure and tissue distribution. *Biochemistry.* 1990:29;5380–5389.
7. Sylvester SR, Morales C, Oko R, Griswold MD. Localization of sulfated glycoprotein-2 (clusterin) on spermatozoa and in the reproductive tract of the male rat. *Biol Reprod.* 1991;45:195–207.
8. Aronow BJ, Lund SD, Brown TL, Harmony JA. Apolipoprotein J expression at fluid-tissue interfaces: Potential role in barrier cytoprotection. *Proc Natl Acad Sci USA.* 1993;90;725–729.
9. Reeder DJ, Stuart WD, Witte DP, Brown TL, Harmony JAK. Local synthesis of apolipoprotein J in human eye. *Exp Eye Res.* 1995;60:495–504.
10. Jones SE, Meerabux JMA, Yeats DA, Neal MJ. Analysis of differentially expressed genes in retinitis pigmentosa retinas, altered expression clusterin mRNA. *FEBS Lett.* 1992;300:279–282.
11. Jomary SE, Neal MJ, Jones SE. Comparison of clusterin gene expression in normal and dystrophic human retinas. *Mol Brain Res.* 1993;20:279–284.
12. Nishida K, Adachi W, Shimizu-Matsumoto A, Kinoshita S, Matsubara K, Okubo K. A gene expresson profile of human corneal epithelium and the isolation of human keratin 12 cDNA. *Invest Ophthalmol Vis Sci.* 1996;37:1800–1809.
13. Tsubota K, Kajiwara K, Ugajin S, Hasegawa T. Conjunctival blush cytology. *Acta Cytol.* 1990;34:233–235.
14. Higashiyama M, Matsumoto K, Hashimoto K, Yoshikawa K. Increased production of transforming growth factor-α in psoriatic epidermis. *J Dermatol.* 1991;18:117–119.
15. Nishida K, Sotozono C, Yokoi N, Kinoshita S. Transforming growth factor-β1, -β2, -β3 mRNA expression in human cornea. *Curr Eye Res.* 1995;14:235–241.
16. Murphy BF, Kirszbaum L, Waiker ID, d'Apice AJF. SP-40,40, a newly identified normal human serum protein found in the SC5b-9 complex of complement and in the immune deposits in glomerulonephritis. *J Clin Invest.* 1988;81:1858–1864.
17. Yamamoto GK, Allansmith MR. Complement in tears from normal humans. *Am J Ophthalmol.* 1979;88:758–763.

90

EFFECTS OF CHRONIC HYPERTONIC STRESS ON REGULATORY VOLUME INCREASE AND Na-K-2Cl COTRANSPORTER EXPRESSION IN CULTURED CORNEAL EPITHELIAL CELLS

Victor N. Bildin,[1] Hua Yang,[2] Jorge Fischbarg,[1] and Peter S. Reinach[2]

[1]College of Physicians and Surgeons
Columbia University
New York, New York
[2]State University of New York
College of Optometry
New York, New York

1. INTRODUCTION

The optical and barrier properties of the corneal epithelium, which are essential for normal vision, depend on the maintenance of epithelial membrane integrity and tight junctional continuity between neighboring cells. Changes in tear film osmolality can occur in daily living (e.g., wind exposure, or swimming in fresh or salt water) and lead to alterations in epithelial cell volume that could affect epithelial topography and tight junction continuity. If these changes are uncompensated, they could compromise the corneal epithelial optical and barrier function. In some other cells, compensation includes reswelling to their isotonic volume through an appropriate regulatory volume increase (RVI). Frequently, this response is dependent on the activity of the bumetanide-sensitive, ouabain-insensitive Na-K-2Cl cotransporter.[1]

The ability of regulatory volume responses to preserve normal corneal epithelial function may be affected by a defect in lacrimal gland function that makes their fluid hypertonic.[2] It is not known whether such hypertonicity could affect the ability of corneal epithelial cells to mediate RVI. For instance, chronic tear defects could prevent a stimulation of Na-K-2Cl cotransporter activity functions necessary to effectively counter increases in tear film osmolarity. If this were to occur, it could help explain some of the symptoms of dry eye patients, who frequently experience recurrent ocular infection, loss of visual acuity, and severe pain.

We have undertaken a study in a cell culture model system to determine if chronic hypertonic stress in vivo may change the ability of corneal epithelial cells to mediate RVI.

With cultured SV40-transformed rabbit and human corneal epithelial cells, we demonstrate that the RVI behavior of rabbit cells is more resistant than that of human cells. Further, our results suggest that the ability to elicit RVI behavior subsequent to hypertonic stress correlates with the amount of Na-K-2Cl cotransporter mRNA present.

2. MATERIALS AND METHODS

2.1. Cell Culture

SV40-virus transformed rabbit corneal epithelial cells were a generous gift from Dr. Araki-Sasaki (Kinki University, Hyogo, Japan). They were cultured in DMEM/F12 as described by Araki-Sasaki et al.[3] For light-scattering measurements, the cells were cultured for 2 days on rectangular 11x22 mm coverslips or in 25 cm^2 culture flasks (mRNA analysis experiments). Hyperosmolarity of the medium (375 mOsm) was achieved by sodium chloride addition. Cell viability was determined with the Sigma MTT cell determination kit and Trypan blue dye exclusion.

2.2. Regulatory Volume Increases

Based on a light-scattering technique, relative cell volume was evaluated as described by Fischbarg et al.[4] Analyses were performed using paired data statistics: Student's two-way t-test for independent means.

2.3. Generation of DNA Probes

^{32}P-labeled, single-stranded DNA probes for Na-K-2Cl cotransporter mRNA with specific activity of 2–4 x 10^9 cpm/μg were generated by asymmetric PCR on corresponding homologous human and rabbit 529-nt DNA fragments. Unreacted nucleotides were removed by spin chromatography through Sephadex G-50. The probes were purified by 1% agarose gel electrophoresis. After identification of radioactive band location by autoradiography, probes were extracted from agarose slices (QIAEXII Agarose Gel Extraction Kit, QIAGEN).

2.4. Nuclease Protection Assay

Cells were lysed in 5 M guanidine thiocyanate (GuSCN)/0.1 M EDTA, and lysates were analyzed immediately or stored at -70°C. Twenty microliters of cell lysate (~100–200 K cells) were mixed with 5 μl of probe contained in 2 x standard saline citrate (SSC)/0.1 M EDTA and incubated at 30°C for 20 h. After hybridization, 2.5 μl of 10 mg/ml proteinase-K (Sigma) and 2.5 μl of 10% sodium dodecyl sulfate (SDS) were added. The mixture was incubated at 37°C for 30 min. Nucleic acids were extracted with an equivalent volume of phenol-chloroform saturated buffer, and isolated by spin chromatography through Sephadex G-50.

An equivalent volume of S1 nuclease buffer containing up to 20 μg denatured salmon sperm DNA and 2000 U/ml of S1 nuclease was added, and the mixture was incubated at 45°C for 30 min. Following digestion and then electrophoresis in a 2% agarose gel, the gels were dried, and the positions of the RNA-DNA duplexes were determined by autoradiography at -70°C, using REFLECTION Film and REFLECTION Intensifying Screens (DuPont NEN). Excised band radioactivity associated with full-length RNA-DNA duplexes was quantified with scintillation counting. Number of mRNA molecules per cell was calculated as described by Haines and Gillespie.[5] All experimental data are expressed as mean + SEM.

3. RESULTS

3.1. Alterations in Volume Regulatory Capacities under Chronic Hypertonic Stress

Fig. 1 shows representative results of experiments in which cell volumes were monitored by light scattering in SV40-transformed corneal epithelial cells (human, HCE, top panel; rabbit, RCE, bottom panel). Responses to changes in osmolality were determined after stressing the cells by culturing them for 48 h in 375 mOsm medium. Subsequent to substituting the isotonic 300 mOsm medium with 375 mOsm medium, there was a rapid decrease in relative cell volume in both cell lines. However, the time course and extent of relative cell volume recovery were markedly different in the two cell lines. In HCE, the RVI recovery response was slow, and after 13 min rose to only 17% of the relative isotonic cell volume. In contrast, in RCE, this response required only 10 min to reach a level that was 69% of the isotonic value. Cells were stressed by hypertonicity during varied periods. Fig. 2 shows the dependence of RVI recovery on the duration of the chronic hypertonic insult. In RCE, after 24 h of exposure, the extent of RVI was unaltered. In marked contrast, for HCE after 24 h of exposure, the recovery decreased linearly to 17 ± 6%, n=6; this level remained unchanged after an additional 24 h. No changes in viability of RCE and HCE cells stressed by hypertonicity up to 48 h were seen (data not shown).

3.2. Changes in Na-K-2Cl Cotransporter mRNA Levels and Spectrum in Response to Hypertonicity

As shown in Fig. 3, we found that in isotonic medium the Na-K-2Cl cotransporter mRNA level is higher in RCE than in HCE cells (4.3 and 2.2 molecules per cell, respectively). As cells were subject to incubation in hypertonic medium, during the first 24 h the mRNA level increased twofold only in RCE cells. In addition, preliminary results of Northern blot analysis (not shown) demonstrate that there are at least three different sizes of mRNAs encoding for the Na-K-2Cl cotransporter protein(s) in both corneal epithelial cell lines. The levels of expression of these putative different isoforms changes also during hypertonic exposure (data not shown).

4. DISCUSSION

We characterized the ability of cultured human and rabbit corneal epithelial cells to undergo RVI subsequent to culture in hypertonic medium for up to 48 h. This was done to develop a model system for simulating a possible change in corneal epithelial function resulting from chronic exposure to hypertonic tears. The rationale for this approach is that in dry eye the tears are frequently reported to be hypertonic.[2] Chronic corneal epithelial cell volume shrinkage could be a contributing factor to some of the symptoms reported in dry eye.

Interestingly, the sensitivity of the RVI behavior to prolonged hypertonicity was markedly different in the two cell lines. In RCE, RVI remained unaffected up to 24 h,

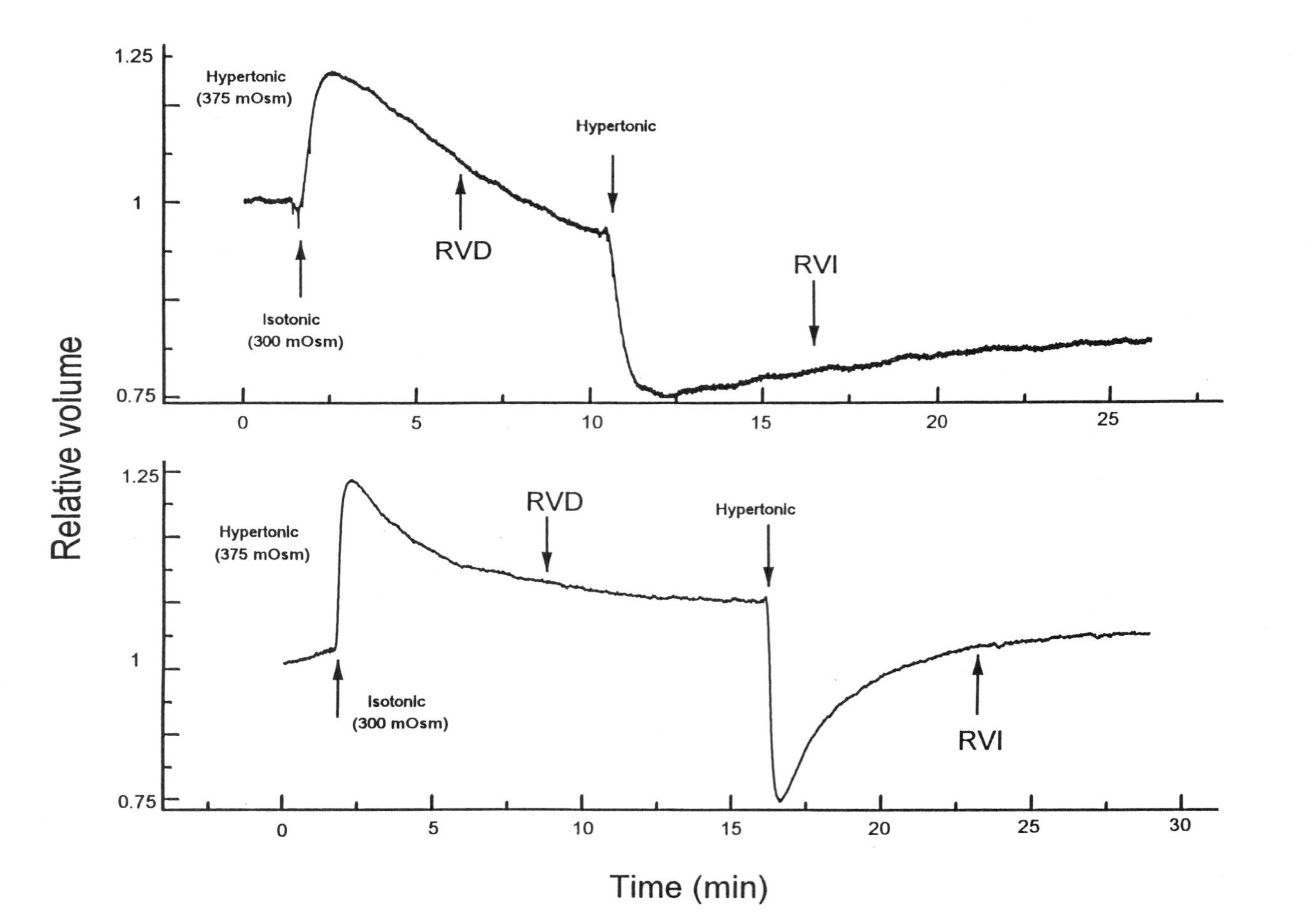

Figure 1. Effect of chronic hypertonic exposure on the capability of HCE cells (top panel) and RCE cells (bottom panel) to undergo regulatory volume increase. Confluent monolayers of cells on 11 x 22 mm rectangular glass coverslips were incubated for 48 h in hypertonic medium. Cells were illuminated with a rectangular cross-section light beam (5 mW helium-neon laser plus expander). The intensity of light scattered by the cells was converted to relative intracellular volume. RVI responses were elicited by a 25% increase in osmolarity, as indicated at the downward-directed arrows.

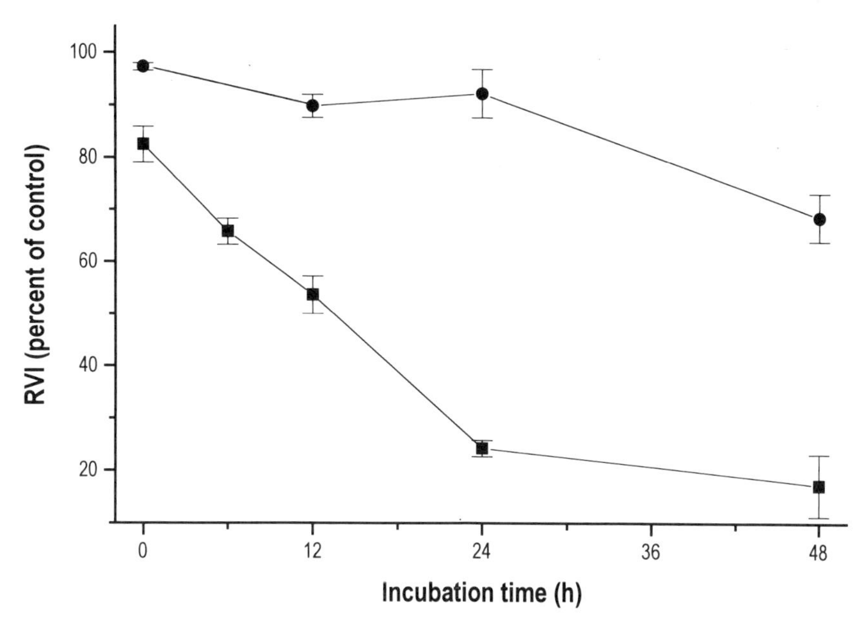

Figure 2. Time dependence of the effect of chronic hyperosmolarity on RVI by RCE and HCE cells. Methodology is the same as described in Fig. 1 legend. The individual responses for the RCE and HCE are shown as solid circles and squares, respectively. Data are mean ± SEM from six different coverslips.

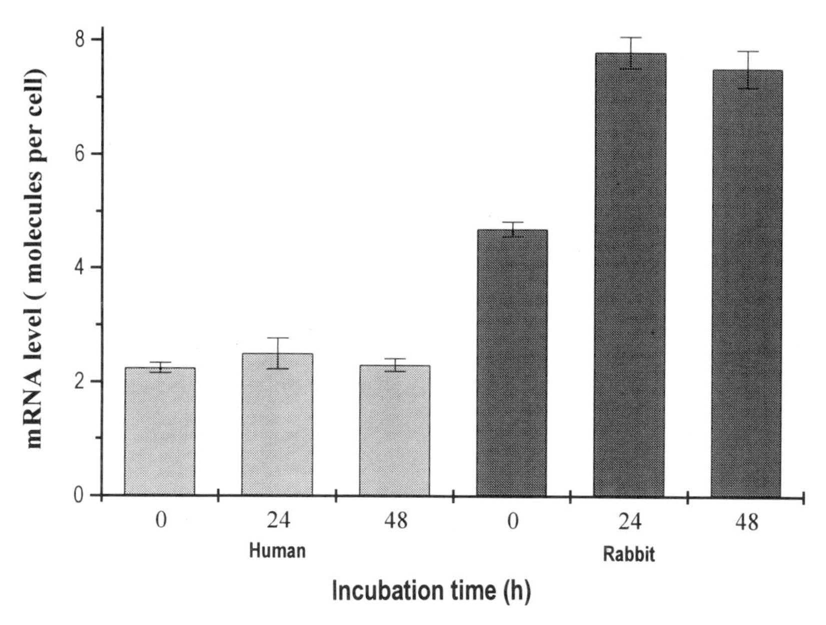

Figure 3. Effect of chronic (24–48 h) hypertonic exposure of HCE and RCE cells on Na-K-2Cl cotransporter mRNA abundance. Cells were incubated in hypertonic medium, lysed in GuSCN, and subjected to lysate nuclease protection analysis. Following digestion, and electrophoresis in a 2% agarose gel, the position of protected RNA-DNA duplexes was determined by autoradiography. Results were quantified by scintillation counting of excised bands corresponding to 529 nt RNA-DNA duplexes. Number of mRNA molecules per cell was calculated and normalized on a DNA content basis.

whereas in HCE, it declined rapidly. In RCE, this resistance correlates with the amount of Na-K-2Cl cotransporter mRNA present. The twofold increase in RCE may be essential for this resistance, because in HCE the mRNA level remained constant even though the RVI capacity fell. It is conceivable that human cells lack (or are depleted of) signaling factors needed to increase bumetanide-sensitive Na-K-2Cl cotransporter mRNA content. Alternatively, the Na-K-2Cl cotransporter(s) expressed by HCE after chronic hypertonicity may be a variant that is unresponsive to signals elicited by such hypertonicity. Therefore, this model system provides a valuable tool for investigating the signaling pathways needed to elicit RVI responses.

We were able to examine the levels of mRNA encoding Na-K-2Cl cotransporter even though such levels are very low. To do this, we developed a novel S1 nuclease protection assay. Its features include the use of asymmetric PCR-generated single-stranded DNA probes labeled to a high specific activity with ^{32}P. Furthermore, hybridization occurs in guanidine thiocyanate cell lysates, obviating the need for membrane transfer from gels. This also has the advantage of allowing very small quantities of RNA to be analyzed. This method of mRNA quantitation is better than routine Northern analysis by a factor or 100 or more.

This is the first report in which the possibility of the existence of several different Na-K-2Cl cotransporter mRNA isoforms in HCE and RCE is mentioned. In this light, the observed increase in mRNA message level in RCE during chronic hypertonic stress suggests a fundamental role for such protein(s) in RVI. Changes in mRNA levels may be due to changes in synthesis rates or message stability, which will be explored using α-amanatin or actinomycin D, inhibitors of RNA polymerase.

Future work may help clarify:

1. whether a family of genes encoding cotransporter proteins exists, or whether a single gene product is being differentially spliced;
2. whether dry eye patients lack the capacity to accumulate enough Na-K-2Cl cotransporter mRNA, which is needed to support RVI behavior during chronic hypertonic stress;
3. whether variations of expression of cotransporter messages occur during ocular development and aging.

ACKNOWLEDGMENTS

Supported by NIH grants EY06178 (JF) and EY04795 (PR), and by RPB, Inc.

REFERENCES

1. O'Donnell ME. Role of Na-K-Cl cotransport in vascular endothelial cell volume regulation. *Am J Physiol.* 1993;264:C1316-C1326.
2. Farris RL. Tear osmolarity - a new gold standard? *Adv Exp Med Biol.* 1994;350:495–503.
3. Araki-Sasaki A, Ohashi Y, Sasabe T. Immortalization of rabbit corneal epithelial cells by a recombinant SV40-adenovirus vector. *Invest Ophthalmol Vis Sci.* 1993;34:2665–2671.
4. Fischbarg J, Li J, Kuang K, Echevarria M, Iserovich P. Determination of volume and water permeability of plated cells from measurements of light scattering. *Am J Physiol.* 1993;265:C1412-C1423.
5. Haines DS, Gillespie DH. RNA abundance measured by a lysate RNase protection assay. *Biotechniques.* 1992;12:736–741.

A UNIFIED THEORY OF THE ROLE OF THE OCULAR SURFACE IN DRY EYE

Michael E. Stern,[1] Roger W. Beuerman,[2] Robert I. Fox,[3] Jianping Gao,[1] Austin K. Mircheff,[4] and Stephen C. Pflugfelder[5]

[1]Allergan, Inc.
Irvine, California
[2]Louisiana State University Eye Center
New Orleans, Louisiana
[3]Scripps Research Foundation
La Jolla, California
[4]University of Southern California
Los Angeles, California
[5]University of Miami
Miami, Florida

1. INTRODUCTION

Dry eye symptoms arise from a series of etiologies and are manifest in different patients with varying severity. The National Eye Institute/Industry Workshop on Clinical Trials in Dry Eyes, under the chairmanship of Dr. Michael A. Lemp, defined specific subtypes of dry eye in order to standardize clinical tests used in diagnosis and design of clinical studies.[1] The use of artificial tears is palliative at best, resulting in a reduction of ocular surface eyelid shear forces and some symptomatic relief. Future research should focus on mechanistic endpoints. What causative factor(s) initiates the sequence of events resulting in the clinical symptoms suffered by the patient?

This review emphasizes observations that the ocular surface (cornea, conjunctiva, accessory lacrimal glands, and meibomian glands), the main lacrimal gland, and the interconnecting reflexive innervation compose a "functional unit" (Fig. 1) whose parts act together as a servomechanism and not in isolation. In the normal individual, when afferent nerves of the ocular surface are stimulated, a reflex results in immediate blinking, withdrawal of the head, and secretion of copious amounts of reflex tears from the main lacrimal gland. These tears contain proteins, mucin, and water. Similarly, in people who face chronic ocular surface irritation due to environmental factors (contact lens, low humidity, wind, etc.), there is chronic stimulation of the lacrimal gland resulting in secretion of "sup-

Lacrimal Gland, Tear Film, and Dry Eye Syndromes 2
edited by Sullivan *et al.*, Plenum Press, New York, 1998

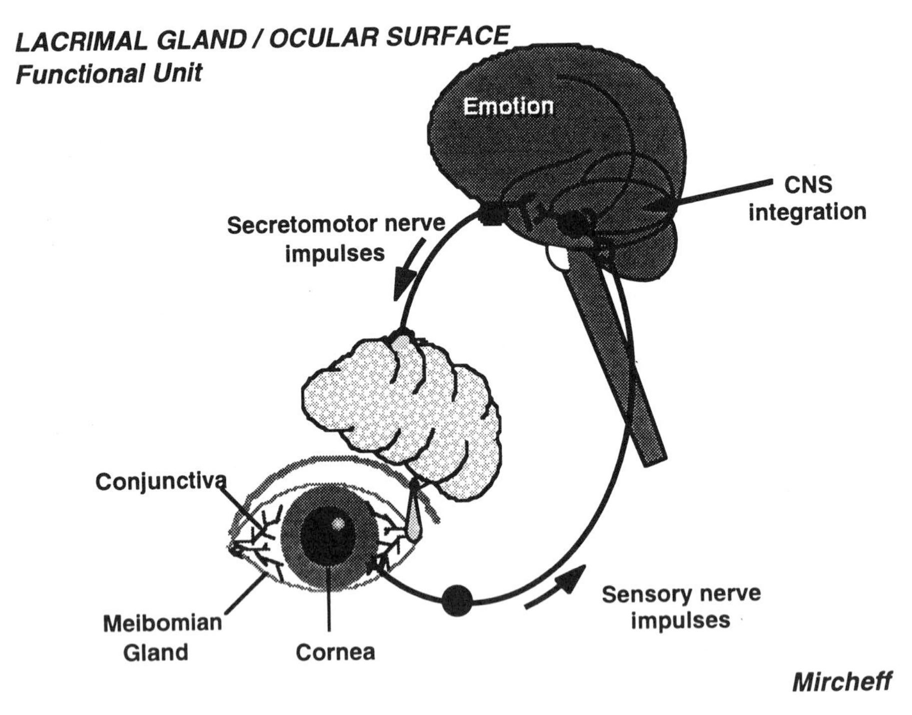

Figure 1. The functional unit comprising the ocular surface, the main lacrimal gland, and the interconnecting innervation.

portive" tears that can maintain and repair the ocular surface. In individuals suffering from dry eye, however, chronic inflammation of the ocular surface as well as of the lacrimal glands can be detected.

This "chronic" inflammation results in inflammatory cytokine secretion from the main lacrimal gland as well as the ocular surface that may interrupt both afferent and efferent arcs of the reflex and therefore impair function. The result of this pathology is a constant ocular surface irritation, which in its most severe form propagates a debilitating disease progression resulting in an inability of the patient to function normally at home or in the workplace.

The alterations in each component of the ocular surface/lacrimal gland reflex will be described.

2. OCULAR SURFACE

The ocular surface is challenged by the shear force across its surface due to blinking,[2] air currents, low humidity-induced desiccation, and foreign bodies (including contact lenses). Additionally, the ocular surface is confronted with several types of bacteria as well as viruses. The ocular surface in normal individuals remains intact and is able to repair the damage produced by these constant insults. Pflugfelder et al.[3] have shown, that diagnostic dyes, rose bengal and fluorescein, do not stain normal conjunctiva or cornea. Nelson et al.,[4] using impression cytology, however have indicated that some transient abnormalities can be found in clinically normal conjunctiva of people living in challenging environments. Patients with Sjögren's syndrome, who demonstrate a severe lack of aqueous tears, stain abundantly in the exposure zone.[4] In normal individuals, minor traumas, such as those already described, are rapidly healed and pose no chronic threat to the ocular surface. This is possibly due to the presence of a trophic surface environment consisting of a normal, non-inflammatory tear film. The tears in the normal individual may vary in quantity. It appears that a chronic alteration in nerve stimulation of the lacrimal gland in a dry eye individual results in inflammation and lymphocytic infiltration of the lacrimal glands. This results in secretion of diminished and altered tears that contain inflammatory cytokines, resulting in an abnormal ocular surface epithelium. The conjunctival and corneal epithelia have also been demonstrated to be competent to secrete IL-1α, TNF-α, IL-6, and IL-8.[5] The question then becomes, what conditions result in the inability of the ocular surface and the lacrimal glands to respond normally to chronic environmental challenges? Although this has not been resolved, several studies have indicated that a dramatic loss in systemic androgens found in a major target population, the peri- and post-menopausal female, results in a loss of support for lacrimal secretory function and production of an anti-inflammatory environment.[6,7]

3. CONJUNCTIVA

The conjunctiva covers the entire ocular surface outside of the cornea. Its surface is composed of a stratified mucus-secreting epithelium and a population of goblet cells also responsible for the mucus secretion. Mucus is one of the main defense mechanisms against various microtrauma. Shear forces applied during blinking (12–15/min) can cause significant trauma to the non-lubricated ocular surface.[2] If superficial trauma is induced by placing a Schirmer test strip or impression cytology membrane on the conjunctival surface, the eye will stain with rose bengal. In the normal eye, staining will no longer be observed after 24 h, indicating that a reparative process actively restores the normal surface barrier. Pflugfelder et al. (personal communications) have developed a model of conjunctival responses to microtrauma in the rabbit using nitrocellulose membranes to remove the superficial two cell layers. Then healing and cellular wound healing behavior are followed. An increase in epithelial proliferation was detected within 1 h and remained elevated for 3 days. Abnormal patterns of expression of various cell markers were detected for 1 week. A marker for basal epithelial cells, cytokeratin 14, was expressed throughout the entire epithelium,[8] and the number of cells staining for the presence of conjunctival mucin was reduced.[9] Increases in the concentrations of mRNA for inflammatory cytokines such as TNF-α, IL1-α, and IL-8 were also detected within conjunctival epithelial cells at the site of the microtrauma.[10] This phenomenon is important in part because of the conjunctival squamous metaplasia seen in moderate to severe dry eye as well as in Sjögren's syndrome. This response is seen as chronic wound healing due to the constant motion of the upper eyelid shear forces generated during blinking. Cytokine synthesis is then initiated in the traumatized corneal and conjunctival epithelium, as well as cytokines present in the lacrimal secretions, in an individual with an unsupported ocular surface (Fig. 1). In Sjögren's syndrome patients, T-cell infiltration of the conjunctiva has been found in both the epithelium and stroma.[11,12] Increased levels of IL-1α, TNF-α, IL-6, IL-8, and IL-10 have been found in the conjunctival epithelium of these patients when compared to control.[5,13] These patients, for the most part, also demonstrated expression of immune activation markers HLA-DR and ICAM-1.[5] The immunomodulatory drug cyclosporine,[13] as well as steroids, have been found to reduce ocular surface rose bengal staining. Additionally, studies in the dry eye dog model have demonstrated that cyclosporine A eliminates both the conjunctival and lacrimal gland lymphocytic infiltrates.[14]

Alterations in the conjunctiva, such as those mentioned, occur as increased tear film abnormalities in people with keratoconjunctivitis sicca (KCS). A chronic inflammatory environment on the ocular surface results in pathologic alterations of the conjunctival epithelium known as squamous metaplasia.[3,15] A decrease in tear fluid secretion has been correlated with an increase in conjunctival rose bengal staining.[4] Patients with Sjögren's syndrome, who are unable to tear even in response to stimulation of the nasal mucosa,[16] have very severe ocular surface irritation. Patients with a decrease in lacrimation also have a decrease in various proteins such as lactoferrin and lysozyme.[17,18] Several other proteins, secreted in tears, that may be trophic to the ocular surface as well as providing an anti-inflammatory environment, are also being investigated.[13,17] It is reasonable to assume that in situations where these proteins are diminished, a pathogenic environment will exist in the ocular surface.

In many types of dry eye, in particular those associated with systemic signs of autoimmune disease, the lacrimal gland becomes infiltrated with lymphocytes. These inflammatory cells adversely affect the function of the lacrimal gland, resulting in altered tear composition and compromise of the ocular surface. The initial glandular dysfunction, however, is most probably caused by a "disconnect" at the neural/glandular interface in the perivascular region. Interruption of the neural signal at this juncture is probably part of the same mechanism that initiates the migration and proliferation of lymphocytes in the lacrimal gland and conjunctiva.

4. OCULAR SURFACE INNERVATION

The ocular surface is exquisitely innervated, with the cornea having a density of free nerve endings approximately 60X that of tooth pulp. Corneal sensation is very acute and is centrally processed and interpreted solely as pain. The conjunctiva does not transmit as acute sensations as does the cornea and is known to feel itch as well as some temperature discrimination. It is well known that corneal stimulation results in a rapid reflex including immediate blinking, profuse reflex tearing, and withdrawal of the head. The neural pathway for this reflex as well as normal tearing have been partially elucidated (Fig. 2). Sensory (afferent) traffic from the cornea and conjunctiva travels down the ophthalmic branch (1) of the trigeminal nerve (V) through the trigeminal ganglion into the spinal trigeminal nucleus located in the brainstem. The initial synapse occurs in this nucleus, and neurons then travel up to the midbrain (pons), or the preganglionic sympathetic neurons in the spinal cord and then the superior cervical ganglion, located in the paravertebral sympathetic chain. Efferent fibers from the pons extend, via the facial (VII) nerve, to the pterygopalatine ganglion located adjacent to the orbit, where they again synapse and then send fibers to the lacrimal gland where they influence the secretomotor function (modulation of water and protein transport). Sympathetic fibers from the superior cervical ganglion also enter the lacrimal gland. Schafer et al.[19] have indicated that parasympathetic neural transmission can be inhibited by cytokines. Therefore, the pro-inflammatory cytokines such as are found in the lacrimal and salivary gland biopsies of patients with Sjögren's syndrome may inhibit neural stimulation of these target tissues.

It is important to note that the control of accessory lacrimal glandular secretion as well as conjunctival goblet cell secretion is only now being investigated. Work by Seiffert

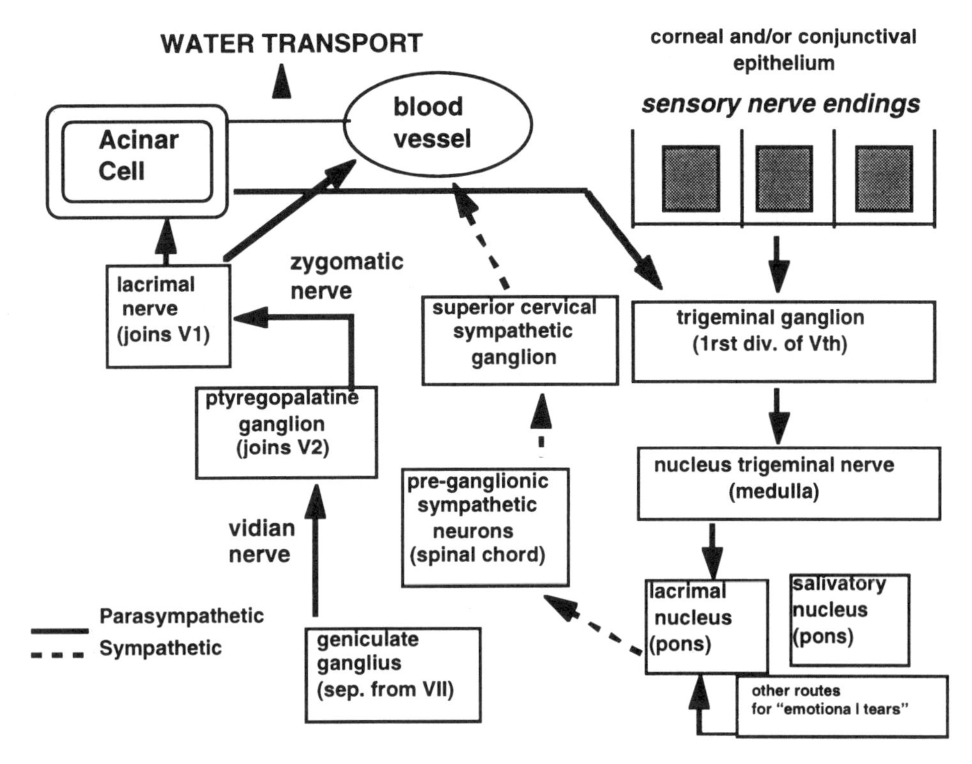

Figure 2. Afferent and efferent paths of lacrimal gland innervation for stimulation of tear flow.

et al.,[20] has demonstrated that the accessory glands are innervated, and Dartt et al.,[21] have also shown that the conjunctival goblet cells are innervated and respond to the presence of vasoactive intestinal peptide (VIP).

5. LACRIMAL GLAND

The lacrimal glands sit at the other end of the neural reflex. The main lacrimal gland resides just superior and temporal to the ocular globe. The accessory glands of Wolfring and Krause reside with the superior bulbar conjunctiva and the upper lid respectively. Although the etiology of dry eye is believed to be multifactorial and can be related to deficiencies in any of the three layers of the tear film, the major cause in Sjögren's syndrome has been reported to be a deficiency in aqueous tear production from the main and accessory lacrimal glands.[1,7] As in the salivary glands of patients with Sjögren's syndrome, as well as the conjunctiva in dogs with KCS,[14] the lacrimal glands of patients with immune-related dry eye have been found to be progressively infiltrated with lymphocytes. Immunohistochemical studies have demonstrated that these infiltrates consist primarily of CD4+ T cells and B cells.[22,23] Classically, this type of lymphocytic accumulation in the interstitium of the lacrimal or salivary gland is thought to result in immune-associated destruction of the epithelial cells in the target tissues, reduce aqueous tear secretion, and subsequently cause dry eye. The possible mechanisms are currently under investigation and discussion. The accumulated evidence indicates that the epithelial cells in the lacrimal and salivary tissues have the potential to be antigen-presenting cells. In vitro, the lacrimal acinar cells have shown the ability to express MHC II following carbachol induction.[24] In vivo, acinar cells in the salivary gland of patients and the lacrimal gland of MRL/lpr mouse model of Sjögren's syndrome strongly express class II antigens.[5,25,26] Additionally, a recent study using PCR-single-strand conformation polymorphism (SSCP) showed that some infiltrating T cells in both lacrimal and salivary glands of Sjögren's patients recognize the shared epitopes on autoantigens, suggesting the importance of restricted epitopes of common autoantigens in the initiation of Sjögren's syndrome.[27] Therefore, it is reasonable to propose that the epithelial cells in inflamed lacrimal or salivary tissues are able to present autoantigens to the cell surface receptors such as T cell antigen receptors. The activated T cells can then secrete inflammatory cytokines such as IL-1β, IL-2, IFN-γ, and TNF-α, which may contribute to a continued local autoimmune stimulation and result in infiltration and proliferation of migrating T-cells within the glands, which, left unchecked, would result in glandular destruction.[28–30] Additionally, these pro-inflammatory cytokines can inhibit neural transmission of parasympathetic pathways and subsequently suppress neural stimulation of the lacrimal gland.[19]

It has become clear that lacrimal gland function is significantly influenced by sex hormones.[31,32] Among these actions discovered during the past decade, androgen has been found to exert essential and specific effects on maintaining the normal glandular function as well as suppressing the inflammation in the lacrimal gland of normal and autoimmune animal models.[32–37] This unique capacity of androgens is initiated through its specific binding to receptors in the acinar nuclei of the lacrimal gland and, in turn, lead to an altered expression of various cytokines and proto-oncogenes in these lacrimal gland epithelial cells.[7,38] The immmunosuppressive activity of androgens in lacrimal gland during Sjögren's syndrome is proposed to be attributed to its ability to induce the accumulation of anti-inflammatory cytokines such as TGF-β.[7,39] Given the critical role that androgen plays in many aspects of lacrimal gland, from anatomy to molecular modulation, it has been hypothesized that a decrease in androgen level below a certain threshold may result in lacrimal atrophy.[6] Apoptosis in the plasma cells of the lacrimal gland interstitium was detected 4 h following withdrawal of androgen in ovariectomized rabbits with atrophic and necrotic changes in the acinar cells occurring over the ensuing several days.[37] The resulting apoptotic fragments are also suggested to be a source of potential autoantigens and could be subsequently presented either by interstitial antigen-presenting cells or acinar cells to CD4 cell antigen receptors to initiate the autoimmune response. Our recent study in KCS dogs indicated that apoptosis plays an important role in dry eye pathogenesis. The data suggest that both the elevated epithelial cell apoptosis and the suppressed lymphocytic apoptosis in the lacrimal and conjunctival tissues of KCS dogs may be involved in the dry eye mechanisms.[40]

6. SUMMARY

It is our belief that the pathology of dry eye occurs when systemic androgen levels fall below the threshold necessary for support of secretory function and generation of an anti-inflammatory environment (Fig. 3). When this occurs, both the lacrimal gland and the ocular surface become irritated and inflamed, and they secrete cytokines that interfere with the normal neural connections that drive the tearing reflex. This leaves the lacrimal gland in an isolated condition, perhaps exacerbating atrophic alterations of the glandular tissue. These changes allow for antigen presentation at the surface of the lacrimal acinar

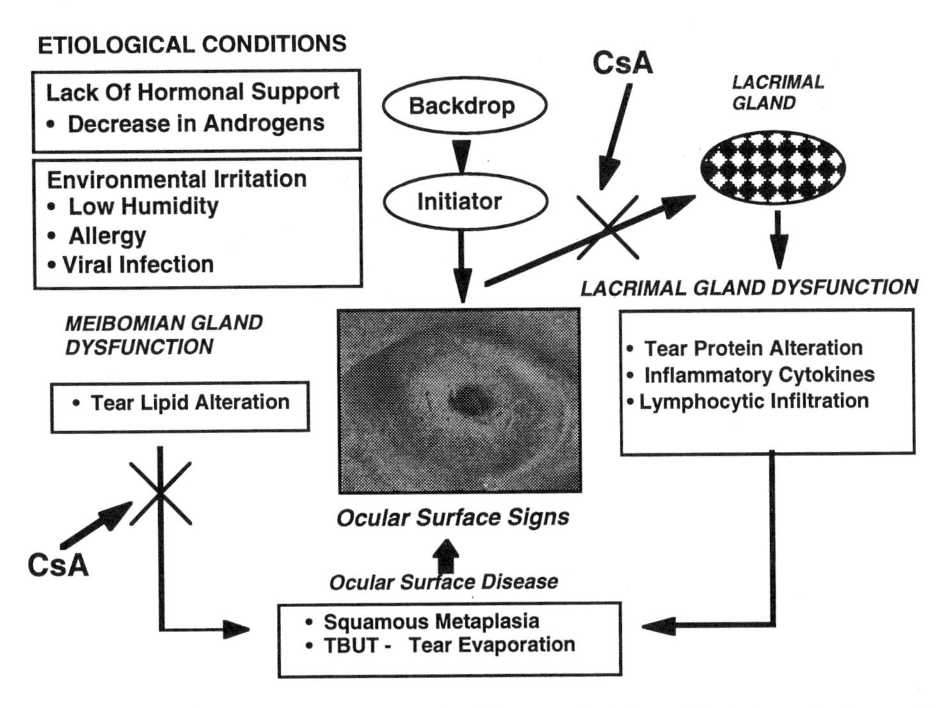

Figure 3. Proposed model of etiology and pathogenesis of dry eye. Included are etiologic factors (background, initiator) and the sequence of events resulting in alterations of the ocular surface. Possible therapeutic interventions (cyclosporine, androgens) are indicated.

cells and increase lymphocytic infiltration of the gland. A similar series of events may be occurring on the ocular surface.

From this hypothesis we conclude:

1. The ocular surface, lacrimal gland, and interconnecting innervation act as an integrated servo-mechanism.
2. Once the lacrimal gland loses its androgen support, it is subject to immune/ neurally mediated dysfunction.
3. The ocular surface is an appropriate target for dry eye therapeutics.

REFERENCES

1. Lemp ME. Report of the National Eye Institute / Workshop on Clinical Trials in Dry Eye. *CLAO J.* 1995;221–232.
2. Kessing AV. A new division of the conjunctiva on the basis of x-ray examination. *Acta Ophthalmol.* 1967;45:680–683.
3. Pflugfelder SC, Tseng SCG, Yoshino K, Monroy D, Felix C, Reis. Correlation of goblet cell density and mucosal epithelial mucin expression with rose bengal staining in patients with ocular irritation. *Ophthalmology*, in press.
4. Nelson JD, Havener VR, Cameron JD. Cellose acetate impression of the ocular surface. *Arch Ophthalmol.* 1983;101:1869–1872.
5. Jones DT, Monroy D, Ji Z, Atherton SS, Pflugfelder SC. Sjîgren's syndrome; cytokine and Epstein-Barr virus gene expression within the conjunctival epithelium. *Invest Ophthalmol Vis Sci.* 1994;35:3493–3503.

6. Mircheff AK, Warren DW, Wood RL. Hormonal support of lacrimal function, primary lacrimal deficiency, autoimmunity, and peripheral tolerance in the lacrimal gland. *Ocul Immunol Inflamm.*, 1996;4:145–172.
7. Sullivan DA, Wickham LA, Krenzer KL, et al. In: Pleyer U, Hartmann C, Sterry W, eds. *Oculodermal Diseases- Immunology of Bullous Oculo-Muco-Cutaneous Disorders.* Buren, The Netherlands: Aeolus Press, 1997 in press.
8. Yen MT, Pflugfelder SC, Crouse CA, Atherton SS. Cytoskeletal antigen expression in ocular mucosa-associated lymphoid tissue. *Invest Ophthalmol Vis Sci.* 1992;33:3235–3243.
9. Huang AJW, Tseng SCG. Development of monoclonal antibodies to rabbit ocular mucin. *Invest Ophthalmol Vis Sci.* 1987;28:1483–1491.
10. Naqui R, Ji Z, Pflugfelder SC. Immune cytokine RNA expression by human conjunctival epithelium after superficial microtrauma. ARVO abstracts. *Invest Ophthalmol Vis Sci.* 1996;37:356.
11. Hikichi T,Yoshida A, Tsubota K. Lymphocytic infiltration of conjunctivia and salivary gland in Sjîgren's syndrome. *Arch Ophthalmol.* 1993;111:21–22.
12. Raphael M, Bellefgih S, Piette, JCH. Conjunctival biopsy in Sjîgren's syndrome; correlations between histologic and immunohistochemical features. *Histopathology.* 1988;13:191–202.
13. Pflugfelder SC, Ji Z, Naqui R. Immune cytokine RNA expression in normal and Sjîgren's syndrome conjunctiva. ARVO Abstracts. *Invest Ophthalmol Vis Sci.* 1996;37:S358.
14. Stern MS, Gelber TA, Gao J, Ghosn CR. The effects of topical cyclosporin A (CsA) on dry eye dogs (KCS). ARVO Abstracts. *Invest Ophthalmol Vis Sci.* 1996;37:S4715.
15. Pflugfelder SC, Huang AJW, Feuer W. Conjunctival cytologic features of primary Sjîgren's syndrome. *Ophthalmology.* 1990;97:985–991.
16. Tsubota K. The importance of the Schirmer test with nasal stimulation. *Am J Ophthalmol.* 1991;111:106–108.
17. Seal DV, Mackie IA. Diagnostic implications of tear protein profiles. *Br J Ophthalmol.* 1984;68:321–324.
18. Danjo Y, Lee M, Horimoto K, Hamano T. Ocular surface damage and tear lactoferrin level in dry eye syndrome. *Acta Ophthalmol.* 1994;72:433–447.
19. Schafer M, Carter L, Stein C. Interleukin 1 beta and corticotropin-releasing factor inhibit pain by releasing opioids from immune cells in inflammed tissue. *Proc Natl* Acad *Sci USA*. 1994;91:4219–4213.
20. Seifert P, Spitznas M. Demonstration of nerve fibers in human accessory lacrimal glands. *Graefes Arch Clin Exp Ophthalmol.* 1994;232:107–114.
21. Dartt DA, Baker AK, Vailant C, Rose PE. Vasoactive intestinal polypeptide stimulation of protein secretion from rat laceimal gland acini. *Am J Physiol.* 1984;247:G502-G509.
22. Pflugfelder SC, Wilhelmus KR, Osato MS, Matoba AY Fond RL. The autoimmune nature of aqueous tear deficiency. *Ophthalmology.* 1986;93:1513–1517.
23. Pepose JS, Akata RF, Pflugfelder SC, Vorgt W. Mononuclear cell phenotypes and immunoglobulin rearrangements in lacrimal gland biopsies from patients with Sjîgren's syndrome. *Ophthalmology.* 1990;97:1599–1605.
24. Mircheff AK, Wood RL, Gierow JP. Traffic of major histocompatibility complex Class II molecules in rabbit lacrimal gland acinar cells. *Invest Ophthalmol Vis Sci.* 1994;35:3943–3915.
25. Fox RI, Bumol T, Fantozzi R, et al. Expression of histocompatibility antigen HLA-DR by salivary gland epithelial cells in Sjîgren's syndrome. *Arthritis Rheum.* 1986;29:1105–1111.
26. Homma M, Sugai S, Tojo T, Miyasaka N, Akizuki M, eds. Sjîgren's syndrome. *State of the Art.* Amsterdam: Kugler Press; 1994.
27. Matsumoto I, Tsubota K, Satake Y, et al. Common T cell receptor clontype in lacrimal glands and labial salivary glands from patients with Sjîgren's syndrome. *J Clin Invest.* 1996;97:1969–1977.
28. Kroemer G, Martinez A. Cytokines and autoimmune diseases. *Clin Immunol Immunopathol.* 1991;61:275–195.
29. Rowe D, Griffiths M, Stewart J, Novick D, Beverly PCL, Isenberg DA. HLA class I and II, interferon, interleukin 2 and interleukin 2 receptor expression on labial biopsy specimens from patients with Sjîgren's syndrome. *Ann Rheum Dis.* 1987;46:580–586.
30. Oxholm P, Daniels TE, Bendtzen K. Cytokine expression in labial salivary glands from patients with primary Sjîgren's syndrome. *Autoimmunity.* 1992;12:185–191.
31. Ahmed SA, Penhale WJ, Talal N. Sex hormones, immune responses and atuoimmune diseases. *Am J Pathol.* 1985;121:531–551.
32. Ahmed SA, Talal N. Sex hormones and the immune system-part 2. Animal data. *Baillieres Clin Rheumatol.* 1990;4:13–31.
33. Sullivan DA, Bloch KJ, Allansmith MR. Hormonal influence on the secretory immune system of the eye: Androgen regulation of secretory component levels in rat tears. *J Immunol.* 1984;132:1130–1135.

34. Vendramini AC, Soo C, Sullivan DA. Testosterone-induced supression of autoimmune disease in lacrimal tissue of a mouse model (NZB/NZW F1) of Sjîgren's syndrome. *Invest Ophthalmol Vis Sci.* 1991;32:3002–3006.
35. Sato EH, Sullivan DA. Comparative influence of steroid hormones and immunosupressive agents on autoimmune expression in lacrimal glands of female mouse model of Sjîgren's syndrome. *Invest Ophthalmol Vis Sci.* 1994;35:2632–2642.
37. Azzarolo AM, Kaswan RL, Mircheff AK, Warren DW. Androgen prevention of lacrimal gland regression after ovariectomy of rabbits. ARVO Abstracts. *Invest Ophthalmol Vis Sci*, 1994;35:S1793.
38. Azzarolo AM, Olsen E, Huang ZM, et al. Rapid onset of cell death in lacrimal glands after ovariectomy. ARVO Abstracts. *Invest Ophthalmol Vis Sci.* 1996;37:S856.
39. Clark JH, Schrader WT, O'Malley Mechanisms of action of steroid hormones. In: Wilson JD, Foster DW, eds. *Willilam Textbook of Endocrinology.* Philadelphia: WB Saunders 1992: 35–90.
40. Huang Z, Gao J, Wickham LA, Sullivan DA Influence of gender and androgen treatment on TGF-β1 mRNA levels in the rat lacrimal gland. ARVO Abstracts. *Invest Ophthalmol Vis Sci.* 1995;35:S991.
41. Gao J, Gelber-Schwalb TA, Addeo JV, Stern ME. Apoptosis in the lacrimal gland and conjunctiva of dry eye dogs. This volume.

92

HUMAN TEAR FILM COMPONENTS BIND *PSEUDOMONAS AERUGINOSA*

Nancy A. McNamara and Suzanne M. J. Fleiszig

University of California, Berkeley
School of Optometry
Berkeley, California

1. INTRODUCTION

Contact lens-induced *Pseudomonas aeruginosa* keratitis is one of the leading causes of vision-threatening corneal disease. To cause corneal infection in vivo, *P aeruginosa* must bind to and colonize the epithelial cell surface. Before interaction can occur with the epithelial cell surface, however, the pathogen must interact with several nonspecific ocular immune components that may help to prevent infection under normal conditions. The tear film contains enzymes such as lysozyme and lactoferrin that provide some amount of antibacterial action. The shearing forces of the blink wipe the ocular surface, enhance the flow of tears across the cornea, and may facilitate removal of pathogens from the eye. Finally, the innermost layer of the human tear film contains mucus, a glycoprotein gel, that improves the spread and stability of aqueous tears, as well as lubricates and protects the underlying corneal epithelium.[1]

In a recent in vitro study, we showed that removal of endogenous mucus from the ocular surface of rat and rabbit corneas with 5% *N*-acetylcysteine caused a significant increase in bacterial adherence to the underlying epithelium.[2] Adding mucin back to the rat ocular surface inhibited bacterial adherence to the cornea, and we were able to demonstrate *P. aeruginosa* binding to rat mucin as direct evidence of the interaction between bacteria and mucin. These findings suggested that adherence of *P. aeruginosa* to rat ocular mucus may act as a defense against infection in the open blinking eye by preventing bacterial adherence to the underlying epithelium and by facilitating bacterial clearance from the anterior ocular surface as the tear film is exchanged with each blink.

In the current study, we investigated the interaction between human ocular tear film components and *P. aeruginosa*. Our preliminary data suggest that tear film components obtained by irrigating the human corneal surface are able to bind *P. aeruginosa*. Preliminary characterization of *P. aeruginosa* binding factors using certain charged molecules

Lacrimal Gland, Tear Film, and Dry Eye Syndromes 2
edited by Sullivan *et al.*, Plenum Press, New York, 1998

and enzymes may indicate the presence of asialo receptors for *P. aeruginosa* in human tears. However, reduced electronegativity produced by the selective removal of negatively charged carbohydrate residues may also influence bacterial binding.

2. METHODS

2.1. Collection and Isolation of Human Tear Film Components

Tear film components (TFCs) were collected from the ocular surface of 16 human subjects using a noninvasive corneal irrigation chamber (NI-CIC) as previously described.[3,4] Briefly, each cornea was irrigated for 30 sec with 10 ml of sterile saline using a metered pump. To remove cells and debris, the eyewash samples were passed through a 3 μm pore-size filter and then washed three times in Hank's balanced salt solution (HBSS) by centrifugation at 5000 rpm for 20 min. The supernatant was dialyzed, lyophilized, and resuspended in 1 ml of HBSS.

2.2. Preparation of Bacteria

Bacteria were grown overnight at 37°C on a Trypticase soy agar (TSA) plate covered with a 12,000 to 14,000 MW pore-size dialysis membrane.[3] The inoculum was prepared by resuspension of bacteria from the membrane into PBS until the appropriate optical density was achieved. A corneal isolate of *P. aeruginosa* (strain 6294, serogroup O6) was used for all experiments.

2.3. Bacterial Adherence Assay

To test bacterial adherence to isolated TFCs, Limbro/Titertek (Flow, McLean, VA) microtiter wells were coated with 100 μl of a 1:25 dilution of the tear film material and incubated overnight at 37°C. The next day each well was washed four times with PBS to remove non-adherent components. Wells were then filled with 30 μl of a bacterial inoculum made up in HBSS to contain 10^8 cfu of *P. aeruginosa* per ml. Following a 30 min incubation at 37°C, the inoculum was discarded and each well was washed 20 times with PBS to remove non-adherent bacteria. Each well was then filled with 300 μl of 0.5% Triton X and incubated at 37°C for 30 min to dislodge adherent bacteria from the well surface. Following vigorous stirring of each well with a sterile pipette, a viable count was performed in duplicate to determine the number of bacteria that had adhered to the ocular material. Control wells were pretreated with HBSS rather than TFCs which does not promote adherence to these wells.[2]

2.4. Treatment of TFCs with Enzymes and Charged Molecules

To study the nature of the interaction between *P. aeruginosa* and human TFCs, we treated TFCs with the following enzymes and charged molecules prior to conducting the bacterial adherence assay: Sodium metaperiodate, which oxidizes vicinal hydroxyl groups on carbohydrates; neuraminidase, which removes sialylated glycoproteins; SP54, a negatively charged molecule that binds and neutralizes positive charges; and protamine sulfate, a positively charged molecule that binds and neutralizes negative charges.

Microtiter wells were pre-incubated with tear film components overnight and then treated for 1 h with 120 μl of a 1:1000 dilution of each enzyme, or for 30 min with 10

Table 1. Median, 25th, and 75th percentile colony forming units of *P. aeruginosa* recovered per well

		Percentiles	
Sample	Median	25th	75th
Control	0	0	0
Tears	10950*	4778	27825
Control	0	0	0
Tears	7500*	4162	12300
Periodate only	0	0	0
Neuraminidase only	0	0	0
Tears + neuraminidase	18150*†	13425	31612
Tears + periodate	23100*†	13650	32738
Control	210	0	600
Tears	4725*	2869	7200
SP54 only	0	0	0
Protamine only	0	0	0
Tears + SP54	5775*	1106	13950
Tears + protamine	42300*†	31125	47738

* = Adherence significantly greater than control wells at the 0.05 alpha level.
† = Adherence significantly greater than wells coated with tears alone at the 0.05 alpha level.

mg/ml of protamine sulfate or 20 mg/ml of SP54. Following treatment, wells were washed ten times with PBS, and bacterial adherence assays were performed. Adherence was compared to untreated wells coated with tear components and control wells pretreated with HBSS rather than the tear film material, and then treated with each enzyme and charged molecule in an identical manner.

3. RESULTS

Non-parametric statistical techniques were employed since the number of colony forming units (cfu) recovered per well produced outlying values that caused some distributions to deviate from normality. Listed in Table 1 are the median and interquartile range of bacterial cfu recovered from six wells coated with either HBSS or TFCs and then either left untreated or treated with an enzyme or charged molecule.

Human ocular TFCs increased the median number of *P. aeruginosa* adhering to microtiter wells compared to controls (10950 vs. 0 cfu, respectively, $p<0.01$), and these results are displayed as a box plot in Fig. 1. Fig. 2 shows that complete deglycosylation of carbohydrate residues present in the TFCs with sodium periodate increased the median number of *P. aeruginosa* binding to wells coated with TFCs 3-fold compared to untreated TFC-coated wells (23100 vs. 7500 cfu, $p<0.01$), while selective removal of negatively charged sialic acid groups with neuraminidase increased binding 2.5-fold (18150 vs. 7500 cfu, $p=0.01$). Finally, Fig. 3 demonstrates the effects of altering surface charge on *P. aeruginosa* binding to TFCs. No bacterial adherence was detected on control microtiter wells treated with protamine or SP54; however, neutralization of negatively charged groups on TFC-coated wells with protamine increased median binding 8-fold compared to wells coated with untreated TFCs (42300 vs. 4725 cfu, $p<0.05$). Alternatively, neutraliza-

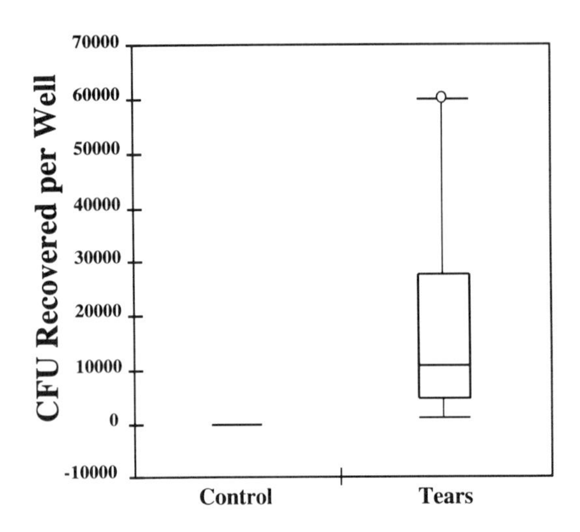

Figure 1. *P. aeruginosa* strain 6294 adherence to human tear film components. Each box encloses the interquartile range of data; a bold line within the box marks the median number of bacteria recovered from 6 wells coated with either HBSS or TFCs. Lines extending from the top and bottom of each box mark the maximum and minimum number of bacteria adhering for each assay. Unfilled circles represent values that fall outside the 99% confidence interval. A single horizontal line indicates that no colony forming units were recovered from the 6 control wells.

tion of positively charged groups on TFC-coated wells with SP54 had no effect on *P. aeruginosa* binding compared to untreated TFCs (5775 vs. 4725 cfu, $p>0.05$).

4. DISCUSSION

Our findings suggest that soluble factors in human eyewash supernatant promote *P. aeruginosa* adherence to microtiter wells that normally resist *P. aeruginosa* binding. The tear film factors used in this study likely contain aqueous, lipid, and mucus tear components, as well as material from the cell surface glycocalyx.

It is possible that *P. aeruginosa* binding to TFCs may involve mucins (the major glycoprotein constituent of mucus), since both removal of carbohydrate residues present

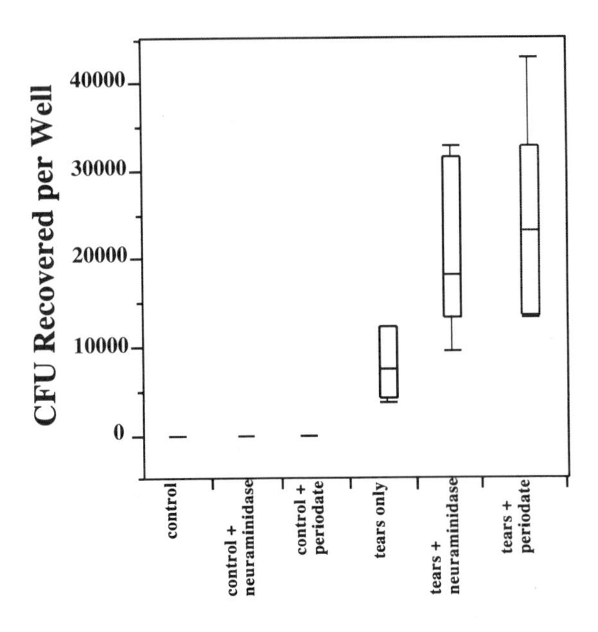

Figure 2. Effect of sodium periodate and neuraminidase on *P. aeruginosa* strain 6294 adherence to human tear film components. (See Fig. 1 legend for interpretation of box plots.)

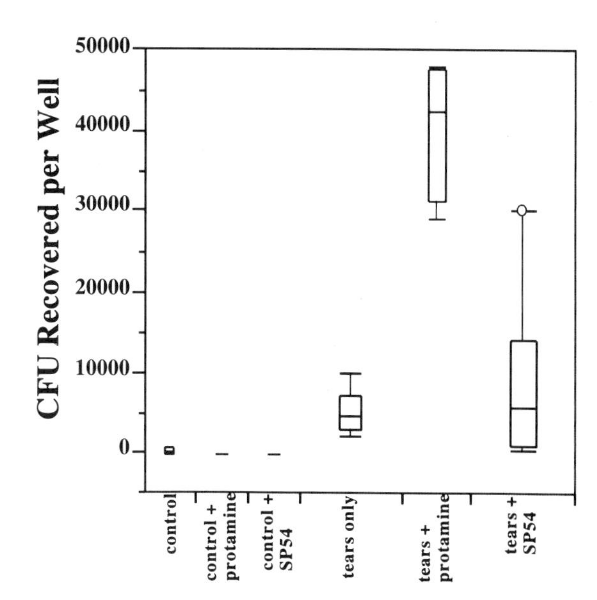

Figure 3. Effect of protamine sulfate (a positively charged molecule) and SP54 (a negatively charged molecule) on *P. aeruginosa* strain 6294 adherence to human tear film components. (See Fig. 1 legend for interpretation of box plots.)

in the tear film and selective removal of negatively charged sialic acid residues increased the interaction between remaining tear film factors and *P. aeruginosa*. Although these results suggest the presence of asialo receptors for *P. aeruginosa* in human tears, the greatest increase in *P. aeruginosa* binding to TFCs was elicited following the neutralization of negative charges present in TFCs using protamine sulfate. Thus, the effect of neuraminidase and periodate on *P. aeruginosa* binding to TFCs may have been the result of reduced electronegativity following the enzymatic treatments. Further investigation is necessary to decipher the effects of altering surface charge and removing carbohydrate moieties from tear film components on *P. aeruginosa* binding.

The intact corneal epithelium is resistant to infection. However, even minor disruption to the mechanical and physiological barrier formed by the corneal epithelium and associated tear film, such as that experienced during contact lens wear, may predispose the eye to pathogen adherence and subsequent infection.[5,6] Our preliminary studies provide evidence that factors other than the corneal epithelial barrier may contribute to nonspecific inhibition of bacterial adherence to the *in vivo* cornea, since bacterial binding by tear film factors is likely to inhibit bacterial adherence to other factors such as the cornea. Bacterial binding to human tear film factors support the data of Miller et al., who reported that tear film factors may promote bacterial adherence to contact lenses.[7] Further studies of the interaction between TFCs and *P. aeruginosa* [1]in contact lens wearers may provide important information about the pathogenesis of contact lens-related infection that could be used in developing strategies to prevent or lessen the risk of contact lens-related ulcerative keratitis.

REFERENCES

1. Holly F, Lemp M. Wettability and wetting of corneal epithelium. *Exp Eye Res*. 1971;11:239–250.
2. Fleiszig S, Zaidi T, Ramphal R, Pier G. Modulation of *Pseudomonas aeruginosa* adherence to the corneal surface by mucus. *Infect Immun*. 1994;62:1799–1804.

3. Fleiszig S, Efron N, Pier G. Extended contact lens wear enhances *Pseudomonas aeruginosa* adherence to human corneal epithelium. *Invest Ophthalmol Vis Sci.* 1992;33:2908–2916.
4. Fullard R, Wilson G. Investigation of sloughed corneal epithelial cells collected by non-invasive irrigation of the corneal surface. *Curr Eye Res.* 1986;5:847–856.
5. Klontz S, Au Y, Misra R. A partial thickness epithelial defect increases the adherence of *Pseudomonas aeruginosa* to the cornea. *Invest Ophthalmol Vis Sci.* 1989;30:1069–1074.
6. Ramphal R, McNiece M, Polack F. Adherence of *Pseudomonas aeruginosa* to the injured cornea: A step in the pathogenesis of corneal infections. *Ann Ophthalmol.* 1981;100:1956–1958.
7. Miller M, Wilson L, Ahearn D. Effects of protein, mucin and human tears on adherence of *Pseudomonas aeruginosa* to hydrophilic contact lenses. *J Clin Microbiol.* 1988;26:513–517.

SMALL-VOLUME ANALYSIS OF RABBIT TEARS AND EFFECTS OF A CORNEAL WOUND ON TEAR PROTEIN SPECTRA

Ray J. Varnell,[1] Dmitri Y. Maitchouk,[1] Roger W. Beuerman,[1]
James E. Carlton,[2] and Anthony Haag[2]

[1]LSU Eye Center
Louisiana State University Medical Center
New Orleans, Louisiana
[2]CORE Laboratory Facility
Louisiana State University Medical Center
New Orleans, Louisiana

1. INTRODUCTION

Tears are a complex fluid composed of secretory products from several sources, such as the orbital lacrimal gland, accessory lacrimal glands, goblet cells, meibomian glands, and perhaps other cells of the conjunctiva and cornea. Previous studies have shown that the lacrimal fluid contains growth factors, as well as a number of other proteins, that may act to protect the ocular surface.[1–3] A compilation of known tear proteins of laboratory animals is shown in Table 1.[4–23]

As emphasized by Farris,[24] there are a large number of proteins in tears. However, most studies have focused on one or two of these proteins, while the majority remain unexplored. Although standard protein separation techniques demonstrate a variety of substances in tears, it has not been possible to determine the biochemical nature and source of all these substances nor have many of the substances been described. The reason is that with standard techniques, it is necessary to develop a specific procedure for each particular protein, using either antibody or sequence analysis, and such analysis is difficult in the absence of some prior knowledge of the specific protein under examination.

However, with mass spectrometry, it is feasible to obtain the mass (or molecular) weights of a large number of proteins simultaneously in a single mass spectrum of one tear sample.[25–27] Mass spectrometry does not require a preassigned bias to focus on a single protein. Thus, qualitative changes in the presence or absence of a large number of tear proteins can be determined in a single analysis. Furthermore, the sensitivity and repro-

Table 1. Proteins reported in tear and lacrimal fluids of various animals

Protein	Species
Albumin	Rabbit, dog[4-6]
Amylase	Guinea pig, rat[7,8]
Angiotensin converting enzyme	Rabbit, guinea pig[7]
β-Hexosaminidase	Guinea pig, rabbit, rat[7,8]
Chondroitin sulfate	Rabbit[9]
Epidermal growth factor	Mouse, rabbit, rat[10-12]
Globulin	Dog[6]
Glycoprotein	Rabbit[13]
Glycosaminoglycans	Rabbit[9]
Hyaluronic acid	Rabbit[9]
IgA	Rabbit[14]
IgG	Rabbit[14]
Kallikreins, plasma	Rabbit[15]
Kallikreins, tissue	Rabbit[15]
Lactate dehydrogenase	Rabbit, guinea pig, rat[5,7,8]
Lysozyme	Dog, rabbit[6,7,16]
Malate dehydrogenase	Rabbit[5]
Peroxidase	Rat, guinea pig, cat[7,8,17]
Plasmin	Rabbit[18]
Plasminogen activator	Rat, guinea pig, rabbit[7,8]
Prostaglandin E_2	Rabbit, guinea pig[7]
Protein content	Dog, rabbit, horse, cow[6,19-21]
Unidentified proteins	Rabbit, mouse[22,23]

ducibility of mass spectrometry are excellent. Most mass spectrometers, however, have definite limitations for tear analysis; large amounts (10 μl or more) of sample are required for partial purification and derivatization because contaminants such as salt and metal ions interfere significantly with the analysis.

Recently, a new type of mass spectrometry instrument, referred to as the matrix-assisted laser desorption/ionization (MALDI) time-of-flight mass spectrometer, has been developed; this instrument is capable of analyzing proteins contained in an unpurified sample as small as 1 μl or less.[28] One drawback to the application of MALDI is the lack of good methods for quantification. Another is that this procedure is best suited for the resolution of proteins with molecular weights between 1,000 and 20,000 mass units. Both these limitations are being investigated and will likely be resolved.

The purpose of our study was to determine the nature of the changes in tear proteins following corneal wounding. Because wounding removes the epithelial barrier that protects the anterior surface of the cornea, the tears penetrate the stroma and become the extracellular fluid surrounding the stromal keratocytes, allowing growth-modulating components in the tears direct access to these cells. One possible result of this contact may be involvement in the reactive fibrosis that occurs after excimer laser ablation of the cornea.[29] For this reason, knowledge of how tear proteins change under various conditions would be valuable to understand normal and pathologic conditions. For instance, growth factors are active at extremely low concentration, and MALDI analysis can be particularly useful in this area as it has a sensitivity as low as 10^{-15} moles.[30]

2. MATERIALS AND METHODS

2.1. Tear Samples

Tear samples were collected from 11 New Zealand white rabbit eyes before and 1, 2, and 3 h after the corneas were wounded. Wounds were made with a 6-mm diameter trephine to a depth of about 100 μm in the center of the cornea. Tears were collected with flame-polished, 10-μl micropipettes, with care taken to avoid touching the ocular surface. The amount of tears collected at each sampling was about 2 μl under non-stimulated conditions before wounding and about 10 μl after wounding. Tears were used fresh or stored frozen at -20°C in sealed microvials.

Substance P, calcitonin gene-related peptide, epidermal growth factor, and fibroblast growth factor (Sigma, St. Louis, MO) were used as standards. Each was reconstituted in 0.9% balanced saline solution (BSS) to concentrations ranging from 10^{-4} M to 10^{-12} M. BSS was used because it is physiologically similar to the aqueous phase of tears.

2.2. MALDI Mass Spectrometry

Tears and standard solutions were diluted 1:100 (v:v) with a saturated solution of α-cyano-4-hydroxycinnamic acid (Aldrich, Milwaukee, WI) with vigorous mixing, after which about 0.5 μl of this mixture was spotted on a 5-mm diameter stainless steel planchet and quickly air dried.[31] The planchet was then loaded into a Voyager™ Biospectrometry™ Workstation (PerSeptive BioSystems, Framingham, MA), where the sample was analyzed. The resulting mass spectrum of each sample was the average of 256 laser pulses and was smoothed using the 19-point Savitsky-Golay method. Distributions of peaks from tears before and after wounding were analyzed statistically.

3. RESULTS

In each eye, changes were seen in the proteins delineated by the mass spectra over time after wounding (Fig. 1). In each spectrum, certain peaks of specific molecular mass were found to be unique, that is, they were seen in one sample at one interval, but not in other samples at any other interval.

Before wounding, 24 specific peaks ranging in mass weight from about 2400 to almost 20,000 Da were found in more than half the samples from the 11 eyes. Because of their high frequency of occurrence, these peaks were followed after wounding. The peak that occurred most frequently had a mass weight of 4783 Da and was seen in all 11 samples prior to wounding. The next most frequently occurring peak had a mass weight of 2418 Da and was seen in 10 of 11 (90.9%) samples before wounding. After wounding, the peak at 4783 Da occurred in 8 of 11 (72.7%) samples at 1 and 2 h and in 9 of 11 (81.8%) samples at 3 h. The peak at 2418 Da occurred in 8 of 11 (72.7%) samples at 1 h, 4 of 11 (36.4%) samples at 2 h, and 6 of 11 (54.5%) samples at 3 h. By comparison, a peak at 10,945 Da that was found in 54.5% of samples prior to wounding did not occur in any samples 1 h after wounding and occurred in only 2 of 11 (18.2%) samples at 2 and 3 h after wounding.

The total number of peaks varied considerably from eye to eye at each interval; before wounding, as many as 60 peaks were observed in one eye and as few as 15 peaks in another eye. However, the average number of peaks seen at each of the four sampling intervals was similar, and the variability was small (Fig. 2). The number of peaks unique to

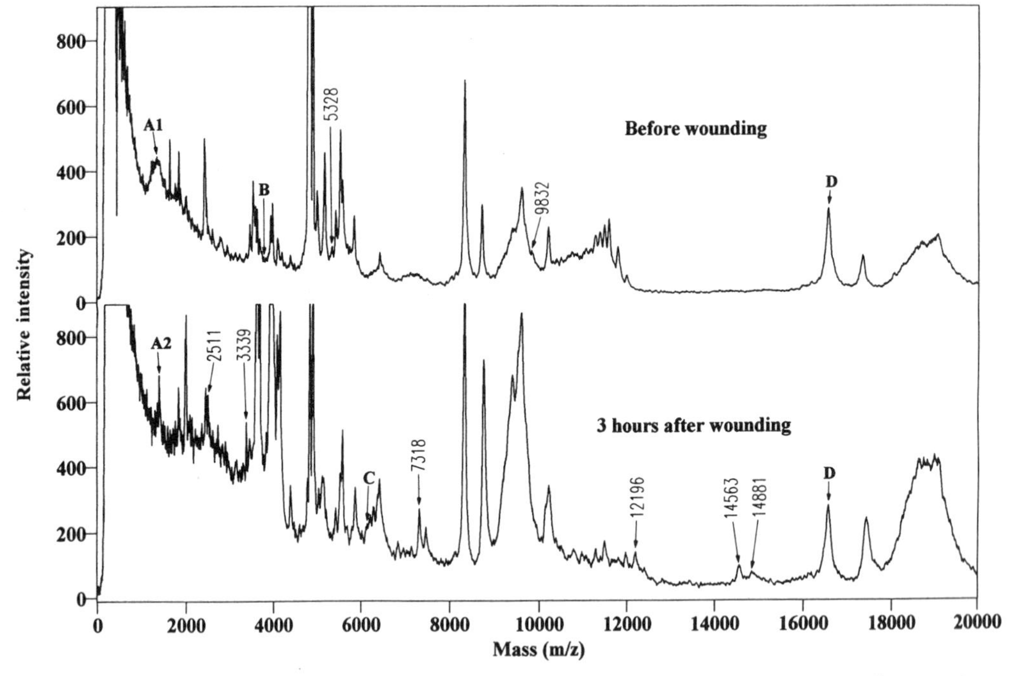

Figure 1. Mass spectra of tears from a single rabbit eye before and 3 h after central corneal wounding. Numbers indicate the mass weight of some peaks unique to each sample. The presence in tears of substance P (A1), with m/z 1348, and its oxidized form (A2) has been verified.[32] Peaks tentatively identified include calcitonin gene-related peptide (B), epidermal growth factor (C), and fibroblast growth factor (D).

each sampling interval increased markedly over time; the number of unique peaks at 3 h after wounding was twice as large as the number before wounding or at 1 h after wounding (Fig. 2). This change suggests the dramatic modulation of the nature of the tear proteins after corneal wounding, including the production of novel tear proteins.

4. DISCUSSION

The changes in the frequency of occurrence of individual peaks, and especially the appearance and disappearance of peaks unique to each sampling period, indicate the po-

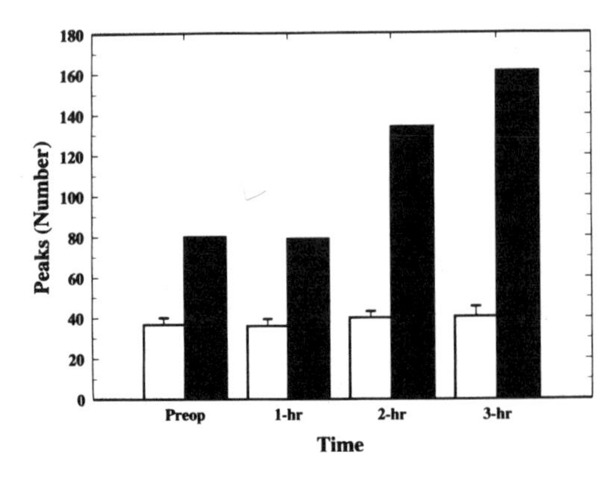

Figure 2. The average (mean ± SE.; n=11) number of tear protein peaks (open bars) observed before and 1, 2, and 3 h after a central corneal wound was made. Solid bars indicate the number of peaks unique to each sampling time in all 11 samples.

tential for new substances to appear in the extracellular environment of keratocytes and epithelial cells. A cell signalling role for these new constituents can be suggested. The interesting aspect of these findings is that the changes in the occurrence of peaks in the mass spectra are the result of changes in production of tear proteins. However, it has often been thought that tears are more aqueous after wounding, and it would be expected that their protein components would actually decrease. Our findings indicate the opposite may be true.

ACKNOWLEDGMENTS

This work was supported by the Department of the Army, Cooperative Agreement DAMD17–93-V-3013 (This does not necessarily reflect the position or the policy of the government, and no official endorsement should be inferred), U.S. Public Health Service grant EY02377 (Core Grant) from the National Eye Institute, National Institutes of Health, and an unrestricted departmental grant from Research to Prevent Blindness, Inc., New York, NY.

REFERENCES

1. van Setten GB, Viinikka L, Tervo T, Pesonen K, Tarkkaren A, Perkeentupa J. Epidermal growth factor is a constant component of normal human tear fluid. *Graefes Arch Clin Exp Ophthalmol.* 1989;22:184–187.
2. Yoshino K, Rahul G, Monroy D, Ji Z, Pflugfelder SC. Production and secretion of transforming growth factor beta (TGFβ) by the human lacrimal gland. *Curr Eye Res.* 1996;15:615–624.
3. Wilson SE, Lloyd SA, Kennedy RH. Basic fibroblast growth factor (FGFb) and epidermal growth factor (EGF) receptor mRNA production in human lacrimal gland. *Invest Ophthalmol Vis Sci.* 1991;32:2816–2820.
4. Eglitis I. The glands. In: Prince JH, ed. *The Rabbit in Eye Research.* Springfield, IL: Charles C Thomas; 1964:38–56.
5. Imayasu M, Moriyama T, Ohashi J-I, Cavanagh HD. Effects of rigid gas permeable contact lens extended wear on rabbit cornea assessed by LDH activity, MDH activity, and albumin levels in tear fluid. *CLAO J.* 1993;19:153–157.
6. Roberts SR, Erickson OF. Dog tear secretion and tear proteins. *J Small Anim Pract.* 1965;3:1–5.
7. Thörig L, van Agtmaal EJ, Glasius E, Tan KL, van Haeringen NJ. Comparison of tears and lacrimal gland fluid in the rabbit and guinea pig. *Curr Eye Res.* 1985;4:913–920.
8. Thörig L, van Haeringen NJ, Wijngaards G. Comparison of enzymes of tears, lacrimal gland fluid and lacrimal gland tissue in the rat. *Exp Eye Res.* 1984;38:605–609.
9. Oya T, Obata H, Miyata K, Tsuru T, Sawa M, Miyauchi S. Quantitative analysis of glycosaminoglycans in tear fluids during corneal epithelial wound healing in rabbits. *J Jpn Ophthalmol Soc.* 1994;98:1049–1055.
10. Tsutsumi O, Tsutsumi A, Oka T. Epidermal growth factor-like, corneal wound healing substances in mouse tears. *J Clin Invest.* 1988;81:1067–1071.
11. Steinemann TL, Thompson HW, Maroney KM, et al. Changes in epithelial epidermal growth factor receptor and lacrimal gland EGF concentration after corneal wounding. ARVO Abstracts. *Invest Ophthalmol Vis Sci.* 1990;31:S55.
12. Tervo K, van Setten GB, Virtanen I, Tarkkanen A, Tervo T. The lacrimal gland is the source of epidermal growth factor in tear fluid: An immunochemical study in rats. ARVO Abstracts. *Invest Ophthalmol Vis Sci.* 1990;31:S45.
13. Iwata S, Kabasawa I. Fractionation and chemical properties of tear mucoids. *Exp Eye Res.* 1971;12:360–367.
14. Sapse AT, Ivanyi J, Bonavida B, Stone W, Kramer R, Sercarz EE. Tears as carriers of antibodies. II. Charge heterogeneity of IgG antibodies in rabbit tears. *Int Arch Allergy Appl Immunol.* 1968;33:598–608.
15. Kuznetsova TP, Chesnokova NB, Paskhina TS. Activity of tissue and plasma kallikreins and content of their precursors in rabbit lacrimal fluid after alkaline burn of the cornea. *Vopr Med Khim.* 1994;40:37–40.

16. Bonavida B, Sapse AT, Sercarz EE. Rabbit tear proteins. I. Detection and quantitation of lysozyme in non-stimulated tears. *Invest Ophthalmol.* 1968;7:435–440.
17. van Haeringen NJ, Ensink FTE, Glasius E. The peroxidase-thiocyanate-hydrogen peroxide system in tear fluid and saliva of different species. *Exp Eye Res.* 1979;28:343–347.
18. Cejková J, Lojda Z, Dropcová S, Kadlecová D. The histochemical pattern of mechanically or chemically injured rabbit cornea after aprotinin treatment: Relationships with the plasmin concentration of the tear fluid. *Histochem J.* 1993;25:438–445.
19. Davidson JH, Blanchard GL, Montgomery PC. Comparisons of tear proteins in the cow, horse, dog and rabbit. *Adv Exp Med Biol.* 1994;350:331–334.
20. Maidment DCJ, Kidder DE, Taylor MN. Electrolyte and protein levels in bovine tears. *Br Vet J.* 1985;141:169–173.
21. Dartt DA, Botelho SY. Protein in rabbit lacrimal gland fluid. *Invest Ophthalmol Vis Sci.* 1979;18:1207–1209.
22. Dartt DA, Knox I, Palau A, Botelho SY. Proteins in fluids from individual orbital glands and in tears. *Invest Ophthalmol Vis Sci.* 1980;19:1342–1347.
23. Matsushima Y, Ikemoto S. Biochemical genetic and sexual variations in mouse tear proteins. *Jpn J Zootech Sci.* 1981;52:775–779.
24. Farris RL. Abnormalities of the tears and treatment of dry eyes. In: Kaufman HE, Barron BA, McDonald MB, Waltman SR, eds. *The Cornea.* New York: Churchill Livingstone; 1988:139–159.
25. Varnell RJ, Carlton JE, Beuerman RW. Tear protein composition determined by laser desorption mass spectrometry. ARVO Abstracts. *Invest Ophthalmol Vis Sci.* 1995;36:S1043.
26. Maitchouk DY, Varnell RJ, Beuerman RW, Yan J. Analyzing proteins in tears from patients with dry eye syndrome, with contact lenses, or undergoing photorefractive keratectomy (PRK). ARVO Abstracts. *Invest Ophthalmol Vis Sci.* 1996;37:S18.
27. Varnell RJ, Maitchouk DY, Beuerman RW, Carlton JE, Haag A. Analysis of rabbit tear proteins using capillary electrophoresis (CE) and matrix-assisted laser desorption/ionization mass spectrometry (MALDI). ARVO Abstracts. *Invest Ophthalmol Vis Sci.* 1996;37:S848.
28. Hillenkamp F, Karas M. Mass spectrometry of peptides and proteins by matrix-assisted ultraviolet laser desorption/ionization. *Methods Enzymol.* 1990;193:280–295.
29. Beuerman RW, McDonald MB, Shofner RS, et al. Quantitative histological studies of primate corneas after excimer laser photorefractive keratectomy. *Arch Ophthalmol.* 1994;112:1103–1110.
30. Beavis RC, Chait BT. Rapid, sensitive analysis of protein mixtures by mass spectrometry. *Proc Natl Acad Sci USA.* 1990;87:6873–6877.
31. Beavis RC, Chaudhary T, Chait BT. α-Cyano-4-hydroxycinnamic acid as a matrix for matrix-assisted laser desorption mass spectrometry. (Letter) *Org Mass Spectrom.* 1992;27:156–158.
32. Varnell RJ, Freeman JY, Maitchouk D, Beuerman RW, Gebhardt BM. Detection of substance P in human tears by laser desorption mass spectrometry and immunoassay. Curr Eye Res, in press.

31-27 kDa CASEINOLYTIC PROTEASE IN HUMAN TEARS

Miki Sakata,[1] Ann R. Beaton,[1] Sonal Sathe,[1] and Robert A. Sack[1,2]

[1]Department of Biological Sciences
State University of New York
State College of Optometry
New York, New York
[2]CRCERT
University of New South Wales
Sydney, Australia

1. INTRODUCTION

Normal wound healing requires the presence and regulation of cell adhesion proteins, proteases, and growth and other factors. Several of the proteases involved in tissue remodeling and wound healing have been detected in tear fluid. Previous studies on ocular pathologies have focused on the plasminogen-plasmin system[1,2] and its role in fibrinolysis, which results in a decrease of epithelial cell adhesion and subsequent destruction of the basement membrane.[3] This study was designed to identify other proteases reported in tear fluid that may impact ocular pathologies and wound healing.

2. METHODS

2.1. Tear Samples

Reflex (R)-0.2~1 ml, open (O)- 2~3 µl, and closed (C)-2~4 µl eye tear samples were collected from a population of normal non-contact lens wearers using disposable microcapillary tubes. Samples were centrifuged, and the supernatant was analyzed by zymography and the pellet subjected to cytological examination.

2.2. Zymography Assays

Caseinolytic activity was analyzed on 10% polyacrylamide-SDS mini gels containing 0.1% casein. Samples were loaded in electrophoresis buffer without reducing agent or boiling. After pre-electrophoresis for 1 h at 40 mA, the gels were run at 25 mA at 4°C. SDS was removed and the protein renatured by incubation in 2.5% (v/v) Triton-X 100 for 1 h at room temperature. Proteinase reactions were carried out in developing buffer (50 mM Tris, pH 7.5, 200 mM NaCl, 5 mM $CaCl_2$, 0.02% Brij 35) for 15–18 h at 37°C. The gel was then stained with Coomassie blue. Enzymatic activities were seen as clear bands against a stained background, indicating that the casein within the SDS gel was digested.

2.3. Immunoprecipitation Reaction

C tears were subjected to zymography before and after immunoadsorption and immunoprecipitation with polyclonal antibodies to polymorphonuclear (PMN) leukocyte elastase or plasmin. Immunoprecipitation was carried out at 4°C overnight with polyclonal antibody. The specific bound protein was eluted from the beads without denaturation in SDS-PAGE loading buffer (no reducing agent) at 50°C for 10 min. The unbound protein and bead washes were examined after acetone precipitation.

3. RESULTS

No detectable caseinolytic activity was seen in the supernatants of R or O tear samples. C tears contained three distinct bands of caseinolytic activity 27–31 kDa in molecular weight. Immunoprecipitation studies showed that this activity was due to PMN leukocyte elastase, with similar levels of activity seen in tears from males and females.[4] Analysis of the pellets from R and O samples revealed a population of desquamated epithelial cells, while the C pellet consisted of a large population of PMN cells with a few epithelial cells. Zymography of a saline wash of the pellets revealed significant elastase activity only in the C pellet wash.

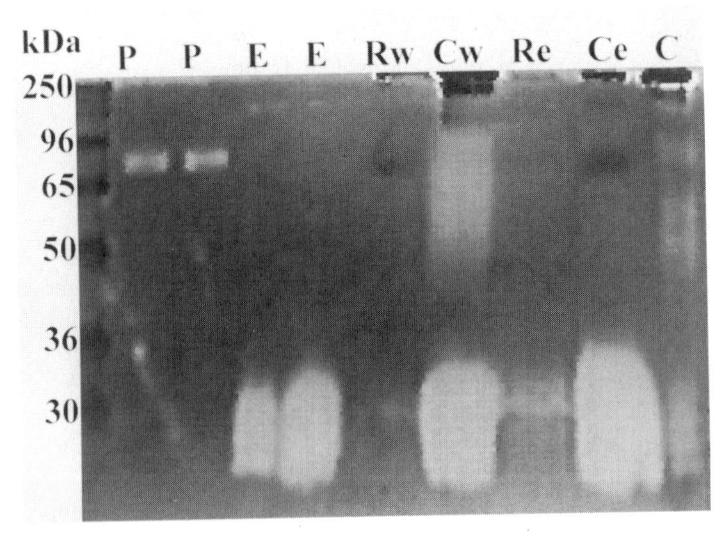

Figure 1. Casein zymography of 20, 10 ng plasmin (P) std; 20, 10 ng elastase (E) std; saline washes (w) and extract (e) of R tear pellet (Rw and Re) and C tear pellet (Cw and Ce); and closed-eye tear supernatant (C) 1 μl.

4. CONCLUSIONS

Although R and O tear supernatants are devoid of caseinolytic activity as measured by zymography, very high levels of activity can be detected in the C tear supernatant. We have shown that this caseinolytic activity is not plasmin as previously thought, but elastase. The elastase is released from functionally active PMN leukocytes recruited and activated in the closed-eye environment. These findings suggest that the external ocular surfaces must have evolved an extremely efficient means of preventing autologous cell damage in the highly reactive closed-eye environment. In other related work, we have documented that this damage may be prevented by a parallel build-up in C fluid of down-regulating antiproteases. The details of this work and related studies will be published in Current Eye Research.

REFERENCES

1. Tervo T, Salonen EM, Vaheri A, et al. Elevation of tear fluid plasmin in corneal disease. *Acta Ophthalmol.* 1988;66:393–399.
2. Van Setten GB, Salonen EM, Vaheri A, et al. Plasmin and plasminogen activator activities in tear fluid during corneal wound healing after anterior keratectomy. *Curr Eye Res.* 1989;8:1293–1298.
3. Liotta LA, Goldfarb RH, Brundage R, Siegel GP, Terranova V, Garbisa S. Effect of plasmin activator (urokinase), plasmin and thrombin on glycoprotein and collagenous components of basement membrane. *Cancer Res.* 1981;41:4629–4636.
4. Hamdi HK, Berjis F, Brown DJ, Kenney MC. Protease activity in normal human tears: Male-female dimorphism. *Curr Eye Res.* 1995;14:1081–1086.

TEAR PROTEIN LEVELS FOLLOWING PUNCTAL PLUGGING

E. Ian Pearce,[1] Alan Tomlinson,[1] Jennifer P. Craig,[1] and Gerald E. Lowther[2]

[1]Department of Vision Sciences
Glasgow Caledonian University
Glasgow, Scotland, United Kingdom
[2]School of Optometry
Indiana University
Bloomington, Indiana

1. INTRODUCTION

Occlusion of the lacrimal drainage system has been advocated in cases of dry eye where other forms of treatment including artificial tears and lid hygiene have proved inadequate. The first method of occlusion suggested was cautery of the tissue around the canaliculus and punctum to prevent tear drainage.[1] The first temporary occlusion technique was described in the city of Glasgow by Foulds[2] in 1961 who advocated the use of gelatin implants. Today collagen plugs are commercially available (Eagle Vision, Memphis, TN; Lacrimedics, Rialto, CA) to temporarily block tear drainage. The efficacy of dissolvable collagen in blocking punctal drainage is less than complete.[3–5] Lowther and Semes[5] found that collagen plugs had no appreciable effect on tear break-up, tear prism height, fluorescein, or rose bengal staining, yet patients reported that dryness-related symptoms were significantly reduced.

Non-dissolvable punctal plugs first suggested by Freeman[6] in 1973 are now available commercially in two different designs. The Punctum Plug (Eagle Vision) was introduced in 1983 and produced in varying sizes to rest at the punctum. The Herrick lacrimal plugs (Lacrimedics) introduced in 1990 also come in a variety of sizes and are designed to be inserted into the horizontal portion of the canaliculus. These non-dissolvable plugs, made from silicone rubber, are more effective at reducing dry eye symptoms.[7]

Studies on the effect of lacrimal occlusion in dry eye patients have concentrated on changes in the biophysics of the tears. Dohlman[8] suggested that an increase in tear volume with punctal occlusion would lead to a decrease in tear osmolality, and this has been found to be the case with electrocautery[9] and punctal plugs.[10] Tsubota and Yamada[11] demonstrated that, although absolute evaporation rate may increase with occlusion, the relative concentration of tears decreases with growing volume. Tomlinson, Craig, and Lowther[12] provided evidence for a decrease in tear production subsequent to occlusion in normal subjects. The beneficial effects of the procedure in dry eyes is illustrated by the decrease in rose bengal staining.[9,10,13] Decreased fluorescein staining, improved Schirmer strip wetting, and reduction in symptoms are also reported.[13]

Lacrimal occlusion has also been used in cases of contact lens-related dry eye. Giovagnoli and Graham[14] demonstrated an average 34.6% increase in contact lens wearing time following inferior punctal occlusion. Slusser[15] found that lacrimal drainage occlusion with non-dissolvable plugs significantly reduces symptoms of dryness, lens awareness, and blurred vision with soft lens wear. A common complication of the use of punctal occlusion in contact lens wearers is intermittent epiphora which is reported initially by 82% of patients.[15] This problem diminishes with time and occurs only rarely after the first few days of occlusion.

The effect of punctal occlusion on the tear composition has not been given much consideration in the literature. The only report to date is by Lowther and Semes[5]; they found that the lactoferrin concentration levels, measured using the Lactoplate test (Eagle Vision), were not different in the test eyes from those in controls. An absorbable intracanalicular collagen implant had been inserted into the upper and lower puncta of the test eye. However, the authors expressed considerable doubt about the efficacy of these plugs in attaining total occlusion.

The purpose of the present study was to determine the effect of complete punctal occlusion using a non-dissolvable punctal plug on tear composition. The samples for composition analysis were taken 1 week following punctal plugging, to allow measurements in a steady-state condition. This avoided any initial perturbation in tear chemistry produced by the insertion procedure.

2. MATERIALS AND METHODS

Nine subjects (three male and six female), mean age 22 (range: 18 to 29 years), were non-contact lens wearers and pathology free. In each subject, all four puncta were occluded using Tapered Shaft Punctal Plugs (Eagle Vision). The dome-shaped head of the plugs allowed visual checks on plug integrity. The patients complained of epiphora initially, but by day 7 overflow from the eye had abated. Tear samples were taken in the unplugged condition and again 7 days after plugging, to reflect the tear composition that would be seen in the long-term use of these devices.

Tear samples were collected in glass microcapillary tubes which had been drawn to a fine tip in a flame and preweighed. The subject was instructed to look up, and the capillary tube was dipped in the lower tear meniscus on the temporal side of the eye. The procedure was carried out by an experienced operative; no sensation was reported by the subject and therefore no reflex tearing was induced. The capillaries were then reweighed and sealed in 0.5 ml microcentrifuge tubes (Eppendorph, Hamburg, Germany) with thermoplastic film (Parafilm, American National Can, Neenah, WI, USA). The microcentrifuge tubes were then stored at –20°C until analysis. Immediately prior to analysis, the samples were warmed to room temperature and flushed from the capillaries with 12 µl of the HPLC analysis buffer.

The HPLC method was based on those of Fullard and Snyder[16] and Boonstra and Kylstra[17] with some modifications. Samples were placed in glass conical 200 µl vials

(Chromacol, Welwyn Garden City, UK) which were mounted in 2 ml vials on plastic three-legged inserts. The tubes were sealed with screw caps and a silicone rubber seal faced with PTFE to stop evaporation A Shimadzu SIL-10Axl autosampler (Shimadzu, Kyoto, Japan) was then used to inject 5 µl of the solution onto the column. The sample needle, unlike those in most autosamplers, makes up part of the flow path, ensuring injection with zero sample loss.

A state-of-the-art TSK-G2000SWXL size exclusion column (Tosoh Corporation, Tokyo, Japan) with a particle size of 5 µm was used. The column separates macromolecules by size, with the smaller molecules eluting last. The proteins were eluted from the column with 0.5 M NaCl, 0.0167 M phosphate buffer, pH 5.28, with 0.5% (w/v) sodium azide as a preservative, using a Shimadzu LC-10AD pump. The eluate was monitored with a Shimadzu SPD-10A ultraviolet spectrophotometric detector at 280 nm. The data from the detector was collected and analyzed on a 386PC running Shimadzu CLASS-VP software version 4.0.

3. RESULTS

The method of tear sample collection used in this study was optimized to avoid reflex lacrimation. The capillary tube was placed into the lower tear meniscus only once, and then withdrawn. As a result, the tear sample obtained varied in size from 0.2 to 1.6 µl with a mean of 0.8 µl. Over this range, the HPLC technique was able to produce an acceptable tear protein profile.

The peaks obtained were identified (Fig. 1) by comparison to known pure proteins (Sigma, Poole, UK), where available. Retention time and overall peak shape were taken into consideration to aid identification. In this way the peaks derived from IgA, lysozyme, and lactoferrin were attributed. The identity of the tear-specific pre-albumin peak was in-

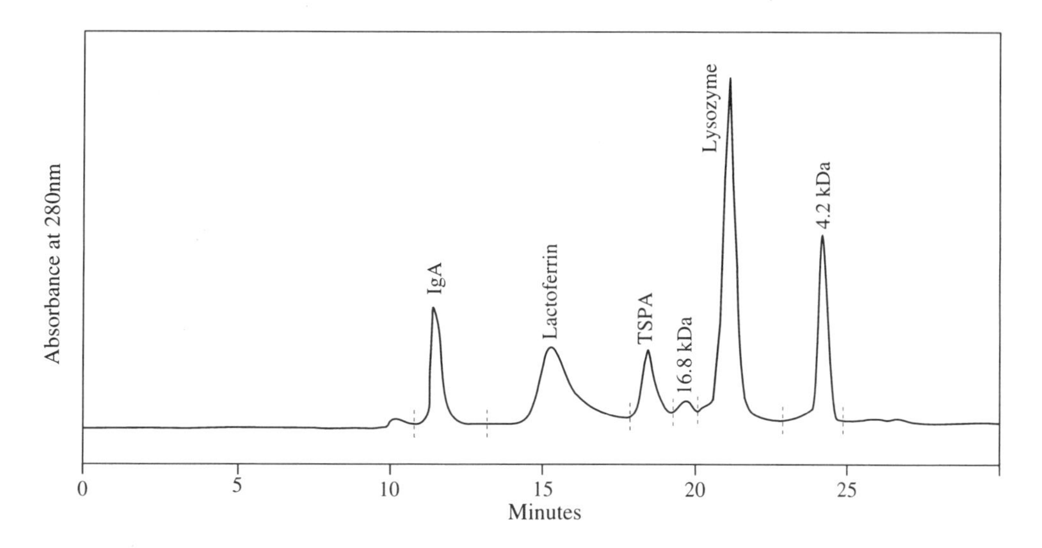

Figure 1. HPLC trace of human tears. The eluate from the column was monitored at 280 nm. Molecules elute in order of size with high-molecular-weight molecules first and smaller molecules later. The identity of the molecules was arrived at by comparison to known standards and also by their apparent molecular weight. Unknown molecules are designated by their apparent molecular weight.

ferred from its apparent molecular weight. Two other peaks were seen in all samples. The apparent molecular weight of these proteins was calculated from their retention time. One peak was at 16.8 kDa, and the other peak represented small molecules with a molecular weight of 4.2 kDa or less.

The separation of the peaks was sufficient such that individual proteins could be quantified from peak areas without contributions from adjacent peaks (Fig. 1). To allow this quantitative analysis, standard curves of varying known amounts of pure lactoferrin, lysozyme, and IgA were prepared. Linear regression analysis was carried out, and the slope and intercept obtained were used for quantification. When the identity of the molecule was not known, or when it was not commercially available, the apparent level was derived by comparison to a bovine serum albumin standard curve.

As the tear samples were of known size, the concentration of each molecule in the tears could be calculated.

Levels of the tear proteins arrived at, both with and without punctal plugs (data not shown), were comparable to those for normal subjects published in the literature.

The levels of individual proteins from each subject plugged and unplugged were compared with a paired sample t-test to overcome inherent inter-subject variation.

It was found that levels of lactoferrin (Fig. 2A) and lysozyme (Fig. 2B) had undergone a significant increase as a result of punctal occlusion ($p = 0.002$ and $p = 0.011$, respectively). The other molecules, IgA ($p = 0.193$), TSPA ($p = 0.715$), the 16.8 kDa ($p = 0.259$), and the 4.2 kDa ($p = 0.981$) proteins showed no significant alteration.

The total protein levels were also calculated by the addition of the individual protein levels seen over the six peaks, with unplugged subjects having a mean tear protein concentration of 5.42 mg/ml (SD ± 1.73) and plugged individuals at 6.29 mg/ml (SD ± 1.78). The paired sample t-test showed no significant change in total tear protein level on plugging ($p = 0.190$).

4. DISCUSSION

Complete punctal occlusion by means of silicone plugs potentially causes a number of changes in the dynamics of tear protein production and elimination. Tear proteins produced in the lacrimal gland have two possible fates. The proteins can, along with the aqueous tears, exit via the puncta. The second possibility is for the proteins to undergo proteolytic degradation.[18] In the eye, tear turnover is approximately 16% of total tear volume per minute.[19] Therefore the half-life of the proteins, as a result of drainage, would be about 4 min, and proteolytic degradation would be of little importance in the turnover of these proteins. When punctal plugs are utilized, aqueous outflow is stopped and the residency time of the tear proteins increases. In this situation, proteolytic degradation could become more important. In the plugged eye, the aqueous component of tears can exit by evaporation or conjunctival adsorption. Macromolecules such as proteins cannot escape by this means. The expected result would be that the concentration of all the proteins increase. This is not seen in the current study, where only lactoferrin and lysozyme concentrations are significantly higher in the plugged state. This could be due to downregulation of production of the other proteins or a differential rate of proteolytic degradation. As total protein levels showed no significant change, differential degradation seems the most likely explanation.

After 7 days, epiphora disappears in most patients, suggesting a feedback mechanism to regulate aqueous tear production.[12] The trigger for this homeostatic response could

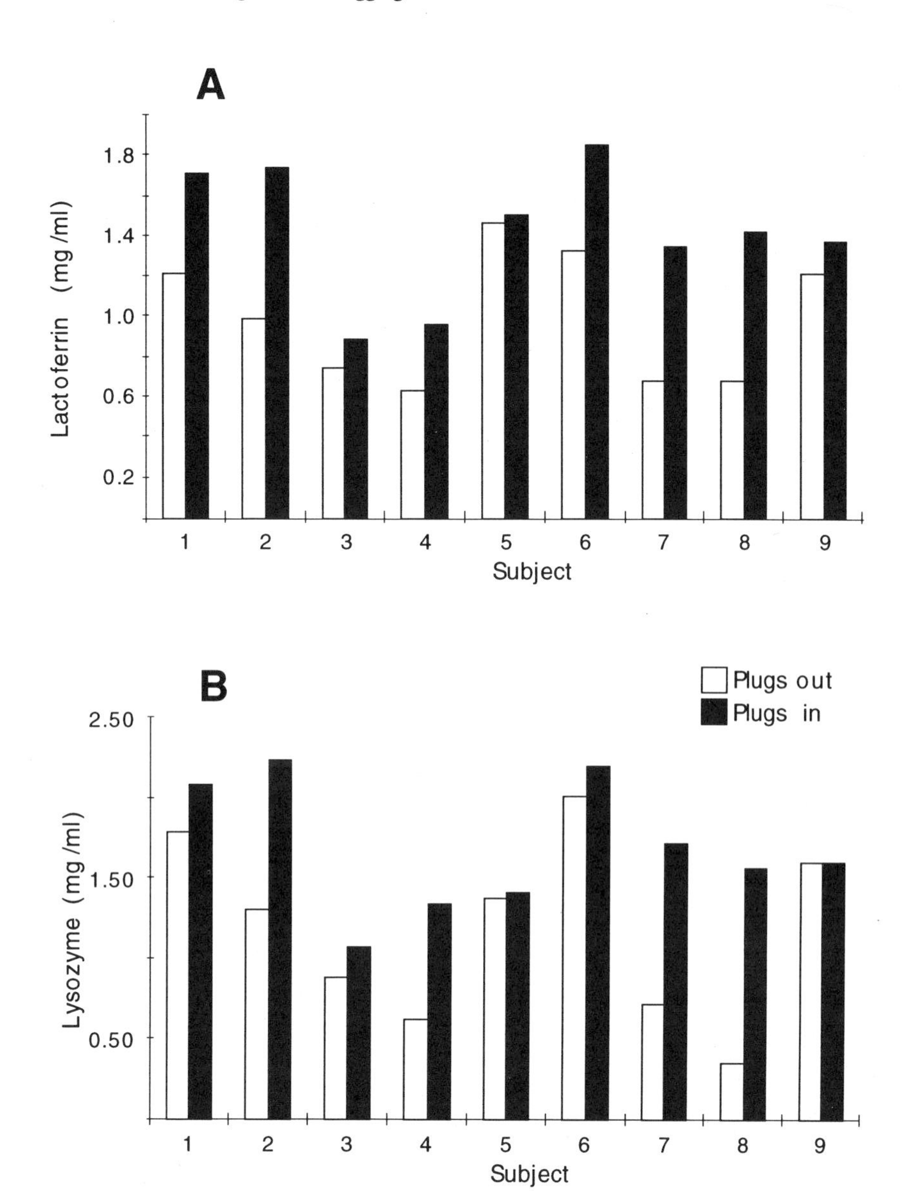

Figure 2. Tear protein concentrations with and without punctal occlusion. Concentrations of lactoferrin (A) and lysozyme (B) for the nine subjects derived from HPLC analysis. Black bars, levels 7 days after punctal occlusion. White bars, levels without punctal plugs.

be the increase in protein degradation products, signalling that tear drainage is impeded and tear proteins are residing in the eye for extended periods of time. This mechanism may be an additional factor in contact lens-induced dry eye when protein-contaminated lenses are persistently worn.

The implications of this study, for the use of punctal plugs, rests on the elevated levels of lysozyme and lactoferrin. An increase in these anti-bacterial molecules can be seen as a desirable side effect in the dry-eye patient. In the contact lens-wearing patient, similar benefits may exist in the borderline dry-eye case. However, changes in tear film composition could have implications for the undesirable interactions that take place between tear fluid proteins and the contact lens. The mixture of denatured and degraded proteins would not be an ideal environment in which to place a contact lens. In this situation there is obvious potential for inflammatory or infectious complications.

ACKNOWLEDGMENTS

The punctal plugs used in this study were donated by CIBA Vision Ophthalmics Inc. and Eagle Vision.

REFERENCES

1. Beetham WP. Filimentary keratitis. *Trans Am Ophthalmol Soc.* 1935;33:413–435.
2. Foulds WS. Intracannicular gelatin implants in the treatment of keratoconjunctivitis sicca. *Br J Ophthalmol.* 1961;45:625–627.
3. Groves SW, Forstat SL, Lear J. Intracannicular collagen implants: A dacryoscintigraphy study. *Invest Ophthalmol Vis Sci.* 1991:32:727.
4. Apte RV. Quantitative analysis of tear outflow reduction due to intracannicular collagen implants using lacrimal scintillography. *Invest Ophthalmol Vis Sci.* 1994;35:1692.
5. Lowther GE, Semes L. Effective of absorbable intracannicular collagen implants in hydrogel contact lens patients with drying symptoms. *International Contact Lens Clinic.* 1995;22:238–243.
6. Freeman JM. The Punctum Plug: Evaluation of a new treatment for the dry eye. *Trans Am Acad Ophthalmol Otolaryngol.* 1973;79:874–879.
7. Slusser TG. A review of hydrogel contact lens-related dry eye and lacrimal drainage occlusiion. *International Contact Lens Clinic.*. 1995;22:250–254.
8. Dolhman CH. Punctal occlusion in keratoconjunctivitis sicca. *Ophthalmology.* 1978;85:1277–1281.
9. Gilbard J. Tear film osmolarity in keratoconjunctivitis sicca. *CLAO J.* 1985;11:243–250.
10. Gilbard J, Rossi SR, Azar DT, Heyda KJ. Effective punctal occlusion by Freeman silicone plug insertion on tear osmolality in dry eye disorders. *CLAO J.* 1989;15:216–218.
11. Tsubota K, Yamada M. Tear evaporation from the ocular surface. *Invest Ophthalmol Vis Sci.* 1992;33:2943–2950.
12. Tomlinson A, Craig JP, Lowther GE. The biophysical role in tear regulation. (This volume).
13. Willis RM, Folberg R, Krachmer JH, Holland EJ. The treatment of aqueous deficient dry eye with removable punctal plugs. *Ophthalmology.* 1987;94:514–518.
14. Giovagnoli D, Graham S. Inferior punctal occlusion with the removable silicone punctal plugs in the treatment of dry eye related contact lens discomfort. *J Am Optom Assoc.* 1992;63:481–485.
15. Slusser TG. Effects of lacrimal drainage occlusion with non dissolvable intracannicular plugs on hydrogel contact lens related dry eye. Thesis, University of Alabama at Birmingham, Birmingham, AL, 1984.
16. Fullard RJ, Tucker DL. Changes in human tear protein-levels with progressively increasing stimulus. *Invest Ophthalmol Vis Sci.* 1991;32:2290–2301.
17. Boonstra A, Kijlstra A. Separation of human tear proteins by high-performance liquid chromatography. *Curr Eye Res.* 1984;3:1461–1469.
18. Hamdi HK, Berjis F, Brown DJ, Kenney MC. Proteinase activity in normal human tears - male-female dimorphism. *Curr Eye Res.* 1995;14:1081–1086.
19. Occhipinti JR, Mosier MA, LaMotte J, Monji GT. Fluorophotometric measurement of human tear turnover rate. *Curr Eye Res.* 1988;7:995–1000.

CHARACTERIZATION OF CELLS SHED FROM THE OCULAR SURFACE IN NORMAL EYES

Carolyn G. Begley,[1] Jie Zhou,[1] and Graeme Wilson[2]

[1]Indiana University
School of Optometry
Bloomington, Indiana
[2]University of Alabama at Birmingham
Birmingham, Alabama

1. INTRODUCTION

The corneal and conjunctival epithelia constitute the ocular surface, which is a dynamic layer under constant renewal. As corneal epithelial cells migrate anteriorly and centripetally,[1,2] the cells terminally differentiate and flatten. Cells reach the surface, tight junctions form between adjacent epithelial cells, and cells apically differentiate. Older surface cells, presumably the large, dark squamous cells viewed by scanning electron microscopy (SEM),[3–5] slough into the tear film as new epithelial cells move anteriorly to replace them. The conjunctiva replaces itself similarly,[6] although the cells may be more cuboidal[7] or polyhedral[8] in shape. The renewal process of the normal ocular surface can be studied by examination of cells shed from the surface.

In 1986, Fullard and Wilson[9] reported a new, non-invasive technique for the collection of cells specifically from the human precorneal tear film. Cell collection by the non-contact corneal irrigation chamber (NC-CIC) yielded squamous, cuboidal, and columnar appearing cells with either green or orange-red cytoplasm following staining with acridine orange dye. Normal cell shedding showed a diurnal variation with irrigation by BTS (basic tear solution), and was increased with irrigation by isotonic NaCl. Trauma to the corneal cells via ultraviolet-B irradiation caused a prolonged acceleration in the shedding rate after a 3 h latent period.[10] There was also an increase in the total number of cells collected in the morning upon eye opening. Wilson et al.[11] showed that a large number of leukocytes was collected in conjunction with (median: 8,154 cells) and without (median: 6,500 cells) extended wear of soft contact lenses.

Cells can also be collected from the ocular surface using contact lens cytology. This technique employs short-term wearing of a contact lens to collect ocular surface cells. Using this method, Zhou et al.[12] showed a strong diurnal variation in the numbers of sloughed cells, with the largest number collected in the morning and evening, and the lowest at mid-day. On average, 16% of these cells were viable, 49% were dead, and 35% showed double staining with two fluorescent probes for intracellular esterase activity (calcein AM) and nucleic acids (ethidium homodimer). Double staining indicated a cell with intact esterase activity, but a permeable cell membrane. In a separate study, Reed and Wilson[13] determined that the size range they found for rabbit corneal epithelial shed cells agreed with that reported for SEM dark cells, whereas human shed cells reflected the full range of epithelial cell sizes.

Both the NC-CIC and contact lens cytology collect surface cells, but contact lens cytology has the advantage of cell collection from a wider area of the ocular surface. The lens comes into contact with corneal and conjunctival epithelia, as well as inflammatory cells present on the surface of the eye. Therefore, the technique has the potential for use in studying ocular surface disease and inflammation. However, the cell types must be distinguished in normal eyes before disease conditions can be effectively studied. In this investigation we characterized the numbers of corneal, conjunctival, and inflammatory cells present over the course of one day using immunohistochemical stains for the three cell types.

2. METHODS

Six subjects between the ages of 22 and 42 participated in the study (two male, four female). All subjects were healthy. Subjects with ocular or systemic disease were excluded. Three subjects were soft contact lens wearers and three had never worn contact lenses.

Ocular surface cells were collected six times over the course of one day from each subject. To collect the cells, subjects wore a NewVues® lens in each eye for 10 min, and then the lens was removed. Lenses were inserted and removed using sterile gloves. Immediately after removal, lenses were washed in culture medium and vortexed to remove adherent cells. Samples were pooled from both eyes of each subject and divided into thirds. Following centrifugation, pellets were resuspended in 1:1 ethanol/acetone at -20°C for 15 min, and pre-incubated in 0.1% BSA/0.1 M PBS (pH 7.2) for 30 min at room temperature. Samples to be processed for keratin staining also included 0.1% Triton X-100 in the pre-incubation mixture. Cell samples were then incubated in the primary antibodies for 24 h at 4°C, either anti-common leukocyte antigen T200 (1:20) (Boehringer Mannheim, Indianapolis, IN) for inflammatory cells, anti-epithelial keratin-AE5 (1:20) (ICN, Costa Mesa, CA) for corneal epithelium, or anti-epithelial keratin-AE1&3 monoclonal mix (1:50) (ICN) for all epithelium. After washing with PBS, all samples were incubated in goat-anti-mouse IgG (H+L)-fluorescein antibody (1:100) (Boehringer Mannheim) overnight at 4°C in the dark. To counterstain the nucleus, cells were incubated for 30 min in 24 μM ethidium homodimer-1 (Molecular Probes, Eugene, OR). Samples were filtered through a 5 μm PCTE* filter membrane (Fisher, Pittsburgh, PA) and viewed by Nikon MICROPHOT-FXA epifluorescent microscopy. Cells on the filters were counted at 250X and photographed at 1000X. In the 7:00 AM samples, the size of each cell was also measured. Cell size was recorded as the longest dimension.

3. RESULTS

All samples contained corneal epithelial cells, as labeled by AE5. AE1&3-positive cells were also present. The AE1&3 antibodies label all epithelial cells and thus include AE5-positive cells. Fig. 1a and b illustrates typical AE5- and AE1&3-positive cells.

All but one sample contained inflammatory cells stained by T200. Morphologically, these cells were either mononuclear and resembled lymphocytes, or multinuclear and re-

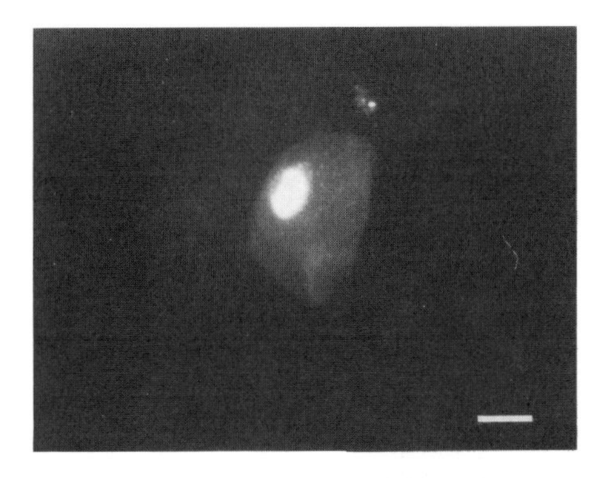

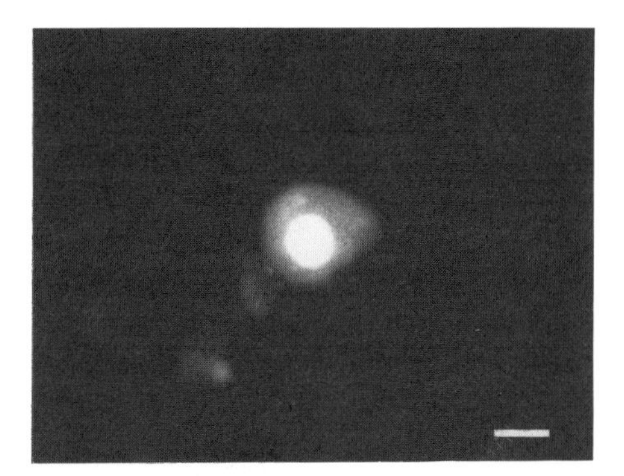

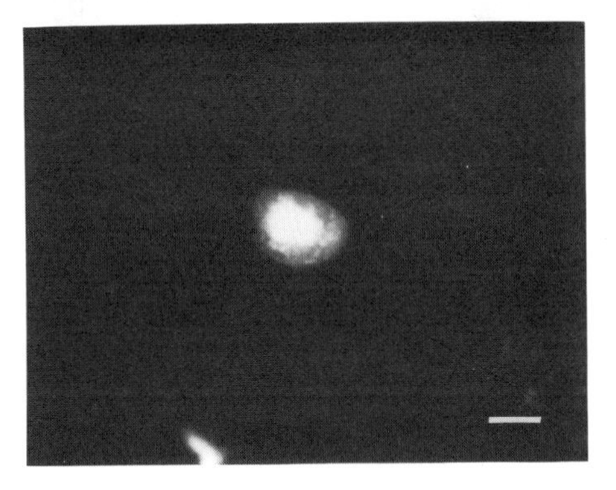

Figure 1. Typical cells obtained by contact lens cytology (bar = 10um). (a) AE5-positive (labels most corneal epithelial cells). (b) AE1&3-positive (labels all epithelial cells). (c) T200 (labels all inflammatory cells).

sembled neutrophils or other polymorphonuclear leukocytes. A mononuclear cell, presumably a lymphocyte, is pictured in Fig. 1c.

Over the period of one day, there was a statistically significant ($p<0.05$) diurnal variation in total cell counts. As Fig. 2 demonstrates, total cell counts were highest in the morning and evening, and lowest at midday. Corneal epithelial (AE5-positive) and inflammatory (T200-positive) cells followed the same pattern, but differences between times of day were not significant. On average, AE5-positive cells were the most predominant cell type during most times of day.

In addition to counting, the longest dimension of each cell was measured for the 7:00 AM collection time. As Fig. 3 illustrates, AE1&3-positive cells range in size from 18 to 58 μm in their longest dimension. These cells show a strong bimodal distribution, with peaks at 26 and 42 μm. AE5-positive cells demonstrate the same bimodal distribution, but the 26 μm peak is much less marked. Fig. 4 represents the size distribution of T200-positive cells from the 7:00 AM samples. In their longest dimension, these cells range from 6–18 μm, with two peaks at 10 and 16 μm.

4. DISCUSSION

The technique of contact lens cytology yields a sampling of ocular surface cells, including corneal and other epithelial and inflammatory cells. In most samples, the majority

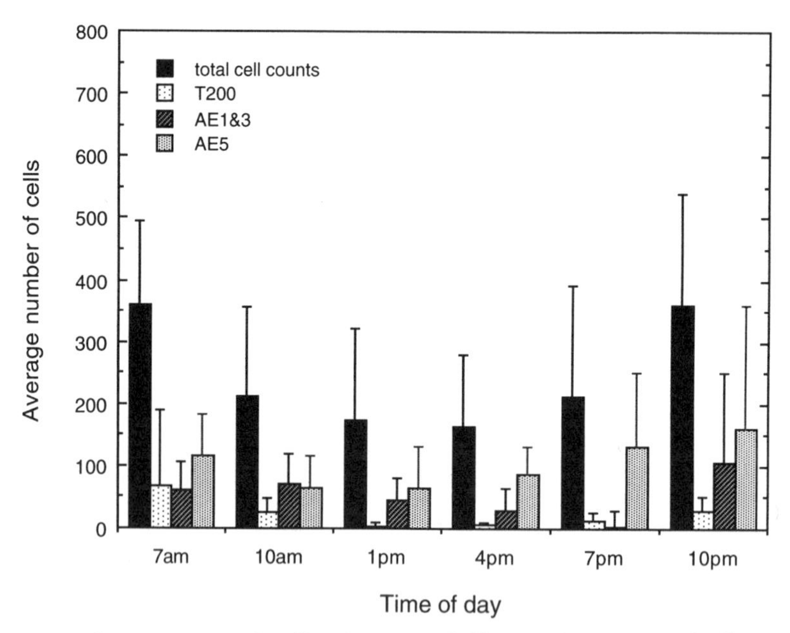

Figure 2. Average cell counts over one day. (Samples were pooled between the eyes, so total cell counts represent the average number of cells derived from both eyes, and all labeled cell types are one-third of the pooled sample.)

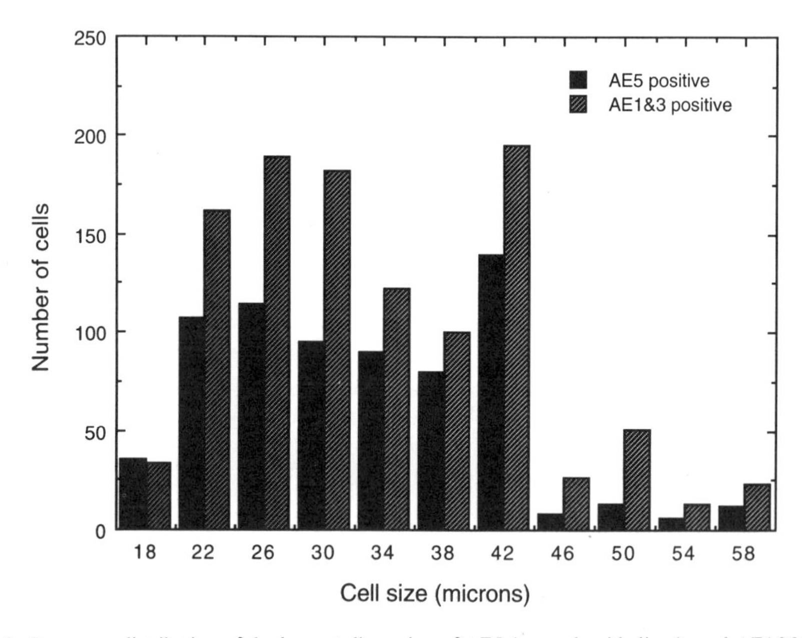

Figure 3. Frequency distribution of the longest dimension of AE5 (corneal epithelium)- and AE1&3 (all epithelium)-positive cells.

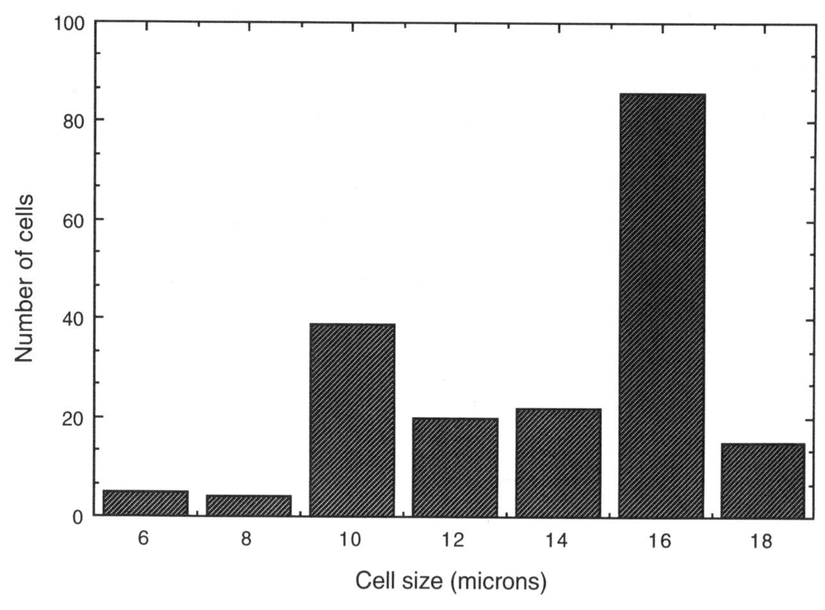

Figure 4. Frequency distribution of the longest dimension of T200-positive (inflammatory) cells.

of epithelial cells collected by contact lens cytology stained positively for AE5, a monoclonal antibody that recognizes a corneal epithelial basic keratin (K3). According to Wiley et al.,[14] AE5 labels all layers of the central corneal epithelium, but is immunohistochemically undetectable in the basal layers of the limbus. During development, K3, the basic 64 kDa keratin, and K12, the acidic 55 kDa keratin, are not expressed by the corneal epithelium until it stratifies.[15] AE5 stains mature corneal epithelium, which includes the corneal cells collected by contact lens cytology under normal conditions.

The hypothesis that large, dark cells viewed by SEM represent those next to shed from the corneal surface[4] is not wholly supported by our data. At the 7:00 AM sampling, we collected AE5-positive cells that ranged widely in size, with peaks at 26 and 42 μm. The 42 μm peak probably corresponds to the large, dark cells described by SEM. We also found many smaller AE5-positive cells in the range of 22 to 26 μm. The presence of these smaller cells suggests that large cell size is not the only determinant for sloughing from the ocular surface.

It is possible that the smaller-sized AE5-positive cells were prematurely shed due to mechanical abrasion by soft contact lenses used in this study. The technique of contact lens cytology involves inserting and removing a soft contact lens over a 10 min period. This artificial contact lens-wearing situation may be sufficiently invasive to dislodge younger cells. In a separate study using this method, Zhou et al.[12] showed that an average of 14% of shed cells were viable, as determined by ethidium dye exclusion and intact esterase activity. Whether these cells were corneal was not determined, but their presence indicates that all cells collected by this method are not terminally differentiated and dying.

AE1&3-positive cells also showed a bimodal size distribution with peaks similar to AE5-positive cells. This is understandable because AE1 and AE3 antibodies react to acidic and basic cytokeratins in most epithelial cells, including corneal epithelium.[16] Therefore, the distribution of these cells was influenced by the distribution of corneal epithelial cells. However, the peak in cell numbers at 26 to 30 μm is larger for AE1&3-posi-

tive cells than AE5-positive cells. This peak may be due to the presence of conjunctival cells, which tend to be smaller[7] than corneal cells. Our samples may also contain other epithelial cell types. In the process of inserting and removing a contact lens, squamous epithelium from the skin of lids could also attach to the lens surface and enter the cell collection.

The T200 antibody labels all cells of bone marrow origin. Almost all our samples contained cells positive for this antibody. In the early-morning samples (7:00 AM), many T200-positive cells resembled lymphocytes or neutrophils morphologically. The bimodal size distribution of the cells supports the presence of these cell types. Lymphocytes and polymorphonuclear cells range in size from 6–18 and 12–15 μm, respectively.[17] In our samples, smaller cells peaking at 10 μm were mononuclear and are thus likely to be lymphocytes, while most of the 16 μm cells were multinuclear and are probably neutrophils. These results support an earlier study that demonstrated the presence of large numbers of cells resembling leukocytes in early-morning eyewash samples.[11]

Later in the day, T200-positive cells were almost all mononuclear. We did not measure their size, but all were small, round cells similar to the cell pictured in Fig. 1c. These were also presumably lymphocytes.

Total cell counts varied diurnally, with the highest number in the morning and evening, and the lowest at midday. This agrees with an earlier study by Zhou et al.[12] Corneal epithelial (AE5-positive) and T200 cells followed a similar diurnal pattern, but the variation was not statistically significant. The reason for this diurnal pattern is unknown. The higher numbers in the morning may reflect a washout of cells after eye closure overnight, but the rise in the evening suggests a cyclical shedding of ocular surface cells.

5. CONCLUSIONS

A variety of ocular surface cells can be sampled by the method of contact lens cytology. Immunohistochemical staining with antibodies specific for corneal and other epithelial cells and inflammatory cells demonstrates a wide range of cell types and sizes. However, further investigation will be necessary to confirm subpopulations of inflammatory cells in normal patients. This technique shows great promise for studying the inflammatory and epithelial components of ocular surface disease.

REFERENCES

1. Thoft RA, Friend J. The X, Y, Z hypothesis of corneal epithelial maintenance. *Invest Ophthalmol Vis Sci.* 1983;24:1442–1443.
2. Lavker RM, Dong G, Cheng SZ, Kudoh K, Cotsarelis G, Sun T-T. Relative proliferative rates of limbal and corneal epithelia. *Invest Ophthalmol Vis Sci.* 1991;32:1864–1875.
3. Hoffmann F. The surface of epithelial cells of the cornea under the scanning electron microscope. *Ophthalmic Res.* 1972;3:207–214.
4. Pfister RR. The normal surface of corneal epithelium: A scanning electron microscopic study. *Invest Ophthalmol.* 1973;12:654–668.
5. Hazlett LD, Wells P, Spann B, Berk RS. Epithelial desquamation in the adult-mouse cornea: A correlative TEM-SEM study. *Ophthalmic Res.* 1980;12:315–323.
6. Setzer PY, Nichols BA, Dawson CR. Unusual structure of rat conjunctival epithelium. *Invest Ophthalmol Vis Sci.* 1987;28:531–537.
7. Fine BS, Yanoff M. *Ocular Histology.* Hagerstown, MD: Harper & Row; 1979:309.

8. Latkovic S, Nilsson SEg. The ultrastructure of the normal conjunctival epithelium of the guinea pig. *Acta Ophthalmol.* 1979;57:123–135.
9. Fullard RJ, Wilson GS. Investigation of sloughed corneal epithelial cells collected by non-invasive irrigation of the corneal surface. *Curr Eye Res.* 1986;5:847–856.
10. Ren H, Wilson G. The effect of ultraviolet-B irradiation on the cell shedding rate of the corneal epithelium. *Acta Ophthalmol.* 1994;72:447–452.
11. Wilson G, O'Leary DJ, Holden BA. Cell contact of tears following overnight wear of a contact lens. *Curr Eye Res.* 1989;8:329–335.
12. Zhou J, Begley CG, Wilson G. Diurnal variation and viability of cells shed from the ocular surface. ARVO Abstracts. *Invest Ophthalmol Vis Sci.* 1996;37:S73.
13. Reed R, Wilson G. The size of shed corneal epithelial cells. *Optom Vis Sci.* 1994;71(12s):157.
14. Wiley L, SundarRaj N, Sun T-T, Thoft RA. Regional herterogeneity in human corneal and limbal epithelia: An immunohistochemical evaluation. *Invest Ophthalmol Vis Sci.* 1991;33:594–602.
15. Rodrigues M, Ben-Zvi A, Krachmer J, Shermer A, Sun T-T. Suprabasal expression of a 64k (no. 3) keratin in developing human corneal epithelium. *Differentiation.* 1987;34:60–67.
16. Tseng SCG, Jarvinen MJ, Nelson WG, Huang JW, Woodcock-Mitchell JW, Sun T-T. Correlation of specific keratins with different types of epithelial differentiation: Monoclonal antibody studies. *Cell.* 1982;30:361–369.
17. Junqueira LC, Carneiro J, Kelley RO. *Histology.* Norwalk, CT: Appleton & Lange; 1995:222.

HEALING EFFECT OF SODIUM-SUCROSE-OCTASULFATE AND EGF ON EPITHELIAL CORNEAL ABRASIONS IN RABBITS

Sven Johansen, Steffen Heegaard, Kirsten Bjerrum, and Jan Ulrik Prause

Eye Pathology Institute
University of Copenhagen, Denmark

1. INTRODUCTION

Our aim in this study was to investigate the possible healing effect of sodium-sucrose-octasulfate (SOS) on epithelial corneal abrasions, and to evaluate any increase in healing rate when combining epidermal growth factor (EGF) with SOS.

2. MATERIALS

2.1. Animals

We used 25 Copenhagen white rabbits (female, 2.5–3 kg).

2.2. Topical Test Solutions

(a) SOS 2% + NaCl 0.6% in phosphate buffer*
(b) SOS 10% in phosphate buffer*
(c) EGF 50 ng/ml in phosphate buffer*
(d) Hydroxypropyl methylcellulose (HPMC) 0.25%, made isotonic with NaCl and preserved with phenylmercury nitrate 0.001%

The test drugs were kindly supplied by BM Research, DK-3500 Vaerloese, Denmark.

* Phosphate buffer: NaH_2PO_4 0.560 g + Na_2HPO_4 0.284 g, NaCl 0.5 g, Na_2EDTA 0.1 g in sterile H_2O, pH 6.2

Lacrimal Gland, Tear Film, and Dry Eye Syndromes 2
edited by Sullivan *et al.*, Plenum Press, New York, 1998

3. METHODS

The study was divided into two parts:

Part I: Corneal abrasions treated with SOS (2 or 10%) vs. vehicle.
Part II: Corneal abrasions treated with SOS (2 or 10%) vs. SOS (2 or 10%) + EGF.

A corneal abrasion was created bilaterally in all animals under topical (cocaine 4%) and general (ketamine 25 mg/kg IM and diazepam 25 mg/kg IM) anesthesia. An area 6 mm in diameter on the central part of both corneas was marked with a trephine. The corneal epithelium was thereafter removed by rubbing for 30 sec, with a swab soaked in *n*-heptanol.

The first rabbit was killed without further pharmacological treatment. The remaining 24 rabbits were treated with test solutions (see Table 1) applied twice a day until the animals were sacrificed. One drop (50 μl) of solution was delivered by a plastic container fitted with a pipette and placed in the inferior fornix.

3.1. Analyses

The corneal re-epithelialization was recorded by corneal staining with fluorescein (1%), which could be observed with a UV lamp, and photographed after 0, 12, 24, 36, 48, and 94 h, or until healed.

At each time of observation, two rabbits were killed with an overdose of pentobarbital (20%) injected in an ear vein. The enucleated eyes were stored in formalin for later hematoxylin-eosin light microscopy histology.

To quantify the observed corneal healing rate, computer planimetry (Leitz texture analyzing system) of the photographs was employed. Decreases in the fluorescein-stained, non-healed areas were expressed as % area healed at 0, 12, 24, 36, 48, 72, and 94 h, or healed.

3.2. Statistics

Differences between the % area healed were determined by paired *t*- and log rank test. A level of $p < 0.05$ was considered significant.

Table 1. Topical applications of test solutions

Part	Treatment	No. of rabbits
I	SOS 2% vs. HPMC 0.25%	6
	SOS 10% vs. HPMC 0.25%	6
II	SOS 2% vs. SOS 2% + EGF 50 ng/ml	6
	SOS 10% vs. SOS 10% + EGF 50 ng/ml	6

SOS, sodium-sucrose-octasulfate; HPMC, hydroxypropyl methylcellulose; EGF, epidermal growth factor.

4. RESULTS

4.1. Slit-Lamp Examination

No morphological differences, as evaluated by slit-lamp examination, were noted between treatments. All corneas healed within 72 h of treatment, regardless of which treatment had been applied.

4.2. Light Microscopy

Histological examination (hematoxylin and eosin) of SOS-, EGF-, and HPMC-treated corneas showed no morphological differences. Re-epithelialization with sliding of the superficial epithelium cells into the denuded area was seen after 12 h in all specimens. No inflammation was observed.

4.3. Computer Planimetry and Statistics

No significant difference in healed epithelial mean area was found between SOS 2% and SOS 10% ($p = 0.60$). In Part I, healed epithelial mean area in % according to time in the period 0–12 h was *higher* after treatment with SOS 10% compared to SOS 2%. In Part I, healed epithelial mean area in % according to time at 12 h was significantly higher after SOS 2% treatment vs. placebo ($p = 0.043$). No difference was found after treatment with SOS 10% vs. placebo. In Part II, no significant difference in healed epithelial area in % according to time after treatment with SOS 2 or 10% +/- EGF was found ($p = 0.32$). Consistently in Parts I and II, no significant differences in the median time to healing were seen.

5. DISCUSSION

Based on the information that healing of epithelial surface may be encouraged by the ability of sucralfate to transport EGF,[1,2] we found it of interest to evaluate the healing effect of SOS in two different concentrations, and to evaluate any increase in healing rate when adding EGF to SOS.

The persulfated disaccharide sucralfate (aluminum sucrose-octasulfate; Caralfate®, Ulcogant®, or Antepsin®) was introduced more than 25 years ago for the treatment of peptic ulcer. The sodium salt of sucrose-octasulfate is SOS. The epithelial healing mechanism induced by sucralfate is incompletely understood; healing may be favored by the ability of SOS to transport EGF, to bind basic fibroblast growth factor (bFGF), and to induce prostaglandin generation and the formation of a protective mucin-like plaque.[3–6] SOS 2% eyedrops significantly improve rose bengal scores in patients suffering from keratoconjunctivitis sicca in primary Sjögren's syndrome.[7,8] SOS 2% eyedrops are effective in treating erosive corneal lesions in keratoconjunctivitis sicca, particularly those associated with Sjögren's syndrome.[9] Sucralfate pellets in rabbit cornea stimulate angiogenesis, and the combination of sucralfate and bFGF induces intense angiogenesis.[3]

EGF is constantly present in human tear fluid.[10–12] The polypeptide growth factor EGF was isolated by Cohen[13] from the submaxillary gland of the mouse. Since then, a large number of growth factors have been identified in almost every ocular tissue, including cornea.[14]

In the present study, no significant difference in healed epithelial mean area between SOS 2% and 10% was found, although the healing rate of SOS 10% in the period 0–12 h is high. SOS in a 2% solution vs. HPMC increases the healing rate by the time point 12 h. An increase in healing rate could perhaps be obtained by increasing the application frequency. Addition of EGF to SOS 2 or 10% did not increase the healing rate. An explanation for the lack of combination effect may be the use of inappropriate EGF concentration.

In conclusion, no significant difference in healed epithelial mean area was found between SOS 2% and 10%. However, at time point 12 h, treatment with SOS 2% was significantly better than placebo, but no difference was found between treatment with SOS 10% and placebo. Addition of EGF did not increase healing rate.

REFERENCES

1. Nexø E, Poulsen SS. Does epidermal growth factor play a role in the action of sulcralfate? *Scand J Gastroenterol.* 1987;12:45.
2. Konturek SJ, Brozozowski T, Bielanski W, Warzecha Z, Drozdowiecz D. Epidermal growth factor in the gastroprotective and ulcer-healing actions of sulcralfate in rats. *Am J Med.* 1989;86:32.
3. Folkman J, Szabo S, Stovroff M, McNeil P, Li W, Shing Y. Duodenal ulcer. *Ann Surg.* 1991;214:414.
4. Nagashima R, Hinohara Y, Hirano T. Selective binding of sucralfate to ulcer lesion: Experiments in rats recieving 14C sucralfate. *Drug Res.* 1980;30:88.
5. Sasaki H, Hinohara Y, Tsunoda Y, Nagashima R. Binding of sucralfate to duodenal ulcer in man. *Scand J Gastroenterol* 1983;83:13.
6. Szabo S. The mode of action of sucralfate: The 1 x 1 x 1 mechanism of action. *Scand J Gastroenterol* . 1991;26 (suppl 185):7.
7. Prause JU. Beneficial effect of sodium sucrose-sulfate on the ocular surface of patients with severe KCS in primary Sjögren's syndrome. *Acta Ophthalmol.* 1991;69:417.
8. Prause JU, Bjerrum K, Johansen S. Effect of sodium sucrose-sulfate on the ocular surface of patients with keratoconjunctivitis sicca in Sjögren's syndrome. *Adv Exp Med Biol.* 1994;350:691.
9. van Bijsterveld OP, van Hemel OLH. Sulcralfate and sodium sucrose sulphate in the treatment of superficial corneal disease in keratoconjunctivitis sicca. *Acta Ophthalmol.* 1992;70:518.
10. Ohashi Y, Motokura M, Mano T, et al. Presence of epidermal growth factor in human tears. *Invest Ophthalmol Vis Sci.* 1989;30:1879.
11. van Setten GB, Tervo T, Tarkkanen A, Pesonen K, Viinikka L, Perheentupa J. Epidermal growth factor is a constant component of human tear fluid. *Graefes Arch Clin Exp Ophthalmol.* 1989;227:184.
12. van Setten GB. Epidermal growth factor in human tear fluid: Increased release but decreased concentrations during reflex tearing. *Curr Eye Res.* 1990;9:79.
13. Cohen S. Isolation of a mouse submaxillary gland protein accelerating incisor eruption and eyelid opening in the new-born animals. *J Biol Chem.* 1962;237:1955.
14. Tripathi BJ, Kwait PS, Tripathi RC. Corneal growth factors: A new generation of ophthalmic pharmaceuticals. *Cornea.* 1990;9:2.

98

A NOVEL APPROACH TO RESOLVE CELLULAR VOLUME RESPONSES TO AN ANISOTONIC CHALLENGE

Pavel Iserovich,[1] Peter S. Reinach,[2] Hua Yang,[2] and Jorge Fischbarg[1]

[1]Department of Ohpthalmology
College of Physicians and Surgeons
Columbia University
New York, New York
[2]State University of New York College of Optometry
New York, New York

1. INTRODUCTION

The ability of cells to regulate their volume in response to a change in external osmolality can be essential to the maintenance of their normal function. In the case of corneal epithelial cells, changes in tear film osmolarity can occur in daily living because of potential exposure to the drying effect of wind, fresh or salt water, etc. Such changes can lead to alterations in epithelial cell volume that affect cell function. For some cells, the inability to undergo volume regulation contributes to rapid cell death.[1]

We have described that SV40-transformed rabbit corneal epithelial cells (tRCE) subject to osmotic stress undergo a regulatory volume response (RVR).[5,6] We report here on a quantitative method to evaluate relative cell volume transients caused by changes in bath osmolality. The method is applied here to separate and quantify osmotic response and regulatory volume decrease (RVD).

2. MATERIALS AND METHODS

2.1. Cell Culture

We used tRCE cultured in DMEM/F12 as described by Araki-Sasaki et al.[2] The cells were subcultured on 11 x 22 mm rectangular glass coverslips until they reached 70–80% confluence (usually 2 days).

Lacrimal Gland, Tear Film, and Dry Eye Syndromes 2
edited by Sullivan *et al.*, Plenum Press, New York, 1998

2.2. Determination of Cell Volume

We employed the low-angle light-scattering technique of Fischbarg et al.[3] to monitor relative cell volume. All experiments were done at 37°C.

2.3. Solutions

The composition of isotonic medium was (in mM): NaCl, 147; KCl, 4.7; $MgCl_2$, 0.4; $CaCl_2$, 1.8; $MgCl_2$, 0.4; glucose, 5.5; HEPES Na, 5.3. A 25% hypotonic solution (225 mOsM) was achieved by appropriately decreasing the [NaCl] or [cyclamic acid sodium salt] (in the case of a Cl^--free solution). NPPB (Sigma) was diluted from a stock solution (0.1 M in DMSO) to a final concentration of 50 μM.

2.4. The Mathematical Procedure

Experimental data for relative cell volume were fitted to an Exponential Associate function (EAf) using ORIGIN™ software (Microcal, Northampton, MA).

$$z(t) = 1 + A \bullet [1 - \exp(-t/\tau_{osm})] + R \bullet [1 - \exp(-t/\tau_{vr})] \tag{1a}$$

where

$$z = \frac{V - b}{V_0 - b}, \tag{1b}$$

z is the relative cell volume, V and V_0 are the time-transient and initial cell volumes, b is the cell osmotic inactive volume, and τ_{osm} and τ_{vr} are characteristic times. To fit, we optimized the parameters: A, amplitude of osmotic response; R, amplitude of the RVR; τ_{osm} and τ_{vr}, characteristic times of the osmotic and RVR, respectively. Fitting was achieved by minimizing χ^2.

Usually, indirect methods are employed for relative cell volume monitoring, so normalization is necessary. However, this is often impossible to do because either the application of an inhibitor or a change in the cell medium composition produces a shift of the baseline (cell changes equilibrium volume and shape, etc.). During the process of normalization, it is not correct to assume that the maximum relative cell volume increase is equal (in the case of a hypotonic challenge) to the theoretical value expected for an ideal osmometer. This assumption is invalid because the initial osmotic transient includes both osmometric swelling as well as the onset of RVD, as exemplified in Fig. 1. In Equation 1, this complex response is deconvoluted by representing it as the sum of two exponential responses. With this approach, it is possible to account for the confounding factors and apply a meaningful normalization. Because now osmometric and RVR are separated the parameter A (amplitude of osmotic response) can be used as a criterion for normalization by making A equal to the theoretical volume change for an ideal osmometer. The normalized volume $z_n(t)$ is therefore:

$$z_n(t) = 1 + [z(t) - z(0)] \cdot \frac{C_0 - C_f}{A \cdot C_f} \tag{2}$$

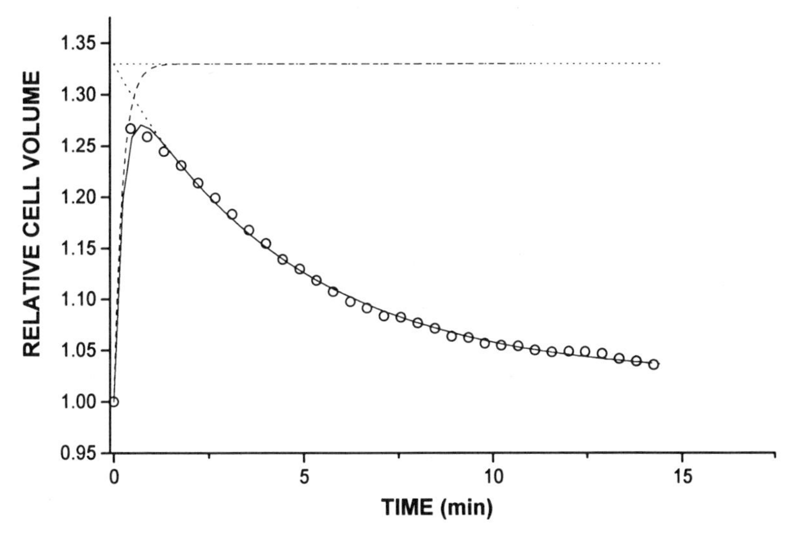

Figure 1. tRCE relative cell volume transient in response to a substitution with a 25% hypotonic solution (o). Two components of the response: osmotic (--) and regulated volume decrease (....). Data normalized to make the amplitude of the osmotic response equal to the theoretical value of 1.33.

where C_0 and C_f are the initial and final osmolarities, and $(C_0\text{-}C_f)/C_f$ is the normalized volume change for an ideal osmometer.

The extent of RVR can be defined as $(R/A)\cdot 100\%$, and the intensity of the RVR and cell osmotic responses can be characterized by evaluating τ_{vr} and τ_{osm}.

To find the relationship of τ_{osm} to the osmotic cell membrane water permeability, we can analyze Equation 6 from[3]

$$\frac{dz}{dt} = k \cdot \left[\frac{1}{z} - \frac{C_f}{C_0} \right] \tag{3}$$

$$k = \frac{P_f \cdot V_w \cdot C_0 \cdot S}{V_0 - b} \tag{4}$$

where P_f is the cell membrane osmotic permeability, V_w is the molar volume of water, and S is the area of the cell membrane. After integration of Equation 3, we have:

$$k \cdot t = \frac{\Delta z \cdot C_0}{C_f} - \left(\frac{C_0}{C_f} \right)^2 \cdot \ln\left(1 - \frac{\Delta z \cdot C_f}{C_0 - C_f} \right) \tag{5}$$

where t is time and $\Delta z = z - 1$. Since the first term on the left side of Equation 5 is less than 0.5, at times larger than k we can neglect this term and simplify Equation 5:

$$k \cdot t = -\left(\frac{C_0}{C_f} \right)^2 \cdot \ln\left(1 - \frac{\Delta z \cdot C_f}{C_0 - C_f} \right) \tag{6}$$

and finally, solving for Δz,

$$\Delta z = \frac{C_0 - C_f}{C_f} \cdot \left[1 - \exp\left(-\left(\frac{C_o}{C_f} \right)^2 k \cdot t \right) \right] \tag{7}$$

Equation 7 is equivalent to the "osmotic" part of Equation 1. We can estimate the cell osmotic permeability using Equations 7, 1, and 4 as

$$P_f = \frac{C_f^{\,2} \cdot (V_0 - b)}{C_0^{\,3} \cdot \tau_{osm} \cdot Vw \cdot A} \tag{8}$$

With regard to the other characteristic time, τ_{vr}, it cannot be simply interpreted because it is a combination of several characteristic times corresponding to the several volume regulatory mechanisms involved.

3. RESULTS

Fig. 1 shows a typical light-scattering measurement prior to and during the hypotonic challenge (shift from 300 to 225 mOsM). The upward deflection is characteristic of, in large part, the osmometric volume increase but could include the initiation of RVD. The reversal and trend towards restoration of the control volume is accounted for by an RVD. The experimental data were fitted and normalized with EAf with very good accuracy.

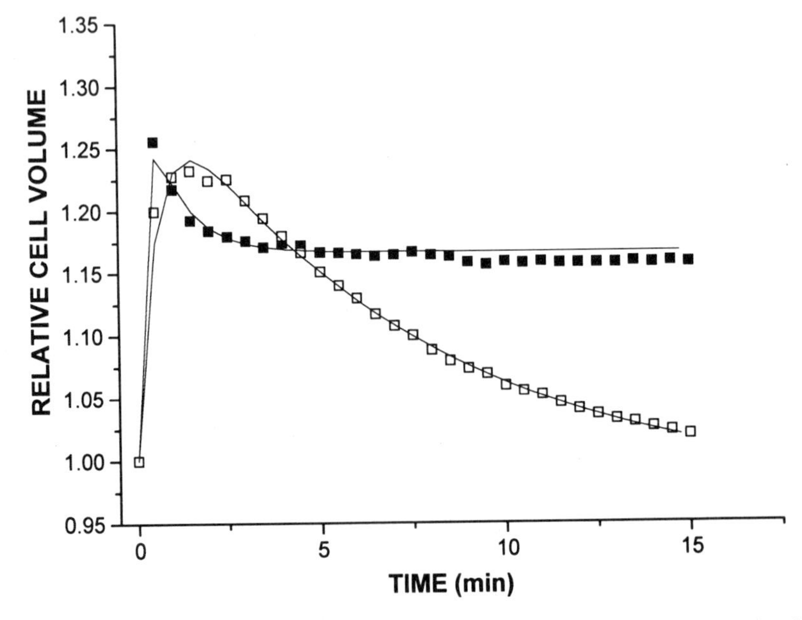

Figure 2. Effect of preincubation with 50 μM NPPB (10 min) on tRCE relative cell volume transient in response to the 25% hypotonic solution challenge. 50 μM NPPB (dark box) ; control (light box). Fitting curves (-). (n=6).

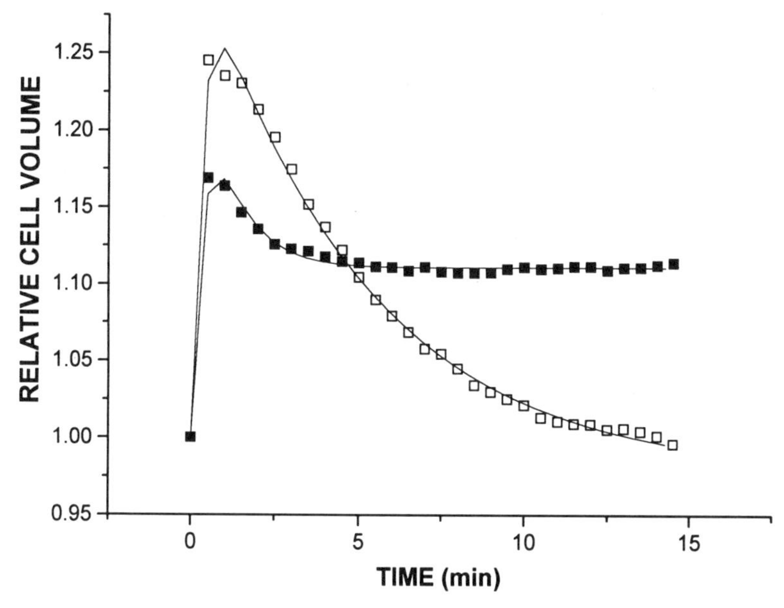

Figure 3. Effect of Cl⁻ substitution with cyclamic acid on tRCE relative cell volume transient in response to the 25% hypotonic solution challenge. Cl⁻-free (dark box) ; control (light box). Fitting curves (-). (*n*=6).

Figs. 2 and 3 show the effects of the putative Cl⁻ channel blocker, NPPB (50 μM) and of the isosmotic replacement of Cl⁻ with cyclamate Ringers, respectively.

Table 1 summarizes the results of the fitting procedure in six experiments.

In both the control and the experimental conditions, we had good fitting accuracy (χ^2 values were small). In addition, the near identity between the effects of Cl⁻ removal and NPPB on RVD validates the idea that NPPB acts as a Cl⁻ channel inhibitor in tRCE cells.

4. DISCUSSION

The relevance of our approach for characterizing RVR is substantiated by the close agreement between the extracted values of the parameters used to characterize the effect of NPPB and Cl⁻-free medium on the light-scattering response to hyposmolality. In both cases, the extent of RVD recovery decreased to ≈ 63% with near identical characteristic times. This agreement is in accord with current indications that NPPB is an inhibitor of Cl⁻

Table 1. Characterization of RVD responses

Parameter	Control	Cl⁻-free	NPPB
χ^2	3 x 10^{-5}	3 x 10^{-5}	2 x 10^{-5}
RVD (%)	108	60	67
τ_{vr} (min)	3.7	0.68	0.98
τ_{osm} (min)	0.49	0.22	0.35

RVD, regulatory volume decrease.

channel activity in a variety of cells. Furthermore, this agreement allows us to suggest that the tRCE cells have two stretch-activated anion channels: one Cl^- selective channel with a low threshold, and another, less selective, anion channel that is permeable for larger organic anions. The latter pathway has a higher stretch-activated threshold whose closing occurs at ≈ 37% cell volume increase. In addition, from the decrease in the values of τ_{vr} we can conclude that there are a large number of those channels in tRCE cells.

The change in τ_{osm} during exposure to either 50 μM NPPB or Cl^- replacement can be interpreted as resulting from increases in cell membrane water permeability. An alternative and more likely explanation is that Cl^- removal caused shrinkage of the cell volume. Additional experiments are needed to distinguish between these alternatives.

It is readily evident from an examination of Equation 1 and Fig. 1 that it would be erroneous to evaluate the extent of the RVD based on an evaluation of the $t_{1/2}$ associated with the RVD response (as is often found in the literature[4]). Instead the evaluation should take into account τ_{vr} and R-to-A ratio.

5. CONCLUSIONS

1. This approach enables a quantitative characterization of the cells' ability to mediate volume regulatory responses to osmotic stress.
2. The experimental data can be fitted with an Exponential Associate function with high accuracy, and can be accurately normalized.
3. In the given example, the contribution of Cl^- channel activity to regulatory volume decrease behavior is more meaningfully quantified as it is isolated from the initial swelling response to hypotonicity.

ACKNOWLEDGMENTS

This study was supported by NIH grants EY06178 and EY04795.

REFERENCES

1. Bortner CD, Cidlowski JA. Absence of volume regulatory mechanisms contributes to the rapid activation of apoptosis in thymocytes. *Am J Physiol.* 1996;271:C950-C961.
2. Araki-Sasaki K, Ohashi Y, Sasabe T. (1993) Immortalization of rabbit corneal epithelial cells by a recombinant SV40-adenovirus vector. *Invest Ophthalmol Vis Sci.* 34:2665–71.
3. Fischbarg J, Li J, Kuang K, Echevarría M, Iserovich P. Determination of volume and water permeability of plated cells from measurements of light scattering. *Am J Physiol.* 1993;265:C1412-C1423.
4. Cornet M, Lambert IH, Hoffmann EK. Relation between cytoskeleton, hypo-osmotic treatment and volume regulation in Ehrlich ascites tumor cells. *J Membr Biol.* 1993;131:55–66.
5. Iserovich P, Wu X, Yang H, Reinach P. Volume regulation by SV40 transformed rabbit corneal epithelial cells. ARVO Abstracts. *Invest Ophthalmol Vis Sci.* 1995;36:S585.
6. Reinach P, Yang H, Araki-Sasaki, and Fischbarg J. Ion transport pathways mediating volume regulatory responses in SV-40 transformed rabbit corneal epithelial cells. ARVO Abstracts. *Invest Ophthalmol Vis Sci.* 1996;37:S1110.

CLASSIFICATION OF ARTIFICIAL TEARS

I: Composition and Properties

J. Murube, A. Paterson, and E. Murube

Department of Ophthalmology
Hospital Ramon y Cajal
Madrid, Spain

Natural tears serve three main functions: they maintain the metabolism of the ocular surface tissues, afford a smooth surface that allows regular light refraction, and lubricate the ocular surface to facilitate blinking. They also have other functions, such as specific and non-specific antimicrobial and buffer effects, etc. Natural tears have a complex composition, water being the major component (98–98.5%), as well as salts, hydrocarbons, proteins, and lipids. There are two types of tear components: structural and active. The structural components of tears are rather abundant, and are the basis of the three-layered architecture of the tear film (lipid, aqueous, and mucinic) because of their physical and chemical properties. The active components of tears are very scarce, and act as nutrients, antimicrobials, enzymes, messengers, inhibitors, etc. Some tear components may serve both types of functions.

As a tear substitute, artificial tears cannot reproduce the natural components exactly, because the structural and active components break down quickly and are not well known. Furthermore, natural tear production is virtually constant, with *nyctemeral* and circumstantial fluctuations, but artificial tears are delivered intermittently. Therefore, not only must they reproduce the metabolic, optical, and physical characteristics of natural tears, but they must also possess additional characteristics. For example, they must last longer than normal tears in the *lacrimal basin*. Even more importantly, artificial tears should provide therapeutic additives to treat primary or secondary damage to the dry eye.

Artificial tears are currently classified as *demulcents* (gums, polyvinyl derivatives, etc.) and emollients (lanolin, vaseline, etc.). This classification was taken chiefly from the effects of the ingredients when used in digestive medicine and dermatology. Mucilaginous gels are *emulcent* when used on the skin and intestinal mucosa, but are mainly *viscosant* and emulsive when used in artificial tears. Lanolin is a dermatological and intestinal emollient, but is mainly a lubricant when used in dry eye. Therefore, the classification as *emulcents* and emollients is useless in *dacryology*.

Lacrimal Gland, Tear Film, and Dry Eye Syndromes 2
edited by Sullivan *et al.*, Plenum Press, New York, 1998

Here we present the most logical classifications of artificial tears in order to facilitate their understanding for clinical use. These classifications are based mainly on chemical composition and biological effects.

Classification by chemical composition includes (1) water, (2) saline solutions, (3) glycerol, monosaccharides, disaccharides, (4) polysaccharides, (5) synthetic polymers, (6) gelatins, (7) biological fluids, and (8) lipids.

Classification by biophysical and biochemical properties includes (1) osmotic pressure, (2) surface tension, (3) lubricity/viscosity, and (4) evaporation.

1. CLASSIFICATION BY CHEMICAL COMPOSITION

1.1. Water

Water makes up 98–98.5% of natural tear.* Since we now know that the lacrimal fluid is secreted continuosly, and its fragmentation into drops is something accidental and occasional, we believe that in the scientific world, when referring to the lacrimal fluid in a generic form, we should say tear and not tears, as we say urine and saliva, and not urines and salivas. In dry-eye drops, water is 97–99%.

1.2. Saline Solutions

These are water solutions of electrolytes, designed to maintain the osmolarity of the nascent aqueous tear (approximately 300 mOsm/l) and the precorneal tear film (approximately 303–310 mOsm/l), or to lower the osmolarity without causing discomfort. Some electrolytes, besides their osmotic purpose, play an important role in corneal epithelium metabolism (K^+, HCO_3^-) and buffer system (HCO_3^-, PO_4^{3-}).

The most widely used electrolytes (salts) are:

Sodium chloride: *Aquasite, Bion Tears, Cellufresh, Celluvisc, Chlorure de sodium 9%, Colircusí Humectante, Dacriosol, Dacrolux, Dropstar TG, Dry Eye Therapy, Dulcilarmes, Hyalistil, Lacril, Lacrimalfa, Lacrisifi, Larmabak, Larmes artificielles, Liquifilm Lágrimas, Liquifilm Tears, Minims Artificial Tears, Ocustil, Refresh, Unilarm, Vidisept N.*

Potassium chloride: *Bion Tears, Cellufresh, Celluvisc, Dacriosol, Dacrolux, Dakrina, Dry Eye Therapy, Dwelle, Estivin II, Hyalistil, Murine Plus, Tears Naturale, Tears Naturale II, Tears Naturale Free.*

Calcium chloride: *Bion Tears, Cellufresh, Celluvisc, Dakrina, Dry Eye Therapy, Dwelle, Lacril.*

Magnesium chloride: *Bion Tears, Cellufresh, Dakrina, Dry Eye Therapy, Dwelle, Lacril.*

Zinc chloride: *Bion Tears, Dry Eye Therapy.*

Sodium bicarbonate: Used to buffer the solutions, but also has an electrolytic effect. *Bion Tears, Lacrimalfa.*

* Since prehistoric times until two centuries ago, humans noticed the tear fluid only when it spills from the eye in drops. So, they gave the fluid a plural name, tears.

Sodium phosphate: *Dry Eye Therapy.*
Sodium phosphate monobasic: *Colircusí Humectante, Dropstar TG, Hyalistil, Ialurex Ipotonico, Lacrimalfa, Lacrisifi, Liquifilm Lágrimas, Liquifilm Tears, Vidisept N.*
Sodium phosphate dibasic: *Colircusí Humectante, Dropstar TG, Hyalistil, Ialurex Ipotonico, Lacrimalfa, Lacrisifi, Larmabak, Liquifilm Lágrimas, Liquifilm Tears, Ocustil, Vidisept N.*
Dipotassium phosphate: *Artelac, Siccaprotect.*
Monopotassium phosphate: *Artelac, Siccaprotect.*
Sodium thiosulfate: Used to protect the active substances from oxidation. *Prefrin, Relief.*
Magnesium sulfate: *Lacrimalfa.*
Calcium carbonate: *DEO ointment.*
Sodium borate: *Cilclar, Lephagel, Oculotect, Tears Naturale Free.*
Sodium citrate: *Dry Eye Therapy.*
Sodium lactate: Used as a nutrient of the corneal and conjunctival epithelium, but also increases osmolarity. *Cellufresh, Celluvisc.*
Boric acid: Used as a preservative, but decreases pH and increases osmolarity of the solution. *Cilclar Oculotect, Protaget SE.*
Hydrochloric acid: Used to decrease pH and increase osmolarity. *Bion Tears, Cellufresh, Refresh, Tears Naturale Free.*
Sodium hydroxide: Used to adjust pH, but increases osmolarity. *Aquasite, Bion Tears, Cellufresh, Lacryvisc, Refresh, Tears Naturale Free, Viscotears, Viscotirs Gel.*

1.3. Glycerol, Monosaccharides, Disaccharides

Glycerol: Used as an emollient and *emulcent* for the skin and intestinal tract; also used as a humectant in dry eye. *Dry Eye Therapy, Moisture Drops.*
Sucrose: Used as a surfactant.
Dextrose: Used as an osmolyte and nutrient. *Hypotears PF, Hypo Tears, Lacril, Murine Plus.*
Sorbitol: Acts as an isotonizing agent. *Artelac, Lacrinorm, Siccagel, Vidisic, Viscotears, Viscotirs Gel.*
Mannitol: Isomeric with sorbitol, used as a lubricant. *Lacryvisc, Thilo-Tears, Dacriogel.*

1.4. Polysaccharides

These include mucilages (gums), dextrans, and mucopolysaccharides.

1.4.1. Mucilages. Mucilages are true or colloidal solutions with viscous and adhesive properties, made of cellulose or vegetable gums. They contain polysaccharides, usually bound to uronic acids. The cellulose mucilages are used extensively in dry eye applications since Swan[1] reported that methylcellulose increases the residence time of artificial tears.

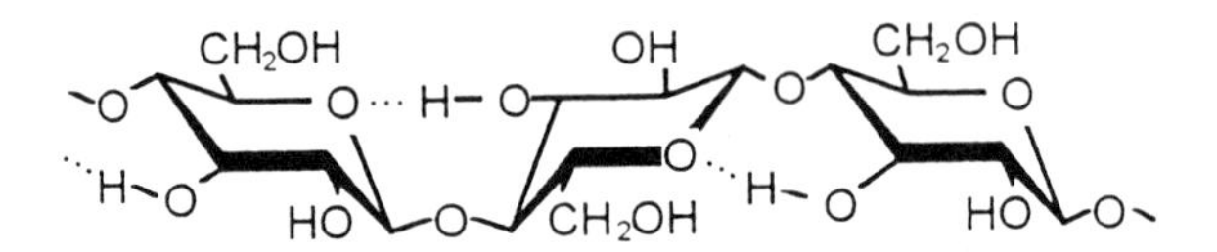

Figure 1. Structural formula of cellulose.

Mucilages considerably increase the viscosity of tears, even at low concentrations (0.1–0.4%), and also provide moderate adsorptive properties to the corneal epithelium. They have little effect on surface tension, little influence on osmotic pressure, and moderate oncotic properties due to the large molecular size.

The most widely used mucilages are cellulose derivatives. The structural formula of cellulose is seen in Figure 1.

Methylcellulose is the methyl derivative of cellulose. The refractive index of the 1% solution is 1.336, very similar to that of natural tear. Table 1 shows the large increase in viscosity of methylcellulose solutions when the concentration is moderately increased.

Methylcellulose is a component of *Lacril, Lacrimart, M-Cel-4, Murocel.*

Hydroxyethyl cellulose is a component of *Barnes-Hind Wetting and Soaking Solution, Degest 2, Minims Artificial Tears, TearGard.*

Hydroxypropyl cellulose: *Lacrisert.*

Hydroxypropyl methylcellulose (Hypromellose): *Artelac, Artelac EDO, Bion Tears, BJ6, Colircusí Humectante, Contactol, Dacriosol, Dacrolux, Duratears (Neth.), Hidroxipropilmetilcelulosa, Hydroxypropyl Methylcellulose Ophthalmic Solution, Hypromellose Eye Drops, Ilube, Isopto Alkaline, Isopto Flucon, Isopto Fluid, Isopto Naturale, Isopto Plain, Isopto Tears, Lacril, Lacrimil, Lacrisifi, Lacrisol, Methocel, Methopt, Metildrops, Moisture Drops, Murine Plus, Neo-Lacrim, Oculotect, Ocutears, Spersatears, Tears Naturale, Tears Naturale Free, Vidisept N, Ultra-Tears, Viscotraan.*

Carboxymethylcellulose (Carmellose): *Celluvisc, Cellufresh.* (Also a component of some artificial salivas.)

Other mucilages that are not cellulose derivatives are used as thickening and suspending agents in fields other than ophthalmology, and are currently under investigation for use as artificial tears. These include the following:

Table 1. Viscosity and surface tension changes of methylcellulose solutions with increasing concentration

Concentration, aqueous solution at 37°C (%)	Viscosity mP (millipoise)	Surface tension (dynes/cm)	Half life in tear film (sec)
0.10	10.1	62.0	52
0.25	10.5	60.0	75
0.33	44.5	58.5	135
0.50	86.6	55.5	155
0.75	700.0	48.5	210
1.00	980.0	46.0	235
2.00	20,600.0	45.0	260

From Aguilar.[2]

- Alginates are mixtures of polyuronic acids extracted from algae. Propylene glycol alginate and sodium alginate are used as thickening and suspending agents in dermal creams and pastes.
- Xanthan gum is a vegetal gum of high molecular weight polysaccharide containing D-glucose, D-mannose, and D-glucuronic acid.
- Guar gum is obtained from the seeds of the leguminous *Cyamopsis psoraloides*, contains a *galactomannan*, and is used in lubricants and as dietary fiber (*Lubo*, *Guarina*).

1.4.2. Dextrans. Dextran (poly-D-glucose) is a branched glucose polymer, slightly acid. Dextran 70 in a 6% solution exerts a high oncotic pressure similar to that of plasma proteins. Dextrans are highly allergenic. The structural formula of dextran is seen in Figure 2.

Dextran 85, 70, 60, or 40 are components of *Bion Tears*, *Dacriosol*, *Dacrolux*, *Dialens*, *Duratears (Neth.)*, *Dwelle*, *Isopto Naturale*, *Moisture Drops*, *Ocutears*, *Tears Naturale*, and *Tears Naturale Free*.

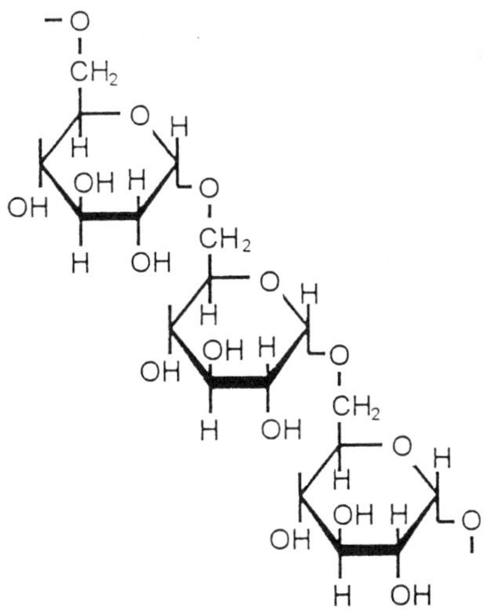

Figure 2. Structural formula of dextran.

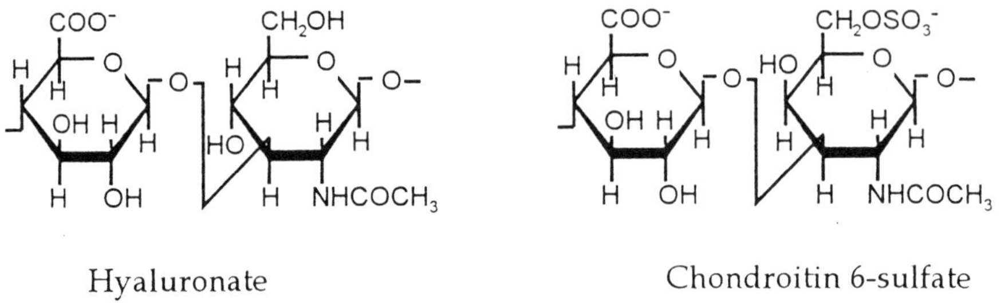

Hyaluronate Chondroitin 6-sulfate

Figure 3. Structural formulas of the repeating disaccharide units of hyaluronate and chondroitin 6-sulfate.

1.4.3. Mucopolysaccharides. Examples are sodium hyaluronate, used in ophthalmic solution 0.2% (*Dropstar TG, Hyalistil, Ialurex Ipotonico, Ocustil*), and sodium chondroitin sulfate A 3% (*Lacrypos*). The structural formulas of the repeating disaccharide units of hyaluronate and chondroitin 6-sulfate are seen in Figure 3.

1.5. Synthetic Polymers

Several water-soluble artificial polymers began to be used in artificial tears in the 60's. They are basically vinyl derivatives or polyethylene glycol, forming a backbone to which different functional groups may be attached to confer the desired mucomimetic properties, i.e., water solubility, good oncotic pressure, good surfactant, good adsorption to the epithelial surface (making it hydrophilic), and tear-film stabilizer.

1.5.1. Vinyl Derivatives. Vinyl is an aliphatic chain with two unsaturated carbons that can polymerize. When a hydrogen atom is substituted by a hydroxylic (-OH), pyrrolidonic (-NOC_4H_6), carboxylic (-COOH), or chloride (Cl^-) group, the results are, respectively, polyvinyl alcohol, polyvinyl pyrrolidone, polyvinyl (or polyacrylic) acid, and polyvinyl chloride.

Polyvinyl alcohol. At a concentration of 1.4%, the alcohol's viscosity is low, but it remains in the *lacrimal basin* for 30 min because it is quite adsorptive. It is included in many drugs for dry eye disorders and contact lens lubricants, such as *Contafilm, Dakrina, Dispatenol, Dwelle, Hypotears, Hypo Tears, Lacrilux, Liquifilm, Liquifilm lágrimas, Liquifresh, Murine, Murine Plus, PMS Artificial Tears, PMS Artificial Tears Plus, Refresh, Siccaprotect, Sno Tears, Teardrops, Tears Plus, Total, Uligin.*

Polyvinyl pyrrolidone (povidone) is a synthetic polymer of 1-vinylpyrrolidin-2-one that works as a non-ionic surfactant. A 3–5% concentration is necessary to increase the

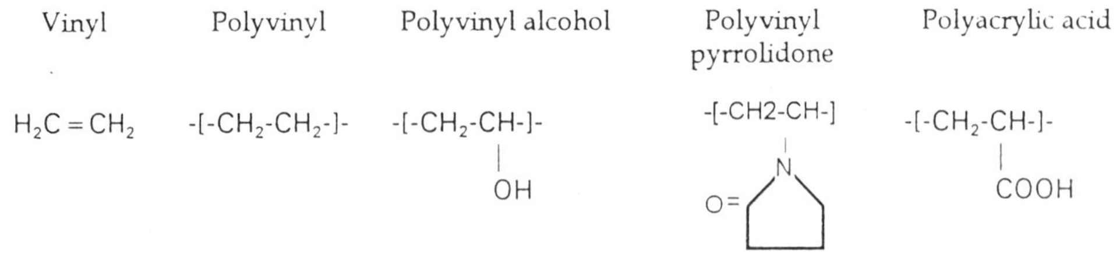

Vinyl	Polyvinyl	Polyvinyl alcohol	Polyvinyl pyrrolidone	Polyacrylic acid
$H_2C=CH_2$	-[-CH_2-CH_2-]-	-[-CH_2-CH-]- (OH)	-[-CH2-CH-] (N-pyrrolidone, O=)	-[-CH_2-CH-]- (COOH)

Figure 4. Vinyl derivatives.

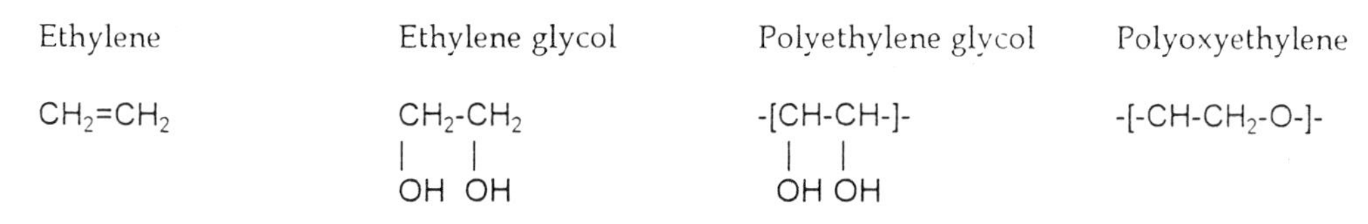

Figure 5. Ethlene glycol derivatives.

viscosity of the solutions. It is a component of artificial tears and contact lens wetting solutions, such as *Adapettes, Barnes-Hind Wetting and Soaking Solution, Bausch & Lomb Sensitive Eyes Lens Lubricant, Bolinan, Dakrina, Dulcilarmes, Murine, Murine Plus, PMS Artificial Tears Plus, Protagens, Protagent SE, Refresh, Teardrops, Tears Plus, Vidisept N.*

Polyacrylic acid. Carbomer is a polymerized carboxyvinylic acid. The most widely used in ophthalmic hydrogels are Carbopol 934 P and Carbopol 940 P. The acrylic acid of carbomer is cross-linked with polyalkenyl ethers of sugars or polyalcohols. It contains 56–68% of carboxylic groups (-COOH).

Carbomers are viscosant agents used in *Dacriogel, Lacrinorm, Lacryvisc, Thilo-Tears, Vidisic, Vidisic EDO, Viscotears,* and *Viscotirs Gel. Siccafluid Gel* is a carbomer without *bencens*, because *bencens* are toxic.

Polycarbophils are polyacrylic acid cross-linked with divinyl glycol. They are used for artificial tears (*Aquasite*), vaginal dryness (*Replens*), and as an intestinal emollient (*Fiberall, Equalactin*).

1.5.2. Ethylene Glycol Derivatives. Polyethylene glycols (macrogols) are condensation polymers of ethylene glycols. As their molecular weight increases, their viscosity increases and their hygroscopicity decreases. They are good non-ionic surfactants and bioadhesives. Macrogol 8000 is a component of *Hypotears*.

Polyoxyethylene-polypropylene glycol (poloxamers, pluronics) are copolymers used as non-ionic surfactants, contact lens cleaners and wetters (*Cilclar, Clerz, Lepha Gel, Pliagel*). They are under investigation for use as artificial tears.

1.5.3. Other Synthetic Polymers. Polysorbates are mixtures of fatty acid esters of sorbitol and its anhydrides copolymerized with approximately 20 moles of ethylene oxide for each mole of sorbitol and its anhydrides. Polysorbate is a non-ionic surfactant macromolecule that decreases surface tension in some lid cleaners and artificial tears, such as cilclar and *Tears Encore*.

Tyloxapol is a polymer of 4-(1,1,3,3-tetramethylbutyl)phenol with ethylene oxide and formaldehyde with non-ionic surfactant and mucolytic properties, used in *Contactol* and *Enuclene*.

1.6. Gelatins

Gelatins are proteins extracted from animal collagen; they contain abundant lysine and little tryptophan. They were used as artificial tears by Gifford et al.[3] over 50 years ago, because in a 5–7% isoosmotic solution they are good surfactant artificial tears. Today, gelatin type A is an ingredient of *Lacril*.

1.7. Biological Fluids

Biological fluids, such as autologous serum, colostrum, saliva, egg whites, and mucins are not commercialized. Mucins are high molecular weight glycoproteins, so they form good colloidal solutions. They decrease surface tension (wetters) and increase viscosity moderately (lubricants). Bovine submaxillary mucin has been used experimentally. Gastric mucin from hog stomach is used in dry mouth patients as artificial saliva (*Saliva Orthana*).

1.8. Lipids

Lipids are organic substances formed by fatty acid esters with glycerol, cholesterol, etc., insoluble in water and soluble in organic solvents (alcohol, ether, benzene, acetone, chloroform). As dry-eye drugs, waxes, which are esters of fatty acids with long-chain alcohols, are frequently used. The most common are petrolatum (paraffin, vaseline, mineral oil) and lanolin. When used for dry eye, they act as a tear lubricant/*viscosant*, but secondarily contribute to creating the *foremost* lipid layer of the tear film. They are formulated as ointments. Eye drops containing lipids have been used only experimentally.

Lanolin:	*Dacriosol, Duratears, Duratears Z, Lacrilube, Lacrilube E.F.P., Lubrifilm Pomada, Tears Gel.*
Paraffin:	A wax-like methylene hydrocarbon present in *Dacriosol ointment, Duolube, Duratears, Duratears Z, Lacri-Lube, Lacrilube, Lacrilube E.F.P., Lubrifilm Pomada, Tears Gel.*
Lecithin is used as a non-ionic surfactant and nutrient in *TearGard*.	
Cholesterol:	Cholesterol 0.05% is being used experimentally with an aqueous solution of 20% oligosaccharide 2-hydroxypropyl-β-cyclodextrin to create a tear substitute that mimics the three main components of natural tears.[4]

2. CLASSIFICATION BY BIOPHYSICAL AND BIOCHEMICAL PROPERTIES

2.1. Osmotic Pressure

When two solutions of different concentrations of a solute (in natural and artificial tears, the solutes are mainly ions and crystalloids) are separated by a membrane (corneal epithelium) that is, for example, permeable to the solvent (aqueous tear) but impermeable to the solutes, there is a net diffusion of the solvent across the membrane towards the side with the higher concentration. This phenomenon is called osmosis, and the pressure to balance this osmotic flow is called osmotic pressure.

Colloidosmotic, or oncotic, pressure is the fraction of the osmotic pressure due to the presence of macromolecules where membranes allow the passage of only small molecules. In natural tear, only a small part of the osmotic pressure is due to oncotic pressure.

The oncotic pressure of natural tear is about 2mmHg, higher than that of the aqueous humor (0.1 mmHg) and smaller than that of plasma (25 mmHg).

The osmolarity of human serum is 290 mOsm/l, equivalent to a 0.9% NaCl aqueous solution. The osmolarity of normal human tear is about 300 mOsm/l when recently produced, and 305–310 mOsm/l in the tear film. NaCl is the most important electrolyte in natural and artificial tears, but K is also necessary as a nutrient, because it plays a role in the metabolism of the corneal epithelium.

Many general collyria and even dry-eye drops (such as *Vidisept N*, which is 320 mOsm/l) are slightly hyperosmolar.

Isoosmolar artificial tears are the most common. The first artificial tears, Cantonnet's *larmes artificielles*,[5] were a 1.4% aqueous NaCl or 2.5% boric acid solution, which tried to maintain the usual osmolarity of the tear film. Most eye drops are serum-isotonic or tear-isotonic: *Adsorbo Tears, Celluvisc, Comfort Tears, Dakrina Dwelle, Isopto Tears, Just Tears, Lacril, Liquifilm Forte, Moisture Drops, Murine (310 mOsm/kg), Neo Tears, Nutra Tear, Refresh, Tears Naturale (290 mOsm/kg), Tears Naturale II (290 mOsm/kg), Tears Plus, Ultra Tears.*

Hypoosmolar artificial tears decrease the osmolarity of the fluid in the lacrimal basin. Mastman et al.[6] reported that dry eye tears are hyperosmolar, and that, in general, the more severe the disease, the higher the tear osmotic pressure. This plays an important disturbing role in dry eye, attracting water from the corneal epithelial cells and interfering with their metabolism, decreasing their vitality, and reducing their microvilli and glycocalyx and mucin layer.

Some artificial tears are hypotonic in order to dilute the hyperosmolar tear film and return it to the osmolarity of normal tears: *Liquifilm, TearGard, Vit-A-Drops. Hypotears* and *Hypotears P.F.* have an osmolarity of 210–230 mOsm/l. A hypoosmolarity of 150 mOsm/l (as Ialurex Ipotonico) is subjectively well accepted by patients, but 75 mOsm/l produces irritation of the eye.[7] The absolute hypoosmolarity (0 mOsm/l) is *distilled water*, which causes itching and swelling of the epithelium. The dilution effect of hypotonic tears lasts only about 20 sec in a normal eye.[8]

The colloidosmotic fraction of artificial tears must be higher than in natural tear, because if the corneal epithelium is damaged (a frequent condition in dry eyes), it allows the passage of many electrolytes but no complex molecules. Therefore, some artificial tears achieve suitable osmotic pressure by decreasing salts and adding dextrose (*Hypotears PF, Lacril, Murine Plus*) or sorbitol (*Vidisic*).

Other artificial tears achieve an oncotic pressure of 40 mmHg or more by including a high concentration of high molecular weight synthetic polymers. For cellulose ethers and synthetic polymers, the necessary concentration results in excessive viscosity. The dextran content in *Dwelle* artificial tears results in high oncotic pressure (65 mmHg) while maintaining low viscosity (27 mP).

Dakrina, Dwelle, and *Hypo Tears* have high oncotic pressure. *Adsorbo Tear* and *Liquifilm Forte* have moderate oncotic pressure. *Liquifilm, Moisture Drops, Nutra Tear, Refresh, Tears Naturale, Tears Naturale II, Tears Naturale Free, TearGard, Tears Plus,* and *Ultra Tears* have low oncotic pressure.

2.2. Surface Tension

When two phases (e.g., liquid-solid, liquid-gas) are in contact with each other, there is a transition region called the interface. The molecules located within the bulk of each phase are, on average, subject to equal forces of attraction in all directions, whereas molecules located at the interface experience imbalanced attractive forces. These short-range intermolecular forces, which include van der Waals forces (in particular, London dispersion forces) and may include hydrogen bonding, are responsible for surface/interfacial tension.

The surface tension of water at 32ºC (the temperature of the corneal surface) is 70.85 dynes/cm, of natural in vitro tears 40–45 dynes/cm, and of natural in situ tears 38 dynes/cm.

When a liquid spreads on a solid, we can distinguish two situations: a partial wetting situation where the liquid remains as a drop with a definite contact angle with respect to the solid surface, and a total wetting situation where the liquid spreads, tending to cover the solid completely. To characterize the wettability of solid surfaces, a useful parameter is the critical surface tension:[9] liquids with a surface tension below the critical surface tension of the solid will present total wetting behavior.

Whereas the surface tension of water is 70.85 dynes/cm and the critical surface tension of the corneal epithelium is 28 dynes/cm,[10] so water on this surface will adopt a drop shape. Natural tears contain glycoproteins and other surfactant macromolecules that decrease the surface tension of water to about 38 dynes/cm. On the other hand, the corneal surface is covered by mucins that increase its critical surface tension to about 38 dynes/cm. The corneal-tear system is therefore close to a total wetting situation. Also contributing to the stability of the tear film, the *foremost* lipid layer of the tear film applies a film pressure over the aqueous tear layer, avoiding the formation of dry spots. This film pressure is 10–12 dynes/cm when the lipid layer is about 160 nm thick.[10] The lipid layer of the tear film contains 90% low polarity lipids, including mainly wax and cholesterol esters and traces of triglycerides, and 10% high polarity lipids, including free fatty acids, aliphatic alcohols, glycolipids, and small amounts of lecithin and other phospholipids. Table 2 lists the surface tension of various artificial tears consitiuents.

In artificial tears, non-ionic synthetic surfactants such as polyvinyl alcohol, polyvinyl pyrrolidone, macrogol, poloxamers, and tyloxapol are used. Other non-ionic surfactants are lecithin (in *Tear Gard*) and polysorbate (in *Tears Encore*). Ionic surfactants, although very active, tend to disperse the lipid layer of the tear film. Some quaternary ammonium antiseptics, such as benzalkonium chloride (in *Tearisol*), cetrimonium chloride (in *Vidisept N*), and cetrimide (in *Artelac)*, are used as cationic surfactants.

Table 2. Surface tension of various artificial tears constituents

Artificial tears constituent	Concentration in water (%)	Surface tension (dynes/cm)
Sodium chloride	0.9	72.2
Gelatin	0.5	54
Methylcellulose	1.0	64
Carboxymethylcellulose	1.0	66
Hydroxyethylcellulose	1.0	60
Hydroxypropylcellulose	1.0	43
Hydroxypropyl methylcellulose	1.0	47
Polyvinyl alcohol	1.4	47
Polyvinyl pyrrolidone	1.0	68
Polyvinylmethyl ether	1.0	42
Bovine submaxillary mucin	0.5	38
Benzalkonium chloride	0.004	39
Chlorobutanol	Saturated	59
Thiomersal	0.004	72.6

Low molecular weight alcohols, such as glycerol (in *Moisture Drops*), and sugars, such as sucrose, are used as humectants.

An appropriate relation must exist between the surface tension of tear and the critical surface tension of the corneal epithelium, since molecules can be adsorbed at the corneal surface. This changes the critical surface tension of the solid, threatening the film stability. However, the complexity of the tear film stability problem depends not only on wetting conditions but also on the adsorption/desorption of molecules on the cornea and bioadhesivity. Some examples of molecular adsorption are the lacrimal mucin adsorbed at the epithelial mucin (glycocalyx) and the macromolecular polymers of artificial tears adsorbed at the corneal epithelium.

An optimum water content exists for maximum bioadhesion.[11] Anionic polymers, such as acrylic acid polymers, have better bioadhesivity than cationic and neutral ones: maximum bioadhesivity is observed at pH 5–6. Adhesivity increases with enhancement of the charge density and simplification of the polymer structure.

2.3. Lubricity/Viscosity

Lubricity and viscosity are not synonymous concepts, but they can be included under the same heading for practical purposes. Lubrication is the creation of a low-viscosity layer between two solid surfaces in order to help them slide against each other. Lubricant tears and ointments are intended to decrease the friction of blinking. Any type of fluid will have a lubricating effect, but mild and moderate *viscosant* tear agents are best for this purpose, as their viscosity also prevents the mechanical removal of the tear film.

Viscosity is related to the internal friction of a fluid, i.e., the frictional force between different layers moving with different velocities. Viscosity is measured in poises (P). The viscosity of water at 32–33ºC (the normal temperature of the anterior surface of the cornea and tear film) is 7.6 mP. The viscosity of normal tears at 32–33ºC is 9 mP. Any collyrium with viscosity above these values may be considered a hyperviscous collyrium because of its increased resistance to elimination due to blinking with respect to natural tear. Isoviscous and mild and moderate hyperviscous collyria are the best ocular lubricants. Highly viscous collyria, though they remain in the *lacrimal basin* for a long time, are not good lubricants because they make blinking difficult and form lumps that disturb vision. Solutions with viscosities above 1000 mP do not mix with natural tear.

Many collyria used for purposes other than artificial tears, such as antibiotic drops, decongestive drops, etc., have a viscosity of 10–15 mP at 32ºC, i.e., 1–6 mP higher than normal tear. They are therefore *hyperviscosant* but do not clinically affect blinking lubrication. Some components of artificial tears are deliberately *viscosants*, such as mucilage solutions and lipids. Other hyperviscous components are used not for their *viscosant* properties but for their surfactant, adsorbent, nutrient, etc. properties. In these cases, viscosity is prevented from increasing too much so that their lubricant effects are not reduced. Some examples are: synthetic polymers (polyvinyls, polyacrylic acid, polyethylene glycols, macrogols, alginates, xanthans, *galactomannans*, dexpanthenol, etc).

Some sample viscosity values are: *Refresh*, 10 mP; *Liquifilm* and *Nutra Tear*, 13 mP; *Dwelle*, *Hypo Tears*, and *Hypo Tears PF*, 26 mP; *Dakrina*, 27 mP; *Cellufresh*, 30 mP; *Tears Naturale II*, *Tears Naturale Free,* and *Tears Plus*, 32 mP; *Tears Naturale*, 44 mP; *Celluvisc*, 2900 mP.

2.4. Evaporation

The evaporation rate of a normal tear film is about 0.1 μl/min, i.e., about 8% of the normal tear production rate (1.2 μl/min). The evaporation rate of the tear film of a dry eye patient is twice that in normal eyes.[12]

The lipid layer does not evaporate at biological temperatures because of its low vapor pressure. Hence, a continuous lipid layer over the aqueous tear film decreases water evaporation. Abnormality of the lipid layer has proven to be one of the most important causes of dry eyes.[13]

Lipid drops have been used since ancient times as a lubricant in xerophthalmos. Now, petrolatum and lanolin are used for the same purpose, but also to rebuild the lipid layer.

REFERENCES

1. Swan KC. Use of methylcellulose in ophthalmology. *Arch Ophthalmol.* 1945;33:378–381.
2. Aguilar Bartolomé JM. El empleo de la metilcelulosa en oftalmología. *Arch Soc Oftalmol Hisp-Amer (Arch Soc Españ Oftalmol).* 1957;17:203–214.
3. Gifford S, Puntenney I, Bellows J. Keratoconjunctivitis sicca. *Arch Ophthalmol.* 1943;30:207–216.
4. Guömundsson OG, Thórisdóttir S, Loftsson T, et al. 2-Hydroxypropyl-β-cyclodextrin (HPβCD) and cholesterol tear substitute. ARVO Abstracts. *Invest Ophthalmol Vis Sci.* 1994;35:S1694.
5. Cantonnet A. Formules de collyres isotoniques aux larmes. *Arch Ophtalmol (Paris).* 1908;28:617–621.
6. Mastman GJ, Baldes EJ, Henderson JW. The total osmotic pressure of tears in normal and various pathologic conditions. *Arch Ophthalmol.* 1961;65:509–513.
7. Gilbard JP, Kenyon KR. Tear diluents in the treatment of keratoconjunctivitis sicca. *Ophthalmology.* 1985;92:646–650.
8. Holly FJ, Lamberts DW. Effect of nonisotonic solutions on tear osmolarity of tear substitutes. *Invest Ophthalmol Vis Sci.* 1981;20:236–245.
9. Zisman WA. Relation of the equilibrium contact angle to liquid and solid constitution. *ACS Adv Chem Series.* 1964;43:1–51.
10. Holly FJ, Lemp MA. Wettability and wetting of corneal epithelium. *Exp Ey Res.* 1971;11:239–250.
11. Chen JL, Cyr GN. In: Manly RS, ed. *Adhesion in Biological Systems.* New York: Academic Press; 1970:163.
12. Rolando M, Refojo MF, Kenyon KR. Increased tear evaporation in eyes with keratoconjunctivitis sicca. *Arch Ophthalmol.* 1983;101:557–558.
13. Shimazaki J, Sakata M, Tsubota K. Ocular surface changes and discomfort in patients with meibomian gland dysfunction. *Arch Ophthalmol.* 1995; 113:1266–1270 .

CLASSIFICATION OF ARTIFICIAL TEARS

II: Additives and Commercial Formulas

J. Murube, A. Murube, and Chen Zhuo

Department of Ophthalmology
Hospital Ramón y Cajal
Madrid, Spain

1. ADDITIVES TO DRY-EYE DROPS

The additives used in artificial tears serve as (1) buffers, (2) preservatives, (3) nutrients, (4) antiseptics and antibiotics, (5) hypersecretors, (6) mucolytics and mucosecretors, (7) anti-inflammatories and immunosuppressors, and (8) adrenergic vasoconstrictors and astringents.

1.1. Buffers

The pH of normal natural tear is 7.4. In normal tear, the buffer system depends mostly on bicarbonates, but proteins, phosphates, ammonium, and other substances also contribute to it.

The pH of artificial tears should reproduce the pH of normal tear, but also maintain the required physical and chemical characteristics of the artificial tears (osmolarity, colloidosmolarity, adsorbance, viscosity, etc.). If the pH of a solution is far from the desired one, it can be adjusted with HCl or NaOH (*Liquifilm Lágrimas, Liquifresh*) and the required buffer. For example, carbomers are acidic in solutions, and therefore NaOH is added to attain a physiological pH; *Tears Naturale* has a pH of 6.91, with HCl and/or NaOH.

A slightly alkaline environment is more compatible with the corneal epithelium than are neutral or acid conditions. Artificial tears with a pH close to 8.5 are the most comfortable for patients with dry eye.[1]

The most frequent buffer systems for maintaining an appropriate pH in artificial tears are:

Lacrimal Gland, Tear Film, and Dry Eye Syndromes 2
edited by Sullivan *et al.*, Plenum Press, New York, 1998

Phosphate:	*Adsorbo Tears, Liquifilm Forte, Neo Tears.*
Phosphate/acetate:	*Liquifilm, Prefrin, Relief.*
Phosphate/citrate:	*Dakrina, Dwelle, Isoptofrin, Isopto Tears, Nutra Tear, Ultra Tears.*
Phosphate/citrate/bicarbonate:	*Murine Plus.*
Citrate:	*Degest 2, Zincfrin.*
Borate:	*Allergy Drops, Cilclar, Lacrinorm, Moisture Drops, Oculotect, Soothe, Tears Naturale II, Tears Naturale Free, Visine Extra.*
Acetate/borate/citrate:	*Lacril.*
Sodium hydroxide:	*Aquasite, Bion Tears, Cellufresh, Lacryvisc, Refresh, Tears Naturale Free. Viscotears*
Some collyria for dry eye are unbuffered:	*Celluvisc, Comfort Tears, Estivin II, Hypo Tears, HypoTears PF, Refresh, Tears Naturale, TearGard, Tears Plus, Vit-A-Drops.*

1.2. Preservatives

Preservatives are ingredients included in preparations to kill or inhibit the growth of microorganisms inadvertently introduced during use. The usual preservatives in ophthalmic solutions are noted here:

1.2.1. Quaternary Ammonium Compounds. Benzalkonium chloride (BAC) is a cationic surfactant that damages the lipid layer in concentrations over 0.01%.[2] It is used as a preservative in *Akwa Tears, Allergy Drops, Comfort Tears, Comfort Eye Drops, Contactol, Dacriosol, Dacrolux, Degest 2, Enuclene, Estivin II, Hypotears, Hypo Tears, Isopto Fluid, Isoptofrin, Isopto Tears, Just Tears, Lacrinorm, Lacrisifi, Lacrisol, Larmabak, Moisture Drops, Murine, Murine Plus, NutraTear, Prefrin, Siccagel, Siccaprotect, Soothe, Tearisol, Tears Naturale, Tears Renewed, Visine Extra, Zincfrin.*

Cetrimonium chloride: *Vidisept N.*

Cetrimide: *Artelac, Vidisic, Viscotears, Viscotirs Gel.*

1.2.2. Mercurials. Thiomersal (sodium ethyl mercurithiosalicylate) causes little epithelial damage, but produces allergic reactions in many people and is therefore used less frequently in ophthalmic preparations: *Adsorbo Tears, Colircusí Humectante, Dacriogel, Dulcilarmes, Hyalistil, Lacryvisc, Liquifilm Forte, NeoTears, Thilo Tears.*
Phenylmercuric borate: *Lacrypos, Larmes artificielles.*

1.2.3. Alcohols. Chlorobutanol: *Lacril, Lacrilube E.F.P., Liquifilm, Liquifilm Lágrimas, Lacrifilm Tears, Tears Plus.*

1.2.4. Esters of Parahydroxybenzoic Acid. Methylparaben and propylparaben: *Duratears, LephaGel , Murocel.*

1.2.5. Other Preservatives. Sodium edetate (EDTA) on its own does not have sufficient antimicrobial strength, but it enhances the quaternary ammonium base's activity. It chelates the calcium required for junction formation, and therefore may be cytotoxic during prolonged use. Used as a preservative in *Adsorbo Tears, Allergy Drops, Aquasite, Colircusí Humectante, Comfort Eye Drops, Comfort Tears, Dacrolux, Dacriosol, Dakrina,*

Degest 2, Dulcilarmes, Duratears (Neth.), Dwelle, Estivin II, Hypo Tears, Isoptofrin, Just Tears, Lacrisifi, Liquifilm Lágrimas, Moisture Drops, Murine, Murine Plus, NutraTear, Prefrin, Soothe, TearGard, Tears Naturale, Vidisept N, Vidisic, Viscotears, Viscotirs Gel, Visine, Visine A.C., Visine Extra, Visine L.R.

Chlorhexidine: *Oculotect.*

Polyquaternium-1 (Polyquad 0.001%): *Dacriosol Collyrium, Tears Naturale II.*

Poly[oxyethylene-(dimethylimino)-ethylene(dimethylimino)-ethylene dichloride] (NPX): *Dwelle.*

Sorbic acid/potassium sorbate are used as preservatives in *Dakrina, Dwelle, TearGard, Vit-A-Drops.*

Boric acid: *Oculotect, Protagent SE.*

To avoid secondary effects, some artificial tears are preservative-free (PF): *AquaSite, Artelac EDO sine, Bion Tears, Celluvisc, Hypo Tears PF, Liquifresh, Moisture Drops, Refresh, Refresh Plus, Tears Naturale Free, Unilarm, Vidisic EDO, Vit-A-Drops.*

1.3. Nutrients

Natural tears provide nutrients to corneal and conjunctival epithelium and improve their metabolism, ameliorating the resistance of the epithelium, enhancing the synthesis of glycocalyx and goblet cell mucin, and facilitating the stability of the aqueous tear film.

In artificial tears, the most important nutrient is water, but since the late eighties[3] they also include dextrose (*Hypo Tears, Hypo Tears PF, Lacril, Murine Plus),* sodium lactate (*Cellufresh, Celluvisc),* sodium citrate *(Dry Eye Therapy).,* vitamin A[4] (*Dakrina, Oculotect, Pomada Epitelizante, Vit-A-Drops*), dexpanthenol, a vitamin B_3-related drug with hyperviscous properties (*Artelac, Dispatenol, Siccaprotect, Uligin*), vitamin B_{12}, (*Isopto*B_{12}*, NutraTear*), vitamin C (*Dakrina*), and vitamin E (*Oculotect*). Other substances are in the experimental stage, such as growth factors,[5–7] hormones,[8] and RGD peptide,[9] which has three amino acids: arginine, glycine, and aspartic acid. These contribute to the binding site of fibronectin to *integrine*, the surface glycoproteins of cell adhesion.

1.4. Antiseptics and Antibiotics

Antiseptics and antibiotics are used to destroy or inhibit microorganisms in the tear film, cornea, and conjunctiva, or to preserve the sterility of artificial tears.

The ocular surface defends itself from microorganisms physically through the integrity of the epithelial cells and tight junctions, and by the cellular and humoral non-specific and specific defense systems. Natural tear normally contain antimicrobial components, such as lysozyme, β-lysine, lactoferrin, peroxidase, phospholipase A_2, and immunoglobulins. The most widely known is lysozyme (muramidase); however, contrary to general opinion, lysozyme is not a natural antibiotic of tear, but a structural protein providing the appropriate surface tension, viscosity, and other rheological properties, with only a secondary—and poor—antibiotic effect.[10]

Dry eyes have fewer defenses against infections than do non-dry eyes due to epithelial damage and decreased tear defenses. Therefore, some dry-eye drops and ointments add antimicrobial ingredients.

Some preservatives included in eye drops (benzalkonium chloride, cetrimide, cetylpyridinium chloride, thiomersal, etc.) have an antiseptic effect on the tear film as well.

Some preparations deliberately include antiseptics in their formulas to act in the *lacrimal basin*, such as the antifungal sodium thiosulfate in *Prefrin* and *Relief.*

1.5. Hypersecretors

Hypersecretor collyria are not actually artificial tears, but real drugs (strictly speaking, drugs are medicines with active pharmacological action). Any hypersecretor collyrium has an excipient that may act as artificial tears because of its aqueous, electrolytic, and macromolecular compounds.

Some dry-eye formulas include ingredients with a secretagogue effect. Vitamin A increases the mucus production of goblet cells, and probably the aqueous and lipid components as well. Bromhexine is a synthetic derivative of the vegetable alkaloid *vasicine*, used as a systemic mucolytic (*Bisolvon*) and as eye drops (*Dakryo Biciron*). Ambroxol is a metabolite of bromhexine and has similar actions and uses. Bromhexine and ambroxol also increase mucus production and quality. Methyl cysteine (*Visclair*) and acetyl cysteine (*Fluimucil*) are mucolytics mostly given orally in respiratory-tract disorders; they probably also increase the quality of nascent mucin, and are therefore occasionally used in *magistral* formulas as a topical medication for dry eye.

Some drugs increase intracellular levels of cyclic adenosine monophosphate (cAMP), stimulating the secretion of the lacrimal glands. In this group vasoactive intestinal peptide, *Colforsin* (*Boforsin, Forscolin*), 3-isobutyl-1-methylxanthine, and other pharmacological agents are under investigation. Pilocarpine is a parasympathomimetic agent, whose muscarinic secretagogue effect must be evaluated against its undesirable miotic effect. In glaucomatous patients, the pilocarpine could compensate for the dryness produced by beta-blockers. *Bethanechol* has the muscarinic action of acetylcholine. Eledoisin (*Eloisín*) is an endecapeptide extracted from the posterior salivary glands of the Mediterranean octopus *Eledone moscata*, which stimulates lacrimal secretion. *Eloisin* eye drops contain 0.04% eledoisin.

1.6. Mucolytics and Mucosecretors

Mucolytics are substances that loosen mucus. The mucolytic action produces and maintains a fluid mucin and, to a certain extent, even reverses its conglutination after it is produced. In dacryology, mucolytics, not lipid tears, could be called emollients because they soften the mucin and make it more fluid.

Part of the action of mucolytic drugs affects the tears already existing in the tear film, but the most important part of the action occurs intracellularly during the mucin formation process, facilitating production and improving quality. Therefore, most mucolytics also act as mucin hypersecretors.

The mucolytic drugs are bromhexine (in *Dakryo Biciron, Ophtosol*), tyloxapol (in *Enuclene*), *N*-acetyl cysteine (*Ilube, Fluimucil ampules*, *Mucomyst*), methyl cysteine (*Visclair*), and sodium 2-mercaptoethanesulfonate (Mesna) (*Ausobronc*, *Mistrabron*, *Mucofluid*, *Mucolene*). Fibrinolysin or plasmin (E.C. 3.4.21.7) are being used experimentally as mucolytics.

1.7. Anti-inflammatories and Immunosuppressors

Some collyria with certain effects in dry eyes include anti-inflammatories, such as the fluorometholone in *Isopto Flucon*.

Collyrium of cyclosporin A, an immunosuppressor, in 2% olive oil is not commercialized, but is often used in *magistral* formulas.

1.8. Adrenergic Vasoconstrictors and Astringents

Tetrahydrozoline chloride:	*Murine Plus, Soothe, Visine Extra.*
Naphazoline:	*Allergy Drops, Comfort Eye Drops, Degest 2, Estivin II.*
Phenylephrine:	*Isoptofrin, Neo-Lacrim, Prefrin, Relief, Zincfrin.*
Zinc sulfate:	*Zincfrin.*

2. COMMERCIAL FORMULAS

The following list has been compiled from information provided by the respective laboratories. Preservatives, buffers, and other ingredients have been omitted in some cases. Water is present in all these products but is not listed among the ingredients here.

Adaptettes (Alcon, Australia)	Povidone, polyoxyethylene, polyoxypropylene. (Contact lens wetters)
Adapt	Now marketed as *Adsorbo Tears*
Adsorbo Tears (Alcon, USA)	Povidone, polyethylene oxide, hydroxyethyl cellulose, thimerosal, EDTA. (Dry eye)
Akwa Tears (Akorn)	Benzalkonium chloride (BAC). (Dry eye)
Allergy Drops (Bausch & Lomb, USA)	Macrogol 300, BAC, EDTA. (Allergic conjunctivitis, redness, dry eye)
AquaSite (CibaVision, USA)	Polyethylene glycol 400 0.2%, Dextran 70 0.1%, polycarbophil, NaCl, EDTA, NaOH. Preservative-free. (Dry eye)
Artelac (Dr. Mann, Germany)	Dexpanthenol 3%, hypromellose (hydroxypropyl methylcellulose) 0.32%, sorbitol 1.5%, Na_2HPO_4 0.462%, NaH_2PO_4 0.1%, cetrimide 0.01%. (Dry eye) pH 7.3, viscosity 10 mPa.s., osmolarity 300 mOsm/kg, density 1.0119, refractive index 1.340.
Artelac EDO sine (Dr. Mann, Germany)	Preservative-free *Artelac*.
Barnes-Hind Wetting Solutions (Barnes-Hind, Australia)	Polyvinyl alcohol. (Hard contact lens wetters)
Bausch & Lomb Sensitive Eyes Lens Lubricant (Bausch & Lomb, Australia)	Povidone, hypromellose. (Soft contact lens wetter)
Bion Tears (Alcon, USA)	Hypromellose 2910 0.3%, dextran 70 0.1%. NaCl, KCl, $MgCl_2$, $CaCl_2.2H_2O$, $ZnCl_2$, $NaHCO_3$, HCl, and/or NaOH. Preservative-free. (Dry eye)
BJ6 (Macarthys, UK; Thornton & Ross, UK)	Hypromellose. (Dry eye)
Cellufresh (Allergan, USA)	Carmellose 0.5%, NaCl, KCl, $MgCl_2$, $CaCl_2$, sodium lactate, HCl, and/or NaOH. (Dry eye) Viscosity 30 mP, pH 6.4–6.6, osmolarity 280–315 mOsm/l.
Celluvisc (Allergan, USA)	Carmellose sodium 1%, NaCl, KCl, $CaCl_2$, $MgCl_2$, sodium lactate, unpreserved and unbuffered. (Dry eye) Viscosity 2900 mP.

Celluvisc (Allergan, Spain)	Carmellose 1%, carmellose sodium 1%, NaCl, KCl, $CaCl_2$, $MgCl_2$, sodium lactate, HCl, and/or NaOH. (Dry eye). pH 6.2–6.8, osmolarity 270–350 mOsm/l.
Chlorure de Sodium 9% Faure (CibaVision, France)	NaCl 9%. (Dry eye, eye surgery)
Cilclar (Ciba Vision, Spain)	Propylene glycol, Polysorbate 20, Miranol 2MCT, PMBH, H_3BO_3, $Na_2B_4O_7$ EDTA (Lid cleaner)
Clerz (Ciba Vision, Australia)	Poloxamer 407. (Contact lens wetter)
Colforsin (USA adopted name)	$C_{22}H_{34}O_7$ (Glaucoma)
Colircusí Humectante (Cusí-Alcon, Spain)	Hypromellose, NaCl, NaH_2PO_4, Na_2HPO_4, thiomersal, EDTA. (Dry eye)
Comfort Eye (Barnes-Hind, USA)	Hydroxyethyl cellulose, BAC, EDTA. (Dry eye, redness)
Comfort Tears (Sola Barnes-Hind, USA)	Hydroxyethyl cellulose, BAC, EDTA (Dry eye)
Contactol (Merck Sharp & Dohme-Chibret, France)	Hypromellose 4000, tyloxapol, BAC. (Dry eye, contact lens and ocular prosthesis cleanser)
Contafilm (Allergan, Germany)	Polyvinyl alcohol. (Dry eye)
Dacriogel (Alcon, Italy)	Mannitol 5%, carbopol 0.3%, thiomersal. (Dry eye)
Dacriosol Collyrium (Alcon, Italy)	Hypromellose 0.3%, dextran 70/75, NaCl, KCl, EDTA, polyquad. (Dry eye)
Dacriosol Eye Drops (Alcon, Italy)	Hypromellose, dextran 70, NaCl, KCl, BAC. (Dry eye)
Dacriosol Eye Ointment (Alcon, Italy)	Liquid paraffin, lanolin, white soft paraffin. (Dry eye)
Dacrolux (Cusí-Alcon, Spain)	Hypromellose 3 mg/g, Dextran 70 1 mg/g, NaCl, KCl, EDTA, BAC, Tween 80 atlas. (Dry eye)
Dakrina (Dakryon, USA)	Polyvinyl alcohols, povidone, vitamin A, vitamin C, KCl, $CaCl_2$, $MgCl_2$, sorbic acid-sorbate, EDTA. (Dry eye) Viscosity 27 mP
Dakryo Biciron (Basotherm, Germany)	Bromhexine hydrochloride. (Dry eye)
Degest 2 (Barnes-Hind)	Hydroxyethyl cellulose, povidone, BAC, EDTA. (Dry eye, redness)
DEO ointment (MacKeen magistral formula)	Calcium carbonate. (Dry eye)
Dialens (Fisons, France)	Dextran. (Dry eye, contact lens wetter)
Dialens (Novopharma, Switz.)	Dextran 60–85. (Dry eye, contact lens wetter)
Dispatenol (Dispersa, Switz.)	Dexphanthenol, polyvinyl alcohol. (Dry eye)
Dry Eye Therapy (Bausch & Lomb)	Glycerin 0.3%, KCl, NaCl, $CaCl_2$, $MgCl_2$, $ZnCl_2$, sodium phosphate, sodium citrate.
Dulcilarmes (Dulcis, Monaco)	Povidone, NaCl, EDTA, thiomersal. (Dry eye)
Dulcilarmes (Allergan, Switz.)	Povidone, NaCl, EDTA. (Dry eye)
Duolube (Bausch & Lomb, Canada)	White soft paraffin, liquid paraffin. (Dry eye)
Duratears (Alcon, Neth.)	Hypromellose, dextran 70, electrolytes, EDTA. (Dry eye)

Duratears ointment (Alcon, USA)	White petrolatum, anhydrous liquid lanolin, mineral oil, methylparaben, propylparaben. (Dry eye)
Duratears (Alcon, France)	Lanolin, vaseline, methylparaben, propylparaben. (Dry eye)
Dwelle (Dakryon, USA)	Polyvinyl alcohols, dextran, KCl, $CaCl_2$, $MgCl_2$, sorbic acid-sorbate, EDTA, NPX. (Dry eye) Viscosity 26–65 mP, tonicity 304 mOsm/kg, oncotic pressure 65 mmHg, surface tension 44 mN/m.
Eloisin (Cusí-Alcon, Spain)	Eledoisin. (Dry eye)
Enuclene (Alcon, Canada)	Tyloxapol, BAC. (Dry eye)
Estivin II (Alcon, USA)	Hypromellose, dextran, naphazoline, KCl, BAC, EDTA. (Dry eye, redness)
Fluimucil (Inpharzam, Germany; Zambon, Spain)	Acetyl cysteine. (Respiratory viscous secretion)
Hidroxipropilmetilcelulosa (Cusí-Alcon, Spain)	Hypromellose. (Dry eye)
Hyalistil (Sifi, Italy)	Sodium hyaluronate 0.2%, NaCl, KCl, Na_2HPO_4, NaH_2PO_4, thiomersal. (Dry eye)
Hypotears (Iolab, USA)	Polyvinyl alcohol 8000, dextrose, BAC, EDTA. (Dry eye)
Hypotears PF (Iolab, USA)	Polyvinyl alcohol, polyethylene glycol 400, dextrose, unpreserved and unbuffered. (Dry eye) Viscosity 26 mP
Hypo Tears (CibaVisión, Italy)	Polyvinyl alcohol 1%, polyethylene glycol 400 1%, dextrose 3.3%, BAC 0.01%, EDTA 0.03%. (Dry eye)
Ialurex Ipotonico (Fidia, Italy)	Hyaluronic acid 0.4%, NaCl, Na_2HPO_4, NaH_2PO_4. (Dry eye) Osmolarity 150 mOsm/l
Ilube (Cusí, UK)	Hypromellose, acetyl cysteine. (Dry eye)
3-Isobutyl-1-methylxanthine (Sigma, USA)	3-Isobutyl-1-methylxanthine. (Apnea, bronchospasm)
Isopto Alkaline (Alcon, UK)	Hypromellose. (Dry eye)
Isopto Flucon (Alcon, Spain)	Hypromellose 0.2%, fluorometholone. (Dry eye, eye inflammation)
Isopto Fluid (Alcon-Thilo, Germany)	Hypromellose 0.5%, BAC 0.01%. (Dry eye)
Isoptofrin (Alcon, USA)	Hypromellose, phenylephrine, BAC, EDTA, phosphate-citrate. (Dry eye, redness)
Isopto Naturale (Alcon-Thilo, Germany)	Hypromellose, dextran. (Dry eye)
Isopto Plain (Alcon, UK)	Hypromellose. (Dry eye)
Isopto Tears (Alcon, USA)	Hypromellose 0.5%, BAC. (Dry eye)
Just Tears (Blairex, USA)	Polyvinyl alcohol 1.4%, BAC, EDTA. (Dry eye)
Lacril (Allergan, USA)	Hypromellose, gelatin type A, polysorbate 80, dextrose, NaCl, KCl, $CaCl_2$, $MgCl_2$, chlorobutanol
Lacri-Lube (Allergan, USA)	White soft paraffin, liquid paraffin. (Dry eye)
Lacri-Lube (Allergan, Eire)	Liquid paraffin, hydrous lanolin. (Dry eye)
Lacrilube (Allergan, S. Afr.)	White soft paraffin, liquid paraffin. (Dry eye)

Lacrilube E.F.P. (Allergan, Spain)	Paraffin 550 mg/g, lanolin 425 mg/g, chlorobutanol 5 mg/g. (Dry eye)
Lacrilube Pomada Oftálmica (Allergan, Spain)	*White soft* paraffin 550 mg/g, liquid paraffin 425 mg/g, lanolin 20 mg/g, chlorobutanol 5 mg/g. (Dry eye)
Lacri-Lube S.O.P. (Allergan, USA)	Mineral oil 42.5%, white petrolatum 55%, lanolin, chlorobutanol. (Dry eye)
Lacrilux (Allergan, Italy)	Polyvinyl alcohol. (Dry eye)
Lacrimalfa (Intes, Italy)	NaCl, $NaHCO_3$, $MgSO_4$, Na_2HPO_4. (Dry eye)
Lacrimart (Baif, Italy)	Methylcellulose. (Dry eye)
Lacrimill (Ottolenghi, Italy)	Hypromellose. (Dry eye)
Lacrinorm (Thea, France)	Carbopol 940, sorbitol 70%, NaOH, BAC. (Dry eye)
Lacrisert (Merck Sharp & Dohme, USA)	Hydroxypropyl cellulose. (Dry eye)
Lacrisifi (Sifi, Italy)	Hypromellose 4000 0.5%, NaCl 0.2%, Na_2HPO_4, NaH_2PO_4, BAC, EDTA. (Dry eye)
Lacrisol (Bruschettini, Italy)	Hypromellose, BAC. (Dry eye)
Lacrypos (Alcon, France)	Sodium chondroitin sulfate. (Dry eye)
Lacrypos P.O.S. (Kayserberg, France)	Disodium chondroitin sulfate 3%, phenylmercuric borate. (Dry eye)
Lacryvisc (Alcon, Spain)	Carbomer 934P 3 mg/g, mannitol, NaOH, thiomersal 0.04 mg/l, NaOH. (Dry eye)
Larmabak (Thea, France)	NaCl 0.9%, NaH_2PO_4, BAC. (Dry eye)
Larmes Artificielles (Cantonnet, 1908; not commercialized)	NaCl 1.4%, or H_3BO_3 2.5%.
Larmes Artificielles Martinet (CibaVision, France)	NaCl 1.4%, phenylmercuric borate. (Dry eye)
LephaGel (Thea, Spain)	Ethoxypolypropylenglycol 5%, Polyglycol 2%, Polyvinyl 0.35%, P-hydroxybenzoate 0.3%, $Na_2B_4O_7$ 0.84% (Lid cleaner)
Liquifilm (Allergan, USA)	Polyvinyl alcohol 1.4%, chlorobutanol, NaCl. (Dry eye, hard contact lens wetter) Viscosity 13 mP
Liquifilm Forte (Allergan, USA)	Polyvinyl alcohol 3%, thimerosal. (Dry eye)
Liquifilm Lágrimas (Allergan, Spain)	Polyvinyl alcohol 14 mg/g, NaCl, Na_2HPO_4, NaH_2PO_4, EDTA, chlorobutanol 5 mg/g, HCl or NaOH. (Dry eye)
Liquifilm Tears (Allergan, USA)	Polyvinyl alcohol, NaCl, Na_2HPO_4, NaH_2PO_4, EDTA, chlorobutanol. (Dry eye)
Lubrifilm Pomada (Cusí-Alcon, Spain)	Petrolatum 0.6 g, white mineral oil 0.3 g, anhydrous lanolin 0.1 g. (Dry eye)
Liquifresh (Allergan, Spain)	Polyvinyl alcohol 5.6 mg, povidone 2.4 g, NaCl, NaOH, HCl. Preservative-free
M-Cel-4 (Charton, Canada)	Methylcellulose. (Dry eye)
Methocel 2% (CIBA Vision, Switz.)	Hydroxypropyl methylcellulose 20 mg/ml, NaCl, BAC. (Contact lens examination)
Methopt (Sigma, Australia)	Hypromellose. (Dry eye)
Metildrops (Oculos, Spain)	Hypromellose. (Dry eye)

Minims Artificial Tears (Smith & Nephew, UK)	Hydroxyethylcellulose, NaCl. (Dry eye)
Moisture Drops (Bausch & Lomb, USA)	Hypromellose, dextran 40, glycerin 0.2%, BAC, EDTA. (Dry eye)
Moisture Drops (Bausch & Lomb, Canada)	Hypromellose, dextran 40, BAC. (Dry eye)
Mucofluid (UCB, Spain)	Mesna. Ampules diluted 1:3 in saline. (Dry eye)
Mucomyst (Apothecon, USA)	Acetyl cysteine sodium diluted 1:10 in saline. (Dry eye)
Murine (Ross, USA)	Polyvinyl alcohol, povidone, Na^+ 130 mEq/l, K^+ 30 mEq/l, dextrose, BAC, EDTA. (Dry eye)
Murine (Abbott, UK)	BAC, EDTA. (Dry eye)
Murine Plus (Abbott, Australia)	Hypromellose, tetrahydrozoline hydrochloride. (Dry eye)
Murine Plus (Ross, USA)	Polyvinyl alcohol, povidone, dextrose, KCl, tetrahydrozoline hydrochloride, BAC, EDTA. (Dry eye, redness)
Murocel (Bausch & Lomb, Canada)	Methylcellulose 1%, methylparaben 0.023%, propylparaben 0.01%. (Dry eye)
Neo Tears (Sola Barnes-Hind, USA)	Hydroxyethyl cellulose, polyvinyl alcohol, thimerosal.
Neo-Lacrim (Alcon Iberhis, Spain)	Hypromellose, phenylephrine. (Dry eye)
NutraTear (Dakryon, USA)	Polyvinyl alcohol, vitamin B_{12} 0.05%, BAC 0.004%, EDTA 0.08%. (Dry eye) Viscosity 13 mP.
Oculotect (Dispersa, Germany)	Hypromellose 0.4%, chlorhexidine diacetate, retinol palmitate, α-tocopherol acetate, H_3BO_3, $Na_2B_4O_7$. (Dry eye)
Ocustil (Sifi, Italy)	hyaluronic acid 1%, Sodium methyl parahydroxybenzoate 0.12%, sodium propyl parahydroxybenzoate 0.022%, Na_2HPO_4 0.004%, NaH_2PO_4 0.056%, NaCl 0.68% (Ocular prosthesis)
Ocutears (Charton, Canada)	Hypromellose, dextran 40. (Dry eye)
Opthosol (Winter, Germany)	Bromhexine hydrochloride. (Dry eye)
Pliagel (Alcon, Australia)	Poloxamer. (Contact lens cleaner)
PMS Artificial Tears (Pharmascience, Canada)	Polyvinyl alcohol. (Dry eye)
PMS Artificial Tears Plus (Pharmascience, Canada)	Polyvinyl alcohol, povidone. (Dry eye)
Pomada Oculos Epitelizante (Frumtost, Spain)	Retinol 10,000 IU, amino acids. (Epitheliopathies)
Prefrin (Allergan, USA)	Polyvinyl alcohol, $Na_2S_2O_3$, phenylephrine, BAC, EDTA. (Dry eye, redness)
Protagent SE. (Alcon-Thilo, Germany)	Povidone, H_3BO_3. (Dry eye)
*Refresh (*Allergan, USA)	Polyvinyl alcohol 1.4%, povidone 0.6%. Unpreserved and unbuffered. (Dry eye)
Refresh (Allergan, Canada)	Polyvinyl alcohol 1.4%, povidone 0.6%, NaCl, HCl, and/or NaOH. (Ocular lubricant) Viscosity 10 mP.

Relief. $Na_2S_2O_3$. *Siccafluid Gel* (Thea, France)	Carbomer. (Dry eye)
Siccagel (Thea, France)	Carbomer 974P 2.5%, Polyvinyl alcohol, sorbitol, sodium acetate, lysine monohydrate, BAC. (Dry eye)
Siccaprotect (Ursapharm, Germany)	Polyvinyl alcohol 1.4%, NaH_2PO_4, Na_2HPO_4, dexpanthenol 3%, BAC 0.005%. (Dry eye)
Sno Tears (Smith & Nephew, UK)	Polyvinyl alcohol. (Dry eye)
Soaclens (Alcon, France, Australia)	(Hard and gas-permeable contact lens wetter)
Soothe (Alcon, USA)	Povidone, macrogol 90, tetrahydrozoline, BAC, EDTA. (Dry eye, redness)
Teardrops (Dispersa, Canada)	Polyvinyl alcohol, povidone. (Dry eye)
Tear Gard (Bioproducts, USA)	Hydroxyethyl cellulose, lecithin, sorbic acid-sorbate, EDTA. (Dry eye)
Tearisol (Inibsa, Spain)	BAC. (Dry eye disorders)
Tears Gel (Alcon, Spain)	Liquid paraffin 0.03 mg/g, liquid anhydrous lanolin 0.03 mg/g. (Dry eye)
Tears Humectante (Alcon, Spain)	Hypromellose 3 mg, dextran 70 1 mg, preservatives. (Dry eye)
Tears Lubricante Pomada Ocular (Alcon, Spain)	Liquid vaseline, lanolin. (Dry eye)
Tears Encore (Alcon, Canada)	Polysorbate 80. (Dry eye)
Tears Naturale (Alcon, USA)	Hypromellose 2910, dextran 70, Na^+ 134 mEq/l, K^+ 16 mEq/l, BAC, EDTA. (Dry eye) pH 6.91, viscosity 44 mP.
Tears Naturale II (Alcon, USA)	Hypromellose, dextran 70, KCl, borate, polyquad. (Dry eye) Viscosity 32 mP.
Tears Naturale Free (Alcon, USA)	Hypromellose 2910 0.3%, dextran 70 0.1%, NaCl, KCl, $Na_2B_4O_7$, HCl, and/or NaOH. (Dry eye) Viscosity 32 mP.
Tears Plus (Allergan, USA)	Polyvinyl alcohol, povidone, chlorobutanol. (Dry eye) Viscosity 32 mP.
Tears Renewed (Akorn)	BAC. (Dry eye)
Thilo-Tears (Alcon-Thilo, Germany)	Carbomer 3mg/g, mannitol 50 mg/g, thiomersal 40μg/g. (Dry eye)
Total (Allergan, Australia)	Polyvinyl alcohol. (Hard contact lens lubricant)
Uligin (Lichtwer, Germany)	Dexpanthenol, polyvinyl alcohol. (Dry eye)
Ultra Tears (Allergan, USA)	Hypromellose, chlorobutanol, BAC.
Unilarm (CibaVision, France)	NaCl 9%. Preservative-free. (Dry eye)
Vasoactive intestinal peptide.	Note: not commercialized for patients. Only used in research.
Vidisept N (Mann, Germany)	Povidone MW 25,000 5%, hypromellose 0.1%, NaCl 0.5%, Na_2HPO_4 1.373%, NaH_2PO_4 0.075%, cetrimonium chloride 0.004%, EDTA 0.01%. (Dry eye) pH 7.3, viscosity 3 mP, osmolarity 320 mOsm/kg, density 1.0188, refractive index 1.344.
Vidisic (Mann, Germany)	Carbopol 940 0.2%, sorbitol 4%, EDTA 0.01%, cetrimide 0.01%. (Dry eye)

Vidisic Gel (Mann, Germany)	Carbomer MW 4,000,000 0.2%, sorbitol 4%, NaOH 0.084%, cetrimide 0.01%, EDTA 0.01%. (Dry eye) pH 7.3, viscosity 4.5 mP, osmolarity 250 mOsm/kg, refractive index 1.338.
Visclair (Sinclair, UK)	Methyl cysteine hydrochloride. (Mucolytic)
Viscotears (CibaVisión, Spain)	Carbomer 940 2 mg/g, sorbitol, NaOH, EDTA, cetrimide 0.1 mg. (Dry eye)
Viscotirs Gel (CibaVision, Italy)	Sorbitol 4%, carbopol 940 0.2%, NaOH 0.084%, cetrimide 0.01%, EDTA 0.01%. (Dry eye)
Viscotraan (Covan, S. Afr.)	Hypromellose. (Dry eye, contact lens lubricant)
Visine Extra (Pfizer, USA)	Macrogol 400, tetrahydrozoline. (Dry eye, redness)
Vit-A-Drops (Vision Pharm, USA)	Polysorbate 80, vitamin A. Unpreserved and unbuffered.
Zincfrin (Alcon, USA)	Polysorbate 80, $ZnSO_4$, phenylephrine, BAC. (Dry eye, redness)

REFERENCES

1. Jones BR, Coop HV. The management of keratoconjunctivitis sicca. *Trans Ophthalmol Soc UK.* 1965;85:379–390.
2. Ubels JL, Williams KK, López Bernal D, Edelhauser HF. Evaluation of effects of a physiologic artificial tear on the corneal epithelial barrier: Electrical resistance and carboxyfluorescein permeability. *Adv Exp Med Biol.* 1994;350:441–452.
3. Holly FJ. Pharmacology of ocular lubricants. In: Onofrey BE (ed). Clinical optometric pharmacology and therapeutics. Philadelphia: Lipincott; 1992. Chapter 57A.
4. Gilbard JP, Huang AJ, Belldegrun R, Lee JS, Rossi SR, Gray KL. Open-label crossover study of vitamin A ointment as a treatment for keratoconjunctivitis sicca. *Ophthalmology.* 1989;96:244–246.
5. Ohashi Y, Motokura M, Kinoshita Y, et al. Presence of epidermal growth factor in human tears. *Invest Ophthalmol Vis Sci.* 1989;30:1879–1887.
6. Danielle S, Gilbard JP, Schepens CL. Treatment of persistent epithelial defects in neurotrophic keratitis with epidermal growth factor: A preliminary open study. *Graefes Arch Clin Exp Ophthalmol.* 1992;230:314–317.
7. van Setten GB, Schultz GS, Macauley S. Growth factors in human tear fluid and in lacrimal glands. *Adv Exp Med Biol.* 1994;350:315–319.
8. Sullivan DA, Ariga H, Vendramini AC, Rocha FJ, Ono M, Sato EF. Androgen-induced suppression of autoimmune disease in lacrimal gland of mouse models of Sjögren's syndrome. *Adv Exp Med Biol.* 1994;350:683–690.
9. Tsubota K. Novedades en el tratamiento del ojo seco. *Int Ophthalmol Clin* 1995;1(Sp. ed):115–128.
10. Murube J. *Dacryologia Basica*, 2nd ed. Madrid: Royper; 1982;480–482, 554–563.

101

THE OCULAR SURFACE, THE TEAR FILM, AND THE WETTABILITY OF CONTACT LENSES

Carol A. Morris,[1] Brien A. Holden,[1] Eric Papas,[1] Hans J. Griesser,[1,2] Shirley Bolis,[1] Philip Anderton,[1,3] and Fiona Carney[1]

[1]Cooperative Research Centre for Eye Research and Technology
Cornea and Contact Lens Research Unit
University of New South Wales
Sydney, NSW, Australia
[2]CSIRO
Division of Chemicals and Polymers
Clayton, Victoria, Australia
[3]School of Optometry
University of New South Wales
Sydney, NSW, Australia

1. INTRODUCTION

The tear film is the interface between the ocular surface and the external environment and, as such, plays several important roles.[1] (i) It forms a refracting thin film that smooths out the irregular corneal surface topography. (ii) It maintains an extracellular environment for the epithelial cells of the cornea and the conjunctiva that is fairly constant in terms of pH, oxygen and carbon dioxide levels, and nutrient and growth factor concentrations. (iii) Tears dilute and wash away noxious stimuli, including bacteria, which are also combated by an elaborate and effective antibacterial system. (iv) The tear film changes its composition in response to physiological stimuli.

The human tear film appears to be reasonably stable and adherent to the cornea, taking up to 60 sec to show the first rupture if the eyes are kept open.[2] However, when a contact lens is inserted, a stable tear film is not maintained over the lens, and the surface actually dries out between blinks.[1] On average, humans blink once every 10 sec, whereas the tear film begins to break up on the surface of a contact lens in 4–6 sec (hard)[3] or 7–9 sec (soft).[4] Lens drying is the single most important factor in deposit formation and influences long-term discomfort during contact lens wear.

Various theories have been advanced to explain this phenomenon. Physical chemistry considerations maintain that a thin film, containing surface-active molecules such as

Lacrimal Gland, Tear Film, and Dry Eye Syndromes 2
edited by Sullivan *et al.*, Plenum Press, New York, 1998

those present in tears, should not rupture so rapidly from a curved surface. In fact, laboratory experiments with lenses show that certain lens surfaces can hold a water film for up to 2 min! Various factors drive this surface-drying phenomenon on-eye. Drainage, due to the tear meniscus which is clearly visible around the lens periphery, certainly has an effect with a rigid lens.[5] The structured layers of the tear film with lens on-eye are not able to build up as they do on the cornea. Another possibility is that the lipid layer is thin or disrupted and not able to prevent the tear film from evaporating.[6] Early binding events of tear components to lenses may also affect wettability.

To explore some of these possibilities, a number of experimental lenses were fabricated and placed on-eye. As the measure of tear break-up, we recorded the time to the appearance of the first dry spot in the central region of the lens, ie, the non-invasive surface drying time (NISDT).

2. EFFECT OF SURFACE CHEMISTRY

A number of thin coatings containing different surface chemical groups were permanently attached to a contact lens material. To maximize any effects, we used a siloxane-based, high Dk, soft material, the chemical nature of which, in the untreated state, is so hydrophobic that tear break-up post blink was virtually instantaneous, ie, the NISDT could not be differentiated from zero. All lenses were of identical base-curve, thickness, and power (-1.00). The method of attachment and the thickness of the coatings were added variables.

Extremely thin, uniform coatings of 5–10 nm thickness and varying hydrophilicities were obtained by glow discharge (gas plasma) deposition and a wet chemistry attachment step. A ten times thicker coating (50 nm) was achieved by an alternative plasma procedure. Much thicker coatings were fabricated by wet chemistry grafting techniques; however, these were shown to be uneven and even patchy in places. Thicknesses of approximately 200 nm up to 1 μm were achieved by this route (see Table 1).

Lenses were placed on-eye in a contralateral study with a HEMA control material (38% water) in the other eye for 4 h. At the end of this time, NISDT was measured, and lenses were removed and analyzed for protein and lipid uptake. This was repeated on 10 subjects (n=10). For protein, lenses were twice extracted in 6 M urea, 0.1% SDS, 0.1 M

Table 1. *In vivo* protein uptake and wetting of different lens surfaces on a siloxane-based material

Lens coating	Thickness	Protein uptake* (μg/lens)	NISDT after 4 h Wear^ (sec)
HEMA (38%)	–	5.7±5.0	5.0±2.0
No coating	–	–⁺	0±0 ⁺
S	5-10 nm	0.7±0.1	4.2±1.1
A	5-10 nm	1.9±2.2	6.0±2.5
V_1	50 nm	3.0±3.0	4.8±1.9
V_2	0.1 μm (patchy)	3.7±1.6	6.1±2.1
P_1	1 μm (patchy)	3.2±1.1	4.6±1.7
P_2	1 μm (patchy)	0.5±0.8	5.9±2.3

*n=3.

^n=10.

⁺, worn only for 10 min; protein uptake ND.

Table 2. Uptake of bovine submaxillary mucin (BSM) on worn and unworn lenses with various surface treatments

Lens coating	μg BSM Bound* (unworn lens)	μg BSM Bound* (4 h worn lens)
HEMA (38%)	0.5	0.4
Uncoated	1.2	–
S	3.4	0.7
A	1.0	0.7
V_1	2.0	0.4
V_2	3.4	0.7
P_1	0.4	0.4
P_2	0.5	0.5

*$n = 4$.

Tris, and 1 mM DTT 30 min at 95°C (90% of protein removed), the combined extract was dialyzed and concentrated, and the protein content was measured by the BCA assay (Pierce, USA). Control unworn lenses were run in each case to correct for any interferences.

All of the lenses took up very little protein during 4 h wear, even compared to the HEMA control, and irrespective of which surface chemistry was present (Table 1). Interestingly, no differences were seen in the NISDTs, regardless of degree of hydrophilicity or thickness of coating.

As a further differentiation, lenses were subjected to a binding assay with ^{125}I-bovine submaxillary mucin (BSM) for 16 h at room temperature, before extensive washing and counting. An identical procedure was carried out on the same lenses after 4 h wear.

Binding of this mucin initially was variable on the different surfaces, and generally higher than the HEMA material with the exception of the two thick surface coatings. After 4 h wear, however, the amount of mucin binding was essentially the same for all lenses, including HEMA (Table 2).

All the coatings markedly altered the lens surface, providing NISDT values equivalent to the HEMA material even with extremely thin coatings (<10 nm) on the unwettable siloxane material. Surprisingly, however, the NISDT and protein and mucin uptake were identical, within experimental error, for all coatings, regardless of hydrophilicity and thickness. The tear film appears to equalize them to a standard low level of wettability. One possible complication is that the surfaces may not have been completely uniform (in the case of wet grafting reactions, they certainly were not). Therefore, the bare polymer patches may be driving this effect, although the size of any bare patches that may have been present would have been well below the spatial resolution (10 nm) of imaging XPS surface analysis.

3. EFFECT OF LIPID

Instability of the tear film on the surface of a lens may be due to excessive depletion of the lipid layer, as a result of deposition of lipid at the material surface. To test the hypothesis that lipid saturation of a surface would help stabilize a tear film, lipophilic, siloxane-based, high Dk material was soaked in either lanolin (Lacri-lube, Allergan) or saline at 34°C for 16 h. Then lenses were washed with sterile saline and rinsed with Instacare (CIBAVision) before being placed on-eye for 4 h.

Table 3. Lipid uptake of lanolin-soaked lenses

Lens	Surface	Lipid in lens after lanolin soak* (μg/lens)	Lipid in lens after wear* (μg/lens)
Siloxane-based	Plasma deposition	74±16	151±38
Siloxane-based	Plasma treatment plus polysaccharide layer	11±2	100±15

*n=2.

Table 4. Wetting of saline- and lanolin-soaked lenses

Lens	Surface	Saline soak* NISDT (sec)	Lanolin soak* NISDT (sec)
Siloxane-based	Plasma deposition	3.6±0.5	2.0±0.8
Siloxane-based	Plasma treatment plus polysaccharide layer	4.7±0.5	5.0±0.8

*n=2.

To verify that indeed there was lipid uptake in the lens, lenses were extracted both before and after wear, and the amount of lipid was determined by the sulfovanillin reaction. As shown in Table 3, lanolin was able to penetrate lenses effectively when they were deposited only with a plasma coating, whereas an additional polysaccharide coating layer presented a greater barrier to this triglyceride. Lenses with both types of coating, however, showed significant further lipid uptake after 4 h wear.

The NISDTs at 4 h wear were similar for the plasma/wet chemistry lenses regardless of a lipid or a saline soak (Table 4). Lenses that were plasma-deposited-only took up substantial amounts of lipid, which resulted in a lower NISDT than the saline-soaked lenses. Overall, the tear film was thinner on the lipid-soaked lenses, and surface drying was simultaneous all over the lens, rather than starting in one discrete spot.

These results show that excessive quantities of this type of lipid on or in a lens surface can destabilize the tear film, presumably by providing the wrong 'base' for a stable tear film.

4. EFFECT OF LENS DIAMETER AND MATERIAL

This line of experimentation derived from an observation by Jean-Pierre Guillon[8] that haptic lenses show a much longer break-up time than do standard intracorneal lenses, suggesting that PMMA can support a long-lasting tear film, and that the meniscus at the edge of the corneal lenses can have a dramatic effect.

These hypotheses were studied in the following manner. PMMA haptics were molded for one eye of each of five subjects. As well as comparing wetting between a 9 mm corneal lens and the 23 mm haptic PMMA lens, the same subjects wore 13.8 mm vs. 18 mm HEMA lenses, and a 13.8 mm vs. an 18 mm high Dk, soft, siloxane-based material. The large soft lenses were specially fabricated. The siloxane-based material was worn uncoated. The NISDT was measured after 1 h wear; the results are compared in Fig. 1.

With soft lenses, however, both HEMA and siloxane-based, the NISDT was essentially the same regardless of lens diameter. Although somewhat variable, the PMMA haptic lenses showed substantially greater tear film stability (16 ± 18 sec) than the smaller PMMA corneal lenses.

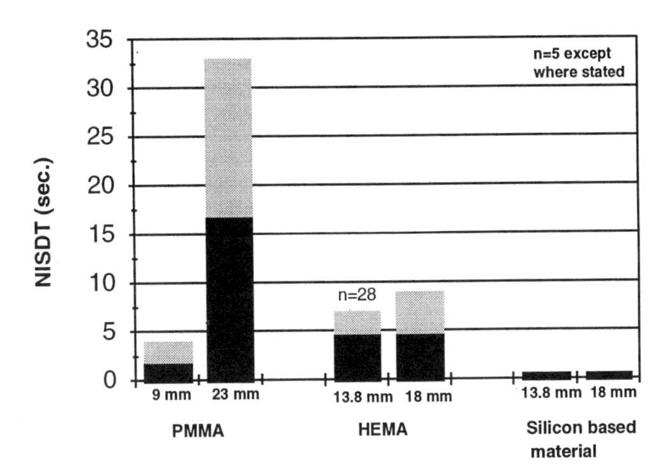

Figure 1. Effect of different materials at two different diameters on non-invasive surface drying time. (Plotted as mean value - dark, plus standard deviation - hatched).

To further pursue the effect of diameter, five additional subjects had a set of five haptic lenses molded on-eye, and these were cut down to a defined smaller diameter between 13 and 17 mm inclusive. Each lens was then placed on-eye, and a HEMA soft lens in the contralateral eye. NISDT was measured after wear for 1 h.

Comparing these intermediate to larger lenses with the 9 and 23 mm lenses from the previous study (Fig. 2), it is clear that diameter has a large influence on the break-up time of tears on this rigid material. Overall, NISDT values for PMMA ranged from 2 sec (9 mm lenses) through 8 to 12 sec (13 to 17 mm lenses) to 17 sec (23 mm lenses). What is also evident is that the soft HEMA material has a substantially lower mean NISDT (5 ± 5 sec) than the PMMA lens (11 ± 10 sec) at the same 17 mm diameter.

The effect of coating with various hydrophilicities on the large haptic lenses was also determined. The PMMA lenses originally molded for the five patients were tested uncoated,

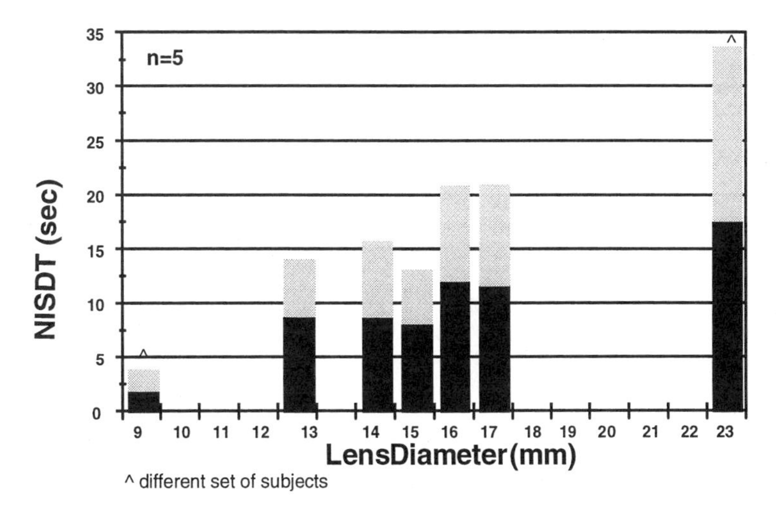

Figure 2. Non-invasive surface drying time of PMMA haptic lenses with varying diameters. (Plotted as mean value - dark, plus standard deviation - hatched).

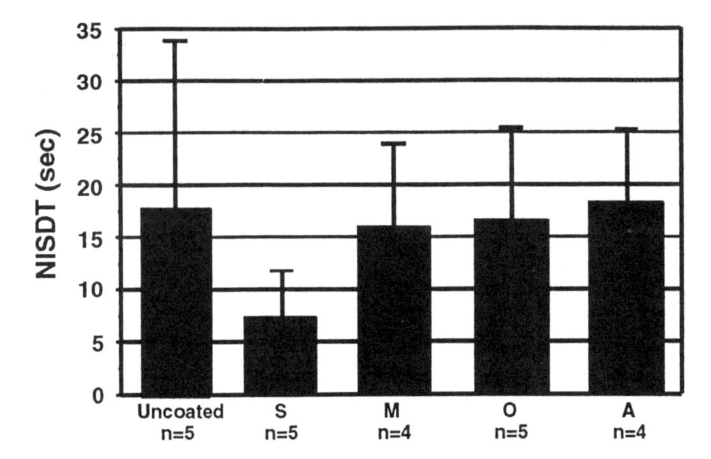

Figure 3. Effect of surface composition on haptic PMMA lenses.

plasma coated only, or plasma coated and further coupled with various thin polysaccharide layers. All five patients wore each surface-treated lens, and NISDTs were compared after 1 h wear to that with the untreated lens. Surface treatment with increasingly hydrophilic coatings did not appear to affect the NISDT of the PMMA haptic lenses in any significant way (Fig. 3), except that, surprisingly, the most hydrophilic coating gave a lower NISDT.

The conclusion we can reach from these studies is that hard materials such as PMMA create inherently better eye tear stability for large diameter lenses than do soft materials. Increasing the size of soft lenses did not improve the surface drying time, whereas for the hard PMMA lenses, the diameter basically determines the surface drying time. It is clear that, in the absence of a highly effective "draining" meniscus at the edge of an intracorneal lens, this material will support a stable tear film. It may be that the intrinsic property of the hard material that affects this phenomenon is the impermeability to water. The tear fluid deposited on these materials after the blink is unable to penetrate the lens and therefore is subject to the drainage forces involved in the pooling of tears at the periphery of the lens. In the case of large haptic lenses, the lens edge is largely covered by the eyelids, with the meniscus eliminated.

REFERENCES

1. Holly FJ. Formation and stability of the tear film. *Int Ophthalmol Clin.* 1978;30:73–96.
2. Mengher LS, Bron AJ, Tonge SR, Gilbert DJ. A non-invasive instrument for clinical assessment of the pre-corneal tear film stability. *Curr Eye Res.* 1985;4:1–7.
3. Guillon M, Guillon JP, Mapstone V, Dwyer S. RGP lenses in vivo wettability. *J Br Contact Lens Assoc.* 1989;6:24–26.
4. Guillon JP, Guillon M, Malgouyres S. Corneal desiccation staining with hydrogel lenses: Tear film and contact lens factors. *Ophthalmic Physiol Opt.* 1990;10:343–350.
5. Fatt I. Observations of tear film break-up on model eyes. *CLAO J.* 1991;17:267–281.
6. Mishima S, Maurice DM. The oily layer of the tear film and evaporation from the corneal surface. *Exp Eye Res.* 1961;1:39–45.
7. Korb DR, Greiner JV, Glonek T. Tear film lipid layer formation: Implications for contact lens wear. *Optom Vis Sci.* 1996;73:189–192.
8. Guillon JP. *Tear Film Structure of the Contact Lens Wearer.* PhD thesis. The City University, London, 1988.

102

CONJUNCTIVAL PERMEABILITY AND ULTRASTRUCTURE

Effects of Benzalkonium Chloride and Artificial Tears

John L. Ubels,[1] Eric M. Woo,[1] William J. Watts,[1] Lisa K. Smith,[1] Uko Zylstra,[1] Jonnell Beaird,[2] and Mitchell D. McCartney[2]

[1]Biology Department
Calvin College
Grand Rapids, Michigan
[2]Alcon Laboratories, Inc.
Fort Worth, Texas

1. INTRODUCTION

The conjunctiva, a mucin-secreting epithelium that covers most of the ocular surface as well as the inner surface of the eyelids, can be adversely affected by many disease processes. It is susceptible to infection and inflammation in response to pathogens and allergens. Conjunctival irritation is also a hallmark of keratoconjunctivitis sicca, with increasing mucin deficiency due to goblet cell loss as the disease progresses. In ocular cicatricial pemphigoid, autoimmunity results in severe conjunctival disease with conjunctival scarring and forniceal shortening.

Conjunctival disease may occur without corneal involvement or prior to development of corneal abnormalities, but the cornea, because of its essential role in transmittance and refraction of light, has been the subject of a greater amount of research. Although the surface area of the conjunctiva in the human is nine times that of the cornea, and in the rabbit 17 times,[1] studies of drug penetration and uptake by the eye have also emphasized the cornea. As regards dry eye syndromes, investigation of effects of ophthalmic preservatives and artificial tear solutions have also focused primarily on the barrier function and ultrastructure of the corneal epithelium.[2–5] In contrast, studies of the conjunctiva in keratoconjunctivitis sicca have usually been limited to clinical observations, rose bengal staining, and measurement of goblet cell density using impression cytology.[6]

The purpose of the present study was to investigate the effects of benzalkonium chloride (BAC) and artificial tear solutions on the rabbit conjunctival epithelium. Experi-

Lacrimal Gland, Tear Film, and Dry Eye Syndromes 2
edited by Sullivan *et al.*, Plenum Press, New York, 1998

ments were designed to measure BAC influence on the barrier function and ultrastructure of the conjunctiva, and to determine whether various artificial tear formulations can provide an environment in which the damaged conjunctival epithelium can recover normal structure and function.

2. METHODS

2.1. Experimental Preparation

The methods used in this study were an adaptation of methods previously reported for studies of the corneal epithelium.[3,5] New Zealand White rabbits (2 kg) were anesthetized with ketamine HCl (30 mg/kg) and xylazine HCl (6 mg/kg). A conjunctival cup was formed around one eye, and two 9–0 polypropylene sutures were placed 4–5 mm apart in the cornea at the lateral limbus. The sutures were used to pull the globe nasally, exposing the lateral bulbar conjunctiva. At the completion of the procedures described below, the rabbits were euthanized with a sodium pentobarbital overdose.

2.2. Carboxyfluorescein Uptake

To determine effects of BAC on the conjunctival epithelium, the conjunctival cup was filled with 0.01%, 0.005%, or 0.001% BAC (Sigma, St. Louis, MO) in balanced salt solution (BSS; Alcon Laboratories, Ft. Worth, TX) for 5 min. The eye was then rinsed with BSS and exposed to 2.7×10^{-3} M 5,6-carboxyfluorescein (CF; Kodak, Rochester, NY) in BSS for 5 min. The eye was again rinsed, and the conjunctiva was marked with a 6 mm trephine. The outlined conjunctival epithelium and substantia propria were then removed from the sclera, weighed, and placed in 20 ml of BSS at 4°C for 48 h. After dialysis of the tissue sample in BSS, the CF concentration of the dialysate was measured by fluorescence using a Series 2 Luminescence Spectrometer (SLM–AMINCO, Rochester, NY), and total uptake of CF per gram of tissue was calculated. For studies of effects of artificial tear solutions on BAC–damaged corneas, the conjunctival cup was filled with an artificial tear solution for 1.5 h prior to CF treatment. The commercially available tear replacements used in this study were: (i) AquaSite (Ciba Vision Ophthalmics, Atlanta, GA), which contains EDTA and NaCl but no other electrolytes; (ii) Bion Tears (Alcon), with an ionic composition similar to tears including HCO_3^-; (iii) Cellufresh (Allergan, Irvine, CA), a lactated Ringer's-based solution, and (iv) Hypotears PF (Iolab, Claremont, CA), which is hypotonic and contains EDTA. Controls were the contralateral conjunctiva exposed to CF without prior BAC or tear solution treatment.

2.3. Electron Microscopy

For studies of effects of BAC and artificial tears on the ultrastructure of the conjunctival epithelium, the conjunctiva was exposed to 0.01% or 0.005% BAC alone or to BAC followed by artificial tears (AquaSite or Bion Tears), as described above. Eyes were also exposed to artificial tears for 1.5 h without prior BAC treatment. After treatment, the rabbit was euthanized and the conjunctiva was fixed in situ using 2% paraformaldehyde and 2.5% glutaraldehyde in a 0.1 M sodium cacodylate buffer. For fixation of the mucin layer, 0.5% cetylpyridinium chloride (CPC) was added to the fixative. A 4x4 mm sample of fixed bulbar conjunctiva attached to the underlying stroma was removed from the eye, processed, and examined by transmission electron microscopy, as previously described.[5]

All tissue samples were taken from the same location on each eye. Protocols were performed on three rabbits per group.

3. RESULTS

3.1. Carboxyfluorescein Uptake

The 6 mm diameter control conjunctival buttons analyzed in the experiments illustrated in Fig. 1 took up 34.5 ± 2.83 nmoles CF/g (mean ± SE) and weighed 17.2 ± 0.49

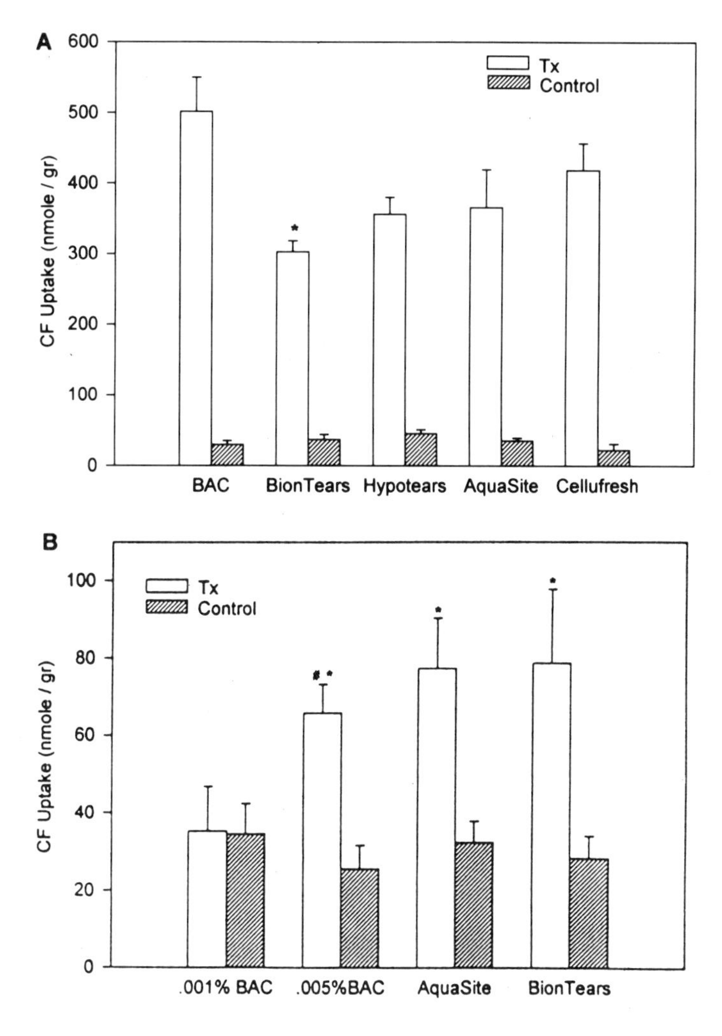

Figure 1. Effect of artificial tears on uptake of carboxyfluorescein (nmoles/gr) by conjunctiva damaged with BAC. (A) Conjunctiva was treated with 0.01% BAC alone or with artificial tears after exposure to BAC (Tx). Values are mean ± SE; n=8. *Significantly different from BAC alone (Kreskal–Wallis ANOVA and Dunn's test, $p \leq 0.05$, data not normally distributed). Control values do not differ (one–way ANOVA). All Tx values differ from control (paired t–test, $p \leq 0.05$). (B) Conjunctiva was treated with 0.001% or 0.005% BAC alone, or with artificial tears after exposure to 0.005% BAC. #Significantly different from 0.001% BAC (t–test, $p \leq 0.05$). *Significantly different from control (paired t–test, $p \leq 0.05$). Controls do not differ (one–way ANOVA).

mg, giving a total uptake of about 0.59 nmoles CF. This amount of CF was taken up over a surface area of 0.28 cm^2 during a 5 min exposure. This yields a control CF uptake rate of 0.42 nmoles/cm^2/min.

Exposure to 0.01% BAC resulted in CF uptake of 501.6 nmoles/g, a 16.95-fold increase in CF uptake compared to paired controls. Exposure to artificial tear solutions following damage of the conjunctiva by 0.01% BAC had little effect on CF uptake. A small, but statistically significant, decrease in CF uptake was observed following treatment with Bion Tears, but none of the four tear solutions studied was able to promote recovery of the epithelial barrier to control levels (Fig. 1A).

Because of the level of damage to the conjunctiva caused by 0.01% BAC, additional experiments were conducted using lower concentrations. After exposure to 0.005% BAC, the uptake of CF was 65.8 nmoles/g, still two-fold higher than paired controls, while exposure to 0.001% BAC had no effect on CF uptake as compared to controls. Treatment with Bion Tears or AquaSite had no effect on CF uptake by corneas damaged with 0.005% BAC (Fig. 1B).

3.2. Ultrastructure

The control conjunctiva had two to four layers of epithelial cells with numerous surface microvilli on the superficial cells. Prominent goblet cells were present, and CPC–fixation revealed a layer of mucin on the surface of all epithelial cells (Fig. 2A and B). Exposure to 0.01% BAC had an inconsistent effect on the conjunctival epithelium. Some areas were unaffected and undistinguishable from control specimens. In regions where damage occurred, it was severe with only a single epithelial cell layer remaining or with complete disruption of the epithelium. Remaining cells were highly vacuolated, with nuclear disruption (Figs. 2C and 3B). BAC at 0.005% also caused non–uniform damage to the conjunctiva, which was not as severe as that observed after 0.01% BAC (Fig. 2D). When corneas were exposed to AquaSite following treatment with 0.01% BAC, in regions with minimal damage by BAC the cells appeared swollen with reduced numbers of microvilli (Fig. 3A). In more severely damaged regions, however, the cells that remained and were subsequently exposed to AquaSite were vacuolated, with numerous intercellular spaces and a lack of microvilli (Fig. 3B). In contrast, cells exposed to Bion Tears appeared normal in minimally damaged areas (Fig. 3C). Even in areas where only one cell layer remained following BAC treatment, epithelial cells exposed to Bion Tears exhibited a more normal cellular morphology with well-defined nuclei and formation of microvilli (Fig. 3D). In keeping with observations on CF uptake, no reestablishment of an intact epithelial barrier during artificial tear treatment was observed. All conjunctivae exposed to Bion Tears or AquaSite without prior exposure to BAC had normal morphology (Fig. 3E and F).

4. DISCUSSION

The most striking observations of this study are the differences between the rabbit conjunctival and corneal epithelia in their permeability to CF, susceptibility to damage by the commonly used preservative benzalkonium chloride, and response to artificial tears. These findings have important implications for our understanding of the pharmacokinetics of topical ophthalmic drugs and the treatment of ocular surface disease. The conjunctiva has a much higher permeability to CF than does the cornea. In the present study, the conjunctival CF uptake rate was 0.42 nmoles/cm^2/min. In our previous studies of the corneal

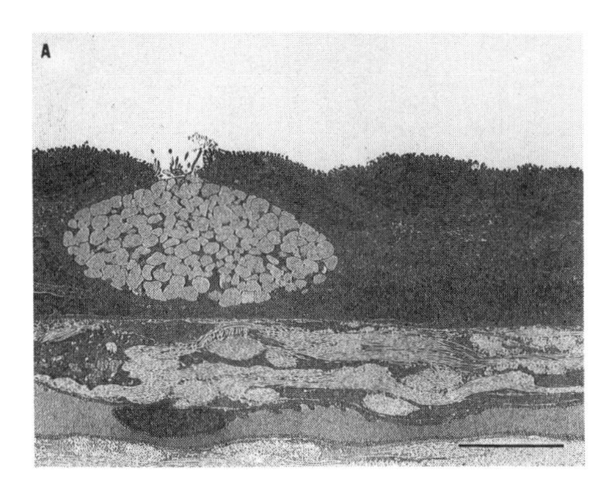

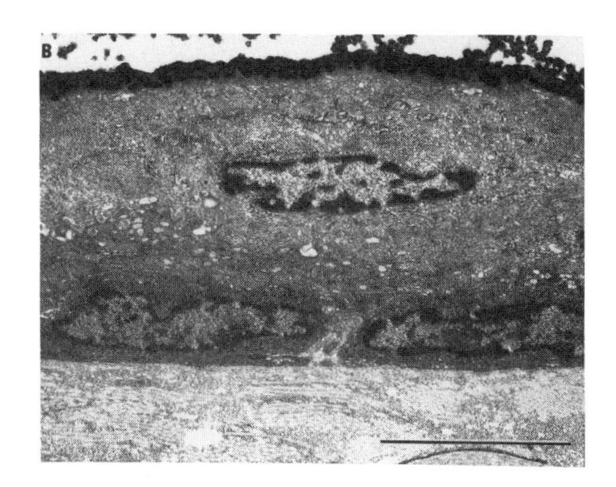

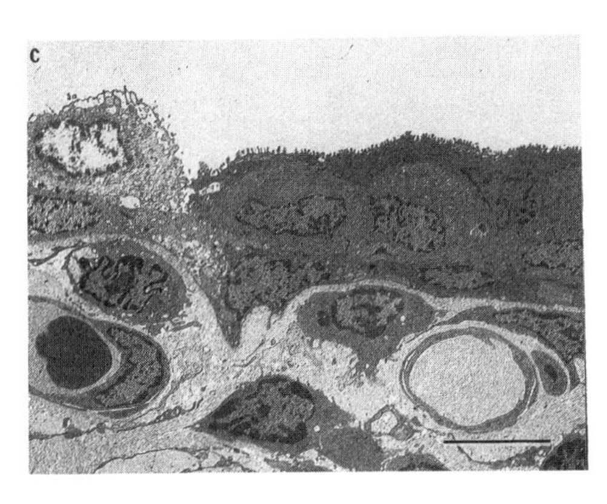

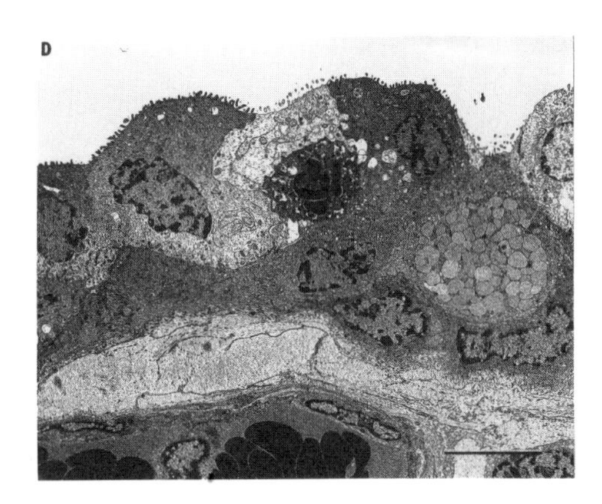

Figure 2. Electron micrographs of conjunctival epithelium. (A) Control conjunctival epithelium. Bar=10 μm (B) Control conjunctival epithelium fixed with CPC to show the mucin layer. Bar=5 μm. (C) Conjunctiva exposed to 0.01% BAC. Note region of sloughing cells and damage to basal layer with adjacent cells of normal appearance. Bar=10 μm. (D) Conjunctiva exposed to 0.005% BAC. Superficial two cell layers are damaged. Bar=10 μm.

epithelium, CF uptake was 3 nmoles/g by a 50 mg cornea with a 1 cm^2 surface area during a 5 min exposure to CF.[3] This gives a corneal CF uptake of 0.03 nmoles/cm^2/min. The permeability of the undamaged conjunctival epithelium to CF is therefore 14 times higher than that of the normal corneal epithelium. This value may be compared to corneal and conjunctival electrical resistance. Klyce and Crosson[7] reported that corneal epithelial resistance is about 15 Kohms/cm^2. In contrast, the resistance of the conjunctival epithelium as measured by Kompela *et al.*[8] is 1.3 Kohms/cm^2, while Shi and Candia[9] reported that R= 1.2 Kohms/cm^2. The resistance of the pericellular shunt in corneal epithelium is therefore about 12 times that of the conjunctival epithelium, which corresponds well to the difference in permeability of these epithelia to CF. This observation agrees with Wang et al.,[10] who reported that conjunctiva is more permeable to beta blockers than is cornea.

The conjunctival epithelium is also more susceptible to damage by BAC than is the corneal epithelium. Exposure to 0.01% BAC caused a 13.5-fold increase in CF uptake (expressed as nmoles CF/g tissue) by conjunctiva, compared to control. In contrast, CF uptake by corneas exposed to 0.01% BAC increased only 6.6-fold in our previous studies.[3] Reductions in BAC concentration resulted in decreased damage to the conjunctival epithelial barrier as compared to control, but it is noted that CF uptake after exposure to

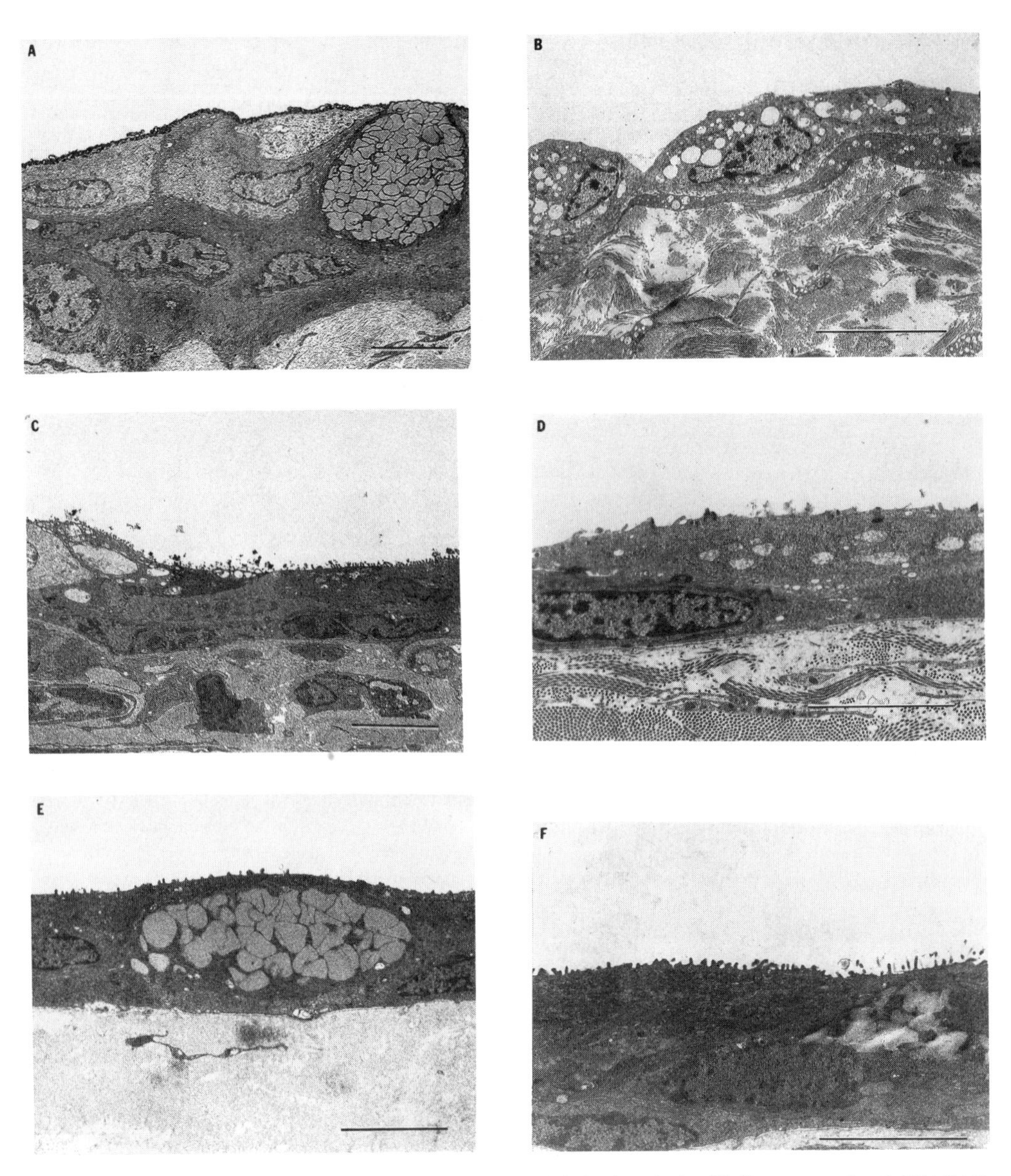

Figure 3. Electron micrographs of conjunctival epithelium. (A) Conjunctival epithelium exposed to 0.01% BAC followed by AquaSite (fixed with CPC). Area is relatively undamaged, but cells are swollen and mucin layer is reduced. Bar=10 μm. (B) Conjunctival epithelium exposed to 0.01% BAC followed by AquaSite. More severely damaged area with a single layer of vacuolated cells and lack of microvilli. Bar=10 μm. (C) Conjunctival epithelium exposed to 0.01% BAC followed by Bion Tears. Area is relatively undamaged, with two to three cell layers that appear healthy with numerous microvilli. Bar=10 μm. (D) Conjunctival epithelium exposed to 0.01% BAC followed by Bion Tears. More severely damaged area. Remaining cells appear healthy and have microvilli. Bar=5 μm. (E) Conjunctiva exposed to Bion Tears without BAC treatment. Bar=10 μm. (F) Conjunctiva exposed to AquaSite without BAC treatment. Bar=5 μm.

0.005% BAC remained 3-fold higher than that observed in cornea after exposure to 0.01% BAC. This difference in the effect of BAC may be related to the higher permeability of the conjunctival epithelium, which would give BAC access to deeper cell layers.

The high level of conjunctival CF uptake after BAC damage may easily be explained by ultrastructural observations. Although damage by 0.01% BAC was not uniform over the entire conjunctival surface, in affected areas damage reached the basal cell layer or even the basement membrane, allowing CF to diffuse easily into the substantia propria. The thickness of the conjunctival epithelium is variable and may be as thin as two to four cell layers in some regions.[11] This explains why the degree of recovery of barrier function during exposure of the damaged conjunctiva to artificial tears is less than that observed previously in our cornea experiments. When the epithelial cells slough to the basal layer of the conjunctival epithelium, the remaining cells, which are less differentiated than the superficial cells, are unable to form new tight junctions. In our previous studies, BAC caused uniform damage to the entire corneal epithelium; however, only two or three cell layers of a total of five to seven cell layers were disrupted, leaving a new superficial layer capable of partial recovery during exposure to Cellufresh and complete recovery of the barrier function during exposure to Bion Tears.[3,5]

Because of the high degree of localized damage caused by BAC, it was difficult to differentiate among various tear formulations in this study. Bion Tears–treated tissue had a slightly lower CF uptake than did conjunctiva exposed to other products, but the response to all solutions was highly variable. Clearly, when cells capable of forming tight junctions are no longer present, no tear formulation will be capable of promoting recovery of barrier function within the time period of our protocol. With electron microscopy, however, it was evident that the remaining cells in areas damaged by BAC and treated with Bion Tears had a more healthy morphology than those exposed to AquaSite. This suggests that with long-term treatment an artificial tear with physiologic composition would be better tolerated and may permit recovery of the damaged conjunctival epithelium, while formulations with ingredients foreign to the tear film, such as EDTA, should be avoided by the severe dry eye patient. Results of clinical trials comparing Bion Tears to other formulations support this idea, since treatment with the physiologic tear substitute was better tolerated by patients and resulted in decreased conjunctival rose bengal staining.[6,12]

In conclusion, this study confirms that significant anatomic and functional differences exist between the rabbit conjunctival and corneal epithelia. The clinical relavance of our rabbit model has not been established; however, the high degree of permeability of the rabbit conjunctiva and its susceptibility to damage by BAC suggest that preservative–free, physiologic artificial tear solutions should be used by patients with severe keratoconjunctivitis sicca and other ocular surface diseases caused by abnormal tear function.

ACKNOWLEDGMENTS

This work was supported by Alcon Laboratories, Inc., Fort Worth, TX, and by the Howard Hughes Medical Institute.

REFERENCES

1. Watsky MA, Jablonski MM. Edelhauser HF. *Curr Eye Res.* Comparison of conjunctival and corneal surface area in rabbit and human. 1988; 7:483–486.

2. Burstein NL, Klyce SD. Elecctrophysiologic and morphologic effects of ophthalmic preparations on rabbit crnea epithelium. *Invest Ophthalmol Vis Sci.* 1977;16:899–911.
3. Lopez Bernal D, Ubels JL. Artificial tear composition and promotion of recovery of the damaged corneal epithelium. *Cornea.* 1993;12:115–120.
4. Berdy GJ, Abelson MB, Smith LM, George MA. Preservative-free artificial tear preparations. *Arch Ophthalmol.* 1992;110:528–532.
5. Ubels JL, McCartney MD, Lantz WK, Beaird J, Dayalan A, Edelhauser HF. Effects of preservative-free artificial tear solutions on corneal epithelial structure and function. *Arch Ophthalmol.* 1995;113:371–378.
6. Nelson JD, Drake MM, Brewer JT, Tuley M. Evaluation of a physiologic tear substitute in patients with keratoconjunctivitis sicca. *Adv Exp Med Biol.* 1994;350:453–457.
7. Klyce SD, Crosson CE. Transport processes across the rabbit corneal epithelium: a review. *Curr Eye Res.* 1985;4:323–331.
8. Kompela UB, Kim KJ, Lee VHL. Active chloride transport in the pigmented rabbit conjunctiva. *Curr Eye Res.* 1993;12:1041–1048.
9. Shi XP, Candia OA. Active sodium and chloride transport across the isolated rabbit conjunctiva. *Curr Eye Res.* 1995;14:927–935.
10. Wang W, Sasaki H, Chien DS, Lee VHL. Lipophilicity influence on conjunctival drug penetration in the pigmented rabbit: a comparison with corneal penetration. *Curr Eye Res.* 1991;10:571–579.
11. Gipson IK. Anatomy of the conjunctiva, cornea and limbus. In: Smolin G, Thoft RA, eds. *The Cornea:: Scientific Foundations and Clinical Practice.* 3rd ed. Boston: Little, Brown; 1994: 3–24.
12. Wolf T, McCulley J, Beasley C, Scoper S, Nelson J. Clinical effects of Bion Tears, Cellufresh, and Aqua-Site in moderate to severe dry eye patients. ARVO Abstracts. *Invest Opthalmol Vis Sci.*1994; 35:S1694.

103

PRESERVATIVE EFFECT ON EPITHELIAL BARRIER FUNCTION MEASURED WITH A NOVEL TECHNIQUE

Jerry R. Paugh,[1] Alan Sasai,[2] and Abhay Joshi[2]

[1]Cooperative Research Centre for Eye Research and Technology
University of New South Wales
Sydney, Australia
[2]Allergan Inc.
Irvine, California

1. INTRODUCTION

In relation to clinical acceptance, the ideal ophthalmic solution is one that requires no preservative. Although unit dose and, more recently, non-preserved multi-dose delivery systems are available, the cost is significant and limits their overall utilization. In addition, poor sanitation and uncertain water supplies in developing nations will necessitate preserved formulations for the foreseeable future. Still not well understood is the degree to which current preservatives are toxic to the ocular surface and a reliable method to quantitate those effects.

Assessment of epithelial barrier function may be an optimum measure of topical solution toxicity for several reasons. Permeability changes are evident prior to gross biomicroscopic observations[1] and provide an objective and quantitative measure of effect. We have recently introduced[2,3] a novel method of permeability measurement in humans that requires only a single drop instillation of sodium fluorescein, and is rapid, simple, and convenient. It also provides concurrent tear turnover data.

Ophthalmic solution preservatives increase permeability of the corneal epithelial barrier. However, these effects have been shown *in vitro,*[4] and in humans under exaggerated conditions of either elevated concentration[5] or dosing.[6] The purpose of this investigation was to measure the effect on the human epithelium of preserved and non-preserved artificial tears in healthy subjects on an acute basis, but with dosing simulating that for moderate to severe dry eye patients. The results suggest that the preservatives tested had minimal effect on barrier function of normal individuals when measured at clinical concentrations.

2. MATERIALS AND METHODS

2.1. Subjects

All subjects were young (18–45 years), non-contact lens wearing individuals in good health and taking no tear-influencing medications such as antihistamines. Sixteen males and 12 females were recruited into the study. The study was conducted according to the tenets of the Declaration of Helsinki; all subjects gave written informed consent after being apprised of the nature of the study and procedures. Preliminary trial sessions were conducted with all subjects to familiarize them with the fluorometry procedures.

2.2. Epithelial Permeability Assessment

The instrument used was an objective scanning fluorometer (Fluorotron® Master, Ocumetrics, Palo Alto, CA) set up for anterior segment fluorescence measurement. The technique measures epithelial permeability in nanometers per second and has been reported in detail elsewhere.[2,3]

Briefly, the drop method involves baseline intrinsic fluorescence measurement, instillation of 2.0 μl of 0.70% sodium fluorescein in borate buffer, and repeated scans to 20 min of combined tear film and corneal fluorescence. The central premise of the epithelial permeability calculation, P_{dc}, is that the area under the axial distance peak (AUC) of the fluorometer is proportional to the mass of fluorescein in the tissue. Thus the individual corneal peak areas are calculated and re-plotted to give a fluorescence vs. time curve. The area under this curve is used to calculate P_{dc} according to the equation:

$$P_{dc} = a_c * q_d / \int a_d \, dt \qquad (1)$$

where a_c represents the final mass of fluorescein in the cornea (distance AUC following irrigation of tear fluorescein), q_d is tear film thickness, assumed to be 8.0 μm in all calculations, and the integral a_d dt is the AUC of the individual tear-cornea distance AUCs plotted against time. Fig. 1 depicts the time AUC. Raw data were extracted from the fluorometer hard drive and areas calculated with Kaleidagraph software (Abelbeck/Synergy Software, Reading, PA).

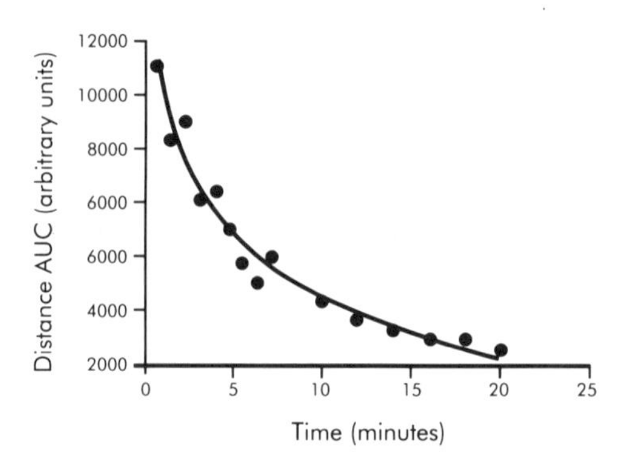

Figure 1. Fluorescein availability (time AUC).

2.3. Study Design

The study was divided into two parts, a pilot investigation to determine technique sensitivity, and a comparison over 1 week (the acute/chronic study), both utilizing the same marketed tears as test formulations. The preserved tear contained polyvinylalcohol and polyethylene glycol as viscous agents, and 0.01% w/v benzalkonium chloride and 0.03% w/v EDTA as preservatives. The unit dose non-preserved tear contained carboxymethylcellulose as the viscous component.

The sensitivity study involved eight subjects who were willing to undergo the sequential drop instillation. Six drops of the preserved artificial tear were instilled in one eye 1 min apart, and the epithelial permeability measured within 2–5 min. Once the first eye had been measured, six drops of a non-preserved saline control were instilled, and the permeability measured in the same manner in the fellow eye. Both subject and investigator were masked, and treatment eyes were selected randomly.

The acute/chronic study (so called because of the short duration—1 week—but with the aggressive dosing necessary to simulate moderate dry eye conditions) was a masked, two-arm parallel acute dosing study wherein subjects were randomly assigned to either preserved or non-preserved artificial tear regimens. The drops were instilled in both eyes a minimum of eight times per day, and both subject and investigators were masked as to drop. Epithelial permeability measurements were made at the same time of day to control for diurnal variation; assessments were made on day 0 (baseline), and days 3 and 7. The day 3 measurement was included since Doughty has suggested a potential adaptation effect of preserved artificial tears on epithelial morphology.[7] No drops were allowed 2 h prior to permeability measurement.

A priori statistical power calculations were made to determine the necessary sample size to achieve ^{3} 80% statistical power (α = 0.05, parallel two-arm design, SE = 0.06 nm/sec based on historical data from this laboratory, δ = 0.20 nm/sec). Fourteen subjects for each arm were required to allow the desired statistical power.

3. RESULTS

3.1. Sensitivity Study

Eight subjects completed the sensitivity study. The P_{dc} data, summarized in Table 1, demonstrate a slight mean increase in permeability for the eye receiving the preserved tear and an average permeability ratio of 1.45. Taken together, these measures suggest that the technique can elucidate barrier function changes, at least under exaggerated dosing. Subject 7 demonstrated a paradoxical effect, wherein obvious irritation and subsequent fluorescein staining were found for the preserved tear, yet the permeability was equal to the non-preserved tear. It is possible that the increased lacrimation from the irritative drop washed the dye out too rapidly for the true permeability to be determined.

3.2. Acute/Chronic Study

Of the 28 subjects recruited, 25 (14 control, 11 preserved tear) completed the study. The P_{dc} values for baseline and days 3 and 7 are presented in Table 2. The baseline P_{dc} values were similar to normative values reported earlier with this technique.[2,3] There were no clinically or statistically significant differences for either the control or the treatment group at days 3 and 7 compared to baseline.

Table 1. Sensitivity study summary

Subject	P_{dc} Preserved tears	P_{dc} Control	Ratio = $\frac{P_{dc} \text{ Preserved}}{P_{dc} \text{ Control}}$	Comments
1	0.23	0.16	1.44	No symptoms or staining
2	0.18	0.14	1.23	No symptoms or staining
3	0.12	0.05	2.40	No symptoms or staining
4	0.15	0.29	0.52	No symptoms or staining
5	0.05	0.10	0.50	No symptoms or staining
6	0.41	0.16	2.56	No symptoms or staining
7*	0.09	0.09	1.00	Irritation and staining
8	0.18	0.09	1.94	Irritation and staining
Mean	0.18	0.14	1.45	
SD	0.11	0.07	0.80	

P_{dc} in nm/sec.
*Paradoxical effect; toxicity obviously present; excessive tearing may have eliminated dye too rapidly.

4. DISCUSSION AND CONCLUSIONS

This investigation endeavored to simulate in healthy normals the typical dosing requirement over 1 week that might be necessary for mild to moderate dry eye sufferers. The aims were to determine the response to a preserved artificial tear and to ascertain whether adaptation occurs in the initial few days. It appears that adaptation does not occur under these conditions and that little or no barrier compromise can be measured. The reason that no measurable effect was observed might be due to one or more of the following: (i) poor sensitivity of the method, (ii) excessive time lag between last instillation and P_{dc} assessment (i.e., the epithelial intercellular adhesion systems were able to re-establish normality prior to measurement, or (iii) preservative effects in the concentrations tested are not harmful to the barrier in healthy eyes.

The sensitivity study demonstrated that barrier effects could be observed in most cases, with the restriction that if significant irritation and subsequent reflex tearing occur, the small amount of instilled fluorescein is washed out too rapidly for an accurate meas-

Table 2. Acute/chronic permeability summary

Day		P_{dc} preserved tears	P_{dc}.control	p Value
Baseline	Mean	0.16	0.19	0.5328
	SD	0.11	0.13	
	n	12	14	
Day 3	Mean	0.15	0.15	0.4115*
	SD	0.08	0.15	
	n	11	14	
Day 7	Mean	0.16	0.16	0.5841*
	SD	0.09	0.13	
	n	11	14	
Difference baseline to day 7	Mean	0.00	–0.03	0.5672
	SD	0.13	0.16	
	n	11	14	

P_{dc} in nm/sec.

urement. The re-healing possibility seems reasonable in view of rabbit epithelial studies showing that, following exfoliation, normal permeability is restored very rapidly (i.e., within minutes).[8] Relative to the current study, the re-healing effect may have occurred prior to permeability measurement, which suggests that only longer-term studies where more permanent ocular surface damage is likely to be present can elucidate the true effect of preservative use.

The third possibility is more intriguing if little effect on barrier function is observed with longer-term studies. Although deleterious effects have been observed in dry eye subjects with these preservatives,[9] new preservatives under development may prove efficacious in preservation of solution sterility while causing minimal changes to the ocular surface.[10]

REFERENCES

1. Green K, Bowman KA, Elijah RD, Mermelstein R, Kilpper RW. Dose-effect response of the rabbit eye to cetylpyridinium chloride. *J Toxicol-Cut Ocular Toxicol.* 1985; 4:13–26.
2. Paugh JR, Joshi A. Novel fluorophotometric methods to evaluate tear flow dynamics in man. ARVO Abstracts. *Invest Ophthalmol Vis Sci.* 1992;33:S950.
3. Joshi A, Maurice DM, Paugh JR. A novel fluorophotometric method to measure epithelial permeability in man. *Invest Ophthalmol Vis Sci.* 1996;37:1008–1016.
4. Tripathi BJ, Tripathi RC. Hydrogen peroxide damage to human corneal epithelial cells *in vitro*. *Arch Ophthalmol.*1989;107:1516–1519.
5. Burstein N. Preservative alteration of corneal permeability in humans and rabbits. *Invest Ophthalmol Vis Sci.* 1984;25:1453–1457.
6. Ramselaar JAM, Boot JP, van Haeringen NJ, Oosterhuis JA. Corneal permeability of ophthalmic solutions containing local anaesthetics and preservatives. *Curr Eye Res.* 1988;7:947–950.
7. Doughty MJ. Twice-daily use of a chlorbutanol-preserved artificial tear on rabbit corneal epithelium assessed by scanning electron microscopy. *Ophthalmic Physiol Opt.* 1992;12:457–466.
8. Wolosin JM. Regeneration of resistance and ion transport in rabbit corneal epithelium after induced surface cell exfoliation. *J Membr Biol.* 1988;104:45–55.
9. Gobbels M, Spitznas M. Corneal epithelial permeability of dry eyes before and after treatment with artificial tears. *Ophthalmology.* 1992;99:873–878.
10. Boets EPM, Kerkmeer MJ, van Best JA. Contact lens care solutions and corneal epithelial barrier function: A fluorophotometric study. *Ophthalmic Res.* 1994;26:129–136.

VIDEO BIOMICROSCOPY OF THE TEAR FILM

P. Noel Dilly,[1] Michel Guillon,[2,3] and Lucia McGrogan[2]

[1]Department of Anatomy and Developmental Biology
St. George's Hospital Medical School
London, United Kingdom
[2]Contact Lens Research Consultants
Westminster, London, England, United Kingdom
[3]Department de Biochimie des Transports Cellulaires
Universite Paris-Sud
Orsay, France

1. INTRODUCTION

The tear film is produced by the lacrimal gland, the accessory lacrimal glands, and the goblet cells of the conjunctiva, with the meibomian glands of the eyelids providing an oily surface layer. The tear film covers the eye and is drained from the surface via two puncta that lead through the canaliculi to the lacrimal sac, which is in the lacrimal fossa. From here, the tears flow in a duct in the bony nasolacrimal canal to the inferior nasal meatus.

Normal non-reflex tearing secretion amounts to something less than 1 μl/min, not even the volume of the canaliculi. This tiny volume could be absorbed by the muscosal surface of the nasolacrimal duct with little actual expulsion into the nose.[1] The pumping capacity of the canaliculi is 100 mm^3/min; secretion in excess of this will overflow onto the cheeks. The unstimulated eye produces less than 1 mm^3/min of aqueous, and the canalicular system is virtually empty at all times.

Jones proposed that the basic supply of lacrimal fluid is from the accessory lacrimal glands.[2] The main lacrimal gland is a sort of emergency organ. The relative lack of symptoms of hypofunction following complete removal of the main lacrimal gland supports this conclusion.

Normally the tear flow is along the margin of the lower lid in the form of a meniscus that is in contact with both the lid margin and the globe. The tears flow medially towards the lacrimal puncta which are situated near the medial end of the upper and lower lids. It is thought that with each blink the upper lid distributes the tear film evenly across the globe. Arlt suggested that the lacrimal sac was compressed by a blink, and his idea was

Lacrimal Gland, Tear Film, and Dry Eye Syndromes 2
edited by Sullivan *et al.*, Plenum Press, New York, 1998

supported by the radiographic studies of Ploman et al.[3–5] The contrary view that the sac was expanded by blinking was proposed in 1853 by Hyrtl[6] and supported some 50 years later by Schirmer.[7]

In his seminal 1957 report, Jones concluded that the positive force in the ampulla and canaliculus is many times greater than the negative force in the tear sac. This has become known as the lacrimal pump mechanism--the pumping being produced by alternating negative and positive pressures in the lacrimal sac.[2]

Several studies have attempted to confirm this hypothesis. Doane used high-speed photography to follow fine carbon lampblack particles as markers in the tear film.[8] He showed that tears moved over the cornea only with blinking. He found that both the punctum and the marginal tears meniscus move medially 2–5 mm with each blink. He demonstrated that without blinking there is no movement of the tear meniscus. In the absence of blinking, there is a steady flow of tears into the sac, which may be seen in the slit lamp microscope after introduction of a carbon particle suspension. The jerky motion of the tear meniscus associated with each blink is called Krebiel flow, named after its discoverer (as reported by Rosengren[9]). Earlier Maurice had shown that the pressure in the lacrimal sac increased on blinking, suggesting that the pressure on the sac increased during the contraction of the eyelid during a blink.[1]

There is a surprisingly stable lipid film that persists even if the meibomian gland orifices are sealed. The lipid layer is 40–100 nm thick. There are about 50–70 meibomian glands arranged along the eyelid margins, with about 10 more glands in the upper lid than in the lower lid. They have a rich parasympathetic and a sparse sympathetic innervation. The control of the lipid secretion is not known, but the glands probably secrete meibomian lipid on the lid margins in minute amounts continuously, and small amounts are drawn into the tear film at each blink.

Much is known about the control of reflex tearing from the lacrimal gland, but we are only slowly progressing towards an understanding of "normal" secretion from the lacrimal gland, the goblet cells of the conjunctiva, and the meibomian glands of the eyelids. Tears are normally stable, and this great excess of secretory tissue is part of the secondary support mechanisms of the tear film.

The upper lid is in forceful and intimate contact with the cornea. The pressure is often great enough to push the anterior surface of the globe in a posterior direction over 1.5 mm.[7] But could some of this movement be caused by the combined retractive forces of the extraocular muscles, especially those involved in Bell's phenomenon? Doane considered that mucus was pushed down by the upper lid and towards the inner canthus during a blink.

Lemp and Weiler used slow-speed 64 fps cinematography to monitor the movement of black polystyrene microspheres approximately 15 μm in diameter (3M's) installed into the sac.[10] The spheres rushed into the puncta immediately on completion of the blink. There was some regurgitation. Krebiel flow was observed. Many of the spheres were taken up in the surface mucin network of the conjunctiva.

2. MATERIALS AND METHODS

2.1. Flow Markers

Calibrated polystyrene microspheres supplied in an aqueous solution were used in the investigation. The microspheres were sterilized using gamma irradiation (Isotron

PLC). The 1–40 μm microspheres were suspended in water and provided in small plastic containers holding 15 ml (2.5% solids). The microspheres were dyed in a variety of colors, red being the easiest to see. The handling of the microspheres involved a sterile technique using microsyringes and fine-gauge (25) needles. Before syringing the beads, we agitated them in their container to achieve a homogeneous suspension.

2.2. Contact Lenses

The contact lenses used in our investigations were Acuvue (Vistakon) and CSIThin (PBH) in a range of base curves 8.30–9.20 mm. The flattest lenses proved uncomfortable with or without microspheres.

2.3. Clinical Material

This initial pilot investigation was carried out on five subjects including one of the authors (PND). Four subjects were social contact lens wearers, and the fifth was a naive

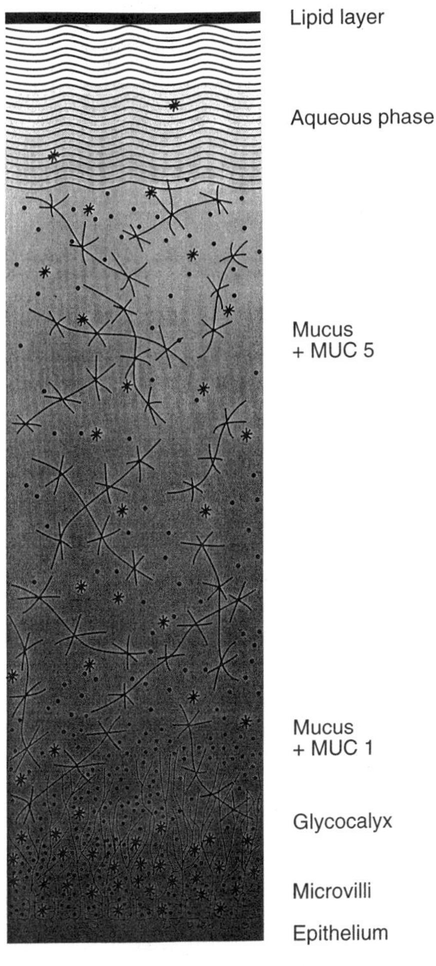

Figure 1.

contact lens wearer. The investigation had received approval from the St. George's Hospital Ethics Committee.

2.4. Clinical Routine

For the experiments without a contact lens, one drop of the aqueous solution containing the beads was expressed from a microsyringe onto the sclera. The drop was aimed to land in the superior conjunctiva just above the superior limbus while the subject looked downwards with head tipped back. This procedure did not produce discomfort or reflex tearing.

For the experiments with a contact lens, one drop of the bead suspension was expressed, using the microsyringe, into the concave surface of the lens immediately before insertion into the eye.

At times, a mixture of beads of different sizes was administered onto the back of the contact lens using this technique. The addition of the beads to the lens produced no change in sensation to the lens wearer.

2.5. Observation Technique

A macro videotape of the eye, using a wide, diffused light source, was taken to establish the presence of microspheres. The technique involves a high-magnification macro video system with a Nikon macro lens.

3. RESULTS

3.1. Without Contact Lens

Particles of all sizes were visible in the tear film; they were rapidly removed via the blinking process. Apart from a few beads attached to the lipid layer, most had been removed within 5 min of insertion. Microspheres were not seen within the mucus layer. Beads attached to the lipid layer were remarkable in that after every blink they returned to the same position relative to the eye underneath.

The marginal flow that occurs in the tear prism between the globe and the lower lid was intermittent. The particles moved at a high speed for about 1–2 sec after each blink. The particles were within the aqueous, and not in contact with the eyes or the lids.

Particles such as mascara or other make-up products that were within the tear prism and attached to the lower lid caused a considerable obstruction to the tear flow. Particles frequently built up behind the obstruction until a particularly strong blink passed them beyond it. Not only were the particles held up, but the aqueous prism appeared larger on the side of the obstruction. Small and large microspheres travelled at very similar speeds.

The microspheres did not cause any symptoms or discomfort in the eye, yet foreign bodies of a similar size that reach the corneal epithelium do cause considerable discomfort.

3.2. With Contact Lens

The pre-lens tear flow remained similar to the pre-ocular tear flow. The lipid layer remained in register; it appeared to be a very stable layer. Marginal flow remained more or less the same as in the non-contact lens wearing eye.

The post-lens tear film microsphere movement was highly influenced by the lens fit. The steep-fitting lens revealed very little movement for any particles behind the lens, even when the lens was moved with a finger. With the flat-fitting lens, the microspheres exhibited large movements in the aqueous under the periphery of the lens; the microspheres were pumped in and out again following each blink. However, where the lens rested on the mucus layer, it pushed the particles into the mucus.

After lens removal, no microspheres were seen in the aqueous, and initially there were a few in the lipid layer. The surprise was the persistence of microspheres deeper within the post-lens tear film. Several days later, some of these microspheres were still in the same place, apparently completely asymptomatic and remaining in the mucus layer.

Very little evidence of casual meibomian gland secretion was observed. When a gland secretes, its globule remains discrete for many blinks before the secretion gets distributed into the heterogeneous lipid layer. In asymptomatic eyes, it is extremely hard to break up the lipid layers.

4. DISCUSSION

The physical properties of the microspheres may not be very important for their behavior in the tear film. Carrington (personal communication) has shown that as soon as charged or hydrophobic or hydrophilic particles enter the tear film, their properties become quite similar, presumably because of a coating or biofilm derived from the tear film.

The corneal epithelium did not stain with fluorescein when the particles were in situ. Sometimes there was a ring of dye around the microspheres. There was no staining with rose bengal. Rose bengal 1% stains dead cells, whereas fluorescein remains extracellular. Fluorescein staining indicates superficial defects in the cell mucus layer, whereas rose bengal and alcian blue indicate severely damaged or dead cells.

The microspheres could be found for several days in the same location. This suggests either a firm binding of the microspheres to the cell surface or that the particles were trapped within the mucus layer that is itself firmly bound to the cell surface. It is unlikely that a cell would survive and not be exfoliated if it had a 6–10 μm particle inside it.

The mucus layer is a stable, tough, viscoelastic covering of the cornea. It is the clinical experience of at least one of us (PND) that only high-velocity foreign bodies are capable of penetrating it. Slower particles are deflected by its surface and remain in the aqueous where they are either washed out by reflex tearing, or end up below the upper lid. It is probably the pressure of the eyelids on the contact lens that pushed the microspheres into the mucus. They were sufficiently deep so as not to be disturbed by normal blinking in eyes after the contact lens had been removed.

We consider the tear film to be an inherently stable structure, capable of rapid secondary reconstruction. The lacrimal gland can produce much more aqueous than is needed to maintain the aqueous layer. There is easy reflex secretion in response to noxious stimuli and emotions.

"Screwing up" the eyes following a painful stimulus probably serves to compress the conjunctival goblet cells, causing them to discharge a replacement layer of mucus. The lipid layer too is capable of rapid replacement from the meibomian glands. It is rarely possible to see more than the occasional single gland discharging in the normal eye. There are vast reserves on each lid. There is, of course, a well- established autonomic supply to the vicinity of the goblet cells that might well control their "normal" secretion (Bron, 1996, personal communication). Much of the maintenance of the tear film is a function of blinking.

The lipids form a heterogeneous sea of varying thickness and immense tensile strength. As our results have shown, the lipid layer has a considerable "memory," with particles descending and returning to their original position after each blink. In the normal dry eye, the stability of the lipid layer far exceeds the normal interval between blinks. Besides its function of preventing evaporation,[1] the lipid layer probably serves to dampen down "waves" caused by the eyelids passing through the aqueous phase of the tear film. An initial attempt at a wave can frequently be seen just after the lid is fully open following a blink. The wave is rapidly surpassed by the lipid layer, thus preventing the visual distortion that would otherwise occur.

Blinking aids drainage and prevents overflow. There is a misconception that the tear film somehow "waters" the ocular surface. The idea that the tear film is an aqueous-dominated, three-layered structure is probably incorrect. Our concept of the tear film is that it is a mucin-dominated gel, hydrated by aqueous fluid, with a lipid layer floating on its surface.

The major structural molecules of the tear film mucus layer are highly glycosylated glycoproteins called mucins. At least 50% of their mass is carbohydrate. The glycoprotein MUC 1 is a membrane-spanning molecule. The entire ocular surface epithelium, ie, corneal and conjunctival, expresses MUC 1. This long filament structure probably extends about 0.5 μm out into the tear film. The conjunctival goblet cells have MUC 5 mucins that contribute the viscous and adhesive properties responsible for tear film stability. The ocular lipids have a very high tensile strength, and reflect this in their "memory" as they return to register after each blink.

What the tears *do* is far more important in the understanding of dry eyes than is the nature of their composition.

The blink is an essential part of tear film stability. Our observations showed that the lid does not contact the globe. Otherwise the microspheres in the mucus would have been moved. Indeed, we would suggest that the lid sweeps over the surface of the mucus in the aqueous phase. This will "flip/flop" the large molecules on the surface of the more mucous phase, keeping it smooth and restoring its refractile properties. The upper lid pushes the lipid layer into forming a thick layer between the closed lids, and our observations have shown that as the lid opens, it is restored to its identical pre-blink position. A major function of the high tensile strength of this layer is to stop the moving lid from inducing waves in the more aqueous surface phase of the mucus layer.

During sleep, there is no blinking and drainage is probably minimal. Sack et al. have shown a great decrease in aqueous volume and dramatic changes in its composition.[11] Lying on one's side and not blinking, it would be expected, even with the minimum tear secretion of 1 mm^3/min, that tears would accumulate in the lower canthus, but this does not happen. Indeed, the eye does not fill. At least an extra 10 mm^3 of aqueous can be added before there is any overflow. It is probable that during sleep and non-blinking, there is another absorptive mechanism, or that secretion ceases. Maurice has proposed a conjunctival site for this absorption.

5. CONCLUSIONS

1. The tear film, especially the lipid and mucus, is an extremely stable structure.
2. The effect of a contact lens is to push particulate matter into the mucus layer, and once it is there it takes a very long time for it to be removed.

3. The mucus layer is thick enough to protect beads of 10 μm from the eyelid without the epithelial cells beneath them being damaged.
4. The epithelial cells remain undamaged, and do not stain, suggesting a powerful protective role for the mucus layer.

REFERENCES

1. Maurice DM. The dynamics and drainage of tears. *Int Ophthalmol Clin.* 1973;13:103–116.
2. Jones LT. Epiphora. II. Its relation to the anatomical structures and surgery of the medial canthal region. *Am J Ophthalmol.* 1957;43:209–212.
3. Arlt 1855 von *Graefes Arch Ophthalmol.* 1:135.
4. Arlt 1863 von *Graefes Arch Ophthalmol.* 9:64.
5. Ploman KC, Ergel A, Knutsson F. 1928 *Acta Ophthalmol.* 6 pg 55 in Duke Elder System of Ophthalmol. 4 pg 430 Mosby
6. Hyrtl 1853 Handbach der Topograph. *Anatomie Wien* 1 130. Ref.: Duke Elder System of Ophthalmol. 4 pg 430 Mosby
7. Schirmer O. Studien zur Physiologie und Pathologie der Tränenabsrbierung und Tränenabfuhr. *Arch Ophthalmol.* 1903;56:197–291.
8. Doane MG. Interaction of eyelids and tears in corneal wetting and the dynamics of the normal human eyeblink. *Am J Ophthalmol.* 1980;89:507–516.
9. Rosengren B. On lacrimal drainage. *Ophthalmologica.* 1973;164:409–421.
10. Lemp MA, Weiler HH. How do tears exit? *Invest Ophthalmol Vis Sci.* 1983;24:619–622.
11. Sack RA, Underwood PA, Tan KO, Sutherland H, Morris CA. Vitronectin, possible contribution to the closed-eye external host-defence mechanism. *Ocular Immunol Inflam.* 1993;1:327–336.

105

PATIENT-DEPENDENT AND MATERIAL-DEPENDENT FACTORS IN CONTACT LENS DEPOSITION PROCESSES

Brian J. Tighe, Lyndon Jones, Kathryn Evans, and Valerie Franklin

Aston Biomaterials Research Unit
Department of Chemical Engineering and Applied Chemistry
Aston University
Birmingham, England, United Kingdom

1. INTRODUCTION

The interaction of polymers with the tears is conveniently studied by observing the nature of the contact lens surface contamination produced during wear. In developing analytical techniques capable of distinguishing between anterior and posterior surface deposition and between the different chemical species involved, the potential is established to use the lens as a probe for differences in patient-to-patient and eye-to-eye tear chemistry. In the broader context, the eye is an ideal body site to study the interactions between a complex biochemical system and an implantation material, both because of the ease of access and the fact that materials can be worn for a set period of time under known conditions before removal for analysis.

In this paper, both aspects are addressed. A comparison is made of two types of neutral hydrophilic functional groups in non-ionic hydrogel contact lenses and between two levels of anionicity in ionic hydrogel contact lenses. Additionally the results are used to demonstrate differences in patient-to-patient and eye-to-eye behavior.

The deposition of contact lenses with substances derived from the tear fluid (commonly referred to as "spoilation" in the contact lens literature) is a well-known clinical complication. Contact lens deposits have been implicated as the causative mechanism behind inflammatory reactions such as giant papillary conjunctivitis (GPC) and acute red eye (ARE) syndrome and eventually produce a deterioration in visual performance. There is, currently, inadequate understanding of the relationship between deposition behavior of lenses and their chemical structure and water content, which for existing materials varies from 38% to 85%. Two principal strategies are available to increase the water content of hydrogels above that of poly(2-hydroxyethyl methacrylate), or polyHEMA, which has a

Lacrimal Gland, Tear Film, and Dry Eye Syndromes 2
edited by Sullivan *et al.*, Plenum Press, New York, 1998

water content of 38%. Small quantities of charged groups such as methacrylic acid or larger amounts of more hydrophilic, neutral groups such as polyvinyl alcohol (PVA) or *N*-vinyl pyrrolidone (NVP) are added to polyHEMA or methyl methacrylate to raise their equilibrium water content to 60% or greater. The US Food and Drug Administration (FDA) currently classifies contact lens materials into four groups, depending upon their charge and water content. Two of these groups are relevant here: ionic materials with water content in excess of 55%, categorized as Group IV, and non-ionic materials with water content in excess of 55%, as Group II.

Control of contact lens deposition currently utilizes two opposing methodologies. An attempt can be made to keep the lenses clean for longer periods of time or the lenses can be routinely discarded before they deposit to the extent where clinical problems occur. The principal factor in preventing all patients immediately switching to disposable lens systems is that currently such lenses are available only in a limited parameter range. Strategies to reduce deposition in non-disposable lenses include optimizing the care regimen and manipulating the lens material such that the polymers used are able to resist deposition to a greater extent. This paper compares two neutral hydrophiles, the hydroxyl group and the pyrrolidone ring, in lenses worn for 6 months, and two levels of ionicity, together with the effect of pyrrolidone ring incorporation in lenses worn for 1-month periods.

2. MATERIALS AND METHODS[1]

2.1. Comparison of FDA Group II Materials

Twelve subjects (six male and six female) were entered in the study. All subjects had <1.00 diopter of astigmatism and had previously worn daily wear, low water content, hydrogel contact lenses without complications for at least 1 year. None of the subjects had previous exposure to either of the test materials. Subjects with eye disease, allergies, insufficient lacrimal secretions, pre-existing ocular infections, or history of problems with lens wear, and pregnant subjects were all excluded.

Two Group II lens materials were chosen to ensure that one of the constituents in each material remained consistent (methyl methacrylate) and that equilibrium water contents were similar. All subjects were fitted by one of the investigators (LJ) with the two test lenses, which were Lunelle ES70, in which *N*-vinyl pyrrolidone is the neutral hydrophile, from Essilor, and Excelens, which is based upon polyvinyl alcohol as the neutral hydrophile, from CIBA-Vision. Lenses were fitted using conventional clinical methods, with care being taken to ensure that both lens types provided adequate movement and corneal coverage. If either lens type fitted inadequately, that subject was excluded from the study. The subjects were divided into two equal groups, each of which wore one of the lens types on a daily wear basis every day for 6 months. Subjects were reviewed in the clinic on initial collection and after 1, 3, and 6 months. At the 6-month visit, a cross-over took place, with all subjects repeating the above visit schedule with the other lens material. The study design used a randomized, double-masked paradigm, in which neither the practitioner nor the subject was aware of the lens type being used at the time of the clinical visits.

Lenses were cleaned at the end of every day with an isopropyl alcohol-based surfactant cleaner (CIBA-Vision Miraflow), rinsed in non-preserved saline, and subsequently disinfected overnight with a two-step 3% hydrogen peroxide system (CIBA-Vision 10:10). In the morning, the lenses were neutralized for at least 20 min with sodium pyruvate before insertion took place. No enzyme tablets or rewetting drops were used at any time during the study.

2.2. Comparison of FDA Group IV Materials

Twenty-one subjects (six male and 15 female) fitting the same patient profiles described under Section 2.1. were entered in the study. All subjects were fitted with the two test lenses, which were Vistakon Surevue (Etafilcon, based on HEMA and methacrylic acid) and Ciba-Vision Focus (Vifilcon, based on HEMA, polyvinyl pyrrolidone, and methacrylic acid). All subjects used Bausch & Lomb's ReNu disinfecting system, utilizing a rub-and-rinse step prior to disinfection. No enzyme tablets were used during the study.

The subjects were divided into two roughly equal groups, each of which wore a pair of lenses for 1 month. Then a cross-over took place. Each test lens was exposed equally for each month. Subjects were reviewed in the clinic on collection and after 2 and 4 weeks. The study design utilized a randomized, double-masked paradigm, in which neither practitioner nor patient was aware of the lens type being used at the time of the clinical visits.

2.3. Clinical and Analytical Protocols

During the scheduled clinic visits, a number of variables were assessed by various clinical monitoring techniques. Such techniques provide a unique opportunity to investigate the interaction of artificial materials with the body. The results are beyond the scope of this paper and will be reported elsewhere.

Previous work in these laboratories has shown that handling hydrogel lenses results in the transfer of skin lipids onto the lens surface, which will significantly influence the degree of deposition subsequently assessed. To overcome this, a specialized collection technique was developed. The subject looked in towards the nose (with the eye wearing the lens to be collected), and the contact lens was gently moved onto the temporal sclera and subsequently removed with sterile, plastic-tipped tweezers. In a separate series of experiments, this technique was shown to prevent lens contamination and to result in no ocular trauma.

At the end of the wearing schedule, each lens was removed with the tweezers and placed in a glass vial containing sterile, non-preserved saline. The vials were then capped, labelled with the subject's initials and study number, and refrigerated prior to spoilation analysis.

2.3.1. Lipid Analysis – Fluorescence Spectrophotofluorimetry. Lipoidal deposition was evaluated using a fluorescence technique on a modified Hitachi F-4500 fluorescence spectrophotometer. This is a non-destructive technique that relies on the low-level fluorescence of lipoidal species following excitation by UV light. This technique is based on the intrinsic fluorescence produced by the conjugated double bonds present in all lipids. Lenses are placed in distilled water in a specially designed cylindrical quartz cell that allows reproducible orientation of the lens to the incident light beam. An incident (excitation) beam wavelength of 360 nm is used, and the height of the resultant fluorescence (emission) peak is monitored at approximately 440 nm. The height of the emission peak provides an estimate of the relative level of deposition.

Baseline fluorescence for the lens materials was evaluated by examining a blank, unworn lens of each material as described above. This background trace was then subtracted from the result achieved with each worn lens to assess accurately the degree of deposited material.

2.3.2. Surface Protein Deposition – Fluorescence Spectrophotofluorimetry. Surface protein deposition was evaluated in a similar manner to that for lipids, using non-destructive fluorescence spectrophotofluorimetry. An incident (excitation) beam wavelength of 280 nm was used, and the height of the resultant fluorescence (emission) peak was monitored at approximately 340 nm. Baseline fluorescence for each material was subtracted as described above.

2.3.3. Total Protein Deposition – Ultraviolet Spectroscopy. Total protein concentration was measured using UV absorbance on a Hitachi U2000 spectrophotometer. The measurement was carried out in matched quartz cells, against an unworn contact lens of the same material and power at 280 nm.

3. RESULTS AND DISCUSSION

The surface lipid and surface protein for each worn lens was determined. The results were first plotted as left eye vs. right eye, in which the value obtained for the right eye of each lens was plotted against that of the lens concurrently worn in the left eye. Values for lipid levels are shown for both FDA Group IV materials and one Group II material in Fig. 1 (Excelens, Etafilcon, and Vifilcon) and for all four materials in Fig. 2. The surface protein levels for all four materials are shown in Fig. 3.

Values for left and right eyes for the same material worn for identical time periods show significant variations, and for protein, these are not appreciably less than the patient-to-patient variations. The material differences are significant, with the two FDA Group IV materials attracting the highest protein levels despite their shorter wear periods, and Etafilcon, which contains the highest level of ionicity, attracting significantly more than the less anionic Vifilcon. Lipid levels are highest for the two *N*-vinyl pyrrolidone-containing materials, Vifilcon and ES 70. The values for ES 70, worn for 6 months, are markedly

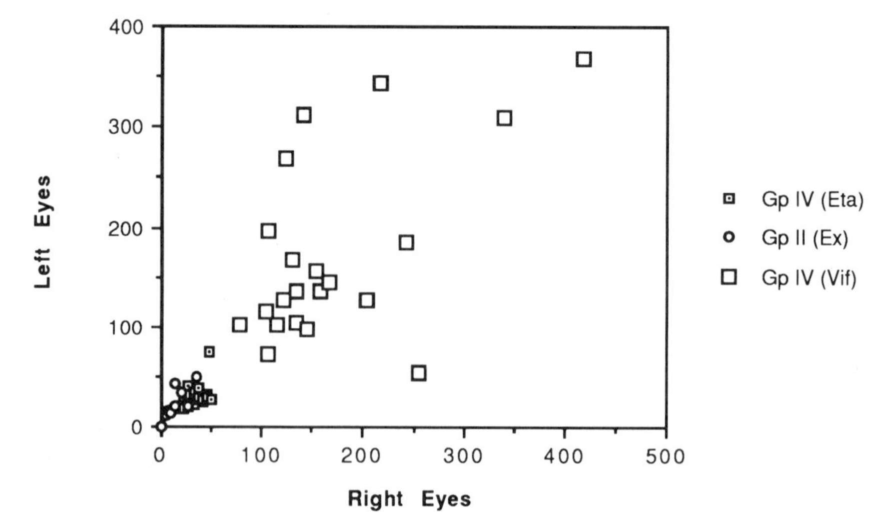

Figure 1. Surface lipid measured by fluorescence showing left eye vs. right eye variations for Excelens, Etafilcon, and Vifilcon.

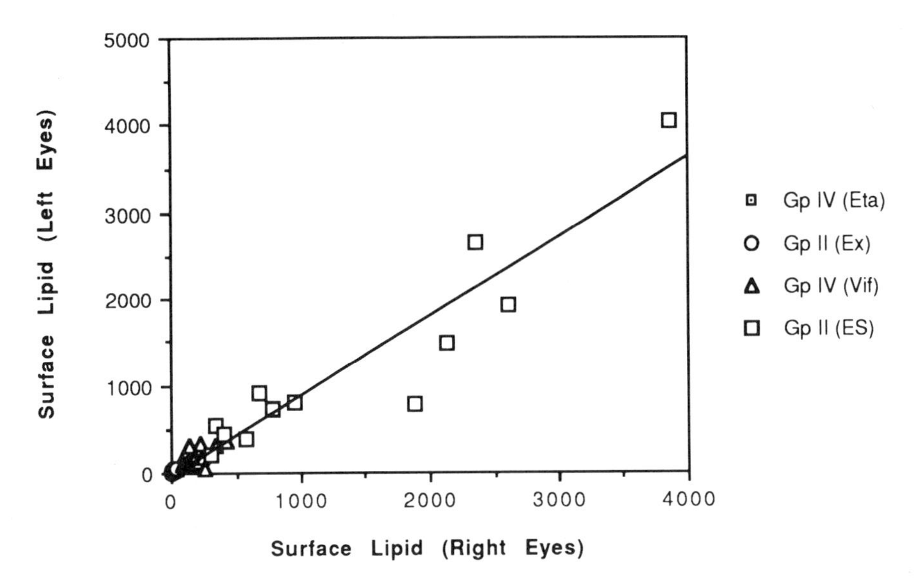

Figure 2. Surface lipid measured by fluorescence showing left eye vs. right eye variations for Excelens, Etafilcon, Vifilcon, and ES 70.

higher than for any of the other materials (Fig. 1 and 2). Left eye-right eye variations are very patient specific, and much greater patient-patient variations occur with lipid than with protein. Separate experiments show that lipid penetrates into the polymer matrix of the hydrogel lenses up to a depth of ~5 μm and that similar, but not identical levels of deposition occur on posterior and anterior surfaces. Experimental values shown here are for anterior surfaces throughout.

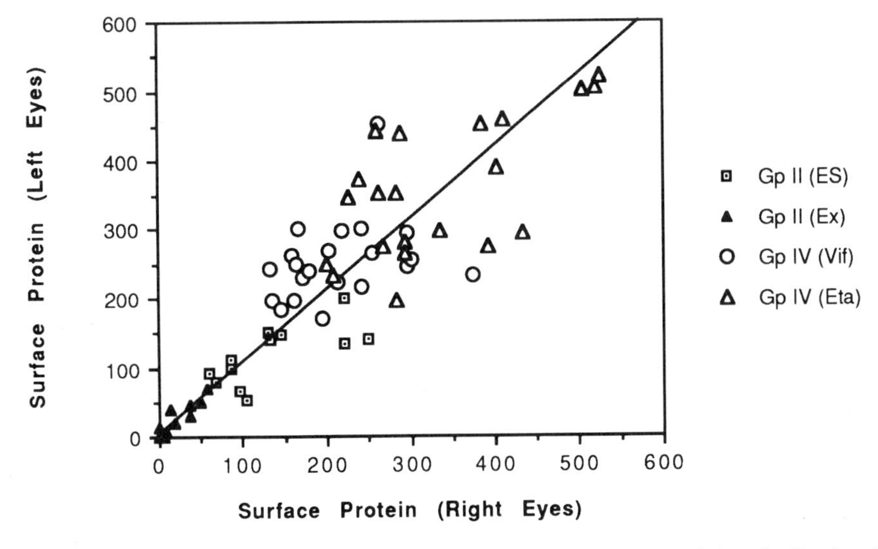

Figure 3. Surface protein measured by fluorescence showing left eye vs. right eye variations for Excelens, Etafilcon, Vifilcon, and ES 70.

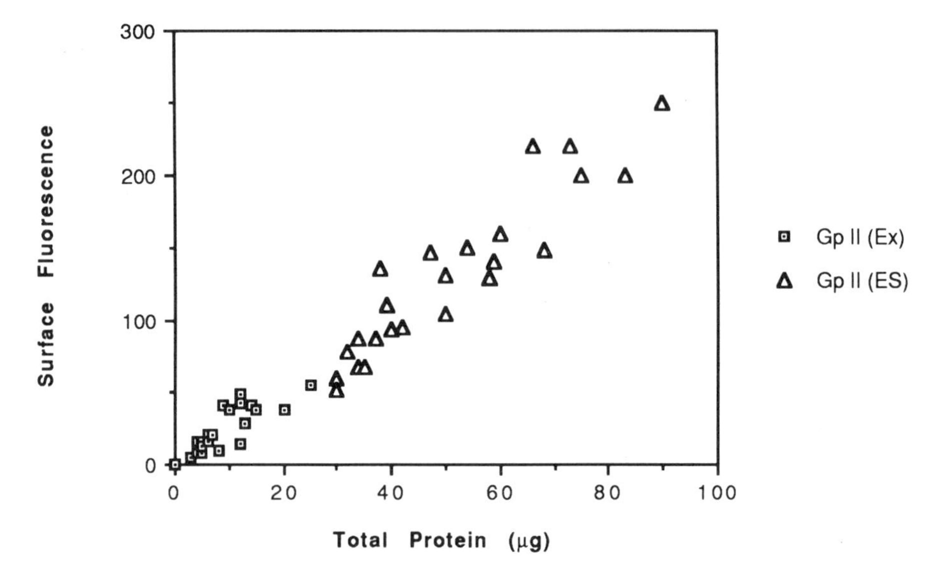

Figure 4. Surface protein measured by fluorescence plotted against total protein (UV) for FDA Group II materials (Excelens and ES 70).

Protein penetration experiments were also carried out; these indicate that no significant protein penetration occurs with either of the FDA Group II materials, Excelens and ES 70. This is also reflected in the substantially linear plot of total protein against surface protein for these materials (Fig. 4).

Protein penetration measurements on both FDA Group IV materials show protein penetration concentrated within 10 µm of both anterior and posterior surfaces. The fact that protein penetrates into the matrix of these materials is illustrated by incorporating re-

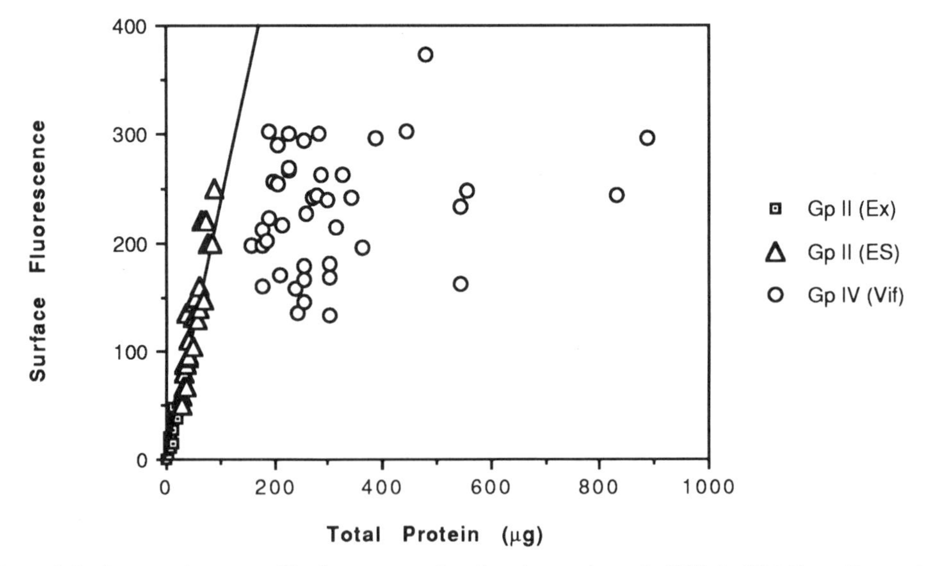

Figure 5. Surface protein measured by fluorescence plotted against total protein (UV) for FDA Group II materials (Excelens and ES 70) together with one Group IV material (Vifilcon).

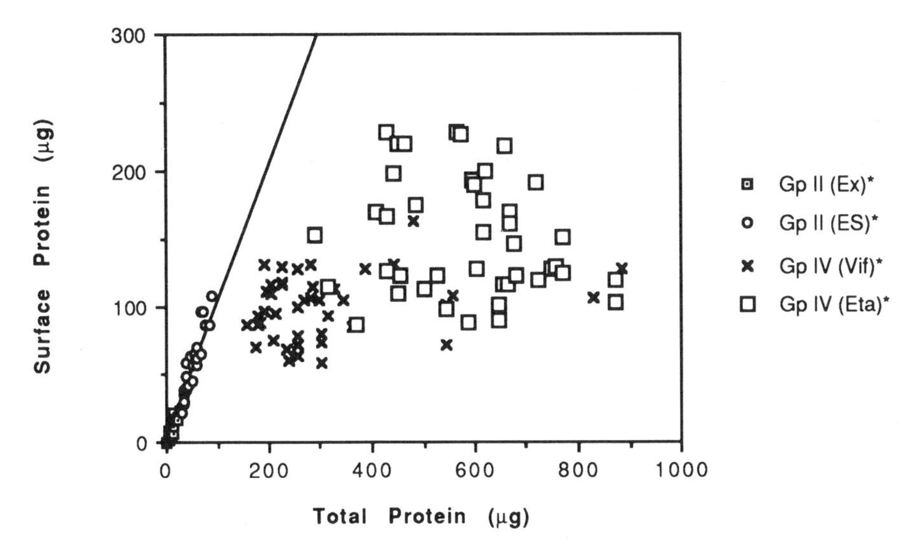

Figure 6. Calibrated plot of anterior plus posterior surface protein against total protein (surface plus matrix) for FDA Group II and Group IV materials (Excelens, ES 70, Vifilcon, and Etafilcon).

sults of surface vs. total protein for Group IV Vifilcon together with those for the two Group II materials in the form of the plot presented in Fig. 4. The combined results are shown in Fig. 5. Here the fact that not all Group IV protein is located on the lens surface is reflected by the fact that the points lie in a cluster to the right of the line associated with the Group II materials. By calibrating the surface protein technique, a plot showing actual protein levels associated with both anterior and posterior surfaces against total protein associated with surface and matrix may be produced. This is shown for all four materials in Fig. 6. Separate experiments established that the protein associated with Group IV lenses is substantially lysozyme and that it is readily extracted from the lens matrices and has intact biological activity. Our results show that satisfactory comparisons of lens deposition can be carried out only if protein and lipid associated with lens storage solutions is assayed in order that losses into solution may be taken into account.

4. CONCLUSIONS

- Both eye-to-eye and patient-to-patient variations are important but are more marked with lipid than with protein.
- Lipid accumulation and penetration occur with all materials tested but is enhanced in materials that contain *N*-vinyl pyrrolidone.
- Both surface-located and matrix-located protein are enhanced by increasing ionicity in FDA Group IV materials.
- Matrix protein is located within 10 µm of the lens surface and consists predominantly of mobile, biologically active lysozyme.

REFERENCE

1. Jones L, Franklin V, Evans K, Tighe BJ. Spoilation and clinical performance of monthly vs three monthly disposable contact lenses. *Optom Vis Sci.* 1996;73:16–21.

EFFECTIVENESS OF BION TEARS, CELLUFRESH, AQUASITE, AND REFRESH PLUS FOR MODERATE TO SEVERE DRY EYE

Peter C. Donshik,[1] J. Daniel Nelson,[2] Mark Abelson,[3] James P. McCulley,[4] Cliff Beasley,[5] and Robert A. Laibovitz[6]

[1]Eye Physician Associates of Hartford
West Hartford, Connecticut
[2]St. Paul Ramsey Medical Center
St. Paul, Minnesota
[3]Ophthalmic Research Associates
N. Andover, Massachusetts
[4]Southwest Medical School
Dallas, Texas
[5]Low Vision Centers of Texas
Fort Worth, Texas
[6]Eye Research Associates
Austin, Texas

1. INTRODUCTION

Keratoconjunctivitis sicca (KCS) is a dry eye syndrome characterized by insufficient aqueous tears to maintain health of the conjunctival and corneal epithelium.[1] KCS can affect individuals of both sexes, most commonly postmenopausal women, and may be associated with considerable ocular morbidity.[2] In the absence of any definitive cure for dry eye disorders, artificial tears remains the only therapeutic choice.

Several different formulations of artificial tear solutions are currently available. Most have short retention times, require frequent dosing, and contain preservatives. Formulations that contain viscosity agents, including hyaluronic acid, have been reported to be effective in treating symptoms and prolonging fluid residence time, but have not been found to improve objective findings.[2,3] Preserved eye drops have been reported to elicit exacerbation of ocular surface disease, while nonpreserved eye drops do not appear to have similar effects.[1,2,4–6] In vitro exposure of damaged corneas to experimental tear solutions with an electrolyte composition similar to human tears and buffered with bicarbonate

Lacrimal Gland, Tear Film, and Dry Eye Syndromes 2
edited by Sullivan *et al.*, Plenum Press, New York, 1998

has been reported to promote the recovery of damaged corneal epithelium to a greater extent than unbuffered formulas or solutions buffered with nonbiologic buffers, such as borate.[6,7]

The evaluation of dry eyes has historically been based on clinical findings such as tear breakup time, Schirmer tear test, and rose bengal and fluorescein staining of the conjunctiva and cornea.[8,9] In recent years, conjunctival impression cytology, a noninvasive technique, has been added to measure conjunctival changes due to KCS. These changes are embodied in the term squamous metaplasia, as evidenced by goblet cell loss, enlargement of nongoblet epithelial cells, increased cellular stratification, keratinization, and loss of surface microvilli.[10–12]

The overall objective of the studies reported here was to evaluate the effectiveness of various physiological tear solutions among patients with moderate to severe dry eye syndrome compared to BION Tears, a preservative-free physiological tear preparation containing bicarbonate.

2. METHODS AND MATERIALS

For inclusion into all three studies, patients were required to have a moderate to severe case of KCS. Schirmer tests of equal to or less than 5 mm of wetting after 5 min, positive rose bengal staining equal to or greater than 3 out of 9, and a conjunctival impression cytology mean score of 3 out of 9 for the bulbar conjunctiva and less than 1 out of 3 for the palpebral conjunctiva were required. Patients were excluded if they wore contact lenses, had ocular surgery within the past 6 months, had active infection or inflammatory disease not related to KCS, or had any corneal abnormalities unrelated to KCS. Study participants could not use any ocular medication during this study. In addition, they could not have taken any concomitant systemic medication for less than 1 month prior to being entered into the study, and their dose regimens could not be changed during the course of the study. All studies were reviewed and approved by an Institutional Review Committee. At the screening visit, the inclusion criteria were verified, an informed consent was obtained, a complete ophthalmic examination was performed, and a complete medical history obtained.

Study 1 consisted of a randomized, open-label, single-center, parallel group study comparing BION Tears (Hydroxypropyl Methylcellulose 0.3%; Dextran 70, 0.1%; electrolytes similar to human tears with bicarbonate as a buffer) to a similar formulation, Formulation 2, which contained the exact same formulation of artificial tears without bicarbonate. Twenty-seven patients with a mean age of 51.8 years (range, 26–72 years) were enrolled. Patients were examined on Days 0, 7, 28, and 56. At follow-up visits on Days 7, 28, and 56, visual acuity, slit lamp examination, and rose bengal staining were performed, and ocular symptoms were recorded. In addition to the examination and recording of ocular symptoms, a Schirmer test and conjunctival impression cytology were also performed at follow-up visits. The ocular symptoms evaluated consisted of ocular discomfort, foreign body sensation, dryness, and photophobia, which were rated on a scale of 0 (absent) to 3 (severe). Itching was also evaluated, on a scale of 0 (no abnormal sensation) to 4 (severe).

Study 2 was a randomized, investigator-masked, multicenter, parallel group study that compared the efficacy of BION Tears (Alcon Laboratories) to Cellufresh (Allergan Pharmaceuticals) and AquaSite (Ciba Vision Ophthalmics) in moderate to severe dry eye patients. Forty-one patients with a mean age of 52 years (range, 30–82 years) were enrolled. Fourteen patients were on BION Tears, 13 on AquaSite (Polyethylene glycol 400, 0.2%; Dextran 70, 0.1%), and 14 on Cellufresh (Carboxymethylcellulose Sodium 0.5%). Following a 1–2 week washout period with Refresh, all patients meeting the inclusion criteria were randomized to one of the three study medications. One week prior to receiving the medication, patients underwent a complete ocular examination with impression cytology. On Day 0, patients received the study medication and then were examined on Days 21, 42, 63, and 84. At each follow-up visit, visual acuity, Schirmer tests, slit lamp examination, fluorescein staining, rose bengal staining, and impression cytology were performed. In addition, ocular signs and symptoms were evaluated. Comfort was assessed utilizing a scale of 0 (none) to 9 (extreme) for burning, stinging, blurriness, filminess, grittiness, dryness, tightness, stickiness, matted, scratchiness, and itchiness, for a total possible score of 0 to 99. Acceptability of the tear substitute was also ascertained, using the sum of the patient's ratings scores from 0 (poor) to 9 (excellent) for the relief of burning, stinging, grittiness, ease of putting in the medication, control over the amount of tears, or the accuracy of getting the tears into the eye, and the overall ease of use. In addition, the lack of messiness and an overall acceptability were also determined. A total possible score ranged from 0 to 81.

Study 3 was a randomized, investigator-masked, multicenter, parallel group study that compared BION Tears to Refresh Plus in patients with moderate to severe KCS. Study procedures were similar to Study 2, except for the rose bengal evaluation. The nasal and temporal conjunctiva were divided into three areas each. Each area was evaluated with a possible rating of 0 to 4 (thus, a possible score ranged from 0 to 24). One hundred and twenty-four subjects with a mean age of 56 years (range, 24–82 years) were enrolled. Sixty-one patients were on BION Tears and 63 on Refresh Plus.

The primary efficacy variables in all of the studies were rose bengal staining and impression cytology. In Studies 2 and 3, comfort and acceptability were also primary efficacy variables. Secondary efficacy variables were ocular signs and symptoms. In Study 1, because of the small patient numbers, only descriptive statistics were calculated for each treatment group, and between-treatment comparisons were based on clinical judgment. No inferential statistical analyses were performed. In Studies 2 and 3, all parameters were analyzed using an analysis of variance model conducted with a 0.05 probability. Standard descriptive statistics were calculated for each variable in the analysis. Statistical analyses consisted of change from baseline score for the worst eye.

3. RESULTS

Table 1 shows the mean rose bengal staining scores for Studies 1, 2, and 3. A consistent improvement in rose bengal staining was seen in all three studies in patients using BION Tears. BION Tears appeared to have a greater clinical benefit (improvement of more than 1 score unit) than formulations without bicarbonate, AquaSite and Cellufresh. Both BION Tears and Refresh Plus were statistically significant in improving rose bengal staining on Day 84 compared to baseline.

Table 2 shows the changes in impression cytology on Days 56 (Study 1) and 84 (Studies 2 and 3). The BION Tears showed an improvement in impression cytology in all three studies. In Study 1, BION Tears appeared to be clinically superior to formulations without bicarbonate, showing an improvement of 0.9 units in BION Tears compared to a worsening of 0.5 units with Formulation 2. In Study 3, BION Tears and Refresh Plus were both statistically significant in improving rose bengal staining on Day 84 compared to baseline.

Table 1. Mean rose bengal staining scores

Study	Treatment	Baseline	Day 56	Day 84
1	BION Tears	4.3	2.5[a]	—
	Formulation 2	3.8	3.0	—
2	BION Tears	4.9	—	3.7[a]
	AquaSite	3.8	—	4.1
	Cellufresh	5.5	—	5.6
3	BION Tears	12.9	—	10.2[b]
	Refresh Plus	12.5	—	8.6[b]

[a]Changes from baseline were clinically significant based on improvement of more than 1 score unit.
[b]Changes from baseline were statistically significant ($p < 0.001$).
Rating scales in Studies 1 and 2 consisted of a sum of scores for the temporal and nasal conjunctiva and cornea regions (range, 0-3 for each region) with a total possible score of 0-9.
Rating scale in Study 3 consisted of a sum of scores for the temporal and nasal conjunctiva divided into 3 regions (range, 0-4 for each of 6 regions) with a total possible score of 0-24.

Table 2. Impression cytology

Study	Treatment	Baseline	Day 56	Day 84
1	BION Tears	6.5	5.6[a]	—
	Formulation 2	5.8	6.3	—
2	BION Tears	5.3	—	5.1
	AquaSite	5.8	—	6.3
	Cellufresh	5.3	—	6.2
3	BION Tears	6.9	—	5.8[b]
	Refresh Plus	6.7	—	5.9[b]

[a]Changes from baseline approached clinical significance based on improvement of almost 1 score unit (0.9).
[b]Changes from baseline were statistically significant ($p < 0.05$).

Table 3. Mean rating of symptoms, study 1

Symptom	Treatment	Baseline	Day 56
Discomfort	BION Tears	1.9	1.0
	Formulation 2	2.1	1.5
Foreign body sensation	BION Tears	1.7	0.6
	Formulation 2	1.6	1.3
Dryness	BION Tears	2.5	1.7
	Formulation 2	2.2	1.8
Photophobia	BION Tears	1.5	1.1
	Formulation 2	1.5	1.4
Itching	BION Tears	0.4	0.6
	Formulation 2	0.8	0.8

Because of small patient numbers, only descriptive statistics were calculated for each treatment group. Between-treatment comparisons were based on clinical judgment. No inferential statistical analyses were performed.

Table 4. Mean rating of symptoms, studies 2 and 3

	Study 2			Study 3		
Symptom	Treatment	Baseline	Day 84	Treatment	Baseline	Day 84
Dryness	BION Tears	2.3	1.9	BION Tears	2.4	1.8[b]
	Cellufresh	2.1	2.2	Refresh Plus	2.3	1.9[b]
	AquaSite	2.0	1.8			
Itching	BION Tears	2.1	1.6	BION Tears	1.6	0.9[a]
	Cellufresh	1.6	1.5	Refresh Plus	1.5	0.9[a]
	AquaSite	1.5	0.7			
Tearing	BION Tears	0.2	0.5[a]	BION Tears	0.4	0.2
	Cellufresh	0.1	0.0	Refresh Plus	0.5	0.2[a]
	AquaSite	0.2	0.1			
Foreign body sensation	BION Tears	1.4	1.0	BION Tears	1.5	0.7[b]
	Cellufresh	0.6	0.9	Refresh Plus	1.5	0.9[b]
	AquaSite	1.2	1.0			

[a]Changes from baseline were statistically significant ($p < 0.05$).
[b]Changes from baseline were statistically significant ($p < 0.01$).

Table 3 shows the mean rating of symptoms in Study 1 comparing BION Tears to Formulation 2 for discomfort, foreign body sensation, dryness, photophobia, and itching. Both study medications had an improvement in all symptoms except itching.

Table 4 reports the mean rating of symptoms in Studies 2 and 3. In Study 2, BION Tears had an improvement in all symptoms except tearing. In Study 3, both BION Tears and Refresh Plus had statistically significant improvements related to dryness, itching, and foreign body sensation.

Fig. 1 shows the mean comfort rating for Study 2.

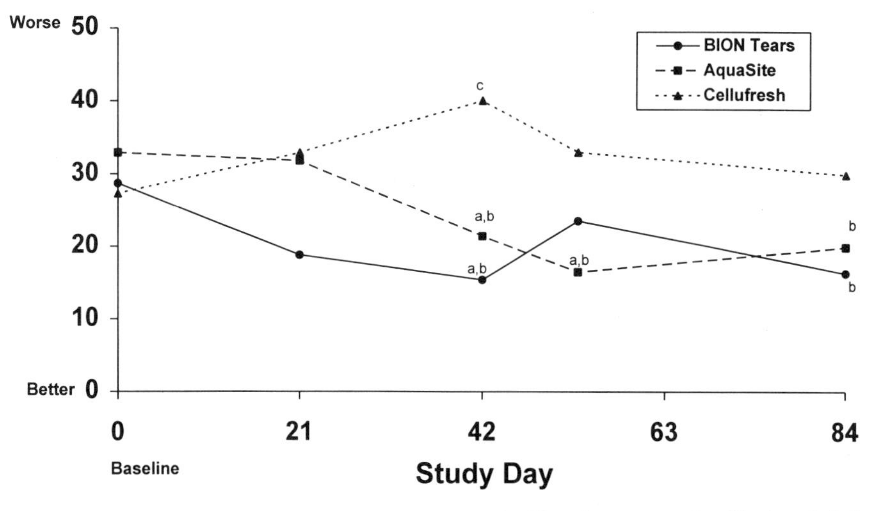

Figure 1. Mean comfort rating, Study 2. a, BION Tears was significantly more comfortable than Cellufresh, as was AquaSite ($p < 0.01$). b, comfort ratings for BION Tears and AquaSite improved significantly relative to baseline ($p < 0.05$). c, comfort ratings for Cellufresh decreased significantly relative to baseline ($p < 0.03$).

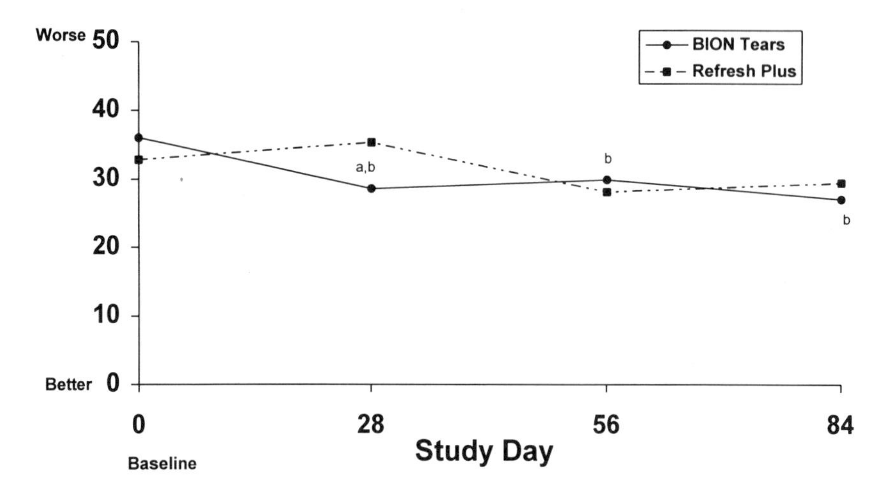

Figure 2. Mean comfort rating, Study 3. a, BION Tears was significantly more comfortable than Refresh Plus at day 28 ($p = 0.02$). b, comfort ratings for BION Tears improved significantly relative to baseline ($p < 0.05$).

Fig. 2 shows the mean comfort rating for Study 3. BION Tears provided the greatest relief from discomfort in Study 1; BION Tears and AquaSite were considerably more comfortable than Cellufresh ($p < 0.01$) in Study 2. In Study 3, BION Tears was significantly more comfortable than Refresh Plus; this difference was statistically significant at Day 28 ($p = 0.02$) and improved significantly relative to baseline ($p < 0.05$).

Fig. 3 shows the mean acceptability rating for Study 2.

Fig. 4 shows the mean acceptability rating for Study 3. In Study 2, BION Tears was significantly more acceptable than AquaSite ($p < 0.05$); however, Cellufresh was found to be slightly more acceptable than BION Tears. Both Cellufresh and BION Tears were significantly more acceptable compared to baselines, while AquaSite was somewhat less ac-

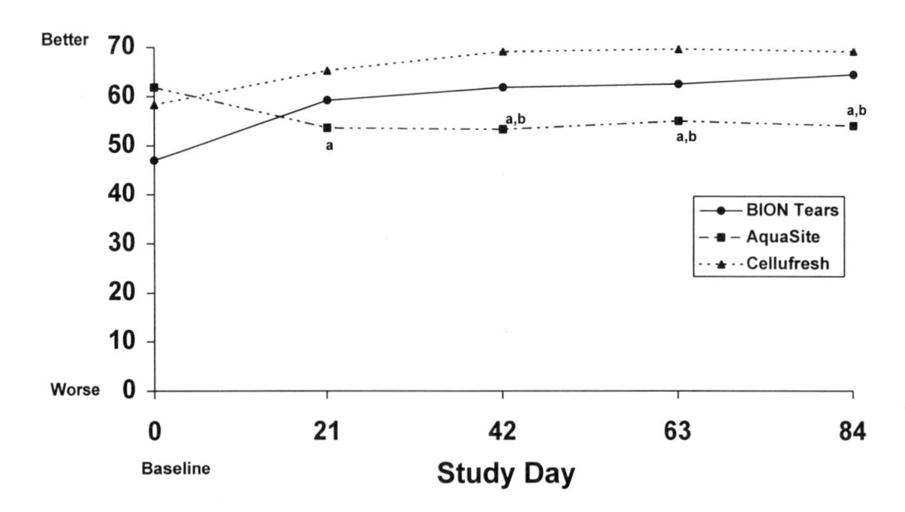

Figure 3. Mean acceptability rating, Study 2. a, BION Tears was significantly more acceptable than AquaSite ($p < 0.05$). b, acceptability ratings for BION Tears improved significantly relative to baseline ($p < 0.04$).

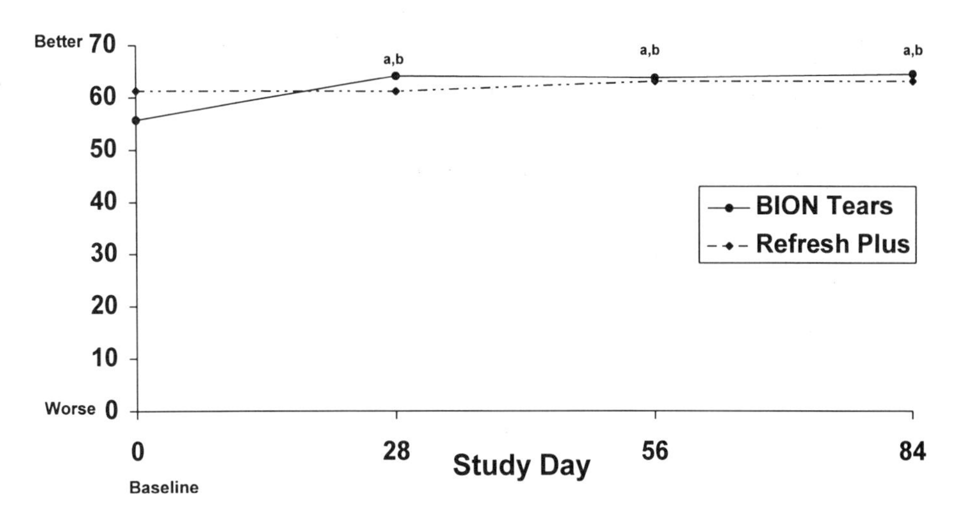

Figure 4. Mean acceptability rating, Study 3. a, BION Tears was significantly more acceptable than Refresh Plus throughout the study ($p < 0.05$). b, acceptability ratings for BION Tears improved significantly relative to baseline ($p < 0.001$).

ceptable compared to baseline. BION Tears was significantly more acceptable than Refresh Plus in Study 3 ($p = 0.05$), with BION Tears showing an improved acceptability compared to baseline throughout the study.

4. CONCLUSION

All solutions tested appeared to be helpful in relieving patient symptoms and comfort, with BION Tears providing the greatest relief from discomfort. With regard to ocular surface disease, both Refresh Plus and BION Tears were significant in decreasing rose bengal staining and improving impression cytology.

ACKNOWLEDGMENT

Research supported by Alcon Laboratories, Fort Worth, TX.

REFERENCES

1. Holly FJ, Lemp MA. Tear physiology and dry eyes. *Surv Ophthalmol.* 1977; 22:69–87.
2. Nelson JD, Farris RL. Sodium hyaluronate and polyvinyl alcohol artificial tear preparations: A comparison in patients with keratoconjunctivitis sicca. *Arch Ophthalmol.* 1988;106:484–487.
3. Stuart JC, Linn JG. Dilute sodium hyaluronate (Healon) in the treatment of ocular surface disorders. *Ann Ophthalmol.* 1985;17:190–192.
4. Lemp MA. Recent developments in dry eye management. *Ophthalmology.* 1987;94:1299–1304.
5. Gobbels M, Spitznas M. Cornea epithelial permeability of dry eyes before and after treatment with artificial tears. *Ophthalmology.* 1992;99:873–887.
6. Gilbard JP, Rossi SR, Gray Heyda K. Ophthalmic solutions, the ocular surface, and a unique therapeutic artificial tear formulation. *Am J Ophthalmol.* 1989;107:348–355.

7. Keller N, Moore D, Carper D, Longwell A. Increased corneal permeability induced by the dual effects of transient tear film acidification and exposure to benzalkonium chloride. *Exp Eye Res.* 1980;30:203–213.
8. Farris RL, Gilbard JP, Stuchell RN, Mandel ID. Diagnostic tests in keratoconjunctivitis sicca. *CLAO J.* 1983;9:23–28.
9. Van Bijsterveld OP. Diagnostic tests in the sicca syndrome. *Arch Ophthalmol.* 1978;96:677–681.
10. Reddy M, Reddy PR, Reddy SC. Conjunctival impression cytology in dry eye states. *Indian J Ophthalmol.* 1991;39:22–24.
11. Bron AJ, Mengher LS. The ocular surface in keratoconjunctivitis sicca. *Eye.* 1989;3:428–437.
12. Rivas L, Oroza MA, Perez-Esteban A, Murube del Castillo J. Morphological changes in ocular surface in dry eyes and other disorders by impression cytology. *Graefes Arch Clin Exp Ophthalmol.* 1992;230:329–334.

107

OCULAR RESIDENCE TIME OF CARBOXYMETHYLCELLULOSE SOLUTIONS

Jerry R. Paugh,[1] Ronald C. Chatelier,[2] and Joseph W. Huff[1]

[1]Cooperative Research Centre for Eye Research and Technology
University of New South Wales
Sydney, Australia
[2]Commonwealth Scientific and Industrial Research Organisation
Division of Chemicals and Polymers
Clayton, Victoria, Australia

1. INTRODUCTION

An important goal of topical ophthalmic solution formulation is to prolong ocular surface contact time, with the expectation of reduced dosages delivered more uniformly. The ideal method of assessing residence time is to covalently bond a tracer to the active ingredient; however, the expense of such labelling is prohibitive for all but the most promising formulations. An alternative is to admix tracers into the formulation, to date with the small molecular weight (MW) tracers sodium fluorescein and technetium-99, a gamma emitter. Both small molecule tracer techniques used to estimate residence time are limited by two factors. First, small molecules penetrate the ocular tissues, confounding elimination; second, they may be eliminated at a rate that differs from that of the polymeric viscous agent. In addition, the gamma scintigraphic method has been criticized for poor sensitivity and spatial resolution,[1] and requires expensive equipment and a radioisotopic pharmacy.

Recently, a new method has been suggested[2,3] that uses a fluorescent dextran of large molecular weight (~70,000) admixed with the topical formulation to measure retention of dry eye formulations. The required instrumentation consists of a biomicroscope modified with an appropriate excitation filter and a photomultiplier/amplification system. The assumption of the technique is that the tracer will admix with the formulation's viscous agents and be eliminated at the same rate. The advantages are accuracy, sensitivity, low cost, and simplicity.

Distinct from measurement methodology, the retention of topical ophthalmic vehicles is complex and is broadly influenced by the comfort, instillation parameters, and formulation characteristics. Comfort can be influenced by vehicle pH, osmolality, and possibly surface tension, but, assuming physiologically acceptable values, is most influenced by viscosity and the perception of "stickiness." A formulation attribute potentially related to comfort is the rheological nature of the topical adjunct, with the idea that a pseudoplastic (i.e., shear thinning) material that is similar to tears may provide optimum comfort,[4] although this premise has not been well investigated.

Assuming acceptable comfort, key formulation variables that are believed to influence retention are viscosity and bioadhesion, with surface tension playing a lesser role, at least with traditional ophthalmic viscous agents.[5] Viscosity is influenced by polymer concentration, branching, and molecular weight,[6] although all residence time studies to date have examined various concentrations of a *given* molecular weight polymer. Viscosity effect on residence time, examined with one molecular weight, has been studied for polyvinylalcohol (PVA) and hydroxypropylmethylcellulose (HPMC) with the sodium fluorescein and gamma scintigraphic methods.[5,7] Only modest improvement in elimination coefficient and area under the concentration vs. time curve (AUC) were found. The reason that viscosity has not shown a greater effect on residence time likely has to do with the limited bioadhesion of traditional viscous agents.

Mucus adhesion is enhanced by polyanionic polymers,[8–10] and recently prolonged residence time has been demonstrated for polyacrylate preparations (Carbopol) using fluorometry[11] and for hyaluronic acid (HA) using gamma scintigraphy.[12] In the case of polyacrylate materials, ionic strength greatly influences rheologic behavior (i.e., increased ionic strength reduces viscosity), which may limit the usefulness of the material for sustained drug delivery. Hyaluronic acid appears to have several desirable attributes (e.g., pseudoplastic rheology, and good bioadhesion) but is costly to produce in commercial quantities. Similar to these materials, carboxymethylcellulose (CMC) is a polyanionic polymer with pseudoplastic rheology and excellent mucoadhesion.[8] Further, CMC is inexpensive and readily available, and thus may offer a unique material to optimize residence time.

This research project involved the study of CMC residence time using the FITC-dextran method in normal subjects. A key aspect was the examination of iso-viscous solutions with low (~25 mPa.s), medium (~100 mPa.s), and high (~250 mPa.s) apparent viscosity, but composed of different molecular weight and concentration CMC polymers. The results suggest that CMC MW at these levels is not the primary determinant of residence time. It appears instead that residence time is a function of increasing viscosity. However, there were significant differences in comfort between the low and high MW formulations, probably related to the greater pseudoplasticity of the high MW solutions.

2. MATERIALS AND METHODS

Eighteen normal subjects were recruited to participate in the study. Subjects had normal ocular surface and tear film characteristics and did not wear contact lenses or take tear-influencing medications. Subjects ranged in age from 22–37 years and provided written consent following description of study procedures.

A maximum of two solutions were instilled at one experimental session. The test eye received a viscous solution of ~25, 100, or 250 mPa.s viscosity, composed of either large MW-low concentration CMC polymer, or lower MW and higher concentration CMC formulation, and the contralateral eye received a borate-buffered saline control. The CMC solutions were of ~90K and ~700K MW pharmaceutical grade polymer (Aqualon® Cellulose Gum, A.C. Hatrick Chemicals, Sydney, Australia) dissolved in high-purity water, adjusted to physiological osmolality with NaCl, and sterilized to preserve MW. The concentrations were (lowest to highest viscosity, all as percent, wt/vol) 3.50, 5.01, 6.21; and 0.40, 0.57, 0.80, for the 90K and 700K CMCs, respectively. Formulations were identified as e.g., "L-25" meaning *L*ow MW CMC of approximately 25 mPa.s viscosity. A Brookfield RVT cone-plate viscometer (Brookfield Engineering, Stoughton, MA) was used for rheological characterization. The fluorescent tracer was FITC-dextran (~72,000 MW, Sigma, St. Louis, MO) in buffered saline, that was sterile filtered and admixed with the CMC formulations and saline control to a final concentration of 0.1% wt/vol.

A Nikon FS-2 photographic biomicroscope (Nikon, Japan) was used as the base instrument for the fluorometer system. It provides excellent optics, variable magnification, and a built-in beam splitter for convenient switching to the photomultiplier from the normal observation arrangement. One filter in the light tower filter wheel was replaced with a custom fluorescein excitation filter (l_{max} 488 nm, FWHM 10 nm, Omega Optics, Brattleboro, VT). The emitted light was transmitted through the standard instrument optics onto a side-on photomultiplier tube (PMT) mounted on the camera port (PMT housing = Oriel #70680; PMT = Oriel #77349, Oriel, Stratford, CT). The emitted light passed through a black metal mask (to reduce stray light) and an emission filter (l_{max} 530 nm, FWHM 30 nm, Omega Optics). The PMT output was directed to an amplifier (#7070, Oriel), and the final signal fed to a standard strip chart recorder. The observation angle was approximately 50 degrees.

2.1. Study Design

The study was a double masked, six-way crossover block design randomized for test formulation and eye. One eye received the test formulation (one of the six CMC mixtures) and the other the saline control. The saline control for the contralateral eye served to reduce inter- and intra-subject variation by compensating for variation in tear turnover rate. A 24 h washout period was enforced to prevent carryover effects. The key parameter was residence time, as indicated by one of several measures. These included gross residence time, as determined by the time in minutes for the fluorescent signal to return to 1.5 times the baseline corneal fluorescence signal, total area under the curve (AUC, a measure of contact time), and the elimination rate constants, k_1 and k_2, which are related to the double exponential nature of the fluorescence decays.

2.2. Power Calculations

Sample size calculations were made by setting Type 1 error $\alpha = 0.05$, Type 2 error $\beta = 0.20$ (i.e., 80% power), two sided test, Students t distribution, and a δ of 15 min retention time. A pooled standard deviation of 10.3 min was selected for the power calculation based on the individual standard deviations from studies with normal subjects using the FITC-dextran retention method. The sample size was adjusted to correct for repeated measures reduction in statistical significance.

Given the above considerations, the calculated sample size was 12 subjects. For convenience and redundancy (necessary in case dropout in any block occurred), 18 subjects in 3 blocks of 6 subjects were enrolled.

2.3. Procedures

Subjects were familiarized with the fluorometry procedure and baseline corneal fluorescence intensity was measured. One 25 μl drop of formulation or control solution was instilled into each eye with a positive displacement pipette according to the randomization schedule. The subject was asked to blink several times and to "roll" the eyes around to distribute the formulation. Fluorescence data from the central precorneal tear film was recorded on a strip chart recorder until fluorescence intensity level of 1.5 times the baseline intrinsic fluorescence was regained. Fluorescent intensity values were obtained as rapidly as possible in the initial 8 min, then about every 2 min throughout the measurement period. The right eye was always measured first to completion, followed immediately by the left eye. A final, post-session slit lamp examination was conducted to monitor anterior segment condition.

Two variables were recorded from each subject, fluorescence intensity, calculated as fluorometer output over time, and comfort, recorded immediately post instillation. Comfort was rated without investigator prompting on a 1–100 point rating scale wherein 1 was defined as "extremely irritating," and 100 was defined as "cannot be felt."

3. RESULTS

3.1. Rheology

Both low and high MW CMC formulations exhibited pseudoplastic behavior, although the high MW exhibited greater shear thinning than low MW. One measure used to describe shear thinning behavior is the "shear thinning index," defined as the ratio of change in viscosity with a 10-fold shift in rpm.[13] For example, the ~100 mPa.s formulations had shear thinning indexes of 1.17 and 1.63 for low and high MW CMCs, respectively (5–50 rpm or shear rates of 19.2 and 192 sec^{-1}). Fig. 1 demonstrates typical shear thinning curves for the two molecular weight CMCs.

3.2. Residence Time

Gross residence time was calculated on a relative basis by subtracting the saline time (in minutes) from the test formulation residence time; mean residence times are summarized in Fig. 2. All formulation residence times were positive, indicating retention greater than saline that is clinically significant. Gross residence time was approximately equal for the low and medium viscosity formulations of both MWs, but increased substantially for the highest viscosity (~250 mPa.s) solutions. The residence time increase from lowest to highest viscosity was statistically significant (paired data Students t test, 2-tailed, $n = 15$; $p = 0.0124$ for H-25 vs. H-250; $p = 0.0411$ for L-25 vs. L-250).

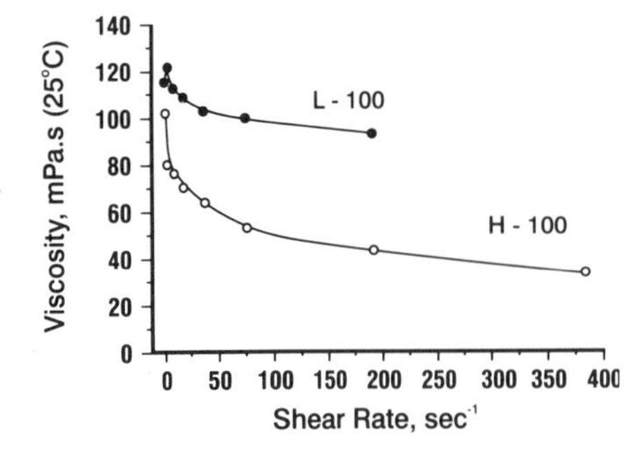

Figure 1. CMC viscosity data, 100 mPa.s formulations. Shear Thinning Index (ratio of viscosity change per 10-fold rpm change) = 1.17 and 1.63 for L-100 and H-100, respectively (5–50 rpm and shear rate change of 19.2–192 sec^{-1}).

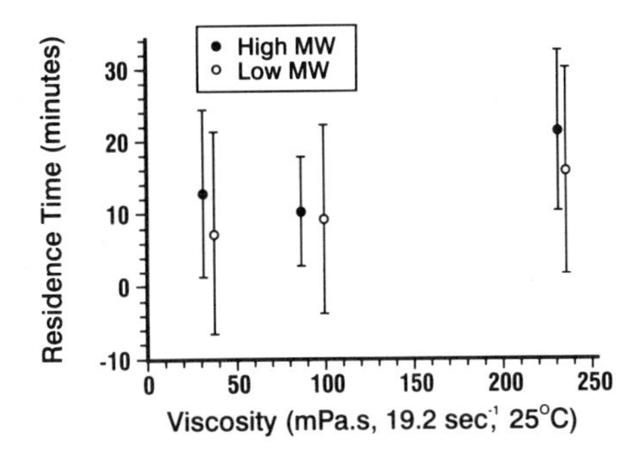

Figure 2. Relative residence time (formulation minus control). Mean ± SD, n = 15 subjects for each determination.

Area under the curve was calculated for formulations and controls (Table 1). The formulation AUC increased from the ~25 mPa.s to the ~250 mPa.s viscosity for both MWs, indicating significant improvement in contact time with increasing viscosity. As with gross residence time, MW differences were not statistically significant (at the 0.05 level) at similar viscosities (i.e., L-25 vs. H-25), but the AUC increase, from lowest to highest viscosity, was statistically significant for each MW (multiple comparisons, Bonferroni test, $p = 0.001$ for the 25–250 mPa.s comparisons, low and high MW CMC).

The decay analysis is preliminary at the time of this writing, although the control and ^{3} 90% of the formulation decays appear to adhere to bi-exponential kinetics. The equation has the form:

$$F(t) = A_1 e^{-k1t} + (1 - A_1)\, e^{-k2t} \tag{1}$$

The initial rate constant, k_1, appears extremely variable and is probably unreliable for statistical analysis; however, the k_2 value (representing the slower decay under physiological conditions) appears realistic and is thus a useful quantitative measure of the decay behavior. A typical plot is depicted in Fig. 3.

3.3. Comfort

Mean comfort scores, calculated on a relative basis (1–100 scale, test formulation minus control score) are summarized in Fig. 4. All high MW formulations were positive, whereas all low MW formulations had negative scores that became more negative with increasing viscosity. These data suggest that the high MW comfort was greater than that for low MW, probably due to the greater pseudoplasticity of the high MW solutions.

Table 1. Mean area under the curve (AUC), fluorescent intensity vs. time

	L-25	H-25	L-100	H-100	L-250	H-250
Formulation AUC	5.45	5.81	7.17	7.17	9.08	10.03
(± SD)	(1.9)	(1.9)	(2.6)	(3.2)	(3.1)	(3.5)
Control AUC	3.69	3.77	3.90	3.36	3.93	3.70
(± SD)	(1.3)	(1.8)	(1.6)	(1.3)	(1.6)	(1.8)

Area in arbitrary units.
"L-25" refers to 90K MW CMC, ~25 mPa.s apparent viscosity formulation, etc.

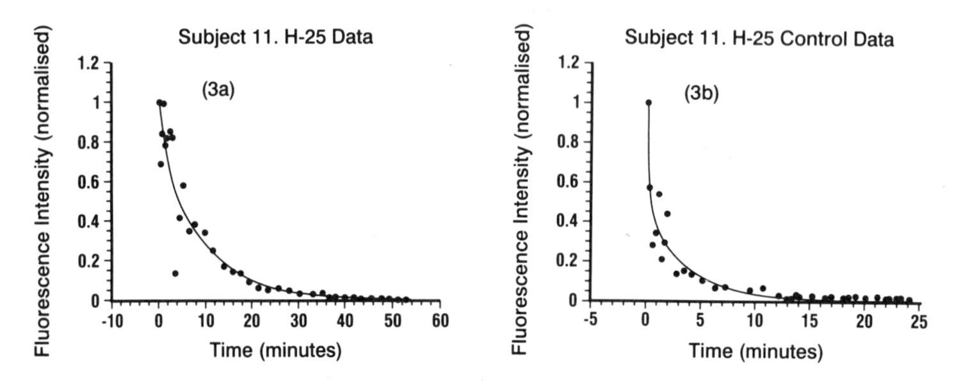

Figure 3. Least squares regression curve fit for one representative subject to the normalized data. A bi-exponential function gave the best fit (see text). (a) H-25 *Formulation*: $k_1 = 0.440$, $k_2 = 0.102$, $\chi^2 = 0.397$, R = 0.950;. (b) H-25 *Control*: $k_1 = 6.0$, $k_2 = 0.276$, $\chi^2 = 0.106$, R = 0.965; elimination coefficients in min^{-1}.

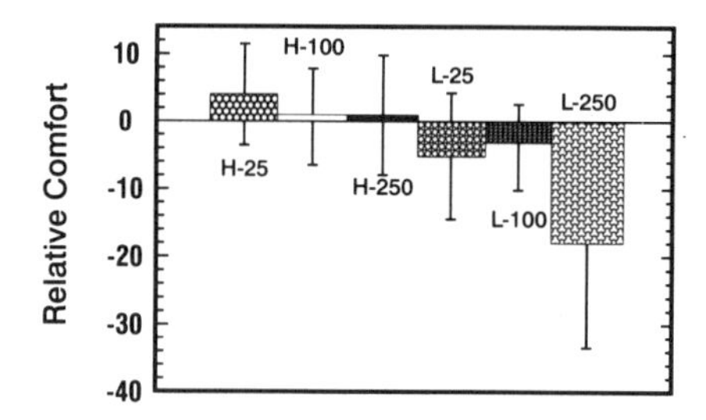

Figure 4. Mean ± SD relative comfort scores (formulation minus control); n = 15 subjects.

4. DISCUSSION AND CONCLUSIONS

This investigation demonstrated by at least two measures, gross residence time and area under the curve, that CMC residence time was on average greater than a saline control, and that these differences were clinically meaningful. Furthermore, the residence time could be extended by increasing the viscosity to ~250 mPa.s from 25 mPa.s, a difference that was statistically significant for both 90K and 700K MW formulations. Residence time was independent of molecular weight, suggesting that enough anionic sites were present on the 90K molecule to promote muco-adhesion.

Another major finding is that topical solution comfort appears to be related to its pseudoplasticity, i.e., the degree of shear thinning. The high MW CMC formulations were all slightly more comfortable than saline (Fig. 4), which itself is generally comfortable; they also demonstrated greater shear thinning compared to the low MW solutions. In contrast, the low MW formulations exhibited less comfort relative to control and worsened as viscosity increased. Shear thinning has also been reported for human tears,[14] and research into optimization of shear thinning properties of palliative formulations warrants more investigation. Shear thinning of tear replacements and comfort drops may enhance comfort by reducing the frictional resistance during blinking, reducing the potential for ocular fatigue.

REFERENCES

1. Maurice DM, Srinivas SP. Use of fluorometry in assessing the efficacy of a cation-sensitive gel as an ophthalmic vehicle: Comparison with scintigraphy. *J Pharm Sci.* 1992;81:615–619.
2. Paugh JR, Meadows DL, Garcia CG, et al. Evaluating the ocular retention time of two different artificial tear polymers using fluorophotometry in patients with dry eye. ARVO Abstracts. *Invest Ophthalmol Vis Sci.* 1994;35:S1694.
3. Meadows DL, Joshi AJ, Paugh JR. Diagnostic apparatus for determining pre-corneal retention time of ophthalmic formulations. US Patent #5,323,775; 1994.
4. Bron AJ, Tiffany JM. Pseudoplastic materials as tear substitutes: An exercise in design. In: van Bijsterveld OP, Lemp MA, Spinelli D, eds. *The Lacrimal System.* Symposium on the Lacrimal System, Singapore, March 17, 1990. Amsterdam: Kugler & Ghedini; 1991;27–33.
5. Ludwig A, van Ooteghem M. The evaluation of viscous ophthalmic vehicles by slit lamp fluorophotometry in humans. *Int J Pharm.* 1989;54:95–102.
6. Bothner H, Wik O. Rheology of hyaluronate. *Acta Otolaryngol Suppl.* 1987;442:25–30.
7. Zaki I, Fitzgerald P, Hardy JG, Wilson CG. A comparison of the effect of viscosity on the precorneal residence of solutions in rabbit and man. *J Pharm Pharmacol.* 1986;38:463–466.
8. Smart JD, Kellaway IW, Worthington HEC. An in vitro investigation of mucosa-adhesive materials for use in controlled drug delivery. *J Pharm Pharmacol.* 1984;36:295–299.
9. Leung SS, Robinson JR. The contribution of anionic polymer structural features to mucoadhesion. *J Cont Release.* 1988;5:223–231.
10. Mortazavi SA, Smart JD. Factors influencing gel-strengthening at the mucoadhesive-mucous interface. *J Pharm Pharmacol.* 1994;46:86–90.
11. Ludwig A, Unlü N, van Ooteghem M. Evaluation of viscous ophthalmic vehicles containing carbomer by slit-lamp fluorophotometry in humans. *Int J Pharm.* 1990;61:15–25.
12. Snibson GR, Greaves JL, Soper ND, Prydal JL, Wilson CG, Bron AJ. Precorneal residence times of sodium hyaluronate solutions studied by quantitative gamma scintigraphy. *Eye.* 1990;4:594–602.
13. ASTM Standard D 2196. Standard test methods for rheological properties of non-Newtonian materials by rotational (Brookfield) viscometer. *Annual Book of ASTM Standards.* Philadelphia: ASTM; 1991.
14. Tiffany JM. The viscosity of human tears. *Int Ophthalmol.* 1991;15:371–376.

108

VITRONECTIN ADSORPTION IN CONTACT LENS SURFACES DURING WEAR

Locus and Significance

Brian J. Tighe,[1] Valerie Franklin,[1] Christopher Graham,[1] Aisling Mann,[1] and Michel Guillon[2]

[1]Aston Biomaterials Research Unit
Aston University
Birmingham, England, United Kingdom
[2]Contact Lens Research Consultants
London, England, United Kingdom
Laboratoire de Biochimie des Transports Cellulaires
Universite de Paris Sud
Orsay, France

1. INTRODUCTION

There has been growing interest in recent years in the factors associated with inflammatory response in the eye. The possibility that different types of synthetic materials may enhance or modulate their response is now being given serious consideration. This has focused greater attention on tear biochemistry and the biochemical changes that result from the interaction of synthetic materials with the ocular environment. One such change involves vitronectin, now well established as an inflammatory marker in various body sites, having an important pro-active role in modulating the conversion of plasminogen to plasmin.[1,2] The activation of plasminogen into plasmin is accepted as the most important mechanism of extracellular proteolysis and is consequently involved in localized inflammatory response and tissue repair.

The plasminogen-plasmin activation process is kept in equilibrium via the antagonistic activities of the plasminogen activator system and plasminogen activator inhibitor (PAI) system. The most significant PA inhibitor in cellular systems is PAI-1, which is known to bind to vitronectin and has been suggested as a key feature in serving to localize PAI-1 to the pericellular compartment where active PA is generated. If vitronectin is localized on a surface adjacent to the cellular site, it removes PAI from the reaction by fix-

Lacrimal Gland, Tear Film, and Dry Eye Syndromes 2
edited by Sullivan *et al.*, Plenum Press, New York, 1998

ing it, creating an imbalance in favor of the PA and production of active plasmin. This results in local upregulation of plasmin formation, thereby providing an important regulatory mechanism in wound repair. We have shown in recent studies that this localization is produced in the eye by the presence of a contact lens.

We have previously demonstrated the value of a cell-based assay for the study of vitronectin in the ocular environment during periods of contact lens wear.[3,4] Our aim was to evaluate the potential influence of the contact lens and lens material on the vitronectin-mediated inflammatory process and the nature of the post-lens micro climate, particularly during closed eye wear. In particular, the location of the adsorbed vitronectin: front vs. back surface, center vs. periphery, and the effects of lens material and of wear regimen were examined. In this paper, we compare the levels of vitronectin adsorption encountered in these various situations and locations with the levels of vitronection obtained by direct adsorption from solution, and the implications of these results in terms of the origin of the vitronectin found on worn contact lenses.

2. METHODOLOGY[3,4]

The potential value of harnessing tissue culture techniques and of using cell adhesion studies to amplify the presence of inflammatory markers arises because of the special role of vitronectin in cell adhesion phenomena. The cell assay involves selective inhibition of the action of individual adhesion proteins by the use of antibody blocking techniques. Comparative studies were set up to obtain ex vivo worn lenses and in vitro spoiled lenses, which were analyzed using the cell assay previously described. The worn lenses were all aseptically removed prior to assay.

3. RESULTS

3.1. Overview of Worn Lens Studies

The results of the worn lens studies, which are summarized in Fig. 1, reveal the following points.

- The concentration of the vitronectin located on the front and back surfaces was compared in the contact lens peripheral region. Significantly greater ($p < 0.001$) levels occurred on the back than the front surface, irrespective of wear modality or lens ionicity.
- Vitronectin concentration at the center and periphery of the lens was determined for both the front and back surfaces. The results for both surfaces revealed a far greater vitronectin concentration at the lens periphery than at the center, especially on the back surface ($p < 0.006$ to $p < 0.001$).
- Comparison of 2 weeks daily wear, averaging 196 h of wear, and 1 week of extended wear, averaging 168 h of wear, revealed a greater vitronectin concentration after extended than daily wear for both Group II and Group IV lenses, and a greater concentration on Group IV than on Group II lenses.

3.2. Overview of in Vitro Studies

A series of in vitro experiments were carried out to establish whether the levels of vitronectin activity observed on posterior and anterior surfaces of Group II and Group IV

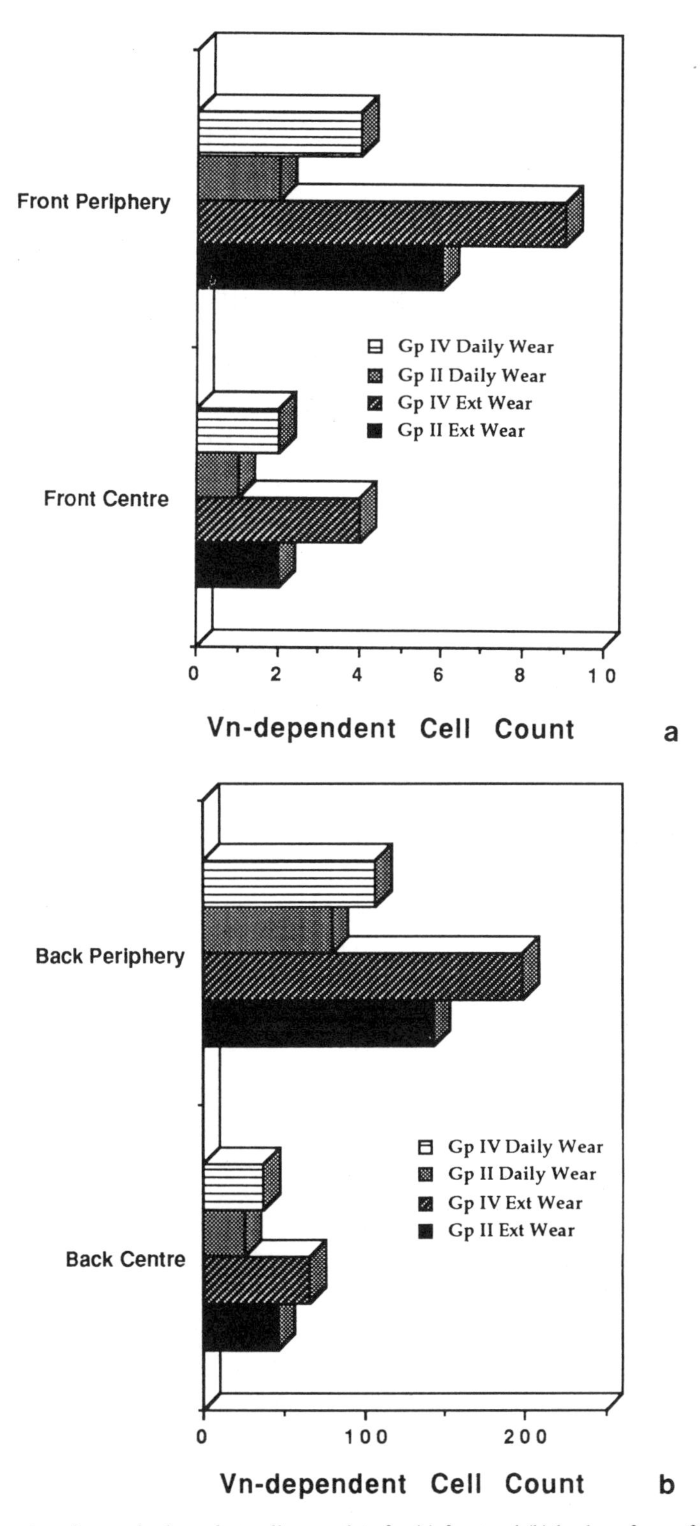

Figure 1. Comparative vitronectin-dependent cell count data for (a) front and (b) back surfaces of FDA Group II and Group IV materials in extended wear and daily wear modalities.

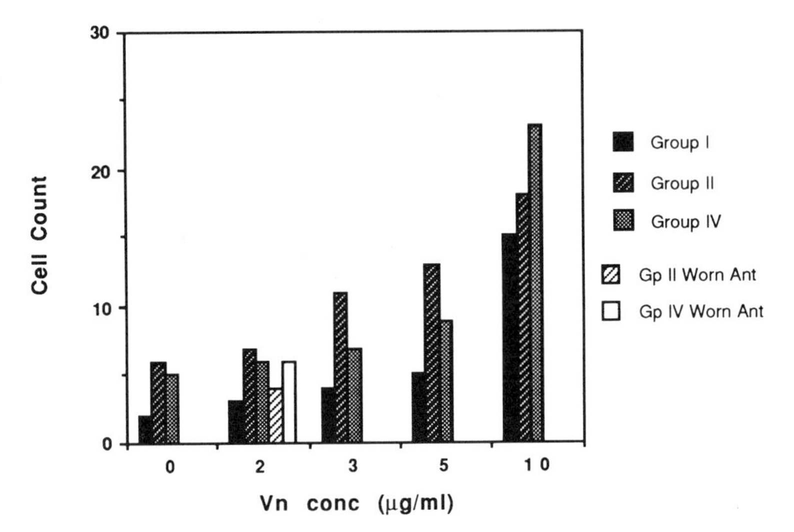

Figure 2. Comparative vitronectin-dependent cell count data for in vitro exposure of FDA Group I, II, and IV materials to vitronectin at a range of concentrations encompassing reported tear levels, in comparison to cell counts on anterior surfaces of ex vivo Group II and IV materials following daily wear.

lenses are consistent with adsorption from known vitronectin concentration levels in open and closed eye situations. The in vitro results are summarized in Fig. 2.

These results reveal a remarkable consistency between the vitronectin levels on the anterior surfaces of ex vivo lenses, which are compared in Fig. 2, with those on in vitro lenses exposed to vitronectin adsorption at tear concentration levels. Vitronectin levels at the periphery of posterior surfaces of ex vivo lenses are, however, so high (Fig. 3) as to be

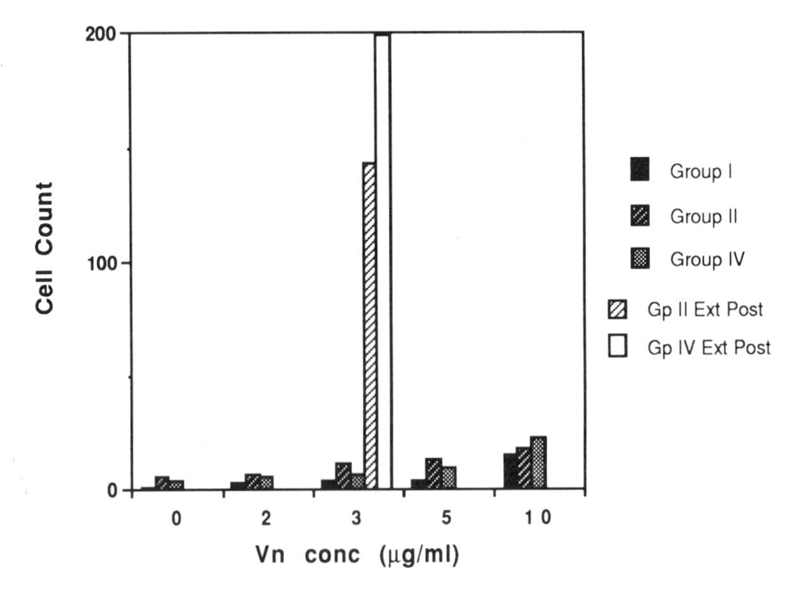

Figure 3. Comparative vitronectin-dependent cell count data for in vitro exposure of FDA Group I, II, and IV materials to vitronectin at a range of concentrations encompassing reported tear levels, in comparison to cell counts on posterior surfaces of ex vivo Group II and IV materials following extended wear.

inconsistent with simple adsorption from vitronectin solution, even at levels vastly above those encountered in the eye. This strongly suggests interaction of the lens and the corneal surface as the source.

4. CONCLUSIONS

Several direct conclusions follow from these results.

- Vitronectin concentrations at the contact lens surface are greater for extended wear than for daily wear contact lenses.
- Vitronectin concentrations at the contact lens posterior surface are vastly greater than at the anterior surface.
- Vitronectin concentrations at the contact lens posterior surface cannot be accounted for by adsorption from tears and require tissue-lens interaction.
- Contact lens posterior surface vitronectin concentrations are markedly dependent upon material type, suggesting the potential for materials effects in lens-induced inflammatory processes.

These conclusions lead to two important implications: (i) The location of vitronectin on the lenses studied confirmed that the post-lens micro climate, particularly towards the periphery, is rich in vitronectin to a degree that is capable of influencing localized plasmin production. This production of a micro-climate, which is influenced by lens material, could provide an important element in the understanding of contact lens induced inflammatory processes. (ii) The contact-induced transfer of vitronectin to the periphery of the posterior surface of the lens suggests the use of this assay as a means of studying the interactive effect of, for example, lens parameters and mechanical properties, on lens-cornea interactions.

REFERENCES

1. Vaheri A, Bizik J, Salonen EM, et al. Regulation of the pericellular activation of plasminogen and its role in tissue-destructive process. *Acta Ophthalmol.* 1992;70:34–41.
2. Sack RA, Underwood PA, Tan KO, Sutherland H, Morris CA. Vitronectin. *Ocular Immunol Inflam.* 1993;1:327–336.
3. Tighe BJ, Fitton JH, Jones L, Guillon M. Victronectin as an inflammatory marker in contact lens wear monitored by a novel on-lens assay. *Optom Vis Sci.* 1993;70:67.
4. Guillon M, Tighe BJ, Franklin VJ, Mann A. The role of adsorbed vitronectin in contact lens involved inflammatory processes. *Optom Vis Sci.* 1994;71:18.

EFFECT OF TEARS AND TEAR RESIDUES ON WORN ETAFILCON AND POLYMACON DISPOSABLE CONTACT LENSES ON THE ADHESION OF *PSEUDOMONAS AERUGINOSA*

Miguel F. Refojo, Fee-Lai Leong, and Marta Portolés

Schepens Eye Research Institute and Department of Ophthalmology
Harvard Medical School
Boston, Massachusetts

1. INTRODUCTION

Pseudomonas aeruginosa (Pa) ulcerative keratitis is the most severe complication of contact lens wear. The highest incidence of infection occurs with extended-wear hydrogel contact lenses, compared with daily-wear hydrogel lenses and with rigid gas-permeable lens wear.[1–6] Corneal infection in contact lens users is believed to be initiated by the adhesion of the microorganism to the lenses.[7–9] The microorganisms can then be transferred to the contact lens-compromised cornea when the immunological defense mechanism of the ocular surface is blocked by the wearing of the lens.[10,11] In normal eyes, without a contact lens, blinking and the normal flow of tears cleanse the cornea of surface debris, desquamated epithelial cells, air polluters, and microorganisms. The mucoid component of the tears inhibits bacterial binding to the cornea surface. Enzymes and immunoglobulins in the tear fluid also protect the cornea against microorganisms.[12] Polymorphonuclear leukocytes (PMN) released from the palpebral conjunctiva to the tears destroy invading microorganisms.[13] However, hydrogel contact lenses that allow minimum tear exchange over the cornea interfere with all these lines of defense of the ocular surface.[14] Also, with some of the available hydrogel lenses during daily-wear, and with all hydrogel lenses under the closed eye during sleep, the cornea becomes more vulnerable to microorganism invasion, due to the hypoxic stress induced by the lenses on the corneal epithelium.[10,15,16]

Pa attaches to specific receptors on the cells and tissues[17] and to non-biological materials, such as the contact lenses, by hydrogen, electrostatic and hydrophobic bonds, and by adhesives secreted by certain bacteria.[18,19] Worn contact lenses are readily coated with tear protein and, with increased wearing time, focal heterogeneous deposits are observed on many lenses.[20] Bacteria attach to new[16,21] and to worn lenses that contain different amounts of deposits, but preferentially to the discrete deposits that occur in some lenses.[22,23] However, although the discrete deposits are not found on disposable contact lenses, a relative high incidence of Pa keratitis is associated with the use of these lenses.[4,5,24,25]

The contribution of tear deposits on the worn lenses to their infectivity is controversial.[16,26,27] Although hydrogel contact lenses of different chemical compositions have been used that coated in vivo with different tear deposits, none of these lenses, provided that they were used in the same conditions, were found to be more prone to induce bacterial keratitis than other lenses. Also, the age of the worn lenses, presumably related to the amount of accumulated tear deposits, was not found to be a significant factor in the incidence of ulcerative keratitis associated with contact lens wear.[2]

Previous studies of Pa adhesion to human worn contact lenses were done with diverse hydrogel lenses collected from the users and stored for different times before they were used for the *in vitro* bacterial adhesion studies.[22,23,28] One of these studies used extended-wear surfilcon A, nonionic hydrogel lenses of 74% water content;[22] the other study included extended-wear non-ionic polymacon (38% water) and lidofilcon B (79% water), and ionic bufilcon A (45% water) and perfilcon A (71% water) lenses.[23] Both studies reported significantly higher bacteria count on the worn than on new lenses. A third study reported significantly fewer Pa adherents on the 1 week worn disposable etafilcon lenses (ionic, 58% water), compared to new lenses of the same kind.[28] In that study, lenses were stored in PBS in the refrigerator for 3 days to 6 weeks before the bacterial adhesion experiments were performed.

In the present study, we repeated the experiments of human worn extended-wear disposable etafilcon lenses,[28] using Pa obtained from a different corneal ulcer, and included in the experiments the extended-wear disposable polymacon lenses. In addition, we compared Pa adhesion to the etafilcon and polymacon lenses, but infecting the lenses immediately after removal from the eye, without rinsing or after a brief rinsing. The *in vitro* bacterial adhesion to the non-rinsed worn lenses is the closest situation to the in *vivo* infection of the lenses.

2. METHODS AND MATERIALS

Pa strain 6294 from a corneal ulcer (Channing Laboratory, Boston, MA) was used in all assays. The bacteria were processed as previously reported.[29]

The disposable etafilcon A hydrophilic contact lenses (ionic, 58% water) are made of a crosslinked copolymer of 2-hydroxyethyl methacrylate with 2% methacrylic acid (Acuvue™, Johnson & Johnson, Vistakon Div., Jacksonville, FL). The disposable polymacon hydrophilic contact lenses (nonionic, 38% water) are made of crosslinked poly(2-hydroxyethyl methacrylate). The polymacon lenses were the blue-tinted SeeQuence® and SeeQuence2® (Bausch & Lomb, Contact Lens Div., Rochester, NY) and the non-tinted SeeQuence® (Bausch & Lomb, Shiba, Minato-ku, Japan). Although all the SeeQuence® lenses are made of the same polymer, they were tested separately because of the possible effect of the method of manufacture and the presence or absence of the blue tint that may affect their surface energy, and hence bacterial adhesion.[30] All the SeeQuence® lenses are spin casted, but the base curves of the SeeQuence2® lenses are, apparently, lathe cut. The Acuvue™ lenses are cast molded.

The lenses were worn for 4 or 5 days and nights, depending on the preference of the subject, but the length of wear was not considered in the analysis of the results. The subjects were informed of the nature of the experiments. The subjects were young adults of both sexes, Caucasians and Asians, and all were experienced lens wearers.

The worn lenses were separated into three groups: *Group 1*, lenses were placed in the Pa suspension [1×10^8 colony forming units (CFU)/ml] immediately after removal from the eye. *Group 2*, lenses were rinsed in 3 aliquots of 10 ml PBS before they were placed in the Pa suspension. *Group 3*, lenses were rinsed as in group 2 and then stored at 5°C in BSS for 2 weeks before they were treated with the Pa suspension. The controls were new lenses from the original container or after rinsing in 3 aliquots of 10 ml PBS, before they were treated in the Pa suspension. The new lenses used for control of group 3 were also stored in BSS in the cold room for 2 weeks.

All the infected lenses were incubated statically in 1 ml of the Pa suspension for 1 h at room temperature, and then rinsed in 5 aliquots of 10 ml PBS (10 dips in each aliquot) to remove non-adherent bacteria.[29] Then the lenses were ground in a tissue grinder, and 3–4 serial dilutions (1:10) were performed. One ml of each dilution was plated in trypticase soy agar. After incubation at 37°C, the CFU/lens were counted.

Comparisons of the CFU/lens on the three groups of lenses and on control lenses were determined by one factor ANOVA (Fisher PLSD and Scheffe F-test) using StatView® computer program (Abacus Concepts, Berkeley, CA). Significance is reported at the 95% level.

The data are represented in box plots using the StatView® computer program that displays graphically the 10th, 25th, 50th, 75th, and 90th percentiles of CFU per lens (Fig. 1). The top of the box represents the 75th and the bottom the 25th percentiles of the bacteria counts on the lenses. The middle 50% of the bacteria counts are contained within the top and bottom of the box. The horizontal line within the box represents the median (50th percentile) of the bacteria counts per lens. When the distribution is symmetric, the median is in the middle of the

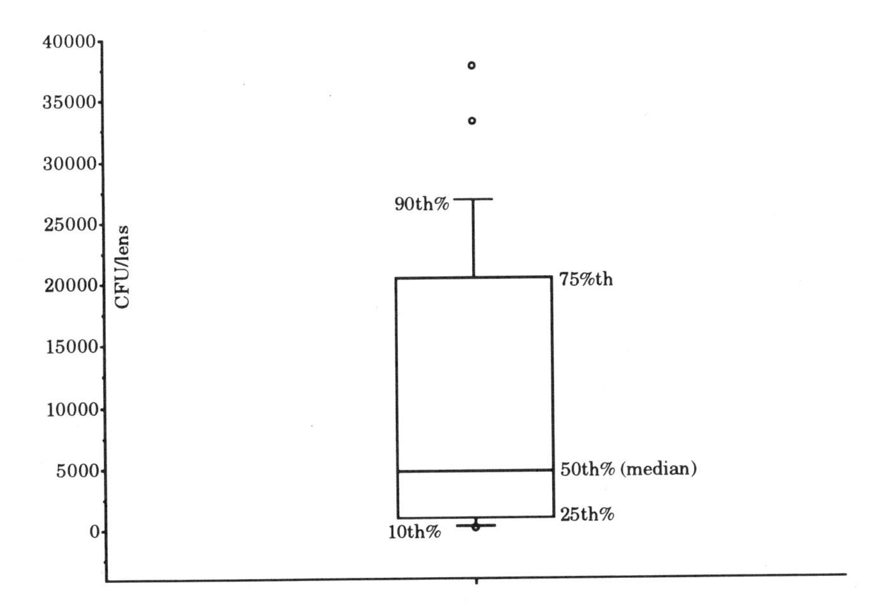

Figure 1. Percentiles of Pa colonies counted on 24 brand-new etafilcon lenses.

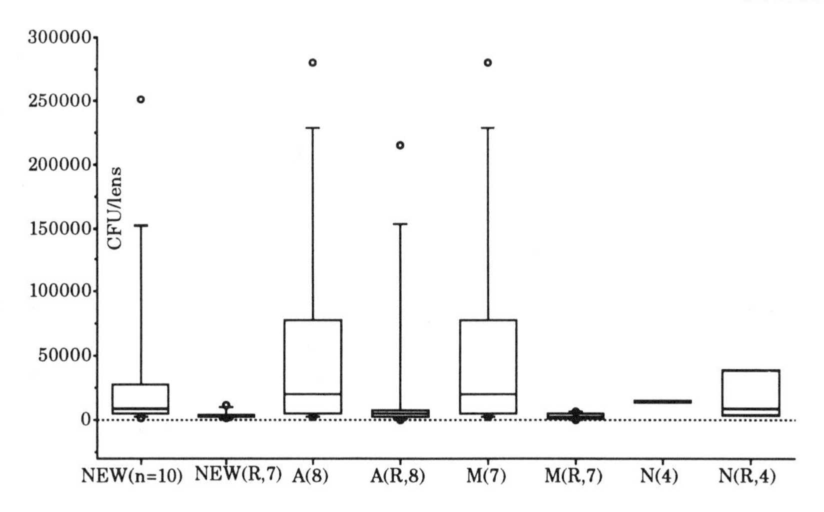

Figure 2. Pa adhesion on new etafilcon lenses and on etafilcon lenses worn by subjects A, M, and N for 4–5 days and nights, before and after (R) rinsing.

box. The line extending from the top of the box (top whisker) represents the percentile of counts between the 75th and the 90th percentiles. The bottom whisker represents the counts betweem the 25th and the 10th percentiles. The circles above and below the whiskers represent the bacteria counts above and below the 90th and 10th percentiles.

3. RESULTS

3.1. Etafilcon

Fig. 2 shows the Pa adhesion on the etafilcon lenses worn by three subjects, before (group 1) and after (group 2) rinsing and on the brand-new lenses, before and after rinsing. There is no significant difference on the bacterial adhesion among any of the groups of lenses.

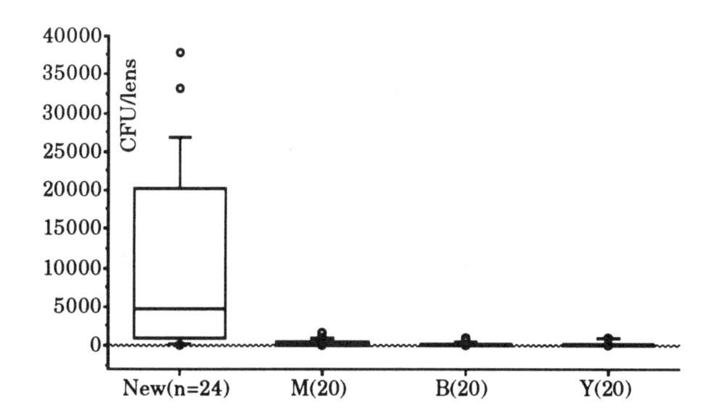

Figure 3. Pa adhesion on etafilcon lenses, new and worn by subjects M, B, and Y for 4 days and nights, and then rinsed in PBS and stored in BSS for 2 weeks.

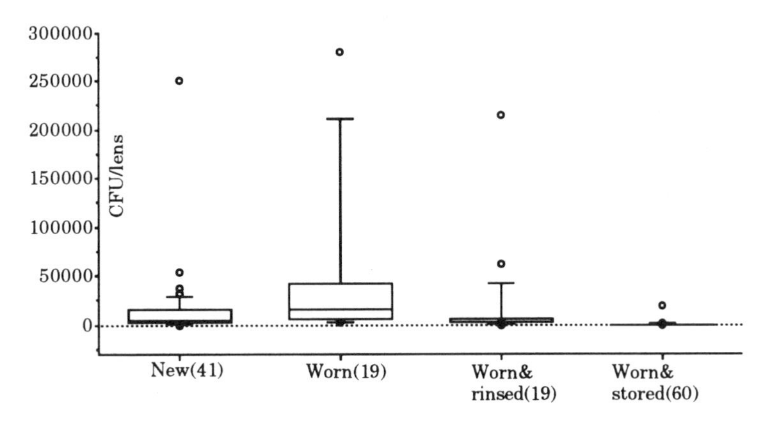

Figure 4. Combined results of Pa adhesion on all the new and worn etafilcon lenses.

Fig. 3 shows the Pa adhesion on etafilcon lenses, new and worn, rinsed and stored at 5°C in BSS for 2 weeks (group 3). There was a significantly lower amount of Pa adherent on the lenses worn by each of the subjects vs. the new lenses (Fisher PLSD and Scheffe F-test).

Fig. 4 shows the combined Pa adhesion to all the etafilcon lenses used in the experiments. The worn and non-rinsed lenses (group 1) had significantly more bacteria than the new lenses, and also more than the lenses of groups 2 and 3 (Fisher PLSD). However, the Scheffe F-test gave only significantly more bacteria on group 1 than on group 3 lenses.

3.2. Polymacon

Figs. 5 and 6 show the Pa adhesion on new, group 1, and group 2 tinted SeeQuence® and SeeQuence2® lenses, respectively. There was no significant difference among any of the SeeQuence® groups of lenses. However, for the SeeQuence2® lenses there were significantly more (Fisher PLSD only) bacteria on group 1 lenses worn by subject C than on the new rinsed lenses, and more than on group 2 lenses worn by subjects C, M, and N.

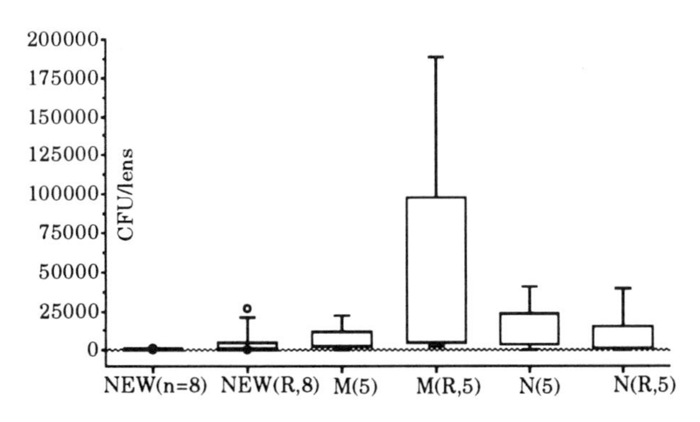

Figure 5. Pa adhesion on SeeQuence® lenses new and worn by subjects M and N for 4 days and nights, before and after (R) rinsing in PBS.

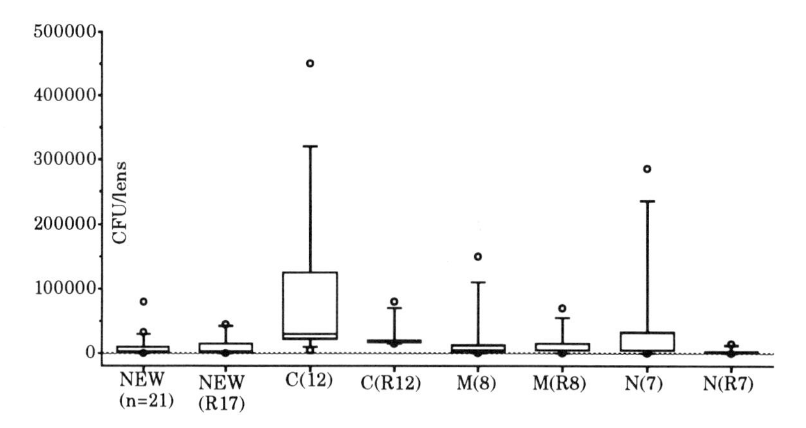

Figure 6. Pa adhesion on SeeQuence2® lenses new and worn by subjects C, M, and N, before and after (R) rinsing in PBS.

Fig. 7 shows the bacteria adherent on the non-tinted SeeQuence® lenses that were worn 4 days and nights and then rinsed in PBS and stored at 5°C in BSS before they were infected with the Pa (group 3). Statistically, there was no difference among any of the groups. There were also no statistical differences among the adherent bacteria on the new, non-rinsed, and rinsed polymacon lenses (tinted SeeQuence® and SeeQuence2®, and non-tinted SeeQuence®).

Fig. 8 contains the combined data of Figs. 5–7, and shows the bacterial adhesion on all the polymacon lenses: ie, new, worn (group 1), worn and rinsed (group 2), and worn, rinsed, and stored (group 3) for all subjects. There were significantly more bacteria on group 1 than on all other groups of lenses including the new lenses (Fisher PLDS). However, the Scheffe F-test gave significantly more Pa only on group 1 than on the new and group 3 lenses.

3.3. Etafilcon vs. Polymacon

Fig. 9 compares the Pa adhesion on the new etafilcon lenses, the new non-tinted polymacon lenses, and the worn rinsed and stored lenses (group 3) of both kinds. There were

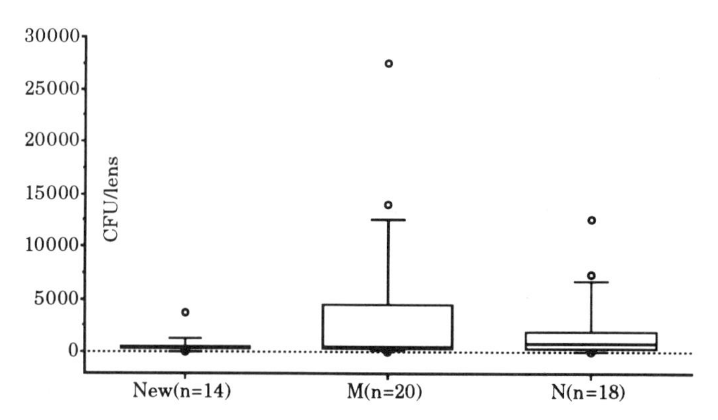

Figure 7. Pa adhesion on non-tinted SeeQuence® lenses new and worn by subjects M and N for 4 days and nights and then rinsed in PBS and stored in BSS.

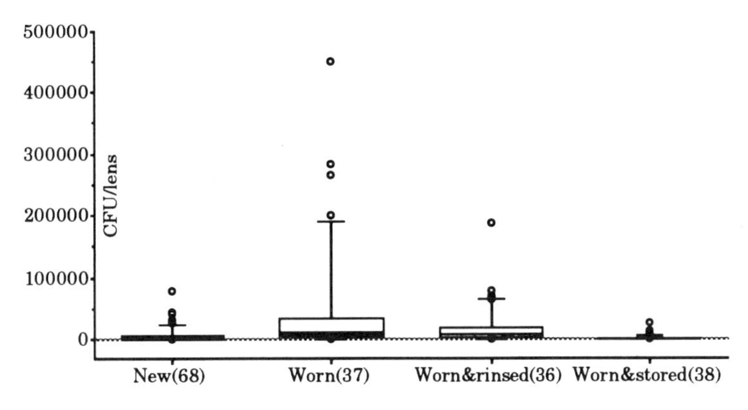

Figure 8. Combined results of Pa adhesion on all the new and worn polymacon lenses.

significantly more Pa attached on the new etafilcon than on the new polymacon lenses, and more on the new etafilcon than on the two kinds of worn lenses (Fisher PLSD and Scheffe F-test).

4. DISCUSSION

The number of Pa colonies found was more variable on the new, non-rinsed etafilcon lenses than on the rinsed lenses of the same kind, and more so than on the new polymacon lenses, either non-rinsed or rinsed. This could be related to residual releasing agents on the cast-molded etafilcon lenses.[30] We also found highly variable bacteria counts on most of the worn, non-rinsed lenses of both kinds (group 1), when worn by the same or different subjects. This variability in bacterial attachment to the lenses is probably related to the different amounts of residual tears on the lenses. As expected, the rinsed

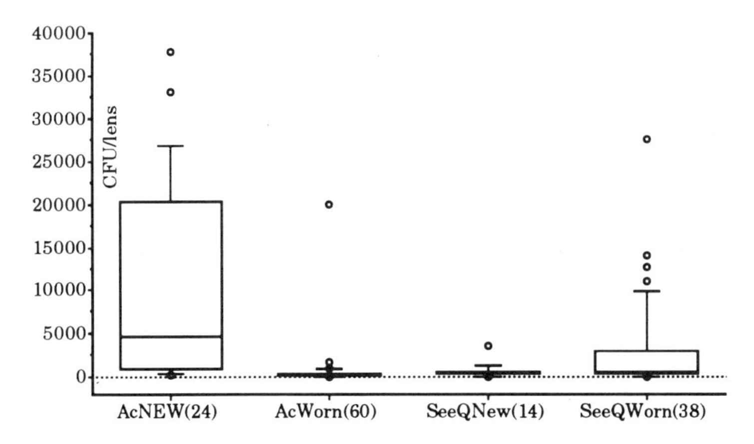

Figure 9. Pa adhesion on new and on worn and stored etafilcon (Ac) lenses and non-tinted polymacon (SeeQ) lenses.

lenses (group 2), and more so the rinsed and BSS-stored lenses (group 3), produced less variability in the bacterial counts. This, we speculate, is due to the removal of the loosely adsorbed tear components by rinsing and storing the lenses in PBS, resulting in a more uniform surface coating on the lenses of groups 2 and 3 than on the worn but non-rinsed lenses of group 1. In our opinion, the in vitro bacterial adhesion on the worn, non-rinsed lenses is the most relevant model to the in vivo bacterial adhesion to any lens.

Fewer Pa colonies were found on the worn and then rinsed and BSS stored-etafilcon lenses than on the brand-new lenses of the same kind that were also rinsed and stored in the same conditions, confirming previous findings with a virulent strain of Pa harvested from a different human corneal ulcer.[28] This is probably due to the high lysozyme bound to the ionic etafilcon lenses.[20] However, residual fabrication contaminants on the etafilcon lenses could also be a factor in the higher bacterial adhesion on the new than on the worn and BSS stored lenses. Under the same conditions, there was no significant difference in the bacterial adhesion on the new, on the worn and rinsed, and on the worn, rinsed, and PBS-stored polymacon lenses.

When all the new SeeQuence® (polymacon) and Acuvue® (etafilcon) lenses, rinsed and non-rinsed, used in the diverse experiments were compared, we found more variability in the bacterial counts on the non-rinsed SeeQuence2 and Acuvue lenses than on the tinted and non-tinted SeeQuence® lenses. This is probably due to residual fabrication contaminants on the SeeQuence2® and Acuvue® lenses. However, we did not find significant difference of bacterial adhesion among any of the lenses. When we compared the bacterial adhesion on the two sets of new lenses used to determine bacterial adhesion on the group 3 lenses, we found significantly fewer bacteria on the new polymacon than on the new etafilcon lenses (Fig. 9). These results indicate that the negative charge on the etafilcon lenses favors Pa adhesion. Nevertheless, with the used and stored lenses there was no significant difference in the amount of bacteria bound on the worn etafilcon and on the new and worn polymacon lenses, indicating that the tear protein bound to the worn ionic lenses had neutralized the negative charge of the brand-new lenses.

Although it has been found in this and previous studies that some specific lenses adhere statistically different amounts of Pa than other lenses,[22,23,28] in clinical practice none of the available hydrogel lenses have been found to cause significantly more infections than other kinds of lenses when worn under the same conditions. Indeed, because highly pathogenic strains of bacteria bind profusely to all kinds of contact lenses,[19,31] all contact lenses, new or old, can induce bacterial keratitis.

Infections occur more frequently with hydrogel contact lenses than with rigid lenses, because the hydrogel lenses interfere more with the immune defense mechanism of the ocular surface than do the more loosely fitted hard lenses. Furthermore, during overnight wear all currently available hydrogel lenses induce corneal hypoxia that opens the epithelial barrier to opportunistic microorganism invasion.

In conclusion, a Pa isolate from a human corneal ulcer was found to adhere profusely on new and on used ionic, high hydration, disposable etafilcon lenses, as well as on non-ionic, low hydration, disposable polymacon lenses. The number of bacteria colonies per lens was very variable, often extending several orders of magnitude, even when the lenses were new or worn in the same conditions by the same person at different times. Corneal hypoxia, bacterial pathogenicity, and interference of the contact lens with the ocular surface immune system seem to be more important in contact lens-associated bacterial keratitis than the specific lenses used or the degree of tears deposited on the lenses.

ACKNOWLEDGMENTS

Supported by NIH grant EY00327. The Vistakon™ lenses were supplied by Dr. George W. Mertz of Johnson & Johnson Vision Products, Inc., and the SeeQuence® lenses were supplied by Dr. Ian Cox of Bausch & Lomb Contact Lens Division.

REFERENCES

1. Poggio EC, Glynn RJ, Schein OD, et al. The incidence of ulcerative keratitis among users of daily-wear and extended-wear soft contact lenses. *N Engl J Med.* 1989;321:779–783.
2. Schein OD, Glynn RJ, Poggio EC, Seddon JM, Kenyon KR, The Microbial Keratitis Study Group. The relative risk of ulcerative keratitis among users of daily-wear and extended-wear soft contact lenses: A case-control study. *N Engl J Med.* 1989;321:773–778.
3. Schein OD, Buehler PO, Stamler JF, Verdier DD, Katz J. The impact of overnight wear on the risk of contact lens-associated ulcerative keratitis. *Arch Ophthalmol.* 1994;112:186–190.
4. Nilsson SE, Montan PG. The hospitalized cases of contact lens induced keratitis in Sweden and their relation to lens type and wear schedule: Results of a three-year retrospective study. *CLAO J.* 1994;20:97–101.
5. Nilsson SE, Montan PG. The annualized incidence of contact lens induced keratitis in Sweden and its relation to lens type and wear schedule: Results of 3-month prospective study. *CLAO J.* 1994;20:225–230.
6. Dart JK, Stapleton F, Minassian D. Contact lenses and other risk factors in microbial keratitis. *Lancet.* 1991;338:650–653.
7. Dart JKG. Predisposing factors in microbial keratitis: The significance of contact lens wear. *Br J Ophthalmol.* 1988;72:926–930.
8. Fletcher EL, Weissman B, Efron N, Fleiszig SMJ, Curcio AJ, Brennan NA. The role of pili in the attachment of *Pseudomonas aeruginosa* to unworn hydrogel contact lenses. *Curr Eye Res.* 1993;12:1067–1071.
9. Fletcher EL, Fleiszig SMJ, Brennan NA. Lipopolysaccharide in adherence of *Pseudomonas aeruginosa* to the cornea and contact lenses. *Invest Ophthalmol Vis Sci.* 1993;34:1930–1936.
10. Lawin-Brüssel CA, Refojo MF, Leong F-L, Hanninen L, Kenyon KR. Time course of experimental *Pseudomonas aeruginosa* keratitis in contact lens overwear. *Arch Ophthalmol.* 1990;108:1012–1019.
11. Lawin-Brüssel CA, Refojo MF, Leong F-L, Kenyon KR. Scanning electron microscopy of the early host inflammatory response in experimental *Pseudomonas* keratitis and contact lens wear. *Cornea.* 1995;14:355–359.
12. Cheng KH, Soanjaard L, Rutten H, Dankert J, Polak BCP, Kijlstra A. Immunoglobulin A antibodies against *Pseudomonas aeruginosa* in the tear fluid of contact lens wearers. *Invest Ophthalmol Vis Sci.* 1996;37:2081–2088.
13. Robb RM, Kubawara T. Corneal wound healing: The movement of polymorphonuclear leukocytes into corneal wounds. *Arch Ophthalmol.* 1962;68:636–642.
14. Polse KA. Tear flow under hydrogel contact lenses. *Invest Ophthalmol Vis Sci.* 1979;18:409–413.
15. Polse KA, Brand RJ, Cohen SR, Guillon M. Hypoxic effects on corneal morphology and function. *Invest Ophthalmol Vis Sci.* 1990;31:1542–1554.
16. Koch JM, Refojo MF, Hanninen LA, Leong F-L, Kenyon KR. Experimental *Pseudomonas aeruginosa* keratitis from extended wear of soft contact lenses. *Arch Ophthalmol.* 1990;108:1453–1459.
17. Hazlett LD, Moon N, Strejc M, Berk RS. Evidence for *N*-acetylmannosamine as an ocular receptor for *P. aeruginosa* adherence to scarified cornea. *Invest Ophthalmol Vis Sci.* 1982;28:1978–1985.
18. Doyle RJ. Strategies in experimental microbial adhesion research. In: Mozes N, Handley PS, Busscher HJ, Rouxhet PG, eds. *Microbial Cell Surface Analysis: Structural and Physicochemical Methods.* New York: VCH Publishers; 1991:291–316.
19. Miller MJ, Ahearn DG. Adherence of *Pseudomonas aeruginosa* to hydrophilic contact lenses and other substrata. *J Clin Microbiol.* 1987;25:1392–1397.
20. Minarik L, Rapp J. Protein deposits on individual hydrophilic contact lenses: Effects of water and ionicity. *CLAO J.* 1989;15:185–188.
21. Duran JA, Refojo MF, Gipson IK, Kenyon KR. *Pseudomonas* attachment to new hydrogel contact lenses. *Arch Ophthalmol.* 1987;105:106–109.
22. Butrus SI, Klotz SA. Contact lens surface deposits increases the adhesion of *Pseudomonas aeruginosa.* *Curr Eye Res.* 1990;9:717–724.

23. Aswad MI, John T, Barza M, Kenyon K, Baum J. Bacterial adherence to extended wear soft contact lenses. *Ophthalmology*. 1990;97:296–302.
24. Buehler PO, Schein OD, Stamler JF, Verdier DD, Katz J. The increased risk of ulcerative keratitis among disposable soft contact lens users. *Arch Ophthalmol*. 1992;110:1555–1558.
25. Matthews TD, Frazer DG, Minassian DC, Radford CF, Dart JKG. Risks of keratitis and patterns of use with disposable contact lenses. *Arch Ophthalmol*. 1992;110:1559–1562.
26. Aswad MI, Barza M, Baum J. Effect of lid closure on contact lens-associated *Pseudomonas* keratitis. *Arch Ophthalmol*. 1989;107:1667–1670.
27. Portolés M, Refojo MF. The role of tear deposits on hydrogel contact lenses induced bacterial keratitis: *Pseudomonas aeruginosa* adhesives. *Adv Exp Med Biol*. 1994; 350:421–426.
28. Boles SF, Refojo MF, Leong FL. Attachment of *Pseudomonas* to human-worn, disposable etafilcon A contact lenses. *Cornea*. 1992;11:47–52.
29. Portolés M, Refojo MF, Leong FL. Poloxamer 407 as a bacterial abhesive for hydrogel contact lenses. *J Biomed Mater Res*. 1994;28:303–309.
30. Grobe III GL, Valint JP Jr, Ammon DM Jr. Surface chemical structure for soft contact lenses as a function of polymer processing. *J Biomed Mater Res*. 1996;32:45–54.
31. Miller MJ, Wilson LA, Ahearn DG. Adherence of *Pseudomonas aeruginosa* to rigid gas-permeable contact lenses. *Arch Ophthalmol*. 1991;109:1447–1448.

110

HYALURONAN IN DRY EYE AND CONTACT LENS WEARERS

Monica Berry,[1] William K. Pastis,[1] Roger B. Ellingham,[1] Louise Frost,[2] Anthony P. Corfield,[1] and David L. Easty[1]

[1]Department of Hospital Medicine
Division of Ophthalmology
University of Bristol
Bristol, United Kingdom
[2]Harold Turner & Son Ophthalmic Opticians
Bristol, United Kingdom

1. INTRODUCTION

Hyaluronan is a ubiquitous glycosaminoglycan, which in the cornea participates in the organization of the extracellular matrix and spacing of collagen fibrils,[1] and modulates the access of macromolecules and enzymes to cells.[2] Through fibronectin receptors, hyaluronan influences adhesion and migration.[3] Hyaluronan is considered a nonspecific response to trauma, and its levels are depressed by corticosteroid therapy.[4]

We investigated the influence of dry eyes and contact lens wear on hyaluronan concentrations in tears of normal volunteers, mild dry eye patients (by clinical or symptomatic criteria), and asymptomatic established wearers of contact lenses.

2. METHODS

2.1. Subjects

Eighteen normal volunteers, 10 men and 8 women, age range 22–42 years, median 32, were recruited from staff and students at the Bristol Eye Hospital.

Fifteen mild dry eye patients by clinical and symptomatic criteria, 4 men and 11 women, age range 32–90 years, median 64, were sampled while taking part in a trial of an artificial tears preparation. Mild dry eye was defined by any two of the symptoms combined with at least two of the test results described in Table 1.

Table 1. Distribution of inclusion criteria in the mild dry eye patients

Symptom	Number of patients with symptom		Test	Number of test-positive patients	
	Both eyes	One eye		Both eyes	One eye
Foreign body sensation	12	0	Break-up time ≤ 10 sec	10	1
Ocular dryness	10	2	Fluorescein staining ≥ 2, scale 0–5	4	2
Burning or pain	10	1	Schirmer 1 without anesthesia ≤ 6 mm in 5 min	10	1
Photophobia	10	0	Lissamine green staining ≥ 4, scale 0–9	0	0

Patients with a combination of two symptoms and at least two positive test results in one eye were included in the mild dry eye group.

Twelve asymptomatic established contact lens wearers, 5 men and 7 women, age range 22–61 years, median 34, donated tears during routine appointments at their optician's practice.

2.2. Sampling of Tears

Unstimulated tears from normals and contact lens wearers were collected in 5 μl capillary tubes (Blaubrand®, BRAND). The capillary gently touched the tear meniscus without touching the eye; and was retracted if the subject felt tearing. Tears were immediately stored on ice in Eppendorf tubes and frozen until tested.

Tears from dry eye patients were collected on Schirmer strips (Cooper Vision) after instillation of a fluorescein drop (Minims), as prescribed in the trial. Tests were performed after a 7 day washout period with nonpreserved saline (Minims), and after 15 and 30 days of use of artificial tears containing carbomer 940 and preserved with either benzalkonium chloride or cetrimide. Schirmer strips were kept frozen until assayed.

2.3. Volume of Tears in Schirmer Strips

An empirical relation between tear volume and wetting of Schirmer strips was established by wetting of strips with known volumes of artificial tear buffer with or without fluorescein, in one bolus, and measuring the wetted distance after 1 and 5 min.

2.3.1. Extraction of Hyaluronan from Strips. Each strip was wetted with a volume of buffer calculated from the above relation, plus 50 μl, and vigorously mixed for 10 min using a rotamixer. Following centrifugation, the supernatant was used in the hyaluronan assay.

2.4. Hyaluronan Assay

The concentration of hyaluronan in tears was assayed using the Pharmacia HA Test (Pharmacia, Uppsala, Sweden), a radiometric assay using ^{125}I-hyaluronan binding proteins isolated from bovine cartilage. All solutions were cold, and incubations were performed at 4°C. Because of very small tear samples, the assay volumes recommended by the manufacturers were halved; all samples were mixed with 50 μl of buffer, and other reagents adjusted accordingly. Hyaluronan concentrations were adjusted by sample volume.

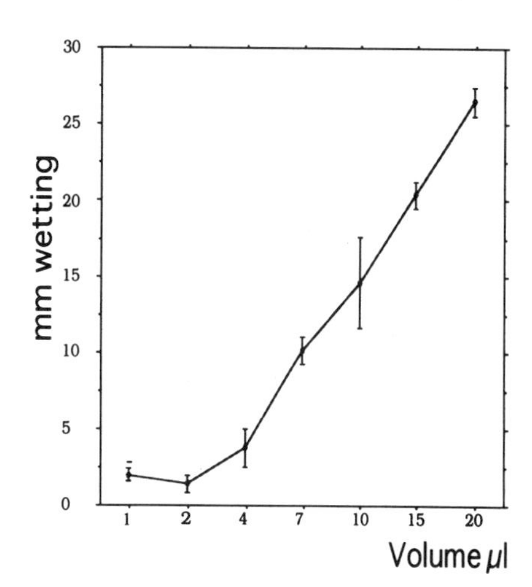

Figure 1. Sample volume and wetting of Schirmer strips. The graph presents means and 95% confidence interval of the mean for 8 determinations. Regression relation is: Wetting = 0.076 + 1.361 * volume, $r^2 = 0.981$.

3. RESULTS

Lissamine green staining and fluorescein staining of the cornea and conjunctiva were the least met criteria of inclusion among the mild dry eye patients (Table 1).

3.1. Volume of Tears in Schirmer Strips

The relation between wetting distance and sample volume is linear for sample volumes larger than 2 µl (Fig. 1). Fluorescein did not affect wetting, and the distance did not increase after 1 min. Tear volumes accumulated in Schirmer strips were larger than capillary collections (Kruskal-Wallis test, p=0.002).

3.2. Hyaluronan Assay

Halving the manufacturer's recommended volumes did not affect the behavior of the assay, as demonstrated by the similarity of standard curves with the manufacturer's published data (Fig. 2). The experimental standard curves fitted semilogarithmic curves with $r^2 \geq 0.9$.

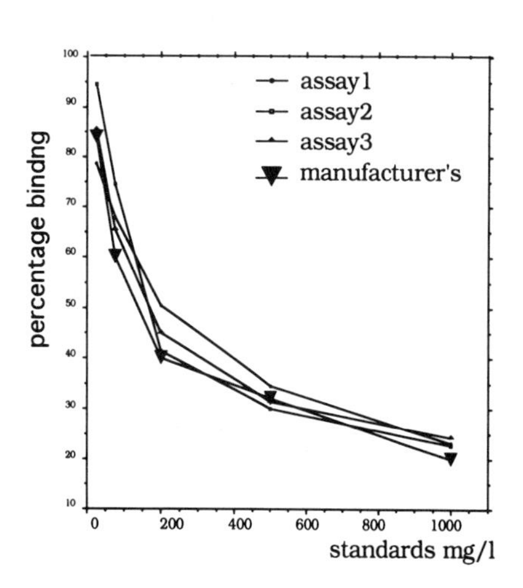

Figure 2. Standard curves of radiometric assays. Standard curves established using half the recommended volumes are stable and similar to the manufacturer's published data.

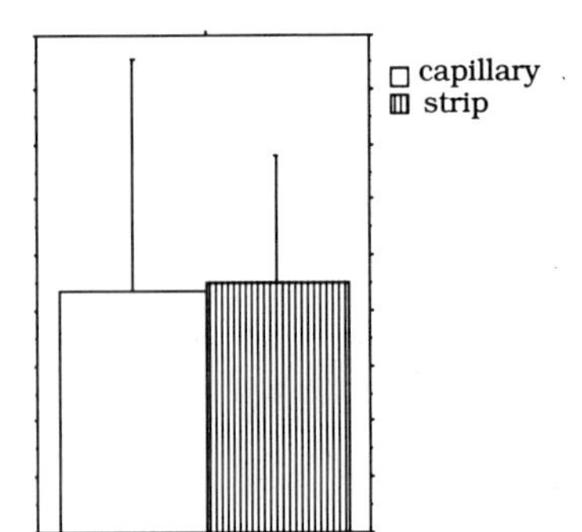

Figure 3. Comparison between hyaluronan (mean ± 95% confidence intervals) measured in a sample with or without Schirmer strip transfer and extraction.

3.2.1. Extraction of Hyaluronan from Schirmer Strips. The efficiency of hyaluronan extraction from Schirmer strips was checked by measuring hyaluronan in aliquots of pooled tears, which were assayed either directly or after having been applied to and extracted from Schirmer strips. Several volumes were compared. Measured hyaluronan concentrations were not statistically different (paired t test, $p=0.49$, Fig. 3).

3.2.2. Hyaluronan Levels in Tears. Wide variation of hyaluronan levels was found in normal eyes ranging from undetectable to 840.6 mg/l, with very low values in most cases (Fig. 4a). In dry eyes, the range is smaller (maximum 83.9 mg/l), while contact lens wearers display the smallest variation (1.4 - 40 mg/l). Both these distributions are skewed to the left (Fig. 4b). In dry eye patients, right and left eyes were not different ($p=0.13$, paired t test), nor did hyaluronan levels change during the use of the artificial tear preparation (Kruskal-Wallis test, $p=0.14$). Among lens wearers, higher levels appeared where there had been a change in lens type in the past.

4. DISCUSSION

The functions ascribed to hyaluronan depend on its concentration, molecular size, and tertiary organization.[2] Hyaluronan binding proteins, such as those used in the assay, are not known to discriminate between molecules with different molecular masses. Changes in permeability of conjunctival or corneal structures may affect both level and molecular size distribution of preocular hyaluronan. Thus, measured concentrations may depend on the method of collection. In normal volunteers, we found much higher levels when the eyes were washed with saline before capillary collection.[5] Even then, individual levels varied by orders of magnitude (12–1900 mg/l).

A wide range of levels in unstimulated normal tears was also found by Fukuda et al.[6] The relation of published mean and range values suggests that very high levels also were observed in a minority of individuals.

We probably underestimated hyaluronan concentration in some of our samples because of the very small volume of tears collected. Schirmer strips collected much larger volumes of preocular fluid from dry eyes than did capillary collections from normal eyes. In dry eyes, tears may have been stimulated by touch, fluorescein, or psychological factors. They failed, however, to exhibit the wide range of normal tears.

Contact lenses are believed to increase evaporation through altered tear lipid. The expectation, therefore, is for increased hyaluronan concentrations in contact lens wearers compared to normal eyes.

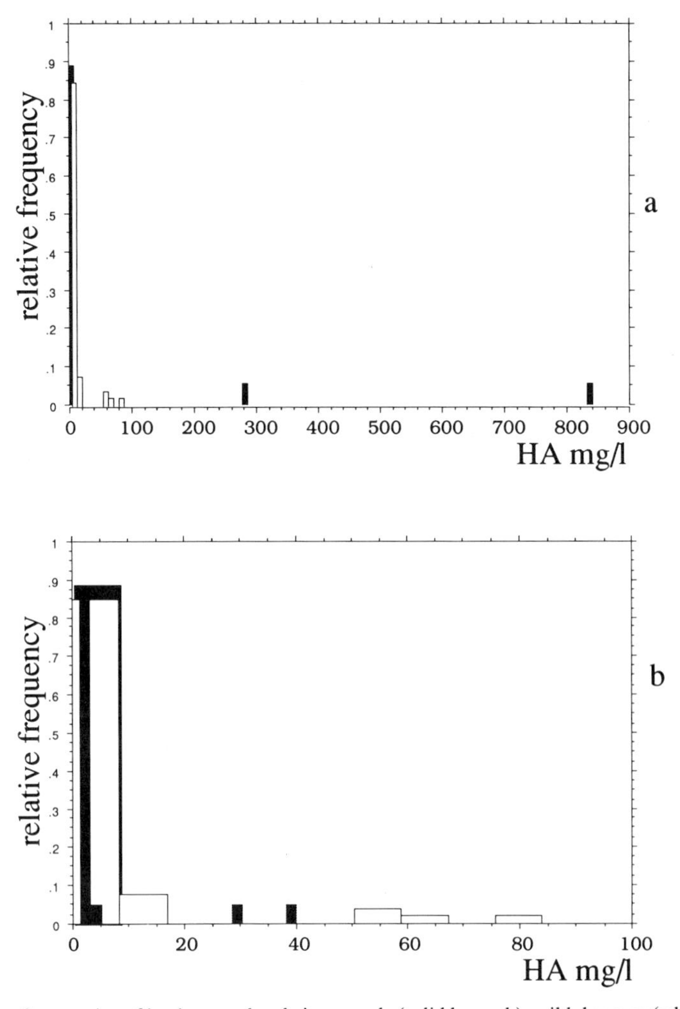

Figure 4. Relative frequencies of hyaluronan levels in normals (solid bars a,b), mild dry eyes (white bars a,b) and contact lens wearers (b, narrow solid bars). All distributions are very asymmetrical, with values clustered at low concentrations.

Decreased hyaluronan concentrations may reflect a metabolic or permeability effect of dry eye pathology or contact lenses on the ocular surface. It remains to be established whether molecular weight distribution of hyaluronan is affected by contact lens wear or surface pathology.

If the very high concentrations of hyaluronan in normal tears represent a response to constant low-grade local insult, the smaller range of values observed in mild dry eyes and in contact lens wearers may point to an inability to respond to a change in local environment.

ACKNOWLEDGMENTS

This work was supported by the Guide Dogs for the Blind Association (MB) and the Wellcome Trust (RBE, APC).

REFERENCES

1. Scott JE. Extracellular matrix, supramolecular organisation and shape. *J Anat.* 1995;187:259–269.
2. Presti D, Scott JE. Hyaluronan-mediated protective effect against cell damage caused by enzymatically produced hydroxyl (OH) radicals is dependent on hyaluronan molecular mass. *Cell Biochem Funct.* 1994;12:281–288.
3. Nakamura M, Nishida T, Hikida M, Ottori T. Combined effects of hyaluronan and fibronectin on corneal epithelial wound closure of rabbit in vivo. *Curr Eye Res.* 1994;13:385–388.
4. Molander N, Ehinger B, Stenevi U, Lindquist U, Lind ML. Corticosteroid suppression of trauma-induced hyaluronan in rabbit cornea and aqueous. *J Refract Surg.* 1995;11:260–266.
5. Frescura M, Berry M, Corfield A, Carrington S, Easty DL. Evidence of hyaluronan in human tears and secretions of conjunctival cultures. *Biochem Soc Trans.* 1994;22:228S.
6. Fukuda M, Miyamoto Y, Miyara Y, Mishima H, Ottori T. Hyaluronic acid concentration in human tear fluids. *Invest Ophthalmol Vis Sci.* 1996;37:S848.

EPIDEMIOLOGY AND CLASSIFICATION OF DRY EYE

Michael A. Lemp

University Ophthalmic Consultants of Washington
Department of Ophthalmology
Georgetown University Medical Center
Washington, DC

1. INTRODUCTION

Dry eye, or keratoconjunctivitis sicca (KCS), is a rubric for a number of clinical disease states characterized by disturbances in the pre-ocular tear film resulting in ocular surface disease.[1] KCS accounts for a high percentage of patient visits to ophthalmologists, and its treatment has given rise to a pharmaceutical market in the US of over $100 million per year (industry estimates). Approximately 7–10 million Americans currently require the use of artificial tear preparations.[2] KCS is associated with a broad spectrum of ocular symptoms ranging from mild transient irritation to persistent irritation, burning, itchiness, redness, pain, ocular fatigue, and vision disturbances.[3,4] In its more severe form, Sjögren's-associated KCS, it is associated with vision-threatening complications including scleritis, corneal ulceration, and secondary bacterial infection.[5] Pathophysiologic factors include deficiencies in aqueous tear production, qualitative changes in tear composition, alterations in the oil secretion of the meibomian glands of the eyelids, blinking abnormalities, and inflammatory changes in the ocular surface.[6]

In the report of the NEI/Industry Workshop on Clinical Trials in Dry Eyes, a new classification system for the various types of dry eye was presented (Fig. 1).[7] The two major categories are *aqueous tear-production-deficient* and *evaporative*. In the aqueous production-deficient category, there is a further breakdown into non-Sjögren's-associated KCS (non-SS-KCS) and Sjögren's-associated KCS (SS-KCS). The leading cause of evaporative tear deficiency is *meibomian gland dysfunction*. Deficiency and/or alterations in lipid secretions by the meibomian glands result in increased evaporation of aqueous tears from the ocular surface and are frequently associated with inflammatory changes of the lid margin.[8–10] The relative frequency of aqueous tear-production deficiency and evaporative KCS is not known. However, both types are commonly seen in clinical ophthalmologic

Lacrimal Gland, Tear Film, and Dry Eye Syndromes 2
edited by Sullivan *et al.*, Plenum Press, New York, 1998

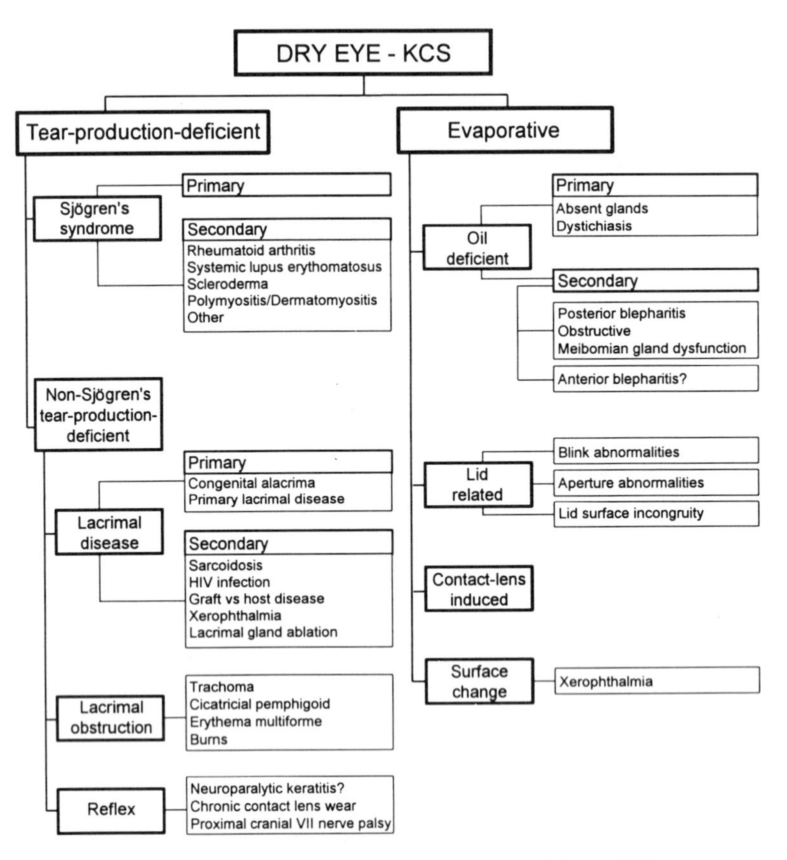

Figure 1. Classification of dry eye. (Adapted from ref. 7.)

practice and account for a large percentage of visits to ophthalmologists throughout the US and the world.[11]

Sjögren's syndrome is a chronic multisystem condition affecting excretory glands in all patients and extra-glandular organs in some (primary SS).[12] About half the cases (secondary SS) are associated with another connective tissue disease, most commonly rheumatoid arthritis.[13] SS is the second most common connective tissue disease (after rheumatoid arthritis). The incidence is unknown, but the prevalence has been estimated to be 1–2% of the general population; 80–90% of the patients are adult women.[12–15] The onset is usually insidious. The major symptoms are dry eyes and dry mouth, with evidence of decreased secretion. There is major salivary gland enlargement in 30–50%. Extra-glandular features include Raynaud's phenomenon, primary biliary cirrhosis, diffuse interstitial lung disease, interstitial nephritis, atrophic gastritis, peripheral neuropathies, and inflammatory vascular disease.[12–15] The prognosis and natural history of SS are not clear.[13] In many patients, symptoms and signs can plateau at a level of moderate severity and do not appear to shorten life span. Some patients with SS, however, can progress to the development of a B-cell lymphoma, which can be fatal. There have been reports that SS can undergo spontaneous remission, but this appears to happen only rarely. In SS-KCS, immunologically mediated inflammatory disease of the ocular surface, such as scleritis and corneal ulceration, can result in loss of vision.[5]

The relationship between SS-KCS and non-SS-KCS is unclear.[16] Whether non-SS-KCS can be a prodromal state to the development of SS-KCS is unknown. In a Swedish study of several hundred seemingly healthy mature adults aged 52–72, approximately one-third of the patients reported symptoms of dry eye and dry mouth.[17] In these subjects, there was a 1.54- to 2.88-fold increase in abnormal exocrine gland function and elevated levels of anti-Ro and anti-La antibodies. This high prevalence of immunologic abnormalities in otherwise seemingly healthy adults in this age group tends to blur the distinction between SS-KCS and non-SS-KCS. There are no good data concerning the clinical course of dry eye states.

2. DIAGNOSIS OF SJÖGREN'S SYNDROME

There are at least five commonly used classification criteria for diagnosing primary Sjögren's syndrome (Fig. 2).[18] Primary SS is diagnosed when all other well-established connective tissue systemic diseases are excluded. If an established systemic disease is present in conjunction with SS, the patient is said to have secondary SS. In a recent European study (Fig. 3),[19] all the initial three criteria were found to have a high specificity (range: 97.9–100%) and similar sensitivity (range: 72.2–57.8%), whereas the California criteria had a substantially lower sensitivity (22.9%). The European criteria also are attractive because they employ criteria currently available in all clinical settings and meet with patient

Copenhagen criteria: Require at least two abnormal objective tests for the function of lacrimal glands and salivary glands (eg, pathological salivary gland biopsy). No subjective complaint or presence of autoantibodies are required.

Japanese criteria: At least two of three items must be fulfilled in patients who have overt dry symptoms of unknown aetiology. (1) KCS diagnosed when Schirmer-1 result and van Bijsterveld score and/or fluorescein staining is abnormal. (2) Pathological lacrimal or salivary gland biopsy. (3) Pathological sialography. Presence of autoantibodies is not required.

Greek criteria: Requires the presence of pathological salivary gland biopsy. In addition, at least two of three items must be fulfilled. (1) KCS diagnosed if subject symptoms and pathological Schirmer-1 or van Bijsterveld score is present. (2) Xerostomia diagnosed if subjective symptoms and decreased stimulated parotic flow rate measured for 5 minutes is present. (3) Parotic gland enlargement either by history or on physical examination. Presence of autoantibodies is not required.

California criteria: Requires the presence of all of the following. (1) KCS diagnosed if subjective symptoms and pathological Schirmer-1 and van Bijsterveld score or fluorescein staining is present. (2) Xerostomia diagnosed if subjective symptoms and decreased stimulated parotic flow rate measured for 3 minutes. (3) Pathological salivary gland biopsy with a focus score of $\geq$2 per 4 mm. (4) Presence of any specific autoantibodies (ANA, IgM-RF, anti-SS-A ab, anti-SS-B ab) in abnormal titre. Lymphoma, sarcoidosis, graft vs host disease and AIDS must be excluded.

European criteria: At least four of six items must be fulfilled. (1) Subjective complaint of dry eyes. (2) Subjective complain of dry mouth. (3) Pathological salivary gland biopsy. (4) Abnormal Schirmer-1 test or van Bijsterveld score (=rose bengal staining). (5) Abnormal unstimulated whole sialometry or sialography

Figure 2. Most commonly used criteria for diagnosing primary Sjögren's syndrome.[17]

1. **Ocular symptoms**
   Definition. A positive response to at least 1 of the following 3 questions:
   a. Have you had daily, persistent, troublesome dry eyes for more than 3 months?
   b. Do you have a recurrent sensation of sand or gravel in the eyes?
   c. Do you use tear substitutes more than 3 times a day?
2. **Oral symptoms**
   Definition. A positive response to at least 1 of the following 3 questions:
   a. Have you had a daily feeling of dry mouth for more than 3 months?
   b. Have you had recurrent or persistently swollen salivary glands as an adult?
   c. Do you frequently drink liquids to aid in swallowing dry foods?
3. **Ocular signs**
   Definition. Objective evidence of ocular involvement, determined on the basis of a positive result on at least 1 of the following 2 tests:
   a. Schirmer-I test (≤5 mm in 5 minutes)
   b. Rose bengal score (≥4, according to the van Bijsterveld scoring system)
4. **Histopathologic features**
   Definition. Focus score ≥1 on minor salivary gland biopsy (focus defined as an agglomeration of at least 50 mononuclear cells; focus score defined as the number of foci 4 mm^2 of glandular tissue)
5. **Salivary gland involvement**
   Definition. Objective evidence of salivary gland involvement, determined on the basis of a positive result on at least 1 of the following 3 tests:
   a. Salivary scintigraphy
   b. Parotid sialography
   c. Unstimulated salivary flow (≤1.5 ml in 15 minutes)
6. **Autoantibodies**
   Definition. Presence of at least 1 of the following serum autoantibodies:
   a. Antibodies to Ro/SS-A or La/SS-B antigens
   b. Antinuclear antibodies
   c. Rheumatoid factor

Figure 3. European criteria for classifying Sjögren's syndrome.[19]

acceptance, as only four of the six criteria must be met. (A labial biopsy is not required.) These criteria were reported to have a sensitivity of 93.5% and a specificity of 94.0% in a secondary care setting.

Tear production decreases with age in otherwise seemingly healthy individuals.[20] Whether this is a purely age-related atrophic glandular phenomenon or an immunologically mediated event associated with pathophysiologic processes common to both SS-KCS and non-SS-KCS is unclear. Recent studies, however, have shed considerable light on factors that may be causative in the development of decreased lacrimal gland activity and the development of Sjögren's syndrome.[21]

The relative roles of viral infections, i.e., Epstein-Barr virus, cytomegalovirus, or herpes virus-6, have been suggested—perhaps these viruses activate or precipitate immune responses in the lacrimal glands, leading to decreased lacrimation.[22–35]

Hormones have been implicated as having a significant effect on tear production.[36,37] The sexual dichotomy manifested in many autoimmune processes, e.g., Sjögren's syndrome, rheumatoid arthritis, systemic lupus erythematosus, thyroid disease, etc., along with considerable experimental evidence from animal studies have led to hypotheses that identify androgens as being particularly important for support of lacrimal gland function and as downregulators of immune activity in the lacrimal glands.[37–40] The role of estrogens in lacrimal secretion and the management of the ocular surface remains somewhat controversial.[41,42]

It has been suggested that a primary lacrimal deficiency, probably the most common cause of tear-production-deficient dry eye, occurs when bio-available androgen levels fall below critical values.[43] Cell death following withdrawal of hormonal support may lead to processing and presentation of parenchymal cell antigens. Normal intracellular traffic patterns may cause acinar cells to secrete autoantigens, leading ultimately to acinar cell death.

Menopausal and post-menopausal women and older men all share a common denominator, i.e., their level of bio-available androgen is low. In KCS-SS, androgen deficiency also plays a major role, particularly as the loss of this hormonal support renders the lacrimal glands more vulnerable to other insult, such as viral infection, which may act as triggering agents in the initiation of an autoimmune cascade.[44]

In a recent study, 26 patients with dry eye and 19 controls were matched approximately by menstrual status, hormone replacement therapy usage, and age.[45] Serum concentrations of androgens, estrogens, LH, FSH, and prolactin were measured. Tear osmolarity was used as an objective index of dry eye severity, and a subjective symptom questionnaire was used. Low testosterone levels were more prevalent in patients and correlated with the subjective severity of the dry eyes.

3. EPIDEMIOLOGIC DATA ON DRY EYE

Anecdotal knowledge indicates that dry eye can affect adults at all ages, particularly elderly people, and that it affects women more often than men. Information about its prevalence is limited. A small study by Jacobsson et al. in Sweden of people aged 55–72 found the prevalence of KCS to be 14.9%; the female:male ratio was 11:8.[46] A small survey of elderly women in a retirement community in Washington, DC, ages 63–92, reported a KCS prevalence of 35%.[47] A recent dry eye prevalence survey (the SEE Project) of 2,514 elderly persons aged 65–84 in Salisbury, Maryland reported a dry eye prevalence of 11.3% based on rose bengal of ^{2}5 and 11.9% based on Schirmer of ^{2}5 mm.[48] The survey found no age, sex, or race differences in prevalence for these definitions of dry eye. A retrospective 15-year study in Minnesota of a county-wide HMO, not yet reported, found an approximate annual incidence of 2–4 per 100,000 for Sjögren's syndrome (personal communication, S. Pillemer). The study is limited by lack of ethnic diversity and the fact that it is limited to cases diagnosed by HMO physicians and therefore excludes undiagnosed cases and carries the likely difficulty of attempting to apply uniform criteria to retrospective diagnostic information prepared by multiple physicians. Hence, information on a key basis for epidemiologic understanding of KCS, the frequency of its occurrence in a large sample of the general population and how that frequency varies by age, sex, and ethnicity, is needed.

4. SCREENING OF KCS BY QUESTIONNAIRE

Several investigative groups have developed and tested symptom questionnaires for KCS screening. Based on a discriminant analysis of 12-item symptom questionnaires completed by 63 women with sicca syndrome and 37 normals, McMonnies and Ho[49] classified the 100 women as either positive or negative for sicca syndrome. Of the 63 sicca subjects, 62 (98%) were positive; of the 37 controls, 36 (97%) were negative. The authors characterized the screening procedure as having 98% sensitivity and 97% specificity; however,

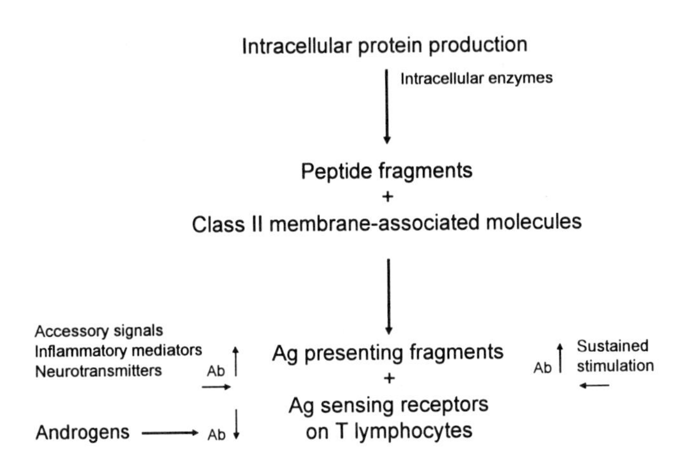

Figure 4. Theory of lymphocytic proliferation in Sjögren's syndrome.

these are inflated estimates, derived as they are from the same data from which the classification procedure was derived. To obtain unbiased estimates of sensitivity and specificity, one would need to apply the authors' classification procedure to an independently tested population.

Jacobsson et al.[46] submitted a two-item screening questionnaire to a random sample of the general population aged 52–72, classifying as positive those who responded "yes" to either question and as negative those who responded "no" to both. One question was about dry eye and the other about dry mouth. Upon clinical examination of a random sample of the respondents, the authors presented data[50] from which sensitivity of 90% and specificity of 40% can be derived.

Vitali et al.[50] tested a 20-item questionnaire (13 on dry eye and 7 on dry mouth) on 240 patients with primary Sjögren's syndrome (33) and 240 controls. Six questions (three on dry eye and three on dry mouth) were selected as most predictive of primary SS. These six questions were then tested on an independent sample of 246 persons with primary SS, 201 with secondary SS, 113 with connective tissue disease but without SS, and 133 controls; all these persons were also tested clinically for Sjögren's syndrome. The authors present an operating characteristic curve that includes a sensitivity of 86% coupled with a specificity of 92%, based on classifying as positive those with one positive response to the dry eye questions and one positive response to the dry mouth questions.

A major finding of the NEI/Industry Workshop was "the paucity of data concerning the frequency of dry eye states in the population and how that frequency varies according to age, sex, and race. There is also virtually no data in reference to risk factors for the development of dry eye conditions. Furthermore, there are no data concerning the natural course and prognosis of these states, or their relationship to each other...."

In an effort to develop an epidemiologic study to answer some of these questions, the National Eye Institute awarded a Planning Grant to the EMMES Corporation of Rockville, Maryland, an epidemiologic study organization, and to University Ophthalmic Consultants of Washington. A formal application has been submitted to the National Institutes of Health to perform a major epidemiologic study directed at answering many of these questions.

DEEP (Dry Eye Epidemiology Projects) consists of two parallel projects, PRIDE (Population Research in Dry Eye), a population-based study of dry eye, and CRISS (Clinical Research in Sjögren's Syndrome), a clinic-based investigation of Sjögren's syndrome (Fig. 5). PRIDE's aims are to estimate the age-, sex-, and race-specific prevalence of various forms of dry eye in a probability sample of residents of Montgomery Country, Maryland, and with persons from that sample as subjects to the case-controlled and longitudinal studies to investigate risk factors, quality of life, clinical course, and prognosis of dry eye. The conditions will be classified according to the NEI/Industry classification. In CRISS, samples of Sjögren's syndrome patients and matched controls from five clinical settings will supplement the cases of Sjögren's syndrome in the population sample for case-controlled and longitudinal studies of primary and secondary Sjögren's syndrome.

The prevalence survey in PRIDE will be conducted in two stages: (1) a screening interview, and (2) an examination of all positive screens and of a sample of negative screens. During the planning phase, we developed and pretested a six-question screening questionnaire to be used in the Dry Eye Prevalence (DEP) survey on 77 dry eye patients (cases) and 80 persons without dry eye (controls). Cases and controls were selected from my ophthalmic practice in Washington, DC, and that of Dr. Daniel Nelson in St. Paul, Minnesota. A staff member of Westat, the proposed DEEP survey research center, experienced in interview training, prepared an interview manual for the questionnaire and trained an experienced interviewer to conduct the interviews. The results of this pre-test indicate that the questionnaire is an effective tool to screen efficiently a population for dry eye and demonstrated a specificity of 94 and a sensitivity of 62.5.

The aims of the case-controlled studies in PRIDE and CRISS, in which persons without dry eye will serve as controls, are to investigate risk factors (medical history, hormonal factors, viral, socio-demographic, and psycho-social variables) for dry eye. Samples of serum of case controls will be collected for future hypothesis testing. The aims of the 6-year longitudinal studies in PRIDE and CRISS are to investigate whether the manifestations of dry eye and quality of life improve or worsen with time and what factors may be prognostic of the condition's course.

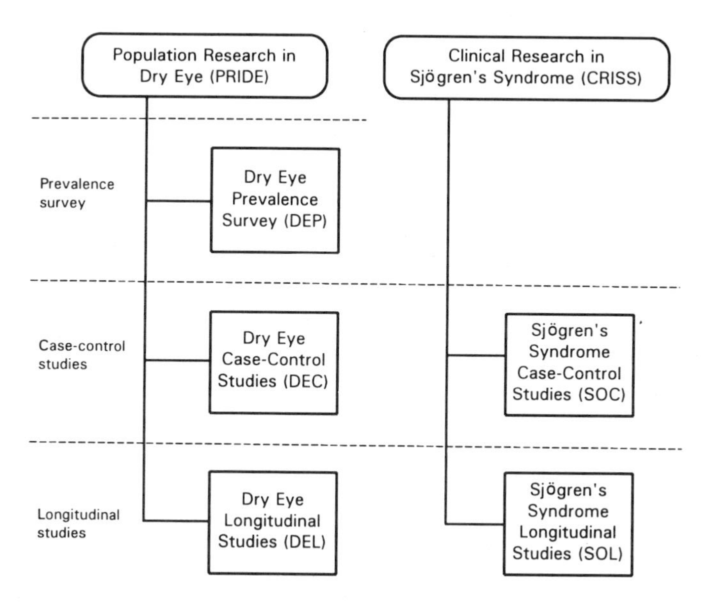

Figure 5. DEEP design schema.

5. DRY EYE PREVALENCE SURVEY (DEP)

The primary goal of the DEP survey is to determine the age-, sex-, and race-specific prevalence of all forms of dry eye and the number of specific forms of the disease among residents of Montgomery County aged 25 to 79. An equally important goal is to identify persons with dry eye for enrollment in case control and longitudinal studies, and persons without dry eye for enrollment as controls in the case control studies. Montgomery County was chosen as a study site in part because although incomes are above average, the population racial make-up is similar to that of the United States. The survey will be conducted in two stages, a screening stage and an examination stage. In the screening stage, a sample of county residents selected by random digit dialing will be administered a telephone screening questionnaire asking about dry eye symptoms. In the examination stage, all positive screenees in a 4.85% sample of negative screenees will be invited for a clinical examination for dry eye. The screening sample will be selected by the survey research center (Westat), and the examinations will be conducted at two examination centers in Montgomery County. It is estimated that approximately 4,000 persons will be examined at one of the PRIDE examination centers.

6. DRY EYE CASE CONTROL STUDIES (DEC)

The goal of the five case control studies, one each of the primary lacrimal disease (PLD), meibomian gland dysfunction (MGD), contact lens induced disease (CLD), primary Sjögren's syndrome (PSS), and secondary Sjögren's syndrome (SSS), is to investigate possible risk factors. Cases for these studies will be those persons of the 1200 examined in DEP who are classified as having one or more of the specified diagnoses of dry eyes. Controls will be drawn from those persons from the 1200 who had screened negative for dry eyes, and who did not have dry eye on examination. Persons who screened positive for dry eye but whose objective tests for the condition were negative will not be used as controls, because their reported screening of having frequent dry eye symptoms indicate they may be borderline cases of dry eye. The risk factors include hormonal factors (pharmacologic, menstrual, reproductive history); disease history (thyroid disease, graft vs. host disease, viral disease [HIV, mononucleosis, mumps, measles]); other medical history (bone marrow transplant, silicone breast implant, dry eye symptoms during pregnancy, irregular sleep cycles); medical history of blood relatives (dry eye, systemic immunologic abnormalities); socio-demographic variables (e.g., educational achievement), and psycho-social variables (information obtained in the quality-of-life interview). Age, sex, and race will be assessed as risk factors in the prevalence study; to avert confounding these factors will be controlled analyses of the case control studies (e.g., Mantel-Haenszel, logistic regression). In addition to the data on the aforementioned potential risk factors obtained by interview, serologic data from immunologic (anti-RO [SSA], anti-La [SSB], anti-nuclear antibodies, and rheumatoid factor) and hormonal assays will be subjected to risk factor analyses (Table 1).

7. DRY EYE LONGITUDINAL STUDIES (DEL)

All persons found to have dry eye on examination in DEP, whether they had screened positive or negative for dry eye, will be invited to enroll in DEL. We estimate

Table 1. PRIDE examination procedures and frequency of follow-up

	Interviews and examinations						
	Base-line	1-year	2-year	3-year	4-year	5-year	6-year
General interview	•	*	o	*	o	*	o
Quality of life	•		o		o		o
Visual acuity	•	*	o	*	o	*	o
Biomicroscopy	•	*	o	*	o	*	o
External examination	•	*	o	*	o	*	o
Meibomian gland assessment	•	*	o	*	o	*	o
Schirmer test	•	*	o	*	o	*	o
Tear breakup time	•	*	o	*	o	*	o
Fluorescein staining	•	*	o	*	o	*	o
Rose bengal staining	•	*	o	*	o	*	o
Tear film osmolarity	•	*	o	*	o	*	o
Lactoferrin	•	*	o	*	o	*	o
Impression cytology	•	*	o	*	o	*	o
Blood drawing	•		*		*		*

• Procedure applied to all participants in DEC, DEL, SOC, and SOL.
o Procedure applied to all participants.
* Procedure applied to all participants with SS in DEL and SOL.
DEC, Dry Eye Case Control Studies. DEL, Dry Eye Longitudinal Studies. SOC, case control studies of risk factors. SOL, longitudinal studies of clinical course and prognosis.

that 90% will agree to participate. There will be a separate longitudinal study for each diagnostic category of dry eye. Persons with multiple dry eye diagnoses will participate in as many studies as they have diagnoses. These studies will go on for 6 years; Sjögren's syndrome patients will be examined annually and all participants at 2-year intervals. Items to be investigated in the study of the clinical course and prognosis include:

- Variables from baseline examination: general interview, clinical examination, impression cytology
- Serologic immune and hormonal levels
- Co-morbidity, especially connective tissue disease
- Changes in the use of drugs for the treatment of dry eyes
- Additional diagnoses of dry eyes; incidence of dry eye diagnostic groups among individuals initially and other dry eye diagnostic groups (e.g., incidence of Sjögren's syndrome among individuals with other types of dry eye)
- Changes in disease severity; ordinal severity scales (e.g., mild, moderate, severe) will be developed during the study
- Quality of life: a global index quality of life as well as particular items from quality-of-life interview
- Seasonal variation of symptoms

PRIDE seeks to answer the following questions:

- What is the age-, sex-, and race-specific prevalence of dry eye in the general population?
- What hormonal, viral, and other immunomodulating factors are associated with the development of dry eye?
- Does dry eye tend to progress in severity, and, if so, at what rate? (Patients identified by the initial survey will be followed over a 6-year period.)

Table 2. CRISS examination procedures and frequency of follow-up

	Interviews and examinations						
	Base-line	1-year	2-year	3-year	4-year	5-year	6-year
General interview	o	o	o	o	o	o	o
Quality of life	o		o		o		o
Visual acuity	o	o	o	o	o	o	o
External examination	o	o	o	o	o	o	o
Schirmer test	o	o	o	o	o	o	o
Tear breakup time	o	o	o	o	o	o	o
Fluorescein staining	o	o	o	o	o	o	o
Rose bengal staining	o	o	o	o	o	o	o
Tear film osmolarity	o	o	o	o	o	o	o
Lactoferrin	o	o	o	o	o	o	o
Impression cytology	o	o	o	o	o	o	o
Blood drawing	o		o		o		o
Biomicroscopy	o	o	o	o	o	o	o
Meibomian gland assessment	o	o	o	o	o	o	o

o Procedure applied to all participants.

- Is dry eye a prodromal state for Sjögren's syndrome?
- Are any pharmaceuticals associated with the development of dry eye?
- Are any co-morbid states associated with the development of dry eye?

8. CLINICAL RESEARCH IN SJÖGREN'S SYNDROME (CRISS)

The CRISS component of DEEP consists of two types of investigations: case control studies of risk factors (SOC), and longitudinal studies of clinical course and prognosis (SOL). In each investigation, two entities will be studied: primary Sjögren's syndrome (PSS), and secondary Sjögren's syndrome (SSS). PSS requires the absence of connective tissue disease (CTD), and SSS requires its presence. In addition to patients with SS, patients without SS will be enrolled as controls in the case control studies; the controls will be without CTD for the PSS and with CTD for SSS. Data collected in SOC and SOL will be pooled with corresponding Sjögren's syndrome data from DEC and DEL and PRIDE. Five sites will take part in this study. Table 2 notes the examinations and frequency of follow-up.

CRISS seeks to answer several important questions about Sjögren's syndrome: its pathogenesis, and the clinical progression of the condition. In this cross-sectional and prospective study, which will be carried out over a 6-year period at five sites in a well-characterized clinical population with Sjögren's syndrome, the following questions will be addressed:

- Is primary Sjögren's syndrome a risk factor for the development of secondary Sjögren's syndrome?
- What are the risk factors (e.g., hormonal, viral, immunologic) associated with primary and secondary Sjögren's syndrome?
- What is the clinical course of primary and secondary Sjögren's syndrome? Do these conditions tend to progress in severity, and if so, at what rate?

- How is quality of life affected by Sjögren's syndrome? Are the effects different for the two types of Sjögren's syndrome (primary and secondary)?

We expect DEEP to produce badly needed information that will provide new research leads and help to develop new strategies for managing dry eye states.

ACKNOWLEDGMENT

I would like to express my sincere appreciation to Mr. Fred Ederer, my colleague and principal investigator for the DEEP studies.

REFERENCES

1. Lemp MA. Basic principles and classification of dry eye disorders. In: Lemp MA, Marquardt R, eds. *The Dry Eye.* New York: Springer Verlag; 1993:101–130.
2. Kaswan R. Cyclosporine drops: A potential breakthrough for dry eyes. In: *Research to Prevent Blindness Science Writers Seminar.* New York: Research to Prevent Blindness; 1989:18–20.
3. Lemp MA. Recent developments in dry eye management. *Ophthalmology.* 1987;94:1299–1304.
4. Toda I, Fujishima H, Tsuboto K. Ocular fatigue is a major symptom of dry eye. *Acta Ophthalmol.* 1993;71:347–352.
5. Lemp MA, Keenan PC. Sjögren's syndrome. In: Fraunfelder FT, Roy FN, eds. *Current Ocular Therapy.* Philadelphia: WB Saunders; 1995:238–239.
6. Holly FJ, Lemp MA. Tear physiology and dry eyes. *Surv Ophthalmol.* 1977;22:69–87.
7. Lemp MA. Report of the National Eye Institute/Industry Workshop on Clinical Trials in Dry Eyes. *CLAO J.* 1995;21:221–232.
8. Mathers WD, Shields WJ, Petroll WM, Jester JV. Meibomian gland dysfunction in chronic blepharitis. *Cornea.* 1991;10:277–285.
9. Mathers WD. Ocular evaporation in meibomian gland dysfunction and dry eye. *Ophthalmology.* 1993;100:347–351.
10. Driver PJ, Lemp MA. Meibomian gland dysfunction. *Surv Ophthalmol.* 1996;40:343–367.
11. American Academy of Ophthalmology. Blepharitis and the dry eye in the adult. Preferred Practice Pattern. San Francisco. 1991.
12. Homma M, Sugai S, Tojo T, Miyaraka N, Arizuki M, eds. *Sjögren's Syndrome - State of the Art.* Amsterdam: Kugler Publications; 1994.
13. Daniels TE. Sjögren's syndrome in a nutshell. Report. Port Washington, NY: Sjögren's Syndrome Foundation; 1990.
14. Talal N, Moutsopoulos HM, Kasson SS, eds. *Sjögren's Syndrome: Clinical and Immunological Aspects.* Berlin: Springer Verlag; 1987.
15. Fox RI, Kanh HI. Pathogenesis of Sjögren's syndrome. *Rheum Dis Clin North Am.* 1992;18:517–538.
16. Tsubota K. Sjögren's syndrome dry eye and non-Sjögren's syndrome dry eye: What are the differences? In: Homma M, Sugai S, Tojo T, Miyasaka N, Akizuki M, eds. *Sjögren's Syndrome - State of the Art.* Amsterdam: Kugler Publications; 1994:27–31.
17. Jacobsson LTH, Hansen BV, Manthorpe R. Association of dry eyes and dry mouth with anti-Ro/SS-A and anti-Ro/SS-B autoantibodies in normal adults. *Arthritis Rheum.* 1992;35:1492–1501.
18. Manthorpe R, Jacobsson LTH. Sjögren's syndrome. *Baillieres Clin Rheumatol.* 1995;9:483–496.
19. Vitali C, Moutsopoulos HM, Bombardiere S. The European Community Study Group on Diagnostic Criteria for Sjögren's Syndrome. Sensivitivity and specificity of tests for ocular and oral involvement in Sjögren's syndrome. *Ann Rheum Dis.* 1994;53:637–647.
20. Mathers WD, Lane JA, Zimmerman MB. Tear film changes associated with normal aging. *Cornea.* 1996;15:229–234.
21. Sullivan DA, Wickham A, Krenzer KL, Rocha AM, Toda I. Aqueous tear deficiency in Sjögren's syndrome: Possible causes and potential treatment. In: Pleyer U, Hartmann C, Sterry W, eds. *Ocular Diseases: Immunology of Bullous Oculo-Muco-Cutaneous Disorders.* Buren The Netherlands: Aeolus Press; 1997:95–152.

22. Fox RI, Saito I. Sjögren's syndrome: Immunologic and neuroendocrine mechanisms. *Adv Exp Med Biol.* 1994;350:609–621.
23. Ahmed SA, Talal N. Sex hormones and the immune system. Part 2. Animal data. *Baillieres Clin Rheumatol.* 1990;4:13–31.
24. Fox RI, Luppi M, Kang HI, Pisa P. Reactivation of Epstein-Barr virus in Sjögren's syndrome. *Springer Semin Immunopathol.* 1991;13:.
25. Sullivan DA, Sato EH. Potential therapeutic approach for the hormonal treatment of lacrimal gland dysfunction in Sjögren's syndrome. *Clin Immunol Immunopathol.* 1992;64:9–16.
26. Fox R. Epstein-Barr virus and human autoimmune diseases: Possibilities and pitfalls. *J Virol Methods.* 1988;21:19–27.
27. Fox RI, Chelton T, Scott S, Benton L, Howell FV, Vaughan FH. Potential role of Epstein-Barr virus in Sjögren's syndrome. *Rheum Dis Clin North Am.* 1987;13:275–292.
28. Burns JC. Persistent cytomegalovirus infection -- the etiology of Sjögren's syndrome. *Med Hypotheses.* 1983;10:451–460.
29. Fox RI, Pearson G, Vaughan JH. Detection of Epstein-Barr virus-associated antigens and DNA in salivary gland biopsies from patients with Sjögren's syndrome. *J Immunol.* 1986;137:3162–3168.
30. Pflugfelder SC, Tseng SCG, Pepose JS, Fletcher MA, Klimas N, Feuer W. Epstein-Barr virus infection and immunological dysfunction in patients with aqueous tear deficiency. *Ophthalmology.* 1990;97:313–323.
31. Pepose JS, Akata RF, Pflugfelder SC, Voigt W. Mononuclear cell phenotypes and immunoglobulin gene rearrangements in lacrimal gland biopsies from patients with Sjögren's syndrome. *Ophthalmology.* 1990;97:1599–1605.
32. Shillitoe EJ, Daniels TE, Whitcher JP, Strand CV, Tolol N, Greenspan JS. Antibody to cytomegalovirus in patients with Sjögren's syndrome as determined by an enzyme-linked immunosorbent assay. *Arthritis Rheum.* 1982;25:260–265.
33. Krueger GRF, Wassermann K, de Clerck LS, et al. Latent herpes virus-6 in salivary and bronchial glands. *Nature.* 1990;336:1255–1256.
34. Mariette X, Gozlan J, Clerc D, Bisson M, Morinet F. Detection of Epstein-Barr virus DNA by in situ hybridization and polymerase chain reaction in salivary gland biopsy specimens from patients with Sjögren's syndrome. *Am J Med.* 1991;90:286–294.
35. McMurray R, Keisler D, Kanuckel K, Izui S, Walker SE. Prolactin influences autoimmune disease activity in the female B/W mouse. *J Immunol.* 1991;147:3780–3787.
36. Ahmed SA, Penhale WJ, Talal N. Sex hormones, immune responses and autoimmune disease. *Am J Pathol.* 1985;121:531–551.
37. Sullivan DA. Possible mechanisms involved in reduced tear secretion in Sjögren's syndrome. In: Homma M, Sugai S, Tojo T, et al., eds. *Sjögren's Syndrome --State of the Art.* Amsterdam: Kugler Publications; 1993:11–19.
38. Sullivan DA. Possible mechanisms involved in the reduced tear secretion in Sjögren's syndrome. In: Homma M, Sugai S, Tojo T, et al., eds. *Sjögren's Syndrome --State of the Art.* [Proceedings of the 4th International Symposium, Tokyo, Japan, August 11–13, 1993.] New York: Kugler Publications; 1994:13–19.
39. Mircheff AK. Understanding the causes of lacrimal insufficiency: Implications for treatment and prevention of dry eye syndrome. In: *Research to Prevent Blindness Science Writers Seminar.* New York: Research to Prevent Blindness; 1993:51–54.
40. Sullivan DA, Allansmith MR. Hormonal influence on the secretory immune system of the eye: Androgen modulation of IgA levels in tears of rats. *J Immunol.* 1985;134:2978–2982.
41. KramerP, Lubkin V, Potter W, Jacobs M, Labay G, Silverman P. Cyclic changes in conjunctival smears from menstruating females. *Ophthalmology.* 1990;97:303–307.
42. Lubkin V, Nash R, Kramer P. The treatment of perimenopausal dry eye syndrome with topical estradiol. ARVO Abstracts. *Invest Ophthalmol Vis Sci.* 1992;33:S1289.
43. Mircheff AK, Warren DW, Wood RL. Hormonal support of lacrimal function, primary lacrimal deficiency, autoimmunity, and peripheral tolerance in the lacrimal gland. *Ocular Immunol Inflamm.* Aeolus Press, Buren (The Netherlands) 1996;4: No 3 pp. 145–172.
44. Sullivan DA, Sato EH. Potential therapeutic approach for the hormonal treatment of lacrimal gland dysfunction in Sjögren's syndrome. *Clin Immunol Immunopathol.* 1992;64:9–16.
45. Mamalis N, Harrison D, Hiura G, et al. Are "dry eyes" a sign of testosterone deficiency in women? Presentation, *10th International Congress of Endocrinology,* San Francisco June 12–15, 1996 Abstract # P3–379, page 849.
46. Jacobsson LTH, Axell TE, Hansen BU. Dry eyes or mouth: An epidemiologic study in Swedish adults, with special reference to primary Sjögren's syndrome. *J Autoimmun.* 1989;2:521–527.

47. Strickland RW, Tesar JT, Beene BH. The frequency of sicca syndrome in an elderly population. *J Rheumatol.* 1987;14:766–771.
48. Schein OD, Munoz B, Tielsch JM. Estimating the prevalence of dry eye among elderly Americans: SEE Project. ARVO Abstracts. *Invest Ophthalmol Vis Sci.* 1996;37:S636.
49. McMonnies CW, Ho A. Patient history in screening for dry eye conditions. *J Am Optom Assoc.* 1987;58:296–301.
50. Vitali C, Bombardieri S, Moutsopoulos HM, et al. Preliminary criteria for the classification of Sjögren's syndrome. *Arthritis Rheum.* 1993;36:340–347.

CANDEES

The Canadian Dry Eye Epidemiology Study

Barbara E. Caffery,[1] Doris Richter,[1] Trefford Simpson,[1] Desmond Fonn,[1]
Michael Doughty,[2] and Keith Gordon[3]

[1]Centre for Contract Lens Research
School of Optometry
University of Waterloo
Waterloo, Ontario, Canada
[2]Glasgow-Caledonia University
Department of Vision Science
Glasgow, Scotland
[3]Alcon, Canada Inc.
Mississauga, Ontario, Canada

1. INTRODUCTION

Although dry eye seems to be a common symptom in eye care practice, very little is known of its true prevalence in a large population. One reason for this lack of knowledge is that until the recent industry-sponsored dry eye meeting at the National Eye Institute, there was no standard definition for dry eye and the diagnostic criteria varied from study to study.

The pupose of our study was to determine the prevalence of dry eye symptoms in a large patient population. There is some literature to suggest that questionnaires have good specificity and sensitivity in the diagnosis of dry eye. A secondary purpose was to determine patient ratings of severity of dryness and to understand the associated patient characteristics for dry eye symptoms.

2. METHODS

Questionnaires were mailed to all optometrists in Canada in October of 1994, together with an explanatory letter. The optometrists were asked to give the survey to 30 consecutive patients. On completion, the forms were mailed to the Centre for Contact

Lacrimal Gland, Tear Film, and Dry Eye Syndromes 2
edited by Sullivan *et al.*, Plenum Press, New York, 1998

Lens Research (CCLR) at the University of Waterloo where they were electronically scanned, and the data was analyzed using STATT Release 3.0b. The forms were received until March 1, 1995.

3. RESULTS

A total of 13,517 questionnaires were returned, which represents 15.6% of the 86,160 that were mailed. This represents 15.7% of the optometrists who were asked to participate.

The patients completing the questionnaires included all age group; 55% of the sample were in the 21 to 50 years age range. The sample comprised 60.7% female patients, and 24.3% of the total were contact lens wearers.

Symptoms of dry eye were reported by 28.7% of the respondents. When asked to grade the symptoms, 7.6% noted constant but moderate symptoms and 1.6% reported severe symptoms which represents a prevalence of 45 in 10,000 for severe symptom. Those reporting severe dry eye were predominantly female by a ratio of 46 to 1.

Of the contact lens-wearing group, 50.1% reported dry eye symptoms which is significantly greater than the non-lens-wearing group (21.7%). By gender, 33% of women and 22% of men reported dry eye symptoms. Patients taking medication and those with allergies were more likely to report dry eye symptoms.

4. CONCLUSIONS

CANDEES represents the most substantial survey to date on the prevalence of dry eye symptoms. In the non-contact lens-wearing population, 21.7% reported dry eye symptoms. The contact lens-wearing group had significantly more subjects reporting dry eye (50.1%). Further work is needed to determine which ocular characteristics best relate to these symptoms.

ACKNOWLEDGMENT

Alcon Canada provided partial funding for this study.

REFERENCES

1. McMonnies CW. Key questions in a dry eye history. *J Am Optom Assoc.* 1986;57:512–517.
2. McMonnies CW, Ho A. Patient history in screening for dry eye conditions. *J Am Optom Assoc.* 1987;58:296–301.
3. McMonnies CW, Ho A. Response to a dry eye questionnaire from a normal population. *J Am Optom Assoc.* 1987;58:588–591.
4. Golding TR, Brennan NA. Survey of optometric opinion on dry eye topics. *Clin Exp Optom.* 1992;75:98–108.
5. Hikichi T, Yoshida A, Fukui Y, et al. Prevalence of dry eye in Japanese eye centres. *Graefes Arch Clin Exp Ophthalmol.* 1995;233:555–558.
6. Lemp MA. Report of National Eye Institute/Industry Workshop on clinical trials in dry eyes. *CLAO J.* 1995;21:221–232.

113

SENSITIVITY AND SPECIFICITY OF A SCREENING QUESTIONNAIRE FOR DRY EYE

Neal L. Oden,[1] David E. Lilienfeld,[1] Michael A. Lemp,[2] J. Daniel Nelson,[3] and Fred Ederer[1]

[1]The EMMES Corporation
Potomac, Maryland
[2]University Ophthalmic Consultants of Washington
Washington, DC
[3]St. Paul-Ramsey Medical Center
St. Paul, Minnesota

1. INTRODUCTION

Dry eye, or keratoconjunctivitis sicca (KCS), is a rubric for a number of clinical disease states characterized by disturbance in the pre-ocular tear film, resulting in ocular surface disease.[1] KCS accounts for a high percentage of patient visits to ophthalmologists, and its treatment has given rise to a large pharmaceutical market. About 7–10 million Americans currently require use of artificial tears preparation.[2]

We have anecdotal knowledge that dry eye affects adults at all ages, particularly elderly people, and that it affects women more often than men, but information about its prevalence from surveys is limited to elderly persons. A small study in Sweden of people aged 55–72 estimated the prevalence of KCS as 14.9%[3]; the female-to-male prevalence ratio was 11:8. A small survey of yet older people (ages 63–92), women in a retirement home in Washington, DC, reported a KCS prevalence of 35%.[4] The SEE project in Salisbury, MD, has reported a dry-eye prevalence of 11.3% based on a rose bengal of ≥5, and of 11.9%, based on a Schirmer of ≤5 mm.[5] Several investigators have reported on the sensitivity and specificity of screening tests for dry eye. Based on a discriminant function analysis of 12-item symptom questionnaires completed by 63 women with and 37 women without sicca syndrome, McMonnies and Ho[6] classified the respondents as either positive or negative for sicca syndrome. Of the 63 sicca subjects, 62 (98%) were positive; of the 37 controls, 36 (97%) were negative. The authors characterized the screening procedure as having 98% sensitivity and 97% specificity; however, these may be biased (inflated) estimates, derived as they are from the same data from which the classification procedure was developed.[7] The Swedish investigators reported a sensitivity of 89% of their screening

Lacrimal Gland, Tear Film, and Dry Eye Syndromes 2
edited by Sullivan *et al.*, Plenum Press, New York, 1998

Table 1. Dry eye screening questionnaire

Do you use:
eye washes?
hot or cold compresses to your eyes?
eye drops?
How often are your eyes:
itchy?
sore (hurting)?
dry (not wet enough)?
scratchy?
gritty or sandy?
burning?
irritated?
watering?
sensitive to light (light hurts your eyes)?
red?
sticky?
achy?
Is your mouth frequently dry?
Do you experience seasonal allergies affecting your eyes?
How often do you wear contact lenses?
Have you ever been told by a physician that you have dry eye?

Note: Questions about frequency were answered on a 4-point scale: Never, Sometimes, Often, Constantly.

questionnaire consisting of two questions, and a specificity of 40%. Vitali et al.,[8] in a European Community study of Sjögren's syndrome, a form of dry eye, obtained a sensitivity of 84% and a specificity of 97% when they took two positive responses to three questions about dry eye symptoms as a "positive" screening criterion for dry eye.

During the planning of the Dry Eye Epidemiology Projects (DEEP), a proposed large epidemiologic study of dry eye that is to include a population survey, we developed a questionnaire for screening the population (Table 1). In DEEP, all persons in a population sample of about 52,000 adults aged 25–79 who screen positive and a 5% sample of those who screen negative are to be clinically examined for dry eye. The screening questionnaire, which can be administered on the telephone in only a few minutes, is intended as a cost-saving measure. In comparison to interviews, clinical examinations for dry eye are expensive. With reasonably high sensitivity and specificity, the survey cost can be markedly reduced (of the 52,000, only about 4,000 may need to be examined clinically) and yet provide unbiased and statistically reliable estimates of prevalence. For reasons of cost, population screening is an essential component of large dry eye prevalence surveys.

In this paper, we present the results of pretesting the screening questionnaire, and the estimates of sensitivity and specificity.

2. METHODS

2.1. Questionnaire, Interviews, and Description of Sample

The questionnaire, containing 19 questions pertaining to dry eye (see Table 1), was pretested in a telephone interview on 77 dry-eye patients (cases) and 79 persons without dry eye (controls). The sex and age distributions of respondents are shown in Table 2. Sex

Table 2. Sex and age distributions of questionnaire respondents

		Sex		Age (yr)					
Sample	Total	M	F	<40	40-49	50-59	60-69	70-79	≥80
Controls	79	31	48	8	15	17	18	14	7
Cases	77	13	64	11	13	14	17	18	4

ratios of controls and cases differed; 83% of cases, but only 61% of controls, were women. Age distributions did not differ significantly ($X^2 = 2.2$, $p = 0.82$). Distribution of primary diagnoses is shown in Table 3. Half the cases and half the controls were selected from the practices of each of two ophthalmologists, one in Washington, DC (Michael A. Lemp, University Ophthalmic Consultants of Washington) and one in St. Paul, MN (J. Daniel Nelson, St. Paul-Ramsey Medical Center). All interviews were conducted on the telephone by a person who was masked as to whether the interviewee was a case or control. The interviewer was trained in conducting the interviews and followed the procedures outlined in an interviewer manual. Nearly all the interviews lasted 3–5 min.

2.2. Statistical Methods

We used stepwise multiple logistic regression, with both forward (p to enter = 0.05) and backward (p to remove = 0.05) elimination to construct our discriminator function. Press and Wilson[9] indicate that logistic regression is superior to parametric discriminant analysis when data are not multivariate normal within classes. Logistic regression produces a predictor equation of the form:

$$\log\left(\frac{p}{1-p}\right) = \beta_1 x_1 + \ldots + \beta_n x_n \tag{1}$$

where $x_1,\ldots,x_n$ are the values of the predictor variables for the observation in question, p is the estimated probability that an observation with these values for predictor variables will be of the dry eye type, and $\beta_1,\ldots,\beta_n$ are fitted constants.

Table 3. Primary dry eye diagnoses of questionnaire respondents

Tear-production-deficient	
Primary Sjögren's syndrome	19
Secondary Sjögren's syndrome	12
Indeterminate Sjögren's syndrome	7
Other	3
Total	41
Evaporative	
Meibomian gland disease	27
Anterior blepharitis	2
Lid-related	3
Other	4
Total	36

The predictor equation having been estimated, a prediction is made about the classification of each observation (case or control), based on the values of the predictor variables. This is done by first choosing a constant K, where $0 \le K \le 1$. If, for a particular observation, the predicted probability p exceeds K, the predicted classification is "dry eye"; otherwise, it is "normal." Of course, the proportion of observations that give rise to predicted classifications of dry eye depends upon the value of K. The comparison between the observed and predicted classifications is summarized by the estimated sensitivity and specificity of the screening test. That is,

$$\text{estimated sensitivity} = \frac{\text{n(both predicted and observed to be dry eye)}}{\text{n(observed to be dry eye)}} \tag{2}$$

$$\text{estimated specificity} = \frac{\text{n(both predicted and observed to be normal}}{\text{n(observed to be normal)}} \tag{3}$$

A function, known as the *receiver operating characteristic* (ROC) curve, displays the various possible estimated sensitivity-specificity pairs as K varies from 0 to 1. The investigator chooses a particular sensitivity-specificity pair from the ROC curve, thus implicitly choosing a value for K. Together with the ß values, this defines the discriminator.

In stepwise regression, data are searched and re-searched for relationships between predictor and predicted variables that appear to be meaningful. Thus, predictors whose relationship to the predicted value is only accidental will too often be incorporated into the regression function.[10] To counter this tendency, we first randomly divided the data into halves, A and B, by randomly assigning to set A half the observations in each of the four strata composed by crossing the two ophthalmologists with the two diagnoses (dry eye/normal). The remainder of the observations were assigned to set B. Then we allowed a variable to be in the prediction equation only if it was selected by the stepwise regression in both halves of the data.

Sensitivity and specificity summarize aspects of a discriminator's probability of misclassification. In general, the probability of misclassification is underestimated by the *resubstitution estimate*.[7] The resubstitution estimate is the proportion of misclassified observations in the same data that were used to estimate the discriminator. We used an unbiased method, *test sample estimation*. Here, we let A be the *training set*, in which we estimated the discriminator. We then applied this discriminator to set B, our *test set*, to calculate unbiased misclassification rates. As a check on the stability of our method, we then switched the roles of sets A and B. Note the reduction in sample size that is the main drawback of the test set method.

We expected to be able to demonstrate a considerable reduction in bias from the use of the *test sample* estimate instead of the resubstitution estimate. However, we shall see below that, for our data, there is very little bias associated with the resubstitution estimate, which, with its larger sample size, is therefore preferred.

We use AB to describe a ROC curve estimate for which the predictor equation is fitted using the A half of the data, and the result is applied to the B half to get sensitivity and specificity values. We refer to types AB and BA estimates as *heterotypic*, types AA and BB as *homotypic*, and type (A+B)(A+B) as *full data*. Homotypic and full data ROC curve estimates are resubstitution estimates. Although only heterotypic estimates are technically test sample estimates, we use training set and test set to refer to halves X and Y of a type XY estimator, regardless of whether X = Y.

In summary, we took the following steps to construct our discriminator function and ROC curves.

(a) *Identification of predictor variables:* Use only those variables independently chosen by the stepwise regression in both sets A and B.

(b) *Construction of ROC curves:* Using only the predictors from step (a), construct the heterotypic ROC curve estimates. Inspect their difference from homotypic estimates to assess the degree of resubstitution bias. Below, this bias will be found to be small, indicating that for this dataset a full data estimate of the ROC curve is preferable because of its larger sample size.

(c) *Construction of discriminator:* Use only the variables of step (a) to construct a full data estimate of the ROC curve.

2.3. Selection of Variables

The following variables were available for analysis (see Table 1 for list of questions):

- Wetting: Applies Washes, Compresses, or Drops to Eye
- Symptoms: Itchy, Sore, Dry, Scratchy, Gritty, Burning, Irritated, Watering, Light-sensitive, Red, Sticky, Achy, Mouth Dry, Allergies
- Patient history: Wears contact lenses; Physician told respondent he or she has dry eye
- Demographic: Age, Sex

For the construction of a discriminator, we use only symptom and demographic variables. The rationale for excluding the wetting and patient history variables from consideration follows.

Wetting. The major objection to the use of wetting variables was that our data are a clinic series, but the discriminant function will be applied to individuals from the general population. It seemed likely that, in the clinic series, many of the dry-eye patients would have been offered wetting agents by their physicians, so that wetting variables would appear to discriminate well between cases and controls. In partial confirmation of this, 95% of dry eye patients in our clinic series used wetting agents, whereas only 49% of normals did. However, when the discriminator is applied to a population-based series, containing many undiagnosed cases, wetting agents may prove to be useless for discrimination.

Even if our data had been population- rather than clinic-based, the manner of incorporation of wetting variables into the prediction equation would be unclear. This is because wetting agents might interact with symptom variables in a way that varies from one individual to the next. Wetting agents might mask symptoms in proactive individuals, but not in non-proactive ones.

Patient History. Clearly, a variable indicating whether an individual was told by a physician that he or she has dry eye is likely to be a good discriminator in a clinic series, but not in a population-based series, which may contain many undiagnosed cases.

Contact Lenses. This variable was intended not as a predictor, but rather to provide information on how well the discriminator functions within classes. Most (92%) individuals in this dataset do not wear contact lenses; the proportion does not depend significantly on case/control status.

2.4. Scoring the Variables

Thus, only the symptom and demographic variables were used in the prediction equation. Symptom variables are ordinal, most on a 4-point scale (Constantly, Often, Sometimes, Never), but two on a 3-point scale (Yes, No, Don't know). We re-coded all symptom variables on a comparable scale according to the following scheme: Constantly = 6; Yes = 6; Often = 4; Don t know = 3; Sometimes = 2; No = 0; Never = 0. In this coding, the numerical values are directly associated with dry eyes for all variables, except "How often are your eyes watering?" which is inversely associated with dry eyes. To achieve consistency, we reversed the coding of this variable.

3. RESULTS

3.1. First Discriminator Function

In our first analysis, we applied stepwise logistic regression on the symptom and demographic variables to sets A and B.

The coefficient *c* in Table 4 is a measure of rank correlation between predicted values and diagnosis. There are three equivalent interpretations of *c*:[11]

1. The area under the ROC curve. A measure of predictive accuracy, this ranges from 0 to 1, with a value of 1/2 when the discriminator cannot distinguish cases from controls, and a value of 1 when a sensitivity and a specificity of 1 are simultaneously attained.
2. The probability that the discriminator, when furnished a pair of observations, one from the case and one from the control group, will be able to correctly assign the observations to the groups on the basis of the relative magnitudes of the predicted values (ties in predicted values are broken by tossing a fair coin).
3. Wilcoxon's rank sum statistic (or equivalently the Mann-Whitney U-test) as a measure of the rank difference between predicted values in the case and control groups.

Table 4. Predictive equations and correlations for the first discriminator [For explanation, see text]

Line	Training set	Predictor Equation	c test set A	c test set B
		Stepwise logistic regressions for variable selection		
1	A	−6.08 + 0.47 dry + 0.76 irritated + 0.60 watering	.90	.90
2	B	−5.92 + 0.90 dry + 1.22 irritated	.87	.94
		Ordinary logistic regressions for assessment of resubstitution bias		
3	A	−3.11 + 0.61 dry + 0.47 irritated	.87	.93
4	B	−5.92 + 0.90 dry + 1.22 irritated	.87	.94
		Ordinary logistic regression for final model		
5	A+B	−4.02 + 0.69 dry + 0.71 irritated	.90 (A+B)	

Because c is a correlation between predicted value and case or control status, it can be calculated both homotypically and heterotypically. Both c values are shown in Table 4, except in line 5, where a full data calculation is appropriate.

All values of c are high, indicating good discriminative ability for all five models.

Table 4 shows the prediction equations used to construct the first discriminator. Lines 1 and 2 of Table 4 show the results of stepwise logistic regression on the two halves of the data. As described in step (a) above, for lines 3–5 we used ordinary logistic regression with only those predictor variables selected in both lines 1 and 2.

Fig. 1 shows the ROC curves constructed from the predictor equations in lines 3 and 4 of Table 4 when applied homotypically and heterotypically to test sets A (solid lines) and B (broken lines). All curves are similar and reveal a strong relationship between observed and predicted.

If resubstitution bias were strong in these data, one would expect to find that a type AA ROC curve estimate would be consistently outside of (i.e., with ostensibly larger sensitivity/specificity values and c values than) a type AB estimate, and similarly for type BB

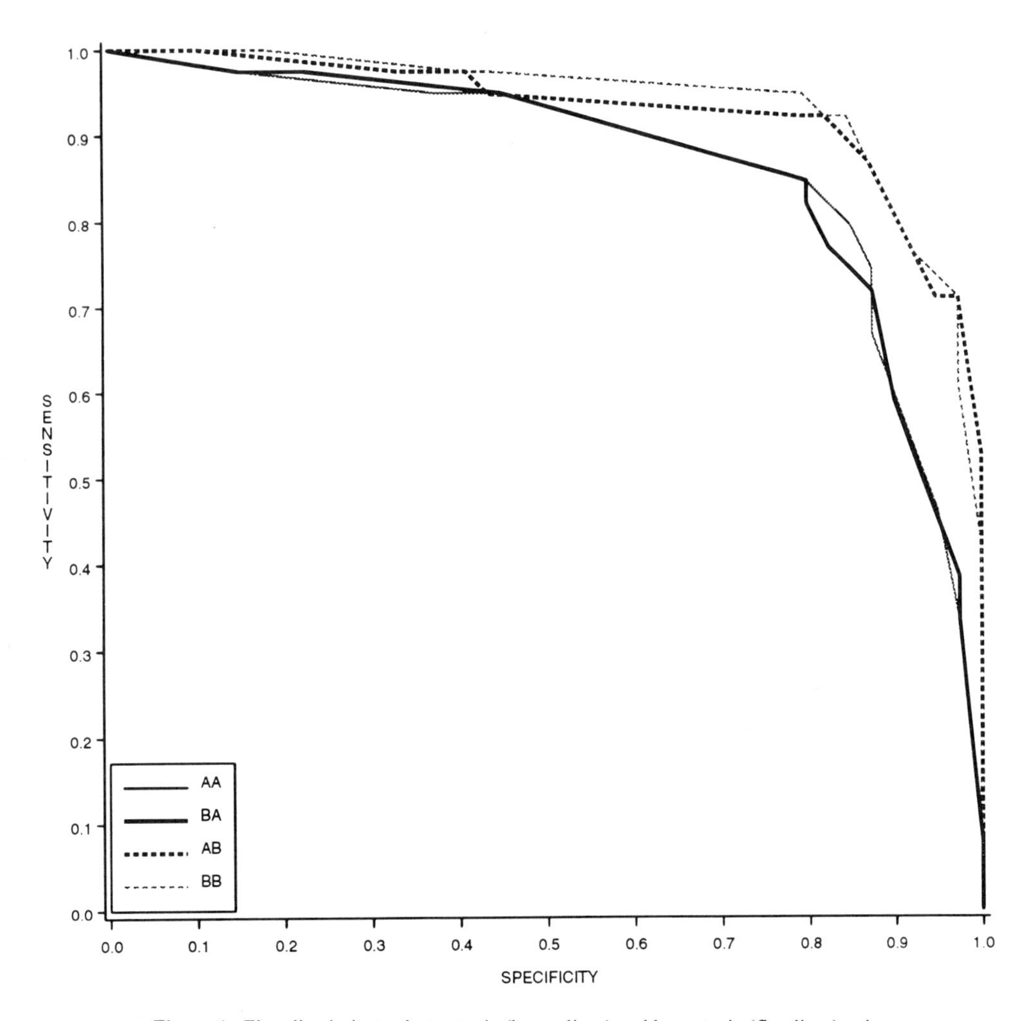

Figure 1. First discriminator heterotypic (heavy lines) and homotypic (fine lines) pairs.

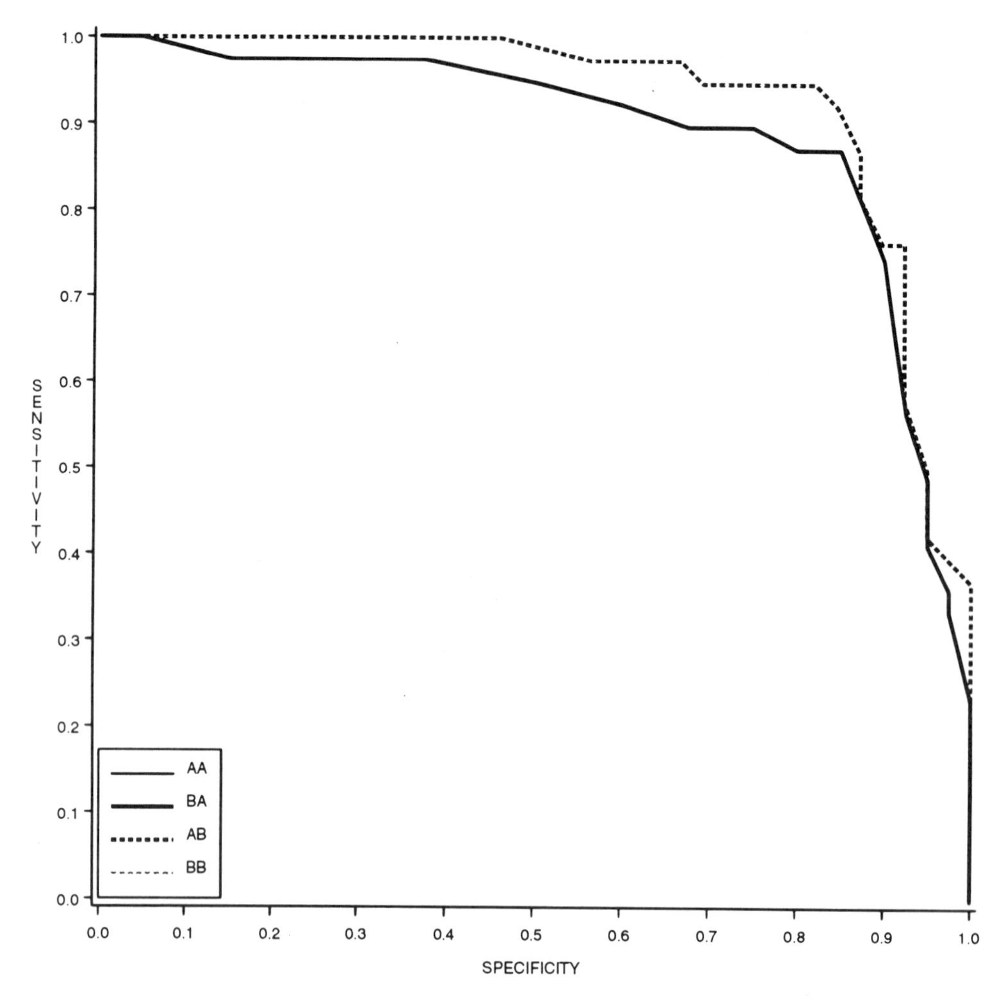

Figure 2. Second discriminator-hetero- and homotypic pairs.

vs. type BA. However, while BB is outside BA (c = 0.94 vs. 0.87), AA is inside AB (c = 0.87 vs. 0.93). In these data, ROC curve differences are dominated by the test set rather than by resubstitution bias or other effects. That is, the solid pair of lines differs from the broken pair more than the two members of each pair differ from each other (Fig. 1), and c values depend more on the choice of test set than on hetero- or homotypicality (Table 4).

Because resubstitution bias is small, a full data estimate, with its larger sample size, is preferred. The full data predictor equation appears in line 5 of Table 4, and the ROC curve derived from it is shown as a solid line in Fig. 3. Table 5 gives the sensitivity/specificity pairs upon which Fig. 3 is based.

Notice that the final linear combination in line 5 of Table 4, because of the near equality of the ß coefficients, is very close to simply using Dry + Irritated. To illustrate more explicitly the dependence of observed dry eye/control status on these two predictor variables, Table 6 shows the number of observed dry eye and normal observations in the data, cross-classified in a 4x4 table. Cells along the NE-SW diagonals of the 4x4 table

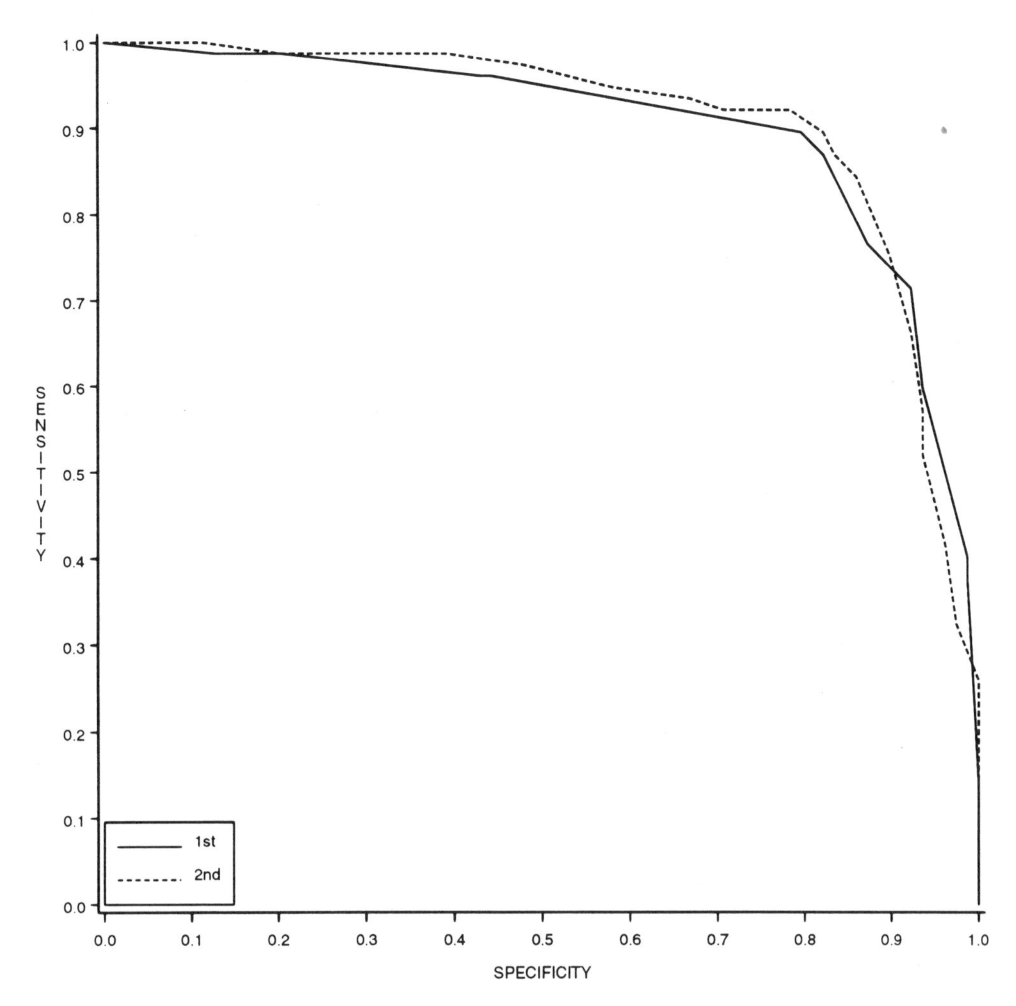

Figure 3. First vs. second discriminator-entire data.

correspond to constant values of the sum Dry + Irritated. As these values of this sum increase (0–12), the ratio of cases to controls increases markedly, in agreement with the logistic regression results.

3.2. Second Discriminator Function

In an attempt to construct a more robust estimator, we calculated, for each subject, 4 variables, each of which summarized the 14 symptom variables: Maximum = the maximum value of the 14 symptom variables; Minimum = the minimum value of the 14 symptom variables; Number = the number of symptom variables with values ≥3; Sum = the sum of symptom variables.

Stepwise logistic regression (not shown) selected Sum in preference to the other 3 predictors in both data halves. Table 7 shows the prediction equations used to construct the second discriminator, which considered for variable selection not only the 14 symptom

Table 5. Sensitivities and specificities of the two discriminators featured in Fig. 3

First discriminator			Second discriminator		
Cut	Sens. (%)	Spec. (%)	Cut	Sens. (%)	Spec. (%)
	100	0		100	0
−4.02	99	13	−3.72	100	11
−2.64	99	20	−3.40	99	20
−2.59	96	43	−2.43	99	39
−1.26	96	44	−2.11	97	48
−1.21	90	80	−1.79	95	58
−1.15	88	81	−1.47	94	67
0.12	87	82	−1.15	92	71
0.17	77	87	−0.83	92	78
0.23	71	92	−0.51	90	82
1.55	60	94	−0.19	87	84
1.61	40	99	0.14	84	86
1.66	38	99	0.46	75	90
2.99	14	100	0.78	66	92
4.42	00	100	1.10	57	94
			1.42	52	94
			1.74	47	95
			2.06	42	96
			2.38	32	97
			2.70	26	100
			5.92	00	100

The columns labelled "Cut" specify the cutpoints giving rise to the sensitivity/specificity pairs. For example, a sensitivity/specificity of (60%, 94%) is obtained when classifying as screen-positive all individuals whose first discriminator score (−4.02 + 0.69 Dry + 0.71 Irritated) exceeds 1.55.

Table 6. Cross-classification of Dry by Irritated by diagnosis in the entire data

	Irritated				
Dry	Never (0)	Sometimes (2)	Often (4)	Constantly (6)	Total
Never (0)					
Control	10	18	1	0	29
Case	1	2	1	0	4
Sometimes (2)					
Control	6	28	4	0	38
Case	0	5	4	2	11
Often (4)					
Control	1	4	4	0	9
Case	0	8	15	1	24
Constantly (6)					
Control	1	1	1	0	3
Case	1	9	18	10	38
Total					
Control	18	51	10	0	79
Case	2	24	38	13	77

Table 7. Predictive equations and correlations for the second discriminator. [For explanation, see text]

Line	Training set	Predictor equation	*c* test set A	*c* test set B
		Stepwise logistic regressions for variable selection		
1	A	−5.38 + 0.19 sum - 0.40 sensitive	.90	.89
2	B	−8.14 + 0.21 sum + 0.52 dry - 0.75 sticky	.87	.95
		Ordinary logistic regressions for assessment of resubstitution bias		
3	A	−5.32 + 0.15 sum	.89	.93
4	B	−6.03 + 0.17 sum	.89	.93
		Ordinary logistic regression for final model		
5	A+B	−5.65 + 0.16 sum	.91 (A+B)	

variables, but also Sum. Lines 1 and 2 of Table 7 show the results of stepwise logistic regression on the two halves of the data. As described in step (a) above, we used ordinary logistic regression with the intersection of the sets of selected variables to construct the models described in lines 3 and 4 of Table 7.

The lack of resubstitution bias and dependence of the ROC curves on the test set is even more pronounced for the second discriminator than for the first. In Fig. 2, AA = BA, and BB = AB. ROC curves and *c* values of lines 3 and 4 of Table 7 depend entirely on the test set, not on whether the *c* values are homotypic or heterotypic. This is because, when applied to the same training set, the predictions made by the two predictor equations are rank-quivalent, and thus generate identical ROC curves. As before, the conclusion is that resubstitution bias is small, so that a full data estimate is preferred. The full data predictor equation is shown in line 5 of Table 4. The ROC curve derived from it is shown as a broken line in Fig. 3.

Do these results generalize to other ROC curves? Because ROC curves are invariant under rank-preserving transformations of the predictor equations giving rise to them, the mapping from predictor equations to ROC curves is many to one. Thus, the ROC curve is more likely to be affected by the test set than by differences between heterotypic and homotypic estimates of the predictor equation. This mitigates the effect of resubstitution bias in ROC curve estimation.

3.3. Comparison of Discriminators

The ROC curves of the first and second discriminators in Fig. 3 hardly differ. Is one discriminator better than the other?

The Pearson correlation coefficients of Sum with the 2 demographic predictors, the 14 symptomatic predictors, and the linear combination 0.69 Dry + 0.72 Irritated (Table 4, line 5) are given in Table 8. The high correlation between Sum and the Dry/Irritated linear combination suggest that their ROC curves are very nearly the same.

A regression of Sum on Dry + Irritated has the following parametric values:

$$\text{Sum} = 12.41 + 3.96^{*}\,(\text{Dry} + \text{Irritated}) + \text{Residuals}$$

Table 8. Pearson correlation coefficients of Sum with other predictor variables

	Sum
Sex	–0.23
Age	–0.03
Watering	0.12
Allergy	0.15
Itchy	0.35
Red	0.43
Sensitive	0.52
Burning	0.56
Mouth dry	0.57
Sticky	0.59
Gritty	0.67
Achy	0.67
Sore	0.68
Scratchy	0.72
Irritated	0.77
Dry	0.77
0.69 Dry + 0.72 Irritated	0.87

By construction, Residuals is uncorrelated with Dry + Irritated. However, a logistic regression of diagnosis (case or control) on Dry + Irritated and Residuals reveals that Residuals is still significantly correlated with diagnosis (p =0.0029). (The association of Dry + Irritated with outcome is significant at p =0.0001.) That the portion of Sum uncorrelated with Dry + Irritated is still associated with outcome suggests that Sum contains elements not in Dry + Irritated that help in the prediction of outcome. Thus, from a statistical viewpoint, Sum appears to be a better predictor. However, it seems likely that the improvement attendant upon using Sum instead of Dry + Irritated is minor compared to the inaccuracies introduced when applying this discriminator, derived from a clinical series, to data randomly sampled from the general population.

3.4. Applying the Discriminator as a Screener for the General Population

The clinical series probably contains few, if any, undiagnosed cases of dry eye, whereas the general population presumably contains many undiagnosed cases of dry eye. Note that individuals previously diagnosed with dry eye will, upon confirmation by their physician, not be screened, but rather, automatically scheduled for a clinical examination. Thus, the population "seen" by the discriminator will contain *no* undiagnosed cases. This means that a previous diagnosis of dry eye is useless as a predictor variable for both the clinical and population series, but for opposite reasons.

As indicated in Section *2.3. Selection of Variables*, the lack of undiagnosed cases in the clinical series also precluded the use of wetting agents as predictive variables in constructions when fitting the discriminator. In DEEP, we plan to remedy inaccuracies that we introduced by constructing the discriminator from a clinical series by re-fitting the discriminator to the first several hundred subjects from whom clinical examination results are available. In this revision fitting, we plan to add wetting-agent questions to our list of potential predictor variables. There is an element of circularity in thus improving the discriminator; the population used for the revision fitting of the discriminator will, to some extent, depend on the definition of the unrevised discriminator. The data will be far richer in screen-positives than a random sample from the population. However, this is also true for the clinic series.

4. SUMMARY

We developed a *Dry Eye Screening Questionnaire* for the *Dry Eye Epidemiology Projects* (DEEP), a proposed large epidemiologic study. All persons who screen positive and a small sample of those who screen negative are to be invited for a diagnostic examination. Containing 19 questions, of which only 14 were used in the analysis, the questionnaire takes only a few minutes to administer on the telephone. To construct a discriminator function and thus a ROC curve, we used stepwise multiple regression on screening responses from a clinic series of 77 cases and 79 controls. Stepwise regression may incorporate into the predictor equation variables whose relation to the predicted is only accidental. Further, misclassification rates are underestimated by the *resubstitution* method, in which the proportion misclassified is obtained from the same dataset in which the discriminator function was fitted. To counter these problems, we randomly divided the data in half. We chose as predictors only those variables (Dry and Irritated) selected by stepwise regression in both data halves. We estimated unbiased misclassification rates using the unbiased *test set* method, in which the discriminator is fitted in one data half, and misclassification rates are calculated in the other half. Comparison of ROC curves arising from resubstitution and test set estimates indicates that resubstitution bias in misclassification rate estimation is negligible in our data. A resubstitution estimate made on the entire data is thus preferred. The resulting sensitivity/specificity values are reasonably high (e.g., 60%/94%), suggesting that the questionnaire will be a useful screening tool in the DEEP study. A second discriminator using the sum of all 14 responses is similar in its misclassification characteristics to the first discriminator. A second potentially significant error, arising from applying results from a clinical series to a general population, will be investigated as survey results in DEEP become available.

ACKNOWLEDGMENTS

The authors acknowledge the participation of Marline Bradford (The EMMES Corporation, Potomac, MD) who conducted the interviews, and of Diane Cadell (Westat, Inc., Rockville, MD) who prepared an interviewer manual and provided the interviewer training. They also thank Sophia Pallas for her consummate word processing.

REFERENCES

1. Lemp MA. Basic principles and classification of dry eye disorders. In: Lemp MA, Marquardt R, eds. *The Dry Eye.* Berlin: Springer Verlag; 1993:101–130.
2. Kaswan R. Cyclosporine drops: A potential breakthrough for dry eyes. In: *Research to Prevent Blindness Science Writers Seminar.* New York: Research to Prevent Blindness; 1989:18–20.
3. Jacobsson LTH, Axell TE, Hansen BU, et al. Dry eyes or mouth - an epidemiological study in Swedish adults, with special reference to primary Sjögren's syndrome. *J Autoimmun.* 1989;2:521–527.

4. Strickland RW, Tesar JT, Beene BH. The frequency of sicca syndrome in an elderly population. *J Rheumatol.* 1987;14:766–771.
5. Schein OD, Munoz B, Tielsch JM, Bandeen-Roche K, West SK. Estimating the prevalence of dry eye among elderly Americans. ARVO Abstracts. *Invest Ophthalmol Vis Sci.* 1996;37:S646.
6. McMonnies CW, Ho A. Patient history in screening for dry eye conditions. *J Am Optom Assoc.* 1987;58:296–301.
7. Breiman L, Friedman JH, Olshen RA, Stone CJ. *Classification and Regression Trees.* New York: Chapman-Hall; 1993;11.
8. Vitali C, Bombardieri S, Moutsopoulos HM, et al. Preliminary criteria for the classification of Sjögren's syndrome: Results of a prospective concerted action supported by the European Community. *Arthritis Rheum.* 1993;36:340–347.
9. Press SJ, Wilson S. Choosing between logistic regression and discriminant analysis. *J Am Statistical Assoc.* 1978;73:699–705.
10. Hosmer DW, Lemeshow S. *Applied Logistic Regression.* New York: Wiley; 1989;111.
11. Hanley JA, McNeil BJ. The meaning and use of the area under a receiver operating characteristic (ROC) curve. *Radiology.* 1982;143:29–36.

114

USE OF A QUESTIONNAIRE FOR THE DIAGNOSIS OF TEAR FILM-RELATED OCULAR SURFACE DISEASE

Maurizio Rolando, Angelo Macrì, Trillo Carlandrea, and Giovanni Calabria

Department of Ophthalmology
University of Genoa
Genoa, Italy

1. INTRODUCTION

Dry eye is defined as a disorder of the tear film, due to tear deficiency and/or tear evaporation that causes damage to the ocular surface and is associated with symptoms of ocular discomfort. This definition clearly underlines that subjective symptoms have a relevant part in the manifestation of the disease. Although collecting ocular surface signs have become rather standardized, symptom collection and their relevance as a diagnostic indicator have not yet been established. Furthermore, the frequent incongruence between the level of ocular surface damage and symptomatic complaints are characteristic of dry eye.

There is wide clinical experience of patients with minimal ocular surface involvement and a high level of complaints as well as a large number of patients with definite typical conjunctival and corneal damage who do not volunteer any symptoms except an undefined feeling of discomfort. These patients, only if asked, will report specific typical symptoms.

Collection of symptoms in dry eye is not straightforward. It is a common clinical experience that dry eye patients will not designate, if not explicitly asked, the occurrence of precise symptoms, but will only make the general complaint of *eye* discomfort. Furthermore, many patients, if not precisely asked, will mismatch symptoms. Burning and pain, and foreign body sensation and itching, are frequently confused. There has been a recent suggestion that a questionnaire about subjective symptoms and history should have a part in the workup of the dry eye suspect and, in general, in the course of ocular surface disease diagnosis.[1]

A validated questionnaire could well be a useful tool for population screening for dry eye, as well as a useful diagnostic parameter in the clinic.

2. MATERIAL AND METHODS

We have developed a questionnaire based on the symptoms and history most frequently reported by dry eye patients. In order to select the three most frequent and specific symptoms for dry eye, 20 patients with tear film-related ocular surface disease and 20 patients with non-tear film-related ocular surface disease were selected. From these patients, on the basis of a list of dry eye symptoms in the ophthalmic literature,[2] the three symptoms with the best sensitivity and specificity in detecting dry eye pathology were selected. These were identified as fundamental symptoms, and the remaining were considered accessory symptoms (Table 1). Our questionnaire was largely derived from the questionnaire of McMonnies and Ho.[3]

For every symptom considered in the questionnaire, a score was given according to an analogic semi-quantitative presentation consisting of a line with a ruler graded zero to nine. The patients, with the help of a staff member, marked the grade of the symptom. In this way a subtotal score for each symptom was obtained. The total score for each patient was calculated by the sum of the individual subtotal scores, with the following exception. The main symptoms were scored on the basis of the sum of the subtotal scores, and the accessory symptoms were scored by the sum of the subtotal scores divided by three.

In sections 3 to 5 of the questionnaire, each positive answer scored 1.

The questionnaire was applied to a population of 268 patients including a group of 112 healthy normals (age, 62 ± 18 years; female/male ratio (f/m), 2.7), a group of 43 ophthalmic patients with no ocular surface diseases (no-OSD) (age, 67 ± 12 years; f/m, 2), a group of 70 patients with non-tear film-related ocular surface diseases (OSD-no KCS) (age, 53 ± 9 years; f/m, 2.05), and a group of 43 patients with tear film-related ocular surface diseases (KCS) (age, 56 ± 13 years; f/m, 6.1).

The sensitivity and specificity of the questionnaire as a whole in separating the KCS patients from the other groups were tested. By studying the area under the ROC (Receiver Operator Characteristics) curves, the score with the best ability to separate dry eyes from normals and other pathologic eyes was identified.

3. RESULTS

Fig. 1 reports the frequency distribution of symptoms in the four groups of subjects. The sensitivity and specificity of the questionnaire in separating dry eye from non-dry eye conditions are sensitivity, 80.8%; specificity, 76.3%, for a diagnostic precision of 78.6%.

Fig. 2 shows the best area under ROC curves indicating the best compromise specificity and sensitivity, obtained when the questionnaire total score of 17 was used as a separator between dry eye and non-dry eye patients.

4. DISCUSSION

The ocular surface has receptors for touch, heat, and pain. Subjective symptoms are the result of activation of these receptors. Systematic and ordered evaluation of the prevalence and the significance of a number of subjective symptoms could be of great help in setting diagnostic networks for ocular surface diseases.

As has been recognized, symptoms are a key basis in diagnosis of keratoconjunctivitis sicca, which has typical subjective signs and typical time relation occurrences. Our

Table 1. Questionnaire for collecting dry eye symptoms and history

1. Main symptoms:
burning
foreign body sensation
need to keep eye closed
2. Accessory symptoms:
heavy eye
photophobia
wet eye feeling (without overflow)
discomfort on lid movement and/or closure
warm eye feeling
feeling his/her own eyes
Time relation:
symptoms appear:
in the morning
during the day
in the evening
symptoms are:
stable
increase with time
3. Environmental stress condition symptoms:
Eyes are bothered by:
cigarette smoke
smoke, or vapors in general
wind
air conditioning
contact lens wear
4. Need for topical therapy:
decongestant
tear substitutes
5. Use of systemic therapy and presence of systemic diseases:
β-blocking agents
diuretics
antihistaminic agents
antispasmodic agents
hormones
oral contraceptives
sleeping pills
antianxiety agents, etc.
mucous tissue dryness
rheumatic disease
thyroid disease
hormone disorders
connective tissue diseases

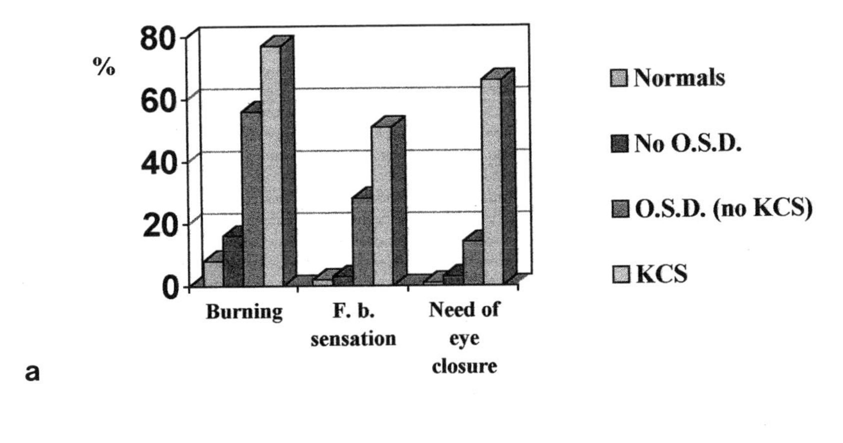

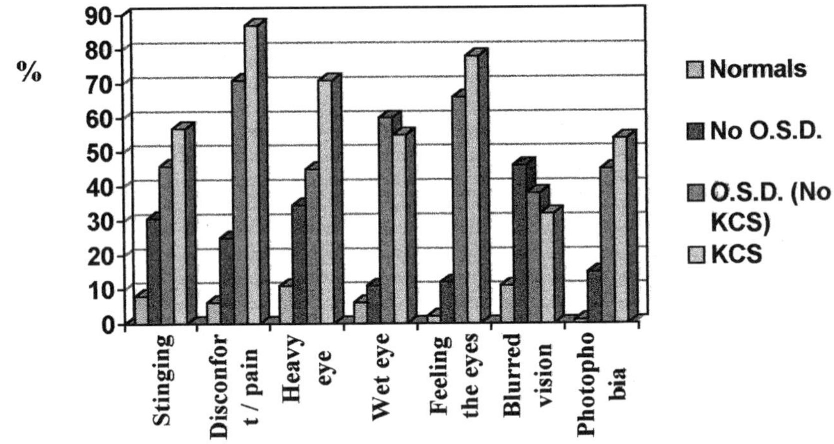

Figure 1. Frequency distribution of (a) main and (b) accessory symptoms.

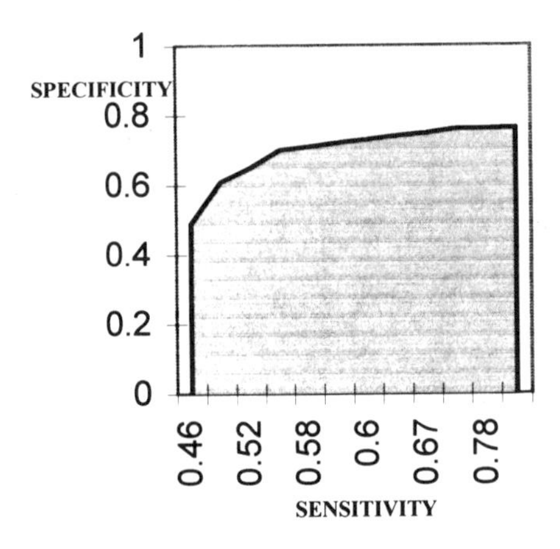

Figure 2. Best area under ROC curves corresponding to the total score of 17.

questionnaire showed an interesting ability to separate dry eyes from non-tear-related ocular surface diseases. We believe that it will prove useful in the clinic, especially in controversial cases where the ocular surface pathology is mixed because of the presence of concurrent diseases, or in early cases where objective signs are not clear.

REFERENCES

1. Lemp MA. Report of the National Eye Institute/Industry Workshop on clinical trials in dry eyes. *CLAO J.* 1995;21:221–232.
2. Liotet S, van Bijsterveld OP, Bletry O, et al. In L'oeil sec. Capter II, Liotet L ed. Paris, Masson; 1987:9–11.
3. McMonnies CW, Ho A. Patient history in screening for dry eye conditions. *J Am Optom Assoc.* 1987;58:296–301.

IMPORTANCE OF CONJUNCTIVAL EPITHELIAL EVALUATION IN THE DIAGNOSTIC DIFFERENTIATION OF DRY EYE FROM DRUG-INDUCED EPITHELIAL KERATOPATHY

Norihiko Yokoi and Shigeru Kinoshita

Department of Ophthalmology
Kyoto Prefectural University of Medicine
Kyoto, Japan

1. INTRODUCTION

Both dry eye and drug-induced epithelial keratopathy[1] often show severe corneal epithelial damage; sometimes the two conditions occur together.[1–3] Dry eye patients in particular are likely to be exposed to the cytotoxic effect of topically applied ocular drugs or preservatives, since frequent instillation is often encouraged and corneal epithelial damage is a predisposing background. It is therefore clinically important to distinguish dry eye-induced corneal epithelial damage from that which is drug-induced.

This study emphasizes the importance of conjunctival epithelial evaluation in differentiating dry eye corneal epithelial damage from drug-induced epithelial keratopathy.

2. SUBJECTS AND METHODS

Representative dry eye cases comprised 16 eyes of eight Sjögren's syndrome patients (one male, seven females; age, 46–62 years), and drug-induced diffuse epithelial keratopathy cases comprised 28 eyes of 16 patients (nine males, seven females; age, 36–85 years). Sjögren's syndrome was diagnosed on the basis of Fox et al.'s criteria[4]; the corneal fluorescein staining ranged from minimal to severe. Drug-induced diffuse epithelial keratopathy was retrospectively diagnosed in reference to the patient history of eyedrop use, including frequency, drug type, and preservative, until the first consultation. The reversibility of corneal damage after treatment was also documented. Charac-

Lacrimal Gland, Tear Film, and Dry Eye Syndromes 2
edited by Sullivan *et al.*, Plenum Press, New York, 1998

teristically, seven eyes of four patients had been exposed to a recently developed antiglaucoma drug (the prostaglandin F2α derivative, isopropyl unoprostone) in combination with a ß-blocker. In the course of our treatment, patients were asked to replace all eyedrops they had been using with preservative-free artificial tears until complete recovery, with no corneal fluorescein staining even after the stoppage of artificial tears. Of the 16 patients with drug-induced epithelial keratopathy, 12 were bilaterally involved and four were unilateral.

To evaluate conjunctival epithelial damage, sulforhodamine B[5] (SRB) and a clinically useful dye, rose bengal[6] (RB), were used (0.5%, 10 μl, and 1%, 2 μl, respectively). These stainings were scored separately (0 to 3 points) at the nasal and temporal bulbar conjunctiva, according to the method of van Biesterveld.[7] The scores were then totaled for each eye.

3. RESULTS AND DISCUSSION

In Sjögren's syndrome patients, severe conjunctival epithelial damage was demonstrated using SRB and RB staining [5.1 ± 1.5 (mean±SD) and 4.9 ± 1.5, respectively]. Drug-induced diffuse epithelial keratopathy patients, in contrast, showed minimal bulbar conjunctival epithelial staining in only two eyes of two patients, using SRB and RB staining (all scores 1); the other patients exhibited no staining. Representative cases are shown in Figs. 1 and 2. There was a significant difference between the two groups in both SRB and RB staining ($p<0.0001$; Mann-Whitney U test). Judging from the results with SRB and RB staining, superficial punctate lesions and epithelial lesions not covered with mucin are formed on the bulbar conjunctiva in Sjögren's syndrome patients, possibly due to a

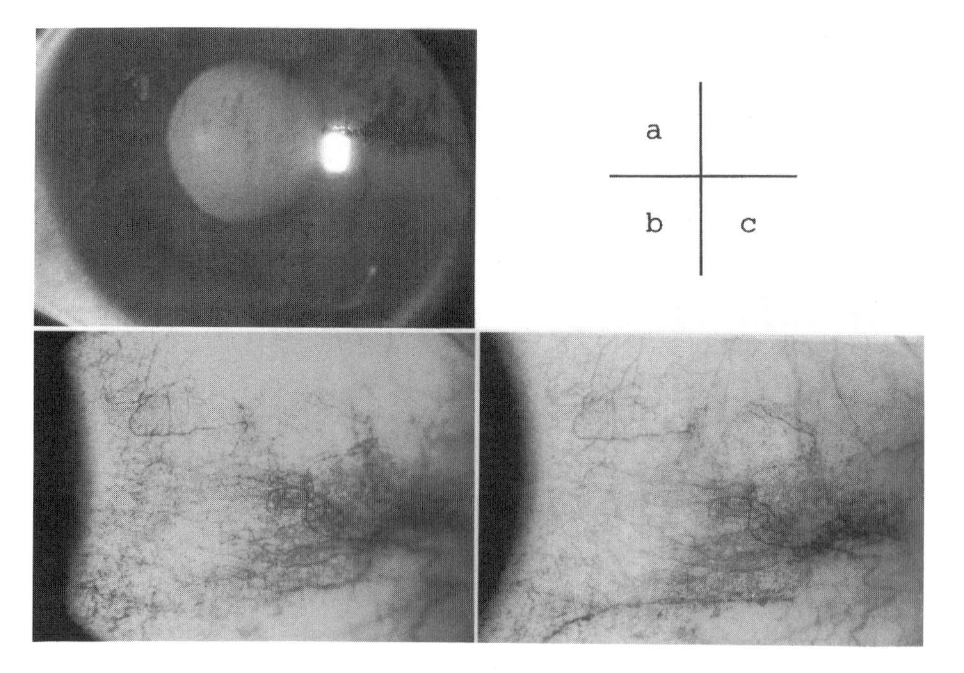

Figure 1. Representative Sjögren's syndrome case. Sulforhodamine B and rose bengal showed moderate staining of conjunctival epithelium. (a) Fluorescein staining of cornea, (b) sulforhodamine B staining of conjunctiva, (c) rose bengal staining of conjunctiva. Only temporal bulbar conjunctiva is shown.

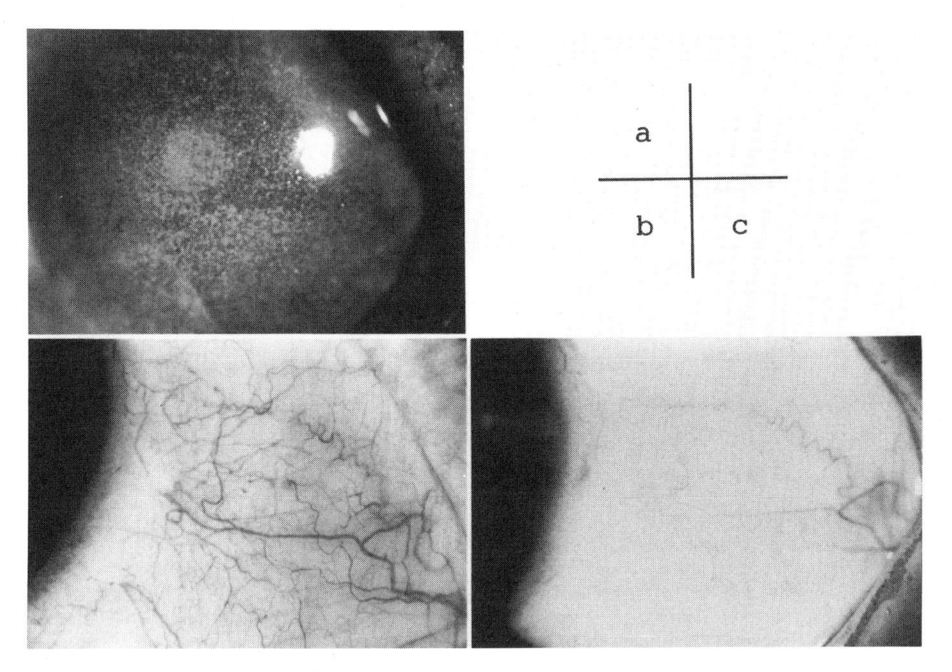

Figure 2. Representative case of drug-induced diffuse epithelial keratopathy. Sulforhodamine B and rose bengal showed no staining of conjunctival epithelium. (a) Fluorescein staining of cornea, (b) sulforhodamine B staining of conjunctiva, (c) rose bengal staining of conjunctiva. Only temporal bulbar conjunctiva is shown.

tear deficiency mechanism. Also, these findings were in marked contrast to those in drug-induced epithelial keratopathy. Considering the well-known fact that conjunctival epithelium is generally involved in dry eye,[7,8] this characteristic staining of Sjögren's syndrome patients can reasonably be applied to dry eye in general. Therefore, differences in bulbar conjunctival staining (Fig. 3) between dry eye and drug-induced epithelial keratopathy are

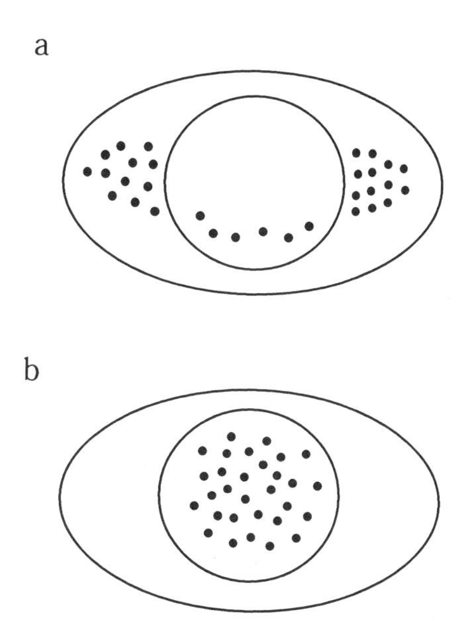

Figure 3. Schema indicating representative patterns of ocular surface epithelial damage in (a) dry eye, and (b) drug-induced diffuse epithelial keratopathy.

expected to be useful in elucidating the etiology of keratopathy. Moreover, the results in the present study encourage us to think that corneal epithelium is damaged predominantly by drug toxicity, whereas bulbar conjunctiva is damaged by dry eye mechanism. On this basis, we propose that leaky barrier function[9] and liability to the meniscus effect[10] in bulbar conjunctival epithelium cause this result. Further investigation is needed to confirm this.

ACKNOWLEDGMENTS

Supported in part by a research grant from Kyoto Foundation for the Promotion of Medical Science, and the intramural research fund of Kyoto Prefectural University of Medicine.

REFERENCES

1. Wilson FM II. Adverse external ocular effects of topical ophthalmic medications. *Surv Ophthalmol.* 1979;24:57–88.
2. Göbbels M, Spitznas M. Influence of artificial tears on corneal epithelium in dry-eye syndrome. *Graefes Arch Clin Exp Ophthalmol.* 1989;227:139–141.
3. Burstein NL. Corneal cytotoxicity of topically applied drugs, vehicles and preservatives. *Surv Ophthalmol.* 1980;25:15–30.
4. Fox RI, Robinson C, Curd J, Kozin F, Howell F. Sjögren syndrome: Proposed criteria for classification. *Arthritis Rheum.* 1986;29:577–585.
5. Eliason JA, Maurice DM. Staining of the conjunctiva and conjunctival tear film. *Br J Ophthalmol.* 1990;74:519–522.
6. Feenstra RPG, Tseng SCG. Comparison of fluorescein and rose bengal staining. *Ophthalmology.* 1992;99:605–617.
7. van Bijsterveld OP. Diagnostic tests in the sicca syndrome. *Arch Ophthalmol.* 1969;82:10–14.
8. Nelson JD, Havener VR, Cameron JD. Cellulose acetate impressions of the ocular surface, dry eye states. *Arch Ophthalmol.* 1983;101:1869–1872.
9. Huang AJW, Tseng SCG, Kenyon KR. Paracellular permeability of corneal and conjunctival epithelia. *Invest Ophthalmol Vis Sci.* 1989;30:684–689.
10. McDonald JE, Brubaker S. Meniscus-induced thinning of tear films. *Am J Ophthalmol.* 1971;72:139–146.

THE SIZE OF CORNEAL EPITHELIAL CELLS COLLECTED BY CONTACT LENS CYTOLOGY FROM DRY EYES

Graeme Wilson[1] and John Laurent[2]

[1]Vision Science Research Center
School of Optometry
The University of Alabama at Birmingham
Birmingham, Alabama
[2]Naval Aerospace Medical Research Laboratory
Pensacola, Florida

1. INTRODUCTION

The success of impression cytology as a technique for collecting conjunctival cells[1] suggested to us that a contact lens could be used to collect cells from the corneal surface.[2] This paper describes the size of corneal epithelial cells collected using contact lens cytology (CLC) from patients with dry eye.

2. MATERIALS AND METHODS

Six dry eye subjects were diagnosed as having moderate KCS by a combination of tests that included tear film break-up time, Schirmer test, and rose bengal staining. Subjective complaints from these subjects ranged from occasional discomfort to severe continuous discomfort necessitating the use of day and night eye lubricants. There were five females and one male, with ages ranging from 30 to 54 years.

Cells were collected from both eyes. The CLC technique of collecting and processing cells has been described in a previous paper.[2] Briefly, a disposable soft contact lens with a water content of 55% or 58% was inserted in the right eye, allowed to stabilize for 2 min, and then removed directly from the cornea without decentration onto the conjunctiva. Cells were irrigated from the back surface of the lens. This procedure was repeated in the same eye. The procedures were then applied to the left eye. The same experimenter inserted and removed all contact lenses.

Table 1. Mean cell length ± SD for both eyes of the six subjects

Subject	Right eye (μm)	Left eye (μm)
1	30.5±9.3 (45)	31.1±10.0 (27)
2	35.9±10.7 (96)	43.2±11.4 (21)
3	34.4±8.6 (47)	40.6±12.6 (21)
4	34.9±8.9 (68)	42.7±17.3 (38)
5	30.7±13.7 (53)	36.1±8.8 (88)
6	34.2±11.5 (89)	33.2±11.3 (138)

The sum of the number of cells obtained from the two collections on each eye is shown in parentheses.

Cells in suspension were stained with acridine orange (Sigma, St. Louis, MO), filtered, and transferred to a microscope slide. The slide was examined on a Leitz microscope equipped with epifluorescence. Cell size was measured using the reticle in the eyepiece of the microscope to determine the longest cell dimension. Only cells with an identifiable nucleus were measured. The statistical significance of cell size differences was tested using the Kolmogorov-Smirnov test.[3]

3. RESULTS

For individual subjects, the mean cell length had a range of 30.5±9.3 μm to 43.2±11.4 μm (Table 1). The grand mean of cell lengths for all 12 eyes was 35.6±4.4 μm. When cell lengths for all 12 eyes were placed in 5 μm bins and plotted as a frequency histogram (Fig. 1), there was a significant difference between the data from dry eyes and normal eyes ($p<0.01$; Kolmogorov-Smirnov). Neither data set was normally distributed ($p<0.01$; Kolmogorov-Smirnov).

4. DISCUSSION

Specular microscopy has been used to describe the size of cells on the epithelial surface. Quantitative studies using this technique have shown an increase in the number of small cells on the corneal surface of subjects with KCS compared to normal subjects,[4] and colonies of small cells on the corneas of subjects with moderate KCS have been reported.[5] There are two reports of larger-than-normal cells on the corneal surface of subjects with KCS.[6,7]

In our study, the size of cells obtained using CLC had a range and size similar to normal subjects. In the sample of 46 normal eyes,[2] the range of mean cell size was from 26.5 to 44.2 μm, with a grand mean of 35.3±5.1 μm. This is similar to the report here for dry eyes of a range from 30.5 to 43.2 μm, with a grand mean of 35.6±4.4 μm. This is a surprising result, because it was expected from specular microscopy studies[4,5] that cells here would be smaller. However, when all the data are considered together (Fig. 1), there is a statistically significant difference between the histograms of the two distributions when the nonparametric Kolmogorov-Smirnov test is applied. This arises in part due to the bimodal nature of the frequency distribution of cells from the normal cornea,[2] so that there is a large difference in the percent of cells in the 30 to 35 μm bin. In addition, the

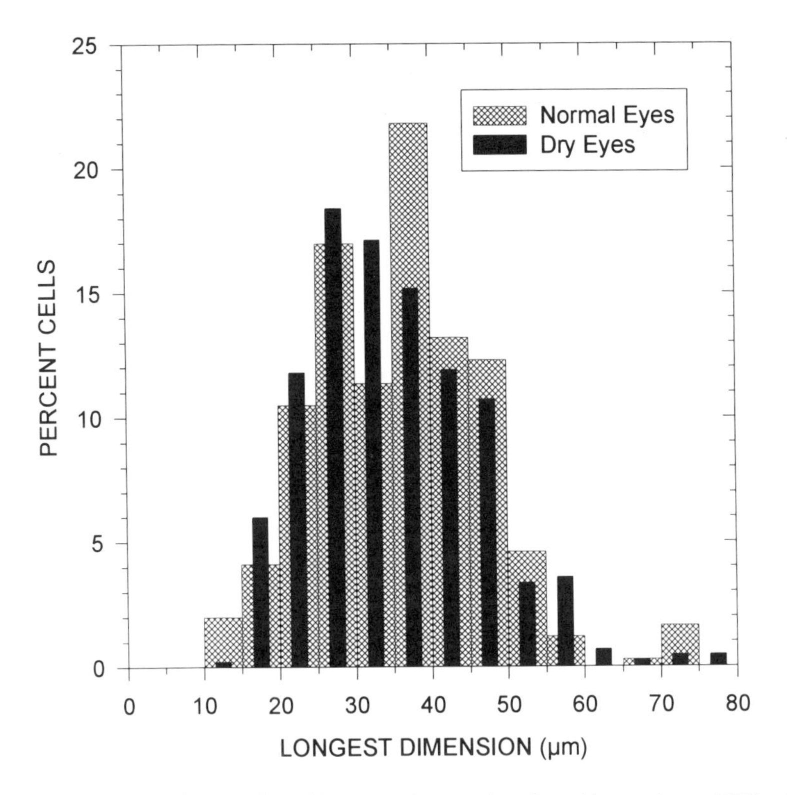

Figure 1. Longest dimension of cells collected by contact lens cytology from 46 normal eyes (3075 cells) and 12 dry eyes (731 cells). The longest dimensions of the cells were placed in 5 μm wide bins to create the frequency histogram.

three bins from 15 to 30 μm all contain a greater percentage of cells in the dry eye histogram, so that cumulatively there are 8.5% more small cells in the dry eye collections (cells less than 35 μm in longest dimension). Therefore, in spite of the similarity of the means, the data support the idea that cells are smaller in dry eye.

It has been hypothesized that the lubricating ability of tears is decreased in dry eye, and, as a consequence, the shear force that the lids apply to the ocular surface during a blink is increased.[8,9] Because the shear force removes cells from the corneal surface, more cells are removed in dry eye when the tears are compromised.[9] In an experimental test of the effect of shear in organ-cultured rabbit eyes,[10] shear did increase the rate of cell shedding, and also reduced the size of cells. It was proposed that the increase in cell shedding rate was due to an increase in the number of apoptotic cells, and not an increase in terminally differentiated cells. Observations of the rabbit cornea[11] suggest that cells shed from the epithelium fall into two categories: terminally differentiated cells and apoptotic cells. In the human eye, when tear composition is inadequate, the increased shear stress accompanying blinking may play a role in creating apoptotic cells.[10]

The bimodal distribution shown for the normal eye in Fig. 1 could be explained by this dual nature of cell shedding. Whether this is a reflection of a move towards apoptotic cells and away from terminally differentiated cells requires further study with specific markers such as TUNEL labeling.

ACKNOWLEDGMENTS

This research was supported by the National Institutes of Health (EY03039) and CIBA Vision. The authors thank Ms. Julia Munroe for technical help and Ms. Katharine Wilson for help with the data collection. Dr. Tom Woolley generously gave time to guide us in the analysis of the data.

REFERENCES

1. Nelson JD, Havener VR, Cameron JD. Cellulose acetate impressions of the ocular surface. *Arch Ophthalmol.* 1983;101:1869–1872.
2. Laurent J, Wilson G. The size of cells collected from normal subjects using contact lens cytology. *Optom Vis Sci.*, in press.
3. Tate M, Clelland R. *Nonparametric and Shortcut Statistics in the Social, Biological, and Medical Sciences.* Danville, IL: Interstate Printers and Publishers; 1959;93–95.
4. Lemp M, Gold J, Wong S, Mahmood M, Guimaraes R. An in vivo study of corneal surface morphologic features in patients with KCS. *Am J Ophthalmol.* 1984;98:426–428.
5. Tsubota K, Yamada M, Naoi S. Specular microscopic observation of human corneal epithelial abnormalities. *Ophthalmology.* 1991;98:184–191.
6. Serdarevic ON, Koester CJ. Colour wide field specular microscopic investigation of corneal surface disorders. *Trans Ophthalmol Soc UK.* 1985;104:439–445.
7. Marechal-Courtois C, Delcourt JC. Wide-field specular microscopy of corneal epithelium. *CLAO J.* 1986;12:165–172.
8. Holly FJ, Holly TF. Advances in ocular tribology. *Adv Exp Med Biol.* 1994;350:275–283.
9. Lemp MA, Mathers WD. Renewal of the corneal epithelium. *CLAO J.* 1991;17:258–266.
10. Ren H, Wilson G. The effect of a shear force on the cell shedding rate of the corneal epithelium. *Acta Ophthalmol*, in press.
11. Ren H, Wilson G. Apoptosis in the corneal epithelium. *Invest Ophthalmol Vis Sci.* 1996;37:1017–1025.

OPTIMUM DRY EYE CLASSIFICATION USING QUESTIONNAIRE RESPONSES

Charles McMonnies,[1] Arthur Ho,[1] and Denis Wakefield[2]

[1]School of Optometry
[2]School of Pathology
University of New South Wales
Kensington, Australia

1. INTRODUCTION

For the diagnosis of dry eye, global criteria are required that recognize a commonality among all forms of dry eye, even though they do not necessarily identify a particular etiology.[1] A dry eye questionnaire[2] can serve as a screening instrument with clinic populations and has been shown to be capable of delivering valid sensitivity and specificity information.[3] In dry eye research a questionnaire can be used to define treatment groups according to symptoms.[4] The feasibility of dry eye related epidemiological surveys of population-based samples may depend on the use of a self-reporting questionnaire.[5] Ideally, classification into dry eye and non-dry eye samples is preferred, but a third group with an equivocal diagnosis is usually present for which the terms questionable[6] and marginal[7] dry eye have been used. The marginal dry eye classification is appropriate for the common presentation of individuals whose tear function is adequate only in favorable conditions and is otherwise deficient in provocative circumstances of air conditioning, central heating, with use of dehydrating medications, following alcohol consumption, and/or with contact lens wear. This investigation is concerned with how dry eye questionnaire responses should be weighted to achieve optimum classification into dry eye, marginal dry eye, and non-dry eye groups.

2. METHOD

Questionnaire responses were analyzed for 50 subjects with a confirmed diagnosis of Sjögren's syndrome and 124 subjects presenting for the correction of refractive error. The criteria for the Sjögren's syndrome diagnosis included anti-nuclear antibody, rheumatoid factor, minor salivary gland biopsy, Schirmer (without anesthetic), and rose bengal. All samples were restricted to females over 45 years old. The original version of the dry eye questionnaire[2] was used to permit valid comparisons with a previous study of this type.[3] Discriminant analysis and logistic regression were used to examine the contributions of predictors and to calculate coefficients for individual questionnaire components to achieve optimized classification. Subsets of the samples were selected randomly so that sensitivity of methods of analysis to variation of responses to the questionnaire within samples could be examined. An additional analysis for which questionnaire responses were weighted according to a scale based on clinical experience[7] was also completed and compared with the other methods of classification.

3. RESULTS

Because variances of the two groups were not equal (Box M analysis M=195.4, $p<0.0001$) and because scores were not normally distributed, discriminant analysis was abandoned in favor of logistic regression. The initial logistic regression analysis involving 75% of each sample (randomly selected) used individual questionnaire components (eg, sensitivity to smog, smoke, air conditioning, etc.) as separate predictor variables and indicated that optimal classification was achieved with seven predictor variables (95% sensitivity and 99% specificity). However, applying the coefficients derived from the initial analysis on the 25% hold-out sample to estimate the 'real world' correctness of classification resulted in a sensitivity of 85% and specificity of 88%. In the second analysis, the individual questionnaire components were combined to calculate a 'frequency' factor, and the result for the initial 75% of the samples was 92% sensitivity and 98% specificity. However, applying the coefficients derived from that analysis on the hold-out sample indicated a 'real world' correctness of classification of 69% sensitivity and 95% specificity. When responses to individual questionnaire components were scored according to a weighted scale based on clinical experience,[7] a sensitivity of 92% and specificity of 93% were achieved.

4. DISCUSSION

Classification based on weighting of responses determined by logistic regression, while objective, can be swayed by sampling variations. The coefficients (or even predictor sets) derived from one sample do not match those obtained from another sample. This loss of precision is illustrated by the inferior classification results obtained when the coefficients derived from the initial 75% samples are applied to the 25% hold-out samples. The classification procedure involving individual questionnaire components being scored according to a weighted scale based on clinical experience is equivalent to a planned analysis, and so is not swayed by sampling variations, but it will be influenced by subjective variability in ascribing weights when the scale is devised. The better results obtained using weighting based on clinical experience (92% sensitivity and 93% specificity) represent a cross validation of a similar study[3] that yielded 98% sensitivity and 97% specificity.

In a clinical setting, clinical experience weighting has the advantage of individual classification score calculation by simple summation of established values. In contrast, logistic regression classification requires more complex calculations involving inverse and logarithmic functions.

It is apparent that questionnaire results can provide provisional disease classification into three groups: normals, dry eye, and a marginal dry eye group for whom an equivocal classification is appropriate. The marginal dry eye group may show greater variability of symptoms and signs if their tear function is adequate in favorable conditions of high humidity and normal blink function, etc. However, the marginal dry eye group is likely to demonstrate increased symptoms and signs in provocative circumstances such as during prolonged intense reading activities, especially in air conditioning, central heating, or natural conditions of low humidity. Other provocative factors may include the use of dehydrating medications, the consumption of alcohol, and the wearing of contact lenses. Fig. 1 shows how the equivocal classification group have scores ranging between 10 and 20. Although the referent score of 14.5 gives a satisfactory sensitivity and specificity for a screening instrument, in practical terms those subjects with scores between 10 and 20 might usefully be excluded from research studies involving responses from groups defined as not having and having a dry eye condition. In some research designs, treating the equivocal classification group of marginal dry eye subjects separately may improve the reliability and validity of outcomes. A dry eye questionnaire is not expected to obviate the need for subsequent examination for signs of dry eye and special tests for dry eye. Thus, the questionnaire is useful in a clinical setting for screening out patients for whom a more thorough assessment for dry eye signs is not warranted. Patients with an equivocal classification based on history warrant further examination. As a screening instrument, reduced specificity is an acceptable outcome if sensitivity is enhanced. In a study of 500 patients presenting for the correction of refractive error, scoring of questionnaire responses according to the scale based on clinical experience resulted in a mean score of 8 (median 7, mode 4). The range was 0 to 32 out of a possible maximum of 45.[7] The mean score would not have been as high if the sample had not included 177 contact lens wearers who reported significantly more dry eye symptoms than non-wearers ($p<0.0001$).[7] In another investigation using the same questionnaire and scoring system, 87% sensitivity and 87% specificity for dry eye diagnosis were achieved for a referent score of greater than or equal to 14.[4]

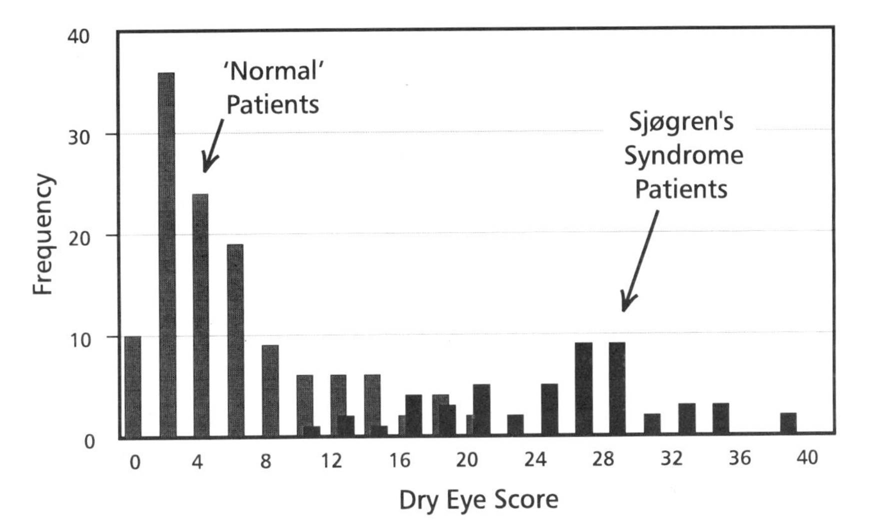

Figure 1. Distribution of scores determined by weighting of responses according to a scale based on clinical experience.

5. CONCLUSIONS

These results confirm the potential for a dry eye questionnaire to be used as a screening instrument with both clinic-based and population-based samples. The intention of improving the precision of this instrument by using logistic regression to optimize the accuracy of weighting of scores for individual questionnaire components has not been successful. Sampling variations reduce sensitivity and specificity results when the weights determined by logistic regression of sample subsets were applied to the balance of the samples studied. These findings indicate that a classification procedure involving individual questionnaire components being scored according to a weighted scale based on clinical experience yields 92% sensitivity and 93% specificity. The referent score on the scale used is 14.5, but overlap of the dry eye and non-dry eye groups between the scores of 10 and 20 define an equivocal classification group with marginal dry eye whose symptoms are provoked by everyday conditions of increased demand on tear function. It is concluded that classification into dry eye, marginal dry eye, and non-dry eye groups is the appropriate application of the questionnaire used.

ACKNOWLEDGMENT

This study was funded by a grant from Allergan Australia.

REFERENCES

1. Lemp MA. Report of the National Eye Institute/Industry Workshop on Clinical Trials in Dry Eyes. *CLAO J.* 1995;21:221–232.
2. McMonnies CW. Key questions in a dry eye history. *J Am Optom Assoc.* 1986;57:512–517.
3. McMonnies CW, Ho A. Patient history in screening for dry eye conditions. *J Am Optom Assoc.* 1987;58:296–301.
4. Golding TR, Brennan NA. Diagnostic accuracy and inter-correlation of clinical tests for dry eye. *Invest Ophthalmol Vis Sci.* 1993;34:823.
5. Bandeen-Roche K, Schein OD, Munoz B, et al. Challenges to defining and quantifying dry eye. ARVO Abstracts. *Invest Ophthalmol Vis Sci.* 1995;36:S862.
6. Seal DV, Mackie IA. The questionably dry eye as a clinical and biochemical entity. In: Holly FJ, Lamberts DW, MacKeen DL, eds. *The Preocular Tear Film in Health, Disease, and Contact Lens Wear.* Lubbock, TX: Dry Eye Institute; 1986:41–51.
7. McMonnies CW, Ho A. Marginal dry eye diagnosis: History versus biomicroscopy. In: Holly FJ, Lamberts DW, MacKeen DL, eds. *The Preocular Tear Film in Health, Disease, and Contact Lens Wear.* Lubbock, TX: Dry Eye Institute; 1986:32–40.

118

DIAGNOSTIC VALUE OF TEAR FILM ABNORMALITIES IN A NEW SYNDROME AFFECTING THE NEUROENDOCRINE AND IMMUNE SYSTEMS

Janos Feher

Institute of Ophthalmology
University of Rome "La Sapienza"
Rome, Italy

1. INTRODUCTION

Keratoconjunctivitis sicca, or dry eye, may be a cardinal sign of Sjögren's syndrome and some other systemic autoimmune diseases characterized by multiple glandular and extraglandular involvement.[1,2]

Dry eye may also be associated with inflammatory diseases of one or more mucous membranes, with sex, thyroid, and adrenal hormonal dysfunctions, and with affective/anxiety disorders, but without any signs of autoimmune disease. The pathogenetic correlation between mucous membrane inflammation and neurohormonal dysfunction suggested that these diseases may form a syndrome.[3] Most of these patients showed abnormal blood cortisol levels and/or absence of the postmeridian decrease of cortisol levels, in good correlation with the degree of mucosal membrane inflammation. Based on these alterations of the pituitary-adrenal axis, the related dry eye was called "stress dry eye."[4]

The diagnosis of dry eye may be difficult, as it has no single characteristic symptom or sign, and the usual diagnostic tests may show false negative results in mild or early cases.[5,6] Recently we elaborated a clinical diagnostic system that allowed us to establish the diagnosis of tear abnormalities much before the development of severe dry eye. The diagnosis of early and mild forms of dry eye called our attention to the associated neurohormonal and immune dysfunctions.[3]

2. MATERIALS AND METHODS

A clinical diagnostic system to evaluate the tear film state was applied routinely on 269 outpatients. The subjective symptoms, the objective signs, and the tear diagnostic tests (Schirmer, breakup time, ferning, rose bengal, and conjunctival brush cytology) were recorded (Table 1). Medical history and current condition of the patients were also noted. Particular attention was paid to gastrointestinal, airway, urogenital, hormonal, and affective-anxiety disorders. Serological tests (rheumatoid factor, antinuclear antibody, SS-A, SS-B antibodies) were also performed to separate dry eye and autoimmune dry eyes.[7]

3. RESULTS

Of the 269 patients, 102 showed tear film abnormalities. They were divided into three main groups (Fig. 1).

Simplex dry eye (with no circulating autoantibodies) was found in 72 cases (65 female, 7 male), aged 19–68 years (mean age 39.6 years). This group was divided into two subgroups: The minor subgroup comprised 6 patients (5 female, 1 male), who were affected by "questionable dry eye," without any other disease. The major subgroup was formed by 66 patients,

Table 1. Clinical diagnosis of keratoconjunctivitis sicca

Subjective Symptoms/History

1) reduced tearing to emotion
2) sensation of dryness
3) foreign body sensation/grittiness
4) soreness/mild pain
5) scratchiness
6) difficulties of opening the eye upon waking
7) burning
8) episodes of excessive tearing
9) sensation of "tired eye"
10) particular sensibility of the ocular surface to topical medication, wind, air conditioning, central heating, smoking, swimming pool, alcohol abuse, and/or light (photophobia)

Each of these symptoms corresponds to 1 score. Diagnosis of "questionable KCS" is established if the total score is higher than 5, but one of the first five symptoms must be positive.

Objective Signs

1) particulate matter on the corneal surface
2) mucus thread at the inner canthus or in the lower fornix
3) deposits at the orifices of the meibomian glands (Mathers' sign)
4) filamentary keratitis
5) thin tear meniscus
6) hyperemia of the bulbar conjunctiva
7) papillary hypertrophy of the tarsal conjunctiva
8) decreased brightness of the conjunctiva
9) frequent blinking
10) tear film irregularities with fluorescein staining

Each of these signs corresponds to 2 scores. Diagnosis of "mild KCS" is established if the total score is higher than 10, but one of the first five signs must be positive.

Diagnostic Tests

1) Schirmer test
2) BUT
3) rose bengal test
4) ferning test
5) conjunctival cytology

Each positive test corresponds to 4 scores. Diagnosis of "severe KCS" is established if the total score is higher than 20, but one of the diagnostic tests must be positive.

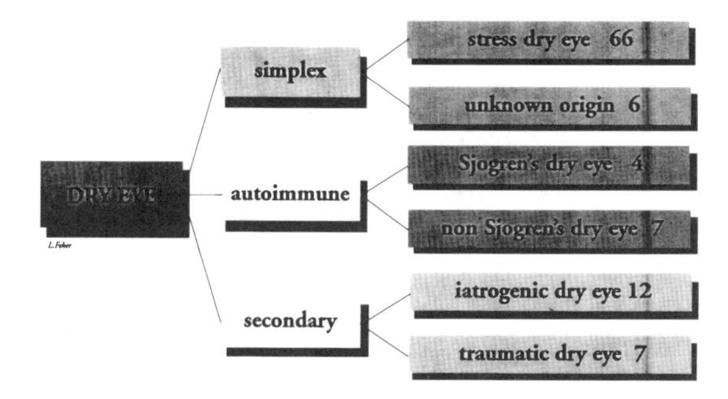

Figure 1. Classification of dry eye syndromes and distribution of our cases.

among them 33 cases of questionable dry eye, 18 cases of mild dry eye, and 15 cases of severe dry eye. All of these patients suffered from chronic or relapsing inflammation of at least one mucous membrane and neurohormonal dysfunction; the breakdown was as follows:

Inflammation of mucous membranes:		*Mental disorders:*	
gastrointestinal	48	unipolar affective	12
airway	32	bipolar affective	4
urogenital	45	anxiety	22
mixed anxiety-depressive	23		
Hormonal dysfunction:			
sex hormonal	52		
thyroid	31		
low blood pressure	62		

Autoimmune dry eye was found in 11 cases (10 female, 1 male), including 3 mild dry eye and 8 severe dry eye. The patients' age distribution was 46–62 years (mean age 50.3 years). The two subgroups, based on the serological tests, were Sjögren's dry eye (SS-A and/or SS-B positive), 4 cases (all female), and non-Sjögren's dry eye (negative to SS antigens), 7 cases (6 female, 1 male).

Secondary dry eye was found in 19 cases (16 male, 3 female), including 9 questionable dry eye, 6 mild dry eye, and 4 severe dry eye. The patients' age distribution was 25–48 years (mean age 36.4 years). Topical or systemic medications in 12 cases and chemical burns or exposure to toxic substances in 7 cases seemed to be responsible for the development of these secondary dry eye syndromes.

4. CONCLUSIONS

A new clinical diagnostic system, based on subjective symptoms, objective signs, and the common diagnostic tests, allowed us to establish the diagnosis of dry eye even if the manifestations of the tear abnormalities were poor. Of the 269 outpatients, 102 showed tear film abnormalities, including 48 cases questionable dry eye, 27 mild dry eye, and 27 severe dry eye. These data show that this clinical diagnostic system is approximately three times more sensitive than the usually applied methods, which can verify only severe cases.

Our findings confirmed the previous observations of the existence of subgroups of simplex dry eye associated with inflammation of the mucous membranes and neuro-endo-

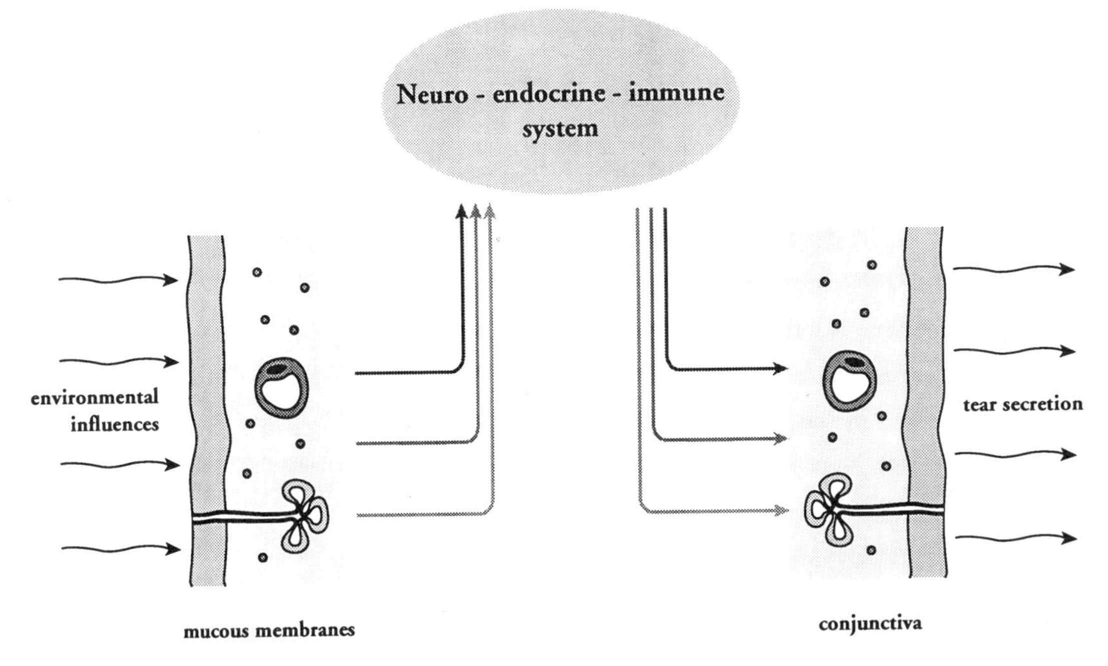

Figure 2. Putative pathomechanism of "stress dry eye."

crine dysfunctions. Tear film abnormalities, even if very slight, may have diagnostic value for these disorders of the neuro-endocrine-immune system.

In addition, these findings suggest that the exogeneous activation of the mucosal immune system affects the neuro-endocrine system, and that the tear film abnormalities may come from the abnormal regulatory influences of neuro-endocrine and immune systems on all mucous membranes (Fig. 2).

1. Mucous membranes transform environmental stimuli into neuronal (neurocrine), hormonal (endocrine), and cellular (cytocrine) signals, which influence the central nervous system, the hypothalamic-pituitary centers, and lymphoid organs, respectively (centripetal arc).
2. The neuro-endocrine-immune system regulates the tear secretion and ocular surface functions by neurocrine, endocrine, and cytocrine signals (centrifugal arc.).
3. This continuous cross-talk between mucous membranes and the neuro-endocrine-immune system, in normal conditions, serves to maintain adequate surface conditions in response to the current environmental influences.
4. In disease states, dysfunction of one or more mucous membranes represents more or less stress to the neuro-endocrine-immune system, which influences the functions of mucous membranes including that of the conjunctiva, resulting in tear film abnormalities, called "stress dry eye."

REFERENCES

1. Sjögren H, Bloch KJ. Keratoconjunctivitis sicca and the Sjögren syndrome. *Surv Ophthalmol.* 1971;16:145–160.

2. Fox RI, Kang HI. Pathogenesis of Sjögren's syndrome. *Rheum Dis Clin North Am.* 1992;18:517–538.
3. Feher J. Tear film abnormalities and mucous membrane diseases associated with neuro-hormonal dysfunctions. In: Hurwitz J, Miglior M, Spinelli D, van Bijsterveld OP. (Eds.) *The Lacrimal System.* 1994, Milano, Kugler & Ghedini Publications; 1995;65–72.
4. Feher J, Papale A, Tomei M, Balacco Gabrieli. C.: Cheratoconjiuntivite secca e malattie delle mucose associate a disfunzioni neurormonali: Una nuova sindroma. Clin Oculist Patol Ocul. 1997;18:17–22.
5. Seal DV, Mackie IA. The questionably dry eye as a clinical and biochemical entity. In Holly FJ, ed. *The Preocular Tear Film in Health, Disease, and Contact Lens Wear.* Lubbock, TX: Dry Eye Institute; 1986:41–51.
6. Goren MB, Goren SB. Diagnostic tests in patients with symptoms of keratoconjunctivitis sicca. *Am J Ophthalmol.* 1988;106:570–574.
7. Tsubota K, Toda I, Yagi Y, Ogawa Y, Ono M, Yoshino K. Three different types of dry eye syndrome. *Cornea.* 1994;13:202–209.

119

LOW-TECH DETECTION OF TEAR FILM-RELATED DISEASES OF THE OCULAR SURFACE

Maurizio Rolando, Salvatore Alongi, Angelo Macrì, Marco Schenone, and Giovanni Calabria

Department of Ophthalmology
University of Genoa, Italy

1. INTRODUCTION

The detection of tear film-related pathology of the ocular surface during routine clinical eye examination, especially in early cases, is not always straightforward. Therefore a number of specific tests such as Schirmer test, fluorescein and rose bengal staining, etc. are usually added to the examination.[1–4] The addition of these specific tests is time consuming, and they are often neglected by the general ophthalmologist. In consequence, the diagnosis of tear film disorders by the general ophthalmologist is not as frequent as it could be.

A number of symptoms and characteristic aspects of the ocular surface in tear film-related diseases could be used in a clinical setting as indicators of a tear film problem or as suggestive of the need for a correct diagnostic procedure. A "conjunctival hyperemia localized inside the interpalpebral fissure," the presence of a "lip-like fold of the conjunctiva at 6'clock,"[5] the presence of an "irregularity of the black line,"[6] and a "rapid deteriorating image at Javal keratometry"[7] are easily collected clinical signs, available to every ophthalmologist in a routine practice.

2. PATIENT SELECTION

A population of patients with ocular surface disease was evaluated. Included were 26 subjects with a diagnosis of keratoconjunctivitis sicca (KCS) assessed on the basis of at least two positive results among the following:

- Schirmer I Test (< 7 mm/5 min)
- typical rose bengal staining of the ocular surface (score > 2, by van Bijsterveld's method)
- BUT (< 10 sec)

Lacrimal Gland, Tear Film, and Dry Eye Syndromes 2
edited by Sullivan *et al.*, Plenum Press, New York, 1998

Also included were 25 subjects with a diagnosis of non-tear film-related ocular surface disease (n-KCS). These patients had typical signs and symptoms of ocular surface disease but negative results to the dry eye tests: Schirmer I test (>10 mm/5 min), rose bengal staining of the ocular surface, and BUT.

3. METHODS

All patients refrained from using any topical therapy for at least 12 h before the visit. They were evaluated by the following means:

- use of a symptoms *questionnaire* (see Table 1)[8]
- search for the presence of a *lip-like fold* at 6 o'clock in the inferior bulbar conjunctiva (Fig. 1)
- study of the distribution of bulbar conjunctiva hyperemia, which was considered positive for dry eye diagnosis when it was *strictly localized* into the interpalpebral area (Fig. 2)
- search for any *irregularity of the black line* when observed after fluorescein staining with a fluorescein dry strip in the inferior fornix at dim Wood light (Fig. 3)
- evaluation of the quality of central pre-corneal tear film surface reflection during *Javal keratometry* (ni-BUT) (which was considered positive when irregular targets appeared within 10 sec)

Schirmer I test, rose bengal staining of the ocular surface, and fluorescein BUT were also performed in every subject.

The sensitivity and specificity of the questionnaire, the conjunctival lip-like fold, the irregularity of the black line, and the observation of tear surface reflection at Javal keratometry, alone and in association, in detecting dry eyes were then evaluated in comparison with the classical methods used for diagnosis of KCS.

4. RESULTS

As a single test, the presence of *irregularity of the black line* had the best compromise between sensitivity and specificity (92% sensitivity and 76% specificity) in diagnosing dry eye conditions.

The associations *Questionnaire + Irregularity of the black line* and *Hyperemia + Irregularity of the black line* showed the best compromise between sensitivity and specificity in diagnosing dry eye conditions.

4.1. Single Test

	KCS	nKCS
Questionnaire: sensitivity 76.93%, specificity 81.20%		
+	20	2
–	6	23
Lip-like fold: sensitivity 65.38%, specificity 72.00%		
+	17	7
–	9	18

Table 1. Questionnaire for collecting dry eye symptoms and history

1. Main symptoms:
    - burning
    - foreign body sensation
    - need to keep eye closed
2. Accessory symptoms:
    - heavy eye
    - photophobia
    - wet eye feeling (without overflow)
    - discomfort on lid movement and/or closure
    - warm eye feeling
    - feeling his/her own eyes

   Time relation:
    - symptoms appear:
        - in the morning
        - during the day
        - in the evening
    - symptoms are:
        - stable
        - increase with time
3. Environmental stress condition symptoms:

   Eyes are bothered by:
    - cigarette smoke
    - smoke or vapors in general
    - wind
    - air conditioning
    - contact lens wear
4. Need for topical therapy:
    - decongestant
    - tear substitutes
5. Use of systemic therapy and presence of systemic diseases:
    - β-blocking agents
    - diuretics
    - antihistaminic agents
    - antispasmodic agents
    - hormones
    - oral contraceptives
    - sleeping pills, antianxiety agents, etc.
    - mucous tissue dryness
    - rheumatic disease
    - thyroid disease
    - hormone disorder
    - connective tissue disease

Localized hyperemia: sensitivity 69.23%, specificity 81.20%

+	18	2
–	8	23

Irregularity of the black line: sensitivity 92.30%, specificity 76.00%

+	24	6
–	2	19

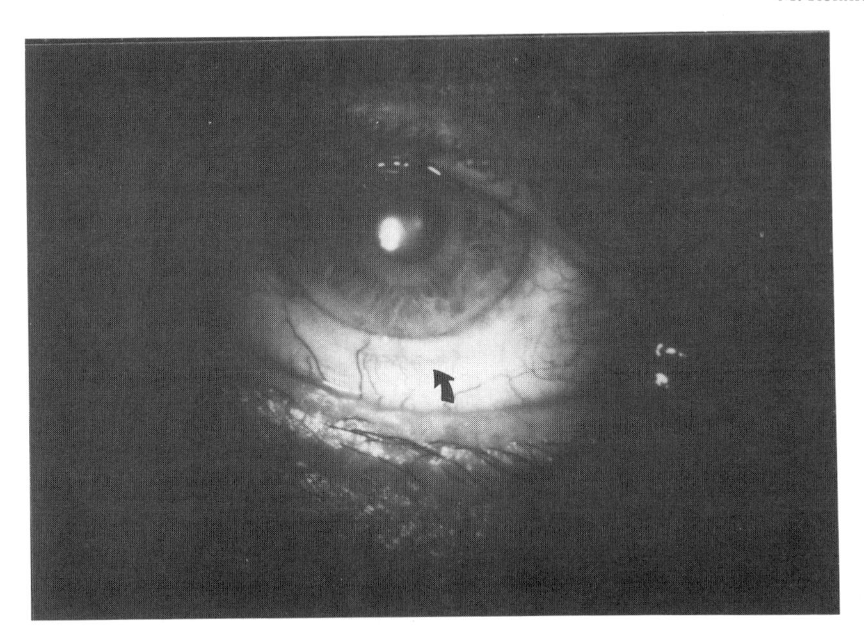

Figure 1. Lip-like fold at 6 o'clock (arrow).

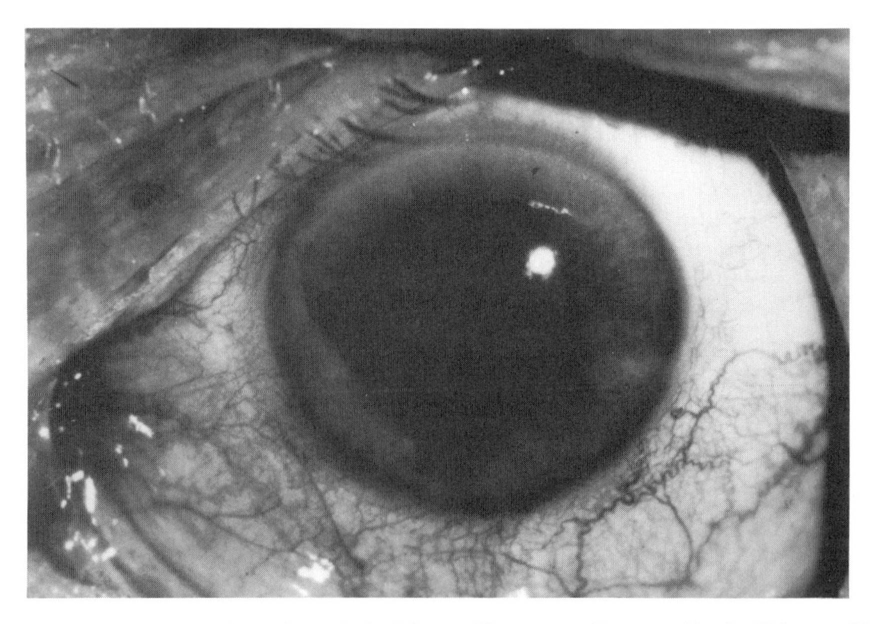

Figure 2. Localized hyperemia in the interpalpebral fissure. The area usually covered by the lid is spared by the hyperemic reaction.

Figure 3. Black-line irregularity after fluorescein staining (arrowhead).

Javal keratometry (ni-BUT): sensitivity 76.93%, specificity 88.00%		
+	18	3
–	8	22

4.2. Two Tests Associated

	KCS	nKCS
Questionnaire + Lip-like fold: sensitivity 76.93%, specificity 76.00%		
++	24	6
–/+-	2	19
Questionnaire + Hyperemia: sensitivity 80.76%, specificity 81.20%		
++	21	2
–/+-	5	23
Questionnaire + Irregularity of black line: sensitivity 96.15%, specificity 84.00%		
++	25	4
–/+-	1	21
Questionnaire + Javal ni-BUT: sensitivity 76.93%, specificity 76.00%		
++	23	4
–/+-	3	21

Lip-like fold + Hyperemia: sensitivity 84.61%, specificity 76.00%		
++	22	6
–/+-	4	19
Lip-like fold + Black line: sensitivity 73.87%, specificity 76.00%		
++	19	6
–/+-	7	19
Lip-like fold + Javal ni-BUT: sensitivity 84.61%, specificity 72.00%		
++	22	7
–/+-	4	18
Lip-like fold + Black line: sensitivity 73.87%, specificity 76.00%		
++	19	6
–/+-	7	19
Hyperemia + Black line: sensitivity 92.30%, specificity 84.00%		
++	24	4
–/+-	2	21
Hyperemia + Javal ni-BUT: sensitivity 88.00%, specificity 84.69%		
++	24	4
–/+-	2	21
Black line + Javal ni-BUT: sensitivity 92.30%, specificity 76.00%		
++	24	6
–/+-	2	19

5. DISCUSSION

Dry eye diagnosis is the result of a composite approach to patient history and subjective and objective symptoms. The diagnosis is often reached after a quite complex, time-consuming sequence of tests. Quick and easy tests are, of course, much preferred by both ophthalmologists and patients, and also are much more likely to be used.

Tests easy to perform during a routine eye examination, such as the *dry eye questionnaire*,[8,9] search for a *lip-like fold at 6 o'clock*, search for *hyperemia localized into the interpalpebral area*, search for *black-line irregularity and irregular targets at Javal keratometry* (ni-BUT), have good sensitivity and specificity in diagnosing dry eye conditions. They are easily learned and mastered by any ophthalmologist; their use will add no substantial extra time to the examination schedule and can provide further information about tear film and ocular surface conditions. Unfortunately no single test can provide a sure diagnosis of dry eye, but it can suggest the need for further investigation, and when necessary, in suspect cases, the diagnosis of KCS can be assessed by performing the classical dry eye tests.

REFERENCES

1. Farris RL, Gilbard JP, Stuchell RN, et al. Diagnostic tests in keratoconjunctivitis sicca. *CLAO J.* 1983;9:23–28.
2. Goren MB, Goren SB. Diagnostic tests in patients with symptoms of keratoconjunctivitis sicca in clinical practice. *Am J Ophthalmol.* 1988;106:570–574.
3. Nelson JD. Diagnosis of keratoconjunctivitis sicca. *Int Ophthalmol Clin.* 1994:34(1):37–56.
4. van Bijsterveld OP. Diagnostic tests in the sicca syndrome. *Arch Ophthalmol.* 1969;82:10–14.
5. Hoh H, Schirra F, Kienecker C, Ruprecht KW. Lid-parallel conjunctival folds (LIPCOF): A definite diagnostic sign of dry eyes. *Ophthalmologe.* 1995;92:802–808.
6. Holly FJ, Lemp MA. Tear physiology and dry eyes. *Surv Ophthalmol.* 1977;22(2):69–87.
7. Mohrman R. The keratometer. In: *System of Ophthalmology*. Duane, ed. 1066. vol 1, chapter 60.
8. Rolando M, Macrì A, Alongi S, Carlandrea T, Calabria G. Use of a questionnaire for the diagnosis of tear film-related ocular surface disease. This volume.
9. McMonnies CW, Ho A. Marginal dry eye diagnosis: History versus biomicroscopy. In: Holly FJ, ed. *The Preocular Tear Film in Health, Disease and Contact Lens Wear.* Lubbock, TX: Dry Eye Institute; 1986:32–40.

USING LIPCOF (LID-PARALLEL CONJUNCTIVAL FOLD) FOR ASSESSING THE DEGREE OF DRY EYE, IT IS ESSENTIAL TO OBSERVE THE EXACT POSITION OF THAT SPECIFIC FOLD

Frank Schirra,[1] Helmut Höh,[2] Christina Kienecker,[1] and Klaus W. Ruprecht[1]

[1]University of Saarland
Hospital and Department of Ophthalmology
Homburg, Germany
[2]Neubrandenburg General Hospital
Department of Ophthalmology
Neubrandenburg, Germany

1. INTRODUCTION

In a previous extensive investigation,[1,2] LIPCOF (stages 0–3; Figs. 1–4) proved a very useful tool for quick and reliable diagnosis of many types of dry eye and for estimating its severity (Figs. 5 and 6). However, it is essential to pay careful attention to the correct spot for examination of the fold (Fig. 7), thus providing the basis for accurate statements.

2. METHODS

The appearance and grade of LIPCOF are correlated with the degree of ocular dryness, thus illustrating the use of LIPCOF for diagnosing this disease. The four-grade LIPCOF scale is derived from a 13-stage scale that is based on Schirmer I test, lysozyme test, tearfilm break-up time, fluorescein and rose bengal staining, and impression cytology.

Lacrimal Gland, Tear Film, and Dry Eye Syndromes 2
edited by Sullivan *et al.*, Plenum Press, New York, 1998

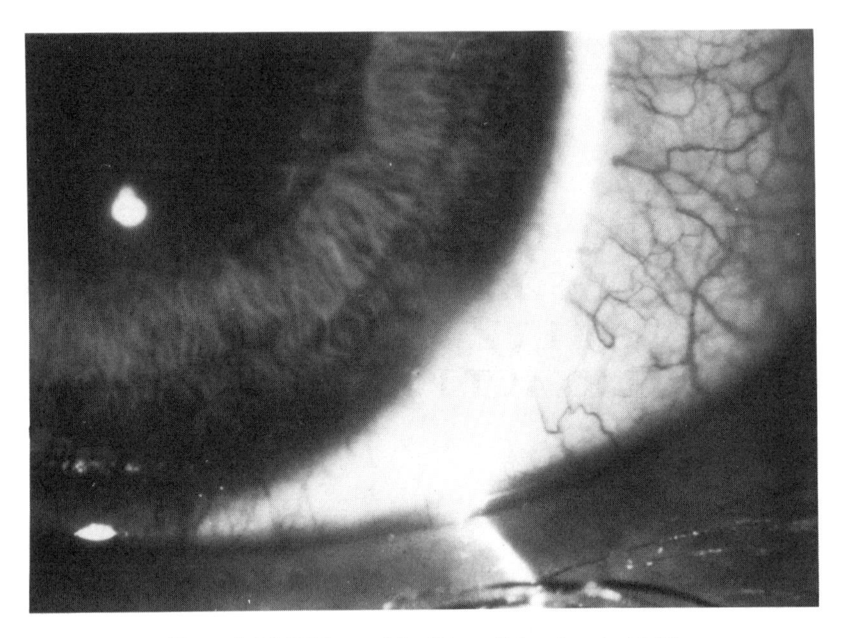

Figure 1. LIPCOF stage 0 (no lid-parallel conjunctival fold).

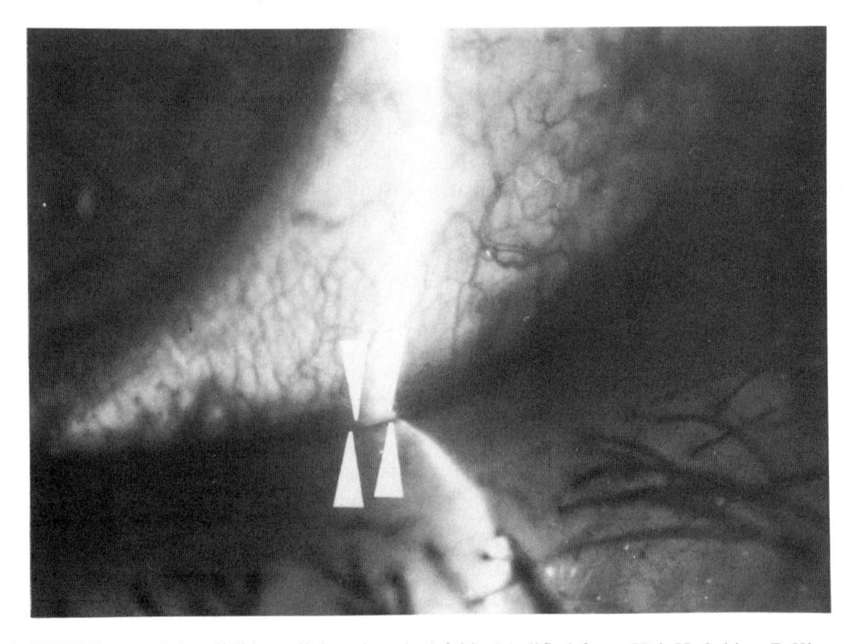

Figure 2. LIPCOF stage 1 (small lid-parallel conjunctival fold). Modified from: Höh H, Schirra F, Kienecker C, Ruprecht KW. Lid-parallel conjunctival fold (LIPCOF) and dry eye: A diagnostic tool for the contactologist. *Contactologia*. 1995;17E:104–117.

Figure 3. LIPCOF stage 2 (medium-sized lid-parallel conjunctival fold). Modified from: Höh H, Schirra F, Kienecker C, Ruprecht KW. Lid-parallel conjunctival fold (LIPCOF) and dry eye: A diagnostic tool for the contactologist. *Contactologia.* 1995;17E:104–117.

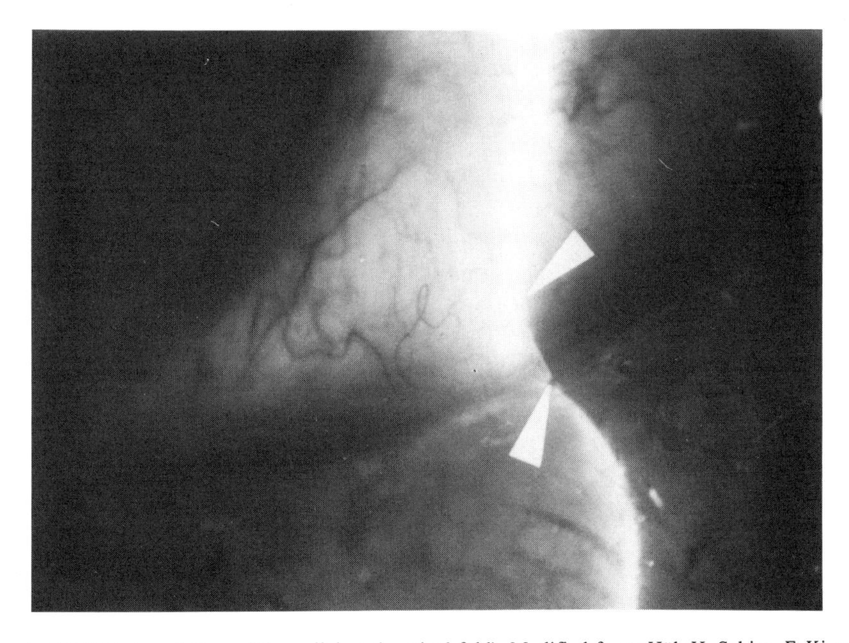

Figure 4. LIPCOF stage 3 (large lid-parallel conjunctival fold). Modified from: Höh H, Schirra F, Kienecker C, Ruprecht KW. Lid-parallel conjunctival fold (LIPCOF) and dry eye: A diagnostic tool for the contactologist. *Contactologia.* 1995;17E:104–117.

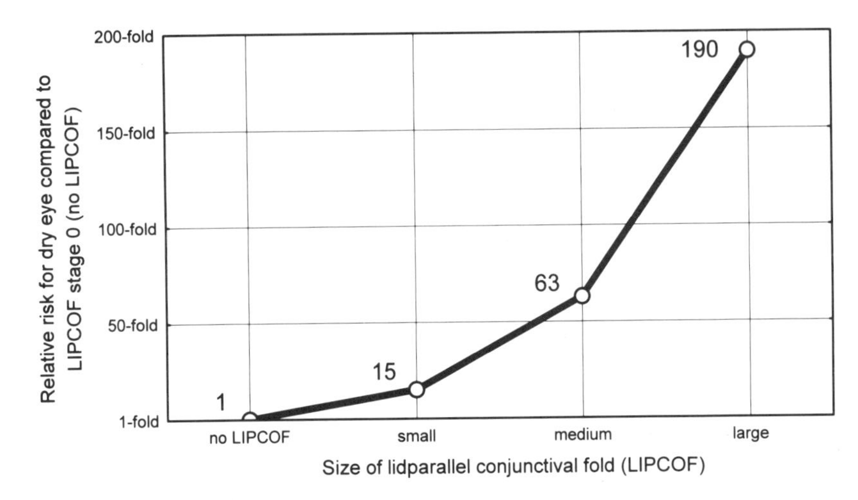

Figure 5. Probability of dry eye depending on LIPCOF size.

3. RESULTS

Depending on its size, LIPCOF changes its morphologic appearance. Small folds are single folds in 91.2%; 73.3% of medium-sized LIPCOFs split into several parallel folds. Among large LIPCOFs, a single fold represents an exception (4.6%).

With regard to dry eye, LIPCOF in the lower temporal quadrant shows a positive predictive value of 93.09%—that is, among all patients with LIPCOF, 93.09% actually do have a dry eye condition. It shows a negative predictive value of 75.95%, which means that among all patients without LIPCOF, 75.95% definitely are free of dry eye. For the other quadrants, both the positive and the negative predictive values are considerably lower (Table 1).

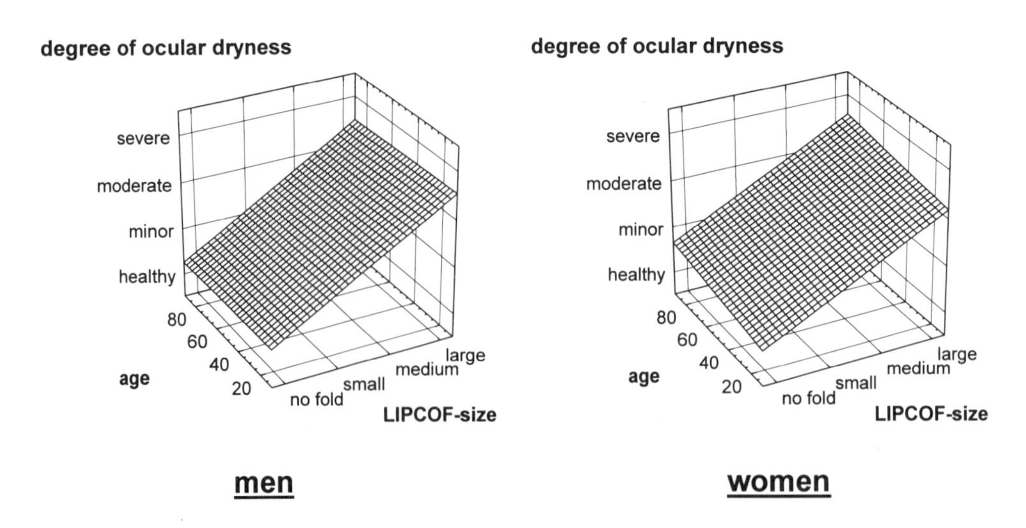

Figure 6. Linear regression of the stage of ocular dryness on age and degree of LIPCOF in males and females. Modified from: Höh H, Schirra F, Kienecker C, Ruprecht KW. Lid-parallel conjunctival fold (LIPCOF) and dry eye: A diagnostic tool for the contactologist. *Contactologia.* 1995;17E:104–117.

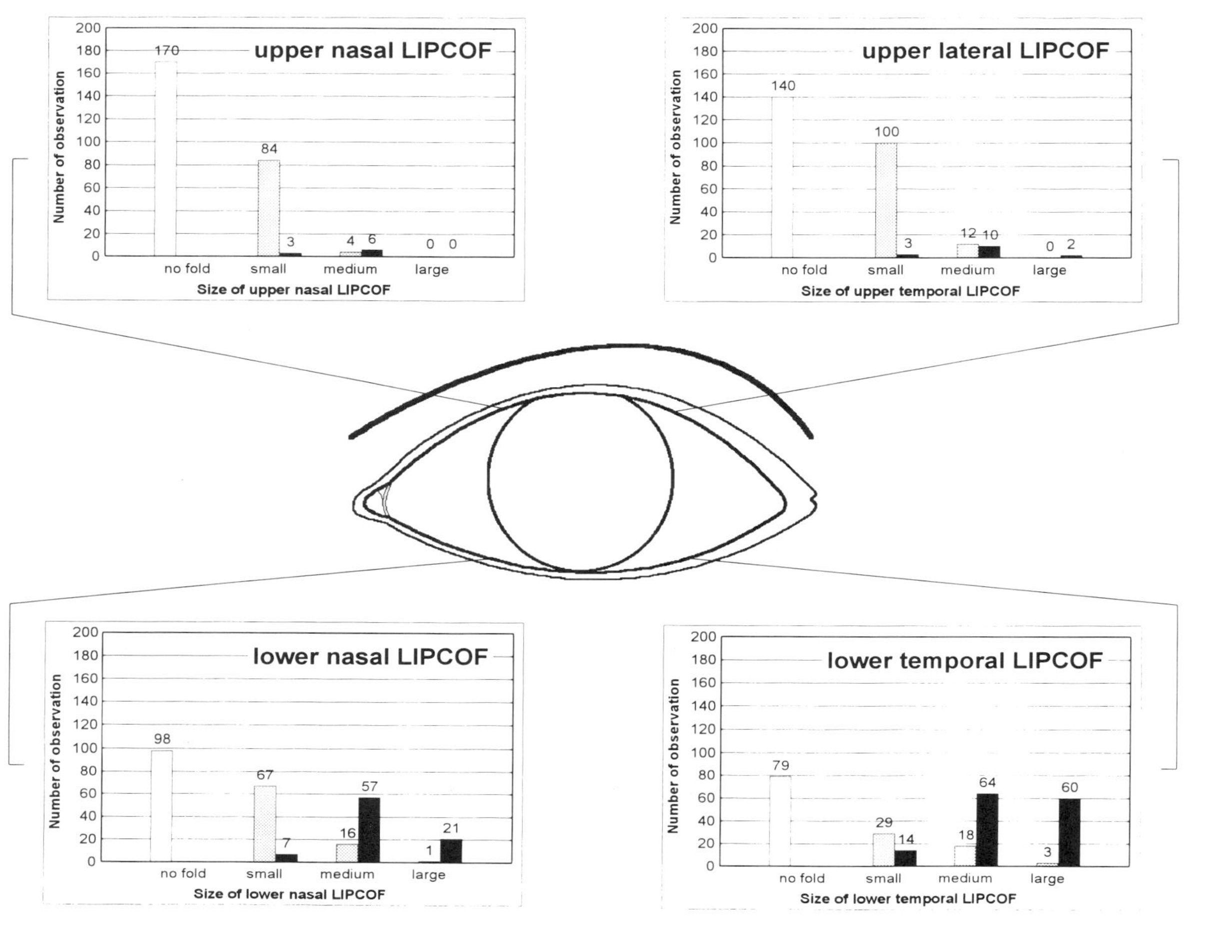

Figure 7. Expression of LIPCOF in the four quadrants of the external eye.

Table 1. Positive and negative predictive values of LIPCOF in the four quadrants of the eye with respect to diagnosing dry eye

	Dry eye		
LIPCOF quadrant	Yes	No	Predictive value
Upper medial			
Yes	92	5	Positive **94.84%** (A/A+B)
No	102	68	Negative **40.00%** (D/C+D)
Upper temporal			
Yes	121	6	Positive **95.28%** (A/A+B)
No	73	67	Negative **47.86%** (D/C+D)
Lower medial			
Yes	162	7	Positive **95.86%** (A/A+B)
No	32	66	Negative **67.35%** (D/C+D)
Lower temporal			
Yes	175	13	Positive **93.09%** (A/A+B)
No	19	60	Negative **75.95%** (D/C+D)

Not considering the particular size of the lower temporal LIPCOF, the model of logistic regression reveals a 44-fold higher risk for dry eye in patients with LIPCOF than for those without it. Considering the particular degree of LIPCOF, the model of logistic regression shows that the risk of having a dry eye condition for patients with LIPCOF stage 1 is 15-fold higher, for patients with LIPCOF stage 2 63-fold higher, and for patients with LIPCOF stage 3 190-fold higher than for patients without lid-parallel conjunctival folds (LIPCOF stage 0) (Fig. 5). In both logistic regression model designs mentioned above, patients' age and sex were non-significant variables, which means that considering these variables does not provide better evaluation of dry eye than not considering them. Investigating how often a certain grade of LIPCOF coincides with the same stage of dry eye, we found that in 52.81%, the lid-parallel conjunctival fold exactly estimates the correct degree of ocular dryness on a four-stage scale. In 35.96%, ocular dryness is over- or underestimated by one grade. Misclassification by two or three grades occurred in 11.23%.

4. CONCLUSION

To avoid serious misclassification of dry eye, it is essential to determine LIPCOF size and quality at the location originally described. Correct estimation of LIPCOF easily leads to expeditious and reliable information about a patient's stage of dry eye.

REFERENCES

1. Höh H, Schirra F, Kienecker C, Ruprecht KW. Lid-parallel conjunctival fold (LIPCOF) and dry eye: A diagnostic tool for the contactologist. *Contactologia.* 1995;17E:104–117.
2. Höh H, Schirra F, Kienecker C, Ruprecht KW. Lidparallele konjunktivale Falten (LIPCOF) sind ein sicheres diagnostisches Zeichen des trockenen Auges. *Ophthalmologe.* 1995;92:802–808.

121

USE OF THE TEARSCOPE PLUS AND ATTACHMENTS IN THE ROUTINE EXAMINATION OF THE MARGINAL DRY EYE CONTACT LENS PATIENT

Jean-Pierre Guillon

Scope Research Limited, and
Dry Eye Symptoms Clinic
Institute of Optometry
London, England, United Kingdom

1. INTRODUCTION

The Tearscope was invented primarily to allow 360° observation of the tear film by specular reflection. The preocular and prelens tear film is viewed with a cold-cathode light source with minimum effect on their structure. Their stability is measured by the noninvasive break-up time (NIBUT).

The redesigned model, Tearscope Plus, affords greater ease of use and is enhanced with attachments that permit a complete examination of the contact lens wearer presenting marginal dry eye symptoms. The attachments include:

- a grid-type insert for XEROSCOPE measurements
- a ring-type insert for PLACIDO disc observation
- a blue and yellow filter set for fluorescein observations
- a slit lamp attachment for hands-free viewing
- a magnifying lens for hand-held operation
- a set-up for dark field examination of contact lenses

The usefulness of the attachments is demonstrated during the routine examination of the marginal dry eye contact lens wearer, both before fitting and at aftercare visits.

Lacrimal Gland, Tear Film, and Dry Eye Syndromes 2
edited by Sullivan *et al.*, Plenum Press, New York, 1998

2. TEARSCOPE PLUS DESIGN

The Tearscope is based on an illuminated, translucent, tubular design which also acts as support for inserts and add-ons.

- The tapered outer casing permits a better fit close to the eye and full corneal specular coverage, avoiding obstruction from the nose (Fig. 1).
- A double concentric illuminating cold-cathode light source situated away from the tear surface limits any evaporation effect.
- The even illumination is achieved by inner reflection from the outer tapered protective casing.
- The light transmission is maximized by the translucence of the inner protective sleeve.
- Any heat generated at the base of the cathode tube is contained and dispersed within the handle away from the subject's eye.
- A slit lamp adapter is available. It secures the Tearscope Plus on the tonometer mount, ready for immediate use and hands-free examination. It can be moved aside for normal slit lamp use.
- A timer in 0.1 sec is provided for break-up time recording, including a lap time function.
- The green numeric display allows easy reading in the dark and in fully illuminated rooms.
- The power unit is available in various voltages for use in the USA, Japan, Europe, etc.
- A magnifying lens permits operation as a portable examination instrument when a biomicroscope is not available.
- A series of removable inserts are available: filters for fluorescein observation, grid patterns for NIBUT measurement, concentric rings for PLACIDO disc mires, a set-up for dark field observation.

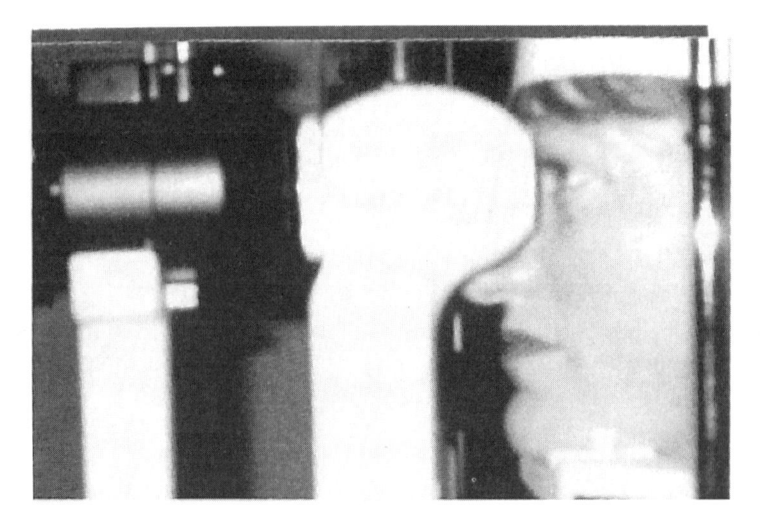

Figure 1. Tearscope Plus in position for slit lamp observation. The Tearscope Plus is aligned between the eye and the microscope. For precise focusing, it can be adjusted horizontally. Binocular observations are made.

3. BIOMICROSCOPE MOUNT

A slit lamp adapter is available. The adapter is secured at the back of the Tearscope Plus and fits on the biomicroscope tonometer mount. In operating position, it is set in line with the observation system and provides hands-free binocular observation of the tear film surface. For precise individual focusing, the Tearscope Plus can be adjusted horizontally on the mount. The timer can be easily operated, and the display is visible.

The Tearscope Plus can be moved aside for normal slit lamp use, ready for immediate use.

4. MAGNIFYING LENS

A magnifying lens has been designed to be in focus with the tear film plane. It can be secured close to the observation aperture at the back of the Tearscope Plus and provides useful magnification when the instrument is hand held. It acts as a focusing lens when the observer is close to the instrument (Fig. 2) and gives increasing magnification when the image is viewed at 30–50 cm (Fig. 3). It is best used with the different inserts and filters. With the magnifying lens and the inserts, the Tearscope Plus becomes a portable examination instrument for contact lens fitting.

5. TEARSCOPE GRIDS

Two removable grid inserts have been designed to fit in the inner illuminated tube. This allows xeroscope-type illumination for the practitioner who wants to determine the noninvasive break-up time of the preocular tear film using this particular method.

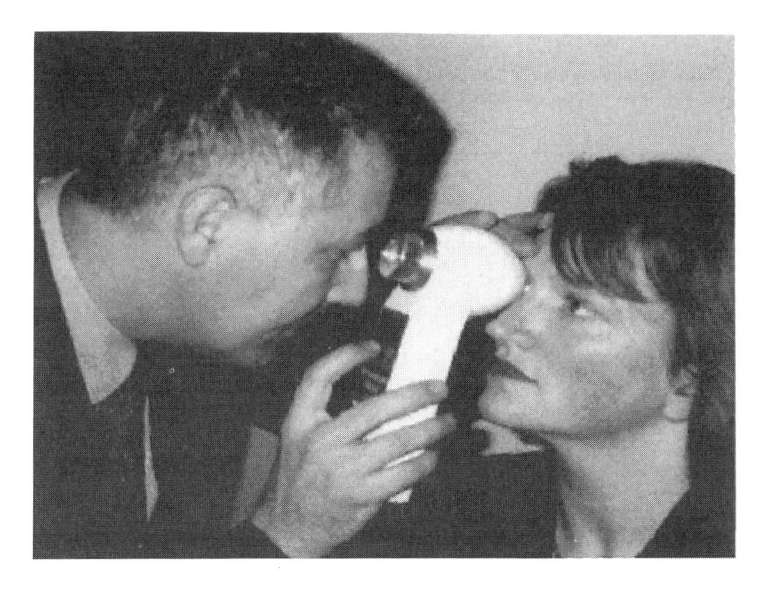

Figure 2. Tearscope Plus for hand-held observation. A magnifying lens is secured at the back of the Tearscope Plus in focus with the tear film plane. It acts as a focusing lens when the observer is close to the instrument.

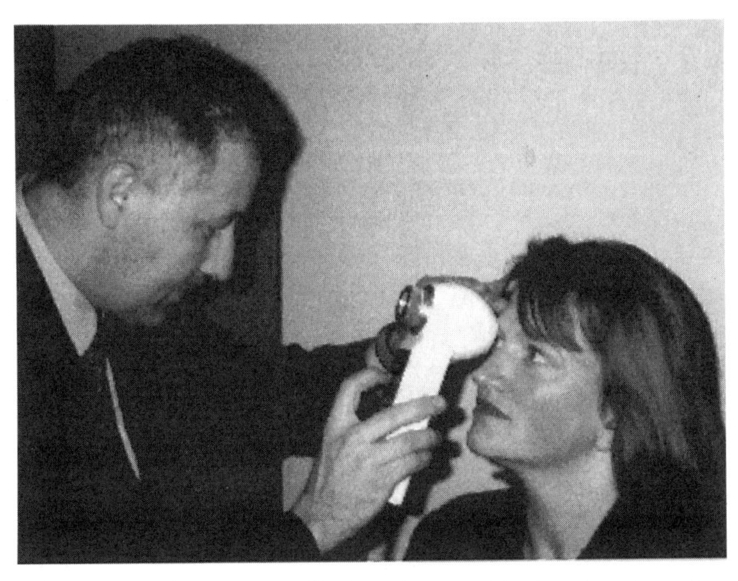

Figure 3. Observation to achieve increased magnification. The magnifying lens gives increasing magnification when the image is viewed at 30–50 cm. It is best used with the filters, grid, and rings systems.

Small peripheral disruptions or a break of the tear film produce a distortion or break of the grid pattern. A coarse grid with a black background and white grid lines is intended for hand-held use (Fig. 4). A fine grid intended for slit lamp observation has a white background and black grid lines. This grid has a dual purpose based on two focusing positions. Position one: the observer can focus on the black grid itself and measure the NIBUT by observing the grid distortion. Position two: by moving slightly away from the subject, the tear film is visible by specular reflection. This allows the combined observation of break-up time and localized tear film structure.

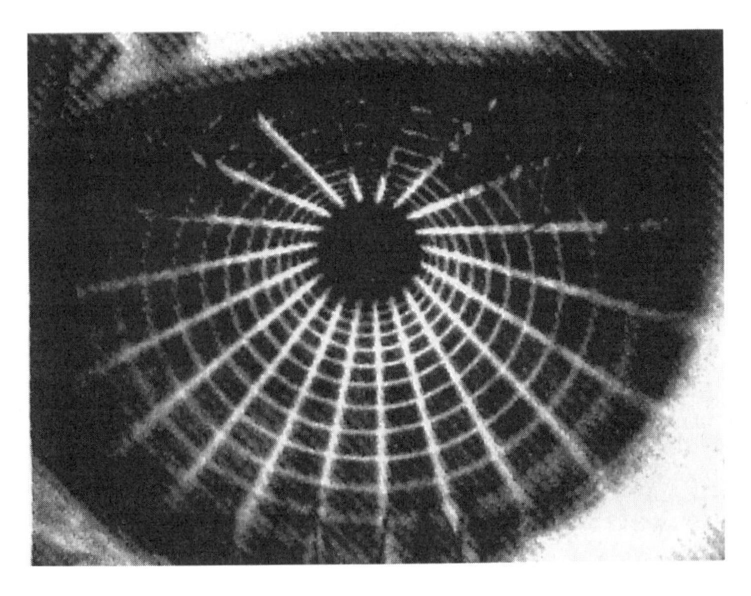

Figure 4. Coarse grid for hand-held observation.

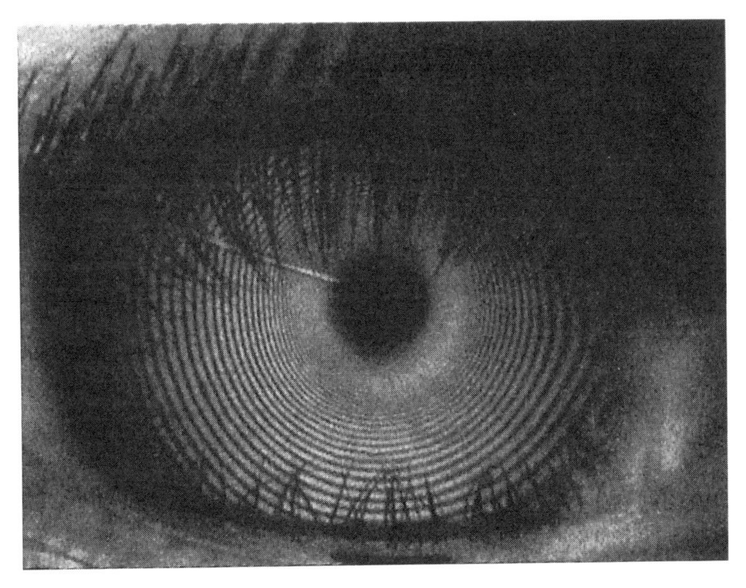

Figure 5. Concentric rings insert in position.

6. TEARSCOPE RINGS

A removable ring insert transforms the Tearscope Plus into an internally illuminated PLACIDO disc by reflecting keratoscope-like rings for the observation of corneal distortion (Fig. 5). This can be applicable for rapid determination of keratoconus, observation of high or irregular astigmatism, and surface distortion of contact lenses. The ring insert is useful to the ophthalmologist or surgeon for quick assessment of corneal distortion during or following graft surgery, to observe the effect of stitches in graft and cataract surgery, and in PRK surgery for the immediate assessment of corneal flattening and possible distortion at the junction area.

7. FILTERS

As a hand-held instrument, the Tearscope Plus is used to assess rigid contact lens centration (Fig. 6) and movement in white light. A blue filter is placed to cover the illuminating light, and a yellow filter is placed in front of the observation aperture. This system has advantages over ultraviolet lamps, as it emits mainly in the visible range and allows fluorescein observation through contact lenses with UV filters. With the use of these filters, the Tearscope Plus allows observation of the preocular fluorescein-stained tear film, its spreading, and break-up. The stained tear prisms are visible for assessment of regularity. RGP lenses fluorescein fitting patterns can also be seen (Fig. 7).

8. DARK FIELD OBSERVATIONS

The Tearscope Plus can now be set up as a perfect dark field examination device for deposits on contact lenses (Fig. 8). A lens in a suitable holder, placed at the back of the

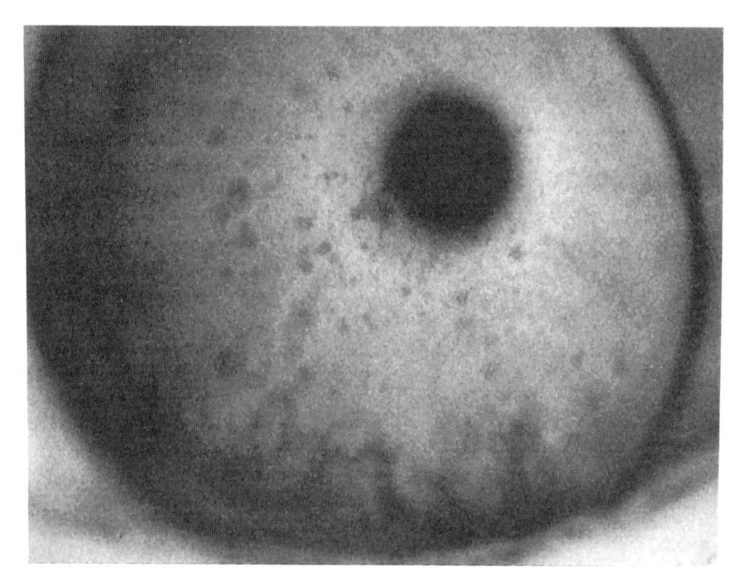

Figure 6. RGP contact lens fit in white light.

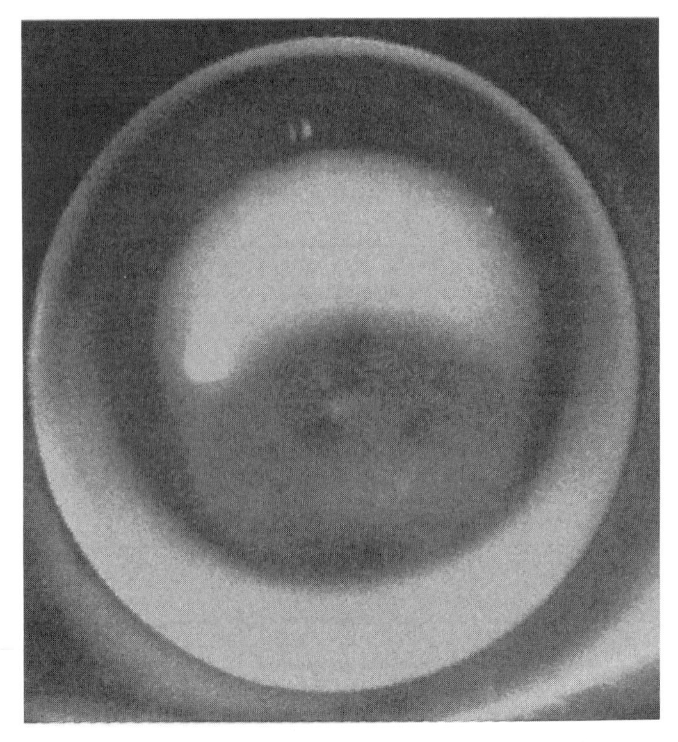

Figure 7. RGP fluorescein fit with use of blue and yellow filters.

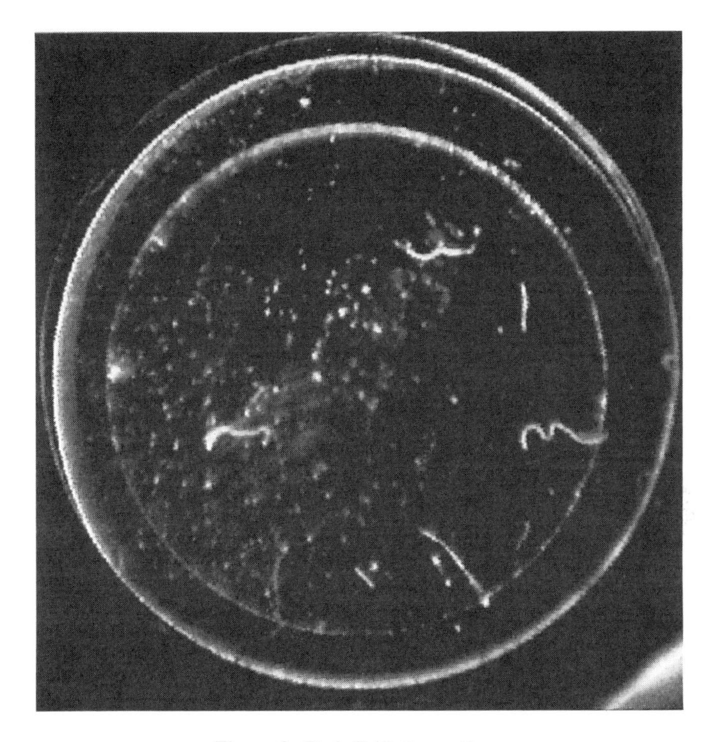

Figure 8. Dark field observation.

Tearscope Plus, is illuminated laterally and viewed against a dark background. Lens deposits and scratches will be highlighted as bright secondary light sources by the 360° oblique and even illumination. The observations are made using the magnification of the biomicroscope but are also possible with the naked eye.

Photographs can be taken by the researcher for future image analysis. Deposits and scratches can be immediately demonstrated to the patient to emphasize the need for adequate cleaning, polishing, or lens replacement.

9. TEARSCOPE ROUTINE FOR THE CONTACT LENS PATIENT

All the new features of the Tearscope Plus can be part of a routine designed to facilitate rapid diagnosis of marginal dry eye cases in contact lens wearers. The current routine makes use of image capture and diagnosis via interactive image database.

The principal observations used for contact lens wearers are as follows.

9.1. Pre-Fitting Stage

Preocular tear film (POTF) observations:

- lipid layer pattern observation and classification.
- lipid layer contamination observation and classification.
- lipid layer spreading observation during blink.
- tear prism observation and classification for regularity on the lid side and on the ocular surface side.

- tear prism observation for continuity in the corneal and conjunctival areas.
- measurement of noninvasive break-up time with and without grid.
- meibomian gland manipulation and observation of lipid secretion spreading at the tear surface.

9.2. Fitting Stage

Soft lens prelens tear film (PLTF) observations:

- lipid layer presence and coverage.
- lipid layer pattern observation and classification.
- lipid layer contamination observation and classification.
- lipid layer spreading observation during blink.
- tear prism observation and classification for regularity along the area of lid contact with the lens.
- assessment of aqueous phase visibility and thickness.
- measurement of noninvasive break-up time (NIBUT) with and without grid.
- assessment of PLTF break-up position and break-up type (spot – band – surface – lid edge).
- assessment of contamination of lens surface.
- assessment of PLTF resurfacing following a blink.
- assessment of artificial tears instillation.

RGP prelens tear film observations:

- lipid layer presence and coverage.
  - lipid islands - 25% cover - 50% cover - 75% cover - full
- lipid layer pattern observation and classification.
- lipid layer contamination observation and classification.
- lipid layer spreading observation during blink.
- tear prism observation and classification for regularity along the area of lid contact with the lens and around the contact lens edge.
- assessment of aqueous phase visibility and thickness.
- measurement of noninvasive break-up time (NIBUT) and non invasive drying up time (NIDUT) with and without grid. The NIBUT is the time taken from the blink to the appearance of the first break in the PLTF. The NIDUT is the time taken by the whole PLTF to leave the lens surface by drainage and evaporation. It is measured from the blink to the appearance of a dry overall contact lens surface and is a measure relating to the volume of tears present in the PLTF.
- assessment of PLTF break-up position and break-up type (spot - band - surface - lid edge).
- assessment of PLTF resurfacing following a blink.
- assessment of POTF at the contact lens edge for 3 and 9 o'clock instability to predict staining.
- assessment of artificial tears instillation.

10. ROUTINE FOR HAND-HELD USE

- ring pattern observation for corneal distortion.
- white light examination for lens centration and movement.

- observation of the PLTF drying sequence.
- measurement of the PLTF noninvasive break-up time (NIBUT) corresponding to the first break in the film.
- measurement of the PLTF noninvasive drying-up time (NIDUT) when the lens surface is devoid of tear cover by drainage and evaporation.
- blue and yellow filter examination of the fluorescein lens fit.
- observation of gross fluorescein corneal staining.

11. AFTER-CARE STAGE

Dark field examination of lens surface deposits and scratches.

12. CONCLUSIONS

The Tearscope Plus now has an improved design, and many useful add-ons are available. It can be mounted on a biomicroscope for hands-free operation or be used as a hand-held device at low magnification observation. It is best used as part of a routine examination that also includes dry eye questionnaire, fluorescein and rose bengal staining, thread test measurements, etc.

It should be a useful tool to help the practitioner in the diagnosis, treatment, and after care of the marginal dry eye contact lens wearer. In addition, it should encourage the non-specialist to tackle and fit marginal dry eye subjects and reduce the drop-out rate in the current contact lens wearing population.

EVALUATION OF SCHIRMER TESTS BY TWO TYPES OF TEAR CLEARANCE TESTS

Masafumi Ono,[1,2] Yukiko Yagi,[1] Eiki Goto,[1] Hao-Yung Yang,[1] and Kazuo Tsubota[1]

[1]Department of Ophthalmology
Tokyo Dental College
Ichikawa, Chiba, Japan
[2]Department of Ophthalmology
Tokai University, School of Medicine
Isehara, Kanagawa, Japan

1. INTRODUCTION

Evaluation of the tear volume measurements is an indispensable part of the examination for the diagnosis of dry eye.[1] The Schirmer I test is the most general tear volume measurement in the clinic, but this test has many modifications and interpretations and may not have enough sensitivity and specificity for the diagnosis of dry eye and keratoconjunctivitis sicca (KCS) of Sjögren's syndrome (SS).[2,3] This clinical study was designed to evaluate the Schirmer test reading using two types of dye dilution tests, tear clearance ratio (TCR) tests 1[4] and 2.

2. METHODS

The study group included 13 age- and sex-matched SS patients (26 eyes). They were diagnosed by modified Fox's criteria[5–7]: Schirmer 1 test <10 mm, focus score ≤1, female, 30–63 years. There were also 10 non-SS dry eye patients (simple dry eye patients: SDE, 20 eyes); Schirmer 1 test <10 mm or fluorescein (FL) or rose bengal (RB) positive staining,[8,9] female, 35–58 years. There were 10 normal volunteers (NL: n=14, 28 eyes); Schirmer 1 test ≤10 mm, no FL and RB staining, including BUT <10 sec, female, 30–60 years. Tear volumes were measured by (i) Schirmer I, (ii) TCR 1, using 10 μl of 0.5% fluorescein and 0.4% oxybuprocaine eyedrop,[3] (iii) TCR 2, using 1 μl of 5% fluorescein eyedrop (Fig. 1), and (iv) the Schirmer test with nasal stimulation (NST).[10] Additionally, SS

The measurement of the Tear Clearance Ratio tests (TCRs)

instillation of dye
(TCR1: 10μl of 0.5%FL with anesthesia,
TCR2: 1μl of 5%FL without anesthesia)

↓

eyes open for 5 min.

↓

Schirmer test

↓

measurement of Schirmer reading
measurement of TCR (dye dilution)

a

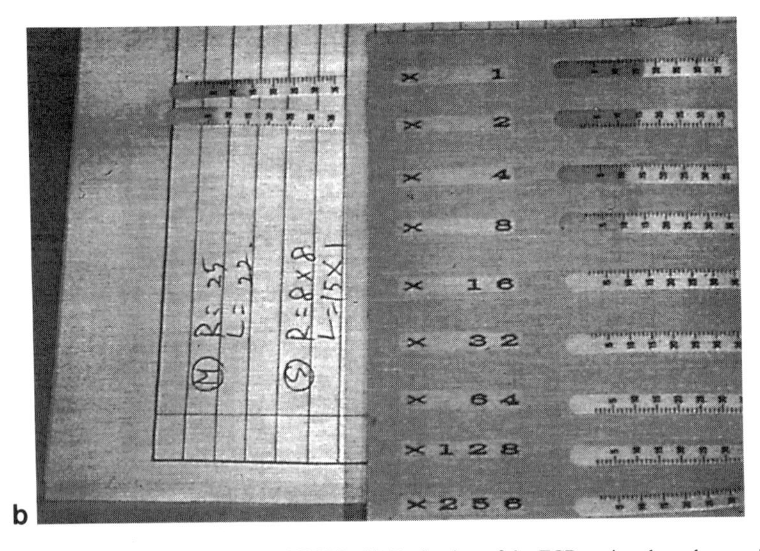

Figure 1. (a) Test methods for TCR1 and TCR2. (b) Evaluation of the TCRs using the color scoring scale.

patients with punctum occlusion had tear volumes measured using these same methods (Fig. 2).

The modified Schirmer reading was increased and the result of TCR was decreased after punctum occlusion by the punctum plug. The loosening of the upper punctum plug produced a decreased modified Schirmer reading and increased the result of TCR. The replacement of the punctum plug restored the results to those of the original condition. (Patient with Sjögren's syndrome, 54 years, female, using TCR1.)

3. RESULTS

There are no significant differences between the Schirmer tests readings of the normals (Fig. 3) and the simple dry eye patients (Fig. 4), except in Schirmer tests with nasal

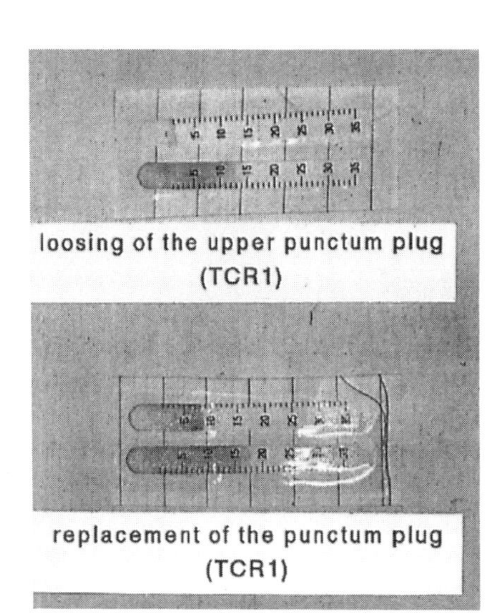

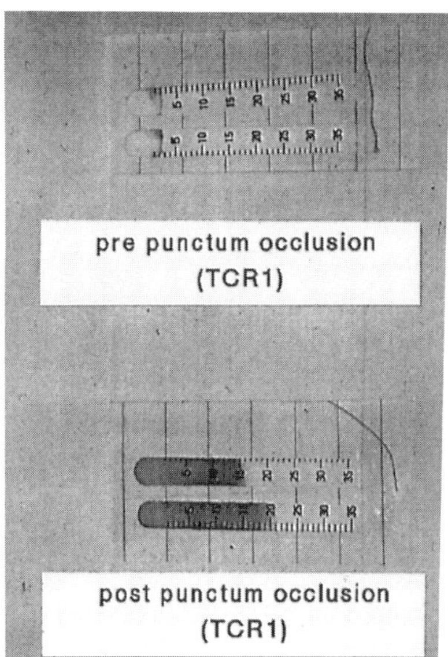

Figure 2. Alteration of the results of Schirmer test and TCR by the punctum plug.

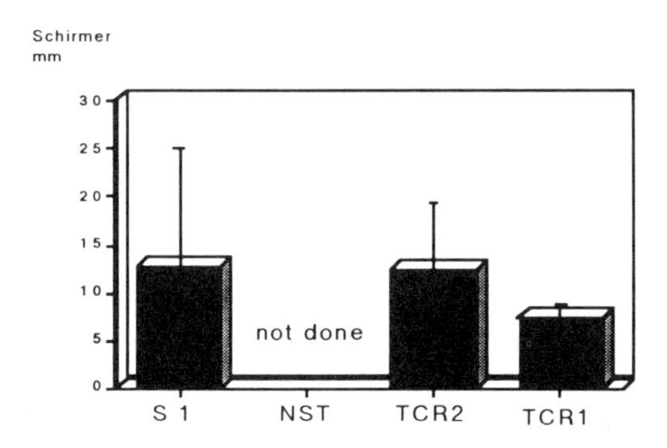

Figure 3. Results of the modified Schirmer tests in normals.

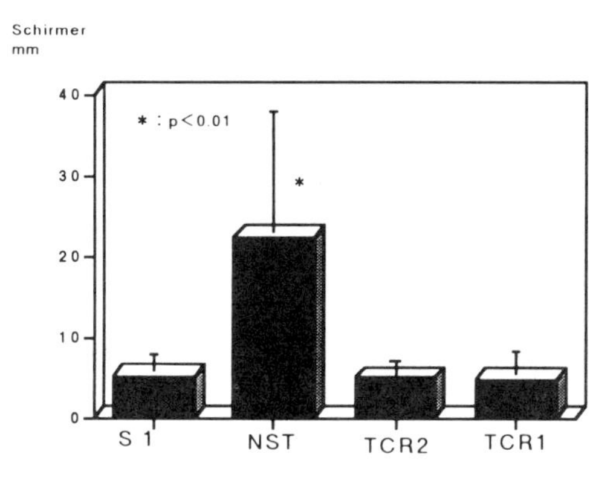

Figure 4. Results of the modified Schirmer tests in simple dry eye patients (non-SS).

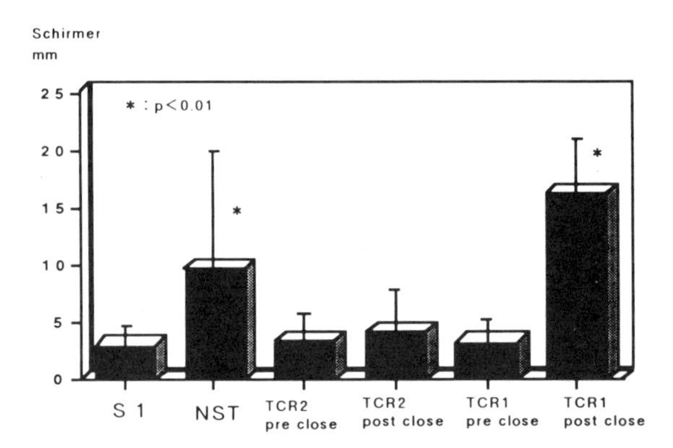

Figure 5. Results of the modified Schirmer tests in Sjögren's syndrome patients.

stimulation. The modified Schirmer test reading with TCR1 was increased after the punctum occlusion. There are no significant differences between the Schirmer test reading with TCR2 pre and post punctum occlusion. The modified Schirmer test reading with NST was higher than the Schirmer 1 test reading in SS; however, this Schirmer test reading of SS was lower than half of the readings of SDE (Fig. 5). There were no significant differences between the results of the ratio of TCR1 and TCR2 (Fig. 6).

4. CONCLUSIONS

Under the low tear drainage condition of SS patients, the modified Schirmer 1 test reading with instillation may show over 10 mm. TCR1 may be a better method for the evaluation of the effects of punctum occlusion in SS. There are no significant differences between the modified Schirmer test reading with TCR2 and the Schirmer 1 test reading. For a more accurate diagnosis of dry eye, the measurement of tear volume by Schirmer 1 test with other co-examinations such as TCR and NST may be important.

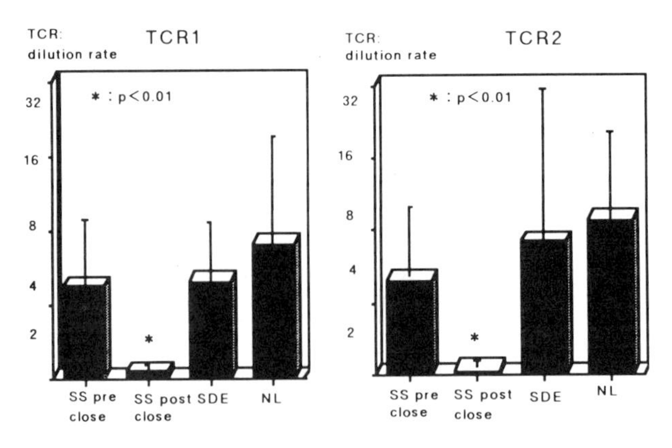

Figure 6. Results of the TCR tests in the three study groups.

REFERENCES

1. Lemp MA. Report of the National Eye Institute/ Industry workshop on clinical trials in dry eye. *CLAO J.* 1996;21:221–231.
2. Sjögren H. Zur Kenntnis der Keratoconjunctivitis Sicca. *Acta Ophthalmol.* 1933;11:1–15.
3. Fox RI, Kang H-I. Pathogenesis of Sjögren's syndrome. *Rheum Dis Clin North Am.* 1992;18:517–538.
4. Ono M, Yoshino K, Ogawa Y, Tsubota K. Tear clearance rate in normal and dry eye patients. ARVO Abstracts. *Invest Ophthalmol Vis Sci.* 1991;32:S2178.
5. Fox RI, Robinson CA, Curd JG, Kozin F, Howell F. Sjögren's syndrome: Proposed criteria for classification. *Arthritis Rheum.* 1986:29:577–585.
6. Pflugfelder SC, Huang AJW, Feuer W, Chuchovski PT, Pepeira IC, Tseng SCG. Conjunctival cytologic features of primary Sjögren's syndrome. *Ophthalmology.* 1990;97:985–991.
7. Ono M, Saito I, Yoshino K, Tsubota K, Moro I. Distribution of IgA1- and IgA2-containing cells in lacrimal glands of patients with Sjögren's syndrome. *Reg Immunol.* 1994;6:218–224.
8. van Bijsterveld OP. Diagnostic tests in the sicca syndrome. *Arch Ophthalmol.* 1969;82:10–14.
9. Toda I, Tsubota K. Practical double vital staining for ocular surface evaluation. *Cornea.* 1993;12:366–368.
10. Tsubota K. The importance of nasal stimulation in Schirmer test. *Am J Ophthalmol.* 1991;111:106–108.

FLUORESCEIN DYE CONCENTRATION AS A FACTOR IN TEAR FILM FLUORESCENCE

Victor M. Finnemore,[1] Donald R. Korb,[1] Jack V. Greiner,[2,3] Thomas Glonek,[4] and John P. Herman[1]

[1]100 Boylston Street
Boston, Massachusetts
[2,3]Schepens Eye Research Institute, and Department of Ophthalmology
Harvard Medical School
Boston, Massachusetts
[4]MR Laboratory
Midwestern University
Chicago, Illinois

1. INTRODUCTION

A large variation in the degree of fluorescence is observed clinically when using fluorescein dye for the evaluation of the ocular surface. Although this variation may not be so important in the case of diagnosing an epithelial defect, such as in a corneal or conjunctival abrasion, it becomes very important when attempting to evaluate the ocular surface in the presence of various degrees of corneal defects that occur in dry eye syndromes. For example, clinically there may be a high degree of variation in fluorescence when using paper impregnated with fluorescein (fluorescein strips) to introduce dye onto the ocular surface, in contrast to the more consistent observations using fluorescein liquid dye, though the appearance is also dependent upon the concentration and volume instilled onto the eye. The degree of fluorescence observed on the ocular surface appears to be concentration dependent, and light of a longer wavelength is emitted with increasing fluorescein concentration.[1] We have observed that increasing the concentration of fluorescein dye to 10% results in nearly 100% quenching of fluorescence for the first few minutes after instillation. Quenching is related to the fluorescein concentration on the ocular surface.[2] The purpose of this study was to evaluate this fluorescein quenching phenomenon in normal eyes and its relationship to the tear film.

2. METHODS

Human subjects (n = 20) aged 25–50 years, with tear film break-up times of >10 sec and Schirmer's test with anesthetic >10 mm at 5 min, were observed at ambient temperature with a relative humidity of 40–45%. Following a complete ophthal-mologic examination, subjects satisfying the following criteria were included in the study: (i) absence of abnormal ocular surface features as determined by biomicro-scopic evaluation, (ii) tear film lipid layer ≥ 90 nm as determined with a custom-built tear film apparatus described previously,[3] (iii) absence of overt ocular disease, and (iv) signed informed consent. Commercially available standard 2% sodium fluorescein (Iolab, Claremont, CA) and 10% Ak-Fluor (Akorn, Abita Springs, LA) dyes were used. An Eppendorf ultramicrodigital pipette was used to instill 1 µl of fluorescein 2% onto the inferior palpebral conjunctiva of the gently depressed lower lid in one eye, and fluorescein 10% in the corresponding position of the contra-lateral eye. Left and right eyes were randomized. Since it has been emphasized that tear flow measurements must be made without causing reflex lacrimation,[4,5] which may result in decreased tear film stability following instillation of fluorescein dye,[6] the following procedure was developed. The 1 µl amount was ejected from the micropipette so that it was held by capillary attraction to the pipette tip. The micropipette tip was then brought near the palpebral conjunctiva of the depressed lower lid until the fluorescein solution was released. The lower eyelid was slowly released, and the subject was asked to blink normally during the course of the experiment.

Fluorescence was measured using the cobalt-blue light on the Haag-Streit slitlamp at the lowest intensity setting. Measurements were recorded 2–3 sec following a blink, to allow the thickness of the precorneal tear film to stabilize. Using a double-masked experimental design, the degree of fluorescence was graded on a scale of 0 to 3+ by two independent examiners as follows: 0, no fluorescence; 1+, mild fluorescence; 2+, moderate fluorescence; 3+, brilliant fluorescence. The degree of fluorescence was recorded at 30 sec intervals for the first 2 min post-instillation and every minute thereafter until the 10 min time point. Observations continued at 5 min intervals until 30 min post-instillation.

3. RESULTS

Immediately after instillation of fluorescein 2% dye, a grade of 3+ fluorescence was attained by all subjects (Fig. 1). With fluorescein 10% dye, no fluorescence was detected

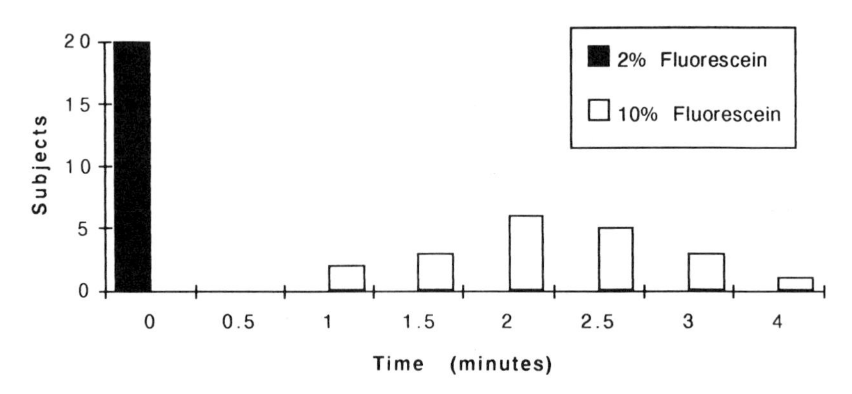

Figure 1. Time to 3+ fluorescence.

immediately. All subjects reached a grade 3+ fluorescence with fluorescein 10% 1–4 min post-instillation, with an average of 2.2 min. The duration of grade 3+ fluorescence was shorter (6.35 ± 1.57 min) for fluorescein 2% than for fluorescein 10% (13.4 ± 4.33 min) once grade 3+ fluorescence was attained (Table 1). Since the degree of fluorescence is dependent not only on the concentration of instilled fluorescein but also on the volume of the tear film layer, such information lends itself to the development of a method to calculate the relative volume of the tear film.

4. DISCUSSION

This study of fluorescence intensity as a function of time and initial fluorescence concentration was undertaken to address the apparent paradox of an initial "hypofluorescence" resulting from 10% fluorescein when compared to the use of a 2% fluorescein solution. This observation may reflect a "self-quenching" phenomenon with concentrated fluorescein solutions, in which fluoresced light from lower layers of the tear film is absorbed by fluorescein dye molecules located in the upper layers of the tear film. This type of reaction is well recognized in the fields of chemical and x-ray spectroscopy, and is sometimes referred to as a "saturation effect." With the use of lower concentrations of fluorescein, insufficient numbers of absorbing molecules would be available in the upper layers of the tear film to quench the light emitted from lower layers, thus allowing the fluoresced light to be detected biomicroscopically. Similarly, the initial quenching effect with a 10% solution would be gradually reduced as the dye is diluted over time by the in-

Table 1. Average tear film fluorescence after fluorescein instillation

Time (min)	2% Fluorescein (mean ± SD)	10% Fluorescein (mean ± SD)
0	3.00 ± 0.00	0
0.5	3.00 ± 0.00	0.36 ± 0.50
1	2.86 ± 0.36	1.14 ± 0.66
1.5	2.57 ± 0.51	1.64 ± 0.93
2	2.36 ± 0.63	2.21 ± 0.80
2.5	2.00 ± 0.78	2.64 ± 0.63
3	1.93 ± 0.92	2.93 ± 0.27
4	1.43 ± 1.02	2.93 ± 0.27
5	0.86 ± 0.86	2.83 ± 0.53
6	0.71 ± 0.83	2.71 ± 0.61
7	0.43 ± 0.76	2.50 ± 0.65
8	0.29 ± 0.61	2.07 ± 0.92
9	0	1.86 ± 0.77
10	0	1.29 ± 0.83
15	0	0.50 ± 0.52
20	0	0.07 ± 0.27
25	0	0
30	0	0

$n = 20$.

Grades of fluorescence: 0, no fluorescence; 1+, mild fluorescence; 2+, moderate fluorescence; 3+, brilliant fluorescence.

flux of "new tears" effectively reducing the concentration of fluorescein molecules and thereby allowing maximum fluorescein levels to be observed.

This apparent "quenching" phenomenon has implications beyond determining the solution of choice for purposes of detecting epithelial defects. Since the degree of fluorescence is concentration-dependent, the dilution of instilled fluorescein by natural tears may allow evaluation of the tear production in normal and dry eyes. That is, the rate of loss of fluorescence would be proportional to the amount of new tears created. Thus, this method would potentially allow distinction of dry (i.e., "aqueous-deficient") vs. normal eyes on a relative basis, as it might be expected that the fluorescence observed in dry eyes would persist for a longer time, since the influx of new tears in these eyes would be slower than in the "normal" eye. Observations with a lacrimal streak dilution test supports this prediction.[7]

Fluorescein 2% achieves grade 3+ at instillation and then declines to 0 over the next 9 min. Instillation of 10% fluorescein, by contrast, begins with a quenched signal exhibiting completely undetectable fluorescence, rising to a value of grade 3+ in an average of 2.2 min and declining to 0 over a more lengthy time course. It has been hypothesized that fluorescein undergoes self-quenching as its concentration increases, due to the absorption of the emitted light from lower layers of the tear film by fluorescein molecules in the upper layers. Since the degree of fluorescence is dependent on concentration and changes are attributable to the dilution of fluorescein by natural interaction of the tear film, quenching may be used to evaluate and determine the relative volume of the liquid phase of the tear film in which the fluorescence becomes soluble. This fluorescence phenomenon may be useful in developing a clinically useful, safe, and accurate test for the quantification of tears relative to corresponding norms and the evaluation of patients with dry eye syndromes.

REFERENCES

1. Romanchuk KG. Fluorescein: Physicochemical factors affecting its fluorescence. *Surv Ophthalmol.* 1982;26:269–283.
2. Wilson G, Horgwei R, Laurent J. Corneal epithelial fluorescein staining. *J Am Optom Assoc.* 1995;66:435–447.
3. Korb DR, Baron DF, Herman JP, et al. Tear film lipid layer thickness as a function of blinking. *Cornea.* 1994;13:354–359.
4. Mishima S, Gasset A, Klyce SH, Baum JL. Determination of tear volume and tear flow. *Invest Ophthalmol.* 1966;5:264–275.
5. Sorenson T. Determination of tear volume and tear flow. ARVO Abstracts. *Invest Ophthalmol.* 1975;25:S43.
6. Mengher LS, Bron AJ, Tonge SR, Gilbert DJ. Effect of fluorescein instillation on precorneal tear film stability. *Curr Eye Res.* 1985;4:9–12.
7. Norn MS. Lacrimal apparatus tests. *Acta Ophthalmol.* 1965;43:557–566.

A NEW METHOD FOR MEASURING HUMAN BASIC TEAR FLUID OSMOLALITY

Pierre L. Emond and Christopher J. Corbett

Advanced Instruments, Inc.
Norwood, Massachusetts

1. INTRODUCTION

Osmolality measures the total number of particles dissolved in a kilogram of water. Since the colligative properties of a solution depend primarily on the number of particles in solution, a measurement of any of these properties can, in principle, be used as the basis for determining osmolality. Freezing (melting) point determinations are a sensitive and well-established method for determining osmolality.

Available melting point osmometers for use with sample volumes of less than 500 nanoliters are labor intensive and require considerable technician training for proper calibration and data collection. Determining the osmolality of tear film samples is one application of small-volume osmometry requiring equipment specifically configured for improved operating convenience in both research and clinical settings.

Over the last decade, tear fluid osmolality measurements based on sample melting point determinations have been described by a number of authors.[1,2,3] Using the same melting point principle but with a markedly different instrumentation approach, we constructed and tested a prototype nanoliter osmometer in our laboratory. The data presented here are the result of a preliminary evaluation of system performance.

2. OBJECTIVES

Our goal was to develop a prototype melting point osmometer with the following features: simplified sample handling and instrumentation operating technique, test sample volume of 500 nanoliters or less, operating range between 260 and 340 mOsm/kg H_2O, and accuracy, precision, and stability of +/- 1% using osmolality test standards.

3. MATERIALS AND METHODS

3.1. Apparatus

The system we developed and tested has four primary components: (i) A personal computer for process control, data collection, and display of results. (ii) A CCD camera, lens system, and video monitor for sample visualization. (iii) A self-contained cooling assembly to freeze and thaw the test sample under computer control. (iv) A mating 500 nanoliter sample holder.

3.2. Procedure

A 500 nanoliter aliquot of test solution is transferred from its storage container to a sample holder using a factory-calibrated micro volume syringe (Microliter #7001, Hamilton, Reno, NV). A new sample holder is used with each determination.

The sample holder is inserted into the system's thermoelectrically driven cooling block. The geometry of the holder, cooling assembly, and CCD camera permits sample visualization during the pre-programmed freeze/thaw cycle initiated by the operator. The sample is fast frozen and then taken through a thawing protocol without operator intervention.

The disappearance of the last ice crystal in the sample during the thawing phase corresponds to its melting point, and this event is noted by the operator using the computer's pointing device. The computer calculates and displays the sample's osmolality based on the temperature at which the last crystal disappeared and the temperature/osmolality calibration equation stored in memory.

The system is self contained and requires no external connections other than 110 volt, 60 Hz electrical power.

3.3. Evaluation Protocol

National Commission for Clinical Laboratory Standards[4,5] guidelines were adapted for evaluation of the nanoliter osmometer as described below.

3.4. Calibration

Instrument calibration was performed on Day 1 of testing using commercially available osmolality standards (Advanced Instruments Clinitrol 290, Cryoscope 621H and 530H). Five measurements were made at each of the following levels using 500 nanoliter aliquots from freshly opened vials of the standards: 275, 290, and 323 mOsm/kg H_2O. A least squares polynomial fit was used to generate a calibration equation which was then programmed into the instrument control's computer software. The calibration of the instrument remained unchanged throughout the subsequent testing.

3.5. Test Samples

On Days 2 through 6 of testing, the calibrated instrument was challenged using freshly prepared salt solutions of known osmolality at the Low, High, and Mid points of the clinical range of interest.

Table 1. Measurements (in mOsm/kg H_2O) of prepared salt standards at three points in clinical range of interest

	Day 2	Day 3	Day 4	Day 5	Day 6	Day 7	Grand value	True osmolality
High								
Mean:	337.3	–	–	337.7	337.6	–	337.5	339.8
SD:	3.4	–	–	2.5	3.8	–	3.2	
CV%:	1.0%	–	–	0.8%	1.1%	–	0.9%	
Mid								
Mean:	301.2	297.2	–	–	296.8	–	298.4	299.6
SD:	2.8	3.1	–	–	1.5	–	3.2	
CV%:	0.9%	1.1%	–	–	0.5%	–	1.1%	
Low								
Mean:	–	259.6	–	–	258.8	259.3	259.3	260.4
SD:	–	1.1	–	–	1.2	1.3	1.2	
CV%:	–	0.4%	–	–	0.5%	0.5%	0.5%	

Daily values, 10 measurements. Grand values, all 30 measurements. CV, coefficient of variation.

The challenge solutions were gravimetrically prepared on Day 1 with reagent-grade NaCl and deionized water to have osmolalities of 260 (Low), 300 (Mid), and 340 (High) mOsm/kg H_2O. Solution osmolality values were confirmed independently using a calibrated cryoscope (Advanced Instruments Model 3D3) prior to use. To assure stability, the challenge solutions were kept in closed containers except during sample loading and stored at 4°C when not in use.

Thirty determinations were made at each of the three challenge levels. At each level, an uninterrupted series of ten determinations were run on each of three different days by one of two operators. Results were analyzed using Microsoft Excel Version 6.0.

4. RESULTS

4.1. Replicate Measurements

The mean ± SD values for all measurements are shown in Table 1.

4.2. Precision and Accuracy

Precision[6,7] of the measurement at each challenge level is characterized by the Standard Deviation and Coefficient of Variation, as noted in Table 1. Measurement accuracy as defined by Systematic Bias (difference between mean of measured and true values), Relative Error (bias as a percent of the true value), and Standard Error of the Mean (SEM) is detailed in Table 2.

Table 2. Accuracy at each challenge level

	Low	Mid	High
Systematic bias	–1.15	–1.22	–2.27
Relative error	–0.4%	–0.4%	–0.7%
Standard error of the mean	0.22	0.58	0.58

4.3. Calibration Stability and Linearity

Calibration stability over a period of 5 days as determined from the standard deviation of the means at 340 mOsm/kg H_2O was within (4σ) 3.72 mOsm/kg H_2O, or 1.1%. Regression statistics for the grand means gave an R^2 of 0.99994713 and a Standard Error of 0.402.

5. CONCLUSIONS

Sample manipulation and instrument operation for measuring tear fluid osmolality have been simplified over prior methods, but collection and loading of the test material into the sample holder needs refinement in order to be applicable in the clinical environment. The operating range of the current instrument is appropriate for tear film analysis at the targeted test volume of 500 nanoliters. A further reduction in sample volume is desirable, while maintaining the other performance characteristics of the system. The measurement accuracy and linearity goals of 1% were achieved with the present system configuration. Precision and calibration stability were good but require further study.

ACKNOWLEDGMENTS

The support and encouragement of J. P. Gilbard, M.D., and D. A. Sullivan, Ph.D., are gratefully acknowledged.

REFERENCES

1. Benjamin WJ, Hill RM. Human tears: Osmotic characteristics. *Invest Ophthalmol.* 1983;29:1624–1626.
2. Gilbard JP, Farris RL, Santamaria J, II. Osmolarity of tear microvolumes in keratoconjunctivitis sicca. *Arch Ophthalmol.* 1978;96:677–681.
3. White KM, et al. Human basic tear fluid osmolality. *Acta Ophthalmol.* 1993;71:524–529.
4. Evaluation of Precision Performance of Clinical Chemistry Devices. 2nd ed., NCCLS Document EP5-T2.
5. Preliminary Evaluation of Clinical Chemistry Methods. NCCLS Document EP 10-T.
6. Skoog DA, Leary JJ. Performance characteristics of instruments; figures of merit. In Principles of Instrumental Analysis. 4th ed., pp. 5–7. Saunders.
7. Kelley WD, Ratliff TA, Nenadic C. Measuring variability. In Basic Statistics for Laboratories, pp. 13–19. New York: Van Nostrand.

MICROPACHOMETRIC QUANTIFICATION OF TEAR ASSESSMENT TESTS

Hans W. Roth[1,2] and Rolf Marquardt[2,3]

[1]Center for Sight
Georgetown University Medical Center
Washington, DC
[2]Insititut für wissenschaftliche Kontaktoptik
Ulm, Germany
[3]University Eye Clinic
University of Ulm, Germany

1. INTRODUCTION

Dry eye presents as a wide distribution of signs and symptoms. It has particularly great clinical significance because of its causal relationship to chronic conjunctivitis.[1] There are multiple causes of dry eye—for example, exogenous factors such as air-conditioned workplaces and long plane flights or auto trips, as well as endogenous factors such as disturbances in salt or hormone concentrations.[2]

The diagnosis of dry eye is especially difficult; many of the tests commonly used have various problems and rarely direct one to an appropriate therapeutic action. Micropachometry is a new method that can indicate disturbances of the corneal physiology. All negative influences from either the surroundings or illnesses result in a decrease in corneal thickness.[3] This method, developed by us, utilizes a commercially available pachometer whose clinical sensitivity has been enhanced to ± 3 μM. This sensitivity results from the combination of modifications to the device and the use of an image projection grid.

In a large clinical study,[3] we demonstrated that a variety of systemic or local diseases result in a thinning of the central corneal thickness (CCT) from the average value of 545 μM. We also reported that any disturbance of the tears results in an initial thinning of the cornea; when the cornea is decompensated, the CCT is significantly greater than normal.

The purpose of the present study was to use the CCT measured by micropachometry as a sensitive marker of tear disturbances in order to assess the clinical usefulness of commonly used tests.

Lacrimal Gland, Tear Film, and Dry Eye Syndromes 2
edited by Sullivan *et al.*, Plenum Press, New York, 1998

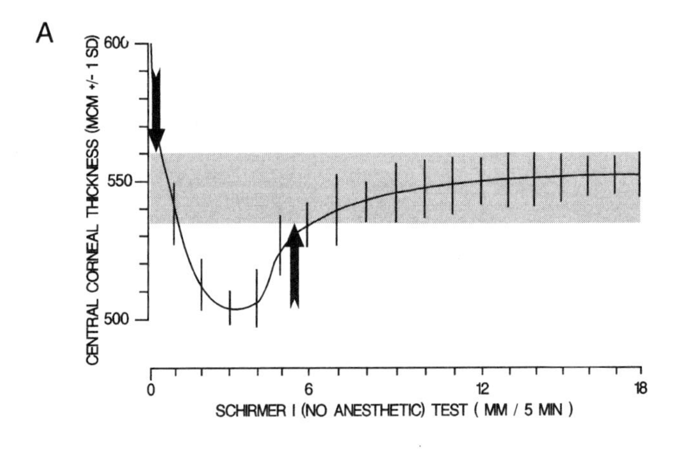

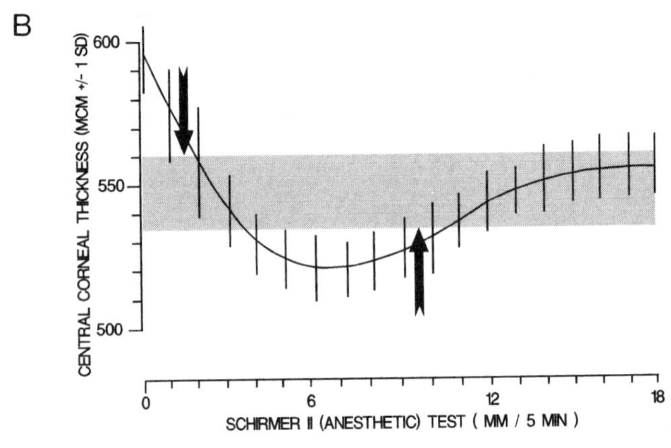

Figure 1. Relationship between central corneal thickness and (A) Schirmer I test data, (B) Schirmer II test data.

2. METHODS

We wanted to determine which of the following tests was the best indicator of a disturbance in corneal physiology: Schirmer tests (I, without anesthetic, and II, after anesthetic), the meniscus test (width of lower meniscus), rose bengal staining, and electrical resistivity of the tears.[4] These tests were carried out on 172 patients with dry eye symptoms.

3. RESULTS

Figures 1–5 list the test results, the shaded areas sign the average pachometric values of the healthy eye.

3.1. Schirmer Tests

The CTT was abnormally thin when Schirmer I or II tests were less than 6 or 9.5 mm/5 min, respectively (Fig. 1). Corneal decompensation as indicated by an abnormally thick CCT became evident when values were less than 0.5 or 1.5 mm/5 min, respectively.

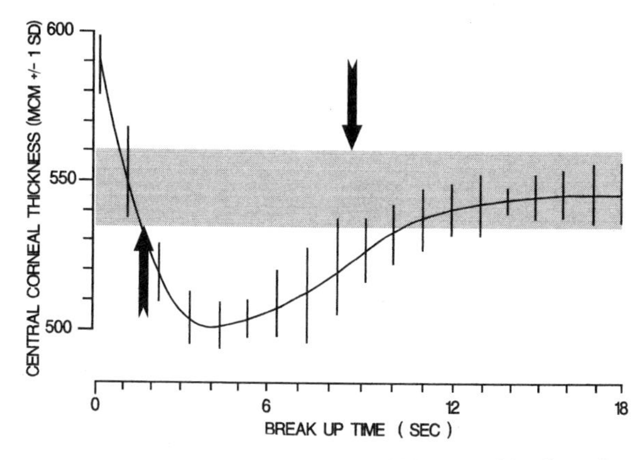

Figure 2. Relationship between central corneal thickness and break-up time values.

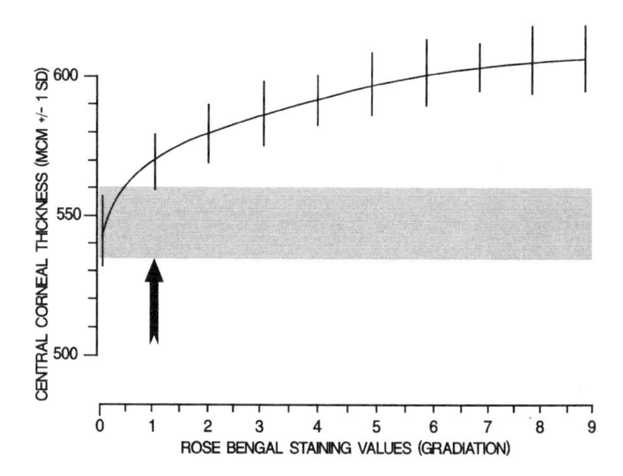

Figure 3. Relationship between central corneal thickness and rose bengal staining gradation.

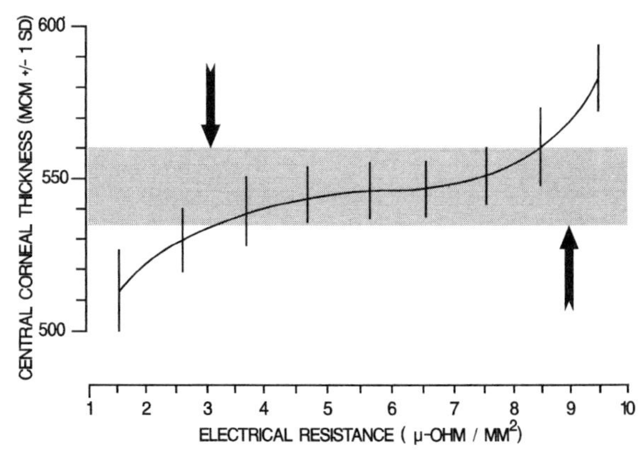

Figure 4. Relationship between central corneal thickness and electrical resistance of the tear liquid.

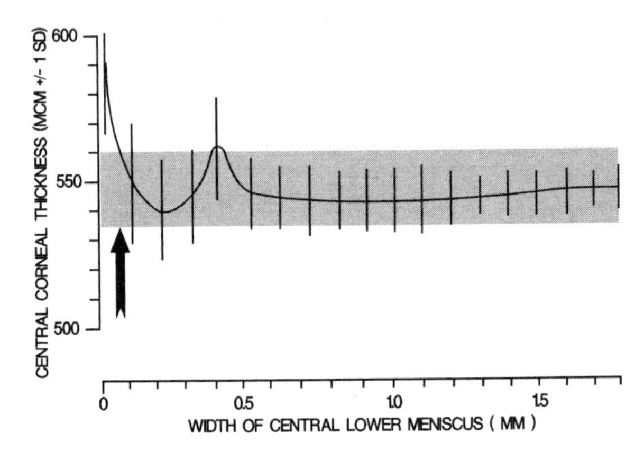

Figure 5. Relationship between central corneal thickness and the width of the central lower tear meniscus.

3.2. Tear Film Breakup Time (BUT)

Thinning of the cornea appeared when the BUT was less than 7.5 sec; decompensation, when the value was less than 1.5 sec (Fig. 2).

3.3. Rose Bengal Staining

Eyes stained with rose bengal usually were decompensated (Fig. 3).

3.4. Tear Resistance

When testing tears for free ions, resistances up to 3 ohms indicated normal tears; values greater than 9 ohms indicated decompensation (Fig. 4).

3.5. Meniscus Test

When the meniscus was visible, there was no detectable change in CCT (Fig. 5).

4. SUMMARY AND CONCLUSION

After reviewing the results of available methods for defining the dry eye with an exact assessment of severity, we found that micropachometry was the most sensitive. The following is a ranking of the other tests according to usefulness: Tear resistance, Schirmer, and BUT. Rose bengal stained eyes were usually decompensated; the meniscus test's primary use was as an "all or nothing" indicator, but recording the widths of extant menisci were unproductive.

REFERENCES

1. Lemp MA, Marquardt R. *The Dry Eye, a Comprehensive Guide*. Heidelberg: Springer Verlag; 1992.
2. Bijsterveld OP, Baardman J. Hornhautdickenmessung bei Patienten mit Keratoconjunctivitis sicca. *Klin Mbl Augenheilkd*. 1990;197:240–243.
3. Roth HW, Roth HW. *Hornhautpachymetrie beim Gesunden, Erkankten und beim Kontaktlinsentraeger*. Stuttgart: Enke Verlag; 1994.
4. Roth HW, MacKeen DL. Micropachometric differentiation of dry eye syndromes. *Adv Exp Med Biol*. 1994;350:505–511.

126

USE OF CORNEAL THICKNESS CHANGES TO COMPARE THE EFFICACY OF CONVENTIONAL EYE DROPS WITH SUPRACUTANEOUS TREATMENT OF DRY EYE

Hans W. Roth

Center for Sight
Georgetown University Medical Center
Washington, DC
Institut für wissenschaftliche Kontaktoptik
Ulm, Germany

1. INTRODUCTION

The dry eye contact lens wearer has many problems. Disturbances in the qualitative or quantitative status of tear production can result in a decrease in both wearing comfort and visual acuity, an increase in glare, and numerous other problems.

A change in either the central or the peripheral corneal thickness (CCT and PCT, respectively), determined by micropachometry, is an excellent indicator of the pathophysiological status in any dry eye condition. As previously reported, these changes mirror decreases in tear secretion and range from a slight corneal thinning seen with mild flow decreases to marked swelling seen with corneal decompensation from drastic tear decrements.[1,2]

2. METHODS

We measured CCT and PCT by micropachometry to assess the efficacy and duration of action of various dry eye treatments, including saline and commercial artificial tears (Lens Fresh) as well as a new modality of treatment termed supracutaneous in which a water-free ointment containing calcium carbonate (DEO™)[3,4] is delivered to the eye by application to the skin of the lower eyelid.

Lacrimal Gland, Tear Film, and Dry Eye Syndromes 2
edited by Sullivan *et al.*, Plenum Press, New York, 1998

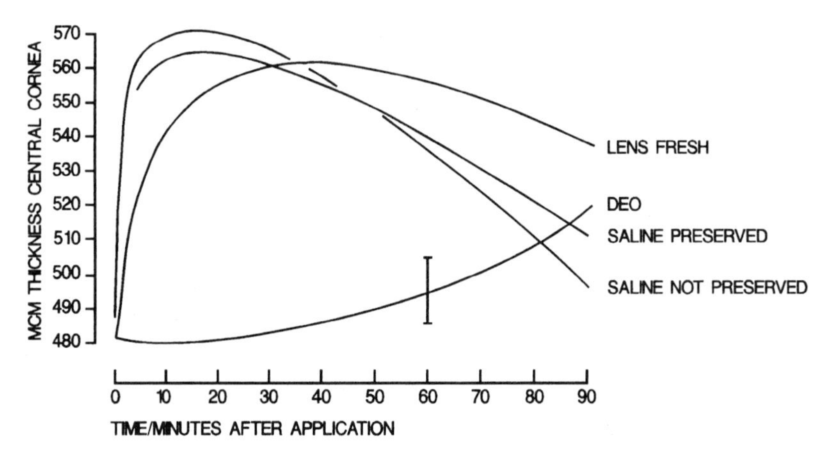

Figure 1. Effect of various treatment modalities on central corneal thickness.

3. RESULTS

Our findings to date have shown that saline and artificial tears normalized the CCT and PCT within a few moments of instillation; the maximum duration of effectiveness was approximately 45 min. The effect of DEO™ was observed 20 to 30 min after application; the duration of effectiveness was many hours (Figs. 1 and 2).

4. DISCUSSION

These results show that the effectiveness of dry eye treatment can be assessed by means of micropachometry. Any normalization of the CCT can be considered indication of improvement of the physiological status of the cornea. These findings were also informative about the effects of the instillation of a conventional artificial tear. In respect to the

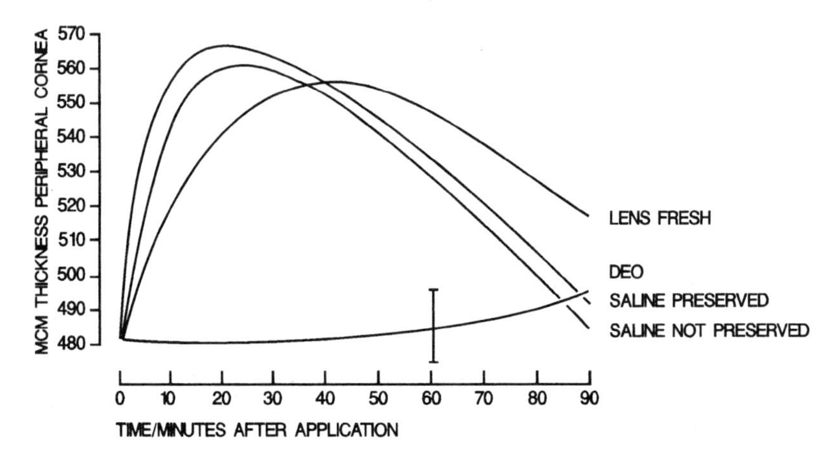

Figure 2. Effect of various treatment modalities on peripheral corneal thickness.

delay between instillation and response, aqueous preparations have the shortest delay; oily or viscous preparations require a longer time. The ideal therapy for dry eye is a combination of an aqueous drop having a rapid initial onset and the delayed onset but protracted effect provided by a supracutaneously applied ointment.[5,6]

5. CONCLUSION

Different dry eye formulations show different effects on the corneal thickness. There are two sets of actions based on normalization of corneal thickness: rapid onset and short duration of effect observed with an instilled aqueous preparation vs. slow onset and prolonged effect from a supracutaneously applied ointment formulation.

REFERENCES

1. Roth HW. Hornhautpachymetrie beim Gesunden, Erkankten und beim Kontaktlinsentraeger. Stuttgart: Enke Verlag; 1994.
2. Roth HW, MacKeen DL. Micropachometric differentiation of dry eye syndromes. Adv Exp Med Biol. 1994;350:505–511.
3. MacKeen DL, Roth HW, Doane MA. Ocular drug delivery by the LLD (lower lid delivery) method. ARVO Abstracts. Invest Ophthalmol Vis Sci. 1996;37:S77.
4. MacKeen DL, Roth HW. Revolutionary method of ocular drug delivery. ARVO Abstracts. Invest Ophthalmol Vis Sci. 1995;36:S166.
5. Roth HW, Vegh M. Diagnosis of dry eye in contactology. IV International Conference of the International Society of Dacryology. Stockholm, June 10, 1996.
6. Roth HW, Vegh M. Dry eye problems in contactology. Congress of the Hungarian Ophthalmological Society, Budapest, August 29, 1996.

127

SJÖGREN'S SYNDROME

Pathogenesis and New Approaches to Therapy

Robert I. Fox

Division of Rheumatology
Scripps Clinic and Research Foundation
La Jolla, California

1. BACKGROUND

During the past decade, several different sets of diagnostic criteria for Sjögren's syndrome (SS) have been proposed. The appearance of minor salivary gland biopsy is the key feature in the San Francisco criteria;[1] evidence for a systemic autoimmune disease as manifest by characteristic autoantibodies or minor salivary gland biopsy is required in the San Diego criteria;[2] and, in contrast, the European (EEC) proposed criteria can be fulfilled without requirement for histologic or serologic abnormality.[3] Also, exclusions to diagnosis of SS differ in the San Diego and EEC criteria. For example, patients with sicca symptoms associated with hepatitis C infection[4–7] are excluded from diagnosis as SS in the San Diego criteria but included in the EEC criteria.

No agreement on a single set of diagnostic criteria has been achieved at several international symposia on SS.[8] Since only about 15% of the patients fulfilling the EEC criteria would be diagnosed as SS using the San Diego criteria,[3] the groups of patients included in various clinical and research studies are very different. This leads to confusion in clinical practice and in the research literature. For example, it is difficult to compare studies published from Europe, where EEC criteria are frequently used, to studies published in US, where other diagnostic criteria are often used. In addition, basic research on pathogenetic mechanisms of SS relies on samples that clinicians provide to researchers. Thus, attempts to find a better animal model or to test a particular molecular hypothesis for SS are difficult when there is no "gold" standard for clinical definition of the disease. This report will use the San Diego criteria for SS, since these patients share common features of minor salivary gland biopsy, autoantibodies, and particular HLA-DR/DQ alleles.[9,10]

2. PATHOGENESIS

2.1. Features of the Salivary Gland Biopsy

A key feature of SS is the appearance of focal lymphocytic infiltrates in the minor salivary gland biopsy (Fig. 1A). A normal salivary gland biopsy lacks these focal infiltrates, (Fig. 1B). Although great attention has been focused on the lymphoid infiltrates it is important to note that residual acinar cells are still present in the SS biopsy (Fig. 1A). A common misconception about SS is that severe "aqueous deficiency" results from the total immune destruction of the gland. Morphometric analysis of SS biopsies shows that only approximately half of the acinar cells are destroyed.[11] The failure of residual acini in SS glands to function adequately may result from interruption of efferent neural signals[12] that would normally trigger increased secretion of water, proteins, and mucin. A failure of neu-

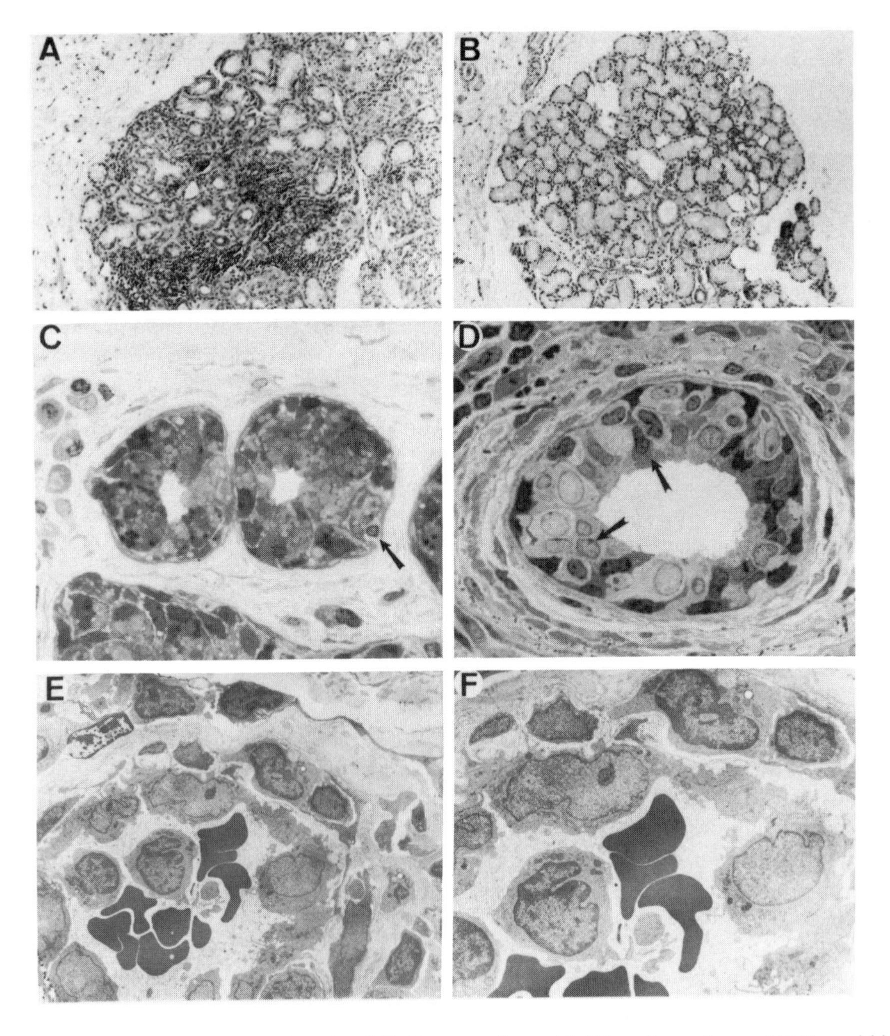

Figure 1. Minor salivary gland biopsies. (A, C-F) From a patient with Sjögren's syndrome. (B) Normal biopsy. Arrows in (D) indicate lymphocytes in direct contact with epithelial cells.

ral innervation of the lacrimal glands occurs at an early stage of pathogenesis in the non-obese diabetic (NOD) mouse.[13]

The destruction of the gland appears to be mediated by direct lymphocyte-glandular contact (Fig. 1C, D). These T-lymphocytes have been shown to exhibit CD4 (+), memory T-cell phenotype[14] and can initiate glandular cell apoptosis through granzyme A and perforin production.[15] The focal lymphocytic infiltrates usually begin around a high endothelial venule (HEV) (Fig. 1 E, F, where the blood vessel is identified by their red blood cells), a vascular structure initially recognized in lymph nodes and known to express adhesive molecules that allow homing of lymphocytes.[16]

2.2. Initiation of Immune Responses in SS: Tolerance and Antigen Presentation

SS provides an excellent model to study the mechanisms for "failure" of tolerance that permit the emergence of T-cells with self reactivity, and how these cells interact with a specific target organ. The first step in the pathogenesis of an autoimmune response is the failure of thymic selection to eliminate any T-cells with potential self reactivity.[17,18] T-cells with an antigen receptor (TCAR) that can bind with high affinity to thymic epithelial cells expressing MHC/self peptide complexes are usually deleted by a process termed "negative selection." Recent studies on thymic selection of T-cells have demonstrated the importance of the balance of fas, fas ligand, bax, and bcl-2 in determining whether nascent T-cells are released into the periphery or undergo an apoptotic death.[19] For example, overproduction of bcl-2 occurs in certain early stages of T-cell development in mice with autoimmune disease, allowing the emergence of autoimmune T-cells.[20] Recently, it has been recognized that certain MHC alleles bind self antigens with a low affinity (termed cryptic antigens) and that T-cells reactive with such self antigens may escape negative selection in the thymus;[21] these studies provide a structural basis for the association of particular MHC alleles, autoantigens, and autoimmune diseases.

After a nascent lymphocyte with potential autoimmune reactivity leaves the thymus and enters the peripheral circulation, it must bind to a HEV to avoid apoptosis.[22] In SS salivary glands, HEV exhibit upregulation of adhesive molecules, including integrins, V-CAMs, selectins, and cadherins,[23] that promote the entry of autoimmune T-cells into the gland and help prevent their apoptosis. T-cells eluted from the SS gland recognize SS-A (Ro) and SS-B (LA) ribonuclear protein antigens;[24] of interest, these nuclear antigens can have aberrant expression on the cell membrane of the salivary gland epithelial cells in SS biopsies.[25] After stimulation by a particular antigen, additional T-cells may be recruited through a process termed antigenic spreading. For example, immunization of mice with a particular peptide of SS-B led to immune responses against different peptides of SS-B and even SS-A peptides.[26]

Recent studies have shown a distinct and different repertoire of T-cell receptor beta gene usage in different patients.[27] However, in the same SS patient, similar TCAR clonotypes were noted in the lacrimal and salivary glands.[28] In comparison, B-cells in the SS salivary gland biopsies of unrelated SS patients have been shown to use a particular sub-subclass of immunoglobulin kappa chain.[29,30]

After entry of the potentially autoimmune lymphocytes into the SS gland, the T-cells must undergo clonal expansion. This process is facilitated by the induction of class II MHC (HLA-DR/DQ) on glandular epithelial cells,[31,32] probably by local production of interferon gamma by infiltrating CD4 T-cells. The steps involved in the assembly of HLA-DR chains and potential autoantigens into a complex are shown schematically in

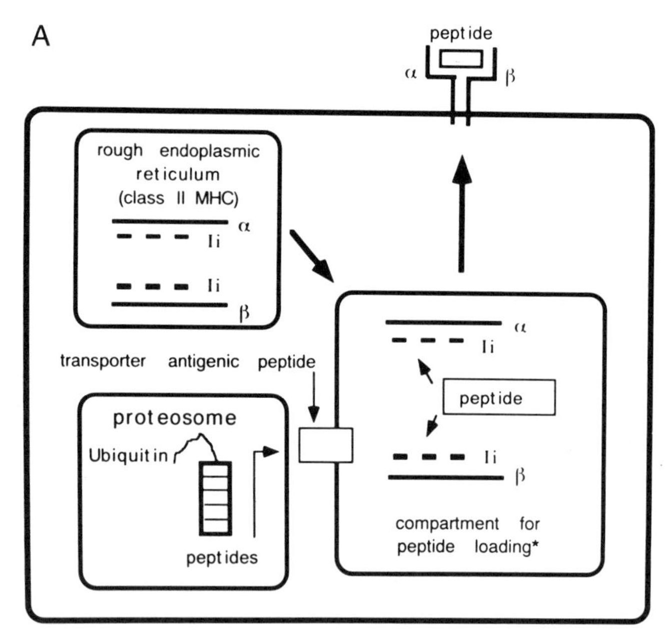

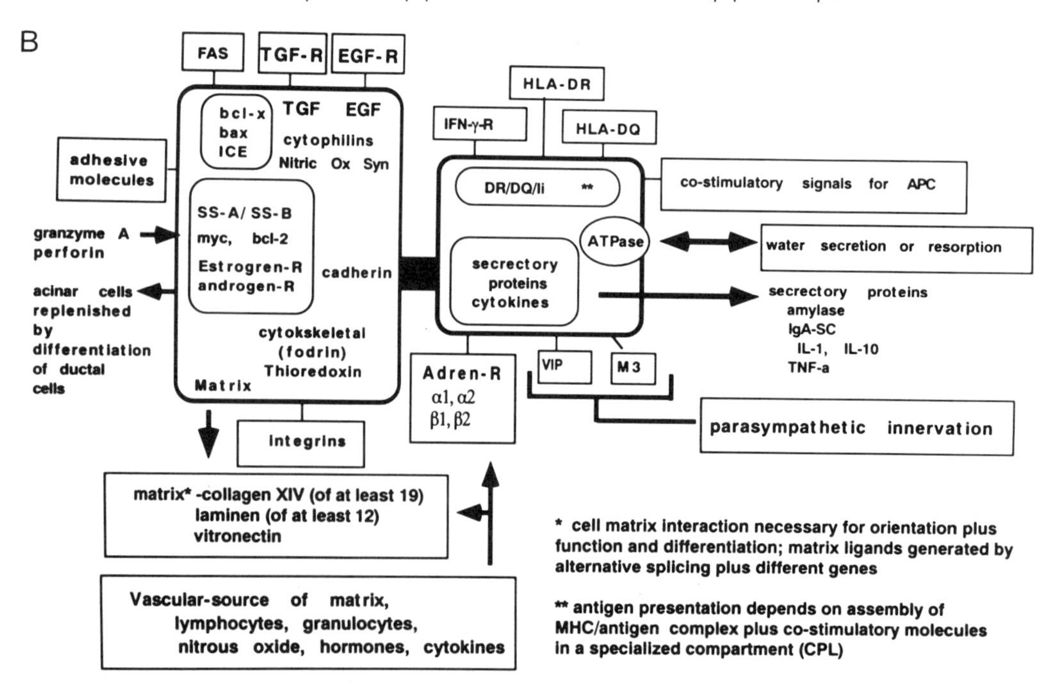

Figure 2. Schematic diagram of antigen presentation in the salivary gland of SS patients. (A) Assembly of class II major histocompatibility complex (MHC) molecules. (B) Numerous factors contribute to antigen presentation, including co-stimulatory molecules, cell matrix, neural stimuli, and cytokines.

Fig. 2. Immediately after synthesis of HLA-DR alpha and beta chains, each chain is bound to a chaperone Ii chain to prevent its binding to self peptides.[33,34] The HLA-DR alpha chain-Ii and HLA-DR beta chain-Ii complexes are eventually transported to a specialized acidic cytoplasmic vesicle termed the compartment for peptide loading, a cytoplasmic vesicle with some similarities to neural vesicles involved in synaptic function.[33,35] The autoimmune "self" peptides compete with Ii chain; if the self peptide has sufficient affinity to displace Ii, a new complex containing HLA-DR/peptide is formed and transported to the cell surface. The resultant HLA-DR positive glandular cells in the salivary or lacrimal glands can "present" autoantigens to the immune T-cells and perpetuate the immune response. A key question is: what are the characteristics of the self peptides that enable them to be "loaded" onto self MHC in the microenvironment of the autoimmune salivary gland? Available evidence suggests that cryptic self peptides (the same peptides bound weakly to MHC antigens in the thymus) can be loaded onto self MHC in the lower pH of the compartment for peptide loading.[34,35] These HLA-DR(+) glandular cells can present these "cryptic" antigens to self reactive T-cells (that escaped negative selection in the thymus).[21]

However, it is important to note that induction of class II MHC (i.e., HLA-DR) antigens on epithelial cells is not sufficient for antigen presentation and that the majority of epithelial cells in SS biopsies that are HLA-DR(+) do not express either DQ or Ii.[31,36] Even if the epithelial cell is able to assemble the MHC/peptide complex, other factors such as co-stimulatory molecules (including I-CAM)[37] are required for antigen presentation to T-cells by the epithelial cells.[38] Upregulation of I-CAM in SS salivary glands has been noted.[39] To summarize, CD4 T-cells with autoreactivity are able to escape "negative selection" in the thymus due to their weak binding of cryptic antigens to self MHC, and the same "autoimmune" T-cells are able to undergo clonal expansion in the peripheral tissues when their weak binding to MHC/peptide is augmented by other co-stimulatory factors in the peripheral tissues such as upregulated I-CAMS, stimulatory cytokines, and growth factors.[34,40]

2.3. Cytokine Production in SS Gland Biopsies

Cytokine production in SS salivary and lacrimal gland biopsies has been demonstrated by use of reverse transcription-PCR and in situ hybridization methods. These studies demonstrate that CD4 T-cells produced predominantly "Th-1"-like cytokines (interferon-gamma, TNF-alpha, IL-2), and that gland epithelial cells produced cytokines, including IL-1, TNF-alpha, and IL-10.[41–45] These cytokines were also detected in SS saliva in increased levels.[41] Several points were of particular interest: (i) IL-4 was not transcribed in samples taken directly from the gland, but was easily detectable after T-cells eluted from the SS salivary gland were cultured in vitro. This indicates that Th-2 cells were present in the SS gland, but were relatively "suppressed" or needed to be induced from uncommitted precursors.[41] (ii) Glandular acinar and ductal cells transcribed pro-inflammatory cytokines such as IL-1 and TNF-alpha, indicating their active role in the perpetuation of the local immune response. In the absence of any T-cells, the salivary gland epithelial cells of NOD/SCID mice express pro-inflammatory cytokines and novel protein antigens.[46] These authors propose that the salivary gland dysplasia may precede lymphocytic infiltrate and play a key role in subsequent stimulation and clonal expansion of T-cells in the SS gland that eventually lead to autoimmune disease. In summary, recent studies in man and in animal models indicate that the glandular epithelial cells are not simple "victims" of autoimmune attack but are active co-participants in pathogenesis.

2.4. Termination of the Immune Response

The persistence of autoimmune T-cell infiltrates in the glands of SS patients suggests the failure of mechanisms to adequately delete these self reactive T-cells. Studies in man and in animal models regarding fas-mediated apoptosis have yielded interesting but confusing results.[47] In animal models such as the MRL/lpr mouse, defects in fas have led to increased longevity of T-cells that could contribute to autoimmunity. However, circulating T-cells in SS patients[48] and in systemic lupus erythematosus (SLE) patients[49] increased rates of fas-mediated apoptotic death in vitro; this finding is in contrast to decreased fas-mediated apoptotic death of T-cells in MRL/lpr mouse.

In comparison to SS peripheral blood,[50] T-cells eluted from salivary gland biopsies of SS patients have a resistance to fas-mediated apoptosis in vitro (Fig. 3). This resistance must be "acquired" in the microenvironment of the gland, since peripheral blood lymphocytes from the same SS patient show normal or accelerated fas-mediated apoptosis after in vitro stimulation with anti-fas antibody (Fig. 3A). This resistance to fas-mediated apoptosis may be due to upregulation of mRNA for bcl-2 and bcl-x (Fig. 3B). Bcl-x has a long and a short form due to splicing variation; the observed bcl-x in Fig. 3B corresponds to the

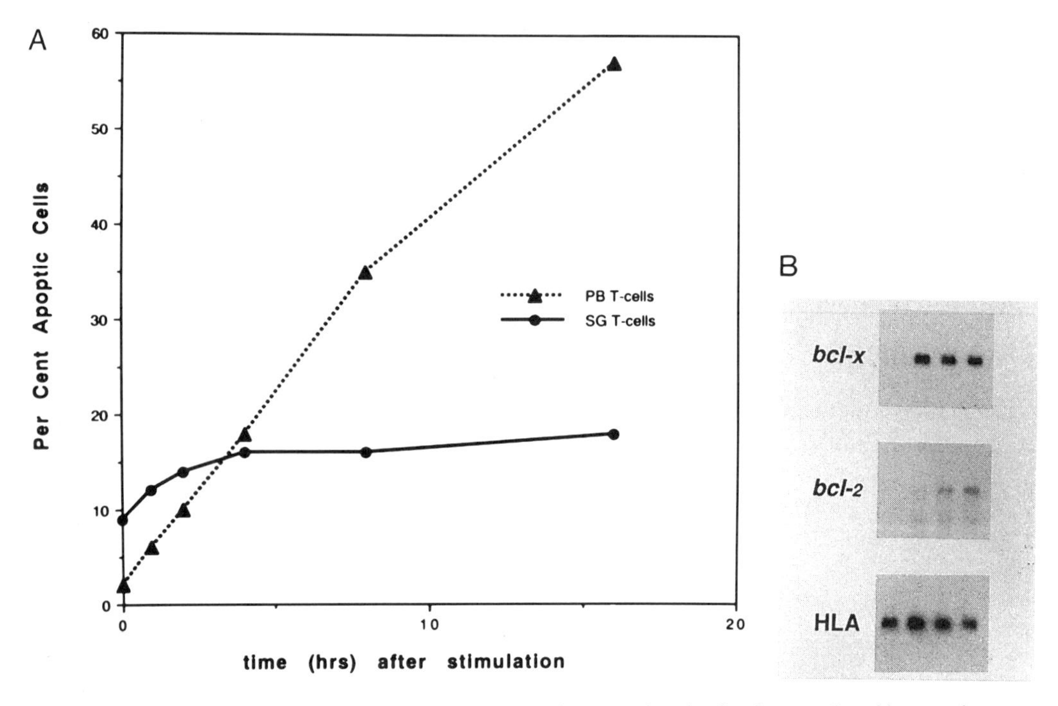

Figure 3. Salivary gland lymphocytes of SS patients have resistance to fas-stimulated apoptosis and increased expression of bcl-2 and bcl-x. (A) Response of peripheral blood and salivary gland T-cells to fas-mediated stimulation. The peripheral blood (PB) and salivary gland (SG) lymphocytes from the same patient were stimulated in vitro with rabbit anti-fas antibody. The percent cells undergoing apoptosis was determined by flow cytometry. (B) Peripheral blood lymphocyte (PBL) and salivary gland lymphocyte (SGL) production of mRNA for bcl-x (top), bcl-2 (middle) and HLA-DR beta chain (bottom). mRNA was extracted from PBL and SGL of SS patients. cDNA was created by reverse transcription, amplified polymerase chain reaction, separated on acrylamide gels and hybridized with specific DNA probes. The left lane in each panel is cDNA derived PBL from SS patients. The other 3 lanes show cDNA derived from SGL eluted from different biopsies of SS patients.

"long" form, which confers resistance to apoptosis. In this regard, resistance to fas-mediated apoptosis in the microenvironment of the salivary gland may be similar to resistance of T-cells that escape "negative selection" in the thymic microenvironment due to upregulation of bcl-2.[20]

An interesting recent observation was that fas ligand helps maintain a site of immune privilege in the eye.[51] Another factor that increases resistance to fas-mediated apoptosis is c-myc, which also is known to be upregulated in SS gland biopsies.[52] If the salivary gland undergoes premature dysplasia in SS as suggested in animal models,[46] then defects in the normal production of fas ligand or other inhibitors of apoptosis in the SS salivary glands may promote a microenvironment conducive to autoimmune disease.

2.5. Cytokine Production May Influence Neural Innervation of the Glands

The morphologic observation of residual glandular acinar tissue in virtually all SS patients leads to the conclusion that the glandular tissue is either not receiving or not responding to "signals" that normally stimulate glandular secretion. Neural signals received from muscarinic receptors and peptidergic receptors are necessary for homeostasis and regeneration of salivary gland epithelial cells[53,54] and influence these responses to other hormonal stimuli.[55] Since the SS patient has symptoms of dryness, the afferent signals are probably reaching the central nervous system and the gland is not responding to efferent signals. In support of this hypothesis, the ability to recover some salivary function in SS patients has been demonstrated by the use of muscarinic M3 agonists such as pilocarpine.[56,57]

Immunohistologic evidence has indicated that efferent neural innervation of the SS gland may be decreased as an early step in pathogenesis.[12] Application of pro-inflammatory cytokines such as IL-1 or TNF-alpha (in amounts similar to the levels found in SS glands or saliva) to nerve cells grown in vitro leads to decrease of neural release of acetylcholine.[58] Of interest, these effects on nerve preparations do not appear to be simple cytokine receptor mediated effects but require a non-neural cell that apparently delivers a slower "toxicity" signal to the nerve acetylcholine.[58] Similar effects of cytokines on neural function have been noted in other experimental models of autoimmunity.[59–63] Of particular interest, pro-inflammatory cytokines expressed in transgenic mice under the control of neural specific promoters may mimic autoimmune disease.[64,65] These results indicate that a common mechanism (i.e., cytokine release by lymphocytes or epithelial cells) may link early neural inhibition, early glandular dysfunction, and clonal expansion of autoimmune T-lymphocytes.

2.6. Role of Extracellular Matrix and Growth Factors

Recent studies have demonstrated the importance of extracellular matrix in normal homeostasis, regeneration, and function of epithelial cells in salivary and lacrimal glands. For example, the differentiation of salivary gland duct cells into acinar cells (including the morphologic appearance of acinar structures and induction of acinar proteins such as amylase) depends on their interaction of cell membrane integrins with their ligands in the cell matrix.[66] In particular, signals from the laminen, collagen IV, and vitronectin contribute to the proliferative and differential of salivary gland ductal epithelial cells.[67,68] Further, response to hormonal signals, cytokines, and growth factors depend on cell-matrix interactions. For example, stimulation of glandular ductal cells by interferon gamma in the presence of matrix leads to differentiation and induction of HLA-DR; however, in the absence of matrix, similar stimulation leads to apoptosis.[69] In lacrimal gland cells grown in vitro, cell-matrix interactions are necessary for secretory responses to muscarinic M3 agonists and are further augmented by novel proteins such as BM180.[70] The necessity for matrix-cell interactions by salivary and lacrimal gland epithelial cells is similar to the requirement for cell-matrix interaction demonstrated in the mammary gland of the lactating rat, where cellular integrins modulate response to hormones, growth factors, and neural signals. If the cell-matrix signals are not "intact," the mammary epithelial cell is triggered into apoptosis pathways.[71–73]

The ocular surface as well as the oral mucosa depend on a continuous supply of growth factors for their homeostasis and regeneration.[74–76] Important members of the epidermal growth factor family (EGF, TGF-α) are produced in the salivary glands. Although these growth factors may be produced and act in an autocrine fashion at mucosal surfaces by epithelial cells, wound healing in the oral cavity is significantly diminished if the parotid gland (a major source of EGF and TGF) is removed and indicates an important role of EGF in saliva in response to "stress" wound healing.[77]

3. FUTURE DIRECTIONS FOR THERAPY

Although SS is often termed a deficiency of aqueous tears, it has become evident that the clinical symptoms depend on the activity of the entire functional unit that includes the glands and the mucosal surfaces which are linked by neural and vascular channels. In a simplest model of SS, several sites for therapy are suggested in Fig. 4. At the mucosa, the surface epithelial cells are maintained by a combination of growth factors produced by the cells themselves (i.e., autocrine), neural efferent fibers (muscarinic and peptidergic stimuli required for homeostasis), and saliva. In addition to "water" content, saliva provides proteins, mucins, and hormones that constitute a hydrated gel that allows both maintenance of lubricating function (i.e., reduction of stress forces associated with mastication) and epithelial cell maintenance. The stimuli from the inflamed mucosal surface is transmitted by afferent nerves and perhaps by other mucosal factors to the central nervous system (CNS). Efferent neural fibers and hormones from the CNS return to the periphery and eventually affect both glandular and vascular function. Changes in vascular endothelial cells and perivascular cells (producing nitric oxide) directly affect the homing of lymphocytes, apoptosis of lymphocytes, and migration of granulocytes into the tissues. Changes in the glandular epithelium create a microenvironment that favors antigen presentation and depresses lymphocytic apoptosis. Thus, a cycle of continuing glandular dysfunction, lymphocytic infiltrates, and mucosal damage eventually reaches a stage where a clinical diagnosis of disease is evident.

Recent advances in the fields of vascular biology (inhibitors and agonists of adhesive molecules and nitric oxide synthetases), endocrinology (development of agonists and inhibitors for production/response to hormones), immunology (immune modifier drugs developed for RA and SLE), and neurobiology (particularly in the field of neural regrowth) may lead to a generation of new therapeutic agents (or better utilization of older drugs) that may be used in combination for future SS therapy.

ACKNOWLEDGMENTS

Supported in part by Grant MO1-RR000833 from the National Institutes of Health and by the Florence, Ramsdell, Kovler, Hennings, and Scripps Charitable Foundations.

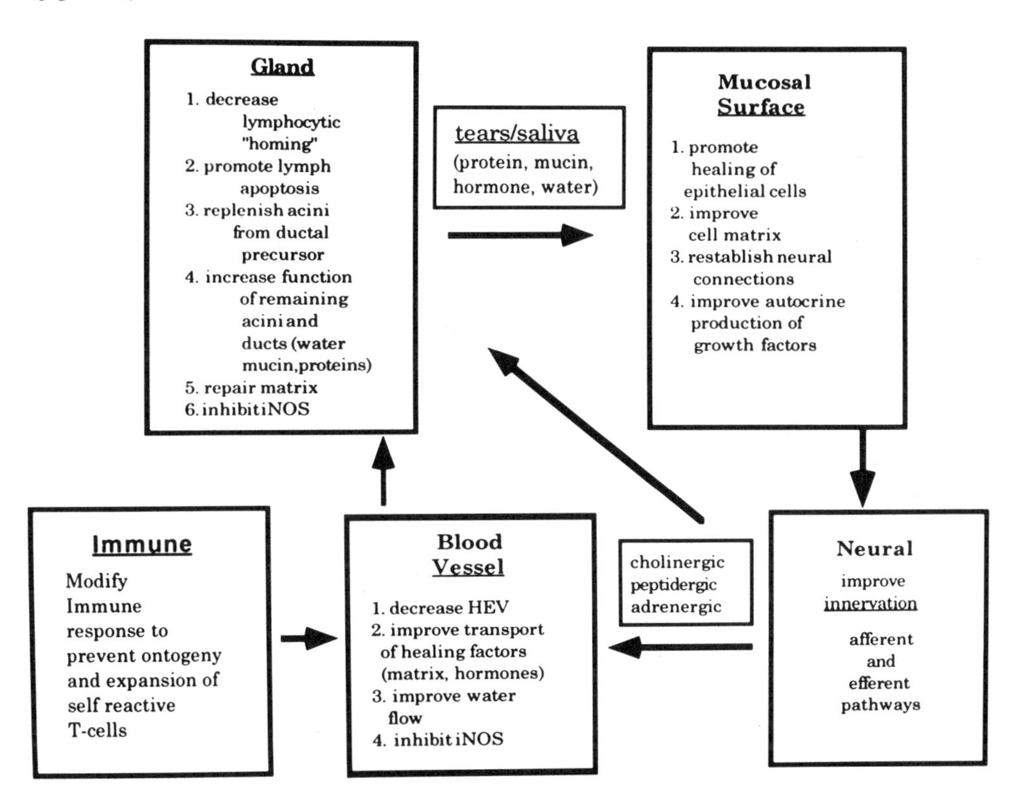

Figure 4. Future directions for therapy of SS. The oral or ocular surface must be considered as part of a functional unit that includes neural innervation, salivary and lacrimal glands, vascular supply, and immune innervation. Immune interactions in the salivary or lacrimal gland can influence neural innervation of the glands. Secretions of the glands contain not only water, but also proteins and mucins. Blood vessels supply not only water and lymphocytes, but also growth factors and adhesive molecules. Each of these sites can be augmented by selective therapy.

REFERENCES

1. Daniels TE, Whitcher JP. Association of patterns of labial salivary gland inflammation with keratoconjunctivitis sicca: Analysis of 618 patients with suspected Sjögren's syndrome. *Arthritis Rheum.* 1994;37:869–877.
2. Fox RI, Robinson CA, Curd JC, Kozin F, Howell FV. Sjögren's syndrome: Proposed criteria for classification. *Arthritis Rheum.* 1986;29:577–585.
3. Vitali C, Moutsopoulos HM, Bombardieri S. The European Community Study Group on diagnostic criteria for Sjögren's syndrome: Sensitivity and specificity of tests for ocular and oral involvement in Sjögren's syndrome. *Ann Rheum Dis.* 1994;53:637–647.
4. Fox RI. Epidemiology, pathogenesis, animal models, and treatment of Sjögren's syndrome. *Curr Opin Rheumatol.* 1994;6:501–508.
5. Haddad J, Deny P, Munz-Gotheil C, et al. Lymphocytic sialadenitis of Sjögren's syndrome associated with chronic hepatitis C virus liver disease. *Lancet.* 1992; 8:321–323.
6. Jorgensen C, Legouffe MC, Perney P, et al. Sicca syndrome associated with hepatitis C virus infection. *Arthritis Rheum.* 1996;39:1166–1171.
7. Mariette X, Zerbib M, Jaccard A, et al. Hepatitis C virus and Sjögren's syndrome. *Arthritis Rheum.* 1993;36:280–281.
8. Fox R, Saito I. Criteria for Diagnosis of Sjögren's Syndrome. *Rheum Dis Clin North Am: Current Controversies in Rheumatology,* in press.

9. Fox RI, Howell FV, Bone RC, Michelson P. Primary Sjögren's syndrome: Clinical and immunopathologic features. *Semin Arthritis Rheum.* 1984;14:77–105.
10. Fox RI, Kang HI, Pathogenesis of Sjögren's syndrome. In: Fox RI, ed. *Clinics of North America*, vol. 18. Philadelphia: W.B. Saunders; 1992:517–538.
11. Andoh Y, Shimura S, Sawai T, Sasaki H, Takishima T, Shirato K. Morphometric analysis of airways in Sjögren's syndrome. *Am Rev Respir Dis.* 1993;148:1358–1362.
12. Konttinen YT, Sorsa T, Hukkanen M, et al. Topology of innervation of labial salivary glands by protein gene product 9.5 and synaptophysin immunoreactive nerves in patients with Sjögren's syndrome. *J Rheumatol.* 1992;19:30–37.
13. Walcott B, Brink P. Age-related decrease in innervation density of the lacrimal gland in mouse models of Sjögren's syndrome. This volume.
14. Adamson TC III, Fox, RI, Frisman DM, Howell FV. Immunohistologic analysis of lymphoid infiltrates in primary Sjögren's syndrome using monoclonal antibodies. *J Immunol.* 1983;130:203–208.
15. Alpert S, Kang HI, Weissman I, Fox RI. Expression of granzyme A in salivary gland biopsies from patients with primary Sjögren's syndrome. *Arthritis Rheum.* 1994;37:1046–1054.
16. Freemont AJ, Jones CJP. Endothelial specialization of salivary gland vessels for accelerated lymphocyte transfer in Sjögren's syndrome. *J Rheumatol.* 1983;10:801–804.
17. Theofilopoulos AN. The basis of autoimmunity: Part I. Mechanisms of aberrant self-recognition. *Immunol Today.* 1995;16:90–98.
18. Theofilopoulos AN. The basis of autoimmunity: Part II. Genetic predisposition. *Immunol Today.* 1995;16:150–159.
19. Bissonnette R, Echeverri F, Mahboub A, Green D. Apoptotic cell death induced by c-myc is inhibited by bcl-2. *Nature.* 1992;359:552–554.
20. Gatenby PA, Irvine M. The bcl-2 proto-oncogene is overexpressed in systemic lupus erythematosus. *J Autoimmun.* 1994;7:623–631.
21. Theofilopoulos A. Regulation of the immune response in autoimmune disease. *Immunol Today*, 1995;16:90–98.
22. Goodnow CC, Adelstein S, Basten A. Peripheral tolerance and homing receptors on endothelial cells. *Science.* 1990;248:1373–1379.
23. Aziz KE, McCluskey PJ, Wakefield D. Expression of selectins (CD62 E,L,P) and cellular adhesion molecules in primary Sjögren's syndrome: Questions to immunoregulation. *Clin Immunol Immunopathol.* 1996;80:55–66.
24. Namekawa T, Kuroda K, Kato T, et al. Identification of SSA reactive T cells in labial salivary glands from patients with Sjögren's syndrome. *J Rheumatol.* 1995;22:2092–2099.
25. de Wilde P, Kater L, Bodeutsch D, van den Hoogen F, van de Putter L, van Venroooij W. Aberrant expression pattern of the SS-B antigen in the labial salivary glands of patients with Sjögren's syndrome. *Arthritis Rheum.* 1996;39:783–791.
26. Toffer F, Gordon T, McCluskey T. Intra and intermolecular spreading of autoimmunity involving the nuclear self antigens LA (SS-B) and Ro (SS-A). *Proc Natl Acad Sci USA.* 1995;92:875–879.
27. Mizushima N, Kohsaka H, Tsubota K, Saito I, Miyasaka N. Diverse T cell receptor beta gene usage by infiltrating T cells in the lacrimal glands of Sjögren's syndrome. *Clin Exp Immunol.* 1995;101:33–38.
28. Matsumoto I, Tsubota K, Nishioka K, Saitoh Y, Sumida T. Common T cell receptor clonotype in lacrimal glands and labial salivary glands from patients with Sjögren's syndrome. *J Clin Invest.* 1996;97:1969–1977.
29. Freimark B, Fantozzi R, Bone R, Bordin G, Fox R. Detection of clonally expanded salivary gland B-lymphocytes but not T-cells in Sjögren's syndrome. *Arthritis Rheum.* 1989;32:859–869.
30. Kipps TJ, Tomhave E, Chen PP, Fox RI. Molecular characterization of a major autoantibody-associated cross-reactive idiotype in Sjögren's syndrome. *J Immunol.* 1989;142:4261–4268.
31. Fox RI, Bumol T, Fantozzi R, Bone R, Schreiber R. Expression of histocompatibility antigen HLA-DR by salivary gland epithelial cells in Sjögren's syndrome. *Arthritis Rheum.* 1986;29:1105–1111.
32. Lindahl G, Hedfors E, Kloreskog L, Forsum U. Epithelial HLA-DR expression and T-cell subsets in salivary glands in Sjögren's syndrome. *Clin Exp Immunol.* 1985;61:475–482.
33. Teyton L. Assembly and transport of major histocompatibility complex molecules. *Nouv Rev Fr Hematol.* 1994;36 Suppl 1:S33–36.
34. Fox R. Mechanisms of assembly of Class II MHC molecules and self peptides in autoimmune disease: Possible model for mechanism of action of antimalarial drugs. *Lupus*, 1996;5:31–39.

35. Sette A, Southwood S, O'Sullivan D, Gaeta F, Sidney J, Grey H. Effect of pH on the MHC Class II peptide interactions. *Proc Natl Acad Sci USA*. 1992;148:844–850.
36. El Kaissouni J, Bene M, Faure G. Investigation of activation markers demonstrates overexpression of secretory component, bcl-2, and ICAM-1 in salivary glands of Sjögren's patients. *Clin Immunol Immunopathol.* 1996;79:236–243.
37. Altmann D, Hogg J, Trowsdale J, Wilkinson D. Cotransfection of ICAM-1 and HLA-DR reconstitutes human antigen-presenting cell function in mouse L cells. *Nature.* 1989;338:512–514.
38. Lotteau V, Teyton L, Peleraux A, et al. Intracellular transport of class II MHC molecules directed by invariant chain. *Nature.* 1990;348:600–606.
39. St. Clair EW, Angellilo JC, Singer KH. Expression of cell-adhesion molecules in the salivary gland microenvironment of Sjögren's syndrome. *Arthritis Rheum.* 1992;35:62–66.
40. Fox R. Sjögren's syndrome: Diagnostic and pathogenetic features. *Curr Opin Rheumatol.*,1996;8:435–438.
41. Fox RI, Kang HI, Ando D, Abrams J, Pisa E. Cytokine mRNA expression in salivary gland biopsies of Sjögren's syndrome. *J Immunol.* 1994;152:5532–5539.
42. Oxholm P, Daniels TE, Bendtzen K. Cytokine expression in labial salivary glands from patients with primary Sjögren's syndrome. *Autoimmunity.* 1992;12:185–191.
43. Jones DT, Monroy D, Ji Z, Atherton SS, Pflugfelder SC. Sjögren's syndrome: Cytokine and Epstein-Barr viral gene expression within the conjunctival epithelium. *Invest Ophthalmol Vis Sci.* 1994;35:3493–3504.
44. Ohyama Y. Cytokine mRNA in salivary glands of patients with Sjögren's syndrome. *Arthritis Rheum.*,1996;39:1376–1384.
45. Boumba D, Skopouli F, Moutsopoulos H. Cytokine mRNA expression in the labial salivary gland tissues from patients with primary Sjögren's syndrome. *Br J Rheumatol.* 1995;34:326–333.
46. Robinson CP, Yamamoto H, Peck AB, Humphreys-Beher MG. Genetically programmed development of salivary gland abnormalities in the NOD (nonobese diabetic)-scid mouse in the absence of detectable lymphocytic infiltration: A potential trigger for sialoadenitis of NOD mice. *Clin Immunol Immunopathol.* 1996;79:50–59.
47. Tan E. Autoimmunity and apoptosis. *J Exp Med.* 1994;179:1083–1086.
48. Ichikawa Y, Yoshida A, Horiki Y, et al. Abnormal expression of apoptosis related antigens, Fas and bcl-2, on circulating T-lymphocyte subsets in primary Sjögren's syndrome. *Clin Exp Rheumatol.* 1995;13:307–313.
49. Emlen W, Niebur J, Kadera R. Accelerated apoptosis of lymphocytes from patients with SLE. *J Immunol.* 1994;152:3685–3692.
50. Ogawa N, Dang H, Kong L, Anaya J, Liu G, Talal N. Lymphocyte apoptosis and apoptosis-associated gene expression in Sjögren's syndrome. *Arthritis Rheum.* 1996;39:1875–1886.
51. Griffith T, Brunner T, Fletcher S, Green D, Ferguson T. Fas ligand-induced apoptosis as a mechanism of immune privilege. *Science.* 1995;270:1189–1192.
52. Skopouli FN, Kousvelari EE, Mertz P, et al. c-myc mRNA expression in minor salivary glands of patients with Sjögren's syndrome. *J Rheumatol.* 1992;19:693–699.
53. Schneyer CA, Humphreys-Beher M, Hall HD. Nerve growth factor effects on inhibition by submandibular gland ablation or autonomic denervation of activity-induced parotid growth. *J Auton Nerv Syst.* 1992;37:137–144.
54. Schneyer CA, Humphreys-Beher MG, Hall HD, Jirakulsomchok D. Mitogenic activity of rat salivary glands after electrical stimulation of parasympathetic nerves. *Am J Physiol.* 1993;264:G935-G938.
55. Schneyer CA, Humphreys-Beher MG, G, Jirakulsomchok D, Hall HD. Effects of autonomic antagonists and growth factors on activity-mediated hyperplasia of rat parotid. *Am J Physiol.* 1993;264:G763-G766.
56. Wiseman LR, Faulds D. Oral pilocarpine: A review of its pharmacological properties and clinical potential in xerostomia. *Drugs.* 1995;49:143–155.
57. Rhodus NL, Schuh MJ. Effects of pilocarpine on salivary flow in patients with Sjögren's syndrome [see comments]. *Oral Surg Oral Med Oral Pathol.* 1991;72:545–549.
58. Soliven B, Wang N. Tumor necrosis factor-alpha regulates nicotinic responses in mixed cultures of sympathetic neurons and nonneuronal cells. *J Neurochem.* 1995;64:883–894.
59. Campbell IL, Hobbs MV, Dockter J, Oldstone MB, Allison J. Islet inflammation and hyperplasia induced by the pancreatic islet-specific overexpression of interleukin-6 in transgenic mice. *Am J Pathol.* 1994;145:157–166.
60. Watkins LR, Maier SF, Goehler LE. Cytokine-to-brain communication: A review and analysis of alternative mechanisms. *Life Sci.* 1995;57:1011–1026.
61. Madden KS, Sanders VM, Felten DL. Catecholamine influences and sympathetic neural modulation of immune responsiveness. *Annu Rev Pharmacol Toxicol.* 1995;35:417–448.

62. Gundersen K, Maehlen J. Nerve-evoked electrical activity regulates molecules and cells with immunological function in rat muscle tissue. *Eur J Neurosci.* 1994;6:1113–1118.
63. Main C, Blennerhassett P, Collins SM. Human recombinant interleukin 1 beta suppresses acetylcholine release from rat myenteric plexus. *Gastroenterology.* 1993;104:1648–1654.
64. Campbell IL. Neuropathogenic actions of cytokines assessed in transgenic mice. *Int J Dev Neurosci.* 1995;13:275–284.
65. Campbell IL, Chiang CS. Cytokine involvement in central nervous system disease: Implications from transgenic mice. *Ann N Y Acad Sci.* 1995;771:301–312.
66. Royce L, Kibbey M, Mertz P, Kleinman H, Baum B. Human neoplastic submandibular intercalated duct cells express an acinar phenotype when cultured on basement membrane matrix. *Differentiation.* 1993;52:247–255.
67. Bernfield M, Banerjee S, Cohn R. Dependence of salivary epithelial morphology and branching morphogenesis upon proteoglycans at the epithelial surface. *J Cell Biol.* 1972;52:674–679.
68. Hoffman M, Kibbey M, Letterio J, Kleinman H. Role of laminen-1 and TGF-beta 3 in acinar differentiation of a human submandibular gland cell line (HSG). *J Cell Sci..* 1996;109:2013–2021.
69. Wu A, Kurrasch R, Katz J, Fox P, Baum B, Atkinson J. Effect of tumor necrosis factor alpha and interferon gamma on the growth of a human salivary gland cell line. *J Cell Physiol.* 1994;161:217–226.
70. Laurie G, Glass D, Ogle R, Stone C, Sluss J, Chen L. BM180: A novel basement membrane protein with a role in stimulus secretion coupling by lacrimal acinar cells. *Am J Physiol.* 1996;39:C1743-C1750.
71. Boudreau N, Sympson CJ, Werb Z, Bissell MJ. Suppression of ICE and apoptosis in mammary epithelial cells by extracellular matrix. *Science.* 1995;267:891–893.
72. Huhtala P, Humphries MJ, McCarthy JB, Tremble PM, Werb Z, Damsky CH. Cooperative signaling by alpha 5 beta 1 and alpha 4 beta 1 integrins regulates metalloproteinase gene expression in fibroblasts adhering to fibronectin. *J Cell Biol.* 1995;129:867–879.
73. Damsky C, Werb Z. Signal transduction by integrin receptors for extracellular matrix; cooperative processing of extracelllular information. *Curr Opin Cell Biol.* 1992;4:772–781.
74. Watanabe K, Nakagawa S, Nishida T. Stimulatory effects of fibronectin and EGF on migration of corneal epithelial cells. *Invest Ophthalmol Vis Sci.* 1987;28:205–211.
75. Maldonado BA, Furcht LT. Epidermal growth factor stimulates integrin-mediated cell migration of cultured human corneal epithelial cells on fibronectin and arginine-glycine-aspartic acid peptide. *Invest Ophthalmol Vis Sci.* 1995;36:2120–2126.
76. Schultz G, Khaw PT, Oxford K, MacCauley S, Van Setten G, Chegini N. Growth factors and ocular wound healing. *Eye.* 1994;8:184–187.
77. Dolce C, Anguita J, Brinkley L, et al. Effects of sialoadenectomy and exogenous EGF on molar drift and orthodontic tooth movement in rats. *Am J Physiol.* 1994;266:E731-E738.

REFLEX TEARING IN DRY EYE NOT ASSOCIATED WITH SJÖGREN'S SYNDROME

Kazuo Tsubota

Department of Ophthalmology and Oral Health Science Center
Tokyo Dental College
Chiba, Japan
and
Department of Ophthalmology
Keio University School of Medicine
Tokyo, Japan

1. INTRODUCTION

Dry eye with tear deficiency is now classified into two categories: associated with Sjögren's syndrome (SS), and not associated with Sjögren's syndrome (non-SS).[1] SS is an autoimmune disorder characterized by lymphocytic infiltration and destruction of exocrine glands.[2] SS dry eye is more severe than non-SS dry eye, with the former showing more vital staining, squamous metaplasia, and lymphocytic infiltration of the conjunctival epithelium (Fig. 1).[3–5]

2. NON-SS DRY EYE

We have previously suggested that reflex tearing can distinguish between different types of dry eye.[6,7] In SS dry eye, the lacrimal gland is destroyed by lymphocytic infiltration, so that it cannot produce any tears,[8] even with nasal stimulation. Non-SS dry eye does not involve such lacrimal gland destruction (Fig. 2), and the gland can produce tears in response to strong stimuli.[8] Since tears include important substances such as EGF, TGF-β, and vitamin A necessary for the healthy proliferation and differentiation of the ocular surface,[9–12] the absence of tears can cause severe cellular damage.[6] The periodic reflex tearing in non-SS dry eye may provide these substances to the ocular surface, enabling normal epithelial proliferation and differentiation. In contrast, the generally worse course of SS dry eye may be caused not by simple desiccation, but by the lack of essential chemi-

Lacrimal Gland, Tear Film, and Dry Eye Syndromes 2
edited by Sullivan *et al.*, Plenum Press, New York, 1998

cals. We have shown that drops made from autologous serum replenish these chemicals and improve SS dry eye.[6,13,14]

Non-SS dry eye has traditionally encompassed everything from severe disease associated with graft vs. host disease (GVHD) to the mild form caused by infrequent blinking with no reflex tearing.[15] Since dry eye in GVHD lacrimal gland destruction is clinically similar to SS dry eye, we group together the two conditions. The other forms of dry eye should be considered as having decreased basic but normal reflex tearing.

2.1. Pathogenesis

Since non-SS dry eye involves at least some supply of essential tear components through reflex tearing, the symptoms are caused by simple desiccation and friction with blinking.[15,16]

The reason for the lacrimal gland malfunction is unknown. Lacrimal gland biopsy shows normal configuration and no lymphocytic infiltration.[17] Since strong stimuli can cause tearing, an abnormality in the neurosensory pathway is suspected. Factors that can shorten tear break-up time (BUT) and exacerbate symptoms include meibomian gland disease (MGD), which decreases the oil supply,[18,19] allergic conjunctivitis, and the decreased blink rate and increased palpebral fissure (both of which enhance tear evaporation) associated with the use of video display terminals.[15]

Non-SS dry eye does sometimes involve a local deficiency of tear components, such as in superior limbic keratitis (SLK). The pathogenesis of SLK is unknown, although it can be considered a local form of SS dry eye accompanied by vital staining and squamous metaplasia.[16]

2.2. Diagnosis

Non-SS and SS dry eye both involve the same symptoms. In Japan, the most frequent complaint is ocular fatigue;[20] other symptoms include burning, irritation, and heavy sensation. There are neither qualitative nor quantitative differences between the presentations of SS and non-SS dry eye.

The only practical means of identifying non-SS dry eye is measurement of reflex tearing;[7,8,20] asking patients whether they can cry is inadequate. We use the Schirmer test with nasal stimulation (Fig. 3) to provide objective, quantitative information. Briefly, the patient is checked with the simple Schirmer test for 5 min without topical anesthesia. Then a small cotton swab is inserted into the nasal cavity toward the entrance of the ethmoid sinus. The Schirmer paper is then placed in the conjunctival sac for 5 min, as with the ordinary Schirmer test, with the swab kept in place.[7,8,20] A wet portion of the paper less than 10 mm is considered low.

It is important to distinguish between non-SS and SS dry eye because the former is associated with less severe ocular surface abnormalities.[4] Patients with non-SS dry eye may be acceptable candidates for refractive surgery since they are capable of producing some tears,[21] which would provide lactoferrin or IgA to prevent infection and growth factors and vitamins to facilitate epithelial wound healing.

2.3. Reflex Tearing

All cases of dry eye can be divided into two categories according to the presence or absence of reflex tearing. Most dry eye patients have the non-SS form (Table 1). Reflex

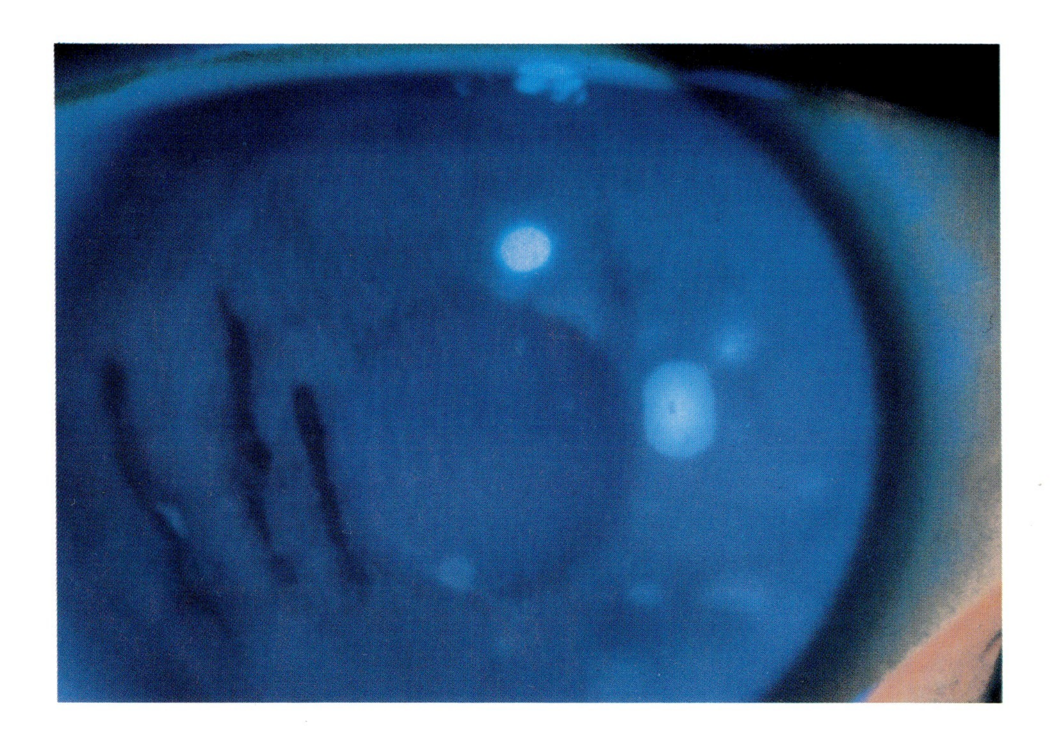

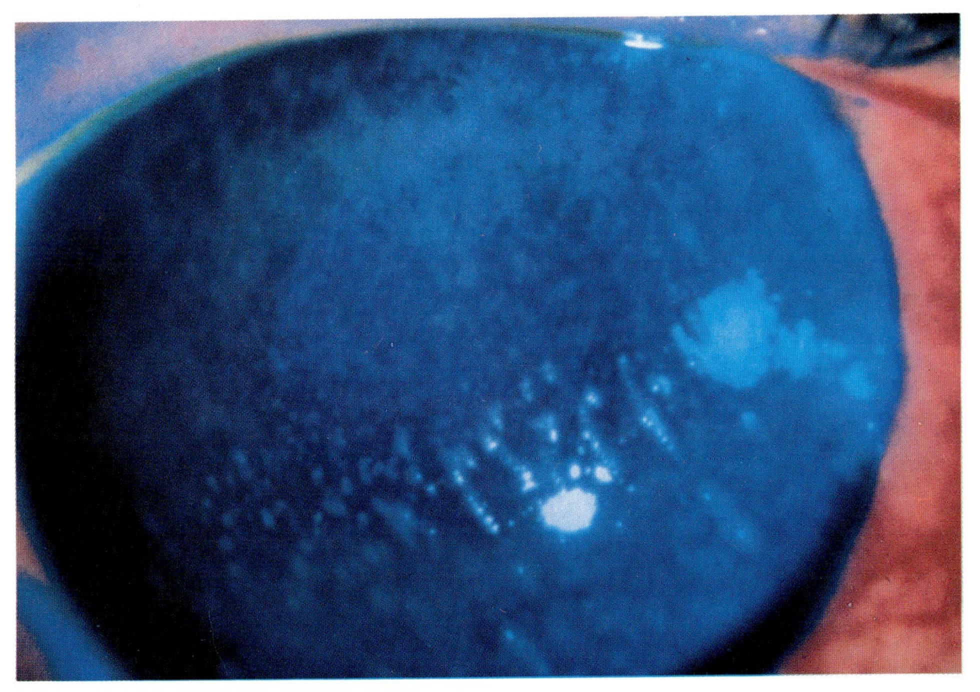

Figure 1. Ocular surface abnormalities of (a) non-SS dry eye and (b) SS dry eye. Note the severe vital staining in SS. (Reprinted with permission from *J Rheumatol.*)

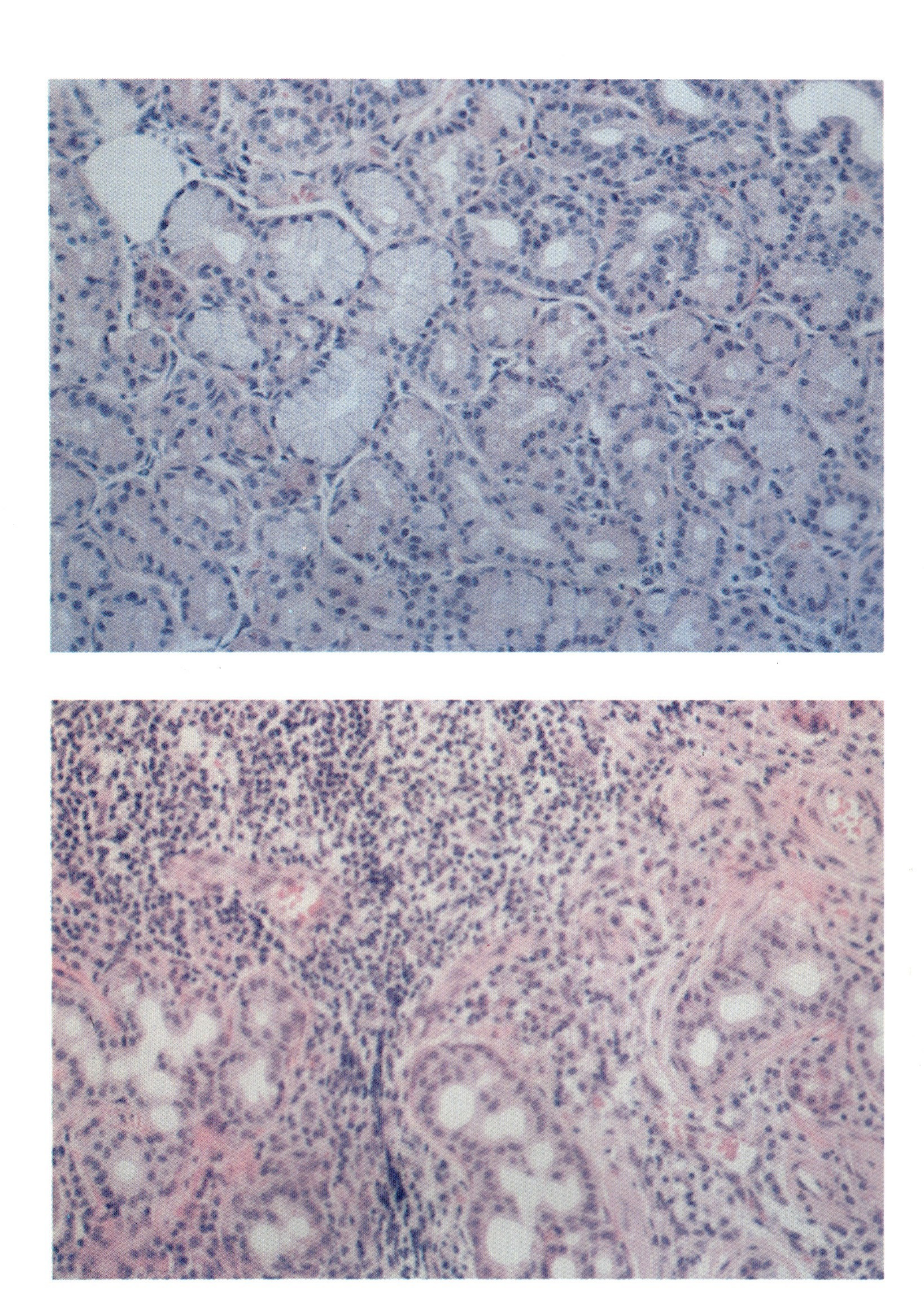

Figure 2. Lacrimal gland biopsy of (a) non-SS dry eye and (b) SS dry eye. Note the lymphocytic infiltration in SS. (Reprinted with permission from *J Rheumatol.*)

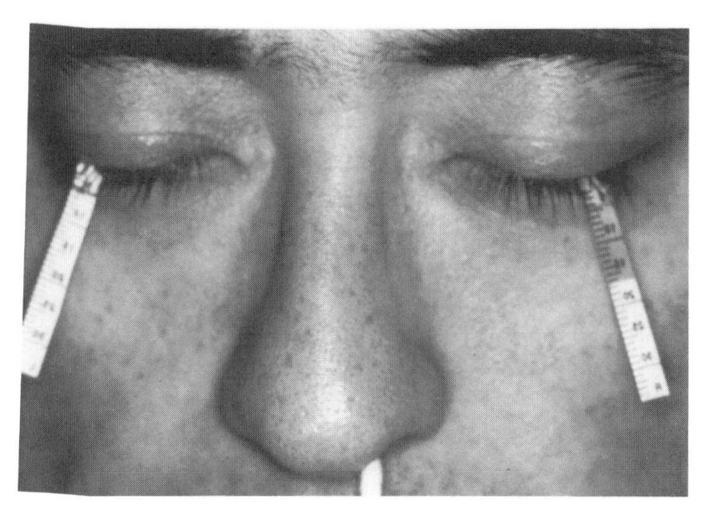

Figure 3. Schirmer test with nasal stimulation. (Reprinted with permission from *J Rheumatol*)

tearing may be a more appropriate criterion for labeling cases of dry eye, since there are actually some patients with SS who do have some reflex tearing, and hence would be said to have non-SS dry eye. This variant of dry eye involves less vital staining and squamous metaplasia than does the usual SS type. Most cases of severe GVHD, end-stage ocular cicatricial pemphigoid, and Stevens-Johnson syndrome are considered SS dry eye, whereas milder cases of these conditions can be categorized as non-SS. Most dry eye associated with contact lens use is 7th nerve palsy, or acquired primary lacrimal gland disorder, non-SS.

2.4. Treatment

Since the squamous metaplasia of the ocular surface epithelium is less severe in non-SS than in SS dry eye, use of autologous tears to supply missing components is not necessary; these substances can be provided by reflex tears. The major treatment goal in this group is prevention of desiccation, which can be achieved with the frequent use of ar-

Table 1. Classification of tear-deficient dry eye by absence or presence of reflex tearing

Disease	Reflex tearing	Type
SS	–	SS
OCP	–	SS
Stevens-Johnson	–	SS
GVHD	–	SS
PLD	+	non-SS
CL-related	+	non-SS
7th nerve palsy	+	non-SS

SS, Sjögren's syndrome. OCP, ocular cicatricial pemphigoid. GVHD, graft vs. host disease. PLD, primary lacrimal gland dysfunction. CL, contact lens.

tificial tears, wearing protective eyeglasses with side panels,[22,23] and treatment of concomitant diseases such as MGD or allergic conjunctivitis.[19,24] Punctal occlusion may be necessary in rare cases, such as if deficiency of tear components is suspected, as in SLK. Then only the upper punctum is closed, because the reflex tears need a place to drain.[16]

3. CONCLUSIONS AND FUTURE WORK

I have proposed that non-SS dry eye be thought of as that form of dry eye in which reflex tearing is maintained (Table 1). This is important because the pathogenesis of ocular surface alterations differs between the two forms of the disease. A supply of tear components via reflex tearing in non-SS dry eye provides a sufficiently stable ocular surface that patients may be able to use contact lenses and undergo refractive surgery. The therapy in that case is targeted simply at relieving discomfort.

It is still unclear why tear production decreases in some patients without apparent destruction of the lacrimal gland. Abnormal neuropeptide production or reception is suspected but has not been studied. Another possibility is a qualitative difference between basic and reflex tears, with the latter perhaps not providing all the necessary substances. Future work on non-SS dry eye should thus concentrate on both quantitative and qualitative tear measurement, as well as on means of stimulating and restoring lacrimal gland function.

ACKNOWLEDGMENT

This study was supported by the Oral Health Science Center of Tokyo Dental College, Chiba, Japan.

REFERENCES

1. Lemp MA. Report of the National Eye Institute/Industry workshop on clinical trials in dry eyes. *CLAO J.* 1995;21:221–232.
2. Fox R, Robinson C, Curd J, Kozin F, Howell F. Sjögren's syndrome: Proposed criteria for classification. *Arthritis Rheum.* 1986;29:577–583.
3. Pflugfelder S, Huang A, Feuer W, Chuchovski P, Pereira I, Tseng S. Conjunctival cytologic features of primary Sjögren's syndrome. *Ophthalmology.* 1990;97:985–993.
4. Tsubota K, Toda I, Yagi Y, Ogawa Y, Ono M, Yoshino K. Three different types of dry eye syndrome. *Cornea.* 1994;13:202–209.
5. Tsubota K. SS dry eye and non-SS dry eye: What are the differences? In: Homma M, Sugai S, Tojo T, Miyasaka N, Akizuki M, eds. *Sjögren's Syndrome.* Amsterdam: Kugler Publications; 1994:27–31.
6. Tsubota K. Dry eye is not caused by simple desiccation, it is rather due to the lack of supply of tear components to ocular surface epithelia. Kyoto Cornea Club, in press.
7. Tsubota K, Xu K, Fujihara T, Katagiri S, Takeuchi T. Decreased reflex tearing is associated with lymphocytes infiltration in lacrimal and salivary glands. *J Rheumatol.* 1996;23:313–320.
8. Tsubota K. The importance of the Schirmer test with nasal stimulation [Letter]. *Am J Ophthalmol.* 1991;111:106–108.
9. Gupta A, Monroy D, Ji Z, Yoshino K, Huang A, Pflugfelder S. Transforming growth factor beta-1 and beta-2 in human tear fluid. *Curr Eye Res.* 1996;15:605–614.
10. Ohashi Y, Motokura M, Kinoshita Y, et al. Presence of epidermal growth factor in human tears. *Invest Ophthalmol Vis Sci.* 1989;30:1879–1887.

11. Ubels J, Loley K, Rismondo V. Retinol secretion by the lacrimal gland. *Invest Ophthalmol Vis Sci.* 1986;27:1261–1269.
12. Wilson S. Lacrimal gland epidermal growth factor production and the ocular surface. *Am J Ophthalmol.* 1991;111:763–765.
13. Tsubota K, Satake Y, Ohyama M, et al. Surgical reconstruction of the ocular surface in advanced ocular cicatricial pemphigoid and Stevens-Johnson syndrome. *Am J Ophthalmol.* 1996;122:38–52.
14. Tsubota K, Satake Y, Shimazaki J. Treatment of severe dry eye. *Lancet.* 1996;348:123.
15. Tsubota K, Nakamori K. Effects of ocular surface area and blink rate on tear dynamics. *Arch Ophthalmol.* 1995;113:155–158.
16. Yang H, Fujishima H, Toda I, Shimazaki J, Tsubota K. Punctal occlusion for the treatment of superior limbic keratoconjunctivitis. *Am J Ophthalmol.*, in press.
17. Xu K, Katagari S, Takeuchi T, Tsubota K. Biopsy of labial salivary glands and lacrimal glands in the diagnosis of Sjogren's syndrome. *J Rheumatol.* 1996;23:76–82.
18. Mathers W. Ocular evaporation in meibomian gland dysfunction and dry eye. *Ophthalmology.* 1993;100:347–351.
19. Shimazaki J, Sakata M, Tsubota K. Ocular surface changes in the patients with meibomian gland dysfunction. *Arch Ophthalmol.* 1995;113:1266–1270.
20. Toda I, Fujishima H, Tsubota K. Ocular fatigue is the major symptom of dry eye. *Acta Ophthalmol.* 1993;71:18–23.
21. Toda I, Yagi Y, Hata S, Itoh S, Tsubota K. Excimer laser photorefractive keratectomy for patients with contact lens intolerance caused by dry eye. *Br J Ophthalmol.* 1996;80:604–609.
22. Tsubota K. The effect of wearing spectacles on the humidity of the eye. *Am J Ophthalmol.* 1989;108:92–93.
23. Tsubota K, Yamada M, Urayama K. Spectacle side panels and moist inserts for the treatment of dry-eye patients. *Cornea.* 1994;13:197–201.
24. Toda I, Shimazaki J, Tsubota K. Dry eye with only decreased tear break-up time is sometimes associated with allergic conjunctivitis. *Ophthalmology.* 1995;102:302–309.

129

CYTOKINE mRNA EXPRESSION IN LABIAL SALIVARY GLANDS AND CYTOKINE SECRETION IN PAROTID SALIVA IN SJÖGREN'S SYNDROME

Philip C. Fox, Margaret M. Grisius, Debra K. Bermudez, and Di Sun

Clinical Investigations and Patient Care Branch
National Institute of Dental Research
National Institutes of Health
Bethesda, Maryland

1. INTRODUCTION

Sjögren's syndrome is an autoimmune exocrinopathy involving primarily the salivary and lacrimal glands. The secondary form of this disorder consists of exocrine involvement (the sicca syndrome) in an individual with another connective tissue disease. The most common associated condition is rheumatoid arthritis. Primary Sjögren's syndrome is characterized by exocrine involvement alone. Although there is more restricted tissue involvement in primary Sjögren's syndrome, this is a systemic condition, as evidenced by prominent serological evidence of autoimmune reactivity, including high titer autoantibodies (anti-nuclear antibodies [ANA], rheumatoid factors [RF], SS-A/Ro, SS-A/La) and hypergammaglobulinemia in the majority of patients. There is also an increased incidence of lymphoma.

Most Sjögren's patients present with complaints of oral and ocular dryness. These symptoms are the result of salivary and lacrimal gland secretory hypofunction. Patients also experience a variety of oral and ocular complications related to decreased gland function. In the oral cavity there is extensive dental decay, increased oral infections, mucosal pain, atrophy and ulceration, and difficulties speaking, chewing, and swallowing.

Characteristic histological changes in the salivary glands include extensive mononuclear cell infiltrates in a peri-ductal, peri-vascular distribution. The infiltrates are predominantly activated T-cells of the CD4+ phenotype. There is also marked acinar cell loss with relative preservation of the ductal structures. At late stages of the disorder, fibrosis and fatty degeneration of the glandular parenchyma are seen.

Lacrimal Gland, Tear Film, and Dry Eye Syndromes 2
edited by Sullivan *et al.*, Plenum Press, New York, 1998

Most research studies concerning mechanisms of pathology in salivary glands in Sjögren's syndrome have focused on the infiltrating cells and their products. Immunohistochemical techniques have been used to demonstrate expression of a number of cytokines by these cells of both the Th1 and Th2 subtypes.[1] These lymphocytes are immunologically very active and have been thought to be the central force in initiating and perpetuating the gland inflammatory reaction. It is only relatively recently that attention has been directed to the possible role of the salivary epithelial cells (acini and ducts) in the autoimmune-mediated destruction of the gland.

There have been reports, using immunohistochemical methods, that duct cells express cytokines. Oxholm et al.[2] and Cauli et al.[3] showed IFN-γ, TNF-α, IL-1α, and IL-6 expression in labial minor gland duct cells in Sjögren's syndrome patient biopsies. Much less activity was noted in acini compared to the ducts and infiltrating T-cells. Fox et al.,[4] using RT-PCR techniques, reported that epithelial cells of enlarged major salivary glands of Sjögren's syndrome patients expressed TNF-α, IL-1α, IL-6, and IL-10. These reports suggest that the epithelial cells in the salivary glands in Sjögren's syndrome are not passive 'victims' of products produced by the infiltrating T-cells, as had been thought previously, but are active participants in the immune process.

Still unknown are the specific cytokines that are produced by each of the cell types found in the salivary glands in Sjögren's syndrome and which are the critical molecules for the perpetuation of the inflammatory process. Problems hampering the search for answers to these questions are insufficient sensitivity of many of the available techniques to detect cytokine expression from individual cells of cell structures (such as immunohistochemistry and in situ hybridization) and an inability to identify products of specific cells with techniques that are extremely sensitive (such as RT-PCR). We have attempted to overcome these problems by using a technique of direct tissue microdissection coupled with RT-PCR amplification of specific cytokine mRNA. This technique allows detection of small amounts of mRNA present in individual gland elements. We have examined acini, ducts, and mononuclear cell infiltrates in labial minor salivary glands of patients with primary Sjögren's syndrome and healthy controls. This allows a description of the cytokines expressed by the different glandular elements in Sjögren's syndrome and by normal salivary glands and will aid in deciphering the immunopathologic mechanisms leading to gland dysfunction.

2. METHODS

2.1. Tissues

Labial minor salivary gland tissues were obtained from patients with primary Sjögren's syndrome and healthy volunteers. Diagnosis of primary Sjögren's syndrome was confirmed in all patients by the findings of symptoms of oral and ocular dryness, keratoconjunctivitis sicca demonstrated by rose bengal staining, a labial minor gland biopsy with a focus score[5] of >1, positive autoimmune serologies (ANA or RF >1:80, or positive anti-SS-A/Ro or anti-SS-B/La), and the absence of another connective tissue disease. Healthy volunteers had no complaints of oral or ocular dryness, no autoimmune serologies, and normal salivary function. All human tissue samples were acquired and utilized in accordance with approvals from the NIDR Human Subject Review Committee. Immediately after removal, specimens were placed in OCT compound (Miles, Elkhart, IN), snap frozen in methyl butane on dry ice, held overnight at -70°C, and then stored in liquid nitrogen until use. After partial rewarming, tissue samples were cut at a 14 μm thickness using aseptic technique, disposable blades, and changing blades between samples. Each section was placed on a non-coated glass slide.

2.2. Microdissection and RNA Extraction

The microdissection technique was as previously described,[6,7] with modifications. After warming tissue sections to room temperature for 1 min, individual acini, ducts, and lymphoid cells were microdissected with a 30 gauge needle while viewing the tissue through a dissecting microscope using a hematoxylin-eosin stained adjacent section as a guide. The dissected cells were immediately placed into an mRNA denaturing solution (Stratagene, La Jolla, CA). This was repeated for five to ten serial tissue sections of each specimen. Five units of human placental RNase inhibitor (Amersham, Arlington Heights, IL) and 5 units of RQ1 RNase-free DNase (Promega, Madison, WI) were added to each sample. Total RNA was extracted from the microdissected tissue samples by the acid-phenol guanidinium isothiocyanate chloroform method[8] using an mRNA isolation kit (Stratagene).

2.3. Primers and Probes

Sense and antisense primers and oligonucleotide probes were obtained from Clonetech Laboratory (Palo Alto, CA). The sequences are shown in Table 1. Probes were labeled with [γ-^{32}P]ATP (Amersham) using a 5' DNA end labeling kit (Stratagene).

2.4. RT-PCR Reactions

2.4.1. RT. Total cellular RNA was converted to cDNA using a first-strand cDNA synthesis kit (Pharmacia Biotech, Piscataway, NJ). The extracted RNA was transferred

Table 1. Human sequences used for RT-PCR and Southern blotting

Gene	RT-PCR product predicted size (bp)	Primer pair upstream/downstream and oligonucleotide probe
IL-2	305	5'-CATTGCACTAAGTCTTGCACTTGTCA-3'
		5'-CGTTGATATTGCTGATTAAGTCCCTG-3'
		Probe:5'TTCTTCTAGACACTGAAGATGTTTCAG-3TTC-3'
IL-4	344	5'-CGGCAACTTTGACCACGGACACAAGTGCGATA-3'
		5'-ACGTACTCTGGTTGGCTTCCTTCACAGGACAG-3'
		Probe: 5'-GTCCTTCTCATGGTGGCTGTAGAACTGCCG-3'
IL-6	628	5'-ATGAACTCCTTCTCCACAAGCGC-3'
		5'-GAAGAGCCCTCAGGCTGGACTG-3'
		Probe:5'-CTCCTCATTGAATCCAGATTGGAAGCATCC-3'
IL-10	328	5'-AAGCTGAGAACCAAGACCCAGACATCAAGGCG-3'
		5'-AGCTATCCCAGAGCCCCAGATCCGATTTTGG-3'
		Probe: 5'-GGCTTTGTAGATGCCTTTCTCTTGGAGCTT-3'
TNF-α	444	5'-GAGTGACAAGCCTGTAGCCCATGTTGTAGCA-3'
		5'-GCAATGATCCCAAAGTAGACCTGCCCAGACT-3'
		Probe: 5'-ATCTCTCAGCTCCACGCCATTGGCCAGGAG-3'
IFN-γ	427	5'-GCATCGTTTTGGGTTCTCTTGGCTGTTACTGC-3'
		5'-CTCCTTTTTCGCTTCCCTGTTTTAGCTGCTGG-3'
		Probe: 5'-TCTGGTCATCTTTAAAGTTTTTAAAAAGTT3'
TGF-β 1	161	5'-GCCCTGGACACCAACTATTGCT-3'
		5'-AGGCTCCAAATGTAGGGGCAGG-3'
		Probe:5'GCGGAAGTCAATGTACAGCTGCCGCACGCA-3'

into a 200 µl RT-PCR reaction tube, and DEPC-treated H_2O was added to bring the RNA solution to 8 µl. This was denatured at 65°C for 9 min, then cooled to 4°C using a DNA Thermal Cycler (Perkin Elmer Cetus, Norwalk, CT; model 9600). RNA then was reverse transcribed using 5 µl of the First-Strand cDNA reaction mixture (consisting of cloned, FPLC-pure Moloney murine leukemia virus reverse transcriptase, RNAguard, RNase/DNase-free BSA, dATP, dCTP, dGTP, and dTTP in aqueous buffer), 1.0 µl of random hexadeoxynucleotides [pd(N)6 primers] at 0.2 µg/µl in aqueous solution, and 1.0 µl of 100 mM DTT. The solution was gently vortexed, spun momentarily at 14,000x*g* to remove the condensation, and incubated at 37°C for 1 h. RT products (cDNA) were stored at -20°C until further use.

2.4.2. PCR. Each 50 Êl PCR reaction mixture was prepared containing 1X PCR buffer (10 mM Tris-HCl, pH 8.3, 1.5 mM $MgCl_2$, 0.001% (w/v) gelatin), 10 pmol of each of the selected primers, 200 µM dNTP, 2 units of thermostable AmpliTaq Gold DNA polymerase (Perkin Elmer Cetus), and 2.5 µl of reverse transcribed cDNA, as described above. A PCR thermal profile of 95°C for 45 sec (denaturation), 65°C for 45 sec (annealing of primers), and 72°C for 2 min (DNA extension) was performed for 35 cycles using the GeneAmp PCR system (Perkin Elmer), with a final step of 72°C for 7 min. The amplified RT-PCR products were visualized and photographed on 1% agarose gels, containing 0.5 µg/ml ethidium bromide in 0.5X TBE buffer, under ultraviolet light. A l-Hind III digest (GIBCO, Grand Island, NY), ϕX174-HaeIII digest (Promega), or 100 bp ladder (GIBCO) was run in parallel as a size marker. The amplified RT-PCR products for human beta actin were examined for each preparation initially, to assure the quality of RNA in the dissection.

2.5. Southern Blotting of PCR Products

Southern blot analyses of the PCR products were performed to establish the specificity of the amplification and to enhance the sensitivity of PCR product detection. The gels containing PCR products were denatured and neutralized according to standard protocols.[9] Southern transfer of the products onto Nytran plus membranes (Schleicher & Schuell, Keene, NH) was made with the capillary blotting method. The membranes were then baked for 1 h at 80°C under vacuum and hybridized using ^{32}P-labeled gene-specific oligonucleotide probes. The resulting blots were subjected to autoradiography with Kodak BioMax films.

2.6. Saliva Collection

Parotid saliva stimulated with repeated applications of 2% citric acid applied to the lateral borders of the tongue was collected with Carlson-Crittenden collectors.[10] Samples were obtained on ice from Sjögren's syndrome patients, healthy volunteers, and patients with primary biliary cirrhosis who had no salivary gland involvement. Samples were stored at -70°C and slowly thawed at 4°C before assay.

2.7. Cytokine Concentration Determination

Cytokine concentrations in saliva and serum were determined using high-sensitivity ELISA kits from R&D (Minneapolis, MN).

Table 2. Cytokine mRNA expression in human labial minor salivary glands of patients with primary Sjögren's syndrome and healthy controls

	IL-2	IL-4	IL-6	IL-10	TNF-α	TGF-β	IFN-γ
Patients							
Acini	+	–	+	+	+	+	+
Ducts	–	–	+	+	+	+	–
Mononuclear foci	+	–	+	+	+	+	+
Controls*							
Acini	+	+	+	+	+	+	–
Ducts	–	–	+	+	+	+	–

*Healthy controls did not have mononuclear foci.
+, present in specimen; –, not detected in specimen.

3. RESULTS

3.1. Cytokine mRNA Expression by Microdissection and RT-PCR

Both the acinar and ductal cells of the labial minor salivary glands demonstrated cytokine mRNA expression. Duct expression was less prominent. mRNA for IL-2, IL-6, IL-10, TNF-α, TGF-β, and IFN-γ were identified in acini of patients with Sjögren's syndrome (Table 2); only TGF-β was found in each specimen examined. The duct cells had a similar complement of cytokine mRNA except that IL-2 and IFN-γ were not found. IL-4 mRNA was not found in any of the Sjögren's syndrome specimens. Infiltrating mononuclear cell foci were dissected and were found to express mRNA for all the above cytokines except IL-4. In the specimens from the healthy volunteers, IL-4 expression was seen in an acinar dissection. IFN-γ mRNA was not found in either ducts or acini. Other cytokines studied were present in the majority of samples from the normal tissue, although IL-2 was not detected in ducts.

3.2. Cytokine Secretion in Parotid Saliva and Serum

Salivary levels of IL-6 were found to be significantly elevated in the Sjögren's syndrome patients when compared to both the healthy controls and primary biliary cirrhosis (PBC) patients (Table 3). Serum IL-6 levels were elevated in both the Sjögren's syndrome and PBC patients compared to controls, although these values were less than those found in the saliva of the same patients (data not shown). Salivary TGF-β concentrations also

Table 3. Parotid salivary cytokine concentrations in primary Sjögren's syndrome and control groups

Group	IL-6 concentration (pg/mL)	TGF-β concentration (pg/mL)
Sjögren's syndrome patients	16.21±4.6 n=31	117.85±20.6 n=41
Healthy controls	0.77±0.2 n=14	6.13±4.3 n=16
Primary biliary cirrhosis patients	1.07±0.4 n=15	None detected n=10

Values are mean ± SEM.

were elevated significantly in the Sjögren's syndrome samples compared to the other two groups (Table 3).

4. DISCUSSION

Use of microdissection coupled with RT-PCR amplification allows one to obtain information on the specific cytokine mRNAs expressed by specific cells in a small tissue sample containing mixed cell types. It is ideal for use in a situation where minimal material is obtained for diagnostic examination, such as labial minor gland biopsies in Sjögren's syndrome. The amount of material obtained routinely precludes fractionation of infiltrating cells from salivary epithelial elements. Further, there are not fully adequate means of separating the acinar and ductal elements. Direct observation and dissection of the glandular elements obviate the need for complex fractionation. Use of RT-PCR techniques with highly specific primers and probes allows identification of mRNA from a small number of cells.[6,7] In the present study, cytokine mRNA could be identified from individual acini, ducts, and infiltrating foci in the labial minor salivary glands of Sjögren's syndrome patients which were obtained by routine biopsy for diagnosis. There was sufficient material for analysis of a number of different cytokines. With more extensive dissection of a specimen, a larger number of mRNAs could be amplified. We are presently examining a number of additional mediators.

Given the small number of samples in the present study, we cannot make any definitive statements about the role of a specific cytokine in the minor salivary glands in Sjögren's syndrome. An interesting finding was that normal minor glands from healthy volunteers expressed a number of cytokines. This emphasizes the immunomodulating roles that these molecules likely play in normal homeostasis and lends a note of caution to findings in tissues from patients. One must consider carefully that the presence of a particular substance in an involved tissue is not proof of a pathogenic role for that molecule. As an example, IL-6 and TGF-β expression were found with about the same frequency in minor glands from Sjögren's syndrome patients and healthy controls. The critical issue in cases such as this is the intensity of expression and the concentration of the cytokine at the tissue site.

The latter issue can be addressed by examining concentrations of these cytokines in the secreted product of these tissues. Volumes of minor salivary gland saliva are too small for analysis by ELISA, so parotid gland saliva was utilized. IL-6 could be found in both patients and healthy controls, confirming the RT-PCR results. However, the IL-6 concentration was much higher in the patient saliva than in the controls. This suggests that IL-6 is both present and elevated in Sjögren's syndrome and may be an important factor in the gland pathology due to its pro-inflammatory activity. A similar argument can be made for the results with TGF-β, although here we anticipate that the TGF-β found is likely a consequence of the chronic inflammation in the salivary glands. This may play a role in the fibrosis seen in later stages of involvement. We are presently looking at salivary concentrations of a number of the other cytokines examined by RT-PCR.

In summary, these results demonstrate extensive cytokine production within the salivary glands in Sjögren's syndrome by the infiltrating lymphocytes *and* the salivary epithelial cells. There is differential expression by cells within a gland. There also is substantial cytokine activity in normal salivary tissues. Further research will delineate the complete profile of cytokines present in salivary glands in Sjögren's syndrome and in healthy controls. This will help in formulation of hypotheses of the mechanisms of tissue destruction in this disorder and in designing new, rational therapies. An exciting possibility is that if specific molecules can be identified solely in patients, definitive diagnosis of this disorder would be possible. By using RT-PCR techniques, this could be accomplished on minute tissue samples such as those obtained by needle aspiration. Continuing studies of this type should advance our understanding of the salivary pathogenesis of this condition and lead to improved diagnosis and treatment.

REFERENCES

1. Fox PC, Speight PM. Current concepts of autoimmune exocrinopathy: Immunologic mechanisms in the salivary pathology of Sjögren's syndrome. *Crit Rev Oral Biol Med.* 1996;7:144–158.
2. Oxholm P, Daniels TE, Bendtzen K. Cytokine expression in labial salivary glands from patients with primary Sjögren's syndrome. *Autoimmunity.* 1992;12:185–191.
3. Cauli A, Yanni G, Pitzalis C, Challacombe S, Panayi GS. Cytokine and adhesion molecule expression in the minor salivary glands of patients with Sjögren's syndrome and chronic sialoadenitis. *Ann Rheum Dis.* 1995;54:209–215.
4. Fox RI, Kang H, Ando D, Abrams J, Pisa E. Cytokine mRNA expression in salivary gland biopsies of Sjögren's syndrome. *J Immunol.* 1994;152:5532–5539.
5. Greenspan JS, Daniels TE, Talal N, Sylvester RA. The histopathology of Sjögren's syndrome in labial salivary gland biopsies. *Oral Surg.* 1974;37:217–229.
6. Emmert-Buck MR, Roth MJ, Zhuang Z, et al. Increased gelatinase A (MMP-2) and cathepsin B activity in invasive tumor regions of human colon cancer samples. *Am J Pathol.* 1994;145:1285–1290.
7. Emmert-Buck M, Vocke CD, Pozzatti RO, et al. Allelic loss on chromosome 8p12–21 in microdissected prostatic intraepithelial neoplasia (PIN). *Cancer Res.* 1995;55:2959–2962.
8. Chomczynski P, Sacchi N. Single-step method of RNA isolation by acid guanidinium thiocyanate-phenol-chloroform extraction. *Anal Biochem.* 1987;162:156–169.
9. Sambrook J, Fritsch EF, Maniatis T. *Molecular Cloning: A Laboratory Manual.* 2nd ed. Cold Spring Harbor, NY: Cold Spring Harbor Laboratory, 1989, pp 6.3–6.60.
10. Fox PC, van der Ven PF, Sonies BC, Weiffenbach JM, Baum BJ. Xerostomia: Evaluation of a symptom with increasing significance. *J Am Dent Assoc.* 1985;110:519–525.

130

AGE-RELATED DECREASE IN INNERVATION DENSITY OF THE LACRIMAL GLAND IN MOUSE MODELS OF SJÖGREN'S SYNDROME

Benjamin Walcott, Nidia Claros, Alpa Patel, and Peter R. Brink

Departments of Neurobiology and Behavior, Physiology, and Biophysics
State University of New York
Stony Brook, New York

1. INTRODUCTION

In many dry eye conditions including Sjögren's syndrome, significant loss of secretory ability can be present even though areas of the lacrimal gland appear morphologically not affected by the disease process.[1] This suggests an interruption(s) in the stimulation/secretion process that has affected the remaining acini, reducing the amount of protein and water they can secrete. The stimulation/secretion process begins with neurons releasing neurotransmitters, at least two of which, acetylcholine and vasoactive intestinal polypeptide (VIP), bind to receptors on the acinar cell surface and activate G protein pathways, increasing cAMP and IP_3 levels.[2] The IP_3 will induce an increase in intracellular free Ca^{++}, which, together with the increased levels of cAMP, will stimulate the secretion of protein. These same second message pathways may also affect anion and cation membrane channels that can alter the movement of water as well as ions such as Ca^{++} across the secretory epithelium into the lumen of the gland.[3,4] We have examined this stimulus/secretion process in the extraorbital lacrimal glands from two mouse models for Sjögren's syndrome,[5] the MRL/MR-*lpr-lpr* [hereafter called MRL] and the (NZB/NZW) F_1 [hereafter called NZB] strains, at different ages, and thus different stages of the disease, and compared them to those of a normal strain of mouse, the Swiss Webster [hereafter called SW] strain, at comparable ages. We have determined the innervation pattern and density in these glands and measured their ability to secrete protein in response to direct exposure to carbachol, an acetylcholine agonist. Our results strongly suggest that, in the NZB strain at least, there are important changes in the stimulus/secretion process associated with the disease that occur *before* significant lymphocytic infiltration that could account for the reduced tear output seen in lacrimal glands from these animals.

Lacrimal Gland, Tear Film, and Dry Eye Syndromes 2
edited by Sullivan *et al.*, Plenum Press, New York, 1998

2. MATERIALS AND METHODS

The NZB and MRL animals were obtained from Jackson Laboratories, and the SWs were bought from Teconic Farms. All animals were female and were maintained in an accredited animal care facility. For immunocytochemistry, animals were killed with an overdose of halothane, and the glands were removed and placed either in freshly prepared 4% buffered paraformaldehyde or in Streck fixative. After 3–4 h of fixation, the tissue was treated at least overnight or for a few days with 30% buffered sucrose, then sectioned at 14 mm on a cryostat. The frozen sections were dried and then stored at -20°C until stained. One primary antiserum was to synaptophysin, a protein associated with synaptic vesicles, which was obtained from Zymed Laboratories; the other antiserum was to VIP, obtained from Incastar. Both primary antisera were used at a dilution as recommended and were applied to the sections overnight. FITC coupled secondary antisera were applied for 1 h, then the slides were washed and coverslipped. They were examined with epifluorescence optics, and the images were captured with a video camera, digitized, and analyzed on a Macintosh computer using NIH-image software. For the quantitative measurements, three different people each analyzed at least 15 images from each preparation independently, and the results were pooled.

For the secretion studies, each fresh gland was weighed and then minced into 5 ml of a standard saline with the addition of glucose.[6] The gland fragments were incubated at 37°C for 15 min with constant agitation by 95% O_2, 5% CO_2 gas mixture, then washed again with fresh saline. This process was repeated three times. Finally, the gland fragments were placed in a fresh aliquot of saline for 10 min, at the end of which time 1 ml of saline was removed, stored, and replaced with 1 ml of fresh media. After another 10 min incubation, 1 ml of media was removed, stored, and replaced with another 1 ml of media, which contained carbachol. After 10 min, 1 ml of media was removed and stored. The total protein concentration of each 1 ml aliquot of media was determined using Coomassie Blue (Pierce). The first two samples indicate the basal level of secretion, and the third aliquot which was in the presence of carbachol was used as a measure of the effectiveness of the given concentration of carbachol to stimulate protein secretion. Each gland was exposed to a single concentration of carbachol so that a number of glands were used to construct a dose response curve.

3. RESULTS

3.1. Innervation Pattern of Control SW Mouse Glands

Immunocytochemical examination of 1 month or 6 month old SW mouse extraorbital lacrimal gland innervation showed a very dense varicose (beaded) pattern of synaptophysin-like immunoreactivity (SLI) uniformly distributed throughout the gland (Fig. 1). Since the antiserum was directed against a synaptic vesicle associated protein, it is likely that these swellings are major neurotransmitter release sites, as are seen in other tissues innervated by the autonomic nervous system. Every acinar cell seemed to have one or several varicosities associated with it, which is a higher density of innervation than is seen in most mammalian lacrimal glands. VIP-like immunoreactivity (VIPLI) had a similar density and distribution to SLI, suggesting that the majority of the nerve fibers were parasympathetic with both acetylcholine and VIP as is found in many mammalian lacrimal glands.[8] A few fibers were seen associated with the ducts that were positive for an antis-

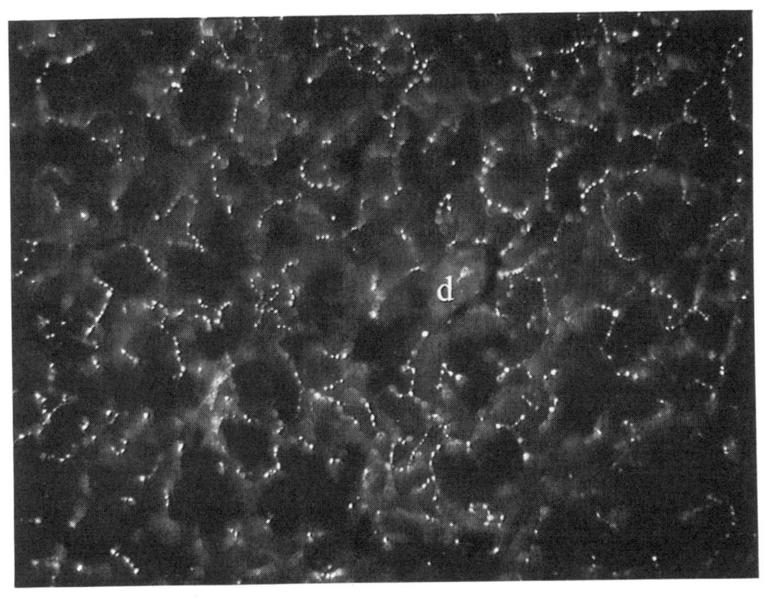

Figure 1. Synaptophysin-like immunoreactivity (SLI) in a normal mouse (SW) extraorbital lacrimal gland. Note the very varicose (beaded) and dense nature of the innervation. d indicates a duct.

erum to Substance P. Given the highly varicose nature of the innervation, we were able to use the computer to count the number of particles and thus determine the particle density per unit area of tissue stained with either synaptophysin or VIP antiserum. The results for both SLI and VIPLI showed no significant difference between SW animals at 1 and at 6 months of age (Fig. 2).

3.2. Innervation of NZB Glands

Young (1 month old) NZB lacrimal glands were indistinguishable from glands from SW control strain. The SLI and VIPLI were both dense and very varicose, and there was a uniform distribution throughout the gland. Beginning at about 3 months of age and continuing on until about 7–8 months, however, there was a decline in the density of the innervation. This decline was not uniform as it seemed to begin in the central region of the gland and proceed outwards. The gland, therefore, from a 4–5 month old animal, would have large regions in the center of the gland with few SLI or VIPLI varicosities or fibers, while the periphery would have a higher density, although less than that seen in young animals. The innervation of the blood vessels between the lobes of the gland appeared not to change significantly, suggesting that the reduction of SLI and VIPLI among the acini may be specific. These observations have been quantified, and the results are shown in Fig. 2. The large variability seen particularly in the 3–5 month period is due in part to the non-uniform distribution of SLI in these glands.

In these NZB animals, there is little to no acinar cell destruction or significant lymphocytic infiltration even at 6 or 7 months of age, so that the entire extraorbital lacrimal gland appears to consist of morphologically normal acinar cells. The fact that the SW strain shows no decrease in innervation over approximately the same time span indicates that the decrease in the NZB is a strain-specific change that could have functional significance.

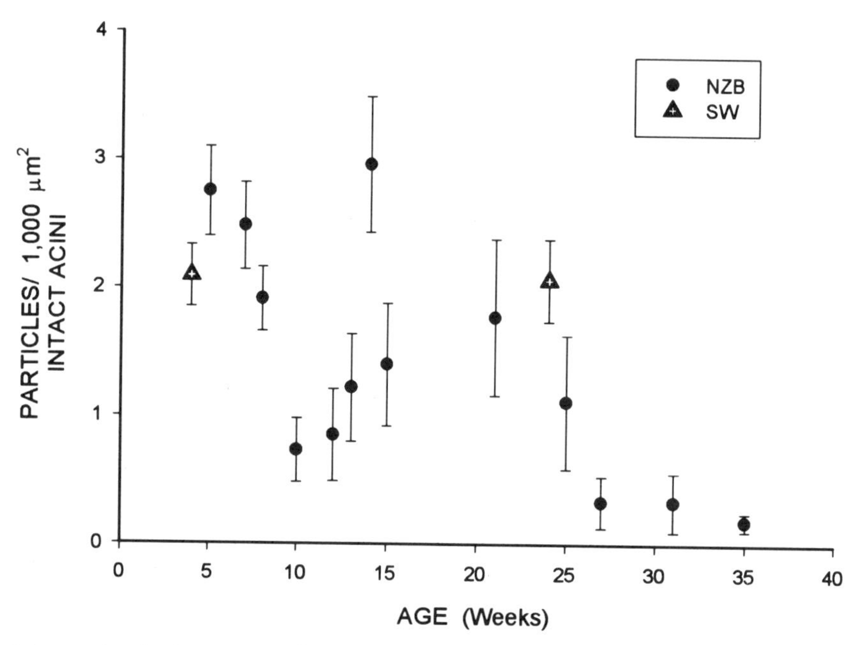

Figure 2. Innervation density as measured by counts of the number of varicosities (particles) that show SLI in the lacrimal glands of NZB mice at different ages. Each round point represents the mean of measurements from at least 30 images from a number of sections from the gland of a single animal; the error bars represent the 99% confidence level. The two triangular points represent similar measurements from SW animals.

To determine if the changes in SLI density in the lacrimal gland are part of a general process of neuropathy, we immunocytochemically examined the Harderian glands from the same animals. The Harderian glands of rodents secrete a very lipid-rich fluid that is deposited on the corneal surface. Little is known about its function, but given the high concentration of lipid in its secretory product, it could be fulfilling some of the role of the meibomian glands in humans. The gland has large secretory acini, many of which contain poryphyrin.[7] In young animals, the gland is highly innervated with many SLI fibers that are associated with the secretory acini. Examination of glands from animals 6–9 months old whose extraorbital lacrimal glands had significant loss of innervation showed little change in the pattern or density of innervation. These results suggest that the changes taking place in the lacrimal gland may be specific to that gland and not other secretory glands in the orbit.

3.3. Secretory Response of NZB Glands

Gland fragments from the SW animals, when stimulated by the direct application of carbachol, released an increased amount of protein into the media that was above a baseline, non-stimulated rate. A minimal response was obtained to carbachol at 0.001 mM and a maximal response at 100 mM. At higher carbachol doses, the amount of stimulated protein secretion was less than the maximal amount. From 0.1 to 10 mM, the response was linear. Stimulation of gland fragments from 1–2 month old NZB animals with carbachol at either 0.1, 1.0, or 10 mM also increased protein secretion above the basal level linearly and at a rate identical to that of the SW animals. The total amount of protein secreted was also similar. When the glands from 6 month old NZB animals were used, the amount of protein secreted in response to the lower doses of carbachol was identical to that of control and young animals. At the high carbachol concentration (10 mM), however, the protein secreted was less than that due to 1.0 mM, which may indicate a change in the sensitivity or number of acetylcholine receptors in the cell surface. However, it does appear that the receptor/second message systems for protein secretion are still functional in these animals at this stage of the disease.

3.4. Innervation Density of MRL Animals

MRL animals were obtained at 1 month old, and many already had significant lymphocytic infiltration of the lacrimal glands. However, when the tissue was examined for SLI or VIPLI, nerve fibers were seen associated with the intact acini. In the areas of infiltration, occasional non-varicose nerve fibers showing SLI were seen but they were rare. As these animals age, the number of intact acini decrease, although even at 6 months, when the animals usually die, there are still a number of areas of intact acini. Examination of the innervation of these remaining intact acini showed a general decrease in the SLI innervation density with age and with the general reduction of number of intact acini. Fig. 3 shows the innervation density for a number of animals at different stages of the disease. Currently it is difficult to quantify the exact extent of acinar cell loss, but estimations from histological preparations show that as acini are reduced in number, the density of innervation on the remaining acini clearly decreased.

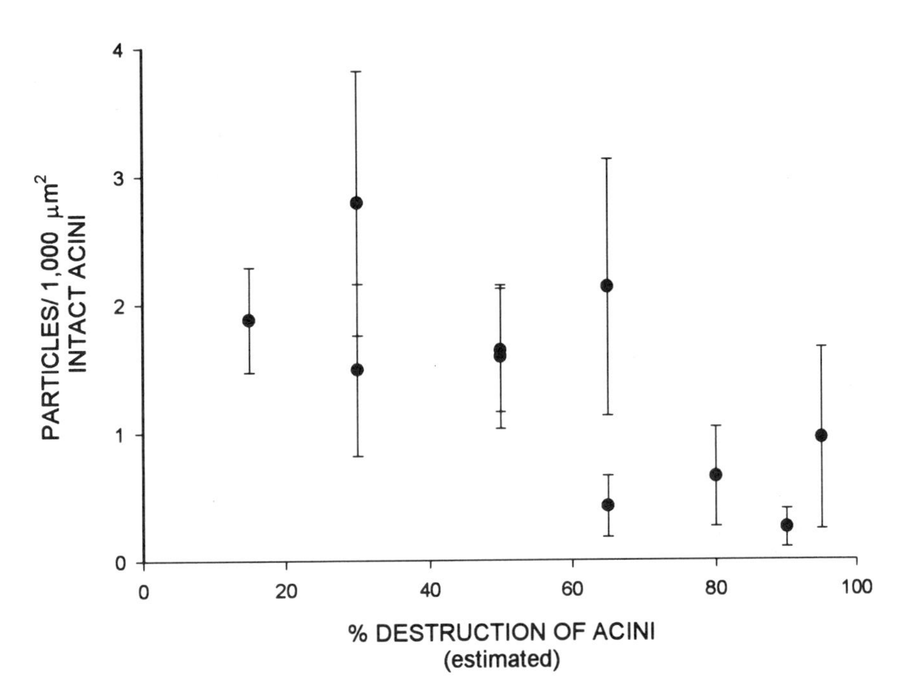

Figure 3. Density of SLI varicosities (particles) per unit area of intact acini in MRL female animals with increasing amounts of acinar tissue loss due to the disease process. Note that there is a reduction in density of innervation to the remaining intact acini as the % destruction increases.

3.5. Secretory Response of MRL Glands

Gland fragments from 1, 2.5, and 5 month old MRL animals were stimulated by the direct application of different concentrations of carbachol, as described in Methods. Even in the older MRL animals, the glands were able to secrete protein in response to carbachol stimulation. In the older animals, the amount of protein per gram of tissue was reduced, presumably due to the presence of large numbers of lymphocytes that did not contribute to the secretory product. With increasing age, however, the sensitivity of the glands to carbachol seemed to increase, which implies some changes in either receptor sensitivity or receptor number, a phenomenon similar to what was seen in the NZB animals.

4. DISCUSSION

The immunocytochemical examination of the NZB animals at different ages showed a progressive decrease in the density of innervation to the acini, so that by about 7–8 months of age very few varicosities showing SLI or VIPLI were present. It seems likely that this decrease is due primarily to the reduction of the parasympathetic cholinergic innervation. The antibody to synaptophysin will bind to any neuron that has synaptic vesicles, and thus will show the entire innervation pattern. VIPLI, however, is usually found in cholinergic neurons in lacrimal gland,[8] and since the pattern of VIPLI is similar in density to that of the SLI, it is reasonable to assume that most of the innervation in these glands is cholinergic.

It is important to note that there is sympathetic innervation to the blood vessels in these same lacrimal glands in these animals and that there was no change in the density of that pattern with age. In dry eye conditions such as Sjögren's syndrome, the manifestation of the disease is a dry ocular surface and the failure of the lacrimal gland to secrete fluid even in response to reflex stimulation.[9] In other dry eye conditions, basal secretion of tear fluids is compromised, but stimulation of the cornea or nasal mucosa will reflexly elicit a normal production of fluid. Based on this distinction, it would seem that the NZB animals are a better model for Sjögren's syndrome than for other dry eye conditions. However what is clear in these animals is that the reduction in cholinergic innervation *precedes* the invasion and replication of large numbers of lymphocytes that has previously been considered the major characteristic of the disease. In fact, one could suggest that the invasion of lymphocytes is the result of a compromised secretory epithelium and that significant changes that affect the secretory ability of the gland have already taken place before the arrival of the lymphocytes. The fundamental issue, therefore, is what causes the loss of innervation in the NZB animals. The fact that the sympathetic innervation to the blood vessels in the glands is normal suggests that the "neuropathy" may be specifically targeted to the parasympathetic fibers and is not a general neural degeneration in the gland. The observation that the parasympathetic innervation to the Harderian glands in the same animal is normal in older animals also suggests that the disease process is at least somewhat specific in its effects. In the MRL extraorbital lacrimal glands, we also see a reduction in the innervation density to the remaining intact acini. In these animals, however, since glandular destruction begins at a very young age, it is hard to tell whether the loss of innervation precedes the lymphocytic invasion or not. There does appear to be a correlation between the extent of loss of acini with the reduction of innervation density to the remaining acini. What causes the reduction in innervation in either strain is at present a mystery.

In both the NZB and MRL extraorbital lacrimal glands, stimulation of gland fragments by the direct application of carbachol caused an increase in the release of protein into the media. This suggests that the receptor/second message/secretory vesicle release pathway is intact. In the 5 month old MRL animals, protein released in response to direct carbachol stimulation increased, and although at least 50% of the acini were destroyed, there were few nerves on those remaining acini and there were large areas of lymphocytic infiltration. However, protein secretion/release is only a small portion of the secretory product. Water is the crucial component of the tears providing lubrication and transport for the soluble proteins needed for a normal corneal surface. Little is known about the transport of water across the lacrimal gland epithelium, but it is likely to be associated with ion channels in the acinar cell membranes. Patch-clamp studies of acinar cells from 6 month old NZB animals show differences in the membrane ion channels that would be consistent with a reduced flow of water across the epithelium.[10] These changes occur before significant lymphocytic infiltration and thus are not likely to be produced by direct action of lymphocytes on the acinar cells. In both strains of mouse, it is likely that the extensive reduction in innervation density and the changes in membrane channels that occur within the first 8 months would significantly reduce the volume of tears the lacrimal gland could produce.

ACKNOWLEDGMENT

Supported by NIH grant EY0940606.

REFERENCES

1. Fox RI. Sjögren's syndrome: Clinical features, pathogenesis, and treatment. This volume.
2. Dartt DA. Regulation of tear secretion. *Adv Exp Med Biol.* 1994;350:1–9.
3. Parod RJ, Putney JW. Stimulation of Ca^{++} efflux from rat lacrimal gland slices by carbachol and epinephrine. *Life Sci.* 1979;25:2211–2215.
4. Saito Y, Kuwahara S. Effect of acetylcholine on the membrane conductance of the intraorbital duct cells of the rat exorbital lacrimal gland. *Adv Exp Med Biol.* 1994;350:87–89.
5. Jabs DA, Prendergast RA. Murine models of Sjögren's syndrome. *Invest Ophthalmol Vis Sci.* 1988;29:1437–1443.
6. Bromberg BB. Autonomic control of lacrimal protein secretion. *Invest Ophthalmol Vis Sci.* 1981;20:110–116.
7. Wetterberg L, Geller E, Yuwiler A. Harderian gland: Development and influence of early hormone treatment on porphyrine content. *Science.* 1970;168:996.
8. Dartt DA, Baker AK, Vaillant C, Rose PE. VIP stimulation of protein secretion from rat lacrimal gland. *Am J Physiol.* 1984;247:G502.
9. Tsubota K. Non-Sjögren's type dry eye. This volume.
10. Brink PA, Peterson E, Banach K, Walcott B. Patch-clamp studies of acinar cells from normal and Sjögren's syndrome mouse model lacrimal glands. This volume.

ABERRANT EXPRESSION AND POTENTIAL FUNCTION FOR PAROTID SECRETORY PROTEIN (PSP) IN THE NOD (NON-OBESE DIABETIC) MOUSE

Christopher P. Robinson,[1] Denise I. Bounous,[2] Connie E. Alford,[3] Ammon B. Peck,[1] and Michael G. Humphreys-Beher[3]

[1]Department of Pathology
[3]Department of Oral Biology
University of Florida
Gainesville, Florida
[2]Department of Pathology
University of Georgia
Athens, Georgia

1. INTRODUCTION

The protein and mucin-rich secretions derived from the salivary, lacrimal, and other minor exocrine tissues are essential for maintaining the health and integrity of the oral and ocular surfaces.[1] For the most part, both tear and saliva secretions serve similar functions and contain many of the same protein constituents, e.g., EGF, NGF, TGF-α, lactoferrin, lysozyme, and immunoglobulins.[1,2] These proteins provide anti-microbial defense mechanisms, growth factors important for mucosal tissue maintenance and regeneration, and mucinous lubricating agents.

Another protein reported to be specific for saliva is parotid secretory protein (PSP). PSP is a 25 kDa, leucine-rich glycoprotein of unknown function that is synthesized in abundant quantities in murine and rat parotid glands and to a lesser extent in the sublingual glands.[3] PSP is also produced in the developing neonatal submandibular gland up to 5 days post-partum, after which time synthesis ceases.[4,5] PSP has been proposed as a marker for tissue-specific protein production of salivary glands[3] and appears coordinately regulated with salivary amylase.

The NOD mouse, a murine model for Type I insulin-dependent diabetes, not only develops an autoimmune-dependent destruction of its pancreatic β cells leading to loss of blood glucose regulation, but also manifests an autoimmune attack against its submandibular and lacrimal glands, resulting in a decline of secretory fluids from both

Lacrimal Gland, Tear Film, and Dry Eye Syndromes 2
edited by Sullivan *et al.*, Plenum Press, New York, 1998

glands.[6] Unique changes in the synthesis of several proteins noted specifically in saliva, including de novo synthesis of PSP as well as decreased concentrations of amylase and EGF, occurs in both NOD and NOD-*scid*, which lack functional T- and B-lymphocytes due to a homologous mutation at a severe combined immuno-deficiency (*scid*) locus.[7] Detection of these changes in NOD-*scid* mice suggest that changes in saliva protein constituents in the NOD strain are independent of the autoimmune destruction of the glands.

The aberrant synthesis of PSP in the submandibular glands of NOD mice led us to examine the question of whether PSP may be abnormally synthesized in other exocrine tissue of NOD mice. In the present report, we show that although PSP is synthesized in the lacrimal glands of NOD mice, it is detected in the lacrimal glands of several other laboratory mouse strains as well. We provide evidence of a potential anti-microbial function for PSP, which may explain its normal synthesis by both the salivary and lacrimal glands. In addition, however, NOD mice were found to have lost the normal regulation of PSP gene transcription in a number of organs/tissues.

2. MATERIALS AND METHODS

BALB/c, CBA/J, and NOD/Uf mice were bred and maintained in the mouse colony at the University of Florida, Gainesville. C3H/HeJ and NOD-*scid* mice were purchased from Jackson Laboratories (Bar Harbor, ME). Binding studies were performed using *Listeria monocytogenes, Streptococcus mutans, Actinobacillus actinomycetemcomitans,* and *Escherichia coli.* Glands were pooled from a minimum of two age- and sex-matched mice of each strain and homogenized in 10 mM Tris buffer (pH 7.4), centrifuged at 500g for 15 min, and assayed for protein concentration. Excised tissues were homogenized in 2 ml of 4 M guanidinium isothiocyanate (GITC), as described by Chirgwin et al.[8] Total tissue RNA (2 µg) was pelleted by centrifugation and reverse transcribed. The resulting cDNA was amplified by PCR using 5'GCAGAGAAACAAGGATCTCG and 3' CACTGGAGAGT AGCCAGCAGG PSP-specific primers. To confirm the identity of specific PCR products, hybridization of digoxigenin-labeled oligonucleotide internal probe (5' AATGCGACCGTTCTTGCC) specific for PSP cDNA was carried out, using the Genius system of Boehringer-Mannheim (Indianapolis, IN).[9] Primers and probes specific for G3PDH were used as positive controls for all PCR reactions.

Gland lysates (30 µg of total protein per lane) were separated on 12% SDS-PAGE gels and Western blotted. Polyclonal rabbit anti-mouse parotid secretory protein (mPSP) IgG antibody, kindly provided by Dr. William Ball (Dept. of Anatomy, Howard University),[10] was incubated with each membrane for 2 h at 25°C. The membranes were incubated with a secondary alkaline phosphatase conjugated goat anti-rabbit immunoglobulin (Sigma) and exposed to substrate.

The bacteria *L. monocytogenes*, *E. coli*, *S. mutans*, and *A. actinomycetemcomitans* were grown in overnight cultures at 37°C. The binding of PSP was performed as described.[11] Specificity of PSP binding was determined by pre-incubation of the cells with 10 µg/ml of unlabeled PSP. Cation-dependent binding was determined by incubation where one or all of the salts were altered in the incubation buffer.

3. RESULTS

We have previously reported[9] a loss of transcriptional regulation of PSP in the submandibular gland of autoimmune NOD mice, which correlates to the appearance of lym-

Table 1. RT-PCR and Southern blot detection of PSP mRNA isolated from NOD, C3H, and BALB/c tissues

Animals	Pancreas	Submandibular	Parotid	Lacrimal	Brain	Liver	Heart
NOD	+	+	+	+	–	–	+
C3H	–	–	+	–	–	–	–
BALB/c	–	–	+	+	–	–	–

Band intensity was scored and ranked by –, no detection; +, weak detection.; or ++, strong detection. G3PDH housekeeping gene was used as a positive control for RNA standardization.

phocytic infiltration. To investigate potential gene dysregulation in other NOD tissues, we screened a panel of NOD, BALB/c, and C3H RNA samples by RT-PCR and Southern probing. Bands corresponding to PSP (785 bp) were observed in RNA from lacrimal, heart, and pancreas of NOD mice (Table 1). Unexpectedly, PSP was additionally detected in the lacrimal glands of BALB/c mice but not in heart or pancreas. The lack of lacrimal PSP expression in C3H mice is in concordance with previous reports by Hjorth et al.[11]

Since PSP transcripts were detected in several NOD tissues, we used Western blot analyses to screen a similar panel of tissues for the presence of PSP. As expected, in C3H mice, PSP was found only in parotid gland lysates. Despite the presence of detectable mRNA transcripts in heart and pancreas of NOD mice, we were unable to detect the presence of PSP at the protein level. However, as depicted in Table 2, PSP bands were detected in the lacrimal glands of BALB/c, NOD, and NOD-*scid*, but not C3H mice. There appeared to be inter-strain variability in the levels of PSP production with BALB/c > NOD > C3H. Relative expression of PSP was independent of age; however, female mice displayed lower concentrations of PSP when compared with syngeneic males. Because C3H mice did express PSP in their parotid glands, it is unknown why they fail to express PSP in their lacrimal glands.

As one of the main functions of both saliva and tears is to serve as a non-immune barrier to bacterial infection from the external environment, PSP was examined for its ability to bind to bacterial pathogens. When radiolabeled PSP was incubated with intact bacteria (10^8 cells) in the presence of NaN3, binding ranged from 18–28% of the radiolabeled protein added (Fig. 1). Pre-incubation of bacteria with unlabeled PSP reduced the level of binding to 3–5% of the total counts added. The addition of NaN_3 to the incubation buffer metabolically kills bacterial cells, thus eliminating the potential for active uptake of PSP rather than strictly binding of PSP to the cell surface. Since PSP is a leucine-rich protein, the potential exists for protein-protein interactions that are dependent on the presence of Zn^{2+}. As shown in Fig. 1, binding to intact bacteria was dependent on the presence of

Table 2. Western blot detection of parotid secretory protein in NOD, C3H, and BALB/c tissues

Animals	Pancreas	Submandibular	Parotid	Lacrimal	Brain	Liver	Heart
8 wk NOD	–	–	+++	++	–	–	–
20 wk NOD	–	++	+++	++	–	–	–
20 wk NOD-*scid*	–	++	+++	+	–	–	–
20 wk C3H	–	–	+++	–	–	–	–
20 wk BALB/c	–	–	+++	++	–	–	–

Positive bands were ranked by band intensity with –, no detection; +, weak detection; ++, intermediate detection; +++, strong detection.

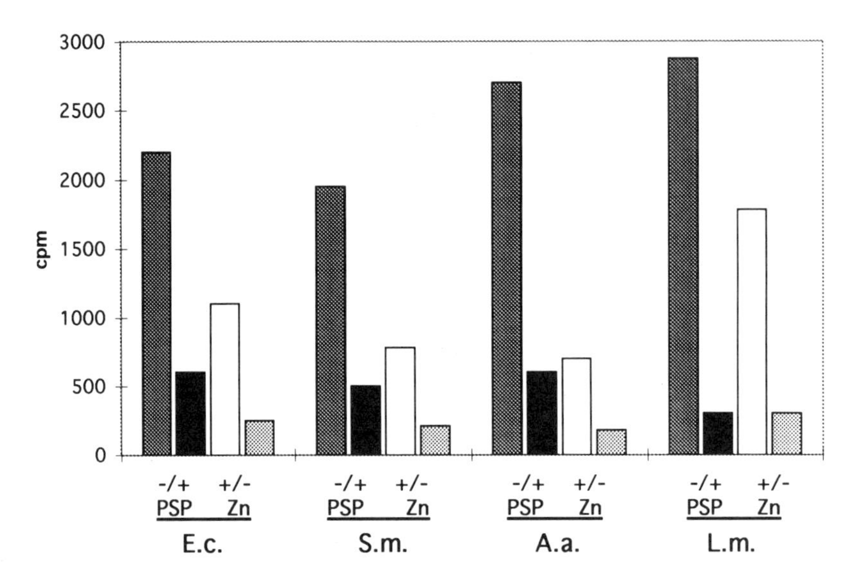

Figure 1. Bacterial binding of PSP. Histogram representing the binding of radiolabeled PSP (10^4 cpm) incubated for 2 h at 37 degrees with 10^8 bacterial cells/ml. Radiolabeled PSP was incubated alone (-) or following the additional pre-incubation with unlabeled PSP (+). Radiolabeled PSP binding was also determined with (+) or without (-) the presence of ZN^{2+}. All values represent mean +/- SEM for 3 separate determinations performed in duplicate. E.c. Escherichi coli., S.m., Streptococcus mutans. A.a., Actinobacillus actinomycetemcomitans. L.m., Listeria monocytogenes.

Zn^{2+}. Maximal binding occurred at 100 μM $ZnCl_2$ for all the bacteria except *A. actinomycetemcomitans,* which showed maximal binding at 1 mM of Zn^{2+} cation.

The open reading frame of the PSP gene of NOD mice was sequenced using cDNA amplified from mRNA isolated from both the lacrimal and parotid glands. As expected, the sequences of the lacrimal and parotid gland PCR products were identical, and revealed a 98% homology with the published murine PSP DNA sequence, with only 14 base pair differences out of 705. Two regions, corresponding to amino acid residues 111–144 and 198–227, contain a few random mutations; however, two additional regions are of particular interest. The first, containing the first 110 amino acid residues, represents a highly conserved region where no amino acid changes were detected between the NOD and the published mouse PSP sequence. The second, containing 7 amino acid residues at 177–183, is a highly substituted region in the NOD PSP gene sequence. These amino acid changes, including a proline substitution at amino acid 178, could potentially cause major structural differences between the two PSP proteins.

4. DISCUSSION

Our results showing PSP production in the lacrimal glands of all normal mice tested, with the exception of C3H, suggest that this secretory protein is not restricted to the parotid and sublingual gland as first reported.[11] The lack of expression in the lacrimal gland of C3H mice supports the findings of Hjorth et al.,[11] but raises further questions about the importance and/or functions of PSP as a secretory protein. Our detection of PSP in the lacrimal glands would indicate that the protein is not involved in saliva-specific functions such as digestion. By far the most widely recognized function of saliva is growth control of microorganisms. The ability of PSP to bind to bacterial surface proteins in a Zn^{2+} dependent manner suggests that the protein may function as another host protein capable of modulating bacterial growth or colonization rates.[12]

The presence of PSP transcripts in the submandibular gland, pancreas, and heart of NOD mice is unique among the strains examined. Despite the presence of aberrant PSP transcription in several tissues of NOD mice possessing exocrine function, protein production was detected only in the submandibular, lacrimal, and parotid glands, and not in the pancreas or heart. Furthermore, non-exocrine tissue (e.g., brain, liver, kidney) did not express PSP mRNA transcripts. One explanation is that only the salivary and lacrimal tissues possess both PSP transcripts and the necessary cytoplasmic factors for proper transcript processing and post-translational modification. Evidence for this was first described by Shaw and Schibler,[13] who found abundant PSP transcription in the developing murine parotid gland at birth, yet translated protein was not detected until 12 days of age. This apparent lag between PSP gene expression and protein translation was attributed to rapid turnover of PSP transcripts in the cytoplasm.

Submandibular glands of neonatal mice produce PSP until about day 5, then gene expression and secretion subside.[5] Under normal conditions, cellular activity necessary for the production of PSP protein might still be present in adult submandibular glands, but PSP transcription would have ceased. Reappearance of PSP mRNA transcripts in the submandibular glands of NOD mice,[14] in addition to the presence of appropriate cytoplasmic factors, could explain why abnormal protein production of PSP is present only in the submandibular gland, but not in the heart or pancreas.

The ectopic expression of PSP in the submandibular gland of NOD mice at the time of initial autoimmune infiltration questions whether PSP is directly involved in the loss of immunological tolerance to these exocrine tissues. Lymphocytic infiltrates are first detected in the submandibular glands of NOD mice at 8 weeks of age. However, it is not until 10–12 weeks of age that lymphocytic populations are detected in the lacrimal tissues.[11] Therefore, it is possible that the loss of immunological tolerance is dependent on an antigen common to both exocrine tissues, such as PSP. Using the NOD-*scid* mouse, we have shown that the aberrant expression of PSP as well as abnormally proteolytic cleavage of PSP during this time is independent of the presence of lymphocytes. It follows that these novel protein changes occurring in the background of autoimmune activity may actually represent a potential triggering mechanism for the loss of exocrine tolerance.

ACKNOWLEDGMENTS

This study was supported in part by NIDR grant DE10515 and funds from the Office for Women's Health Policy. We thank Mr. Micah Kerr and Ms. Elizabeth Bowen for their excellent technical assistance and Ms. Janet Cornelius and Dr. Jeffrey Anderson for helpful discussions.

REFERENCES

1. Fox PC. Saliva composition and its importance in dental health. *Comp Contin Dent Educ* Suppl 1. 1989;13:5456–5460.

2. Zelles T, Purushotham KR, Macauley SP, Oxford GE, Humphreys-Beher MG. Saliva and growth factors: The fountain of youth resides in us all. *J Dent Res.* 1995;74:1826–1832.
3. Larsen HJ, Broderson CH, Hjorth JP. High-level salivary gland expression in transgenic mice. *Transgenic Res.* 1994; 3:311–316.
4. Ball WD, Hand AR, Moreira JE, Iverson JM, Robinovitch MR. The B1-immunoreactive proteins of the perinatal submandibular gland: Similarity to the major parotid gland protein RPSP. *Crit Rev Oral Biol Med.* 1993;4:517–524.
5. Mirels L, Ball WD. Neonatal rat submandibular gland protein SMG-A and parotid secretory protein and alternatively regulated members of a salivary protein multigene family. *J Biol Chem.* 1992;267:2679–2687.
6. Humphreys-Beher MG, Hu Y, Nakagawa Y, Wang PL, Purshotham KR. *Adv Exp Med Biol.* 1994;350:631–635.
7. Schultz LD, Schweitzer PA, Christianson SW, et al. Multiple defects in innate and adaptive immunologic function in NOD/LtSz-*scid* mice. *J Immunol.* 1995;154:180–191.
8. Chirgwin JM, Przybyla AE, MacDonald RJ, Rutter WJ. Isolation of biologically active ribonucleic acid from sources enriched in ribonuclease. *Biochemistry.* 1979;18:5294–5299.
9. Robinson CP, Bounous DE, Alford CE, et al. Expression of parotid secretory protein in murine lacrimal tissues and its possible function as a bacterial binding protein in exocrine secretions. *Am J Physiol.* 1997;272:G863-G871.
10. Mirels L, Girard L, Ball WD. Molecular cloning of developmentally regulated neonatal rat submandibular gland proteins. *Crit Rev Oral Biol Med.* 1993;4:525–529.
11. Hjorth JP, Owerbach D. Inheritance of a parotid secretory protein in mice and its use in determining salivary amylase quantitative variants. *Genetics.* 1980;95:129–140.
12. Isberg RR. Discrimination between intracellular uptake and surface adhesion of bacterial pathogens. *Science.* 1991;252:934–938.
13. Shaw P, Schibler U. Structure and expression of the parotid secretory protein gene of the mouse. *J Mol Biol.* 1986;192:567–576.
14. Robinson CP, Yamamoto H, Peck AB, Humphreys-Beher MG. Genetically programmed development of salivary gland abnormalities in the NOD-*scid* mouse: A potential trigger for sialoadenitis of NOD mice. *Clin Immunol Immunopathol.* 1996;78:50–59.

SURVEY OF CANINE TEAR DEFICIENCY IN VETERINARY PRACTICE

Renee Kaswan,[1] Christopher Pappas, Jr.,[2] Keith Wall,[2] and Susan G. Hirsh[1]

[1]University of Georgia College of Veterinary Medicine
Athens, Georgia
[2]Schering-Plough Animal Health
Union, New Jersey

1. INTRODUCTION

Keratoconjunctivitis sicca (KCS, dry eye) is a common cause of ocular morbidity and blindness in dogs. Typical clinical signs are mucoid to mucopurulent discharge, rubbing and pawing eyes, and progressive corneal scarring. Canine KCS can result from a number of causes including trauma, canine distemper virus, and sulfonamide toxicity, but most cases are thought to have an autoimmune pathogenesis.[1–4] Like human dry eye, canine dry eye may occur as an isolated disorder or in association with a number of autoimmune diseases, such as atopy, hyperadrenocorticism, diabetes mellitus, and hypothyroidism.[4,5]

Early reports on canine tear deficiency and KCS suggested that aging females[6,7] were particularly predisposed. Later studies[8] on the interaction of age and gender on the incidence of tear deficiency confirmed that the likelihood of tear deficiencies increases with age. However, female predisposition was not apparent when data analysis was controlled for frequency of neutering. Apparently, loss of sex hormones following castration or ovariohysterectomy predisposed older dogs of either sex to tear deficiency; the predisposition of females resulted from an increased neuter frequency in female dogs.

An increased incidence of tear deficiency (as KCS) was reported among West Highland White Terriers in the UK[3,6] and among cavalier King Charles Spaniels, Lhasa Apsos, English Bulldogs, and Shih Tzus in the United States.[1,9] Breed predispositions for a number of ocular disorders, including KCS, were recently reviewed.[10] Dog breeds likely to develop dry eye included the American Cocker Spaniel, Dachshund, English Bulldog, Lhasa Apso, Shih Tzu, Miniature Schnauzer, Pekingese, and West Highland White Terrier.[1,10]

Although awareness of canine KCS has increased among the veterinary community, the absolute incidence of the condition remains unknown. The study presented here was

Lacrimal Gland, Tear Film, and Dry Eye Syndromes 2
edited by Sullivan *et al.*, Plenum Press, New York, 1998

designed and conducted by Schering-Plough Animal Health to assess the prevalence of tear deficiency and dry eye among dogs presenting with ocular complaints to general veterinary practices. The influence and interaction of animal age, breed, gender, neuter status, and concurrent medical conditions upon the incidence of tear deficiency were considered.

2. MATERIALS AND METHODS

2.1. Study Design

Forty-eight collaborating United States veterinarians performed 1-min Schirmer Tear Tests (STT; Schering-Plough Animal Health, Kennilworth, NJ) bilaterally without anesthetic on all canine patients presented for ophthalmic diagnosis between March and May, 1996. A standardized history was completed for each patient; case history data included age, sex, breed, presenting complaint, ophthalmologic history and/or conditions, and concurrent medical conditions and medications. Dogs receiving drugs known to suppress tear production (e.g., atropine) at the time of examination were excluded from the study.

Following elimination of non-qualifying dogs, 460 canine ophthalmology cases were available for data analysis. Patients were initially categorized with regard to tear production as follows: 'tear deficient' (STT value <10 mm/min for one or both eyes); 'marginal' tear production (STT ≥ 10 mm bilaterally, but <14 mm/min in either or both eyes); 'normal' tear production (STT value ≥14 mm/min bilaterally); prior KCS diagnosis, with concurrent cyclosporine treatment (Optimmune®, Schering-Plough, or other ophthalmic cyclosporine).

2.2. Statistical Analyses

The proportion of dogs with marginal or deficient tearing was compared to the proportion with normal tearing within four sex groups (male, castrated male, female, ovariohysterectomized female) by testing the significance of difference between two proportions as follows: all males vs. all females; intact males vs. castrated males, intact males vs. ovariohysterectomized females; intact males vs. intact females; castrated males vs. ovariohysterectomized females; intact females vs. ovariohysterectomized females; neutered dogs (male or female) vs. intact dogs. The differences in proportions of tear deficient and normally tearing dogs were also tested for significance within seven age groups (1 year or less, 1–2 years, 2–4 years, 4–7 years, 7–10 years, 10–15 years, 15–17 years) and for sex groups among age groups. The relative proportion of tear deficient animals among the 30 most commonly reported breeds was similarly analyzed, as were the types and frequency of coexisting conditions among tear deficient dogs.

Mean STT's of qualifying eyes from tear deficient animals not receiving cyclosporine were compared to those of previously diagnosed KCS patients receiving cyclosporine via a two-sample T-test, one way analysis of variance, and least significant difference (LSD) comparison of means. Sex, age, and breed distributions of cyclosporine treated vs. untreated tear deficient dogs were compared via a two-sample Kolmogorov-Smirnov test. Breed distributions among tear deficient and normally tearing dogs were compared in the same manner. Ages of tear deficient and normal dogs were compared via a one-way analysis of variance followed by LSD comparison of means.

3. RESULTS

3.1. Population Description and Contributing Factors

Four hundred sixty dogs qualified for analysis: 161 (35%) were classified as tear deficient (including 59 dogs previously diagnosed with KCS); 58 (13%) were classified with marginal tear production; 241 (52%) had normal tear production. Fifty-nine dogs previously diagnosed with KCS were receiving ophthalmic cyclosporine at the time of examination; 15 of these had normal STT values. Statistical analyses revealed significantly higher ($p \leq 0.001$) STT values (mean of qualifying eyes) among tear deficient dogs concurrently receiving ophthalmic cyclosporine (11.0 mm/min, SD 6.6) compared to tear deficient dogs not receiving cyclosporine (8.1 mm/min, SD 3.3). Animals receiving cyclosporine did not differ significantly in either sex or age distribution from animals not treated with cyclosporine.

Stepwise regression analysis and F-tests indicated that sex (including neuter status), age, and breed all contributed significantly to a regression model (Cp=3.0) for diagnosis (normal or non-normal tear production). Kruskal-Wallis one-way analysis of variance confirmed that the factors sex (p=0.0088), age (p=0.0086), and breed ($p \leq 0.0001$) all significantly affected diagnosis.

3.2. Age and Sex

The mean age of tear deficient dogs (8.5 years±4.0) was significantly greater ($p \leq 0.0001$) than that of normally tearing animals (6.5 years±4.4). Dogs 10 years of age and older were significantly (p=0.008) more likely to exhibit deficient tearing, regardless of sex. In contrast, younger animals (4 years and less) of any sex group were significantly less likely ($p \leq 0.0001$) to be tear deficient. The prevalence of KCS increased with age (Fig. 1), without regard to gender or neuter status (p=0.005).

Without regard to tear production or gender, neutered animals accounted for 70% of surveyed dogs (Fig. 2). Females were 56% of the total survey population (46% ovariohysterectomized, 10% intact) and 44% were males (24% castrated, 20% intact). Among ab-

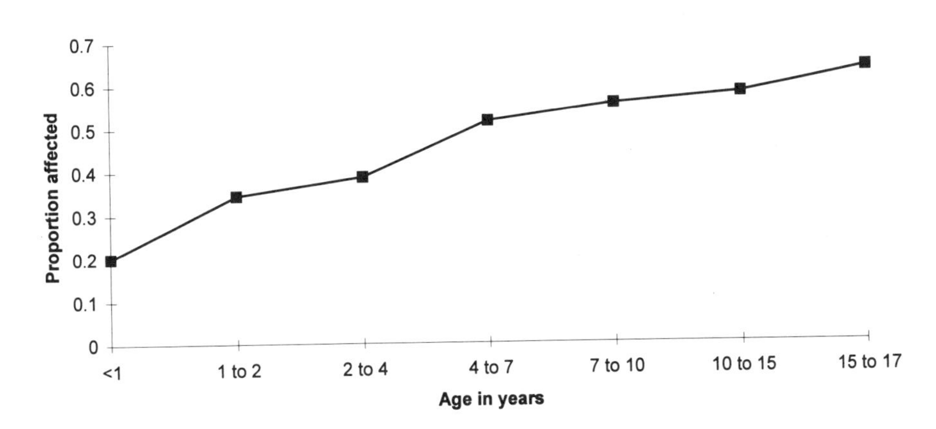

Figure 1. Change in prevalence of tear deficiency with age among survey dogs. Prevalence of tear deficiency increased with age (p=0.005), regardless of sex or neuter status.

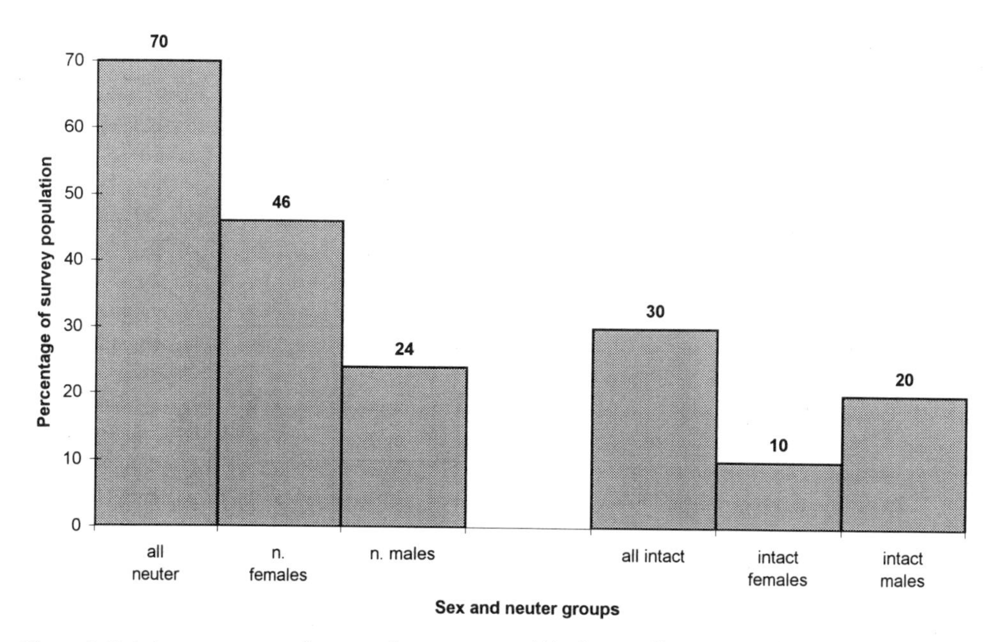

Figure 2. Relative percentages of sexes and neuter status within the overall survey population. Neutered animals (46% female, 24% male) were 70% of the sample population.

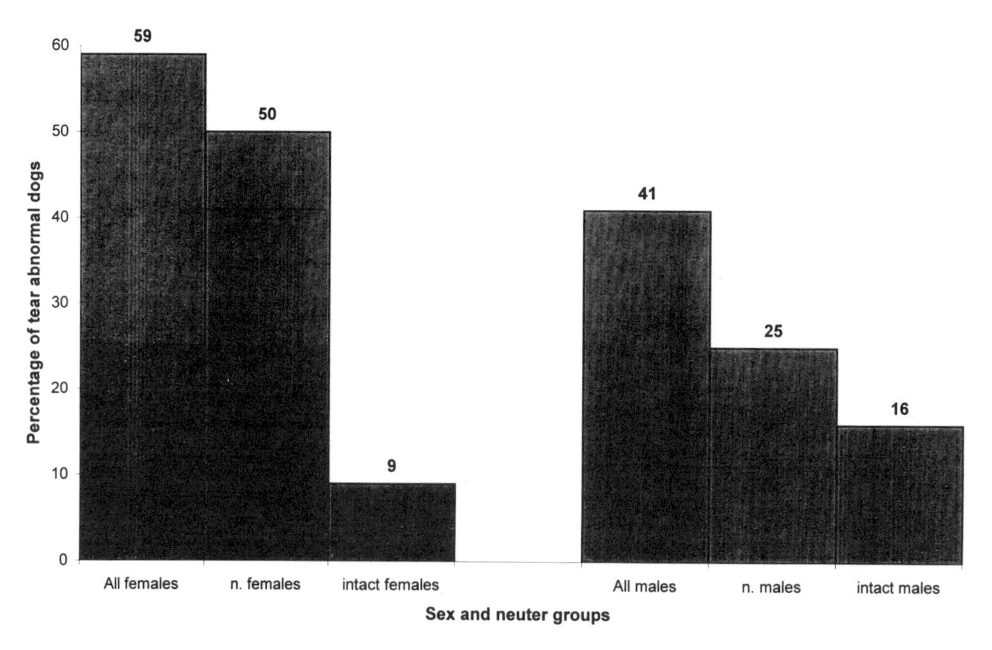

Figure 3. Relative percentages of sexes and neutered animals among tear abnormal (tear deficient + marginal tear production) dogs surveyed. Neutered males and neutered females together were 75% of tear abnormal dogs; only 25% (9% female, 16% male) were intact animals. Neutered animals of either sex were statistically inclined to be tear deficient. When neuter status was ignored, no female predisposition was observed; the overall proportion of tear deficient females was not significantly greater than that of males.

normal (tear deficient + marginal tear production) dogs (Fig. 3), 59% were female (50% ovariohysterectomized, 9% intact) and 41% male (25% castrated, 16% intact). In comparison, 46% of normally tearing dogs (Fig. 4) were male (22% castrated, 24% intact) and 54% were female (42% ovariohysterectomized, 12% intact).

Neutered dogs, regardless of age, were more likely to be marginal or tear deficient than intact dogs (p=0.016). No statistically significant differences were observed between the proportions of abnormal intact males and intact females, or between castrated males and neutered females. Ignoring age and neuter status, the overall proportions of abnormal females and males did not differ significantly (p=0.263). Castrated males (p=0.048) and ovariohysterectomized females (p=0.023) both occurred in statistically higher proportions than intact males within the abnormal group. When age is not considered, the proportion of tear deficient intact females is not statistically different from that of ovariohysterectomized females. However, among animals over 5 years of age, the relative proportion of tear deficient ovariohysterectomized females was significantly greater (p=0.048). Other age-diagnosis relationships were unaffected by age partitioning of sexes.

3.3. Breed Predisposition

Cocker Spaniels, Lhasa Apsos, Shih Tzus, mixed breeds, Bulldogs, Miniature Poodles, and Miniature Schnauzers accounted for 74.5% of tear abnormal dogs (Table 1). However, of these breeds only Cocker Spaniels, Lhasa Apsos, Shih Tzus, Bulldogs, and Schnauzers were statistically ($p \leq 0.0001$) more likely to be tear deficient. Mixed breed dogs were statistically less likely ($p \leq 0.0001$) to be tear deficient. A two sample Kolmogorov-Smirnov test corroborated ($p \leq 0.0001$) different breed distributions for normal and tear deficient dogs.

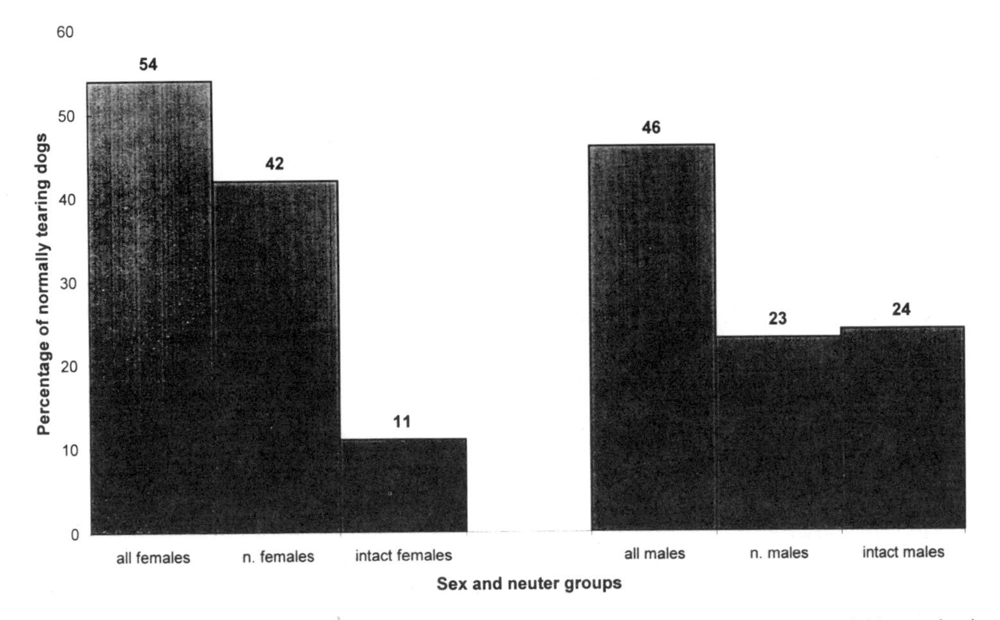

Figure 4. Relative percentages of sexes and neuter status among normally tearing dogs surveyed. Neutered animals were 64% (42% female, 22% male) of normal dogs; 36% (12% female, 24% male) were intact dogs.

Table 1. Thirty breeds most commonly presented among survey animals and their relative prevalence among study groups

Dog breed	% of deficient dogs	% of normal dogs	% of survey
American Cocker Spaniel*	20.6	12.1	15.8
Lhasa Apso*	12.7	2.8	7.1
Shih Tzu*	11.5	6.0	8.4
Mixed breed	11.5	20.9	16.8
English Bulldog*	9.1	1.4	4.7
Miniature Poodle	5.5	12.6	9.5
Miniature Schnauzer*	3.6	1.9	2.6
Dachshund	1.8	2.3	2.1
Yorkshire Terrier	1.8	0.9	1.3
Labrador Retriever	1.8	3.9	3.9
Basset Hound	1.8	0.9	1.3
West Highland White Terrier	1.8	0.5	1.1
Chow Chow	1.8	0.5	1.1
Shar Pei	1.8	<0.05	0.7
Toy Poodle	1.8	1.4	1.6
Golden Retriever	1.2	4.7	3.2
German Shepherd	1.2	3.3	2.4
Basenji	1.2	<0.05	2.6
Siberian Husky	1.2	<0.05	0.5
Chinese Pug	1.2	1.4	1.3
Beagle	1.2	2.3	1.8
Boston Terrier	1.2	2.3	1.8
Border Collie	0.6	1.4	1.1
Shetland Sheepdog	0.6	3.7	2.4
Bichon Frise	0.6	1.4	1.1
Springer Spaniel	0.6	2.8	1.8
Rottweiller	<0.05	2.7	1.6
Brittany Spaniel	<0.05	1.4	0.8
Maltese	<0.05	0.9	0.5
Chihuahua	<0.05	1.9	1.1

*Breeds demonstrating statistically significant ($p \leq 0.05$) greater proportion of tear-deficient than of normal individuals.

3.4. Presenting and Coexisting Conditions

The most common presenting complaints were conjunctivitis, ocular discharge, ocular inflammation, and dried periocular mucus ("crusts"). Coexisting conditions included atopy, hypothyroidism, autoimmune mediated dermatitis, otitis externa, hyperadrenocorticism, renal and/or liver disease, and diabetes (Table 2), but only dermatitis ($p \leq 0.0001$), hypothyroidism ($p \leq 0.0001$), otitis externa ($p = 0.0014$), hyperadrenocorticism ($p \leq 0.0001$), and pannus ($p \leq 0.0001$) were statistically more likely among tear deficient dogs.

4. DISCUSSION

The diagnosis of canine KCS requires a STT ≤10 mm/min (<14 mm/min being marginal[4]) in the presence of clinical manifestations, the hallmark being mucoid to mucopurulent discharge. Corneal vascularization, inflammation, ulceration, or pigment deposition

Table 2. Common concurrent medical conditions among surveyed tear deficient and normal dogs

Condition	Proportion of affected dogs that are also tear deficient	Proportion of affected dogs that tear normally	p^*
Hypothyroidism	0.619	0.381	≤0.0001
Atopy (or allergy)	0.487	0.514	=0.66
Dermatitis	0.667	0.333	≤0.0001
Otitis externa	0.529	0.491	=0.0014
Hyperadrenocorticism	0.750	0.250	≤0.0001
Pannus	0.800	0.200	≤0.0001

*Considered significant at ²0.05; highly significant at ²0.01.

occur with chronicity. Identification of pre-symptomatic animals or those likely to develop KCS presents an advantage in preventing the ocular morbidity associated with advanced canine KCS. Thus, the authors feel that administration of Schirmer Tear Tests during routine anterior segment ophthalmologic examinations is valuable. The percentage of survey dogs (35%) identified with deficient tearing and/or diagnosed with KCS was much higher than previous estimates,[1,8] probably a consequence of the survey's design. Routine administration of a Schirmer Tear Test, even in the absence of other clinical signs, revealed tear deficiencies that had not yet resulted in other clinical signs. The increased occurrence of marginal tear deficiency in breed, sex, and age categories predisposed to KCS suggests that these dogs are developing KCS and should be monitored closely or treated with ophthalmic cyclosporine. The efficacy of ophthalmic cyclosporine in treating canine KCS is well-established;[2–4] therefore, significantly greater STT values observed among cyclosporine-treated dogs were not unexpected.

In agreement with Kaswan and associates,[8] neutered animals of either gender were significantly more likely to develop tear deficiencies and KCS (Figs. 2–4). Previous reports suggesting a female predisposition[6] to tear deficiency likely resulted from owners' preferential neutering of females. Unlike earlier reports, however, the overall proportion of tear deficient females (ignoring neuter status) in the current study was not significantly greater than that of males. This may have resulted from one or both of two differences in the gender status of the sample population. First, the percentage of neutered animals in the study population had generally increased from 22%[8] to 70%, a trend that may have contributed to the increased incidence of tear deficiency. Additionally, within the current survey's population, females were neutered only 1.5 times as often as males; earlier reports indicated that the females were neutered 3.4 times as frequently as males.[8] This reduced the iatrogenic 'gender gap' within the population. Intact dogs of either gender presented with deficient tearing significantly less frequently, an observation that underscores the importance of sex hormones in the etiology of KCS.[11–16]

Another factor significantly affecting the probability of tear deficiency is age. In people, tearing declines with age,[17–21] while the prevalence of general autoimmune disease and autoimmune KCS increases. Previous researchers have suggested that canine KCS was autoimmune in origin;[22] others reported decreased tear production among older dogs.[8] In the present study, dogs aged 10 years and over were significantly more likely to be tear deficient; the mean age of tear deficient dogs was significantly greater than that of normally tearing dogs. Several immune-mediated conditions (hypothyroidism, hyperadrenocorticism, and immune-mediated dermatologic conditions) occurred in statistically greater frequency among tear deficient dogs. Increased prevalence of dry eye among aged and autoimmune-reactive individuals is compatible with current theories of an autoimmune etiology for canine KCS.[3–5,22,23]

Survey results were also consistent with previous reports of KCS breed predisposition.[1,10] Nearly 75% of tear deficient dogs were one of the following: Cocker Spaniel, Lhasa Apso, Shih Tzu, Bulldog, Miniature Poodle, Miniature Schnauzer, or mixed breed. Interestingly, comparisons of proportions indicated that mixed breeds were significantly less likely to be tear deficient.

5. SUMMARY AND RECOMMENDATIONS

The prevalence of tear deficiency among a survey population of 460 dogs presented in general practice (35%) was substantially higher than expected, likely due to routine Schirmer Tear Test screening of the survey animals. Regression analyses indicated that animal sex/neuter status, age, and breed all influenced the occurrence of tear deficiency. With respect to age and breed predisposition, statistical analyses of survey results were consistent with earlier reports. Increased prevalence of dry eye occurred among aged and autoimmune suspect individuals, an observation compatible with current theories of an autoimmune etiology for canine KCS. Predisposed breeds included American Cocker Spaniels, Lhasa Apsos, Shih Tzus, Bulldogs, and Miniature Schnauzers. With regard to gender and neuter status, neutered dogs of either gender were statistically more likely to be tear deficient. Unlike previous reports, however, no overall female predisposition was observed. This was attributed to the survey population's sex composition; neutered males and hysterectomized females occurred in more nearly equal numbers than in previous reports. Moreover, the percentage of neutered animals was more than three times that of earlier studies. The authors propose that routine administration of the Schirmer Tear Test to aged, neutered dogs, predisposed breeds, and dogs already affected by other autoimmune disorders would permit early identification of canine dry eye.

REFERENCES

1. Kaswan RL, Salisbury MA. A new perspective on canine keratoconjunctivitis sicca: Treatment with ophthalmic cyclosporine. *Vet Clin North Am Small Anim Pract.* 1990;20:583–613.
2. Bounous D, Carmichael KP, Kaswan RL, Hirsh S, Stiles J. Effects of ophthalmic cyclosporine on lacrimal gland pathology and function in dogs with keratoconjunctivitis sicca. *Prog Vet Comp Ophthalmol Vet Comp Ophthalmol.* 1995;5:5–12.
3. Sansom J, Barnett KC, Neumann W, et al. Treatment of keratoconjunctivitis sicca in dogs with cyclosporine ophthalmic ointment: A European clinical field trial. *Vet Rec.* 1995;137:504–507.
4. Kaswan RL, Bounous D, Hirsh SG. Diagnosis and management of keratoconjunctivitis sicca. *Vet Med* 1995;90:539–560.
5. Quimby FW, Schwartz RS, Poskitt T, Lewis RM. A disorder of dogs resembling Sjögren's syndrome. *Clin Immunol Immunopathol.* 1979;12:471–476.
6. Sansom J, Barnett KC. Keratoconjunctivitis sicca in the dogs: A review of 200 cases. *J Small Anim Pract.* 1985;26:121–131.
7. Aguirre GD, Rubin VMD, Harvey CE. Keratoconjunctivitis sicca in dogs. *J Am Vet Med Assoc.* 1971;158:1566–1579.
8. Kaswan RL, Salisbury MA, Lothrup CD. Interaction of age and gender on occurrence of canine keratoconjunctivitis sicca. *Prog Vet Comp Ophthalmol Vet Comp Ophthalmol.* 1991;1:93–97.
9. Gelatt KN. *Veterinary Ophthalmology*, 2nd ed. Philadelphia: Lea & Febiger; 1991;281–289.

10. Whitley RD, McLaughlin SA, Gilger BC. Update on eye disorders among purebred dogs. *Vet Med.* 1995;90:575–592.
11. Hiroko A, Edwards J, Sullivan DA. Androgen control of autoimmune expression in lacrimal glands of MRL/Mp-lpr/lpr mice. *Clin Immunol Immunopathol.* 1989;53:499–508.
12. Gao J, Lambert RW, Wickham A, Sullivan DA. Androgen regulation of secretory component mRNA levels in the rat lacrimal gland. *Adv Exp Med Biol.* 1994;350:219–224.
13. Rocha FJ, Kelleher RS, Edwards JA, Pena JDO, Ono M, Sullivan DA. Binding characteristics, immunocytochemical location and hormonal regulation of androgen receptors in lacrimal tissue. *Adv Exp Med Biol.* 1994;350:157–160.
14. Warren DW, Azzarolo AM, Becker L, Bjerrum K, Kaswan RL, Mircheff A. Effects of dihydrotestosterone and prolactin on lacrimal gland function. *Adv Exp Med Biol.* 1994;350:99–104.
15. Sullivan DA, Ariga H, Vendramini AC, Rocha FJ, Ono M, Sato EH. Androgen-induced suppression of autoimmune disease in lacrimal glands of mouse models of Sjögren's syndrome. *Adv Exp Med Biol.* 1994;350:683–690.
16. Ono M, Rocha FJ, Sato EH, Sullivan DA. Distribution and endocrine regulation of androgen receptors in lacrimal glands of the MRL/lpr mouse model of Sjögren's syndrome. In: Homma M, Sugai S, Tojo T, et al., eds. *Sjögren's Syndrome-- State of the Art.* Amsterdam: Kugler Publications; 1994:261–263.
17. de Rotth A. On the hypofunction of the lacrimal gland. *Am J Ophthalmol.* 1941;24:20–25.
18. Holm S. Keratoconjunctivitis sicca and the sicca syndrome. *Acta Ophthalmol.* 1949(suppl 33):1–6.
19. Henderson JW, Prough WA. Influence of age and sex on flow of tears. *Arch Ophthalmol.* 1950;43:224–231.
20. Norn MS. Tear secretion in normal eyes. *Acta Ophthalmol.* 1965;43:567–573.
21. McGill JI, Liakos GM, Goulding N. Normal tear protein profiles and age-related changes. *Br J Ophthalmol.* 1984;99:888–890.
22. Kaswan RL, Martin CL, Dawe DL. Keratoconjunctivitis sicca: Immunological evaluation of 62 canine cases. *Am J Vet Res.* 1985;46:376–383.
23. Kaswan RL, Martin CL, Dawe DL. Rheumatoid factor determination in 50 dogs with keratoconjunctivitis sicca. *J Am Vet Med Assoc.* 1983;183:1073–1075.

LACRIMATION AND SALIVATION ARE NOT RELATED TO LYMPHOCYTIC INFILTRATION IN LACRIMAL AND SALIVARY GLANDS IN MRL lpr/lpr MICE

Hiromi Fujita,[1] Tsutomu Fujihara,[1] Tsutomu Takeuchi,[2] Ichiro Saito,[3] and Kazuo Tsubota[1]

[1]Department of Ophthalmology
Ichikawa General Hospital
Tokyo Dental College
Ichikawa, Chiba, Japan
[2]Department of 2nd Internal Medicine
Saitama Medical Center
Saitama Medical School
Kawagoe, Saitama, Japan
[3]Department of Pathology
Tokushima University School of Dentistry
Tokushima, Japan

1. INTRODUCTION

Sjögren's syndrome (SS) is an autoimmune disease characterized by lymphocytic infiltration into salivary and lacrimal glands leading to xerostomia and keratoconjunctivitis sicca.[1–5] Lymphocytic infiltration into these glands has been demonstrated in several kinds of autoimmune mice, such as MRL/MP–+/+ (MRL/+), MRL/MP~lpr/lpr (MRL/lpr), NZB, NZB/W F1 hybrid, and NOD.[6–8] Among these strains, MRL/lpr mice may be one of the best models for Sjögren's syndrome, since marked infiltration of lymphocytes has been reported in lacrimal and salivary glands, which is indistinguishable from human disease.[6] The infiltrated lymphocytes in the glands of MRL/lpr mice were largely $CD4^+$ T-cells with lesser numbers of $CD8^+$ T-cells and B-cells,[8–11] but $CD4^-$ $CD8^-$ double-negative T-cells were rarely detected.

Although the histologic findings in MRL/lpr mice are similar to human disease, the secretory function of the glands has been shown to be nearly normal as judged by Schirmer's test of tear secretion. Given the inadequacy of Schirmer's test as a sensitive measure of ocular wetting in small animals,[6] we measured tear secretion with cotton thread, which is thought to be more sensitive. We also analyzed the salivary secretion as well as the histologic features of the glands in MRL/lpr mice. We will also discuss the apparent lack of correlation between the infiltration of lymphocytes and the impaired secretory function of lacrimal and salivary glands in MRL/lpr mice.

2. MATERIALS AND METHODS

2.1. Comparison of Methods to Measure Volume

The cotton thread was purchased from SHOWA YAKUHIN KAKO (Tokyo). Whatman #1 filter paper was cut into 0.5 x 5.0 mm strips. The indicated volume of PBS, pH 7.2, was absorbed by the strips, and then the length of wetting was measured at 10X magnification using a micrometer on a dissecting microscope. As shown in Fig. 1, the cotton thread was more sensitive than the Schirmer's test paper to measure small volumes. We concluded that the cotton thread is much more suitable to detect small differences in volume of tear secretion of small animals.

2.2. Mice

Female mice, 23–30 weeks old, MRL/lpr, MRL/+, and BALB/c, were purchased from Japan SLC and Japan CLEA. At the indicated times after birth, lacrimal and salivary glands were removed, and specimens for light microscopy were fixed in 4% phosphate-buffered formaldehyde (pH 7.2) and stained with hematoxylin and eosin (H&E). Lymphocytic infiltration into the exocrine glands was graded.[12] Histologic sections of the glands were scored from 0 to IV based on the presence or absence of foci of 50 or more mononuclear inflammatory cells: grade 0 = no inflammatory cells; grade I = inflammation without

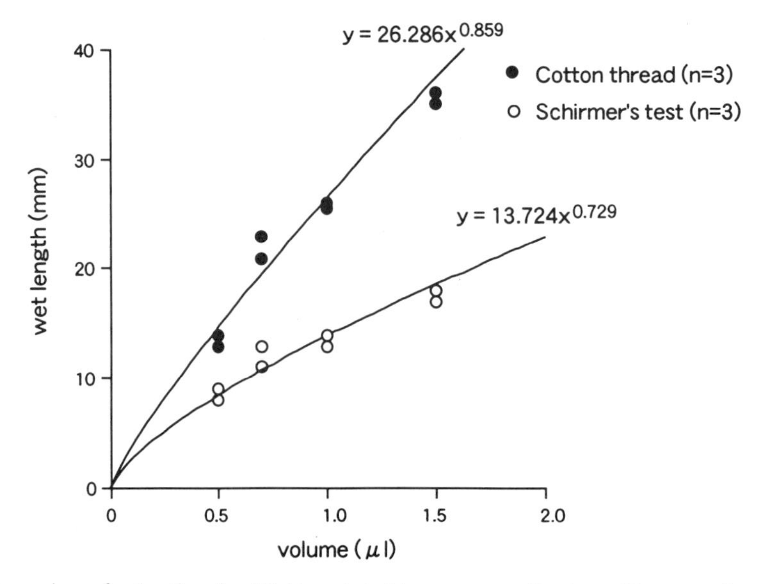

Figure 1. Comparison of cotton thread and Schirmer's test to measure small volume. The cotton thread was more sensitive.

any focus; grade II = the presence of at least one focus; grade III = multiple foci; and grade IV = multiple foci plus evidence of gland destruction.

2.3. Measurement of Tear and Salivary Secretion

Experimental groups of animals were injected intraperitoneally with 0.05 mg of pilocarpine/100 g of body weight in saline, under anesthetization with sodium pentobarbital. Tear secretion volume was measured with cotton thread, which was placed under the lower lid of each eye near the medial canthus. After 5 min, the cotton thread was removed, and the length of soaked areas was measured. These operations were repeated for 30 min. Salivary secretion volume was measured with a capillary tube for 30 min.

2.4. Immunohistologic Analysis

Frozen sections of lacrimal and salivary glands were analyzed as described.[8] Briefly, frozen sections were fixed in acetone for 10 min, rinsed in PBS, and incubated with PBS containing 10% normal goat serum for 20 min. They were then incubated with PE conjugated anti-CD4 or fluorescein (FITC) conjugated anti-CD8 monoclonal antibody (PharMingen) for 1 h, followed by extensive washing with PBS.

3. RESULTS

3.1. Histologic Analysis

Representative pictures of the lacrimal gland in a female MRL/lpr mouse are shown in Fig. 2. Severe inflammatory lesions were observed in lacrimal and salivary glands in

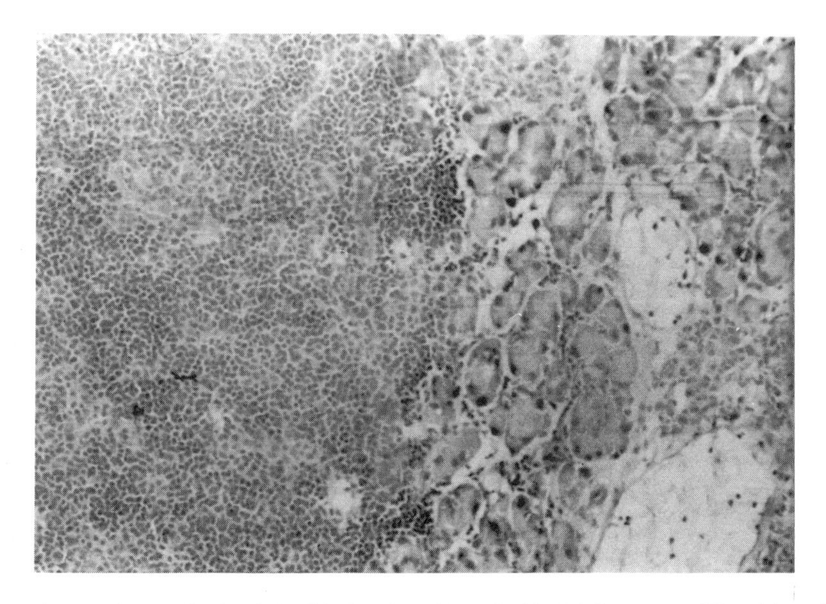

Figure 2. Histopathology of lacrimal gland lesions in a 30-week-old MRL/lpr mouse. HE staining of salivary glands from MRL/lpr mice shows inflammatory infiltrates. (Original magnification X200)

Table 1. Histopathologic grading of lacrimal and salivary gland lesions

	Strain	No.	Focus scale score 0	I	II	III	IV
LG	MRL/lpr	5	0	0	0	1	4
	MRL/+	4	0	0	0	3	1
	BALB/c	5	5	0	0	0	0
SG	MRL/lpr	5	0	0	0	0	5
	MRL/+	4	0	0	0	3	1
	BALB/c	5	5	0	0	0	0

LG, lacrimal glands. SG, salivary glands.

MRL/+ and MRL/lpr mice at 23–30 weeks after birth.[13,14] As summarized in Table 1, MRL/lpr and MRL/+ mice had grade IV and grade III lesions in the lacrimal and salivary glands, respectively, as reported previously. In contrast, BALB/c mice had no inflammatory lesions.

3.2. Measurement of Tear and Salivary Secretion

As shown in Fig. 3, there are no significant differences in the stimulated tear secretion between BALB/c and MRL/lpr or MRL/+ mice. However, the stimulated salivary secretion of MRL/lpr mice was significantly reduced by 10% in comparison with BALB/c (Fig. 4).

3.3. Immunohistochemical Analysis

Immunohistologic analysis revealed that most of the infiltrating lymphocytes were $CD4^+$ T-cells, and a lesser proportion were $CD8^+$ T-cells, as reported.[14] When we analyzed the distribution, we found that $CD4^+$ T-cells were predominantly localized in inflammatory sites, but $CD8^+$ T-cells were detected not only in inflammatory sites but also in sites

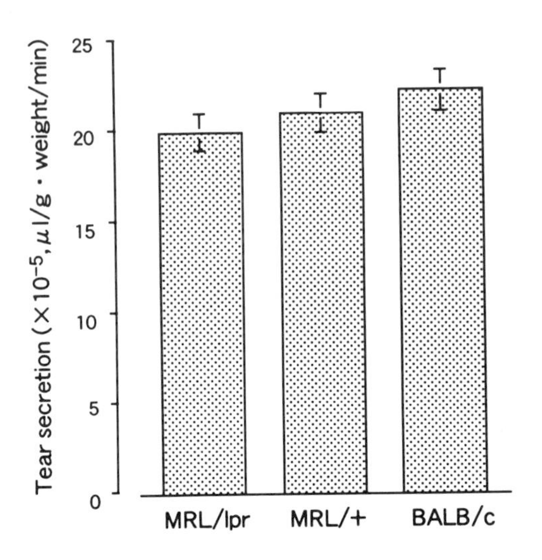

Figure 3. Tear secretion after stimulating with pilocarpine in MRL/lpr, MRL/+, and BALB/c mice. Each value represents the mean ±SE of 8–10 eyes. There was no significant difference between BALB/c and MRL/lpr or MRL/+ mice.

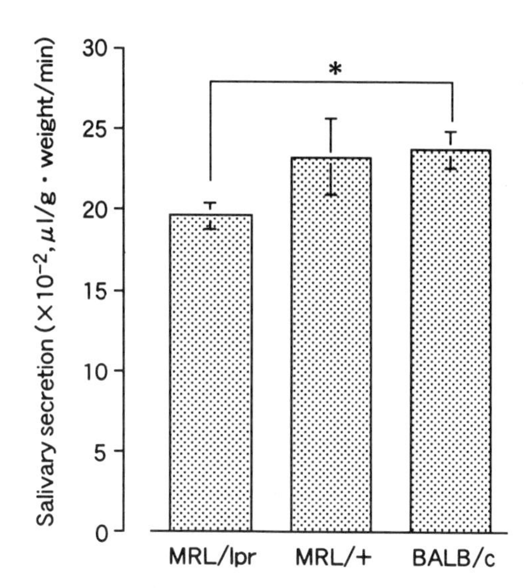

Figure 4. Salivary secretion after stimulation with pilocarpine in MRL/lpr, MRL/+, and BALB/c mice. Each value represents the mean ±SE of 4–5 animals. The salivary secretion in MRL/lpr mice was significantly reduced in comparison with BALB/c mice. (* : $p<0.05$, ANOVA test)

close to acini and ducts in both lacrimal and salivary glands of MRL/lpr and MRL/+ mice (Table 2; Figs. 5–8).

4. DISCUSSION

Sjögren's syndrome is an autoimmune disease characterized by lymphocytic infiltration and destruction of lacrimal and salivary glands resulting in dry eye and dry mouth. Immunohistochemical analysis have revealed that most infiltrated lymphocytes are $CD4^+$ T-cells. However, the mechanism of destruction of the exocrine glands is still unclear.

In this study, we found that cotton thread is more sensitive than Schrimer's test paper when measuring a small volume. By this method we confirmed that the tear secretion is not significantly reduced, even when stimulated with pilocarpine to induce maximal secretion, in MRL/lpr mice. In addition, we measured the volume of salivary secretion for a longer time than previous reports and found a significant 10% decrease in salivary secretion of MRL/lpr mice in comparison with BALB/c mice.

Immunohistologic analysis showed similar findings in the lacrimal and salivary glands of MRL/lpr and MRL/+ mice, suggesting that the grade of infiltration of lympho-

Table 2. Immunohistologic analysis of lacrimal and salivary glands

	Strain	Secretion	CD4	CD8
LG	MRL/lpr	→	F	F, A, D
	MRL/+	→	F	F, A, D
	BALB/c	→	–	–
SG	MRL/lpr	↓	F	F, A, D
	MRL/+	→	F	F, A, D
	BALB/c	→	–	–

LG, lacrimal glands. SG, salivary glands. F, focus. A, acini. D, ducts.

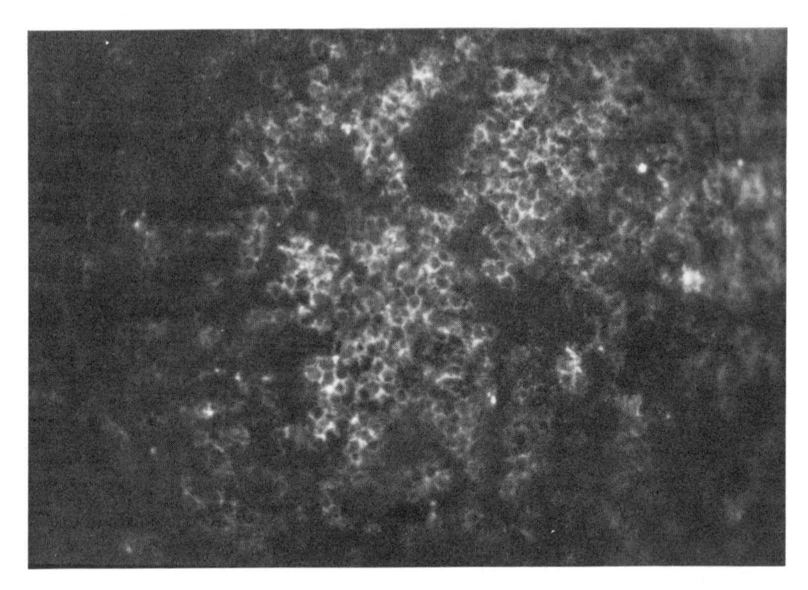

Figure 5. Immunohistochemical detection of CD4 in the lacrimal gland in MRL/lpr mice. The $CD4^+$ T lymphocytes were located in the focus region. (Original magnification X400)

cytes, or predominance of $CD4^+$ T-cells in the foci, does not correlate with the functional impairment of the secretion. Taken together, these data suggest that the grade of inflammatory lesion in the gland does not correlate with the functioning of the gland in mice and that the distribution of $CD4^+$ and $CD8^+$ T lymphocytes in the gland can not explain the dysfunction of the gland.

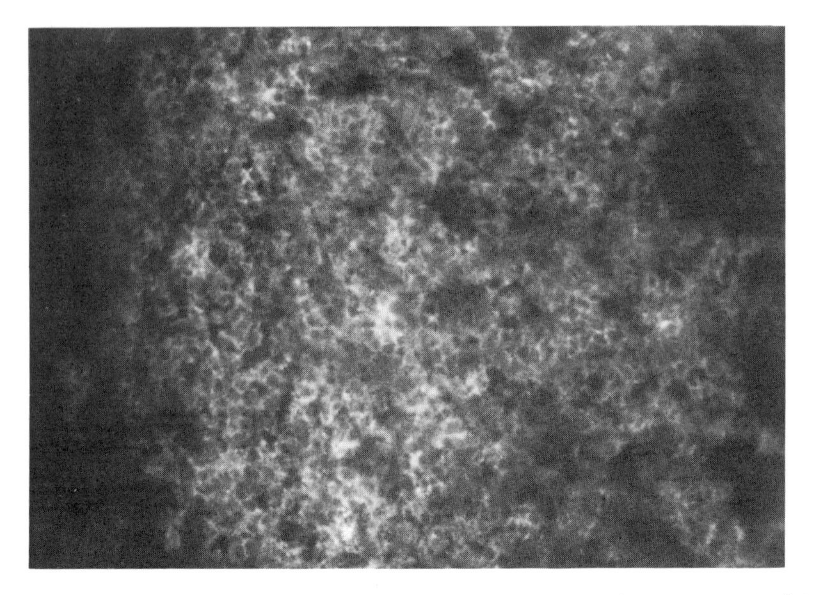

Figure 6. Immunohistochemical detection of CD4 in the salivary gland in MRL/lpr mice. The $CD4^+$ T lymphocytes were located in the focus region. (Original magnification X400)

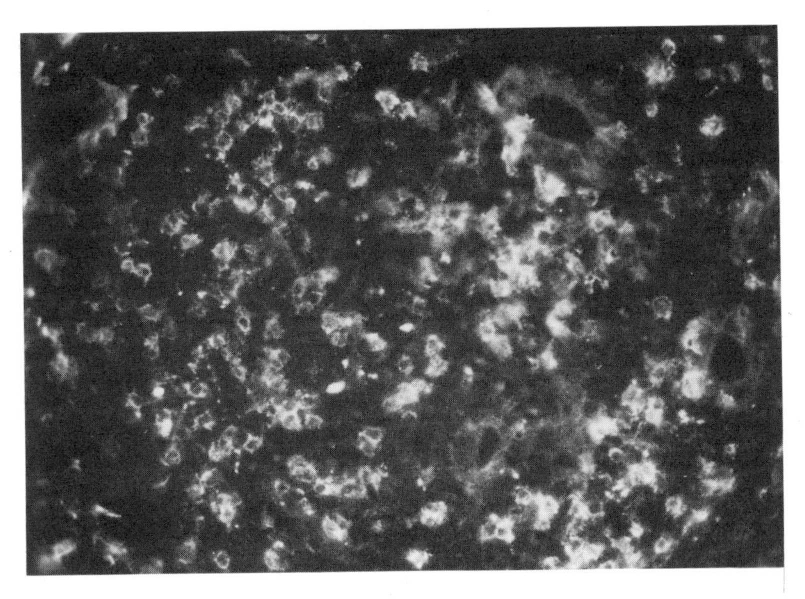

Figure 7. Immunohistochemical detection of CD8 in the lacrimal gland in MRL/lpr mice. The $CD8^+$ T lymphocytes were observed in the focus region and around both the acini and ducts in MRL/lpr mice. (Original magnification X400)

In non-obese diabetic (NOD) mice[15,16] with lymphocytic infiltration in lacrimal and salivary glands, secretory dysfunctions were observed.[17] However, the grade of lymphocytic infiltration in both types of glands in NOD mice is less than in MRL/lpr mice.[6] The grade of lymphocytic infiltration may not be sufficient to explain the glandular dysfunction. There may be other factors causing the dysfunction in the exocrine glands. Further

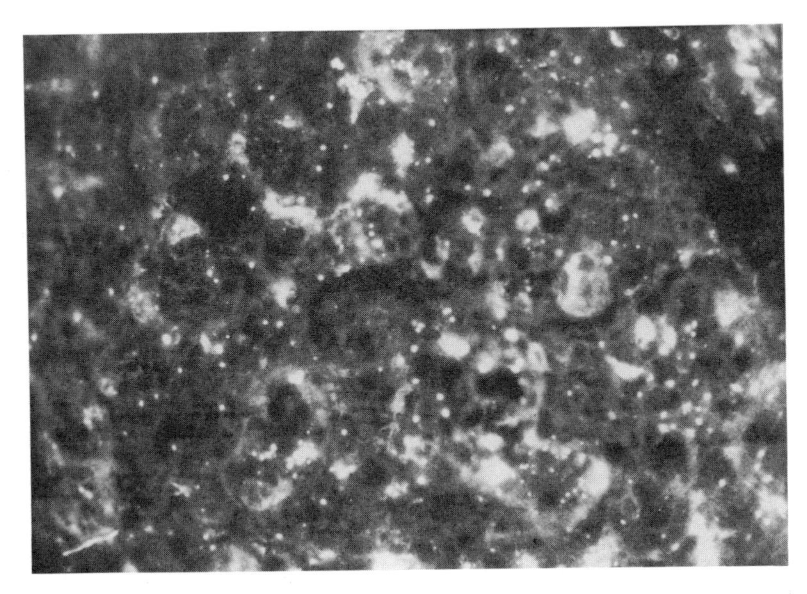

Figure 8. Immunohistochemical detection of CD8 in the salivary gland in MRL/lpr mice. The $CD8^+$ T lymphocytes were observed around both the acini and ducts in MRL/lpr mice. (Original magnification X400).

studies, including measurement of tear and salivary secretion volume and analysis of T-cell receptors and various cell adhesion molecules in the lacrimal and salivary glands of various autoimmune mice, are now in progress in our laboratories.

REFERENCES

1. Fox RI, Robinson CA, Curd JG, Kozin F, Howell FV. Sjîgren's syndrome: Proposed criteria for classification. *Arthritis Rheum.* 1986;29:577.
2. Talal N, Moutaspoulos HM, Kassan SS, eds. *Sjîgren's Syndrome.* New York: Springer-Verlag; 1987.
3. Fox RI, Saito I. Sjögren's syndrome: Immunologic and neuroendocrine mechanisms. *Adv Exp Med Biol.* 1994;350:609.
4. Oxholm P. Primary Sjîgren's syndrome: Clinical and laboratory markers of disease activity. *Semin Arthritis Rheum.* 1992;22:114.
5. Talal N. Sjîgren's syndrome and connective tissue diseases associated with other immunologic disorders. In: McCarty DA, ed. *Arthritis and Allied Conditions: A Textbook of Rheumatology.* Philadelphia: Lea & Febiger; 1989:1197.
6. Jabs DA, Prendergast RA. Murine models of Sjögren's syndrome. *Invest Ophthalmol Vis Sci.* 1988;29:1437.
7. Eisenbarth G. Insulin dependent diabetes mellitus: A chronic autoimmune disease. *N Engl J Med.* 1986;314:1360.
8. Jabs DA, Prendergast RA. Reactive lymphocytes in lacrimal gland and vasculitic renal lesions of autoimmune MRL/lpr mice express L3T4. *J Exp Med.* 1987;166:1198.
9. Theofilopoulos AN, Eisenberg RA, Bourdon M, Growel JS Jr, Dixon FJ. Distribution of lymphocytes identified by surface markers in murine strains with systemic lupus erythematosus-like syndromes. *J Exp Med.* 1979;149:516.
10. Lewis DE, Giorgi JV, Warner NL. Flow cytometry analysis of T-cells and continuous T-cell lines from autoimmune MRL/lpr mice. *Nature.* 1981;289:298.
11. Wofsy D, Hardy RR, Seaman WE. The proliferating cells in autoimmune MRL/lpr mice lack L3T4, an antigen on 'helper' T cells that is involved in the response to class II major histocompatibility antigens. *J Immunol.* 1984;132:2686.
12. Jabs DA, Lee B, Burek CL, et al. Cyclosporine therapy suppresses ocular and lacrimal gland disease in MRL/Mp-lpr/lpr mice. *Invest Ophthalmol Vis Sci.* 1996;37:377–383.
13. Jonsson R, Tarkowski A, Backman K, Holmdahl R, Klareskog L. Sialadenitis in the MRL-1 mouse: Morphological and immunohistochemical characterization of resident and infiltrating cells. *Immunology.* 1987;60:611.
14. Hoffman RW, Alspaugh MA, Waggie KS, Dutham JB, Walker SE. Sjogren's syndrome in MRL-1 and MRL/n mice. *Arthritis Rheum.* 1984;27:157.
15. Asamoto H, Oishi M, Akagawa Y, Tochino Y. Histologic and immunologic changes in the thymus and other organs in NOD mice. In: Seiichiro T, Tochino Y, Noraka K, eds. *Insulitis and Type 1 Diabetes.* Tokyo: Academic Press; 1986.
16. Hu Y, Nakagawa Y, Purushotham KR, Humphreys-Beher MG. Functional changes in salivary glands of autoimmune disease-prone NOD mice. *Am J Physiol.* 1992;263:E607.
17. Humphreys-Beher MG, Hu Y, Nakagawa Y, Wang PL, Purushotham KR. Utilization of the non-obese diabetic (NOD) mouse as an animal model for the study of secondary Sjögren's syndrome. *Adv Exp Med Biol.* 1994;350:631.

THE SJÖGREN SYNDROME AND TEAR FUNCTION PROFILE

O. Paul van Bijsterveld

Oogcentrum Houten
Houten, The Netherlands

1. INTRODUCTION

As early as 1992, Fleming reported the presence of lysozyme in the tear fluid.[1] Other proteins have also been reported, notably lactoferrin[2] and tear-specific prealbumin.[3,4] All these proteins are major components of the lacrimal fluid. Their presence under both normal and pathological conditions have been well documented.[5–9] The tear proteins are synthesized by the lacrimal gland[10] and, in degenerations of the tear gland, the concentration of tear proteins in the lacrimal fluid decreases.

In most cases, the cause of diminished tear gland function is uncertain, as biopsy of the tear gland is not commonly performed. Less frequently, degeneration of the tear gland is the result of autoimmune inflammatory reactions such as can be observed in Sjögren's syndrome; one characteristic of this condition is lymphocytic cell infiltration into the lacrimal gland, which ultimately destroys its epithelial cells.

The diagnosis of Sjögren's syndrome as an underlying disease of keratoconjunctivitis sicca (KCS) is important, as this syndrome can lead to many other ocular and also nonocular medical complications. As the ophthalmological routine investigation of tear gland function is not particularly elaborate, it would be of considerable interest if differences could be demonstrated in the profile of the common parameters of tear function in patients with KCS associated with Sjögren's syndrome and in those without such association.

2. PATIENTS, MATERIALS, AND METHODS

Two groups were examined. One group consisted of 30 patients, ranging in age from 37 to 86 years, with clinical and laboratory evidence of KCS associated with Sjögren's syndrome (KCS-SJ). The second group consisted of 30 patients, ranging in age from 39 to 79 years, with KCS but without evidence of Sjögren's syndrome or other systemic disease

Lacrimal Gland, Tear Film, and Dry Eye Syndromes 2
edited by Sullivan *et al.*, Plenum Press, New York, 1998

(KCS-NSJ). An effort was made to stratify both groups according to sex, age, and tear fluid lysozyme concentration.

The tear fluid lysozyme concentration was determined by the agar diffusion method.[11] Hen egg-white lysozyme (HEL) standards were used to express the lysozyme concentration in micrograms per milliliter HEL equivalent.[12] The lactoferrin concentration was determined by the radial immunodiffusion method, using the Lactoplate[9] (JDC, Culemborg, The Netherlands). The diameters of the precipitation rings were read after 72 h.

For the measurement of tear flow, the Schirmer I test was used in a routine fashion, using sterile filter paper strips measuring 5 x 35 mm (Whatman No. 41). The wetting of the filter paper strip was measured after 5 min. The rose bengal score was determined as described earlier.[7] The tear film break-up time was performed according to the method of Norn.[13] All tests were performed in the unanesthetized eye.

To achieve normal or near-normal distribution for statistical analysis, some of the data were transformed. For lysozyme and lactoferrin, the square roots of the diameter lysis c.q. precipitation were used. For the Schirmer test values, the logarithms were used. As the data for both the rose bengal score and the tear film break-up time were compatible with normal distribution, these were not transformed. The analysis of variance technique was used throughout as the statistical test.

3. RESULTS

Table 1 shows the patient characteristics, distributed according to sex, age, and tear fluid lysozyme concentration. No significant difference between the KCS-SJ and the KCS-NSJ groups could be demonstrated for any of these parameters.

With regard to the lactoferrin tear fluid concentration, a significant difference could be demonstrated between the groups. In the KCS-SJ group, the lactoferrin concentration in the tear fluid was lower than in the KCS-NSJ group. Also, the break-up time was significantly lower and the Schirmer values almost significantly lower in the KCS-SJ group (Table 2).

Although the lactoferrin tear fluid concentration and the tear film break-up time values were significantly lower in the KCS-SJ group and the Schirmer values nearly so, none of these tests alone could differentiate well between patients with and without Sjögren's syndrome.

For lactoferrin, if balanced, the probability of misclassification at the limit of 521 μg/ml is 41%; for the break-up time at the limit of 3.2 sec, this value is 53%, and for the

Table 1. Distribution of patients with KCS associated (KCS-SJ) and not associated (KCS-NSJ) with Sjögren's syndrome

Parameter	KCS-SJ	KCS-NSJ	Significance of difference between groups
Average age (yr)	62	60	NS
Sex (no.)			NS
M	4	5	
F	26	25	
Average lysozyme (HEL) concentration (μg/ml)	681	779	NS

NS, not significant.

Table 2. Lactoferrin tear concentration, Schirmer values, and tear film break-up time in patients with KCS with (KCS-SJ) and without (KCS-NSJ) association with Sjögren's syndrome

Parameter	KCS-SJ	KCS-NSJ	Significance of difference between groups
Lactoferrin concentration (μg/ml)	478	799	$p = 0.05$
Median Schirmer values (mm)	4	7	$p = 0.10$
Tear film break-up time (sec)	3	5	$p = 0.025$

Schirmer test at the limit of 5.5 mm wetting of the filter paper strip, it was no better. This means that, in practice, one fails trying to diagnose a patient with Sjögren's syndrome with any of the individual parameters at the stated limits.

As the values of lactoferrin, break-up time, and Schirmer test were lower in the group of KCS-SJ patients, combining the values of these tests, as in Hotelling's multivariate T 2 analysis, could provide a clinically useful method to defferentiate between patients with KCS associated with Sjögren's syndrome and those that are not.

For any individual patient, the discriminant function value (DFV) can be arrived at by using the formula: DFV = (-0.083 * the mm diameter of lactoferrin precipitation) + (-0.0219 * the mm wetting of the filter paper strip in 5 min) + (-0.1414 * the break-up time in sec). The higher the value, the more likely the KCS is associated with Sjögren's disease, and vice versa.

Table 3 shows the probabilities of misclassification at various DFV's. If for a certain patient a value of -0.78 is calculated, the KCS is likely associated with Sjögren's syndrome. Consulting Table 3 shows that in 20%, i.e., one in every five cases, this diagnosis would be wrong. A low value, say -2.41, is indicative of KCS not associated with Sjögren's disease, but in 3% of the cases this is not so.

These probabilities of misclassification indicate that calculating the DFV from the diameter precipitation of the lactoferrin test, the millimeters of wetting of the filter paper strip of the Schirmer test, and the tear film break-up time in seconds provides hardly a powerful diagnostic criterion to differentiate between these two group of patients. However, it does allow for a more rational policy of referral for the medical diagnosis of Sjögren's syndrome.

Table 3. Probability of misclassifying a patient with KCS with (KCS-SJ) and without (KCS-NSJ) associated Sjögren's syndrome at various discriminant function values

KCS-SJ probability of misclassification (%)	Discriminant function value equal or higher/lower than	KCS-NSJ probability of misclassification (%)
0	−0.40	100
20	−0.79	73
37	−1.19	33
53	−1.59	20
67	−1.99	3
73	−2.39	3
90	−2.79	0
100	−3.19	0

4. DISCUSSION

To make the two groups comparable, the patients were stratified according to sex, age, and one important tear function parameter—tear fluid lysozyme concentration. This parameter was chosen as we have most experience with this test. If tear fluid lactoferrin concentration had been chosen, the DFV would have been calculated from the diameter of lysozyme lysis, the Schirmer test values, and the tear film break-up time, with similar results.

It is rather surprising that the profile of the tear function parameters in patients with KCS associated with Sjögren's syndrome differs from patients with KCS not associated with any systemic disease, as in both categories the lacrimal gland is damaged, in the first group because of lymphocytic cell infiltration and in the second group presumably because of age-related involution. Yet there is some histopathological and immunochemical support for these clinical findings.

Gillette et al.[14] demonstrated that lysozyme is produced in 30% to 50% of acinar and ductal cells, whereas lactoferrin is secreted in 90% of acinar cells but not in ductal cells. In Sjögren's syndrome the acinar cells are destroyed early,[15] so differences in the relation of protein concentration in the tear fluid can be expected: a more pronounced decrease of lactoferrin, with little or no difference in the lysozyme tear fluid concentration.

It is also in the acinar epithelial cells that tear fluid is produced; therefore it is hardly surprising that the Schirmer values were lower in patients with KCS associated with Sjögren's syndrome; the decrease almost reached statistical significance. An explanation for the significantly decreased tear film break-up time is more difficult to provide because of the complex nature of this physicochemical phenomenon.

REFERENCES

1. Fleming A. On a remarkable bacteriolytic element found in tissues and secretions. *Proc R Soc Lond B Biol Sci.* 1922;93:306–317.
2. Broekhuyse RM. Tear lactoferrin: A bacteriostatic and complexing protein. *Invest Ophthalmol.* 1974;13:550–554.
3. Josephson AS, Wald A. Enhancement of lysozyme activity by anodal tear protein. *Proc Soc Exp Biol Med.* 1969;131:677–679.
4. Janssen PT, van Bijsterveld OP. The relations between tear fluid concentrations of lysozyme, tear-specific prealbumin and lactoferrin. *Exp Eye Res.* 1983;36:773–779.
5. Regan EF. The lysozyme content of tears. *Am J Ophthalmol.* 1950;33:600–605.
6. Bonavida B, Sapse AT. Human tear lysozyme: II. Quantitative determination with standard Schirmer strips. *Am J Ophthalmol.* 1968;66:70–76.
7. van Bijsterveld OP. Diagnostic tests in the sicca syndrome. *Arch Ophthalmol.* 1969;82:10–14.
8. Janssen PT, van Bijsterveld OP. A simple test for lacrimal gland function: A tear lactoferrin assay by radial immunodiffusion. *Graefes Arch Clin Exp Ophthalmol.* 1983;220:171–174.
9. Boersma MHG, van Bijsterveld OP. The lactoferrin test for the diagnosis of keratoconjunctivitis sicca in clinical practice. *Ann Ophthalmol.* 1987;19:152–154.
10. Janssen PT, van Bijsterveld OP. Origin and biosynthesis of human tear fluid proteins. *Invest Ophthalmol Vis Sci.* 1983;24:623–630.
11. van Bijsterveld OP. Standardization of the lysozyme test for a commercially available medium: Its use for the diagnosis of the sicca syndrome. *Arch Ophthalmol.* 1974;91:432–434.
12. van Bijsterveld OP, Mansour KH. Lysozyme in reflex and "basic" secretion. *Graefes Arch Clin Exp Ophthalmol.* 1983;221:130–132.
13. Norn MS. Desiccation of the precorneal tear film. I. Corneal wetting time. *Acta Ophthalmol.* 1969;47:865–880.
14. Gillette TE, Allansmith MR, Greiner JV, Janusz M. Histologic and immunohistologic comparison of main and accessory lacrimal tissue. *Am J Ophthalmol.* 1980;89:724–730.
15. Chomette G, Auriol M, Liotet S. Ultrastructural study of the lacrimal gland in a case of Sjögren's syndrome. *Scand J Rheumatol Suppl.* 1986;61:71–75.

135

CYTOKINES MAY PROVE USEFUL IN THE TREATMENT OF SJÖGRENS SYNDROME (SS) DRY EYE

Norman Talal,[1] Toru Nakabayashi,[1] John J. Letterio,[2] Liping Kong,[1] Andrew G. Geiser,[3] and Howard Dang[1]

[1]Department of Medicine
Division of Clinical Immunology
University of Texas Health Science Center
San Antonio, Texas
[2]National Institutes of Health
National Cancer Institute
Laboratory of Chemoprevention
Bethesda, Maryland
[3]Lilly Research Laboratories
Lilly Corporate Center
Indianapolis, Indiana

1. INTRODUCTION

Patients with Sjögren's syndrome (SS) often complain of xerophthalmia and xerostomia. The loss of tear and saliva production in SS is secondary to lymphocytic infiltration and destruction of the lacrimal and salivary glands. Neither the reason for lymphocyte accumulation nor the mechanism for acinar destruction is known. In this and other autoimmune diseases, research has focused on specialized functions of immune cells and their products (e.g., cytokines). Particular attention has centered on the interaction between MHC molecules and T-cell receptors and on specific autoantibody and autoantigen interactions. This has led to a large body of information on immune repertoire, but offers few clues for treatment. Our recent investigations in SS have focused more on generalized functions that immune cells share with other cells. Chief among these is the important homeostatic mechanism called apoptosis, or programmed cell death (PCD).

Lacrimal Gland, Tear Film, and Dry Eye Syndromes 2
edited by Sullivan *et al.*, Plenum Press, New York, 1998

2. APOPTOSIS (PROGRAMMED CELL DEATH, PCD) AND AUTOIMMUNE DISEASE

In biology, the death of a cell is not left to chance but is energy-dependent and elegantly directed as a vital part of the cell's molecular machinery. Apoptosis, or PCD, is currently a "hot area" in medical science, important in AIDS, cancer, autoimmune diseases such as systemic lupus erythematosus (SLE), and degenerative neurologic diseases such as Alzheimer's. The creation of the immunodeficiency in AIDS may in part be due to apoptotic killing of helper T (Th)-cells. In cancer, some anti-cancer drugs appear to work by inducing apoptosis in malignant cells. Moreover, PCD is regulated by oncogenes such as myc, p53, and bcl-2.

Apoptosis was of interest initially to embryologists studying cell deletion during fetal development. For example, apoptosis of fetal interdigital webs is responsible for the development of digits. PCD is responsible for normal involutional processes including ovarian follicular atresia. Many animal cells undergo apoptotic cell death when exposed to viruses. The cell, once aware of the virus, initiates its own suicide mechanism, blocking cellular production of additional viral particles.

For almost half a century, research in autoimmunity has been guided by the forbidden clone hypothesis of Sir MacFarlane Burnet. He proposed that lymphocytes with antigen-binding receptors for self fail to be eliminated from the body and, therefore, escape from tolerance. They persist, leading to autoantibody formation and tissue destruction. We now know that thymocyte elimination occurs through a process of negative selection mediated by apoptosis, or PCD. B-cells undergo negative selection by PCD in germinal centers unless positively selected by antigen rescue. Thus, positive and negative selection operates throughout the life of lymphocytes. Apoptosis is a major mechanism controlling lymphocyte balance through cell turnover and renewal. Immune cells are now studied not only for their MHC immunoglobulin and their T-cell receptors, but also for their ability to undergo apoptosis. The status of apoptosis has become important in human and murine autoimmune diseases.

Nucleosome bodies produced during apoptosis may become immunogenic[1] and induce a variety of autoantibodies to nucleoproteins (U1 RNP, Sm, Ro/SSA, histone). If this is the case, then accelerated apoptosis in human autoimmune rheumatic disorders, as recently demonstrated, could explain the enigmatic nature of the immunogens in rheumatic diseases and help resolve the controversy of true autoantigen vs. molecular mimicry.[2]

Fas/APO-1 is a transmembrane protein that mediates PCD. The relationship between PCD and autoimmunity first became apparent when a mutated *Fas*/APO-1 gene was found in genetically autoimmune susceptible MRL/*lpr* mice.[3] MRL mice homozygous for *lpr* (lymphoproliferation) or *gld* (generalized lymphoproliferative disease) develop a systemic autoimmune disease resembling SLE, SS, and rheumatoid arthritis (RA). There is autoantibody formation and an accumulation of non-malignant T-cells in the spleen and lymph nodes of these animals. Because of the genetic mutation, there is little or no *Fas* expressed in MRL/*lpr* mice. Similiarly, *gld* strains have defects in expressing *Fas* ligand. Lymphocytes fail to undergo PCD and thus accumulate in tissues, compromising organ function. Another autoimmune mouse strain homozygous for the *me*^{v} defect lacksaphophatase needed for the *Fas* death signal.[4,5]

Together, this shows that *Fas* is important in the development of autoimmune diseases.

3. SS PATIENTS

The problem of lacrimal and salivary gland destruction in SS patients can be broken down into two separate questions of lymphocyte biology: (i) What attracts immune cells to these sites in the first place? (ii) Once immune cells enter these target organs and commence the destructive process, why don't they ever leave? Lymphocytes will diminish if steroids or immunosuppressive drugs are given to the patient. Left untreated, however, the process is relentless.

The mechanisms responsible for the attraction of lymphocytes into a target organ is unknown, but genetic, viral, and hormonal factors may all be involved.[6,7]

It is generally assumed that this process represents an immune response to a viral or autoantigen.[8,9]

Our studies on apoptosis in salivary glands of SS patients[10] reveal that infiltrating lymphocytes are *Fas*- positive, thereby showing a commitment to PCD. However, these cells are also bcl-2-positive, suggesting that this suppressive oncogene may block PCD in the gland. This phenomenon of "blocked apoptosis" is most true of lymphocytes in the dense periductal nodules present in the SS lesion. Thus, the lesions of SS are analogous to the lesions of autoimmune MRL/*lpr* mice in that they arise as a consequence of impaired PCD. Lymphocytes fail to die by PCD because of a genetic defect in the mice and a regulatory defect in the SS patients.

A second finding in our studies suggests that the cause of dryness in SS is due to apoptotic death of acinar epithelial cells. Salivary gland biopsies in SS patients show that both *Fas* and *Fas* ligand are expressed by these acinar cells which then promote the death sequence by their interaction (Fig. 1).

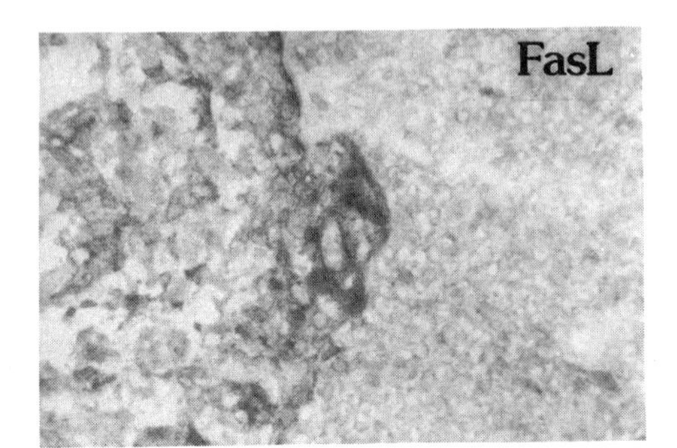

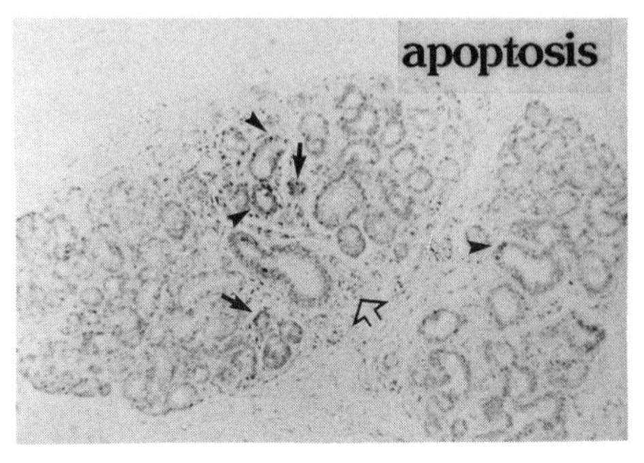

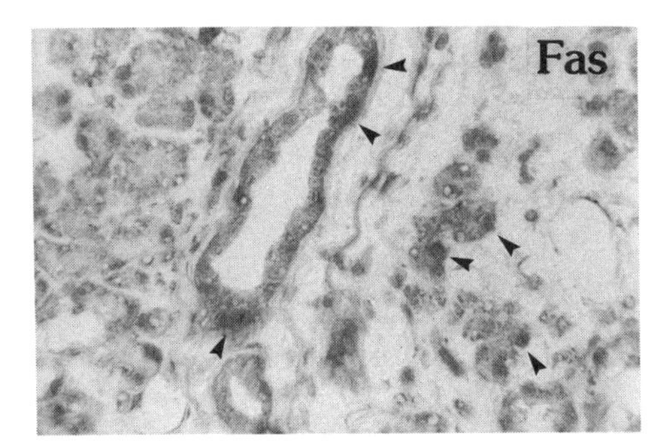

Figure 1. In situ staining for DNA fragmentation (ie, apoptosis), *Fas*, and *Fas*L in an SS salivary gland. Note that acinar cells (arrows) and epithelial cells (arrowheads) stain positive for apoptosis, and the infiltrating lymphocytes do not (open arrow). *Fas* L is seen on an epimyoepithelial island. (magnification, 109X)

4. TGF-β KNOCK-OUT MICE

A new mouse model for SS has appeared in the TGF-β1 (-/-) "knock-out" mouse.[11,12] TGF-β1 is a multipotent cytokine that has predominantly suppressive effects on the immune and inflammatory response. It is also an inducer of apoptosis in many different cells. Mice bearing the TGF-β1 null mutation (-/-) develop lymphoid infiltrates in the heart, lungs, lacrimal and salivary glands, and other organs similar to those seen in the pseudolymphoma of SS.[13,14]

We studied sera from TGF-β1 (-/-) mice and found elevated autoantibody levels to dsDNA, ssDNA, and Sm ribonucleoprotein.[12] No antibodies to Ro/SSA or La/SSB and no IgM rheumatoid factor were found. Serum autoantibodies were predominantly IgG and were specific as shown by ELISA inhibition studies. Antinuclear antibody patterns on Western blots varied from one mouse to the next, indicating a random process responsible for the diversity. Wild-type and heterozygote mice had no autoantibodies. Immunoglobulin glomerular deposits were found in TGF-β1 (-/-) mice, indicating that these autoantibodies may be pathogenic.

Treatment of TGF-β1 (-/-) mice with dexamethasone or TGF-β1 (-/-) failed to suppress autoantibody production. These mice represent an overlap combining the autoimmune serology of SLE with the tissue infiltrates of SS. These results support the concept that TGF-β1 is an important, naturally occurring immunosuppressive cytokine whose absence can lead to a systemic autoimmune disease.

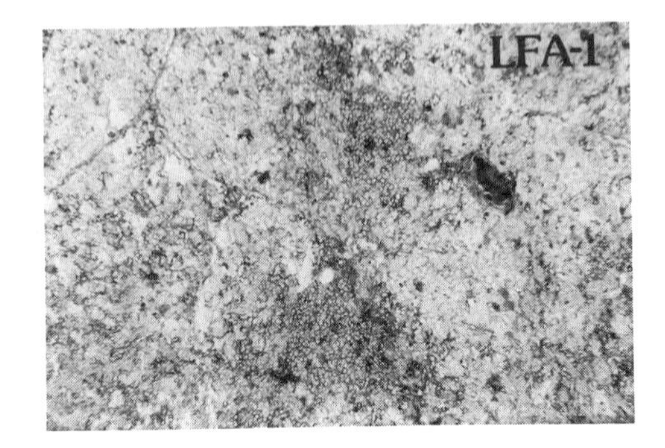

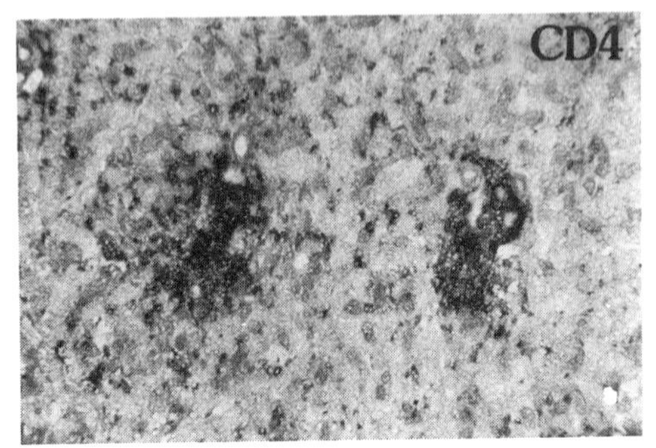

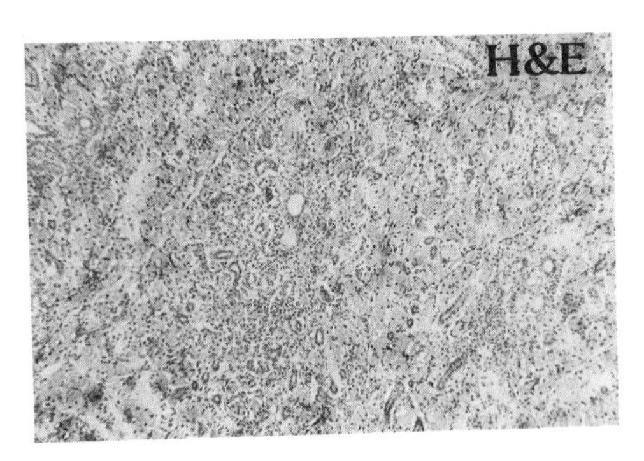

Figure 2. Immunohistochemical staining for CD4 and LFA-1 of a salivary gland from a TGF-β1 (-/-) mouse. Hematoxylin and eosin staining shows the typical mononuclear cells infiltrating the salivary gland of the TGF-β1 (-/-) mouse. (magnification, 109X)

Analysis of the salivary glands from TGF-β1 (-/-) mice show enhanced expression of MHC class II antigen, VCAM-1, and ICAM-1 expression (Fig. 2). The infiltrating lymphocytes in these mice express LFA-1, indicating their activated state. Furthermore, local production of IL-α, IL-1β, IL-2, IL-4, IL-6, IL-10, and γ-interferon was seen. These features resemble the features in SS patients.[15–17]

5. NEW STRATEGIES FOR TREATING AUTOIMMUNE DISEASE

Today much interest in the treatment of autoimmune disease centers on cytokines and their receptors. We know in great detail the factors that promote or inhibit immunity and autoimmunity, creating or failing to create a harmonious regulatory balance. Molecular targeting at the gene, mRNA, or protein level can influence this balance and restore immunoregulation in autoimmune states.

There are now four circumstances in which failure of apoptosis is associated with autoimmune disease: (i) MRL/*lpr* mice; (ii) tissue infiltrates of SS patients; (iii) bcl-2 transgenic mice where the suppressor oncogene results in long-lived B-cells that produce SLE-like autoantibodies and immune complex glomerulonephritis;[18] and (iv) TGF-β1 "knock-out" mice where the absence of this apoptosis-inducing cytokine keeps cells in continuous cycling, resulting in autoaggressive tissue lymphoid infiltrates resembling those seen in SS and SLE-like serum autoantibodies.

SS may soon be understood on the basis of inappropriate lymphocyte cell survival related to cytokine dysregulation and the function of oncogenes and autogenes.[19,20] Cytokine therapy to restore immunoregulatory balance could be an appropriate new treatment for immune-based dry eye syndromes.

REFERENCES

1. Casciola-Rosen LA, Anhalt G, Rosen A. Autoantigens targeted in systemic lupus erythematosus are clustered in two populations of surface structures on apoptotic keratinocytes. *J Exp Med.* 1994;179:1317–1330.
2. Ogawa N, Dang H, Talal N. Apoptosis and autoimmunity. *J Autoimmun.* 1995;8:1–20.
3. Watanabe-Fukunaga R, Brannan CI, Copeland NG, Jenkins NA, Nagata S. Lymphoproliferative disorder in mice explained by defects in *fas* antigen that mediates apoptosis. *Nature.* 1992;356:314–317.
4. Su X, Zhou T, Yang PA, Wang Z, Mountz JD. Hematopoietic cell protein-tyrosine phosphatase-deficient motheaten mice exhibit T cell apoptosis defect. *J Immunol.* 1996;156:4198–4208.
5. Mountz JD, Gause WC, Jonsson R. Murine models for systemic lupus erythematosus and Sjögren's syndrome. *Curr Opin Rheumatol.* 1991;3:736–756.
6. Talal N. Sjögren's syndrome: Historical overview and clinical spectrum of disease. *Rheum Dis Clin North Am.* 1992;18:1–9.
7. Talal N. In: Roitt IM, Delves PJ, eds. *Encyclopedia of Immunology.* Sex hormones and autoimmunity. London: Academic Press;1992;vol. 3:195–198.
8. DeKeyser F, Hoch SO, Takei M, et al. Cross-reactivity of B/B' subunit of the Sm ribonucleoprotein autoantigen with proline-rich polypeptides. *Clin Immunol Immunopathol.* 1992, 62:285–290.
9. Talal N, Mountz J. Viewpoint. Retrovirus, apoptosis, and autogenes. *Immunol Today.* 1993;14:532–536.
10. Kong L, Ogawa N, Nakabayashi T, et al. *Fas* and *Fas* ligand expression in salivary glands of patients with primary Sjögren's syndrome. *Arthritis Rheum..* 1997. 40:87–997.
11. Yaswen L, Kulkarni AB, Frederickson T, et al. Autoimmune manifestations in the TGF-β1 knockout mouse. *Blood.* 1995;87:1439–1445.
12. Dang H, Geiser A, Letterio J, Nakabayashi T, Kong L, Talal N. SLE-like autoantibodies and Sjögren's syndrome-like lymphoproliferation in TGF-β1 "knock-out" mice. *J Immunol.* 1995;155:3205–3212.
13. Shull MM, Ormsby I, Kier AB, et al. Targeted disruption of the mouse TGF-β1 gene results in multifocal inflammatory disease. *Nature.* 1992;359:693–699.

14. Kulkarni AB, Huh C-G, Becker D, et al. Transforming growth factor-β1 null mutation in mice causes excessive inflammatory response and early death. *Proc Natl Acad Sci USA*. 1993;88:2918–2921.
15. Ogawa N, Dang H, Lazaridis K, McGuff HS, Aufdemorte TB, Talal N. Analysis of TGF β and other cytokines in autoimmune exocrinopathy (Sjögren's syndrome). *J Interferon Cytokine Res*. 1995;15:759–767.
16. Fox RI, Bumol T, Fantozzi R, Bone R, Schreiber R. Expression of histocompatibility antigen HLA-DR by salivary gland epithelial cells in Sjögren's syndrome. *Arthritis Rheum*. 1986;29:1105–1111.
17. Fox R, Kang H-I, Ando D, Abrams J, Pisa E. Cytokine mRNA expression in salivary gland biopsies of Sjögren's syndrome. *J Immunol*. 1994;152:5532–5539.
18. Strasser A, Senga W, Vaux DL, Bath ML, Adams CS, Harris AW. Enforced bcl-2 expression in B-lymphoid cells prolongs antibody responses and elicits autoimmune disease. *Proc Natl Acad Sci USA*. 1991;88:8661–8665.
19. Mountz JD, Steinberg AD, Klinman DM, Smith HR, Mushinski JF. Autoimmunity and increased *c-myb* transcription. *Science*. 1984;266:1087–1092.
20. Mountz JD, Talal N. Retroviruses, apoptosis and autogenes. *Immunol Today*. 1993;14:532–536.

THE NOW AND FUTURE THERAPY OF THE NON-SJÖGREN'S DRY EYE

Gary N. Foulks

Department of Ophthalmology
Eye and Ear Institute
University of Pittsburgh
Pittsburgh, Pennsylvania

1. INTRODUCTION

The National Eye Institute/Industry Workshop Classification for Dry Eye was developed to coordinate clinical research in dry eye, but also provides a framework to consider the various therapeutic options for managing dry eye disease.[1] This classification differentiates between evaporative and tear-deficient dry eye, although it is important to recognize that both mechanisms for development of dry eye may occur in any given patient. Traditional therapy for the dry eye has concentrated on tear replacement with a variety of topical solutions. As our understanding of the structure and function of the tearfilm has improved, so has our knowledge of the important interactions this tearfilm has with the eyelid and the ocular surface.[2] We are able to appreciate the variety of ocular conditions that can produce or aggravate dry eye, and with this appreciation we can develop a practical strategy for its management.[3] This discussion proposes some therapeutic principles for the treatment of dry eye, and summarizes the things we presently do to treat dry eye, and then identifies new strategies that should soon be available for dry eye therapy. An important caveat before any treatment is employed, however, is that the patient must be fully educated as to the nature and characteristics of his/her dry eye disorder in order to assure understanding of the goals of therapy and to stimulate compliance with the recommended treatment.

2. TREATMENT PRINCIPLE #1

Treatment Principle Number One is to treat coexistent or ancillary conditions that provoke or aggravate dry eye. As summarized in the recent NEI/Industry Workshop Classification, a number of conditions contribute to increased tear evaporation or instability of the tearfilm, and these features should be treated if there is any hope of controlling the symptoms of dry eye. Meibomian gland dysfunction certainly contributes to tear instabil-

Lacrimal Gland, Tear Film, and Dry Eye Syndromes 2
edited by Sullivan *et al.*, Plenum Press, New York, 1998

ity and ocular surface desiccation.[4,5] The present mainstay of therapy for meibomian gland inspissation is intensive lid hygiene with application of hot compresses and subsequent mechanical massage or expression of the meibomian glands. Dry eye patients are at increased risk for bacterial blepharitis, and treatment of any associated infectious blepharitis should include intensive hygienic lid massage and topical antibiotic treatment with agents that are not toxic to the ocular surface (eg, erythromycin, bacitracin, sulfacetamide). Often systemic therapy with oral tetracycline or doxycycline is indicated to improve the character of the meibomian secretion.[6]

Eyelid dysfunction with incomplete lid closure or incomplete blink should be identified early in the course of therapy, as it may require eyelid surgery, such as lateral canthoplasty, to correct deficiencies in blink or an abnormally large evaporative surface.[7] When the ancillary problems have been corrected, attention can then effectively be directed to the inadequacy of tear secretion and the abnormalities of the ocular surface.

A search for concurrent use of systemic or topical medications that can decrease tear production is necessary, since the list of medications capable of compromising tear production is large and includes antihistamines, drugs with anti-cholinergic effects, and some anti-glaucoma medications.[8]

3. TREATMENT PRINCIPLE #2

Treatment Principle Number Two is to supplement the tears in eyes with decreased lacrimal production. There are well-defined differences between Sjögren's syndrome and non-Sjögren's syndrome associated tear insufficiency, [4] but the approach to tear supplementation is generally the same. In the past, a wide variety of formulations have been used including crystalloid, colloid, and polymer preparations. A useful categorization of these options has been collated by Juan Murube; it includes classification by function (osmotic, oncotic, surfactant, adsorbent, lubricant/viscosant, mucolytic, or lipid layer reconstructive), but can equally well be characterized by chemical composition (saline, lipid, monosaccharide, disaccharide, polysaccharide, gelatin, synthetic polymer, or organic fluid).[9] Similarly, the additives to these formulations include buffers, preservatives, antiseptics, astringents, vasoconstrictors, nutrients, and growth factors. Whatever the formulation, the goal has generally been to improve lubrication of the ocular surface and increase tear volume or tear retention time.

With the identification of the importance of hypertonicity of the tearfilm in pathogenesis of the ocular surface disorder in dry eye disease by Farris and Gilbard in the early 1970s, attempts to correct the hyperosmolality of the tearfilm led to application of hypotonic preparations to decrease tonicity of the tearfilm and to maintain integrity of the ocular surface.[10–12] The difficulty in reducing tear tonicity has been documented by who noted that topical instillation has a very short-term effect in lowering tonicity.[13] Nonetheless, the clinical studies with hypotonic tear supplements (Hypotears and TheraTears; ATF) have suggested that tonicity reduction can improve integrity of the ocular surface after 6 weeks of continuous therapy. Although further clinical trials are needed to verify the benefit of TheraTears, the initial clinical experience has been favorable.[14,15]

4. TREATMENT PRINCIPLE #3

Treatment Principle Number Three is to conserve the tearfilm. Generally this has meant attempts to decrease evaporation by the use of spectacles with side panels (either with or without humidification)[16,17] or by punctal occlusion. A number of methods for punctal occlusion are available, including the transient effect of biologic adhesive or collagen inserts, or the more permanent punctal plugs or canalicular inserts. Finally, thermal cautery, electrocautery, or laser occlusion is most definitive.[18] Temporary occlusion with dissolvable punctal inserts is advisable prior to permanent occlusion, to exclude that unusual patient who will experience epiphora, or overflow tearing after punctal occlusion.

5. TREATMENT PRINCIPLE #4

Treatment Principle Number Four is to heal the ocular surface. This is attempted by addition of various nutrients to the tear supplement. The retinoids have had mixed success in healing the surface, and although at least one large clinical trial showed no statistically significant response to retinoic acid in a broad range of dry eye patients,[19] there are documented cases of improvement in some patients with severe dry eye and reversal of keratinization of the conjunctiva.[20] The use of topical fibronectin has not been successful in a large multicenter clinical trial,[21] but some encouraging data are appearing with the use of synthetic RGP peptides20 and further investigation of these formulations is needed. Vitamin supplementation has been advocated and there is some anecdotal success, but prospective clinical trials are needed to prove efficacy.[22,23]

6. TREATMENT PRINCIPLE #5

Treatment Principle Number Five is to stimulate tear secretion. The future of dry eye therapy will probably include the use of tear stimulants, but to date the systemic side effects of therapy and occasional serious adverse events have limited this type of treatment. Systemic bromhexine has been used in Europe with efficacy in increasing tear production and relieving symptoms in a selected cohort of dry eye patients.[24] The findings that 3-isobutyl-1-methylxanthine (IBMX) stimulates tear secretion offer further options for therapy with derivatives or congeners of this chemical agent.[25]

The findings of increased tear secretion with topical cyclosporine therapy in canine dry eye[26] have not been confirmed in human trials. Although some clinical trials have demonstrated increased tear secretion by Schirmer I testing in humans during topical therapy,[27,28] other studies failed to find a stimulation of tear secretion.[29,30]

7. TREATMENT PRINCIPLE #6

Treatment Principle Number Six is to control inflammation. The work of Pflugfelder,[31] Tsubota,[4] and others in better characterizing the inflammatory nature of Sjögren's syndrome dry eye suggests that selective anti-inflammatory therapy should be effective in control of ocular surface disease. Certainly there is evidence that topical cyclosporine alters the inflammatory cell expression in the conjunctiva of patients with Sjögren's syndrome.[32] Although the recent clinical trials of topical cyclosporine have been complicated by difficulties with the delivery vehicle,[27,28] there is nonetheless evidence of clinical benefit in symptoms, punctate keratopathy, and mucus clumping of dry eye.[28] A newly formulated emulsion delivery system for cyclosporine is presently in clinical trials in the United States; results will be described later in this meeting.

8. TREATMENT PRINCIPLE #7

Treatment Principle Number Seven is to provide hormonal support to the lacrimal glands, meibomian glands, and ocular surface. The recent work of Sullivan,[33] Mircheff,[34] and associates demonstrate the importance of hormonal support of the lacrimal gland both in maintenance of function and in resistance to immune insult in the lacrimal glands. The role of androgenic hormone seems critical and suggests that future therapy of lacrimal dysfunction and dry eye may be not only supportive, but possibly prevent progressive disease. The finding that androgenic hormones also have a role in the maintenance of meibomian gland function suggests a further therapeutic benefit of androgenic hormonal support.[33] The development of non-virilizing androgenic agents is feasible and could provide the substratum needed for maintenance of lacrimal gland function. Reports of the beneficial effect of topical estradiol on the ocular surface of patients with dry eye finds continued support in recent clinical trials.[35] It is possible that a combination of estrogenic and androgenic hormones might demonstrate an additive salutory effect in dry eye patients.

The treatment principles for dry eye enumerated here are summarized in Table 1.

9. FUTURE DIRECTIONS IN THERAPY

The future continues to provide new options for therapy of the dry eye and we can expect to see advances in tear stimulants and agents to control inflammatory disease that causes or complicates dry eye disease. This prospect, along with better modalities and strategies to maintain lacrimal gland function, suggests that we may be able to treat dry eye disease better—or even to prevent it.

Table 1. Treatment principles for dry eye

Principle #1	Control coexistent conditions
	blepharitis, meibomitis
	eyelid abnormalities
	medications altering tear secretion
Principle #2	Supplement the tearfilm
	crystalloid
	colloid
	polymer
Principle #3	Conserve the tearfilm
	spectacles
	punctal occlusion
Principle #4	Heal the ocular surface
	? retinoids
	RGP peptides
	? vitamins
Principle #5	Stimulate tear secretion
	bromhexine
	? IBMX
Principle #6	Control inflammation
	cyclosporine
Principle #7	Provide hormonal support
	androgen
	estrogen

REFERENCES

1. Lemp MA. Report of the National Eye Institute/Industry Workshop on Clinical Trials in Dry Eyes. *CLAO J.* 1995;21:221–232.
2. Gilbard JP. Dry eye disorders. In: Albert DM, Jakobiec FA, eds. *Principles and Practice of Ophthalmology*. Philadelphia: Saunders; 1994:257–276.
3. Tsubota K, Toda I, Yagi Y, et al. Three different types of dry eye syndrome. *Cornea.* 1994;13:202–209.
4. Tsubota K. SS dry eye and non-SS dry eye: What are the differences? In: Homma M, Sugai S, Tojo T, Miyasaka N, Akizuki M, eds. *Sjögren's Syndrome: State of the Art.* Amsterdam: Kugler; 1994:21–26.
5. Mathers WD. Ocular evaporation in meibomian gland dysfunction and dry eye. *Ophthalmology.* 1993;100:347–351.
6. Foulks GN. Anterior segment inflammation and skin disease. *Int Ophthalmol Clin.* 1983;23:81–95.
7. Lemp MA. General measures in management of the dry eye. *Int Ophthalmol Clin.* 1987;27:36–43.
8. *Update on Dry Eye Syndromes*. Otsuka American Pharmaceutical, Inc. Key West, FL. Monograph. 1996.
9. Murube J. History of dakryology. In: Holly FJ, Lamberts DW, MacKeen DL. eds. *The Preocular Tearfilm in Health, Disease, and Contact Lens Wear.* Lubbock, TX: Dry Eye Institute. 1986;3–29.
10. Gilbard JP, Farris RL. Tear osmolarity and ocular surface disease in keratoconjunctivitis sicca. *Arch Ophthalmol.* 1979;97:1642–1646.
11. Farris RL, Stuchell RN, Mandell ID. Tear osmolarity variation in the dry eye. *Trans Am Ophthalmol Soc.* 1986;84:250–268.
12. Farris RL. Tear osmolarity: A new gold standard? *Adv Exp Med Biol.* 1994;350:495–503.
13. Holly FJ, Lamberts DW. Effect of non-isotonic solutions on tearfilm osmolality. *Invest Ophthalmol Vis Sci.* 1981;20:236–245.
14. Gilbard JP, Rossi SR. An electrolyte-based solution that increases corneal glycogen and conjunctival goblet-cell density in a rabbit model for keratoconjunctivitis sicca. *Ophthalmology.* 1992;99:600–604.
15. Gilbard JP. The tearfilm: Pharmacological approach and effects. *Adv Exp Biol Med.* 1994;350:377–384.
16. Hart DE, Simko M, Harris E. How to produce moisture chamber glasses for the dry eye patient. *J Am Optom Assoc.* 1994;65:517–522.
17. Tsubota K, Yamada M, Urayama K. Spectacle side panels and moist inserts for the treatment of dry-eye patients. *Cornea.* 1994;13:197–201.
18. Lamberts DW. Punctal occlusion. *Int Ophthalmol Clin.* 1994;134:145–150.
19. Soong HK, Martin NF, Wagoner MD, et al. Topical retinoid therapy for squamous metaplasia of various ocular surface disorders. A multicenter, placebo-controlled double-masked study. *Ophthalmology.* 1988;95:1442–1446.
20. Tsubota K. New approaches to dry-eye therapy. *Int Ophthalmol Clin.* 1994;34:115–128.
21. Nelson JD, Gordon JF. Topical fibronectin in the treatment of keratoconjunctivitis sicca. Chiron Keratoconjunctivitis Sicca Study Group. *Am J Ophthalmol.* 1992;114:441–447.
22. Holly FJ. Vitamins and polymers in the treatment of ocular surface disease. *Contact Lens Spectrum.* 1990;May:37–42.
23. Holly FJ. Diagnostic methods and treatment modalities of dry eye conditions. *Int Ophthalmol.* 1993;17:113–125.
24. Avisar R, Savir H. Our further experience with bromhexine in keratoconjunctivitis sicca. *Ann Ophthalmol.* 1988;20:382.
25. Gilbard JP, Rossi SR, Heyda K, Dartt DA. Stimulation of tear secretion and treatment of dry-eye disease with 3-isobutyl-1-methylxanthine. *Arch Ophthalmol.* 1991;109:672–676.
26. Kaswan RL, Salisbury MA, Ward DA. Spontaneous canine keratoconjunctivitis sicca, a useful model for human keratoconjunctivitis sicca: Treatment with cyclosporine eye drops. *Arch Ophthalmol.* 1989;107:1210–1216.
27. Laibovitz RA, Solch S, Andriano K, O'Connell M, Silverman MH. Pilot trial of cyclosporin 1% ophthalmic ointment in the treatment of keratoconjunctivitis sicca. *Cornea.* 1993;12:315–323.
28. Foulks GN, Pflugfelder SC, Lemp MA, Stern KI, Burk CT, Reis BL. A randomized double-masked clinical trial to assess efficacy and safety of 1% cyclosporin containing ophthalmic ointment versus placebo in patients with keratoconjunctivitis sicca associated with Sjögren's syndrome. Arvo Abstracts. *Invest Ophthalmol Vis Sci.* 1996;37:S646.
29. Gunduz K, Ozdemir O. Topical cyclosporin treatment of keratoconjunctivitis sicca in secondary Sjögren's syndrome. *Acta Ophthalmol.* 1994;72:438–442.
30. Helms HA, Rapoza P, Stern KI, Burk CT, Rosenthal A, Reis BL. A randomized double-masked clinical trial to assess safety and comparative efficacy of three concentrations of a cyclosporin containing ophthal-

mic ointment versus placebo in patients with keratoconjunctivitis sicca. Arvo Abstracts. *Invest Ophthalmol Vis Sci.* 1996;37:S646.

31. Pflugfelder SC, Huang AJW, Feuer W, et al. Conjunctival cytologic features of primary Sjögren's syndrome. *Ophthalmology.* 1990;97:985–993.
32. Power WJ, Mullaney P, Farrell M, Collum LM. Effect of topical cyclosporin A on conjunctival T cells in patients with secondary Sjögren's syndrome. *Cornea.* 1993;12:507–511.
33. Sullivan DA, Wickham LA, Krenzer KL, Rocha EM, Toda I. Aqueous tear deficiency in Sjögren's syndrome: Possible causes and potential treatment. In: Pleyer U, Hartman C, Steery W, eds. *Oculodermal Diseases- Immunology of Bullous Oculo-Muco-Cutaneous Disorders.* Buren; The Netherlands: Aeolus Press, 1997:95–152.
34. Mircheff AK, Warren DW, Wood RL. Hormonal support of lacrimal function, primary lacrimal deficiency, autoimmunity, and peripheral tolerance in the lacrimal gland. *Ocular Immunol Inflam.* 1996:145–172.
35. Lubkin V. Drugs for topical applications of sex steroid in the treatment of dry eye syndrome and methods of preparation and application. US Patent 5,041,434. August 20, 1991.

DRY EYE TREATMENT WITH EYE DROPS THAT STIMULATE MUCIN PRODUCTION

Takashi Hamano

Department of Ophthalmology
Osaka University Medical School
Suita, Osaka, Japan

1. INTRODUCTION

The main treatment curently available for dry eye is eye drops. Despite a large variety of eye drop formulations, they can only provide lubrication and increased tear film viscosity. These formulations can be characterized as first-generation eye drops. A more effective treatment would be possible with formulations that could stimulate aqueous or mucous secretion. We are currently developing formulations that directly stimulate mucin production. Such formulations would be second-generation eye drops.

I chose to develop a mucin stimulant based on experience in my outpatient clinic. Sometimes over 90% of the clinic patients are dry eye. Despite having many different types of dry eye, they all have a very short tear breakup time. It occurs in less than a few seconds.

Based on seeing thousands of dry eye patients, I was convinced that one of the best treatments would be to develop a stimulant of mucin production. This would stabilize the tear film. This is necessary to more effectively treat dry eye. The idea of developing stimulants of mucin production is based on therapies used in the treatment of gastritis and ulcers; some of those drugs stimulate mucin secretion. We started by screening these drugs, one of which is gefarnate. Here I would like to present the results of two experiments with gefarnate. In the first experiment, the effect of gefarnate on conjunctival goblet cells was determined. In the second experiment, its effects in a dry eye animal model were determined.

2. EFFECT OF GEFARNATE ON CONJUNCTIVAL GOBLET CELLS

2.1. Materials and Methods

Sixteen Japanese albino rabbits were divided into four groups. One group was untreated. One group received the vehicle. The remaining two groups received eye drops con-

Lacrimal Gland, Tear Film, and Dry Eye Syndromes 2
edited by Sullivan *et al.*, Plenum Press, New York, 1998

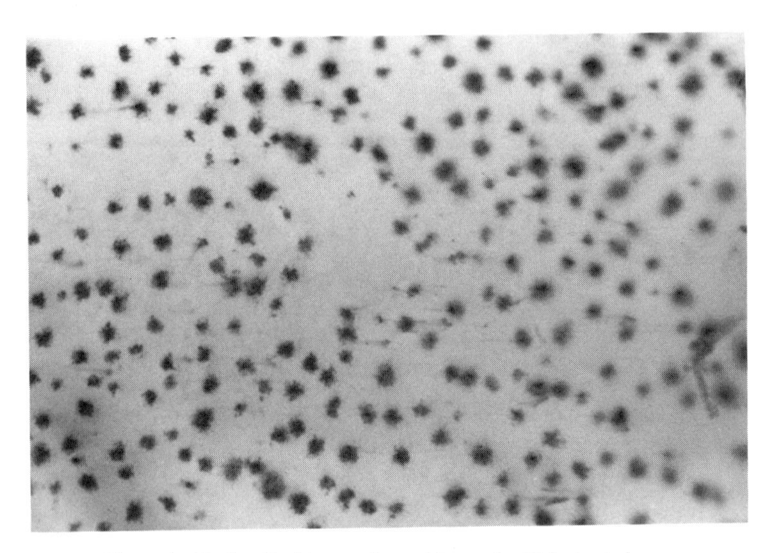

Figure 1. Conjunctival impression cytology of vehicle-treated eye.

taining 0.1% or 1.0% gefarnate, respectively. The drops were given in both eyes six times a day for 7 days. On the eighth day, impression cytology was done on the conjunctiva. At three points randomly selected in each specimen, the number of goblet cells were counted.

2.2. Results

Fig. 1 shows the conjunctival impression cytology after vehicle treatment, and Fig. 2 after treatment with 1.0% gefarnate eye drops. Fig. 3 shows the density of conjunctival goblet cells in the four groups. In the 1.0% treated group, there was a significant increase of goblet cells; the 0.1% treated group showed no increase.

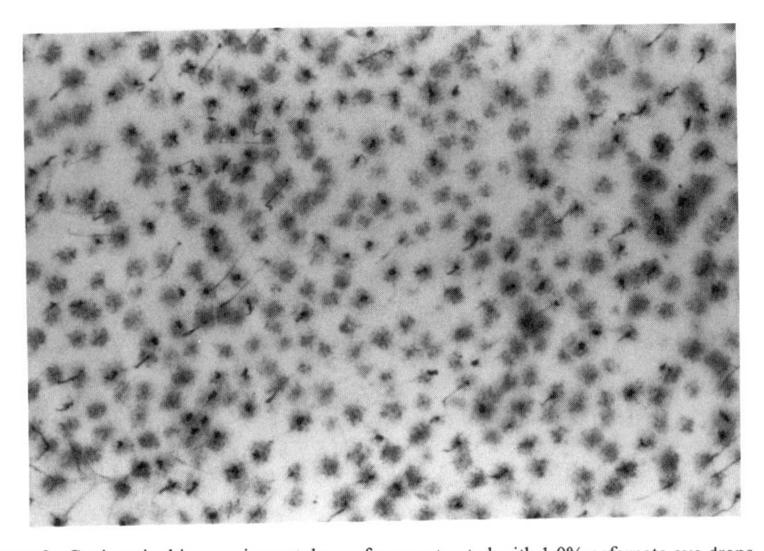

Figure 2. Conjunctival impression cytology of an eye treated with 1.0% gefarnate eye drops.

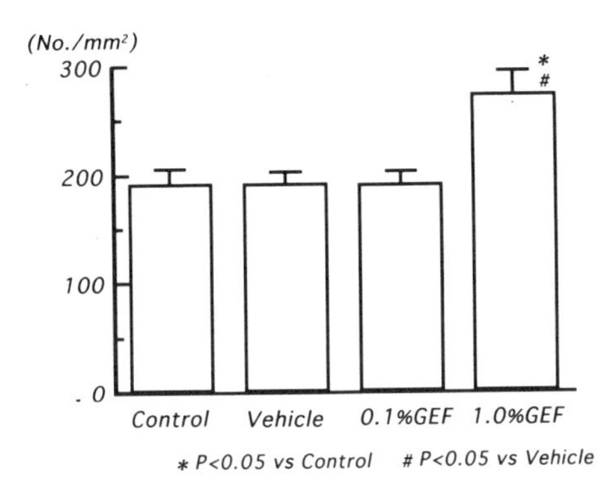

Figure 3. Density of conjunctival goblet cells.

3. EFFECT OF GEFARNATE AGAINST CORNEAL DESICCATION

3.1. Materials and Methods

For the dry eye animal model, 12 Japanese albino rabbits were used. They were divided into five groups: untreated control, treated with vehicle, treated with eye drops containing 0.1%, 0.3%, or 1.0% gefarnate. Rabbits were anesthetized with 25% carbamic acid ethyl ester, and their eyes were held open for 2 h with a speculum. Eye drops were instilled when desiccation was just starting. After 2 h of eye opening, all rabbits were sacrificed with an overdose of sodium pentobarbital. The whole globe was enucleated and stained with methylene blue. Then each cornea was dissected. Methylene blue from the cornea was extracted with acetone/sodium sulfate solution for 16 h. Its absorbance was measured at 660 nm with a spectrophotometer, and the amount of methylene blue was calculated.

3.2. Results

Fig. 4 shows the results of methylene blue quantitative analysis. In the 0.3% and 1.0% treated groups, there is significantly less staining, indicating that desiccation of the cornea caused less damage in these treated groups.

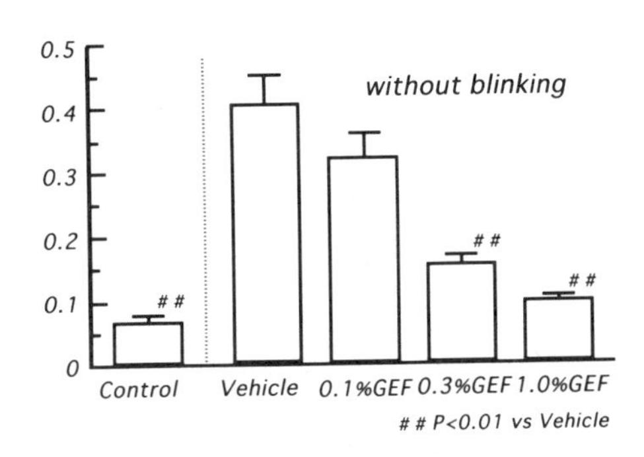

Figure 4. Absorbance of methylene blue extracted from rabbit corneas at 660 nm.

4. DISCUSSION

Although we do not yet have proof, we think that the results of these experiments suggest that gefarnate-containing eye drops stimulate the expression of transmembrane mucin and stimulate mucin production enough to decrease corneal epithelial damage in dry eye. Therefore, we think that gefarnate-containing eye drops are a candidate for clinical use in the treatment of dry eye.

138

A DOSE-RANGING CLINICAL TRIAL TO ASSESS THE SAFETY AND EFFICACY OF CYCLOSPORINE OPHTHALMIC EMULSION IN PATIENTS WITH KERATOCONJUNCTIVITIS SICCA

Joseph Tauber,[1] for the Cyclosporine Study Group*

[1]4400 Broadway
Kansas City, Missouri

*Cyclosporine Study Group:

Peter Donshik, MD
University of Connecticut
West Hartford, CT

Gary Foulks, MD
Duke University
Durham, NC

Harold Helms, MD
Birmingham, AL

Robert Laibovitz, MD
Eye Research Associates
Austin, TX

Marta Lopatynsky, MD
Contemporary Eye Associates
Bayonne, NJ

Earl Nelson, MD
Eye Surgery Center of Louisiana
New Orleans, LA

Peter Rapoza, MD
Cornea Consultants
Boston, MA

Dara Stevenson, MD
New Orleans, LA

Joseph Tauber, MD
Kansas City, MO

Allergan Ophthalmology Clinical Research:
Brenda Reis, PhD
Caroline Burk, PharmD
Katherine Stern, MS

1. KERATOCONJUNCTIVITIS SICCA: EPIDEMIOLOGY AND ETIOLOGY

Keratoconjunctivitis sicca (KCS), or dry eye, is a common diagnosis, affecting approximately 30 million individuals worldwide. Patients with KCS experience symptoms associated with a chronic ocular irritation and often complain of a dry, gritty feeling or the

Lacrimal Gland, Tear Film, and Dry Eye Syndromes 2
edited by Sullivan *et al.*, Plenum Press, New York, 1998

sensation of a foreign body in their eyes. These symptoms are primarily the result of an unstable ocular tear film and an abnormal ocular surface that has a diminished ability to respond to environmental challenges.

The etiology of KCS is varied and multifactorial. The disease may be the result of age-related decreases in systemic androgen support to the lacrimal gland or of systemic autoimmune diseases such as Sjögren's syndrome (an autoimmune disease that includes the triad of KCS, deficient salivation, and a connective tissue disease, such as rheumatoid arthritis). Recently, a growing body of research suggests that KCS is the result of an underlying cytokine and receptor mediated inflammatory process.[1–4]

2. TREATMENT OF KCS AND THE ROLE OF OPHTHALMIC CYCLOSPORINE

Until the middle to late 1980s, the treatment strategies for KCS were symptomatic and palliative in nature, consisting primarily of tear replacement, tear preservation, and parasympathomimetic tear stimulation. These treatment modalities provided partial to complete relief of the symptoms but did not address the underlying pathologic mechanism of KCS. Given the evidence that there may be an immune component to the etiology of KCS, agents that inhibit the underlying inflammatory process may prove beneficial in correcting the disorder.

In 1970, cyclosporines were first discovered in a mixture of metabolites produced by newly identified strains of *Fungi Imperfecti* isolated from soil samples. By 1972, scientists found that the cyclosporines had immunosuppressive properties. Researchers then purified cyclosporine from the 25 or more metabolites and conducted further studies to elucidate the pharmacology and toxicology of this compound. By 1977, their results indicated that this compound could specifically and reversibly inhibit immunocompetent T lymphocytes. In 1983, cyclosporine was approved by the FDA to prevent rejection in patients receiving organ transplantation.

The beneficial effects of cyclosporine on the immune response in transplantation patients, coupled with the fact that cyclosporine had been used successfully for nearly a decade in the treatment of canine KCS, prompted researchers to evaluate the compound in a number of human ocular autoimmune disorders.[5] Results of these studies have established the efficacy of ophthalmic cyclosporine in a variety of conditions, including KCS, corneal transplantation, vernal keratoconjunctivitis, and ligneous conjunctivitis.

The mechanism of action of cyclosporine in the treatment of patients with KCS may be twofold. Cyclosporine's primary mechanism of action is as an immune modulator that prevents T-cell-mediated immunoreactivity. Patients with KCS have significantly increased numbers of cytokine-secreting T-cells in their conjunctiva.[6,7] These cytokines may be responsible for the localized inflammation and irritation associated with KCS.

Cyclosporine has a local immunoregulatory effect in the conjunctiva that is most likely mediated through the local suppression of the T-cell population. In addition, cyclosporine may act as a lacrimomimetic agent. Based on observational reports of post-transplant patients, the effects of cyclosporine on tear production was evaluated in a cohort of patients before and after kidney transplantation. All patients received oral cyclosporine following the transplant procedure to prevent organ rejection. Tear production was increased significantly following transplantation, as compared to pre-transplant/pre-cyclosporine levels.[8]

In summary, cyclosporine ameliorates the chronic ocular surface inflammatory disease seen in patients with KCS, in part by suppressing the production of soluble mediators of inflammation in the tear film, and relieves the symptoms of dry eye by stimulating tear production. Therefore, this study was designed to assess the efficacy, tolerability, and safety of four concentrations of cyclosporine ophthalmic emulsion in the treatment of patients with KCS.

3. CYCLOSPORINE DOSE-RANGING STUDY IN PATIENTS WITH KCS

This double-masked, dose-ranging study tested 0.05%, 0.1%, 0.2%, and 0.4% cyclosporine ophthalmic emulsion in patients with moderate to severe KCS with or without Sjögren's syndrome. Patients used Refresh® between four to eight times daily for 2 weeks before being randomized to receive either a cyclosporine-containing emulsion or vehicle dosed twice daily for 12 weeks. Refresh® was used during the 12-week treatment period as needed, but not to exceed eight times daily. A 4-week post-treatment observation period followed, during which only Refresh® was used. Evaluations occurred at Weeks -2, 0, 1, 4, 8, 12, 14 (post-treatment Week 2), and 16 (post-treatment Week 4). Objective ophthalmic evaluation included Schirmer Tear Test, conjunctival cultures, brush cytology, tear protein profiles, intraocular pressure, visual acuity, and biomicroscopy evaluation of superficial punctate keratitis, rose bengal stain, tear break-up time, and tear film debris. Subjective assessments of functional disability due to KCS and the severity of symptoms of discomfort were also evaluated. Additional measures included trough and peak blood cyclosporine levels, chemistry and hematology panels, and the use of concomitant Refresh®.

Of the 162 patients enrolled in the study, 133 completed the trial and were considered evaluable. In general, results found the mean improvement-from-baseline in objective ophthalmic evaluations to be greater in patients randomized to the cyclosporine groups than in those who received vehicle, although the differences between the two groups were not statistically significant. Pairwise analysis of the treatment groups revealed that, at several timepoints, patients receiving cyclosporine exhibited significantly greater improvements in Schirmer Tear Test, superficial punctate keratitis, and symptoms of discomfort as compared to those randomized to the vehicle.

Clinically significant change-from-baseline improvement favored the 0.1% concentration of cyclosporine, although a clear dose-response effect was not demonstrated among the active treatment groups. Although not to the same extent as patients receiving active treatment, those randomized to vehicle did experience improvement in some objective signs and in symptoms of discomfort. The greater than expected results seen in patients treated with vehicle may be attributed to the oil-containing emulsion, which has been shown to possess a residence time of up to 3 h in the eye.

Peak whole blood cyclosporine levels were less than 0.2 ng/mL, approximately 100-fold lower than the 20 ng/mL level used to monitor patients receiving systemic cyclosporine. No serious treatment-related adverse events were reported. No cyclosporine-related increases in the number of organisms cultured during treatment were noted, and no ocular infections were reported.

4. CONCLUSIONS

Topical cyclosporine represents a significant step forward in the treatment of patients with KCS. Previous agents were strictly palliative in nature, but cyclosporine ophthalmic emulsion may represent the first therapeutic agent with a mechanism of action targeted at the underlying immune-mediated inflammatory response that occurs in patients with KCS. Results of this study found cyclosporine ophthalmic emulsion to be effective, well-tolerated, and safe. The absence of a clear dose-response among the concentrations evaluated suggests that a threshold effect was attained at the lower concentrations. We conclude that cyclosporine ophthalmic emulsion produces clinically significant change-from-baseline improvements in both the signs and symptoms of KCS and that further investigation of this promising agent is warranted.

REFERENCES

1. Kaswan RL, Martin CL, Doran CC. Rheumatoid factor determination in 50 dogs with keratoconjunctivitis sicca. *J Am Vet Med Assoc.* 1983;183:1073–1075.
2. Kaswan RL, Martin CL, Chapman WL. Keratoconjunctivitis sicca: Histopathologic study of nictitating membrane and lacrimal glands for 28 canine cases. *Am J Vet Res.* 1984;45:112–118.
3. Kaswan RL, Martin CL, Dawe DL. Keratoconjunctivitis sicca: Immunological evaluation of 62 canine cases. *Am J Vet Res.* 1985;46:376–383.
4. Stern M, Beuerman R, Fox R, Mircheff A, Pflugfelder S. A unified theory of the role of the ocular surface in dry eye. This volume.
5. Hess AD. Mechanisms of action of cyclosporine: Consideration for the treatment of autoimmune diseases. *Clin Immunol Immunopathol.* 1993;68:220–228.
6. Soukiasian SH, Rice B, Foster CS, Lee SJ. The T-cell receptor in normal and inflamed human conjunctiva. *Invest Ophthalmol Vis Sci.* 1992;33:453–459.
7. Power WJ, Mullaney P, Farrell M, Collum LM. Effect of topical cyclosporin A on conjunctival T-cells in patients with secondary Sjögren's syndrome. *Cornea.* 1993;12:507–511.
8. Green K, Palmer SL, Bowen PA. Systemic cyclosporine increases tear flow in renal allograft recipients. *Invest Ophthalmol Vis Sci.* 1994;35:S1963.

ORAL PILOCARPINE FOR SYMPTOMATIC RELIEF OF DRY MOUTH AND DRY EYES IN PATIENTS WITH SJÖGRENS SYNDROME

A. S. Papas,[1] M. M. Fernandez,[1] R. A. Castano,[1] S. C. Gallagher,[2] M. Trivedi,[2] and R. C. Shrotriya[2]

[1]Tufts University School of Dental Medicine
Boston, Massachusetts
[2]MGI PHARMA, Inc.
Minnetonka, Minnesota

1. INTRODUCTION

1.1. Sjögrens Syndrome

Sjögrens syndrome (SS) is an immunologic disorder characterized by progressive destruction of the exocrine glands leading to mucosal and conjunctival dryness.[1] Virtually any organ system of the body may be affected in the SS patient, but the disease progress is usually most striking in the salivary and lacrimal glands, where there is a progressive mononuclear cell infiltrate that generally leads to scarring.[2] Loss of exocrine function accompanying these tissue changes are responsible for many of the manifestations of SS, including the clinical symptoms of xerostomia and keratoconjunctivitis sicca (KCS).[2,3] Patients complain of severe dryness of the mouth and gritty, dry, burning, or itching sensations in the eyes.[2] Lack of saliva may cause oral discomfort, pain, inflammation, mucositis, dysgeusia, candidiasis, increased caries, and periodontal disease. This may lead to poor nutrition and difficulty with social interactions.

Current treatment of SS is geared toward symptomatic relief of mucosal dryness and meticulous oral hygiene, and includes artificial tears, ophthalmologic lubricating ointments, nasal sprays of normal saline, moisturizing skin lotions, frequent sipping of water, artificial saliva preparations, and oral fluoride treatments.[2] Treatment may also be directed to extraglandular manifestations of joint pains and organ involvement.

Lacrimal Gland, Tear Film, and Dry Eye Syndromes 2
edited by Sullivan *et al.*, Plenum Press, New York, 1998

1.2. Pilocarpine HCl

Pilocarpine is the chief alkaloid obtained from the leaflets of South African shrubs of the genus *Pilocarpus*. This alkaloid, in a dose-dependent fashion, can cause pharmacologic stimulation of exocrine glands in man, producing, among other actions, symptoms such as diaphoresis, salivation, lacrimation, gastric secretion, and pancreatic secretion.[3] It is through stimulation of exocrine secretions that pilocarpine produces increased salivary and lacrimal responses, which may, in turn, alleviate symptoms of dry mouth and dry eye.

Pilocarpine hydrochloride (Salagen® tablets, SF MGI PHARMA, Inc., a brand of pilocarpine HCl) has been shown to be effective and safe for treatment of xerostomia resulting from radiation damage to salivary glands.[4–7] Mechanistically it exerts its beneficial effect through cholinergic stimulation of salivary secretions from residual major and minor functional salivary gland tissue.

2. STUDY DESIGN

The objective of the current study was to determine the efficacy and safety of pilocarpine in the treatment of xerostomia and KCS associated with SS. The study was a double-blind, randomized, placebo-controlled trial conducted at multiple clinical sites. Subjects were randomized to one of two treatment groups for a period of 12 weeks: One group received a placebo, the other group received active treatment with 5 mg and 7.5 mg pilocarpine. Both groups were on a q.i.d. regimen. At Week 6, subjects had their dose escalated: placebo to placebo, and active treatment with 5.0 mg pilocarpine HCl increased to 7.5 mg pilocarpine HCl. If a subject was unable to tolerate the escalated dose, (s)he could return to the previous dose for the remainder of the study.

3. MATERIALS AND METHODS

3.1. Clinical Methods

3.1.1. Serological Screening Failures. To qualify for the study, patients were required to test positive for serological marker(s) (SS-A, SS-B, rheumatoid factor, and/or ANA) or have a positive labial biopsy. The baseline values for subjects who qualified were compared to baseline values for those with negative serology with respect to saliva output and perceptions of dryness at screening.

3.1.2. Dry Mouth Assessments. Five-minute collections of unstimulated (resting) whole saliva were obtained predose and at 30, 45, and 60 min postdose and weighed. Subjects ingested nothing by mouth for 60 min prior to collection. During sample collection, subjects were placed in a quiet room and asked to expectorate their saliva into a collection cup as frequently as would be comfortable over a 5-min period. Their heads were tilted slightly forward, as they were not allowed to swallow during this interval. Sialometry samples were collected upon study admission and at Week 6 and Week 12.

Global assessment of dry mouth symptoms was completed by each subject at Week 6 and Week 12 using a 100 mm visual analog scale (VAS).

3.1.3. Safety Assessments. Safety assessments included electrocardiograms (prior to study and at end of study), physical examination (prior to study and at end of study), vital signs (predose, Week 6, and Week 12), laboratory evaluations (predose and Week 12), and adverse experiences. Adverse experiences were reviewed from subjects' diaries throughout the study.

3.2. Statistical Methods

Statistical analyses of an intent-to-treat cohort (i.e., a cohort of all patients who received at least one dose of study medication and had at least one efficacy assessment postdose) were performed for dry mouth and dry eye assessments. Separate analyses were done for Week 6 and Week 12 results. In addition, a subset of the Week 6 and the Week 12 analyses, an endpoint analysis, was conducted that included the Week 6 results for subjects who did not complete the study and Week 12 results for all other subjects.

3.2.1. Serological Screening Failures. Salivary flow scores for Sjögren's subjects *vs.* non-Sjögren's subjects (screening failures) were compared using a t-test. VAS scores were compared for the two groups using Mann-Whitney U test for independent samples.

3.2.2. Dry Mouth Assessments. For each subject, the total change from baseline in saliva weight was calculated by summing the differences in saliva weights from predose to 30 min postdose, 45 min postdose, and 60 min postdose for each collection day. The mean total change between treatment groups was compared using a t-test. A t-test was also used to compare the mean VAS scores.

3.2.3. Dry Eye Assessments. A t-test was used to compare the mean VAS scores between treatment groups.

3.2.4. Safety Assessments. No statistical analyses were done for safety parameters.

4. RESULTS

Sixty subjects were enrolled at a single study site, of whom 58 completed all 12 weeks. Results for these 60 subjects are presented below. This population represented 23% (60/250) of this 11-site multicenter trial.

4.1. Serological Screening Failures

The salivary flow measurements were significantly ($p<0.01$) lower for the Sjögren's subjects enrolled into the study (0.837±1.226) than for the non-Sjögren's subjects (1.827±1.974). There was no statistically significant difference between the two groups in subjects' perceptions of dryness (VAS).

4.2. Demographics

Table 1 presents a summary of demographics for enrolled subjects.

Table 1. Subject demographics

Demographic	Pilocarpine (n = 30)	Placebo (n = 30)
Age (yr)		
Mean (SD)	55.8 (13.8)	61.2 (13.7)
Range	27-77	30-85
Gender		
Female	26	30
Male	4	0
Race		
Caucasian	29	29
Asian	0	1
Black	1	0

Table 2. Global dry mouth assessment

Analysis timepoints	Pilocarpine		Placebo		
	n	% Responded[b]	n	% Responded[b]	p Value
Week 6	28	46.4	28	17.9	0.0164
Week 12	26	65.4	27	25.9	0.0016
Endpoint	28	64.3	28	25.0	0.0012

[a]Includes patients who received at least one dose of study drug and had at least one efficacy assessment postdose.
[b]A patient marking the VAS at a point >55 mm, which indicated improvement, was considered a responder.

4.3. Efficacy Results

Global dry mouth assessment was statistically significant ($p \leq 0.05$) between pilocarpine and placebo for the analysis at Week 6, Week 12, and endpoint (Table 2).

Global dry eye assessment was statistically significant ($p \leq 0.05$) between pilcarpine and placebo for the analysis at Week 12 and endpoint (Table 3).

The objective measurement of change from predose in saliva weight was statistically significant ($p \leq 0.05$) in favor of pilocarpine over placebo following the first dose of study drug (admission) and at Week 6 and Week 12. Results are shown in Table 4 and Fig. 1.

Table 3. Global dry eye assessment

Analysis timepoints	Pilocarpine		Placebo		
	n	% Responded[b]	n	% Responded[b]	p Value
Week 6	28	35.7	28	28.6	0.5686
Week 12	26	53.9	27	25.9	0.0308
Endpoint	28	50.0	28	25.0	0.0456

[a]Includes patients who received at least one dose of study drug and had at least one efficacy assessment postdose.
[b]A patient marking the VAS at a point >55 mm, which indicated improvement, was considered a responder.

Table 4. Total change from baseline in saliva weight (grams)

Analysis timepoints	Pilocarpine		Placebo		
	n	Mean (SD)	*n*	Mean (SD)	*p* Value
Admission	30	0.4 (0.42)	30	−0.1 (0.31)	.0001
Week 6	28	0.6 (1.20)	28	0.0 (0.16)	.0156
Week 12	26	0.7 (0.79)	27	−0.0 (0.21)	.0002
Endpoint	28	0.7 (0.76)	28	−0.0 (0.21)	.0001

4.4. Safety Results

The most common adverse experience reported was sweating (73% of pilocarpine patients), which is consistent with the current labeling for Salagen® Tablets. No clinically significant adverse experiences and no serious drug-related adverse experiences were reported.

5. DISCUSSION

Except for the side effect of sweating, oral pilocarpine was well tolerated by the patients in the trial. We found that it takes at least 3 months for the salivary glands to hypertrophy sufficiently to show a statistically significant difference both objectively in the flow rate and subjectively as on the questionnair's global assessment of dry mouth and dry eye. We expect that the subjects who continued on the medication for 1 year will have an even better response.

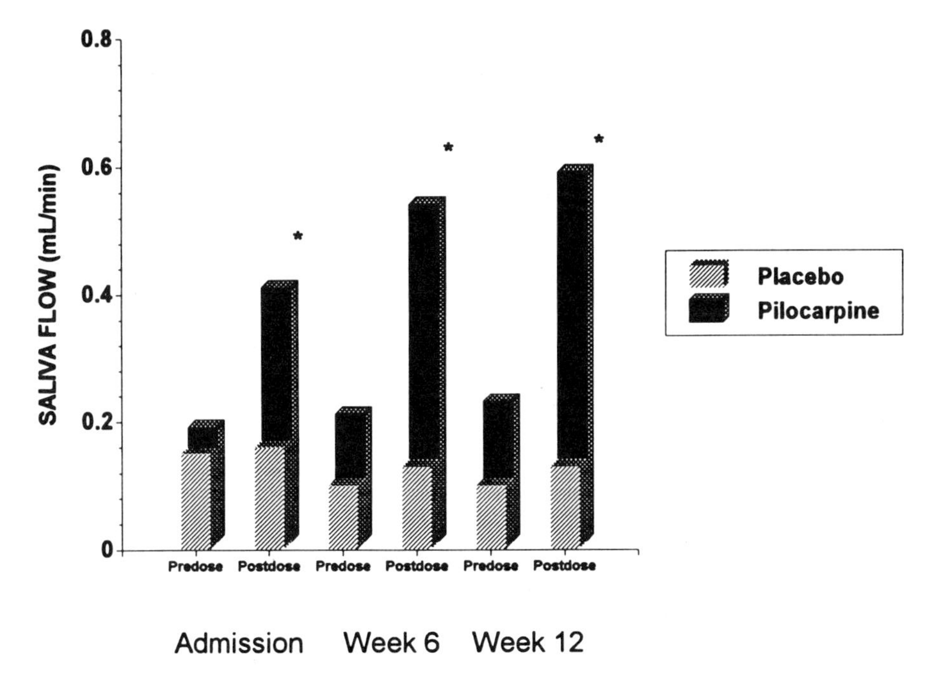

Figure 1. Saliva flow at various timepoints. *Statistically significant ($p \leq 0.05$) difference between pilocarpine and placebo.

6. CONCLUSIONS

- Global dry mouth assessment was significantly ($p \leq 0.05$) improved following pilocarpine treatment as compared with placebo for analysis at Week 6, Week 12, and endpoint.
- Global dry eye assessment was significantly ($p \leq 0.05$) improved following pilocarpine treatment as compared with placebo for analysis at Week 12 and endpoint.
- Increase from predose in salivary flow rate was significantly ($p \leq 0.05$) higher following pilocarpine treatment as compared with placebo at admission, Week 6, and Week 12.

REFERENCES

1. Moutsopoulos HM. Sjögren's syndrome. In: Isselbacher J, Braunwald E, Wilson JD, Martin JB, Fauci AS, Petersdorf RG, Root R, eds. *Harrison's Principles of Internal Medicine*, 12th ed. New York: McGraw-Hill; 1991:1449–1450.
2. Fox RI, Izuno G, Simon RA, Poceta S, Stewart R, Friedlaender MH. Sjögren's syndrome: A guide for patients. *Natl Sjögren's Syndrome Assoc.* 1990; (September 16):2:30.
3. Taylor P. Cholinergic agonists. In: Gilman AG, Goodman LS, Rall TW, Murad F, eds. *Goodman and Gilman's The Pharmacologic Basis of Therapeutics*, 8th ed. New York: Macmillan; 1990:122–130.
4. Fox PC, *et al.* Pilocarpine treatment of salivary gland hypofunction and dry mouth (xerostomia). *Arch Intern Med.* 1991;151:1149–1152.
5. Greenspan D, Daniels TE. The use of pilocarpine in post-radiation xerostomia. *Cancer.* 1987;59:1123–1125.
6. Johnson JT, *et al.* Oral pilocarpine for post-radiation xerostomia in patients with head and neck cancer. *N Engl J Med.* 1993;329:390–395.
7. LeVeque FG, *et al.* A multicenter, randomized, double-blind, placebo-controlled, dose-titration study of oral pilocarpine for treatment of radiation-induced xerostomia in head and neck cancer patients. *J Clin Oncol.* 1993;11:1124–1131.

ORAL PILOCARPINE FOR SYMPTOMATIC RELIEF OF KERATOCONJUNCTIVITIS SICCA IN PATIENTS WITH SJÖGREN'S SYNDROME

J. D. Nelson, M. Friedlaender, R. P. Yeatts, R. Yee, M. McDermott, S. Orlin, S. C. Gallagher, R. C. Shrotriya, and
the MGI PHARMA Sjögren's Syndrome Study Group

Ramsey Clinic & St. Paul Ramsey Medical Center
Scripps Clinic & Research Foundation
Bowman Gray School of Medicine
University of Texas Health Science Center-Houston
Kresge Eye Institute
Scheie Eye Institute
and MGI PHARMA, Inc.

1. SJÖGREN'S SYNDROME AND KCS

Sjogren's syndrome (SS) was first described by Henrik Sjögren, who documented association of rheumatoid arthritis and systemic immunopathy characterized by symptoms of dryness, especially KCS and xerostomia.

1.1. Classification of Sjögren's Syndrome

Although there is considerable controversy regarding criteria for classification it is generally agreed that SS may be classified as primary or secondary.[1,2] Primary SS consists of the features of KCS, xerostomia, the presence of autoantibodies (anti-IgA and anti-SS-A/B), and a positive focus score on minor salivary gland biopsy. Secondary SS consists of the features of primary SS in conjunction with overt clinical manifestations of immunoinflammatory connective tissue disorder(s).[3,4] Of these disorders, rheumatoid arthritis is the most common.[5]

1.2. Symptoms of Sjögren's Syndrome

Although virtually any organ system of the body may be affected, SS patients most often notice profound symptoms of dryness due to hypofunction of the salivary and lacrimal glands.[6] Patients complain of gritty, dry, or burning sensations in the eyes, and oral dryness, pain, irritation, mucositis, taste aberration, increased caries, and periodontal disease.[7] This devastating combination of symptoms can lead to a poorer quality of life characterized by decreased activity and difficulty with social interaction.[8]

2. ORAL PILOCARPINE FOR SYMPTOMS OF DRYNESS

Pilocarpine (P) is a cholinergic agonist that binds to muscarinic receptors and stimulates exocrine glands, eg, salivary and lacrimal glands. Pilocarpine hydrochloride tablets (SF Salagen® Tablets, MGI PHARMA) have been shown to provide clinically significant symptomatic relief of xerostomia from salivary gland hypofunction caused by radiotherapy for head and neck cancer.[9–11]

2.1. Clinical Trial Objectives

Two prospective, randomized, double-masked, placebo-controlled, multicenter clinical trials were conducted to determine the efficacy and safety of oral P as treatment for symptoms of xerostomia and KCS in patients with SS.

2.2. Clinical Trial Designs

The two multicenter trials included a total of 629 patients (oral P, 376; placebo, 253) in the intent-to-treat cohort (all patients who received at least one dose of study drug and had at least one efficacy assessment post-dose). Key efficacy parameters were global improvement of symptoms of dry mouth and dry eyes and increased salivary flow.

2.2.1. Dosage and Regimen. Study P92–01 was a fixed-dose study with 3 arms: P 10 mg/d, P 20 mg/d, and placebo. Study P92–02 was a 2-arm dose-adjustment study that compared P 20–30 mg/d to placebo. Dosing in both studies was four times a day for 12 weeks.

2.2.2. Diagnostic Criteria. Diagnostic criteria for identifying patients with KCS and xerostomia included baseline assessment of dry mouth symptoms, dry eye symptoms, and salivary flow. Diagnostic criteria for SS included the presence of autoantibodies to SS-A, and/or SS-B, and/or rheumatoid factor, and/or ANA. If subjects failed serology, a labial biopsy was performed and sent to a central lab (focus scores >1 focus/4 mm^2).

2.3. Clinical Trial Methods

Criteria for response included assessment of key symptoms associated with KCS and xerostomia by a questionnaire using a visual analog scale (VAS), a change in the quantity of tear and saliva substitute use, global assessment of KCS and xerostomia, and measurement of salivary flow. Assessments were made at Week 6 and Week 12 or at endpoint if patients discontinued (Table 1).

2.3.1. Global Improvement Assessment. In a global xerostomia and KCS assessment using a VAS questionnaire (100 mm scale), patients were asked to rate their present condition of dry mouth and dry eye compared with their condition at the start of the study. For

Table 1. Subsets of evaluated dryness symptoms (Studies P92–01 and P92–02)*

Mouth (Xerostomia)	Eyes (KCS)	Other
speaking	comfort	dry skin
eating	itching	vaginal dryness
sleeping	sense of foreign body	nasal dryness
oral comfort	light sensitivity	mucus production*
swallowing	blurring of vision*	
denture discomfort*	tiredness	
reduced sialogue use	redness	
	discharge or drainage*	
	matting or sticking	
	ability to focus*	
	difficulty driving at night*	
	reduced artificial tear use	

*Some subsets were included in only one study.

the primary efficacy variable of global improvement, responders were defined as those patients whose scores improved (>55 mm). Additional efficacy variables, i.e., subsets of dryness symptoms (Table 2), were also measured via a 100 mm VAS. Change from baseline was computed by subtracting the baseline score from each available post-baseline score (Week 6, Week 12, endpoint). Patients having an improvement (increase) of ≥25 mm were classified as responders.

2.3.2. Salivary Flow Assessment. Whole salivary flow was measured by collecting four 5-min saliva samples at each visit (one sample at pre-dose and three samples at timed intervals post-dose). Each collection was recorded as grams of saliva per 5 min. For analysis, these 5-min values were divided by 5 to obtain mean flow per minute. Total post-dose increase in salivary flow for the observed period was estimated by computing Area Under the Curve (AUC) using a trapezoidal rule.

2.4. Efficacy Results at Endpoint

Highly significant results were seen in both studies.

2.4.1. Global Improvement Results. In Study P92-01, significant improvement ($p \leq 0.0001$) was seen with oral pilocarpine in the global mouth assessment in an overall com-

Table 2. Efficacy summary (intent-to-treat cohort evaluated at endpoint)

	Study P92-01				Study P92-02		
	Oral P 10 mg/d	Oral P 20 mg/d	Placebo	*p*-Value overall ≤	Oral P 20-30 mg/d	Placebo	*p*-value pairwise ≤
Improved mouth global*	39.09	61.34	31.09	0.0001	57.85	27.64	0.0001
Improved eye global*	31.07	39.81	25.89	0.0259	48.36	24.59	0.0001
Salivary flow mean increase AUC	0.11	0.26	0.04	0.0001	0.22	0.01	0.0001

*% responders (improvement score of >55 mm on 100 mm VAS).
AUC, estimate of overall improvement of salivary flow over observed study period.

parison of P 10 mg/d, P 20 mg/d, and placebo. Overall comparison favored pilocarpine in the global eye assessment ($p < 0.05$). Additionally, there was a statistically significant decrease in the use of artificial saliva ($p < 0.05$).

In Study P92–02, significant improvement ($p \leq 0.0001$) was seen with oral pilocarpine in global mouth and global eye assessments in pairwise comparisons of P 20–30 mg/d and placebo. Additionally, there was a highly significant decrease in the use of artificial saliva ($p \leq 0.0001$) and artificial tears ($p < 0.01$).

2.4.2. Salivary Flow Results. Salivary flow mean increases were measured as AUC, defined as the estimated increase in saliva during the observed period. For Study P92–01, the AUC values of 0.04 for placebo and 0.26 for P 20 mg/d indicated a 650% overall improvement in salivary flow over the observed period. Similarly, for Study P92–02, the AUC values of 0.01 for placebo and 0.22 for P 20–30 mg/d indicated a 2200% (or 22-fold) overall improvement of salivary flow during the observed period. Salivary flow results for both studies were highly significant ($p \leq 0.0001$) (Table 2).

2.5. Safety Results

In both clinical trials, safety was assessed through reported adverse experiences, physical examination, lab analyses, ECG, and vital signs obtained periodically throughout the study.

2.5.1. Adverse Events. No serious drug-related adverse events were reported. The most common adverse experiences were sweating, headache, urinary frequency, nausea, dyspepsia, flushing, dizziness, and rhinitis. The side effect most often reported was sweating (Table 3).

2.5.2. Withdrawal from Studies. The incidence of withdrawal due to sweating was comparable (3.1% in both studies) at doses ranging from 20–30 mg/d. Of the 373 patients in Study P92–01, 16 patients (placebo, 5/125, or 4.0%; P 10 mg/d, 6/121, or 5.0%; P 20 mg/d = 5/127, or 3.9%) withdrew due to an event(s) with a potential relationship to study drug. In Study P92–02, 17 patients of the 256 enrolled (placebo, 6/128, or 4.7%; P, 11/128, or 8.6%) withdrew from the study due to an event(s) with a potential relationship

Table 3. Summary of safety (incidence of most common adverse experiences)

	Study P92-01 (% of patients)			Study P92-02 (% of patients)	
Adverse side effect	Placebo	Oral P 10 mg/d	Oral P 20 mg/d	Placebo	Oral P 20-30 mg/d
Sweating	7	11	43	7	64
Headache	25	21	15	13	16
Urinary freq.	2	11	9	6	15
Nausea	9	12	12	9	13
Dyspepsia	8	7	6	7	12
Flushing	2	2	10	3	9
Dizziness	8	5	10	6	4
Rhinitis	6	7	10	9	6
Withdrawals due to sweating	≤1	0	3	0	3

to study drug. The term "potential relationship to study drug" includes any event classified by the investigators as having a possible relationship to study drug, a probable relationship to study drug, or for which the relationship was unknown. A review of the incidence of adverse experiences for these studies indicates that doses up to P 30 mg/d in a q.i.d. regimen were well tolerated.

2.6. Conclusions

Oral pilocarpine (20–30 mg/d) is safe and effective in relieving symptoms of xerostomia and KCS in patients with SS. Significant improvements in these symptoms were seen in the pilocarpine-treated group vs. the placebo group. Dose adjustment (20–30 mg/d) may be desirable for maximum benefits with minimal side effects. The most common adverse effect was sweating, in addition to other minor cholinergic effects. No serious drug-related event occurred during these studies.

REFERENCES

1. Fox RI. Vth International Symposium on Sjögren's Syndrome: Clinical aspects and therapy. *Clin Rheumatol.* 1995;14(Suppl.1):17–19.
2. Fox RI, Robinson CA, Curd JG, et al. Sjögren's syndrome: Proposed criteria for classification. *Arthritis Rheum.* 1986;29:577–585.
3. Lemp MA. Report of the National Eye Institute/Industry Workshop on Clinical Trials in Dry Eyes. *CLAO J.* 1995;21:221–232.
4. Oxholm P, Asmussen K. Classification of disease manifestations in primary Sjögren's syndrome: Present status and a new proposal. *Clin Rheumatol.* 1995;14(Suppl.1):3–7.
5. Lemp MA. Management of the dry-eye patient. *Int Ophthalmol Clin.* 1994;34(1):101–113.
6. Fox RI, Izuno G, Simon RA, et al. *Sjögren's Syndrome: A Guide for the Patients.* National Sjögren's Syndrome Association. 1990:2–30.
7. Nelson JD. Dry eye: Etiology, diagnosis, and management." *Pharmacy Times.* 1996 September;66–78.
8. Moutsopoulos HM. Sjögren's syndrome. In: Isselbacher KJ, Braunwald E, Wilson JD., et al., eds. *Harrison's Principles of Internal Medicine*, 12th ed. New York: McGraw-Hill; 1991;1449–1450.
9. LeVeque FG, Montgomery M, Potter D, et al. A multicenter, randomized, double-blind, placebo-controlled, dose-titration study of oral pilocarpine for treatment of radiation-induced xerostomia in head and neck cancer patients. *J Clin Oncol.* 1993;11:1124–1131.
10. Johnson JT, Ferretti GA, Nethery WJ, et al. Oral pilocarpine for post-radiation xerostomia in patients with head and neck cancer. *N Engl J Med.* 1993;329:390–395.
11. Rieke JW, Hafermann MD, Johnson IT, et al. Oral pilocarpine for radiation-induced xerostomia: Integrated efficacy and safety results from two prospective randomized clinical trials. *Int J Radiat Oncol Biol Phys.* 1995;31:661–669.

SUPRACUTANEOUS TREATMENT OF DRY EYE PATIENTS WITH CALCIUM CARBONATE

Donald L. MacKeen,[1,2] Hans Walter Roth,[2,3] Marshall G. Doane,[4] and Patricia D. MacKeen[1]

[1]MacKeen Consultants Ltd.
Bethesda, Maryland
[2]Center for Sight
Georgetown University Medical Center
Washington, DC
[3]Institut für wissenschaftliche Kontaktoptik
Ulm, Germany
[4]Schepens Eye Research Institute
Boston, Massachusetts

1. BACKGROUND

1.1. Problems of Conventional Ocular Drug Delivery

The general mode of treatment of ocular conditions has not changed in thousands of years, ie, it has required the instillation of aqueous preparations or ointments onto the ocular surface or into the inferior cul de sac. These methods have many inherent problems: instillation difficulties, reflex washout from the solution's temperature and pH, eventual irritation from preservative ingredients required by multiple-dose aqueous preparations, wash-out of resident tear biomolecules, difficulty in resuspension of insoluble components, marked absorption across the large, permeable conjunctival surface, short duration of action, and vision blurring from ointments. There have been numerous attempts to mitigate some of these problems, e.g., addition of viscosity-increasing agents, creation of slowly soluble inserts, aerosolized delivery methods, single-use unpreserved solutions, etc. None, however, has addressed all of these problems.

1.2. Supracutaneous Drug Delivery

Supracutaneous drug delivery is an entirely novel, patented (US 5,366,739) method for patient self-treatment that eliminates the aforementioned problems. A drug, either particulate or non-particulate, is incorporated into a water-free, preservative-free hydrophobic ointment. This is applied supracutaneously by spreading it in a thin layer onto the skin of the lateral lower eyelid. Subsequently both vehicle and active component enter the eye; this has been shown by the response of dry eye patients[1,2] or by markers such as pilocarpine[3] or fluorescein.[2]

1.3. Formulation Rationale for Calcium Carbonate Ointment

Lacrimal gland fluid appears to be the only source of ionic calcium needed for the physiology of the avascular cornea. A certain minimal level must be maintained on the ocular surface for basic actions such as cell-to-cell adhesion. There have been many reports of the frequency of association between meibomian dysfunction and dry eyes and of the production of free fatty acids (FFA) during these conditions.[4] FFA are irritating, surface-active, probably a cause of chronic reflex tearing (with potential lacrimal gland exhaustion), and they act as a sink for tear ionic calcium by reacting to form an insoluble calcium soap. Because ionic calcium in tear fluid requires an active process in the lacrimal glands, its concentration could decrease to non-physiological levels because of lacrimal gland exhaustion plus its physiological sequestration as insoluble calcium fatty acid salts. We formulated calcium carbonate in a water-free ointment (US Pat. 5,290,572), because unlike soluble acid calcium chloride, calcium carbonate is relatively neutral, can act as a depot of ionic calcium to react with excess FFA, and can act as a source of ionic calcium when the concentration is below physiological needs.

2. MATERIALS AND METHODS

2.1. Preparations Used

Movement of particulate and non-particulate substances were observed by the use of several ointments such as Zinc Oxide (ZnO) Ointment USP and fluorescein sodium 10% in a glycerol-petrolatum vehicle. Sections of human lower eyelids had been examined in an attempt to detect the source of the supracutaneous ointment movement. Calcium carbonate 10% w/w in a glycerolpetrolatum (DEO™) ointment was used in the treatment of dry eyed patients.

2.2. Sites of Application

Although not considered an appropriate ophthalmic formulation, ZnO ointment USP was utilized because it could be readily observed during video-taping. Two spots of ZnO ointment were placed on the skin of the lateral lower lid to assess speed and direction of movement over the skin surface.

Fig. 1 shows the recommended site for application of a suitable ointment intended for supracutaneous administration—the lower lid in a swath about 3 mm wide between 6 o'clock and the lateral canthus. Application to the skin of the temporal canthus works equally well, but is not recommended for clinical practice because patients often place the ointment too high where it moves onto the upper lid. To visualize its movement, a minimal volume of fluorescein ointment, approximately 5 mg, was smeared with a cotton-tipped applicator to the recommended lower lid site.

Calcium carbonate ointment was tested on dry eye patients who had not responded adequately to conventional treatment[1,2] and to 24 gel and rigid gas-permeable contact lens

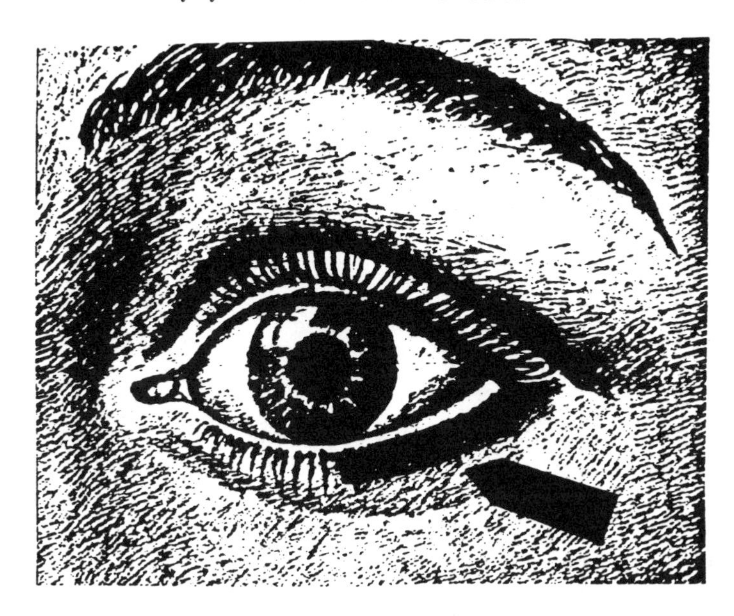

Figure 1. Arrow points to shaded area recommended for ointment application for supracutaneous drug delivery (ref. 5).

wearers who developed dry eye symptoms after 3 or more months of previously successful contact lens wear.[6] The most common complaint was a decrease in daily wearing times. These patients had not responded adequately to conventional instillation of artificial tears.

3. RESULTS

3.1. Movement of Zinc Oxide Ointment

Examination of the path in time showed a vertical movement from the spot of ZnO ointment placed nearer the center of the lid and a temporal movement from the spot placed nearer the canthus. After 20 min or more, one could see the whiteness of the ointment over the lateral half of the lower lid margin and on the edge of the *upper* lid.[6]

3.2. Entrance of Fluorescein into the Tears

The first appearance of fluorescence was a very narrow section, ca. 0.5 mm, of the lower meniscus at 4 o'clock. By 1 min, the width of the streak had increased five-fold. By 1.5 min, the fluorescent streak extended from 5 o'clock to the temporal canthus. By 2 min, the streak extended from 7 o'clock to a section of the upper meniscus. At this time, the fluorescence enhanced visualization of the irregularities of the temporal bulbar conjunctiva. At 6.5 and 7 min, there was increasing fluorescence from the preocular tear film. Fluorescence was observable for more than 15 min following application.[2]

3.3. Probable Motive Force for Supracutaneous Movement

Examination of coronal sections of the lower lid showed the muscles of Riolan immediately subjacent to the lid skin. These muscles appeared somewhat vertically oriented.

Table 1. Response following 1 to 10 weeks of treatment with DEO™

	Weeks of treatment with DEO									
Responses*	1	2	3	4	5	6	7	8	9	10
Improvement in signs and symptoms	40	10	2	2	2	2	1	–	–	–
Minimal improvement in signs and symptoms	58	–	–	–	–	–	–	–	–	–
Improvement in signs	40	10	8	1	–	–	–	–	–	–
No improvement in signs or symptoms	–	–	–	–	–	–	–	–	–	2
Worsened status	1	–	–	–	–	–	–	–	–	–

*Patients may have exhibited one or more changes; thus the total may not equal 60.
Reprinted with permission of publishers of ref. 1.

Unlike the deeper situated orbicularis orbis muscles, these fiber bundles appear situated close enough to the ointment application site to enable the supracutaneous movement observed.[2,3]

3.4. Treatment of Dry Eye Patients

DEO™ was used in the treatment of dry-eyed patients who obtained inadequate relief from conventional treatments.[1] It resulted in a statistically significant increase in meniscal height and BUT and an improvement in rose bengal staining and symptoms (Tables 1 and 2). Many patients found single daily applications effective in treating signs and symptoms.

3.5. Treatment of Dry Eye Contact Lens Wearers

Dry-eyed contact lens wearers benefited from the application of DEO™. Although it might seem irrational to apply an ointment during contact lens wear, it proved most effective. One patient inadvertently applied the ointment directly to her lenses; simple removal and normal cleaning corrected the situation. In a study of 24 patients, all achieved significantly (<0.01) longer daily wearing times following application of DEO two or three times daily.[6]

3.6. Pharmacological Response to Pilocarpine

The pharmacological response to a single application of pilocarpine in a glycerol-petrolatum vehicle applied to the lower lid skin was shown to be effective. Patients stud-

Table 2. Results termed total success, partial success, or failure following 1 to 10 weeks of treatment with DEO™

	Total success		Partial success		Failure	
Age (yr)	Male	Female	Male	Female	Male	Female
20-39	2	4	0	0	0	0
40-59	4	6	3	5	1	0
60-79	8	12	2	8	0	1
80	2	2	0	0	0	0

Reprinted with permission of publishers of ref. 1.

ied were ocular hypertensive persons without field loss; the average IOP was 29 mm Hg. After 2 h, the average IOP was 21 mm Hg. The pressure reached a plateau of approx. 18 mm Hg at 6 to 8 h and gradually increased to an average of 24 mm Hg at 20 h.[6]

4. DISCUSSION

Supracutaneous ophthalmic drug delivery is a simple, effective method of protracted ocular drug delivery. The only requirement for its effective use, regardless of the active ingredient, is contact between the lower lid and the bulb; this is necessary for delivery of the vehicle and drug from the placement site into the lipid of the lower meniscus. The delivery appears to be as follows: the subcutaneous musculature of the lower lid must move the ointment toward and over the lid margin; the meniscal surface of the lower margin has been reported[7] to possess the thickest layer of lipids. Based on our results, there is a ready admixture of endogenous tear lipids and exogenous lipid vehicle plus drug following the supracutaneous movement. This mixture spontaneously spreads across the aqueous phase of the preocular tear film on successive blinks.

The movement of the lipid toward the lid margin appears to be slow, based on the time delay between placement and presence, as judged by either the appearance of fluorescence or the pharmacological response seen with pilocarpine. The observed motion of the lower lid during normal blinks is lateral, ie, nasally during closing and temporally during the opening phase. Judging from the findings, there is a concomitant, previously unnoticed vertical movement of the lid skin from the action of the subjacent musculature. It is this natural movement that appears to gradually move portions of the lipid vehicle over the skin to mix with the lipid layer of the preocular tear film. Based on patient comments regarding improved visual acuity, the slow introduction of a neutral lipid enhances the optical characteristics of the abnormal precorneal tear film and does not adversely affect the normal tear film.

The initial site of action of supracutaneously delivered drugs is in the interpalpebral space; once in the aqueous phase, the secondary site of action is the conjunctival surface. This appears to be the desired sequence of action for most ophthalmic drugs: either the interpalpebral space for the treatment of dry eyes, or the penetration into the globe across the corneal surface. The protracted delivery of the drug to this space optimizes entrance through the desired location. The secondary location to which this method delivers the drug, ie, the conjunctiva, is in contrast to that delivered by conventional instillation. Moreover, this method, which delivers over a long period of time, should not disturb the tear volume, not wash off normal tear biopolymers such as mucin, etc., and not result in a sensitization to preservative ingredients because none is needed in this water-free ointment preparation.

Particulate or soluble drugs are appropriate for supracutaneous delivery. Particulate drugs such as calcium carbonate can neutalize free fatty acids within the lipid medium and, because of hydrophilicity, move into the subjacent aqueous medium of the tear film to provide a source of calcium and bicarbonate ions for normal physiology. Readily soluble drugs move readily into the aqueous portion of the tear film as evidenced by fluorescein sodium and can interact as dictated by their structure on the surface of the cornea (and subsequently the conjunctival surface once in the aqueous phase) or penetrate the ocular surface to effect an action within the tissue or within the eye.

Patient compliance is enhanced. The application of ointment is simple. It can be done without contortions or concern about injuring the eye or lids. The ointment is merely smeared over the lid surface. There is no concern regarding resuspension of particulate material such as with the steroids, because the ointment base maintains the suspended state.

5. CONCLUSION

Supracutaneous delivery to the eye eliminates all of the problems historically associated with conventional methods of dry eye treatment. This method is appropriate for any drug regardless of its relative solubility in water. Calcium carbonate ointment in the glyceryl-petrolatum vehicle (Dry Eye Ointment [DEO^{TM}]) has been found effective when administered supracutaneously in the treatment of dry eye conditions.

REFERENCES

1. MacKeen DL, Roth HW. A preliminary report on a new method for the treatment of dry eye. *Pract Optom.* 1995;6:132–134.
2. MacKeen DL, Roth HW, Doane MG. Ocular drug delivery by the LLD (lower lid delivery) method. ARVO Abstracts. *Invest Ophthalmol Vis Sci.* 1996;37:S77.
3. MacKeen DL, Roth HW. Revolutionary method of ocular drug delivery. ARVO Abstracts. *Invest Ophthalmol Vis Sci.* 1995;36:S166.
4. Dougherty LM, Osgood JK, McCulley JP. The role of wax and sterol esters of meibomian secretions in chronic blepharitis. *Invest Ophthalmol Vis Sci.* 1989;30:1958–1961.
5. Murube del Castillo J. *Dacriologica Basica.* Las Palmas: Sociedad Española de Oftalmologia. 1981;192.
6. MacKeen PD, Roth HW, MacKeen DL. Treatment of dry eyed contact lens wearers with DEO. ARVO Abstracts. *Invest Ophthalmol Vis Sci.* 1996;37:S72.
7. Doane MG. Contact lenses and lipid layers in the tear film. ARVO Abstract. *Invest Ophthalmol Vis Sci.* 1989;30:481.

PRECLINICAL SAFETY STUDIES OF CYCLOSPORINE OPHTHALMIC EMULSION

O. Angelov, A. Wiese, Y. Yuan, J. Andersen, A. Acheampong, and B. Brar

Allergan
Irvine, California

1. INTRODUCTION

Cyclosporine A (CsA) is an immunomodulator that inhibits T-cell mediated immunoreactivity.[1–4] In addition, data indicate that CsA may act as an anti-inflammatory, due to its phosphatase inhibition activity. Because of these actions, CsA ophthalmic emulsion formulation was developed by Allergan for treatment of keratoconjunctivitis sicca. Since the safety profile of this new formulation is very important for the clinical success of the drug, long-term safety studies were carried out as described below.

2. MATERIALS AND METHODS

2.1. Three-Month Ocular Study in New Zealand White Rabbits

Groups of ten rabbits/sex were treated with 0.05%, 0.2%, or 0.4% cyclosporine ophthalmic emulsion or the vehicle of 0.4% cyclosporine ophthalmic emulsion, three times daily, one drop in the left eye, for 3 months. The right eye served as an untreated control. Eight animals/sex/group were sacrificed after 3 months of treatment, and two animals/sex/group were allowed to recover for 1 month, after which they were sacrificed.

The parameters measured were: daily clinical observations; weekly body weight; weekly gross ocular observations; slit lamp biomicroscopy, ophthalmoscopy, hematology, and serum chemistry measured pretest, at the end of 1 month, 3 months, and recovery; end-of-study blood drug concentration (AUC, C_{max}, and C_{min}) using liquid chromatography-tandem mass spectrometry; gross observations at necropsy; organ weights; and histological evaluation of all tissues and organs.

Lacrimal Gland, Tear Film, and Dry Eye Syndromes 2
edited by Sullivan *et al.*, Plenum Press, New York, 1998

2.2. Six-Month Ocular Study in New Zealand White Rabbits

Groups of 15 rabbits/sex were treated with 0.05%, 0.2%, or 0.4% cyclosporine ophthalmic emulsion or vehicle of 0.2% cyclosporine ophthalmic emulsion three times/day, or 0.4% cyclosporine ophthalmic emulsion or the vehicle of 0.4% cyclosporine ophthalmic emulsion six times/day, one drop in the left eye, for 6 months. The right eye served as an untreated control. Ten animals/sex/group were sacrificed after 6 months of treatment, and five animals/sex/group were allowed to recover for 2 months, after which they were sacrificed.

The parameters measured were: daily clinical observations; weekly body weight; weekly gross ocular observations; slit lamp biomicroscopy and ophthalmoscopy pretest and at the end of 1 month, 3 months, treatment, and recovery; hematology and serum chemistry pretest, at the end of 3 months, treatment, and recovery; end-of-study blood drug concentration (AUC, C_{max}, and C_{min}) using liquid chromatography-tandem mass spectrometry; gross observations at necropsy; organ weights; and histological evaluation of all tissues and organs.

2.3. One-Year Ocular Study in Beagle Dogs

Groups of six beagle dogs/sex were treated with 0.4% cyclosporine ophthalmic emulsion or the vehicle of 0.4% cyclosporine ophthalmic emulsion six times/day, or 0.1% or 0.2% cyclosporine ophthalmic emulsion three times/day, one drop in the left eye, for 1year. The right eye served as an untreated control. Four animals/sex/group were sacrificed after 1 year of treatment, and two animals/sex/group were allowed to recover for 1 month, after which they were sacrificed.

The parameters measured were: daily clinical observations; weekly physical examinations; body weight and food consumption weekly during the first 3 months, monthly thereafter; weekly gross ocular observations; slit lamp biomicroscopy and ophthalmoscopy pretest and at the end of 1, 3, 6, 9, and 12 months and recovery; blood pressure and ECG pretest, at mid-study, and at end-of-study; hematology and serum chemistry pretest and at the end of 3, 6, 9, and 12 months and recovery; urine analysis pretest and at the end of 3, 6, and 12 months and recovery; blood drug concentration (AUC, C_{max}, and C_{min}) using liquid chromatography-tandem mass spectrometry during week 1 and at the end of study; gross observations at necropsy; organ weights; and histological evaluation of all tissues and organs.

This study was conducted at Corning Hazleton Laboratories in Vienna, VA.

3. RESULTS

3.1. Three-Month Ocular Study in New Zealand White Rabbits

Transient slight ocular discomfort and transient slight conjunctival hyperemia were observed in all treatment groups, including the vehicle group. There were no systemic adverse effects, and no compound-related histological changes in the eye or in any organs or tissues. Blood concentrations after the 0.05% treatment were generally below the limit of quantitation (0.1 ng/ml). The mean C_{max} values after 3 months of dosing with 0.2% and 0.4% cyclosporine ophthalmic emulsion were 1.48 and 0.721 ng/ml, respectively. (See Fig. 1.) AUC values for the 0.2% and 0.4% cyclosporine ophthalmic emulsion treatment groups were 4.52 and 4.28 ng•h/ml, respectively.

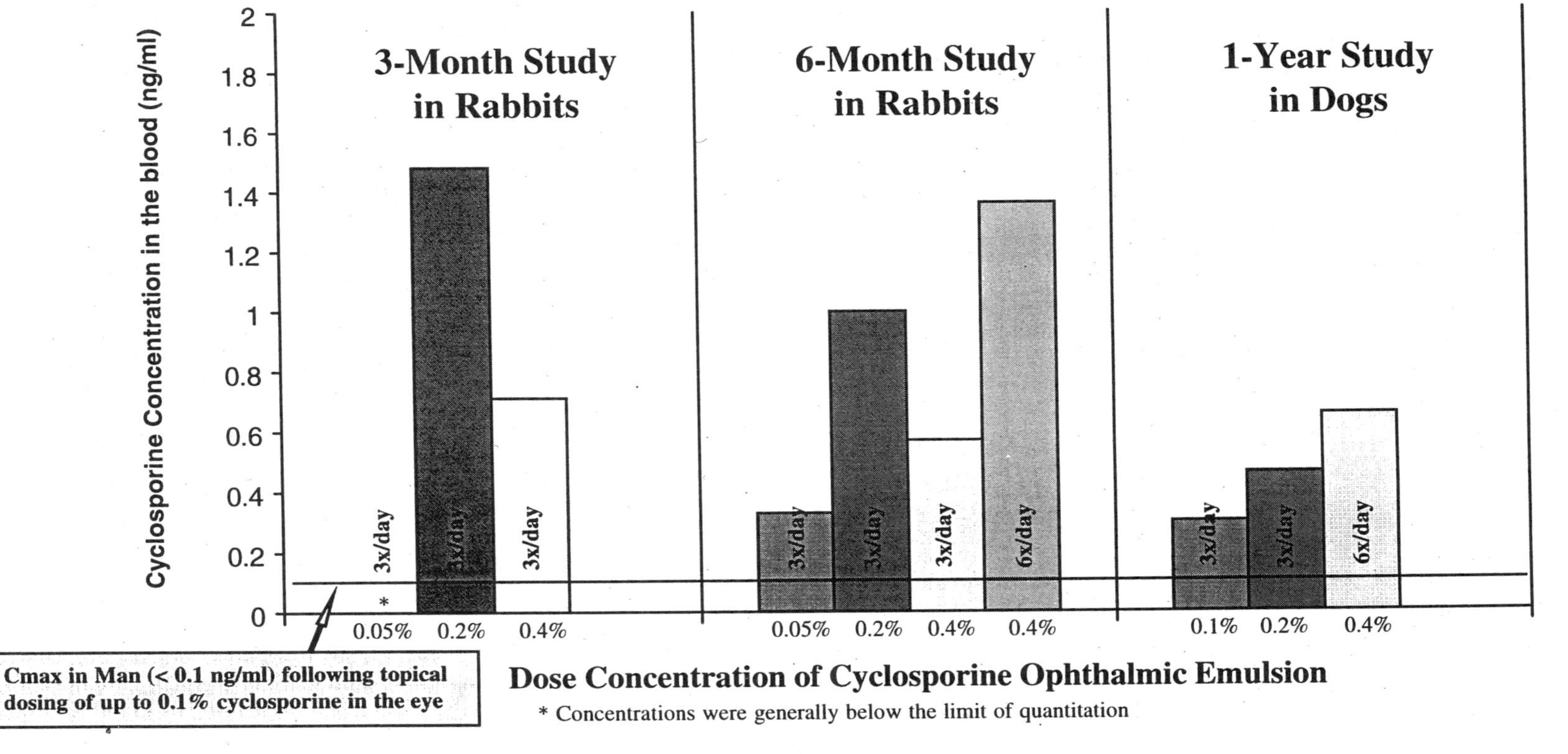

Figure 1. Mean peak blood cyclosporine concentration (C_{max}).

3.2. Six-Month Ocular Study in New Zealand White Rabbits

Transient slight ocular discomfort and transient slight conjunctival hyperemia were observed in all treatment groups, including the vehicle groups. There were no systemic adverse effects, and no compound-related histological changes in the eye or in any organs or tissues. The mean C_{max} values after 6 months of dosing with cyclosporine ophthalmic emulsion at concentrations of 0.05% TID, 0.2% TID, and 0.4% TID and six times daily were 0.328, 0.997, 0.570, and 1.36 ng/ml, respectively. (See Fig. 1.) AUC values for these groups were 3.48, 9.25, 6.85, and 16.7 ng•h/ml, respectively.

3.3. One-Year Ocular Study in Beagle Dogs

No ocular or systemic adverse effects were observed in any treatment group. There was a tendency toward an increased incidence of slight lacrimation of the treated eye in the group treated with 0.4% cyclosporine, which corresponds to the pharmacologic effect of this drug. There were no compound-related histological changes in the eye or in any organs or tissues. The mean C_{max} values after 1 year of dosing with cyclosporine ophthalmic emulsion at concentrations of 0.1% TID, 0.2% TID, and 0.4% six times daily were 0.299, 0.459, and 0.654 ng/ml, respectively. (See Fig. 1.) AUC values for these groups were 2.35, 3.39, and 9.55 ng•h/ml, respectively. There was no accumulation during this prolonged period of treatment.

4. DISCUSSION

In these animal studies, the concentrations and frequency of administration of cyclosporine ophthalmic emulsion represented an exaggerated dosing regimen compared with the proposed clinical regimen, which is once or twice daily administration of the maximum concentration of 0.1%. There were no clinical signs of toxicity in the eyes or systemically, and the histological examination of all tissues and organs did not reveal any compound-related findings.

The liquid chromatography-tandem mass spectrometry analytical method that was used to analyze cyclosporine concentrations in the blood is a significantly more sensitive method, with a limit of detection of 0.1 ng/ml, than the widely used radioimmunoassays and HPLC methods for cyclosporine, which have a limit of detection of 20 ng/ml. In the rabbit study and the dog study using concentrations of up to 0.4% cyclosporine ophthalmic emulsion six times daily, the majority of the individual peak blood drug concentrations were generally less than 1.5 and 1.2 ng/ml, respectively. This is at least 10 times higher than the peak blood drug concentrations in patients where the blood concentrations following ocularly administered 0.05 or 0.1% cyclosporine BID are less than 0.1 ng/ml.

5. CONCLUSION

The preclinical safety data in rabbits and dogs showed that cyclosporine ophthalmic formulations at concentrations up to 0.4% dosed as much as six times daily for up to 6 months in rabbits and up to 1 year in dogs did not show any topical or systemic toxicity. The systemic exposure to the cyclosporine formulation after ocular administration is very minimal. Therefore, cyclosporine ophthalmic formulation, at the proposed concentration, is judged to be safe for long-term application in humans.

REFERENCES

1. Borel JF, Feurer C, Gubler HU, Stahelin H. Biological effect of cyclosporin A: a new antilymphocytic agent. *Agents Actions*. 1976;6:468–475.
2. Borel JF. Cyclosporine: historical perspectives. *Transplant Proc*. 1983;15:3–13.
3. Shevach EM. The effects of cyclosporine A on the immune system. *Annu Rev Immunol*. 1985;3:397–423.
4. Kahan BD, Bach JF (eds). Proceedings of the Second International Congress on Cyclosporine. *Transplant Proc*. 1988;20, Suppl. 3:1–1131.

143

CONJUNCTIVAL IMPRESSION CYTOLOGY FROM DOGS WITH KERATOCONJUNCTIVITIS SICCA

Pre- and Post-Treatment with Topical Cyclosporine

Denise I. Bounous,[1] Kathleen L. Krenzer,[2] Renee L. Kaswan,[3] and Susan G. Hirsh[3]

[1]Department of Pathology
College of Veterinary Medicine
University of Georgia
Athens, Georgia
[2]New England College of Optometry
Boston, Massachusetts
[3]College of Veterinary Medicine
University of Georgia
Athens, Georgia

1. INTRODUCTION

Keratoconjunctivitis sicca (KCS), or dry eye, is a major cause of ocular morbidity in humans and dogs.[1,2] Although the cause of KCS in dogs often cannot be determined, as in non-Sjögren's dry eyes, histopathologic and serologic studies suggest that canine KCS frequently results from immune-mediated lacrimal gland destruction or dysfunction. Tear film deficiencies cause changes in the conjunctival epithelium that may account for many of the clinical symptoms and the corneal morbidity in these cases. The normal conjunctival surface is composed of nonkeratinized, stratified, squamous epithelium with goblet cells. These cells and the products from them, including mucin, are critical for ocular surface integrity. Squamous metaplasia from cuboidal epithelium to increasing keratinization and goblet cell loss is associated with KCS in patients with Sjögren's syndrome and with ocular problems in postmenopausal women.[2–4] Methods for staging squamous metaplasia by impression cytology have been established for use in humans.[2,5–7] Although KCS in dogs and response to treatment with cyclosporine have been characterized histologically by biopsy of lacrimal glands, there are no reports on conjunctival cytology from dogs with

KCS.[1,8] Clinical trials in dogs and humans have supported the validity of treating KCS with cyclosporine. In this project we characterized conjunctival cytology from dogs with KCS before and after treatment with cyclosporine in order to determine the effect of topical cyclosporine treatment on the conjunctival epithelium.

2. MATERIALS AND METHODS

Impression cytology was obtained from the conjunctiva of 15 dogs of various breeds. All dogs were diagnosed with idiopathic chronic KCS based on a Schirmer Tear Test (STT) (Alcon, Fort Worth, TX) $\leq$ 10 mm/min in one or both eyes and signs of conjunctivitis with or without keratitis, and no history of a precipitating factor such as distemper virus infection or sulfonamide treatment. Routine baseline hematology and chemistry panels were also performed to assess the dogs for pre-existing conditions. All dogs were housed in an animal care facility in accordance with the guidelines for animal care and use at The University of Georgia. Specimens were taken prior to and 6 weeks after twice-daily treatment with 0.2% topical cyclosporine (Allergan, Irvine, CA). Because different cytological changes have been observed in specimens from various sites of the bulbar conjunctiva in humans, collections were made from four areas.[7] Impression cytology from the dorsal nasal, dorsonasocentral, dorsocentrotemporal, and temporal limbus were collected onto nitrocellulose filter discs.[9] The preparations were stained with hematoxylin and evaluated in a masked fashion by two authors (DB, KK). Cytological characteristics of the preparations were graded by presence of goblet cells, degree of squamous metaplasia (Table 1, Fig. 1), and degree of inflammation. Data were collected per epithelial cell morphology and degree of inflammation for each quadrant. Individual scores from each quadrant of each dog were evaluated for any significant differences. The scores from the four conjunctival areas of each dog were then added together to acquire pre- and post-treatment total scores of morphology and of inflammation. These numerical scores were compared using a paired sample t-test.

3. RESULTS

No significant differences could be detected in specimens from different individual conjunctival areas. There was a numerical decrease in the total morphology score (total scores from all four areas), corresponding to improvement in epithelial cell morphology, of specimens from 12 of the 15 dogs (Fig. 2). Of these specimens, 4 scores were significantly improved at $p \leq 0.05$, and 1 was significant at $p < 0.10$. Specimens from 3 of 15 dogs had a numerical increase in scores, corresponding to a worsening in the epithelial cell morphology. Of these 3 specimens, 2 were significantly worse at $p \leq 0.05$. Inflammatory cells consisted primarily of neutrophils. After treatment with cyclosporine, specimens

Table 1. Scoring system for epithelial cells and inflammation in conjunctival cytology

Morphology	Grade	Inflammation	Grade
Goblet cells present	0	No inflammation	0
Cuboidal epithelial cells present	0	Trace inflammation	1
Mild squamous metaplasia	1	Focal inflammation	2
Moderate squamous metaplasia	2	Diffuse inflammation	3
Marked keratinization	3		

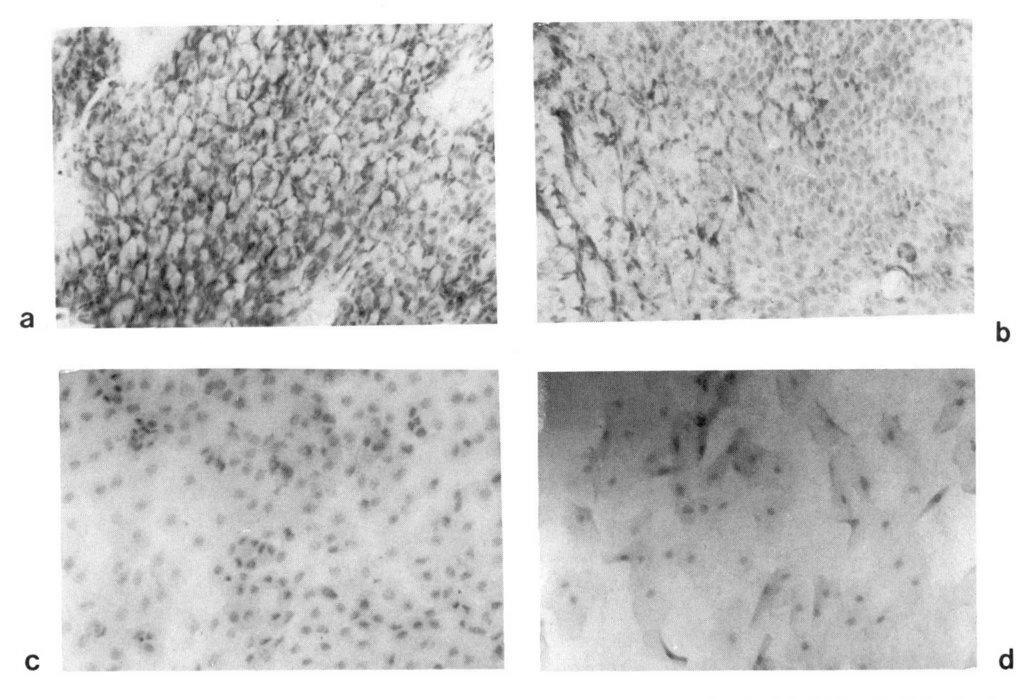

Figure 1. Cell types observed in canine conjunctival cytology. (a) Goblet cells. (b) Cuboidal epithelial cells. (c) Mild squamous metaplasia. (d) Marked squamous metaplasia with keratinization.

from 11 of the 15 dogs showed a decrease in the neutrophilic inflammation, and specimens from 4 dogs had an increase in inflammatory cells. Inflammation in 1 post-treatment specimen was significantly decreased and in 2 specimens was significantly increased.

4. CONCLUSIONS

Impression cytology is a well-established method for evaluating ocular surface disease in humans.[4,6,8–10] However, this is the first report of impression cytology in dogs. Our

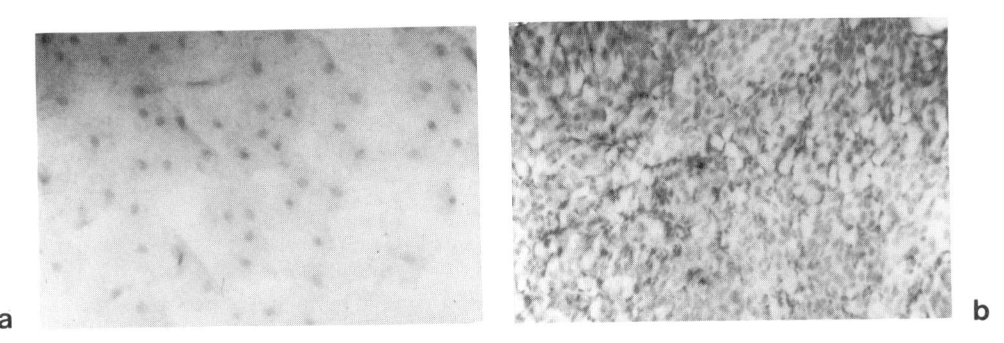

Figure 2. (a) Pre-treatment cytology from dog with KCS showing marked squamous metaplasia and goblet cell loss. (b) Post-treatment cytology of same dog after 6 weeks of topical cyclosporine, showing recovery of goblet cells and more cuboidal epithelial cells.

findings indicate that impression cytology is an appropriate method for use in dogs and may serve as a valid comparison for human studies.

In general, pre-treatment impression cytology specimens exhibited moderate squamous metaplasia; there was improvement to more cuboidal epithelium in the 6 weeks post-treatment specimens. The inflammatory cell infiltrate in most specimens consisted primarily of neutrophils. At 6 weeks post-treatment, improvement in the degree of inflammation was not significant. Our results suggest that the degree of squamous metaplasia of conjunctival epithelial cells, as determined by impression cytology, may be significantly lessened with topical cyclosporine treatment.

REFERENCES

1. Bounous DI, Kaswan RL, Hirsch S, Carmichael KP, Stiles J. Effects of ophthalmic cyclosporine on lacrimal gland pathology and function in dogs with keratoconjunctivitis sicca. *Vet Comp Ophthalmol.* 1995;5:5–12.
2. Budak K, Kaur AC, Gunduz K, Kilelioglu G, Mizrah B, Ekinci C. Conjunctival impression cytology in postmenopausal women. *Ann Ophthalmol.* 1996;28:250–253.
3. Hess AH. Mechanisms of action of cyclosporine: Considerations for the treatment of autoimmune diseases. *Clin Immunol Immunopathol.* 1993;68:220–228.
4. Takamura T, Takano H, Shinozaki K, Tsubota K, Kobayashi T. The effect of vitamin A on squamous metaplasia with Sjögren's syndrome. In: Homma M, Sugai S, Tojo T, Miyasaka N, Akizuki M, eds. *Proceedings of the Fourth International Symposium on Sjögren's Syndrome.* Amsterdam: Kugler Publications; 1993:253–255.
5. Kaswan RL, Martin CL, Chapman WL. Keratoconjunctivitis sicca: Histopathologic study of nictitating membrane and lacrimal gland from 28 dogs. *J Vet Res.* 1984;45:112–118.
6. Nelson JD, Drake MM, Brewer JT, Tuley M. Evaluation of a physiological tear substitute in patients with keratoconjunctivitis sicca. *Adv Exp Med Biol.* 1994;350:453–457.
7. Tseng SCG. Staging of conjunctival squamous metaplasia by impression cytology. *Ophthalmology.* 1985;92:728–733.
8. Kaswan RL, Salisbury MA. Spontaneous canine keratoconjunctivitis sicca, a useful model for human keratoconjunctivitis sicca: Treatment with cyclosporine eye drops. *Arch Ophthalmol.* 1989;107:1210–1216.
9. Krenzer KL, Freddo TF. Patterns of cytokeratin expression in impression cytology specimens from normal human conjunctival epithelium. *Adv Exp Med Biol.* 1994;350:289–293.
10. Liotet S, Kogbe O, Wattiaux MJ. Sjögren's syndrome diagnosis: A comparison of conjunctival and gingival impressions and salivary gland biopsy. *Adv Exp Med Biol.* 1994;350:661–665.

CYCLOSPORINE DISTRIBUTION INTO THE CONJUNCTIVA, CORNEA, LACRIMAL GLAND, AND SYSTEMIC BLOOD FOLLOWING TOPICAL DOSING OF CYCLOSPORINE TO RABBIT, DOG, AND HUMAN EYES

Andrew Acheampong, Martha Shackleton, Steve Lam, Patrick Rudewicz, and Diane Tang-Liu

Allergan
Irvine, California

1. INTRODUCTION

Cyclosporine is an immune modulator that inhibits T-lymphocyte-mediated immunoreactivity. Allergan is currently evaluating the clinical efficacy of 0.05%-0.4% cyclosporine emulsion for the treatment of immuno-inflammatory eye diseases, such as keratoconjunctivitis sicca, or dry eye syndrome. Topical ocular application of cyclosporine, formulated as 2% cyclosporine in olive oil, 0.2% cyclosporine in corn oil ointment (Schering-Plough), or 0.2% cyclosporine emulsion (Allergan), was found to reduce ocular surface inflammation and improve lacrimal gland secretion in dogs with KCS.[1–3]

The aim of the present research was to determine the ocular tissue distribution of cyclosporine in rabbits and dogs, and to compare tissue concentrations in rabbits, dogs, and humans after topical administration. Determination of relationships between the ocular tissue drug concentrations and efficacy is important for optimizing delivery of pharmacologically active concentrations in the target ocular surface tissues, providing support to the local mechanism of action, and optimizing dosing regimen.

2. METHODS

2.1. Animal Studies

[Mebmt -^{3}H]-cyclosporin-A was prepared by Amersham (UK) with radiochemical purity greater than 98%. Female New Zealand white rabbits (2–3 kg) received a single 50

Lacrimal Gland, Tear Film, and Dry Eye Syndromes 2
edited by Sullivan *et al.*, Plenum Press, New York, 1998

μl dose of 0.2% ^{3}H-cyclosporine formulation (~1 mCi/ml) into the lower conjunctival cul-de-sac of the left eye. Male beagle dogs (10–13 kg) received a 35 μl dose of 0.2% ^{3}H-cyclosporine emulsion (~1 mCi/ml) into the lower conjunctival cul-de-sac, twice daily for 7 days. Ocular tissues and systemic blood were also collected at selected time points over a 96-h period postdose. Two dogs or four rabbits were used per time point. The rabbit experiments were conducted according to USDA and Allergan ACUC guidelines. The dog study was conducted at Huntingdon Life Sciences. Tissue radioactivity concentrations were expressed as ng equivalents (eq) of cyclosporine per gram of tissue, using the specific activity of the dose formulation.

2.2. Human Range-Finding Study

One hundred sixty-two human subjects with KCS received an eyedrop of vehicle or 0.05%, 0.1%, 0.2%, or 0.4% cyclosporine emulsion twice daily for 12 weeks. Blood samples were collected from all subjects at morning troughs after 1, 4, and 12 weeks of dosing. In addition, blood samples were collected from selected subjects at 1, 2, and 4 h after the last dose at week 12. Cyclosporin A (CsA) concentrations in blood samples were measured by a validated liquid chromatography-tandem mass spectrometry (LC/MS/MS) method with Cyclosporin G as the internal standard. The lower limit of quantitation of the blood assay was 0.1 ng/ml.

3. RESULTS AND DISCUSSION

Figs. 1 and 2 depict the time course of cyclosporine in tears, ocular surface tissues, and orbital lacrimal gland of rabbits and dogs after eyedrop instillation of 0.2% ^{3}H-cyclosporine emulsion. Significant cyclosporine concentrations (C_{max}, ~1000 ng/g) were found in the conjunctiva and cornea, the target tissues for CsA reduction of ocular surface inflammation. The 0.2% emulsion provided approximately 7-fold higher cyclosporine concentrations in the rabbit cornea and conjunctiva than those for 0.2% cyclosporine in pure castor oil.[4] The lacrimal gland C_{max} was several-fold that of blood (~1 ng-eq/g), especially in the dog.

The ocular absorption and disposition of cyclosporine in rabbits and dogs were characterized by rapid absorption into ocular and extraocular tissues, reservoir effect of the cornea, relatively low intraocular tissue concentrations, and a long terminal elimination half-life of 20–44 h in most ocular tissues (Figs. 1 and 2). Similar ocular distribution characteristics were noted in previous rabbit and human studies.[4–7]

Table 1 shows less than 0.2 ng/ml blood concentrations in humans following multiple topical instillation of 0.05%, 0.1%, 0.2%, and 0.4% cyclosporine ophthalmic emulsion over a 12-week period of dosing. The systemic blood CsA concentrations in humans after topical CsA doses of the emulsions were much lower than the blood trough concentrations of 20–100 ng/ml used for monitoring the safety of patients receiving systemic cyclosporine therapy.

4. CONCLUSIONS

Topically applied cyclosporine emulsion can produce significant concentrations in the cornea and conjunctiva to exert a local immunomodulatory effect. The ocular distribu-

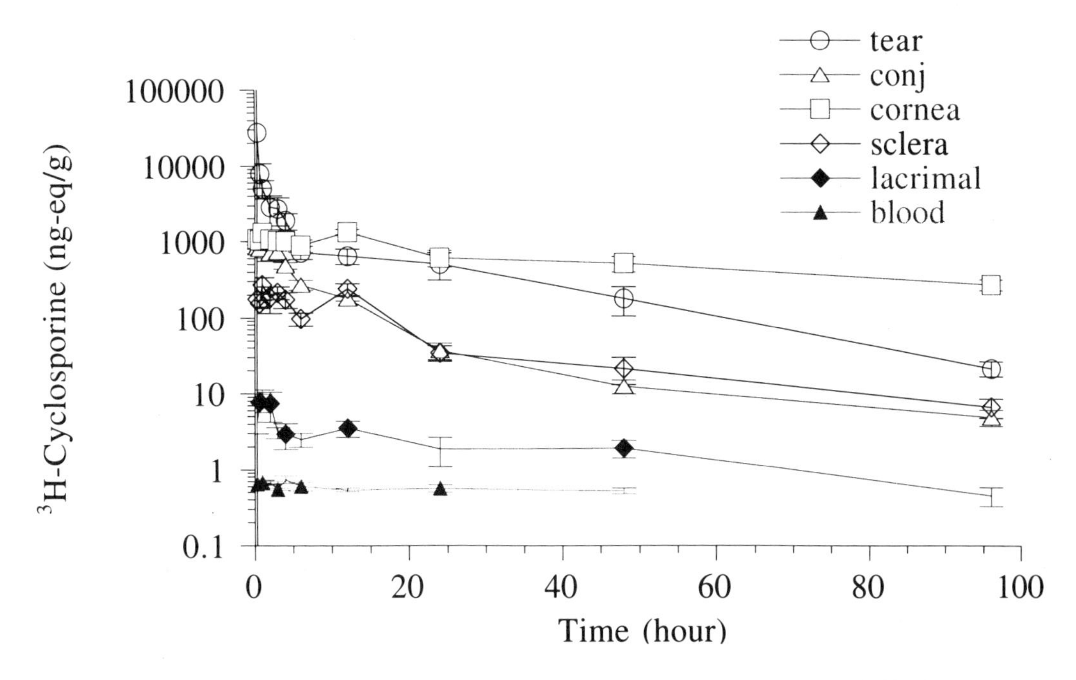

Figure 1. Total radioactivity concentrations (mean ± SEM) in rabbit eyes and systemic blood.

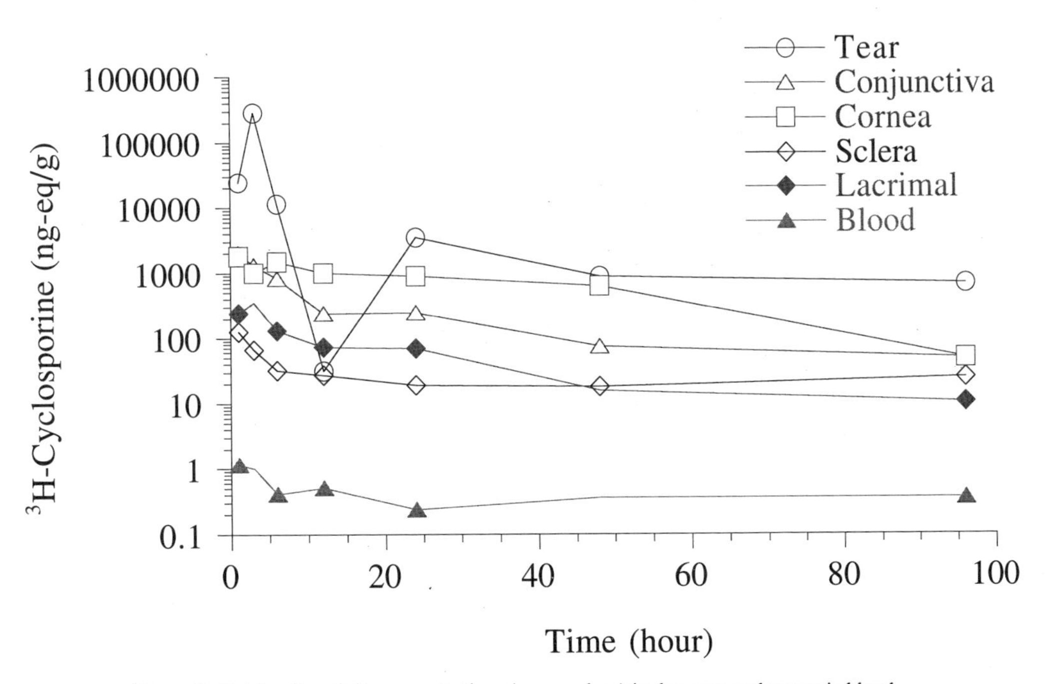

Figure 2. Total radioactivity concentrations (mean values) in dog eyes and systemic blood.

Table 1. Human blood trough and maximum cyclosporin A concentrations over 12 weeks

Cyclosporine emulsion	Range of blood cyclosporine A concentration (ng/ml)	
	Trough level	Maximum level
0.05%	<0.1	<0.1
0.1%	<0.1 to 0.102	<0.1
0.2%	<0.1 to 0.108	<0.1 to 0.144
0.4%	<0.1 to 0.157	<0.1 to 0.158

tion of cyclosporine after topical application of CsA emulsion was generally similar in rabbit and dog. In the rabbits dosed with 0.2% emulsion, the C_{max} tissue distribution was: tears > cornea > conjunctiva > lacrimal gland > blood.

Systemic blood cyclosporine concentrations following topical application of cyclosporine emulsion were very low in rabbits, dogs, and humans, obviating concerns about systemic toxicity or systemic mechanism of action. The human blood cyclosporin A concentrations were less than 0.2 ng/ml, much lower than the blood trough concentrations of 20–100 ng/ml used for monitoring the safety of patients receiving systemic cyclosporine therapy.

REFERENCES

1. Kaswan RL. Spontaneous canine keratoconjunctivitis, a useful model for human keratoconjunctivits sicca: Treatment with cyclosporine eyedrops. *Arch Ophthalmol.* 1989;107:1210–1216.
2. The 0.2% cyclosporine ointment is marketed by Schering-Plough for treatment of keratoconjunctivitis sicca in dogs.
3. Stern ME, Gelber TA, Gao J, Ghosn CR. The effects of topical cyclosporine A (CsA) on dry eye dogs (KCS). ARVO Abstracts. *Invest Ophthalmol Vis Sci.* 1996;37:S1026.
4. Acheampong A, Tang-Liu D, Shackleton M, Lam S, Angelov O, Ding S. Ocular absorption of cyclosporine from an aqueous emulsion: Comparison to other eyedrop formulations. ARVO Abstracts. *Invest Ophthalmol Vis Sci.* 1996;37:S1026.
5. Wiederholt M, Kossendrup D, Shulz W, Hoffman F. Pharmacokinetics of topical cyclosporin A in the rabbit eye. *Invest Ophthalmol Vis Sci.* 1986;27:519–524.
6. Kaswan RL. Intraocular penetration of topically applied cyclosporine. *Transplant Proc.* 1988;20 (Suppl 2):650–655.
7. Ben Ezra D, Mafzir G, de Courten C, Timonen P. Ocular penetration of cyclosporin A. III. The human eye. *Br J Ophthalmol.* 1990;74:350–352.

145

ESTROGEN THERAPY IN KERATOCONJUNCTIVITIS SICCA

Jinus Akramian,[1] Andreas Wedrich,[1] Johannes Nepp,[1] and Michael Sator[2]

[1]Department B of Ophthalmology
[2]Department of Gynecology
University of Vienna, Austria

1. INTRODUCTION

Recently, attempts have been made to prove the presence of hormone receptors on the eye surface; this would mean that they are able to react to hormone therapy. The absorption of percutaneously or vaginally applied estradiol has already been proved.[1,2]

With a pronounced decrease of the estrogen level over a long period of time, as is the case in postmenopause, the maturation process ceases, resulting in a thin epithelium prone to inflammatory changes, such as atrophic vaginitis. This atrophy can be treated with estradiol vaginal cream. After such therapy, the vaginal mucosa shows cyclic changes again.[3,4]

We have chosen to treat our menopausal patients with topically applied estradiol ointment. However, the absorption of estradiol in the conjunctiva has not yet been proved, so we embarked on a placebo-controlled trial to determine whether the conjunctiva is able to absorb and react to topically applied estradiol.

2. PATIENTS AND METHODS

In a placebo-controlled, randomized, double-blind pilot study, we enrolled 22 women 45–65 years of age (mean, 57 years). They had varying degrees of ocular irritation, burning, foreign body sensation, and pain characteristic of keratoconjunctivitis sicca. Patients with arthritis, allergic reaction, refractive error, or other systemic complaints were excluded from the study. Informed consent was obtained from all subjects.

Schirmer test, tear film breakup time (BUT) without anesthesia, rose bengal staining, and subjective assessment of dry eye symptoms with a standardized questionnaire were performed before therapy and after 1 week of therapy. The tests were separated by a 10-min interval, to minimize reflex tearing and ocular surface changes as a consequence of the testing.

Lacrimal Gland, Tear Film, and Dry Eye Syndromes 2
edited by Sullivan *et al.*, Plenum Press, New York, 1998

Schirmer 1 test was carried out with sterile filter paper strips in a routine fashion. Rose bengal 1% was used to examine the degree of staining. The placebo and the estradiol ointments were prepared by a pharmacist at our clinic, and ointments were randomized. The randomization was sealed in an envelope, and the encoded information was disclosed at the end of the study.

3. RESULTS

One patient of each group interrupted the therapy, one because of an acute infection and the other because of blurry vision caused by the ointment.

The data were analyzed by the Student paired t-test.

The mean value for the Schirmer tests changed from 6.1 to 8.9 mm in the hormone-treated group and from 8.25 to 8.2 mm in the placebo group (Fig. 1). This improvement in the hormone-treated group was statistically significant ($p = 0.01$). The change in the placebo group was not statistically significant.

The mean value for BUT increased from 11.3 to 15.7 sec in the hormone-treated group and from 13 to 15 sec in the placebo group (Fig. 2). There was only a slight improvement in the hormone-treated group but it was statistically significant ($p=0.05$); the change in the placebo group was not statistically significant.

The rose bengal score decreased from 1.7 to 1.35 in the hormone-treated group and from 1.55 to 0.89 in the placebo group (Fig. 3). The change was not statistically significant in either group.

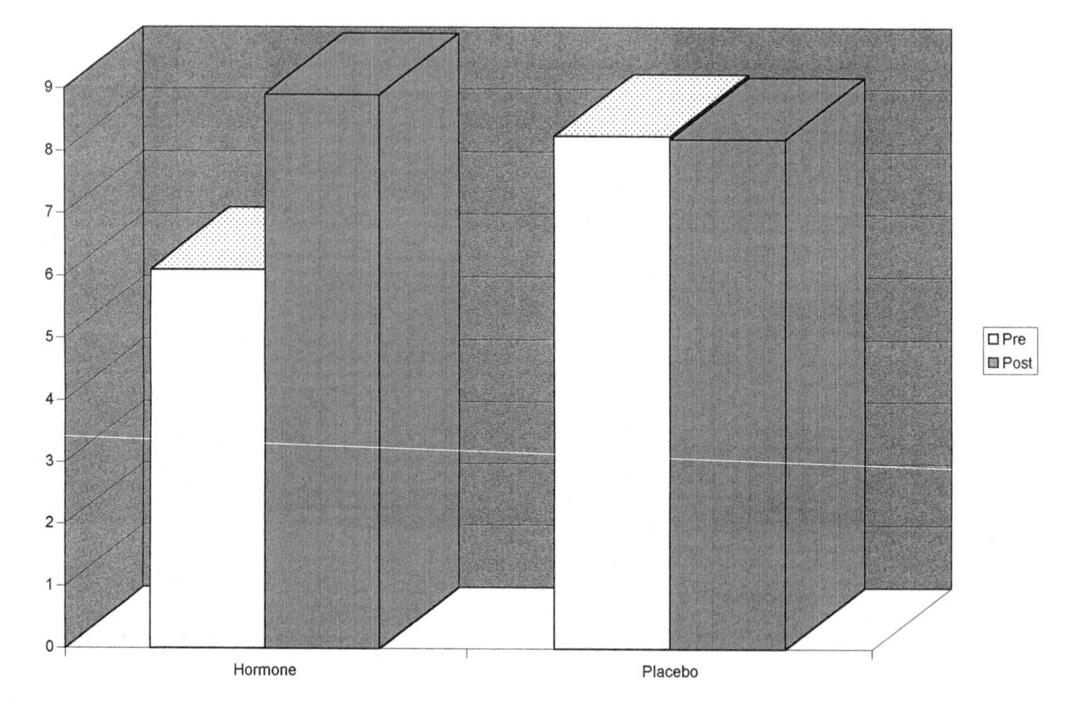

Figure 1. Mean Schirmer test value before and after 1 week of placebo or estradiol ointment therapy.

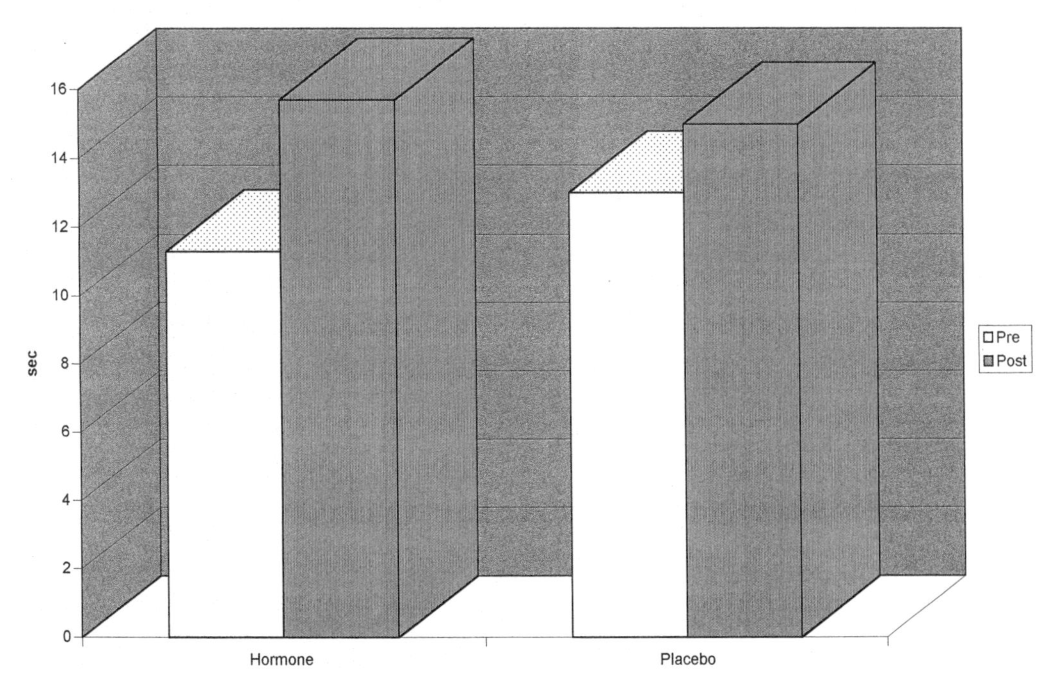

Figure 2. Mean BUT value before and after 1 week of placebo or estradiol ointment therapy.

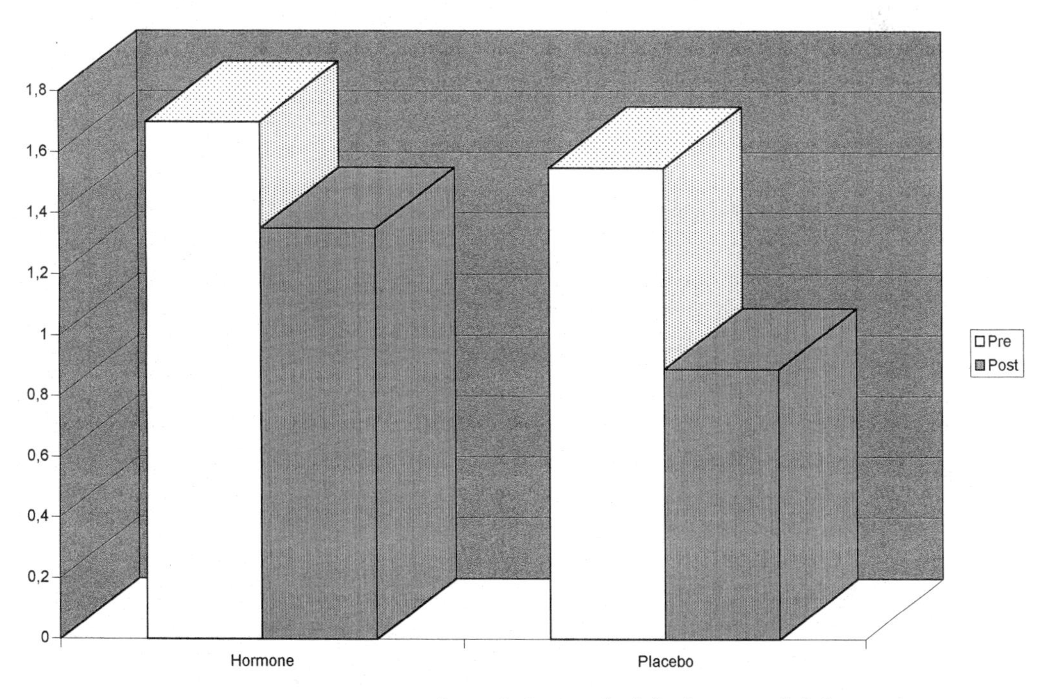

Figure 3. Mean rose bengal staining score before and after 1 week of placebo or estradiol ointment therapy.

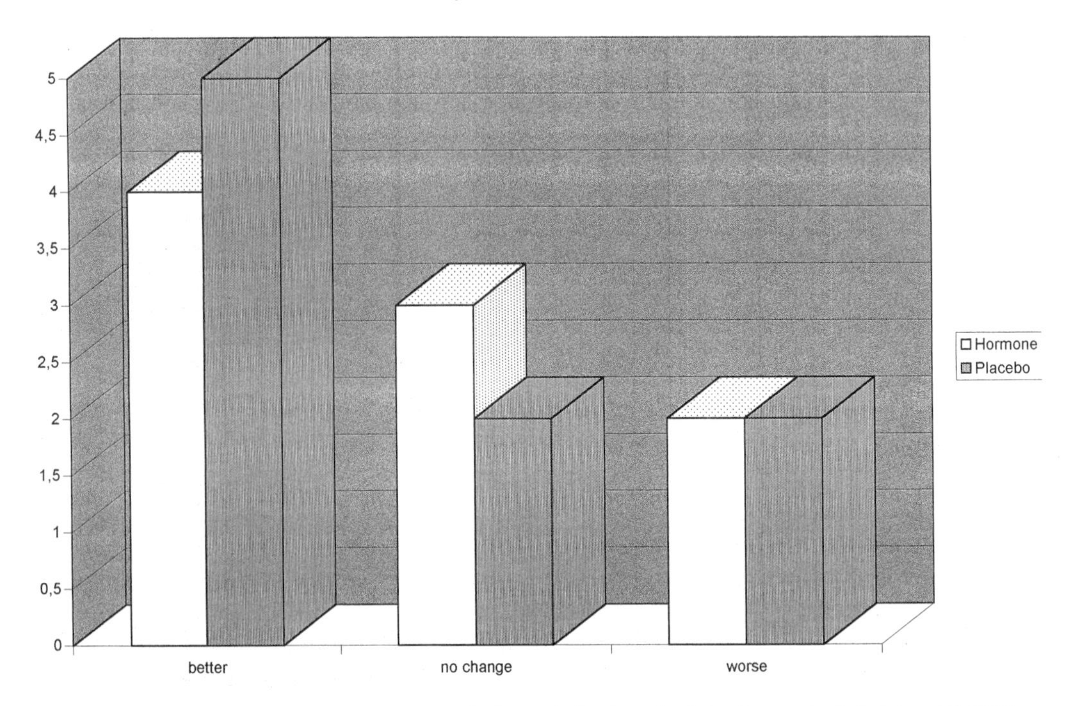

Figure 4. Mean subjective assessment responses before and after 1 week of placebo or estradiol ointment therapy.

In both groups, seven patients subjectively felt better or the same after the treatment (Fig. 4).

4. CONCLUSION

After only 1 week of topically applied estradiol, the values for the Schirmer test and breakup time increased. The changes in rose bengal staining seem to be too small to be detected in our small group of patients. Based on these encouraging preliminary results, a prospective, randomized, double-blind study with a larger number of patients is currently in progress. The absorption of locally applied estradiol in the conjunctiva has yet to be proved.

ACKNOWLEDGMENT

We are grateful to Dr. Schneider for statistical analysis of the data.

REFERENCES

1. Jemec GBE, Serup J. Short-term effects of topical 17β-oestradiol on human post-menopausal skin. *Maturitas.* 1989;11:229–234.

2. Elkik F, Gompel A, Mercier-Bodard C, et al. Effects of percutaneous estradiol and conjugated estrogens on the level of plasma proteins and triglycerides in postmenopausal women. *Am J Obstet Gynecol.* 1982;8:888–892.
3. Vavilis D, Maloutas S, Nasioutziki M, Boni E, Bontis J. Conjunctiva is an estrogen-sensitive epithelium. *Acta Obstet Gynecol Scand.* 1995;74:799–802.
4. Kramer P, Lubkin V, Potter W, et al. Cyclic changes in conjunctival smears from menstruating females. *Ophthalmology*. 1990;97:303–307.

DRY EYE TREATMENT WITH ACUPUNCTURE

A Prospective, Randomized, Double-Masked Study

J. Nepp, A. Wedrich, J. Akramian, A. Derbolav, C. Mudrich, E. Ries, and J. Schauersberger

University Eye Clinic of Vienna
Vienna, Austria

1. INTRODUCTION

Dry eye is defined as a disorder of the tear film due to tear deficiency or excessive tear evaporation, which causes damage to the interpalpebral ocular surface and is associated with symptoms of ocular discomfort.[1] Tear film homeostasis is based on an adequate, proper function and balance of at least three systems that produce the lipid, aqueous, and mucus layers of the tear film. Previous studies give evidence that humoral factors—for instance, vasointestinal active peptide, endocrine factors (estrogen, androgens), and the vegetative nerve system—are part of the complex regulatory mechanism of these systems.[2]

Acupuncture, an ancient Chinese method, has been practiced in Europe especially since the late sixties. Since then, it has been the object of critical investigations regarding its effect and its mechanisms of action. The irritation of histologically specific skin points has been shown to enhance the emission of transmitter substances by the central nervous system in the experimental animal.[3,4] Acupuncture effects are based on secretion of transmitters (e.g., endorphins, dymorphins, serotonin, kinins) within neural synapses in the spinal cord and hypothalamus,[5] which cause vasodilatation[6,7] and relaxation of muscles.[8] Other effects are presumed, but unproven, such as immunologic, psychologic, and vegetative effects.[9] In a previous preliminary study, needle acupuncture had a beneficial effect in patients with dry eye syndrome.[10,11] Nevertheless, acupuncture may also have a placebo effect.[12]

We performed this prospective study to answer two questions: (i) Is there any objective effect of acupuncture? (ii) If so, is there any difference between the two methods - needle vs. laser acupuncture?

2. METHODS

Patients with Schirmer test below 5 mm and BUT below 5 sec were included in this study. A total of 114 patients participated. The age range was 20–84 years (mean, 57.6 years); 87% female, 13% male. After gaining informed consent, acupuncture treatment was performed in a randomized, double-blind fashion with two identical-appearing soft laser devices, one emitting infrared light: verum laser (4 mW, 780 nm, gallium-aluminum-arsenit), the other without any function (falsum laser); 32 patients received verum laser, and 30 received falsum laser. Treatment was performed once a week for 10 weeks; the duration of point stimulation was 20 sec with the laser and 30 min for needle acupuncture.

The points were selected based on a preceding ophthalmologic and general questionnaire. There were local points (GB1, UB2, ST5, Ex2YinTang), specific points for the eyes and mucosa (LI4, SI3, Li3, Kd6, TH5), and individual points according to present complaints. There was a session once a week for 10 weeks. Before and 1 week after the last treatment, a slit lamp examination was performed. For evaluation of the therapy effect, we used the results of Schirmer test, break-up time (BUT) of the tear film, and drop-frequency protocol.

After completion of the study, the identity of the two lasers was uncovered.

The results of the laser-treated patients were compared with a third group, 30 patients treated with needle acupuncture. A fourth group of 22 patients, treated with artificial tears only, served as control.

For statistical analysis, the multiple Tukey-Kramers test was used.

3. RESULTS

We observed the patients for 3 months. The results of Schirmer test ($p<0.01$), BUT ($p<0.001$), and drop frequency ($p<0.001$) in the four groups are shown in Tables 1–3 and Fig. 2. There were statistically significant differences in the Tukey-Kramers multiple-test between the results of verum laser and falsum laser and between needles acupuncture and the controls ($p<0.01$). However, we found no statistically significant difference between verum laser and needle acupuncture or between falsum laser and the control group ($p<0.05$).

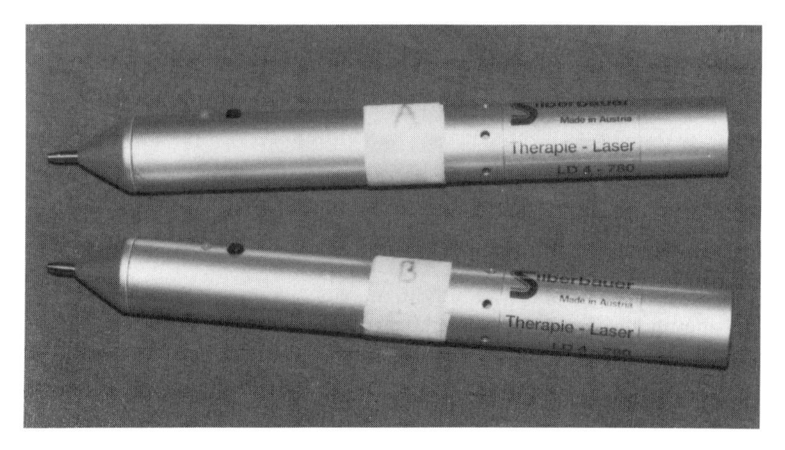

Figure 1. The two laser instruments.

Table 1. Schirmer II test results

	Pat.	%	Kind of therapy	
Normal- >10 mm/5 min	16	26.7	Needle-acupuncture	n:60
	23	35.9	Verum-Laser	n: 64
	14	23.3	Falsum-Laser	n: 60
	7	15.9	Control	n: 44
Low - pathologic (<10 mm/5 min)	20	33.3	Needle-acupuncture	n: 60
	13	20.3	Verum-Laser	n: 64
	6	10	Falsum-Laser	n: 60
	4	9.1	Control	n: 44
Severe - pathologic (<5 mm/5 min)	24	40	Needle-acupuncture	n: 60
	28	43.8	Verum-Laser	n: 64
	40	66.7	Falsum-Laser	n: 60
	33	75	Control	n: 44

Table 2. Break-up time test results

BUT	Pat. n	%	Kind of therapy	
Normal (>10 sec)	44	73.3	Needle-acupuncture	n: 60
	23	35.9	Verum-Laser	n: 64
	3	5.0	Falsum-Laser	n: 60
	4	9.1	Control	n: 44
Low - pathologic (5-9 sec)	12	20	Needle-acupuncture	n: 60
	27	42.2	Verum-Laser	n: 64
	14	23.3	Falsum-Laser	n: 60
	9	20.45	Control	n: 44
Severe - pathologic (0-4 sec)	4	6.7	Needle-acupuncture	n: 60
	14	21.9	Verum-Laser	n: 64
	43	71.7	Falsum-Laser	n: 60
	31	70.4	Control	n: 44

Table 3. Frequency of artificial tears

Drop-frequency	Pat. n.	%	Kind of therapy	
No use of artificial tears	34	56.7	Needle-acupuncture	n:60
	20	31.0	Verum-Laser	n: 64
	2	3.3	Falsum-Laser	n: 60
	0	0	Control	n: 42
Less use of artificial tears (1-3/day)	20	33.3	Needle-acupuncture	n: 60
	23	35.9	Verum-Laser	n: 64
	18	30	Falsum-Laser	n: 60
	4	29.5	Control	n: 42
4-10x use of artificial tears / day	6	10	Needle-acupuncture	n: 60
	21	32.8	Verum-Laser	n: 64
	38	63.4	Falsum-Laser	n: 60
	28	66.7	Control	n: 42
>10x use of artificial tears / day	0	0	Needle-acupuncture	n: 60
	0	0	Verum-Laser	n: 64
	2	3.3	Falsum-Laser	n: 60
	10	23.8	Control	n: 42

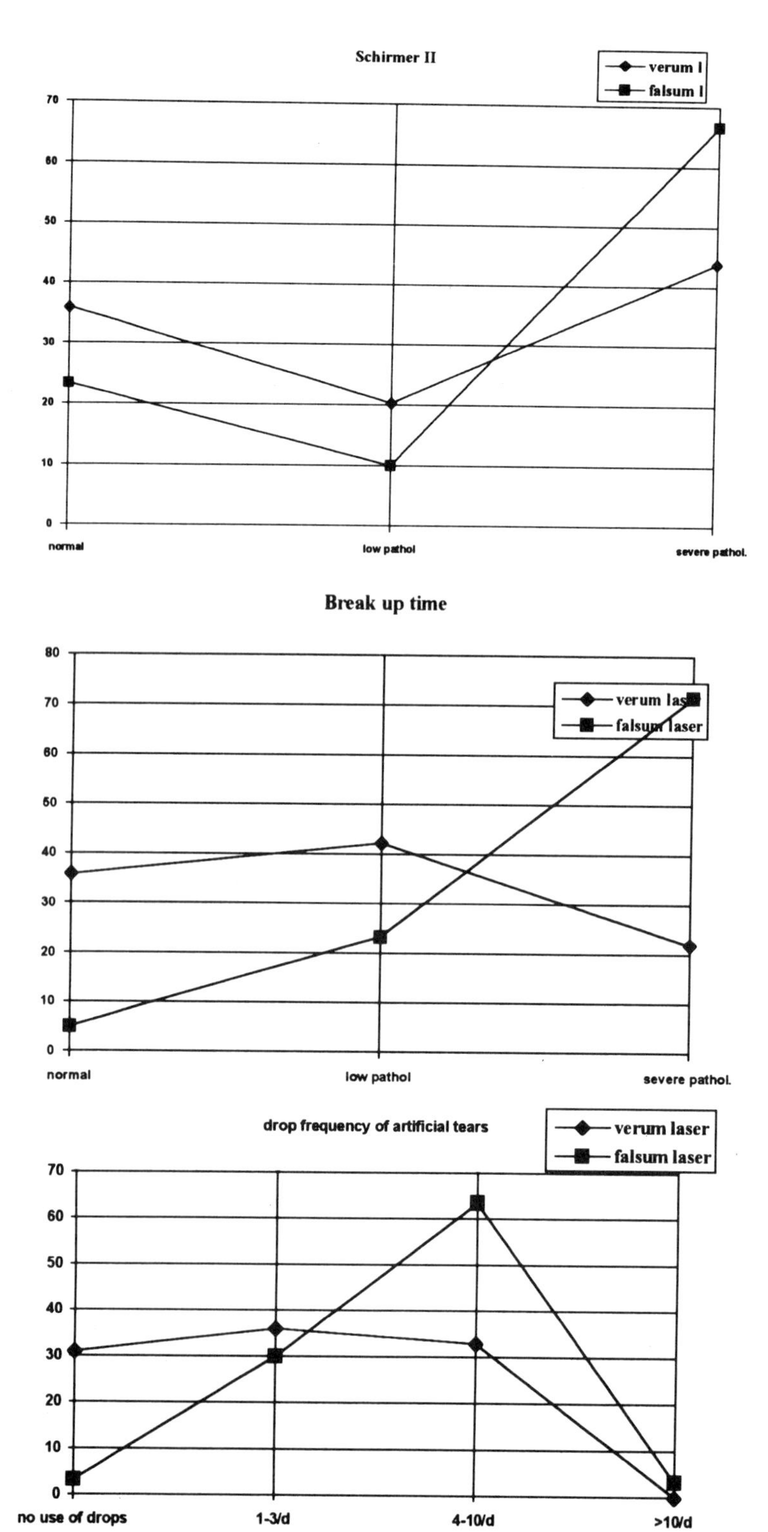

Figure 2. Schirmer II and BUT test results, and frequency of artificial tears.

4. DISCUSSION

Conjunctivitis sicca is well known as a dry-eye symptom, associated with many other symptoms, especially psychovegetative disorders.[13] Head recognized that irritations of organic dysfunctions may be transmitted to the skin within the same segment.[14] Kecik investigated an irritating therapy within the same segment. In his series, low-energy laser treatment was compared with short-wave diathermy on 10 patients with dry eyes.[15] The eye region was stimulated once a day for 10 days. The Schirmer test increased 2–6 mm for both methods, and there was no significant difference between the two methods.

Acupuncture is presumed to affect organs and functional systems within the same segment, according to Head‘s theory, as well as along muscle lines[16] and along vegetative nerves.[17] Acupuncture is supposed to affect psychic tension,[18] and Erb has shown that patients with dry eyes suffer from psychological disorders: anxiety, nervousness, and social-emotional lability.[19] Earlier pilot studies[10,11,20] showed a good effect of needle acupuncture in patients with dry eyes. Ocular symptoms were reduced in 91% of the patients; 59% were symptom free. The subjective improvement was accompanied by an improvement of BUT and drop frequency. Yet a kind of placebo effect, or suggestion, may also play a role in acupuncture treatment.[12] To the real effects, we performed a randomized, double-masked study in patients with dry eyes using an infrared, low-energy laser as the stimulating tool. Laser may influence disorders of the autonomous nervous system similar to needle acupuncture.[21–23]

We found no statistically significant difference between verum laser and needle acupuncture. Therefore, we suggest that both stimulation techniques have a similar effect.

5. CONCLUSION

Based on our results of Schirmer test, BUT, and drop frequency, we found an objective effect of acupuncture in the treatment of dry eyes. There was a statistically significant difference between the verum laser group and the falsum laser group, as well as between the needle group and the control group. There was no statistically significant difference between the needle acupuncture group and the verum laser group. The mechanism of action is still speculative.

ACKNOWLEDGMENT

We thank Mr. Dipl. Ing. Silberbauer for the development of the placebo-laser for this study.

REFERENCES

1. Lemp MA. Report of the National Eye Institute/Industry Workshop on Clinical Trials in Dry Eyes. *CLAO J.* 1995:21:221–232.
2. Dickhaus H, Pauser G, Zimmermann M. Hemmung im Rückenmark, ein neurophysiologische Wirkmechanismus bei Hypalgesie durch Stimulationsakup. *Wiener KlinWochenschr.* 1978;90:59–64.
3. Heine H. Morphologie der Ohrakupunkturpunkte. *Deutsche Z Akupunktur.* 1993;36(10):99–103.
4. Heine H. Zur Morphologie der Akupunkturpunkte. *Deutsche Z Akupunktur.* 1987;30(4):75–79.

5. Riederer P, Tenk H, Werner H, Bischko J, Rett A, Krisper H. Manipulation of Neurotransmitters by *Acupuncture. J Neural Transm.* 1975;37:81–94. Transmittersubstanzen im Hypothalamus.
6. Kitzinger E. Temperaturreaktion an den Akren der Finger bei Akupunktur. *Akupunktur, Theorie u. Praxis.* 1991;19,220–223.
7. Trnavsky G. Venenverschlußrheographische Durchblutungsmessung zur Kontrolle von Durchblutungsveränderungen nach perkutaner elektrischer Reizung von Akupunkturpunkten. *Deutsche Z Akupunktur.* 1980;23:64–68.
8. Bergsmann O. Über muskuläre Resonanz und Dämpfungsphänomene bei Akupunktur. *Deutsche Z Akupunktur.* 1985;28:59–69.
9. Bischko J. *Einführung in die Akupunktur.* 13.Aufl. HaugVlg. 1983:51–74.
10. Nepp J, Wedrich A, Akramian J, Strenn K, Velikay M. Conjunctivitis sicca. *Deutsche Z Akupunktur.* 1993;2;36:26–37.
11. Wedrich A, Nepp J, Akramian J, Strenn K, Velikay M. Akupunktur und Conjunctivitis sicca - Erste Ergebnisse. *Spektrum Augenheilkd.* 1993;7/6:267–271.
12. Loma Linda. Acupuncture: The position paper of the National Council Against Health Fraud. *Clin J Pain.* 1991;7, 2:162–166.
13. Murube del Castillo J. Der Begriff des trockenen Auges. In: Marquart R, Lemp MA eds. *Das trockene Auge.* Berlin: Springer 1991;5–6.
14. Head H. Die Sensibilitätsstörung der Haut bei Visceralerkrankungen. Berlin: Hirschwald. 1898.
15. Kecik T, Switka-Wieclavska I, Portacha L, Ciszewska J. The efficiency of short wave diathermy and laser stimulation of the lacrimal gland in the treatment of dry eye syndrome. *Adv Exp Med Biol.* 1994;350:601–604.
16. Bergsmann O, Meng A. Akupunktur und Bewegungsapparat: Versuch einer Synthese. Heidelberg Haug. 1982;10:154–58.
17. Feyrter F. Über die Pathologie der vegetativen nervösen Peripherie und ihrer ganglionären Regulationsstätten. In: Bachmann G, ed. *Akupunktur, eine Ordnungstherapie.* Ulm, Haug. 1959;(1):20.
18. Bischko J. Einführung in die Akupunktur. Heidelberg, Haug. 1970;56.
19. Erb C, Horn A, Gunther A, Saal JG, Thiel HJ. Psychosomatische Aspekte bei Patienten mit primärer Keratoconjunctivitis sicca. *Klin Monatsbl Augenheilkd.* 1996;208:96–99.
20. Wedrich A, Nepp J, Strenn K, Akramian J, Velikay M. Akupunktur bei Conjunctivitis sicca- mittelfristige Ergebnisse. Spectrum Augenheilkd. 1996;10/2:66–70.
21. Schjelderup W. Historische und theoretische Grundlage der Laser-Therapie. In: Bahn J, ed. *Laser- und Infrarotstrahlen in der Akupunktur.* Heidelberg, Haug Verlag. 1982:88–91.
22. Krötlinger M. On the use of laser in acupuncture. *Am J Acupuncture.* 1980;5:297–311.
23. Karu T. Photobiological fundamentals of low-power laser therapy. *IEEE J QuantumElektronics.* 1987;QE23(10):1703–1717.

PUNCTAL OCCLUSION FOR THE TREATMENT OF DRY EYE

Mitchell H. Friedlaender and Robert I. Fox

Divisions of Ophthalmology and Rheumatology
Scripps Clinic and Research Foundation
La Jolla, California

1. INTRODUCTION

Punctal occlusion is thought to be highly effective in the treatment of severe dry eye, often producing dramatic improvement of signs and symptoms. Its role in mild or moderate dry eye is less clear, and sometimes complications, including epiphora, may occur. There are several methods for performing temporary and permanent punctal occlusion. We have performed these techniques for many years and present our experience here with these useful procedures. We have also developed some guidelines for the timely use of punctal occlusion for dry eye patients.

2. MATERIALS AND METHODS

A retrospective chart review from 1987 to 1996 was undertaken to determine the effects of punctal occlusion on the signs and symptoms of dry eye. Patients were evaluated for Sjögren's syndrome and other systemic disease. Our criteria for Sjögren's syndrome are rather strict; they require the presence of dry eye, dry mouth, and a connective tissue disease. The severity of the dry eye was evaluated using three variables: the symptom of dryness (the primary variable), Schirmer test (a secondary variable), and rose bengal staining (a secondary variable). Patients were followed for 6 months to 9 years after punctal occlusion.

Temporary punctal occlusion was performed under topical proparacaine anesthesia by applying a heated platinum spatula to the lower punctum (Fig. 1). Permanent punctal occlusion was performed under local xylocaine anesthesia by dilating the punctum and applying electrocautery or hyfrecation to the canaliculus (Fig. 2).

Lacrimal Gland, Tear Film, and Dry Eye Syndromes 2
edited by Sullivan *et al.*, Plenum Press, New York, 1998

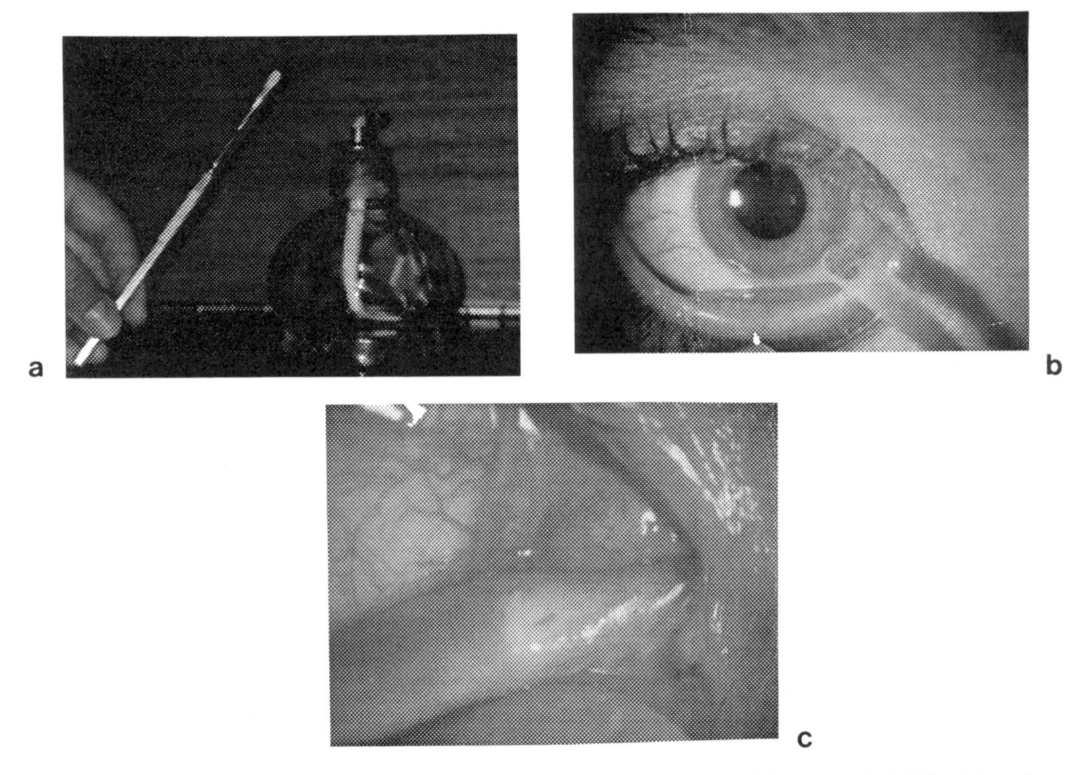

Figure 1. Temporary punctal occlusion: (a) Heating platinum spatula. (b) Cauterizing punctum. (c) Blanching of punctum after cautery.

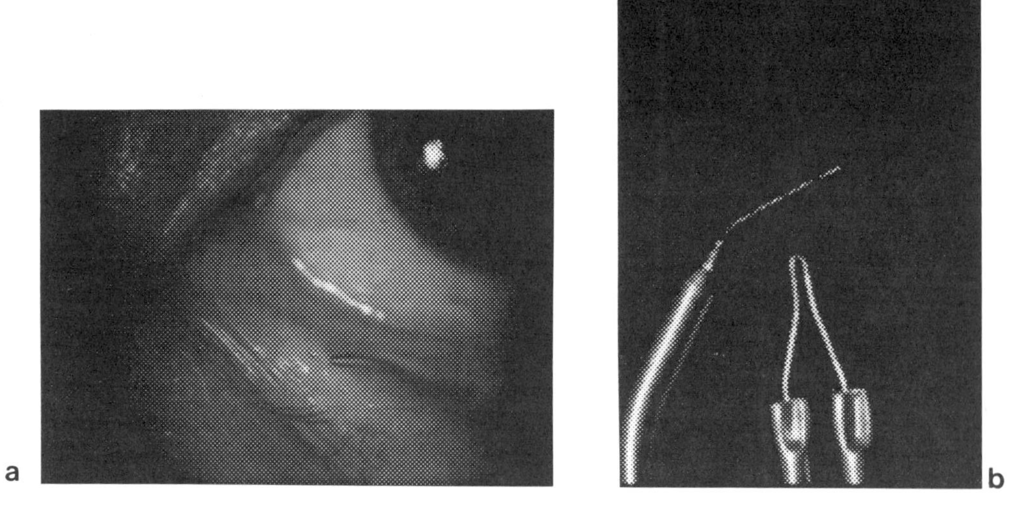

Figure 2. Permanent punctal occlusion: (a) Dilating punctum. (b) Hyfrecator and cautery.

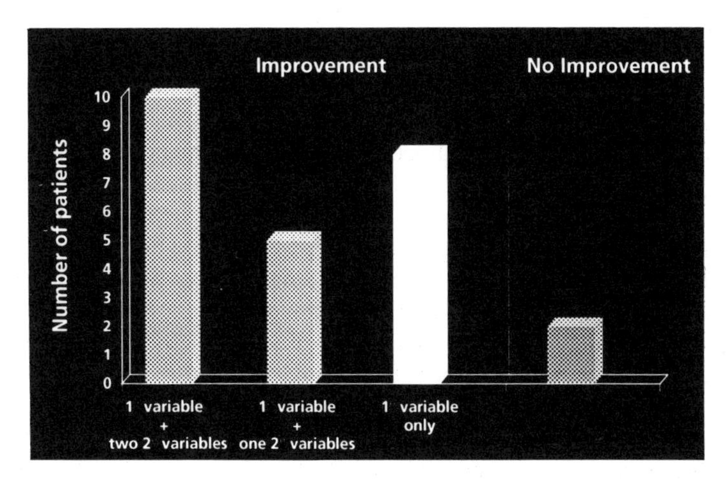

Figure 3. Improvement of signs and symptoms after temporary punctal occlusion.

3. RESULTS

Based on clinical impression and the presence of one or more variables, 386 patients were considered to have dry eye. Of the 73 of these patients who underwent punctal occlusion, 48 carried a diagnosis of Sjögren's syndrome. All 73 patients who underwent punctal occlusion had temporary punctal occlusion, and 32 of these subsequently underwent permanent punctal occlusion. Among all patients who had *only* temporary punctal occlusion, 23/25 had improvement of at least one variable (Fig. 3). Among the patients who underwent permanent punctal occlusion, 30/32 had improvement of at least one variable (Fig. 4). Three patients had troublesome epiphora after permanent occlusion of both upper and lower puncta. Two of these were non-Sjögren's patients and one was a questionable Sjögren's patient with negative serological tests.

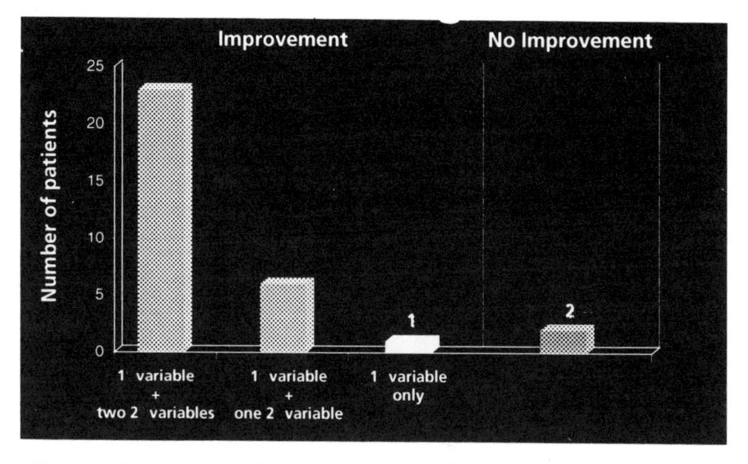

Figure 4. Improvement of signs and symptoms after permanent punctal occlusion.

4. DISCUSSION

Punctal occlusion is helpful in the treatment of dry eyes in both Sjögren's and non-Sjögren's patients. It seems reasonable to reserve punctal occlusion for the most severe dry eye patients. In our study, punctal occlusion was recommended for only 20% of dry eye patients. Those who were comfortable using frequent (even hourly) artificial tears were not considered candidates for punctal occlusion. Most patients (66%) who underwent punctal occlusion had Sjögren's syndrome. Many of the 34% who did not have Sjögren's syndrome had one or more diagnostic creteria for Sjögren's syndrome, and several will no doubt one day carry a diagnosis of Sjögren's syndrome.

Both temporary and permanent punctal occlusion usually led to improvement in symptoms, Schirmer test, and rose bengal staining. Only four patients who underwent punctal occlusion (5.5%) had no improvement.

We believe punctal occlusion should be done in stages. All candidates for punctal occlusion should have temporary punctal occlusion first. If there is improvement, it may be reasonable to perform permanent punctal occlusion. Even when great care is taken to select appropriate patients for punctal occlusion, epiphora can occur as a complication. In this study, three patients had bothersome epiphora after closure of all four puncta. This represents 4% of all patients receiving punctal occlusion, and 12.5% of those who had permanent punctal occlusion. Interestingly, two of the patients with epiphora did not have Sjögren's syndrome, and one had possible Sjögren's syndrome.

5. CONCLUSIONS

i. Punctal occlusion is a useful treatment for severe dry eye, but it should be reserved for a minority of cases.
ii. Caution should be exercised when performing permanent punctal occlusion, especially in non-Sjögren's patients.
iii. Temporary punctal occlusion should be done before attempting permanent punctal occlusion.
iv. If permanent punctal occlusion is necessary, the upper puncta should be left open to avoid persistent epiphora.

KERATOPROSTHESIS IN PEMPHIGOID AND STEVENS-JOHNSON SYNDROME

Claes H. Dohlman and Hisao Terada

Department of Ophthalmology
Harvard Medical School
and Massachusetts Eye and Ear Infirmary
Boston, Massachusetts

1. INTRODUCTION

In pemphigoid and Stevens-Johnson syndrome, inflammation of the ocular surface epithelium can lead to conjunctival shrinkage and symblepharon formation, as well as corneal vascularization, opacity, and reduction of vision. In the end stages, tear access to the fornices is blocked, resulting in complete dryness and keratinization of the entire ocular surface. In these severe cases with only hand movements or light perception vision, standard corneal transplantation is almost always unsuccessful.[1] Keratoprosthesis has been tried in the past, but the complication rate has been formidable (for review, see ref. 2). More recently, a number of modifications have improved the prognosis, and the outcome in a small number of cases has been published.[3] This series, now substantially expanded, is described here more fully.

2. PATIENTS, MATERIALS, AND METHODS

Seventeen eyes with pemphigoid (Table 1) and 7 eyes with Stevens-Johnson syndrome (Table 2) were operated with keratoprosthesis. Two types of collarbutton-shaped prostheses of methylmethacrylate were used, one type for the still-wet eye and the other type for a through-the-lid placement in the end-stage dry eye. The designs,[4] as well as the surgical techniques,[5] have been described previously.

In addition, a glaucoma valve shunt (Ahmed) was implanted in most cases, either simultaneously or secondarily. Intraocular pressure was estimated with finger palpation; disc appearance and visual fields were followed.

The keratoprostheses were temporarily covered with conjunctiva (Type I) or lid skin (Type II), and these were opened after 1–2 months. Subsequently topical anti-collagenase

Lacrimal Gland, Tear Film, and Dry Eye Syndromes 2
edited by Sullivan *et al.*, Plenum Press, New York, 1998

Table 1. Keratoprosthesis in pemphigoid patients

Patient no.	Age at oper. (yr)	Months postop	Type K Pro	Shunt	Years blind pre-op	Preop VA	Best postop VA	Months ≥ 20/200 postop	Present VA	Severe complications
1	77	72	II	No	4	HM	20/20	3	NLP	Endophthalmitis
2	74	60	II	Yes (2°)	4	HM	20/20	48	CF	Glaucoma
3	79	60	II	No	9	HM	CF	0	NLP	Loosening of back plate; ARMD
4	75	48	II	Yes (2°)	2	HM	20/25	24	LP	Retinal detachment
5	49	24	I	No	6	LP	20/25	12	20/25	Glaucoma
6	76	30	II	Yes	3	HM	20/50	29	20/50	–
7	59	25	II	Yes	2	LP	20/20	24	20/20	–
8	80	25	II	Yes	1	HM	20/30	25	20/30	–
9	64	21	II	Yes	4	HM	20/60	6	20/60	–
10	55	18	II	Yes	?	LP	20/60	18	20/60	–
11	72	21	II	Yes	?	HM	20/30	15	20/80	–
12	78	19	I	Yes	8	HM	20/20	15	20/20	–
13	57	13	I	Yes	2	LP	20/60	12	HM	Retro membrane hemorrhage
14	61	9	I	Yes	1	LP	HM	0	HM	End-stage glaucoma
15	81	9	II	Yes	8	LP	20/30	9	20/30	–
16	91	5	II	Yes	2	LP	20/80	3	20/80	–
17	74	5	II	Yes	2	LP	20/20	3	LP	Endophthalmitis

VA, visual acuity. HM, hand movements. LP, light perception. NLP, no light perception. CF, counting fingers. ARMD, age-related macular degenertion.

Table 2. Keratoprosthesis in Stevens-Johnson syndrome patients

Patient no.	Age at Oper. (yr)	Months postop	Type K Pro	Shunt	Years blind pre-op	Preop VA	Best postop VA	Months ≥ 20/200 Postop	Present VA	Severe complications
1	26	65	I	Yes (2°)	3	HM	20/40	24	HM	Glaucoma
2	30	38	II	Yes (2°)	3	LP	20/20	38	20/20	Glaucoma
3	40	36	I	Yes (2°)	4	LP	20/25	12	HM	Glaucoma
4	24	32	I	No	5	HM	20/40	13	NLP	Endophthalmitis
5	61	24	II	Yes	10	HM	20/50	2	NLP	Endophthalmitis
6	68	24	II	Yes	3	HM	20/50	12	NLP?	Infections Retinal detachment
7	33	21	I	Yes	2	LP	20/30	12	NLP	Endophthalmitis

VA, visual acuity. HM, hand movements. LP, light perception. NLP, no light perception.

medication (medroxyprogesterone 1% suspension) and topical antibiotics such as Ocuflox or Polytrim were prescribed for life, at least twice daily.[4]

If skin retraction after Type II or corneal necrosis after Type I occurred, it was repaired as previously described.[5]

3. RESULTS

The tables summarize the patient characteristics, visual acuity, and severe complications. The complications of infection and glaucoma will be detailed in separate publications.

4. DISCUSSION

It is clear from the results presented in the tables that the prognosis for sustained good postoperative vision is much better in ocular pemphigoid than in Stevens-Johnson syndrome. The latter category had a very high rate of explosive endophthalmitis, which seemed to be related to the unpredictable necrosis and melting of the corneal tissue, or to the lid skin retraction around the prosthesis, allowing microbial penetration into the eye. This tissue breakdown seems to parallel the chronic inflammation of the surface epithelia in this syndrome and the frequent flare-up after any surgical intervention. It is also possible that the holes in the plastic support plates of the prosthesis, which are expected to aid in the circulation of fluid and nutrients, may in fact do more harm than good by shortening the route for invading bacteria. Revision of the prosthesis design to eliminate the holes is now in progress.

In pemphigoid, the ocular structures are much more quiet than in Stevens-Johnson syndrome and are less apt to respond to surgery with violent inflammation. As can be seen in Table 1, only five eyes out of 17 had a serious complication, aside from glaucoma. One of them resulted from an unscrewing of a loosely fitted back plate, a maloccurrence that can not be repeated thanks to design change. Two infections occurred, one in a patient (our first through-the-lid prosthesis) who could not be retrieved for return visit in time, an event underscoring the importance of regular follow-up. Minor revisions and YAG laser membraneotomies were frequently needed; they seemed to be the price for the visual improvement. One difficulty with an elderly patient with pemphigoid is that the fundus is not visible preoperatively, and, therefore, posterior segment health cannot be taken for granted. One patient with no history of glaucoma was found postoperatively to have end-stage optic nerve damage and only a temporal visual field remnant. Another patient turned out to have macular degeneration to a maximal degree, again unanticipated preoperatively.

Despite these uncertainties, end-stage pemphigoid seems to be a very reasonable indication for keratoprosthesis. In elderly, bilaterally blind patients with vision in the range of hand movements or light perception only, and no other known contraindications, it is most reasonable to proceed with a keratoprosthesis (plus glaucoma shunt and cataract extraction or vitrectomy) in the less-seeing eye.

5. SUMMARY

Keratoprosthesis was implanted in 17 eyes with ocular pemphigoid and 7 eyes with Stevens-Johnson syndrome and followed for 6 months to 6 years. Visual outcomes and complications were compared. Pemphigoid was found to have a much better prognosis than did Stevens-Johnson syndrome. Keratoprosthesis is now firmly indicated in elderly pemphigoid patients whose vision in both eyes has become reduced to hand movements or less due to a corneal opacity.

ACKNOWLEDGMENTS

This work was sponsored by Mr. Saad Abdulaziz Abdulmohsin Al-Rashed, Kuwait.

REFERENCES

1. Tugal-Tutkin I, Akova YA, Foster CS. Penetrating keratoplasty in cicatrizing conjunctival diseases. *Ophthalmology*. 1995;102:576–586.
2. Dohlman CH. Keratoprosthesis. In: Krachmer JJ, Mannis MJ, Holland EJ, eds. *The Cornea*. Philadelphia: Mosby; 1996:1855–1863.
3. Dohlman CH, Doane MG. Keratoprosthesis in end-stage dry eye. *Adv Exp Med Biol*. 1994;350:561–564.
4. Dohlman CH, Doane MG. Some factors influencing outcome after keratoprosthesis surgery. *Cornea*. 1994;13:214–218.
5. Dohlman CH, Waller S, Netland PA. Keratoprosthesis surgery. In: Lindquist TD, Lindstrom RL, eds. *Ophthalmic Surgery Update #4*, Mosby-Yearbook, Inc. Chicago, 1996;V-L-O-V-L-24.

PORTABLE DEVICE FOR PROGRAMMABLE, AUTOMATIC, OR ON-DEMAND DELIVERY OF ARTIFICIAL TEARS

Vo Van Toi[1] and P. A. Grounauer[2]

[1]Tufts University
Electrical Engineering and Computer Science Department
Biomedical Engineering Laboratory
Medford, Massachusetts
[2]Hôpital Ophtalmique Universitaire de Lausanne
Switzerland

1. INTRODUCTION

Efficient delivery of artificial tears is crucial in the treatment of dry eyes. Usually, eyedrops are commercially available in plastic bottles that allow manual administration by the patient. This method is simple but has significant drawbacks; one is that the drops are too large and only 1% to 10% of the administered dose is absorbed into the eyes.[1] It is often tedious for the patient to dispense the drops at the prescribed quantity and timing. Self-administration of eyedrops is particularly difficult for the elderly, children, and patients suffering from various impairments. The irregularity in patient compliance greatly compromises the treatment's efficacy.

As there is a substantial market for topical eye therapy, considerable resources have been invested in research to find ways to administer medications effectively, to prolong a medication's effects, and to improve a patient's compliance in using the medication. Several investigations have been carried out to design devices for delivery of eye medication. These devices allow patients to self-administer medications in an easier way. They include devices placed in the conjunctival sac,[2] externally worn devices mounted on an eyeglass frame with a tubing tip in direct contact with the eyelids,[3] semi-implants put partially under the skin by surgery,[4] and full implants inserted entirely under the skin.[5] Some common side effects result from these devices being invasive, impractical, prone to contamination, and unsafe.

The study presented here offers a solution to this problem by proposing (a) a novel topical delivery method that consists of automatically ejecting, from nearby nozzles, a single microdroplet of medication into the patient's eyes, and (b) a new, battery-powered device to perform this function, the characteristics of which include reliability, safety, portability, and low cost. Most importantly, this method requires no major effort on the part of the patient. The novel concept of the device is that it ensures delivery of a medication without contamination or altering its characteristics. It can be used for patients of all ages and will assist medical treatment for out-patients as well as hospitalized patients who need constant or long-term treatment. Therefore, it is useful for sufferers of dry eye, eye infections, glaucoma, and other eye diseases.

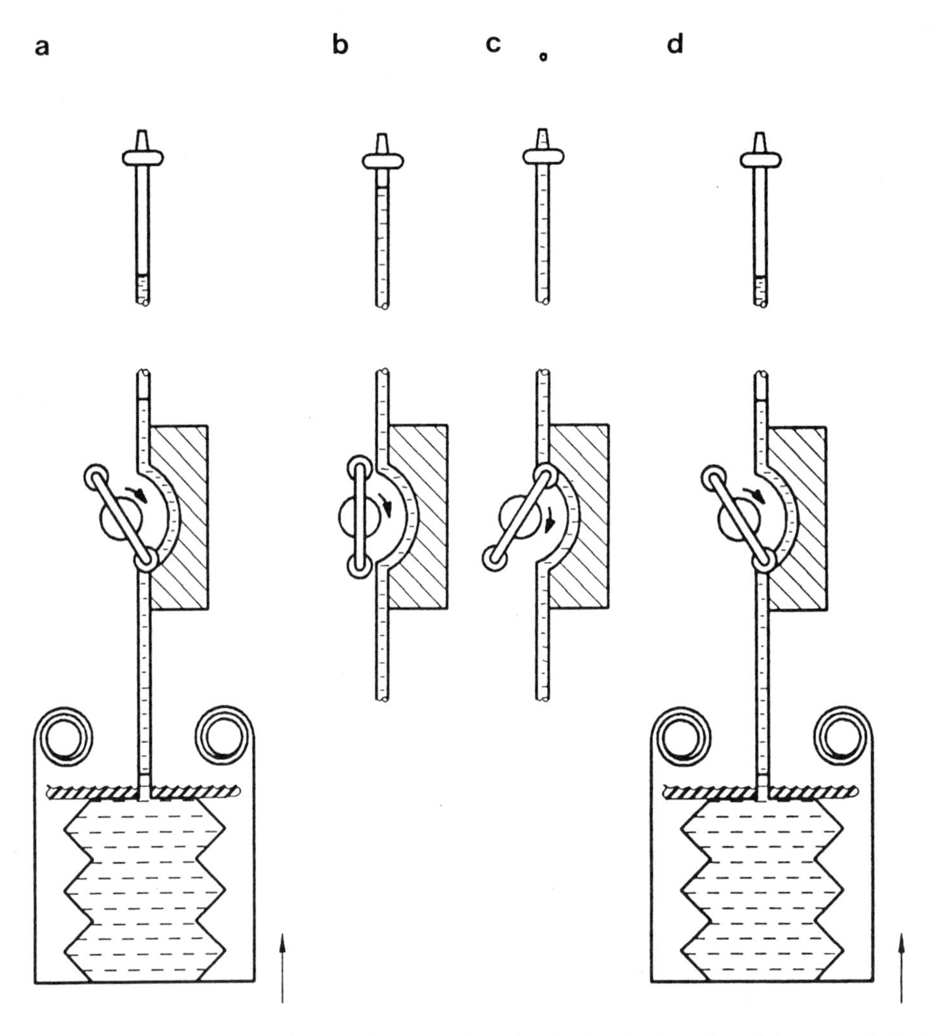

Figure 1. The four phases involved in the formation of a microdroplet via the action of the patented ejection mechanism. (a) Resting: the rotating wiper acts as a closed valve, preventing fluid pressure from reaching the nozzle. (b) Priming: the wiper moves to an open position, allowing fluid pressure to pass to the nozzle. (c) Ejecting: the wiper snaps the valve closed, completing the formation of the microdroplet. (d) Clearing: the wiper continues to rotate against the flow, creating a back pressure, drawing fluid away from the nozzle's end. This action keeps the nozzles clean, and prevents the liquid from drying and subsequent clogging.

2. PRINCIPLE

The system consists of (a) a medication reservoir connected by flexible tubing to miniature nozzles, (b) a rotating wiper that acts as a valve in the tubing, and (c) a spring that exerts continuous pressure on the liquid in the reservoir.

The functioning principle of the device is shown in Fig. 1, illustrating the different phases of the drop-ejecting action when the wiper performs a whole cycle. In the first (resting) phase, the rotating wiper acts as a closed valve that prevents pressurized fluid from reaching the nozzle (Fig. 1a). In the second (priming) phase, the rotating wiper acts as an open valve that allows just enough fluid to reach the nozzle and form a droplet at the nozzle's tip (Fig. 1b). In the third (ejecting) phase, the rotating wiper acts again as a closed valve; however, the droplet, which has accumulated sufficient momentum, leaves the nozzle to reach the patient's eye (Fig. 1c). In the fourth (clearing) phase, as the wiper continues to rotate back to the position indicated in the first phase, the fluid is drawn inward from the nozzle's end, which prevents the fluid from drying and clogging the nozzle (Fig. 1d).

In summary, the system produces a positive pressure to form the droplet followed by a negative pressure to sever the droplet from the nozzle. This permits the controlled formation of minuscule droplets. Additionally, the negative pressure draws the remaining solution back into the tube, which keeps the nozzle clean, and prevents liquid from drying and clogging the device. The action of advancing and reversing is the key feature of the invention, which differentiates it from peristaltic pumps and other atomizing devices.

3. DESIGN

The engineering model, which has been built, is called the AMIE (Automatically Medicated Instrument for Eyedrops) and consists of a pocket case and a headset connected by a slender (3 mm diameter), flexible tubing (Fig. 2).

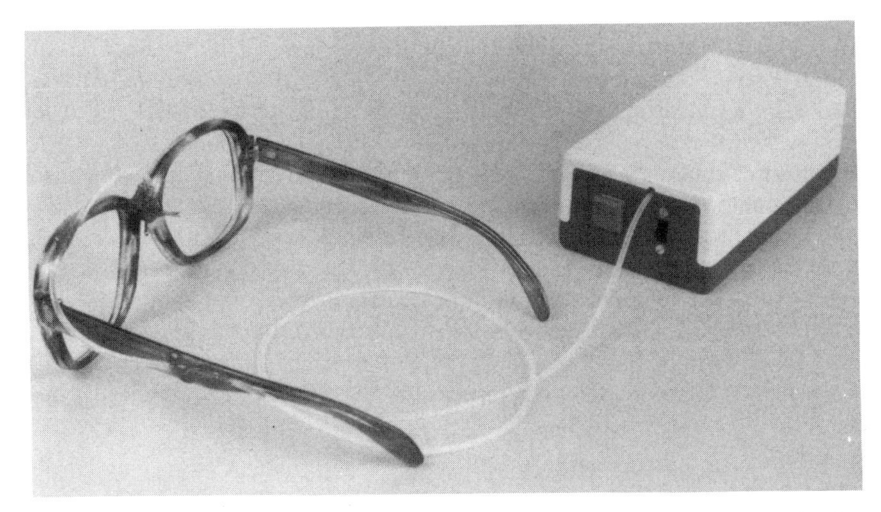

Figure 2. AMIE (Automatically Medicated Instrument for Eyedrops) can be mounted on any eyeglass frame. Two nozzles are attached to the bridge. On the box, one switch is for power on/off, and another (the larger) is for manual ejection, although AMIE will administer microdroplets automatically at a programmable rate monitored by a microcomputer.

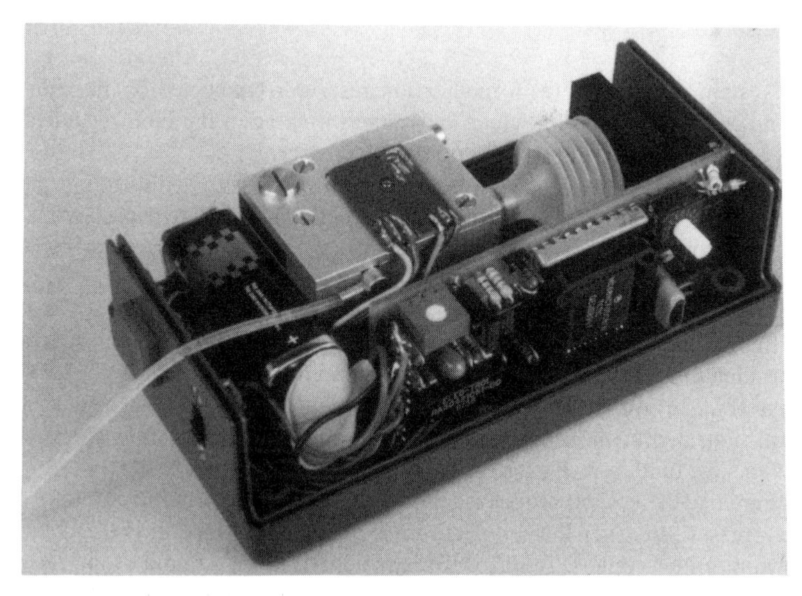

Figure 3. The case contains the reservoir, ejection mechanism, compression springs, 9 volt battery, and electronic controls. The microcomputer monitors the device functions in programmable, automatic, or manual mode.

The case consists of a 120 x 40 x 65 mm box made to fit into a shirt pocket, and contains a flexible reservoir, two compression springs, an ejection mechanism, an electronic control board, and a 9 volt battery. The electronic control contains an 8 bit, single-chip microcontroller and support electronics that allows the patented ejection mechanism to be opened and closed at different speeds and frequencies. This unique design permits the user to preset the droplet volume, ejection rate, and velocity of the ejected droplet. The automatic control can be overridden temporarily by pressing a button that allows the patient to obtain supplementary drops at will (Fig. 3).

An important feature of the AMIE is that the reservoir, tubing, and nozzles are connected and constitute a fluid cartridge. The medication is stored inside the unit separate from the driving mechanism. This prevents the medication's characteristics from being altered, preserves medication sterility, and makes replacing it convenient.

The headset consists of an eyeglass frame fitted with a pair of ejector nozzles attached to the frame bridge. The tubing and nozzles are small enough to mount easily and discreetly, either provisionally or permanently, on the patient's own eyeglasses or on an eye shield. The nozzles are aimed toward the nasal corners of the eyes, and their distance to the eyes is sufficient to prevent them from touching the eyes even upon sudden jarring.

When the power is on, the device automatically delivers eyedrops at preset intervals; however, a patient can get additional drops at will by simply pressing a button. The volume of the ejected microdroplets is adjustable from 1 to 5 μl, and the ejected dosage is regulated from 2 sec to several hours.

4. DISCUSSION

We are proposing a new tropical delivery method for eye conditions and an instrument that automatically ejects smaller droplets at a more frequent rate from miniature nozzles located a comfortable distance from the eyes. As this method requires no major effort from patients, it will greatly improve their compliance, and thus ensure the efficacy of the treatment. Medication enters the eye in a single droplet instead of a sprayed mist; it neither wets the patient's eyelids nor floods the patient's eyes. Therefore, it prevents waste of medication, reduces medical costs, and augments patient comfort. As the droplet impinges the eyeball, instinctive blinking will help to spread the medication over the eye. By adjusting the drop volume and ejection pace, the medication concentration in the eyes can be maintained at a steady and precise level, thus increasing the effectiveness and minimizing any side effects of the medication. Since there is no contact between the device and the eyes, the problems of eye irritation and medication contamination are avoided.[6–8]

An engineering model has been built and tested for reliability and practicality. The fluid cartridge unit keeps the medication at room temperature and separate from the driving mechanism. The cartridge can be sterilized for reuse, be pre-filled, and is disposable. Although designed for one day's use, testing during three consecutive days of operation revealed that the sterility was well preserved. This constitutes an important advantage of the device, because it can be used without the need of adding a preservative to the solution.

The momentum of the droplet can be controlled by varying the reservoir pressure and wiper speed. The impact energy generated by the droplet hitting the eye is approximately 0.04 mJ, well below the bearable level of 0.1 mJ.[6] We have also found that blinking did not interfere with the ejection; drops can still penetrate through the palpebral slits when the eyelids are closed. More importantly, the manual function mode of the device helps dispense additional drops as needed.

The design is highly efficient, as battery energy is used only to open and close the valve; all the mechanical energy used to eject the drop is derived from the spring compressing the reservoir. In the engineering model, the average power consumption was 45 μW. Therefore, the battery lasts many months.

Since development of the AMIE, two other devices based on the well-known principles of ink-jet printers have been proposed.[9,10] One approach uses a thermal mechanism, and the other a piezoelectric mechanism. The thermal mechanism functions by locally heating the medication to 400°C. The heat pulse vaporizes a small bubble and ejects the medication out of an orifice. This device raises concerns that the heating process will alter the medication characteristics and produce a real risk of clogging the nozzles and injuring the patient.[11] It also uses excessive energy, which requires a bulky battery. The piezoelectric device uses a piezoelectric crystal as a droplet generator, and a voltage pulse applied to the crystal induces a mechanical pulse and ejects a droplet. A high voltage (over 100 V) is required to drive the crystal, which adds bulky circuitry and the risk of electrical discharge to the user.[11] Both these approaches require unsightly ejector pumps mounted beside the eyes. Additionally, these mechanisms require intimate contact of the ejector mechanism with the medication, leading to contamination and cleaning problems. These drawbacks are avoided with the low-voltage, electronic-powered circuit AMIE, which provides uncontaminated eyedrops at room temperature.

Since the method is non-invasive and requires no action on the part of the patient to ensure accurate medication delivery, the AMIE can be used for patients of all ages and will assist medical treatment for out-patients as well as hospitalized patients who need continuous or long-term care. For many patients who normally require assistance, nursing and hospital care costs can be reduced.

In addition, the AMIE can be used as a tool for clinical, physiological, and pharmaceutical research, such as determining the efficacy of new medicines and the systemic effects caused by medication absorption. The most direct thrust of the AMIE is to help patients who suffer from dry eye syndrome due to congenital or acquired diseases, pollution, or excessive eye strain as in the case of video-display-terminal (VDT) users.[12] For VDT users, the AMIE could be a useful industrial hygiene tool as it continually maintains the eyes' moisture.[13]

ACKNOWLEDGMENTS

We thank Professor J. Bille, Dr. M. Doane, Mr. Durwood Marshall, Mr. Jianguo Sun, Mr. Khalil Zaman, and Mr. Xiao Dong Zhang for their assistance.

REFERENCES

1. Oleijnik O. Conventional systems in ophthalmic drug delivery. In Ophthalmic Drug Delivery, Ashim K. Mitra, editor. Marcel Dekker, Inc., 1993, pp. 177–198.
2. Pavan-Langston D. Potential use of ocular-insert therapy in dry eyes. *Int Ophthalmol Clin.* 1973;13:231–238.
3. Doane MG. Method of ophthalmic fluid delivery. *Int Ophthalmol Clin.* 1979;20:93–101.
4. Dohlman CH, Doane MG, Reshmi CS. Mobile infusion pumps for continuous delivery of fluid and therapic agents to the eye. *Ann Ophthalmol.* 1971;3:126–128.
5. Rehkopf PG, Goldbert DB. Implantable closed delivery system. *Invest Ophthalmol Vis Sci.* (supp), 1979:20:145.
6. Toi VV, Zhang XD. Eye medication delivery system. *Invest Ophthalmol Vis Sci.* (supp), 1990;31:404.
7. Toi VV, Grounauer PA. Automatic delivery of eye medication by droplet ejection. *Invest Ophthalmol Vis Sci.* (supp), 1992;33:1012.
8. Toi VV. Eye medication delivery system. US Patent No. 5,171,306, 1992.
9. Bertera J. A prototype tear jet irrigates dry eyes. *Invest Ophthalmol Vis Sci.* (supp), 1994;35:2040.
10. Bertera J. A prototype piezo jetpump irrigation for dry eye. *Invest Ophthalmol Vis Sci.* (supp), 1995;36:4599.
11. Germano P. Micro-pompe pour lunettes médicales. Technical Report, Institut de Microtechnique, Ecole Polytechnique Fédérale de Lausanne, 1989.
12. Tsubota K, Nakamori K. Dry eyes and video display terminals [letter]. *N Engl J Med.* 1993;328:548.
13. Toi VV, Dumery B. Relieving dry eye symptoms for computer users, Investigative Ophthalmology and Visual Science (supp) 1997;38:653.

CONTRIBUTORS

INDEX